Charophytes of Europe

Hendrik Schubert · Irmgard Blindow · Emile Nat ·
Heiko Korsch · Thomas Gregor · Luc Denys ·
Nick Stewart · Klaus van de Weyer ·
Roman Romanov · Michelle T. Casanova
Editors

Charophytes of Europe

Springer

Editors
Hendrik Schubert
Institute of Biosciences
University of Rostock
Rostock, Germany

Irmgard Blindow
Biological Station of Hiddensee
University of Greifswald
Kloster, Germany

Emile Nat
Dutch Information Center on Charophytes
Naturalis Biodiversity Center
Leiden, The Netherlands

Heiko Korsch
Jena, Germany

Luc Denys
Research Institute for Nature and Forest
Brussels, Belgium

Thomas Gregor
Senckenberg Nature Research Society
Frankfurt am Main, Germany

Klaus van de Weyer
lanaplan GbR
Nettetal, Nordrhein-Westfalen, Germany

Nick Stewart
Glastonbury, UK

Michelle T. Casanova
The Natural History Museum
London, UK

Roman Romanov
Dobra Voda, Montenegro

ISBN 978-3-031-31897-9 ISBN 978-3-031-31898-6 (eBook)
https://doi.org/10.1007/978-3-031-31898-6

Illustration by: Heiko Korsch, Themar, Thüringen, Germany

This Springer imprint is published by the registered company Springer Nature Switzerland AG
The registered company address is: Gewerbestrasse 11, 6330 Cham, Switzerland

Paper in this product is recyclable.

Foreword

I am delighted to welcome *Charophytes of Europe*, and what could be more timely. The charophytes are such an important group of green algae, surviving > 500 million years of geological history, providing evolutionary insight into the origin of land plants and now, with the planetary crisis, impacted by the degradation and loss of wetlands across the world.

As a diehard marine phycologist, I have only recently come to studying charophytes. Initiated by Michelle T. Casanova into the 'art' of charophyte collecting with an improvised grapple (a repurposed kitchen whisk weighted with a broken plant pot), studying them is opening up a fascinating world of these organisms and their habitats. But my interest in their taxonomy came indirectly through research on the structural colour in *Nitella*, a phenomenon caused by the helicoidal architecture of the cell and hypothesised to provide protection of UV light. Naturally, my immediate thought was what species are we working on.

Now we have this wonderful book on the charophytes, an authoritative text that is a testament to the charaphytologist community. It cannot have been easy to come to a consensus on the final number of species, given the 'battle of concepts' and the different schools on what constitutes a charophyte species. However, led by Hendrik Schubert and a taxonomy supported by a robust phylogenetic framework, 68 authors, from 29 countries spanning five continents have managed to reach a consensus here. What a fantastic example of what can be achieved by such leadership and collaboration.

It is brilliant to be able to delve into this comprehensive book and find a wealth of information and insight into so many aspects of the charophytes, from ecology to palaeontology, ontogenesis to oospores, and the detailed descriptions of the *Characeae* species recorded from Europe. The book is beautifully illustrated with a range of line drawings of species, colour photographs, micrographs and electron microscopy images. For each of the taxa of *Characeae* in Europe, there are nomenclature and taxonomic details, and sections covering morphology, phenology and reproduction, habitat conditions and ecology, variability, identification and risk of confusion, status and threat, and distribution.

The inclusion of distribution maps for all species, with coloured dots indicating the last time species were recorded at each site, must be a wake-up call to everyone. With so many records of species last recorded at least 70 years ago over swathes of some areas, the impact of environmental change and human pressure on a wide range of species is clear. It is imperative that we look after our water bodies and charophyte habitats, and this informative book can only help in raising awareness of the charophytes to a wider audience. Here, also, the authors are to be congratulated for including the first European Red List for this group of invaluable algae.

What a superb book. As the authors hope, I am sure that this will be a useful tool and a common standard for charaphytologists, which is what the authors wish for.

August 2022

Prof. Juliet Brodie
Department of Life Sciences
Division of Algae, Fungi and Plants
The Natural History Museum
London, UK
j.brodie@nhm.ac.uk

Contents

Part III Glossary and Index

Part I
General Section

Chapter 1
Compiling What We Know About European Charophytes—An Introduction

Hendrik Schubert, Irmgard Blindow, Michelle T. Casanova, Luc Denys, Thomas Gregor, Heiko Korsch, Emile Nat, Roman Romanov, Nick Stewart, and Klaus van de Weyer

Although introductory chapters are often ill-fated, editors are obliged to write one in the hope that it will be read (and perhaps even cited) by someone, given that its

H. Schubert (✉)
Institut für Biowissenschaften, Universität Rostock, Albert-Einstein-Straße 3, 18051 Rostock, Germany
e-mail: hendrik.schubert@uni-rostock.de

I. Blindow
Universität Greifswald, Biologische Station Hiddensee, 18565 Kloster, Germany
e-mail: blindi@uni-greifswald.de

M. T. Casanova
273 Casanova Rd, 3351 Westmere, Victoria, Australia
e-mail: mt.casanova@federation.edu.au

L. Denys
VAC Herman Teirlinck, Research Institute for Nature and Forest, Havenlaan 88 Bus 73, 1000 Brussel, Belgium
e-mail: luc.denys@inbo.be

T. Gregor
Department of Botany and Molecular Evolution, Senckenberg Research Institute and Natural History Museum Frankfurt, Senckenberganlage 25, 60325 Frankfurt am Main, Germany
e-mail: thomas.gregor@senckenberg.de

H. Korsch
Am Werragrund 1, 98660 Themar, Germany
e-mail: heiko.korsch@t-online.de

E. Nat
Dutch Information Center on Charophytes, Naturalis Biodiversity Center, P.O. Box 9517, 2300 RA Leiden, The Netherlands
e-mail: e.nat@kranswieren.nl

H. Schubert et al. (eds.), *Charophytes of Europe*,
https://doi.org/10.1007/978-3-031-31898-6_1

title is sufficiently alluring and that its contents add something more than just the structure and intention of the book. May this chapter find some readers….

This book originated from discomfort. After several regional overviews of charophytes were published in national languages other than English in the last three decades (e.g. Krause, 1997; van Raam, 1998; Cirujano et al., 2008; Bazzichelli and Abdelahad, 2009; Mouronval et al., 2015; Arbeitsgruppe Characeen Deutschlands, 2016; Borisova et al., 2016), many botanists and limnologists yearned for a comprehensive treatise spanning the entire European flora. This reflected the fact that these regional publications were much more than just accounts of the floristic diversity and distribution of charophytes, but also exemplified different and developing taxonomic viewpoints at a time where applied ecology and conservation called, more and more, for a robust common framework.

Most readers will probably be well aware of the problems arising in this field where charophytes are concerned. Two conflicting taxonomic concepts persist: the 'macro-species' concept lumping widely varying morphotypes into the smallest number of entities possible advocated by Wood (1965) and the older 'micro-species' approach clinging to a much more detailed separation (see Chap. 5). Whereas some of us may willingly accept the micro-species approach as the legitimate consequence of a long European research tradition, this is certainly not the case world-wide. Wherever a non-native species appears in Europe, taxonomic problems arise. In such cases, Wood (1965) will always be the primary reference and given the still uncharted variability of many non-European species, incertitude lingers. Moreover, the microspecies concept on its own is not trouble-free. As our knowledge increases, especially with respect to distribution and ecology, assigning "atypical morphotypes" often becomes a matter of debate. If a given morphotype exhibits a distinct distribution pattern, the question of its taxonomic rank becomes inevitable and needs to be solved. The "quick fix" use of molecular genetics has stirred up debates and sometimes even spurred conflicts. Although adding information, it cannot serve as a panacea (see Chaps. 5 and 6). The rank of a taxon is made by a kind of agreement among scientists, and other scientists may or may not agree with their arguments. Whether a group of similar individuals warrants a certain taxonomic rank remains a reasoned decision, but arguments may not necessarily be met with general agreement. Consequently, all authors of this book thought it wise to found such decisions for as much as possible on a broad base of expert opinions to avoid dissent afterwards (see Schubert, 2014).

R. Romanov
Nišice Ujtin Potok B.B, Dobra Voda, 85356 Bar, Montenegro
e-mail: streptophytes@gmail.com

N. Stewart
4 Hexton Road, BA6 8HL Glastonbury, UK
e-mail: nfstewart@freeuk.com

K. van de Weyer
lanaplan, Lobbericher Str. 5, 41334 Nettetal, Germany
e-mail: klaus.vdweyer@lanaplan.de

So, instead of just compiling regional and national surveys, the idea of a truly European Flora of charophytes arose. Of course, with the last one dating from the late 1990s (Krause, 1997), one might wonder whether this was already necessary only two decades later? Obviously, the answer was a clear "Yes". Not just because of language issues or geographical scope, but also because of the wealth of new information collected since the works of Werner Krause and Joop van Raam motivated botanists once again to dive into charophytology. Field work, experiments, morphological observations, systematic and taxonomic studies changed established opinions and led to fresh delineations, new species, the solving of nomenclatural issues and reconsideration of distribution records, giving ample material for this book.

Moreover, it allowed us to consider the entire European continent as a whole for the first time. While former charophyte monographs largely focused on Central, Western and Northern Europe, we could now include Eastern Europe in much more detail, with consequences for both taxonomic coverage and biogeography.

In all this, the heritage of the late Joop van Raam (17.08.1941–01.10.2011; Nat, 2012) requires special mention. Unable to finish his life-long project of a microspecies-based world monograph of charophytes, he provided the editors with all his material shortly before passing away. This awesome collection of distribution data, ecological and nomenclatural information accompanied by original opinions and brilliant systematic insights provided a most welcome starting point for a European treatise. Also heartily acknowledged is Uwe Raabe's contribution who, unfortunately, felt obliged to decline editorship for health reasons. However, he helped to launch the project with much enthusiasm and shared his in-depth knowledge throughout the preparatory stage.

European charophytologists are well organised. Besides national and regional working groups and networks, there is an active pan-European consortium (Group of European Charophytologists, GEC, which first met in Lausanne in 1987), now embedded in an equally active global association (International Research Group on Charophytes, IRGC, active since 1989). Both regularly organise conferences and foster lively scientific discussions. These networks allowed the large group of authors and correspondents to gather securing the necessary basis for broadly acceptable taxonomic decisions.

Coordinating the activity of so many authors became quite an endeavour. Several in-person editorial meetings (Schmöckwitz, 2016, Scotlandwell, 2017, Leiden, 2017, Wandlitz, 2019) were necessary to get the work started, set goals and outline book contents and structure. Two more extensive workshops (Fig. 1.1), the first one in Vienna, organised by Karl-Georg Bernhardt and the second one in St. Petersburg, organised by Roman Romanov, allowed the authors to arrive at common agreements on matters of taxonomy and species delineation. Then, CoVid 19 barged in. Scheduled identification workshops and planned fellowships had to be cancelled, access to herbaria became problematic and editorial meetings had to be done online. These obstacles, and especially the lack of face-to-face discussions, delayed the finishing of manuscripts by almost two years. Production, unfortunately, also took another one and a half years.

Fig. 1.1 Impressions from the author workshops: Upper panels: debates and discussions during field work at the first author workshop in Vienna (20th–22nd of April 2018). Middle and lower panels: exploring unexpected treasures in the herbarium, taking samples beneath the ice while freezing and debating characters of the *Hartmania* group during the second author workshop in St. Petersburg (24th–29th of March 2019)

It is our aspiration that the final product of 70 authors will now be accepted as a reference work that reflects and evaluates the current knowledge on charophytes in Europe with respect to their taxonomy, ecology and distribution. For this purpose, and following a checklist of accepted taxa known to occur in Europe (Chap. 2), reviews on ontogenesis and systematics (Chaps. 3 and 6) provide the background for the taxonomic concepts introduced in Chap. 5. Oospores, although part of the charophyte life cycle, and therefore ontogenesis, are dealt with separately (Chap. 4) because of their outstanding importance for environmental monitoring, palaeontology and, increasingly, species identification. Chap. 7 focuses on the broad range of niches occupied by charophytes and their ecological role in aquatic ecosystems, which is complemented by a synthesis of their palaeontological record (Chap. 8). Chap. 9 covers the current threats faced by charophytes and their ranking in national and regional Red Lists, reflecting their broad range of ecological strategies and acclimation capabilities. It also includes the first European Red List for this group.

The main part of the book consists of a taxonomic treatment, introduced by a determination key (Chap. 10) and followed by detailed descriptions of all European species in alphabetical order (Chaps. 16–78) covering morphological, autecological and distributional aspects and briefly discussing their relationships to other taxa. This part finishes with several taxa of uncertain affiliation and an addendum of species that were probably recorded erroneously in Europe (Chaps. 78–86). In addition, characteristics of the genera dealt with (Chap. 13) and detailed treatments of species groups that are difficult to distinguish (Chaps. 14 and 15).

While working on the manuscript, it became evident that a clear-cut terminology was essential, provided here in a comprehensive glossary (Chap. 87). In spite of the huge effort by John Bruinsma to provide this, there are probably still some inconsistencies left in the use of certain terms. We call for some mercy if readers stumble on those we may have overlooked. Because there were still some shortcomings in the terminology published so far and to assist our readership, we decided to include a dedicated glossary. This may help to prevent future misunderstandings and confusion.

This book was prepared without any external support. This applies not only to the authors, but especially to the many fieldworkers and enthusiasts who provided distribution data and took part in the many preceding discussions and identification workshops. We tried our best to acknowledge these contributions, but are well aware that there is always someone neglected or forgotten. Please forgive us, if this is the case.

We hope that this book will be a useful tool, perhaps even a common standard, in the forthcoming years and wish all readers a great deal of success with their related projects.

References

Arbeitsgruppe Characeen Deutschlands. (Hrsg.), (2016). *Armleuchteralgen - Die Characeen Deutschlands*. Springer-Verlag. [in German].

Bazzichelli, G., & Abdelahad, N. (2009). *Alghe d'acqua dolce d'Italia. Flora analitica delle Caroficee*. Sapienza Università di Roma - Ministero dell'Ambiente e della Tutela del Territorio e del Mare, Roma. [in Italian].

Borisova, O. V., Palamar-Mordvintseva, G. M., & Tsarenko, P. M. [Борисова, О. В., Г. М. Паламар-Мордвинцева & П. М. Царенко]. (2016). *Флора водоростей України. Том 12. Харофітові водорості. Вип. 2. [Flora algarum Ucrainicae. Charophyta, 12(2)]*. Kholodny Institute of Botany of the National Academy of Science of Ukraine, Kyiv. [In Ukrainian].

Cirujano, S., Cambra, J., Sánchez Castillo, P. M., Meco, A., & Flor Arnau, N. (2008). *Flora ibérica. Algas continentales. Carófitos (Characeae)*. Real Jardín Botánico, Madrid. [in Spanish].

Krause, W. (1997). Charales (Charophyceae). Süßwasserflora von Mitteleuropa, vol. 18. Gustav Fischer Verlag. [in German].

Mouronval, J. B., Baudouin, S., Borel, N., Soulié-Märsche, I., Klesczewski, M., & Grillas, P. (2015). *Guide des Characées de France méditerranéenne*. Office National de la Chasse et Faune Sauvage. [in French].
Nat, E. (2012). In Memoriam Joop van Raam (*17-08-1941, †01-10-2011) & Bibliografie. Rostocker Meeresbiologische Beiträge 24: 2–5 (K. van de Weyer, Trans.). [in German].
Schubert, H. (2014). Phylogeny and taxonomy of charophytes, good news from a battlefield of concepts. *Journal of Phycology, 50*(5), 773–775.
van Raam, J. C. (1998). *Handboek Kranswieren*. Chara Boek. [in Dutch].
Wood, R. D. (1965). *A revision of the Characeae*. Cramer Verlag.

Chapter 2
Checklist of European Charophytes

Thomas Gregor

We recognise 63 European species of *Characeae*. This is substantially more than the 50 corresponding taxa covered by Krause (1997) in his treatment for Central Europe, an increase of 26%.

The increase is mostly due to the expansion of the covered area. Ten taxa occurring in the southern, southeastern or northern part of Europe were not in Krause's geographical focus: *Chara altaica*, *C. corfuensis*, *C. globata*, *C. hellenica* (described in 2010), *C. oedophylla*. *C. oryzae* (described in 2021), *C. uzbekistanica, C. zeylanica* (first European record in 2019), *Nitella palhinae* and *N. wahlbergiana*. The description of *Lamprothamnium sardoum* has been published in 2022 (Casanova and Becker, 2022).

In 5 cases names have changed for nomenclatural reasons: *Chara polyacantha* is treated as *Chara aculeolata*, *Chara intermedia* as *C. papillosa*, *Chara rudis* as *C. subspinosa*, *Chara delicatula* as *C. virgata*, *Nitella batrachosperma* as *N. confervacea*.

In 11 cases we changed names for taxonomic reasons: European plants thought to belong to *Chara denudata*, a taxon described from South Africa, are now considered to belong to *Chara dissoluta*. We consider, albeit with doubts, *Chara kokeilii* to belong to *Chara tenuispina*, *Chara crassicaulis* to be part of *C. vulgaris, Chara muscosa* to be part of *C. contraria*, and *Chara ohridiana* to belong to *Chara dissoluta*, *Lamprothamnium hansenii* to belong to *L. papulosum*. We split *Tolypella* into *Tolypella* and *Sphaerochara*: *Tolypella canadensis* becomes *Sphaerochara canadensis*, *Tolypella intricata* changes to *Sphaerochara intricata* and *Tolypella prolifera* into *Sphaerochara prolifera*.

In 6 cases the delimitation of species changed: We include *Chara desmacantha* into *Chara curta*, the latter considered by Krause as a variety of *Chara aspera*. *Chara*

T. Gregor (✉)
Department of Botany and Molecular Evolution, Senckenberg Research Institute and Natural History Museum Frankfurt, Senckenberganlage 25, 60325 Frankfurt am Main, Germany
e-mail: thomas.gregor@senckenberg.de

H. Schubert et al. (eds.), *Charophytes of Europe*,
https://doi.org/10.1007/978-3-031-31898-6_2

fibrosa was recognised by Krause as occurring extremely rarely in rice fields in Italy. There are now more records of such a rice field *Chara*; alas we are unable to name theses neophytes from south-east Asia with certainty and use the name *Chara fibrosa* s.l. We separate Krause's *Chara gymnophylla* into the aulacanthous *C. gymnophylla* and the tylacanthous *C. squamosa*, the latter being the older name. The extinct *Chara filarszkyana*, a taxon similar to *Chara tenuispina*, is treated as a taxon of uncertain affiliation. We separate *Chara affinis* (= *C. baltica* var. *affinis*) from *C. hispida*. Plants from the Mediterranean formally considered to belong to *Chara baltica* are distinct from this taxon and named informally as "Mediterranean Chara".

European taxa of *Characeae*:

Accepted names:

Chara aculeolata Kütz.
Chara affinis Blindow, H. Schub. & N. F. Stewart
Chara altaica A. Braun
Chara aspera Willd.
Chara baltica (Hartm.) Bruzelius
Chara baueri A. Braun
Chara braunii C. C. Gmelin
Chara canescens Desv. & Loisel.
Chara connivens Salzm. ex A. Braun
Chara contraria A. Braun ex Kütz.
Chara corfuensis J. Groves ex Fil.
Chara curta Nolte ex Kütz.
Chara dissoluta A. Braun ex Leonh.
Chara fibrosa C. Agardh ex Bruzelius s. l.
Chara filiformis Hertzsch ex A. Braun
Chara fragifera Durieu
Chara galioides DC.
Chara globata Mig.
Chara globularis Thuill.
Chara gymnophylla (A. Braun) A. Braun
Chara hellenica Langangen
Chara hispida L.
Chara horrida Wallman ex Wahlst.
Chara imperfecta A. Braun
Chara kirghisorum Less.
Chara neglecta Hollerb.
Chara oedophylla Feldm.-Maz.
Chara oryzae R. E. Romanov, Nikulin & A. Gontcharov
Chara papillosa Kütz.
Chara squamosa Desf.
Chara strigosa A. Braun
Chara subspinosa Rupr.
Chara tenuispina A. Braun

Chara tomentosa L.
Chara uzbekistanica Hollerb.
Chara virgata Kütz.
Chara vulgaris L.
Chara zeylanica Willd.
Lamprothamnium papulosum (Wallr.) J. Groves
Lamprothamnium sardoum Ralf Becker
Lychnothamnus barbatus (Meyen) Leonh.
Nitella capillaris (Krock.) J. Groves & Bull.-Webst.
Nitella confervacea (Bréb.) A. Braun ex Leonh.
Nitella dixonii H. Groves & J. Groves
Nitella flexilis (L.) C. Agardh
Nitella gracilis (Smith) C. Agardh
Nitella hyalina (DC. in Lam. & DC) C. Agardh
Nitella mucronata (A. Braun) Miq.
Nitella opaca (C. Agardh ex Bruzelius) C. Agardh
Nitella ornithopoda A. Braun
Nitella syncarpa (Thuill.) Chev.
Nitella tenuissima (Desv.) Kütz.
Nitella translucens (Pers.) C. Agardh
Nitella wahlbergiana Wallman
Nitellopsis obtusa (Desv.) J. Groves
Sphaerochara canadensis (Sawa) Soulié-Märsche
Sphaerochara intricata (Trentep. ex Roth) Feist-Castel & N. Grambart
Sphaerochara prolifera (Ziz ex A. Braun) Soulié-Märsche
Tolypella glomerata (Desv.) Leonh.
Tolypella hispanica Nordst. ex Allen
Tolypella nidifica (O. F. Müll.) A. Braun
Tolypella normaniana (Nordst.) Nordst.
Tolypella salina Corill.

Taxa of uncertain affiliation.
Chara filarszkyana R. E. Romanov & Mesterházy
Mediterranean Chara
Nitella palhinae (A. G. Cunha) R. E. Romanov
Nitella opaca var. *attenuata* H. Groves & J. Groves

References

Becker, R., Schubert H., & Nowak, P. (2021). *Chara zeylanica* J.G. Klein ex Willd. (*Charophyceae, Charales, Characeae*): First European record from the island of Sardinia, Italy. *Plants, 10*, 2069. https://doi.org/10.3390/plants10102069

Casanova, M. T., & Becker, R. (2022). *Lamprothamnium sardoum* sp. nov. (*Characeae, Streptophyta*): A new species of *Lamprothamnium* for Europe. *Phytotaxa, 567*(3), 269–277. https://doi.org/10.11646/phytotaxa.567.3.6

Krause, W. (1997). *Charales* (*Charophyceae*) Süßwasserflora von Mitteleuropa. In: H. Ettl, G. Gärtner, H. Heynig, & D. Mollenhauer (Eds.), Süßwasserflora von Mitteleuropa 18. Fischer, Jena.

Langangen, A. (2010). *Chara hellenica* spec. nov. and other charophytes collected in Corfu and Amorgos islands (Greece) in 2008. *Studia Botanica Hungarica, 41*, 21–30.

Romanov, R. E., Yu. Nikulin, A., Yu. Nikulin, V., & Gontcharov, A. A. (2021). New species *Chara oryzae* and a new section *Corillionia* of *Chara* (*Charales, Charophyceae*) from European Mediterranean rice fields. *European Journal of Phycology*. https://doi.org/10.1080/09670262.2021.1987527

Chapter 3
Ontogenesis and Life Cycle of Charophytes

Hendrik Schubert, Anja Holzhausen, Petra Nowak, and Michelle T. Casanova

3.1 Introduction

The development of the *Characeae*, especially the formation of their unique reproductive organs, had already been studied extensively at the end of the nineteenth century (for an overview see Migula, 1889–1897). The material available is correspondingly extensive and detailed. Therefore, this chapter does not even attempt to make an all-encompassing account of the individual development of the *Characeae*.[1] Rather, it will be a comprehensive outline of the basic principles of ontogenetic processes that result in a charophyte morphology. Through this form of presentation, the reader should understand the development of relevant features and be able to make a weighting of the importance of individual deviations in cortication, appearance of gametangia etc.—which are characters used for species delineation. For this,

[1] This was already impossible in 1922. Oltmanns (1922) already stated laconically: "*Wir können hier nur das Wichtigste bringen, den Kleinkram, der reichlich vorhanden, übergehen wir*" (c.f. Oltmanns, 1922, [we can only present the most important, we skip the odds and ends that are abundantly available]).

H. Schubert (✉) · P. Nowak
Institut für Biowissenschaften, Universität Rostock, Albert-Einstein-Straße 3, 18051 Rostock, Germany
e-mail: hendrik.schubert@uni-rostock.de

P. Nowak
e-mail: petra.nowak@uni-rostock.de

A. Holzhausen
Institute of Agricultural and Nutritional Sciences, Martin-Luther-University Halle-Wittenberg, Betty-Heimann-Str. 5, 06120 Halle (Saale), Germany
e-mail: anja.holzhausen@landw.uni-halle.de

M. T. Casanova
273 Casanova Road, 3351Westmere, Victoria, Australia
e-mail: mt.casanova@federation.edu.au

H. Schubert et al. (eds.), *Charophytes of Europe*,
https://doi.org/10.1007/978-3-031-31898-6_3

all illustrations should be seen as idealised concepts of ontogenetic patterns, they do not represent drawings of microphotographs, which can be found in several publications (e.g. Pickett-Heaps, 1975). Special features of the individual taxa as well as open questions and debates are contained in the form of footnotes and references to further literature.

3.2 General Morphology

The vegetative thallus of charophytes consists of shoots (axes), rhizoids and whorls of branchlets, in most cases adventitious shoots (lateral branches) originate from the main axis nodes. Complete specimens still retain the first growth of uniseriate cells, called "protonema" (sometimes also "prochara" or "proembryo", note that the relation to protonemata of mosses is not clear!), which originates directly from the oospore and from which the first and only whorl-node of the main shoot originates (Fig. 3.1).

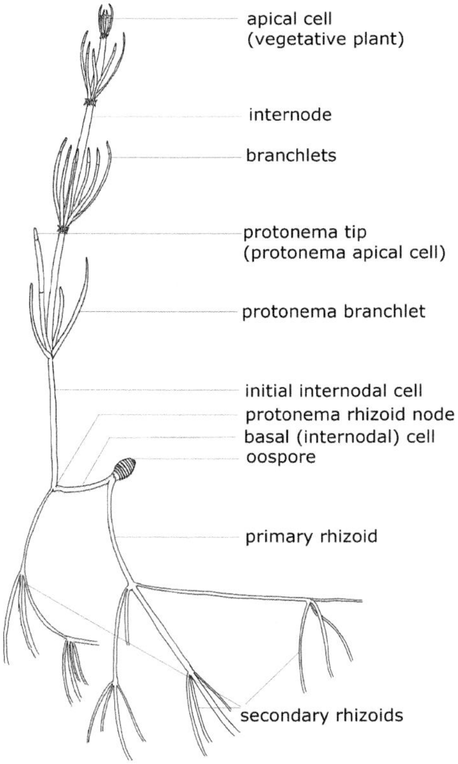

Fig. 3.1 General morphology of a complete charophyte, consisting of oospore, protonema with primary rhizoid and vegetative plant. © Anja Holzhausen

The main axis is divided into internodes and nodes. Whorls of branchlets, as well as lateral branches and adventitious shoots, always arise at the nodes. The branchlets themselves are in turn divided into branchlet nodes and branchlet internodes. Bract cells originate from branchlet nodes; bract cells may grow into short side-branchlets, developing the reproductive organs (gametangia). Bracteoles and bractlets, being homologous to oogonia (bracteoles) or replacing antheridia in dioecious species (bractlets), are found exclusively on gametangial short branches.

Axis cortication originates primarily from the axis nodes and consists of rows of alternating cortex nodal cells and cortex internodal cells. Cortex nodal cells can develop peripheral cells (in the cortex plane) as well as spine-cells (perpendicular to the cortex plane). Rhizoids, which anchor the plant in the sediment can originate from several places. The primary rhizoid is one of the filaments formed by the germinating oospore. In addition, the first protonemal node ("rhizoid node") also develops a cluster of secondary rhizoids. Additional secondary rhizoids can also be formed by the shoot nodes of the *Chara* plant.

A stringent and unambiguous description of the different morphological characters of charophytes might be possible by a further development of the system introduced by Chadefaud (1979). Unfortunately, his notation is complicated and, in addition, a number of open questions still need to be clarified before universal applicability is given. Figure 3.2 presents an attempt at an overview of cell-descendency, which may serve as a basis for such a notification. However, it also reveals some gaps of interpretation which needs to be studied in more detail.

3.3 Germination and Formation of Protonema

Characeae are regarded as haplonts, with the fertilised egg being the only diploid state throughout the life cycle. The only exception to this is probably the parthenogenetic race of *C. canescens* which, according to Ernst (1918), never does undertake meiosis.[2]

[2] This was either rejected or ignored by many later authors, still referring to *C. canescens* as a unique example of generative parthenogenesis (e.g. Bower, 1923). Still the issue is not yet settled unanimously (see Haig, 2010).

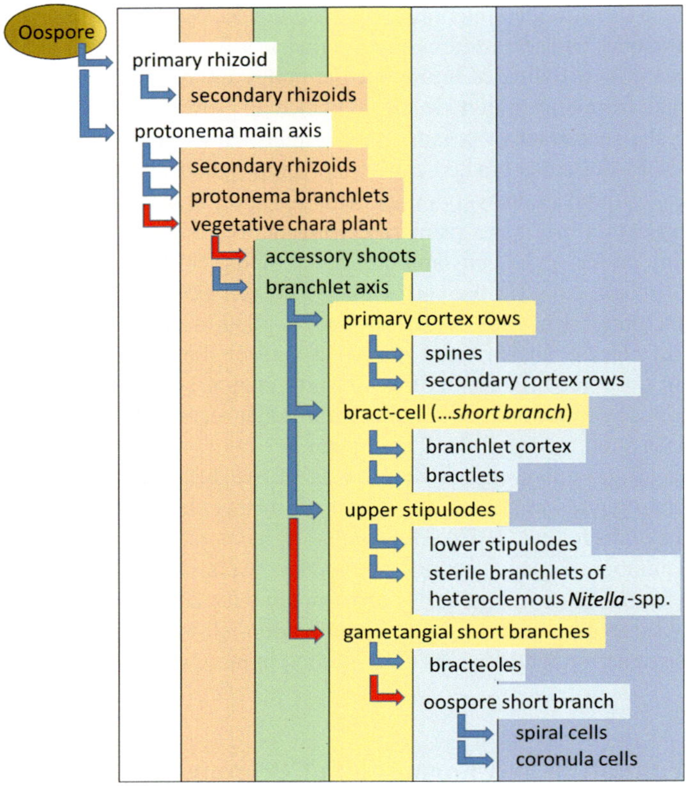

Fig. 3.2 Cell-descendency scheme. Ontogenesis of charophytes follow a strict pattern of meiosis products of an apical cell. Shown here is an overview of the homologous organs of a charophyte (indicated by different background colours). Each new generation of apical cells, derived from lateral formation of nodal cells, is seen as homologues as first proposed by Kuczewski (1906). Red arrows indicate descendencies bound to first laterals only. However, reproductive organs are still hard to interpret. Whether or not the turning cells are pauperising laterals or consist of a series of internodal and nodal cells is not clear as well as the interpretation of antheridial cell lines. One may assume that the divisions giving rise to shield cells are transversal division of an apical cell, similar to the ones of the oospore and main axis nodes, which do not change their status and are in fact just multiplications of apical cells. Accepting this, formation of manubrias and capitulae could be interpreted as a "normal" pattern, resulting in internode (manubrium) and node (capitulum), the latter dissecting laterals (secondary capitulae) which are the origin of spermatogenous threads. However, there are alternative explanations as, e.g. series of limb cells or even direct conversion (Kuczewski, 1906) a concept still to be proven. © Hendrik Schubert

The Charophyte Meiosis Debate

There is still uncertainty at which stage of the life cycle meiosis takes place. Three different stages of the life cycle have been reported for being the place of meiosis: (A) oospore germination, (B) formation of gametes[3] and (C) in the protonemal branchlet whorl node.

(A) The recent standard interpretation has been that meiosis takes place during germination of oospores as was reported by Oehlkers (1916). In investigating germination of *Chara vulgaris* (*C. foetida*) he concluded that meiosis should have taken place during the first nucleus divisions inside the closed oospore, resulting in four nuclei out of which three degenerate. However, his conclusion is an indirect one, because the only successful chromosome counts have been done during telophase I and II (and subsequent divisions) of the gametophyte plant or in gametogenesis, which still leaves opportunity for speculation, taking the arguments of Haig (2010) into account. Haig (2010) discussed in detail the problems arising from applying generation cycle concepts derived from higher plants to algae and raised doubt about the robustness of generalisations made by reviewing examples of "unusual life cycles reported in charophyte algae".

(B) Tuttle (1924, 1926) reported a reduction in chromosome number during division of the primary oocyte, resulting in three sterile polar bodies (interpreted as Braun's "Wendungszellen", Braun, 1853), and the large oocyte. Similarly, he found a reduction in chromosome number in formation of the antheridial filaments. However, his concept was rejected by Guerlesquin and Noor (1982) for several reasons. Firstly, the identity of his organism was not clear; there was a precise description of the tetrascutate character of the antheridia, but, the only herbarium specimen that has been traced to his work so far is described as *Nitella microcarpa* by Wood (1954, p. 67)—which is a species having octoscutate antheridia. Secondly, his observations are in contrast to numerous other reports (e.g. Debski, 1897, Karling, 1926; Gonçalves da Cunha, 1942; Lindenbein, 1927; Mendes, 1946, Shen, 1967a, 1967b; Walther, 1929), in which reduction of chromosome number during antheridia formation was not found. The case is still open because the argument that meiosis takes place at different stages in different taxa cannot be excluded.

(C) Another candidate for the site of meiosis is during the formation of the apical cell of the *Chara* thallus in the first branchlet whorl of the protonema as reported by Gonçalves da Cunha (1936, 1942). These are

[3] With respect to this, the genera of charopyhtes may even behave differently. Goetz (1899) reported a kind of (partial?) nucleus division in developing oocytes of *Nitella flexilis* and *N.opaca* ("Ausscheidung von Kernsubstanz") but couldn't observe this in *Chara vulgaris*.

still the only reports of protonemal chromosome numbers, but his observations were rejected by Mendes (1946) who wrote in a laconic footnote: "this cannot be taken into consideration as this author probably has not seen real chromosomes" (c.f. Mendes, 1946). This is not a very strong counterargument as these authors differ in their counts of chromosome numbers (Mendes 1946, $n = 14$ and Gonçalves da Cunha 1942, $n = 16$). Thus this option must also be retained as a potential alternative to the recent standard interpretation.

In addition, the argument of Haig (2010) that different life cycle modes could coexist in charophyte algae, does not restrict these differences to a certain taxonomic rank, so differences among genera as well as between species groups (e.g. octo- and tetrascutate ones) could also be possible.

Inside the oospore, the nucleus moves first to the apex of the oocyte, where it performs a series of nuclear divisions resulting in a mononuclear apical cell and three rapidly degrading nuclei that remain within in the starch-filled oocyte (Oehlkers, 1916; Fig. 3.3). The exact timing of these events is not known, the formation of this mononuclear apical cell might terminate primary dormancy or, alternatively, might represent the first step of germination. Nevertheless, the first visible sign of germination is an increase in volume of this apical cell, which creates fissures in the oospore wall. The apical cell now undertakes a longitudinal division, and the two daughter cells grow out of the oospore and become the apical cell of the primary rhizoid and the apical cell of the protonema, respectively.

The rhizoidal apical cell forms a richly branched primary rhizoid, whose structure has not been studied in detail yet. By a series of cross sections, the protonemal apical cell divides into a filament consisting of a basal cell, a rhizoid node, an initial internode cell, an axis node and a protonemal extension of a series of 3–4 cells.

In the rhizoid node several longitudinal divisions take place. The resulting lateral cells grow into rhizoids, each of which can, in turn, branch and anchor the plant in the sediment.

The initial internodal cell between the rhizoid node and the axis node of the protonema does not divide further. The development of the protonemal branchlet node initially proceeds in the same way as described below for the axis nodes of the mature plant: a sequence of longitudinal divisions results in the formation of a cell plate of about 8 cells; the 6 lateral cells of this plate become transformed to protonemal branchlet apical cells except for the one first formed, which becomes the apical cell of the main axis of the *Chara* plant (de Bary, 1875; Kamieński, 1876; Pringsheim, 1863).

Interestingly, this pattern seems to be different among species. Ernst (1901) reported that for *Nitella syncarpa* the axis originates from the protonemal rhizoidal node.

The protonemal branchlet apical cell undergoes a series of divisions, invariably forming an ecorticated branchlet consisting of nodal and internodal cells. As the

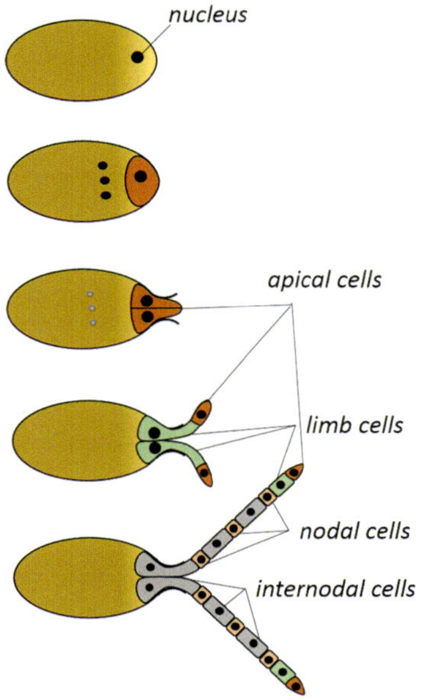

Fig. 3.3 Germination and development of primary rhizoid and protonema. © Hendrik Schubert

branchlet is an organ of limited growth only, the apical cell ceases activity after a couple of divisions and becomes the terminal cell of the protonemal branchlet.

Following all of these divisions the respective daughter cells do not separate completely. They are still connected by plasmodesmata. These connections, resembling the sieve-plates of higher plants, ensure the supply of photosynthetic products and oxygen to the rhizoids, and also transport nutrients from the rhizoids into the photosynthetic active area of the thallus. This also explains why some species can germinate and grow in complete darkness for a while before they exhaust the oospore reserves, as reported by de Winton et al. (2004).

3.4 Shoot Development

As outlined above, the apical cell that develops from the first lateral cell of the protonemal branchlet node (or rhizoidal node in the case of *N. syncarpa*) gives rise to the main axis development. In a series of transverse divisions, a filament of omnipotent cells is formed (Fig. 3.4). Each of these cells perform just one further transverse division. As a result of this secondary division, the initial row of equal omnipotent cells transforms to an alternating pattern of vegetative internodal cells and omnipotent nodal cells, with the active apical cell at the top.

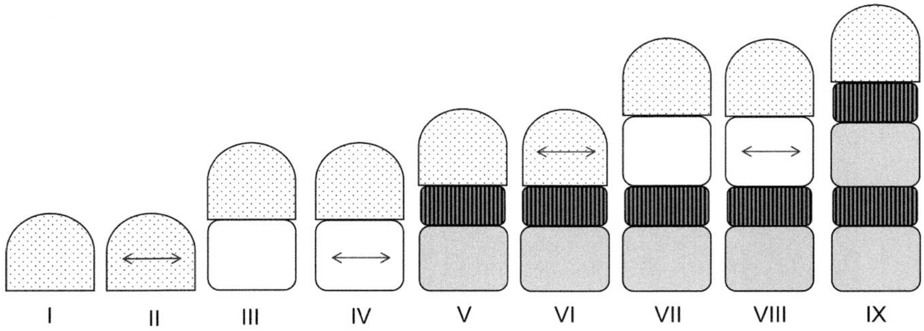

Fig. 3.4 Shoot development. The apical cell (dotted, I) cuts off (II) a limb cell (white, III) which further divides (V) into a nodal cell (hatched) and an internodal cell (grey). This pattern continuously repeats (VI–IX) in axes of unlimited growth. © Anja Holzhausen

The omnipotent nodal cell now initiates the formation of a branchlet node. For this, first a longitudinal division creates two equal cells, each of these cells rapidly undertakes a series of 3–4 unequal longitudinal divisions which finally separates them completely from the surrounding water, turning them to internal cells without any direct contact to the ambient environment (Fig. 3.5). At the end of this stage, the branchlet node consists of 2 central cells, surrounded by 6–8 lateral cells.

These lateral cells are essentially similar to the apical cell, but the number of divisions they can perform is limited. So, instead of a new axis, a branchlet of limited growth will be formed by these branchlet apical cells.

The branchlet apical cells now develop in a similar way to the axis apical cell. A series of transverse divisions result in a filament of equal cells each of which, except for the terminal one, undergo exactly one unequal transverse division (Fig. 3.6). After

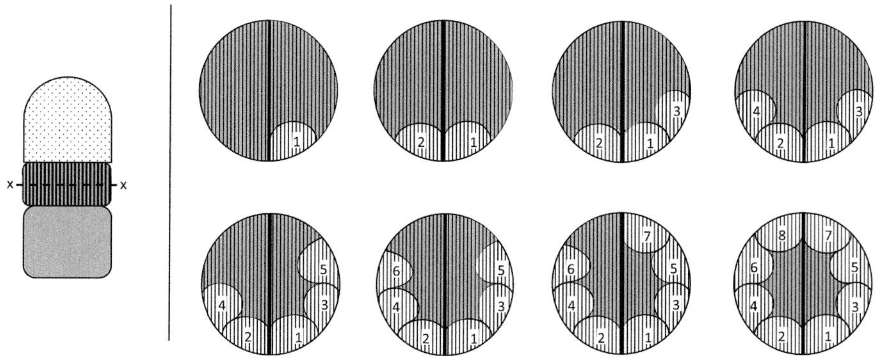

Fig. 3.5 Differentiation sequence of axis nodes. Axis nodal cell in xx-cross view (right panels): after first the longitudinal division of the nodal cell into a pair of central nodal cells, laterals (numbered 1–8) are cut successively until the central nodal cells are completely surrounded by them. The pattern of formation of laterals is alternating between the right and left central nodal cell. © Anja Holzhausen

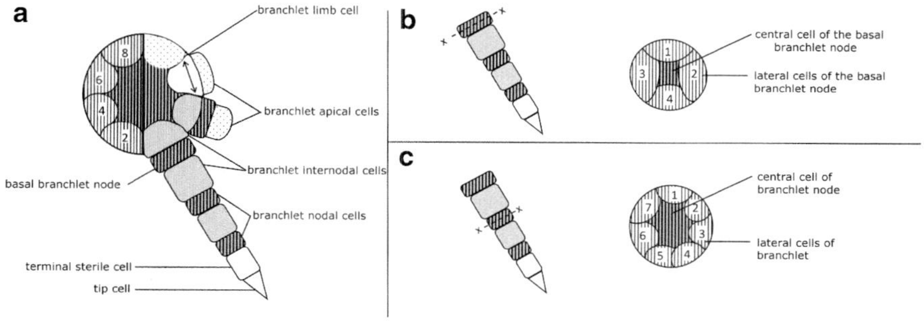

Fig. 3.6 Branchlet formation. **a** Stages of branchlet formation (only shown for the right nodal central cells) clockwise from top: lateral not yet differentiated, still resting as apical cell; first limb cell cut; limb cell differentiated into an internodal and nodal cell; fully developed branchlet, apical cell stalled divisions, becoming the tip cell, below them undifferentiated, sterile cell(s); **b** Cross section of a basal branchlet node; **c** Cross section of a non-basal branchlet node. © Anja Holzhausen

this, the branchlet consists of an alternating pattern of branchlet internodal cells, unable to perform further divisions, and branchlet nodal cells. The nodal cells give rise to the development of branchlet nodes and can further differentiate. The apical part of a branchlet in most charophyte species does not develop further, so beneath the apical cell a series of 2–5 sterile cells remain. These protonemal branchlets are always ecorticate.

3.5 Cortication

Some members of the genus *Chara* develop a cortex on the axis and, except for some species, also on some of the branchlet segments. In the development of axis cortication, the lowermost branchlet nodal cell (basal node) is the starting point for the development of axis cortication. In contrast to the axis nodes, branchlet nodes directly cut off laterals by unequal longitudinal segmentation (Figs. 3.7 and 3.8). The previously described equal division of the initial nodal cell that is exhibited by the axis nodes has not been observed at branchlet nodes. After dividing to produce four lateral cells, the remaining central cell of the branchlet node ceases further development and becomes inactive. Cortex cells can be formed above and below the axis nodes. The lateral cell that faces towards the apex of the axis is the starting point for the formation of the cortex lobe of the upper internode; the lateral cell that faces the rhizoids becomes the cortex lobe of the internode beneath the axis node. The laterals to the left and the right will be dealt with in Sect. 3.6.

The upper and the lower laterals are different with respect to their further development. Prior to the transformation into a cortex apical cell, the upper lateral performs a longitudinal division. One of these cells remains omnipotent and may develop later into rhizoids, lateral axes or "bare-footed branches" (originally called "nacktfüssige

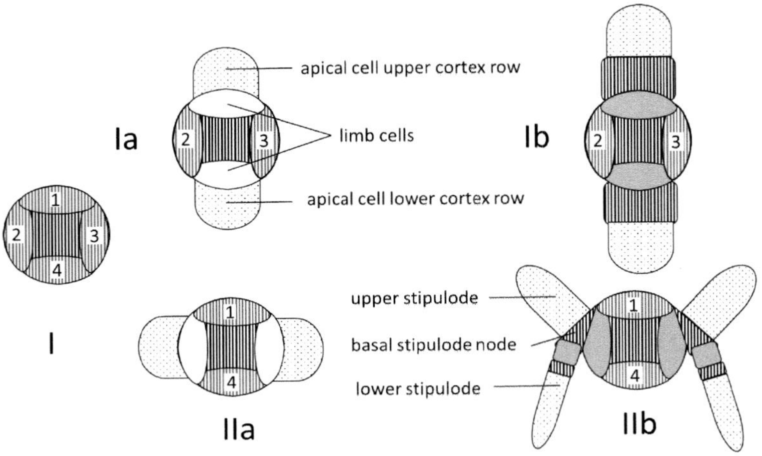

Fig. 3.7 Formation of axis cortex and stipulodes. Four laterals are dissected from the basal branchlet node (I, see also fig. 3.5b) leaving a central basal branchlet node in the centre. Axis cortication will be initiated by laterals 1 and 4 whereas laterals 2 and 3 will develop to stipulodes. Laterals 1 and 4 convert to apical cells and cut off limb cells, (Ia), developing to main axis cortical nodes and internodes (Ib). An exception from this is the first (*Chara*) or the first two (*Nitella*) branchlet basal node(s) where lateral 1 becomes the apical cell for branches, which are (secondary) axes of unlimited growth. Consequently there is one primary cortex row missing in the upward direction, compared with the number of downward rows. Laterals 2 and 3 also convert to apical cells (IIa), developing the upper stipulodes (IIb) by leaving a limb cell behind, which further differentiates into an internodal and nodal cell. According to Kuczewski (1906) the nodal cell below the upper stipulode cuts off laterals, which again convert to apical cells growing to the lower stipulodes (IIb). Please note that these drawings are purely schematic ones, detailed illustrations of natural shape and arrangements can be found in Kuczewski (1906). © Hendrik Schubert (redrawn and amended from Schubert et al. (2016) on a previous version prepared by Anja Holzhausen)

Zweige" by Kuczewski, 1906). This cell can be compared with the first lateral of the protonemal branchlet node, which gives rise to the axis. The second cell becomes a common cortex apical cell like the lower lateral cell of the basal branchlet node. These cortex apical cells now follow the same pattern—first building a filament by equal transverse division, then each of those cells dividing unequally once, resulting in an alternating pattern of cortex internodal and cortex nodal cells. Cortex internodal cells, as in all internodal cells, elongate, but do not differentiate further. In common with all internodal cells, they become polynucleate. The cortex nodal cells start a series of unequal divisions, giving rise to two "peripheral cells" to the left and the right of the cortex nodal cell in line with the cortex plane and one (to several) perpendicular to the cortex plane. The peripheral cells, which can be compared with the laterals of the shoot and branchlet nodes, sometimes remain so small that they cannot be detected without a microscope. In such a case a haplostichous cortex will be formed. In the case where they become elongate halfway along the cortex internodal cell, an alternating pattern of peripheral cells from the left and the right neighbouring cortex

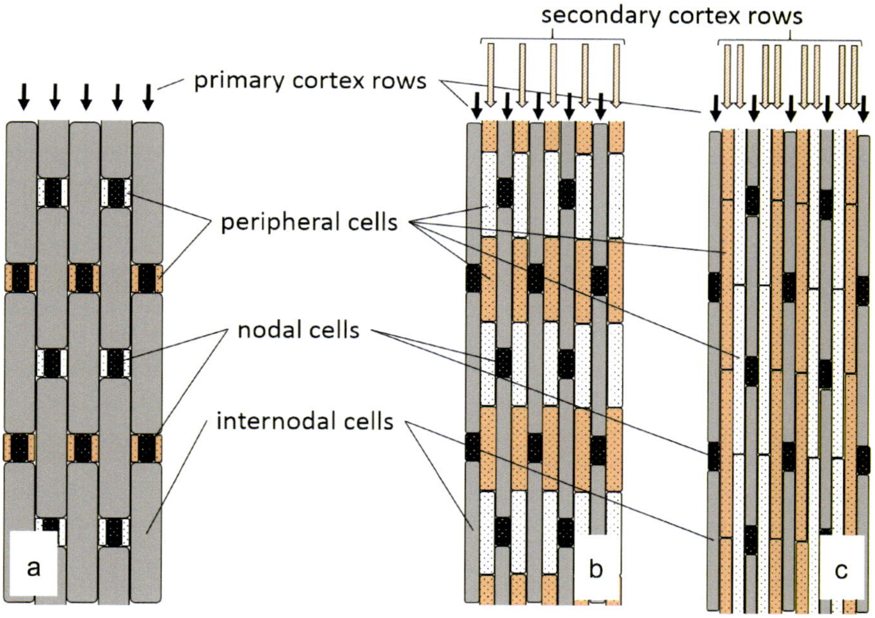

Fig. 3.8 Cortication. **a** Haplostichous—peripheral cells are inconspicuous and too small for the formation of a cortex row, consequently the cortex consists of primary rows only; **b** Diplostichous—primary rows are supplemented by secondary rows of peripheral cells, originating from the left and right nodal cells in an alternating pattern; **c** Triplostichous—each primary cortex row develops a complete row of peripheral cells to the left and the right. © Hendrik Schubert (redrawn and amended from Schubert et al. (2016) on a previous version prepared by Anja Holzhausen)

filaments forms together a single "secondary cortex row"—resulting in a diplostichous cortex. If the peripheral cells fully develop along the full length of an internode, a triplostichous cortex will be the result. Secondary cortex rows never bear any spine-cells, because the nodal cells from which spine-cells originate are absent in secondary cortical rows. Besides these peripheral cells, the cortex node laterals that are perpendicular to the cortex plane can form spine-cells. How far these laterals can grow is species specific and not directly linked to the growth of the peripheral cells. In *C. canescens* for example, the peripheral cells stay tiny, whereas the perpendicular cells develop to long spine-cells. On the other hand triplostichous species, e.g., *Chara globularis*, do not develop visible spine-cells from the perpendicular cells, but do develop secondary cortex rows along the entire length of the internode.

The genera *Lamprothamnium, Nitella, Nitellopsis, Sphaerochara* and *Tolypella* do not possess a cortex.

3.6 Stipulodes

Stipulodes are formed from the left and right lateral of the basal branchlet node. Each of these two laterals divide once, resulting in a very short filament consisting of an apical cell and a still omnipotent one beneath (Fig. 3.7). The apical cell elongates to form the upper stipulode and ceases further growth or development. The lower cell of the filament divides at least once, in the way all nodal cells usually do, producing one or several laterals which then elongate to form the lower stipulodes. The upper and lower stipulodes thus differ in the sequence of development.

It is not completely clear whether or not the typical unequal division of the filament cell occurs prior to the cutting off of the stipulode laterals. Kuczewski (1906) speaks of a direct transformation in his text, but clearly shows the existence of an internodal cell in his drawings and the respective legend, so he might just have been unaware of the general pattern observed.

The genera *Tolypella*, *Sphaerochara* and *Nitellopsis*, unlike *Lychnothamnus*, *Lamprothamnium* and most *Chara* species, do not form stipulodes. For *Nitella* the conditions are more complicated. Ernst (1904) investigated the formation of the short branches of the heteroclemous species *Nitella hyalina* and could prove that they are equivalent to stipulodes.[4] Accordingly, he calls them "stipular leaves". The recently described species *Nitella fuiana* (Chen et al., 2018) illustrates the principal ability in the genus for the development of stipulodes in homoeoclemous species as well.[5]

3.7 Lateral Axes or Branches

Lateral axes or "side shoots" develop from the cell[6] located between the basal branchlet internode and the central cell pair of the axis node. In the genus *Chara*, it is usually only at the base of the first branchlet a lateral axis can form, whereas for *Nitella*, cells at the base of the the first and the second branchlet are capable of

[4] This raises the question of whether or not this is just a homology or, alternatively, the short branches of the heteroclemic *Nitella* species should be regarded as stipulodes. Oltmanns (1974) goes one step further and suggests that the short branches may be a homology to the cortex, illustrating the necessity of the meticulous (but also fatiguing the reader) cell formulas of Ernst's school. Unfortunately, the work of Ernst (1904), though often cited, is rarely reflected in content (for example, by Fritsch, 1977). The question cannot be satisfyingly answered yet because there is neither a clear definition of the term "stipulode" nor an agreement about whether cell line descent or the function should be given priority in the designation.

[5] In some heteroclemous species of *Nitella* only a single-celled secondary branchlet forms in some whorls (e.g. *Nitella lhotzkyi*), which can be regarded as equivalent to a stipulode. Consequently the question whether or not *Nitella fuiana* is a homoeoclemous species or rather belongs to the heteroclemous section Decandollea should be investigated.

[6] The relationships are somewhat intricate—usually this cell should be an internodal cell that cannot perform any further divisions. Kuczewski (1906) calls this cell neutral only as "x"; from which the primary internodal cell of a branchlet develops. An interpretation of this pattern, deviating from the normal course of development, was not found in the literature.

developing lateral axes. The difference is marginal, because in *Nitella* it is still at the first branchlet, but both the left and the right cell of the central cell pair of the axis node, can give rise to the development of lateral axes. For this purpose, a so-called lateral axis origin cell ("Achselspross Urzelle") is divided off at the apical part of the cell, still being connected to one of the central cells of the axis node. The remaining residual cell, an internodal cell, has a connection to the shoot internode only on the side on which it sprouts. The lateral axis origin cell now takes the role of an apical cell in further development. The development of a side shoot is thus the same as the development of the main shoot. The only difference is the development at the basal node of the lateral axis. The basal nodal cell divides, as in the axis nodes, initially by longitudinal division, in which the resulting central cells subsequently constrict lateral cells. However, according to Kuczewski (1906), in contrast to the conditions at the shoot node, only four lateral cells are formed at the basal node of a lateral axis. In addition, Kuczewski (1906) found in several *Chara* species that two of these lateral cells usually become so-called accessory axes. These accessory axes usually remain very small and can only be seen under a microscope. The other two lateral cells of the basal node of a lateral axis also remain capable of dividing. Given the right conditions, they may become rhizoids, shoot-borne protonemata[7] or so-called "nacktfüssige Zweige" ("bare-footed branches").[8]

3.8 Branchlet Cortication

Branchlet cortication, if any, originates from the branchlet nodes. The basal branchlet node does not contribute to branchlet cortication, as it is responsible for shoot cortication instead,[9,10] The branchlet nodal cell divides unequally until it is surrounded by

[7] "Right conditions" in this context means, for example, that rhizoids are formed only when this node is covered by sediment. Axis-borne protonemata and bare-footed branches are only formed when the axis-apical cell has been damaged or removed (Kuczewski, 1906). The admissibility of the use of the term "protonema" in this context is unclear, as long as we remain uncertain about the location of reduction division (meiosis).

[8] Bare-footed branches are lateral axes without cortication that are formed at the lowest internode. Their basal node does not develop cortex apical cells at all. The next node develops normal cortication at its apical part whereas the downward cortex rows are spreading and not in contact to the lower internode (Pringsheim, 1863).

[9] An exception, according to Witt (1906), is *Chara tomentosa*. In this species, the basal branchlet node also contributes to branchlet cortication so *Chara tomentosa* is the only known charophyte species with ascending and descending cortex cells at the lowermost branchlet section. This unique character, which is easily observed with the use of a hand lens, allows for reliable determination of even poorly developed specimens.

[10] This also seems to prevent the formation of gametangia at the basal branchlet node, at least in species with corticated basal branchlet segments. However, Yuanjie et al. (1991) described a species (*Chara pseudocanescens*) with corticated basal segments and gametangia at the basal node. Because at least the upper lateral of the basal node can perform longitudinal division, this may

a complete ring of lateral cells.[11] After these divisions are completed, the branchlet nodal cell is in contact with the upper and lower internode and the laterals only (Fig. 3.6c). The lateral cells now transform into apical cells as usual. The first limb cell originating from the apical cell differentiates into an internodal cell, remaining in the branchlet node, and a branchlet cortex nodal cell. After this, the apical cell converts into a bract-cell sensu Wood (1965) also called "leaflet" by, e.g. Pickett-Heaps (1975) (Fig. 3.9). The branchlet cortex nodal cell now produces three lateral cells, which, together with the nodal cell, cover the ascending and descending branchlet internodes. However, there are often deviations from this general scheme. Kuczewski (1906) observed for instance the formation of four instead of three lateral cells, as well as cases in which only two lateral cells take over the ascending cortication, whereas the descending cortex is formed solely by the nodal cell. Since the number of lateral cells varies widely, branchlet cortication often is irregular.[12] The terminal segment in all species of *Chara* is always ecorticated. At the uppermost branchlet node, directly adjacent to the ecorticated terminal segment, the limb cell does not divide into an internodal and nodal cell. Here the limb cell, after a longitudinal division, grows directly into two descending cortical rows.

3.9 Terminology "Leaflets"

Unfortunately, the terms used to describe the morphology of the *Characeae* have not been used consistently. Even older authors frequently used a large number of synonyms in the same publication, so reference to them is not always helpful. In many cases, however, it is possible to supplement the various types of nodal and internodal cells with the corresponding prefixes. The term "leaflets" is different because of confusion that has resulted in the interpretation of species descriptions. The term "branchlet" is used here for lateral organs of limited growth originating from axis nodes. An axis node itself is characterised by the equal division of its nodal cell prior the cutting off of lateral cells. So in an axis node two nodal central cells are always present. The term "branchlet" is thus synonymous with the "Blättchen" (e.g. Braun, 1853), the "primäre Blättchen" (e.g. Kuczewski, 1906), the "branchlets" or "laterals of limited growth" (e.g. Wood, 1965) or the "leaves" and "laterals" of, e.g., Fritsch (1977).

Bract-cells, on the other hand, originate exclusively from branchlet nodes. They represent a transformation of an apical cell, which descends directly from a branchlet

---

explain gametangia formation. In any case, gametangia formation of this species as well as for Gymnobasalia should be studied in more detail.

[11] The number of lateral cells formed by the branchlet nodal cell is apparently not fixed. Kuczewski (1906) observed between 5 and 7 lateral cells on one and the same branchlet.

[12] This is just a statement, no decent investigation about the use of this character (haplo-, diplo-, triplostichous branchlet cortication) for species delineation is known to the authors. However, the character is mentioned at least in Wood (1965) and should be investigated for European species in more detail.

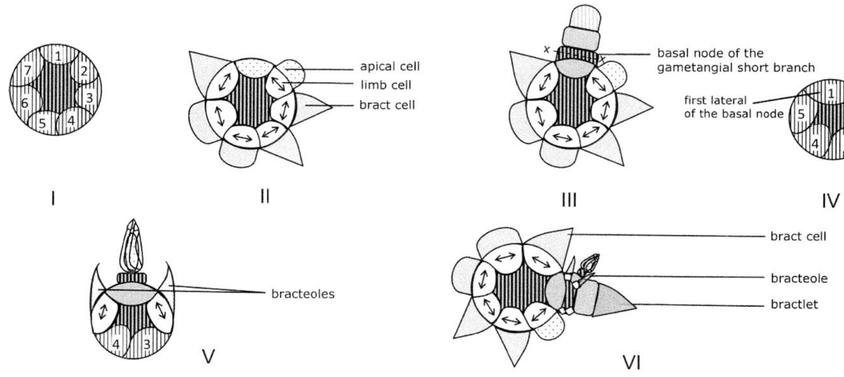

Fig. 3.9 Bract-cells, bracteoles and bractlets. Bract cells originate directly from laterals dissected by the branchlet nodal cell (I). These laterals become apical cells, releasing a limb cell beneath and growing to either to bract cells (II) or developing further to gametangial short branches (III). At the basal node of the gametangial short branch, laterals are dissected (IV), the first lateral may become an apical cell itself, giving rise to oogonium formation (see Fig. 3.11); the two neighbouring laterals developing into bracteols (V). In monoecious female plants, the apical cell of the gametangial short branch ceases growth after releasing another limb cell and becomes a bractlet, replacing the antheridium (VI). Please note that this simplified scheme species-specific is modified largely. In species with corticated branches the limb cell beneath the bract cell differentiates into an internodal cell and a nodal cell, the latter giving rise to the formation of branchlet cortication. © Anja Holzhausen

node lateral cell. Prior to the transformation to a bract-cell, whether or not this branchlet node lateral cell produces a limb cell is decisive for the existence of branchlet cortication. The term "bract-cell" or "-leaflet" is thus used in the sense of Wood (1965) and is synonymous with "Foliola" (Braun, 1853), "Blättchen" (Braun, 1853), "Blättchen 2. Ordnung" (Kuczewski, 1906) and "leaflets" (Pickett-Heaps, 1975). Bract cells are, as far as they have been examined, homologous to gametangial short branches.

Bracteoles arise from the antheridial nodal cell and thus are homologous to the oogonia. The term, introduced by Braun (1853), is also used by Wood (1965), and is synonymous with "Vorblättchen" (Braun, 1853).

Bractlets are the most problematic term because they are ontogenetically identical to bract-cells. Like bract-cells, they originate from a transformations of an apical cell, which itself descends directly from a branchlet node lateral cell. The difference between a bract-cell and a bractlet is for a bractlet, transformation of the apical cell takes place after a series of divisions, leading to the formation of a gametangial short branch. Bractlets replace the antheridium and therefore occur only when antheridia and oogonia are not formed on the same gametangial short branch. Being homologous to the antheridium they are a good character to distinguish between di- and monoecious species (Wille, 1897) and can be used for identification of sejoined gametangial arrangements as well.

3.10 Gametangial Short Branches

Gametangial short branches are formed by apical cells which descend directly from a branchlet node lateral cell. The apical cell performs a series of divisions, resulting in a row of two internodal cells and 2 nodal cells. The apical cell, called the vertex cell, divides by two consecutive longitudinal divisions into four quadrants, which divide by subsequent transverse division into eight octants.[13] Each of these octant cells now divides into 3 cells: (1) the shield cell (scute), which closes the antheridium to the outside; (2) the handle cell (manubrium) connecting the shield cell to the next cell in line which is (3) the capitulum. From the inner cell, the capitulum, threads of spermatogenic cells arise. The spermatogenic cells each convert to spermatozoids which are released by separation of the shield cells.[14]

During this development, the internodal cell beneath the antheridium extends into the space inside the antheridium within the lower four shield cells. It is often named "flask cell" (Flaschenzelle) or "foot cell" (Fusszelle) by several authors (Fig. 3.10). The arrangement of gametangia is different in different genera. In ecorticated genera as, e.g. *Nitella* spp., lateral cells grow into branchlet segments. Their gametangial short branches are thus always in the middle of these branching points on top of the underlying branchlet node (Oltmanns, 1922).

The oogonium develops from the lower nodal cell of the gametangial short branch. This nodal cell divides into 3–4 lateral cells in the same way as described above for the branchlet node formation. One[15] of these lateral cells is the starting point for the formation of the oogonium (oogonium mother cell), the others becoming bracteoles. Therefore bracteoles are not homologous to bract-cells, they originate from a later cell division and thus belong to a "higher order" than bract-cells.[16] The oogonium mother cell behaves like an apical cell. It divides into a limb cell that divides into an internodal cell and a nodal cell as usual. The internodal cell becomes the stem cell of the oogonium and, like all internodal cells, does not further differentiate. However, in some species it can elongate and produce so-called "stalked" or "stipitate" oogonia. The nodal cell cuts off several lateral cells in quick succession, which convert into

---

[13] In tetrascutate species like, e.g. *Chara zeylanica*, this division step is omitted, so that the antheridium has only four shield cells (Griffin and Proctor 1964).

[14] At first glance, the complicated structure of antheridiae was (and is) interpreted by many authors as a "complex structure" and compared to the propagating organs of higher plants. Goebel (1930), on the other hand, came to the astonishing conclusion in his investigations on the ontogenesis of antheridiae: he showed that antheridia represent only a higher order of branching. Following his interpretation, the spermatogenic threads are just rows of unicellular gametangia, as typical of green algae. Fritsch (1977) recognised this interpretation as "ingenious", unfortunately the insight of both authors is rarely acknowledged irrespective of being often cited.

[15] For the genus *Chara* it is always the upward-oriented lateral cell which gives rise to oogonium formation. For *Nitella, Sphaerochara* and *Tolypella*, often several laterals develop oogonia whereas for *Lychnothamnus* there is no conclusive analysis available so far (see Migula, 1889–1897).

[16] Branchlets would have been "primary laterals" according to this accounting. Primary cortex rows of both axes and branchlets, together with the upper stipulodes and side-shoots would be "secondary laterals". Secondary cortex rows, lower stipulodes, bract-cells and antheridiae would respresent "tertiary laterals" whereas oogonia and bracteoles are "quaternary laterals" (see Kuczewski, 1906).

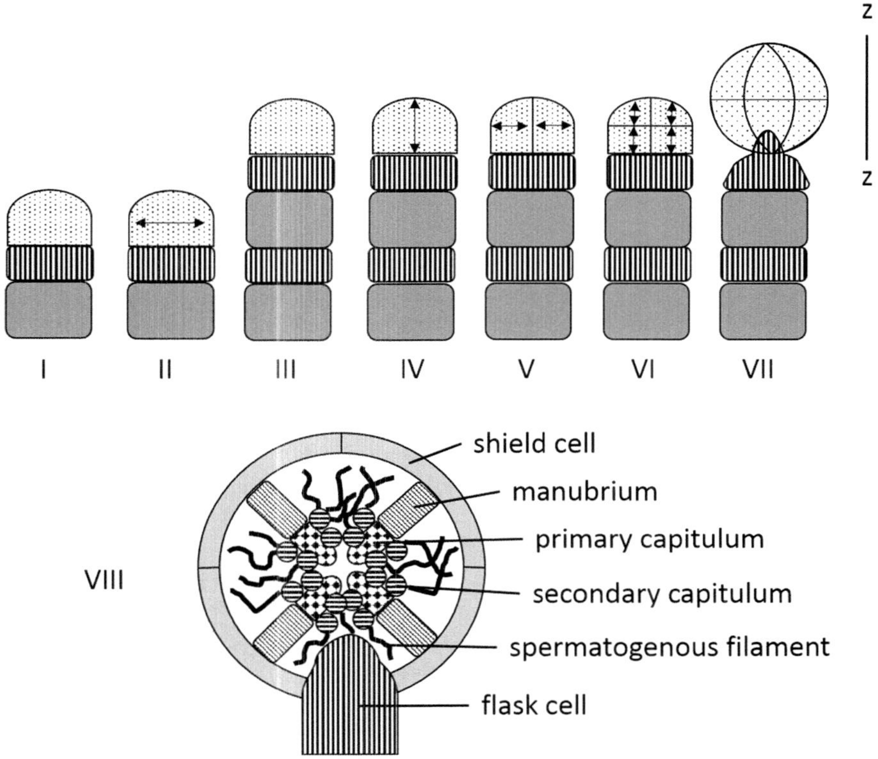

Fig. 3.10 Antheridium formation. (I) The first lateral of a branchlet node developed into an gametangial short-branch apical cell, releasing a limb cell which divides into an internodal and nodal cell. This first nodal cell of the gametangial short branch in monoecious species is the origin of oogoniae (see Fig. 3.11) whereas in dioecious, ecorticated species the apical cell of the gametangial short branch directly develops into an antheridium after releasing the first limb cell (see Picket-Heaps, 1975: Fig. 7.26). The example shown here, typical for monoecious species, releases another limb cell (II, III) differentiating into an internodal cell and a nodal cell, which becomes at a later stage the flask cell (VIII). The apical cell itself now divides into octants by a series of three periclinal divisions, turning the division plane 90° after each step (IV–VII). All of these octant cells are "apical cells" still, further differentiating into an outer shield cell, a manubrium and a primary capitulum, surrounded by secondary capitula, from which the spermatogenic filaments develop (VIII, showing a cross-cut in z-z direction). These filaments are unique structures for charophytes, consisting of a row of identical cells, each of them developing an antherozoid which gets released by breaking up the shield cells, probably by excessive growth of the flask cell. © Hendrik Schubert (redrawn and amended from Schubert et al. (2016) on a previous version prepared by Anja Holzhausen)

the envelope apical cell that gives rise to the formation of the spiral cells, enveloping the oocyte.[17] Whether or not the general pattern of division of charophyte organs

[17] The spiral cells wrap the oocyte clockwise in recent *Characeae* as well as in *Eochara*; Paleotaxa from the Devonian as e.g. *Trochiliscus* have anticlockwise spiral cells instead (Graham and Wilcox, 2000, see also Chap. 8).

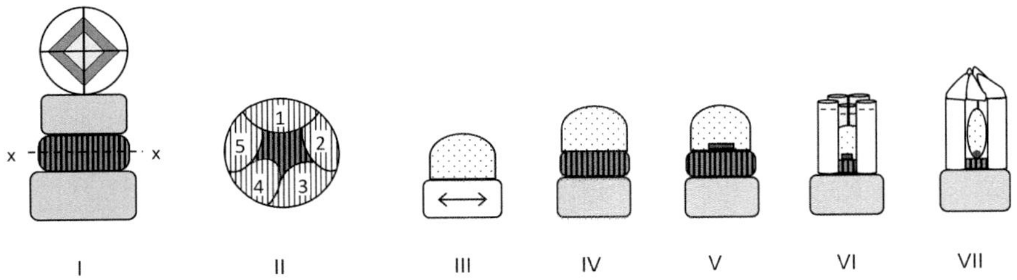

Fig. 3.11 Oogonium formation. At the first node of a gametangial short branch (I) five laterals are dissected from the nodal cell (II, xx-view). The first lateral becomes an apical cell and cuts off a limb cell (III). This limb cell follows the general pattern by dividing unequally into an internodal and a nodal cell (IV). The internodal cell develops into the stalk cell of the oogonium and may elongate, resulting in a "stalked oogonium". The apical cell itself now cuts off the turning cell (V), simultaneously the nodal cell dissects 5 laterals. These laterals all become apical cells, performing one (genus *Chara*, shown here, as well as the genera *Lamprothamnium*, *Nitellopsis* and *Lychnothamnus*) or two (genera *Nitella*, *Sphaerochara* and *Tolypella*) divisions, differentiating into the large spiral cells and with coronula cells on top, together enveloping the oocyte and the turning cell(s) (VI–VII). © Anja Holzhausen

is followed here again is not fully clear. The apical cell cuts off a limb cell, which becomes the spiral cell. However, whether or not the difference between the genera *Nitella, Sphaerochara* and *Tolypella* with two rows of coronula cells and *Chara* with just one can be interpreted by completing/lacking limb-cell differentiation has not yet been sufficiently investigated.

The apical cell which gave rise to the formation of the oogonium mother cell is now largely surrounded by the enveloping cells. It cuts off a "Wendungszelle",[18] which can be considered an internodal cell and this develops into the oocyte. The oocyte itself begins to swell as a result of the storage of reserve substances. The growth of the oocyte is largely synchronous with that of the enveloping cells, leaving only a small channel open between the coronula cells, through which fertilisation occurs (Fig. 3.11). After fertilisation, the oocyte forms the impermeable oospore wall (complex multi-layered oospore wall sensu Leitch (1989), for details as well as formation of gyrogonites see Chap. 4, this volume) and the nucleus migrates from the basal part of the oospore to the apical area, from which, after a period of rest, germination begins.

The relative position of oogonium and antheridium is consistent among genera for monoecious species. In *Chara*, the antheridium always is situated adaxially (i.e. between the branchlet and axis) whereas the oogonium arises from the cell on the branchlet side. In *Lamprothamnium* the order is reversed. In *Nitella* paired oogonia can occur, analogous to lateral axis formation.

---

[18] In *Nitella*, three instead of one such "Wendungszellen" are observed. Initially they surround the oocyte on three sides, but get forced into a basal position by the onset of rapid growth of the oocyte. The term "Wendungszelle" (meaning turning cell) has been coined by A. Braun (1853) because of theoretical reflections about their function to redirect growth towards the oocyte—an interpretation already rejected by Goebel (1902).

In dioecious species the location of gametangia is unchanged; the antheridium in female plants is replaced by a bractlet that supplements the two bracteoles. The presence or absence of this bractlet between the two bracteoles allows the possibility of a clear distinction between specimens of monoecious species (e.g., *Chara globularis*) having just lost antheridia and dioecious species (e.g., *Chara connivens*).

3.11 Rhizoids and Bulbils

The rhizoids of charophytes are unicellular threads, which serve both for anchoring as well as nutrient uptake. The tip cell of each rhizoid has gravity receptors that provide positive gravitropy (Braun and Limbach, 2006; Kiss and Staehelin, 1993). Although a large number of studies on the development of rhizoids have been published, the question of whether the division mode of the rhizoid cells corresponds to that of the shoot is still not satisfactorily answered. A detailed description of the development of rhizoids and bulbils can be found in Giesenhagen (1896); in the following account, only the essential steps are presented.

The rhizoid cells have a typical S-shaped cell wall at their contact surface. A small nodal cell is located at the foot end of this S-shaped contact zone. This nodal cell via two successive divisions can form a cell plate giving rise to the formation of secondary rhizoids. In *Chara aspera* some of these cells can be transformed into rhizoidal bulbils, storage organs for vegetative reproduction. These organs are often described as "single-celled bulbils", in reality they consist of a large vegetative storage cell with a complex of nodal cells on top, which can grow into a protonema, developing as described above. In *Chara fragifera*, the formation of bulbils follows the same scheme in principle, but the number of starch-filled storage cells is larger. Whether or not *C. fragifera* is able to develop shoot bulbils in addition cannot be answered with certainty—the specimen reproduced by Giesenhagen (1896: Fig. 19) shows an attachment of a secondary rhizoid which may originate from a shoot node.

Shoot bulbils develop on subterranean shoot nodes; they correspond to extremely shortened whorls of branchlets, in which the internodal cells especially are converted to storage compartments filled with starch and consequently extremely swollen. In contrast to rhizoidal bulbils, shoot bulbils "germinate" (i.e. produce new axes) directly without formation of a protonema.

References

Bower, F. O. (1923). *Botany of the living plant* (p. 545). Macmillan.

Braun, A. (1853). Über die Richtungsverhältnisse der Saftströme in den Zellen der Characeen. Bericht über die zur Bekanntmachung geeigneten Verhandlungen der Königl. *Preuss. Akademie der Wissenschaften zu Berlin, 1853*, 45–76.

Braun, M., & Limbach, C. (2006). Rhizoids and protonemata of characeen algae: Model cells for research on polarized growth and plant gravity sensing. *Protoplasma, 229*, 133–142.

Chadefaud, M. (1979). L'évolution de la structure cladomienne chel les charales et les céramiales. *Rev. Algol. NS, 14*, 253–273.
Chen, K., Qing, R., Fu, H., Ren, L., Liu, K., Cao, S., & Lan, L. (2018). Identification of a new characteristic species of Nitella (Charophyceae) from southwestern China based on morphology and molecular phylogeny. *Plant Systematics and Evolution, 304*, 123–132.
de Bary, A. (1875). Zur Keimungsgeschichte der Charen. *Botanische Zeitung, 33*, 377–420.
de Winton, M. D., Casanova, M. T., & Clayton, J. S. (2004). Charophyte germination and establishment under low irradiance. *Aquatic Botany, 79*, 175–187.
Debski, B. (1897). Beobachtungen über Kerntheilung bei *Chara fragilis*. *Jahrbücher Für Wissenschaftliche Botanik, 30*, 227–248.
Ernst, A. (1918). *Bastardisierung als Ursache der Apogamie im Pflanzenreich*. Gustav Fischer.
Ernst, A. (1901). Ueber Pseudo-Hermaphroditismus und andere Missbildungen der Oogonien von *Nitella syncarpa* (Thuill.) Kützing. *Flora, 88*, 1–36.
Ernst, A. (1904). Die Stipularblätter von *Nitella hyalina* (DC). *Vierteljahrsschr. Naturf. Ges. Zürich, 49*, 64–114.
Fritsch, F. E. (1977). *The structure and reproduction of the Algae* (pp. 447–469). Cambridge University Press.
Giesenhagen, K. (1896). Untersuchungen über die Characeen. *Flora, 82*, 381–433.
Goebel, K. (1902). Morphologische und biologische Bemerkungen. 11. Ueber Homologien in der Entwickelung männlicher und weiblicher Geschlechtsorgane. *Flora, 90*, 279–333.
Goebel, K. (1930). Die Deutung der Characeen-Antheridien. Ein Versuch. *Flora, 124*, 491–498.
Goetz, G. (1899). Ueber die Entwicklung der Eiknospe bei den Characeen. *Botanische Zeitung*, 57, 1–13.
Gonçalves da Cunha, A. (1942). *Contribuïção para o estudo dos Caröfitos portugueses*. Dissertation Univ. Lisboa, Lisboa.
Gonçalves da Cunha, A. (1936). Sur la position de la meiose dans le cycle evolutif de Characées. *Bulletin de la Société Portugaise des Sciences Naturelles, 12*, 175–188.
Graham, L.E. & Wilcox, L.W., 2000. Green algae V—Charophyceans. In *Algae* (pp. 494–543). Prentice Hall.
Griffin, D. G., & Proctor, V. W. (1964). A population study of *Chara zeylanica* in Texas, Oklahoma and New Mexico. *American Journal of Botany, 51*, 120–124.
Guerlesquin, M., & Noor, N. N. (1982). La meiose et sa place dans le cycle des Charophytes: Thèses en présence et incertitudes. *Cryptogamie, Algologie, 3*, 323–328.
Haig, D. (2010). What do we know about charophyte (Streptophyta) life cycles? *Journal of Phycology, 46*, 860–867.
Kamieński, F. (1876). Kilka spostrzezen nad rozwojem ramienicowatych (Characeae). *Rozprawy i sprawozdania z posiedzen Wydzialu Matematyczno-Przyrodniczego Akademii Umiejetnosci, 3*, 210–240.
Karling, J. S. (1926). Nuclear and cell division in *Nitella* and *Chara*. *Bulletin of the Torrey Botanical Club, 53*, 319–379.
Kiss, J. Z., & Staehelin, L. A. (1993). Structural polarity in the *Chara* rhizoid: A reevaluation. *American Journal of Botany, 80*, 273–282.
Kuczewski, O. (1906). Morphologische und biologische Untersuchungen an *Chara delicatula* f. *bulbillifera* A. Braun. *Arbeiten aus dem Laboratorium für Allgemeine Botanik und Pflanzenphysiologie der Universität Zürich, 5*, 1–51.
Leitch, A. R. (1989). Formation and ultrastructure of a complex, multilayered wall around the oospore of *Chara* and *Lamprothamnium* (Characeae). *European Journal of Phycology, 24*, 229–236.
Lindenbein, W. (1927). Beitrag zur Cytologie der Charales. *Planta, 4*, 436–466.
Mendes, E. J. (1946). Mitosis in the spermatogenous threads of *Chara vulgaris* L. var. *longibracteata* Kütz. *Portugaliae Acta Biologica, 1*, 251–264.
Migula, W. (1889–1897). Die Characeen Deutschlands, Oesterreichs und der Schweiz. In L. Rabenhorst, L. (Hrsg.) *Kryptogamenflora* (Band V, 2. Auflage). Eduard Kummer.

Oehlkers, F. (1916). Beitrag zur Kenntnis der Kernteilungen bei den Characeen. *Berichte der Deutschen Botanischen Gesellschaft, 34*, 223–227.
Oltmanns, F. (1922). *Morphologie und Biologie der Algen. 2. Auflage. Erster Band Chrysophyceae – Chlorophyceae* (pp. 437–459). Gustav Fischer.
Oltmanns, F. (1974). *Morphologie und Biologie der Algen. Erster Band Chrysophyceae – Chlorophyceae.* A. Asher & Co.
Pickett-Heaps, J. D. (1975). *Green algae.* Sinauer.
Pringsheim, N. (1863). Ueber die Vorkeime und die nacktfüssigen Zweige der Charen. *Jahrbücher für Wissenschaftliche Botanik, 3*, 294–324.
Schubert, H., Holzhausen, A., & Nowak, P. (2016). Individualentwicklung der Characeen. In Arbeitsgruppe Characeen Deutschlands (Eds.) *Armleuchteralgen - Die Characeen Deutschlands* (pp. 57–78). Springer.
Shen, E. Y. F. (1967a). Microspectrophotometric analysis of nuclear DNA in *Chara zeylanica. Journal of Cell Biology, 35*, 377–384.
Shen, E. Y. F. (1967b). The amount of nuclear DNA in *Chara zeylanica* measured by microspectrophotometry. *Taiwania, 13*, 111–114.
Tuttle, A. H. (1924). The reproductive cycle of the Characeae. *Science, 60*, 412–413.
Tuttle, A. H. (1926). The location of the reduction division in a Charophyta. *University of California Publications in Botany, 13*, 227–235.
Walther, E. (1929). Entwicklungsgeschichtliche und zytologische Untersuchungen an einigen Nitellen. *Arch. Julius Klaus-Stiftung, 4*, 23–121.
Wille, N. (1897). Characeae. In A. Engler (Hrsg) *Die natürlichen Pflanzenfamilien* (S 161–175). Engelmann.
Witt, A. (1906). *Beiträge zur Kenntnis von Chara ceratophylla Wallroth und Chara crinita* Wallroth. Dissertation Univ. Zürich, Zürich.
Wood, R. D. (1965). *A Revison of the Characeae. First Part: Monograph of the Characeae.* Cramer.
Wood, R. D. (1954). The Characeae of Lake Mattamuskeet, North Carolina. *Journal of Elisha Mitchell Scientific Society, 70*, 60–74.
Yuanjie, L., Shulian, X., & Lichuan, Q. (1991). 山西轮藻属新植物. *Acta Hydrobiologica Sinica, 15*, 350–355 [Some new taxa of Chara from Shanxi, Chinese, with English abstract].

Chapter 4
Extant Oospores

Anja Holzhausen, Nick Stewart, Carl D. Sayer, Ben Goldsmith, and Michelle T. Casanova

4.1 Introduction

Oospores represent the diploid zygote of the charophyte plant. They are the product of fertilisation of the egg-cell within the oogonium, on female or monoecious thalli, by sperm-cells from male or monoecious thalli. The complex oospore wall is derived from both the developing zygote and the maternally produced helical enveloping cells (Leitch, 1989). This has often been referred to as the 'membrane', however, the complexity of the structure and its formation from both the developing zygote and maternal cells makes the word 'wall' more appropriate (e.g. Leitch, 1989). Within the multilayered oospore wall, along with the fertilised nucleus, are membrane-bound starch grains and lipids within the cytoplasm. Oospores range from c. 180–1100 μm long, and can be round, oval, ovate, obovate, elliptical, to almost rectangular in profile, and are either terete, orbicular or somewhat triangular in cross-section.

A. Holzhausen (✉)
Martin Luther University Halle Wittenberg, Institute of Agricultural and Nutritional Sciences, Betty-Heimann-Str. 5, 06120 Halle (Saale), Germany
e-mail: anja.holzhausen@landw.uni-halle.de

N. Stewart
4 Hexton Road, Glastonbury BA6 8HL, United Kingdom
e-mail: nfstewart@freeuk.com

C. D. Sayer · B. Goldsmith
Environmental Change Research Centre, Department of Geography, University College London, Gower Street, London WC1E 6BT, UK
e-mail: c.sayer@ucl.ac.uk

B. Goldsmith
e-mail: ben@goldsmithecology.co.uk

M. T. Casanova
273 Casanova Rd, Westmere, Victoria 3351, Australia
e-mail: m.casanova@nhm.ac.uk

H. Schubert et al. (eds.), *Charophytes of Europe*,
https://doi.org/10.1007/978-3-031-31898-6_4

Oospore morphology varies among genera and species in the *Characeae* and can be taxonomically important. Oospores are easily distinguished from other biological propagules on the basis of their rigid multilayered cell wall and the pattern of helical sutures on the surface. No other spores or seeds are the same.

The multilayered oospore wall is constructed from a variety of materials, including sporopollenin (the polymerised compound of carotenoids and carotenoid acids (Cronk, 2009) that was initially discovered in angiosperm pollen grains; Wallace et al., 2011). It is usually pigmented, appearing variously coloured, often yellow, brown or black. Some species have a white calcified covering on the oospore which is referred to as a gyrogonite (see Chap. 8 Palaeontology). The gyrogonite is deposited by the enveloping helical cells and retains the helical sutures characteristic of the underlying oospore. The structure and composition of the oospore make it particularly resistant to degradation and decay, so that oospores can remain viable for long periods of time, and withstand a variety of environmental conditions (e.g., drought and desiccation, low-oxygen environments, external chemicals). Similarly, the zygote cell, with its stores of starch, can be physiologically dormant for long periods of time.

Decay-resistance and dormancy are key features of oospores that render them useful for a number of reasons. Firstly, because oospores and gyrogonites are frequently found in the sediment of lakes, ponds and lagoons there is considerable potential for their use in palaeoecological studies (García, 1994). Preservation of oospores is mostly restricted to Late Quaternary sediments, but remarkably, calcified gyrogonites afford evidence of former charophyte existence as far back as the Late Silurian period (~425 million years before present; Soulié-Märsche and García, 2015). Further, due to their ability to remain dormant and hence disperse through time, as well as through space, oospores are important in the field of resurrection ecology as well as in practical wetland restoration.

Thus, the fact that oospores can be identified and distinguished on the basis of their morphology means they can be used for a range of purposes:

- To identify past charophyte communities to inform ecological restoration
- To inform resurrection ecology studies including the potential to return locally, regionally, and nationally extinct species
- To reconstruct former environmental conditions in studies of environmental change
- To identify the occurrence of extant charophytes at sites even when living charophyte plants are absent (e.g., in drought, or out-of-season).

Whereas charophyte oospores are clearly a highly useful resource, there are limitations to their widespread application in environmental and ecological studies, and these need to be overcome. Firstly, an agreed taxonomic species concept is needed for charophytes in general. The characterisation of the morphology of oospores of a 'species' relies on the standardisation of what a 'species' is (see Chap. 5). Comprehensive sampling throughout a 'species' range is also required for determination of the limits of variation in oospore morphology. Better documentation of the diagnostic features of oospores, especially in relation to type material, and finally, the provision of metadata about how the morphology has been examined: oospore origin, storage, treatment for imaging, cleaning techniques and imaging techniques need to be described. Ideally, a set of standard methods will be developed for reliable comparison of the oospores within and among extant, sub-fossil and fossil species.

4.2 The History

Oospores have been recognised as useful for distinguishing charophyte species since the very beginning of taxonomic studies (e.g., Braun, 1849; Allen, 1888; Holtz, 1891; Migula, 1889–1897; Groves and Bullock-Webster, 1920, 1924; Corillion, 1957; Wood, 1965). Although Braun (1849) was the earliest author to mention oospore characteristics as a tool for species delimitation, Braun and Nordstedt (1882) were among the first to use descriptions of oospores as keying characters and Nordstedt later produced a comprehensive description of many of them (Nordstedt, 1889a, 1889b). Allen (1888) provided a determination key for *Nitella* species using oospore characteristics (along with morphological characters) and Groves and Bullock-Webster (1920, 1924) made the first oospore key for European species in the UK and Ireland. Even in these early days there were opposing views among experts about the reliability of characters of extant oospore species, mirroring the disagreements about what constituted a 'species' a 'variety' or a 'form'. This has continued into the twentieth century: Allen (1937) stated that 'two of the most promising fields' were the study of oospores and fossils, whereas Zaneveld (1940) was critical of this. There was contradictory data and discrepancy among authors, perhaps due to differences in the collection, storage, treatment and preparation of oospores for examination, if not the differences in equipment or skill possessed by the experts. Later studies by Williams (1959) and Cấceres (1975) used stable oospore features to distinguish among species where thallus morphology was not informative. A step forward was the work by Haas (1994) in which a determination key for oospores of 22 species in Europe was provided, along with the allocation to species-groups for species with inconspicuous or overlapping characters.

4.3 Gross Morphological Features: Size and Shape

The most easily seen and measured characteristics of oospores are their colour, length, width, number of striae, width of the fossae, the presence of ridges, flanges, claws, a cage or crests (Fig. 4.1), and the presence of a gyrogonite. The works of Braun and Nordstedt (1882), Migula (1889–1897), Groves and Bullock-Webster (1920, 1924), Corillion (1957), Wood (1959, 1965), Krause (1997) and Bailly and Schaefer (2010) provide these data for many taxa. Within Europe, oospore descriptions were provided for most species from Central Europe (Kützing, 1849; Holtz, 1891; Migula,

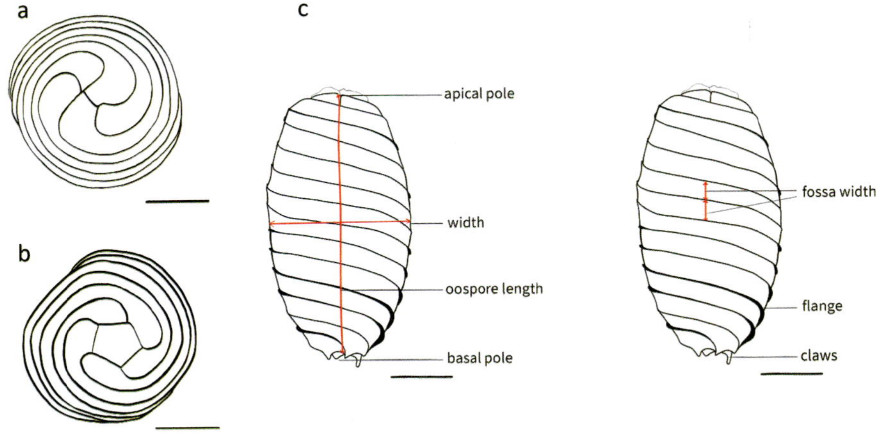

Fig. 4.1 Gross morphological characteristics of *Chara* oospores in **a** Apical, **b** Basal and **c** Lateral view (*Chara baltica*, Darss Zingster Bodden chain). Drawings: Anja Holzhausen

1889–1897; Holtz, 1899; Pankow, 1990; Haas, 1994; Krause, 1997; Schubert and Blindow, 2003), the Netherlands (Raam et al., 1998), France (Corillion, 1957; Bailly and Schaefer, 2010), the United Kingdom and Ireland (Groves and Bullock-Webster, 1920, 1924; Allen, 1950), Spain (Cirujano et al., 2008), Norway (Langangen, 2007), Sweden (Wallman, 1853), Czech Republic (Caisová and Gąbka, 2009). Differences in the distribution of species means that not all species have been treated, or treated equally. There are numerous descriptions of cosmopolitan species such as *Chara aspera*, *C. contraria*, *C. hispida*, *Nitella flexilis* and *N. mucronata,* but few or no descriptions for *C. mucronata*, *C. corfuensis*, *C. hellenica*, *Sphaerochara canadensis (=Tolypella canadensis)* or *Lamprothamnium succinctum*. Additional data is required for these, and other species.

Similar terms are used to describe or define oospore morphology in most of the literature, however, different authors can vary somewhat in their definitions: until recently there has been little consistency (Soulié-Märsche and García, 2015). Whether the size is measured dry or wet, how the striae are counted (at the edge, or across the centre), what is called a flange and what is called a ridge have all been done differently. In this review we will use the definitions provided by Soulié-Märsche and García (2015) based on the definitions by Horn af Rantzien (1956) and Horn af Rantzien (1959), as far as possible. A glossary of oospore terms is also provided (see Part III Glossary and Index).

The gross measurements most often used for oospore description are:

- Oospore length: this is the distance between the top and bottom of the oospore, or the longest polar axis (LPA). It is measured from the top of the oospore excluding flanges and crests, to the base, excluding claws and cage.

- Oospore width: this is the measurement of the widest part of the oospore, perpendicular to the length, the longest equatorial distance (LED). It is measured on the body of the oospore excluding flanges and ridges.
- The ratio of these length and width can be calculated providing an isopolarity index (ISI = LPA/LED × 100). This gives an idea of sphericity or elongation. The actual shape of the oospore is best determined from glossaries of leaf-shapes.
- The number of striae (stripes, ridges or 'spirals') visible in lateral view. These can most easily be counted by the number of crossed striae at the median axis, or at the edges of the oospore. The count on one side can differ from the other side (e.g., 12 or 13) because of the view of a partial ridge at the top or bottom. Given the variation within populations of oospores, this inaccuracy is not likely to be important. In some species the helical ridges are extended down the sides of the basal cell of the oogonium into 'claws'. Sometimes the space between those claws is filled in with a thinner wall (secondary wall layers on the basal cell), forming a 'cage' at the base of the oospore.
- The expression of the striae can be different both within and between species. The expression of ridges or thickened ridges (pachygyra) can be measured and the presence of wing-like extensions to the ridges (flanges) can also be measured.
- Fossa width: This is the space between striae, ridges or flanges and should be measured at the middle of the oospore. It varies among the upper, middle and lower part of a single oospore, thus an average fossa width can be calculated.
- At times the flanges can be extended into a 'crest' at the apex of the oospore, particularly in the genus *Nitella*. The development and height of this crest can be described and measured.
- The impression of the basal cell(s) can be a single pentagonal cell (tribe *Chareae*) or 1–3 square, rectangle, triangle or rounded cells (tribe *Nitelleae*).
- The gyrogonite and its characteristics (see Soulié-Märsche and García, 2015). Where gyrogonites are present and not removed for processing they can possess important taxonomic features.

For taxonomic work and comparison of species it is important that measurements are made on several oospores. A minimum of ten oospores, and usually many more (depending on the statistical variance in features) should be examined, and these should come from different parts of the specimen or population. This is not always possible, especially with the description of new species, or from old type material where it would be negligent to obtain many oospores.

The gross morphology of oospores of the same species can vary within a population, and also among populations (Casanova, 1997; Holzhausen et al., 2015; Brzozowski et al., 2019; Holzhausen et al., 2022, 2023), which is why most treatments express ranges of length and width and number of striae for a species (see Sect. 4.7, "Sources of Variability").

4.4 Gross Morphological Features: Colour

Although oospore colour is considered a consistent character, especially in distinguishing some closely related species (e.g. *C. vulgaris* and *C. contraria*), the determination of colour has been uniquely subjective. Descriptions are given under different terms such as 'ochre', 'wine-red', 'brown', 'chestnut' and 'black', but the interpretation of colour has depended on the light used for examination, pre-treatments and the experiences of the author. For example, a nineteenth century description of a colour 'the same as walnut-ink' might be outside the experience of a researcher working in the twenty-first century. It is recommended to use an official colour chart (e.g. based on the German RAL system) for a more objective description of colour in oospores. This system has the advantage of being well-established and used in industry and commerce (since 1979) as well as able to be enlarged to accommodate additional colours if needed. The colours are described by name, number and RGB ratios and printed guides are readily available. Complete charts can be found online (e.g. RAL Deutsches Institut für Gütesicherung und Kennzeichnung e.V.). The colour of oospores can be best determined by using dim lamp radiation. High external or internal lamp radiation and determination only from images could create misinterpretation.

4.5 Fine Features: Ornamentation

The fine detail of oospore walls was obvious to researchers from the middle of the nineteenth century. Nordstedt (1889b) documented some of the variation in the genus *Nitella*, and the presence of lumps and cavities on the oospore walls of some species of *Chara* were also noted. The details of the ornamentation or patterns present on the oospore wall, as seen via light microscopy, were sometimes equivocal, and can still be misinterpreted today. However, in many cases the detailed ornamentation could be seen and the structures illustrated (Fig. 4.2).

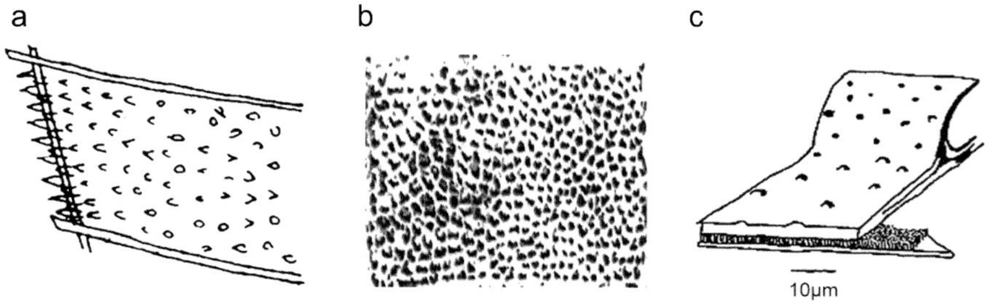

Fig. 4.2 Pen and pencil drawn images of oospores by **a** Braun and Nordstedt (1882); **b** Groves and Bullock-Webster (1920); **c** Williams (1959)

During the middle of the twentieth century new ways of imaging and photographing oospores using the electron microscope provided the impetus for new interpretation and understanding of the value of fine features of the oospore wall. Early use of SEM by John and Moore (1987), Soulié-Märsche (1989) and John et al. (1990) provided the stimulus to further investigation by Casanova (1991), Casanova (1997) and Sakayama and co-workers (Sakayama et al., 2005; Sakayama, 2008; Kato et al., 2010) among others. The reporting of comparative nucleotide analysis by Sakayama and co-workers provided an additional body of evidence that oospore features were diagnostic at the species level in the genus *Nitella*. These studies also revealed a new suite of characters for assessment. John and Moore (1987) discovered the presence of a 'ribbon' attached to the striae of some species of *Chara* and Sakayama et al. (2005) revealed the taxonomic significance of the inner-wall layers. Although scanning electron microscopy can provide unequivocal images of the three-dimensional ornamental features of the oospore wall, once the features are recognised, they can also be distinguished via light microscopy (Fig. 4.3).

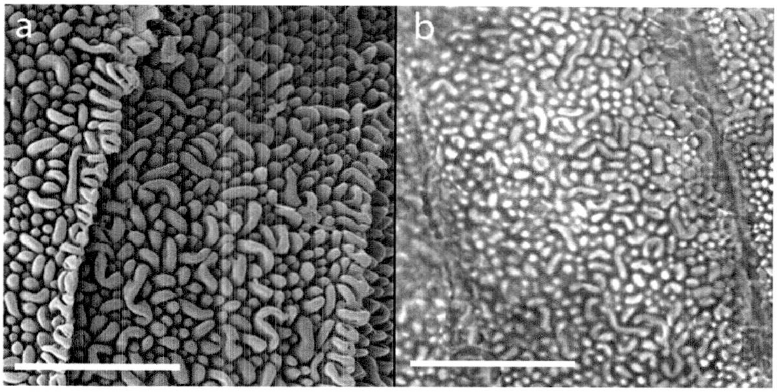

Fig. 4.3 Ornamentation on the oospore wall viewed via **a** SEM and **b** Light microscopy. *Nitella paludigena* Casanova type material from Brawley's Swamp, South Australia, M. T. Casanova p. 810. Scale = 20 μm. Images: M. T. Casanova

The current 'gold-standard' of description of oospore ornamentation requires the following six images (Fig. 4.4):

- The overall view of the oospore in side-view, imaged with a scale bar of c. 100 μm (50–300 μm) so that the length, width, shape and other gross features can be measured and described (Fig. 4.4a).
- A view of the base of the oospore so that the impression of the basal cell(s) (basal plate) can be measured and described (Fig. 4.4b).
- A view of the width of the fossa, including a ridge or flange, usually with the ridges perpendicular to the long axis of the image in landscape view (Fig. 4.4c).
- A view of the fine-detail (construction) of the fossa wall with a scale bar of 2–10 μm (Fig. 4.4d).

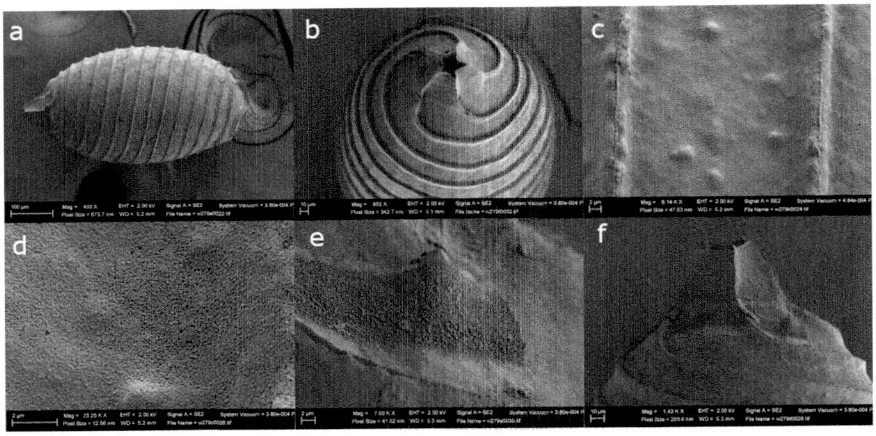

Fig. 4.4 The range of oospore images considered 'gold-standard' for description of oospore features. Specimen: *Chara connivens*, Slapton Ley, UK, Aug. 1899 D. G. Claridge [BM000640161]. **a** Oospore in side view, **b** Basal view of oospore, **c** View of fossa and ridges, **d** Detailed view of fossa, **e** View of ribbon, **f** View of basal cage. Images: M. T. Casanova

- A view of the flanges or ribbon structures associated with the striae or ridges with a scale bar of 2–10 μm (Fig. 4.4e).
- A view of the basal claws and cage complex where present (Fig. 4.4f).

It is not always possible or appropriate to obtain all these images: e.g., there may be no flanges or ribbon, or there may be too few oospores to mount so that the base is visible. It is also likely that other features of taxonomic importance will be discovered in time.

It is important in the provision of images for taxonomic studies to deposit the specimen in a recognised herbarium, or at least designate its taxon and origin. This allows specimens to be checked in the future. In ideal situations the measurements should be made on a variety of oospores from the one plant or population (depending on the purpose of the study). However, this is not always possible or desirable. It can be expensive to use scanning electron microscopes, and time for preparation and viewing is always limited. Oospores from type material should be examined only with permission, and judiciously. One of the advantages of working with oospores is that they are not destroyed by imaging, but can be stored with the specimen and examined again if necessary.

The description of oospore ornamentation has usually followed the terminology applied to the sculpturing on pollen grains (Faegri and Iversen, 1975). In the glossary of this book we provide a pictorial glossary for the standardisation of descriptive terms.

4.6 Taxonomic Significance of Oospore Features

Oospores of different tribes in family *Characeae* can be easily distinguished with the aid of a light microscope, on the basis of their gross morphology. Sometimes their features are visible by eye. Oospores in the tribe *Chareae* have an elongate ellipsoidal or oval shape, whereas in the tribe *Nitelleae* they are either globose or rounded. At the genus level within the *Chareae*, oospores of *Chara* have various shapes in side-view, *Lamprothamnium* (usually elliptical), *Lychnothamnus* (obovate) and *Nitellopsis* (ovate) are all somewhat similar, but differ in relation to their gyrogonites where they are present (Soulié-Märsche and García, 2015; Sinkevičiene, 2019). Within the *Nitelleae,* oospores of *Sphaerochara* and *Tolypella* are terete, and those of *Nitella* are, in cross-section, either ellipsoidal (Haas, 1994; Krause, 1997) or triangular (Casanova pers. obs.). Species in *Sphaerochara* also possess gyrogonites (Soulié-Märsche and García, 2015).

The use of oospores for differentiation of species has been mainly restricted to the genera *Nitella* and *Tolypella*, where a considerable amount of information on the morphometry and ornamentation of the oospore wall exists (Ray et al., 2001; Urbaniak and Blaženčić, 2012; Mouronval et al., 2015). In contrast to species in the *Nitelleae*, the number of detailed descriptions for species of *Chara*, especially for the oospore wall ornamentation, is very small (e.g. John et al., 1990; Casanova, 1997). SEM studies on the oospores of species of *Chara* in Europe exist only for Poland (Boszke and Bociąg, 2008; Urbaniak, 2009, 2011), Sweden (Ray et al., 2001; Mandal et al., 2008), the Balkan region (Mandal et al., 2002; Urbaniak and Blaženčić, 2012), Germany (Holzhausen et al., 2015, 2023) and specimens from the UK and Ireland that are deposited in the Natural History Museum London (Casanova and Wilbraham, 2019). Oospore ornamentation could eventually be important to differentiate even closely related species (e.g. *Chara vulgaris* and *C. contraria* ; *C. globularis* and *C. aspera*, Fig. 4.5).

4.7 Sources of Variability

When the data for oospores of European species is examined, there is great variability within accepted species, not only the gross morphology, but also in the description of ornamentation. This variability can be allocated to different causes. These include: storage conditions, the number of oospores used in the investigation, the equipment used for measurement, the kind of pre-treatment oospores have been exposed to and subjective interpretation of qualitative characters. Furthermore, there can be genetic variation among populations, environmental influences or differences in the developmental stage of the oospore (Casanova, 1997). There are few studies that reveal the particular causes of variation so further investigation is needed.

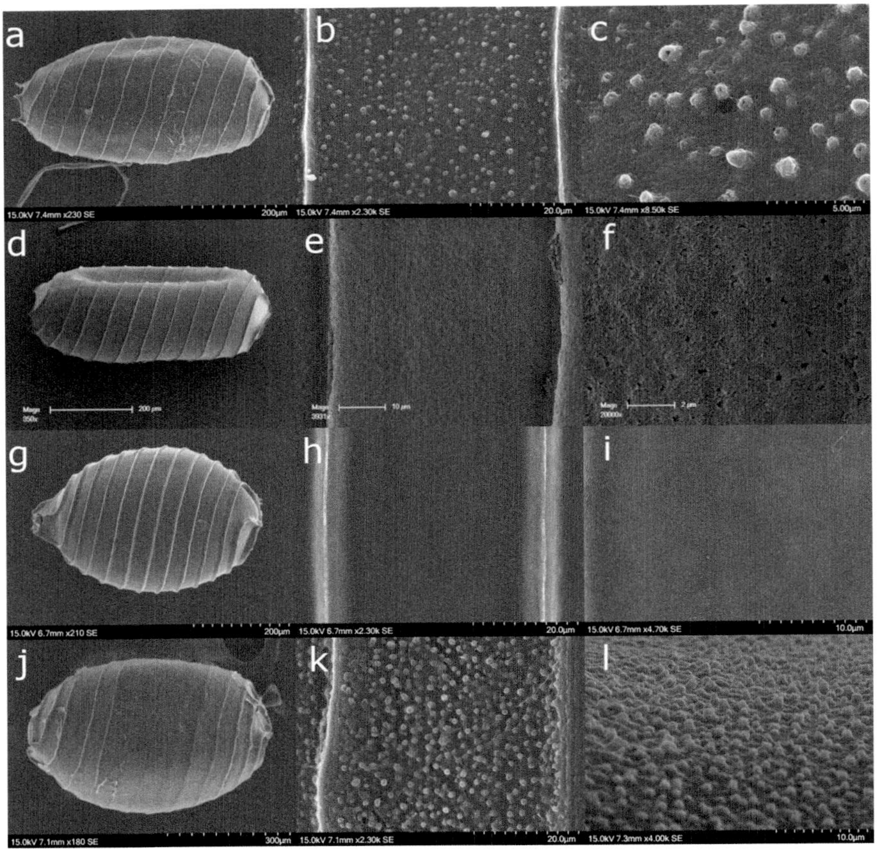

Fig. 4.5 Comparative morphology of four common European species. **a–c** *Chara vulgaris* Col? Neuf Moulin, France, Jul. 1841 [PC0577005]; **d–f** *Chara contraria*, A. Braun, Carlesruhe, viii-1829 [GFW]; **g–i** *Chara aspera* A. Bennett, Hickling Broad, UK, 21-vii-1883 [BM013739063]; **j–l** *Chara globularis*, E. G. Camus 756, Flue de Moreau, 11-vi-1862, France [PC0576748]. In the first column the oospore is depicted with its base to the left and apex to the right (scale = 200 μm), in the second column the fossa wall and ridges are depicted (scale = 20 μm), in the third row the fine detail of the oospore wall is depicted (scale = 5 μm). Images: M. T. Casanova

Storage

It is not known if storage affects the morphology of oospores, certainly plants collected and stored dry as herbarium specimens retain the same oospore features as those collected recently (Fig. 4.6). However, storage conditions are very important for the viability of oospores. In contrast to the existence of numerous studies and species-specific protocols for 'correct' storage conditions for terrestrial and aquatic vascular plant seeds (Hay and Muir, 2000; Probert and Hay, 2000), little is known about the consequences of storage conditions on charophyte oospores. In general, the

Fig. 4.6 Oospores of *Chara benthamii* A. Braun obtained from pressed specimens in herbaria in UK and Australia. **a** Singapore Botanic Gardens Lake, 1896, T. B. Blow; **b** Singapore Botanic Gardens Lake, 1908, Anon.; **c** Singapore Botanic Gardens Lake, 1922, Holtum 8389; **d** Moreton Bay, c. 1849, C. Stuart; **e** Perth, 1960, R. D. Wood; **f** Kalgon Creek, Newman, 2004, Mickle. Images: M. T. Casanova

storage of oospores is only considered in relation to their future use. For oospores used in germination experiments or cytological studies it is crucial that viability is maintained, whereas the storage conditions of oospores used for morphological and taxonomic studies must ensure that their characteristics are not changed during storage.

Proctor (1967) found that storage of dry oospores at low temperatures was best for preserving oospore viability. In his study there was little difference in the viability of oospores stored dry at 3 °C in the dark (refrigerator) compared to those stored dry at − 20 °C in the dark (freezer) over the evaluation period of 4 years. In comparison, storage of oospores either wet at 3 °C, or at 24 °C in dim light resulted in decreasing viability over time. Casanova and Brock (1996) stored oospores over 18 months in both light and dark, room temperature, cold (refrigerator) and freezing, with both intermittently wet and dry pre-treatments. They found that different species responded differently to storage in their rate (%) and speed of germination, and also that none of the oospores survived freezing either wet or dry. Similar results were found for oospores from Germany and Austria; oospores of *Chara canescens* that were stored for six years under wet conditions in the dark germinated at high rates, whereas oospores stored in sediments (in a freezer) did not germinate (Holzhausen pers. obs.). Oospores from sediments that were exposed to temperatures over 30°C in sediment also did not germinate (Holzhausen pers. obs.). In some studies researchers have found very low germination rates of oospores freshly collected from plants, and clearly some sort of innate dormancy exists for many species (Forsberg, 1965; Shen, 1966; Proctor, 1967; Holzhausen, 2016). The influence of storage conditions on germination success, rate and timing is a subject for future investigation.

Moisture

Quantitative measurements of oospores can vary in relation to moisture. Blume et al. (2009) found that oospores can differ in gross morphology depending on whether they are wet or dry. Dry oospores were on average 5% shorter and 27% narrower than wet oospores. In studies where measurements are given, whether they were measured wet or dry should always be stated to allow comparability between studies. This lack of information has contributed to the variability in reported measurements in the past. Oospores that have been stored in alcohol should be considered 'dry' because of the dehydrating effects of the alcohol. Oospores that are measured with the aid of scanning electron microscopy will always be 'dry' due to the effects of the dehydration required for the vacuum in SEM.

Pretreatments

Unless oospores are 'freely released' from the plant, by the natural separation and decomposition of the enveloping cells, the oospore will retain the remnants of those cells on its external surface. Sometimes a gyrogonite is also present. These structures need to be removed for examination and measurement of the oospore. Enveloping cells can be either chemically or mechanically removed. John and Moore (1987) and John et al. (1990) trialled various chemical methods for removing the cells (acids, bleach, acetolysis) and found that the expression of oospore ornamentation did not differ significantly among methods (David John, pers. com.). Casanova (1991), Casanova (1997) used acetolysis at first, and afterwards a method for cleaning diatom frustules involving a strong detergent and sonication (Crawford et al., 2001). Holzhausen et al. (2015) used low concentrated acetic acid with or without sonication. The images obtained using all these methods provide the same results in relation to oospore wall morphology. Sterilisation pre-treatment, such as soaking in sodium hypochlorite can gradually alter the pigmentation and characteristics of the oospores (Fig. 4.7). Although different authors could induce germination after sterilisation of oospores with e.g., sodium hypochlorite or soaps (e.g. Sakayama, 2008; Sato et al., 2014; Holzhausen et al., 2022), a decline in viability could not be excluded and should be investigated further.

Equipment and observer

The microscope used and the methods of measuring can introduce error or variance into measurement of oospore or gyrogonite characteristics (Soulié-Märsche and Joseph, 1991). Different lighting can influence the perception of colour, light microscopy and scanning electron microscopy can allow different interpretation of oospore features.

The 3-dimensional structure of oospore ornamentation can be misinterpreted with light microscopy unless the user is very skilled (Leitch et al., 1990). The colour is only visible with dim light, but some of the fine features of the oospore wall are only visible with the aid of scanning electron microscopy (SEM). With SEM, preparation of the stubs and thickness of the coating can modify the view of the oospore. Some published images show 'charging' (parallel light or dark patches across the image

Fig. 4.7 Comparison of untreated and treated oospores of *Chara vulgaris*. Oospores were treated with Eau de Javel (2.6 g.L^{-1} sodium hypochlorite). The upper row **a–c** Shows the untreated oospores, the lower row the treated ones: **d** 10s, **e** 90s, **f** 300s. Images: Jasmina Gebert and Anja Holzhausen

due to insufficient attachment of the oospore to the stub, insufficient coating, or very thin and pointed structures), and sometimes the glue that is used to attach the oospore to the stub can spread and cover important features (Fig. 4.8).

With the use of uniform and standardised treatment and terms, as is suggested here for colour, errors could be minimised. The expression of flanges, for example, can be measured directly. Moreover, in cases of non-graphical representation, readers can compare numerical values. Categorisation based on numerical values could be used reduce the sometimes confusing use of adjectives to describe the expression of striae (i.e. wide, narrow), flanges (i.e. long, short), or membrane ornamentation e.g. grains (dense or sparse). Soulié-Märsche and Joseph (1991) illustrate the errors that can be introduced via insufficient sampling and different means of measurement.

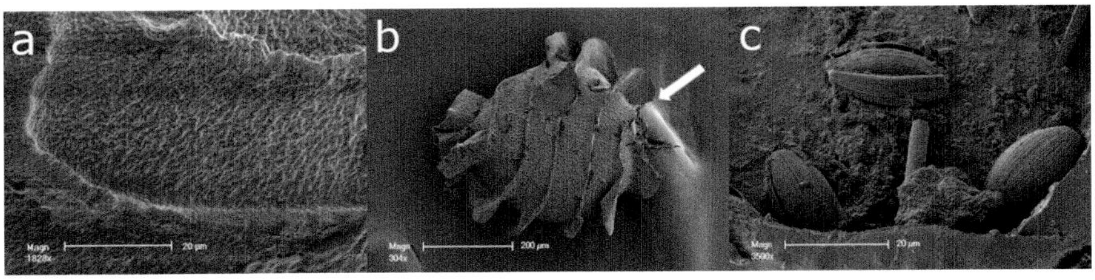

Fig. 4.8 An example of 'charging' on an oospore image. **a** The actual surface is blurred or highlighted by uneven electron reflection due to preparation or mounting. **b** An example where the sticky carbon tab has adhered to the oospore (arrowed). **c** An example of debris, including diatom cells on the surface of an oospore. Images: M. T. Casanova

Maturity and ripening

During development the oogonium and oospore undergo complex post-fertilisation processes involving the accumulation of starch and deposition of multiple layers in the oospore wall. A detailed description is given in Chap. 3 'Ontogenesis'. For some *Nitella* species there is a gradual colour change in the oospore wall from transparent, through yellow and orange to dark brown (Casanova, 1991). In the genus *Chara* the spiral cells are initially orange, and the colour gradually fades as the oospore wall is deposited. The oospore wall becomes visible through the enveloping cells as it changes from transparent to brown or black (Fig. 4.9). The final stages in the development of the oogonium can be the deposition of flanges, fine-features on the oospore wall, or the outermost wall-layers (Casanova, 1991). In many cases these sources of error can be overcome by choosing the most mature oospore visible on the specimen, or allowing oospores to be freely-released in culture. This is also the reason that, where possible, more than one oospore from each plant and more than one specimen of each species should be examined.

Habitat and geographic distribution

There are studies that report that the number of striae on oospores varies with light availability (Ernst, 1918). In species that occur in temporary wetlands, the

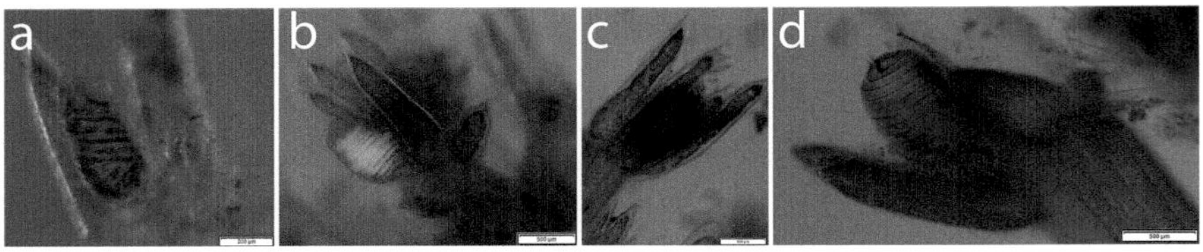

Fig. 4.9 An example of sequential changes in the transparency and colour of the oogonium to oospore formation and maturation of *Chara tomentosa*, Darss Zingster Bodden chain. **a** Early stage of oogonia with orange spiral cells. **b** Early stage of oogonia with partly green coloured spiral cells. **c** Fertilised stage. Spiral cells changed from orange to brown. **d** Oospore with calcification. Photographs: Anja Holzhausen

first oospore is sometimes much larger than oospores produced later in the life-cycle (Casanova, 1993), however it has been hypothesised that species in temporary wetlands have a greater degree of uniformity among oospore sizes and shapes, than those that occur in permanent wetlands (Casanova, 1991), perhaps as a consequence of the dependence on a minimum oospore size for successful germination and establishment in temporary wetlands. Similarly, species with very small oospores (c. 200 μm) have less variation in size than species with very large oospores. (Casanova, 1991). The environmental influence on gross morphology of oospores and gyrogonites has been rarely considered for European species so far. Nevertheless, Sanjuan and Martín-Closas (2015), Sanjuan et al. (2017) and Brzozowski et al. (2019) showed that temperature and light, including depth distribution, affects the gross morphology of *Lychnothamnus barbatus*, *Chara vulgaris* and *Chara globularis* gyrogonites. Additionally, studies on extant oospores of *Chara canescens* and *Chara braunii* have shown that differences between oospore sizes exist between field-collected oospores and oospores that grew under laboratory conditions using varying salinities or light regimes (Holzhausen pers. obs.). These effects have to be considered in further comparative studies.

In many cases differences detected among regionally distinct populations of the same species (Haas, 1994; Ray et al., 2001; Vedder, 2003, 2004; Blume et al., 2009; Urbaniak, 2009; Holzhausen et al., 2015) are not a consequence of taxonomic confusion. Such differences could be the consequence of different growing conditions, but there are few studies on this; or a consequence of genetic variation indicating the existence of short-range or regional evolution, endemicity and cryptic speciation. At present only gross morphological features have been recorded to vary due to environmental controls. The acquisition of additional information on environmental factors when undertaking oospore or edaphic studies could improve the interpretation of oospore differences in gross morphology studies.

4.8 Preparation Methods

Sampling

Examination of oospores is valuable in all taxonomic studies and many ecological ones. When a thallus is collected, if oospores are present, the maturity should be determined. A simple method for fresh collections is to shake the thallus carefully in a bucket of water, then sieve the water (200 – 500 μm sieve for *Chara*; 64–200 μm sieve for *Nitella* and *Tolypella*) to collect oospores that fall from the thallus. If there are no oospores, or they appear too immature for examination, the thallus can be maintained in a simple, window-sill culture to see if oospores will mature or be released. Although some charophyte thalli will survive in any water, the establishment of persistent cultures is often hampered by epiphytic growth of (cyano-) bacteria or fungi and a failure to optimise species-specific conditions for life cycle completion under laboratory conditions. In recent years, an enormous number of studies using various substrates, media compositions, light regimes and culture vessels have been undertaken (e.g. Kuczewski, 1906; Ernst, 1918; Karling, 1924; Forsberg, 1965; Andrews, 1984). Synthetic media as well as water from the collection site at the time of collection can be used. Sometimes it is enough to examine the collecting bag to find oospores released from the thallus. When working with herbarium specimens (dry sheets, or alcohol-preserved) the specimen should be examined for the most mature oospores. Sometimes they will be loose on the sheet or bottom of the jar, but more often they are still attached to the thallus. It is sometimes possible for a loose oospore on a herbarium sheet to belong to a different specimen or species, so care should be taken if these are sampled. Each herbarium will be able to provide guidelines for sampling, but in general the following should be observed. Where there are abundant oospores usually fewer than 5 should be used for imaging. Where there are 8–10 oospores only three should be removed. Where there are five then only two should be used, and where there are three, then a single oospore then all should be used. If there are fewer then all should be left on the plant, and other specimens used, unless there are exceptional circumstances (e.g. the oospore is loose on the sheet and could be lost or misplaced). Notice of removal of oospores should be attached to the specimen, so that future workers will know that samples have been taken, and who to contact for access to the data. The placement and arrangement of oospores on the herbarium specimens should also be noted (e.g. clustered, singular, at the base of the whorl, on the branchlet nodes etc.).

Cleaning

Acetolysis involves the use of boiling concentrated sulphuric acid and acetic anhydride. This removes the organic matter adhering to the oospore very well, but it can also result in the degradation of the oospore. In more recent studies simply soaking the oospores in a strong detergent, then removing the enveloping cells with a set of fine forceps and needles, and then applying sonication for 20–30 s, produced excellent results in many cases (Casanova, 2009, 2013). Freely released oospores can be examined without any particular preparation, but oospores can also have debris and dirt, sometimes epiphytic diatoms and bacteria present on the surface. If necessary, debris can be removed using the former cleaning methods. A very dilute acid (hydrochloric, acetic, citric or nitric acid) can be used to remove gyrogonites and calcium carbonate adhering to the surface of *Chara, Lamprothamnium* and *Sphaerochara* oospores, and can be used in conjunction with manual cleaning or sonication. If the acid is not dilute and the cleaning is not careful, superficial characters can be removed from oospores (especially cages, ribbons, flanges and claws). Oospore walls can sometimes uncoil along the sutures if cleaned too aggressively, or if they are immature.

Handling

Oospores can be remarkably robust, however, their small size creates limitations in handling. Handling can be done with the use of fine forceps or with a pipette. In a mixture of oospores and soil or debris from living charophytes, the easiest way to select individual oospores is to allow them to fall or sediment down in a petri-dish. They can be picked out individually or sucked out with a pipette. Storage in a microcentrifuge tube is ideal because it concentrates the oospores at the base of the tube. Liquid (usually ethanol, unless viability is required or distilled water for maintaining viability) can be introduced into the tube, and the oospores sucked up and delivered to a slide, stub or petri dish for examination. This provides an efficient way of moving the oospores from one cleaning solution to another. When oospores are placed on stubs in a drop of ethanol, with carbon sticky tabs as an adhesive, the ethanol evaporates, and the oospore can be moved into a good position for scanning with the aid of a fine needle. After imaging, the oospore can be removed by applying ethanol, which loosens the oospore from the carbon sticky tab, and it can be picked up with forceps or a pipette and stored for future examination. It is important that the container be sufficiently well-labelled for the specimen to be identified in the future, if needed. Oospores can be stored in ethanol, but if the ethanol evaporates the oospores are not usually damaged. Dry oospores can last on specimens and in storage for centuries (Fig. 4.10). If the viability is to be maintained, oospores should be stored wet or dry. See Sect. 4.7 on storage.

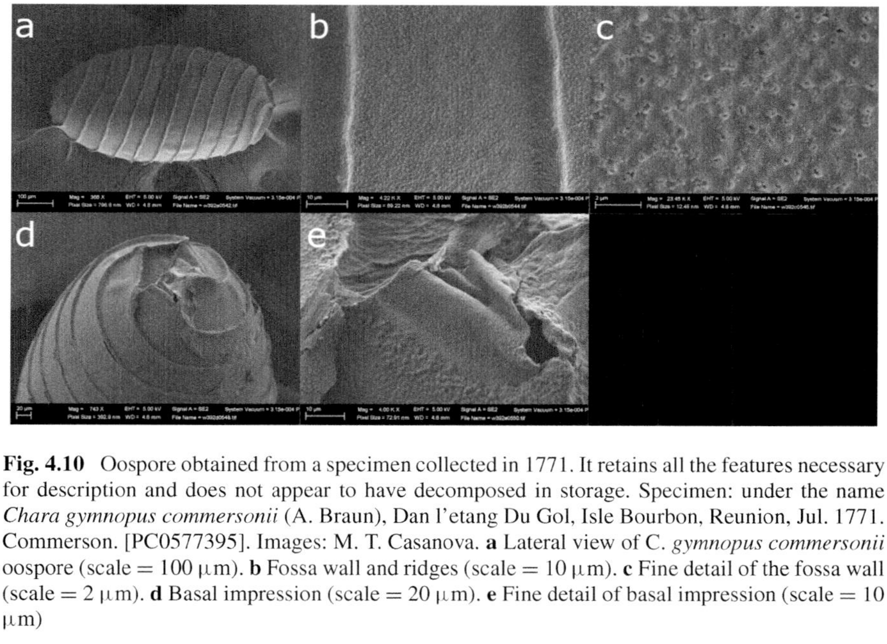

Fig. 4.10 Oospore obtained from a specimen collected in 1771. It retains all the features necessary for description and does not appear to have decomposed in storage. Specimen: under the name *Chara gymnopus commersonii* (A. Braun), Dan l'etang Du Gol, Isle Bourbon, Reunion, Jul. 1771. Commerson. [PC0577395]. Images: M. T. Casanova. **a** Lateral view of C. *gymnopus commersonii* oospore (scale = 100 μm). **b** Fossa wall and ridges (scale = 10 μm). **c** Fine detail of the fossa wall (scale = 2 μm). **d** Basal impression (scale = 20 μm). **e** Fine detail of basal impression (scale = 10 μm)

4.9 Reliability of Characters

Overall shape

The overall shape of oospores is a reliable character for many taxa. It is certainly valuable at the generic level, and often for species, although there are instances where it has been shown to vary (Casanova, 1991 for *Chara australis*). There are many instances where it is constant within a population and over space and time (Fig. 4.6). In the example given (non-European) oospores of the species *Chara benthamii* were removed from specimens collected in the same site (Singapore Botanic Garden) on three dates: 1896, 1908 and 1922, and from specimens collected in Brisbane, Australia, Perth and the Pilbara region of Western Australia.

In each case a mature oospore was obtained from dried material, cleaned (using various methods described above), and imaged (using two different SEM machines). Oospore shape was consistently ovate and any deviation of that shape is likely due to contraction under the vacuum (Fig. 4.6b), or damage in handling (Fig. 4.6e). The remarkable similarity among these oospores can only be appreciated in comparison with those of other species (e.g., Fig. 4.5). In *C. benthamii* the presence of a 'shoulder' or flattening of the apex, and elongation towards the base of the oospores are considered useful characters.

Traditional characters

The length, width and number of striae are the most easily measured and commonly given characters in any oospore analysis. However, there is evidence that these characters vary greatly for species, among populations, as well as within populations, and might not be the best characters to use in the absence of a description of environmental growth conditions. However, the combination of traditional characters with additional morphological characters, such as colour, the presence of a shoulder, the fossa width or thickened apical striae can be used in multivariate analysis. The requirement for these additional details could hamper the ease of use of oospores for species delimitation, but the judicious use of both measurements and descriptive characters, along with vegetative morphology can improve insights obtained from regional studies (Holzhausen et al., 2015).

Expression of flanges and ribbons

These characters can be reliable when they are present. Sakayama et al. (2009) used characters on the ribbon, along with genetic characters, to support the description of *Chara leptospora,* a new species related to *Chara globularis*. However, as for basal cages, the preservation of flanges, and especially ribbons, is dependent on the way the oospores have been cleaned and handled. See Sect. 4.8, "handling". Similarly, their presence can make cleaning by hand difficult, and can result in insufficient imaging of the fossa width (Fig. 4.11).

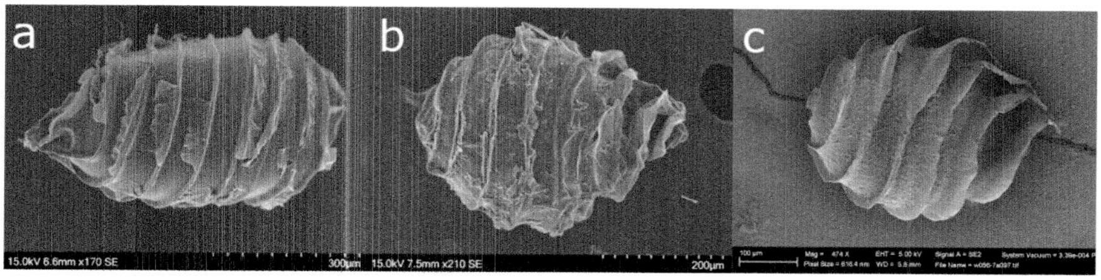

Fig. 4.11 The expression of ribbons and flanges on oospores. The oospore of a specimen of *Chara* sp. that was collected in 1880 (BM000771926) and decalcified with dilute hydrochloric acid. The ribbons (partial wall deposition on the inner edges of the enveloping cells) are well preserved, but not complete. They differ from flanges in being much less robust; **b** Flanges on the oospore of a specimen of *Nitella opaca* (Valliant PC0741129) that have made it difficult to clean. This specimen was undated but probably collected in the early 1700s, so special care was used in its cleaning, which has unfortunately failed to remove debris and cellular material from the oospore, and resulted in a poor specimen for examining features of the oospore wall; **c** A recent collection of *Nitella opaca* from Ireland (Casanova and Brodie W055) that was freely released from the plant, after the enveloping cells had decomposed, allowing the flanges to be well-preserved and the oospore wall easily examined. Images: M. T. Casanova

Fine detail of the oospore wall

The fine detail of oospore wall ornamentation has been recognised as a very useful character for species delineation in the Nitelleae. However, the major issue with the use of oospore ornamentation in the genus *Chara* has been the lack of studies demonstrating firstly that the oospore wall is ornamented, or has structural variation, and secondly that the characteristics are constant within a species. The following example of *Chara vulgaris* from different places in Europe indicates that there is a degree of constancy to ornamentation and gross morphology of the oospores from different regions, provided that the taxonomic limits of 'species' are consistently applied (Fig. 4.12). The oospores all have 11–12 striae except the one from Russia (c. 10), and they all have an oval shape with a barely detectable 'shoulder'. In some the development of claws and a cage are apparent, however these features can be formed late in development and can be missing in oospores that are not fully developed, from sediment samples or too aggressively cleaned. One character that they all have in common is the development of a verrucate oospore wall (Fig. 4.13). The density and the degree of porosity of the oospore wall appears to vary somewhat among specimens, but despite the locality and timing of collection (UK to Russia, Sweden to Greece; from 1841 to the present day), the SEM machine used, as well as the variety and form designated, they are all remarkably similar to each other.

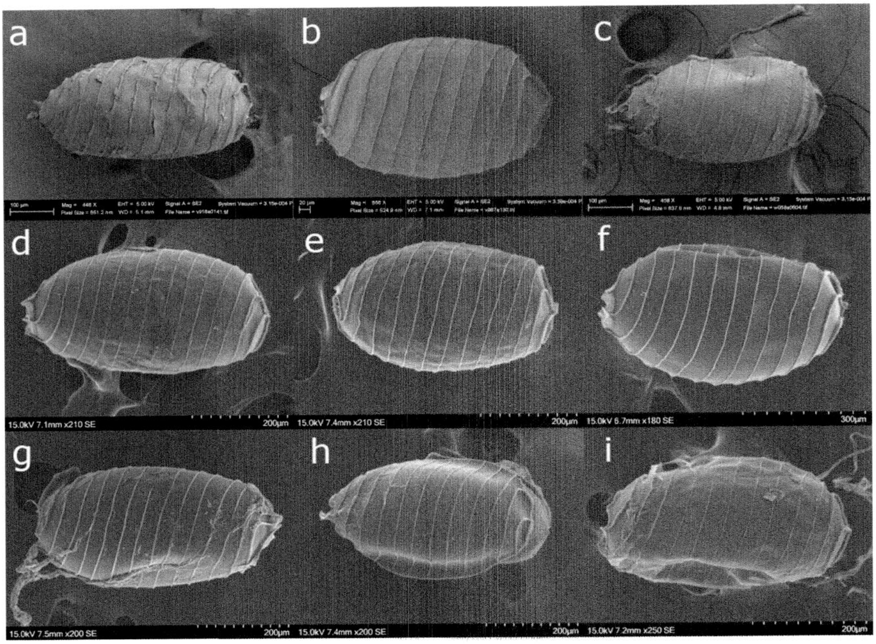

Fig. 4.12 Gross morphology of oospores of different specimens of *Chara vulgaris* L. from thalli of different varieties and forms (all consistent with *C. vulgaris*) collected at different times, in different regions of Europe. **a** UK, specimen: *C. vulgaris* var. *vulgaris* Cotswold Water Park, 19-iii-2019 M. T. Casanova and N. F. Stewart [BM]; **b** Russia, specimen: *C. vulgaris* var. *vulgaris*; Loc? 1975, Col? [IBIW11761]; **c** Ireland, specimen: *C. vulgaris* var. *vulgaris*, Poulnabrone Gorge, 18-v-2019, M. T. Casanova and J. Brodie [BM]; **d** France, specimen: *C. vulgaris* var. *papillata*, Basin de l'Abierou, 1841, A. Weddell [PC0576799]; **e** France, specimen: *C. vulgaris* var. *denta*, Dans le Fosses du Marai de Neufontaine, 1841, A. Weddell {PC0575797]; **f** France, specimen: *C. vulgaris* var. *hispidula*, Mares du Usilles, 1841, A. Weddell [PC0576792]; **g** Greece, specimen: *C. vulgaris* var. *subhispida*, Kephisso, 1847, Col? [PC0577079]; **h** Sweden, specimen: *C. vulgaris* var. *subinermis f. reflexa*, Kristianstad, 6-ix-1895, L. J. Wahlstedt 70 [PC0577076]; **i** Germany, specimen: *C. vulgaris* var. *subinermis f. gracilis*, Draegerbrush, Callies 1897, Dr Rabenhorst [PC0576995]. Images: M. T. Casanova

4.10 Natural Oospore Longevity and Wetland Restoration

Oospores are a key means of dispersal for charophytes through space and time. Early studies demonstrated that oospores retained viability after passing through the digestive system of ducks, suggesting that waterfowl are important long-distance dispersal agents for charophytes (Proctor, 1962; Bonis and Grillas, 2002; Charalambidou and Santamaria, 2005). However, one of the most remarkable features of charophyte oospores is their ability to disperse through time. Charophytes build up substantial oospore propagule banks as a means of surviving unfavourable conditions (Bonis and Grillas, 2002). Following disturbance events in aquatic habitats charophytes

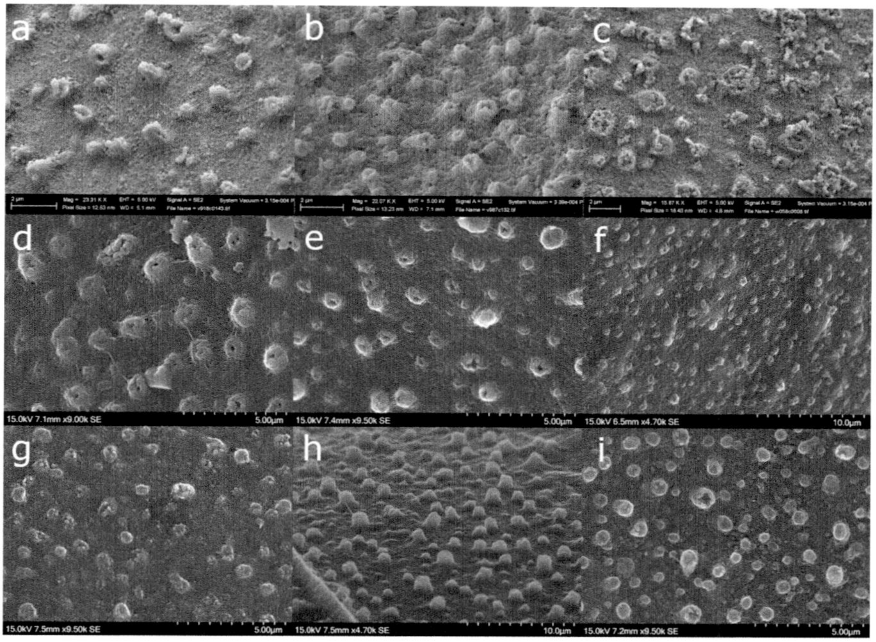

Fig. 4.13 Fine detail of the oospore wall for different specimens of *Chara vulgaris* L. from thalli of different varieties and forms (all consistent with *C. vulgaris*) collected at different times, in different regions of Europe. **a** UK, specimen: *C. vulgaris* var. *vulgaris* Cotswold Water Park, 19-iii-2019 M. T. Casanova and N. F. Stewart [BM]; **b** Russia, specimen: *C. vulgaris* var. *vulgaris*; Loc? 1975, Col? [IBIW11761]; **c** Ireland, specimen: *C. vulgaris* var. *vulgaris*, Poulnabrone Gorge, 18-v-2019, M. T. Casanova and J. Brodie [BM]; **d** France, specimen: *C. vulgaris* var. *papillata*, Basin de l'Abierou, 1841, A. Weddell [PC0576799]; **e** France, specimen: *C. vulgaris* var. *denta*, Dans le Fosses du Marai de Neufontaine, 1841, A. Weddell {PC0575797]; **f** France, specimen: *C. vulgaris* var. *hispidula*, Mares du Usilles, 1841, A. Weddell [PC0576792]; **g** Greece, specimen*:* *C. vulgaris* var. subhispida, Kephisso, 1847, Col? [PC0577079]; **h** Sweden, specimen: *C. vulgaris* var. *subinermis* f. *reflexa*, Kristianstad, 6-ix-1895, L. J. Wahlstedt 70 [PC0577076]; **i** Germany, specimen: *C. vulgaris* var. *subinermis* f. *gracilis*, Draegerbrush, Callies 1897, Dr Rabenhorst [PC0576995]. Images: M. T. Casanova

have often been observed to flourish, presumably due to long-lived oospores which allow quick re-establishment (Wade, 1990).

The duration of oospore viability is, however, relatively poorly understood, although for some species decadal to centennial-scale viability has been documented (Wade and Edwards, 1980; Simons et al., 1994; Beltman and Allegrini, 1997; Tanaka et al., 2003; Rodrigo et al., 2010; Stobbe et al., 2014). For a small lake in the Trans-Urals region of Russia, based on germination experiments conducted on well-dated sediment core sections, Stobbe et al. (2014), showed viability of *N. mucronata* oospores in sediments dating to 300+ years old and attributed that longevity to anaerobic conditions in the sediments, coupled with relatively constant moisture levels and low temperatures. It remains likely, however, that millennial-scale viability is possible for charophyte oospores wherever conditions are ideal, and more research is needed in this respect.

The complex of factors influencing the viability and re-activation of oospore banks as well as optimal growth and reproduction conditions, are not completely understood. Nevertheless, charophyte oospores, and the buried wetland sediments they persist in, clearly afford incredible opportunities for aquatic ecosystem restoration (Steinhardt and Selig, 2007, 2011; Steinhardt et al., 2009; Sayer et al., 2012; Rodrigo and Alonso-Guillén, 2013; Blindow et al., 2016, 2021; Stragauskaitė et al., 2021).

In some studies, nutrient reduction and consequent improved habitat conditions (especially increased water clarity) has been shown to recover charophyte populations (Rodrigo et al., 2010; Baastrup-Spohr et al., 2013; van de Weyer et al., 2014). Further, following sediment removal in lakes, ponds and ditches, charophytes frequently return and dominate, albeit sometimes for 1–2 years only, again almost certainly due to germinating oospores (Wade and Edwards, 1980; Wade, 1990; Nishihiro and Washitani, 2007). In Dutch turf ponds, re-excavation in the 1990s returned charophyte taxa not seen in the ponds for twenty years (Beltman and Allegrini, 1997). In addition, at one eastern English pond, restored by sediment removal, a species presumed extinct in the British Isles since the 1950s (*Nitella capillaris*) was found, suggesting that exceptionally rare species can sometimes be resurrected if appropriate conditions are created (Hawkins, 2019). In a different English study the re-excavation of ponds lost to agricultural land reclamation (so-called "ghost ponds") showed that Characeaen oospores can remain viable even when buried underneath cropped fields (Alderton et al., 2017). In this study at least five charophyte species (*Chara* and *Nitella*) were shown to have centennial-scale viability in tank experiments using mud excavated from exposed pond deposits. The oospores and the ponds were quite literally buried alive!

The long-term success of efforts to restore charophyte populations to wetlands requires optimal conditions for the subsequent and constant accumulation of viable oospores in sediments (especially for annual species) or the ability of reproduction via fragmented thalli for species with perennial life cycles. Translocation of charophyte thalli into new sites to establish for bioremediation, has also been trialled (Marquardt and Schubert, 2009) and suggests potential for using charophytes to 'kick-start' restoration processes as shown by van de Weyer et al. (2014).

4.11 Development of an Oospore Database—A Protocol for the Future

The previous sections delineate the challenges of using oospores for species determination and eco-physiological studies. The first thing that is needed is a consensus species concept, and species descriptions that encompass the range of vegetative morphology recognised for each species. That work is being undertaken as part of the preparation of this volume. Once a consensus species concept is developed, oospores can be allocated to species, and their characters, and range of characters can be determined. At the moment, even where there is a good, and agreed species concept, the use of oospores is hampered by both overlapping characters among species and broad within-species ranges. In addition, previous studies have used erroneous species concepts (e.g. Wood, 1965), data on pre-treatments and sample-sizes can be missing, or a lack of agreement on which morphological data are important. The creation of a (digital) database of oospore characteristics could have a number of advantages, depending on its purpose. It is proposed here to create a uniform European oospore identification database in order:

- to analyse both interspecific and intraspecific variability with the aim of species delimitation, and identification of regional differences between geographically separated areas,
- to provide a comprehensive digital platform to facilitate easy access to and comparison with existing data,
- to increase the comparability of oospore morphometric data from different authors by a standardised approach to the description of oospore characteristics,
- to identify environmental factors that influence the formation of oospore characteristics,
- and to facilitate the establishment of integrated herbarium databases, that combine morphological plant and oospore features with genetic species information and preservation of specimens for further examination

Researchers at Rostock University (M. Feike, H. Schubert) and Greifswald University (Blindow et al., 2004; Vedder, 2004; Blume et al., 2009) started the documentation of oospore features and preservation of related herbarium specimens. This preliminary protocol included the main characteristics of length, width, number of striae, colour and shape.

In the last few years, this protocol was expanded by adding information from oospores from sediment analyses, environmental conditions, pre-treatments, SEM images of Baltic species, additional characteristics such as fossa width and the expansion of replicates per population (A. Holzhausen). The spectrum is, so far, dominated by species of the genus *Chara*, but does also contain *Lychnothamnus barbatus* and a few species of the genera *Tolypella*, *Nitella* and *Lamprothamnium*.

In general, the integrative approach is based on a combination of the main aspects: preparation of herbarium specimens, fresh plant material, plant associated oospores (Anja Holzhausen) and genetic analyses of herbarium material (Petra Nowak). This provides for subsequent examination, and enables the correlation of conspicuously deviating oospore morphometries with corresponding plant characteristics and sequence data. Additional information on habitat-specific parameters can allow the correlation of physico-chemical parameters with oospore characteristics focusing on the elucidation of the influences that lead to important intraspecific variability in separate populations. Information on the origin of the material (fresh plant material, herbarium material, or sediment extraction) and the removal of debris or sterile cells (e.g. with acetic acid or by feeding activity of *Gammarus tigrinus*) will facilitate comparability of data from different authors.

Based on existing literature, the following oospore parameters are considered to be vital information:

qualitative parameter
- colour
- presence of a shoulder
- presence of thickened apical striae that mainly exist within diplostichous species of *Chara*
- presence of appendages (claws or cage)

quantitative parameter
- length,
- width,
- length of the fossae (at least 3 or 4)
- diameter of the basal impression
- and, if applicable, length of appendages

The database does not require access to SEM of all species, but if that is possible then the (optional)
- measurements of the striae and fossa
- measurement of the basal cell(s) impressions
- and a description of the fine detail of the oospore wall ornamentation or construction, and the flanges or ribbons, are also useful.

References

Alderton, E., Sayer, C. D., Davies, R., Lambert, S. J., & Axmacher, J. C. (2017). Buried alive: Aquatic plants survive in 'ghost ponds' under agricultural fields. *Biological Conservation, 212 (Part A)*, 105–110. https://doi.org/10.1016/j.biocon.2017.06.004.

Allen, G. O. (1937). Note on the outer covering of charophyte fruits. *Journal of Botany, British and Foreign, 75*, 153–155.

Allen, G. O. (1950). *British stoneworts (Charophyta). Arbroath, The Haslemere Natural History Society*. Buncle, Arbroath.

Allen, T. F. (1888). *The Characeae of America.*

Baastrup-Spohr, L., Lønsmann Iversen, L., Dahl-Nielsen, J., & Sand-Jensen, K. (2013). Seventy years of change in the abundance of Danish charophytes. *Freshwater Biology, 58*, 1682–1693. https://doi.org/10.1111/fwb.12159.

Bailly, G., & Schaefer, O. (2010). *Guide illustré des Characées du nord-est de la France.* Conservatoire Botanique National de Franche-Comté, Besancon.

Beltman, B., & Allegrini, C. (1997). Restoration of lost aquatic plant communities: new habitats for *Chara. Netherlands Journal of Aquatic Ecology, 30*(4), 331–337. https://doi.org/10.1007/BF02085876.

Blindow, I., Carlsson, M., & van de Weyer, K. (2021). Re-establishment techniques and transplantations of charophytes to support threatened species. *Plants, 10*. https://doi.org/10.3390/plants 10091830.

Blindow, I., Dahlke, S., Dewart, A., Flügge, S., Hendreschke, M., Kerkow, A., & Meyer, J. (2016). Long-term and interannual changes of submerged macrophytes and their associated diaspore reservoir in a shallow southern Baltic Sea bay: Influence of eutrophication and climate. *Hydrobiologia, 778*(1), 121–136. https://doi.org/10.1007/s10750-016-2655-4.

Blindow, I., Rost, E., Jachner, S., Frey, S., & Dilger, M. (2004). Entwicklung eines Schlüssels zur Bestimmung von Characeen-Oosporen—eine Projektbeschreibung—Development of a determination key for charophyte oospores—a project description. *Rostocker Meeresbiologische Beiträge, 13*, 27–34.

Blume, M., Blindow, I., & Dahlke, S. (2009). Oospore variation in closely related *Chara* taxa. *Journal of Phycology, 45*, 995–1002. https://doi.org/10.1111/j.1529-8817.2009.00725.x.

Bonis, A., & Grillas, P. (2002). Deposition, germination and spatio-temporal patterns of charophyte propagule banks: a review. *Aquatic Botany 72*(3–4), 235–248. https://doi.org/10.1016/S0304-3770(01)00203-0.

Boszke, P., & Bociąg, K. (2008). Morphological variation of oospores in the population of *Chara rudis* A. Braun in a mesotrophic lake. *Polish Journal of Ecology, 56*(1), 139–147.

Braun, A. (1849). Chareæ Australis et Antarticæ, or observations on the Chareæ of Australia and the Southern circumpolar regions. *Hooker's Journal of Botany & Kew Garden Miscellany, 1*.

Braun, A., & Nordstedt, C. F. O. (1882). Fragmente einer Monographie der Characcen. Nach den hinterlassenen Manuskripte A. Braun's, herausgegeben von Dr. O. Nordstedt. *Abhandlungen der Königlichen Akademie der Wissenschaftlichen Berlin für das Jahr 1882*, 211 pp.

Brzozowski, M., Palomares Cabanilles, M., Kowalewski, G. & Pełechaty, M. (2019). Environmental factors responsible for the gyrogonite formation by an endangered macroalga, *Lychnothamnus barbatus*, a fertility indicator of past and present lacustrine ecosystems. *Limnologica, 77*. https://doi.org/10.1016/j.limno.2019.125686.

Cáceres, E. J. (1975). Novedades carológicas argentinas. I. Una nueva especie de *Nitella* y tres adiciones al género para la flora argentina. *Kurtziana, 8*, 105–125.

Caisová, L., & Gąbka, M. (2009). Charophytes (Characeae, Charophyta) in the Czech Republic: taxonomy, autecology and distribution. *Fottea, 9*(1), 1–43.

Casanova, M. T. (1991). An SEM study of developmental variation in oospore wall ornamentation of three *Nitella* species. *Phycologia, 30*(3), 237–242. https://doi.org/10.2216/i0031-8884-30-3-237.1.

Casanova, M. T. (1993). *The ecology of charophytes in temporary and permanent wetlands: an Australian perspective*. Dissertation, University of New England.

Casanova, M. T. (1997). Oospore variation in three species of *Chara* (Charales, Chlorophyta). *Phycologia, 36(4)*, 274–280. https://doi.org/10.2216/i0031-8884-36-4-274.1.

Casanova, M. T. (2009). An overview of *Nitella* (Characeae, Charophyceae) in Australia. *Australian Systematic Botany, 22*, 193–218. https://doi.org/10.1071/SB08039.

Casanova, M. T. (2013). *Lamprothamnium* in Australia (Characeae, Charophyceae). *Australian Systematic Botany, 26(4), 268–290*. https://doi.org/10.1071/sb13026.

Casanova, M. T. & Brock, M. A. (1996). Can oospore germination pattems explain charophyte distribution in permanent and temporary wetlands? *Aquatic Botany 54(4)*, 297–312. https://doi.org/10.1016/0304-3770(96)01032-7.

Casanova, M. T. & Wilbraham, J. (2019). Dataset: charophytes of Britain and Ireland. Resource: Charophytes of Britain and Ireland. Natural History Museum Data Portal (data.nhm.ac.uk). https://doi.org/10.5519/0028430.

Charalambidou, I., & Santamaria, L. (2005). Field evidence for the potential of waterbirds as dispersers of aquatic organisms. *Wetlands, 25*(2), 252–258. https://doi.org/10.1672/2.

Cirujano, S., Chambra, J., Sánchez Castillo, P. M., Meco, A., & Flor Arnau, N. (2008). *Flora ibérica Algas Continentales. Carófitos (Characeae)*. Real Jardín Botánico, CSIC.

Corillion, R. (1957). Les Charophycees de France et d' Europe Occidentale. *Bulletin de la Société Scientifique de Bretagne, 32* (Hors série I & II), 499 pp.

Crawford, S. A., Higgins, M. J., Mulvaney, P., Wetherbee, R. (2001). Nanostructure of the diatom frustule as revealed by atomic force and scanning electron microscopy. *Journal of Phycology, 37*, 543–554. https://doi.org/10.1046/j.1529-8817.2001.037004543.x.

Cronk, Q. C. B. (2009). *The molecular organography of plants*. Oxford University Press. https://doi.org/10.1093/acprof:oso/9780199550357.001.1.

Ernst, A. (1918). *Bastardierung als Ursache der Apogamie im Pflanzenreich*. Gustav Fischer Verlag.

Faegri, K., & Iversen, J. (1975). *Textbook of pollen analysis (3d ed.)*. Munksgaard, 295 pp.

Forsberg, C. (1965). Sterile germination of oospores of *Chara* and Seeds of *Najas marina*. *Physiologlia Plantarum, 18*, 128–137. https://doi.org/10.1111/j.1399-3054.1965.tb06875.x.

García, A. (1994). Charophyta: Their use in paleolimnology. *Journal of Paleolimnology, 10*(1), 43–52. https://doi.org/10.1007/bf00683145.

Groves, J. & Bullock-Webster, G. R. (1920). *The British Charophyta Volume I. Nitelleae*. Ray Society, London, 405 pp.

Groves, J. & Bullock-Webster, G. R. (1924). *The British Charophyta. Volume II. Chareæ*. Ray Society, London.

Haas, J. N. (1994). First identification key for charophyte oospores from central Europe. *European Journal of Phycology, 29(4)*, 227–235. https://doi.org/10.1080/09670269400650681.

Hawkins, J. (2019). The discovery and conservation or rare stoneworts in Suffolk's farmland ponds. *Transactions of the Suffolk Naturalists Society, 55*, 109–125.

Hay, F. R., & Muir, J. S. K. (2000). Low temperature survival of slender naiad (*Najas flexilis*) seeds. *Cryo Letters, 21*(5), 271–278. PMID: 12148032.

Holtz, L. (1891). Die Characeen Neuvorpommerns mit der Insel Rügen und der Insel Usedom. *Mittheilungen aus dem naturwissenschaftlichen Vereine von Neu-Vorpommern und Rügen, 23*, 99–156.

Holtz, L. (1899). Die Characeen der Regierungsbezirke Stettin und Köslin. *Mittheilungen des naturwissenschaftlichen Vereines für Neuvorpommern und Rügen, 31*, 101–189.

Holzhausen, A., Nowak, P., Niedrig, C., Feike, M., & Schubert, H. (2015). Morphometry of *Chara aspera*, *C. canescens*, *C. baltica* var. *baltica*, *C. baltica* var. *liljebladii* and *C. intermedia* oospores: local variation versus taxonomic differences. *Aquatic Botany, 120* (Part A), 60–66. https://doi.org/10.1016/j.aquabot.2014.07.001.

Holzhausen, A., Stingl, N., Rieth, S., Kühn, C., Schubert, H., & Rensing, S. A. (2022). Establishment and optimization of a new model organism to study early land plant evolution: germination, cultivation and oospore variation of *Chara braunii* Gmelin, 1826. *Frontiers in Plant Science, 13*, 987741. https://doi.org/10.3389/fpls.2022.987741.

Holzhausen, A., Nowak, P., Ballot, A., Becker, R., Gebert, J., Gregor, T., Karol, K. G., Lambert, E., Pérez, W., Raabe, U., Schneider, S. C., Stewart, N., van de Weyer, K., Wilde, V., & Schubert, H. (2023). Plastid DNA sequences and oospore characters of some European taxa of *Tolypella* section *Tolypella* (Characeae) identify five clusters, including one new cryptic *Tolypella* taxon from Sardinia, but they do not coincide with current morphological descriptions. *Frontiers in Plant Science, 14*, 1096181. https://doi.org/10.3389/fpls.2023.1096181.

Holzhausen, A. (2016). *Resilienz aquatischer Ökosysteme. Vitalitätsabschätzungen von Diasporenbanken*. Dissertation, University Rostock. https://doi.org/10.18453/rosdok_id00002020

Holzhausen, A., Porsche, C., & Schubert, H. (2017). Viability assessment and estimation of the germination potential of charophyte oospores: Testing for site and species specificity. *Botany Letters, 165*(1), 147–158. https://doi.org/10.1080/23818107.2017.1393460.

Horn af Rantzien, H. (1956). Morphological terminology relating to female charophyte gametangia and fructifications. *Botaniska Notiser, 109*, 212–259.

Horn af Rantzien, H. (1959). Recent charophyte fructifications and their relations to fossil charophyte gyrogonites. *Arkiv für Botanik, 2(4)*, 165–332.

John, D. M. & Moore, J. A. (1987). An SEM study of the oospore of some *Nitella* species (Charales, Chlorophyta) with descriptions of wall ornamentation and an assessment of its taxonomic importance. *Phycologia, 26*(3), 334–355. https://doi.org/10.2216/i0031-8884-26-3-334.1.

John, D. M., Moore, J. A., & Green, D. R. (1990). Preliminary observations on the structure and ornamentation of the oosporangial wall in *Chara* (Charales, Chlorophyta). *British Phycological Journal, 25*(1), 1–24. https://doi.org/10.1080/00071619000650011.

Kato, S., Sakayama, H., Morishima, H., Sano, S., Oomori, Y., Kato, N., Ito, M., Kasai, F., Watanabe, M. M., & Nozaki, H. (2010). Morphology and molecular phylogeny of *Chara altaica* (Charales, Charophyceae), a monoecious species of the section *Desvauxia*. *Cytologia, 75*(2), 211–220. https://doi.org/10.1508/CYTOLOGIA.75.211.

Krause, W. (1997). Charales (Charophyceae). In H. Ettl, G. Gärtner, H. Heynig, & D. Mollenhauer (Eds.), *Süßwasserflora von Mitteleuropa (vol.18)*. Gustav Fischer Verlag, 201 pp.

Kützing, F. T. (1849). *Species Algarum*, F. A. Brockhaus, Lipsiae.

Langangen, A. (2007). *Charophytes of the Nordic countries*. Saeculum ANS, 102 pp.

Leitch, A. R. (1989). Formation and ultrastructure of a complex, multilayered wall around the oospore of *Chara* and *Lamprothamnium* (Characeae). *British Phycological Journal, 24*(3), 229–236.

Leitch, A. R., John, D. M., & Moore, J. A. (1990). The oosporangium of the *Characeae* (Chlorophyta, Charales). *Progress in Phycological Research*, 7, 213–268.

Mandal, D. K., Blaženčić, J. M., & Ray, S. (2002). SEM study of compound oospore wall ornamentation of some members of Charales from Yugoslavia, Croatia and Slovenia. *Archives of Biological Sciences, 54*(1–2), 29–34. https://doi.org/10.2298/ABS0202028M.

Mandal, D. K., Ojha, S. N., & Ray, S. (2008). Species-specificity of oospore dimensions in three ecorticate species of *Chara* (Charales, Charophyceae). *Aquatic Botany, 88(2)*, 181–183. https://doi.org/10.1016/j.aquabot.2007.09.006.

Marquardt, R., & Schubert, H. (2009). Photosynthetic characterisation of *Chara vulgaris* in bioremediation ponds. *Charophytes, 2(1)*, 1–8.

Migula, W. (1889–1897). Fünfter Band: Die Characeen. In: Fischer, A. E., Fischer, F., Hauck, G., Limpricht, C., Luerssen, W., Migula, H., Rehm, P., Richter & G. Winter (eds), *Dr L Rabenhorst's Kryptogamenflora von Deutschland, Oesterreich und der Schweiz*, ed. 2. Eduard Kummer, Leipzig [in German].

Mouronval, J.-B., Baudouin, S., Borel, N., Soulié-Märsche, I., x Klesczewski, I., & Grillas, P. (2015). *Guide des Characées de France méditerranéenne*. Office national de la Chasse et de la Faune Sauvage, 211 pp.

Nishihiro, J., & Washitani, I. (2007). Restoration of lakeshore vegetation using sediment seed banks; studies and practices in lake Kasumigaura, Japan. *Global Environmental Research, 11*, 171–177.

Nordstedt, C. F. O. (1889a). Über einige Characeen aus Spanien. *Acta Universitatis Lundensis, 25*, 17–22.

Nordstedt, O. (1889b). *De Algis et Characeis*. Lunds Universitets års-skrift. *Acta Universitatis Lundensis, 25.*

Pankow, H. (1990). *Ostsee – Algenflora*. Gustav Fischer Verlag.

Probert, R. J., & Hay, F. R. (2000). Keeping seeds alive. In M. Black & J. D. Bewley (Eds.), *Seed technology and its biological basis*. Sheffield Academic Press.

Proctor, V. W. (1962). Viability of *Chara* oospores taken from migratory water birds. *Ecology, 43*(3), 528–529.

Proctor, V. W. (1967). Storage and Germination of *Chara* Oospores. *Journal of Phycology, 3*(2), 90–92. https://doi.org/10.1111/j.1529-8817.1967.tb04638.x

Raam, J. V., Maier, E. X., Bruinsma, J., Simons, J., & Stegenga, H. (1998). *Handboek kranswieren*. Chara boek. Hilversum.

Ray, S., Pekkari, S., & Snoeijs, P. (2001). Oospore dimensions and wall ornamentation patterns in Swedish charophytes. *Nordic Journal of Botany, 21*(2), 207–224. https://doi.org/10.1111/j.1756-1051.2001.tb01359.x.

Rodrigo, M. A., & Alonso-Guillén, J. L. (2013). Assessing the potential of Albufera de València lagoon sediments for the restoration of charophyte meadows. *Ecological Engineering, 60*, 445–452. https://doi.org/10.1016/j.ecoleng.2013.09.041.

Rodrigo, M. A., Alonso-Guillén, J. L., & Soulié-Märsche, I. (2010). Reconstruction of the former charophyte community out of the fructifications identified in Albufera de València lagoon sediments. *Aquatic Botany, 92*(1), 14–22. https://doi.org/10.1016/j.aquabot.2009.09.002.

Sakayama, H. (2008). Taxonomy of *Nitella* (Charales, Charophyceae) based on comparative morphology of oospores and multiple DNA marker phylogeny using cultured material. *Phycological Research, 56*(3), 202–215. https://doi.org/10.1111/j.1440-1835.2008.00502.x.

Sakayama, H., Kasai, F., Nozaki, H., Watanabe, M. M., Kawachi, M., Shigyo, M., Nishihiro, J., Washitani, I., Krienitz, L., & Ito, A. M. (2009). Taxonomic reexamination of *Chara globularis* (Charales, Charophyceae) from Japan based on oospore morphology and *rbc*L gene sequences, and the description of *C. leptospora* sp. nov. *Journal of Phycology, 45*(4), 917–927. https://doi.org/10.1111/j.1529-8817.2009.00700.x.

Sakayama, H., Miyaji, K., Nagumo, T., Kato, M., Hara, Y., & Nozaki, H. (2005). Taxonomic reexamination of 17 species of *Nitella* subgenus *Tieffallenia* (Charales, Charophyceae) based on internal morphology of the oospore wall and multiple DNA marker sequences. *Journal of Phycology, 41*(1), 195–211. https://doi.org/10.1111/j.1529-8817.2005.04133.x.

Sanjuan, J., & Martín-Closas, C. (2015). Gyrogonite polymorphism in two European Charophyte biozone index species. *Palaeontology, 1*(3), 223–236. https://doi.org/10.1002/spp2.1013.

Sanjuan, J., Vicente, A., Flor-Arnau, N., Monléon, T., Cambra, J., & Martín-Closas, C. (2017). Effects of light and temperature on *Chara vulgaris* (Charophyceae) gyrogonite productivity and polymorphism—palaeoenvironmental implications. *Phycologia, 56*(2).

Sato, M., Sakayama, H., Sato, M., Ito, M., & Sekimoto, H. (2014). Characterization of sexual reproductive processes in *Chara braunii* (Charales, Charophyceae). *Phycological Research, 62*(3), 214–221. https://doi.org/10.1111/pre.12056.

Sayer, C. D., Bennion, H., Davidson, T. A., Burgess, A., Clarke, G., Hoare, D., Frings, P., & Hatton-Ellis, T. (2012). The application of palaeolimnology to evidence-based lake management and conservation: examples from UK lakes. *Aquatic Conservation: Marine and Freshwater Ecosystems, 22*, 165–180. https://doi.org/10.1002/aqc.2221.

Schubert, H., & Blindow, I. (2003). *Charophytes of the Baltic Sea*. Gantner Verlag.

Shen, E. Y. F. (1966). Oospore germination in two species of *Chara*. *Taiwania, 12*(1), 39–46. https://doi.org/10.6165/tai.1966.12.39.

Simons, J., Ohm, M., Daalder, R., Boers, P., & Rip, W. (1994). Restoration of Botshol (The Netherlands) by reduction of external nutrient load: recovery of a characean community, dominated by *Chara connivens*. *Hydrobiologia, 275*, 243–253. https://doi.org/10.1007/BF00026715.

Sinkevičienė, Z. (2019). Peculiarities of *Chara filiformis* (Charales, Charophyceae) distribution and oospore sizes in Lithuania. *Webbia, 74*(1), 133–138. https://doi.org/10.1080/00837792.2019.1607501.

Soulié-Märsche, I. (1989). *Étude comparée de gyrogonites de Charophytes actuelles et fossiles et phylogénie des genres actuels*. Imprimerie des Tilleuls.

Soulié-Märsche, I., & García, A. (2015). Gyrogonites and oospores, complementary viewpoints to improve the study of the charophytes (Charales). *Aquatic Botany, 120 (Part A)*, 7–17. https://doi.org/10.1016/j.aquabot.2014.06.003.

Soulié-Märsche, I., & Joseph, C. (1991). Fiabilité des données morphométriquesen micropaleontologie: exemple des Charophytes. *Geobios, 24*, 113–123. https://doi.org/10.1016/0899-5362(91)90083-B.

Steinhardt, T., Karez, R., Selig, U., & Schubert, H. (2009). The German procedure for the assessment of ecological status in relation to the biological quality element "Macroalgae & Angiosperms" pursuant to the European Water Framework Directive (WFD) for inner coastal waters of the Baltic Sea. *Rostocker Meeresbiologische Beiträge, 22*, 7–42.

Steinhardt, T., & Selig, U. (2007). Spatial distribution patterns and relationship between recent vegetation and diaspore bank of a brackish coastal lagoon on the southern Baltic Sea. *Estuarine, Coastal and Shelf Science, 74*(1–2), 205–214. https://doi.org/10.1016/j.ecss.2007.04.004.
Steinhardt, T., & Selig, U. (2011). Influence of salinity and sediment resuspension on macrophyte germination in coastal lakes. *Journal of Limnology, 70*(1), 11–20. https://doi.org/10.3274/JL11-70-1-03.
Stobbe, A., Gregor, T., & Röpke, A. (2014). Long-lived banks of oospores in lake sediments from the Trans-Urals (Russia) indicated by germination in over 300 years old radiocarbon dated sediments. *Aquatic Botany, 119*, 84–90. https://doi.org/10.1016/j.aquabot.2014.07.004.
Stragauskaitė, V., Bučas, M., & Martin, G. (2021). Distribution of charophyte oospores in the Curonian lagoon and their relationship to environmental forcing. *Water, 13*(2). https://doi.org/10.3390/w13020117.
Tanaka, N., Momohara, A., Sakayama, H., & Uehara, K. (2003). *Chara* species that emerged from 40-year-old sediments from Lake Teganuma, central Japan. *Nat. Hist. Res, 7*(2), 101–105.
Urbaniak, J. (2009). Oospore variation in *Nitella gracilis* and *Nitella mucronata* (Charales, Charophyceae) from Poland. *Biologia, 64*(2), 252–260. https://doi.org/10.2478/s11756-009-0049-z.
Urbaniak, J. (2011). A SEM and light microscopy study of the oospore wall ornamentation in Polish charophytes (Charales, Charophyceae)—genus *Chara*. *Nova Hedwigia, 93*(1–2), 1–28. https://doi.org/10.1127/0029-5035/2011/0093-0001.
Urbaniak, J., & Blaženčić, J. (2012). SEM study of oospore characteristics in endemic and endangered Balkan Charophytes. *Cryptogamie, Algologie, 33*(3), 277–288. https://doi.org/10.7872/crya.v33.iss3.2012.277.
van de Weyer, K., Sümer, G., Hueppe, H., & Petruck, A. (2014). Das Konzept PHOENIX See: Nachhaltiges Management von Makrophyten-Massenentwicklungen durch eine Kombination nährstoffarmer Standortbedingungen und Bepflanzung mit Armleuchteralgen. *Korrespondenz Wasserwirtschaft, 7*, 23–27.
Vedder, F. (2003). *Morphologie und Taxonomie rezenter und subfossiler Characeen-Oosporen aus der Ostsee*. Greifswald, Ernst-Moritz-Arndt-University Greifswald. Diploma.
Vedder, F. (2004). Morphologie und Taxonomie rezenter und subfossiler Characeen-Oosporen aus der Ostsee—Morphology and taxonomy of recent and subfossil oospores of charophytes out of the Baltic Sea. *Rostocker Meeresbiologische Beiträge, 13*, 43–54.
Wade, P. M. (1990). The Colonisation of disturbed freshwater habitats by *Characeae*. *Folia Geobotanica et Phytotaxonomica, 25*, 275–278. https://doi.org/10.1007/BF02913027.
Wade, P. M., & Edwards, R. W. (1980). The effect of channel maintenance on the aquatic macrophytes of the drainage channels of the monmouthshire levels, South Wales, 1840–1976. *Aquatic Botany, 8*, 307–322. https://doi.org/10.1016/0304-3770(80)90061-3.
Wallace, S., Fleming, A., Wellman, C. H., & Beerling, D. J. (2011). Evolutionary development of the plant and spore wall. *AoB Plants, 2011*, 1–18. https://doi.org/10.1093/aobpla/plr027.
Wallman, J. (1853). *Försök till en systematisk uppställning af växtfamiljen*. P. A. Norstedt & Söner, 114 pp.
Williams, M. B. (1959). A revision of *Nitella cristata* Braun (Characeae) and its allies. Part II. Taxonomy. *Proceedings of the Linnean Society of New South Wales, 84*, 346–356.
Wood, R. D. (1959). Gametangial constants of extant charophyta for use in micropaleobotany. *Journal of Paleontology, 33*(1), 186–194.
Wood, R. D. (1965). Monograph of the characeae. A revision of the Characeae. In I. K. Wood, & R. D. Wood (Eds.), *A Revision of the Characeae*. J Cramer.
Zaneveld, J. S. (1940). *The Charophyta of Malaysia and adjacent countries*. Dissertation, University Leiden, 223pp.

Chapter 5
Taxonomic Concepts and Taxonomic Evolution in European *Characeae*

Hendrik Schubert, Karl-Georg Bernhardt, Michelle T. Casanova, Thomas Gregor, Wolf-Henning Kusber, and Barbara Turner

5.1 Introduction

A monograph containing a taxonomic treatment requires description of the entities dealt with. Whereas until the middle of the last century charophyte monographs (e.g. Migula, 1897; Wood, 1965) gave long lists of herbarium specimens studied and related to the respective taxon, you'll not find such lists in this, nor in many of the monographs published during the last 50 years (e.g. Hollerbach and Krassavina, 1983, Krause, 1997). Therefore, this book shall not just reflect the knowledge gathered by the authors; it shall allow the reader to perform such specimen-to-taxon

H. Schubert (✉)
Institut für Biowissenschaften, Universität Rostock, Albert-Einsteinstraße 3, 18051 Rostock, Germany
e-mail: hendrik.schubert@uni-rostock.de

K.-G. Bernhardt · B. Turner
University of Natural Resources and Life Sciences, Vienna, Gregor-Mendel-Straße 33, 1180 Vienna, Austria
e-mail: karl-georg.bernhardt@boku.ac.at

B. Turner
e-mail: barbara.turner@boku.ac.at

M. T. Casanova
The Natural History Museum, London, UK
e-mail: m.casanova@nhm.ac.uk

T. Gregor
Department of Botany and Molecular Evolution, Senckenberg Research Institute and Natural History Museum, Senckenberganlage 25, 60325 Frankfurt am Main, Germany

W.-H. Kusber
Botanischer Garten Berlin, Freie Universität Berlin, Königin-Luise-Straße 6-8, 14195 Berlin, Germany
e-mail: w.h.kusber@bgbm.org

H. Schubert et al. (eds.), *Charophytes of Europe*,
https://doi.org/10.1007/978-3-031-31898-6_5

sorting exercises (determinations) themselves. Conceptually the basis of species determination starts with an individual. Several individuals that share enough characters can be grouped into 'species', distinct from other 'species'. A description of those distinguishing characters should allow for indisputable identification to 'species' of any newly sampled specimen. It is obvious that to allow distinction, that such a concept includes characters shared by the groups as well as characters distinguishing among them. However, identification of characters that are uniformly expressed by all members of one group, but also reliably distinguish among groups is a challenge. Although "species" is generally accepted as the basic entity for describing the overwhelming diversity to which life has evolved, there is still an ongoing debate in biology about "species concepts". When this monograph was being prepared there were lively debates about "species" delineation. All of the arguments discussed turned out to be non-decisive by themselves—in fact, in each case scientists resolved on *agreed* concepts for species, instead of being completely objective. This illustrated once again that not only "species are hypotheses" (Pante et al., 2015) but, moreover, species are man-made agreements, based on a certain concept which needs (A) to be defined accurately and (B) which has been developed with a well-articulated objective.

It is therefore no surprise that a number of "species concepts" have been developed in the past that reflect diverse aims as well as a diversity of techniques and selection of discriminative characters. Mayden (1997) listed 22 "species concepts", some of which are only slightly different from others, but all designed for general acceptance. Moreover, Mayr (2000) demanded a clear distinction between the use of the term "species" as a taxonomic rank in the Linnaen hierarchy and the use for a species taxon. This argument may sound a bit sophisticated at first, but illustrates the motivation of this chapter. A taxonomic rank in Linnaen hierarchy is something easily accepted as being virtual whereas a species taxon itself is often, as shown by Mayr (2000) believed to be a real entity. Cracraft (2000) exemplified this brilliantly by describing the following situation: "...*We could put 20 biologists in a room and point to a squirrel, and everyone would probably agree that, yes, this is a real entity – a squirrel. But one cannot do that with a species. Biologists do not see a species, yet that does not prevent us from inferring they are discrete, real entities........*" (cf. Cracraft, 2000, p. 11). He goes on to give good advice: "....*One of the things that helps us with that inference is a definition, which forms the basis of individuating what we think are real entities called species...but....definitions do not necessarily make things real...*" (cf. Cracraft, 2000, p. 11).

For these reasons, an overview of the "species concept", as agreed by the authors of this monograph, will be explained, starting with a brief description of the history of charophyte-taxonomy.

5.2 The Naming of the European *Characeae*

The history of European charophyte species is one of discovery and increasing recognition of diversity. Linnaeus (1753) introduced the genus *Chara* with 4 species in "Cryptogamia Algæ" (Tables 5.1 and 5.2). Today, of these four species only *Chara tomentosa* L. ('densely hairy' Chara) is still recognised in its original sense. Linnaeus' *Chara vulgaris* ('common' Chara) incorporated smooth *Chara* species, *Chara* species with spine-cells were *Chara hispida* ('bristly' Chara) and the fourth species: *Chara flexilis* ('flexible' Chara) is now recognised as *Nitella* species. About 200 years later, Wood (1960) clarified typification of the Linnaean names, and an ambiguity concerning *Chara hispida* has been recently resolved by Gregor et al. (2014). *Chara vulgaris* was out of use for a long time and replaced by *Chara foetida* ('smelly' Chara) as the identity of *C. vulgaris* had been obscure.

A few species were added in the second half of the eighteenth century (Fig. 5.1). Müller (1777) described *Conferva nidifica* (= *Tolypella nidifica*). Thuillier (1799) added 5 species but their interpretation is hampered by a lack of original material. Of the Thuillier species only *Chara globularis* and *Chara syncarpa* (= *Nitella syncarpa*) are currently recognised. *Chara capillacea* is a synonym of *Chara globularis ; Chara batrachosperma* is synonymous with *Chara vulgaris;* and *Chara funicularis* is one of the "swords of Damocles" waiting to menace an established name if original material is found. Desfontaines (1799) described *Chara squamosa* from today's Algeria of which *Chara conimbrigensis* is a later synonym.

In the early decades of the nineteenth century description of European *Characeae* accelerated and culminated between 1830 and 1850 (Fig. 5.1). It started with Roth's (1800) description of *Chara intricata* (= *Sphaerochara intricata*). Persoon (1807) was the first after Linnaeus to provide a worldwide treatment of *Characeae.* In this case with 14 species, 8 of them from Linnaeus (1753) and Thuillier (1799) and only 4 from outside Europe. *Chara translucens* (= *Nitella translucens*) is still in use today but as original material is missing the interpretation is not without doubt. In 1809, Desvaux (1809) published *Chara tenuissima* (= *Nitella tenuissima*) and Willdenow (1809) *Chara aspera*. Nicaise Auguste Desvaux (1784–1856) was familiar with *Characeae* and "un travail plus considérable … sur le genre entier" (Loiseleur Deslongchamps, 1810) but never published this work. In Loiseleur Deslongchamps (1810) Desvaux published 6 new species and two of them are still in use: *Chara glomerata* (= *Tolypella glomerata*) and *Chara obtusa* (= *Nitellopsis obtusa*). *Chara fragilis* was used for a long time for today's *Chara globularis* but it is an illegitimate name as specimens of the earlier *Chara vulgaris* were cited. Desvaux's *Chara delicatula,* was an invalid synonym of *Chara aspera,* rendering the much used name *Chara delicatula* C. Agardh, 1824 invalid. *Chara canescens* was published with the addition of "nobis" which implies "Desv. & Loisel." but is often ascribed to Loiseleur Deslongchamps alone. Smith (1810) published *Chara gracilis* (= *Nitella gracilis*), de Candolle (1813) *Chara galioides*, Krocker (1814) *Chara capillaris* (= *Nitella capillaris*) and again de Candolle (1815) *Chara hyalina* (= *Nitella hyalina*).

Table 5.1 Number of currently recognised European species of *Characeae*; species currently in other genera in brackets

	Chara	*Lamprothamnium*	*Lychnothamnus*	*Nitella*	*Nitellopsis*	*Tolypella*	Scope	Sum
Linnaeus (1753)	3			(1)			World	4
Persoon (1807)	5			(2)			World	7
Bruzelius (1824)	7			(6)		(1)	World	14
Agardh (1824)	7			6		(2)	World	15
Wallroth (1833)	7	(1)	(1)	(3)			Germany	12
Braun (1834)	10		(1)	(8)	(1)	(2)	World	22
Braun (1835)	12	(1)	(1)	(6)	(1)	(2)	World	23
Kützing (1845)	12 (1)		(1)	6	(1)	(3)	Germany	24
Ruprecht (1845)	8			(4)	(1)	(1)	Russia	14
Rabenhorst (1847)	10 (1)		(1)	5	(1)	(2)	Germany	20
Kützing (1849)	16	(1)	(1)	8	(1)	(3)	World	30
Wallman (1853)	16	(1)	(1)	10	(1)	(3)	World	32
Wahlstedt (1862)	10			5	(1)	(2)	Scandinavia	18
Braun (1867)	22	(1)	(1)	10	(1)	(4)	Europe	39
Braun and Nordstedt (1882)	18	1	1	11	(1)	4	World	36
Migula (1889, 1897)	24	1	1	12	(1)	6	Europe	45
Reyes-Prosper (1910)	15	1		8	(1)	2	Spain	27
Hy (1913)	15	1	1	11	1	3	France	32
Groves and Bullock-Webster (1920, 1924)	16	1		9	1	4	Britain and Ireland	31
Migula (1925)	20	1	1	10	(1)	4	Germany	37
Corillion (1957)	18	1	1	11	1	5	Western Europe	37
Wood (1965)	29	2	1	10	1	5	World	48
Hollerbach and Krassavina (1983)	26	1	1	11	1	6	USSR	46
Moore (1986)	11	1		7	1	3	Britain & Ireland	23
Krause (1997)	24	1	1	11	1	8	Europe	46
Bryant et al. (2002)	17	1		9	1	4	Britain & Ireland	32
Schubert and Blindow (2004)	12	1		8	1	1	Baltic Sea	23
Langangen (2007)	19	1		11	1	5	Nordic Countries	37
Cirujano et al. (2008)	11	1		9	1	4	Iberian Peninsula	26
Arbeitsgruppe Characeen Deutschlands (2016)	20	1	1	10	1	4	Germany	37

Table 5.2 Authors of currently accepted European *Characeae* species

	Number of species	Year(s) of publication
C. Linnaeus (1707–1778)	4	1753
O. F. Müller (1730–1784	1	1777
A. J. Krocker (1744–1823)	1	1814
R. L. Desfontaines (1750–1833)	1	1799
J. L. Thuillier (1757–1822)	2	1799
A. W. Roth (1757–1834)	1	1800
J. E. Smith (1759–1828)	1	1810
C. H. Persoon (1761–1836)	1	1807
C. C. Gmelin (1762–1832)	1	1826
C. L. Willdenow (1765–1812)	1	1809
J. L. A. Loiseleur-Deslongchamps (1774–1849)	1	1810
A. P. de Candolle (1778–1841)	2	1813, 1815
N. A. Desvaux (1784–1856)	3	1809, 1810
C. J. Hartman (1790–1849)	1	1820
K. F. W. Wallroth (1792–1857)	1	1833
J. H. Wallman (1792–1853)	1	1853
M. C. Durieu de Maisonneuve (1796–1878)	2	1848/1849, 1859
L. A. de Brébisson (1798–1872)	1	1849
A. S. Bruzelius (1799–1865)	1	1824
F. J. F. Meyen (1804–1840)	1	1827
A. K. H. Braun (1805–1877)	10	1834, 1835, 1849, 1855, 1863, 1864, 1878
F. G. Kützing (1807–1893)	5	1832, 1834, 1845, 1849
C. F. Lessing (1809–1862)	1	1835
F. J. Ruprecht (1814–1870)	1	1845
L. J. Wahlstedt (1836–1917)	1	1862
C. F. O. Nordstedt (1838–1924)	1	1868
T. F. Allen (1837–1902)	1	1888
N. Filarszky (1858–1941)	1	1930
J. Groves (1858–1933)	1	1937
E. F. A. W. Migula (1863–1938)	1	1904
A. Gonçalves da Cunha (1902–1986)	2	1935
M. M. Hollerbach (1907–1989)	2	1960, 1981
R. J. Corillion (1908–1997)	1	1960
G. Feldmann-Mazoyer (1910–1994)	1	1945
T. Sawa (1929–2013)	1	1973
A. Langangen (1942–)	1	2010
R. Becker (1960–)	1	2022

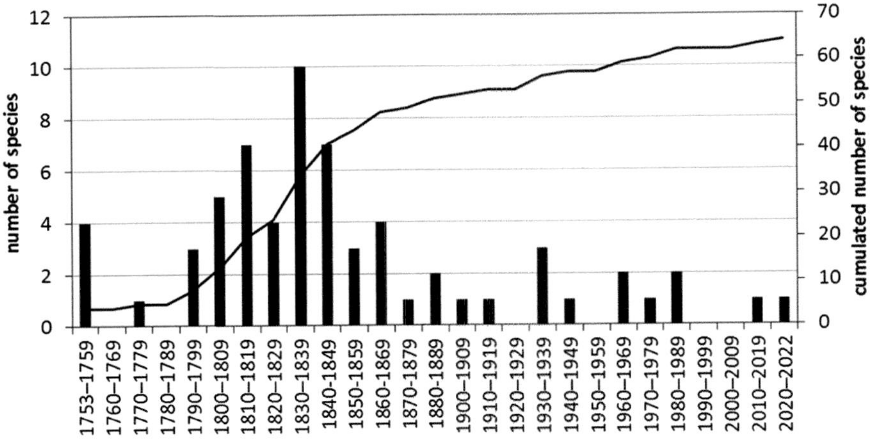

Fig. 5.1 The description of European *Characeae* between 1753 and 2022. Bars: number of species described per period; line: cumulative number of species

Hartman (1820) published *Chara hispida* var. *baltica* Hartm. (= *Chara baltica* (Hartm.) Bruz.). Bruzelius (1824) gave a worldwide overview of the *Characeae* and published *Chara opaca* Bruz. (= *Nitella opaca* (Bruz.) C. Agardh). He divided *Chara* in 4 tribes: Tribus *Charae vulgaris*; Tribus *Charae Hispidae*; Tribus *Charae Tomentosae*: and Tribus *Charae Flexilis*. Agardh (1824) distinguished the genus *Nitella* and repeated, more or less, Bruzelius 'worldwide' overview. Gmelin (1826) described *Chara braunii* Gmelin and Meyen (1827) and *Chara barbata* Meyen (= *Lychnothamnus barbatus* (Meyen) Leonh.).

Sixteen new taxa were added between 1830 and 1849: In 1832, *Chara aculeolata* was described by F. T. Kützing in Reichenbach (1830–1832). Wallroth (1833) published *Chara papulosa* (*Lamprothamnium papulosum*) in an account of German *Characeae* containing 11 species and several varieties. Beginning in the 1830s, Alexander Braun (1805–1877) was for 5 decades the prominent figure in European *Characeae*. A sketch in the unfinished Flora Badensis Alsatica of C. C. Gmelin was his first work (Braun, 1834). This "esquisse monografique" covered the whole world and recognised 23 Chara species with several varieties. *Chara mucronata* (= *Nitella mucronata*) and *Chara prolifera* (= *Sphaerochara prolifera*) were European species new to science in that work and are still accepted. In the same year he gave a lecture about *Chara*-species which was printed in the magazine Flora (Braun, 1835). Here he recognised 30 species in 3 sections: *Chara* Sect. *Charae epigynae* with today's species of *Nitella, Nitellopsis Sphaerochara* and *Tolypella*, *Chara* Sect. *Charae pleurogynae* with *Lamprothamnium* (originally *Lamprothamnus*) and *Lychnothamnus* and Chara Sect. Charae hypogynae with today's *Chara* species. The number of extra-European species was still just four. Species that were new to science in that work and are still accepted were *Chara connivens* and *Chara tenuispina*. The infraspecific differentiation was modest but complicated. He had an additional 9 "Abarten" (varieties) and 1 form. *Chara ceratophylla* Wallr. was divided into two "Hauptformen" (main forms) and *Chara tomentosa* belonged to one of them. His concept was rather modern in dividing species into geographical subspecies in *Chara coronata* Bisch. (= *Chara braunii*) and *Chara crinita* Wallr. (= *Chara canescens*). Braun did not always respect the law of priority. Kützing (1834) added *Chara papillosa* and *Chara virgata*. The latter species was already known under the illegitimate name *Chara delicatula*. Lessing (1835) described *Chara kirghisorum* from the eastern border of Europe. *Chara contraria* Kützing (1845) was described surprisingly late for such a common species, so it is possible that there are older names for this taxon. Kützing (1849) divided *Characeae* into three genera: *Nitella*, *Charopsis* (with *Chara braunii* and *Chara* (*Lychnothamnus*) *barbata*) and *Chara*. Ruprecht (1845) introduced *Chara* section *Lychnothamnus* and described *Chara subspinosa*. Durieu de Maisonneuve (1846–1855) published *Chara imperfecta* in 1848 or 1849. Rabenhorst (1847) gave an overview on German species of *Characeae*, Kützing (1849) produced a worldwide overview with numerous extra-European species, wherein *Chara aspera* var. *crassior* (= the somewhat controversial *Chara curta*) was described. Braun (1849) published a thorough account of the Swiss species with *Chara strigosa* and *Chara baueri*, the latter a replacement name for the illegitimate *Chara scoparia* Rchb. He used a rich infraspecific differentiation incompatible with currently recognised ranks. Brébisson (1849) published *Nitella gracilis* var. *confervacea* (= *Nitella confervacea*). This species was already known to Reichenbach (1830) under the misapplied name *Chara batrachosperma*.

In the second half of the nineteenth century 9 more species were added to the list of European *Characeae* recognised today. Wallman (1853) gave an extended worldwide overview (Table 5.3) on the *Characeae* with 12 new European species but only *Nitella wahlbergiana* is accepted today. *Chara filiformis* was described by A. Braun in Herzsch (1855). Durieu de Maissonneuve (1859) described *Chara fragifera*, Wahlstedt (1862) provided us with *Chara horrida*, A. Braun in Leonhardi (1863) described *Nitella ornithopoda* and A. Braun in Leonhardi (1864) recognised *Chara dissoluta*. Braun (1867) gave a concise overview of the European *Characeae* with 41 species, and *Nitella normanniana* was described by Nordstedt in Norman (1868). Shortly before his death A. Braun in Ascherson (1878) introduced the elusive *Chara succincta* (= *Lamprothamnium succinctum*) from an Egyptian oasis. Allen, in North America (1888), published *Tolypella hispanica*. Five years after Braun's death Nordstedt published "Monographie der Characeen" based on Braun's publications and his legacy (Braun and Nordstedt, 1883). It contains a rich infraspecific differentiation, e.g. *Chara foetida* (= *Chara vulgaris*) is divided into 9 subspecies and many more forms and varieties are mentioned. The nominate concept (i.e. where the type variety retains the name of the species) was not adopted so it is often uncertain which of the infraspecific taxa contains the type. Between 1889 and 1897 Migula published "Die Characeen Deutschlands, Österreichs und der Schweiz" (Migula, 1897) in which, for example *C. foetida* (= *C. vulgaris*) is divided into four "Reihen" (sequences) with 46, 13, 9 and 2 forms respectively.

Table 5.3 Monographic treatments of European *Characeae*

	1753	1807	1824a	1824b	1833	1834	1835	1845a	1845b	1847	1849	1853	1862	1867	1882	1889	1910	1913	1920	1925	1957	1965	1983	1986	1997	2002	2004	2007	2008	2016
	G	G	G	G	D	G	G	D	R	D	G	G	Sc	Eu	G	Eu	Sp	F	BI	D	W	G	UR	BI	Eu	BI	BS	NC	IP	D
Chara aculeolata								C			C	C	C S	C S	C S	C S		C	C	C S	C	C S	C S		C S	C		C S		C
C. aspera			C	C	C S	C	C S	C S	C S	C	C	C	C	C	C	C	C	C	C	C	C	C	C	C	C	C	C	C	C	C
C. baltica			C	C		C	C S	C		C	C	C	C	C	C S	C		C	C		C	C	C	C	C	C	C	C		C
C. baueri						C S	C S	C S		C S	C S	C S		C S	C S	C S	C S	C S		C S		C	C		C			C		C
C. braunii					C S	C S	C S	●	C S	●	C S	C		C S	C S	C S			C	C S	C	C	C	C	C	C		C	C	C
C. canescens			C S	C S	C S	C S	C S	C S	C S	C S	C S	C S	C S	C S	C S	C S	C S	C S	C	C S	C	C	C	C	C	C	C	C	C	C
C. conimbrigensis																					C	C								
C. connivens							C				C	C		C	C	C	C	C	C	C	C	C	C	C	C	C	C	C	C	C
C. contraria								C		C	C	C	C	C	C	C	C	C	C	C	C	C	C		C	C	C	C		C
C. corfuensis																						C								
C. curta																C			C S			C	C S		C S	C		C		
C. dissoluta														C		C			C	C		C	C S		C S	C S		C S		C
C. filiformis														C S		C S				C S		C	C		C			C		C
C. fragifera														C	C	C	C	C	C		C	C	C	C	C	C			C	
C. galioides														C	C	C	C	C		C	C	C	C		C				C	
C. globata																						C	C S							
C. globularis		C			C S	C S	C S	C S	C S	C S	C S	C S	C S	C S	C S	C S	C S	C S	C S	C S	C S	C	C S	C	C	C	C	C	C S	C
C. hellenica																														
C. hispida	C	C	C	C	C	C	C	C	C S	C S	C	C	C	C	C	C	C	C S	C	C	C	C	C	C	C	C	C	C	C	C
C. horrida													C	C		C	C			C		C	C		C		C	C		C
C. imperfecta														C	C	C	C	C			C	C			C				C	
C. kirghisorum												C			C							C	C							

(continued)

Table 5.3 (continued)

	1753	1807	1824a	1824b	1833	1834	1835	1845a	1845b	1847	1849	1853	1862	1867	1882	1889	1910	1913	1920	1925	1957	1965	1983	1986	1997	2002	2004	2007	2008	2016
C. oedophylla																						C								
C. papillosa								C			C	C		C S	C S	C S	C S			C S		C	C S	C S	C S	C S	C S	C S		C
C. squamosa		C	C	C		C	C				C S	C	C S	C S		C	C S			C S		C	C S		C S					
C. strigosa											C	C		C	C	C		C		C	C	C	C		C			C S		C
C. subspinosa														C S		C S			C	C S		C	C S		C S	C S		C S		C
C. tenuispina							C	C		C	C	C		C	C	C				C	C	C	C		C					C
C. tomentosa	C	C	C	C	C S	C S	C S	C	C	C S	C	C	C S	C S	C S	C S	C S	C S	C	C S	C	C	C	C	C	C	C	C	C	C
C. uzbekistanica																							C							
C. virgata									C S							C S			C S	C S	C S	C	C S		C S	C	C	C		C
C. vulgaris	C	C	C	C	C	C S	C S	C	C	C S	C	C S	C S	C S	C S	C S	C S	C	C	C S	C	C	C	C	C	C	C	C	C	C
Lamprothamnium papulosum					●		●				●	●		●	La S	La S	La S	La S	La	La S	La	La	La	La	La	La	La	La	La	La
La. succinctum																						La								
Lychnothamnus barbatus					●	●	●	●		●	●	●		●	Ly	Ly		Ly		Ly	Ly	Ly	Ly		Ly					Ly
Nitella capillaris			●	N S		●			●			N S		N S	N S	N S	N S	N S	N	N S	N	N S	N	N	N	N		N		N
N. confervacea											N S	N S	N S	N	N S	N	N S	N	N S	N	N S	N	N	N	N S	N	N	N S	N	N
N. flexilis	●	●	●	N	●	●	●	N		N	N	N	N	N	N	N	N	N	N	N	N	N	N	N	N	N	N	N	N	N
N. gracilis			●	N	●	●	●	N	●	N	N	N	N	N	N	N		N	N	N	N	N	N	N	N	N	N	N	N	N
N. hyalina			●	N		●	●	N	●	N	N	N		N	N	N	N	N	N	N	N	N	N	N	N	N	N	N	N	N
N. mucronata						●	●	N	●		N	N	N		N	N	N	N	N	N	N	N	N		N	N	N	N	N	N
N. opaca			●	N		●						N	N	N	N	N		N	N	N	N		N		N	N	N	N		N

(continued)

Table 5.3 (continued)

	1753	1807	1824a	1824b	1833	1834	1835	1845a	1845b	1847	1849	1853	1862	1867	1882	1889	1910	1913	1920	1925	1957	1965	1983	1986	1997	2002	2004	2007	2008	2016
N. ornithopoda														N	N	N		N			N	N	N		N				N	
N. syncarpa								N		N	N		N	N	N	N	N	N		N	N	<u>N</u>	N		N		N	N	N	N
N. tenuissima						●	●	N		N	N	N		N	N	N	N	N	N	N	N	<u>N</u>	N	N	N	N		N	N	N
N. translucens		●	●	N	●	●	●				N	N	N	N	N	N	N	N	N	N	N	<u>N</u>	N	N	N	N		N	N	N
N. wahlbergiana												N	N			N											N	N		
Nitellopsis obtusa						●	●	●	●	●	●	●		●	●	●	●	Ni S	Ni	●	Ni	<u>Ni</u>	Ni	Ni	Ni	Ni	Ni	Ni	Ni	Ni
Sphaerochara canadensis																									T			T		
Tolypella glomerata											●	●		●	T	T	T	T	T	T	T	T	T		T	T		T	T	T
T. hispanica																T	T			T	T		T		T				T	
S. intricata				●		●	●	●			●	●		●	T	T		T	T		T	<u>T</u>	T	T	T	T		T		T
T. nidifica			●	●				●	●	●	●			●	T	T			T	T	T	<u>T</u>	T	T	T	T	T	T		T
T. normaniana																T				T		T	T		T			T		
S. prolifera						●	●	●		●		●		●	T	T		T	T		T	T	T	T	T	T			T	T
T. salina																									T				T	

Black dot: Homo- or heterotypic synonym in another genus; addition "S": heterotypic synonym in the same genus; *underlined* entry: macrospecies in Wood (1965). Authors: 1753-Linnaeus; 1807-Persoon; 1824a-Bruzelius; 1824b-Agardh; 1833-Wallroth; 1834-Braun; 1835-Braun; 1845a-Kützing; 1845b-Ruprecht; 1847-Rabenhorst; 1849-Kützing; 1853-Wallman; 1862-Wahlstedt; 1867-Braun; 1882-Braun and Nordstedt; 1889-Migula (1889–1897); 1910-Reyes Prosper; 1913-Hy; 1920-Groves and Bullock-Webster (1920 and 1924); 1925-Migula; 1957-Corillion; 1965-Wood; 1983-Hollerbach and Krassavina; 1986-Moore; 1997-Krause; 2002-Bryant et al.; 2004-Schubert and Blindow; 2007-Langangen; 2008-Cirujano and et al.; 2016-AG Characeen Deutschlands
Geographic coverage abbreviations: G-Global, D-Germany, R-Russia, Sc-Scandinavia, Eu-Europe, Sp-Spain, F-France, W-Western Europe, UR-USSR, BI-Britain and Ireland, BS-Baltic Sea, NC-Nordic Countries, IP-Iberian Peninsula
Genus abbreviations: C-*Chara*, La-*Lamprothamnium*, N-*Nitella*, T-*Tolypella*

In the first half of the twentieth century few species were added to the European charophyte flora, but a couple of geographical surveys appeared. Migula (1904) described *Chara globata*. Reyes-Prospér (1910) gave an overview of Spain's *Characeae* with 26 currently recognised species, Hy (1913) recognised 31 species for France, Groves and Bullock-Webster (1920, 1924) recognised 30 species for Britain and Ireland and Migula (1925) recognised 40 species for Germany. Groves and Bullock-Webster (1920, 1924) applied rigorously the accepted nomenclatural rules and changed the names of a couple of species. Nándor Filarszky worked on a world-wide treatment after having described *Chara mucronata* (Filarszky, 1930) and dealing with the Hungarian species (Filarszky, 1937) but this was never published. A few later additions to the European charophyte flora included *Chara conimbrigensis* and *Tolypella palhinhae* (= *Nitella palhinhae*) by da Cunha (1935), *Chara corfuensis* by Groves in Filarszky (1937) and *Chara oedophylla* by Feldmann (1945).

After 1950 only 4 species[1] were added. Corillion (1957) published a thorough account of the *Characeae* of Western Europe without describing new species, although in 1960 he described *Tolypella salina* (Corillion, 1960). Hollerbach (1960) described *Chara uzbekistanica*. Sawa (1973) published *Tolypella canadensis* (= *Sphaerochara canadensis*). Hollerbach (1981) described *Chara neglecta*. Hollerbach and Krassavina (1983) gave an overview on the *Characeae* of the USSR, Moore (1986) provided a treatment for Great Britain and Ireland with a rather wide species concept and only 22 species (largely following Wood, 1965). Krause (1997) published an influential work about European *Characeae* with 46 species. Bryant et al. (2002) published another species account for the British Isles this time with 31 species. Schubert and Blindow (2004) gave an overview of the Baltic Sea with 22 species, Langangen (2007) for the Nordic Countries (Iceland, Scandinavia) with 36 species and Cirujano et al. (2008) with 25 species of the Iberian Peninsula. At the time of this review the latest member of family *Characeae* recognised for Europe was published by Langangen (2010) with *Chara hellenica*. Joop van Raam (1941–2011) prepared a thorough worldwide database for *Characeae* which covered numerous taxonomic and nomenclatural aspects. Conclusions from this database remained unpublished but parts were integrated in AlgaeBase (https://www.algaebase.org).

5.3 Taxonomic Concepts

5.3.1 Why Do It? Motivations for Grouping Individuals

The aim of this paragraph is to make our knowledge of species and what separates them easier to understand.

A discussion about taxonomic concepts can help us to understand the progression from taxonomy as a basic scientific concept to the use of that concept in applied

[1] *Chara oryzae* was published by Romanov et al. (2021) and is not treated in this chapter.

science. Whereas the naming of groups of individuals as taxa is considered to be basic science, the reasons to identify similar entities in ecology, plant geography, biomonitoring and biodiversity repositories are manifold, the common ground is to communicate any result by using existing, unanimously recognised, scientific names for entities that behave the same, look the same and have the same ecological role.

Taxonomy can be defined as circumscribing and classifying taxa. Taxa can be defined as groups of organisms, grouped by characteristics. These characters can be morphological, molecular or other traits. An integrative approach combining all criteria would be the best way to handle taxonomic information.

There are two crucial inputs into taxonomic work. Scientific names, agreed on since Linnaeus, are mononominals (genus names and higher rank names), binominals (species name as a genus name with species epithet) and trinominals (species name plus one infraspecific epithet, e.g. in varieties). Scientific names can help us to handle taxa in scientific communication. Scientific names are applied with reference to the International Code of Nomenclature for algae, fungi, and plants, ICN (Turland et al., 2018). The Code of Nomenclature is like a toolkit which provides rules for the correct application of scientific names. However, the ICN lacks taxonomical evaluation, i.e. there is no information about which scientific name should be accepted, only whether a scientific name is nomenclaturally correct or not. The main criteria of the ICN are whether or not a name was validly and effectively published and, if so, whether it is regarded as valid. For the correct application of a scientific name it must be linked to a nomenclatural type. Normally the herbarium specimen designated or used by the author who firstly validly described a taxon is the nomenclatural type. In many groups of plants old names lack types. Often original specimens are lost, destroyed, in bad condition or even represent previously described taxa. For these reasons typification is an ongoing task which can lead to different treatments in monographs (such as Wood, 1965; Krause, 1997; Arbeitsgruppe Characeen Deutschlands, 2016, this volume). The ICN provides measures to conserve or reject names or types if there is controversy.

A good example is *Chara hispida* L. which turned out to be problematic, its Linnaean type is a specimen of what is currently recognised as *Chara aspera*. Thus, *Chara* specialists applied to conserve the name *Chara hispida* with a new type according to Art. 14 of ICN (Gregor et al., 2014). This proposal was discussed and voted on by the Nomenclatural Committee for Algae (and was subsequently listed in the Annex of the ICN (Wiersema et al., 2020)). *Chara hispida* might be a good example of how the 'real *Chara hispida*' and what is published about the taxon *Chara hispida* may differ. For validity the ICN requires a description or a diagnosis, and for that to be effectively published. It is up to the author of a taxon to decide how meaningful the description or diagnosis is. Normally a description or diagnosis gives criteria but this does not automatically imply that individual plants or specimens can unequivocally be assigned to this species easily. This leads us to the question of species concepts. Species concepts are what authors of taxa have in their mind when describing species new to science. Species concepts, roughly spoken, provide the criteria for an assemblage of similar individuals to make a species (see below). The question what makes a species is an ongoing academic discourse, see Zachos (2016). The main differences between species concepts will be outlined in the next chapters.

In the *Characeae* many species have been described in the nineteenth century. Authors at this time were rarely explicit about their species concepts. This means that only implicit information can be deduced about their concepts.

The existence of different species concepts is why the naming of species can seem chaotic to non-taxonomists. However, there are ways in which we can identify the taxonomic concept used by particular authors (Berendsohn, 1995; Geoffroy and Berendsohn, 2003). The first mention of a name should also mention the publication or the author who created the taxon concept. For most charophyte species the names used by different authors are probably congruent, but there are always exceptions. Using an agreed taxonomic framework is most important when named taxa are associated with factual data (e.g. ecological parameters, distribution). Confusion can occur when taxa are misidentified, or later split, so retaining a link between the taxon concept (as given by the name and author) and the factual data is important. A good example of the way data and names are linked is shown in the recent Monograph of *Characeae* in Germany (Arbeitsgruppe Characeen Deutschlands, 2016).

5.3.2 Horizontal Versus Vertical Concepts—From "Sorting" to "Identification of Relatives"

Classification is as old as humankind. Organismal systems are much older than our understanding of evolution. A system of organisms, as a result of taxonomic work, helps us to navigate through biodiversity and to 'sort' individuals, i.e. identify and classify them.

From Linnean times the main criteria for descriptions and/or diagnoses were morphological. The hierarchical classification systems that have been set up on the basis of morphological research could consist of polyphyletic clades if there is convergent evolution of morphology.

Extant taxa, their subdivisions (i.e. varieties or forms), populations and individuals are living synchronously. Taxonomic treatments as well as identifications are like a horizontal cross-section through the Tree of Life. The main traits for horizontal concepts are morphological features, geographical occurrence and environmental parameters. These traits can be handled relatively easy if the source of information is known. Horizontal taxonomic concepts need rich descriptions and references to other taxa to be transparent and to allow individual specimens to be identified. However, often there are knowledge gaps for characters and parameters, as well as our knowledge of which characters are variable and which are stable.

With the advent of molecular methods it has become possible to find natural, evolutionarily related or monophyletic groups of taxa. This evolutionary perspective is called a "vertical concept". All biological species are products of evolution in space and time. All described taxa represent those species more or less adequately.

A lineage on any taxonomic rank is a vertical branch in the Tree of Life. This branch could be isolated for a long time or may develop by dividing into two branches over a long time. In the case of division the genetic diversification might not be synchronous with morphological diversification. This can lead to conflicting morphological versus molecular characters.

Generally, all evidence based approaches in phycology have resulted in both morphologically overestimated taxa, and morphologically underestimated taxa (i.e. the number of taxa based on morphology can be either fewer or more than the taxa based on phylogeny). However, the main challenge nowadays is to align horizontal and vertical taxon concepts. Vertical concepts have the advantage of resulting in natural groupings; horizontal concepts have the advantage of being linked to historical data from centuries of empirical investigation. Science can use the best parts of all the different approaches, as long as the taxonomic concepts are clearly defined.

5.3.3 Selected Horizontal Concepts

5.3.3.1 Why Horizontal Concepts?

As outlined above, the different "species concepts" are based on different approaches as well as aiming for different outcomes. When sorting individuals into categories that allow for all kinds of biogeographical mapping, a horizontal concept, which does not require any information about phylogenetic relationship or evolutionary background would be sufficient. The following introduces these horizontal concepts by presenting three purely theory-driven basic forms, followed by a fourth section, where the application of horizontal concepts is demonstrated.

5.3.3.2 Morphological (Phenetic) Species Concept

The morphological species concept is likely to be the first species concept in history, as it sorts individuals to species on the basis of their appearance. The primary sorting criterion is morphological similarity, defined by characters that are shared by all individuals belonging to the respective species, but not with other species. The main problem is "choosing the appropriate characters". It becomes obvious that some easily recognisable characters are not generally applicable—e.g. size changes with growth and even the size when "fully mature" depends on growth conditions. Several attempts have been made to develop principles for choosing the best descriptive characters. An early attempt was made by Jungius (1747), who formulated principles which at least in part can be still found in the International Code of Nomenclature for algae, fungi and plants (Turland et al., 2018). However, the morphological "species concept" is actually a description of the nature of the characters used rather than a concept, because it does not define the limits of species delineation. It also assumes that character expression is discontinuous, allowing for separation of distinct groups along these discontinuities. Such an attempt becomes complicated when distinguishing closely related entities or, what can be even more problematic, simple structured entities with a restricted set of characters e.g., unicells. On the other hand, it can be applied to organisms reproducing vegetatively, parthenogenetically or requiring decades before becoming reproductive, as is often the case for botanical subjects. The main problem, however—having no objective criterion for definition of what a "species" is in terms of morphological similarity/dissimilarity—results in a situation that allows any morphological difference to be used to distinguish a new taxon at species level. In these cases the final acceptance of a new species is made by consensus. Consequently, the type-method (i.e. reference to a single 'typical' specimen) developed as an instrument for independent preservation of opinions. The type specimen, which is selected as a reference point when a plant species is first named, allows for later comparison with other specimens and the description of new species by comparison of all the characters of a specimen whether or not they were referred to in the original description.

5.3.3.3 Biological Species Concept

The biological species concept attempts to introduce an objective criterion for ranking taxa at the level of species. It can facilitate the decision whether or not discrete populations belong to the same species in the Linnaean hierarchy. In one of its latest versions, Mayr (2000) defined them as follows: "*biological species... [are] ...groups of interbreeding natural populations that are reproductively isolated from other such groups*." (cf. Mayr, 2000, p. 17).[2] In brief, this concept is based on the criterion of interfertility and consequently offers not only objectivity, but also links horizontal and vertical species concepts.

In the fully developed version Mayr (2000) refers also to several problems arising from this restriction to interfertility as the decisive character for ranking. Hybridisation, for example, may occur without harming the species status—it is regarded as a "mistake" of an individual. On the other hand spatially or chronologically isolated populations may belong to the same species as long as they have the capacity to interbreed when isolation is terminated.

Naturally, the biological species concept cannot be applied to species that only reproduce asexually or parthenogenetically. Another problem is the way Mayr (2000) deals with the problem of "hybridisation"—taking it as a simple "mistake" by individuals does not reflect the large variety of mechanisms to prevent such a mistake, ranging from offspring sterility to temporal differences in reproductive activity (Fig. 5.2).

It must be accepted that the biological species concept is the best known, often the only concept known to biologists. Probably this is due to the fact that it presents

---

[2] This is just the first sentence of a much longer paragraph, which essence is only given in brief in the following. For anyone critically reflecting about the biological species concept reading the full article cited is highly recommended.

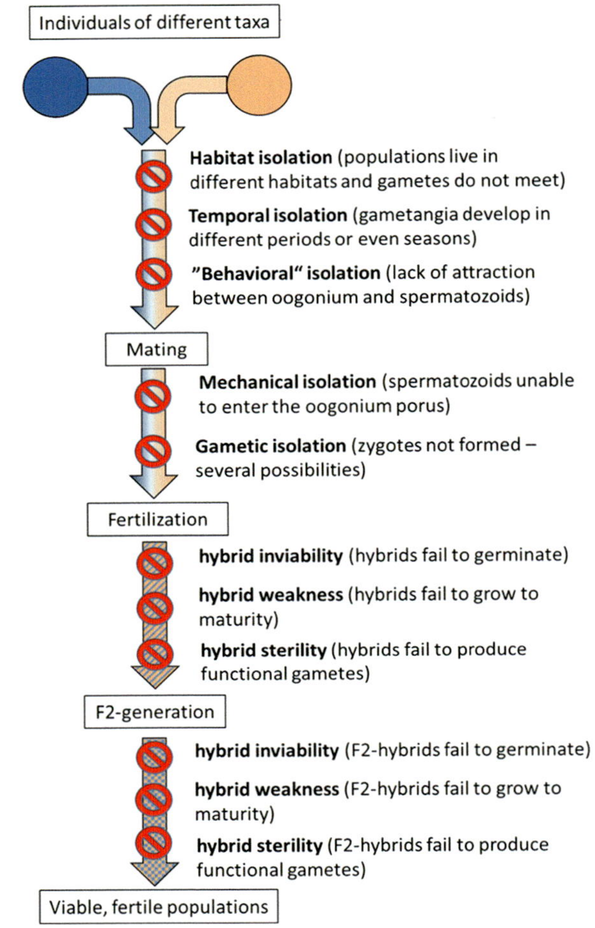

Fig. 5.2 Reproductive barriers in pre- and postzygotic development. Figure re-drawn from Turnbridge and Fitzpatrick (2017) with additions sensu Chen and Lin (2016)

an objective criterion for ranking species taxa and, moreover, this criterion, at first, sounds absolutely logical.

5.3.3.4 Ecological Species Concept

Probably the least known of the horizontal species concepts for a botanist, the ecological species concept, refers to niche occupation. It was defined as "a lineage (or a closely related set of lineages) which occupies an adaptive zone minimally different from that of any other lineage in its range and which evolves separately from all lineages outside its range" (cf. Van Valen, 1976). Reflecting in fact the practice for delineation of asexually, morphologically simple organisms as, e.g. bacteria, where the morphological as well as the biological concept were not applicable, the concept raised an ongoing debate about its validity especially after the influential review of Andersson (1990), who claimed that only this concept provides explanation of phenetic differentiation of lineages and, in agreement with Van Valen (1976), stated that "species are maintained for the most part ecologically, not reproductively" (cf. Van Valen, 1976, p. 235). This review was answered by a number of critical comments (e.g. Bremer and Eriksson, 1992; Grant, 1992) but their criticisms are more focused on Andersson's conclusions about limitations of the biological species concept instead of criticising the ecological species concept per se. The main problem of the ecological species concept is the vague definition of the "adaptive zone"—a term which can be taken as a synonym of the "ecological niche" sensu Hutchinson (1957). As outlined in a more recent review (Holt, 2009) the niche concept is still a valuable tool in ecology, but applying it to taxonomy results in several problems. At first sight, niche occupation could be taken as an objective criterion, similar to interfertility of the biological concept. But on the other hand, niche dimensions are not defined and consequently can be increased subjectively until the result confirms the pre-set expectations, not to mention the problems arising from differences between the fundamental and realised niche of a taxon.

5.3.3.5 Synthesis: Achieving an Agreement by Pluralistic Concepts

As outlined above, there exist three main approaches to rank a given taxon to the species level in the Linnaean hierarchy without information about its phylogeny—by morphology, by interfertility and by niche occupation. All of them have their pros and cons. Morphology is, as long as ultrastructural characters are not required, easy to assess, but the choice of decisive characters can be subjective, and character expression can be altered by several factors e.g. habitat condition, ontogenesis, sex; and may change substantially within the life cycle. Interfertility has the potential to serve as an objective criterion, but cannot be applied to all taxa reproducing asexually or parthenogenetically. Ecological niches too have the potential to serve as an objective criterion, but there is an indefinite number of niche dimensions, resulting in subjectivity again. Applied in practice, none of these three "concepts" solve all aspects of the "species-rank-problem" in full. The taxonomic practitioners consequently developed their own, individual concepts, many of which merge aspects of the above basic concepts, until agreement was reached. In a simplified sense, it could be stated that botany largely refers to the morphological concept, amending it by elements of the biological concept, if required. This is the case for polymorphic or dimorphic "species" where, irrespective of profound morphological differences, gametophytes and sporophytes or male and female individuals are merged to one "species". On other occasions arguments derived from the biological species concept are rejected e.g., *Chara globularis*, which is still amalgamated as a morphospecies irrespective of the results of Proctor (1971) who demonstrated the existence of fertility barriers. The same holds true for microbiology (based on the ecological species concept) and zoology (based on the biological species concept). Both disciplines merge their basic concept with elements of the others until the results, a so-called "pluralistic concept", are commonly accepted. In zoology, for example, tiger and lion are able to interbreed and also the offspring (tigon or liger) has been proven to be interfertile (producing a "litigon"). An ecological argument added to the basic biological species concept, that fertile offspring must be produced in nature, meets the expectations of all who want these taxa to be regarded as separate species.

5.3.4 *Vertical Concepts*

5.3.4.1 Motivation for Developing Vertical Concepts

Vertical species concepts try to circumscribe species based on lineages with consideration of their historical dimension. Since it is hard to reconstruct the real evolutionary history of a species, vertical concepts are quite theoretical concepts and mainly used for phylogenetic reconstructions. Like the horizontal species concepts several different vertical species concepts exist. They all define a species based on its evolutionary history but they differ in the way in which lineages/branches/tips are considered species.

The first vertical species concept was Hennig's Cladistic species concept (Hennig, 1950, 1966). Hennig agreed with Mayr that species are reproductive communities but this criterion alone was less important for him. Therefore, he proposed that if tokogenetic (i.e. the ancestor-decendent) relationships between individuals no longer exist, the individual lineages have split into two species and the stem species no longer exists. That means that according to Hennig (1950) there are only monophyletic species. Ridley has a very similar definition of a species: "*A species is that set of organisms between two speciation events, or between one speciation event and one extinction event, or that are descended from a speciation event*" (cf. Ridley, 1989).

5.3.4.2 Phylogenetic Species Concept(s)

Cladistic analyses of infraspecific groups result in artefactual patterns that look like, but fail to meet the assumptions of phylogenies (Davis and Nixon, 1992; Nixon and Wheeler, 1992; Wheeler, 1999). Therefore, several authors formulated independently new versions of phylogenetic species concepts that were consistent with phylogenetic theory and independent of cladistic analysis. The various phylogenetic species concepts (de Queiroz, 1998) may be classified (following Baum and Donoghue, 1995) into two major types: character-based and lineage-based concepts. The character-based concepts are the phylogenetic concepts of Mishler and Theriot (2000) and Wheeler and Platnick (2000). These concepts are based on shared derived characters (synapomorphies). The lineage-based concept is represented by Wiley's evolutionary species concept (Wiley, 1978).

Mishler and Theriot (2000) refer to a species as "*the least inclusive taxon recognized in a formal phylogenetic classification*", in their version of a phylogenetic species concept. They define a species as a monophyletic group of organisms, recognised as lineages on the basis of shared, derived characters and ranked as species because of causal factors that maintain the lineages as the smallest important monophyletic group recognised in a formal classification.

Wheeler and Platnick (2000) define a species as "*the smallest aggregation of (sexual) populations or (asexual) lineages diagnosable by a unique combination of character states*". According to this concept species are not necessarily monophyletic.

Wiley's evolutionary species concept can be summarised as a combination of a phylogenetic species concept with a biological species concept. An evolutionary species is a lineage of interbreeding organisms, reproductively isolated from other lineages that have a beginning, an end, and a distinct evolutionary trajectory (Wiley, 1978). The beginning of a species' existence is speciation, as a population becomes reproductively isolated from a parent population. The end of a species occurs either with extinction or with the branching of the species into one or more descendants.

5.3.4.3 Support by Genetic Data?—Challenges and Limitations

As for all concepts the big challenge is to choose the appropriate characters for species identification. In the case of phylogenetic species delimitation, the chosen characters should reflect the evolutionary history of a species and put it in relationship with its closest relatives. The data (genetic markers) used have to be autoapomorphic (derived trait unique for a species/taxon) for each species to be characterised as an individual species, and synapomorphic (derived trait that is shared between sister-taxa and their direct progenitor) for species to be placed correctly in the phylogenetic tree. The issue arising here is that for reconstructing the "real" phylogeny of a species, all descendants of a progenitor are needed for the investigation to make correct assumptions. This is unfortunately not always possible because descendants might not be found (because they are very rare or even have died out) or are sometimes not recognised as belonging to this relationship and therefore not considered in investigations. As well as this, evolution of a species is not always linear but often more reticulate, especially in plants. So, when using different genetic markers each of these markers might show a slightly different phylogeny because each gene/genetic region has its own historic lineage. These points make it clear that phylogenetic reconstructions are not always straight forward and that results have to be viewed critically to draw reasonable conclusions.

5.4 Charophyte Taxonomy

5.4.1 The Botanical Taxonomic Concept and the Problems Arising from Applying This Higher-Plant Derived Concept to Algae

The science of botanical taxonomy was initially developed to classify flowering plants. Historically the taxonomy of charophytes relied on an almost entirely morphological species concept as described above (Nowak et al., 2016). Most of the characters used for species delineation in charophytes are vegetative morphological characters with few or weak links to reproductive traits. In contrast to lower-plants, higher-plants have a broader set of characters related to reproduction (floral structures) which are much more stable and better conserved than vegetative characters. As aquatic organisms, charophytes are highly dependent on their surrounding environment. This can lead to high levels of phenotypic variance due to differing ecological conditions (Casanova, 2005). High variability within species paired with apparent low variability among species makes the charophytes a taxonomically difficult group (Ahmadi et al., 2012). These difficulties in species delineation have led to two competing concepts about how much phenotypic variation is accepted within species as taxonomic units especially in the genus *Chara* (Nowak et al., 2016). Both of these concepts use similar vegetative characters for species diagnosis, but they differ in their conclusions regarding the taxonomic relevance of these traits (Nowak et al., 2016). Reproductive features (e.g. dioecism versus monoecism) in addition to the phenetic characters can be used for species delineation.

Apart from the difficulties for determination at species level there are also some poorly resolved questions about the higher level taxonomy of charophytes, as today, only parts of a previously much larger group, are present. This means that while reconstructing the evolutionary history of plants one will come to points where linkage groups are missing because they have gone extinct. From the top down in the hierarchical classification of charophytes, they belong to the Chloroplastida (green plants), within them to the Streptophyta and Phragmoplastophyta (Nishiyama et al., 2018; Wickett et al., 2014). The relationship between charophytes and Embryophyta (land plants) is still debated as the morphological, anatomical and physiological characters show a slightly different pattern to the molecular data. However, one can agree that charophytes and land plants share a common ancestor.

Today the charophytes are represented by one family—*Characeae* which can be subdivided into two tribes—*Chareae* and *Nitelleae* (McCourt et al., 1996). Tribe *Chareae* includes four genera (*Chara, Lamprothaminum, Nitellopsis* and *Lychnothamnus*), tribe *Nitelleae* includes three genera (*Nitella, Tolypella* and *Sphaerochara*). This classification of *Characeae* into two tribes and seven genera is more or less accepted by most Charophyte-specialists although not unambiguously because based on morphology, chromosome studies and molecular data not all tribes and genera represent monophyletic groups (Caisová and Gąbka, 2009 and references therein). The recognition of *Sphaerochara* is new, but well-supported by morphological, genetic and palaeontological data (Soulié-Märsche and García, 2015).

5.4.2 Historical Development of "Classical" Charophyte Concepts

As mentioned above (see Sect. 5.2) the starting point for delineation of 'species' in the family *Characeae* is Linnaeus' *Species Plantarum* (1753) (Horn af Rantzien and Olsen, 1949). The morphological features used to distinguish the four species he recognised were the presence, shape and clustering of the spine cells (*aculeis caulinis*), the presence of bract cells internal to the branchlets (*frondibus interne dentatis*), and the transparency and flexibility of the thallus. Gradually a more complete description of the morphological characters, and a separate language, or glossary of terms, for those characters, became necessary to distinguish different species and genera. The number of new species increased rapidly and Agardh (1820) introduced a second genus: *Nitella*, for specimens without cortication; Leonhardi (1863) separated *Tolypella* from that group. Subsequently erected genera that are still recognised today are *Lychnothamnus, Lamprothamnium* (= *Lamprothamnus* nom illeg.) and *Nitellopsis,* the most recent being *Sphaerochara.* During the days of colonialism and exploration, many more species were discovered and documented from outside Europe. In some cases the new specimens had characters that were similar to those in the European flora (e.g. *Chara globularis*), and were either given that name, or designated as a variety or form of the familiar species. However, many others were recognised as distinctly different and described as new species. It must be remembered that before the 1860s many of the charophytologists based their species concepts on the idea that God created all species in the space of a few days, and there was no idea of evolution or speciation. Additionally, many of the specimens received from far-flung places were dried, or in poor condition (Nordstedt, 1918), and such specimens were always put into the context of species already known (e.g. *having the appearance of a small Nitella gracilis*). The work of describing species was about documenting the fixed number of 'kinds' created by God, and not about understanding or explaining diversity.

The early, often regional, works culminated in the production of a monograph, compiled initially by Alexander Braun, and posthumously completed from Braun's notes together with his own observations by Otto Nordstedt (Braun and Nordstedt, 1883). This was a substantial body of work, the culmination of Braun's life-time of collecting, examining and describing specimens from many parts of the world. Although Braun had a tendency to allocate new names where the prior one did not suit, his morphological species concept for the European continent was reasonably broad, and it appears he was content to indicate how particular specimens varied from the type while still including them in the species (sometimes as varieties or forms).

In the latter part of the nineteenth century and up to the middle of the twentieth century there appeared to be a tendency for experts to work in isolation on their own regional florae (see Sect. 5.2). Migula's approach of compiling a broader overview was to recognise and describe the plethora of variation, providing many of the same names (e.g. *subhispida, elongata*) for forms in different species (Migula, 1897, 1925). This resulted in more than 20 different compound names in many common species, and these were used on specimen labels into the twentieth century by collectors such as Nordstedt, Wahlstedt and Olsson. It is likely that these 'forms' were not construed as a taxonomic distinction (in the sense that they represented distinct entities), rather, they were an added description of the morphology of individual specimens (as we might today say e.g. '*a specimen of Chara vulgaris, but with long internodes and longer bract-cells than normal….*'). It was during this time that many gatherings were made and deposited in herbaria, and the morphological species concept was solidified. Given the variety of treatments in different countries, the tendency for authors to publish treatments in their own language, and for people in their own countries, it's not surprising that there was duplication of effort, and application of different names (species, varieties and forms) in different places for what would eventually be seen as the same species. In effect, at this time '*One man's form was another man's species and vice-versa*' (Anon., 1965). The development of the International Code of Botanical Nomenclature (first called a code after the Stockholm Botanical Congress in 1952, now called the International Code of Nomenclature for algae, fungi, and plants), and the sensible aspiration that the same species be called the same name throughout its range, invited comprehensive standardisation of the species concept in the latter part of the twentieth century.

5.4.3 "Ruling the World". A First Attempt at a Global Overview by Wood, Strengths and Weaknesses of the Macrospecies Concept

At about this time Dick Wood (Richard Dawson Wood 1918–1977) from Rhode Island in the United States came onto the charophyte scene. It seems that he started work on the *Characeae* in North Carolina prior to 1943, claiming authorship of a section on *Characeae* in Whitford's *Freshwater Algae of North Carolina* (Whitford, 1943; Wood, 1965, p. 851). His first solo publication on charophytes was in 1947 (Wood, 1947), and from then, he was author of several taxonomic and ecological works. He visited Guy Oldfield Allen in Britain during WWII when he was a member of the USA armed forces. From the correspondence between them (in BM), it would be fair to say that Wood was somewhat self-assured and thick-skinned, and Allen was not impressed by his approach to the systematics of the *Characeae*. It is apparent that Wood wanted to standardise nomenclature and taxonomy in the

Characeae, in line with the ICBN (1952) from his frequent references to it (Wood, 1965). Wood seems to have focused on two lines of evidence in relation to the definition of 'species' in family *Characeae*: firstly, the morphological changes that occur in corticated species under culture (e.g. Corillion, 1957); and secondly, the results of breeding studies undertaken by Macdonald and Hotchkiss (1956). Wood interpreted the changes in morphology as a kind of plasticity, whereby a single species could display a diversity of characters during growth. He misinterpreted the successful production of oospores (between a specimen identified as *Protochara australis* and a specimen identified as *Chara australis*) as being between monoecious and dioecious entities (calling them *C. australis* and *C. corallina* respectively: Wood, 1965, p. 27). In fact, both plants in Macdonald and Hotchkiss' study were dioecious, and both appear to be (in subsequent examination of the deposited specimens (AD)) *Chara australis*. He supported his contentions with his, unsuccessful, 'punched-card' investigation of herbarium specimens (Wood (1957): an early form of multivariate sorting) saying that 'it was found that within a given species certain characters occurred in all possible combinations but that some combinations occurred more frequently than others'. Wood's very comprehensive, but flawed, work resulted in his revision in two parts, consisting of the *Monograph of the Characeae* (Wood, 1965), and with Kozo Imahori, the *Iconograph of the Characeae* (Wood and Imahori, 1964). In this some hundreds of species of charophytes were amalgamated to just 19 species of *Chara*, 3 of *Lamprothamnium*, 1 of *Lychnothamnus*, 3 of *Nitellopsis*, 53 of *Nitella* and 2 of *Tolypella* (Doty, 1966). These species included 101 varieties and 395 forms. There was a somewhat luke-warm enthusiasm expressed about the publication of the *Monograph* (Anon., 1965; Doty, 1966; Moul, 1966), and most reviewers commented on the addition of the '*Microspecies Index*' in the last pages of the Monograph. It was as if (to quote Vernon Proctor) Wood wrote his 'Bible', then at the end said 'if all this is wrong then you can use *these* species names'; effectively rendering both the new nomenclature and the new combinations in the Microspecies Index invalid. The correspondence of genuine experts in the *Characeae* generally expressed some doubt about the usefulness of the revision (Allen letters, Williams letters), and in Europe it was never embraced. However, newcomers to charophytology generally attempted to use Wood's taxonomy, at least at first (e.g. Casanova and Brock, 1990, c.f. Casanova, 1997). In contrast, Imahori's illustrations and descriptions of individual specimens (often type material) remain useful to this day (Wood and Imahori, 1964).

A post-script to Wood's revision is that it stimulated further research into species concepts and morphological variation in the *Characeae* (e.g. Proctor, 1971). It even appears that in the light of those further studies (and further travel, discussion and examination of charophytes) Wood expressed the idea that the amalgamation of some monoecious and dioecious species might have been unjustified (Wood and Mason, 1977, p. 109). However, he also stated that (with reference to recognising such taxa) 'such a move would be premature at this stage' (Wood and Mason, 1977, p. 110).

5.4.4 *Calming Down the Battle of Concepts: Respect for Paleo-Data, Implementing Genetic Evidence but Still Largely Ignoring the Biological Concept*

It does appear that in English-speaking countries at least, Wood's revision failed to stimulate an inclusion of charophyte research and ecology in general ecological studies. Instead newcomers found charophytes difficult to identify, and found determining the occurrence of species in their own regions difficult. Even where regional florae were produced, a revision by Wood seemed to inhibit charophyte research by neophytes. However, even while Wood was working on his *Revision* others were asking important questions, and finding answers.

Vernon Proctor (who, like Wood, had been in the USA armed forces) started working on the dispersal of aquatic life (crustaceans, bryophytes and charophytes) in the 1950s, which led him to investigate dormancy in oospores. He was interested in the 'big questions': i.e. "*What species exist, where do they occur, and why is this so?*". Given that charophytes could be widely dispersed (Proctor, 1962), he investigated whether individual charophyte species were 'cosmopolitan' as asserted by Wood, by undertaking breeding studies, and investigating the biological species concept in the genus *Chara*. His early conclusion ([that each charophyte] *species consists of a complex of incipient, sibling species that are largely restricted to individual continents, if not local areas, or, in some instances, even single bodies of water*) was expanded in a review (Proctor, 1975). He worked on different sections of the genus over time (*Charopsis*: Proctor, 1970; *Grovesia*: Proctor, 1971; *Willdenowia*: Proctor and Wiman, 1971) and this led to a series of hypotheses or conjectures for future research (Proctor, 1980). These are worth repeating here (Table 5.4), as they are still relevant to current taxonomic and ecological research. It should be noted that Proctor (1980) did not put these forward as contentions, rather as conjectures to be examined and tested. Some of them (i.e. that Wood's concept of Section *Grovesia* is polyphyletic) have been shown to be true. Conjectures 8 and 9 appear to have been disproved on the basis of recent phylogenies. For others there is evidence both for and against. The main value of Proctor's (1980) summary paper has been to show a way forward in taxonomic research in the *Characeae*: always question the basis of the taxonomic approach, and find evidence for or against an hypothesis using the best available tools.

Proctor's breeding studies were not widely replicated in other labs. The difficulty of keeping and monitoring large numbers of clones for crossing was inhibitory for most charophytologists (who then, like now, are usually poorly funded). Instead new methods were developed, including Scanning Electron Microscopy (SEM) of oospore walls (John and Moore, 1987; John et al., 1990) and analysis of gene sequences. These methods involved investigation of hypotheses (e.g. *that Wood's species concepts encompass more than one species*), and in some cases supported Proctor's (1980) conjectures and provided evidence against Wood's (1962) species concept.

Table 5.4 Conjectures developed by Proctor (1980) concerning the nature of charophyte species and their biogeography

Conjecture subject	Statement
Capacity to form hybrids and how species evolve in *Characeae*	1. Hybrids between subsections of *Chara* are unknown and respective components of morphologically similar, dioecious-monoecious combinations are always reproductively isolated
Biogeography and sexual system	2. Dioecious charophytes are less vagile [i.e. likely to spread widely] than monoecious and consequently better indicators of evolutionary tracks
Biogeography and sexual system	3. Most charophytes are endemic to a single land mass but especially dioecious species
Biogeography and sexual system	4. Islands and especially remote oceanic islands, should support predominantly monoecious forms rather than dioecious
How species evolve in *Characeae*	5. Evolution has proceeded consistently from dioecism to monoecism within the Charophyta
Biogeography and sexual system	6. Dioecious representatives of a subsection normally occupy but a single landmass, adjacent landmasses, or landmasses previously in contact
Morphological species markers	7. *C. pseudohydropitys* is conspecific with the earlier named *C. hydropithys* which has the potential to produce both single ranked stipulodes and double throughout its range
Morphological subgeneric markers	8. No phylogenetically sound bases exist for separation of the genus into haplostephanous and diplostephanous subgenera
Phylogeny	9. Members of subsections *Willdenowia* and *Agardhia* are more closely related to one another, phylogenetically, than to those of any other section
Phylogeny	10. Section *Grovesia* is polyphyletic in origin, does not accurately reflect the evolutionary history of the genus and must be viewed as an artificial assemblage
Biogeography and sexual system, and how species evolve in *Characeae*	11. The first 6 conjectures also apply for *Nitella*

5.4.5 The Problem of Geographic Isolation—Biogeographic Issues Leading to Well-Funded "Splitting Attempts"

One of the historical artefacts of taxonomic investigation into the family *Characeae*, is that the first species were described from Europe (Sweden, Germany, UK), and so the European names were retained by Wood (1965) and others for many of the species in Europe, although the only place Wood was given much credence was in the UK (Moore, 1986). In other places (e.g. Japan and Australia) previously described species were placed under European names (Wood, 1965, 1972). It was very apparent in places outside Europe that Wood's species concept was not useful, especially

Table 5.5 Additional conjectures developed since 1980

Conjecture subject	Statement
Biogeography	1. Charophyte species distribution is controlled primarily by biological interactions and secondarily, water chemical and physical characteristics (Proctor, 1999)
Morphological species markers	2. Charophyte morphology is plastic in relation to the presence of herbivores (Proctor, 1992)
Morphological species markers	3. Specimens of *Nitella* and *Chara* with different construction of the oospore wall represent different species (Sakayama, Casanova)
Genetic species markers	4. Specimens of *Nitella* and *Chara* with significantly different gene sequences represent different species (Sakayama, Karol, McCourt, Meiers, etc.)
Morphological species markers	5. In section *Agardhia* characters of the basal branchlet segment can be indicative of speciation (Casanova, 2013)
Species markers	6. If oospores are different **and/or** gene sequences are different **and** distribution is different, specimens should be recognised as different species

when compared with predecessors' treatments (Braun and Nordstedt, 1883; Nordstedt, 1918; Imahori, 1954). This provided impetus to use the most recent molecular and morphological methods to test species concepts (Casanova, 2005; Sakayama et al., 2004a, b, 2005, 2010). Additionally, Proctor continued working on the causes of the distribution of different species, developing hypotheses about morphological adaptations and biological interactions (Proctor, 1990, 1999). These studies provide a new set of conjectures to be tested (Table 5.5), some of which are contrary (e.g. 2 and 5).

5.4.6 *Proof of Concept by Means of Systematics*

In Summary it can be stated that the recent taxonomic approach for charophytes arose from a purely morphological concept and has now developed as an agreement between scientists which, merging elements of horizontal as well as vertical concepts, is pluralistic by nature. Consequently, decisions about the rank of a specific taxon in Linnaean hierarchy, especially with regard to the species rank, are not made with a single set of arguments. In some cases, sexuality is regarded a good distinguishing character, making it impossible to discriminate between some species when infertile without genetic analysis (e.g. *Nitella flexilis/Nitella opaca*); in other cases (*Chara canescens*) bisexual and parthenogenetic individuals are lumped to one single species. In both cases, the arguments for the decision are based on phylogeny so far as it is known (e.g. Schaible et al., 2009a, 2009b). However, phylogenetic approaches lack any kind of definition about the distance required for ranking a certain genotype a "species". In fact, some of the decisions are application-driven as, e.g., in the *Chara hispida*-cluster ("*Hartmania* group", see also Schubert et al., 2016 and Chap. 15, this volume). This cluster consists of a number of morphologically distinct entities, which are linked by various intermediates and with only a few genetic differences detected so far (Nowak et al., 2016; Schneider et al., 2016). On the other hand, at least some of these taxa have indicative value, exhibiting clear biogeographical patterns (e.g. Nowak et al., 2017). For that reason, they are kept at species level, at least until further progress is made in elucidating the genetic basis of ontogenetic character expression. The set of markers currently in use represent proxies only, none of them is decisive for the expression of characters used for species delineation as, e.g. cortication, stipulode formation and all others. Consequently, the absence of genetic difference is seen as a weak argument for combining morphologically distinct taxa, whereas strong genetic differences are regarded as a good support for splitting morphologically similar taxa into different species (e.g. Nowak et al., 2016).

5.5 Consequences for This Volume

Readers will find several cases where we deviate in taxonomy and/or nomenclature from the latest overview of European *Characeae* by Krause (1997). Some cases concerning the German charophyte flora have already been discussed in Arbeitsgruppe Characeen Deutschlands (2016) and need not be repeated. Further changes are listed below:

We regard *Chara baltica* var. *affinis* as different from *Chara baltica* and *Chara hispida*. It may merit species rank and is consequently being treated here as *Chara affinis*.

We regard plants from the Mediterranean formerly attributed to *Chara baltica* as different from *Chara baltica* and *Chara hispida*. These plants may also merit species rank and we treat them as "Mediterranean Chara".

Chara corfuensis was not mentioned by Krause (1997) because it was perhaps unknown to him. We regard it as a "good" species.

Krause (1997) considered *Chara curta* (albeit with reservations) as a variety of *Chara aspera*. We consider it a synonym of the younger name *Chara desmacantha* which was used by Krause (1997) and the legal name for triplostichous dioecious specimens with bundled spine-cells.

We accept *Chara dissoluta* as an European species, *Chara denudata* described from South Africa is genetically clearly different. We treat *Chara dissoluta* var. *ohridiana* a synonym of *Chara dissoluta* in contrast to Krause (1997) who wanted to grant it species rank.

Chara gymnophylla is split into the aulacanthous *Chara gymnophylla* and the tylacanthous *Chara squamosa*.

Chara hellenica was described in 2000 and was the only addition (Fig. 5.1) to the European charophyte species after Krause (1997) until *Lamprothamnium sardoum* was published in 2022 (Casanova and Becker, 2022).

The occurrence of *Chara kirghisorum* in Europe was only recently definitely established (Romanov, this volume).

We consider *Chara kokeilii* a morphotype of *Chara tenuispina*.

We accept *Chara mucronata* and publish the new name *Chara filarszkyana*.

Chara neglecta was known for some time from Europe (e.g. Hollerbach and Krassavina, 1983), but was not mentioned by Krause (1997).

The occurrence of the mainly North-African *Chara oedophylla* in Europe was unknown to Krause (1997). There are a few records in Spain (Comelles, 1981; Cirujano et al., 2008), Southern France (Soulié-Märsche, 2003; Mouronval et al., 2015), (Romanov et al., 2019) and Southeastern Europe (Soulié-Märsche et al., this volume) Sicily.

Although the occurrence of *Chara uzbekistanica* in Europe was already published by Zhivoglyad and Krivonosov (1982) this was unknown to Krause (1997). There are several localities in the Volga floodplain and Volga Delta (Romanov and Zhakova, this book).

We accept *Nitella palhinhae* as a species, this taxon was not mentioned by Krause (1997).

We follow Soulié-Märsche (1989) in her approach to assign species of *Tolypella* sect. *Acutifolia* T. F. Allen, with short, acuminate end cells to the genus *Sphaerochara* Mädler. This refers to *Sphaerochara canadensis*, *S. intricata* and *S. prolifera*.

The next chapter will give an insight into the recent status of molecular techniques for supporting these taxonomic decisions as well as for unravelling the systematics of charophytes.

References

Agardh, C. A. (1820). *Species Algarum 1*. Officina Berlingiana, Lundae.

Agardh, C. A. (1824). *Systema Algarum*. Literis Berlingianis, Lundae.

Ahmadi, A., Riahi, H., Sheidai, M., & van Raam, J. (2012). Numerical taxonomy of some charophytes in Iran. *Nordic Journal of Botany, 30*, 206–214.

Allen, T. F. (1888). *The Characeae of America*, 1. S.E. Cassino.

Andersson, L. (1990). The driving force; species concepts and ecology. *Taxon, 39*, 275–382.

Anon. (1965). Review of the 'Revison of the Characeae'. *Taxon, 14*, 240–242.

Ascherson, P. (1878). Noch einige Bemerkungen über die orientalischen Schismus-Formen und über Pflanzen der kleinen Oase. *Österrreichische Botanische Zeitschrift, 28*, 254–257. [In German].

Baum, D. A., & Donoghue, M. J. (1995). Choosing among alternative "phylogenetic" species concepts. *Systematic Botany, 20*, 560–573.

Berendsohn, W. G. (1995). The concept of "potential taxa" in databases. *Taxon, 44*, 207–212.

Braun, A., & Nordstedt, C. F. O. (1882). Fragmente einer Monographie der Characeen. Nach den hinterlassenen Manuscripten A. Braun's. *Abhandlungen der Königlichen Akademie der Wissenschaften zu Berlin für das, 1882*, 1–211. [In German].

Braun, A. (1834). Esquisse monographique du genre *Chara*. *Annales des Sciences Naturelles, Série 2 Botanique, 1*, 349–357.

Braun, A. (1835). Uebersicht der genauer bekannten *Chara*-Arten. *Flora, 18*, 49–73. [In German].

Braun, A. (1849). Übersicht der Schweizerischen Characeen. Ein Beitrag zur Flora der Schweiz. *Neue Denkschriften der Allgemeinen Schweizerischen Gesellschaft für die gesammten Naturwissenschaften, 10*(3), 1–23. [In German].
Braun, A., 1867. *Conspectus systematicus Characearum europaearum*. [No publisher].
Bremer, B., & Eriksson, O. (1992). Ecological species concepts: A reply to Andersson. *Taxon, 41*, 307–309.
Bruzelius, A. (1824). *Observationes in genus Charae, quae venia ampliss.* Berlingiana, Londoni Gothorum.
Bryant, J. A., Stewart, N. F., & Stace, C. A. (2002). A checklist of Characeae of the British Isles. *Watsonia, 22*, 203–208.
Caisová, L., & Gąbka, M. (2009). Charophytes (Characeae, Charophyta) in the Czech Republic: Taxonomy, autecology and distribution. *Fottea, 9*, 1–43.
Casanova, M. T. & Becker, R. (2022). *Lamprothamnium sardoum* sp. nov. (Characeae, Streptophyta): A new species of *Lamprothamnium* for Europe. *Phytotaxa, 567*(3), 269–277.
Casanova, M. T. & Brock, M. A. (1990). Charophyte germination and establishment from the seed bank of an Australian Temporary Lake. *Aquatic Botany, 36*, 247–254.
Casanova, M. T. (1997). Oospore variation in three species of *Chara* (Charales, Chlorophyta). *Phycologia, 36*, 274–280.
Casanova, M. T. (2005). An overview of *Chara* L. in Australia (Characeae, Charophyta). *Australian Systematic Botany, 18*, 25–39.
Casanova, M. T. (2013). Review of the species concepts *Chara fibrosa* and *C. flaccida* (Characeae, Charophyceae). *Australian Systematic Botany, 26*, 291–297.
Chen, C. E. Z., & Lin, H.-X. (2016). Evolution and molecular control of hybrid incompatibility in plants. *Frontiers in Plant Science, 7*, 1208. https://doi.org/10.3389/fpls.2016.01208
Cirujano, S., Cambra, J., Sanchez Castillo, P. M., Meco, A., & Flor Arnau, N. (2008). *Flora iberica. Algas continentales. Carófitos (Characeae)*. Real Jardín Botánico. [In Spanish].
Comelles, M. (1981) Contribució al coneixement de las caroficies d'Espanya. *Collecteana Botanica, 12*, 97–103.
Corillion, R. (1957). *Les Charophycées de France et d'Europe Occidentale*. Bretonne. [In French].
Corillion, R. (1960). *Tolypella salina*, sp. nov., Charophycée nouvelle des marais de Croix-de-Vie (Vende). *Revue Algologique, Nouvelle Série, 5*(3), 198–207, pl. 21–22. [In French].
Cracraft, J. (2000). Species concepts in theoretical and applied biology: a debate with consequences. In Q. D. Wheeler & R. Meier (Eds.), *Species concepts and phylogenetic theory* (pp. 3–14). Columbia Press.
da Cunha, A. G. (1935). Quelques espèces nouvelles de Characées de Portugal. *Bulletin de la Société Portugaise des Sciences Naturelles, 12*(8), 41–58, pl. 3–5.
Davis, J. I., & Nixon, K. C. (1992). Populations, genetic variation, and the delimitation of phylogenetic species. *Systematic Biology, 41*, 421–435.
de Brébisson, A. (1849). *Flore de la Normandie, Phanérogames et cryptogames semi-vasculaires*, ed. 2. A. Hardel, Caen & Derache. [In French].
de Candolle, A. P. (1813). *Catalogus plantarum horti botanici Monspeliensis, addito observationum circa species novas aut non setis cognita fasciculo*. J. Martel, Monspelii. / Cat. Pl. Horti Monsp.
de Candolle, A. P. (1815). Description succincte des plantes qui croisent naturellement en France. In J. B. A. P. de Lamarck & A. P. de Candolle (Eds.), *Flore Française ou description succincte de toutes les plantes qui croissent naturellement en France* (5 & 6, ed. 3). Desray. [In French].
De Queiroz, K. (1998). The general lineage concept of species, species criteria, and the process of speciation. In D. J. Howard & S. H. Berlocher (Eds.), *Endless forms: Species and speciation* (pp. 57–75). Oxford University Press.
Desfontaines, R. (1799). *Flora Atlantica sive historia plantarum quae in Atlante, agro Tunetano et in Algeriensi crescunt 2*. Blanchon.
Desvaux, N. A. (1809). Essai sur la géographie botanique du Haut-Poitou (Département de la Vienne). *Journal de Botanique, 2*, 290–318. [In French].

Doty, M. S. (1966). Book review: A revision of the Characeae. by Richard D. Wood; Kozo Imahori. *Bulletin of the Torrey Botanical Club, 93*, 357–359.

Durieu de Maisonneuve, M. C. (1846–1855). *Exploration scientifique de l'Algérie pendant les années 1840, 1841, 1842*. Imprimerie Royale. [In French].

Durieu de Maisonneuve, M. C. (1859). Note sur une nouvelle espèce du genre *Chara* (*Ch. fragifera* DR.). *Bulletin de la Société Botanique de France, 6*, 179–186. [In French].

Feldmann, G. (1945). Deux nouvelles espèces de *Chara* de l'Afrique du Nord. *Bulletin de la Société d'histoire Naturelle d'Afrique du Nord, 36*, 168–173. [in French].

Filarszky, N. (1930). A budai höforrások nyílt vizeinek Chara-féléi és néhany más érdekesebb, ritkább algafaja. Matematikai és természettudományi értesitö á M. Tud. *Akadémia III. Osztályának Folyóirata, 47*, 652–676. [In Hungarian].

Filarszky, F. (1937). Idegenföldi charafélék határozása. (Determinatio characearum exoticarum). *Magyar Tudománios Akadémia Matematikai és Természettudományi Értesitöje, 55*, 476–497, 6 pl. [In Hungarian].

Geoffroy, M., & Berendsohn, W. G. (2003). The concept problem in taxonomy: importance, components, approaches. In W. B. Berendsohn (Ed.), *MoReTax Handling factual information linked to taxonomic concepts in biology* (pp. 5–14). Schriftenreihe für Vegetationskunde 39. Bundesamt für Naturschutz.

Gmelin, C. C. (1826). *Flora Badensis alsatica et confinium regionum cis et transrhenana plantes a lacu Bodamico usque ad confluentem Mosellae et Rheni sponte nascentes exhibens secundum systema sexuale cum iconibus ad naturam delineatis 4, Supplement*. A. Müller, Carlsruhe.

Grant, V. (1992). Comments on the ecological species concept. *Taxon, 41*, 310–312.

Gregor, T., Blindow, I., Raabe, U., Schubert, H., & Stewart, N. (2014). (2303) Proposal to conserve the name *Chara hispida* (Characeae) with a conserved type. *Taxon, 63*, 933–934.

Groves, J., & Bullock-Webster, G. R. (1920). *The British Charophyta 1*. Ray Society.

Groves, J., & Bullock-Webster, G. R. (1924). *The British Charophyta 2*. Ray Society.

Hartman, C. J. (1820). *Handbok i Skandinaviens flora, innefattande sveriges och norriges vexter, till och med mossorna: Med inledning, af handlande grunderna for botanikens studium, samt tvanne plancher*. Z. Haeggstrom. [In Swedish].

Hennig, W. (1950). *Grundzüge einer Theorie der phylogenetischen Systematik*. Deutscher Zentralverlag. [In German].

Hennig, W. (1966). *Phylogenetic systematics*. University of Illinois Press.

Hertzsch, H. (1855). Zur *Chara filiformis* Hertzsch. *Hedwigia, 1*, 81–82. [In German].

Hollerbach, M. M., & Krassavina, L. K. (1983). The charophytes—Charophyta. In: M. M. Hollerbach (Eds.), *The identification manual of freshwater algae of the USSR 14*. Science. [In Russian].

Hollerbach, M. M. (1960). De speciebus novis generis *Chara* e sectione Haplostephanae Triplostichae. Botanicheskie materialy otdela sporovyh rastenii. *Botanicheskogo Instituta Imeni v. l. Komarova Akademii Nauk SSSR, 13*, 101–107.

Hollerbach, M. M. (1981). De species nova *Chara neglecta* Hollerb. (Charophyta). *Novitates Systematicae Plantarum non Vascularium, 18*, 3–9.

Holt, R. D. (2009). Bringing the Hutchinsonian niche into the 21st century: Ecological and evolutionary perspectives. *PNAS, 106*, 19659–19665.

Horn af Rantzien, H., & Olsen, S. (1949). A suggested starting point for the nomenclature of Charophyta. *Svensk Botanisk Tidskrift, 43*, 98–106.

Hutchinson, G. E. (1957). Concluding remarks. *Cold Spring Harbor Symp, 22*, 415–427.

Hy, F. (1913). Les Characées de France. *Mémoires de la Société Botanique de France, 26*(1–47), 1–3. [In French].

ICBN (1952). International Code of Botanical Nomenclature, 1952 - Stockholm Code. https://iapt-taxon.org/historic/1952.htm

Imahori, K. (1954). *Ecology, Phytogeography and Taxonomy of the Japanese Charophyta*. Koeltz Scientific Publishers.

John, D. M., & Moore, J. A. (1987). An SEM study of the oospore of some *Nitella* species (Charales, Chlorophyta) with description of wall ornamentation and an assessment of its taxonomic importance. *Phycologia, 23*, 334–354.

John, D. M., Moore, J., & Green, D. (1990). Preliminary observations on the structure and ornamentation of the oosporangial wall in *Chara* (Charales, Clorophyta). *British Phycological Journal, 25*, 1–24.

Jungius, J. (1747). *Opuscula Botanico-Physica*. Typis Ottonis.

Krause, W. (1997). Charales (Charophyceae). In H. Ettl, G. Gärtner, H. Heynig, & D. Mollenhauer (Eds.), *Süßwasserflora von Mitteleuropa 18*. Gustav Fischer. [In German].

Krocker, A. J. (1814). *Florae Silesiacae renovatae, emendatae, prius ultra nongentas, circa mille, nunc ultra ter mille plantas, silesiae indigenas, auctae etc. 3*. Prelo Bieliof, Wratislaviae.

Kützing, F. T. (1834). Beschreibung einiger neuer Arten der Gattung Chara. *Flora, 17*, 705–707. [In German].

Kützing, F. T. (1845). *Phycologia Germanica, d. i. Deutschlands Algen in bündigen Beschreibungen. Nebst einer Anleitung zum Untersuchen und Bestimmen dieser Gewächse für Anfänger*. W. Köhne. [In German].

Kützing, F. T. (1849). *Species Algarum*. F.A. Brockhaus.

Langangen, A. (2007). *Charophytes of the Nordic countries*. Saeculum ANS.

Langangen, A. (2010). *Chara hellenica* spec.nov. and other charophytes collected in Corfu and Amorgos islands (Greece) in 2008. *Studia Botanica Hungarica, 41*, 21–30.

Lessing, C. F. (1835). Beitrag zur Flora des südlichen Urals und der Steppen. *Linnaea, 9*, 145–213. [In German].

Linnaeus, C. (1753). *Species plantarum, exhibentes plantas rite cognitas, ad genera relatas, cum Differentiis Specificis, Nominibus Trivialibus, Synonymis Selectis, Locis Natalibus, Secundum systema sexuale digestas 2*. Laurentius Salvius, Holmiae.

Loiseleur Deslongchamps, L. A. (1810). *Notice sur les plantes a ajouter a la flore de france (Flora Gallica); avec quelques corrections et observations*. Auteur, Migneret & Gabon, Paris. [In French].

Macdonald, M. B., & Hotchkiss, A. T. (1956). An estipulodic form of *Chara australis* R. Br. (= *Protochara australis* Woms. And Ophel). *Proceedings of the Linnean Society of New South Wales, 53*, 274–284.

Mayden, R. L. (1997). A hierarchy of species concepts: The denouement in the saga of the species problem. In M. A. Claridge, H. A. Dawah, & M. R. Wilson (Eds.), *Species: The units of diversity* (pp. 381–424). Chapman & Hall.

Mayr, E. (2000). The biological species concept. In Q. D. Wheeler & R. Meier (Eds.), *Species concepts and phylogenetic theory* (pp. 17–29). Columbia Press.

McCourt, M. R., Karol, K. G., Guerlesquin, M., & Feist, M. (1996). Phylogeny of extant genera in the family Characeae (Charales, Charophyceae) based on rbcL sequences and morphology. *American Journal of Botany, 83*, 125–131.

Meyen, F. J. F. (1827). Beobachtungen und Bemerkungen über die Gattung *Chara*. *Linnaea, 2*(55–81), 2–3. [In German].

Migula, W. (1897). Die Characeen Deutschlands, Oesterreichs und der Schweiz. Unter Berücksichtigung aller Arten Europas. In A. Fischer, E. Fischer, F. Hauck, G. Limpricht, C. Luerssen, W. Migula, H. Rehm, P. Richter, & G. Winter (Eds.), *Dr. L. Rabenhorst's Kryptogamenflora von Deutschland, Oesterreich und der Schweiz* 5, ed. 2. Eduard Kummer. [In German].

Migula, W. (1904). Characeae Rossicae ex Herbarion Horti Petropolitani. Trudy Imperatorskago S.-Peterburgskago Botaniceskago Sada. *Acta Horti Petropolitani, 23*, 533–539.

Migula, W. (1925). Charophyta – (Charales). In A. Pascher (Ed.), *Die Süßwasser-Flora Deutschlands, Österreichs und der Schweiz* (pp. 207–243). Gustav Fischer. [In German].

Mishler, B. D., & Theriot, E. C. (2000). The phylogenetic species concept (sensu Mishler and Theriot): Monophyly, apomorphy, and phylogenetic species concepts. In Q. D. Wheeler & R. Meier (Eds.), *Species concepts and phylogenetic theory* (pp. 44–54). Columbia University Press.

Moore, J. M. (1986). *Charophytes of Great Britain and Ireland*. B.S.B.I Handbook No. 5. Botanical Society of the British Isles, London.
Moul, E. T. (1966). Book review: A revision of the Characeae: Iconograph of the Characeae. Part 2. by Richard D. Wood; Kozo Imahori Source: *Bulletin of the Torrey Botanical Club, 93*, 143.
Mouronval, J. B., Baudouin, S., Borel, N., Soulié-Märsche, I., Klesczewski, M., & Grillas, P. (2015). *Guide des Characées de France méditerranéenne*. Office national de la Chasse et de la Faune Sauvage.
Müller, O. F. (1777). *Icones Plantarum sponte nascentium in regnis Daniae et Norvegiae, in ducatibus Slesvici et Holsatiae, et in comitatibus Oldenburgi et Delmenhorstiae: ad illustrandum opus de iisdem Plantis, Regio jussu exarandum Florae Danicae nomine inscriptum 5(13)*. København.
Nishiyama, T., Sakayama, H., de Vries, J., Buschmann, H., Saint-Marcoux, D., et al. (2018). The *Chara* genome: Secondary complexity and implications for plant terrestrialization. *Cell, 174*, 448–464.
Nixon, K. C., & Wheeler, Q. D. (1992). Extinction and the origin of species. In: M. J. Novacek & Q. D. Wheeler (Eds.), *Extinction and phylogeny* (pp. 119–143). Columbia University Press.
Nordstedt, C. F. O. (1918). Australasian Characeae. *Proceedings of the Royal Society of Victoria, 31*, 1–6.
Nordstedt, C .F. O. (1883). Fragmente einer Monographie der Characeen. Nach den hinterlassenen Manuskripten A. Brauns herausgegeben von Dr. O. Nordstedt. *Abhandlungen der Königlichen Akademie der Wissenschaften zu Berlin, 1882*, 1–221. [In German].
Norman, J. M. (1868). Specialia loca natalia plantarum nonnullarum vascularum & Characearum & Lichenum in Agro Arctico Norvegiae confiniisque Kongel. *Det Kongelige Norske Videnskabers Selskabs Skrifter i det 19de aarhundrede, 5*, 243–378.
Nowak, P., Schubert, H., & Schaible, R. (2016). Molecular evaluation of the validity of the morphological characters of three Swedish *Chara* sections: Chara, Grovesia, and Desvauxia (Charales, Charophyceae). *Aquatic Botany, 134*, 113–119.
Nowak, P., Schubert, H., Holzhausen, A., Sommer, V., & Schaible, R. (2017). Morphological adaptations of *Chara baltica* and *Chara liljebladii* (Characeae) under different light conditions. *Botany Letters, 165*, 67–75.
Pante, E., Puillandre, N., Viricel, A., Arnaud-Haond, S., Aurelle, D., Castelin, M., Chenuil, A., Destombe, C., Forcioli, D., Valero, M., Viard, F., & Samadi, S. (2015). Species are hypotheses: Avoid connectivity assessments based on pillars of sand. *Molecular Ecology, 24*(3), 525–544.
Persoon, C. H. (1807). *Synopsis plantarum seu enchiridium Botanicum complectens enumerationem systematicam specierum hucusque cognitarum 2*. Treuttel et Würtz, Paris, Cottam, Tübingen.
Proctor, V. W. (1962). Viability of *Chara* oospores taken from migratory water birds. *Ecology, 43*, 528–529.
Proctor, V. W. (1970). Taxonomy of *Chara braunii*: An experimental approach. *Journal of Phycology, 6*, 317–321.
Proctor, V. W. (1971). *Chara globularis* Thuillier (= *C. fragilis* Desvaux): Breeding patterns within a cosmopolitan complex. *Limnology and Oceanography, 16*, 422–436.
Proctor, V. W. (1975). The nature of charophyte species. *Phycologia, 14*, 97–113.
Proctor, V. W. (1980). Historical biogeography of *Chara* (Charophyta): An appraisal of the Braun-Wood classification plus a falisifiable alternative for future consideration. *Journal of Phycology, 16*, 218–133.
Proctor, V. W. (1990). Characeae of Llano Estacado (Texas and adjacent New Mexico) playas. *Journal of Biogeography, 17*(75), 84.
Proctor, V. W. (1992). Abstract in the Proceedings of the Charophyte Symposium at Madison Wisconsin.
Proctor, V. W. (1999). Charophytivory, playas y papalotes, a local paradigm of global relevance. *Australian Journal of Botany, 47*, 399–406.
Proctor, V. W., & Wiman, F. H. (1971). An experimental approach to the systematics of the monoecious-conjoined members of the genus *Chara*, series *Gymnobasalia*. *American Journal of Botany, 58*, 885–893.

Rabenhorst, L. (1847). *Deutschlands Kryptogamen-Flora oder Handbuch zur Bestimmung der kryptogamischen Gewächse Deutschlands, der Schweiz, des Lombardisch-Venetianischen Königreichs und Istrien*. Eduard Kummer. [In German].

Reichenbach, H. G. L. (1830–1832). *Flora Germanica excursoria ex affinitate regni vegetabilis naturali disposita sive principia synopseos plantarum in Germania terrisque in Europa media adjacentibus sponte nascentium cultarumque frequentius. Insunt plantae Acroblastae et Phylloblastae*. C. Cnobloch, Lipsiae.

Reyes-Prósper, E. (1910). *Carofitas de España*. Madrid, Artística Española. [In Spanish].

Ridley, M. (1989). The cladistic solution to the species problem. *Biology and Philosophy, 4*, 1–16.

Romanov, R., Napolitano, T., van de Weyer, K., & Troia, A. (2019). New records and observations to the Characean flora of Sicily (Italia). *Webbia, 74*, 111–119.

Roth, A. G. (1800). *Catalecta Botanica quibus plantae novae et minus cognitae describuntur atque illustrantur 2*. Fr. Gleditschiana.

Ruprecht, F. J. (1845). Distributio cryptogamarum vascularium in Imperio Rossico. *Beiträge der Pflanzenkunde des Russischen Reiches, 3*, 1–56.

Sakayama, H., Hara, Y., Arai, S., Sato, H. & Nozaki, H. (2004a). Phyologenetic analyses of *Nitella* subgenus Tieffallenia (Charales, Charophyceae) using ribosomal DNA internal transcribed spaces sequences. *Phycologia, 43*, 672–681.

Sakayama, H., Hara, Y. & Nozaki, H. (2004b). Taxonomic re-examination of six species of *Nitella* (Charales, Charophyceae) from Asia and phylogenetic relationships within the genus based on rbcL and atpB gene sequences. *Phycologia, 43*, 91–104.

Sakayama, H., Miyaji, K., Nagumo, T., Kato, M., Hara, Y., & Nozaki, H. (2005). Taxonomic reexamination of 17 species of *Nitella* subgenus *Tieffallenia* (Charales, Charophyceae) based on internal morphology of the oospore wall and multiple DNA marker sequences. *Journal of Phycology, 41*, 195–211.

Sakayama, H., Kasai, F., Nozaki, H., Watanabe, M. M., Kawachi, M., Shigyo, M., Nishihiro, J., Washitani, I., Kreinitz, L., & Ito, M. (2010). Taxonomic reexamination of *Chara globularis* (Charales, Charophyceae) from Japan based on oospore morphology and rbcl gene sequences, and the description of *C. leptospora* sp. nov. *Journal of Phycology, 45*, 917–927.

Sawa, T. (1973). Two new species of Tolypella (Characeae) from North America. *Journal of Phycology, 9*(4), 472–482.

Schaible, R., Bergmann, I., Boegle, M., Schoor, A., & Schubert, H. (2009a). Genetic characterisation of sexually and parthenogenetically reproductive populations of *Chara canescens* (Charophyceae) using AFLP, rbcL, and SNP markers. *Phycologia, 48*(2), 105–117.

Schaible, R., Bergmann, I., & Schubert, H. (2009b). A survey of sexually reproducing female and male populations of *Chara canescens* (Charophyta) in the National Park Neusiedler See-Seewinkel (Austria). *Cryptogamie, Algol., 30*(4), 279–294.

Schneider, S. C., Nowak, P., von Ammon, U., & Ballot, A. (2016). Species differentiation in the genus *Chara* (Charophyceae): Considerable phenotypic plasticity occurs within homogenous genetic groups. *European Journal of Phycology, 51*, 1–12.

Schubert, H., & Blindow, I. (2004). *Charophytes of the Baltic Sea*. A.R.G. Gantner Verlag.

Schubert, H., Blindow, I., & van de Weyer, K. (2016). *Chara hispida*. In AG Characeen Deutschlands (Hrsg.) *Armleuchteralgen Die Characeen Deutschlands*. Springer. [In German].

Smith, J. E. (1810). *English botany; or, coloured figures of british plants, with their essential characters, synonyms, and places of growth: to which will be added, occasional remarks 30*. R. Taylor & C.

Soulié-Märsche, I. (1989). *Etude comparée de gyrogonites de charophytes actuelles et fossiles et phylogénie des genres actuels*. Imprimerie des Tilleuls.

Soulié-Märsche, I. (2003). *Les Characées de trois lacs du Centre Var (Bonne Cougne - Redon - Gavoti). Rapport d'expertise pour le site FR 9301621 du programme NATURA 2000.*

Soulié-Märsche, I., & García, A. (2015). Gyrogonites and oospores, complementary viewpoints to improve the study of the charophytes (Charales). *Aquatic Botany, 120*, 7–17.

Thuillier, J. L. (1799). *La flore des environs de Paris, ou distribution méthodique des plantes qui y croissent naturellement, faite d'après le systême de Linnée; avec le nom et la description de chacune en latin et en français; l'indication de leur floraison, de la couleur des leurs fleurs, et la citation des auteurs qui les ont mieux décrites ou ont donné les meilleures figures, ed. 2.* Auteur et H.L. Peronneau.
Turland, N. J., Wiersema, J. H., Barrie, F. R., Greuter, W., Hawksworth, D. L., Herendeen, P. S., Knapp, S., Kusber, W. -H., Li, D. -Z., Marhold, K., May, T. W., McNeill, J., Monro, A. M., Prado, J., Price, M. J., & Smith, G. F. (Eds.). (2018). *International code of nomenclature for algae, fungi, and plants (Shenzhen Code) adopted by the Nineteenth International Botanical Congress Shenzhen, China, July 2017. Regnum Vegetabile 159.* Koeltz Botanical Books, Glashütten.
Turnbridge, N., & Fitzpatrick, K. (2017). Chapter 24: The origin of species. In L. A. Urry, M. L. Cain, S. A. Wassermann, P. V. Minorsky, J. B. Reece (Eds.), *Campbell biology.* Pearson.
van Valen, L. (1976). Ecological species, multispecies and oaks. *Taxon, 25,* 233–239.
von Leonhardi, H. (1863). Die böhmischen Characeen. *Lotos, 13*(55–62), 69–80. [In German].
von Leonhardi, H. (1864). Die bisher bekannten österreichischen Armleuchter-Gewächse besprochen vom morphogenetischen Standpuncte. *Verhandlungen Des Naturforschenden Vereins in Brünn, Abhandlungen, 2,* 122–224. [In German].
Wahlstedt, L. J. (1862). *Bidrag till kännedomen om de skandinaviska arterna af växt-familjen Characeae.* N.P. Lundbergh, Lund (in Swedish).
Wallman, J. (1853). *Försök till en systematisk uppställning af växtfamiljen Characeae.* P.A. Norstedt & Söner. [In Swedish].
Wallroth, F. W. (1833). Flora cryptogamia Germaniae. Pars posterior continens Algas et Fungos. In M. J. Bluff & C. A. Fingerhuth (Eds.), *Compendium florae Germaniae 4.* J.L. Schrag.
Wheeler, Q. D., & Platnick, N. I. (2000). The phylogenetic species concept (sensu Wheeler and Platnick). In Q. D. Wheeler & R. Meier (Eds.), *Species concepts and phylogenetic theory* (pp. 45–69). Columbia University Press.
Wheeler, Q. D. (1999). Why the phylogenetic species concept? Elementary. *Journal of Nematology, 31,* 134–141.
Wickett, N. J., Mirarab, S., Nguyen, N., Wamow, T., Carpenter, E., et al. (2014). Phylotranscriptomic analysis of the origin an early diversification of land plants. *PNAS, 111,* 4859–4868.
Wiersema, J. H., Turland, N. J., Barrie, F. R., Greuter, W., Hawksworth, D. L., Herendeen, P. S., Knapp, S., Kusber, W.-H., Li, D.-Z., Marhold, K., May, T. W., McNeill, J., Monro, A. M., Prado, J., Price, M. J. & Smith, G.F. (Eds.). (2020). 2018+ [continuously updated]: *International Code of Nomenclature for algae, fungi, and plants (Shenzhen Code) adopted by the Nineteenth International Botanical Congress Shenzhen, China, July 2017: Appendices I–VII.* Online at https://naturalhistory2.si.edu/botany/codes-proposals/. Accessed 17 April 2020
Wiley, E. O. (1978). The evolutionary species concept reconsidered. *Systematic Biology, 27,* 17–26.
Willdenow, C. L. (1809). Fünf neue Pflanzen Deutschlands. *Magazin für die Neuesten Entdeckungen in der Gesammten Naturkunde, Gesellschaft Naturforschender Freunde zu Berlin, 3,* 296–299. [In German].
Whitford, L. (1943). The freshwater algae of North Carolina. *Journal of the Elisha Mitchell Scientific Society, 59,* 131–170.
Wood, R. D. (1957). Hand-sorted punched cards in taxonomic research. *Brittonia, 9*(2), 65–69.
Wood, R. D. (1960). Linnaean types for Characeae. *Transactions of the American Microscopical Society, 79,* 219–226.
Wood, R. D. (1965). *A revision of the Characeae. First Part: Monograph of the Characeae.* Cramer.
Wood, R. D., & Imahori, K. (1964). *A revision of the Characeae 2: Iconograph of the Characeae.* J. Cramer.
Wood, R. D. (1947). The Characeae of the Put-in-Bay region, Lake Erie (Ohio). *The Ohio Journal of Science, 46*(6), 240–258.
Wood, R. D. (1962). New combinations and taxa in the revision of Characeae. *Taxon, 11,* 7–25.

Wood, R. D. (1972). *Characeae of Australia*. Verlag von J. Cramer
Wood, R. D., & Mason, R. (1977). Characeae of New Zealand. *New Zealand Journal of Botany, 15*, 87–180.
Zachos, F. E. (2016). *Species concepts in biology*. Springer.
Zhivoglyad, A. F., & Krivonosov, G. A. (1982). On the specific composition and the productivity of the Charophyta at the lower reaches of the Volga Delta and Northern Caspy. *Botanicheskii Zhurnal (Moskow & Leningrad), 67*(5), 672–674. [In Russian].

Chapter 6
Systematics and Phylogeny of European *Characeae*

Petra Nowak, Kenneth G. Karol, and Susanne C. Schneider

Charophyceae are one of six freshwater algal lineages that, with the embryophytes (land plants), comprise the Charophyta (streptophyta). Their phylogenetic position in relation to land plants is a subject of ongoing change based on integrative approaches combining anatomy, ecology, genomics, morphology, ontogeny, and phylogeny. The Charophyceae are divided into the genera: *Chara, Lamprothamnium, Lychnothamnus, Nitellopsis, Nitella, Tolypella* and *Sphaerochara,* although some phylogenetic studies indicate that *Lamprothamnium* may be placed within the genus *Chara.* The species concept is, however, still discussed between, on the one hand, supporters of the microspecies concept established by Braun (1883) and Migula (1900), and, on the other hand, the macrospecies concept suggested by R. D. Wood. This monograph follows the microspecies concept and recognises 67 microspecies in Europe. Following Wood (1962), these 67 microspecies would belong to just 24 macrospecies. Here, we summarise the current knowledge on the systematics of the family *Characeae,* and analyse which of the morphologically described species in this flora currently are supported by phylogenetic sequence analyses. We found that not all morphologically described species were consistent with genetic analyses. Consistency between morphological traits and gene sequences varied among the genera. Whereas molecular phylogenetic analyses for the genera *Nitella, Sphaerochara,* and

P. Nowak (✉)
Universität Rostock, Institut für Biowissenschaften, Albert-Einstein-Straße 3, 18051 Rostock, Germany
e-mail: petra.nowak@uni-rostock.de

K. G. Karol
New York Botanical Garden, 2900 Southern Blvd., Bronx, NY 10458, USA
e-mail: kkarol@nybg.org

S. C. Schneider
Norwegian Institute for Water Research, Økernveien 94, 0579 Oslo, Norway
e-mail: susi.schneider@niva.no

H. Schubert et al. (eds.), *Charophytes of Europe*,
https://doi.org/10.1007/978-3-031-31898-6_6

Tolypella strongly support the microspecies concept, high morphological variability in the genus *Chara* is not reflected in molecular phylogenetic analyses.

6.1 Charophyceae

Modern systematics reflects evolutionary processes in the ecosystem and combines results from numerous disciplines. This applies in particular to:

(a) Taxonomy, which deals with the classification and nomenclature of organisms;
(b) Phylogeny, which investigates the evolutionary history of organisms and provides the basis for understanding the natural system;
(c) Ecophysiology, which examines the function of organisms in interaction with their environment.

How are the results from these various fields of science reflected in the systematics of the Charophyceae? One of the important questions of evolution is linked to the origin, diversification, and classification of Charophyceae: the origin of land plants. The "conquerors" of land had to master a multitude of challenges. Compared to aquatic habitats, terrestrial systems exhibit extreme temperature fluctuations, intensive UV radiation, and the danger of dehydration. The ancestors of land plants had to adapt to these new challenges, developing a genetic and ecophysiological "toolbox" to cope with these complex challenges (Floyd and Bowman, 2007; Vries and Archibald, 2018).

One of the first decisive steps in the evolution of green algae, and later land plants, was the establishment of a symbiotic relationship between an unknown eukaryotic original flagellate, single-celled protist (host) and a free-living cyanobacterium that carried out photosynthesis in the host cell. From this cyanobacterium primary plastids evolved. This first phototrophic eukaryote is considered the ancestor of all green algae. Molecular findings date the formation of photosynthesising eukaryotes to 1.5 billion years ago (Yoon et al., 2004). Other algae groups that can be traced back to such a symbiosis are red algae and glaucophytes (Leliaert et al., 2011). Subsequent (secondary) endosymbioses led to further green algae groups such as euglenophytes (Leliaert et al., 2012).

Some descendants of the first phototrophic eukaryotes conquered the land by adapting in a lichen-like community to life outside the water, probably more than a billion years ago (Heckman et al., 2001; Vries and Archibald, 2018). The green algae from this lichen-like symbioses belonged to the Chlorophyta. Today we know that about 1200–725 million years ago, a second group of green algae, the Charophyta (streptophyta), split from the Chlorophyta (Stewart and Mattox, 1975; Wodniok et al., 2011) and occupied predominately freshwater habitats. Unfortunately, the names Charophyta and streptophyta are used controversially. Bremer et al. (1987) circumscribed the streptophyta to include some green algae and land plants, while Jeffrey (1982) used the name more restrictively—based on the twisted (strepto) nature of their sperm—to include only the Charophyceae and land plants. As our understanding of relationships of other lineages of green algae grew, so did the circumscription of the Charophyta/streptophyta (Lewis and McCourt, 2004). Though both names are used interchangeably, the name streptophyta is in common usage. For this reason, we here use streptophyta, and also because this minimises potential confusion between Charophyta and Charophyceae.

Table 6.1 and Fig. 6.1 give an overview of major lineages of Chlorophyta and all known classes of streptophyta. All green plants[1] (green algae and land plants) are called Chlorobionta or Viridiplantae (Judd et al., 2008). Latest phylogenomic analyses proposed the Prasinodermophyta as the earliest diverging lineage within the Viridiplantae, diverging before the split of the Chlorophyta and streptophyta (Li et al., 2020). Within the streptophyta, seven classes are distinguished, among them the class Charophyceae (to be considered here in detail) as well as the embryophytes, which comprises the land plants (hornworts, liverworts, mosses, lycophytes, ferns and seed plants). Although the monophyly of embryophytes (land plants) and tracheophytes (vascular plants: lycophytes, ferns and seed plants) is generally accepted, various hypotheses about the relationships between bryophytes (hornworts, liverworts and mosses) and tracheophytes have gained acceptance; with the fundamental distinction that bryophytes are considered either monophyletic or a paraphyletic grade[2] (e.g. Puttick et al., 2018; Su et al., 2021). The monophyly of mosses and liverworts is unambiguous (Wickett et al., 2014). The position of hornworts, however, depends on the phylogenetic strategy used. The more traditional view is that hornworts are sister to the tracheophytes ("hornworts—Tracheophyta", Qiu et al., 2006). Wickett et al. (2014) suggested originally to consider hornworts as a sister to all other embryophytes ("hornwort—sister"). Recent analyses support for the monophyly of bryophytes (Su et al., 2021; Fig. 6.1 illustrates this hypothesis). However, the position of the hornworts has not been clarified, and only further analyses can help to unravel the topology of the embryophytes.

The monophyletic streptophyta contains various life forms and shows a general increase in the complexity of the thallus up to a cormus (Graham et al., 2000). Single-celled flagellates of the Mesostigmatophyceae (*Mesostigma*) (Marin and Melkonian, 1999), sarcinoid packets of non-motile singled-cells of the Chlorokybophyceae (*Chlorokybus*) (Casper, 1997) and the simple unbranched filaments of the Klebsormidiophyceae (*Klebsormidium*) (Lokhorst, 1996) are the earliest diverging classes of the streptophyta (Leliaert et al., 2011). The Charophyceae, Zygnematophyceae, and Coleochaetophyceae, together with the land plants are united by the presence of a phragmoplast; a microtubular array that helps to establish the new cell

[1] Various names have been proposed for this lineage comprising the green algae and land plants: Viridiplantae or Viridaeplantae (Cavalier-Smith 1981, 1998), Chlorobiota or Chlorobionta (Jeffrey 1971, 1982), Chloroplastida (Adl et al., 2005), or simply green plants (Sluiman et al., 1983) or green lineage.

[2] A paraphyletic grade is an assemblage of any size and systematic rank that originated from a single common ancestor, but does not—as opposed to a monophyletic group—contain all descendants from this ancestor. The perhaps most compelling example for a paraphyletic group are the Reptilia (turtles, tuataras, lizards and snakes, crocodiles plus dinosaur-like reptiles), the lineage which also gave rise to the birds.

Table 6.1 Overview of the main green algae and land plants lineages (*Chlorobionta* Jeffrey, *Viridiplantae* Cavalier-Smith). Modified according to Lewis and McCourt (2004), Leliaert et al. (2012) and Li et al. (2020)

Division *Prasinodermophyta* B.Marin & Melkonian	Division *Charophyta* Migula (streptophyta Cavalier-Smith)
Class *Prasinodermophyceae* B.Marin & Melkonian	Class *Mesostigmatophyceae* Marin & Melkonian
Order *Prasinodermales* B.Marin & Melkonian	Order *Mesostigmatales* Marin & Melkonian
Class *Palmophyllophyceae* Leliaert et al.	Class *Chlorokybophyceae* Irisarri et al.
Order *Palmophyllales* Zechman et al.	Order *Chlorokybales* Irisarri et al.
Order *Prasinococcales* Guillou et al.	Class *Klebsormidiophyceae* C.Hoek, D.G.Mann & H.M.Jahns
	Order *Klebsormidales* K.D.Stewart & K.R.Mattox
Division *Chlorophyta* Reichenbach	Class *Zygnematophyceae* Round
Class *Chlorophyceae* Wille	Order *Zygnematales* Bessey
Order *Chlamydomonadales* F.E.Fritsch	Order *Desmidiales* Bessey
Order *Sphaeropleales* Luerssen	Class *Coleochaetophyceae* C.Jeffrey
Order *Oedogoniales* Heering	Order *Coleochaetales* Chadefaud
Order *Chaetophorales* Wille	Class *Charophyceae* Rabenhorst
Class *Ulvophyceae* K.R.Mattox & K.D.Stewart	Order *Charales* Dumortier
Order *Ulotrichales* Borzì	Family *Characeae* S.F.Gray
Order *Ulvales* Blackman & Tansley	"Tribe" *Chareae* Ganterer

(continued)

Table 6.1 (continued)

Order *Cladophorales* Haeckel	Genus *Chara* Linnaeus
Order *Trentepohliales* Chadefaud	Genus *Lamprothamnium* J. Groves
Class *Trebouxiophyceae* Friedl	Genus *Lychnothamnus* (Ruprecht) A. Braun
Order *Trebouxiales* Friedl	Genus *Nitellopsis* Hy
Order *Microthamniales* M.Melkonian	"Tribe" *Nitelleae* (Leonh.) Zaneveld
Order *Prasiolales* Schaffner	Genus *Nitella* C. Agardh
Order *Chlorellales* Bold & M.J.Wynne	Genus *Sphaerochara* Mädler
Class *Mamiellophyceae* Marin & Melkonian	Genus *Tolypella* A. Braun
Order *Mamiellales* Moestrup	Class *Embryophyceae*
Class *Pyramimonadophyceae* Moestrup & Daugbjerg	
Order *Pyramimonadales* Chadefaud	

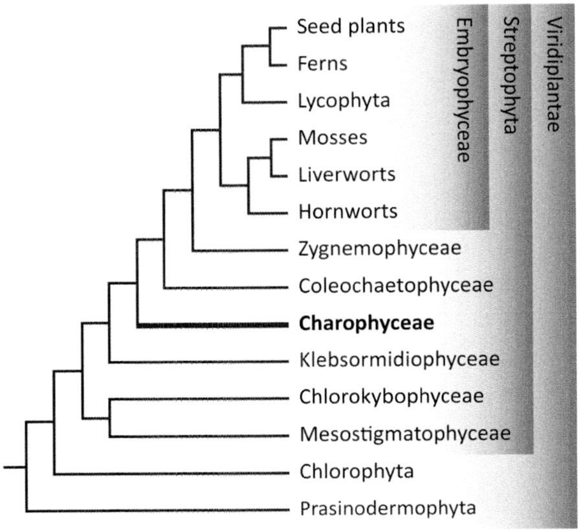

Fig. 6.1 Phylogenetic relationships of major lineages of *Charophyta/Streptophyta*. Modified according to Lewis and McCourt (2004), Leliaert et al. (2012), Li et al. (2020) and Su et al. (2021)

wall after division (Nishiyama et al., 2018; Pickett-Heaps, 1975). In the past, all three of these were treated as candidates for the closest relatives of land plants. Due to the high morphological complexity and their 'plant-like' body plan, the Charophyceae were early regarded as the predecessors of land plants (Pringsheim, 1862) and single-gene analyses supported this idea (Karol et al., 2001; McCourt et al., 2004). Later, especially anatomical and ontogenetic characters (Cook, 2004; Graham et al., 2000; Huss and Kranz, 1997; Kranz et al., 1995), substantiated by multigene analyses (Finet et al., 2010), supported the Coleochaetophyceae as the closest relatives of land plants. However, analyses of chloroplast and nuclear genes supported the Zygnematophyceae or a clade of Coleochaetophyceae and Zygnematophyceae as the closest relatives of land plants (Turmel and Lemieux, 2018; Turmel et al., 2007, 2013; Wodniok et al., 2011). Extant Zygnematophyceae have a simple body plan and grow either as a filamentous thallus (Zygnematales, Conjugatales), as elongated single cells, or as cells consisting of two symmetrical semi-cells (Desmidiales, Desmids). Comparative phylogenomic analyses of *Chara braunii* with selected plant and algal species (Nishiyama et al., 2018) supported the Zygnematophyceae as the sister group of land plants (though a genome representing Coleochaetophyceae was not available) and showed that many of the plant-like characters of the Charophyceae had developed previously in an unknown hypothetical common ancestor.

6.2 Genetic Approaches Used for *Characeae*

Advances in the methodology of molecular phylogenetics have led to a widespread use of molecular markers for studying the relationships among members of the streptophyta including the Charophyceae. DNA Fingerprinting techniques such as AFLPs or microsatellites, and the sequencing of plastid, mitochondrial, and nuclear markers have been used to clarify charophyte phylogeny during the last decades.

Mannschreck et al. (2002) used **AFLP** (amplified fragment length polymorphism) analyses to assess genetic diversity within and among *Chara* species. AFLP is a PCR-based (polymerase chain reaction) technique and uses selective amplification of a subset of digested DNA fragments to generate and compare unique fingerprints for genomes of interest. The technique combines the advantages of needing no prior information regarding the target genome with the reproducibility and sensitivity to detect polymorphisms at the DNA sequence level. AFLP analyses were successfully used by Mannschreck et al. (2002) to assess the relationship of species of the genus *Chara* and by Boegle et al. (2010a, 2010b) to delineate several diplostichous taxa belonging to the *C. hispida* species complex.

Another DNA fingerprint technique uses **microsatellites** or single sequence repeats (SSR), which have been widely used in plant genetic studies. Microsatellites are simple sequence tandem repeats (SSTRs) that usually occur in non-coding regions of the genome. Conserved flanking regions are identified on each side of a repeat unit consisting of "unordered" DNA. These flanking regions are critical and allow the development of locus-specific primers to amplify the microsatellites with PCR. Microsatellite analysis has been used only to a very restricted extent within the *Characeae*. Schaible et al. (2009) used this technique to distinguish parthenogenetic and dioecious populations of *Chara canescens*.

In recent decades, sequences of short DNA fragments have been increasingly used for the diagnosis of taxa, which has improved the objectivity and efficiency of species identification and enabled both inter- and intra-specific phylogenetic analyses. DNA-based species identification using a universal **DNA barcode** consisting of standard DNA regions was first established in animals (Hebert et al., 2003) and later also for plants, fungi and other eukaryotes (Hollingsworth et al., 2009a, 2009b; Saunders and McDevit, 2012; Schoch et al., 2012). The idea of using molecular markers for species identification and differentiation has been used for decades in prokaryotes (Tindall et al., 2010), with the 16S rRNA gene as the most common marker. For eukaryotes, the DNA barcode was developed most successfully for animals, using a portion of the mitochondrial gene cytochrome *c* oxidase subunit 1 (*cox1*/*CO1*), as the standardised tool for molecular taxonomy and identification (Ratnasingham and Hebert, 2007). Mitochondrial DNA of plants evolves much slower than its animal counterpart, and consequently other markers were chosen for genetic analyses (Hollingsworth et al., 2009b). For plants, chloroplast Rubisco large subunit (*rbc*L) or Maturase K (*mat*K) served as barcode markers and their value in evolutionary, systematic, and population studies of photosynthetic organisms increased rapidly (Hall et al., 2010). Also for *Charophyceae*, chloroplast markers are widely used for species identification and evolutionary phylogenetic analyses. Nearly 30 years ago, McCourt et al. (1996) reconstructed the phylogeny of *Characeae* genera by using the sequence data of *rbc*L. Since then, numerous studies have been published focusing on the systematics and taxonomy of *Charophyceae* based on single and combined chloroplast genes (e.g. McCourt et al., 1999; Sakayama et al. 2004b; Pérez et al., 2014; Schneider et al., 2016). Additionally, several authors used parts of the ribosomal DNA in phylogenetic studies on charophycean algae (Meiers et al., 1997, 1999; Ragan et al., 1994; Sakayama et al., 2004a; Wilcox et al., 1993; Zechman et al., 1990). Ribosomal gene sequences (18S, 5.8S and 28S) are relatively conserved with phylogenetic information similar to plastid gene-based studies, usually with less resolution and support, whereas the nuclear internal transcribed spacer (nrITS1 and nrITS2) sequences exhibit relatively high genetic variability. The ribosomal DNA sequences complement plastid genes in recent studies and have shown a broad phylogenetic utility across the characean genera. Especially the combination of plastid genes with the nuclear ITS regions resulted in higher resolution among some species of *Chara* (Nowak et al., 2016; Schneider et al., 2015), *Nitella* (Sakayama, 2008) and *Tolypella* (Pérez et al., 2017).

6.3 Systematics of *Characeae*

The first scientist providing a valid description of the genus *Chara* was none other than the founder of the binary nomenclature, Carl von Linné, in his famous work "Species Plantarum" (1753). He suggested *Chara* to be a green algae. In the following years, many botanists were occupied with the "stoneworts", but did not recognise their true character and classified them into various plant groups. For example, not only were they placed with the mosses or ferns but also with the vascular plants. Migula (1900) summarised this development in the preface of his work concerning the *Characeae* in the context of Rabenhorst's cryptogamous flora and pointed to the outstanding role of Alexander Braun. In several fundamental works on the history of the *Characeae*, Alexander Braun had created the basis for systematics of this unique group of green algae. In the early decades of the twentieth century, scientists tended to split the *Characeae* into many species. This contributed, on the one hand, to a quite exact knowledge of morphological forms, but on the other hand, also lead to an extremely unnatural fragmentation. The monograph by Wood and Imahori (1965), which attempted to summarise many taxa at the species level, deviated drastically from this approach. However, some morphologically clearly distinct species were merged, and for this reason, the work of Wood and Imahori was criticised immediately after its publication (e.g. McCracken et al., 1966; Sarma and Khan, 1967). Until today, science has not come to a conclusion if the microspecies concept (splitting taxa into many entities which sometimes are difficult to distinguish), or the macrospecies concept (merging microspecies into larger morphologically variable groups) should preferably be used for the *Characeae* (Schubert, 2014). For details on the species concepts see Chap. 5 in this monograph.

Currently a pluralistic way is used for the systematics of the *Characeae* which combines elements of the three fundamental species concepts (biological = reproductive isolation; phenetic = morphology; ecological = niche structure) with modern phylogenetic analyses. The difference between the macro- and the microspecies concept is mainly the weighting of individual characteristics; i.e. both concepts have different points of view for the determination of limits between species and what is to be regarded within a species as a site- (variety) or population-specific (subspecies) individual expression of characteristics (see Chap. 5, Taxonomic concepts and taxonomic evolution in European *Characeae*). With the development of molecular genetic methods, it has become possible to test species concepts by determining ancestral relationships—the systematics of species. Genetic similarities are used as a tool for tracing the ancestral evolutionary history. Within this context it is assumed that populations without reproductive barriers exhibit a high degree of genetic similarity. Consequently, the genetic distances between populations will increase with their time of isolation. Similar to morphological species descriptions, however, this approach also suffers from some uncertainties, such as for example: How much genetic variability is "sufficient" to recognise two species, or how much intraspecific variation is "allowed"?

For these reasons, it is difficult to specify the number of extant *Characeae* species. There is no final agreement of the species concept because, in view of the high morphological variability in some groups, it has not yet been possible to reconcile traditional morphological species delimitations with the results of modern phylogenetic analyses. This monograph recognises 67 (excluding addendum) species of *Characeae* for Europe, if following the microspecies concept. According to the macrospecies concept of Wood (1962) and Wood and Imahori (1965), this would result in just 24 species.

6.4 Molecular Identification and Present Phylogenetic Status

Fossil records of the class *Charophyceae* described a large number of families and genera belonging to two different orders (see Chap. 8, The fossil record and evolution of charophytes from European data). In contrast, extant *Charophyceae* form a single family, the *Characeae*, of the order *Charales*. Wood and Imahori (1965) proposed a phylogeny for the family *Characeae* based on morphological studies, which generally agree with analyses of fossil characters. Traditionally, six genera were distinguished within the family *Characeae* (*Chara*, *Lamprothamnium*, *Lychnothamnus*, *Nitellopsis*, *Nitella*, *Tolypella*). However, based on morphological characters of the oospores and gyrogonites (Feist and Grambast-Fessard, 1991; Martín-Closas et al., 2018; Soulié-Märsche, 1989) supported by ultrastructural (Vouilloud et al., 2015) and genetic analyses (Pérez et al., 2014), we in this monograph split the genus *Tolypella* into two genera: *Tolypella* and *Sphaerochara*. *Sphaerochara* is equivalent to *Tolypella* section *Rothia* sensu Wood (1962). A detailed explanation of the background for this splitting is given in Chap. 8. Palaeontological and molecular data corresponded well and demonstrated that *Sphaerochara* represents the oldest genus among the living charophytes. Fossil records documented *Sphaerochara* as early as the Late Jurassic (157.3–152.1 million years ago, Martín-Closas et al., 2018).

Based on morphological criteria, the genera of the *Characeae* are classified into two groups ("tribes"): *Nitelleae* and *Chareae*. This differentiation is based on branching characteristics and on the coronal region of the oogonia. The branchlets of the tribe *Nitelleae* (*Nitella*, *Sphaerochara*, and *Tolypella*) are divided, and the oogonia are characterised by a small coronula of 10 cells (two tiers of five cells) (Wood and Imahori, 1965). In contrast, the tribe *Chareae* (*Chara*, *Lamprothamnium*, *Lychnothamnus*, and *Nitellopsis*) is characterised by undivided, or simple, branchlets, and the coronae of the oogonia is composed of a single tier of five cells. This morphological system, suggesting that the genera *Nitella*, *Sphaerochara*, and *Tolypella* shared a common ancestor and are the earliest diverging group of extant charophytes, has been contradicted by fossil records (e.g., Martín-Closas et al., 2018) and phylogenetic analysis. Phylogenetic studies based on ribosomal and plastid sequences (Karol et al., 2001; McCourt et al., 1996, 1999; Meiers et al., 1999; Pérez et al., 2014; Sakayama, 2008; Schneider et al., 2016) supported monophyly of the tribe *Chareae*. The tribe *Nitelleae*, however, formed a paraphyletic grade (Fig. 6.2) with the clade *Sphaerochara* & *Tolypella* found sister to the remainder of the family.

Morphologically the genera *Tolypella* and *Sphaerochara* are characterised by pinnate branching, and the branchlets often form compact "birds nest" clusters. Species of the genus *Tolypella* have branchlet end cells that are allantoid, rounded at the apex and the coronula is usually deciduous. In contrast, *Sphaerochara* species have small, conical and acute branchlet end cells and a persistent coronula. In this monograph, five *Tolypella* and three *Sphaerochara* species are recognised for Europe based on morphological characteristics. Genetically, *Sphaerochara intricata* (=*Tolypella intricata*), *S. prolifera* (=*Tolypella prolifera*), and *S. canadensis* (=*Tolypella canadensis*) form a distinctive clade (Fig. 6.2) and can be separated from each other and from other species based on *rbc*L sequence data. The monophyletic clade of *Tolypella* revealed a more complex picture (Fig. 6.2). *Tolypella nidifica* and *T. salina* have identical *rbc*L gene sequences (Fig. 6.2) despite morphological differences; e.g., antheridia size or oospore features, especially wall ornamentation pattern (Lambert, 2013; Mouronval et al., 2015). In contrast, the species *T. glomerata*, *T. normaniana* and the dioecious *T. hispanica* showed genetic variability and morphotypes of these species can be found in different clades (Fig. 6.2). Other chloroplast genes (*atp*B, *psb*C) revealed comparable results (unpublished data). Consequently, morphological traits do not match with genetic differences in the genus *Tolypella*. In this European flora, we have decided to keep all "traditional" *Tolypella* species, and await further studies (multigene phylogenetic and ecophysiology analyses) of the *Tolypella* taxa to clarify potential misidentifications, species status and identify morphological characters that would be helpful to characterise genetic clades.

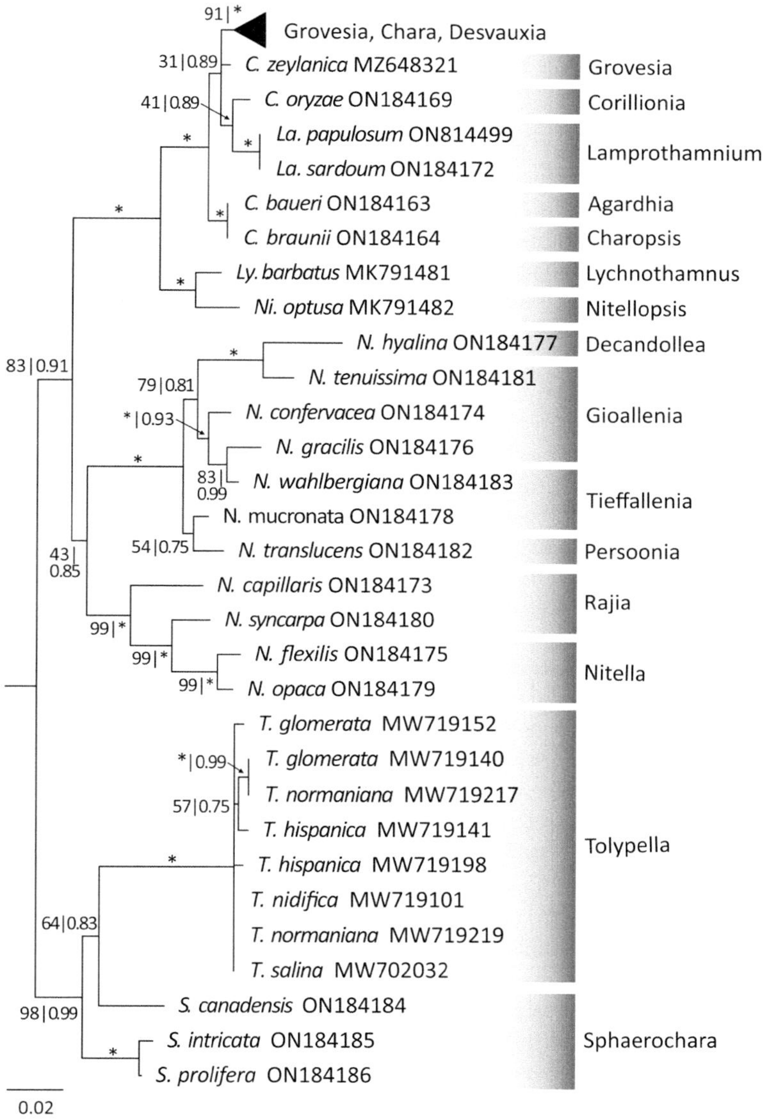

Fig. 6.2 Maximum likelihood and Baysian inferrence (GTR + I + G) phylogeny of *Characeae* based on *rbc*L (1041 bp). Numbers show bootstraps values (1000 replicates) and posterior probabilities (two parallel MCMC runs, 3 million generations) calculated with MEGA X and MrBayes. Stars indicate branches that received 100% bootstraps value and 1.00 posterior probability support. Sections (Wood and Imahori, 1965) are indicated with grey shades. The GenBank accession numbers were given after the species names

The genus *Nitella* is morphologically quite variable and includes plants with loosely arranged undifferentiated branchlets and others with complex branchlet patterns that can form compact fertile heads. The genus *Nitella* is morphologically divided into three subgenera: *Hyella*, *Nitella*, and *Tieffallenia*. This differentiation is based on the number and morphology of dactyls cells; cells that follow the last branchlet furcation (see Ontogenesis Chap. 3). Forms with single-celled dactyls are classified into the subgenus *Nitella* (1). Subgenus *Hyella* (2) has multi-celled (2–5) dactyls that generally terminate with allantoid-shaped cells. The subgenus *Tieffallenia* (3) also has multi-celled dactyls, but their number is typically restricted to two or rarely three, and they terminate with reduced, acute-shaped cells. According to current knowledge, only members of the subgenera *Nitella* (Sect. *Rajia*, *Nitella*) and *Tieffallenia* (sect. *Gioallenia*, *Tieffallenia*, *Decandollea*, *Persoonia*) occur in Europe and both were recovered with strong support in the *rbc*L-based sequence phylogeny (Fig. 6.2). Morphological studies of the plant and oospore morphology as well as multigene analyses (Sakayama, 2008; Sakayama et al., 2002) support the traditional classification of species according to Migula (1900) or Braun (1883). Of the 14 *Nitella* species presented in this monograph, eleven are well supported by *rbc*L sequence data, and they formed distinct genetic entities. The other three species (*N. dixonii*, *N. ornithopoda* and *N. palhinhae*) were not included in the phylogenetic analyses, which does not allow a statement about their species status at this point.

The vegetative characters of the genus *Nitellopsis* are so distinct from any other charophyte that Hy (1889) and Migula (1900) placed it in a separate genus. The main axis and branchlets are ecorticated and stipulodes are absent or rare. Branchlets with 1–2 nodes produce usually 1–2 very long bract-cells, often superficially resembling branchlets of *Nitella*. Nevertheless, the coronae of the oogonia with a single tier of five cells allies *Nitellopsis* in the tribe *Chareae* (Fig. 6.2). Fossil analyses of the gyrogonites of *Nitellopsis* (Grambast and Soulié-Märsche, 1972) and of vegetative characters (Guerlesquin, 1967) suggested an affinity with large-fruited charophytes of the lower Tertiary. Plants of the genus *Nitellopsis* are surviving representatives of an ancient type with *Nitellopsis obtusa* as something simple and archaic (Grambast and Soulié-Märsche, 1972; Soulié-Märsche, 1989). Gene sequence analyses confirm the position of the genus and show that *Nitellopsis* and *Lychnothamnus* formed a strongly supported monophyletic sister clade to *Chara* and *Lamprothamnium* (Fig. 6.2).

Plants of the genus *Lychnothamnus* are monoecious, and exhibit a main axis with imperfect or entirely absent cortex. The stipulodes are 2 times as many as the branchlets and each oogonium is flanked by two lateral antheridia. The genus comprises a single species, *Lychnothamnus barbatus*, and gene sequencing also shows a clear separation of this species (Figs. 6.2 and 6.3). The sister genera *Lychnothamnus* and *Nitellopsis* are native in Europe and were recently discovered in North America (Karol et al., 2017; Larkin et al., 2018). While *Nitellopsis obtusa* is perceived as aggressively invasive, *Lychnothamnus barbatus* shows no known invasive trends and may be a previously undiscovered native species in North America. Phylogenetic analyses placed North America *Lychnothamnus* and *Nitellopsis* specimens with those from Europe (Karol et al., 2017; Larkin et al., 2018) and supports their monophyletic status.

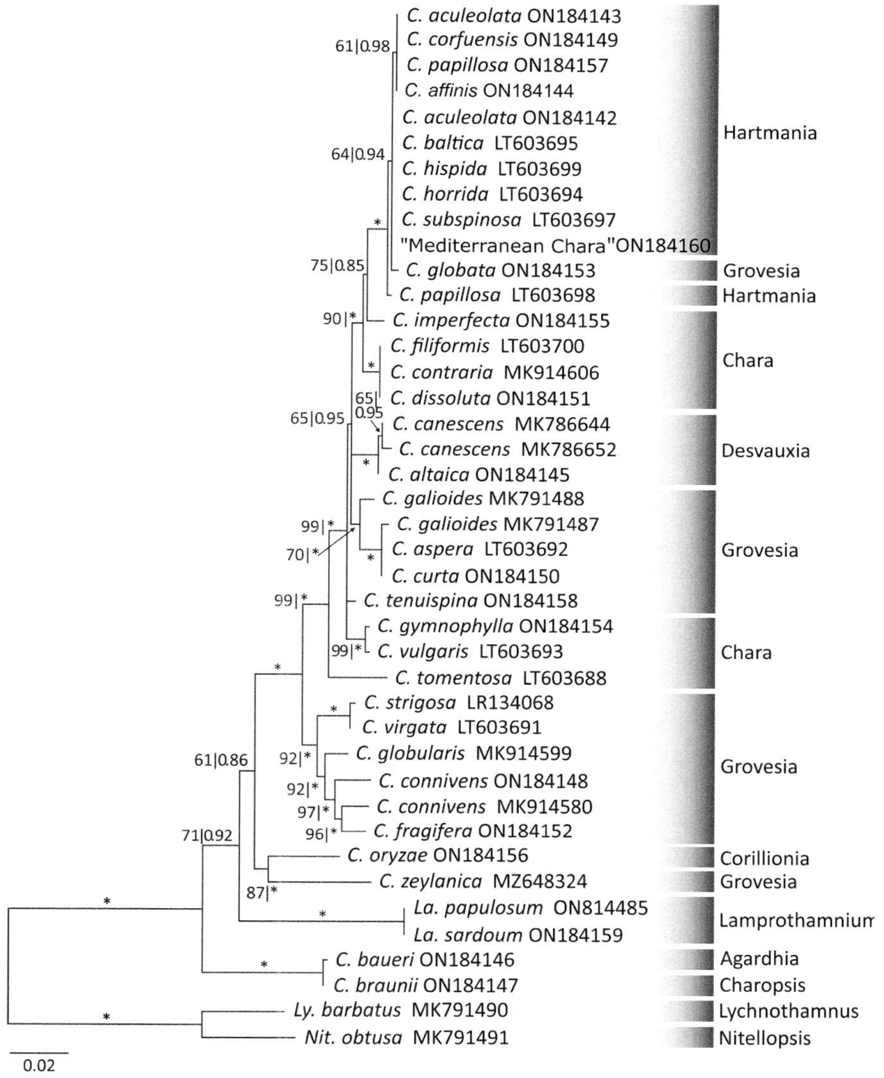

Fig. 6.3 Maximum likelihood and Baysian inferrence (GTR + I + G) phylogeny of *Chareae* based on *mat*K (1045 bp). Numbers show bootstraps values (1000 replicates) and posterior probabilities (two parallel MCMC runs, 3 million generations) calculated with MEGA X and MrBayes. Stars indicate branches that received 100% bootstraps value and 1.00 posterior probability support. Subsections of *Chara* (Wood and Imahori, 1965) are indicated with grey shades. The GenBank accession numbers were given after the species names

The genus *Lamprothamnium* has an ecorticated main axis and branchlets, and stipulodes in one tier. Plants of this genus are monoecious with conjoined gametangia and generally oogonia develop below the antheridia. The status of *Lamprothamnium* had been considerably debated in the past (e.g., Wood and Imahori, 1965) and some phylogenetic studies place it within *Chara*, though only weakly supported. Morphologically the genus *Lamprothamnium* has similarities with species of *Chara* subgenus *Charopsis*; both are totally ecorticated and both have stipulodes in one tier, though stipulodes in *Lamprothamnium* are arranged opposite each branchlet, while stipulodes in *Chara* are arranged alternate to the branchlets. Wood and Imahori (1965) proposed that *Lamprothamnium* could have developed from within the *Charopsis* subsection relatively recently. AFLP, ribosomal and (some) plastid sequences weakly placed *Lamprothamnium* within the genus *Chara*, which partly confirms this conjecture (Mannschreck, 2003; McCourt et al., 1996; Meiers et al., 1999; Pérez et al., 2014). Extant and fossilised oospore characters such as a single-celled basal plate, a closed apex and a peripheral depression not possessed by members of the genus *Chara* (Martín-Closas and Schudack, 1991) support the status of *Lamprothamnium* as a separate genus. Because we do not yet have a sufficient number of *Lamprothamnium* samples for detailed phylogenetic analyses, this monograph continues to recognise *Lamprothamnium* at the genus level until more data become available. Two *Lamprothamnium* species that occur in Europe (*L. papulosum* and *L. sardoum*), notably, shared identical *rbc*L and *matK* sequences (Figs. 6.2 and 6.3). Detailed analyses of additional samples along with more variable markers will be necessary to resolve these uncertainties. *Lamprothamnium succinctum* was not included in this phylogenetic study. It is a very rare charophyte species and not currently recognised in Europe. This species was reported from the Canary Island ("Charco de Maspalomas", Gonzalez Henriquez and Betancort Villalba, 1995), but its status is unknown.

Vegetative characters of the stipulodes and the cortex are generally used to differentiate species of the genus ***Chara***. Traditionally (Braun, 1883; Wood and Imahori, 1965; Zaneveld, 1940), the genus *Chara* has been divided into two subgenera, primarily based on the presence of either one row of stipulodes (haplostephanous: subgenus *Charopsis* sensu Wood and Imahori, 1965) or two rows of stipulodes (diplostephanous: subgenus Chara sensu Wood and Imahori, 1965).

Chara is morphologically the most variable genus of this monograph, with 40 species differentiated using morphological characters alone. We analysed *matK* sequence data of 32 taxa, and found that not all morphological species formed distinct genetic clusters. For 8 species (*C. fibrosa* sensu lato, *C. hellenica, C. kirghisorum, C. neglecta, C. oedophylla, C. squamosa, C. uzbekistanica, C. filarszkyana* nom. nov.), we lack genetic data for phylogenetic analyses.

Plants of the subgenus *Charopsis* exhibit stipulodes in one tier and are delimited into entirely ecorticated forms and forms with corticated axes (subsection *Agardhia*). The question of treating Kützing's "*Charopsis*" as a separate genus has been repeatedly discussed in recent decades. Corillion et al. (1959) argued with ecological, cytological, and anatomical evidence for raising *Charopsis* to genus rank. Wood and Imahori (1965) interpreted vegetative characters used for species delineation as more comparable to those of other subgenera. Even today, vegetative peculiarities give rise to repeated discussions about re-establishing *Charopsis* to generic rank. Unfortunately, results using DNA sequence data with respect to this question are currently inconsistent. Depending on the genetic marker (plastid genes, ribosomal DNA) and even the phylogenetic method used (distance, parsimony, Bayesian, or likelihood analyses), the position of *Charopsis* species varies (Mannschreck, 2003; McCourt et al., 1996; Meiers et al., 1999; Pérez et al., 2014). For example, single gene phylogenetic analyses placed *Charopsis* sister to a clade including the genus *Lamprothamnium* plus subgenus *Chara* (Figs. 6.2 and 6.3, Meiers et al., 1999). However, phylogenetic analyses of combined genetic markers indicate that *Charopsis* is a member of the genus *Chara* to the exclusion of *Lamprothamnium* (e.g., Pérez et al., 2014). A final answer to this question could not yet be found here and there is still need for clarification. The European species within *Charopsis*, i.e., *Chara braunii* (sect. *Charopsis*) and *C. baueri* (subsections *Agardhia*, *Braunia*), are clearly separated from other *Chara* species. However, they are only differentiated by a single base pair substitution (Fig. 6.3). As discussed above under *Lamprothamnium*, additional taxa including *C. fibrosa*, *C. pelosiana* and *C. uzbekistanica* (sect. *Charopsis*, subsec. *Agardhia*) along with more variable molecular markers are necessary.

Traditionally, stem cortication has been one of the most important morphological traits for identifying relationships between sections in the subgenus *Chara* (e.g. Braun, 1883; Wood and Imahori, 1965; Zaneveld, 1940). Sections *Desvauxia*, *Chara* and *Grovesia* have haplo-, diplo- and triplostichous cortex, respectively. Wood and Imahori (1965) continued to delimit the latter two sections further into subsections. The main difference between the subsections *Chara* and *Hartmania* (sect. *Chara*) is the internode diameter. *Hartmania* specimens are described as "tall and stout", whereas *Chara* individuals are, in contrast, "slender and rather small". The distinction between subsections *Grovesia* and *Willdenowia* (sect. *Grovesia*) is based on branchlet parameters. However, gene sequence analyses (plastid) do not support the categorisation into these different sections or subsections and thus show that the type of cortication is not phylogenetically informative at this level (Fig. 6.3). These results support earlier studies (Forsberg, 1963; McCracken et al., 1966; Meiers et al., 1999; Proctor and Wiman, 1971; Proctor, 1971a, 1971b) that dispute the phylogenetic significance assigned to stipulodes and cortication in determining the evolutionary history of subgenera and subsections. Homoplasy[3] seems to limit the utility of these characters in interpreting the phylogenetic history of this genus.

---

[3] Homoplasy is a shared character between two or more taxa that did not arise from a common ancestor. A homoplasy is the opposite of a homology, where a common ancestor provided the genes that gave rise to the trait in two or more taxa. Often, a homoplasy will occur when two unrelated species independently evolve or gain a comparable trait.

The *mat*K sequence data (Fig. 6.3) grouped 33 species of the subgenus *Chara*, into 15 well-defined groups (i.e., monophyletic morphospecies or groups of morphospecies) with considerable morphological variation within genetically homogeneous groups. This included features commonly used for *Chara* determination: length and number of spine cells, length of stipulodes and bract cells, cortication (tylacanthous, isostichous, aulacanthous and absent cortication). However, there were also substantial genetic differences among morphologically similar species (e.g., *C. virgata*—*C. globularis*—*C. connivens*—*C. fragifera*). Members of the prominent group of the subsection *Hartmania* (robust charophytes, diplostephanous, and diplostichous cortex) form a genetic group of nine morphospecies (Fig. 6.3) with only little genetic variation. In addition, the groups that were differentiated within these nine morphospecies are not consistent with morphological species boundaries. *Chara baltica*, *C. hispida* or *C. aculeolata* cannot be separated from each other with different plastid genes either (e.g. Nowak et al., 2016; Schneider et al., 2015, 2016). However, AFLP analyses indeed revealed a clear separation of *C. hispida* and *C. papillosa* according to their morphological traits (Mannschreck, 2003; Mannschreck et al., 2002). Based on ribosomal 18S DNA and their ITS-1 regions the same held true for the separation of *C. baltica* and *C. horrida* from *C. hispida*, *C. subspinosa* and *C. papillosa*; two different genetic clades can be observed (Nowak et al., 2016). Thus, phylogenetic results are partly contradictory, depending on the methods used. This prevents a final decision on whether or not members of the *Hartmania* subsection are independent species. A similar situation holds true for the taxa *C. contraria*, *C. filiformis* and *C. dissoluta*, which are morphologically different but show identical *mat*K sequences (Fig. 6.3). Cultivation experiments have shown that *C. contraria* exhibits high morphological variability, while *C. filiformis* remains morphologically stable (Teppke, 2016). We can observe this kind of assemblages within the section *Grovesia* as well: *C. virgata* and *C. strigosa* are morphologically very different, mainly due to the presence or absence of spines, but genetically they are very similar. Figure 6.3 used two specimens with minor differences, but Schneider et al. (2016) revealed that both species may also have identical *mat*K sequence data. However, these genetically very similar or even identical taxa are considered different species in this European flora due to morphological and/or ecological differences. In contrast, recent studies on *C. connivens* show genetic differences between the dioecious freshwater *C. connivens* or Mediterranean specimens and its brackish equivalents collected at the Baltic Sea (Fig. 6.3, Trbojević et al., 2020, unpublished data) Thus, even within the genus *Chara*, morphologically similar forms may be genetically distinct. Moreover, the status within the genetic cluster around *C. aspera* has not yet been clarified. *Chara curta* cannot be genetically separated from *C. aspera*. In contrast, *C. galioides* shows high genetic variability and occurs in two distinct clusters, one of which is identical with *C. aspera,* while the other cluster is genetically distinct (Fig. 6.3). This means that *C. galioides* exists as a "true" species, but the current morphological description is insufficient to correctly differentiate "true" *C. galioides*, which is genetically distinct, from forms that genetically cluster with *C. aspera*. This includes the species description in this monograph, because we currently do not know of any morphological trait that could consistently separate the "true" *C. galioides*, which forms its own cluster, from specimens that are a morphological variant of *C. aspera*. Until suitable traits are found, DNA barcoding can help to separate *C. galioides* from *C. aspera* individuals that look similar to "*C. galioides*".

The section *Willdenowia* is represented by a single species in Europe: *Chara zeylanica*. Specimens of *C. zeylanica* were observed only recently on the island of Sardinia in 2020 (Becker et al., 2021) and can be easily recognised with *mat*K sequences (Fig. 6.3). Sister to *C. zeylanica* is *C. oryzae*. This is a new species observed in rice fields in S-Europe, for which Romanov et al. (2022) have described a new section (*Corillionia*) because the morphological traits (triplostichous cortex, totally corticated branchlets, haplostephanous stipulodes) do not fit to any other section of *Chara*.

In conclusion, the considerable morphological plasticity found in genus *Chara* is often not reflected in molecular phylogenetic studies. Genetic inferences show extensive morphological variation within genetically homogeneous groups. Thus far, DNA barcode/phylogenetic analyses are not able to separate for example *C. hispida* from *C. subspinosa* and *C. papillosa*, nor *C. baltica* from *C. horrida*. If we continue to recognise these taxa as separate species then they are certainly more closely related to each other than are other species assemblages across genus *Chara*. Thus, species of the *Hartmania* group; are very closely related with each other, and the same was found for species within the subsections *Chara* (*C. contraria—C filiformis—C. dissoluta*) and *Grovesia* (*C. virgata—C. strigosa*). All members of these groups could be interpreted as apparently phylogenetically very young, and we may be, consequently, witnessing ongoing speciation. However, the observed morphological differences may also simply be a consequence of habitat. The morphologically based groups (subgenera, sections and subsections) as developed by Wood and Imahori (1965) are useful for identification, but do not necessarily correspond to natural systematic units.

Remarks and Acknowledgment The authorities of the charophyte species discussed here can be found in the corresponding species chapters of this monograph. Nadia Abdelahad, Ralf Becker, Roland Bengtsson, Linnea Bergdahl, Irmgard Blindow, John Bruinsma, Anders Hargeby, Martin Huber, Mauro Iberite, Volker Krautkrämer, Tina Kyrkander, Anders Langangen, Jean-Baptiste Mouronval, Silke Oldorff, Uwe Raabe, Lothar Rathai, Stefan Rensing, Hendrik Schubert, James Shilland, Nick Stewart and Klaus van de Weyer are gratefully acknowledged for providing the samples used for phylogenetic analyses.

References

Adl, S. M., Simpson, A. G. B., Farmer, M. A., Andersen, R. A., Anderson, R. O., Barta, J. R., Bowser, S. S., Brugerolle, G., Fensome, R. A., Fredericq, S., et al. (2005). The new higher level classification of eukaryotes with emphasis on the taxonomy of protists. *The Journal of Eukaryotic Microbiology, 52*, 399–451.

Becker, R., Schubert, H., & Nowak, P. (2021). *Chara zeylanica* J.G.Klein ex Willd. (Charophyceae, Charales, Characeae): First European record from the Island of Sardinia, Italy. *Plants, 10*(10), 2069. https://doi.org/10.3390/plants10102069

Boegle, M. G., Schneider, S. C., Melzer, A., & Schubert, H. (2010a). Distinguishing *Chara baltica, C. horrida* and *C. liljebladii*—conflicting results from analysis of morphology and genetics. *Charophytes, 2*, 53–58.

Boegle, M. G., Schneider, S. C., Schubert, H., & Melzer, A. (2010b). *Chara baltica* Bruzelius 1824 and *Chara intermedia* A. Braun 1859—Distinct species or habitat specific modifications? *Aquatic Botany, 93*, 195–201.

Braun, A. (1883). *Fragmente einer Monographie der Characeen.—Nach den hinterlassenen Manuscripten A. Braun's herausgegeben von Hrn. Dr. Otto Nordstedt. Abhandlungen der Königlich Akademie der Wissenschaften zu Berlin.* Königliche Akademie der Wissenschaften, Berlin [in German].

Bremer, K., Humphries, C. J., Mishler, B. D., & Churchill, S. P. (1987). On cladistic relationships in green plants. *Taxon, 36*, 339–349.

Casper, S. J. (1997). Glanz und Elend der algologischen Systematik am Beispiel von *Chlorokybus atmophyticus* Geitler—unzeitgemäße Betrachtungen In Memoriam Alfred Rieth (22. 11.1911–27. 03.1997). *Archiv Für Protistenkunde, 148*, 521–535. [in German].

Cavalier-Smith, T. (1981). Eukaryote kingdoms: Seven or nine? *Bio Systems, 14*, 461–481.

Cavalier-Smith, T. (1998). A revised six-kingdom system of life. *Biological Reviews of the Cambridge Philosophical Society, 73*, 203–266.

Cook, M. E. (2004). Cytokinesis in *Coleochaete orbicularis* (Charophyceae): An ancestral mechanism inherited by plants. *American Journal of Botany, 91*, 313–320.

Corillion, R., Gille, C., & Guerlesquin, M. (1959). Critères cytologiques, anatomiques et édologiques en faveur du maintien d'un genre *Charopsis* chez les Charophycées. *Bulletin De La Société Scientifique De Bretagne, 34*, 65–71. [in French].

de Vries, J., & Archibald, J. M. (2018). Plant evolution: Landmarks on the path to terrestrial life. *New Phytologist, 217*, 1428-1434.

Feist, M., & Grambast-Fessard, N. (1991). The genus concept in Charophyta: Evidence from Palaeozoic to recent. In R. Riding (Ed.), *Calcareous algae and stromatolites* (pp. 189–203). Springer.

Finet, C., Timme, R. E., Delwiche, C. F., & Marlétaz, F. (2010). Multigene phylogeny of the green lineage reveals the origin and diversification of land plants. *Current Biology, 20*, 2217–2222.

Floyd, S. K., & Bowman, J. L. (2007). The ancestral developmental tool kit of land plants. *International Journal of Plant Sciences, 168*, 1–35.

Forsberg, C. (1963). Some remarks on Wood's revision of Characeae. *Taxon, 12*, 141–144.

Gonzalez Henriquez, N., & Betancort Villalba, M. J. (1995). Notas sorbe las Characeae de "El Charco de Maspalomas" (Gran Canaria, Islas Canarias). *Botánica Macaronésica, 21*, 37–42. [in in Spain].

Graham, L. E., Cook, M. E., & Busse, J. S. (2000). The origin of plants: Body plan changes contributing to a major evolutionary radiation. *Proceedings of the National Academy of Sciences, 97*, 4535–4540.

Grambast, L., & Soulié-Märsche, I. (1972). Sur l'ancienneté et la diversification des *Nitellopsis* (Charophytes). *Paléobiologie Continentale, 3*, 1–14. [in French].

Guerlesquin, M. (1967). *Recherches caryotypiques et cytotaxinomiques chez les Charophycées d'Europe occidentale et d'Afrique du Nord.* Jouve. [in French].

Hall, J. D., Fučíková, K., Lo, C., & Lewis, L. A. (2010). An assessment of proposed DNA barcodes in freshwater green algae. *Cryptogamie, Algologie, 31*, 529–555.
Hebert, P. D. N., Ratnasingham, S., & deWaard, J. R. (2003). Barcoding animal life: Cytochrome *c* oxidase subunit 1 divergences among closely related species. *Proceedings. Biological Sciences, 270*(Suppl 1), S96–S99.
Heckman, D. S., Geiser, D. M., Eidell, B. R., Stauffer, R. L., Kardos, N. L., & Hedges, S. B. (2001). Molecular evidence for the early colonization of land by fungi and plants. *Science, 293*, 1129–1133.
Hollingsworth, M. L., Clark, A. A., Forrest, L. L., Richardson, J., Pennington, R. T., Long, D. G., Cowan, R., Chase, M. W., Gaudeul, M., & Hollingsworth, P. M. (2009a). Selecting barcoding loci for plants: Evaluation of seven candidate loci with species-level sampling in three divergent groups of land plants. *Molecular Ecology Resources, 9*, 439–457.
Hollingsworth, P. M., Forrest, L. L., Spouge, J. L., Hajibabaei, M., Ratnasingham, S., van der Bank, M., Chase, M. W., Cowan, R. S., Erickson, D. L., Fazekas, A. J., et al. (2009b). A DNA barcode for land plants. *Proceedings of the National Academy of Sciences of the United States of America, 106*, 12794–12797.
Huss, V. A. R., & Kranz, H. D. (1997). Charophyte evolution and the origin of land plants. *Plant Systematics and Evolution, 11*, 103–114.
Hy, M. (1889). Sur les modes de ramification et cortication dans la famille des Characées, et les caractéres qu´ils petent fournir á la classification. *Bulletin de la Société Botanique de France, 36*, 393–398. [in French].
Jeffrey, C. (1971). Thallophytes and kingdoms: A critique. *Kew Bulletin, 25*, 291.
Jeffrey, C. (1982). Kingdoms, codes and classification. *Kew Bulletin, 37*, 403.
Judd, W. S., Campbell, C. S., Kellogg, E. A., Stevens, P. F., & Donoghue, M. J. (2008). *Plant systematics: A phylogenetic approach.* Sinauer Associates Inc.
Karol, K. G., McCourt, R. M., Cimino, M. T., & Delwiche, C. F. (2001). The closest living relatives of land plants. *Science, 294*, 2351–2353.
Karol, K. G., Skawinski, P. M., McCourt, R. M., Nault, M. E., Evans, R., Barton, M. E., Berg, M. S., Perleberg, D. J., & Hall, J. D. (2017). First discovery of the charophycean green alga *Lychnothamnus barbatus* (Charophyceae) extant in the New World. *American Journal of Botany, 104*, 1108–1116.
Kranz, H. D., Miks, D., Siegler, M. L., Capesius, I., Sensen, C. W., & Huss, V. A. (1995). The origin of land plants: Phylogenetic relationships among charophytes, bryophytes, and vascular plants inferred from complete small-subunit ribosomal RNA gene sequences. *Journal of Molecular Evolution, 41*, 74–84.
Lambert, E. (2013). *Tolypella salina R. cor., sur la façade atlantique. Contribution à l'amélioration des connaissances.* Université Catholique de l'Ouest [in French]. https://www.fcbn.fr/sites/fcbn.fr/files/ressource_telechargeable/130117_rapport_tolypella_salina_lambert_vdef.pdf
Larkin, D. J., Monfils, A. K., Boissezon, A., Sleith, R. S., Skawinski, P. M., Welling, C. H., Cahill, B. C., & Karol, K. G. (2018). Biology, ecology, and management of starry stonewort (*Nitellopsis obtusa*; Characeae): A red-listed Eurasian green alga invasive in North America. *Aquatic Botany, 148*, 15–24.
Leliaert, F., Smith, D. R., Moreau, H., Herron, M. D., Verbruggen, H., Delwiche, C. F., & de Clerck, O. (2012). Phylogeny and molecular evolution of the green algae. *Critical Reviews in Plant Sciences, 31*, 1–46.
Leliaert, F., Verbruggen, H., & Zechman, F. W. (2011). Into the deep: New discoveries at the base of the green plant phylogeny. *BioEssays: News and Reviews in Molecular Cellular and Developmental Biology, 33*, 683–692.
Lewis, L. A., & McCourt, R. M. (2004). Green algae and the origin of land plants. *American Journal of Botany, 91*, 1535–1556.
Li, L., Wang, S., Wang, H., Sahu, S. K., Marin, B., Li, H., Xu, Y., Liang, H., Li, Z., Cheng, S., et al. (2020). The genome of *Prasinoderma coloniale* unveils the existence of a third phylum within green plants. *Nature Ecology and Evolution, 4*, 1280.

Lokhorst, G. M. (1996). Comparative taxonomic studies on the genus *Klebsormidium* (Charophyceae) in Europe. *Cryptogamic Studies, 5*. Gustav Fischer.
Mannschreck, B. (2003). *Genetische und morphologische Differenzierung ausgewählter Arten der Gattung Chara* (177 pp.). Dissertation, Technische Universität München [in German].
Mannschreck, B., Fink, T., & Melzer, A. (2002). Biosystematics of selected *Chara* species (Charophyta) using amplified fragment length polymorphism. *Phycologia, 41*, 657–666.
Marin, B., & Melkonian, M. (1999). Mesostigmatophyceae, a new class of streptophyte green algae revealed by SSU rRNA sequence comparisons. *Protist, 150*, 399–417.
Martín-Closas, C., & Schudack, M. E. (1991). Phylogenetic analysis and systematization of post-paleozoic Charophytes. Bulletin de la Société Botanique de France. *Actualités Botaniques, 138*, 53–71.
Martín-Closas, C., Vicente, A., Pérez-Cano, J., Sanjuan, J., & Bover-Arnal, T. (2018). On the earliest occurrence of *Tolypella* section Tolypella in the fossil record and the age of major clades in extant Characeae. *Botany Letters, 165*, 23–33.
McCourt, R. M., Delwiche, C. F., & Karol, K. G. (2004). Charophyte algae and land plant origins. *Trends in Ecology and Evolution, 19*, 661–666.
McCourt, R. M., Karol, K. G., Casanova, M. T., & Feist, M. (1999). Monophyly of genera and species of Characeae based on *rbc*L sequences, with special reference to Australian and European *Lychnothamnus barbatus* (Characeae: Charophyceae). *Australian Journal of Botany, 47*, 361–369.
McCourt, R., Karol, K. G., Guerlesquin, M., & Feist, M. (1996). Phylogeny of extant genera in the family Characeae (Division Charophyta) based on *rbc*L sequences and morphology. *American Journal of Botany, 83*, 125–131.
McCracken, M., Proctor, V. W., & Hotchkiss, A. (1966). Attempted hybridization between monoecious and dioecious clones of *Chara*. *American Journal of Botany, 53*, 937–940.
Meiers, S. T., Proctor, V. W., & Chapman, R. L. (1999). Phylogeny and biogeography of *Chara* (Charophyta) inferred from 18S rDNA sequences. *Australian Journal of Botany, 47*, 347–360.
Meiers, S. T., Rootes, W., Proctor, V. W., & Chapman, R. L. (1997). Phylogeny of the Characeae (charophyta) inferred from organismal and molecular characters. *Archiv Für Protistenkunde, 148*, 308–317.
Migula, W. (1900). *Die Characeen Deutschlands, Österreichs und der Schweiz* (2nd ed.). E. Kummer. [in German].
Mouronval, J.-B., Baudouin, S., Borel, N., Soulié-Märsche, I., Klesczewski, M., & Grillas, P. (2015). *Guide des Characées de France méditerranéenne*. Office National de la Chasse et de la Faune Sauvage. [in French].
Nishiyama, T., Sakayama, H., de Vries, J., Buschmann, H., Saint-Marcoux, D., Ullrich, K. K., Haas, F. B., Vanderstraeten, L., Becker, D., Lang, D., et al. (2018). The *Chara* genome: Secondary complexity and implications for plant terrestrialization. *Cell, 174*, 448–464.e24.
Nowak, P., Schubert, H., & Schaible, R. (2016). Molecular evaluation of the validity of the morphological characters of three Swedish *Chara* sections: Chara, Grovesia, and Desvauxia (Charales, Charophyceae). *Aquatic Botany, 134*, 113–119.
Pérez, W., Casanova, M. T., Hall, J. D., McCourt, R. M., & Karol, K. G. (2017). Phylogenetic congruence of ribosomal operon and plastid gene sequences for the Characeae with an emphasis on *Tolypella* (Characeae, Charophyceae). *Phycologia, 56*, 230–237.
Pérez, W., Hall, J., McCourt, R., & Karol, K. G. (2014). Phylogeny of North America *Tolypella* (Charophyceae, Charophyta) based on plastid DNA Sequences with a description of *Tolypella ramosissima* sp. nov. *Journal of Phycology, 50*, 776–789.
Pickett-Heaps, J. D. (1975). *Green algae: Structure, reproduction and evolution in selected genera*. Sinauer, Sunderland, Mass.
Pringsheim, M. (1862). On the pro-embryos of the Charae. *The Annals and Magazine of Natural History, 59*, 321–326.

Proctor, V. W. (1971a). *Chara globularis* Thuillier (= *C. fragilis* Desvaux): Breeding patterns within a cosmopolitan complex. *Limnology and Oceanography, 16*, 422–436.
Proctor, V. W. (1971b). Taxonomic significance of monoecism and dioecism in the genus *Chara*. *Phycologia, 10*, 299–307.
Proctor, V. W., & Wiman, F. (1971). An experimental approach to the systematics of the monoecious-conjoined members of the genus *Chara*, series Gymnobasalia. *American Journal of Botany, 58*, 885–893.
Puttick, M. N., Morris, J. L., Williams, T. A., Cox, C. J., Edwards, D., Kenrick, P., Pressel, S., Wellman, C. H., Schneider, H., Pisani, D., & Donoghue, P. C. J. (2018). The interrelationships of land plants and the nature of the ancestral embryophyte. *Current Biology, 28*, 733–745.e2.
Qiu, Y.-L., Li, L., Wang, B., Chen, Z., Knoop, V., Groth-Malonek, M., Dombrovska, O., Lee, J., Kent, L., Rest, J., et al. (2006). The deepest divergences in land plants inferred from phylogenomic evidence. *Proceedings of the National Academy of Sciences of the United States of America, 103*, 15511–15516.
Ragan, M. A., Parsons, T. J., Sawa, T., & Straus, N. A. (1994). 18S ribosomal DNA sequences indicate a monophyletic origin of Charophyceae. *Journal of Phycology, 30*, 490–500.
Ratnasingham, S., & Hebert, P. D. N. (2007). bold: The barcode of life data system (http://www.barcodinglife.org). *Molecular Ecology Notes, 7*(3), 355–364. https://doi.org/10.1111/j.1471-8286.2007.01678.x
Romanov, R. E., Nikulin, A. Y., Nikulin, V. Y., & Gontcharov, A. A. (2022). New species *Chara oryzae* and a new section *Corillionia* of *Chara* (Charales, Charophyceae) from European Mediterranean rice fields. *European Journal of Phycology, 57*(3), 328–342. https://doi.org/10.1080/09670262.2021.1987527
Sakayama, H. (2008). Taxonomy of *Nitella* (Charales, Charophyceae) based on comparative morphology of oospores and multiple DNA marker phylogeny using cultured material. *Phycological Research, 56*, 202–215.
Sakayama, H., Hara, Y., Arai, S., Sato, H., & Nozaki, H. (2004a). Phylogenetic analyses of *Nitella* subgenus Tieffallenia (Charales, Charophyceae) using nuclear ribosomal DNA internal transcribed spacer sequences. *Phycologia, 43*, 672–681.
Sakayama, H., Hara, Y., & Nozaki, H. (2004b). Taxonomic re-examination of six species of Nitella (Charales, Charophyceae) from Asia, and phylogenetic relationships within the genus based on rbcL and atpB gene sequences. *Phycologia, 43*(1), 91–104. https://doi.org/10.2216/i0031-8884-43-1-91.1
Sakayama, H., Nozaki, H., Kasaki, H., & Hara, Y. (2002). Taxonomic re-examination of *Nitella* (Charales, Charophyceae) from Japan, based on microscopical studies of oospore wall ornamentation and *rbc*L gene sequences. *Phycologia, 41*, 397–408.
Sarma, Y., & Khan, M. (1967). Dioecism and monoecism as taxonomic criteria in Charophyta. *Current Science, 36*, 243–247.
Saunders, G. W., & McDevit, D. C. (2012). Methods for DNA barcoding photosynthetic protists emphasizing the macroalgae and diatoms. In W. Kress & D. Erickson (Eds.), *DNA barcodes. Methods in molecular biology*, vol 858. Humana Press. https://doi.org/10.1007/978-1-61779-591-6_10
Schaible, R., Bergmann, I., Boegle, M. G., Schoor, A., & Schubert, H. (2009). Genetic characterisation of sexually and parthenogenetically reproductive populations of *Chara canescens* (Charophyceae) using AFLP, *rbc*L, and SNP markers. *Phycologia, 48*, 105–117.
Schneider, S. C., Nowak, P., von Ammon, U., & Ballot, A. (2016). Species differentiation in the genus Chara (Charophyceae): Considerable phenotypic plasticity occurs within homogenous genetic groups. *European Journal of Phycology, 51*, 282–293.
Schneider, S. C., Rodrigues, A., Moe, T. F., Ballot, A., & de Clerck, O. (2015). DNA barcoding the genus *Chara*: Molecular evidence recovers fewer taxa than the classical morphological approach. *Journal of Phycology, 51*, 367–380.

Schoch, C. L., Seifert, K. A., Huhndorf, S., Robert, V., Spouge, J. L., Levesque, C. A., & Chen, W. (2012). Nuclear ribosomal internal transcribed spacer (ITS) region as a universal DNA barcode marker for fungi. *Proceedings of the National Academy of Sciences of the United States of America, 109*, 6241–6246.

Schubert, H. (2014). Phylogeny and taxonomy of charophytes, good news from a battlefield of concepts. *Journal of Phycology, 50*, 773–775.

Sluiman, H. J., Roberts, K. R., Stewart, K. D., & Mattox, K. R. (1983). Comparative cytology and taxonomy of the Ulvophyceae. IV. Mitosis and cytokinesis in Ulothrix (Chlorophyta). *Journal of Phycology, 16*, 537–545.

Soulié-Märsche, I. (1989). *Etude comparée de gyrogonues de Charophytes actuelles et fossiles et Phylogéie des genres actuels* (166 pp.). Dissertation, l'Université des Sciences et Techniques du Languedoc [in French].

Stewart, K. D., & Mattox, K. R. (1975). Comparative cytology, evolution and classification of the green algae with some consideration of the origin of other organisms with chlorophylls A and B. *Botanical Review, 41*, 104–135.

Su, D., Yang, L., Shi, X., Ma, X., Zhou, X., Hedges, S. B., & Zhong, B. (2021). Large-scale phylogenomic analyses reveal the monophyly of bryophytes and Neoproterozoic origin of land plants. *Molecular Biology and Evolution, 38*, 3332–3344.

Teppke, M. (2016). Chapter 12.10 *Chara filiformis*. In Arbeitsgruppe Characeen Deutschlands (Ed.), *Armleuchteralgen. Die Characeen Deutschlands* (pp. 292–299). Springer [in German].

Tindall, B. J., Rosselló-Móra, R., Busse, H.-J., Ludwig, W., & Kämpfer, P. (2010). Notes on the characterization of prokaryote strains for taxonomic purposes. *International Journal of Systematic and Evolutionary Microbiology, 60*, 249–266.

Trbojević, I., Marković, A., Blaženčić, J., Subakov Simić, G., Nowak, P., Ballot, A., & Schneider, S. (2020). Genetic and morphological variation in *Chara contraria* and a taxon morphologically resembling *Chara connivens*. *Botany Letters, 38*, 1–14.

Turmel, M., & Lemieux, C. (2018). Evolution of the plastid genome in green algae. In S.-M. Chaw & R. K. Jansen (Eds.), *Plastid genome evolution* (pp. 157–193). Academic Press.

Turmel, M., Otis, C., & Lemieux, C. (2013). Tracing the evolution of streptophyte algae and their mitochondrial genome. *Genome Biology and Evolution, 5*, 1817–1835.

Turmel, M., Pombert, J.-F., Charlebois, P., Otis, C., & Lemieux, C. (2007). The green algal ancestry of land plants as revealed by the chloroplast genome. *International Journal of Plant Sciences, 168*, 679–689.

von Linné, C. (1753). *Species plantarum: vol. 1.* Holmiae [Stockholm]: Impensis Laurentii Salvii [in Latin].

Vouilloud, A. A., Leonardi, P. I., & Cáceres, E. J. (2015). Mixed evolutionary traits of *Tolypella* (section Rothia, Charales) compared with *Chara* and *Nitella* shown by ultrastructure of vegetative internodal cells. *Aquatic Botany, 120*, 67–72.

Wickett, N. J., Mirarab, S., Nguyen, N., Warnow, T., Carpenter, E., Matasci, N., Ayyampalayam, S., Barker, M. S., Burleigh, J. G., Gitzendanner, M. A., et al. (2014). Phylotranscriptomic analysis of the origin and early diversification of land plants. *Proceedings of the National Academy of Sciences of the United States of America, 111*, E4859–E4868.

Wilcox, L. W., Fuerst, P. A., & Floyd, G. L. (1993). Phylogenetic relationships of four charophycean green algae inferred from complete nuclear-encoded small subunit rRNA gene sequences. *American Journal of Botany, 80*, 1028–1033.

Wodniok, S., Brinkmann, H., Glöckner, G., Heidel, A. J., Philippe, H., Melkonian, M., & Becker, B. (2011). Origin of land plants: Do conjugating green algae hold the key? *BMC Evolutionary Biology, 11*, 104.

Wood, R. D., & Imahori, K. (1965). *A revision of the Characeae. First Part: Monograph of the Characeae.* Cramer.

Wood, R. D. (1962). New combinations and taxa in the revision of Characeae. *Taxon, 11*, 7–25.

Yoon, H. S., Hackett, J. D., Ciniglia, C., Pinto, G., & Bhattacharya, D. (2004). A molecular timeline for the origin of photosynthetic eukaryotes. *Molecular Biology and Evolution, 21*, 809–818.

Zaneveld, J. S. (1940). The Charophyta of Malaysia and adjacent countries. *Blumea, 4*, 1–223.

Zechman, F. W., Theriot, E. C., Zimmer, E. A., & Chapman, R. L. (1990). Phylogeny of the Ulvophyceae (Chlorophyta): Cladistic analysis of nuclear-encoded rRNA sequence data. *Journal of Phycology, 26*, 700–710.

Chapter 7
Ecology of Charophytes: Niche Dimensions of Pioneers and Habitat Engineers

Andrzej Pukacz, Takashi Asaeda, Irmgard Blindow, Norma Catarina Bueno, Michelle T. Casanova, Anne Herbst, Mariusz Pełechaty, Viraj Ranawaka, and Hendrik Schubert

7.1 Introduction

Charophytes, as structurally relatively simple organisms, and at the same time evolutionarily old and positioned among the closest relatives of the first land plants (Karol et al., 2001), have long aroused scientific and practical interest. Their large internodal cells, which are easily accessible for, e.g., electrophysiological techniques, made these macroalgae popular model organisms for studying the physiological aspects of terrestrial acclimation, salinity tolerance and membrane processes (Beilby & Casanova, 2014; Braun et al., 2007). Charophytes are also important bioindicators of water quality and water regime (for review see Doege et al., 2016); and their capacity to absorb pollutants from freshwater makes them useful for bioremediation (Kalin et al., 2002a, 2002b; Marquardt and Schubert, 2009). Concerns about the declining state of water resources have led to intensive study of both these aspects over the few

Andrzej Pukacz: Deceased

A. Pukacz (Deceased)
Collegium Polonicum, Adam Mickiewicz University Poznań, ul. Kościuszki 1, 69-100 Słubice, Poland

T. Asaeda · V. Ranawaka
Department of Environmental Science, Saitama University, 255 Shimo-Okubo, 338-8570 Sakura Ward, Saitama, Japan
e-mail: asaeda@mail.saitama-u.ac.jp

V. Ranawaka
e-mail: vranawaka@hotmail.com

I. Blindow
Universität Greifswald, Biologische Station Hiddensee, 18565 Kloster, Germany
e-mail: blindi@uni-greifswald.de

H. Schubert et al. (eds.), *Charophytes of Europe*,
https://doi.org/10.1007/978-3-031-31898-6_7

last decades. The past decades, charophyte-based bioindication systems have been developed by several authors, serving as a tool for environmental monitoring (e.g. Burns et al., 1999; Gutowski et al., 1998; Jäger, 2000; Kohler, 1982; Krause, 1981; Melzer, 1994; Selig et al., 2009; van Raam, 1998).

However, the bioindication protocols developed so far have been shown to be pretty region-specific, irrespective of basing on similar taxa. Intercalibration between "regional schemes" is possible, but complicated (Schneider, 2007), not only because of the problematic taxonomy in family *Characeae*, but also in relation to data management (Doege et al., 2016). This is highlighted by the fact that some species that are considered to be rare and threatened in their native range have successfully spread to other regions (Karol et al., 2017), raising the common questions about effective ecological niche dimensions of non-native species (Alix et al., 2017).

The interaction between charophytes and their environment is mutual. Charophytes are not only affected by environmental conditions, but dense charophyte vegetation can substantially change its environment. Charophytes can be a key element of aquatic ecosystems, able to create and maintain clear-water conditions by a network of biological interactions (Blindow et al., 2002; Casanova et al., 2003a; Meurer and Bueno, 2012; van den Berg et al., 1998a).

Many past reviews of charophyte ecology and niche dimensions have focused largely on charophyte distribution in relation to water chemistry (nutrients, salinity) and physical properties (pH, temperature, e.g. Corillion, 1957; García and Chivas, 2006; Rey-Boissezon and Auderset Joye, 2015). This review presents a more holistic approach to charophyte ecology, including aspects of biological interactions, relations to sediment properties, availability of diaspores and phenology. All stages in the charophyte life cycle from the dispersal and germination of viable oospores, through requirements for establishment and growth, interactions with the biotic and

N. C. Bueno
Universitário Cascavel, Universidade Estadual do Oeste do Paraná, Rua Universitária 1619, CEP 85819-100 Cascavel Paraná, Brazil
e-mail: normacatarina@hotmail.com

M. T. Casanova
273 Casanova Rd, 3351 Westmere, Victoria, Australia
e-mail: m.casanova@nhm.ac.uk

A. Herbst
Faculty of Interdisziplinary Research, Department Maritime Systems, University of Rostock, Albert-Einstein-Str. 21, 18059 Rostock, Germany
e-mail: anne.herbst@uni-rostock.dede

M. Pełechaty
Faculty of Biology, Department of Hydrobiology, Adam Mickiewicz University, ul. Uniwersytetu Poznańskiego 6, 61-614 Poznań, Poland
e-mail: marpel@amu.edu.pl

H. Schubert (✉)
Institut für Biowissenschaften, Universität Rostock, Albert-Einstein-Straße 3, 18051 Rostock, Germany
e-mail: hendrik.schubert@uni-rostock.de

abiotic environment, to the formation of a diaspore reservoir (oospores and vegetative propagules) will be reviewed.

7.2 Arriving at the New Town: Distribution Pathways of Charophytes

Charophytes occur on every continent, including Antarctica from which, however, only a single *Nitella* sp. record is known yet.

Charophytes prefer freshwater or brackish water habitats rather than open marine situations, and can be found in a wide variety of standing and slow-moving waters (Martin et al., 2004). Some species are widespread (e.g. *Nitella hyalina* (DC.) C. Agardh 1824, *Chara zeylanica* Willd. 1805, *Chara globularis* Thuill. 1799), but many charophytes have a limited geographical distribution. Charophytes are usually the first macrophytes to arrive in newly created ponds and lakes, leading to some species being coined as "permanent pioneers" (Krause and Walter, 1985), (Fig. 7.1).

Besides anthropogenic dispersal pathways as, e.g., via recreational fishing, shipping, bathing and diving (Karol and Sleith, 2017), there are two important natural

Fig. 7.1 Typical habitats of pioneer charophytes. **a** Field pond, **b** Ditch by a fish pond, **c** Puddles in a pasture. Photographs: A. Pukacz

processes that can explain the success of charophytes in rapid colonisation, re-establishment and long-distance dispersal: transport and in situ survival of oospores and bulbils. Charophyte oospores range in size from ~180 to >1000 μm (Haas, 1994; Krause, 1997; Wood, 1959) and are the primary agent of long and short-distance dispersal. Bulbils, which are not resistant to desiccation and consequently less likely to be successfully dispersed overland, are thought to be important for local persistence and dispersal within water bodies (Bonis and Grillas, 2002; Brochet et al., 2010; van den Berg et al., 2001).

Where there is variation in the production of oospores by individual charophyte species, those that produce large numbers are more likely to be widely dispersed and locally persistent (Baastrup-Spohr et al., 2015). Within lakes or wetlands, wind can play an important role in dispersal via wind-induced currents (Soomers et al., 2010). Also, charophyte oospores are very light, and wind dispersal could be significant when wetlands dry (Bakker et al., 2013). There is a body of evidence supporting the role of water birds as principal vectors in long-distance dispersal (Olsen, 1944; Proctor, 1959, 1962). Because of the lack of structures facilitating attachment, oospores are more likely to be transported internally in the birds alimentary tracts than externally. Smaller-sized propagules are more likely than large ones to survive gut passage in birds and to germinate afterwards (Brochet et al., 2010; Charalambidou and Santamaria, 2005; Figuerola et al., 2010; Green et al., 2002; Pollux, 2011). Enhanced germination after gut passage was also observed for *Chara* spp. (Brochet et al., 2010; Figuerola et al., 2010). However, as Proctor (1980) pointed out, although water birds can readily distribute charophyte oospores, there are numerous examples of short-range endemism that are not explained by dispersal capacity alone.

Species of *Chara* display a divergent biogeographic pattern (Proctor, 1980), i.e. some species are restricted to a single continent whereas others are distributed across large, supracontinental areas or are close to be cosmopolitan. Proctor (1980) hypothesised that the sexual reproductive system of charophytes has an impact on their biogeography, contending that cosmopolitan *Chara* species are monoecious (male and female gametangia on the same plant), whereas dioecious species (separate male and female plants) are confined to smaller areas. Dioecious species depend on the arrival of more than one oospore, whereas monoecious species can, in theory, establish a population from a single oospore. Until recently this fits with the distribution of all dioecious charophytes (e.g. Casanova, 2014; García and Casanova, 2003). However, irrespective of being broadly accepted, the evidence may not be as strong as assumed. Where the genetic identity of individual taxa has been determined, only a few taxa are found on more than one continent for both monoecious as well as dioecious species. Of course, it is striking that only the apomictic form of the single charophyte able to reproduce parthenogenetically, the dioecious *Chara canescens* Loisel. 1810, is found all over Europe and even in the Southern Hemisphere (Casanova and Nicol, 2009; Schaible et al., 2012), whereas sexually reproducing dioecious populations of *Chara canescens* are very rare (Schaible et al., 2009, 2011). On the other hand Proctor et al. (1971) demonstrated reduced interfertility between individuals of cosmopolitan (monoecious) species originating from different continents. Consequently, Proctor's hypothesis (Proctor, 1980) should be critically re-examined in the light of the increasing availability of distribution data.

7.3 Establishing Stands: Breaking Dormancy and Induction of Germination

Charophytes have three options to establish stands: (A) vegetative growth from omnipotent nodal cells (Skurzyński and Bociąg, 2011), (B) vegetative growth from bulbils (Asaeda et al., 2007; Wang et al., 2015) or, (C) germination of sexually produced oospores. Charophytes exhibit apical dominance, so lower nodal cells are activated when the apical cell is removed or damaged (Kuczewski, 1906; Pringsheim, 1863; Schubert et al., 2016). Bulbils—irrespective of their origin—are omnipotent organs too, but they contain starch storage cells, which enhances the capacity to endure periods of unsuitable conditions and improves the competitive ability of the vegetative offspring. Bulbil formation was studied intensively by Giesenhagen (1896). However, the mechanisms for activation of bulbil complexes and induction of bulbil formation have been little studied (c.f. Casanova, 1994). In laboratory incubation, bulbils usually germinate instantaneously and do not require pre-treatment to achieve high germination rates (Nowak et al., 2017a; Steinhardt and Selig, 2011). This raises the question of how germination of bulbils is suppressed under natural conditions, preventing germination before separation from the parent plant or its demise (Giesenhagen, 1896).

The regenerative options for charophytes (basal growth, bulbil activation and oospore germination) are targeting on different time frames. Nodal cells allow for immediate replacement of damaged axes, whereas bulbils support some growth even if nutrient and light availability is poor, and survival of short periods with adverse conditions, e.g. winters (Giesenhagen, 1896; van den Berg et al., 2001).

Oospores allow for an even longer perspective. They are protected by a multilayered oospore wall (Leitch et al., 1990) from chemical and physical hazards, and contain large amounts of storage compounds. They display a high degree of primary and secondary dormancy so that, contrary to bulbils, only part of the available pool germinates at any one time if conditions are suitable (Asaeda et al., 2007; Casanova and Brock, 1990, 1996; van den Berg et al., 2001; Vleeshouwers et al., 1995; Wang et al., 2008). Primary dormancy has been related to the concentration of abscisic acid in the oospore (Penfield and King, 2009; Sabbatini et al., 1987; Sederias and Coleman, 2007). After the period of primary dormancy, the oospore becomes susceptible to triggers that induce germination. Oospores can enter into a secondary dormancy if the germination "window" is not available in any one year (Sokol and Stross, 1986). Studies have shown that oospores can survive in sediments for more than 40 years (Alderton et al., 2017; Rodrigo et al., 2010). Oospores were even found to germinate from sediments dated at hundreds of years (Stobbe et al., 2014). Such oospores represent a persistent, dormant propagule bank in lakes and wetlands that is rarely totally depleted (van den Berg et al., 2001).

Hilhorst (1993) developed a hypothetical physiological model of dormancy for terrestrial plants that probably can be applied for charophytes as well (Holzhausen et al., 2017). This model presents a combined model for the regulation of secondary dormancy by temperature and germination induction by the availability of nitrate, light and gibberelic synthesis. It implies that dormancy and germination are directly correlated with the quantity of phytochrome receptors as stated for charophytes as well (Sokol and Stross, 1992).

Most existing studies on dormancy concentrate on individual or multi-factor analyses of triggers for secondary dormancy or germination induction. Stratification periods under low temperatures (e.g. Casanova and Brock, 1996; Forsberg, 1965) as well as osmotic shock (e.g. Lambert, 2013; Lambert et al., 2013; Proctor, 1967) can stimulate germination through synthesis of gibberellins (Hilhorst, 1993; Penfield and King, 2009; Vleeshouwers et al., 1995) when irradiance conditions are favourable (Kalin and Smith, 2007; Matheson et al., 2005; Sabbatini et al., 1987; Sederias and Coleman, 2007, 2009; van den Berg et al., 1998a). The specific conditions required for breaking dormancy and induction of germination, differ among species (for a detailed discussion see: Holzhausen et al., 2017). Temperature regime, osmotic conditions (incl. drying), organic material, sediment redox as well as irradiance conditions required for activation of oospores are related to the ecological conditions the species prefer, including a more or less pronounced seasonality (Bonis and Lepart, 1994; Casanova and Brock, 1996; Forsberg, 1965). Exposure to light (de Winton et al., 2004) as well as seasonality (Sokol and Stross, 1986) can be necessary for germination of some species whereas others are reported to be insensitive to some of the factors included in Hilhorst's (1993) model (e.g. de Winton et al., 2004; Shen, 1966).

7.4 Growing to Fertility—Abiotic Requirements for a Happy Charophyte Life

Most studies on the growth conditions for charophytes concentrate on more readily measured physical and chemical characteristics of water bodies, such as light, nutrients, pH and conductivity. These have been considered to be the most important determinants of charophyte distribution and diversity, both within and among aquatic environments (Blindow et al., 2014; van den Berg et al., 1998a). For some of these characteristics there are clear tolerance limits for individual species (e.g. Corillion, 1957; García and Chivas, 2006; Rey-Boissezon and Auderset Joye, 2015). As a consequence of this apparent sensitivity, particularly to nutrient status, the occurrence of charophytes has widely been applied as an indication of poor "trophic status" of waterbodies (Doege et al., 2016; Krause, 1981; Melzer, 1994; Schneider, 2007; Schwarz and Hawes, 1997). However, some characteristics have received insufficient attention so far as well as interactions among the physical and chemical characteristics of water (e.g. Domin et al., 2004; Schubert et al., 2017a).

Charophytes can take up nutrients through their photosynthetic thalli as well as through the rhizoids. They can accumulate nitrogen, primarily as ammonium-N, and phosphates (Beilby and Casanova, 2014). In general, charophytes are dominant constituents of submerged plant communities at low nutrient concentrations, where they can form extensive underwater meadows ("beds"). As nutrient concentrations increase, charophyte dominance declines and angiosperm abundance increases. Under high nutrient conditions charophytes can completely disappear. This was initially explained as a toxic effect of phosphorus (Forsberg, 1965), but later studies showed that *Chara* species maintained high growth rates at elevated phosphorus concentrations if light was not limiting (Blindow, 1988; Henricsson, 1976). In fact, a number of charophytes can be found at total phosphorus concentrations exceeding 0.4–0.5 mg L^{-1} e.g. *Nitellopsis obtusa* (Desv.) J. Groves 1919 (Doege et al., 2016). Nevertheless, mean values of total phosphorus are often lower than 0.1 mg L^{-1} in sites with abundant charophytes. Charophytes occurring in high nutrient waters are restricted to shallow sites where light is not a limiting factor (Blindow, 1992).

Charophytes as members of the "green chloroplast lineage" have limited options to maximize efficient use of light by expanding their spectral range of light absorption (Schubert, 1996a, 1996b). Still they are able to exist with as little as 1.5% of the light that reaches the water surface. Consequently, charophytes can be found in oligotrophic lakes (Fig. 7.2) down to as much as 65 m (Kufel and Kufel, 2002), and in shallow turbid water where the Secchi depth is less than 5 cm (Casanova and Porter, 2013). Light-use efficiency also allows persistence during the winter with thick ice cover (Herbst et al., 2018; Pełechaty et al., 2017; Pukacz et al., 2016a). Some charophyte species experience permanent shading as a result of the mineral encrustation on the surface of their thalli (Marquardt and Schubert, 2009; Martin et al., 2004; Raven et al., 1986) as well as within dense monospecific vegetation. Conversely some charophyte species in shallow waters need protection from high light intensity and UV-B radiation. These species can produce high concentrations of carotenoids, which protect the photosynthesis apparatus (Schagerl and Pichler, 2000) as well as the genetic material of sensitive organs (in particular antheridia, Wang et al., 2015) from excessive irradiance.

In addition to acclimation by means of changes in pigment content and composition, charophytes adapt to light conditions by a high degree of morphological plasticity. Similar to embryophytes, the effects of light attenuation in the water column, along with depth, result in elongation and spreading of branchlets (Blindow and Schütte, 2007; Nowak et al., 2017b; Pukacz et al., 2014; Schneider et al., 2015) as well as increased variability in biomass production (Pełechaty et al., 2013; Pukacz et al., 2016a).

Not all species react equally to worsening light conditions. Species with large thalli (>1 mm internode diameter) are the first to decline, whereas the smaller ones (<1 mm internode diameter) can still survive in shallow areas irrespective of higher turbidity (Blindow, 1992). Some species with large thalli, but sufficient morphological plasticity to allow for condensed growth forms (e.g. *Chara aculeolata* Kütz. in

Fig. 7.2 Charophytes growing at 15 m depth in an ultraoligotrophic, volcanogenic lake in the Andes. Photographs: H. Schubert

Rchb. 1832 and *Chara tomentosa* L. 1753) can "move" to the shallow sites (e.g. Krause, 1997; Urbaniak and Gąbka, 2014). Similarly, there are species that only persist as scattered individuals in shallow water, which never form extensive "beds" (Casanova and Brock, 1996).

Often regarded as organisms preferring calcium-rich, hard and alkaline waters (Fig. 7.3) in the past, pH tolerance of charophytes is generally wide and differs largely among taxa. While *Chara, Nitellopsis, Sphaerochara, Tolypella* and *Lychnothamnus* usually occur in more alkaline environments with pH values between 6.0 and 9.0, most *Nitella* species tend to occur in slightly acidic to neutral soft waters (e.g. Doege et al., 2016; Martin et al., 2004). Still, especially within the genera *Chara* and *Nitella*, species-specific differences with respect to both pH and Ca-content are large and preference ranges of some species (e.g. *Chara braunii* Gmellin 1826 and *Nitella flexilis* (L.) C. Agardh, 1824) may overlap (Schubert et al., 2017b). The primary mechanism of carbon assimilation in charophytes is the conversion of bicarbonates into CO_2 that diffuses into the charophyte cell, additionally resulting in $CaCO_3$ precipitation onto the surface of the charophyte thallus (Raven et al., 1986). This results in the characteristic strong encrustation of charophytes in Ca-rich freshwaters, where these macroalgae contribute to calcite precipitation and storage in bottom sediments (as summarized by Pełechaty et al., 2013). Calcite precipitation has significant ecological consequences if there is co-precipitation of phosphorus, allowing charophyte beds to act as nutrient sinks (Kufel and Kufel, 2002). In addition, this

Fig. 7.3 *Chara strigosa* A. Braun 1847 in an alpine hard-water lake with calcite deposition. Photograph: A. Pukacz

process of biomineralization results in decalcification of the water (Pełechaty et al., 2013).

Salinity is one of the major factors determining occurrence ranges of aquatic species (Haller et al., 1974; Schubert et al., 2017a; Telesh et al., 2013). Different salinity tolerance capabilities are observed among charophytes (Bisson and Kirst, 1995; Schubert and Blindow, 2004). In brackish water charophytes require the capacity to regulate osmotic processes and ion concentrations in line with external salinity variation. Salinity-tolerant charophytes regulate their turgor pressure in response to osmotic pressure of the medium by accumulation of Na^+, K^+ and Cl^- ions and sucrose. In contrast, obligate freshwater charophytes do not regulate their turgor pressure, but adjust internal osmotic pressure via vacuolar K^+ concentration (Beilby and Casanova, 2014). In general, four groups of charophytes can be distinguished in relation to habitat salinity (as summarized by Martin et al., 2004): freshwater species (e.g. many *Nitella* species), oligohaline halotolerant species (e.g. *Nitellopsis obtusa*, up to 5 psu), mesohaline species (e.g. *Chara canescens*, *Chara aspera* Willd. 1809, up to 15 psu) and euryhaline species (e.g. *Lamprothamnium* species, 20–70 psu).

Most charophytes seem to tolerate variable water temperatures. Their tolerance can be species-specific (Rojo et al., 2017), and this is largely reflected in geographic distribution (Khan, 1991). Temperature tolerance/requirement is a crucial factor for altitudinal distribution (e.g. Calero et al., 2017; Haas, 1994). Most species require warmer waters, which partly may explain diversity and abundance decreases at high altitudes and, to a lesser degree, high latitudes (Schubert et al., 2015). Nevertheless, there is a small group of species (e.g. *Sphaerochara canadensis* Soulié-Märsche, comb. nov., *Nitella opaca* (ex Bruzelius) C. Agardh 1924, *Nitella hookeri* A. Braun 1849, *Chara strigosa*) that seems to prefer lower water temperatures (Casanova et al., 2007; Krause, 1997). Temperature also seems to be very important for the growth and development of charophytes, affecting the maturation of gametangia (Calero et al., 2017) and dormancy of oospores (Holzinger and Pichrtová, 2016).

Although charophytes can be found on different types of substratum (peat, sand, clay and even rock), many species prefer a flat and stable substrate (Andrews et al., 1984; Sanderson et al., 2008). Fewer species are common in running waters (Krause, 1997), and those that occur there are often found in back-waters, or slow moving parts of the stream. Bathymetry is one of the factors affecting species diversity of charophytes (Nowak et al., 2017a; Pukacz et al., 2016b) as well as their spatial distribution within water bodies (Pełechaty et al., 2004) due to the low strength of rhizoids. For this reason, the larger charophytes are confined to areas with low to moderate mechanical stress (Henricson et al., 2006). In contrast, there are some small charophytes whose well-developed rhizoids (e.g. *Chara aspera* or *Chara virgata* Kütz. 1834), facilitate their occurrence in very shallow, steep and wave-exposed sites. There have been few studies on charophyte occurrence in relation to substratum type. Since charophytes lack the complex root and hollow stem systems of vascular macrophytes, they cannot facilitate oxygen release into the rhizosphere (Matheson et al., 2005), which probably affects their capacity to grow on and absorb nutrients from anoxic soils. However, several species are known for abundant growth on anoxic mud and even toxic sediments (Herbst et al., 2019). It must be stated that, except for descriptive studies, relationships between charophyte growth and substratum type are still largely unexplored (e.g. Casanova et al., 2003a; Kufel and Kufel, 2002).

7.5 Nasty Neighbours and Fancy Friends—Interactions Between Charophytes and Their Biotic Environment

Charophytes are not alone in aquatic environments, they contribute to food webs and habitat provision and they coexist, compete with and facilitate or hinder the existence of other organisms. In the geological past, first in the Devonian (~400 Mio years bp, Schudack, 2016), charophytes dominated the submerged vegetation in shallow freshwater and probably many brackish water ecosystems (Riveline et al., 1996; Schudack, 1991). Since the angiosperms showed up on the "stage" around 200–250 million years ago (Morris et al., 2018), interactions with these "newcomers" have possibly been one of the most important issues in the life of the phylogenetically far older (Domozych et al., 2016) charophytes.

Charophytes today occur mainly in mesotrophic ecosystems. Competition for light is assumed to restrict the "bottom-dwelling" charophytes from turbid water, where angiosperms can form canopies near the water surface, enabling them to "escape" from the low light availability in the water column (Blindow, 1992). Charophytes have been shown to produce (Wium-Andersen et al., 1982) and excrete (Blindow and

Hootsmans, 1991) allelopathic substances that reduce the densities of both phytoplankton and epiphyton in the laboratory (Berger and Schagerl, 2004; Blindow and Hootsmans, 1991; Mulderij et al., 2007; Pakdel et al., 2013), but less is known about the importance of this mechanism in natural ecosystems (Hilt and Gross, 2008).

In oligo- to mesotrophic hardwater ecosystems, however, charophytes can form a dense and often wintergreen vegetation. Efficient bicarbonate assimilation (van den Berg et al., 2002) as well as long-term immobilisation of nutrients by uptake and co-precipitation of phosphorus with carbonates (Kufel and Kufel, 2002; Siong and Asaeda, 2006) renders them superior competitors to angiosperms and phytoplankton (Blindow et al., 2014). Charophytes function as key organisms in such "Chara-lakes" (Samuelsson, 1925) and exert a major impact on the whole trophic web. They offer food and shelter from predation, for zooplankton, macroinvertebrates and fish larvae (Jones and Sayer, 2003), though low densities of zooplankton are common in charophyte-dominated lakes possibly due to low food availability and/or high densities of smaller fish (Blindow et al., 2000, 2014). Charophytes, especially when growing in dense meadows, can influence the abundance of invertebrate-eating as well as piscivorous fish, since these are favoured by high food availability and high water clarity (Hargeby et al., 1994, 2005; Schulze et al., 2006). Numbers of both breeding and resting waterfowl have been shown to follow changes of charophyte abundance and thus availability of both plant and invertebrate food in shallow lakes (Hargeby et al., 1994; Milberg et al., 2002).

By a number of feedback-mechanisms, charophytes can improve their own light availability and, as "ecosystem engineers", stabilise their dominance in shallow aquatic ecosystems: Reduction of nutrient availability, reduction of mechanical wave action and thereby decreased sediment resuspension (Casanova et al., 2003a; van den Berg 1998a, 1998b) and possibly allelopathy (Blindow and Hootsmans, 1991) are such mechanisms. These also impact on higher trophic levels such as increased predation efficiency of piscivorous fish and higher zooplankton density, mechanisms that "cascade down" the food web and ultimately cause reduced phytoplankton densities (Scheffer et al., 1993). Charophyte-dominated lakes thus belong to the first (Blindow et al., 1993; Scheffer et al., 1993) and most convincing (Capon et al., 2015) examples of ecosystems for which the existence of alternative stable states has been suggested. Charophytes are assumed to stabilise their environment more efficiently than angiosperms (Blindow et al., 2014). While a positive impact on zooplankton is assumed to be the most important feedback mechanism in angiosperm-dominated lakes, charophytes are considered to promote the clearwater state mainly by nutrient immobilization, reduced sediment resuspension and effects on the benthic food web (Fig. 7.4).

Recent research distinguishes among three alternative states in freshwater lakes: a phytoplankton-dominated "turbid" state, a rather unstable "crashing" state dominated by submerged angiosperms, and a stable, "clear" state dominated by charophytes (Sayer et al., 2010). Therefore, charophytes are regarded as especially suitable for improving water quality as well as bathing and fishing conditions of shallow aquatic ecosystems (Bakker et al., 2013). In order to improve water quality, management aims at restoring charophyte vegetation. Charophytes can be planted as green plants

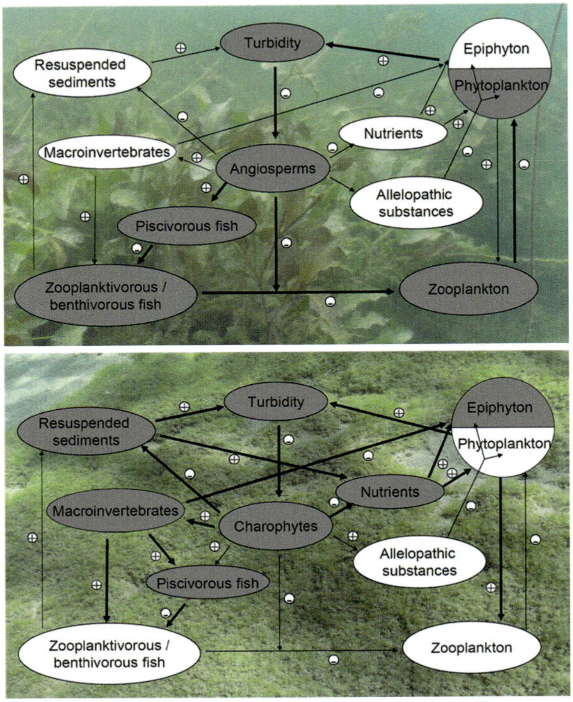

Fig. 7.4 Feedback-mechanisms stabilizing the clearwater state in lakes dominated by canopy-forming angiosperms (upper panel) and "bed-forming" charophytes (lower panel). The qualitative effect of each route in the diagram can be assessed by adding the signs along the way. Dark areas and thick arrows indicate major feedback mechanisms. Light areas and thin arrows indicate feedback mechanisms of minor or unknown importance. After Scheffer et al. (1993) and Blindow et al. (2014), modified

and/or by the import of sediment that contains charophyte oospores (Blindow et al., 2021; Hilt et al., 2006; van de Weyer et al., 2017).

While charophytes favour—and are favoured by—"friendly" organisms such as zooplankton, piscivorous fish and most plant-associated macroinvertebrates, herbivorous animals can harm charophytes and act as "nasty neighbours". Grazing by fish, beetles, amphipods and other invertebrates is thought to have an important role in the distribution and abundance of grazing-sensitive/insensitive species (Proctor, 1999). Grazing by waterfowl seems not to have an overall strong impact on charophytes (Hidding et al., 2010; Rip et al., 2006) and only occasionally damages these macrophytes severely (Matuszak et al., 2012).

Sediment-disturbing fish such as carp or bream, as well as crustaceans and especially invasive or stocked crayfish, can cause substantial declines in charophyte vegetation (Bakker et al., 2013; van der Wal et al., 2013). A strong impact has also been observed from herbivorous fish (Jellyman et al., 2009). Charophytes can be sensitive to grazing, tolerant of grazing, or adapted to avoid grazing (e.g. Proctor, 1999). Grazing-sensitive species will only exist in the absence of high grazing pressure (e.g. via depth distribution or temporal habitat separation, such as in temporary water bodies). Some morphological characteristics such as cortication, spine-cells, dense alopecuroid growth or mucus formation could be interpreted as grazing defences, but there are alternative interpretations (Subramanian, 2018). As it has been found that it is advantageous to reproduce frequently to recolonise lost habitats under high grazing pressure (Idestam-Almquist, 2000) this could also be a mechanism for grazing tolerance.

7.6 Living at the Edge—Persisting Under Inconvenient Conditions

Charophytes developed different adaptation mechanisms that have helped them to survive for more than 400 million years. One of the main challenges was to keep up with changing environmental conditions.

Among charophytes there are some perennial species e.g. *Nitellopsis obtusa, Chara tomentosa, Chara australis* R. Brown 1810 and *Chara subspinosa* Rupr. 1846, that are able to withstand strong competition with embryophytes by features typical of K-selected species i.e. few, large offspring and a perennial, polycarpic life-history (Blindow et al., 2002; Luther, 1951). Other charophytes, with features consistent with r-selected species (many, small offspring and an annual, monocarpic life-history), can be regarded as typical pioneer plants and survival specialists, often being well adapted to variable and ephemeral habitats (Migula 1889–1897). The production of numerous oospores (Fig. 7.5), which can persist for decades or even centuries (Stobbe et al., 2014), within just a few weeks, allows them to continuously re-occupy habitats, which offer favourable growth conditions just for short periods, such as flood ponds, ephemeral wetlands or arctic springs (Casanova and Porter, 2013; Langangen, 2000; Raabe, 2009).

The ability to survive under inconvenient conditions is largely related to light adaptation (Asaeda et al., 2007; Blindow et al., 2003; Sato et al., 2014). In low-light conditions charophyte thalli elongate (Nowak et al., 2017b), and the density of gametangia decreases when passing optimum irradiance (e.g. Blindow et al., 2003; Sato et al., 2014). *Chara fibrosa* ex Bruzelius 1824 and *Nitella hyalina*, for instance exhibit a more elongated habitus in deep water compared to shallower water (Asaeda et al., 2007). However, some species show different morphological and reproductive responses to reduced light. In an experimental study, high-light conditions promoted vegetative growth of *Chara braunii* whereas low light induced sexual reproduction (Sato et al., 2014). If light availability is high, vegetative growth is more profitable than expending energy for the formation of gametangia, whereas sexual reproduction is stimulated when light conditions are insufficient for length growth. In addition to pairs of gametangia, even single antheridia are produced because energy investment in antheridia is lower than that of oogonia production. Oogonia and antheridia are produced to form dormant oospores and ensure the survival in the sediment until light conditions improve.

A similar strategy is reflected by the morphological and reproductive variability of *Lychnothamnus barbatus* (Brzozowski and Pełechaty, 2020). At shallow sites thallus

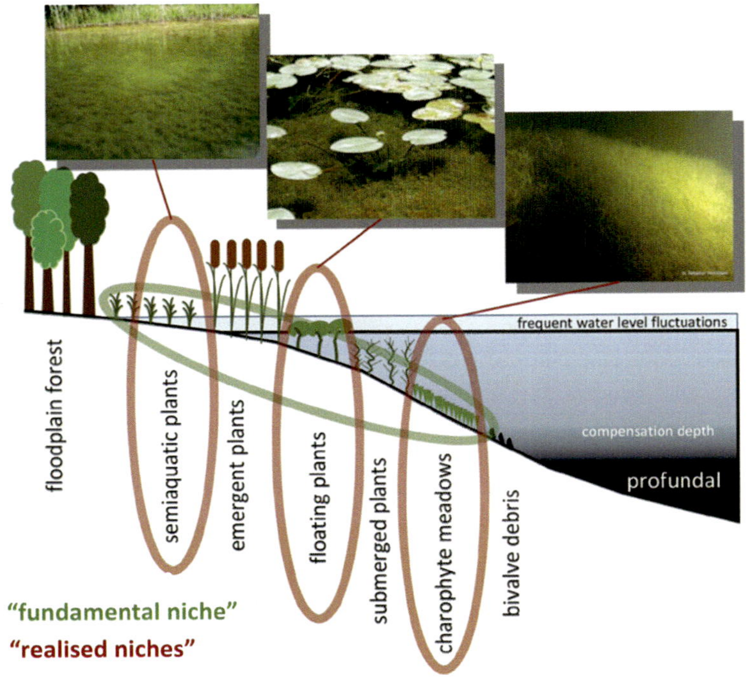

Fig. 7.5 Fundamental and realised niche of charophytes in a deep lake. Charophytes are able to grow from the surface down to depths where irradiance limits the existence of photoautotrophs (compensation depth). In the absence of embryophytes, charophyte meadows may occur from the compensation depth up to the shoreline. However, under eutrophic conditions, when embryophytes are favoured, charophyte meadows become squeezed between the irradiance limit for embryophytes and the compensation depth. In the remaining part of the littoral zone, charophytes are restricted to a few niches where they grow underneath embryophytes (semiaquatic plant belt and floating plant belt) or where embryophytes cannot stand repeated mechanical disturbance by, e.g. grazing or wave action (semiaquatic belt). Drawing: H. Schubert; left and middle photographs: A. Pukacz; right photograph: S. Wesołowski

weight and branching varied, while individuals from deep sites were longer and more similar but produced more oogonia.

In shallow water, charophytes have to cope with high solar radiation, including UV radiation. Charophytes are sensitive to UV-B radiation (de Bakker et al., 2001; Rae et al., 2001). The decline of maximum quantum yield (Fv/Fm) at a high ratio of UV radiation compared to PAR exposure is substantially stronger in charophytes compared to vascular plants (Rae et al., 2001). Charophytes exposed to strong UV also tend to reproduce more by means of bulbils than by antheridia and oogonia and are reduced in length. Smaller stature is advantageous to restrict UV-B exposure (de Bakker et al., 2001).

Moreover, analysis of plants growing at different depths showed that photosynthetic inhibition by UV radiation was more pronounced for deeper growing species.

Species rarely exposed to high UV in their natural environment, such as *C. corallina* Klein ex Willd. 1805, experience more damage compared to shallower growing species (e.g. *C. fibrosa*). Consequently, the intensity of UV radiation determines the upper depth limit (Rae et al., 2001).

However, charophytes can adjust to excessively high or low light by altering their pigment composition (Blindow et al., 2003; Küster et al., 2000, 2004; Schneider et al., 2015; Schwarz et al., 1996). In deeper water where light intensity is low, the chlorophyll a to chlorophyll b ratio increases due to enhanced production of antenna pigments (Schagerl and Pichler, 2000; Schwarz et al., 1996). With increasing irradiance, the chlorophyll a to carotenoid ratio ratio gradually decreases, improving the light-protecting capacity offered by protective carotenoids (Küster et al., 2004; Schagerl and Pichler, 2000; Schneider et al., 2006). Extreme environmental conditions, such as excessive solar radiation cause pigment degradation, which reduces the light absorbing potential of both photosystems (Geissler et al., 2009).

In addition, exposure to high light induces oxidative stress in plants resulting in enhanced generation of reactive oxygen species (ROS) in the chloroplast (Fig. 7.6) (Asada, 2006; Latowski et al., 2011). ROS are by-products of various metabolic pathways localised in different cellular compartments, specifically in chloroplasts, mitochondria, peroxisomes, and plasma membranes, depending on the type of stressors (Choudhury et al., 2017; Nakano and Asada, 1981; Sharma et al., 2012). However, charophytes have developed an antioxidant defense system that includes catalase (CAT), ascorbate peroxidase (APX), and guaiacol peroxidase (POD) in order to scavenge and counter the adverse effects of ROS (Atapaththu et al., 2016). Oxidative stress occurs when the level of ROS exceeds the amount of ROS controllable by natural defense mechanisms (Murata et al., 2007; Sharma et al., 2012) and results in damage to proteins, lipids, and DNA, ultimately ending in cell death (Asada, 2006; Sharma et al., 2012).

Under favorable conditions, calcite encrustation on the plant surface enables charophytes growing in very dense carpets to continue photosynthesis even if dissolved carbon dioxide becomes depleted (McConnaughey, 1991; Smith and Walker, 1980). Most charophyte species are able to use bicarbonate (HCO_3^-), allowing photosynthesis to continue at relatively high pH (>9.0), where bicarbonate is the predominant inorganic carbon source (Sand-Jensen et al., 2018; Van den Berg et al., 2002). Without encrustation, the production of organic matter (CH_2O) releases hydroxyl ions ($HCO_3^- + H_2O \rightarrow CH_2O + OH^- + O_2$), driving pH up and above pH 10.0, which will inhibit cell function and photosynthesis (Sand-Jensen, 1983). However, calcite encrustation counteracts this pH rise as $Ca^{2+} + 2HCO_3^- \rightarrow CaCO_3 + CH_2O + O_2$, and maintains bicarbonate concentration. Without encrustation the molar relationship between carbon assimilation and oxygen production is 1:1, while with encrustation, it becomes 2:1 (Sand-Jensen et al., 2018). Therefore, the extent of encrustation depends on pH, bicarbonate and calcium concentrations (McConnaughey and Whelan, 1997).

At slightly acidic to neutral pH, a high fraction of carbon dioxide is present and calcification rate is likely low, while with increasing pH, and increasing importance

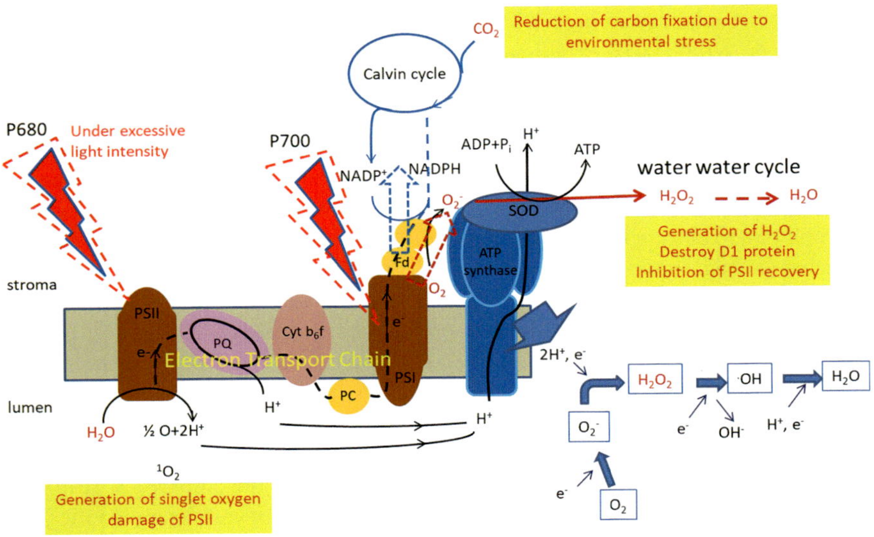

Fig. 7.6 Effects of solar radiation on the chloroplast and metabolic pathways: under excessive light, reactive oxygen species are generated ($^1O_2$ and H_2O_2). Abbreviations: plastoquinone, PQ; cytochrome b_6f, Cyt b_6f; plastocyanin, PC; flavoprotein-ferredoxin, Fd; Cu–Zn superoxide dismutase, SOD. Figure: Takashi Asaeda

of bicarbonate and carbonate, calcification becomes more intense (Sand-Jensen et al., 2018).

The above-mentioned mechanisms probably also allowed charophytes to cope with other environmental extremes, e.g. extreme salinity. Unusual salinity conditions whether by concentration, variability in time as well as ionic composition (e.g. Lambert, 2013), create niches where some charophyte species are superior competitors to embryophytes. The same holds for habitats with either unusually high (Saber et al., 2017) or low (Langangen, 2007) temperatures.

An extreme example of charophyte persistence is their occurrence in a residual mining pond, close to a potash mining heap in Teutschenthal near Mansfeld region, Germany (Fig. 7.7C). The turbid water in this pond has a high ion concentration resulting in a total conductivity of 5.0 mS cm^{-1}. Nevertheless, it is inhabited by *C. tomentosa* (Herbst and Schubert, 2018). Charophytes are also found in different other mine waste waters, e. g. gold and uranium tailings in Canada (Kalin and Smith, 2016; Kalin et al., 2005). In such polluted environments, charophytes can resist high concentrations of heavy metals and other potentially toxic elements without losing photosynthetic capacity (Herbst et al., 2019). Most of the pollutants are deposited in encrustations, whereas concentration within the cells stay low (Herbst et al., 2018, 2019; Laffont-Schwob et al., 2015).

The clue for such a broad tolerance range within a very small group of organisms might be their simple organisation, allowing efficient growth within a short favourable period conditions, combined with extremely persistent diaspores.

Fig. 7.7 Examples of extreme environments with charophytes. **a** *C. hispida* L. 1753 in Noirmoutier, Atlantic coast, France. **b** *C. vulgaris* L. 1753 in a rice field near El Parmar, Spain. **c** *C. tomentosa* in a residual mining pond in front of a potash mining heap in Teutschenthal, Germany. **d** *Chara* meadow growing under ice in Lake Krüselin, Germany. Photographs: a—Hendrik Schubert; b, c, d.1—Anne Herbst; d.2—Petra Nowak

In newly dug or cleared ponds and ditches charophytes can be the first macroscopic plants to arrive, and some species are characteristic of ephemeral water bodies that dry up completely during summer (Bicudo and Bueno, 2013; Casanova and Brock, 1990; Casanova et al., 2003b). However, in the long run, such ephemeral species are outcompeted as soon as nutrient levels and a certain stability of water level allow for the growth of rooted embryophytes (e.g. Blindow, 1992). In such systems charophytes are restricted to mesohabitats where the relatively slow growing embryophytes are hampered by physical disturbance as, e.g., cattle grazing close to the reed belt or ice scraping (Blindow et al., 2016; Idestam-Almqvist, 2000). Any serious disturbance of such systems as, e.g., drawdown of regulated lakes or vegetation clearing (Becker, 2011) creates an opportunity for charophytes to re-establish and to replenish their diaspore reservoir until being outcompeted again ("permanent pioneers", Krause and Walter, 1985). In freshwater as well as coastal systems, the numbers of charophyte diaspores outnumbers all other taxonomic groups by far, but the oospore bank can also contain species which remained undetected for decades (Blindow et al., 2016; Nowak et al., 2017a; Olsen, 1944; Steinhardt and Selig, 2007).

7.7 Summary and Knowledge Gaps

From an ecologists' point of view the available data show that occurrence or absence of charophytes is not just a reflection of water properties such as nutrient availability or ion composition. Charophytes, like all organisms, have a complex niche structure with a multitude of potential limitations. Moreover, as an ancient, but still evolving group of organisms consisting of many cosmopolitan species, it is not surprising that niches are highly species-specific. There may be similarities within some groups with respect to preference ranges, but there is little in common with respect to "ecology of charophytes" at family level (Doege et al., 2016; Schubert et al., 2017c). The fact that many charophyte species dominate the macrophytobenthos of clear-water ecosystems makes them good bioindicators; however, they are not physiologically restricted to such conditions, but are just more competitive under low-nutrient-clear-water conditions than angiosperms. Most, if not all species can also be found under eutrophic turbid water conditions when competitors are missing. The reasons for reduced competition are manifold: unusual ionic composition (e.g. Krawczyk et al., 2016; Rybak et al., 2017), low-frequency physical disturbance (e.g. ice scraping: Blindow et al., 2016; Idestam-Almqvist, 2000); extreme osmotic regimes (e.g. Lambert, 2013); irregular pond-bed drawdown (e.g. Lambert et al., 2013), just to name a few. All of them favour r-selected organisms, able to colonise fast and to resist even prolonged periods of harsh conditions with resting stages that persist decades or longer in dormancy. On the other hand not all of the species are such "pioneers". Several species develop dense vegetation in shallow lakes and brackish lagoons, where they are superior competitors due to a number of feedback mechanisms, which include both abiotic impacts and complex biological interactions (Blindow et al., 2014). This life form is equivalent to that of K-strategists and differs from the r-selected "pioneers", as some of these "competitors", such as *Nitellopsis obtusa* and *Chara tomentosa*, only produce low numbers of oospores and therefore have a limited dispersal ability.

It is when charophytes just do not "behave" as expected. This is the case, for example, when species such as *Chara hispida* and *C. aspera* "switch niches" and build up a dense, persistent vegetation in shallow lakes, as well as act as "pioneers" in nearby small water bodies (e.g. Blindow et al., 1993), or when *Nitellopsis obtusa*, dioecious with low oospore production, becomes an invasive "nuisance plant" in North America (Alix et al., 2017). Another unusual case is when *Sphaerochara intricata* (Trentep.l ex Roth) Feist-Castel & N. Grambart. "spontaneously" shows up in habitats that are "only briefly aquatic" (Moore, 1986) such as mud trampled by livestock or lorry tracks in a forest, while several *Nitella* species with rich oospore production "insist on remaining rare" and do not colonise any of the numerous apparently suitable habitats in their neighbourhood (Blindow, 2008; Moore, 1986). In such case it becomes obvious that in addition to the ongoing work on taxonomic and evolutionary aspects of charophytes, ecological studies are likely to produce exciting insights into ecological theory (e.g. Capon et al., 2015) and management.

There is still a large gap in knowledge about growth, oospore production, germination and dispersal for example, as well as mechanisms facilitating persistence and (re)establishment of charophyte vegetation, which should be filled not only for scientific and management reasons, but also as this knowledge is badly needed to protect this ancient group of permanent pioneers and ecosystem engineers.

Remarks This chapter is based on the content originally published in: Schubert H., I. Blindow, N. C., Bueno, M. T. Casanova, M. Pełechaty & A. Pukacz, 2018. Ecology of charophytes—permanent pioneers and ecosystem engineers. Perspectives in Phycology 5(1): 61–74. https://doi.org/10.1127/pip/2018/0080; https://www.schweizerbart.de/journals/pip.

References

Alderton, E., Sayer, C. D., Davies, R., Lambert, S. J., & Axmacher, J. C. (2017). Buried alive: Aquatic plants survive in 'ghost ponds' under agricultural fields. *Biological Conservation, 212*, 105–110.

Alix, M. S., Scribailo, R. W., & Weliczko, C. W. (2017). *Nitellopsis obtusa* (Desv.) J. Groves, 1919 (Charophyta: Characeae): New records from southern Michigan, USA with notes on environmental parameters known to influence its distribution. *BioInvasions Records, 6*(4), 311–319.

Andrews, M., Davidson, I. R., Andrews, M. E., & Raven, J. A. (1984). Growth of *Chara hispida* L. Apical growth and basal decay. *Journal of Ecology, 72*, 873–884.

Asada, K. (2006). Production and scavenging of reactive oxygen species in chloroplasts and their functions. *Plant Physiology, 141*, 391–396.

Asaeda, T., Rajapakse, L., & Sanderson, B. (2007). Morphological and reproductive acclimations to growth of two charophyte species in shallow and deep water. *Aquatic Botany, 86*, 393–401.

Atapaththu, K. S., Rashid, M. H., & Asaeda, T. (2016). Growth and oxidative stress of Brittlewort (*Nitella pseudoflabellata*) in response to Cesium exposure. *Bulletin of Environmental Contamination and Toxicology, 96*(3), 347–353.

Baastrup-Spohr, L., Iversen, L. L., Borum, J., & Sand-Jensen, K. (2015). Niche specialization and functional traits regulate the rarity of charophytes in the Nordic countries. *Aquatic Conservation, 25*, 609–621.
Bakker, E. S., Sarneel, J. M., Gulati, R. D., Liu, Z., & van Donk, E. (2013). Restoring macrophyte diversity in shallow temperate lakes: Biotic versus abiotic constraints. *Hydrobiologia, 710*, 23–37.
Becker, R. (2011). Dynamik und Konstanz der Grabenvegetation in den "Bornhorster Huntewiesen" (Stadt Oldenburg, Oldb.) zwischen 1982 und 2010. *Drosera, 2011*, 11–34.
Beilby, M. J., & Casanova, M. T. (2014). *The physiology of Characean cells*. Springer Verlag.
Berger, J., & Schagerl, M. (2004). Allelopathic activity of Characeae. *Biologia, 59*, 9–15.
Bicudo, C. E. M., & Bueno, N. C. (2013). Characeae biomass: Is the subject exhausted? In: M. D. Matovic (Ed.), *Biomass now-sustainable growth and use* (pp. 523–540). InTech, Croacia.
Bisson, M. A., & Kirst, G. O. (1995). Osmotic acclimation and turgor pressure regulation in algae. *Naturwissenschaften, 82*, 461–471.
Blindow, I., & Hootsmans, M. J. M. (1991). Allelopathic effects from *Chara* spp. on two species of unicellular green algae. In: M. J. M. Hootsmans, & J. E. Vermaat (Eds.), *Macrophytes, a Key to Understanding Changes Caused by Eutrophication in shallow freshwater ecosystems. IHE Reports Series* (Vol. 21, pp. 139–144). Delft.
Blindow, I. (2008). *Åtgärdsprogram för bevarande av hotade kransalger. V. Tuvsträfse och barklöst sträfse 2008–2011*. Naturvårdsverket, Stockholm.
Blindow, I. (1988). Phosphorus toxicity in *Chara*. *Aquatic Botany, 32*, 393–395.
Blindow, I. (1992). Decline of Charophyta during eutrophication: A comparison to angiosperms. *Freshwater Biology, 28*, 9–14.
Blindow, I., Andersson, G., Hargeby, A., & Johansson, S. (1993). Long-term pattern of alternative stable states in two shallow eutrophic lakes. *Freshwater Biology, 30*, 159–167.
Blindow, I., Carlsson, M., & van deWeyer, K. (2021). Re-establishment techniques and transplantations of charophytes to support threatened species. *Plants, 10*, 1830.
Blindow, I., Dahlke, S., Dewart, A., Flügge, S., Hendreschke, M., Kerkow, A., & Meyer, J. (2016). Composition and diaspore reservoir of submerged macrophytes in a shallow brackish water bay of the southern Baltic Sea—influence of eutrophication and climate. *Hydrobiologia, 778*, 121–136.
Blindow, I., Dietrich, J., Mollmann, N., & Schubert, H. (2003). Growth, photosynthesis and fertility of *Chara aspera* under different light and salinity conditions. *Aquatic Botany, 76*, 213–234.
Blindow, I., Hargeby, A., & Andersson, G. (2002). Seasonal changes of mechanisms maintaining clear water in a shallow lake with abundant *Chara* vegetation. *Aquatic Botany, 72*, 315–334.
Blindow, I., Hargeby, A., & Hilt, S. (2014). Facilitation of clear-water conditions in shallow lakes by macrophytes: Differences between charophyte and angiosperm dominance. *Hydrobiologia, 737*, 99–110.
Blindow, I., Hargeby, A., Wagner, B. M. A., & Andersson, G. (2000). How important is the crustacean plankton for the maintenance of water clarity in shallow lakes with abundant submerged vegetation? *Freshwater Biology, 44*, 185–197.
Blindow, I., & Schütte, M. (2007). Elongation and mat formation of *Chara aspera* under different light and salinity conditions. *Hydrobiologia, 584*, 69–76.
Bonis, A., & Grillas, P. (2002). Deposition, germination and spatio-temporal patterns of charophytes propagule banks: A review. *Aquatic Botany, 73*, 235–348.
Bonis, A., & Lepart, J. (1994). Vertical structure of seed banks and the impact of depth of burial on recruitment in two temporary marshes. *Vegetatio, 112*, 127–139.
Braun, M., Foissner, I., Lühring, H., Schubert, H., & Thiel, G. (2007). Characean green algae: Still a valid model system to examine fundamental principles in plants. *Progress in Botany, 68*, 193–220.
Brochet, A. L., Guillemain, M., Fritz, H., Gauthier-Clerc, M., & Green, A. J. (2010). Plant dispersal by teal (*Anas crecca*) in the Camargue: Duck guts are more important than their feet. *Freshwater Biology, 55*, 1262–1273.

Brzozowski, M., & Pełechaty, M. (2020). Broad morpological and reproductive variability of the endangered macroalga *Lychnothamnus barbatus* in the depth gradient. *Aquatic Botany, 165*, 1–8.

Burns, N. M., Rutherford, J. C., & Clayton, J. S. (1999). A monitoring and classification system for New Zealand lakes and reservoirs. *Lake and Reservoir Management, 15*, 255–271.

Calero, S., Auderset Joye, D., Rey-Boissezo, A., & Rodrigo, M. A. (2017). Time and heat for sexual reproduction: Comparing the phenology of *Chara hispida* of two populations at different latitudes. *Aquatic Botany, 136*, 71–81.

Capon, S. J., Lynch, A. J. J., Bond, N., Chessman, B. C., Davis, J., Davidson, N., Finlayson, M., Gell, P. A., Hohnberg, D., Humphrey, C., Kingsford, R. T., Nielsen, D., Thomson, J. R., Ward, K., & Mac Nally, R. (2015). Regime shifts, thresholds and multiple stable states in freshwater ecosystems: A critical appraisal of the evidence. *Science of the Total Environment, 534*, 122–130.

Casanova, M. T. (1994). Vegetative and reproductive responses of charophytes to water-level fluctuations in permanent and temporary wetlands in Australia. *Australian Journal of Marine and Freshwater Research, 45*, 1409–1419.

Casanova, M. T. (2014). *Lamprothamnium* in Australia (Characeae, Charophyceae). *Australian Systematic Botany, 26*, 268–290.

Casanova, M. T., & Brock, M. A. (1990). Charophyte germination and establishment from the seed bank of an Australian temporary lake. *Aquatic Botany, 36*, 247–254.

Casanova, M. T., & Brock, M. A. (1996). Can oospore germination patterns explain charophyte distribution in permanent and temporary wetlands? *Aquatic Botany, 54*, 297–312.

Casanova, M. T., de Winton, M. D., & Clayton, J. S. (2003a). Do charophytes clear turbid water? *Verhandlungen Internationale Vereiningung für Theroretische und Angewandte Limnologie, 26*, 1440–1443.

Casanova, M. T., de Winton, M. D., Karol, K. G., & Clayton, J. S. (2007). *Nitella hookeri* A. Braun (Characeae, Charophyceae) in New Zealand and Australia: Implications for endemism, speciation and biogeography. *Charophytes, 1*, 2–18.

Casanova, M. T., García, A., & Porter, J. L. (2003b). Charophyte rediscoveries in Australia, what and why? *Acta Micropalaeontologica Sinica, 20*, 129–138.

Casanova, M. T., & Nicol, J. M. (2009). *Chara canescens* (Characeae, Charophyceae) in the Southern Hemisphere. *Charophytes, 1*, 55–60.

Casanova, M. T., & Porter, J. L. (2013). Two new species of *Nitella* (Characeae, Charophyceae) from arid-zone claypan wetlands in Australia. *Muelleria, 31*, 53–59.

Charalambidou, I., & Santamaria, L. (2005). Field evidence for the potential of waterbirds as dispersers of aquatic organisms. *Wetlands, 25*, 252–258.

Choudhury, F. K., Rivero, R. M., Blumwald, E., & Mittler, R. (2017). Reactive oxygen species, abiotic stress and stress combination. *The Plant Journal, 90*, 856–867.

Corillion, R. (1957). Les Charophycées de France et d'Europe Occidentale. *Bulletin de la Société Scientifique de Bretagne, 32*, 1–499.

de Bakker, N. V. J., van Beem, A. P., van de Staaij, J. W. M., Rozema, J., & Aerts, R. (2001). Effects of UV-B radiation on a charophycean alga, *Chara aspera*. *Plant Ecology, 154*, 219–236.

De Winton, M. D., Casanova, M. T., & Clayton, J. S. (2004). Charophyte germination and establishment under low irradiance. *Aquatic Botany, 79*, 175–187.

Doege, A., van de Weyer, K., Becker, R., & Schubert, H. (2016). Bioindikation mit Characeen. In: Arbeitsgruppe Characeen Deutschlands (Hrsg.), *Armleuchteralgen - Die Characeen Deutschlands* (pp. 97–138). Springer Verlag.

Domin, A., Schubert, H., Krause, J. C., & Schíewer, U. (2004). Modelling of pristine depth limits for macrophyte growth in the southern Baltic Sea. *Hydrobiologia, 514*, 29–39.

Domozych, D. S., Popper, Z. A., & Sørensen, I. (2016). Charophytes: Evolutionary giants and emerging model organisms. *Frontiers in Plant Science, 7*, 1470.

Figuerola, J., Charalambidou, I., Santamaria, L., & Green, A. J. (2010). Internal dispersal of seeds by waterfowl: Effect of seed size on gut passage time and germination patterns. *Naturwissenschaften, 97*, 555–565.

Forsberg, C. (1965). Sterile germination of oospores of *Chara* and seeds of *Najas marina*. *Physiologia Plantarum, 18*, 128–137.
García, A., & Casanova, M. T. (2003). *Lamprothamnium heraldii* sp. nov. (Charales, Charophyta) from Australia: The first dioecious representative of the genus. *Phycologia, 42*, 622–628.
García, A., & Chivas, A. R. (2006). Diversity and ecology of extant and quaternary Australian charophytes. *Cryptogamie, Algologie, 27*, 323–340.
Geissler, N., Hussin, S., & Koyro, H. W. (2009). Interactive effects of NaCl salinity and elevated atmospheric CO_2 concentration on growth, photosynthesis, water relations and chemical composition of the potential cash crop halophyte *Aster tripolium* L. *Environmental and Experimental Botany, 65*, 220–231.
Giesenhagen, K. (1896). Untersuchungen über die Characeen. *Flora, 82*, 381–433.
Green, A. J., Figuerola, J., & Sánchez, M. I. (2002). Implications of waterbird ecology for the dispersal of aquatic organisms. *Acta Oecologica, 23*, 177–189.
Gutowski, A., Hofman, G., Leukart, P., Melzer, A., Mollenhauer, M., Schmedtje, U., Schneider, S., & Tremp, H. (1998). Trophiekartierung von aufwuchs- und makrophytendominierten Fließgewässern. *Informationsberichte Des Bayerischen Landesamtes Für Wasserwirtschaft, 4*(98), 1–501.
Haas, J. N. (1994). First identification key for charophyte oospores from central Europe. *European Journal of Phycology, 29*, 227–235.
Haller, W. T., Sutton, D. L., & Barlowe, W. C. (1974). Effects of salinity on growth of several aquatic macrophytes. *Ecology, 55*, 891–894.
Hargeby, A., Andersson, G., Blindow, I., & Johansson, S. (1994). Trophic web structure in a shallow eutrophic lake during a dominance shift from phytoplankton to submerged macrophytes. *Hydrobiologia, 279*(280), 83–90.
Hargeby, A., Blom, H., Blindow, I., & Andersson, G. (2005). Increased growth and recruitment of piscivorous perch, *Perca fluviatilis*, during a transient phase of expanding submerged vegetation in a shallow lake. *Freshwater Biology, 50*, 2053–2062.
Henricson, C., Sandberg-Kilpi, E., & Munsterhjelm, R. (2006). Experimental studies on the impact of turbulence, turbidity and sedimentation on *Chara tomentosa* L. *Cryptogamie, Algologie, 27*, 419–434.
Henricsson, M. (1976). *Nutritional studies of Chara globularis Thuill., Chara zeylanica* Willd., and *Chara haitensis* Turpin (51 pp.). Ph.D. Thesis, University of Uppsala, Uppsala.
Herbst, A., Patzelt, L., Schoebe, S., Schubert, H., & von Tümpling, W. (2019). Bioremediation approach using charophytes—preliminary laboratory and field studies of mine drainage water from the Mansfeld Region, Germany. *Environmental Science and Pollution Research, 26*, 34983–34992.
Herbst, A., & Schubert, H. (2018). Age and site-specific pattern on encrustation of charophytes. *Botanical Studies, 59*(1), 31.
Herbst, A., von Tümpling, W., & Schubert, H. (2018). The seasonal effects on the encrustation of charophytes in two hard-water lakes. *Journal of Phycology, 54*, 630–637.
Hidding, B., Bakker, E. S., Keuper, F., de Boer, T., de Vries, P. P., & Nolet, B. A. (2010). Differences in tolerance of pondweeds and charophytes to vertebrate herbivores in a shallow Baltic estuary. *Aquatic Botany, 93*, 123–128.
Hilhorst, H. W. M. (1993). New aspects of seed dormancy. In *Proceedings of the Fourth International Workshop on Seeds: Basic and Applied Aspects of Seed Biology* (Vol. II, pp. 571–579).
Hilt, S., & Gross, E. M. (2008). Can allelopathically active submerged macrophytes stabilise clear-water states in shallow lakes? *Basic and Applied Ecology, 9*, 422–432.
Hilt, S., Gross, E. M., Hupfer, M., Morscheid, H., Mählmann, J., Melzer, A., Poltz, J., Sandrock, S., Scharf, E.-M., Schneider, S., & van de Weyer, K. (2006). Restoration of submerged vegetation in shallow eutrophic lakes—guideline and state of the art in Germany. *Limnologica, 36*, 155–171.
Holzhausen, A., Porsche, C., & Schubert, H. (2017). Viability assessment and estimation of the germination potential of charophyte oospores: Testing for site and species specificity. *Botany Letters, 165*, 147–158.

Holzinger, A., & Pichrtová, M. (2016). Abiotic stress tolerance of charophyte green algae: New challenges for omics techniques. *Frontiers in Plant Science, 7*, 678.

Idestam-Almqvist, J. (2000). Dynamics of submerged aquatic vegetation on shallow soft bottoms in the Baltic Sea. *Journal of Vegetation Science, 11*, 425–432.

Jäger, D. (2000). Beiträge zur Characeen-Flora Vorarlbergs (Österreich). *Berichte des Naturwissenschaftlich-medizinischen Vereins in Innsbruck, 87*, 67–85.

Jellyman, D., Walsh, J., de Winton, M., & Sutherland, D. (2009). *A review of the potential to re-establish macrophyte beds in Te Waihora (Lake Ellesmere)* (49 pp.). Environment Canterbury Report No. R09/38 ISBN 978-1-86937-968-1.

Jones, J. I., & Sayer, C. (2003). Does fish–invertebrate–periphyton cascade precipitate plant loss in shallow lakes? *Ecology, 84*, 2155–2167.

Kalin, M., Kießig, G., & Küchler, A. (2002a). Ecological water treatment processes for underground uranium mine water: progress after 3 years of operating a constructed wetland. In *Proceedings of the Uranium Mining and Hydrogeology III Conference* (pp. 591–600), Freiberg, Germany, September 15–21.

Kalin, M., Smith, M. P., & Wittrup, M. B. (2002b). Ecosystem restoration incorporating minerotrophic ecology and Stoneworts that accumulate 226Ra. In: *Proceedings of the Uranium Mining and Hydrogeology III Conference*, Freiberg, Germany, September 15–21: 499–508.

Kalin, M., & Smith, M. P. (2016). Characean biomass in alkaline mine waste water: a natural water treatment. In *7th International Symposium on Extant and Fossil Charophytes*, Astana, Kazakhstan, August 31–Sept 2, 2016.

Kalin, M., & Smith, M. P. (2007). Germination of *Chara vulgaris* and *Nitella flexilis* oospores: What are the relevant factors triggering germination? *Aquatic Botany, 87*, 235–241.

Kalin, M., Wheeler, W. N., & Meinrath, G. (2005). The removal of uranium from mining waste water using algal/microbial biomass. *Journal of Environmental Radioactivity, 78*(2), 151–177.

Karol, K. G., McCourt, R. M., Cimino, M. T., & Delwiche, C. F. (2001). The closest living relatives of land plants. *Science, 294*, 2352–2353.

Karol, K. G., Skawinski, P. M., McCourt, R. M., Nault, M. E., Evans, R., Barton, M. E., Berg, M. S., Perleberg, D. J., & Hall, J. D. (2017). First discovery of the charophycean green alga *Lychnothamnus barbatus* (Charophyceae) extant in the New World. *American Journal of Botany, 104*(7), 1108–1116.

Karol, K. G., & Sleith, R. S. (2017). Discovery of the oldest record of *Nitellopsis obtusa* (Charophyceae, Charophyta) in North America. *Journal of Phycology, 53*, 1106–1108.

Khan, M. (1991). Charophytes in time and space: Zonal distribution patterns. *Bulletin de la Société Botanique de France, 138*, 33–45.

Kohler, A. (1982). Wasserpflanzen als Belastungsindikatoren. *Decheniana Beih, 26*, 31–42.

Krause, W. (1997). *Charales (Charophycae). Süsswasserflora von Mitteleuropa, Band 18*. Gustav Fischer Verlag Jena.

Krause, W. (1981). Characeen als Bioindikatoren für den Gewässerzustand. *Limnologica (Berlin), 13*, 399–418.

Krause, W., & Walter, E. (1985). Die Characeen der Teiche in Oberfranken. *Berichte der Bayerischen Botanischen Gesellschaft zur Erforschung der heimischen Flora, 65*, 51–58.

Krawczyk, R., Lis, Ł., & Urbaniak, J. (2016). Water parameters and species composition of macrophytes in reclamation lakes in the area of a former sulphur borehole mine (SE Poland). *Annales Universitatis Mariae Curie-Sklodowska, Lublin – Polonia. Sectio C, 71*, 27–40.

Kuczewski, O. (1906). Morphologische und biologische Untersuchungen an *Chara delicatula f. bulbillifera* A. *Braun. Arbeiten aus dem Laboratorium für Allgemeine Botanik und Pflanzenphysiologie der Universität Zürich, 5*, 1–51.

Kufel, L., & Kufel, I. (2002). *Chara* beds acting as nutrient sinks in shallow lakes: A review. *Aquatic Botany, 72*, 249–260.

Küster, A., Schaible, R., & Schubert, H. (2004). Light acclimation of photosynthesis in three charophyte species. *Aquatic Botany, 79*, 111–124.

Küster, A., Schaible, R., & Schubert, H. (2000). Light acclimation of the charophyte *Lamprothamnium papulosum*. *Aquatic Botany, 68*, 205–216.

Laffont-Schwob, I., Triboit, F., Prudent, P., Soulié-Märsche, I., Rabier, J., Despréaux, M., & Thiéry, A. (2015). Trace metal extraction and biomass production by spontaneous vegetation in temporary Mediterranean stormwater highway retention ponds: Freshwater macroalgae (*Chara* spp.) vs. cattails (*Typha* spp.). *Ecological Engineering, 81*, 173–181.

Lambert, E. (2013). *Tolypella salina* R. cor., *sur la façade atlantique. Contribution à l'amélioration des connaissances.* Rapport, Université Catholique de l'Ouest, http://www.fcbn.fr/sites/fcbn.fr/files/ressource_telechargeable/130117_rapport_tolypella_salina_lambert_vdef.pdf.

Lambert, E., Desmots, D., LeBail, J., Mouronval, J.-B., & Felzines, J.-C. (2013). *Tolypella salina* R. Cor. on the French Atlantic coast: Biology and ecology. *Acta Botanica Gallica, 160*, 107–119.

Langangen, A. (2007). *Charophytes of the Nordic countries.* Saeculum ANS.

Langangen, A. (2000). Charophytes from the warm springs of Svalbard. *Polar Research, 19*, 143–153.

Latowski, D., Kuczyńska, P., & Strzałka, K. (2011). Xanthophyll cycle—a mechanism protecting plants against oxidative stress. *Redox Report, 16*, 78–90.

Leitch, A. R., John, D. M., & Moore, J. A. (1990). The oosporangium of the Characeae (Chlorophyta Charales). *Progress in Phycological Research, 7*, 213–268.

Luther, H. (1951). Verbreitung und Ökologie der höheren Wasserpflanzen im Brackwasser der Ekenäs Gegend in Südfinnland. II Spezieller Teil. *Acta Botanica Fennica, 50*, 1–370.

Marquardt, R., & Schubert, H. (2009). Photosynthetic characterisation of *Chara vulgaris* in bioremediation ponds. *Charophytes, 2*(1), 1–8.

Martin, G., K. Torn, I. Blindow, H. Schubert, R. Munsterhjelm, & Henricson, C. (2004). Introduction to charophytes. In: H. Schubert & I. Blindow (eds.) *Charophytes of the Baltic Sea (pp. 3–14).* The Baltic Marine Biologists Publications No. 19, Gantner Verlag.

Matheson, F. E., de Winton, M. D., Clayton, J. S., Edwards, T. M., & Mathieson, T. J. (2005). Responses of vascular (*Egeria densa*) and non-vascular (*Chara globularis*) submerged plants and oospores to contrasting sediment types. *Aquatic Botany, 83*, 141–153.

Matuszak, A., Mörtl, M., Quillfeldt, P., & Bauer, H. G. (2012). Exclosure study on the exploitation of macrophytes by summering and moulting waterbirds at Lower Lake Constance. *Hydrobiologia, 697*, 31–44.

McConnaughey, T. (1991). Calcification in *Chara corallina*: CO_2 hydroxylation generates protons for bicarbonate assimilation. *Limnology and Oceanography, 36*, 619–628.

McConnaughey, T. A., & Whelan, J. F. (1997). Calcification generates protons for nutrient and bicarbonate uptake. *Earth Science Reviews, 42*, 95–117.

Melzer, A. (1994). Möglichkeiten einer Bioindikation durch submerse Makrophyten—Beispiele aus Bayern. *Beiträge zur Angewandten Gewässerökologie Norddeutschlands, 1*, 92–102.

Meurer, T., & Bueno, N. C. (2012). The genera *Chara* and *Nitella* (Chlorophyta, Characeae) in the subtropical Itaipu Reservoir, Brazil. *Brazilian Journal of Botany, 35*(2), 219–232.

Migula, W. (1889–1897). *Die Characeen Deutschlands, Oesterreichs und der Schweiz.* Eduard Kummer, Leipzig

Milberg, P., Gezelius, L., Blindow, I., Nilsson, L., & Tyrberg, T. (2002). Long-term variation in waterfowl community composition during autumn in Lake Tåkern, southern Sweden. *Ornis Fennica, 79*, 72–81.

Moore, J. A. (1986). *Charophytes of Great Britain and Ireland.* BSBI Handbook No. 5, Botanical Society of the British Isles

Morris, J. L., Puttick, M. N., Clark, J. W., Edwards, D., Kenrick, P., Pressel, S., Wellman, C. H., Yang, Z., Schneider, H., & Donoghue, P. C. J. (2018). Timescale of early land plant evolution. *Proceedings of the National Academy of Sciences, 115*(10), 1–10.

Mulderij, G., van Nes, E., & van Donk, E. (2007). Macrophyte—phytoplankton interactions: The relative importance of allelopathy versus other factors. *Ecological Modelling, 204*, 85–92.

Murata, N., Takahashi, S., Nishiyama, Y., & Allakhverdiev, S. I. (2007). Photoinhibition of photosystem II under environmental stress. *Biochimica et Biophysica Acta—Bioenergetics, 1767*, 414–421.

Nakano, Y., & Asada, K. (1981). Hydrogen peroxide is scavenged by ascorbate-specific peroxidase in spinach chloroplasts. *Plant and Cell Physiology, 22*, 867–880.

Nowak, P., Schubert, H., Holzhausen, A., Sommer, V., & Schaible, R. (2017b). Morphological adaptations of *Chara baltica* and *Chara liljebladii* (Characeae) under different light conditions. *Botany Letters, 165*(7), 67–75.

Nowak, P., Steinhardt, T., von Ammon, U., Rohde, H., Schoor, A., Holzhausen, A., Schaible, R., & Schubert, H. (2017a). Diaspore bank analysis of Baltic coastal waters. *Botany Letters, 165*(7), 1–15.

Olsen, S. (1944). *Danish Charophyta, chorological, ecological and biological investigations*. Det Kongelige Danske Videnskabernes Selskab.

Pakdel, F. M., Sim, L., Beardall, J., & Davis, J. (2013). Allelopathic inhibition of microalgae by the freshwater stonewort, *Chara australis*, and a submerged angiosperm, *Potamogeton crispus*. *Aquatic Botany, 110*, 24–30.

Pełechaty, M., Pukacz, A., & Pełechata, A. (2004). Co-occurrence of two stoneworts of reverse ecological spectra in the same lake ecosystem. Habitat requirements of *Chara delicatula* Agardh and *Chara globularis* Thuillier in the context of bioindication. *Polish Journal of Environmental Studies, 13*(5), 551–556.

Pełechaty, M., Brzozowski, M., & Pietruczuk, K. (2017). Overwintering and gyrogonite formation by the rare and endangered indicative macroalga *Lychnothamnus barbatus* (Meyen) Leonh. in eutrophic conditions. *Aquatic Botany, 139*, 19–24.

Pełechaty, M., Pukacz, A., Apolinarska, K., Pełechata, A., & Siepak, M. (2013). The significance of *Chara* vegetation in the precipitation of lacustrine calcium carbonate. *Sedimentology, 60*, 1017–1035.

Penfield, S., & King, J. (2009). Towards a systems biology approach to understanding seed dormancy and germination. *Proceedings of the Royal Society B - Biological Sciences, 276*, 3561–3569.

Pollux, B. J. A. (2011). The experimental study of seed dispersal by fish (ichthyochory). *Freshwater Biology, 56*, 197–212.

Pringsheim, N. (1863). Ueber die Vorkeime und die nacktfüssigen Zweige der Charen. *Jahrbücher für wissenschaftliche Botanik, 3*, 294–324.

Proctor, V. W. (1959). Dispersal of fresh-water algae by migratory birds. *Science, 130*, 623–624.

Proctor, V. W. (1962). Viability of *Chara* oospores taken from migratory water birds. *Ecology, 43*, 528–529.

Proctor, V. W. (1967). Storage and germination of *Chara* oospores. *Journal of Phycology, 3*, 90–92.

Proctor, V. W. (1980). Historical biogeography of *Chara* (Charophyta): An appraisal of the Braun-Wood classification plus a falsifiable alternative for future consideration. *Journal of Phycology, 16*, 218–233.

Proctor, V. W. (1999). Charophytivory, Playas y Papalotes, a local paradigm of global relevance. *Australian Journal of Botany, 47*, 399–406.

Proctor, V. W., Griffin, D. G., & Hotchkiss, A. T. (1971). A Synopsis of the Genus *Chara*, Series Gymnobasalia (Subsection Willdenowia RDW). *American Journal of Botany, 58*(10), 894–901.

Pukacz, A., Pełechaty, M., Schubert, H., Blindow, I., & Raabe, U. (2016b). *Ochrona jezior ramienicowych Ziemi Lubuskiej i Brandenburgii (Schutz von Characeen-Seenim Lebuser Land und in Brandenburg)*. Polsko-Niemiecki Instytut Badawczy w Collegium Polonicum w Słubicach. VEGA sp.z.o.o.

Pukacz, A., Pełechaty, M., & Frankowski, M. (2016a). Depth-dependence and monthly variability of charophyte biomass production: Consequences for the precipitation of calcium carbonate in a shallow *Chara*-lake. *Environmental Science and Pollution Research, 23*, 22433–22442.

Pukacz, A., Pełechaty, M., Frankowski, M., Kowalski, A., & Zwijacz-Koszałka, K. (2014). Seasonality of water chemistry, carbonate production, and biometric features of two species of *Chara* in a shallow clear water lake. *The Scientific World Journal, 167631*, 1–11.

Raabe, U. (2009). *Chara baueri* rediscovered in Germany—plus additional notes on Gustav Heinrich Bauer (1794–1888) and his herbarium. *IRGC-News, 20*, 13–16.

Rae, R., Hanelt, D., & Hawes, I. (2001). Sensitivity of freshwater macrophytes to UV radiation: relationship to depth zonation in an oligotrophic New Zealand lake. *Marine and Freshwater Research, 52*, 1023–1030.
Raven, J. A., Smith, F. A., & Walker, N. A. (1986). Biomineralization in the Charophyceae sensu lato. In B. S. C. Leadbeater & R. Riding (Eds.), *Biomineralization in lower plants and animals* (pp. 550–557). Clarendon.
Rey-Boissezon, A., & Auderset Joye, D. (2015). Habitat requirements of charophytes—evidence of species discrimination through distribution analysis. *Aquatic Botany, 120*, 84–91.
Rip, W. J., Rawee, N., & de Jong, A. (2006). Alternation between clear, high-vegetation and turbid, low vegetation states in a shallow lake: The role of birds. *Aquatic Botany, 85*, 184–190.
Riveline, J., Berger, J.-P., Feist, M., Martin-Closas, M., Schudack, M., & Soulie-Märsche, I. (1996). European Mesozoic—Cenozoic charophyte biozonation. *Bulletin de la Société Géologique de France, 167*, 453–468.
Rodrigo, M. A., Alonso-Guillén, J. L., & Soulié-Märsche, I. (2010). Reconstruction of the former charophyte community out of the fructifications identified in Albufera de València lagoon sediments. *Aquatic Botany, 92*, 14–22.
Rojo, C., Carramiñana, M., Cócera, D., Roberts, G. P., Puche, E., Calero, S., & Rodrigo, M. A. (2017). Different responses of coexisting *Chara* species to foreseeable Mediterranean temperature and salinity increases. *Aquatic Botany, 138*, 53–63.
Rybak, M., Kołodziejczyk, A., Joniak, T., Ratajczyk, I., & Gąbka, M. (2017). Bioaccumulation and toxicity studies of macroalgae (Charophyceae) treated with aluminium: Experimental studies in the context of lake restoration. *Ecotoxicology and Environmental Safety, 145*, 359–366.
Sabbatini, M. R., Argüello, J. A., Fernández, O. A., & Bottini, R. A. (1987). Dormancy and growth-inhibitor levels in oospores of *Chara contraria* A.Braun ex Kütz. (Charophyta). *Aquatic Botany, 28*, 189–194.
Saber, A. A., Ballot, A., Schneider, S. C., & Cantonati, M. (2017). Morphological and molecular features of a *Chara vulgaris* population from desert springs on the Sinai Peninsula (Springs of Moses, Egypt). *Botany Letters, 165*, 77–89.
Samuelsson, G. (1925). Untersuchungen über die höhere Wasserflora von Dalarne. *Svenska Växtsociologiska Sällskap Handlingar, 9*, 1–37.
Sanderson, B. G., Asaeda, T., Lalith Rajapakse, L., & Redden, A. M. (2008). Mechanisms affecting biomass and distribution of charophytes and *Najas marina* in Myall Lake, New South Wales, Australia. *Hydrobiologia, 608*, 99–119.
Sand-Jensen, K. A. J. (1983). Photosynthetic carbon sources of stream macrophytes. *Journal of Experimental Botany, 34*(2), 198–210.
Sand-Jensen, K., Jensen, R. S., Gomes, M., Kristensen, E., Martinsen, K. T., Kragh, T., Baastrup-Spohr, L., & Borum, J. (2018). Photosynthesis and calcification of charophytes. *Aquatic Botany, 149*, 46–51.
Sato, M., Sakayama, H., Sato, M., Ito, M., & Sekimoto, H. (2014). Characterization of sexual reproductive processes in *Chara braunii* (Charales, Charophyceae). *Phycological Research, 62*, 214–221.
Sayer, C. D., Davidson, T. A., & Jones, J. I. (2010). Seasonal dynamics of macrophytes and phytoplankton in shallow lakes: A eutrophication-driven pathway from plants to plankton? *Freshwater Biology, 55*, 500–513.
Schagerl, M., & Pichler, C. (2000). Pigment composition of freshwater Charophyceae. *Aquatic Botany, 67*, 117–129.
Schaible, R., Bergmann, I., Boegle, M., Schoor, A., & Schubert, H. (2009). Genetic characterisation of sexually and parthenogenetically reproductive populations of *Chara canescens* (Charophyceae) using AFLP, rbcL, and SNP markers. *Phycologia, 48*(2), 105–117.
Schaible, R., Bergmann, I., & Schubert, H. (2011). Genetic Structure of sympatric sexually and parthenogenetically reproducing population of *Chara canescens* (Charophyta). *International Scholarly Research Notices, Ecology, 501838*, 1–13.
Schaible, R., Gerloff-Elias, A., Colchero, F., & Schubert, H. (2012). Two parthenogenetic populations of *Chara canescens* differ in their capacity to acclimate to irradiance and salinity. *Oecologia, 168*, 343–353.

Scheffer, M., Hosper, S. H., Meijer, M. L., Moss, B., & Jeppesen, E. (1993). Alternative equilibria in shallow lakes. *Trends in Ecology & Evolution, 8*, 275–279.

Schneider, S., Ziegler, C., & Melzer, A. (2006). Growth towards light as an adaptation to high light conditions in *Chara* branches. *New Phytologist, 172*, 83–91.

Schneider, S. (2007). Macrophyte trophic indicator values from a European perspective. *Limnologica, 37*, 281–289.

Schneider, S. C., Garcia, A., Carles Martín-Closas, C., & Chivas, A. R. (2015). The role of charophytes (Charales) in past and present environments: An overview. *Aquatic Botany, 120*, 2–6.

Schubert, H., & Blindow, I. (2004). *Charophytes of the Baltic Sea*. Gantner, Ruggell.

Schubert, H., Holzhausen, A., & Nowak, P. (2016). Individualentwicklung der Characeen, In: A. G. Characeen Deutschlands (Hrsg.), *Armleuchteralgen - Die Characeen Deutschlands* (pp. 57–78). Springer Verlag

Schubert, H., Telesh, I., Nikinmaa, N., & Skarlato, S. (2017b). Physiological adaptations. In: P. Snoeijs-Leijonmalm, H. Schubert, & T. Radziejewska (Eds.), *Biological Oceanography of the Baltic Sea* (pp. 255–280). Springer.

Schubert, H. (1996a). *Ökophysiologie der Lichtanpassung des Phytoplanktons eutropher Flachgewässer*. Habilitation, Univ. Rostock.

Schubert, H. (1996b). Starklichtanpassung—Strategien von Grünalgen und Cyanobakterien. *Nova Acta Leopoldina (supplement), 14*, 109–124.

Schubert, H., Blindow, I., Schories, D., Mages, M., von Tümpling, W., & Woelfl, S. (2017c). Biogeography of Chilean Charophytes—determined by climate or by water chemistry? *Botany Letters, 165*, 129–145.

Schubert, H., Marquardt, R., Schories, D., & Blindow, I. (2015). Biogeography of Chilean charophytes. *Aquatic Botany, 120*, 129–141.

Schubert, H., Schories, D., Schneider, B., & Selig, U. (2017a). Brackish water as an environment. In P. Snoeijs-Leijonmalm, H. Schubert, & T. Radziejewska (Eds.), *Biological Oceanography of the Baltic Sea* (pp. 3–22). Springer.

Schudack, M. (2016). Vorkommen fossiler Charophyten in Deutschland. In: AG Characeen Deutschlands (Hrsg.), *Armleuchteralgen - Die Characeen Deutschlands* (pp. 573–603). Springer Verlag.

Schudack, M. (1991). Eine Charophyten-Biozonierung für den Zeitraum Oberjura bis Berriasium in Westeuropa und ihr Vergleich mit Sequenzstratigraphie und eustatischer Meeresspiegelkurve. *Berliner Geowissenschaftliche Abhandlungen, 134*, 311–332.

Schulze, T., Baade, U., Dörner, H., Eckmann, R., HaertelBorer, S. S., Hölker, F., & Mehner, T. (2006). Interactions of residential piscivores with an introduced new predator type in a mesotrophic lake. *Canadian Journal of Fisheries and Aquatic Sciences, 63*, 2202–2212.

Schwarz, A.-M., & Hawes, I. (1997). Effects of changing water clarity on characean biomass and species composition in a large oligotrophic lake. *Aquatic Botany, 56*, 169–181.

Schwarz, A.-M., Hawes, I., & Howard-Williams, C. (1996). The role of photosynthesis/light relationships in determining lower depth limits of Characeae in South Island, New Zealand lakes. *Freshwater Biology, 35*, 69–79.

Sederias, J., & Coleman, B. (2007). The interaction of light and low temperature on breaking the dormancy of *Chara vulgaris* oospores. *Aquatic Botany, 87*, 229–234.

Sederias, J., & Coleman, B. (2009). Inhibition of *Chara vulgaris* oospore germination by sulfidic sediments. *Aquatic Botany, 91*, 273–278.

Selig, U., Steinhardt, T., & Schubert, H. (2009). Interannual variability of submerged vegetation in a brackish coastal lagoon on the southern Baltic Sea. *Ekologia, 28*, 412–423.

Sharma, P., Jha, A. B., Dubey, R. S., & Pessarakli, M. (2012). Reactive oxygen species, oxidative damage, and antioxidative defense mechanism in plants under stressful conditions. *Journal of Botany, 217037*, 1–26.

Shen, E. Y. F. (1966). Oospore germination in two species of *Chara*. *Taiwania, 12*, 39–46.

Siong, K., & Asaeda, T. (2006). Does calcite encrustation in *Chara* provide a phosphorus nutrient sink? *Journal of Environmental Quality, 35*, 490–494.

Skurzyński, P., & Bociąg, K. (2011). Vegetative propagation of *Chara rudis* (Characeae, Chlorophyta). *Phycologia, 50*(2), 194–201.
Smith, F. A., & Walker, N. A. (1980). Photosynthesis by aquatic plants: Effects of unstirred layers in relation to assimilation of CO_2 and HCO_3^- and to carbon isotopic discrimination. *New Phytologist, 86*, 245–259.
Sokol, R. C., & Stross, R. G. (1986). Annual germination window in oospores of *Nitella furcata* (Charophyceae). *Journal of Phycology, 22*, 403–406.
Sokol, R. C., & Stross, R. G. (1992). Phytochrome-mediated germination of very sensitive oospores. *Plant Physiology, 100*(3), 1132–1136.
Soomers, H., Winkel, D. N., Du, Y., & Wassen, M. J. (2010). The dispersal and deposition of hydrochorous plant seeds in drainage ditches. *Freshwater Biology, 55*, 2032–2046.
Steinhardt, T., & Selig, U. (2007). Spatial distribution patterns and relationship between recent vegetation and diaspore bank of a brackish coastal lagoon on the southern Baltic Sea. *Estuarine, Coastal and Shelf Science, 74*, 205–214.
Steinhardt, T., & Selig, U. (2011). Influence of salinity and sediment resuspension on macrophyte germination in coastal lakes. *Journal of Limnology, 70*(1), 11–20.
Stobbe, A., Gregor, T., & Röpke, A. (2014). Long-lived banks of oospores in lake sediments from the Trans-Urals (Russia) indicated by germination in over 300 years old radiocarbon dated sediments. *Aquatic Botany, 119*, 84–90.
Subramanian, D. (2018). *Indian Charophyta*. MJP Publishers.
Telesh, I., Schubert, H., & Skarlato, S. (2013). Life in the salinity gradient: Discovering mechanisms behind a new biodiversity pattern. *Estuarine, Coastal and Shelf Science, 135C*, 317–327.
Urbaniak, J., & Gąbka, M. (2014). *Polish Charophytes—an illustrated guide to identification*. Wydawnictwo Uniwersytetu Przyrodniczego we Wrocławiu, Wrocław.
van den Berg, M. S., Coops, H., Meijer, M.-L., Scheffer, M., & Simons, J. (1998a). Clear water associated with a dense *Chara* vegetation in the shallow and turbid Lake Veluwemeer, The Netherlands. In: E. Jeppesen, M. Søndergaard, M. Søndergaard, & K. Christoffersen (Eds), *The structuring role of submerged macrophytes in lakes* (pp. 339–352). Springer.
van Raam, J. (1998). *Handboek Kranswieren*. Charaboek Hilversum.
van de Weyer, K., Meis, S., & Sümer, G. (2017). Entwicklung von Flora und Vegetation im Phoenix See (Dortmund)—fünf Jahre nach Anpflanzungen mit Armleuchteralgen. *Rostocker Meeresbiologische Beiträge, 27*, 7–18.
van den Berg, M. S., Coops, H., & Simons, J. (2001). Propagule bank buildup of *Chara aspera* and its significance for colonization of a shallow lake. *Hydrobiologia, 462*, 9–17.
van den Berg, M. S., Coops, H., Simons, J., & Pilon, J. (2002). A comparative study of the use of inorganic carbon resources by *Chara aspera* and *Potamogeton pectinatus*. *Aquatic Botany, 72*, 219–233.
van den Berg, M. S., Scheffer, M., & Coops, H. (1998b). The role of Characean algae in the management of eutrophic shallow lakes. *Journal of Phycology, 34*, 750–756.
van der Wal, J. E. M., Dorenbosch, M., Immers, A. K., Forteza, C. V. J., Geurts, J. M., Peeters, E. T. H. M., Koese, B., & Bakker, E. S. (2013). Invasive crayfish threaten the development of submerged macrophytes in lake restoration. *PLoS ONE, 8*(10), e78579.
Vleeshouwers, L. M., Bouwmeester, H. J., & Karssen, C. M. (1995). Redefining seed dormancy: An attempt to integrate physiologie and ecology. *Journal of Ecology, 83*, 1031–1037.
Wang, H., Liu, C., & Yu, D. (2015). Morphological and reproductive differences among three charophyte species in response to variation in water depth. *Aquatic Biology, 24*, 91–100.
Wang, H., Yu, D., & Xiao, K. (2008). The interactive effects of irradiance and photoperiod on *Chara vulgaris* L.: Concerted responses in morphology, physiology and reproduction. *Hydrobiologia, 61*, 33–41.
Wium-Andersen, S., Anthoni, U., Christophersen, C., & Houen, G. (1982). Allelopathic effects on phytoplankton by substances isolated from aquatic macrophytes (Charales). *Oikos, 39*, 187–190.
Wood, R. D. (1959). Gametangial constants of extant Charophyta for use in micropaleobotany. *Journal of Paleontology, 33*, 186–194.

Chapter 8
The Fossil Record and Evolution of Charophytes from European Data

Carles Martín-Closas, Ingeborg Soulié-Märsche, and Josep Sanjuan

8.1 Introduction

Charophytes are among the oldest non-marine plants to be represented in the fossil record thanks to their propensity to calcify. Their fructifications, called gyrogonites and utricles, are microfossils ranging in size from 200 μm to 2 mm that are frequently found in ancient lacustrine deposits, from the Upper Silurian to recent times, representing one of the most complete fossil records of a non-marine plant group.

Palaeontologists' interest in charophytes began with the discovery of *Gyrogonites medicaginula* by Lamarck (1804). This first description of a fossil charophyte species introduced the term gyrogonite for all calcified charophyte oospores. The study of charophyte evolution is based on a large body of palaeontological data, enabling characterisation of their morphological evolution and providing a good basis to support testable hypotheses about their phylogeny, palaeoecology and biogeography. Palaeoecological reconstructions of fossil charophytes are generally obtained by indirect methods based on sedimentology, geochemistry and taphonomy, since the only direct method is by analogy with the ecology of living species in the most modern floras (Neogene and Quaternary). When possible, other groups of organisms such as freshwater molluscs and ostracods are also used for comparison.

One of the most important applications of fossil charophytes is their use to determine the relative age of rock successions in the continental domain. In summary,

C. Martín-Closas (✉) · J. Sanjuan
Departament de Dinàmica de la Terra i de l'Oceà, Facultat de Ciències de la Terra, Universitat de Barcelona, 08028 Barcelona, Catalonia, Spain
e-mail: cmartinclosas@ub.edu

I. Soulié-Märsche
Institut des Sciences de l'Évolution (ISEM), Université de Montpellier, Cc 061, Place Eugène Bataillon, 34095 Montpellier Cedex 05, France

H. Schubert et al. (eds.), *Charophytes of Europe*,
https://doi.org/10.1007/978-3-031-31898-6_8

the study of fossil charophytes represents a substantial contribution to knowledge of aquatic plant evolution throughout the Earth's history.

8.2 The Origin of Charophytes and Their Evolution During the Palaeozoic

It is thought that charophytes had originated at least by the beginning of the Palaeozoic. The first charophytes were probably not calcified and have not left a direct fossil record. However, by the mid-Ordovician, about 470 Ma (million years ago), there are remains of highly evolved land plants such as trilete spores of mosses, which were certainly preceded by charophytes (Gray et al., 1982).

There is still much debate about the putative ancestor of land plants and in consequence, about the group closest to the *Charophyta*. Depending on the authors, the *Zygnematophyta* (the conjugates) or the Coleochaetophyta are considered the most closely related to charophytes (Timme et al., 2012). In comparison to any of the other groups of algae ancestral to land plants, charophytes represent a further degree of structural organisation, which enabled them to attain, for the first time in the history of life, the status of an aquatic macrophyte in the non-marine realm. Moreover, physiological innovation enabled the newly originated charophytes to colonise a broader range of environments, ranging from brackish lagoons to oligohaline mires.

The earliest occurrence of charophyte remains in the fossil record is the fructification of *Primochara* from the Upper Silurian of Podolia (Ukraine), ca. 430 Ma, reported by Ischenko and Ischenko (1982). By the Upper Silurian, fossil charophytes display two types of fructifications, utricles and gyrogonites, which have been classified within two extinct orders, *Sycidiales* and *Moellerinales*, respectively (Feist et al., 2005a, 2005b; Lu et al., 1996).

Utricles of the *Sycidiales* show enormous variation in their Bauplan (generalised structural body plan, perhaps indicating that this was a polyphyletic group. It includes four families, *Chovanellaceae*, *Sycidiaceae*, *Pinnoputamenaceae* and *Trochiliscaceae* (Fig. 8.1).

Uncoiled utricles of the *Chovanellaceae* look like a specialised whorl of bract cells, which surround the oospore. The family *Sycidiaceae* is represented by the cosmopolitan genus *Sycidium* (Fig. 8.2), which was only related to the charophytes long after its discovery. In this genus, the utricle is formed by a basal set of polygonal cells, probably a modified flattened node, from which a whorl of up to 22 vertical branches, regularly segmented in successive prismatic cells, covered the oospore, leaving a large germination pore at the apex. Another family included within the *Sycidiales* is the interesting family *Pinnoputamenaceae*, which displays utricles built of ramified and slightly coiled bract cells, which may hold several antheridia, similar to modern ones, in the vegetative framework (Feist and Feist, 1997). These fossils

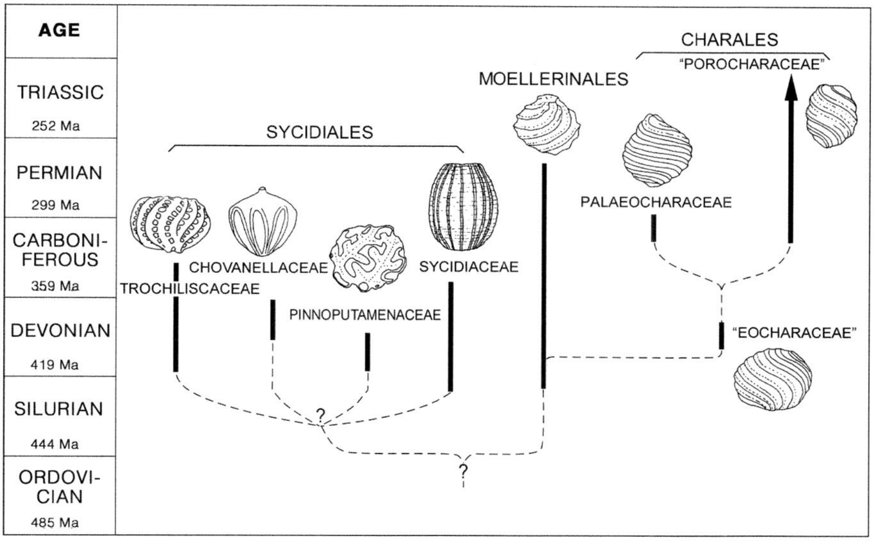

Fig. 8.1 Diagram of the evolution of Palaeozoic charophytes. Modified from Martín-Closas and Soulié-Märsche (2016)

indicate that in contrast to the oogonia, the antheridia present a highly conservative structure, remaining almost unchanged since the dawn of charophyte evolution. The utricles of the *Trochiliscaceae*, for instance in the genera *Karpinskya* and *Trochiliscus*, include about 18 segmented cells, coiled anticlockwise from the base to the apex, laterally appearing to coil to the right (Fig. 8.2).

The order *Sycidiales* developed mainly during the Devonian and became extinct by the beginning of the Carboniferous, lasting in total for about 100 million years. The reasons for its extinction remain unknown.

Representatives of the order *Moellerinales* (for instance the genus *Moellerina*) were the first charophytes to develop gyrogonites (Fig. 8.2). Thus, in comparison with the coeval *Sycidiales*, the fructifications of *Moellerinales* were free of any vegetative coating built by bract cells around the oospore. These first gyrogonites were formed by a variable number of helical cells (6–12), leaving a pore in the apex for germination. The gyrogonite cells were also coiled anticlockwise, i.e. to the right in lateral view, starting from the base up to the apex of the oospore. In contrast to the complex utricle structures of the *Sycidiales*, the *Moellerinales* show quite a homogenous gyrogonite structure over a very long fossil record that lasted for ca. 170 million years, from the Upper Silurian to the Upper Permian. It is not well known what the evolutionary relationship might be between the *Moellerinales* and the *Sycidiales*.

The order *Charales* is defined in a palaeontological sense by gyrogonites that show helical cells coiled clockwise from the base to the apex, i.e. to the left in lateral view (Fig. 8.2). Its first representatives occurred about 393 million years ago, in

Fig. 8.2 Examples of Palaeozoic charophytes. **a-c** Anticlockwise spiralled fructifications. **a** *Karpinskya laticostata*, Middle Devonian, North America. Utricle with apical cells, **b** *Trochiliscus podolicus* Devonian–Carboniferous, North America. Utricle with apical neck, **c** *Moellerina greenei*, Middle Devonian, North America. Flattened gyrogonite with spiral cells turning anticlockwise, **d-f** Gyrogonites with spiral cells turning clockwise, **d** *Eochara wickendenii*, Middle Devonian, North America. Lateral view, **e**, **f** *Palaeochara acadica*, Permian, North America, **e** Lateral view, **f** Basal view showing 6 spiral cells, **g–i** *Sycidium lagenarium*, Middle Devonian, Xichong, China (NIGPAS collection, Nanjing), **g** Lateral view, **h** Apical view, **i** Basal view. Reproduced from Martín-Closas and Soulié-Märsche (2016) with permission

the Middle Devonian. There is some consensus for the idea that *Charales* derived from *Moellerinales*, which would turn the latter taxon into a paraphyletic group. The first charalean gyrogonites belong to the genus *Eochara* (Fig. 8.2). They were still formed by a large and variable number of between 8 and 13 spiral cells, and show a large apical pore for germination. During the Carboniferous, the number of spiral cells in the charalean gyrogonite fell and then remained relatively invariable. These forms have been identified as belonging to two families: the *Palaeocharaceae* (e.g. genus *Palaeochara*), with gyrogonites built by 6–7 cells (Fig. 8.2), and the *Porocharaceae* (e.g. genus *Porochara*), bearing gyrogonites always with 5 spiral cells. Both *Palaeocharaceae* and *Porocharaceae* retained the ancestral apical pore as a germination opening.

From the Middle Devonian to the latest Permian, the two oppositely coiled charophyte orders *Moellerinales* and *Charales* have been reported from North America, Europe and Asia. However, even if they coexisted in time, they are not found together in the same assemblages (Soulié-Märsche, 2004). These two modes of coiling probably represented two equally effective options to organise protective cells around the oospore, without any further adaptive, environmental or palaeogeographic meaning. The coiling of cells around the oospore would endow it with increased resistance against internal pressures and enable it to accumulate more nourishment (starch) to ensure the growth of the young plant after germination (Martín-Closas et al., 1999a).

Vegetative remains of Palaeozoic charophytes are infrequent in the fossil record. One of the best-known thalli is that of *Palaeonitella*, from the famous fossil site of Rhynie in the Devonian of Scotland, where it occurs together with the remains of *Rhynia,* one of the oldest fossils of a terrestrial vascular plant. The material was preserved by silicification, with the silica originating from a nearby geothermal source. As the name suggests, the organisation of *Palaeonitella* is reminiscent of extant *Nitella* (Kelman et al., 2004; Kidston and Lang, 1921). Other, quite different Palaeozoic vegetative remains have been described as *Octochara* and *Hexachara* from the Devonian of South Africa and were found preserved as modified organic remains (Gess and Hiller, 1995). Initially, the thalli of Palaeozoic charophytes were ecorticate, whereas corticated forms did not appear until the end of this Era, in the Permian. In this respect, it can be assumed that the Palaeozoic calcified fructifications were attached to uncalcified thalli for dozens of millions of years.

By the nineteenth century, one of the most important discussions among charophyte palaeontologists concerned the ecology of Palaeozoic groups. As with the hypothesis that chlorophytes originated in the sea, a marine origin was also proposed for charophytes. This idea was based on the occurrence of calcified charophyte fructifications associated with marine invertebrates such as echinoderms and brachiopods in Palaeozoic rocks (Racki, 1982). Subsequently, however, more accurate taphonomic and palaeoecological analyses have challenged this assumption. Fructifications found in marine rocks often show evidence of transport such as erosion and breaking. Furthermore, co-occurrences of fossil thalli and fructifications have only been found in non-marine deposits, suggesting that fructifications found in marine rocks were transported seawards from the continental domain (Martín-Closas, 2003).

8.3 Triassic to Middle Jurassic: A Time for Porocharacean Charophytes and the Rise of Characeans

The Palaeozoic came to an end with a global biotic crisis at the Permian–Triassic boundary (P/T), which is thought to have caused the extinction of about 70–90% of marine organisms. For plants in the terrestrial domain, the crisis was not as severe and had already begun by the end of the Palaeozoic, during the Permian. This event also represented a significant change for charophytes, as the anticlockwise-turning gyrogonites of the *Moellerinales* disappeared completely during the Late Permian (Soulié-Märsche, 2004). As a result, only the *Charales*, represented by the *Porocharaceae*, survived past the Palaeozoic–Mesozoic boundary and continued into the Triassic. From the beginning of the Triassic, which is now established as occurring 252 Ma, all charophyte gyrogonites consistently display five cells turning clockwise around the oospore (Grambast, 1962, 1974). This group has been characterised as a monophyletic clade, called the *Quinquespiralia*, by Martín-Closas and Schudack (1991).

The *Porocharaceae* were the first representatives of the *Quinquespiralia*, and constitute a paraphyletic group, since they are the ancestral stock from which all post-Palaeozoic charophyte families evolved. Porocharacean gyrogonites dominated charophyte assemblages from the Carboniferous to the Upper Jurassic. Later, they became less abundant and were outcompeted by other charophyte groups until becoming extinct soon after the Cretaceous–Cenozoic boundary.

In addition to the constant number of 5 cells turning clockwise, gyrogonites of the *Porocharaceae* are also characterised by an apical germination pore, which was perhaps covered by a non-calcified structure similar to the deciduous coronula. The base of this type of gyrogonite displays the calcified remains of the oosphere sister cells, known in the palaeontological literature as the basal plate. This basal structure can be either unicellular, i.e. formed by a single pentagonal plug (corresponding to one sister cell of the oosphere), or multipartite, formed by two sub-pentagonal cells, which represent two calcified cells out of three sister cells of the oosphere. During the Mesozoic, porocharacean species with the two types of basal plate coexisted and displayed similar gyrogonite morphologies.

Porocharaceans with a unicellular basal plate are already represented by the end of the Palaeozoic. Significant genera in this group are the Permian genus *Leonardosia* (Souza de Faria et al., 2013), the Permian and Triassic genus *Stomochara* and the Jurassic and Cretaceous genus *Feistiella*, which is the last genus of the *Porocharaceae*, surviving briefly after the Cretaceous–Cenozoic boundary (Bignot and Grambast, 1969; Vicente et al., 2019). The *Porocharaceae* with a multipartite basal plate are represented in the Triassic and Jurassic by gyrogonites such as *Stellatochara* and *Stenochara* (Horn af Rantzien, 1954), followed by *Porochara* and *Latochara*, which developed mainly in the Jurassic and Cretaceous (Schudack, 1986).

This difference in basal plate also exists in the living *Characeae* and was reported in the nineteenth century by Braun (1853) and studied in detail by Goetz (1899). From a biological point of view, the two alternative states represent different processes of cell division during female gametogenesis and make it possible to distinguish between two distinct groups of living *Characeae* (Tribes, according to Wood 1962), the *Chareae* and the *Nitelleae*, the latter now being considered paraphyletic. However, the species of genus *Sphaerochara* Mädler (=*Tolypella* section *Rothia* Wood), usually ranged within the *Nitelleae*, do not match this classification as they are characterised by a single basal plate. The biological significance of single and multiple basal plates has still not been satisfactorily explained (Michaux-Ferrière and Soulié-Märsche, 1987).

In light of the evolutionary significance of the basal plate, two alternative phylogenetic hypotheses have been advanced: (1) The first was proposed by Soulié-Märsche (1979, 1989) and Martín-Closas and Schudack (1991) and is based on the homology of the basal plate (Fig. 8.3), resulting in the definition of two post-Palaeozoic groups within the *Quinquespiralia*: the *Monoplacata*, charophytes with a unicellular basal plate, and the *Polyplacata*, formed by charophytes with composite basal plates. (2) The second hypothesis is based on molecular phylogenies from the extant *Characeae* (e.g. McCourt et al., 1996; Pérez et al., 2014) and does not consider the basal structure a homologous character for post-Palaeozoic charophytes since it changed at least twice, once between *Tolypella* and *Sphaerochara* and the second time from *Nitella* to the *Chareae* (Fig. 8.4).

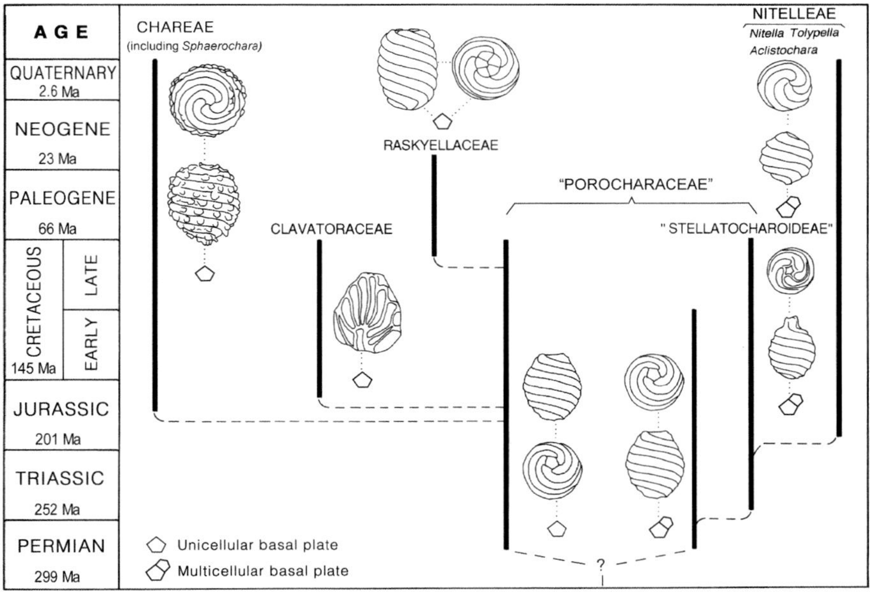

Fig. 8.3 Diagram of the evolution of post-Palaeozoic charophytes according to the hypothesis of the homology of basal plates. Modified from Martín-Closas and Soulié-Märsche (2016)

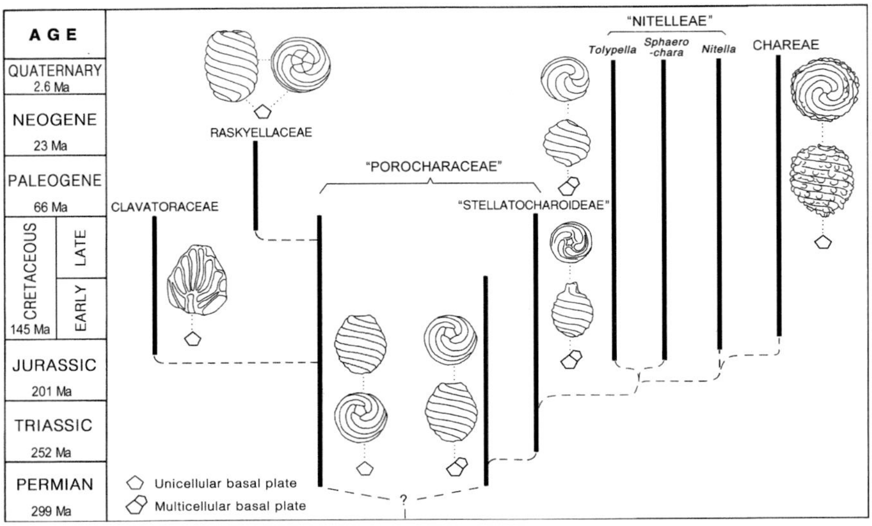

Fig. 8.4 Diagram of the evolution of post-Palaeozoic charophytes based on molecular phylogenies

Whatever the merit of these hypotheses, several charophyte families evolved from the ancestral paraphyletic stock of the *Porocharaceae* (Grambast, 1974) during the Mesozoic: the *Characeae*, the *Clavatoraceae* and the *Raskyellaceae*.

The vegetative structure of porocharaceans is poorly known but was possibly very diverse. The oldest corticated thallus known in the fossil record belongs to a Permian porocharacean, *Leonardosia* (Souza-Faria and Ricardi-Branco, 1999), and corticated thalli are usually associated with porocharacean fructifications.

From the Triassic to the Middle Jurassic, the *Porocharaceae* occupied the entire range of ecological niches available for charophytes, from brackish lagoons to freshwater lakes and fluvial ponds (e.g. Tiss et al., 2019). Later, during the Upper Jurassic and Cretaceous, they occupied a more restricted habitat in brackish water, probably due to the selective pressure of early *Characeae* and *Clavatoraceae* in freshwater habitats (Climent-Domènech et al., 2009; Martín-Closas, 2003).

The paraphyletic family *Porocharaceae* is still enigmatic. Its relatively uninformative gyrogonite morphology renders it difficult to distinguish meaningful species, and the group has remained poorly studied until now. However, in Europe it has been used successfully for biostratigraphic studies, mainly of Triassic and Jurassic deposits (Bilan, 1988; Schudack, 1993), and it has also provided significant palaeoenvironmental information for the entire Mesozoic (e.g. Mojon, 1989; Schudack et al., 1998; Tiss et al., 2019).

The first family to diverge from the *Porocharaceae* was the *Characeae*. Gyrogonites of *Aclistochara* display the oldest known closed apex, which is a character related to the *Characeae* (Feist et al., 1991). The genus *Aclistochara* occurred by the Late Triassic, ca. 237 Ma, and expanded worldwide during the Jurassic. The gyrogonites of this genus show a deep periapical depression, which is somehow reminiscent of the apex of extant *Lamprothamnium* gyrogonites. However, both genera differ in the features of their basal plates, which are unicellular in *Lamprothamnium* and multipartite in *Aclistochara*.

By the Middle Jurassic, a new type of characean gyrogonite appeared and is ranged in the genus *Mesochara* (Tiss et al., 2019), from "Mesozoic *Chara*". This genus, with its unmodified characean apex and unicellular basal plate, is reminiscent of extant *Chara* (Soulié-Märsche, 1989). It has been considered as representing the oldest fossil ancestor of the extant *Chareae* (Martín-Closas et al., 2018). Assemblages composed of diverse porocharacean gyrogonites and the early characeans *Aclistochara* and *Mesochara* are abundant worldwide in Upper Jurassic assemblages. Also during the Middle Jurassic the first *Clavatoraceae* appeared (Trabelsi et al., 2024).

8.4 Charophyte Diversification in the Upper Jurassic and Lower Cretaceous

In the Upper Jurassic, starting about 161.5 Ma, charophyte vegetations shifted significantly with the diversification of the *Characeae* and the *Clavatoraceae* (Figs. 8.3 and 8.4). This diversification continued during the Early Cretaceous and has been related to the worldwide expansion of non-marine wetlands (Martín-Closas, 2003).

The *Clavatoraceae* represents a significant family of tropical charophytes that thrived from the Middle Jurassic until the end of the Cretaceous, i.e. for more than 100 million years. However, the main development of this family occurred during the Early Cretaceous (Grambast, 1974). The clavatoracean fructification displays a thick and well-calcified coat, called the utricle (Fig. 8.5d–h), which surrounds the central, bottle-shaped oospore with five clockwise coiled spiral cells and a unicellular basal plate. Depending on the subfamily, this inner oospore may be calcified or non-calcified. The utricle is complex in structure. Within the *Clavatoraceae*, three monophyletic subfamilies have been characterised on the basis of the arrangement of utricle cells (Martín-Closas, 1996), two of which are very abundant in the fossil record. The subfamily *Atopocharoidae* (e.g. genus *Atopochara*) is characterised by a triradiate symmetry, whereas the subfamily *Clavatoroidae* (e.g. genus *Clavator*) displays bilaterally symmetrical utricles (Fig. 8.5). Many species of this family evolved into gradualistic lineages marked by progressive changes in utricle morphology over time (Fig. 8.6), which increases their great value for biostratigraphic purposes (Grambast, 1974).

Silicified clavatoracean thalli are sometimes exceptionally well-preserved (Harris, 1939; Musacchio, 1971; Peck, 1957) and calcified thalli are also recorded. For instance, in central Spain the calcified fossil thallus described as *Clavatoraxis* has been attributed to this family (Martín-Closas and Diéguez, 1998). This is a corticated thallus heavily coated with rosette-like structures made of radially-arranged spine cells. Other clavatoraceans had *Chara*-like thalli, known in the palaeontological

Fig. 8.5 Examples of Mesozoic charophytes of the families *Porocharaceae* and *Clavatoraceae*. **a** *Latochara tenuicostata*, Tithonian, Oléron Island, France, Lateral view of gyrogonite showing short apical neck and small apical pore, **b**, **c**. *Porochara douzensis*, Berriasian, Cherves de Cognac, France, **b** Lateral view of gyrogonite, **c**. Apical view showing large apical pore, **d–f** *Clavator grovesii jiuquanensis*, Late Barremian–Early Aptian, Pas du Frou, Chartreuse, France, bilaterally symmetrical utricle, **d** Abaxial view, **e** Lateral view, **f** Adaxial view, **g–i** *Atopochara trivolvis triquetra*, Late Barremian–Early Aptian, Pas du Frou, Chartreuse, France, Utricle with triradiate symmetry, **g** Lateral view, **h** Basal view, **i** Apical view with apical pore. Reproduced from Martín-Closas and Soulié-Märsche (2016) with permission

literature as *Charaxis*. For example, utricles of the clavatoracean genus *Echinochara* have been found attached to an exceptionally well-preserved *Charaxis* thallus in the Lower Cretaceous of the Iberian Chain, Eastern Spain (Pérez-Cano et al., 2020).

Most of the ca. 25 clavatoracean species originated on and inhabited the islands of a large archipelago located in the central Tethys sea (mainly today's Europe and

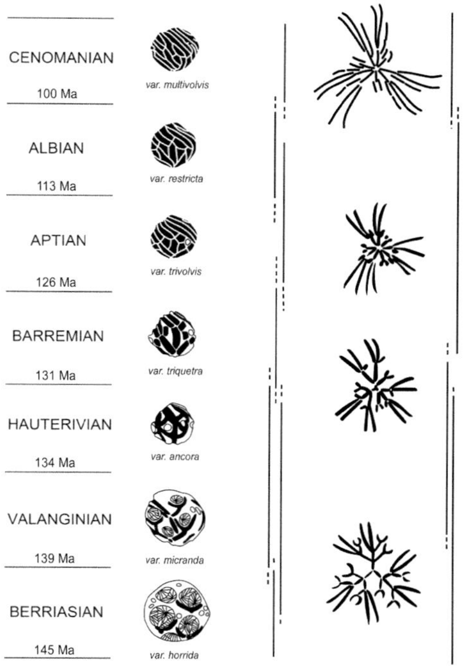

Fig. 8.6 Evolutionary lineage of *Atopochara trivolvis*. *Left*: geological age (stages) given in million years ago (Ma), *Centre*: lateral view of utricles showing a gradualistic evolution; *Right*: schemes of vegetative structure of utricles, folded down on their base; vertical lines showing the duration of each structure. Modified from Grambast (1974), Martín-Closas (1996). Reproduced from Martín-Closas and Soulié-Märsche (2016) with permission

North Africa), which had a tropical and subtropical climate during the Early Cretaceous. In this domain, the *Clavatoraceae* occupied most of the habitats available for charophytes, such as freshwater alkaline lakes, fluvial ponds and even brackish environments (e.g. Vicente and Martín-Closas, 2013). Some species were capable of long-distance migration and expanded from the Tethyan archipelago to East Asia (e.g. China and Japan) and South America (e.g. Argentina and Brazil) in at most a few million years, probably transported in the digestive tracts of dinosaurs and early birds (Martín-Closas, 2015; Martín-Closas and Wang, 2008; Musacchio, 2000). Abundance, relatively rapid evolutionary rates and biogeographic ranges reaching even cosmopolitanism, render the *Clavatoraceae* an extremely useful tool for geological correlation of Lower Cretaceous sedimentary sequences (Grambast, 1974; Riveline et al., 1996; Schudack, 1993).

Although the utricles of the *Clavatoraceae* were represented until near the Cretaceous–Cenozoic boundary, this family had begun a long-lasting decline by the end of the Early Cretaceous ca. 100 million years ago, an event that has been related to the radiation of aquatic angiosperms (Martín-Closas, 2003). For more than 300 million years, charophytes had been the only macrophytes in non-marine aquatic environments, but the diversification of aquatic angiosperms in the Early Cretaceous produced an abrupt change in hydrophytic communities without any intermediary step, such as vegetation composed of free-sporing plants (e.g. ferns) or 'gymnosperms' (e.g. conifers or cycadales), as was the rule in terrestrial habitats (Martín-Closas, 2003). The last relicts of the *Clavatoraceae* were two species from the latest Cretaceous, one belonging to the genus *Clavator* in South Europe (Grambast, 1974), and the other to *Atopochara* in North China and Mongolia (Kyansep-Romashkina, 1975; Li et al., 2019).

In addition to the *Clavatoraceae*, the Upper Jurassic and Lower Cretaceous also yielded the first occurrence of important modern genera, i.e. *Sphaerochara*, *Tolypella* and *Nitella*. The oldest well-documented record of the genus *Sphaerochara* occurs in the Kimmeridgian of Germany (Schudack, 1993). As shown by Soulié-Märsche (1989) and Soulié-Märsche and García (2015), the overall morphology and unicellular basal plate of the gyrogonites of *Sphaerochara* are identical to the extant gyrogonites of *Tolypella* sect. *Rothia* Wood (=section *Acutifolia* Allen) and provide strong arguments to support the synonymy of both taxa, with *Sphaerochara* having nomenclatural priority. Hence, *Sphaerochara* also applies to the living species of section *Rothia* Wood, whereas *Tolypella* s. s. (*Tolypella* section *Tolypella* Wood = section *Obtusifolia* Allen) remains a separate genus (Fig. 8.4). This proposal is in part supported by the results from molecular phylogenies that suggest the two sections separated at the base of the genus (Pérez et al., 2014).

The oldest gyrogonites corresponding in size, shape and multipartite basal plate to the oospores of extant *Tolypella* s. s. were reported by Martín-Closas et al. (2018) in the Lower Cretaceous of Catalonia (Spain). The genus *Tolypella* has a relatively discontinuous fossil record. Besides this first occurrence, it re-appears again in the uppermost Cretaceous of Argentina (Uliana and Musacchio, 1978) and North America (Feist and Brouwers, 1990), and finally in the Oligocene of the Paris Basin (France) and up to the Oligocene–Miocene boundary in the Rhein Graben, Germany (Schwarz, 1985, 1997). This latter occurrence, about 23 million years ago, represents the youngest known record of a calcified gyrogonite of *Tolypella*. Later, this genus lost its capacity to calcify (Soulié-Märsche, 1979, 1989).

The genus *Nitella* appears first in the fossil record during the Early Cretaceous, with exceptionally well-preserved silicified oospores, known as *Nitellites sahnii*, from India (Horn af Rantzien, 1957; Feist et al., 1991). An exceptionally rare occurrence of silicified oospores of *Nitella* documents this genus again in the Upper Cretaceous of Argentina (Musacchio, 2010). This genus was probably much more common than suggested by these sparse data since its fossil record is strongly biased by the lack of calcification.

8.5 The Radiation of the *Characeae* from the Upper Cretaceous to the Eocene

The sedimentary deposits from the early Late Cretaceous (i.e. from the Cenomanian to the Coniacian, in absolute time values 100.5–86.3 Ma) are very poor in fossil charophyte remains. By that time, the Earth had one of the highest relative sea levels in geological history, and almost all lowlands were flooded by the oceans. An exceptionally rich record of charophytes from that time has been unveiled in the Songliao Basin, north China, by Li et al. (2020), whereas in Europe only very limited sites with poorly diverse charophytes occur. Afterwards, in the Santonian to Maastrichtian, 86.3–66.0 million years ago, charophyte assemblages were dominated worldwide by characeans, associated with only a few relict porocharaceans, always displaying a unicellular basal plate (genus *Feistiella*), and even rarer clavatoraceans. In addition, a new fossil family emerged, the *Raskyellaceae* (genus *Saportanella*). Their gyrogonites display an apex hermetically sealed by a 5-celled operculum, a character most probably derived from the porocharaceans bearing a unicellular basal plate, such as *Feistiella*.

Some of the most abundant and widespread characean genera from the Late Cretaceous belonged to the genera *Microchara* (with extremely small gyrogonites, some of them barely reaching 250 μm), *Peckichara* (with heavily ornamented gyrogonites and periapical modifications) and *Platychara* (with flattened gyrogonites). By the same time, ca. 75 million years ago, gyrogonites had appeared in Eurasia and America corresponding to the extant genus *Lychnothamnus* (Vicente et al., 2020). Originally, these were classified within the fossil genus *Amblyochara*, until the gyrogonites of the extant *Lychnothamnus* were described. Both types proved basically identical and the genus *Amblyochara* was put into synonymy with the genus *Lychnothamnus* (Soulié-Märsche, 1989).

The study of fossil thalli from Late Cretaceous charophytes is starting. Limestones deposited in ancient lakes of Southern Europe have so far yielded the best-known record of these thalli. For instance, a thallus classified within the genus *Clavatoraxis* has been found with anatomically connected *Microchara* gyrogonites in the latest Cretaceous of the Pyrenees (Villalba-Breva and Martín-Closas, 2011).

To date, few data are available about the phylogeny of Late Cretaceous characeans, but it is clear that they present a number of gyrogonite and thallus morphologies that are unknown in extant species. This morphological diversity represents a second radiation of the characeans, particularly of the *Chareae*, well separated from the first, small radiation that had occurred in the Late Jurassic and Early Cretaceous (Grambast, 1974; Martín-Closas et al., 2018). The Late Cretaceous species provide sufficient evolutionary events to construct biostratigraphic zonations, which are useful for characterising the age of rocks on all continents where a charophyte record is available (e.g. Riveline et al., 1996 for Europe and Li et al., 2019 for China). These biozonations were the first to have been correlated with the Geological Time Scale through magnetostratigraphic studies (Galbrun et al., 1993).

The palaeoecology of European Upper Cretaceous charophytes has been explored in recent studies. Freshwater and alkaline lakes were the environments where charophyte assemblages showed the highest number of species. In Europe, for instance, the genera *Peckichara*, *Microchara* and *Platychara* shared their shallowest habitats with the last clavatoraceans belonging to the genus *Clavator* (Villalba-Breva and Martín-Closas, 2013). Ponds in fluvial floodplains showed impoverished assemblages of a few species, usually with very small-sized gyrogonites belonging to *Microchara* (Vicente et al., 2016). Finally, brackish water bodies worldwide were usually dominated by the porocharacean *Feistiella*. The first occurrences of the genus *Lamprothamnium* are also documented from the latest Cretaceous of Argentina and China (Musacchio and Moroni, 1983 and Li et al., 2016, respectively), although this genus occurred at that time in a wider range of habitats than at present.

It is well known that at the end of the Cretaceous, 66 million years ago, most marine biota underwent global extinction. In terrestrial environments, this crisis represented the end of the dinosaurs and some plant families. However, it was not as significant for charophytes. The last few remaining species of the *Porocharaceae* and *Clavatoraceae* vanished shortly after the Cretaceous–Cenozoic boundary (Bignot and Grambast, 1969; Vicente et al., 2019). The *Characeae* experienced a reduced extinction limited to the specific level, and after the biotic crisis, in the Palaeocene, this family continued to diversify (Vicente et al., 2019).

The maximum number of fossil species belonging to the family *Characeae* was reached during the Eocene, between 33.9 and 56 million years ago, in the context of a global greenhouse climate (Riveline, 1986). Eocene gyrogonites are often heavily calcified and ornamented with a diversity of tubercles, granules and crests (e.g. Grambast, 1977). Significant Eocene genera were *Peckichara*, *Harrisichara*, *Maedleriella*, *Nodosochara*, *Gyrogona* and the typical gyrogonites of *Nitellopsis*, *Lychnothamnus* and *Chara* s. s (Fig. 8.7). A new genus of the *Raskyellaceae*, the genus *Raskyella*, appeared in the Early Eocene and became more widespread in Europe and North Africa during the Middle Eocene (Grambast, 1959; Martín-Closas et al., 1999b). Some genera, such as *Harrisichara*, showed a wide range of ecological tolerances, from freshwater lakes to brackish habitats, while others such as *Nodosochara* appear to have been restricted to freshwater lakes and ponds (Sanjuan and Martín-Closas, 2012). This diversity of *Characeae* in the Eocene means that they too constitute a very important tool for biostratigraphy (e.g. Anadón and Feist, 1981; Feist-Castel, 1977). Recently, part of the Eocene charophyte biozonation has been correlated with the Geological Time Scale with the help of magnetostratigraphy (Sanjuan et al., 2014).

A significant change occurred in the Oligocene, about 34 Ma, when charophyte diversity was drastically reduced and many characteristic Eocene genera such as *Maedleriella*, *Harrisichara* and *Grovesichara* disappeared (Riveline, 1986). This decline has been related to increasing global cooling, aridification and the trend towards icehouse conditions from the Oligocene onwards. This climatic turnover, known internationally as the "Grande Coupure" (Stehlin, 1910), had a significant influence on many terrestrial groups such as mammals, and is now thought to be related to the origin of the permanently cold Antarctic circumpolar current (Zachos et al., 2001).

Fig. 8.7 Examples of Cenozoic charophytes from the families *Raskyellaceae* and *Characeae*. **a**–**c** *Raskyella peckii*, Middle Eocene, Mallorca (Spain), **a** Lateral view, **b** Apex with five-celled operculum, **c** Base, **d**–**f** *Harrisichara lineata*, Upper Eocene of the Ebro basin (Spain), **d** Apex with small comma-shaped apical nodules, **e** Lateral view with characteristic midcellular crest and base prolongated into a short column, **f** Base. **g**–**i** *Maedleriella mangenotii*, Middle Eocene, Mallorca (Spain), **g, h** Lateral view showing diverse types of tubercle, **i** Base with pentagonal basal plate visible from the exterior and also ornamented with a tubercle, **j-l** *Lychnothamnus barbatus megalicarpus*, Miocene, Cerdanya (Catalonia, Spain), **j** Lateral view with bicarinate spiral cell sutures, **k** Apex with periapical thinning of spiral cells, **i** Base with star-shaped basal funnel. Reproduced from Martín-Closas and Soulié-Märsche (2016) with permission

8.6 Neogene and Quaternary Charophytes and the Origin of Modern Floras

From the Miocene onwards, starting 23 million years ago, most of the charophytes found in the fossil record belong to modern genera. The only exception is the occurrence at the base of the Miocene of a new species of the *Raskyellaceae*, *Rantzieniella nitida*. This species is useful for characterising the basal stage of the Miocene, called the Aquitanian. All other extant genera, with the exception of *Nitella* and *Tolypella*, are represented in the fossil record by their gyrogonites. In this respect, it should be recalled that the last gyrogonites of *Tolypella* with a multipartite basal plate are recorded at the Oligocene–Miocene boundary (Schwarz, 1985, 1997).

In the Miocene, the genus *Nitellopsis*, represented by *N. merianii* and several other species, was abundant and has been found throughout Eurasia (Grambast and Soulié-Märsche, 1972; Heer, 1854; Sanjuan and Alqudah, 2018; Soulié-Märsche et al., 2002). *Lychnothamnus barbatus* also included several subspecies or varieties. In addition, the populations of *Sphaerochara* were very abundant and showed ornamented gyrogonites which are unknown today (Sanjuan and Soulié-Märsche, 2020). *Lamprothamnium* is represented in Europe by *L. papulosum* and was limited to brackish water with changing salinity. There were several species of *Chara*, but in contrast to today's situation, they were less abundant. The special character of some of the *Chara* species from the Miocene is the presence of ornamentation, composed of small tubercles on the spiral cells of the gyrogonite. This type of specimen is extremely rare in present-day populations. Most Miocene gyrogonites are very similar to those from modern species, and in some cases identical, such as in *Lychnothamnus*, *Lamprothamnium* and some *Chara*. The Pliocene has so far delivered only a few localities with charophytes, where *Nitellopsis* and *Lychnothamnus* appear to be dominant (Kovar-Edder et al., 2003).

The Pleistocene glacial intervals repeatedly interrupted the charophyte fossil record in Europe and North Africa. In contrast, charophytes are very well represented in interglacial lakes (Demirci et al., 2023). As these were pioneering plants, charophyte gyrogonites and oospores are especially abundant at the base of these limnic deposits, representing colonisation of the lake bottom after glacial times. As these lakes became more mature and richer in nutrients, charophytes were often substituted by aquatic angiosperms. During the Holocene, European lakes often evolved into peat mires, becoming rich in organic remains and acidified. Under such conditions, *Chara*-dominated vegetation was substituted by *Nitella* (Haas, 1999). As a result of the evolution of charophytes during the Cenozoic, and especially in the Neogene and Quaternary, there is a notable reduction in charophyte species richness and biodiversity even at the genus level.

Initial studies of Quaternary charophytes (*Characeae*) attempted to draw parallels between the taxonomy of fossil charophytes and extant species. For instance, Krassavina (1971) suggested these similarities when describing *Chara palaeofragilis* and *Chara palaeohispida* in the Pleistocene. However, in many cases, modern research on living charophyte species has enabled recognition of specific gyrogonite morphologies of extant species (García, 1999; García and Chivas, 2004; Soulié-Märsche, 1989; Soulié-Märsche et al., 2008). These data form the cornerstone for determination of fossil gyrogonites that may be directly related to modern species. By analogy, these findings also make it possible to determine precise palaeoecological and palaeolimnological characteristics of ancient lakes (García, 1994). Furthermore, this research provides a solid basis for comparing the first occurrences of extant species and genera in the fossil record and allows for a comparison with the results of molecular phylogenies (e.g. Martín-Closas et al., 2018). Modern charophyte flora is the result of evolution over a long period of time, and it can be assumed that all modern species appeared in the fossil record during the Neogene and Quaternary and have existed at least since the beginning of the Holocene (10,000 years ago).

8.7 Overview

The evolution of charophytes shows three main steps, which correspond to global biotic crises of the biosphere, especially in the non-marine domain:

(1) The Permian–Triassic crisis ("P/T crisis"), which occurred about 252 million years ago and represented the end of the highly diverse charophyte fructification morphologies of the Palaeozoic. As a result, only the *Quinquespiralia* survived, i.e. charophytes with gyrogonites showing 5 spiral cells turning clockwise from the base to the apex (to the left in lateral view). At the beginning of the Mesozoic (Triassic and Jurassic), it was mainly representatives of the paraphyletic family *Porocharaceae* that evolved and occupied most environments, while in the Middle Jurassic, biodiversity was increased by the appearance of the first *Chareae* and the *Clavatoraceae*, a charophyte group characterised by the oospore being protected within an utricle.

(2) The passage from the Early to Late Cretaceous, about 100.5 million years ago, heralded the substitution of the *Clavatoraceae*-rich floras by floras dominated by the *Characeae*, which were radiating. This change was especially marked in the European and North African domain, but has also been recognised elsewhere. In contrast, the Cretaceous–Palaeogene boundary ("K/Pg crisis") at the end of the Mesozoic only brought minor changes in charophyte biodiversity. During the first part of the Cenozoic, in the Palaeocene and mainly in the Eocene, *Characeae* biodiversity reached its peak, with 25 genera characterised by their gyrogonites.

(3) A new turnover in charophyte evolution was the "Grande Coupure", which brought a significant reduction in charophyte biodiversity. This crisis occurred at the beginning of the Oligocene, about 34 million years ago, and was related to a global cooling and aridification event. Since then, charophyte biodiversity has steadily dwindled to the present seven genera. Besides the species that are represented by calcified oospores, there were probably abundant species similar to the extant *Nitella* and *Tolypella* s.s. that never calcified.

Charophyte evolution has enabled the development of a well-supported biozonation encompassing the entire time span from the Silurian to the Quaternary (Riveline et al., 1996). Biozones are characterised by index species that provide a tool for rock age determination. In recent years, some of these biozones have been correlated to the global Geological Time Scale by means of magnetostratigraphy. Although most of these biozones have been defined in Europe, some of them can be applied worldwide and contribute to the role of charophytes as index fossils.

8.8 Summary

The charophyte fossil record provides sufficient data to document the history of the group from the Silurian to the present. Fossil charophyte remains include not only their calcified fructifications, i.e. utricles and gyrogonites, but also their thalli. Diversity in the Bauplan of charophyte fructifications reached its peak in the Palaeozoic, with the coexistence of three charophyte orders, *Sycidiales*, *Moellerinales* and *Charales*. The extinction of the two first orders at the end of the Palaeozoic led to a single type of fructification, the gyrogonite of the *Porocharaceae*, with five spiral cells turning clockwise. This paraphyletic group represents the ancestor of all post-Palaeozoic charophytes. It rapidly diversified into different genera represented by fructifications with two types of basal plate (i.e. the oosphere sister cells), either unicellular or composite. This fundamental difference is thought to represent two alternative types of female gametogenesis.

Charophytes with a composite basal plate developed diverse gyrogonite morphologies at the base of the Mesozoic (Triassic–Jurassic), which are classified in the family *Porocharaceae*. They are considered to represent the ancestors of characeans, with *Aclistochara*, a Mesozoic fossil genus, and extant *Tolypella* s.s. and *Nitella*, which are recorded in the Early Cretaceous. The first *Characeae* with a single-celled basal plate occurred in the Middle Jurassic with *Mesochara*. In the Late Jurassic, the oldest known *Sphaerochara* appeared (identical to *Tolypella* sect. *Rothia*), together with the first *Clavatoraceae*. This fossil family is characterised by oospores surrounded by a complex utricle. Clavatoraceans dominated the tropical and subtropical wetlands on the Tethyan islands (present-day Europe and North Africa) during the Early Cretaceous and also developed cosmopolitan species in the genera *Atopochara* and *Clavator*. As a result, the formerly abundant *Porocharaceae* became less diverse and were relegated to brackish coastal lagoons. After a period with a very poor charophyte fossil record at the beginning of the Late Cretaceous (Cenomanian to Coniacian), the *Characeae* underwent significant radiation, leading to the development of many new genera including the first record of *Chara*, *Lamprothamnium* and *Lychnothamnus*. By the beginning of the Cenozoic, the last few remaining species of clavatoraceans and porocharaceans became extinct and the *Chareae* continued to expand. The genus *Nitellopsis* is documented at the very beginning of the Cenozoic. The *Characeae* reached their maximum diversification in the Middle Eocene, with gyrogonites that are quite different from modern ones. These morphological types regressed in the Early Oligocene, when global climate cooling and aridification reduced gyrogonite morphologies to the seven modern genera. Hence, the Neogene to recent charophyte flora is an impoverished remnant of the plants that flourished in the geological past.

References

Anadón, P., & Feist, M. (1981). Charophytes et biostratigraphie du Paléogène inférieur du Bassin de l'Ebre Oriental. *Palaeontographica Abt B, 178*(4–6), 143–168.

Bignot, G., & Grambast, L. (1969). Sur la position stratigraphique et les Charophytes de la Formation de Kozina (Slovénie, Yougoslavie). *Comptes Rendus de l'Académie des Des Sciences Série D, 269*, 689–692.

Bilan, W. (1988). The epicontinental Triassic charophytes of Poland. *Acta Paleobotanica, 28*, 63–161.

Braun, A. (1853). Über die Richtungsverhältnisse der Saftströme in den Zellen der Characeen. *Verhandlungen der Akademie der Wissenschaften Berlin, 1853*, 45–76.

Climent-Domènech, H., MartÍn-Closas, C., & Salas, R. (2009). Charophyte-rich microfacies in the Barremian of the Eastern Iberian Chain (Spain). *Facies, 55*, 387–400.

Demirci, E., Sanjuan, J., & Tonoglu, C. (2023). Early Pleistocene charophyte flora from Dursunlu (Irgin Basin, Turkey): Palaeoecological implications. *Review of Palaeobotany and Palynology, 311*, 104848. http://doi.org/10.1016/j.revpalbo.2023.104848

Feist, M., & Brouwers, E. M. (1990). A new *Tolypella* from the Ocean Point dinosaur locality, North Slope, Alaska, and the Late Cretaceous to Paleocene nitelloid charophytes. *US Geological Survey Bulletin, 1990-F*, 1–7.

Feist, M., Bhatia, S. B., Yadagiri, P. (1991). On the oldest representative of the family Characeae and its relationships with the Porocharaceae. *Bulletin de la Société Botanique de France 138 Actualités Botaniques, 1*, 25–32.

Feist, M., Grambast-Fessard, N., Guerlesquin, M., Karol, K., Lu, H., McCourt, R. M., Wang, Q., & Shenzen, Z. (2005a). *Charophyta. Treatise on Invertebrate Paleontology. Part B., Protoctista 1, Bd. 1.* The Geological Society of America, Boulder.

Feist, M., & Feist, R. (1997). Oldest record of a bisexual plant. *Nature, 385*, 401.

Feist, M., Liu, J., & Tafforeau, P. (2005b). New insights into Paleozoic charophyte morphology and phylogeny. *American Journal of Science, 92*, 1152–1160.

Feist-Castel, M. (1977). Evolution of the charophyte floras in the Upper Eocene and Lower Oligocene of the Isle of Wight. *Palaeontology, 20*, 143–157.

Galbrun, B., Feist, M., Colombo, F., Rocchia, R., & Tambareau, Y. (1993). Magnetostratigraphy and biostratigraphy of Cretaceous-Tertiary deposits, Ager Basin, Province of Lérida, Spain. *Palaeogeography, Palaeoclimatology, Palaeoecology 102*, 41–52.

García, A. (1994). Charophyta: Their use in paleolimnology. *Journal of Paleolimnology, 10*(1), 43–52.

García, A. (1999). Quaternary charophytes from Salina del Bebedero, Argentina: Their relation with extant taxa and palaeolimnological significance. *Journal of Paleolimnology, 21*(3), 307–323.

García, A., & Chivas, A. (2004). Quaternary and extant euryhaline *Lamprothamnium* Groves (Charales) from Australia: Gyrogonite morphology and paleolimnological significance. *Journal of Paleolimnology, 31*(3), 321–341.

Gess, R. W., & Hiller, N. (1995). Late Devonian charophytes from the Witteberg Group, South Africa. *Review of Palaeobotany and Palynology, 89*, 417–428.

Goetz, G. (1899). Ueber die Entwickelung der Eiknospe bei den Characeen. *Botanische Zeitschrift, 57*(1), 1–13.

Grambast, L. (1977). Étude sur les charophytes tertiaires d'Europe occidentale II. Espèces nouvelles de l'Éocène inférieur. *Paléobiologie Continentale, 8*(1), 1–27.
Grambast, L. (1959). *Extension chronologique des genres chez les Charoideae*. Société des Éditions Technip.
Grambast, L. (1962). Classification de l'embranchement des Charophytes. *Naturalia Monspeliensia, 14*, 63–86.
Grambast, L. (1974). Phylogeny of the Charophyta. *Taxon, 23*, 463–481.
Grambast, L., & Soulié-Märsche, I. (1972). Sur l'ancienneté et la diversification des *Nitellopsis* (Charophytes). *Paléobiologie Continentale, III*(3), 1–14.
Gray, J., Massa, D., & Boucot, A. J. (1982). Caradocian land plant microfossils from Libya. *Geology, 10*, 197–201.
Haas, J. N. (1999). Charophyte population dynamics during the Late Quaternary at Lake Bibersee, Switzerland. *Australian Journal of Botany, 47*(3), 315–324.
Harris, T. M. (1939). *British Purbeck Charophyta*. British Museum Natural History.
Heer, O. (1854). Discussion sur l'identité du *Chara meriani* et *helicteres*. *Bulletin de La Société Vaudoise des Sciences Naturelles, III*, 278–281.
Horn af Rantzien, H. (1954). Middle Triassic Charophyta of South Sweden. *Opera Botanica, 1*(2), 1–75.
Horn af Rantzien, H. (1957). Nitellaceous Charophyte gyrogonites in the Rajmahal series (Upper Gondwana) of India. *Stockholm Contributions in Geology, 1*(1), 1–29.
Ishchenko, T. A., & Ishchenko, A. A. (1982). Charophytes found in the Upper Silurian of Podolia. In Y. V. Teslenko (Hrsg.) *Systematics and evolution of fossil plants* (S 21–32). Nauk Dumka, Kiev.
Kelman, R., Feist, M., Nigel, H. A. T., & Hass, H. (2004). Charophyte algae from Rhynie chert. *Transactions of the Royal Society of Edinburgh: Earth Sciences, 94*, 445–455.
Kidston, R., & Lang, W. H. (1921). On old red sandstone plants showing structure from the Rhynie Chert Bed, Aberdeenshire. Part V. *The Thallophyta. Transactions of the Royal Society Edinburgh, 52*(4), 855–902.
Kovar-Eder, J., Schwarz, J., & Wojcicki, J. J. (2003). The predominantly aquatic flora from Pellendorf, Lower Austria, Late Miocene, Pannonian—a systematic study. *Acta Palaeobotanica, 42*(2), 125–151.
Krassavina, L. (1971). *Charophyta e sedimentis quaternariis partis australis depressioniis sibiriae occidentalis* (S 88–103). Academia Scientiarum URSS, Leningrad.
Kyansep-Romashkina, N. P. (1975). Some charophytes from the Upper Jurassic and the Cretaceous of Mongolia. The joint Soviet-Mongolian Paleontological Expedition Transactions. In Kramarenko, N. N. (Ed.), *Fossil fauna and flora of Mongolia* (S 181–204). Nauka, Moskow.
Lamarck, J. B. (1804). Suite des mémoires sur les fossiles des environs de Paris. *Annales du Muséum d'Histoire Naturelle Paris, 5*, 349–357.
Li, S., Wang, Q., Zhang, H., Lu, H., & Martín-Closas, C. (2016). Charophytes from the Cretaceous-Paleogene transition in the Pingyi Basin (Eastern China) and their Eurasian correlation. *Cretaceous Research, 59*, 179–200.
Li, S., Wang, Q., Zhang, H., Wan, X., & Martín-Closas, C. (2019). Charophytes from the Cretaceous-Paleocene boundary in the Songliao Basin (north-eastern China): A Chinese biozonation and its calibration to the Geomagnetic Polarity Time Scale. *Papers in Palaeontology, 5*(1), 47–81.
Li, S., Wang, Q., Zhang, H., Zhang, Y., Wan, X., & Martín-Closas, C. (2020). Filling a gap in the evolution of charophytes during the Turonian to Santonian: Implications for modern physiognomy. *Review of Palaeobotany and Palynology, 274*, 104154.
Lu, H. N., Soulié-Märsche, I., & Wang, Q. F. (1996). Evolution and Classification of Palaeozoic charophytes. *Acta Micropalaeontologica Sinica, 13*, 1–12.
Martín-Closas, C., & Schudack, M. E. (1991). Phylogenetic analysis and systematisation of post-palaeozoic charophytes. *Revue de la Société Botanique de France, Actualités Botaniques, 138*(1), 53–71.

Martín-Closas, C., Bosch, R., & Serra-Kiel, J. (1999a). Biomechanics and evolution of spiralization in charophyte fructifications. In M. H. Kurmann, & A. R. Hemsley (Eds.), *The evolution of plant architecture* (S 399–421). Royal Botanic Gardens Kew, London.

Martín-Closas, C., & Soulié-Märsche, I. (2016). Ursprung und Evolution der Charophyten. In Arbeitsgruppe Characeen Deutschlands (Eds.), *Armleuchteralgen. Die Characeen Deutschlands* (pp. 29–49). Springer, Berlin.

Martín-Closas, C. (1996). A phylogenetic system of Clavatoraceae (Charophyta). *Review of Palaeobotany and Palynology, 94*, 259–293.

Martín-Closas, C. (2003). The fossil record and evolution of freshwater plants. A review. *Geologica Acta, 1*, 315–338.

Martín-Closas, C. (2015). Cosmopolitanism in Northern Hemisphere Cretaceous Charophyta (Clavatoroidae). *Palaeogeography, Palaeoclimatology, Palaeoecology, 438*, 9–23.

Martín-Closas, C., & Diéguez, C. (1998). Charophytes from the Lower Cretaceous of the Iberian Ranges (Spain). *Palaeontology, 41*, 1133–1152.

Martín-Closas, C., Serra-Kiel, J., Busquets, P., & Ramos-Guerrero, E. (1999b). New correlation between charophyte and larger foraminifera biozones (Middle Eocene, Southeastern Pyrenees). *Geobios, 32*(1), 5–18.

Martín-Closas, C., Vicente, A., Pérez-Cano, J., Sanjuan, J., & Bover-Arnal, T. (2018). On the earliest occurrence of *Tolypella* section *Tolypella* in the fossil record and the age of major clades in extant Characeae. *Botany Letters, 165*, 23–33.

Martín-Closas, C., & Wang, Q. (2008). Historical biogeography of the lineage *Atopochara trivolvis* PECK 1941 (Cretaceous Charophyta). *Palaeogeography, Palaeoclimatology, Palaeoecology, 260*, 435–451.

McCourt, R. M., Karol, K. G., Guerlesquin, M., & Feist M. (1996). Phylogeny of extant genera in the family Characeae (Charales, Charophyceae) based on rbcL sequences and morphology. *American Journal of Botany, 83*, 125–131.

Michaux-Ferrière, N., & Soulié-Märsche, I. (1987). The quantities of DNA in the vegetative nuclei of *Chara vulgaris* and *Tolypella glomerata* (Charophyta). *Phycologia, 26*, 435–442.

Mojon, P. O. (1989). Polymorphisme écophénotypique et paléoécologique des Porocharacées (Charophytes) du Crétacé Basal (Berriasien) du Jura Franco Suisse. *Revue de Paléobiologie, 2*, 505–524.

Musacchio, E. A., & Moroni, A. M. (1983). Charophyta y ostracoda no marinos Eoterciarios de la formación El Carrizo en la provincia de Rio Negro, Argentina. *Ameghiniana, 20*, 21–33.

Musacchio, E. A. (1971). Charophytas de la Formación La Amarga (Cretácico inferior), Provincia Neuquén, Argentina. *Revista del Museo de La Plata, 6*, 19–38.

Musacchio, E. A. (2000). Biostratigraphy and biogeography of Cretaceous Charophytes from South America. *Cretaceous Research, 21*, 211–220.

Musacchio, E. A. (2010). Upper Cretaceous *Lychnothamnus*, *Nitella* and *Tolypella* (Charophyta) from Zampal, Argentina. *Cretaceous Research, 31*, 461–472.

Peck, R. E. (1957). North American Charophyta. *Geological Survey Professional Paper 294, A*, 1–44.

Pérez, W., Hall, J. D., McCourt, R. M., & Karol, K. G. (2014). Phylogeny of North American *Tolypella* (Charophyceae, Charophyta) based on plastid DNA sequences with a description of *Tolypella ramosissima* sp. nov. *Journal of Phycology, 50*, 776–789.

Pérez-Cano, J., Bover-Arnal, T., & Martín-Closas, C. (2020). Barremian charophytes from the Maestrat Basin (Iberian chain). *Cretaceous Research, 115*, 104544.

Racki, G. (1982). Ecology of the primitive charophyte algae; a critical review. *Neues Jahrbuch für Geologie und Paläontologie, Abhandlungen, 162*, 388–399.

Riveline, J. (1986). *Les Charophytes du Paléogène et du Miocène Inférieur d'Europe* (Occidentale). Cahiers de Paléontologie. C.N.R.S.

Riveline, J., Berger, J. P., Feist, M., Martín-Closas, C., Schudack, M., & Soulié-Märsche, I. (1996). European Mesozoic-Cenozoic Charophyte Biozonation. *Bulletin de la Société Géologique de France, 167*, 453–468.

Sanjuan, J., & Alqudah, M. (2018). Charophyte flora from Miocene deposits of Zahle (Beeka Valley, Lebanon). Biostratigraphic, Palaeoenvironemental and Palaeobiogeographical implications. *Geodiversitas, 40*(10), 195–209.

Sanjuan, J., & Martín-Closas, C. (2012). Charophyte palaeoecology in the Upper Eocene of the Eastern Ebro basin (Catalonia, Spain). Biostratigraphic implications. *Palaeogeography, Palaeoclimatology, Palaeoecology, 365–366*, 247–262.

Sanjuan, J., Martín-Closas, C., Costa, E., Barberà, X., & Garcés, M. (2014). Calibration of Eocene-Oligocene charophyte biozones in the Eastern Ebro Basin (Catalonia, Spain). *Stratigraphy, 11*(1), 61–81.

Sanjuan, J., & Soulié-Märsche, I. (2020). New charophyte assemblage from middle Miocene lacustrine deposits of Moneva (Ebro Basin, Spain). *Geobios, 79*, 59–90.

Schudack, M. E. (1986). Zur Nomenklatur der Gattungen *Porochara* Mädler (syn. *Musacchiella* Feist & Grambast 1984) und *Feistiella* n. gen. (Charophyta). *Paläontologische Zeitschrift, 60*, 21–27.

Schudack, M. E. (1993). Die Charophyten im Oberjura und Unterkreide Westeuropas. Mit einer phylogenetischer Analyse der Gesamtgruppe. *Berliner Geowissenschaftliche Abhandlungen A, 8*, 1–209.

Schudack, M. E., Turner, C. E., & Peterson, F. (1998). Biostratigraphy, paleoecology and biogeography of charophytes and ostracodes from the Upper Jurassic Morrison Formation, Western interior, USA. *Modern Geology, 22*, 379–414.

Schwarz, J. (1985). Revision der Charophyten-Flora der Süßwasserschichten und des Kalktertiärs im Mainzer Becken (Ober-Oligozän—Unter-Miozän. *Mainzer Geowissenschaftliche Mitteilungen, 14*, 7–98.

Schwarz, J. (1997). Charophyten aus dem Tertiär des Oberrheingrabens (Mitteleozän-Untermiozän). *Palaeontographica. Abt. B, 243*, 1–84.

Soulié-Märsche, I. (1979). Origine et Evolution des genres actuels des Characées. *Bulletin du Centre de Recherches Explor.-Prod. Elf-Aquitaine, 3*(2), 821–831.

Soulié-Märsche, I. (1989). *Étude comparée de gyrogonites de charophytes actuelles et fossiles et phylogénie des genres actuels*. Imprimerie des Tilleuls, Millau.

Soulié-Märsche, I. (2004). Charophyte gyrogonites, the result of enantioselective influence 250 Million years ago. In G. Palyi, C. Zucchi, & Caglioti, (Eds.), *Progress in biological chirality* (Chap. 30, pp. 365–375). Elsevier, Oxford.

Soulié-Märsche, I., Benammi, M., & Gemayel, P. (2002). Biogeography of living and fossil *Nitellopsis* (Charophyta) in relationship to new finds from Morocco. *Journal of Biogeography, 29*, 1703–1711.

Soulié-Märsche, I., Benkaddour, A., Elkhiati, N., Gemayel, P., & Ramdani, M. (2008). Charophytes, indicateurs de paléo-bathymétrie du lac Tigalmamine (Moyen Atlas, Maroc). *Geobios, 41*, 435–444.

Soulié-Märsche, I., & García, A. (2015). Gyrogonites and oospores, complementary viewpoints to improve the study of the charophytes (Charales). *Aquatic Botany, 120*, 7–17.

Souza de Faria, R., Ricardi-Branco, F., & Christiano de Souza, I. (2013) Permian *Leonardosia* organic oospores from Southern Brazil. *Palaeontology, 56*, 797-805.

Souza-Faria, R., & Ricardi-Branco, F. (1999). *Leonardosia langei* Sommer (Charophyta, Porocharaceae) from Corumbataí Formation (Guadalupian), Piracicaba, SP, Brazil: First record of an antheridium and of corticated thalli. *Ameghiniana, 46*, 49–57.

Stehlin, H. G. (1910). Remarques sur les faunules de Mammifères des couches eocènes et oligocenes du Bassin de Paris. *Bulletin de la Société Géologique de France, 4*, 488–520.

Timme, R. E., Bachvaroff, T. R., & Delwiche, C. F. (2012). Broad phylogenetic sampling and the sister lineage of land plants. *PLoS ONE, 7*, e29696.

Tiss, L., Trabelsi, K., Kamoun, F., Soussi, M., Houla, Y., Sames, B., & Martín-Closas, C. (2019). Middle Jurassic charophytes from southern Tunisia: Implications on evolution and paleobiogeography. *Review of Palaeobotany and Palynology, 263*, 65–84.

Trabelsi, K., Sames, B., & Martín-Closas, C. (2024). First occurrence of family Clavatoraceae (fossil Charophyta) in the Middle Jurassic (Bathonian) of France. Papers in Palaeontology, 2024, e1548.
Uliana, M. A., & Musacchio, E. A. (1978). Microfosiles calcareos no-marinos del Cretacico superior en Zampal, Provincia de Mendoza, Argentina. *Ameghiniana, 15*, 111–135.
Vicente, A., Csiki-Sava, Z., & Martín-Closas, C. (2019). European charophyte evolution across the Cretaceous-Paleogene boundary. *Palaeogeography, Palaeoclimatology, Palaeoecology, 533*, 109244.
Vicente, A., Expósito, M., Sanjuan, J., & Martín-Closas, C. (2016). Small sized charophyte gyrogonites in the Maastrichtian of Coll de Nargó, Eastern Pyrenees: An adaptation to temporary floodplain ponds. *Cretaceous Research, 57*, 443–456.
Vicente, A., & Martín-Closas, C. (2013). Lower Cretaceous charophytes from the Serranía de Cuenca, Iberian chain: Taxonomy, biostratigraphy and palaeoecology. *Cretaceous Research, 40*, 227–242.
Vicente, A., Sanjuan, J., Eaton, J., & Villanueva-Amadoz, U. (2020). Oldest report of North American *Lychnothamnus* and their implications on its ecological evolution and regional distribution. *Aquatic Botany, 167*, 103271. https://doi.org/10.1016/j.aquabot.2020.103271
Villalba-Breva, S., & Martín-Closas, C. (2011). A characean thallus with attached gyrogonites and associated fossil charophytes from the Maastrichtian of the Eastern Pyrenees (Catalonia, Spain). *Journal of Phycology, 47*, 131–143.
Villalba-Breva, S., & Martín-Closas, C. (2013). Upper Cretaceous palaeogeography of the Central Southern Pyrenean Basins (Catalonia, Spain) from microfacies analysis and charophyte biostratigraphy. *Facies, 59*, 319–345.
Zachos, J., Pagani, M., Sloan, L., Thomas, E., & Billups, K. (2001). Trends, rhythms, and aberrations in global climate 65 Million years ago to present. *Science, 292*(5517), 686–693.

Chapter 9
Provisional Red List of European Charophytes

Nick Stewart

9.1 Introduction

In most countries charophyte conservation and protection has relied heavily on the conservation actions undertaken for other wildlife groups, including for birds and for general aquatic habitats. Because sites of interest for charophytes often have a shared interest with other groups, many sites have received some form of protection by national legislative frameworks or international treaties (although all too often this protection is nominal and damage continues to occur).

However, actions directed at the protection of individual charophyte species have been more limited. Different countries have approached this in various ways, including legal protection of certain species, drawing up of Red Lists and, in a few cases, the development of action programmes directed at the conservation of individual species.

A commonly used method for assessing priorities in species conservation is by drawing up a Red List, aimed at identifying the species most at risk of decline and extinction. Many European countries have drawn up such lists (see Fig. 9.1 and Table 9.1).

In most cases these have followed the standard format developed by the International Union for the Conservation of Nature (IUCN) (2012) but a few countries have developed their own parallel systems (e.g. Germany, Czech Republic, Russian Federation, Ukraine). This chapter is an attempt to produce an overall assessment for European charophytes based on the IUCN system. Although such an assessment has been under consideration for many years, the information gathered together for this book and by individual countries as part of Red Listing and other conservation-priority processes has made it possible to make an authoritative assessment of all but a few European charophytes. This has been considerably aided by communication with a number of experts with a knowledge of particular species or geographic areas.

N. Stewart (✉)
4 Hexton Road, Glastonbury BA6 8HL, UK
e-mail: nfstewart@freeuk.com

H. Schubert et al. (eds.), *Charophytes of Europe*,
https://doi.org/10.1007/978-3-031-31898-6_9

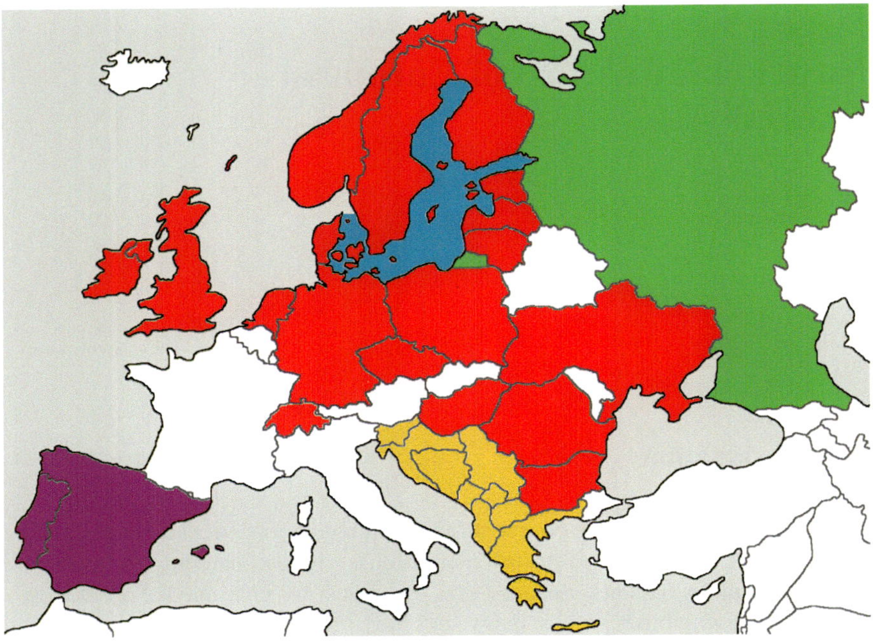

Fig. 9.1 Map showing European countries and regions with Red Lists. Blue: regional Red List (only brackish-water tolerant species of the Baltic Sea coast are considered); Orange and purple: part of regional Red Lists; Green: Red List also covers non-European parts, but regional lists exist in addition; Red: national Red List. For data sources see Table 9.1

Table 9.1 National and regional Red Lists analysed

Country/region	Authors
Albania	Blaženčić et al. (2006)
Baltic Sea	Helsinki Commission (HELCOM) (2013)
Bosnia Herzegovina	Blaženčić et al. (2006)
Britain	Joint Nature Conservation Committee (2004)
Bulgaria	Temniskova et al. (2008)
Croatia	Blaženčić et al. (2006)
Czech Republic	Caisová and Gąbka (2009)
Denmark	Baastrup-Spohr et al. (2013)
Estonia	Nature Conservation Committee (1998)
Finland	Hyvärinen et al. (2019)
Germany	Korsch et al. (2013)
Greece	Blaženčić et al. (2006)

(continued)

Table 9.1 (continued)

Country/region	Authors
Hungary	Németh (2005)
Ireland	Stewart and Church (1992)
Latvia	Ministru kabineta (2000)
Lithuania	Lietuvos Respublikos Aplinkos Ministerija (2021)
Montenegro	Blaženčić et al. (2006)
Netherlands	van Raam (2000)
Norway (excluding Svalbard)	Norwegian Biodiversity Information Centre (2021)
North Macedonia	Blaženčić et al. (2006)
Poland	Siemińska et al. (2006)
Portugal	Cirjuano et al. (2007)
Romania	Blaženčić et al. (2006)
Russia	Trutnev and Kamelin (2008)[a]
Serbia & Kosova	Blaženčić et al. (2006)
Slovenia	Blaženčić et al. (2006)
Spain	Cirjuano et al. (2007)
Switzerland	Auderset Joye and Schwarzer (2012)
Sweden	Artdatabanken (2020)
Ukraine	Palamar-Mordvintseva and Tsarenko (2004)

[a]Given the huge area covered, regional lists for the European part of Russia are more applicable (an overview of these can be found at e.g.: http://www.biodat.ru/db/rb/reg.htm)

9.2 The IUCN System

The IUCN system uses a number of categories to describe different degrees of threat. These are listed in Table 9.2, together with their standard abbreviations.

Table 9.2 Red list categories and abbreviations

Extinct		EX
Extinct in the wild		EW
Regionally extinct		RE
Critically endangered	} Threatened	CR
Endangered	} Threatened	EN
Vulnerable	} Threatened	VU
Near threatened		NT
Least concern		LC
Data deficient		DD
Not evaluated		NE

Each of the threatened categories have quantitative criteria that are used in the assessment and these are summarised in Fig. 9.2. It should be noted that these categories are designed for all types of organisms, including animals and plants and, as discussed further below, only a few have relevance to charophytes. Guidelines on how the criteria should be interpreted, including definitions of the various terms are set out at some length in IUCN (2022). The following notes discuss how these relate to charophyte ecology and state of knowledge and, in particular, how the criteria have been interpreted in this review.

9.3 Some Definitions

Several terms are used in the IUCN system and it is worth summarising these as they are sometimes employed differently from normal usage.

Population: The sum total of all plants of a species at all sites.

Subpopulation: "Geographically or otherwise distinct groups in the population between which there is little demographic or genetic exchange (typically one successful migrant individual or gamete per year or less)" (IUCN, 2022). With charophytes, it is usually difficult to judge how isolated the parts of the total population are. The potential for long distance transport by birds is fairly high as oospores can survive passing through a bird's gut and remain viable (Charalambidou and Santamaria, 2005; Proctor, 1962). An extreme example is the occurrence of *Chara aspera* and *Chara canescens* on Svalbard some 900 km from the nearest other locations in northern Norway but connected by a regular goose migratory route (Langangan et al., 2019). Another example is *Chara canescens* on the east coast of Britain where a number of colonies have appeared briefly and then disappeared because of habitat instability. Here it is suspected that oospores are being brought in at intervals by birds from the Baltic Sea. The frequency of actual oospore transport is unknown but records suggest that successful colonisations occur here only at intervals averaging more than a decade.

Shorter distance transport of a few kilometres probably occurs more often, as for example with the rapid colonisation of mineral extraction sites soon after the workings have been abandoned. Although birds are still likely to be the most frequent vector, transport by animals and humans are also known to play a part. However, it is difficult to assess how frequently and over what distances such transport occurs on anything more than an occasional basis. Generally in this assessment, colonies in separate water bodies more than 10 km apart are considered to be separate subpopulations, unless there is some indication or evidence of genetic exchange.

Location: "A geographically or ecologically distinct area in which a single threatening event can rapidly affect all individuals of the taxon present" (IUCN, 2022). In general this has been interpreted as a single water body, except:

A. Population size reduction. Population reduction (measured over the longer of 10 years or 3 generations) based on any of A1 to A4	Critically Endangered	Endangered	Vulnerable
A1	≥ 90%	≥ 70%	≥ 50%
A2, A3 & A4	≥ 80%	≥ 50%	≥ 30%

A1 Population reduction observed, estimated, inferred, or suspected in the past where the causes of the reduction are clearly reversible AND understood AND have ceased.

A2 Population reduction observed, estimated, inferred, or suspected in the past where the causes of reduction may not have ceased OR may not be understood OR may not be reversible.

A3 Population reduction projected, inferred or suspected to be met in the future (up to a maximum of 100 years) *[(a) cannot be used for A3]*.

A4 An observed, estimated, inferred, projected or suspected population reduction where the time period must include both the past and the future (up to a max. of 100 years in future), and where the causes of reduction may not have ceased OR may not be understood OR may not be reversible.

based on any of the following:

(a) direct observation *[except A3]*

(b) an index of abundance appropriate to the taxon

(c) a decline in area of occupancy (AOO), extent of occurrence (EOO) and/or habitat quality

(d) actual or potential levels of exploitation

(e) effects of introduced taxa, hybridization, pathogens, pollutants, competitors or parasites.

B. Geographic range in the form of either B1 (extent of occurrence) AND/OR B2 (area of occupancy)	Critically Endangered	Endangered	Vulnerable
B1. Extent of occurrence (EOO)	< 100 km²	< 5,000 km²	< 20,000 km²
B2. Area of occupancy (AOO)	< 10 km²	< 500 km²	< 2,000 km²
AND at least 2 of the following 3 conditions:			
(a) Severely fragmented OR Number of locations	= 1	≤ 5	≤ 10
(b) Continuing decline observed, estimated, inferred or projected in any of: (i) extent of occurrence; (ii) area of occupancy; (iii) area, extent and/or quality of habitat; (iv) number of locations or subpopulations; (v) number of mature individuals			
(c) Extreme fluctuations in any of: (i) extent of occurrence; (ii) area of occupancy; (iii) number of locations or subpopulations; (iv) number of mature individuals			

C. Small population size and decline	Critically Endangered	Endangered	Vulnerable
Number of mature individuals	< 250	< 2,500	< 10,000
AND at least one of C1 or C2			
C1. An observed, estimated or projected continuing decline of at least (up to a max. of 100 years in future):	25% in 3 years or 1 generation (whichever is longer)	20% in 5 years or 2 generations (whichever is longer)	10% in 10 years or 3 generations (whichever is longer)
C2. An observed, estimated, projected or inferred continuing decline AND at least 1 of the following 3 conditions:			
(a) (i) Number of mature individuals in each subpopulation	≤ 50	≤ 250	≤ 1,000
(ii) % of mature individuals in one subpopulation =	90–100%	95–100%	100%
(b) Extreme fluctuations in the number of mature individuals			

D. Very small or restricted population	Critically Endangered	Endangered	Vulnerable
D. Number of mature individuals	< 50	< 250	D1. < 1,000
D2. *Only applies to the VU category* Restricted area of occupancy or number of locations with a plausible future threat that could drive the taxon to CR or EX in a very short time.	-	-	D2. typically: AOO < 20 km² or number of locations ≤ 5

E. Quantitative Analysis	Critically Endangered	Endangered	Vulnerable
Indicating the probability of extinction in the wild to be:	≥ 50% in 10 years or 3 generations, whichever is longer (100 years max.)	≥ 20% in 20 years or 5 generations, whichever is longer (100 years max.)	≥ 10% in 100 years

Fig. 9.2 Summary of IUCN Red List criteria reproduced from IUCN (2012)

– where water bodies are interconnected such as by a river, canal or ditch system. Such water bodies are at risk from a single pollution event or ongoing chronic pollution and are therefore considered a single location.

– where water bodies fall within a single management unit, such as a farm complex, nature reserve etc. Here a change of management regime could potentially affect the whole complex of water bodies even though they may not be hydrologically connected. These are therefore considered a single location.

- conversely, in very large water bodies such as the Baltic Sea and the Black Sea, judgement has been used over the extent that a threat event relevant to the species is likely to have an impact. For example, the entire Gulf of Bothnia could be interpreted as a single location because of changes to the vegetation of the east coast due to rising land levels, if this was affecting a particular species in an interconnected way. Nutrient enrichment is also an issue affecting substantial parts of the Gulf of Bothnia but this comes from multiple sources and the impacts of the different sources are separable, even if their cumulative impact is significant. This differs from more enclosed land-locked lakes such as Lake Constance where flushing through of nutrients is constrained and therefore this would be regarded as a single location.

Climate change and its consequences could affect otherwise isolated water bodies over potentially huge areas. For example, rises in sea level could affect the salinity balance of numerous coastal sites for species requiring a particular range of brackish water. However, we have not used this to justify that they should all be regarded as a single location as this does not seem to be the intent set out in IUCN (2022).

In many cases "locations" and "subpopulations" will be the same, for example where hydrological connection is also linked to genetic connection.

Extent of Occurrence: Essentially the range of a species (in this case within Europe), including areas of unsuitable habitat such as sea. It is calculated by drawing a polygon around the known sites, with the polygon not including any concave sections.

Area of Occupancy: The sum of the area occupied by colonies of the species. In the IUCN calculation of this feature, there is a minimum area corresponding to each location of 4 km^2. This applies even if the site is a single pond and the rest of the surrounding 4 km^2 is unsuitable habitat. This is because the thresholds for this feature in Criterion B (<10 km^2 for CR, <500 km^2 for EN, <2000 km^2 for VU) would not be surpassed by hundreds of small ponds if the actual area of occupied ponds was used. However, using the 4 km^2 minimum, this means that, for a species of small water bodies, the thresholds mentioned above would equate to 2 locations for CR, 125 locations for EN and 500 locations for VU. These are the maximum numbers of locations and for species of larger water bodies the thresholds could correspond to considerably fewer locations if many sites are larger than 4 km^2.

Generation: Generations are used in several criteria in the context of setting a time period for quantified declines. In all cases these are defined as declines of X% within a period of "10 years or three generations, whichever is the longer" (IUCN, 2012). Generation length is defined as "the average age of parents of the current cohort (i.e. newborn individuals in the population)". This is problematic for perennial charophytes which, in favourable conditions, can grow continuously at the stem tips while senescing and decaying at the base of the beds. This means that in stable systems, plants within the beds may be of indefinite age, perhaps decades or even centuries. On the other hand, all species can reach full height and sexual maturity within a year and in optimal growing conditions *Chara* beds can recover within a year or two. It therefore seems best to use the 10 year time period as the basis for assessing quantified declines.

Nevertheless, this approach differs from that taken by bryologists (Bergamini et al., 2019) who have adopted a 4-tier approach;

- short-lived species (ephemeral colonists, fugitives, annual shuttles): generation length = 1–5 years; 3 generations = 10 years
- medium-lived species (colonists, short-lived shuttles): generation length = 6–10 years; 3 generations = 20 years
- long-lived species (perennial stayers, long-lived shuttles, dominants): generation length = 11–25 years; 3 generations = 50 years
- species that are never or only very rarely found with sporophytes: generation length = 33 years: 3 generations = 100 years.

In this classification, a few charophytes may fall within the short-lived category but many could qualify as medium-lived and the more perennial dominants (e.g. species in the *Hartmania* group) as long-lived species. However, most aquatic vascular plant groups have been assessed against the 10 year time span, even though many have the capacity for infinite perennation by vegetative spread and/or fragmentation (Richard Lansdown pers. comm.). It follows from this, that Red List statuses also need to be reviewed at least every ten years.

Severely fragmented distributions: "The phrase 'severely fragmented' refers to the situation in which increased extinction risks to the taxon results from the fact that most of its individuals are found in small and relatively isolated subpopulations" (IUCN, 2022). This is used in sub-criteria B1a and B2a as one of the risk factors to be taken into consideration and, in order to qualify under these subcriteria, more than half of the total population needs to be in small and isolated patches.

In the discussion around this subject in the IUCN guidelines (IUCN, 2022), there are two themes to consider; inherent viability of small populations and the ability to recolonise after a catastrophic event, whether natural or human-generated. There is little evidence of the viability of small populations being an issue for charophytes in terms of inbreeding and reduced genetic diversity. However, particularly for species of small water bodies, there is often significant isolation between locations. The potential for long-distance transport by birds is discussed above under the definition of subpopulations. However, many species seem to have rather low mobility, preferring a "sit-and-wait" strategy, waiting for suitable habitat conditions to come to them rather than moving rapidly to new locations, even when lots of suitable habitat appears to be available. An extreme example of this can be seen at Otmoor, Oxfordshire, England where a group of arable fields were bought by a bird conservation charity (RSPB) who raised the water table and dug numerous scrapes and ditches. *Sphaerochara intricata* appeared in three different places, all where an old palaeochannel ran through the fields and it has continued to show no inclination to move into the newly created habitats. It is probable that the oospores successfully survived over 10 years of arable cultivation waiting for suitable conditions to return.

Bryologists, who have similar issues of uncertainty over demographic isolation have adopted a view that a distance of between 100 and 1000 km (depending on the species) represents severe fragmentation (Hallingbäck et al., 2000). A similar approach has been adopted here.

In this assessment we have only used the severely fragmented subcriterion where the fragmentation has been the result of or accelerated by human activities. Several species have scattered distributions due to habitat rarity. For example, *Chara dissoluta* is a species particularly occurring in deep water in highly calcareous lakes and has a disjunct distribution in several parts of Europe. There is probably little demographic movement between subpopulations, even within distribution areas, particularly as there is little opportunity for bird-grazing of a deep water species with no inherent buoyancy. However, prior to modern pollution impacts the subpopulations of this species probably remained stable and viable in these sites for many centuries or, one would guess, millennia. Such examples would not qualify for "severely fragmented" in this assessment.

This sub-criterion has been invoked here in two particular situations;

- to species of small water bodies such as ponds and ditches in urban or agricultural settings. These are habitats that have declined in number and quality, particularly as farming methods have become more intensive and human population pressures have increased in recent decades and these processes are continuing. Often such species are reliant on continuing management and are therefore vulnerable to unpredictable changes. Examples include *Chara baueri, Sphaerochara intricata* and *Sphaerochara prolifera.*
- to species of brackish water along the Mediterranean and southern European parts of the Atlantic coasts. These are habitats that are particularly under pressure from urban, industrial and tourist development which have resulted in significant habitat fragmentation and deterioration. Examples include *Chara corfuensis*, *Tolypella salina* and *Lamprothamnium sardoum.*

Extreme fluctuations: "For the 'extreme fluctuations' sub-criterion to be invoked, populations would need to fluctuate by at least tenfold." (IUCN, 2022). Where there is some dispersal of even small amounts of oospores or plant fragments between subpopulations then the fluctuation is calculated on the total population within the connected subpopulations. However, if the subpopulations are totally isolated, then the fluctuation applies to individual subpopulations and the subcriterion applies if most subpopulations have fluctuations of an order of magnitude (IUCN, 2022). This distinction is important as there may be individual sites within a subpopulation which have extreme fluctuations but if these are asynchronous then the overall fluctuation in the total subpopulation may be much less.

Each species needs to be considered on a case by case basis but for a number of the rarer charophytes the degree of isolation between subpopulations seems to be high, particularly for species usually found in small water bodies and consequently with small populations.

The IUCN guidelines (IUCN, 2022) indicate that "assessors must be reasonably certain that fluctuations in the number of mature individuals represent changes in the total population, rather than simply flux of individuals between different life stages." Some charophytes are be able to survive periods when the habitat is in an unsuitable successional state and there are examples where a species appears from time to time in a location for a year or two following periodic clearances of, for example, a pond or ditch. Because of the long durability of charophyte oospores, such management interventions can be intermittent and irregular events with gaps of several decades. This could be interpreted as analogous to the example of where this subcriterion should not be applied, given in IUCN (2022) of terrestrial plants that require fire episodes to instigate germination of new cohorts of the plant. However, there are a number of key differences which justify the use of this subcriterion for charophytes;

- for the species under consideration, the disturbance events are nowadays rarely natural and direct human-generated management intervention is normally needed.
- although germination of oospores is known to be possible after several decades of dormancy, rather little is known about factors that favour long-dormancy and whether all oospores are capable of this.
- knowledge of the factors that favour germination of the dormant oospores are also often poorly known. Management when conditions are not optimal could deplete the oospore resource without recruitment. For example, recent experimental management of *Nitella capillaris* in Suffolk, England suggests that pond clearance in early autumn has a much better possibility of success than the same clearance in late autumn (Hawkins, 2019; J. Hawkins pers. comm.).

9.4 Assessment of Extinction and Site Loss

Most, if not all, charophytes have at least some oospores that can remain viable for decades. This is based on numerous examples of sites where plants have appeared after pond clearance or other disturbance after periods of many years when the successional state has been unsuitable for the growth of living plants. In the case of rare species there are often no nearby colonies from which new colonisations could have occurred. Experiments with oospores taken from levels in peat cores over a millennium old have also resulted in some successful germination (Carl Sayer pers. comm.).

This long oospore viability can make it difficult to judge conclusively that a colony has been lost or is merely dormant while habitat conditions remain unsuitable. The situation is complicated by the rather limited knowledge of what factors favour oospore survival and whether there is a gradual viability depletion rate over time, whether only a proportion of oospores have long term viability from the beginning and what conditions favour germination of long-dormant spores. This creates a "Schrödinger's Stonewort" situation whereby the only way of proving that a species is extinct or potentially present is to restore the water body to a suitable state or to sample the sediment to hunt for oospores. Since in most cases the exact location of historic records is not known in sufficient detail to test this, the status of the species can often remain with some doubt.

This creates an anomaly in the IUCN criteria which are meant to assess the risk of absolute extinction but are designed around the assessment of functional extinction (i.e. when no growing plants are visible). This, of course, is not a problem that is unique to charophytes; many plants can build up a reserve of viable diaspores in the soil which only germinate in particular conditions. However, the potential for long-term survival of oospore viability makes the situation more pronounced in charophytes as a group. Nevertheless, the only practical way of assessing the status of species remains through the assessment of functional extinction/site loss.

In this review we have assessed species on a case by case basis, taking into account that some regions are less well studied but also taking into account our knowledge from areas that are much better studied. We have also taken note of information from other sources which indicate deterioration of the habitat at particular sites (such as increased nutrient levels). In general terms, sites not seen for 20 years are considered to have been lost with greater leeway given in regions where coverage of charophyte survey is limited. Greater leeway has also been given where a species has not been seen at its last known site(s) and there is a possibility that it could still be present. Here the preference is to hold the species in the threatened category "Critically Endangered" to encourage further survey and/or management, rather than declaring it "Regionally Extinct".

Where the species has been recorded in locations within the last 20 years but not refound in subsequent survey(s), a judgement has been made whether this has been due to temporary lack of suitable habitat conditions or because the habitat has deteriorated significantly. In the latter situation, the location is assumed to have been lost.

9.5 Assessment of Population Sizes

Several IUCN criteria relate to numbers of individuals which are assessed in terms of mature individuals. In most cases this is impossible to assess. It is very common for charophytes to occur in stands which may extend for a few square metres up to hundreds, or even thousands, of square metres. Even in smaller patches it is rarely possible to discern how many stem branches originate from individual rhizoid bases. For more perennial stands, vegetative spread is also common with shoots falling over, particularly in periods of low active growth such as the winter, and then new shoots developing from the lower stem nodes. In the deeper water of lakes there is also the problem of actually inspecting the extent of populations, particularly when visibility is reduced by turbidity or humic colouring. Even with aid of snorkelling, diving or underwater drones, the sites where subpopulation numbers may be under pressure are also the ones where turbidity is commonly an issue.

Nevertheless, there are some cases in small water bodies, such as ponds and ditches, where plants occur as countable tufts identifiable as individual plants. However, subpopulations of such species also tend to have more volatile numbers due to successional changes or the occurrence of disturbance events. So in some years there may be solid patches of plants which are difficult to enumerate. This problem of differentiation of individuals is not, of course, unique to charophytes. Bryologists, for example, have suggested that patches of up to 1 square metre should be assessed as individuals (Bergamini et al., 2019). A similar approach could be applied to charophytes; it is suggested that an individual be equated to patches with a radius equal to the average height of plants in that subpopulation (i.e. the area in which a single stem could fall over into) rather than an absolute area for all species.

However, the issues over the definition of individuals are probably unimportant. The thresholds for numbers of individuals are such that any charophyte species likely to qualify as threatened on population size in criteria C and D1 is much more likely to qualify at a higher level of threat under other criteria, principally criterion B.

Use of numbers of mature individuals are more likely to be possible when relative change is being considered as in criterion Bbv "Continuing decline … in number of mature individuals" and Bciv "Extreme fluctuations … in number of mature individuals". Here it is possible to use other measures without knowing the absolute numbers involved. These include population extent, population density or frequency of presence in a grid based matrix.

9.6 Overview of the Use of the IUCN Criteria for Charophytes

Although there are five IUCN criteria used to assess the Red List status of species, most of these are impossible to use for European charophytes. Although many charophyte species are known to be significantly declining, the level of detail required to provide quantitative evidence is normally lacking. Our knowledge of the distribution of European charophytes has advanced very considerably in recent decades. Even in regions that are less well studied, there is often sufficient information from neighbouring regions to be able to extrapolate that similar trends are occurring in those areas. However, the distribution of charologists even in more studied areas are generally too sparse to keep an eye on the state of all of the sites on the 10 year basis required by the criteria (see paragraph 9.3 "Generations") even for the rarer species. This means that it is necessary to pool data from a longer period (20 or 30 years) and compare with data beyond this (e.g. comparing numbers of locations confirmed since the year 2000 with those recorded last recorded between 1980 and 2000). However, many changes have occurred, and often accelerated, within these time periods which will not therefore show in this pooled data.

The historical information is also extremely patchy. Some countries have quite good historic data but for many the historical record is sparse or only locally more detailed. Even in the countries with better historic data, there are often gaps in the historic data due to the intermittent availability of charologists.

As a result we have not been able to show quantified declines greater than 30% in 10 years (or extrapolated equivalent) required to qualify as threatened under criterion

A, except in species that are so rare that they qualify at a higher level under other criteria. As discussed above (see paragraph 9.5), criteria related to numbers of mature individuals (C and D1) are also more or less impossible to apply except again for species that are so rare that they qualify at a higher level under other criteria. Criterion E is based on statistical analysis which relies on data not available for charophytes.

This means that the assessment of the threat statuses of charophytes relies, rather unsatisfactorily, almost entirely on criteria B and D2. i.e.

Restricted Extent of Occurrence (B1) or Area of Occupation (B2) combined with two out of three by:

(a) few locations or severely fragmented distribution
(b) continuing decline
(c) extreme fluctuations.

or
Five or fewer locations and with a plausible threat (D2).

Even amongst these the often widely dispersed but rare distributions of many charophytes means that criterion B1 is rarely applicable.

The need to rely on just these criteria has, we believe, resulted in taxa being assigned to Near Threatened or even Least Concern because a higher level of threat could not be justified based on the available data, even when we are certain that there is a problem.

9.7 Provisional Red List Assessment

9.7.1 Widespread Species Considered not Threatened in Many or Most Countries

The following are widespread species that are frequent in many countries, although they may be regarded as threatened in a few. They are therefore classified as Least Concern (LC):

Chara aspera
Chara contraria
Chara globularis
Chara virgata
Chara vulgaris
Nitella flexilis
Nitella opaca.

9.7.2 Species Threatened in Many Areas but with Some Remaining Strongholds (Least Concern)

Table 9.3 provides a commented overview about species considered threatened in many countries but still with sufficient populations in some countries to be secure for the present (Least Concern). If the situation changes in these strongholds, this should prompt a re-evaluation of the status.

Table 9.3 Widespread species considered to be of "Least Concern"

Taxon	Comments
Chara aculeolata	LC at least in Ireland, Britain, Iberia, Estonia and Lithuania
Chara baltica	Bulk of the population in the Baltic Sea where LC. Probably European near-endemic, with otherwise only some sites in Greenland
Chara braunii	Threatened in many northern countries, although stable in Sweden (NT) and possibly increasing in Belgium (L. Denys pers. comm.). But LC in several southern countries. Nevertheless, there have been strong population reductions in the Alpine region, Italy, parts of France and the Balkans (R. Becker pers. comm.)
Chara canescens	LC in the Baltic Sea and Iberia. This species mostly reproduces asexually by parthenogenesis but there are a few sites where it reproduces sexually. We have not assessed asexual and sexual populations separately but further research is needed to understand the interaction between the two reproductive types. Sexual populations, if assessed on their own, would qualify for a threat status based on few sites and an ongoing decline (B2ab(ii, iii, iv))
Chara connivens	Fairly frequent in Spain, Portugal and parts of France. Also LC in the Baltic Sea
Chara curta	Fairly frequent in Ireland and scattered in Britain. Also possibly overlooked in some other countries because of inclusion in *Chara aspera*
Chara filiformis	Stronghold south-east of the Baltic Sea from Germany to the Baltic States where still locally quite widespread and abundant. European endemic
Chara fragifera	Still locally frequent in Iberia (LC) and western France
Chara galioides	Locally frequent along the western and central Mediterranean coast and inland in Spain (where LC)
Chara gymnophylla	Strongholds in Spain (where LC), Italy and Greece. Nevertheless, there have been strong declines in Italy (where there has been confusion with *Chara squamosa*), Czechia and elsewhere (R.Becker pers. comm.)
Chara hispida	Threatened in many eastern countries but frequent in the west (e.g. LC in Spain, Netherlands, Ireland, Britain & Sweden)
Chara papillosa	This species is listed as at least NT in most countries with Red Lists. However, there appear to be strongholds in the regions south of the Baltic Sea and around the Alps
Chara strigosa	Stronghold in Fennoscandia, Russia and Lithuania where NT or LC

(continued)

Table 9.3 (continued)

Taxon	Comments
Chara subspinosa	Good populations in Ireland, the Baltic States and Russia but at least NT in other available Red Lists
Chara tomentosa	LC in the Baltic and surrounding countries
Nitella capillaris	Threatened in many countries but LC in Netherlands and also still a number of sites in northern France
Nitella confervacea	Some reasonable populations in France, Ireland and the Rhine valley (although there are strong pressures in the latter and it is Red Listed in Germany). Also NT in several other countries (Britain, Sweden, Finland) and possibly under-recorded due to its size
Nitella gracilis	Reasonable numbers of localities in France, Poland and NT in southern Sweden
Nitella hyalina	Although threatened in many countries, it has a stronghold in France where it seems to be fairly secure (I. Soulié-Märche pers. comm.). It also seems to be secure in the Netherlands, northern Germany and in parts of the Mediterranean region (R.Becker pers. comm.). It has a broad ecological range, including some tolerance of enrichment and may benefit from climate change (K. van de Weyer pers. comm.)
Nitella mucronata	LC and increasing in Ireland, Britain, Netherlands and Germany. Also recently found to be more frequent in Sweden than previously thought (I. Blindow pers. comm.)
Nitella syncarpa	This species is still locally frequent in its stronghold in central Europe. However, it is declining and is Red Listed in most countries with Red Lists, even within its stronghold area
Nitella tenuissima	Listed as threatened in all available Red Lists, including being Regionally Extinct in several countries (Belgium, Sweden, and Latvia). However, still fairly frequent in France (I. Soulié-Märche pers. comm.) and Spain (R. Becker pers. comm.) (although classified as VU in the latter)
Nitella translucens	Reasonable populations in the Atlantic fringe (Ireland, Britain, France, Spain, Portugal) and in the region from Belgium to Southern Sweden
Nitella wahlbergiana	Although quite restricted in distribution, it is LC in Sweden, Finland and Russia and threats are thought to be low
Nitellopsis obtusa	Fairly frequent in northern central Europe; LC in Germany, Netherlands, Lithuania and Estonia and NT in several others. Appears to be increasing in some areas
Sphaerochara canadensis	A rare species, sparsely distributed in arctic regions but with rather limited threats and little indication of decline. NT in Norway, Sweden and Finland
Tolypella glomerata	LC in Ireland, Britain, Germany, Spain and locally frequent in France and Belgium
Tolypella hispanica	Quite restricted distribution and vulnerable habitat but quite a few localities in Iberia (LC), southern France and Sardinia. Outside Europe known from Siberia, Iran and Pakistan
Tolypella nidifica	Frequent in the Baltic Sea (LC) and there are several records in Spain and Sardinia

9.7.3 Short List (Including Candidates for NT)

Chara affinis **VU (European and Globally)** [B2ab(ii, iii, iv)]

Rationale: This is a rare taxon with only 10 known sites which has only here been raised to species level. It has been known (at varietal level) in Britain and the Channel Islands (Iles Anglo-Normande) for over a century and here it has been lost from five of its seven sites, two within the last twenty years. Elsewhere its distribution is less well studied and this assessment is therefore based on extrapolation from the situation in Britain. It has been assessed as Vulnerable due to limited Area of Occupancy, fewer than 10 locations and continuing decline. Although it is currently confirmed from only 5 sites (which would qualify it for Endangered) there are other former sites in Belgium and the Netherlands which may be this species. The latter records have previously been overlooked and they suggest that further survey is needed in these areas.

Chara altaica **EN** [B1&B2ab(iii)]

Rationale: Known currently from 3 locations in Europe (Russia), where it is being impacted by the effects of nutrient enrichment and long-lasting drought. The European sites are outliers from its centre of distribution in southern Siberia and central Asia but the European subpopulations are substantially distant from its nearest other locations.

Chara baueri **EN (European and Globally)** [B2ab(iii, iv)]

Rationale: This is a well-studied species and, although the number of known sites has increased in recent years, these are likely to be pre-existing sites that have been overlooked. It is currently known from 14 small ponds in 7 widely scattered locations in central and eastern Europe. Most sites are unprotected and it was lost from another site with three ponds through drainage in 2009/2010. Further sites have been lost historically due to urban spread, drainage and agricultural improvement. AOO = 28 km^2. Recent decline 18% in the number of subpopulations. Outside Europe it is known only from a single site in Kazakhstan.

Chara corfuensis **EN (European and Globally)** [B2ab(iii, iv)]

Rationale: This species is known from 9 sites in the Balkan countries. It has been lost from at least two other sites, one of these since 2004 and there are ongoing threats from nutrient enrichment at others. There is a good subpopulation at only one location and all of the other subpopulations are small. Its main habitat (brackish waters on the Mediterranean coast) is under continued pressure from urban, industrial and tourist developments and this has resulted in increasing isolation of suitable habitat. AOO = 36 km^2 with severely fragmented habitat and few locations. Recent decline 10% in number of locations. This species is endemic to Europe.

Chara dissoluta **NT (European and Globally)** [B2ab(iii, iv)]

Rationale: This species is listed on the Red Lists for Ireland (Rare under the old IUCN criteria), Denmark (EN) and the Balkans (CR) but hesitation over the distinction of this as a species and its taxonomic limits have resulted in it being placed as DD or not assessed by other countries with Red Lists. It appears to be extant in around 28 locations but has been lost from around 17 other sites throughout its range. There have also been declines in extent in several large sites where it is locally still present. However, it is not clear over what time period these losses have occurred and the data is insufficient to calculate the recent rate of decline. Nevertheless it has an AOO of <20,000 km^2 (exact AOO is difficult to calculate because of the large size of some sites), a continuing decline and a disjunct distribution. However, the latter seems to be a natural phenomenon, rather than human-generated and therefore does not the fit the subcriterion relating to severely fragmented distribution (see discussion in paragraph 9.3). Possibly endemic to Europe as records from south Asia and north Africa are in doubt. *Chara dissoluta* var. *ohridana* is believed to be the only charophyte to have appeared on a postage stamp (Fig. 9.3).

Fig. 9.3 North Macedonian postage stamp

Chara filarszkyana **EX (European and Globally)**

Rationale: Only known from hot springs in the Budapest area in Hungary where it was last seen in 1926. The habitat has been much modified since that time and is almost certainly no longer suitable. European endemic.

Chara globata **NT** [B2ab(iii)]

Rationale: Currently known from around 20 sites in the Black Sea to Caspian Sea region. There is insufficient baseline information to know if and how much this species is declining but there are known losses outside Europe as well as one historic European site which has not been relocated. Although most sites are larger water bodies with some stability, it is threatened by alterations to hydrological regimes, including excessive water abstraction. As a species of arid and semi-arid climates this is likely to become an increasing concern with climate change.

Chara hellenica **CR (European and Globally)** [B1&2ab(iii)]

Rationale: This species is only known from a single site in Corfu (Greece) which has been badly damaged by nutrient enrichment. It was last seen in 2008 and may already be extinct. European endemic.

Chara horrida **NT (European and Globally)** [A3c]

Rationale: This species is restricted to the Baltic Sea and is classified as NT in the Baltic Red List. It seems to be increasing in the central part of the Baltic Sea but it has severely declined in the southern Baltic. If the latter decline continues it will result in a considerable reduction in the EOO, enough for it qualify as VU. European endemic.

Chara imperfecta **NT (European and Globally)** [B2ab(ii, iii, iv)].

A rare species with fewer than 40 known sites globally. It has been lost from several sites in western France although the records here are not recent. Its main stronghold is in Spain where recent information suggests that subpopulations are fairly secure (I. Soulié-Märche pers. comm.). Nevertheless, it is a very restricted species and is listed as VU in Iberia. Outside Europe it is known only from a few sites in Morocco and Algeria.

Chara kirghisorum **VU** [B2ab(iii)]

Rationale: This is a rare Eurasian species with fewer than 20 sites in the world. There are 8 recent records from the European part of its range but it has been lost from another site. Sites are being impacted by nutrient enrichment causing an ongoing decline in water quality. AOO < 50 km^2.

Chara neglecta **NT** [D2]

Rationale: A rare Eurasian species restricted to Ukraine, southern Russia and outside Europe in Azerbaijan and south-western Siberia and with a few mostly unconfirmed records elsewhere in Asia (R. Romanov pers. comm.). In total, fewer than 25 sites are known in its total range. There have been several sites recently discovered in Europe, all of them in good condition. Human pressures are low but sites could be lost because of climate change.

Chara oedophylla **NT** [B2ab(iii, iv?)]

Rationale: A rare species known from 8 ponds, lake edges and irrigation channels in Spain, France, Sicily and Russia. It appears to be at the edge of its range and it is more frequent in North Africa, particularly Tunisia. At present the European populations seem to be stable but this needs to be kept under review.

Chara oryzae **VU** [D2]

Rationale: This is a species of rice fields, a habitat that could suggest that this is an introduced species. However, it is not yet known anywhere else in the world and therefore needs to be allocated a threat status. It is only known from 3 sites in France and Spain but there is no information about whether populations are stable or declining.

Chara squamosa **DD**

Rationale: This appears to be a rare species with less than 35 localities globally, scattered around both sides of the Mediterranean. Of these around 12 are from Europe and many of these are historic records. However, this is a poorly understood species that has been confused with *Chara gymnophylla* and it is too early to assess its Red List status.

Chara tenuispina **EN** [B2ab(i, ii, iii, iv)]

Rationale: Listed as EN or CR in all of its countries that have Red Lists. It has been recorded from a total of 44 past and present sites but it is currently known from only just over half of these, with an increasingly fragmented distribution spread over 8 countries. It is primarily a species of small water bodies and is continuing to decline due to drainage and successional changes.

Chara uzbekistanica **DD**

This species has quite a restricted distribution in the Volga Delta area and has been suggested for inclusion in the Red List for the Russian Federation. However, it seems to be locally frequent in this core area although there are identification issues and some records may be errors for *Chara connivens* (R. Romanov pers. comm.). Further work is needed to assess its distribution and potential threats.

" Mediterranean Chara" **DD**

Rationale: This is an uncertain taxon whose status as a species is unclear and this has hampered a full understanding of its distribution and ecology. It appears to be a rare species and as a brackish species of coastal wetlands along the Mediterranean coast it is subject to significant pressures from development and drainage. However, further information is needed before an assessment can be made.

Lamprothamnium papulosum **NT** [A2ac]

Rationale: Although quite widespread around the coast of Europe, this species has declined considerably throughout the majority of its range (except some stable populations on the western coast of Ireland and Scotland and there are a number of locations in Sardinia). It is a well-documented species and is listed as at least Vulnerable in nearly all of its countries/regions with Red Lists, excepting Britain where it is Near Threatened. Information from other countries that do not have Red Lists suggest similar declines throughout. Although there has been a steady attrition of sites over the last century, it is not thought that the rate of decline has reached the 30% per decade threshold required to qualify under criterion A and it is not yet sufficiently rare to qualify under criterion B.

Lamprothamnium sardoum **CR (European and Globally)** [B1&2ab(iii)]

Rationale: This is a newly discovered species, only known from a single site in Sardinia where the population is small. As a newly discovered species, it is not known if there have been other populations but there are considerable ongoing pressures on coastal wetlands in this area which has fragmented other suitable habitat. Although the location is a Natura 2000 site, this does not provide particular security from these pressures and there are concerns about drainage, land intensification and scrub encroachment.

Lychnothamnus barbatus **EN** [B2ab(i, ii, iii, iv)]

Rationale: Recorded from 34 sites since 1980, but lost from a further 63 since 1900. Since 1980 there have been further losses but intensive surveys have resulted in a few new sites being found. Although there has been a steady attrition of sites over the last century which is continuing, it is not thought that the rate of decline has reached the 30% per decade threshold required to qualify under criterion A. On the other hand, it has a small Area of Occupation and many populations fluctuate significantly due to nutrient enrichment together with unfavourable fish management, recreation and land use changes. In addition it is a rare species globally with less than 60 sites worldwide and it also belongs to a monospecific genus.

Nitella dixonii **CR (European and Globally)** [B2ab(ii, iv)]

Rationale: Known from fewer than 10 localities in Spain and Portugal but not seen in most of these since the 19th or early 20th Century. There is only one recent record from Spain in 2006 but further investigation of its Portuguese sites is needed. European endemic.

Nitella ornithopoda **CR (European and Globally)** [B2ab(ii, iv)]

Rationale: Known from fewer than 15 localities in France, Spain and Portugal but most of these are from the 19th and early 20th Century. There is only one recent record from Spain in 2009 with another Spanish site last seen in 1987. Endemic to Europe and Macaronesia (Azores).

Nitella palhinhae **DD (European and Globally)**

Rationale: This species is known only from its type locality in Portugal where it was seen in the 1930s. However, it is a poorly understood species that needs further investigation before its threat status can be assessed. European endemic.

Sphaerochara intricata **VU** [B2ab(i, ii, iii, iv)]

Rationale: This species is recorded from many countries but is threatened in all with the exception of LC in Netherlands. Many sites are small water bodies with small populations and the estimated AOO is less than 2000 km^2. There have been significant declines in many countries including some Regional Extinctions and this is ongoing. The habitat has also become severely fragmented by agricultural improvement, drainage and infilling. However, although subpopulation numbers fluctuate, it is able to withstand periods when the habitat is successionally unfavourable as dormant oospores.

Sphaerochara prolifera **VU** [B2ab(i, ii, iii, iv)]

Rationale: Similar to the preceding species, this is recorded from many countries but is threatened in all with the exception of LC in Britain. Many sites are small water bodies with small populations and the estimated AOO is less than 2000 km^2. There have been significant declines in many countries including some Regional Extinctions and this is ongoing. The habitat has also become severely fragmented by agricultural improvement and artificial enrichment. However, although subpopulation numbers fluctuate, it is able to withstand periods when the habitat is successionally unfavourable as dormant oospores.

Tolypella normaniana **NT (European and Globally)** [B2ab(iii)]

Rationale: Restricted to Norway where it is assessed as NT. It is restricted to fewer than 10 locations and although sites have been damaged by infilling for roads and industrial use, the subpopulations seem to be stable. European endemic.

Tolypella salina **EN (European and Globally)** [B2ab(iii, iv)]

Rationale: This species is restricted to Spain (CR) and France (legally protected) with one location in Greece (CR), one in Sicily and nine in Sardinia. There are fewer than 50 sites in total (AOO 200 km^2) and many are threatened by modifications to the water regime (and hence to the salinity levels) and by drainage, including several sites that have been lost. The distribution is severely fragmented by coastal development and the inland brackish sites by agricultural pressures. European endemic.

9.7.4 Doubtfully European Species and Probable Introductions (Not Assessed)

The following species have doubtfully been recorded from Europe. They have not been assessed for threat status until they are confirmed as European species.

Chara hydropithys

Lamprothamnium succinctum

Nitella acuminata

Nitella brachyteles

The following species have been recorded from Europe but only as introductions and are therefore not considered for a threat assessment.

Chara fibrosa

Chara socotrensis

Chara zeylanica

In addition, *Chara oryzae* is also thought possibly to be an introduced species in Europe. However it is not currently known elsewhere in the world and it has therefore been given a threat status above.

Acknowledgements This chapter would not have been possible without the help of many people, including the authors of chapters in this book and of the various Red Lists that have been compiled across Europe. Additional information has been provided by Nadia Abdelahad, Ralf Becker, Jelena Blaženčić, Irmgard Blindow, Aurélie Boissezon, John Bruinsma, Michelle Casanova, Elena Chemeris, Luc Denys, Thomas Gregor, Heiko Korsch, Elisabeth Lambert, Anders Langangen, Richard Lansdown, Aleksandra Marković, Emile Nat, Eva Papastergiadou, Marius Pełechaty, Roman Romanov, Susi Schneider, Hendrik Schubert, Zofija Sinkevičienė, Ingeborg Soulié-Märsche, Kaire Torn, Angelo Troia, Klaus van de Weyer and Egita Zviedre. I would particularly like to thank Hendrik Schubert for his help in finalising the text and to Irmgard Blindow, Ralf Becker, Ingeborg Soulié-Märsche, Roman Romanov and Klaus van de Weyer for their responses to my multiple questions.

References

Artdatabanken, S. L. U. (2020). *Rödlistade arter i Sverige 2020*. SLU, Uppsala.

Auderset Joye, D., & Schwarzer, A. (2012). *Rote Liste der Armleuchteralgen. Gefährdeta Arten der Schweiz, Stand 2010*. Bundesamt für Umwelt BAFU and LEBA, Universität Genf, Bern (72 pp.).

Baastrup-Spohr, L., Dahl-Nielsen, J., & Sand-Jensen, K. (2013). Kransnålalger Rummer Mange Truede Arter. *Urt, 37*, 66–70.

Bergamini, A., Bisang, I., Hodgetts, N., Lockhart, N., van Rooy, J., & Hallingbäck, T. (2019) Recommendations for the use of critical terms when applying IUCN red-listing criteria to bryophytes. *Lindbergia, 42*, linbg.01117.

Blaženčić, J., Stevanović, B., Blaženčić, Z., & Stevanović, V. (2006). Red data list of Charophytes in the Balkans. *Biodiversity Conservation, 15*, 3445–3457.

Caisová, L., & Gąbka, M. (2009). Charophytes (Characeae, Charophyta) in the Czech Republic: Taxonomy, autecology and distribution. *Fottea, 9*(1), 1–43.

Charalambidou, I., & Santamaria, L. (2005). Field evidence for the potential of waterbirds as dispersers of aquatic organisms. *Wetlands, 25*(2), 252–258.

Cirjuano, S., Cambra, J., Sánchez Castillo, P.M., Meco, A., & Flor Arnau, N. (2007). *Flora Ibérica. Algas continentale. Carofitos (Characeae)*. Real Jardin Botanico, Madrid.

Hallingbäck, T. Hodgetts, N., Raeymaekers, G., Schumacker, R., Sérgio, C., Söderström, L., Stewart, N. & Váňa, J. (2000). Guidelines for the application of the 1994 IUCN threat categories to bryophytes. In T. Hallingbäck, & N. Hodgetts (Eds.), *Mosses, liverworts and hornworts. Status survey and conservation action plan for bryophytes* (pp. 71–76). IUCN.

Hawkins, J. (2019). The discovery and conservation of rare stoneworts in Suffolk's farmland ponds. *Suffolk Natural History, 55*, 109–125.

HELCOM. (2013). Red List of Baltic Sea species in danger of becoming extinct. *Baltic Sea Environmental Proceedings, 140*, 1–106.

Hyvärinen, E., Juslén, A., Kemppainen, E., Uddström, A. & Liukko, U.-M. (2019). *Suomen lajien uhanalaisuus. Punainen kirja 2019*. Ympäristöministeriö & Suomen ympäristökeskus, Helsinki [in Finnish].

IUCN (2012). *IUCN red list categories and criteria, version 3.1* (2nd edn). International Union for the Conservation of Nature (IUCN), Gland. https://portals.iucn.org/linrary/node/10315

IUCN (2022). *Guidelines for using of IUCN red list categories and criteria*. Version 15.1. IUCN Standards and Petitions Subcommittee, Gland. http://www.iucnredlist.org/resources/redlistguidelines

Joint Nature Conservation Committee. (2004). *Conservation designations of UK taxa*. https://hub.jncc.gov.uk/assets/478f7160-967b-4366-acdf-8941fd33850b

Korsch, H., Doege, A., Raabe, U., & van de Weyer, K. (2013). Rote Liste der Armleuchteralgen (Charophyceae) Deutschlands 3. Fassung, Stand: Dezember 2012. *Hausskenechtia Beiheft 17*, 1–32.

Langangan, A., Ballot, A., Nowak, P., & Schneider, S. C. (2019). Charophytes in warm springs on Svalbard (Spitsbergen): DNA barcoding identifies *Chara aspera* and *Chara canescens* with unusual morphological traits. *Botany Letters, 167*, 179–186.

Lietuvos Respublikos Aplinkos Ministerija. (2021). *Lietuvos Raudonoji Knyga*. Aplinkos ministerija, Vilnius. https://am.lrv.lt/uploads/am/documents/files/Raudonoji%20knyga/Raudonoji_knyga_2021_WEB.pdf

Ministru kabineta. (2000). *Noteikumi par īpaši aizsargājamo sugu un ierobežoti izmantojamo īpaši aizsargājamo sugu sarakstu*. Noteikumi Nr. 396: prot. Nr. 54 9.§, Rīgā.

Nature Conservation Committee of Estonian Academy of Sciences. (1998). *Eesti Punane Raamat*. http://www.zbi.ee/punane/liigid/vetikad_e.html

Németh, J. (2005). Red list of algae in Hungary. *Acta Botanica Hungarica, 47*, 379–417.

Norwegian Biodiversity Information Centre. (2021). *Norsk rødliste for arter 2021*. https://www.biodiversity.no/Pages/135380/Norwegian_Red_List_for_Species

Palamar-Mordvintseva, G. M., & Tsarenko, P. (2004). Red list of Charales of the Ukraine. *International Journal on Algae, 6*(4), 305–318.

Proctor, V. W. (1962). Viability of *Chara* oospores taken from migratory water birds. *Ecology, 43*(3), 528–529.

van Raam, J. (2000). Rode Lijst 2000 voor de Nederlandse kranswieren. *Nieuwsbrief Kranswieren, 11*, 7–8.

Siemińska, J., Bąk, M., Dziedzic, J., Gąbka, M., Gregorowicz, P., Mrozińska, T., Pelechaty, M., P.M., Pliński, M., & Witkowski, A. (2006) Red List of the algae in Poland. In Z. Mirek, K. Zarzycki, W. Wojewoda, & Z. Szelag (Eds.), *Red list of plants and fungi in Poland*. W. Szafer Institute of Botany, Kraków.

Stewart, N. F., & Church, J. (1992). *Red data books of Britain and Ireland: Stoneworts*. Joint Nature Conservation Committee.

Temniskova, D., Stoyneva, M. P., & Kijakov, I. K. (2008). Red list of the Bulgarian algae. I. Macroalgae. *Phytologia Balcanica, 14*(2), 193–206.

Trutnev, Yu. P., & Kamelin, R. V. (Eds.) (2008). *Krasnaya kniga Rossijskoj Federatsii (rasteniya i griby) Red Data Book of the Russian Federation (plants and fungus)* (855 pp.). Tovaritshestvo nautchnykh izdanij KMK.

Chapter 10
Determination Key *Characeae*

Klaus van de Weyer, Roman Romanov, and Irmgard Blindow

10.1 Introduction

All stoneworts have the same basic structure, similar to a horsetail (*Equisetum*). The thallus of charophytes consists of regularly repeating nodes and internodes (axis and branchlets). The branchlets are arranged in whorls around nodes of the axis. The axis can be corticated (Fig. 10i-1) or ecorticated (Fig. 10i-2). Some species have a reduced cortication.

In the genera *Chara*, *Nitellopsis*, *Lamprothamnium* and *Lychnothamnus*, the branchlets are undivided (Fig. 10i-3). The branchlets consist of several cells. Small cells (bract-cells) are arranged around the nodes between the single branchlet segments (Fig. 10i-4) up to the tip, but many species have nodeless end part of branchlet. Species of *Chara*, *Lychnothamnus* and *Lamprothamnium* have several bract-cells on each node, *Nitellopsis* has one.

The branchlets of *Nitella* can be simple, but are mostly divided once (Fig. 10i-5) or several times into similar branchlet segments of more or less equal length (Fig. 10i-6). *Nitella translucens* appears to have undivided branchlets because of a small crown of dactyls at the branchlets tips (Fig. 10i-7). In *Sphaerochara* and *Tolypella* the branchlets are divided into unequally long rays (Fig. 10i-8).

The branchlets can be corticated in case of *Chara* only (Figs. 10i-1 and 10i-9) or ecorticated (Fig. 10i-5). In some species the cortication of the branchlets is reduced to initials only (Fig. 10i-29). In *Nitella*, the dactyl (end segment of the branchlet) can

K. van de Weyer (✉)
lanaplan, Lobbericher Str. 5, 41334 Nettetal, Germany
e-mail: klaus.vdweyer@lanaplan.de

R. Romanov
Nišice Ujtin Potok b.b., Dobra Voda, Bar 85356, Montenegro

I. Blindow
Biologische Station Hiddensee, Universität Greifswald, 18565 Kloster, Germany
e-mail: blindi@uni-greifswald.de

H. Schubert et al. (eds.), *Charophytes of Europe*,
https://doi.org/10.1007/978-3-031-31898-6_10

be single-celled or consist of two or more cells (Fig. 10i-11), of which the outermost forms an acute apex and often is mucronate, i.e., considerably smaller and thinner than the penultimate branchlet cell (Fig. 10i-12). The number of dactyl cells and the shape of the end and penultimate branchlet cells are important for species delineation in *Nitella*. In *Tolypella* the branchlet tips are obtuse (Fig. 10i-13), while *Sphaerochara* is characterised by an acute top segment of the branchlets (Fig. 10i-12).

Fig. 10i-1 (DJ)

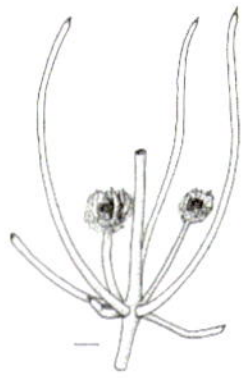

Fig. 10i-2 (DJ)

Fig. 10i-3 (DJ)

Fig. 10i-4 (DJ)

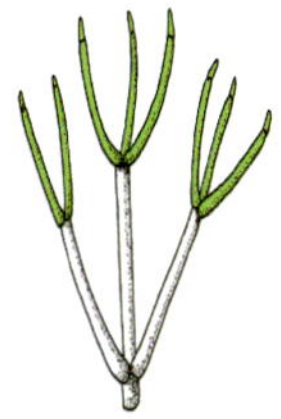

Fig. 10i-5 (DJ)

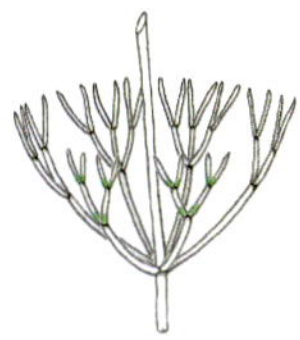

Fig. 10i-6 (DJ)

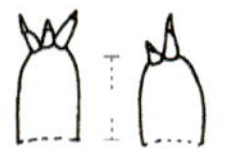

Fig. 10i-7 (DW)

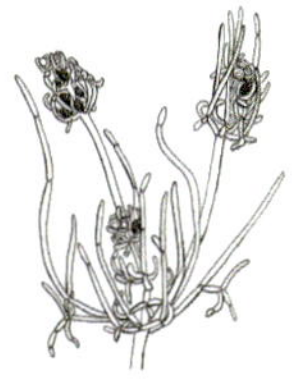

Fig. 10i-8 (DW)

Fig. 10i-9 (DJ)

Fig. 10i-10 (DJ)

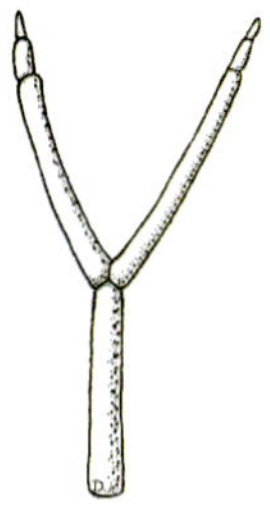

Fig. 10i-11 (DJ)

Fig 10i-12 (DW)

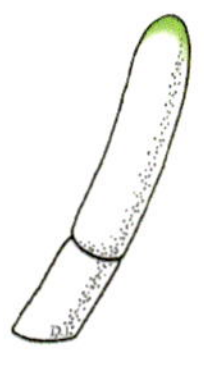

Fig. 10i-13 (DJ)

Fig. 10i-14 (DJ)

Fig. 10i-15 (DJ)

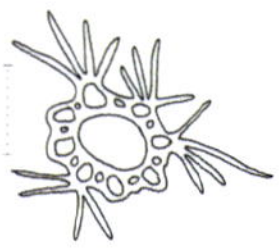

Fig. 10i-16 (cross section, DW)

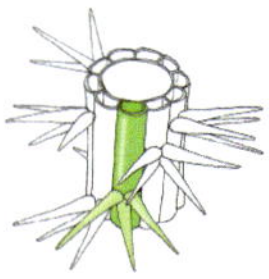

Fig. 10i-17 (DJ)

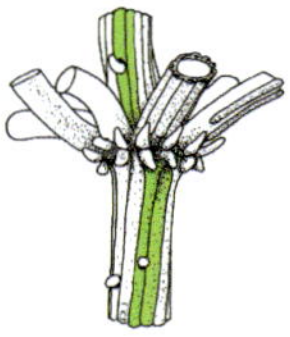

Fig. 10i-18 (DJ)

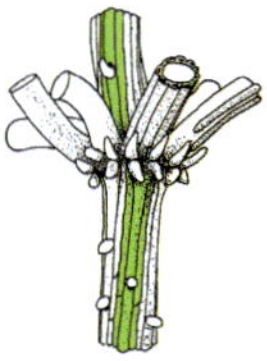

Fig. 10i-19 (DJ)

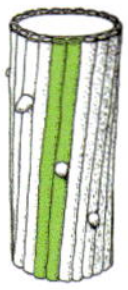

Fig. 10i-20 (DJ)

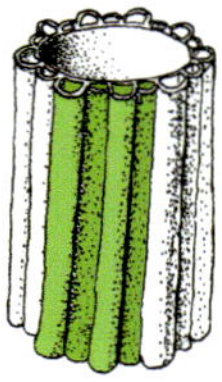

Fig. 10i-21 (DJ)

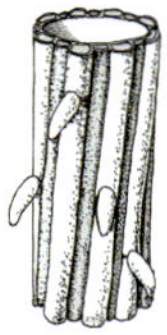

Fig. 10i-22 (DJ)

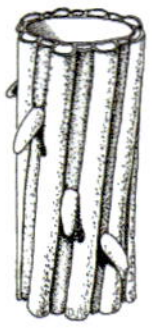

Fig. 10i-23 (DJ)

Fig. 10i-24 (DJ)

In the genera *Chara*, *Lamprothamnium* and *Lychnothamnus*, rows of small cells (stipulodes) are attached to the axis nodes (Fig. 10i-14). The number and shape of these stipulodes are of diagnostic importance. In the genera *Nitella* and *Nitellopsis* cells from axial nodes can protrude resulting in their appearance resembling stipulodes.

In many *Chara* species, the central internode cell is surrounded by layer of several rows of cells forming an axial cortex. The number of axial cortex cell rows is a diagnostic character as is the occurrence, number, shape and position of spine-cells on the cortex cells. The spine-cells may be single (Fig. 10i-15) or arranged in bunches of different habit (Fig. 10i-16). If the number of cortex cell rows corresponds to the number of branchlets, the axial cortex is called haplostichous (Fig. 10i-17). Diplostichous cortex has twice as many cortex cell rows as the number of branchlets (Fig. 10i-18), triplostichous cortex has three times as many (Fig. 10i-19). In haplostichous cortex, spine-cells, if present, are attached to every cortex cell row, in diplostichous to every second cell row and in triplostichous species to every third cortex cell row. Intermediate variants between these three types are not infrequent.

The cortex cell rows can be of equal thickness (isostichous cortex, Fig. 10i-20) or consist of alternating thicker and thinner rows of cortex cells (heterostichous cortex, Fig. 10i-21). Among heterostichous cortex, the primary cortex rows can be thicker than the secondary rows of cortex cells and thus protrude resulting in placement of spine-cells at the ridges (tylacanthous cortex, Fig. 10i-22). Alternatively, the secondary cortex cell rows can be thicker and more prominent (aulacanthous cortex, Fig. 10.23), and spine-cells are situated in furrows. These characters should always be checked at the upper internodes. If it is not clearly recognisable whether a plant is tylacanthous or aulacanthous, it is advisable to make cross sections of the axis

(Fig. 10i-16) and check more plants. Plants heavily encrusted with lime should be placed in a solution consisting of vinegar and tap water (1:1 or 1:2) for about 30 min and then washed with tap water.

Bract-cells arise from the peripheral cells of the branchlet nodes. The bracteoles are formed from the antheridial nodal cell below the oogonium. Adaxial and abaxial bract-cells can be equally long (i.e. verticillate, Fig. 10i-24) or of different lengths. This character and the length of the adaxial bract-cells are used for identification, e.g. within the *Hartmania* group of *Chara*.

The male gametangia (antheridia, Fig. 10i-25) are normally reddish or yellowish with a shape similar to a football, while female gametangia (oogonia, Fig. 10i-26) are brownish or greenish in most species and have an ellipsoidal outline. The size of the antheridia is a diagnostic character for the separation of some species (i.e. *Chara aspera–C. galoides*). Gametangia diameters are around 1.3 times larger when measured on wet material compared to dry. The antheridium of most species is composed of eight triangular shield-like cells (octoscutate, Fig. 10i-27); in *C. zeylanica*, it is composed or four rhomboidal cells (tetrascutate, Fig. 10i-28).

Fig. 10i-25 (DJ)

Fig. 10i-26 (DJ)

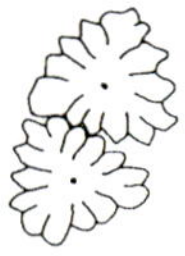
Fig. 10i-27 (ISM)

Fig. 10i-28 (ISM)

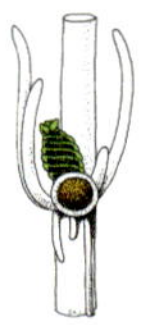
Fig. 10i-29 (DJ)

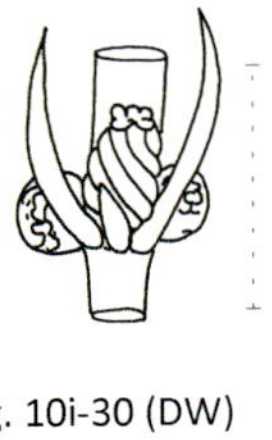
Fig. 10i-30 (DW)

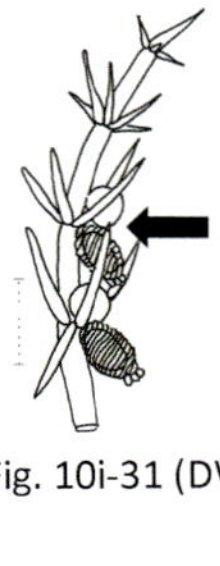
Fig. 10i-31 (DW)

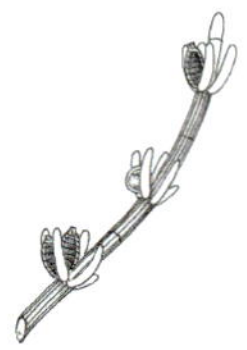
Fig. 10i-32 (DJ)

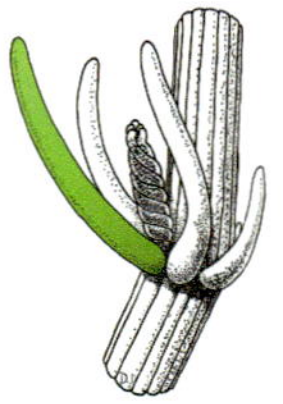
Fig. 10i-33 (DJ)

While some species are dioecious, the majority of the species are monoecious (Fig. 10i-29). Among these, the position of the gametangia is a diagnostic criterion for genera and species. In *Chara* the antheridium is located below the oogonium (Fig. 10i-29). In *Lychnothamnus* the antheridium is next to the oogonium (Fig. 10i-30), while in *Nitella* and *Lamprothamnium*, the antheridium is located above the oogonium (Fig. 10i-31), or antheridium surrounded by oogonia in *Tolypella* and *Sphaerochara*. The antheridia and oogonia of most monoecious species are conjoined, i.e., located on the same branchlet node (Figs. 10i-29, 10i-30, 10i-31). *Chara oedophylla*, rarely *C. gymnophylla* and *C. squamosa* have sejoined gametangia, i.e. the antheridium and oogonium are located on separate branchlet nodes of the same plant (Fig. 10i-32). An unequal ripening of gametangia such as protandry (antheridia ripen earlier than oogonia) can result in presence of solitary gametangia at nodes and could be source of misinterpretation. The bractlet (Fig. 10i-33) forms instead of the antheridium in female plants of dioecious species of *Chara*. Monoecious species of *Chara* have empty spaces below the oogonium and between the bracteoles after disintegration of the antheridium.

Fig. 10i-34 (DJ)

Fig. 10i-35 (DJ)

Fig. 10i-36 (DJ)

Fig. 10i-37 (MC)

Fig. 10i-38 (MC)

The spiral cells of the oogonium form a coronula on the top which consists of 5 cells in one single row in *Chara*, *Nitellopsis* and *Lamprothamnium* (Fig. 10i-34), and 10 cells in two rows in *Nitella*, *Sphaerochara* and *Tolypella* (Fig. 10i-35) at the top of five spiral cells. Sometimes the coronula can be deciduous (Fig. 10i-36).

After fructification the oogonium ripens into an oospore loosing the coronula cells and outer parts of spiral cells and developing a thick wall. The length, width, number of striae (Fig. 10i-36) and the colour of the oospores are of diagnostic importance. The oospores in *Nitella* are laterally compressed (Fig. 10i-37, basal pole view), while the oospores in all other genera are terete (laterally not compressed, Fig. 10i-38).

In *Nitella*, *Sphaerochara* and *Tolypella*, the surface structure of the oospore wall is a diagnostic character and can be studied with light or scanning electron microscopy (smooth, reticulate, granulate, Figs. 10i-39, 10i-40, 10i-41, 10i-42 and 10i-43). The appearance of oospore surface of the same species can be different between light or scanning electron microscopy observations.

Fig. 10i-39 smooth structure of the oospore wall, *Tolypella salina*, SEM (MC)

Fig. 10i-40 smooth structure of the oospore wall, *Tolypella salina*, light microscope (KW)

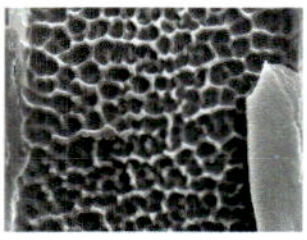

Fig. 10i-41 reticulate structure of the oospore wall, *Nitella mucronata*, SEM (RR)

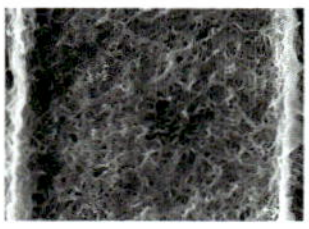

Fig. 10i-42 spongy structure of the oospore wall, *Nitella wahlbergiana*, SEM (RR)

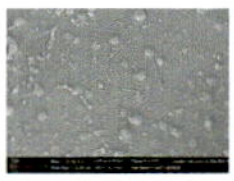

Fig. 10i-43 slightly granulate structure of the oospore wall, *Sphaerochara intricata*, SEM (MC)

Some stoneworts reproduce or hibernate vegetatively by bulbils containing starch. Bulbils may be single-celled like in *Chara aspera* (Fig. 10i-44) and *Lamprothamnium* or multicellular. *Chara fragifera* has strawberry/raspberry-like multicellular rhizoidal bulbils (Fig. 10i-45). The nodal bulbils of *Nitellopsis* are star-shaped (Fig. 10i-46). Some species of *Chara* and *Nitella* have enlarged basal stem nodes which serve as overwintering organs (hibernacles).

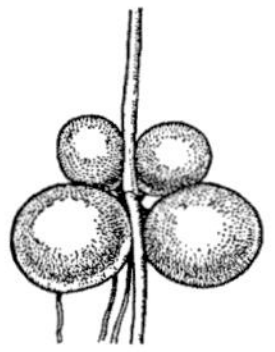

Fig. 10i-44 (RR)

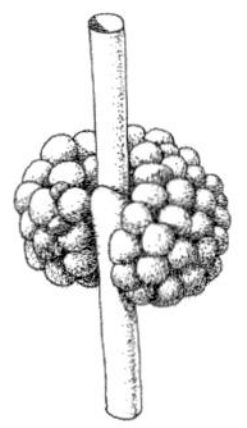

Fig. 10i-45 (DJ)

Fig. 10i-46 (DJ)

10.2 How to Use This Key

This key includes all characean species recorded from Europe. Taxa on an intraspecific level are only considered occasionally. Species without confirmed European records are not included in the determination steps, but some information is given in the key. Within the genus *Chara*, robust, diplostichous species belonging to the so-called *Hartmania* complex can be hard to separate from each other. Please use the matrix key (see Chap. 15) alternatively or additionally for determination! In the Mediterranean area, some additional *Hartmania* species can be expected, which only can be recognised by means of the matrix key.

Especially for beginners, determination of many characean species is a challenge. Within Europe, a number of national and regional determination keys are available, often included in a characean flora (e.g., Bailly and Schaeffer, 2010; Bazzichelli and Abdelahad, 2009; Blindow et al., 2007; Bruinsma et al., 2018; Bryant and Stewart, 2011; Cirujano et al., 2008; Hollerbach and Krassavina, 1983; Krause, 1997; Langangen, 2007; Langangen et al., 2002; Moore, 1986; Mouronval et al., 2015; Schubert and Blindow, 2004; Stewart, s.d.; Urbaniak and Gąbka, 2014; van de Weyer, 2016; van de Weyer and Schmidt, 2018; van Raam, 1998). It is recommended to (additionally) use these keys which only contain the species occurring in the region in question.

Detailed information on characters of different species is also available in the single species chapters (see Chaps. 16–86), in different characean floras which are available on a national (e.g., Bazzichelli and Abdelahad, 2009; Becker et al.,

2016; Corillion, 1957; Groves and Bullock-Webster, 1920, 1924; Langangen, 2007; Migula, 1889–1897), regional (Muller et al., 2017, Schubert and Blindow, 2004) or global (Wood, 1965; Wood and Imahori, 1964) scale. This information has also been used to develop this European key together with a number of publications on single, often rare species (Abdelahad and Piccoli, 2017; Blaženčić and Stevanović, 2018; Casanova, 2013; Krause, 1990; Langangen, 2010, 2012; Romanov et al., 2015, 2022; Soulié-Märsche and Muller, 2016; Troia et al., 2018). Several publications provide information about oospore and gyrogonite characters of single species, part of which also discuss the applicability of these characters for taxonomy (e.g., Holzhausen et al., 2015; John and Moore, 1987; Ray et al., 2001; Soulié-Märsche, 1999; Urbaniak et al., 2012).

10.3 Preservation

Plants can be stored in alcohol or pressed (herbarium specimen) (Moore, 1986).

For herbarium specimens, place the plants on a sheet of white, preferably acid-free paper of appropriate size which will be a permanent part of pressed specimen. The plants with rough texture need to be spread out manually. Overhanging parts are folded over.

Plants that collapse outside the water are placed in a basin filled with water. The specimens are cleaned if necessary. A white, thick paper is placed under the plants. After carefully arranging the plants, the paper with the plants is slowly lifted out of the basin, and covered by a second white paper or tissue. Then the sample with the two white sheets is pressed between a newspaper. As an alternative to white paper, baking paper has also proved successful, from which the dried plants can be easily removed.

The plants are then placed in a plant press including the labelling (see below). Depending on the amount of water, it may be necessary to change the newspaper one or several times. The dry plants are fastened on the thick paper by glue or paper strips.

Always put only plants of one species and one locality on one sheet. The following information is suitable for labelling:

1. Species with author's name
2. Locality (country, state, district, geographical name, grid map or GPS data)
3. Habit (ecological designation)
4. Date of collection
5. Collector ("leg.")
6. Determinator ("det.")
7. Name of herbarium and consecutive no.

10.4 Key to the Genera

1	At least some branchlets divided (Fig. 10-1) (*Nitellopsis obtusa* has long bract cells which might be mistaken for a branching)	2
1*	Branchlets undivided for naked eye	4
2	Branchlets once or repeatedly divided into similar branchlet segments (Fig. 10-1), oospore laterally compressed	*Nitella*
2*	Branchlets divided into unequally long branchlet segments, oospore terete, laterally not compressed	3
3	Top cell of the branchlets acute (Fig. 10-5)	*Sphaerochara*
3*	Top cell of the branchlets obtuse (Fig. 10-6)	*Tolypella*
4	Stipulodes absent	5
4*	Stipulodes present (Fig. 10-7, 10-8 and 10-9) (rarely rudimentary)	8
5	With long bract-cells at the branchlet nodes (Fig. 10-2), with star-shaped bulbils (Fig. 10-3), dioecious, all branchlets undivided, large	*Nitellopsis obtusa*
5*	Without bract-cells, no star-shaped bulbils, monoecious. Sterile branchlets undivided, fertile branchlets divided	6
6	Medium-sized to large, often longer than 1 m, axis diameter 1–2 mm, occasionally less, sterile undivided branchlets with a crown of mainly bicellular dactyls on top (Fig. 10-4), freshwater, throughout Europe	*Nitella translucens*
6*	Small to medium-sized, axis diameter < 1 mm, no dactyl crown on top of the branchlets, Northern regions only	7
7	Top segment of the branchlets acute, freshwater (Fig. 10-5)	*Sphaerochara canadensis*
7*	Top segment of the branchlets obtuse (Fig. 10-6), estuarine habitats	*Tolypella normaniana*
8	Stipulodes in two rows (Fig. 10-7), sometimes small and papillate (Fig. 10-8), rarely rudimentary, plants in most cases with cortex	*Chara*
8*	Stipulodes in one row (Fig. 10-9), cortex lacking or rudimentary	9
9	Antheridium below oogonium (Fig. 10-10)	*Chara*
9*	Antheridium next to or above oogonium (Fig. 10-11)	10
10	Apical branchlets funnel-shaped, not fox-tail like, sometimes axis partially with cortex, antheridium next to the oogonium, freshwater	*Lychnothamnus barbatus*
10*	Apical whorls bushy, fox-tail like (Fig. 10-12), plants completely ecorticated, antheridium above the oogonium (Fig. 10-11), brackish water	*Lamprothamnium*

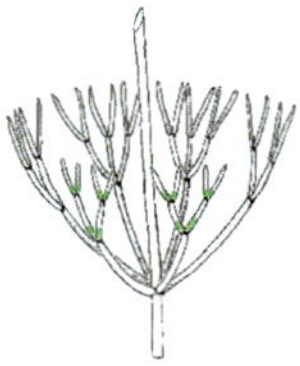
Fig. 10-1 (DJ)

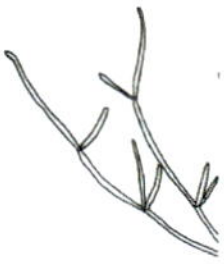
Fig. 10-2 (DW)

Fig. 10-3 (DJ)

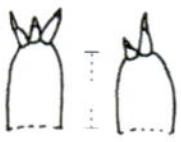
Fig. 10-4 (DW)

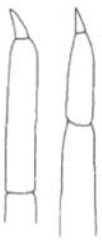
Fig 10-5 (RR)

Fig. 10-6 (DJ)

Fig. 10-7 (DW)

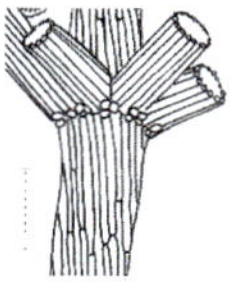
Fig. 10-8 (DW)

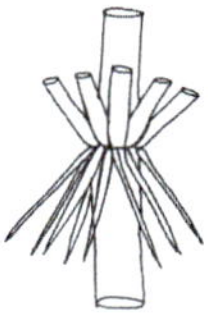
Fig. 10-9 (DW)

Fig. 10-10 (DW)

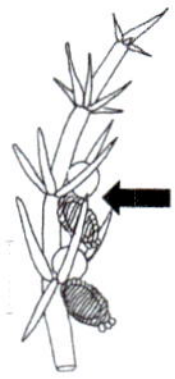
Fig. 10-11 (DW)

Fig. 10-12 (DW)

10.5 Chara

1	Stipulodes in one row (Fig. 10-13)	2
1*	Stipulodes in two rows (Figs. 10-7, 10-8)	6
2	Axis and branchlets corticated	*C. uzbekistanica* The stipulodes of *C. uzbekistanica* were described as haplostephanous, but restudy of its type specimen revealed diplostephanous stipulodes with rudimentary cells in lower row. They could be easily missed at low magnification or in pressed state
2*	At least parts of the axis and branchlets ecorticated	3

(continued)

(continued)

3	Axis and branchlets ecorticated (Fig. 10-16)	*C. braunii* *C. socotrensis*, rare in Europe and only known from the UK, has reduced stipulodes in contrast to *C. braunii* which has well developed stipulodes. *C. braunii* has a corona (=several tiny cells at the tips of the branchlets, see Fig. 10.14) which lacks in *C. socotrensis*. *C. socotrensis* has partly undifferentiated branchlets in contrast to *C. braunii* which has branchlets completely differentiated into nodes and internodes
3*	Axis corticated, branchlets ecorticated (Fig. 10-17) or with imperfect cortex (Figs. 10-15 and 10-18)	4
4	Branchlets with abbreviated segments and mainly incomplete cortex (Fig. 10-18), Mediterranean only	*C. oryzae*
4*	Branchlets completely ecorticated (Fig. 10-17)	5
5	Stem cortex triplostichous, branchlets terminated with a tiny corona (group of terminal cells on the branchlet-tip) (Fig. 10-14)	*C. baueri*
5*	Stem cortex diplostichous, branchlets terminated with an end-cell surrounded by a group of bract-cells, both well developed (Fig. 10-19)	C. subsection *Agardhia* (*fibrosa* s.l.)
6	Axial cortex incomplete, varying from nearly inconspicuous or poorly developed primary rows only to primary rows contacting at the centre of the internode with short undeveloped secondary rows and empty spaces between primary rows	7
6*	Axis completely corticated (Fig. 10-15)	8
7	Dioecious	*C. imperfecta*
7*	Monoecious	*C. dissoluta*

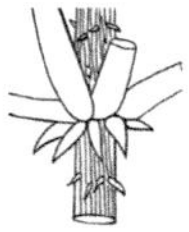

Fig. 10-13 (DW)

Fig. 10-14 (RR)

Fig. 10-15 (DJ)

Fig. 10-16 (DW)

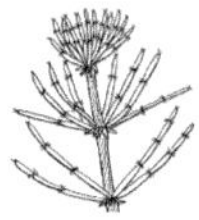

Fig. 10-17 (DW)

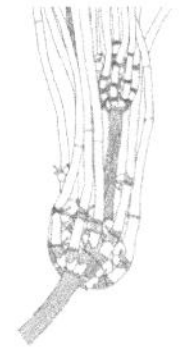

Fig. 10-18 (RR)

Fig. 10-19 (RR)

8	Haplostichous (Fig. 10-20), dioecious	*C. canescens*
8*	Not or only partially haplostichous, monoecious or dioecious	9
9	Diplostichous (Fig. 10-21) or triplostichous (Figs. 10-22 and 10-23)	10
9*	Haplo-diplostichous to diplostichous with secondary rows overriding each other, monoecious, only few locations in South-Eastern Europe	*C. altaica*
10	Triplostichous (Figs. 10-22 and 10-23) (*C. strigosa* is triplo-/diplostichous)	11
10*	Diplostichous (Fig. 10-21) (*C. globata*, *C. neglecta* and *C. baltica* may be di-/triplostichous)	21
11	Without spine-cells (Fig. 10-22) or spine-cells papillate (Fig. 10-23)	12
11*	Spine-cells long (Fig. 10-24), single or in bunches, sometimes lacking on older internodes	16
12	Monoecious	13
12*	Dioecious	14
13	Both rows of stipulodes inconspicuous (Fig. 10-22), papillate, without spine cells	*C. globularis* There are transition forms between *C. globularis* and *C. virgata*
13*	Upper row of the stipulodes prolonged, acute, lower papillate, with papillate spine cells (Fig. 10-23)	*C. virgata*
14	Plants collapsing when removed from the water, multicellular rhizoidal bulbils (strawberry/raspberry-like)	*C. fragifera*
14*	Plants not collapsing when removed from the water, bulbils sphaerical or absent	15

(continued)

(continued)

15	Heterostichous tylacanthous to isostichous, branchlets of male plants not incurved, stipulodes acute, with unicellular sphaerical bulbils (Fig. 10-25)	*C. aspera* var. *subinermis*
15*	Isostichous, branchlets of male plants incurved forming a globose head (Fig. 10-26), stipulodes papillate, rarely with multicellular nodal bulbils, no sphaerical bulbils	*C. connivens*

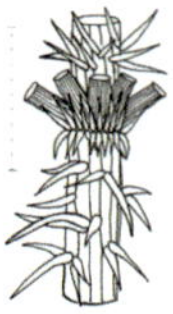

Fig. 10-20 (DW)

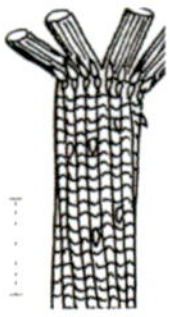

Fig. 10-21 (DW)

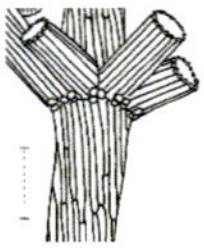

Fig. 10-22 (DW)

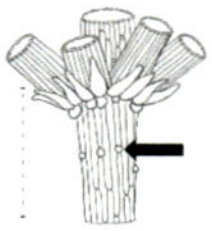

Fig. 10-23 (DW)

Fig. 10-24 (DW)

Fig. 10-25 (RR)

Fig. 10-26 (KW)

16	Dioecious	17
16*	Monoecious	19
17	Spine-cells dense, in bunches of 2–3 (Fig. 10-27)	*C. curta*
17*	Spine-cells single (Fig. 10-28)	18
18	Diameter of antheridia > 800 μm (fresh)/700 μm (dry), male plants robust compared to female ones, brackish water along the Mediterranean, the Atlantic and the Black Sea, without sphaerical bulbils	*C. galioides* *C. neglecta* has an irregular stem cortex and is never completely triplostichous. In SE Europe only
18*	Diameter of antheridia < 800 μm (fresh)/700 (dry) μm, without sexual dimorphism, fresh and brackish water, throughout Europe, often with unicellular sphaerical bulbils on the rhizoids (Fig. 10-25)	*C. aspera*
19	Spine-cells robust, in bunches of 2–5 (Fig. 10-27), axial cortex irregular, partly diplostichous, tylacanthous	*C. strigosa*
19*	Spine-cells slender, solitary (Fig. 10-28), cortex regularly triplostichous	20
20	Lowermost branchlet segment corticated antheridial shields clearly triangular, antheridia octoscutate (Fig. 10-29)	*C. tenuispina* *C. filarszkyana* has translucent ecorticated branchlets. *C. filarszkyana* is insufficiently known. It has disappeared from its single locality in Hungary

(continued)

(continued)

20*	Lowermost branchlet segment always ecorticated, antheridial shields rhomboidal, antheridia tetrascutate (Fig. 10-30)	*C. zeylanica*
21	Axis diameter usually less than 1 mm, plants mostly delicate, medium sized, spine-cells single	22
21*	Axis diameter usually 1–5 mm, plants robust, usually large, spine-cells single or in bunches	30
22	Aulacanthous (Fig. 10-31)	23
22*	Tylacanthous (Fig. 10-32)	26
23	Monoecious	24
23*	Dioecious, male plants robust compared to female ones, diplo- (-triplostichous)	*C. neglecta*

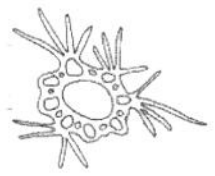

Fig. 10-27 (DW)

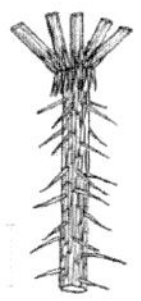

Fig. 10-28 (DW)

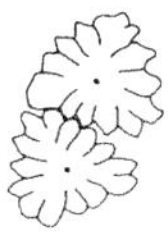

Fig. 10-29 (ISM)

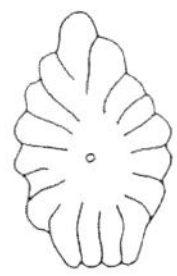

Fig. 10-30 (ISM)

Fig. 10-31 (DW)

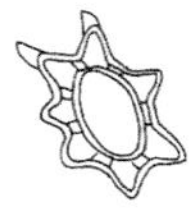

Fig. 10-32 (DW)

24	Branchlets always with bract-cells between ecorticated segments, ecorticated (Fig. 10-33) or only 1–2 basal segments corticated	*C. gymnophylla*
24*	Branchlets with 1–3 corticated segments or entirely corticated, no bract-cells between ecorticated segments (Fig. 10-35)	25
25	All gametangia conjoined (Fig. 10-36), bract-cells not swollen, diameter of the antheridia 250–500 μm, throughout Europe	*C. vulgaris*
25*	Most gametangia sejoined (Fig. 10-37), bract-cells swollen, diameter of the antheridia 560–840 μm, Mediterranean and only one locality in South-Eastern steppe	*C. oedophylla*
26	Dark green, often diplo-triplostichous, antheridia large, diameter (450–)500–800 μm, coastal waters	*C. baltica*
26*	Fresh to light green, not dark green, regularly diplostichous, antheridia diameter 250–610 μm, freshwater, occasionally in brackish lagoons with low salinity	27

(continued)

(continued)

27	Branchlets 0.5–3 cm long, not extraordinary short in comparison with internodes (Fig. 10-34 and 10-35)	28
27*	Branchlets up to 1 (–2) cm long, extremely short compared to the internodes, with 1 or 2 nodes only (Fig. 10-38)	29
28	Always with bract-cells and gametangia between ecorticated segments (Fig. 10-33), branchlets entirely ecorticated or with 1–2 corticated basal segments, antheridia large, diameter 360–610 μm, restricted to the Mediterranean	*C. squamosa*
28*	Bract-cells and gametangia between or at the top of corticated segments (Fig. 10-35), not between ecorticated segments, branchlets with 1–3 (–8) ecorticated segments, antheridia small, diameter 250–400 μm, throughout Europe	*C. contraria*
29	Dioecious, antheridia large, diameter 370–590 μm, Southern Russia	*C. kirghisorum*
29*	Monoecious (Fig. 10-36), antheridia small, up to 400 μm, Northern and Central Europe	*C. filiformis*
30	1–3 top segments of the branchlets, spine-cells and bract-cells swollen (Fig. 10-39), axis of the living plants brownish, apical parts reddish, tylacanthous (Figs. 10-32 and 10-41), dioecious	*C. tomentosa*
30*	Spine-cells and bract-cells not swollen, axis and apical parts of living plants green-greyish to yellow-brownish, tylacanthous (Fig. 10-41) or aulacanthous (Fig. 10-40), monoecious or dioecious	31 (use also *Hartmania* matrix key)

Fig. 10-33 (DJ)

Fig. 10-34 (DJ)

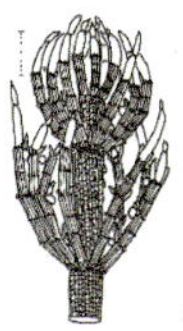

Fig. 10-35 (DW)

Fig. 10-36 (DJ)

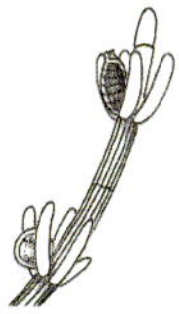

Fig. 10-37 (DJ)

Fig. 10-38 (DW)

Fig. 10-39 (DJ)

31	Aulacanthous (Fig. 10-40) to isostichous	32
31*	Tylacanthous (Fig. 10-41) to isostichous	34

(continued)

(continued)

32	Stipulodes very dense, irregular, in more than two rows (Fig. 10-42), restricted to coastal sites	*C. horrida*
32*	Stipulodes in two regular rows (Fig. 10-43), occasionally with extra stipulodes, freshwater or inland brackish water sites, only occasionally in coastal lagoons with low salinity	33
33	Isostichous (Fig. 10-44) to heterostichous, spine-cells distinctly protruding from the cortex and pointing at different directions, inclined and sticking out from the stem (Fig. 10-45)	*C. hispida*
33*	Clearly heterostichous, spine-cells in antiparallel direction appressed in the cortex furrows or y-shaped and protruding, but in this case appressed to the stem (Fig. 10-46)	*C. subspinosa*
34	Dioecious, rare, Mediterranean only	*C. hellenica*
34*	Monoecious (Fig. 10-36)	35
35	Dark green, often diplo-triplostichous, antheridia large, diameter (450–)500–800 μm, restricted to coastal waters of the North, Baltic and Black Seas	*C. baltica*
35*	Light, greyish or yellowish green, antheridia smaller, diameter rarely exceeding 550 μm, freshwater or inland brackish water, only occasionally in coastal lagoons	36
36	Spine-cells single or in bunches of 2–3	37
36*	Spine-cells dense, in bunches of more than 3, occasionally 2–3, never exclusively single, brush-like pattern	*C. aculeolata*
37	Upper whorls usually with evidently arcuate branchlets, whorls dense and compact, forming heads, bract-cells long and verticillate, squarrose, spine-cells normally shorter than the axis diameter, semiarid regions of southern and eastern Europe	*C. globata*
37*	Different	38
38	Spine-cells shorter than or as long as axis diameter, stipulodes short (to long), throughout Europe	*C. papillosa*
38*	Spine-cells longer than axis diameter, stipulodes long, only recorded from Western Europe (UK, France, Benelux)	*C. affinis*

In the Mediterranean region, use the *Hartmania* matrix key (see Chap. 15) for determination of tylacanthous taxa belonging to the *Hartmania* group, as the presence of additional taxa (*C. corfuensis*, "Mediterranean Chara") complicates the determination in this region.

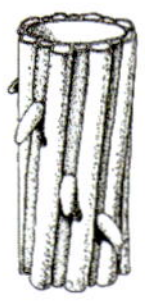

Fig. 10-40 (DJ)

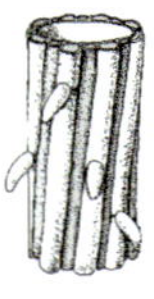

Fig. 10-41 (DJ)

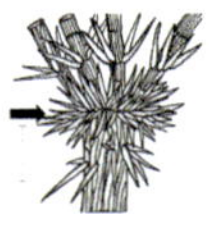

Fig. 10-42 (DW)

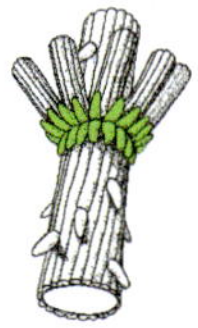

Fig. 10-43 (DJ)

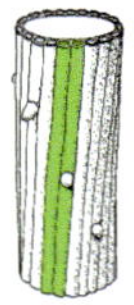

Fig. 10-44 (DJ)

Fig. 10-45 (KW)

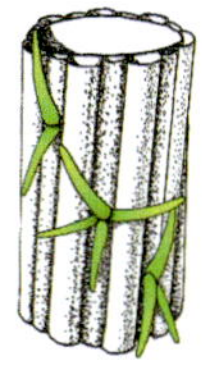

Fig. 10-46 (DJ)

10.6 Lamprothamnium

1	Gametangia only at the branchlet-nodes (Fig. 10-47), most common species of the genus in Europe	*L. papulosum*
1*	Additional gametangia at the base of the whorls	*L. sardoum*
		The ornamentation of the oospore-membrane of *L. succinctum* is densely verrucate in contrast to *L. sardoum* which has a smooth ornamentation of the oospore-membrane. There is no confirmed record of *L. succinctum* in Europe

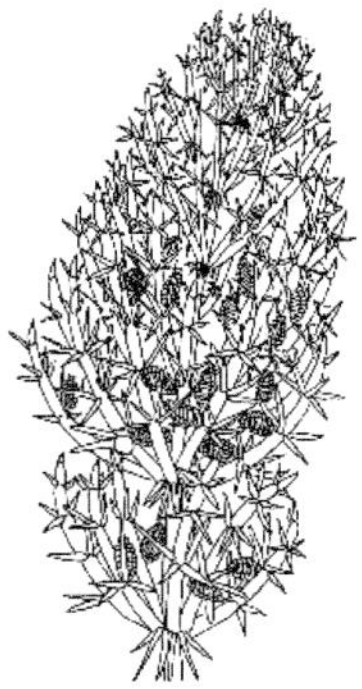

Fig. 10-47 (DJ)

10.7 Nitella

1	At least some branchlets appearing undivided for naked eye, but with a tiny terminal corona, dactyls short, 2–3 celled of 0.2–0.3 mm length (Fig. 10-48), large plants	*N. translucens* *N. brachyteles* has longer secondary branchlet segments and a fibrous oospore-membrane, while *N. translucens* has very short secondary branchlet segments and a reticulate oospore-membrane. *N. brachyteles* is known only from South Europe
1*	Branchlets divided or undivided, without terminal corona	2

(continued)

(continued)

2	Short 1–2 furcate accessory branchlets inside and below of each whorl of long 2–3 furcate branchlets (Fig. 10-49)	*N. hyalina*
2*	Uniform branchlets within one whorl	3
3	Dactyls single-celled, length of oospores 280–600 μm, width 280–525 μm, coronula detaching early in most cases	4
3*	Dactyls 2–5-celled (Fig. 10-50), length of oospores 180–350 μm, width 170–350 μm, coronula persistent	7
4	Gametangia with mucus (Fig. 10-51), plants dioecious	5
4*	Gametangia without mucus, plants monoecious or dioecious	6
5	Fertile female branchlets divided (Fig. 10-52), oospore brown, with flanged ridges, oospore-membrane granulate	*N. capillaris*
5*	Fertile female branchlets not divided, oospore black, with low ridges, oospore-membrane smooth	*N. syncarpa*

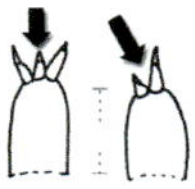

Fig. 10-48 (DW)

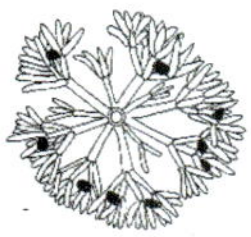

Fig. 10-49 (DW)

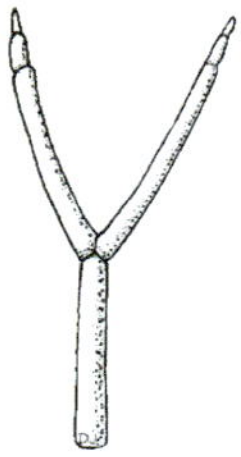

Fig. 10-50 (DJ)

Fig. 10-51 (KW)

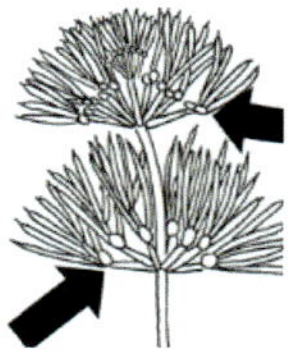

Fig. 10-52 (DW)

6	Dioecious	*N. opaca*
6*	Monoecious (Fig. 10-53)	*N. flexilis* Both species can not be separated from each other when sterile *N. acuminata* is separated from *N. flexilis* and *N. opaca* by very long dactyls with distinctive acuminate tips and a persistent coronula of the oogonium. The oospore ornamentation of *N. acuminata* is somewhat roughened, and sparsely covered with flaccid dentate projections
7	Branchlets single-furcate only (Fig. 10-60), single locality in SW Europe	*N. palhinhae*
7*	Branchlets 1–3-furcate, 1–2, 2–3 or 1–3 may occur within the same plant	8
8	Dactyls 2-celled (Figs. 10-54 and 10-55)	9
8*	Dactyls 2–3 or (2–)3–4(–5)-celled	13
9	Terminal dactyl cell mucronate, i.e., with base much narrower than the preceding cell (Fig. 10-55)	10
9*	Terminal dactyl cell not mucronate, its base approximately of the same diameter as the preceding cell (Fig. 10-54)	11
10	Mainly longer than 10 cm, axis diameter > 0.5 mm, secondary branchlet segments not shorter than others, fertile whorls not condensed, oospore-membrane reticulate (Fig. 10-56), throughout Europe	*N. mucronata*
10*	Short, rarely longer than 10 cm, axis diameter <0.5 mm, secondary branchlet segments shorter than others, fertile whorls condensed, oospore-membrane spongy (Fig. 10-57), appearing as irregularly granulate or 3D-finely-reticulate in light microscope, Northern Europe	*N. wahlbergiana*

Fig. 10-53 (KW)

Fig. 10-54 (DJ)

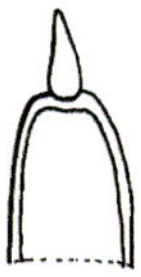
Fig. 10-55 (DW)

Fig. 10-56 (RR)

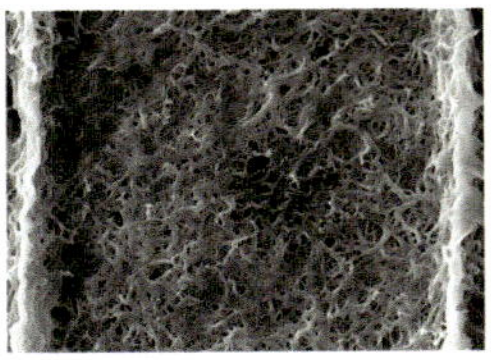
Fig. 10-57 (RR)

11	Gametangia mainly located at the second branchlet divisions, branchlets with central secondary branchlet segment, branchlets forming dense, circular to globular whorls appearing like a string of pearls on the thin axis (Fig. 10-58), oospore-membrane reticulate	*N. tenuissima*
11*	Gametangia not mainly located on the second branchlet divisions, central secondary branchlet segment absent, plant not appearing like a string of pearls on the thin axis	12
12	Gametangia can be located on all branchlet divisions (but are often restricted to one), oospore-membrane spongy (Fig. 10-61), not evidently reticulate in light microscopy	*N. gracilis*
12*	Gametangia preferably located at the first branchlet divisions (Fig. 10-59), oospore-membrane granulate (Fig. 10-62)	*N. confervacea*
13	Dactyls (2–3)4(5)-celled	14
13*	Dactyls 2–3 or 3-celled	15
14	Monoecious, oospore-membrane finely spongy, South-West Europe	*N. ornithopoda*
14*	Dioecious, oospore-membrane roughly agglutinated spongy, South-West Europe	*N. dixonii*
15	Terminal dactyl cell with base approximately of the same diameter as the preceding cell (Fig. 10-54), branchlets regular, oospore-membrane spongy (Fig. 10-61), not evidently reticulate in light microscopy, delicate plants	*N. gracilis*
15*	Terminal dactyl cell mainly mucronate (Fig. 10-55), branchlets irregular, oospore-membrane reticulate (Fig. 10-56), robust plants	*N. mucronata*

Fig. 10-58 (DW)

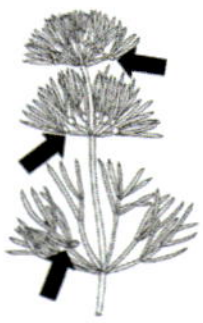

Fig. 10-59 (DW)

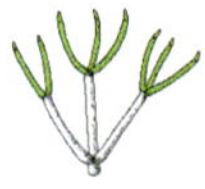

Fig. 10-60 (DJ)

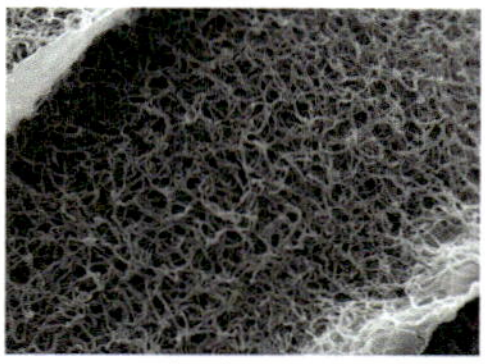

Fig. 10-61 (RR)

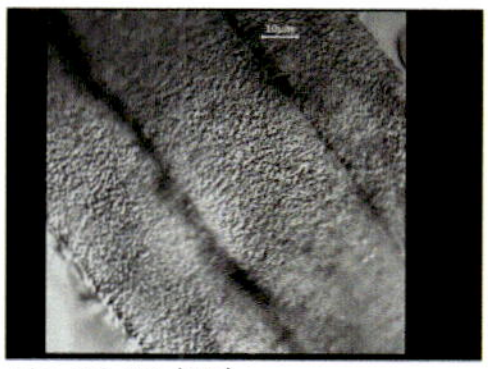

Fig. 10-62 (LD)

10.8 Sphaerochara

1	Sterile branchlets divided (Fig. 10-63), oospore-membrane (slightly) granulate (Fig. 10-65)	*Sphaerochara intricata*
1*	Sterile branchlets undivided (Fig. 10-64), oospore-membrane smooth (Fig. 10-66)	2
2	Robust, encrusted plants, ephemeral, not reliably known north of 59° N, antheridium diameter 300 μm	*Sphaerochara prolifera*
2*	Slender, unencrusted plants resembling *Nitella* in appearance; perennial, often sterile, permanent water bodies, cold water species, not reliably known south of 63° N, antheridium diameter 400–500 μm	*Sphaerochara canadensis*

Fig. 10-63 (DW)

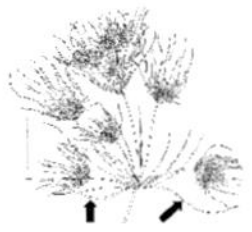

Fig. 10-64 (DW)

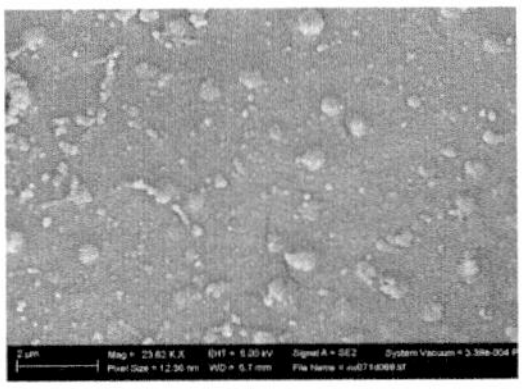

Fig. 10-65 (MC)

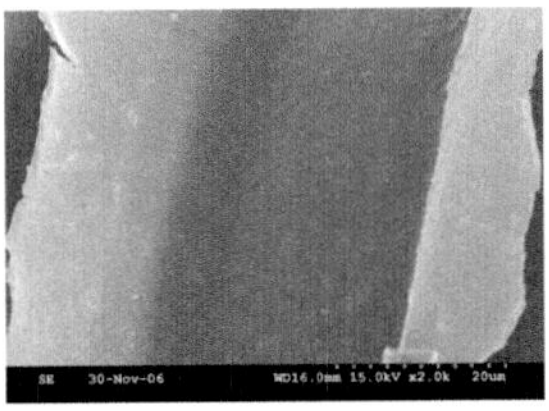

Fig. 10-66 (SC)

10.9 Tolypella

1	Dioecious, restricted to Southern Europe	*Tolypella hispanica*
1*	Monoecious (Fig. 10-67)	2
2	Without sterile branchlets (only protonemal whorls have sterile branchlets), northern tidal flats, in estuaries, not reliably known south of 64° N	*Tolypella normaniana*
2*	With sterile branchlets	3
3	Plant dull/matt (not shiny) because of lime encrustation, oospores brown, oospore-membrane regular/irregular reticulate (Fig. 10-56), fresh and brackish water throughout Europe	*Tolypella glomerata*
3*	Plant shiny and semitransparent, without encrustation, oospore-membrane smooth (Figs. 10-70 and 10-71) or smooth with irregular pits (Figs. 10-68 and 10-69), brackish water	4
4	Oospores brown to black, oospore-membrane variable from smooth with irregular pits (Figs. 10-68 and 10-69) to granulate, throughout Europe	*Tolypella nidifica*
4*	Oospore-membrane smooth (Figs. 10-70 and 10-71), Southern Europe	*Tolypella salina*

	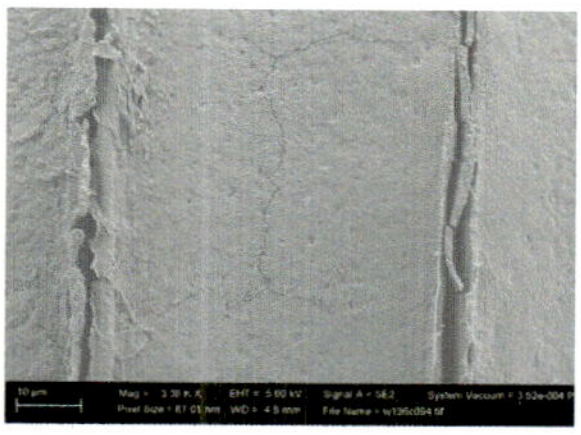	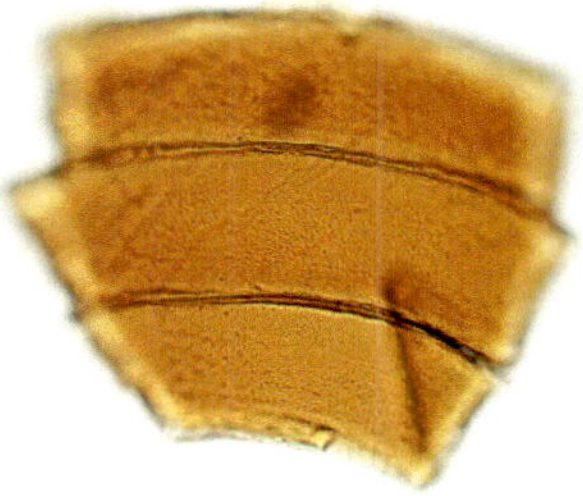
Fig. 10-67 (DW)	Fig. 10-68, SEM (MC)	Fig. 10-69, light microscope (RR)

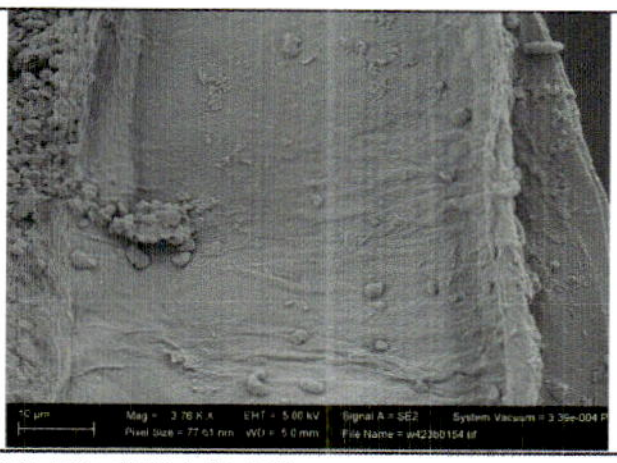	
Fig. 10-70, SEM (MC)	Fig. 10-71, light microscope (KW)

Acknowledgements We would like to thank Nadia Abdelahad, Dietmar Jäger, Ingeborg Soulié-Märsche and Zofija Sinkevičienė for comments.

Abbreviations

SEM Scanning electron microscopy

Drawings/Photos

DJ Dietmar Jäger
DW Dagmar Wassong (in: van de Weyer and Schmidt, 2018)
ISM Ingeborg Soulié-Märsche (in: Soulié-Märsche, 1999)
KW Klaus van de Weyer
LD Luc Denys
MC Michelle T. Casanova

RR Roman Romanov (Figs. 10i-44 and 10.25 modified from Hollerbach and Krassavina, 1983)
SC Santos Cirujano Bracamonte
WK Werner Krause

References

Abdelahad, N., & Piccoli, F. (2017). Report on Charophytes from rice fields in northern Italy including the alien species *Chara fibrosa* ssp. benthamii. *Plant Biosystems—An International Journal Dealing with all Aspects of Plant Biology*. https://doi.org/10.1080/11263504.2017.1320310

Bailly, G., & Schaeffer, O. (2010). Guide illustré des Characées du nord-est de la France. Conservatoire Botanique National de Franche-Comté, Besançon. 96 p.

Bazzichelli, G., & Abdelahad, N. (2009). *Alghe d'aqua dolce d'Italia* (p. 89). Università degli Studi di Roma la Sapienza.

Becker, R., et al. (2016). Beschreibung der Characeen-Arten Deutschlands. In: Arbeitsgruppe Characeen Deutschlands (Hrsg.): Armleuchteralgen - Die Characeen Deutschlands (pp. 209–572). Springer Spektrum. ISBN 978-3-662-47796-0

Blaženčić, J., & Stevanović, B. (2018). Rediscovery of the endemic species *Chara rohlenae* Vilh. 1912 (Characeae)—believed extinct—on the Balkan Peninsula. *Botanica Serbia, 42*, 109–115. http://botanicaserbica.bio.bg.ac.rs/arhiva/pdf/2018_42_1_706_full.pdf

Blindow, I., Krause, W., Ljungstrand, E., & Koistinen, M. (2007). Bestämningsnyckel för kransalger i Sverige. *Svensk Botanisk Tidskrift, 101*, 165–220.

Bruinsma, J., Denys, L., Krause, W., Mes, R., Nat, E., & van Raam, J. 2018. Determinatietabel van Kranswieren van de Benelux: 112 pp., Stichting Jeugdondsuitgeverij. ISBN 9789051070606.

Bryant, J. A., & Stewart, N. (2011). Order Charales. In: D. M. John, B. A. Whitton, & A. J. Brook (Eds.), *The freshwater algal flora of the British Isles* (pp. 742–765). Natural History Museum, Cambridge University Press.

Casanova, M. (2013). Review of the species concepts *Chara fibrosa* and *C. flaccida* (Characeae, Charophyceae). *Australian Systematic Botany, 26*, 291–297.

Cirujano, S., Cambra, J., Sánchez Castillo, P. M., Meco, A., & Flor Arnau, N. (2008). *Flora ibérica. Algas continentales.Carófitos (Characeae)* (p. 132). Real Jardín Botánico.

Corillion, R. (1957). Les Charophycées de France et d'Europe Occidentale. Bulletin de la Société Scientifique de Bretagne 32, fasc. h.s. 1 & 2. also: Traveaux du Laboratoire de Botanique de la Faculté des Sciences d'Angers, fasc. 11 & 1.

Groves, J., & Bullock-Webster, G. R. (1920, 1924). *The British Charophyta, Part I & II*. Ray Society.

Hollerbach, M. M., & Krassavina, L. K. (1983). *Determination of the freshwater algae of the USSR. Vol. 14–Charophyta* (190 pp., 72 pl.). Leningrad.

Holzhausen, A., Nowak, P., Niedrig, C., Feike, M., & Schubert, H. (2015). Morphometry of *Chara aspera, C. canescens, C. baltica* var. *baltica, C. baltica* var. *liljebladii* and *C. intermedia* oospores: Local variation versus taxonomic differences. *Aquatic Botany, 120*, 60–66.

John, D. M., & Moore, J. A. (1987). An SEM study of the oospore of some *Nitella* species (Charales, Chlorophyta) with descriptions of wall ornamentation and an assessment of its taxonomic importance. *Phycologia, 26*, 344–355.

Krause, W. (1990). Zum Neufund der *Chara kokeilii* A. Br. Im Skutariesee. *Carinthia II, 180*(100), 705–713.

Krause, W. (1997). *Charales (Charophyceae). Süßwasserflora von Mitteleuropa, Bd. 18*. G Fischer, Jena/Stuttgart/Lübeck/Ulm.

Langangen, A. (2007). *Charophytes of the Nordic countries* (101 pp.). Saeculum.

Langangen, A. (2010). *Chara hellenica* spec. nov. and other Charophytes collected in Corfu and Amorgos islands (Greece). *Studia Botanica Hungarica., 41*, 21–30.
Langangen, A. (2012). Charophytes (Charales) from Crete (Greece) collected in 2010. *Flora Mediterranea*, 25–32. https://doi.org/10.7320/FlMedit22.025
Langangen, A., Koistinen, M., & Blindow, I. (2002). The charophytes of Finland. *Memoranda Societatis pro Fauna Et Flora Fennica, 78*, 17–48.
Migula, W. (1889–1897). *Die Characeen Deutschlands, Oesterreichs und der Schweiz*. In: Rabenhorst L (Hrsg) Kryptogamenflora, 2. Aufl. Bd. V. Kummer, Leipzig.
Moore, J. A. (1986). *Charophytes of Great Britain and Ireland*. BSBI Handbook 5.
Mouronval, J. B., Baudouin, S., Borel, N., Soulié-Märsche, I., Klesczewski, M., & Grillas, P. (2015). *Guide des Characées de France méditerranéenne*. Office national de la chasse et de la faune sauvage.
Muller, S. D., Rhazi, L., & Soulie-Märsche, I. (2017). Diversity and distribution of Characeae in the Maghreb (Algeria, Morocco, Tunisia). *Crytogamie Algologie, 38*(3), 201–251.
Ray, S., Pekkari, S., & Snoeijs, P. (2001). Oospore dimensions and wall ornamentation patterns in Swedish charophytes. *Nordic Journal of Botany, 21*, 207–224.
Romanov, R. E., Gontcharov, A. A., & Barinova, S. S. (2015). *Chara globata* Mig (Streptophyta: Charales): Rare species revised. *Fottea, 15*, 39–50.
Romanov, R. E., Nikulin, A. Y., Nikulin, V. Y., & Gontcharov, A. A. (2022). New species *Chara oryzae* and a new section *Corillionia* of *Chara* (Charales, Charophyceae) from European Mediterranean rice fields. *European Journal of Phycology, 57*(3), 328–342. https://doi.org/10.1080/09670262.2021.1987527
Schubert, H., & Blindow, I. (2004). *Charophytes of the Baltic Sea* (326 pp.). Ganter [The Baltic Marine. Biologists Publication 19].
Soulié-Märsche, I. (1999). Extant Gyrogonite populations of *Chara zeylanica* and *Chara haitensis*: Implications for taxonomy and Palaeoecology. *Australian Journal of Botany, 47*, 371–382.
Soulié-Märsche, I., & Muller, S. D. (2016). New finds of *Chara oedophylla* Feldmann in Tunisia: Significance of sejoined gametangia. *Botanica Serbia, 40*, 229–236.
Stewart, N. (s.d.). Key to the stoneworts of the Norfolk Broads (8 pp.). Unpublished workshop notes for the Broads Authority, UK.
Troia, A., Napolitano, T., van de Weyer, K., & Romanov, R. E. (2018). New records of Characeae for Sicily (Italy). In *22nd Meeting of the Group of European Charophytologists (GEC)*, Palermo, Italy, 17–21 September 2018 (p. 36). https://www.unipapress.it/it/category/atti-e-convegni_2/scienze-biologiche_27/
Urbaniak, J., Langangen, A., & van Raam, J. (2012). Oospore wall ornamentation in the genus *Tolypella* (Charales, Charophyceae). *Journal of Phycology, 48*, 1538–1545.
Urbaniak, J., & Gąbka, M. (2014). *Polish Charophytes*. Universytet Przyrodiczy we Wroclawiu.
van de Weyer, K. (2016). Bestimmungsschlüssel. In Arbeitsgruppe Characeen Deutschlands (Hrsg.), *Armleuchteralgen - Die Characeen Deutschlands* (pp.193–208). Springer Spektrum. ISBN 978-3-662-47796-0
van de Weyer, K., & Schmidt, C. (2018). Bestimmungsschlüssel für die aquatischen Makrophyten (Gefäßpflanzen, Armleuchteralgen und Moose) in Deutschland: Band 1: Bestimmungsschlüssel. 2., überarbeitete Auflage. *Fachbeiträge des LfU Brandenburg, 119*, 180 pp.
van Raam, J. (1998). *Handboek Kranswieren*. Chara boek.
Wood, R. D., & Imahori, K. (1964). *A revision of the Characeae. Iconograph of the Characeae* (XXV + 7 pp., icones 1–395). Cramer.
Wood, R. D. (1965). *A revision of the Characeae* (p. 904). Cramer Verlag.

Part II
Taxonomic Section

Chapter 11
Introduction to the Species Chapters

Irmgard Blindow, Michelle T. Casanova, Luc Denys, Thomas Gregor, Heiko Korsch, Emile Nat, Roman Romanov, Hendrik Schubert, Nick Stewart, and Klaus van de Weyer

This chapter gives an introduction to the comprehensive species-specific information available for all European charophytes. All chapters follow the taxonomic concept outlined in detail in Chap. 5, for which the determination key (Chap. 10) was developed, and the checklist (Chap. 2), which includes all taxa retained by the authors of this book.

I. Blindow (✉)
Biologische Station Hiddensee, Universität Greifswald, 18565 Kloster, Germany
e-mail: blindi@uni-greifswald.de

M. T. Casanova
273 Casanova Rd, 3351 Westmere, Victoria, Australia
e-mail: mt.casanova@federation.edu.au

L. Denys
Research Institute for Nature and Forest (INBO), VAC Herman Teirlinck Havenlaan 88 Bus 73, 1000 Brussels, Belgium
e-mail: luc.denys@inbo.be

T. Gregor
Department of Botany and Molecular Evolution, Senckenberg Research Institute and Natural History Museum Frankfurt, Senckenberganlage 25, 60325 Frankfurt am Main, Germany
e-mail: thomas.gregor@senckenberg.de

H. Korsch
Am Werragrund 1, 98660 Themar, Germany
e-mail: heiko.korsch@t-online.de

E. Nat
Dutch Information Center on Charophytes, Naturalis Biodiversity Center, P.O. Box 9517, 2300 RA Leiden, The Netherlands
e-mail: e.nat@kranswieren.nl

R. Romanov
Nišice Ujtin Potok b.b., Dobra Voda, 85356 Bar, Montenegro
e-mail: streptophytes@gmail.com

H. Schubert et al. (eds.), *Charophytes of Europe*,
https://doi.org/10.1007/978-3-031-31898-6_11

Criteria to separate the genera are provided and discussed in Chap. 13 "characteristics of the genera". While a number of species can be identified rather easily (see Chap. 10), others are hard to distinguish even for experienced botanists. This is especially true for species belonging to the subsection *Agardhia* and the *Hartmania* group within the *Chara* genus. Both groups are therefore described separately in Chaps. 14 and 15. There the reader will also find details about the character combinations required for identification to species level within these sections. In addition, careful reading of the morphological descriptions and the section "Identification and risk of confusion" is particularly advised for taxa belonging to these groups.

The single species chapters include all charophyte taxa at species level recorded from Europe, in alphabetic order by their accepted names. For some taxa, the taxonomic status still has to be clarified; these are treated as "taxa with uncertain affiliation". These taxa should (and will) be investigated in more detail during the coming years. As we believe that there are valid arguments to treat them as separate entities, they are also included in the determination key (Chap. 10). Some additional taxa are not included in the key, but only described in the addendum. They refer to published records of species for which the occurrence in Europe could not be verified by herbarium material or later field investigations. Both taxa with "uncertain affiliation" and those in the "addendum" follow at the end of the species chapters.

Apart from some "taxa with uncertain affiliation" and "addendum" chapters, which may be rather short, all species chapters have the same structure: They start with their formal "**Taxonomy**", limited to names which were widely used until recently or that are currently considered valid. Scientific names are mentioned with their author(s), including common synonyms, type material and eventual remarks on taxonomy and nomenclature. Special attention is paid to nomenclatural issues and correct typification, for which Thomas Gregor and Roman Romanov performed in-depth checks. Local names are given for different European languages, if available. The following "**Morphology**" section gives detailed information based on literature and personal observations. It describes the general habitus of the species and then gives information on size, axis, cortex, spine-cells, stipulodes, branchlets, bract-cells, bracteoles, and sexuality. Gametangia are described with the dimensions of oogonia, oospores, and antheridia summarised in separate tables, if available. In the details about oogonia, we use the term "no. of helical stripes" and avoid the term "no. of

H. Schubert
Institut für Biowissenschaften, Universität Rostock, Albert-Einstein-Straße 3, 18059 Rostock, Germany
e-mail: hendrik.schubert@uni-rostock.de

N. Stewart
Glastonbury 4 Hexton Road, BA6 8HL, UK
e-mail: nfstewart@freeuk.com

K. van de Weyer
lanaplan, Lobbericher Str. 5, 41334 Nettetal, Germany
e-mail: klaus.vdweyer@lanaplan.de

convolutions". The latter term is used in many publications, but may lead to misunderstanding, as one might relate it to the number of convolutions per spiral cell, i.e., 1/5 of the number of helical stripes (see glossary). For a number of species, light microscope or SEM images show the surface structure of the oospore membrane. Most of these images were provided by Michelle T. Casanova and Roman Romanov. If present, nodal and rhizoidal bulbils are also described. Chromosome numbers are only given for a few species, when they are important for species delineation (e.g., *Nitella opaca* and *N. flexilis*). Most recent literature sources do not provide such information, and the often wide ranges mentioned in older publications may be due to methodology. Corillion (1975), for example, gave n = 14, 24 and 42 for *Chara virgata*. Obviously, some older authors were not aware that the large internodal cells of charophytes are polynucleate (see Chap. 3). "**Phenology and reproduction**" includes information about the life cycle, the time of the year when gametangia and oospores can be found, and hibernation.

The section "**Habitat conditions and ecology**" includes the usual habitat types, substrate, nutrient conditions, water and sediment chemistry (e.g. pH, calcium concentration, salinity and conductivity), the observed depth ranges, thermal and altitudinal ranges and commonly associated macrophytes. The section "**Variability**" presents the range of morphological characters and the most common varieties and/or forms including problematic morphotypes. Hints and difficulties in distinguishing the species from other, similar species are given under the heading "**Identification and risk of confusion**". In addition to the determination key (see Chap. 10), this section will give valuable information concerning species identification.

The section "**Distribution**" describes the global distribution of the taxon and gives a detailed description and a map of its occurrence in Europe. These maps show former and recent records of the species (for detailed information, see Chap. 12) and were compiled by Heiko Korsch, who meticulously gathered all available records. Without the help of many field workers and colleagues to check herbarium records these maps would not have their present quality. Heiko Korsch dedicated Chap. 12 to all these volunteers, acknowledging their valuable contribution.

As submerged macrophytes, charophytes are sometimes hard to investigate. Especially in larger and deeper water bodies, methods such as scuba diving and the use of special sampling equipment from a boat are necessary to record the occurring taxa completely (van de Weyer, 2007). Because of the highly varying phenology of single species, such investigations should be done preferably in different seasons. Several annual species such as *Sphaerochara intricata* and *Nitella capillaris* survive a long time in the sediments by means of oospores and are only found as "green plants" in certain years.

The number of times a certain species was recorded not only reflects its frequency, but also the intensity of charophyte studies. Charophytes were rather intensively collected between 1820 and 1900. During this time period, however, investigations were not systematic and rarely gave much information on habitat characteristics. Herbarium collections are essential, because taxonomic concepts and diagnostic criteria changed substantially through time. Moreover, preserved material allows researchers to check and expand on published information. Overall, our knowledge about the occurrence of charophytes in Europe during the nineteenth century is rather poor. During the twentieth century, a number of monographs of charophytes were published in different European countries (see Chap. 1 for some examples).

Many charophytes have severely declined during the last decennia as a result of increasing eutrophication pressure. Amongst others, this has led to the particular consideration of '*Chara* vegetations' in the European Union's Habitats Directive and their use as indicators of ecological status in the Water Framework Directive. Increased attention towards applied aspects has also resulted in more systematic investigations of the distribution, ecology and taxonomy of charophytes in many European countries during the last decades, and finally, substantially improved our knowledge about this group. Nevertheless, large differences still remain among regions, and the distribution maps not only reflect species occurrence, but also this imbalance of coverage. The section "**Status and threats**", summarises conservation issues relating to the species in different European countries. All "**References**" at the end of each species chapter are based on a comprehensive reference collection compiled and provided by Thomas Gregor, who deserves a special acknowledgment.

References

Corillion, R. (1975). *Flore et végétation du Massif Armoricain. T. IV. Flore des Charophytes (Characées) du Massif Armoricain et des contrées voisines d'Europe occidentale*. Jouve, Paris [in French].

van de Weyer, K. (2007). *Die Bedeutung von Tauchuntersuchungen bei der Erfassung von Makrophyten in Seen und Fließgewässern*. Tagungsbericht Dresden 2006. Deutsche Gesellschaft für Limnologie (DGL), Werder [in German].

Chapter 12
Introduction to the Distribution Maps

Heiko Korsch

The distribution maps of the Charophytes use the grid-system of the Atlas Florae Europapaeae, amended by grid-cells for Cyprus, for some Greek islands in the Aegean Sea and at the eastern boarder of Europe. The grid-cells have an average area of about 2500 km^2. The dots are placed in the centre of the grid-cell.

The background-map is that of openstreetmap (Fig. 12.1).

The data collection was organised country-wise, coordinated by Heiko Korsch. For every country, I tried to find a country-contributor who collected as many charophyte records as possible from literature, herbaria and field investigations. For some countries, e.g., Poland, Russia and Ukraine, almost all records have been verified, whereas for other countries, such a thorough verification could not be performed. For some countries, unfortunately, a contributor could not be found. In such cases I summarised as many published records as possible. Moldova is the only country where almost no data are available. Unfortunately, the web page "laji.fi" which is similar to species recording pages from Norway and Sweden (see webpages mentioned below) was not available in time to be considered. So, it was not possible to use the additional distribution data from there.

All data were collected in tables, and to all the AFE-grid-number was added. Afterwards they were imported into a geographic information system. Please note that the density of records is quite different within Europe. In some countries there are many actual mapping activities, in others almost none. For that reason it cannot be concluded that dots showing (only) historical records mean that the species has disappeared from there.

The number of charophyte species per AFE-gridcell in the map therefore depends mainly on two reasons, i.e., species richness and intensity of mapping. Which of it is more important differs from country to country.

H. Korsch (✉)
Themar, Germany
e-mail: heiko.korsch@t-online.de

H. Schubert et al. (eds.), *Charophytes of Europe*,
https://doi.org/10.1007/978-3-031-31898-6_12

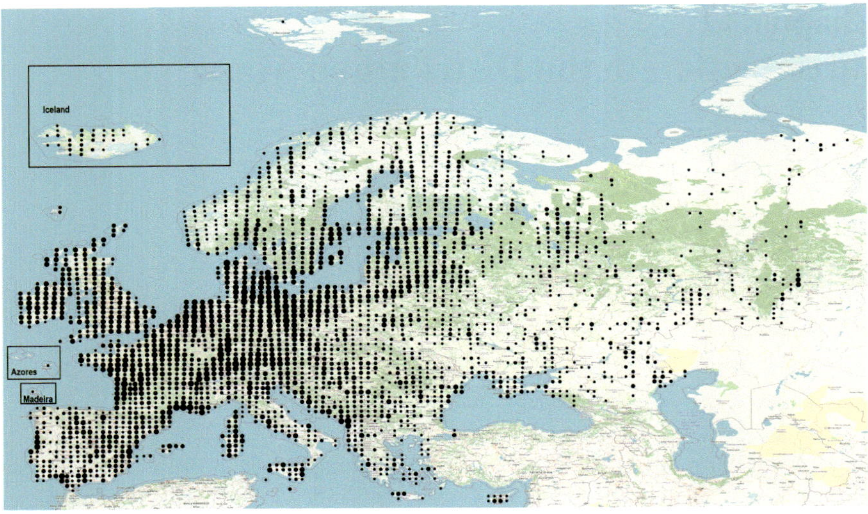

Fig. 12.1 Charophyte diversity. Count of charophyte species over the whole time period per AFE-gridcell symbolized by 6 different sizes of points. The smallest points are records of 1 or 2 different species; the biggest points are records of more than 20 species, the others are intermediate

Black dots are records from 1980 onwards, red dots last record from 1950 to 1979 and yellow dots last record before 1950; doubtful published records are marked with "?"

Country-contributors

Albania: Kashta, Lefter
Austria: Hohla, Michael; Pall, Karin
Belarus: Vishnyakov, Vasily; Romanov, Roman
Belgium: Denys, Luc
Bosnia-Herzegovina: Blaženčić, Jelena
Bulgaria: scarce published records, compiled by Korsch, Heiko
Croatia: Stankovic, Igor
Cyprus: Papastergiadou, Eva
Czech: published records compiled by Korsch, Heiko
Denmark: Båstrup-Spohr, Lars
Estonia: Torn, Kaire
Färöer: records from Langangen (2007), compiled by Korsch, Heiko
Finland: Koistinen, Marja
France: Gourvil, Johan
Germany: Korsch, Heiko
Great Britain: Stewart, Nick
Greece: Papastergiadou, Eva
Hungary: Mesterházy, Attila

Iceland: Hrafnsdottir, Thora
Ireland: Stewart, Nick
Italy: Abdelahad, Nadia
Latvia: Zviedre, Egita
Lithuania: Sinkevičienė, Zofija
Luxembourg: Nat, Emile
Malta: scarce published records, compiled by Korsch, Heiko
Macedonia: Blaženčić, Jelena; Trajanovska, Sonja
Moldova: almost no data available
Montenegro: Blaženčić, Jelena
Netherlands: Nat, Emile
Norway: Data from "Artsdatabanken.no"; mainly supervised by Langangen, Anders
Poland: Urbaniak, Jacek; Gąbka, Maciej
Portugal: Rodrigo, Maria; additional published records, compiled by Korsch, Heiko
Romania: published records (mainly from Caraus, 2017), compiled by Korsch, Heiko
Russia: Romanov, Roman; Chemeris, Elena
Serbia & Kosovo: Blaženčić, Jelena; Marković, Aleksandra
Slovakia: scarce published records, compiled by Korsch, Heiko
Slovenia: Blaženčić, Jelena; additional published records, compiled by Korsch, Heiko
Spain: Rodrigo, Maria
Sweden: Data from Artdatabanken.se downloaded, additional historical records from Langangen (2007), compiled by Korsch, Heiko
Switzerland: Sager, Lionel; Jutzi, Michael
Turkey (European part): scarce published records, compiled by Korsch, Heiko
Ukraine: Borisova, Olena; Romanov, Roman

The distribution maps could not have been produced without the contribution of a huge number of people and some organisations who collected the data. I would like to thank and honour all of these contributors. I tried not to forget anybody in the list below, but this seems almost unavoidable because of the huge number of people. In that case I would like to apologize.

List of contributors for the distribution data; surname, initial(s) of first name(s), (country, city) are given. (Org.) = Organisation

Abbink, C. (the Netherlands, Vriezenveen)
Abts, U. (Germany, Krefeld)
Achterkamp, B. (the Netherlands, Culemborg)
Adolphy, K. (Germany, Erkrath)
Albrecht, J. (Germany, Bielefeld)
Alexander, P. (Germany, Kassel)
Almer, W. (Germany, Regensburg)
Altenburg & Wymenga, (Org.) (the Netherlands, Veenwouden)

Alterra, (Org.) (the Netherlands, Wageningen)
Altrock, M. (Germany, Schinkel)
Amesz, M. (the Netherlands, Amsterdam)
Andeweg, R. (the Netherlands, Rotterdam)
Annema, M. (the Netherlands, Rotterdam)
Aptroot, A. (the Netherlands, Heemskerk)
Aqualysis, (Org.) (the Netherlands, Zwolle)
Arendt, K. (Germany, Gerswalde)
Arendt, W. (Germany, Rees)
Arenhövel, C. (Germany, Weimar)
Arneth, S. (Germany, Deggendorf)
Arp, W. (Germany, Berlin)
Arts, G. (the Netherlands, Wageningen)
Auerswald, J. (Germany, Dreba)
Baade, H. (Germany, Altenburg)
Baart, I. (Austria, Mödling)
Baartmans, M. (the Netherlands, Bennekom)
Bakker, K. (the Netherlands, Geleen)
Bakker, P. (the Netherlands, Hilversum)
Bakker, T. (the Netherlands, Breda)
Ballintijn, K. (the Netherlands, Purmerend)
Balster, W. (the Netherlands, Leiden)
Balters, H. (Germany, Westhausen)
Bardehle, A. (Germany, Essen)
Barendse, R. (Belgium, Genk)
Barth, U. (Germany, Tann)
Baum, S. (Germany, Hinte)
Beck, M. (Germany, Karlsruhe)
Becker, R. (Germany, Oldenburg)
Bednár, F. (Slovakia, Ružomberok)
Behlert, R. (Germany, Haltern)
Bekel, L. (Germany, Frechen)
Bellstedt, R. (Germany, Gotha)
Beltz, A. (Germany, Erndtebrück)
Benigni, E. (Italy, Arpa Piemonte)
Benkert, D. (Germany, Potsdam)
Berg, C. (Germany, Rostock)
Berger, D. (Germany, Jena)
Beringen, R. (the Netherlands, Nijmegen)
Bernhardt, K.G. (Austria, Wien)
Best, E. (the Netherlands, Amsterdam)
Beutler, D. (Germany, Beeskow)
Biemans, C. (the Netherlands, Amsterdam)
Bier, A. (Germany, München)
Biewald, I. (Germany, Jena)

Bijl, R. (the Netherlands, –)
Bikker, H. (the Netherlands, Druten)
Binjamin, F. (France, Vincennes)
Bintzer, W. (Germany, Clausthal-Zellerfeld)
Blaženčić, J. (Serbia, Belgrade)
Blindow, I. (Germany, Kloster)
Blischke, H. (Germany, Hannover)
Blümel, C. (Germany, Rostock)
Blüml, V. (Germany, Osnabrück)
Bockhacker, K. (Germany, Jena)
Boedeltje, G. (the Netherlands, Lochem)
Boersma, W. (the Netherlands, –)
Böger, K. (Germany, Darmstadt)
Böhme, B. (Germany, Annaburg)
Böhnert, W. (Germany, Kurort Hartha)
Böhnig, T. (Germany, Jenalöbnitz)
Bolander, E. (Germany, Isny)
Bolbrinker, P. (Germany, Mölln)
Bolier, A. (the Netherlands, –)
Bolpagni, R. (Italy, Emilia-Romagna, Parma)
Bonder, M. (the Netherlands, Maastricht)
Bönsel, D. (Germany, Eberstadt)
Boonstra, H. (the Netherlands, Leeuwarden)
Borisova, O. V. (Ukraine, Kyiv)
Borkowsky, O. (Germany, Braunschweig)
Borowski, A. (Germany, Altenburg)
Bosma, H. (the Netherlands, Leeuwarden)
Botanical Society of Britain and Ireland (Org.) (Britain and Ireland)
Botta, P. (Italy, Arpa Piemonte)
Bottalico, A. (Italy, Puglia, Bari)
Bouman, A. (the Netherlands, Weesp)
Brabantse Delta, (Org.) (the Netherlands, Breda)
Brand, J. (Germany, Wildeshausen)
Braun, A. (Germany, Berlin)
Braun, W. (Germany, Karlsfeld)
Braun-Lüllemann, J. (Germany, Hohengandern)
Bräutigam, S. (Germany, Görlitz)
Bredenbeek, J. (the Netherlands, Hasselt)
Breitfeld, M. (Germany, Markneukirchen)
Brekelmans, F. (the Netherlands, Utrecht)
Bremer, G. (Germany, Münster)
Brick, H. (Germany, Kitzingen)
Brijker, H. (the Netherlands, –)
Bröckel, A. (Germany, Bayreuth)
Brockhaus, T. (Germany, Chemnitz)

Brouwer, E. (the Netherlands, Nijmegen)
Bruin, K. (the Netherlands, Den Hoorn, Texel)
Bruinsma, J. (the Netherlands, Breugel)
Brullo, S. (Italy, Sicily)
Bruns, E. (Germany, Hildesheim)
Brux, H. (Germany, Oldenburg)
Buchholz, A. (Germany, Rottenburg)
Bukowsky, N. (Germany, Templin)
Bülow, B. v. (Germany, Haltern)
Bureau Everts & De Vries, (Org.) (the Netherlands, Groningen)
Bureau Waardenburg, (Org.) (the Netherlands, Culemborg)
Buro Bakker, (Org.) (the Netherlands, Assen)
Buro Elodea, (Org.) (the Netherlands, Drachten)
Büscher, D. (Germany, Dortmund)
Büscher-Wenst, E. (Germany, Königslutter)
Busse, J. (the Netherlands, –)
Busse, W. (the Netherlands, –)
Calise, C. (Italy, Liguria, Genova)
Calvo, S. (Italy, Sicily)
Caspari, S. (Germany, St. Wendel)
Caviglia, F. (Italy, Arpa Piemonte)
Chemeris, E. V. (Russia, Yaroslavl)
Christiansen, U. (Germany, Worms)
Christmann, K-H. (Germany, Düsseldorf)
Ciongwa, P. (Germany, Northeim)
Cisaro, C. (Italy, Arpa Piemonte)
Claasen, T. (the Netherlands, Leeuwarden)
Comin, S. (Italy, Friuli-Venezia-Giulia)
Conservatoire Botanique National Alpin & Contributors, (Org.) (France, Gap)
Conservatoire Botanique National de Bailleul & Contributors, (Org.) (France, Bailleul)
Conservatoire Botanique National de Brest & Contributors, (Org.) (France, Brest)
Conservatoire Botanique National de Corse & Contributors, (Org.) (France, Corte)
Conservatoire Botanique National de Franche - Comté & Contributors, (Org.) (France, Besançon)
Conservatoire Botanique National des Pyrénées et de Midi-Pyrénées & Contributors, (Org.) (France, Bagnères-de-Bigorre)
Conservatoire Botanique National du Bassin-Parisien & Contributors, (Org.) (France, Paris)
Conservatoire Botanique National du Massif central & Contributors, (Org.) (France, Chavaniac-Lafayette)
Conservatoire Botanique National méditerranéen de Porquerolles & Contributors, (Org.) (France, Hyères)
Conservatoire Botanique National Sud-Atlantique & Contributors, (Org.) (France, Audenge)

Cordes, U. (Germany, Anröchte)
Cornips, N. (the Netherlands, Nijmegen)
Costa, A. (Italy, Liguria, Genova)
Csiky, J. (Hungary, Pécs)
Čturtlikova, M. (Czechia, –)
Daemen, N. (the Netherlands, Woerden)
Dagnino, D. (Italy, Liguria, Genova)
Dahlke, S. (Germany, Greifswald)
Damino, R. (Italy, Sicily)
Damstra, Y. (the Netherlands, Apeldoorn)
De Boer, D. (the Netherlands, Utrecht)
De Boer, E. (the Netherlands, Zwolle)
De Bonte, A. (the Netherlands, Dalfsen)
De Bruin, A. (the Netherlands, Nijmegen)
De Bruyn, U. (Germany, Oldenburg)
De Goede, T. (the Netherlands, –)
De Goeij, S. (the Netherlands, Nijmegen)
De Groot, T. (the Netherlands, –)
De Jong, T. (the Netherlands, Culemborg)
De Koning, H. (the Netherlands, –)
De Lange, L. (the Netherlands, –)
De Vries, D. (the Netherlands, Groningen)
De Vries, H. (the Netherlands, Amsterdam)
De Vries, P. (the Netherlands, –)
De Zwart, A. (the Netherlands, Wilhelminadorp)
Dell'Uomo, A. (Italy, Marche)
Den Hartog, C. (the Netherlands, –)
Denys, L. (Belgium, Brussel)
Derks, P.J.T. (the Netherlands, –)
Desmots, D. (France, Noirmoutier-en-l'Ile)
Dethlefs, M. (Germany, Hermannsburg)
Dethloff, H.-J. (Germany, Trier)
Deutsch, A. (Germany, Bielefeld)
Deventer, M. (Germany, Viersen)
Dickorè, W. B. (Germany, Göttingen)
Didion, A. (Germany, Homburg/Saar)
Diekjobst, H. (Germany, Oberhambach)
Dienemann, H. (Germany, Leipzig)
Dienst, M. (Germany, Konstanz)
Diepolder, U. (Germany, Hohenau)
Dietze, H. (Germany, Uenglingen)
Diewald, W. (Germany, Straubing)
Dijkhuis, E. (the Netherlands, Nijmegen)
Dijkstra, B. (the Netherlands, Alkmaar)
Dilger, M. (Germany, Dresden)

Doef, R. (the Netherlands, Andijk)
Doege, A. (Germany, Triebischtal)
Doege, D. (Germany, Triebischtal)
Doll, R. (Germany, Greifswald)
Donath, F. (Germany, Oldenburg)
Dormels, A. (Germany, Nettetal)
Douwes, R. (the Netherlands, Hilversum)
Drews, H. (Germany, Kronshagen)
Drok, W.J. (the Netherlands, Arnhem)
Drozdowski, I. (Austria, Perchtoldsdorf)
Ducheck, M. (Germany, Creusen)
Duikteam Scuba Services, (Org.) (the Netherlands, Bennekom)
Duistermaat, L. (the Netherlands, Leiden)
Düll, R. (Germany, Bad Münstereifel)
Dullies, F. (Germany, Zwickau)
Dunger, I. (Germany, Görlitz)
Eberl, T. (Austria, Salzburg)
Eckartz-Nolden, G. (Germany, Bonn)
Eckstein, J. (Germany, Göttingen)
Eenshuistra, P. (the Netherlands, Venlo)
Egelmeers, J. (the Netherlands, Eindhoven)
Elfrink, W. (the Netherlands, –)
Ellermann, S. (Germany, Halle/Saale)
Elvio, F. (Italy, Arpa Lombardia)
Engelbach, G. (Germany, Battenberg)
Erzberger, P. (Germany, Berlin)
Eßer, (Germany, Mönchengladbach)
Ettrich, L. (Germany, Mettmann)
Everards, K. (the Netherlands, Utrecht)
Eysink, F. (the Netherlands, Witharen)
Fabricius, K. (Germany, Schinkel)
Falci, A. (Italy, Sicily)
Farenholtz, K. (Germany, Porta Westfalica)
Fechtler, T. (Germany, Göttingen)
Feder, J. (Germany, Bremen)
Federici, G. (Italy, Lombardia, Bergamo)
Feenstra, M. (the Netherlands, –)
Feldhaus, D. (Germany, Halle)
Fels, J.-G. (Germany, Bösel)
Ferber, D. (Germany, St. Augustin)
Ferrara, L. (Italy, Arpa Piemonte)
Ferraro, V. (Italy, Sicily)
Fickel, H. (Germany, Mühlhausen)
Fickel, U. (Germany, Mühlhausen)
Fischer, C. (Germany, Dannenberg)

Fittje, J. (Germany, Oldenburg)
FLORON, (Org.) (the Netherlands, Nijmegen)
Foerster, E. (Germany, Kleve)
Fortino, D. (Italy, Arpa Lombardia)
Francescato, C. (Italy, Friuli-Venezia-Giulia)
Frank, D. (Germany, Halle/Saale)
Franke, T. (Germany, Hemhofen-Zeckern)
Franken, F. (the Netherlands, Nijmegen)
Freese, E. (Germany, Oldenburg)
Freitag, H. (Germany, –)
Frenken, T. (the Netherlands, 's Hertogenbosch)
Frenzel, B. (Germany, Apenrade (DK)
Freund, H.-J. (Germany, Bocholt)
Freymann, F. (Germany, Halle/Saale)
Friedrich, G. (Germany, Krefeld)
Friese, M. (Germany, Oberförstchen)
Fritsch, R. (Germany, Idar-Oberstein)
Fritz, R. (Germany, Hohenheim)
Fuchs, R. (Germany, Bochum)
Funcke, J. (Germany, Hannover)
Fürhaupter, K. (Germany, Schönkirchen)
Furnari, F. (Italy, Sicily)
Gaertner, M. (Germany, Görlitz)
Galm, M. (Germany, Bad Tölz)
Gandolfi, G. (Italy, Arpa Piemonte)
Ganterer, U. (Austria, Pörtschach)
Garibodi, L. (Italy, Arpa Lombardia)
Garniel, A. (Germany, Kiel)
Garve, E. (Germany, Sarstedt)
Gastaldi, E. (Italy, –)
Gausmann, P. (Germany, Herne)
Gebauer, P. (Germany, Klitten/Dürrbach)
Gebhardt, W. (Germany, Marktleuthen)
Geerlings, J. (Germany, Düsseldorf)
Gehroldt, E. (Germany, Gera)
Geiger, A. (Germany, Recklinghausen)
Gerdol, R. (Italy, –)
Gerken, R. (Germany, Celle)
Gerlach, A. (Germany, Clausthal-Zellerfeld)
Gerth, A. (Germany, Jena)
Geyer, H.-J. (Germany, Arnsberg)
Giardini, M. (Italy, Lazio, Roma)
Giesen, T. (the Netherlands, Ulft)
Gilli, C. (Austria, Wien)
Giordano, L. (Italy, –)

Glatfeld, M. (Germany, Bielefeld)
Gminder, A. (Germany, Jenalöbnitz)
Göbel, E. (Germany, Schleid)
Göckeritz, J. (Germany, Gera)
Göding, H. (Germany, Dingolfing)
Golde, A. (Germany, Freiberg)
Görner, M. (Germany, Jena)
Gorny, M. (Germany, Wunsiedel)
Goudsmits, K. (the Netherlands, Utrecht)
Gourvil, J. (France, Vincennes)
Grasselt, A. (Germany, Dresden)
Grätz, C. (Germany, Drieschnitz)
Grebe, T. (Germany, Oer-Erkenschwick)
Gregor, T. (Germany, Schlitz)
Grein, G. (Germany, Hildesheim)
Greiner, R. (Germany, Jena)
Groot, J. (the Netherlands, Broek op Langedijk)
Gross, E. (Italy, –)
Grote, S. (Germany, Braunschweig)
Grotenhuis, H. (the Netherlands, –)
Grünberg, H. (Germany, Goßwitz)
Grundmann, H. (Germany, Chemnitz)
Grunert, H. (Germany, Braunschweig)
Grunewald, R. (Germany, Rostock)
Grüttner, A. (Germany, Halle/Saale)
Guala, P. (Italy, Arpa Piemonte)
Guderitz, T. (Germany, Dresden)
Günther, K.-F. (Germany, Jena)
Gutte, P. (Germany, Leipzig)
Guttman, S. (Germany, Bernburg)
Gysels, J. (Belgium, Mechelen)
Haacks, M. (Germany, Hamburg)
Haese, U. (Germany, Aachen)
Hager, A. (Germany, Heuchelheim)
Hahn, S. (Germany, Demitz-Thumitz)
Hamann, U. (Germany, Schinkel)
Hänel, K. (Germany, Neuwürschnitz)
Hanenburg, J. (the Netherlands, –)
Hanspach, D. (Germany, Ortrand)
Hardtke, H.-J. (Germany, Possendorf)
Harlacher, R. (Germany, Maisach)
Hartenauer, K. (Germany, Halle/Saale)
Hauswirth, L. (Germany, Lippstadt)
Heckmanns, G. (Germany, Krefeld)
Heinemann, K. (Germany, Potsdam)

Heinrich, W. (Germany, Jena)
Heinrichs, J. (Germany, Göttingen)
Heinzel, K. (Germany, Schönkirchen)
Helmecke, K. (Germany, Wittersroda)
Helming, W. (Germany, Münster)
Hendriks, B. (the Netherlands, –)
Hendriks, J. (the Netherlands, –)
Hendriks, J. (the Netherlands, –)
Hendrix, M. (the Netherlands, –)
Henker, H. (Germany, Neukloster)
Hentschel, J. (Germany, Jena)
Hericks, F. (Germany, Saterland)
Herkommer, U. (Germany, Neu-Ulm)
Hermans, T. (the Netherlands, –)
Herwanger, H. (Germany, Bad Waldsee)
Herzberg, I. (Germany, –)
Hesse, J. (Germany, Recklinghausen)
Hibma, J. (the Netherlands, Dokkum)
Hilbig, W. (Germany, Petershausen)
Hinrichsen, A. (Germany, Rangsdorf)
Hirsch, G. (Germany, Jena)
Hoegen, B. (the Netherlands, Leeuwarden)
Hoentjen, B. (the Netherlands, –)
Hoesch, (Germany, –)
Hofbauer, M. (Austria, Wien)
Höglinger, F. (Austria, Lenzing)
Hohla, M. (Austria, Obernberg am Inn)
Hokken, M. (the Netherlands, Lelystad)
Holfeld, H. (Germany, Burghaun)
Höllgärtner, M. (Germany, Jockgrim)
Holsteijn, H. (the Netherlands, Groningen)
Homm, T. (Germany, Oldenburg)
Hooft-van Huijsduijnen, M. (the Netherlands, Zwolle)
Hoogheemraadschap Hollands Noorderkwartier, (Org.) (the Netherlands, Heerhugowaard)
Hoogheemraadschap van Rijnland, (Org.) (the Netherlands, Leiden)
Höpstein, G. (Germany, Bad Blankenburg)
Hospers, A. (the Netherlands, Assen)
Hübner, G. (Germany, Lautertal)
Hübner, T. (Germany, Haltern)
Hübschen, J. (Germany, Coesfeld-Lette)
Hull, H. van't (Germany, Husum)
Huls, J. (the Netherlands, Uithoorn)
Humberg, B. (Germany, Winsen/Luhe)
Husner, A. (Germany, Düsseldorf)

Iberite, M. (Italy, Lazio, Roma)
Ilardi, V. (Italy, Sicily)
Illig, H. (Germany, Luckau)
Inberg, H. (the Netherlands, Culemborg)
Irmscher, B. (Germany, Chemnitz)
Itjeshorst, W. (Germany, Wesel)
Jage, H. (Germany, Kemberg)
Jagel, A. (Germany, Bochum)
Jager, H. (the Netherlands, –)
Jäger, D. (Austria, Hohenems)
Janse, J. (the Netherlands, Nijmegen)
Jansen, A. (the Netherlands, –)
Jedzig, A. (Germany, Friedersdorf OL)
Jenemann, K. (Germany, Dresden)
Jeschke, L. (Germany, Greifswald)
Jessat, M. (Germany, Altenburg)
John, H. (Germany, Halle/Saale)
Jorda, C. (Germany, Schwäbisch Gmünd)
Jütte, M. (Germany, Marsberg-Padberg)
Kabus, T. (Germany, Seddin)
Kaiser, R. (Austria, Salzburg)
Kaiser, B. (Germany, Perl)
Kaiser, T. (Germany, Beedenbostel)
Kambergs, P. (Germany, Düsseldorf)
Kämpfe, S. (Germany, Weimar)
Kaplan, K. (Germany, Bad Bentheim)
Kashta, L. (Albania, Tirana)
Kasparz, S. (Germany, Lübben)
Keil, P. (Germany, Oberhausen)
Kempe, K. (Germany, Finsterwalde)
Kerkhof, D. (the Netherlands, Ridderkerk)
Keulen, S. (the Netherlands, –)
Kiechle, J. (Germany, Konstanz)
Kieckbusch, J. J. (Germany, Felm)
Kierkels, G. (the Netherlands, Baarle-Nassau)
Kiesewetter, H. (Germany, Crivitz)
Kirch, C. (Germany, Hannover)
Kirchner-Hessler, R. (Germany, Freiburg)
Kirste, A. (Germany, Königsbrück)
Kison, H.-U. (Germany, Quedlinburg)
Kläge, C. (Germany, Luckau)
Kleesadl, G. (Austria, Linz)
Kleinknecht, U. (Germany, Leipzig)
Klemm, G. (Germany, Berlin)
Klenke, F. (Germany, Bobritzsch)

Klotz, S. (Germany, Halle/Saale)
Klug, W. (Germany, Gotha)
Knappe, J. (Germany, Marburg)
Koch, E. (Germany, Gottmadingen)
Köck, U.-V. (Germany, Halle/Saale)
Koeman en Bijkerk, (Org.) (the Netherlands, Haren)
Kohl, M. (Germany, Kleingeschwenda)
Kohler, U. (Germany, Landsberg)
Köhler, H. (Germany, Leipzig)
Köhler, U. (Germany, Rositz)
Kölbel, A. (Germany, Kiel)
Koltzenburg, M. (Germany, Tübingen)
Kolvoort, W. (the Netherlands, –)
König, A. (Germany, Frankfurt)
König, O. (Germany, Essen)
Kordges, T. (Germany, Essen)
Korsch, H. (Germany, Jena)
Korte, E. (Germany, Riedstadt)
Koslowski, I. (Germany, Gladbeck)
Köstler, H. (Germany, Berlin)
Krausch, H.-D. (Germany, Potsdam)
Krause, W. (Germany, Aulendorf)
Krautkrämer, V. (Germany, Lippstadt)
Krebs, G. (Germany, Thale)
Krewel, E. (Germany, Bonn)
Krisai, R. (Austria, Braunau)
Krisch, H. (Germany, Greifswald)
Kroning, P. (Germany, Düsseldorf)
Kroy, J. (Germany, Templin)
Kruit, L. (the Netherlands, Nijmegen)
Krüner, U. (Germany, Mönchengladbach)
Kruspe, R. (Germany, Ottendorf-Okrilla)
Kubosch, R. (Germany, Siegen)
Kühn, I. (Germany, Halle/Saale)
Kulbrock, P. (Germany, Bielefeld)
Kummer, V. (Germany, Potsdam)
Kunzendorff, E. (Germany, Osterode)
Kurstjens, G. (the Netherlands, –)
Kurz, A. (Germany, Schleusingen)
Kusel-Fetzmann, E. (Austria, Mödling)
Küster, A. (Germany, Markkleeberg)
Lakmann, G. (Germany, Delbrück)
Lakušić, D. (Serbia, Belgrade)
Lambert, B. (France, Angers)
Lambert, E. (France, Angers)

Lambert, F. (the Netherlands, Eindhoven)
Lämmel, D. (Germany, Ratzeburg)
Landeck, I. (Germany, Finsterwalde)
Landelijk Informatiecentrum voor Kranswieren, (Org.) (the Netherlands, Leiden)
Landmann, M. (Germany, Oldenburg)
Lang, W. (Germany, Erpolzheim)
Langangen, A. (Norway, Oslo)
Langbehn, H. (Germany, Celle)
Langbroek, M. (the Netherlands, Katwijk)
Langbroek, W. (the Netherlands, Katwijk)
Langeweg, S. (the Netherlands, Heerhugowaard)
Langner, N. (Germany, Königswartha)
Lanjouw, B. (the Netherlands, Rottum)
Lansing, P. (the Netherlands, 's Hertogenbosch)
Lasen, C. (Italy, Veneto, Arson di Feltre)
Latini, M. (Italy, Lazio, Roma)
Leeuwerke, G. (the Netherlands, –)
Lefebvre, P. (Italy, Arpa Piemonte)
Leischner, P. (Germany, Erfurt)
Leithe, F. (Austria, Wien)
Lenglachner, F. (Austria, Salzburg)
Lenski, I. (Germany, Marburg)
Lenzen, S. (Germany, Mönchengladbach)
Leupold, P. (Germany, Hemhofen-Zeckern)
Leurs, W. (the Netherlands, Oldeberkoop)
Lienenbecker, H. (Germany, Steinhagen)
Lieneweg, H. (Germany, Halle/Saale)
Linke, C. (Germany, Kargow)
Lippold, H. (Germany, Jena)
Livretti, F. (Italy, Sardinia)
Löbnitz, G. (Germany, Erfurt)
Löbnitz, M. (Germany, Erfurt)
Loeb, R. (the Netherlands, Nijmegen)
Loos, G.-H. (Germany, Kamen)
Luckwaldt, G. v. (Germany, Hameln)
Lüdicke, T. (Germany, Eberswalde)
Lueg, H. (Germany, Freiberg)
Lugmair, A. (Austria, Alkoven)
Luijten, L. (the Netherlands, Groningen)
Lukács, B.A. (Hungary, Debrecen)
Lünterbusch, C. (Germany, Ahaus)
Luwe, M. (Germany, Kempen)
Madsack, G. (Germany, Hannover)
Mäemets, H. (Estonia, Tartu)
Maes, B. (the Netherlands, Utrecht)

Mandos, M. (the Netherlands, Waalwijk)
Manitz, H. (Germany, Jena)
Marcel, H. (the Netherlands, Utrecht)
Markgraf-Maué, K. (Germany, Kranenburg)
Marković, A. (Serbia, Belgrade)
Marrone, F. (Italy, Sicily)
Martin, C. (Germany, Schinkel)
Martins, L. (Germany, Magdeburg)
Mauersberger, H. (Germany, Brandenburg)
Mauersberger, R. (Germany, Templin)
Mause, R. (Germany, Düren)
Mayer, A. (Germany, Eichenau)
Mayer, G. (Germany, Albstadt)
Mayr, B. (Germany, Donauwörth)
Meço, M. (Albania, Tirana)
Meier, G. (Germany, Gera)
Meierott, L. (Germany, Gerbrunn)
Meijer, S. (the Netherlands, Zwolle)
Meijer, W. (the Netherlands, –)
Meineke, T. (Germany, Ebergötzen)
Melzer, A. (Germany, Iffeldorf)
Merluzzi, P. (Italy, Friuli-Venezia-Giulia)
Mes, R. (the Netherlands, Leiderdorp)
Mescheder, C. (Germany, Berlin)
Mesterházy, A. (Hungary, Debrecen)
Mesters, C. (the Netherlands, –)
Meyer, T. (Germany, Kiel)
Meyer-Spethmann, U. (Germany, Nordhorn)
Mezzena, R. (Italy, Friuli-Venezia-Giulia)
Michels, C. (Germany, Recklinghausen)
Miers, S. (Germany, Hannover)
Mietz, O. (Germany, –)
Millozza, A. (Italy, Lazio, Roma)
Mohl, R. (Germany, Jülich)
Molineri, P. (Italy, Piemonte?)
Molla, O. (Albania, Pogradec)
Möllenkamp, I. (Germany, Bramsche)
Moroni, R. (Italy, –)
Mouronval, J.B (France, Camargue)
Mrkvicka, A. (Austria, Perchtoldsdorf)
Mrotzek, A. (Germany, Greifswald)
Mulder, R. (the Netherlands, Leeuwarden)
Mullaj, A. (Albania, Tirana)
Müller, F. (Germany, Dresden)
Müther, K. (Germany, –)

Napolitano, T. (Italy, Sicily)
Naselli-Flores, L. (Italy, Sicily)
Nat, E. (the Netherlands, Amsterdam)
National Parks and Wildlife Service (Org.) (Ireland, Dublin)
Natuur en Milieufederatie Drenthe, (Org.) (Netherlands, Assen)
Natuurmonumenten, (Org.) (the Netherlands, 's Graveland)
Neikes, N. (Germany, Nettetal)
Neumann, J. (Germany, Ottendorf-Okrilla)
Nicola, A. (Italy, Arpa Piemonte)
Niederbichler, C. (Germany, Utting)
Niemeijer, I. (the Netherlands, Ooij)
Nienhaus, I. (Germany, Bonn)
Nijboer, R. (the Netherlands, Zwolle)
Nikolaus, H. (Germany, Danndorf)
Ochse, M. (Germany, Kleve)
Ochsmann, J. (Germany, Göttingen)
Oddo, E. (Italy, Sicily)
Odé, B. (the Netherlands, Nijmegen)
Olberding, P. (Germany, Holdorf)
Oldorff, S. (Germany, Stechlin)
Ondersteijn, R. (the Netherlands, –)
Oonk, M. (the Netherlands, 's Hertogenbosch)
Oosterveld, E. (the Netherlands, Veenwouden)
Oppel, T. (Germany, Gotha)
Oriolo, G. (Italy, Friuli-Venezia-Giulia)
Ostrow, H. (Germany, Grub am Forst)
Otto, A. (Germany, Hannover)
Otto, H.-W. (Germany, Bischofswerda)
Otto, R. (Germany, Gundelsheim)
Oving, B. (the Netherlands, –)
Ozinga, W. (the Netherlands, Assen)
Packet, J. (Belgium, Brussel)
Pall, K. (Austria, Wien)
Pandolfi, F. (Italy, Arpa Lombardia)
Pannocchia, M. (Italy, Arpa Piemonte)
Panzeca, P. (Italy, Sicily)
Papajewski, W. (Germany, Bochum)
Paproth, R. (Germany, Havelberg)
Pardey, A. (Germany, Essen)
Pasch, D. (Germany, Nettersheim)
Pätzold, F. (Germany, Hügelsheim)
Paul, M. (Germany, Radebeul)
Peeters, G. (the Netherlands, –)
Perçini, D. (Albania, Pogradec)
Perdeck, M. (the Netherlands, –)

Petersen, J. (Germany, Hildesheim)
Peterson, J. (Germany, Halle/Saale)
Petrick, W. (Germany, Luckau)
Petzold, F. (Germany, Berlin)
Petzold, S. (Germany, Braunschweig)
Peukert, M. (Germany, Frankfurt/Main)
Piccoli, F. (Italy, Emilia- Romagna, Ferrara)
Pick, T. (the Netherlands, Uithoorn)
Pietsch, W. (Germany, Dresden)
Pilsl, P. (Austria, Salzburg)
Pizzo, V. (Italy, Arpa Piemonte)
Pizzuti, R. (Italy, Friuli-Venezia-Giulia)
Plantinga, J.-E. (the Netherlands, Zeist)
Ploß, E. (Germany, Tann)
Poelgeest, A. (the Netherlands, –)
Polli, E. (Italy, Friuli-Venezia-Giulia)
Pot, R. (the Netherlands, Oosterhesselen)
Provincie Noord-Brabant, (Org.) (the Netherlands, 's Hertogenbosch)
Provincie Utrecht, (Org.) (the Netherlands, Utrecht)
Pudwill, R. (Germany, Gifhorn)
Pukacz, A. (Germany, Slubize (PL)
Pusch, J. (Germany, Bad Frankenhausen)
Put, E. (the Netherlands, –)
Quirini, C. (Germany, Bielefeld)
Raabe, U. (Germany, Marl)
Raangs, K. (the Netherlands, Groningen)
Raape, C. (Germany, Bottrop)
Rabe, H. (Germany, Moschwitz)
Radkowitsch, A. (Germany, Pforzheim)
Raith, H. (Germany, Halle/Saale)
Ratai, L. (Germany, Feldberg)
Ratsch, A. (Germany, Berlin)
Rätzel, S. (Germany, Frankfurt/Oder)
Rau, H. (Germany, Münster-Roxel)
Rechinger, K. (Austria, Wien)
Reich, D. (Austria, Wien)
Reijerse, F. (the Netherlands, Nijmegen)
Reimann, M. (Germany, Heilbronn)
Reuter, D. (Germany, Hermsdorf)
Ristow, M. (Germany, Berlin)
Ritter, B. (Germany, Arnsberg)
RIZA, (Org.) (the Netherlands, Lelystad)
Rode, P. (Germany, Stadtroda)
Rödel, D. (Germany, Osnabrück)
Roden, C. R. (Ireland, Co. Galway)

Rohner, M.-S. (Germany, Berlin)
Romanov, R. E. (Montenegro, Dobra Voda)
Rönicke, H. (Germany, Magdeburg)
Rosati, G. (Italy, Lazio?)
Rosenbauer, S. (Germany, Korntal-Münchingen)
Rosenkranz, F. (Germany, Dessau)
Rossenaar, A.-J. (the Netherlands, Amsterdam)
Rossi, A. (Italy, Arpa Piemonte)
Rößler, M. (Germany, Leipzig)
Roth, K. (Germany, Maibach)
Rothe, J. (Germany, Jena)
Rothschuh, L. (Germany, Krefeld)
Rott, E. (Austria, Innsbruck)
Rott, E. (Italy, –)
Rovers, J. (the Netherlands, –)
Rozema, W. (the Netherlands, –)
Rudat, C. (Germany, Greifswald)
Runge, F. (Germany, Münster-Kinderhaus)
Ruocco, M. (Italy, Emilia-Romagna)
Rüther-Lülfsmann, P. (Germany, Bielefeld)
Ruthsatz, B. (Germany, Trier)
Ruygrok, T. (the Netherlands, 's Hertogenbosch)
Rydl, V. (Germany, Köln)
Sackwitz, P. (Germany, Dettingen)
Samietz, R. (Germany, Gotha)
Samuels, J. (the Netherlands, Etten-Leur)
Sander, R. (Austria, St. Pölten)
Sänger, H. (Germany, Chrimitschau)
Santisi, S. (Italy, –)
Sauberer, N. (Austria, Wien)
Sauter, A. (Austria, Salzburg)
Schaarschmidt, J. (Germany, Plauen)
Schacherer, A. (Germany, Langenhagen)
Schaible, R. (Germany, Rostock)
Schäpers, J. (Germany, Recklinghausen)
Scharbert, A. (Germany, Winnekendonk)
Schätzle, F. (Germany, Weida)
Schaubel, K. (Germany, Imsbach (Pfalz)
Schauer, T. (Germany, München)
Scheepens, M. (the Netherlands, –)
Scheers, K. (Belgium, Brussel)
Scheiberlich, G. (the Netherlands, Wageningen)
Schiedermayer, K. (Austria, Kirchdorf am der Krems)
Schlesinger, S. (Germany, Teningen)
Schlüter, R. (Germany, Recklinghausen)

Schlüter, U. (Germany, Schwerin)
Schmid, M. (Germany, Benningen)
Schmidt, C. (Germany, Münster)
Schmidt, D. (Germany, Potsdam)
Schmidt, H.-C. (Germany, Weida)
Schmidt, O. (Germany, Kaiserslautern)
Schmieder, K. (Germany, Hohenheim)
Schmitz, U. (Germany, Düsseldorf)
Schnabel, A. (Germany, Soest)
Schnabel, H. (Germany, Wittichenau)
Schneider, S. (Germany, Oslo)
Schoknecht, T. (Germany, Potsdam)
Schön, M. (Germany, Weiden)
Schönfelder, I. (Germany, Jena)
Schöpe, D. (Germany, Dresden)
Schorer, A. (Germany, München)
Schossig, A. (Germany, Görlitz)
Schraetz, E. (Germany, Krefeld)
Schröder, U. (Germany, Neustadt/Orla)
Schubert, H. (Germany, Rostock)
Schubert, K. (Germany, Sömmerda)
Schulte-Bocholt, A. (Germany, Gelsenkirchen)
Schumacher, W. (Germany, Mechernich)
Schunselaar, J. (the Netherlands, Enschede)
Schürer, I. (Germany, Schlettau/Dörfel)
Schuster, C. (Germany, Goldbach)
Schütz, P. (Germany, Essen)
Schütz, W. (Germany, Kiel)
Schütze, P. (Germany, Großpostwitz)
Schwamb, H. (Germany, Ludwigshafen?)
Schwarz, R. (Germany, Forst)
Schwarzer, A. (Germany, Löhnberg)
Schwieger, F. (Germany, Braunschweig)
Schwienbacher, E. (Italy, –)
Seegers, J. (the Netherlands, Utrecht)
Seelen, L. (the Netherlands, Haaren)
Seitz, B. (Germany, Berlin)
Sennert, G. (Germany, Grefrath)
Seuffert, A. (Germany, Greifswald)
Simon, B. (Germany, Plossig)
Simons, J. (the Netherlands, Amstelveen)
Sinning, F. (Germany, Edewecht)
Slingerland, P. (the Netherlands, Alkmaar)
Slobboda, S. (Germany, Dresden)
Sluschny, H. (Germany, Schwerin)

Smeekens, V. (Belgium, Brussel)
Sonneborn, I. (Germany, Bielefeld)
Sonneborn, W. (Germany, Bielefeld)
Sonnenburg, F. (Germany, Solingen)
Soulié-Märsche, I. (France, Montpellier)
Spaargaren, J.-J. (the Netherlands, Almere)
Specken, B. (the Netherlands, Amsterdam)
Spieß, H.-J. (Germany, Kratzeburg)
Spiess, R. (Germany, Serrahn)
Spin, J. (the Netherlands, Zwolle)
Spooren, M. (the Netherlands, –)
Spranger, E. (Germany, Bayreuth)
Sprick, P. (Germany, Hannover)
Springer, S. (Germany, München)
Spronk, J. (the Netherlands, Eindhoven)
Sprunkel, E. (Germany, Köln)
Spruyt, T. (the Netherlands, Heiloo)
Stapf, O. (Austria, Innsbruck)
Stark, H. (Germany, Ratingen)
Stegeman-Broos, E. (the Netherlands, Almelo)
Stelzer, D. (Germany, Riemerling)
Stephan, B. (Germany, Altenberge)
Stiller, N. (Germany, Stendal)
Stockmayr, S. (Austria, Stammersdorf/Wien)
Stöhr, O. (Austria, Debant)
Stolwijk, P. (the Netherlands, Enschede)
Stolz, F. (Austria, Innsbruck)
Stouthamer, W. (the Netherlands, Groningen)
Strumpf, K. (Germany, Altenburg)
Stuhr, J. (Germany, Kiel)
Stumpel, A. (the Netherlands, –)
Sumser, H. (Germany, Köln)
Süß, R. (Germany, Tannroda)
Swolfs, A. (the Netherlands, Heiloo)
Tamis, W. (the Netherlands, Leiden)
Tara, K. (Germany, Bochum)
Targan, H. (Germany, Nürnberg)
Täuber, T. (Germany, Northeim)
Täuscher, L. (Germany, Berlin)
Taux, K. (Germany, Oldenburg)
Teodorescu, E.C. (Romania, Bukarest)
Teppke, M. (Germany, Rethwisch)
Teunissen, W. (the Netherlands, Leiden)
Thaler, F. (Austria, Wien)
Thannhauser, M. (the Netherlands, Leeuwarden)

Thiel, H. (Germany, Rosdorf)
Thomas, W. (Germany, Münster)
Tigges, P. (Germany, Weilerswist)
Timmermann, T. (Germany, Berlin)
Tippmann, H. (Germany, Herbertingen)
Tolman, M. (the Netherlands, Groningen)
Tomasella, M. (Italy, Friuli-Venezia-Giulia)
Töppich, D. (Germany, Ullersdorf)
Torn, K. (Estonia, Tallinn)
Trapp, S. (Germany, Wörthsee)
Trbojević, I. (Serbia, Belgrade)
Troelstra, W. (the Netherlands, –)
Troia, A. (Italy, Sicily, Palermo)
Trompetter, R. (the Netherlands, Leiden)
Turcato, C. (Italy, Liguria, Genova)
Uhlig, H. (Germany, Weixdorf)
Uhlmann, D. (Germany, Dresden)
Ulmer, A. (Germany, Coburg)
Unruh, M. (Germany, Zeitz)
Vahle, H.-C. (Germany, Witten)
Van Beers, P. (the Netherlands, Bladel)
Van Berkel, W. (the Netherlands, Kaatsheuvel)
Van Bodegom, P. (the Netherlands, Leiden)
Van Boekel, W. (the Netherlands, Boijl)
Van Buggenum, H. (the Netherlands, Nijmegen)
Van de Haterd, R. (the Netherlands, Culemborg)
Van de Werf, B. (the Netherlands, –)
Van de Weyer, K. (Germany, Nettetal)
Van den Berg, M. (the Netherlands, Harderwijk)
Van den Dool, E. (the Netherlands, Utrecht)
Van den Nieuwenhuijzen, A. (the Netherlands, –)
Van der Burg, R. (the Netherlands, –)
Van der Burgh, L. (the Netherlands, Lelystad)
Van der Haar, P. (the Netherlands, Deventer)
Van der Ham, N. (the Netherlands, Alkmaar)
Van der Heide, Y. (the Netherlands, Veenwouden)
Van der Linden, J. (the Netherlands, Tilburg)
Van der Meijden, R. (the Netherlands, Oostvoorne)
Van der Schans, T. (the Netherlands, –)
Van der Slikke, W. (the Netherlands, Delft)
Van der Veen, K. (the Netherlands, Veenwouden)
Van der Ven, W. (the Netherlands, Eindhoven)
Van der Waaij, M. (the Netherlands, –)
Van der Werff, A. (the Netherlands, –)
Van der Wiel, P. (the Netherlands, –)

Van Dijk, E. (the Netherlands, –)
Van Dort, K. (the Netherlands, Wageningen)
van Geest, G. (the Netherlands, Delft)
Van Gennip, B. (the Netherlands, –)
Van Haaren, T. (the Netherlands, Amsterdam)
Van Hagen, H. (the Netherlands, Driebergen-Zeist)
Van Hoorn, M. (the Netherlands, –)
Van Kalmthout, G. (the Netherlands, –)
Van Kapel, H. (the Netherlands, –)
Van Kessel, J. (the Netherlands, –)
Van Kessel, K. (the Netherlands, –)
Van Moorsel, R. (the Netherlands, Nijmegen)
Van Oers, K. (the Netherlands, Wageningen)
van Raam, J. (the Netherlands, Hilversum)
Van Ruth, P. (the Netherlands, Oirschot)
Van Schie, M. (the Netherlands, Lelystad)
Van Tweel, L. (the Netherlands, Nijmegen)
Van Vliet, B. (the Netherlands, –)
Van Westen, M. (the Netherlands, –)
Van Westreenen, F. (the Netherlands, –)
Van Wijkel, A.-M. (the Netherlands, –)
Van Wijngaarden, W. (the Netherlands, Middelburg)
Vanberg, C. (Germany, Bad Breisig)
Vazzola, S. (Italy, Arpa Piemonte)
Vendrig, K. (the Netherlands, Almere)
Verbeek, M. (the Netherlands, 's Hertogenbosch)
Verlaan, E. (the Netherlands, Spijkernisse)
Verrijdt, T. (the Netherlands, –)
Vielberth, J. (Austria, Graz)
Vishnyakov, V. S. (Russia, Yaroslavl)
Visser, G. (the Netherlands, Midsland, Terschelling)
Vöge, H. (Germany, Hamburg)
Vöge, M. (Germany, Hamburg)
Vogel, A. (Germany, Lippstadt)
Vogel, B. (Germany, Leipzig)
Vogel, J. (Germany, Görlitz)
Vogel, P. (Germany, Karlsruhe)
Vogelzang, P. (the Netherlands, –)
Voggesberger, M. (Germany, Stuttgart)
Vohlmeister, V. (Germany, Bad Tölz)
Völkl, W. (Germany, Seybothenreuth)
Volkmann, H. (Germany, Eisleben)
Vollmer, I. (Germany, Altenkirchen)
Voorn, P. (the Netherlands, 's Graveland)
Voß, K. (Germany, Kiel)

Wagemakers, A. (the Netherlands, Sprang-Capelle)
Wagenknecht, J. (Germany, Eckental)
Wagner, A. (Germany, Unterammergau)
Wagner, H.-G. (Germany, Cottbus)
Wagner, I. (Germany, Unterammergau)
Wahrenburg, P. (Germany, Rastede)
Walkowski, U. (Germany, Osnabrück)
Walter, E. (Germany, Bayreuth)
Walter, B. (Germany, Borgholzhausen)
Walter, J. (Germany, Kiel)
Waltje, H. (the Netherlands, Drachten)
Warthemann, G. (Germany, Dessau)
Waterschap Friesland, (Org.) (the Netherlands, Leeuwarden)
Waterschap Groot Salland, (Org.) (the Netherlands, Zwolle)
Waterschap Hollandse Delta, (Org.) (the Netherlands, Ridderkerk)
Wauer, H. (Germany, Kürten)
Weber, D. (Germany, Bad Berka)
Weber, G. (Germany, Bochum)
Weber, I. (Germany, Marburg)
Weddeling, K. (Germany, Bonn)
Weeda, E. (the Netherlands, Zwolle)
Wegener, K.-A. (Germany, Greifswald)
Weingart, C. (Germany, Landau)
Weiss, G. (Germany, Rangsdorf)
Weiss, J. (Germany, Lüdinghausen)
Weiß, J. (Germany, Leipzig)
Welgraven, G. (the Netherlands, Heiloo)
Wenst, M. (Germany, Glentorf)
Werner, K. (Germany, Halle/Saale)
Werstat, C. (Germany, Letzlingen)
Westermann, K. (Germany, Rheinhausen)
Westermann, S. (Germany, Rheinhausen)
Westhus, W. (Germany, Jena)
Wiehler, K. (Germany, Minden)
Wieland, A. (the Netherlands, Goes)
Wimmer, W. (Germany, Salzgitter)
Winter, H. (Germany, –)
Wirtz, E. (the Netherlands, Tienhoven)
Wittjen, K. (Germany, Münster)
Wittmann, H. (Austria, Salzburg)
Wittwer, S. (Germany, Radebeul)
Wohlfahrt, U. (Germany, Vechta)
Wohlgemuth, J. O. (Germany, Beedenbostel)
Woike, S. (Germany, Haan)
Wolff, P. (Germany, Saarbrücken)

Wolgarten, H. (Germany, Herzogenrath)
Wolkerstorfer, C. (Austria, Linz-Ebelsberg)
Wollert, H. (Germany, Teterow)
Wortmann, D. (Germany, Rommerskirchen)
Wosch, C. (Germany, Dippoldiswalde)
Woschèe, R. (Germany, Neunburg)
Wucherpfennig, D. (Germany, Crailsheim)
Xylander, H. (Germany, Görlitz)
Yousef, M. (Germany, Rostock)
Zahlheimer, W. (Germany, Landshut)
Zander, B. (Germany, Oldenburg)
Zeigerer, A. (Germany, Erfurt)
Zeneli, V. (Albania, Shkodra)
Zhakova, L. V. (Russia, St. Petersburg)
Zimmermann, K. (Germany, Leverkusen)
Zimmermann, W. (Germany, Weimar)
Zoetelief, H. (the Netherlands, –)
Zonderwijk, M. (the Netherlands, Leiden)
Zühlke, J. (Germany, Koblenz)
Zündorf, H.-J. (Germany, Jena)
Zuyderduyn, C. (the Netherlands, Noordwijk)

References

Caraus, I. (2017). Algae of Romania. A distributional checklist of actual algae. Version 2.4. *Studii si stiintifice Cercetari Biologie*, *7*, 1–1002, 1 map.
Langangen, A. (2007). *Charophytes of the Nordic countries* (102 pp.). Saeculum ANS.

Chapter 13
Characteristics of the Genera in Family *Characeae*

Michelle T. Casanova

Chara species are often corticated (i.e., with a cortex) with a layer of linear cells on the internodes and/or branchlets, giving them a striped appearance. If a cortex is present, there can be spine cells arising from the nodes in the cortical cells. Species of *Chara* have the capacity to produce one or two rows of stipulodes below the whorls of branchlets. The branchlets in *Chara* are not forked, they are essentially monopodial branchlets, although they can have bract-cells arising from the branchlet nodes. In *Chara* when the gametangia occur together at the branchlet nodes, the oogonium is above the antheridium. There can also be bracteoles and bractlets (arising from the basal cell complex of the gametangia). The oogonium has five coronula cells, and the oospore is terete and elongate in profile.

Nitella species are never corticated, always ecorticate in all parts. They have no stipulodes, but there can be one or more whorls of branchlets at the same node. The branchlets are furcate or forked, with two or more, more-or-less equally long secondary (or tertiary etc.) branchlet segments. The terminal branchlet segments are called dactyls, and they can consist of a single cell, or two or more cells. In *Nitella* when the gametangia occur together the antheridia are terminal on the branchlets or within the furcations, and the oogonia are adjacent to the nodes. The oogonium has two rows of five cells (i.e., ten cells) in the coronula, and the oospore is in the shape of a flattened sphere or oval.

Lamprothamnium species are never corticated, always ecorticate. They have a single row of stipulodes. They have monopodial branchlets and their bract-cells are usually spreading and can be at right-angles to the branchlet. In *Lamprothamnium,* when the gametangia occur together, the antheridium is above the oogonium. The oogonium has five coronula cells, and the oospore is terete and elongate in profile.

Tolypella species are never corticated, always ecorticate in all parts. The protonemal filament can be retained and obvious on the growing plant. They have

M. T. Casanova (✉)
273 Casanova Rd, Westmere, VIC 3351, Australia
e-mail: m.casanova@nhm.ac.uk; mt.casanova@federation.edu.au

H. Schubert et al. (eds.), *Charophytes of Europe*,
https://doi.org/10.1007/978-3-031-31898-6_13

no stipulodes, but there can be one or more whorls of branchlets at the same node. The branchlets consist of monopodial rays, with secondary rays, tertiary rays (etc.) developing at the nodes of the primary ray. The end cells on the rays are rounded, and obtuse. The gametangia occur at the base of the whorls, and at the nodes of the rays, and are often have stipes, or pedicels connecting them to the node. The coronula cells on the oogonium are frequently deciduous, falling away as the oogonium matures. Oospores have the impression of 2–3 basal cells.

In this treatment we recognise extant species of *Sphaerochara* (= *Tolypella* sect. *Rothia* R. D. Wood = *Tolypella* sect. *Acutifolia* T. F. Allen = *Tolypella* sect. *Conoideae* J. Groves & Bullock-Webster). In common with the genus *Tolypella*, *Sphaerochara* species are never corticated, always ecorticate in all parts. The protonemal filament can be retained and obvious on the growing plant. They have no stipulodes, but there can be one or more whorls of branchlets at the same node. The branchlets consist of monopodial rays, with secondary rays, tertiary rays (etc.) developing at the nodes of the primary ray. In contrast with *Tolypella* the end cells on the rays are short, conical and acute or mucronate. The coronula cells on the oogonium are persistent and remain on the maturing oogonium. Oospores have the impression of a single basal cell.

Lychnothamnus and *Nitellopsis* are both monospecific genera in Europe, so the species description is also the genus description.

Chapter 14
Introduction to *Chara* Subsection *Agardhia*

Michelle T. Casanova

Wood erected Section *Agardhia*, with subsection *Agardhia*, subsection *Braunia* and subsection *Wallmania*. This chapter treats the *Agardhia* group of species (sect. *Agardhia* Wood) which constitutes most of the taxa amalgamated into *Chara fibrosa* by Wood (1962). *Chara fibrosa* was the earliest name for species with a single row of stipulodes and ecorticate branchlets, and the group is most abundant and diverse in the Oceania-Australia-Asian regions. At least 20 dioecious and monoecious taxa have been described in Australia alone (Braun and Nordstedt, 1882; Wood, 1965). A single outlying taxon occurs elsewhere (i.e., *C. curtissii* T.F. Allen ex Rob. in North America [Meiers et al., 1999]), and species that can be allocated to this group have been detected in Europe (Abdelahad and Piccoli 2017; Avetta, 1898; Bergmeier and Abrahamczyk, 2008; Langangen, 2000; Soulié-Märsche et al., 2013). Taxa in the *Agardhia* group share the following characters: They have corticated axes (haplostichous, diplostichous or occasionally triplostichous) and ecorticate branchlets, they have a single row of stipulodes (1x, 2x or numbers in-between), they have a cluster of bract-cells subtending the final branchlet segment, and they do not have gyrogonites. Oospores can be brown (e.g., *C. wightii*) or black (e.g., *C. benthamii*). They are often small and delicate, and in Europe members of this group are always monoecious. Their axes are usually somewhat flexible, with a diameter usually less than 0.5 mm. Spine cells are always single. Specimens could be confused with *Chara baueri* (in fact European material was amalgamated with *C. baueri* by Wood (1962) as *C. baueri* f. *pelosiana*, but *C. baueri* is distinguished by the presence of a corona of cells at the apex of the branchlets, and unilateral bract-cells). Gymnophyllous species (e.g., *Chara gymnophylla*) that are related to *Chara vulgaris* or *C. contraria* can be distinguished from members of subsection *Agardhia* on the basis of two rows of stipulodes in those species, and the characteristic branchlet tips (i.e., a single cell, rather than a segment subtended by verticillate bract-cells).

M. T. Casanova (✉)
273 Casanova Rd, Westmere, VIC 3351, Australia
e-mail: m.casanova@nhm.ac.uk; mt.casanova@federation.edu.au

H. Schubert et al. (eds.), *Charophytes of Europe*,
https://doi.org/10.1007/978-3-031-31898-6_14

The systematic treatment of Zaneveld (1940) is often referred to for designation of species in this group by European authors, because Wood (1965) amalgamated so many distinct taxa that it is difficult to discern the history or typification of what we currently recognise as separate species. However, Zaneveld (1940) worked mainly on south-east Asian species and revised *C. fibrosa* to include three subspecies (subsp. *benthamii*, subsp. *gymnopitys* A. Braun and subsp. *flaccida* A. Braun) separated on the basis of the ratio of stipulodes to branchlets and colour of the oospores. Recent genetic analysis (K. G. Karol pers. comm.) and morphological analyses (Casanova, 1997, 2013) indicate that many more taxa should be recognised, and an investigation into the typification of *C. fibrosa* resulted in clarification of the species concept for *C. fibrosa* (= *C. flaccida* var. *gaudichaudii* A. Braun *nom. inval.*) and *C. wightii* (A. Braun) Casanova (= *C. flaccida* var. *wightii* A. Braun) (Casanova, 2013).

The European material (Table 14.1) that has been determined as belonging to this group has been described or identified as *C. pelosiana* (Avetta, 1898) and *C. fibrosa* (Bergmeier and Abrahamczyk, 2008; Langangen, 2000) or *C. fibrosa* var. *benthamii* (Abdelahad and Piccoli, 2017; Soulié-Märsche et al., 2013). A Lectotype for *C. pelosiana* was designated in 2000 (in O), and its features described (Langangen, 2000).

In relation to morphological characters that can be taxonomically useful within this group several things should be noted. Some characteristics can be considered unreliable (e.g., plant height, degree of calcification, number and length of spine cells), since they vary in relation to water depth, water chemistry or the presence of herbivores (Proctor, 1999). Other, more cryptic characters (such as the length of the basal branchlet cell and the placement of gametangia) have been shown to be useful (Casanova, 2013). Oospore shape and dimensions are also thought to be consistent within species, whereas the number of striae can sometimes vary. Some features are difficult to discern on old, pressed type material (which could be why Langangen's [2000] measurements differ from the type description [Table 14.1]).

Braun and Nordstedt (1882) were able to examine the species known in the group at that time (*C. myriophylla* F. Muell. ex A. Braun, *C. mollusca* A. Braun, *C. benthamii*, *C. leptopitys* A. Braun, *C. gymnopitys*, *C. flaccida nom. inval.* [= *C. fibrosa* and *C. wightii* (A. Braun) Casanova], *C. griffithii nom. inval.* [= *C. erythrogyna* Griffith], *C. psilopitys nom. inval.* [= *C. drummondii* A. Braun and *C. weddellii* A. Braun], *C. thwaitesii* A. Braun, *C. dichopitys* A. Braun *nom. inval.* [= *C. hookeri* A. Braun and *C. preissii* A. Braun]) and various varieties and forms within these species, along with a few other species no longer included in section *Agardhia* (*C. braunii* Gmel., *C. muelleri* A. Braun, *C. hydropithys* Rchb.). Zaneveld's (1940) revision of *C. fibrosa* resulted in *C. fibrosa* subsp. *flaccida* Zaneveld *nom. inval.* (= *C. fibrosa*) with 1x stipulodes and brown oospores, *C. fibrosa* subsp. *benthamii* with stipulodes as numerous as the branchlets plus 1–3 more, and black oospores, and *C. fibrosa* subsp. *gymnopitys* with stipulodes usually 2x the number of branchlets and black oospores. Examination of the type material of *C. benthamii*, *C. wightii*, *C. fibrosa* and *C. gymnopitys* as well as many more recent specimens (including collections from Guam, the type locality of *C. fibrosa*) confirms, as stated by Zaneveld (1940)

Table 14.1 Characteristics of the specimens described in *Chara* subsection *Agardhia* in Europe, excluding material collected in Greece, with the addition of the descriptions of *Chara benthamii* (Type: Wilford 238, BM), C. fibrosa (Type: C. Gaudichaud 1819, PC, GOET) and C. wightii (Type: R. Wight 133. BM)

References	Avetta (1898) (*C. pelosiana*)	Langangen (2000) *C. pelosiana* designated Lectotype (O)	Soulié-Märsche et al. (2013) (*C. fibrosa* subsp. *benthamii*)	Abdelahad and Piccoli (2017) (*C. fibrosa* subsp. *benthamii*)	*Chara benthamii* Braun and Nordstedt (1882)	*Chara fibrosa* type material and material from Guam (Casanova, 2013)	*Chara wightii* type material (Casanova, 2013)
Height (mm)	70–80, hardly branched	Unable to be assessed	400–500	100–150	120–200	c. 300	150–250
Incrustation	Incrusted	Heavy	Banded	Slightly	Not encrusted	Not encrusted	Slightly
Internode width (μm)	500	400	Not stated	320–400	400–550, 570[a]	400–500	700
Internode length (mm)	50–70 mm	10	30	Not stated	9–35[b]	10–30	Not stated
Cortex	2x	2x	2x	2x	2x, uneven	2x	2x
Iso-heterostichous?	Almost isostichous	Tylacanthous	Isostichous	Slightly tylacanthous	Tylacanthous, to somewhat isostichous[a]	Isostichous	Tylacanthous
Spine cells	Solitary, verticillate	Not stated	Solitary, verticillate	Solitary, verticillate	Solitary, sparse	Solitary, obscure	Solitary, shortly conical
Spine cell length (μm)	1000	900	1000–1600	300–500	60–800	<50	165
stipulodes	1 tier, 1x, equal in length to basal branchlet cell	1 tier, 1–2 per branchlet	1 tier, 1x about as long as the basal branchlet cell? (Fig. 2)	1 tier, 1x, equal in length to basal branchlet cell (Figs. 1, 2)	1 tier, 1x plus 1–3 extra[a], 1/2 to 1/3 the length of the basal branchlet cell	1 tier, 1x	1 tier, 1x, as long as basal branchlet cell
Number of branchlets	10–12	6–8	10	9–12	8–10 rarely 11, 12[a]	c. 10	10–11
Branchlet cortex	None	None	None	None	None	None	None
Branchlet length (mm)	5–7	4.5	Not stated	Not stated	8–11[b]	12–22	8–14

(continued)

Table 14.1 (continued)

References	Avetta (1898) (*C. pelosiana*)	Langangen (2000) *C. pelosiana* designated Lectotype (O)	Soulié-Märsche et al. (2013) (*C. fibrosa* subsp. *benthamii*)	Abdelahad and Piccoli (2017) (*C. fibrosa* subsp. *benthamii*)	*Chara benthamii* Braun and Nordstedt (1882)	*Chara fibrosa* type material and material from Guam (Casanova, 2013)	*Chara wightii* type material (Casanova, 2013)
Number of branchlet segments	3–4, uniform in length	3–4, rarely 5	3–4, rarely 2 to 6	3–4, rarely 6	4–6, basal longest	4, uniform in length	5–6, basal segment longest
Number of bract-cells	6–8 verticillate	4–5, verticillate	4–5, verticillate, (Fig. 1)	4–5, verticillate (Fig. 1, 2)	4–6, 2 on sterile whorls, mostly verticillate	2–4 verticillate	4–6 verticillate
Bracteole length (μm)	c.800, double the oogonium length	Up to 1000, double the oogonium length	Up to 1000, double the oogonium length	About as long as the oogonium (Figs. 1, 2)	2–3x longer than the oogonium	1000, up to 2x longer than oogonium	2500
Gametangia	Monoecious conjoined	Monoecious conjoined	Monoecious conjoined	Monoecious Conjoined or sejoined (possibly antheridia missing due to maturity)	Monoecious conjoined	Monoecious conjoined	Monoecious, conjoined, single and geminate
Oogonia length (μm)	Not stated	420–650 c. coronula	c. 600	540–600	500[b]	500	495
Oogonia width (μm)	Not stated	270–380	440	340–390	380[b]	350	375
Antheridia diameter (μm)	Not stated	240–250	300	280–320	360	300	500
Oospore length (μm)	415–440	370–400	400–500	350–390	740	410	470
Oospore width (μm)	240–256	250	250–290	240–316	480	280	310

(continued)

Table 14.1 (continued)

References	Avetta (1898) (*C. pelosiana*)	Langangen (2000) *C. pelosiana* designated Lectotype (O)	Soulié-Märsche et al. (2013) (*C. fibrosa* subsp. *benthamii*)	Abdelahad and Piccoli (2017) (*C. fibrosa* subsp. *benthamii*)	*Chara benthamii* Braun and Nordstedt (1882)	*Chara fibrosa* type material and material from Guam (Casanova, 2013)	*Chara wightii* type material (Casanova, 2013)
Number of striae	Not stated	Not stated	7–9	7–8	8[b]	9	9–10
Colour of oospores	Not stated	Brown	Brown (Fig. 2)	Brown (Fig. 4)	Black	Brown	Brown

[a] From Zaneveld, 1940 where different; [b] From examination of type material

that the oospores of *C. fibrosa* and *C. wightii* are brown, whereas *C. benthamii* and *C. gymnopitys* have black oospores (Table 14.1).

It has been suggested that specimens within this group were introduced to Europe with the culture of rice, since they occur in rice fields (Soulié-Märsche et al., 2013). In the absence of a comprehensive morphological and genetic analysis of all the species in subsection *Agardhia*, throughout the range of the subsection, it is difficult to decide if the specimens occurring in Europe are native or exotic in origin. Additionally, we cannot decide if *C. pelosiana* is the same as the specimens described more recently from France, Greece and Italy without access to the type material as well as additional gatherings. However, it is clear from their descriptions that none of the material can be ascribed to *C. benthamii* or *C. gymnopitys* because the oospores are brown to golden-brown, rather than black, and there are 1 × stipulodes, rather than 1 × plus 1 to 3 extra (in *C. benthamii*) or up to 2x (as in *C. gymnopitys*). Similarly, it appears that despite having brown oospores, the specimens are not *C. fibrosa* since that species differs markedly in appearance and ecology, and in relation to the length of stipulodes and branchlet cells. Although *C. wightii* has a similar appearance to European material (Casanova, 2013), the regular occurrence of geminate gametangia and large antheridia in that species rules it out for European material. Given the similarities described in the publications concerning the European specimens (Table 14.1, with the exception of some of the measurements by Langangen 2000), they might all be the same species, and that species has a valid and legitimately published name: *Chara pelosiana*. If, after comprehensive investigation, it turns out that *C. pelosiana* is the same as a species occurring in rice-fields elsewhere in the world, it might be appropriate to change the name, depending on nomenclatural precedence.

References

Abdelahad, N., & Piccoli, F. (2017). Report on Charophytes from rice fields in northern Italy including the alien species *Chara fibrosa* ssp. *benthamii*. *Plant Biosystems, 152*, 599–603.

Avetta, C. (1898). Nuova specie di *Chara* (*Chara pelosiana* mihi). *Malpighia, 12*, 229–235. [in Italian].

Bergmeier, E., & Abrahamczyk, S. (2008). Current and historical diversity and new records of wetland plants in Crete, Greece. *Willdenowia, 38*, 433–453.

Braun, A., & Nordstedt, C. F. O. (1882). Fragmente einer Monographie der Characeen. Nach den hinterlassen Manuscripten A. Braun's herausgegeben von Dr. Otto Nordstedt. *Abhandlungen der Königlichen Akademie der Wissenschaften zu Berlin für das, 1882*, 1–211. [in German].

Casanova, M. T. (1997). Oospore variation in three species of *Chara* (Charales, Chlorophyta). *Phycologia, 36*, 274–280.

Casanova, M. T. (2013). Review of the species concepts *Chara fibrosa* and *C. flaccida* (Characeae, Charophyceae). *Australian Systematic Botany, 26*, 291–297.

Langangen, A. (2000). *Chara fibrosa* Agardh ex Bruzelius, a Charophyte new to the European Flora. *Allionia, 37*, 249–252.

Meiers, S. T., Proctor, V. W., & Chapman, R. L. (1999). Phylogeny and biogeography of *Chara* (Charophyta) inferred from 18S rDNA sequences. *Australian Journal of Botany, 47*, 347–360.

Proctor, V. W. (1999). Charophytivory, playas y papalotes, a local paradigm of global relevance. *Australian Journal of Botany, 47*, 399–406.

Soulié-Märsche, I., Triboit, F., Despréaux, M., Rey-Boissezon, A., Laffont-Schwob, I., & Thiéry, A. (2013). Evidence of *Chara fibrosa* Agardh ex Bruzelius, an alien species in South France. *Acta Botanica Gallica: Botany Letters, 160*, 157–163.
Wood, R. D. (1962). New combinations and taxa in the revision of Characeae. *Taxon, 11*, 7–25.
Wood, R. D. (1965). *Monograph of the Characeae*. J. Cramer.
Zaneveld, J. S. (1940). The Charophyta of Malaysia and adjacent countries. *Blumea, 4*, 1–224.

Chapter 15
Introduction to the *Hartmania* Group (*Chara* Subsection *Hartmania*)

Irmgard Blindow, Michelle T. Casanova, and Klaus van de Weyer

The *Hartmania* group was initially designated a subsection of the genus *Chara* in the subgenus and section *Chara*, containing a single species (*Chara hispida*) and was named in honour of Carl Johan Hartman (Wood, 1965: xiii). Wood (1965) described the subsection as "species in subgenus *Chara* that are similar to *Chara hispida*" and separated it from other subsections by the presence of "usually fasciculate spine cells" (Wood, 1965: 65). It is now recognised that the group contains several species that occur in Europe: *Chara aculeolata, C. affinis, C. baltica, C. corfuensis, C. globata, C. hellenica, C. hispida, C. horrida, C. papillosa,* and *C. subspinosa.* There is also a taxon with uncertain affiliation ("Mediterranean Chara") (Table 15.1).

Taxa in the *Hartmania* group share the following characters: They have corticated axes (diplostichous, more rarely triplostichous) and corticated branchlets, they have two rows of twinned stipulodes. They are often coarsely textured, and, except for *Chara hellenica*, are monoecious. Their axes are robust, with a diameter exceeding 0.5 mm. Spine-cells can be single but are often fasciculate. The few exceptions are slender *Chara affinis* which can be confused with *C. contraria* var. *hispidula*, slender *C. papillosa* which can be confused with *C. contraria*, slender, single-spined *C. hispida* which can be similar to *C. vulgaris* and small individuals of *C. baltica* with a diplo- to triplostichous cortex, which are hard to distinguish from *C. aspera* when sterile.

I. Blindow (✉)
Biologische Station Hiddensee, Universität Greifswald, 18565 Kloster, Germany
e-mail: blindi@uni-greifswald.de

M. T. Casanova
273 Casanova Rd, Westmere, VIC 3351, Australia
e-mail: mt.casanova@federation.edu.au

K. van de Weyer
lanaplan, Lobbericher Str. 5, D-41334 Nettetal, Germany
e-mail: klaus.vdweyer@lanaplan.de

H. Schubert et al. (eds.), *Charophytes of Europe*,
https://doi.org/10.1007/978-3-031-31898-6_15

Table 15.1 Overview over taxa/clusters in the *Hartmania* group

Cluster	Taxa
1	*Chara hellenica*
2	*Chara hispida,* including *C. equisetina* *Chara subspinosa* (= *C. rudis*)
3	*Chara baltica,* including var. *liljebladii*, var. *densa* and var. *rigida* *Chara horrida*
4	*Chara aculeolata* (= *C. polyacantha*) *Chara papillosa* (= *C. intermedia*), including *C. visianii* *Chara corfuensis* *Chara globata* *Chara affinis* (= *C. baltica* var. *affinis*) "Mediterranean Chara" (= *C. baltica* from the Mediterranean region; taxon with uncertain affiliation)

Within the *Hartmania* group, however, determination is difficult even for experienced botanists. The group forms a "species complex" sensu Proctor (1971a) of similar, but reproductively isolated and morphologically distinct entities. Single taxa are highly variable with a wide range of characters, which connects them to "adjacent" taxa. On the whole, the species complex forms a continuum rather than clearly separated clusters or easily defined species. As a rule, "typical" specimens of a certain taxon are "rather" easy to determine, but individual gatherings often show one or several "unusual" characters and can even appear as intermediates between two taxa.

Current analyses of genetic variation generally provide few characters to distinguish entities within this species complex. Habitat salinity seems to be an important criterion on which we can form sensible groupings (Nowak and Schubert, 2019). This is not surprising, as occurrence at higher salinity requires physiological mechanisms for turgor regulation, which are lacking in typical freshwater species (Winter and Kirst, 1991).

Morphologically, within the *Hartmania* group, several clusters of species can be separated despite of the high degree of overlap in characters (Table 15.1). The first cluster consists of a single species (*Chara hellenica*), which is strongly tylacanthous, recorded only from its type locality on Corfu, and the only dioecious taxon within the group.

The second cluster combines the aulacanthous (to isostichous) inland species, mainly or only occurring in freshwater and variously designated as *Chara subspinosa, C. hispida* and *C. hispida* f. *equisetina*. These have each been described as separate species, varieties or forms in the past and after evaluation are here designated *C. hispida* (which includes *C. equisetina*) and *C. subspinosa*.

The third cluster consists of typically brackish water taxa which are restricted to coastal brackish environments: *C. baltica* and *C. horrida*. *Chara baltica* var. *rigida*, which is known only from one inland, near-coastal brackish water site in Great Britain, is amalgamated with *C. baltica*. All these taxa are separated from other taxa of the *Hartmania* group by their larger antheridia.

The fourth cluster combines tylacanthous to isostichous monoecious taxa from freshwater and inland brackish water sites. *Chara visianii*, which has been described as a separate species (Blaženčić and Randjelović, 1994) is here included within *C. papillosa*, as it seems morphologically and ecologically inseparable from that species. *Chara dominii* is included in *C. globata*. This cluster also contains two taxa, up to now included in *C. baltica* (*Chara baltica* var. *affinis* and *C. baltica* recorded from the Mediterranean area), but morphologically separated from *C. baltica*. They are treated as a separate species (*C. affinis*) or as "Taxon with uncertain affiliation" ("Mediterranean Chara"), respectively.

Calcium encrustation is a typical character of all freshwater collections belonging to the *Hartmania* group (which all are hardwater plants). In the Baltic Sea, however, encrustation is weak overall despite similar Ca concentrations and pH values to those in freshwater calcium-rich lakes. High Mg/Ca ratios in the Baltic might be preventing this encrustation (Herbst et al., 2018). Encrustation is therefore explained by habitat conditions rather than by some (genetically manifested) physiological behaviour of the plants; however, as such encrustations are indirectly connected to the different salinity tolerances of each taxon (and thereby to genetically manifested characters), they can be used as a criterion to separate the different taxa.

Monographs rarely give information about how many individuals were measured from how many populations. Different monographs often give very similar or even identical values for measured characters, possibly due to authors copying from each other. Information on methods is often lacking or insufficient. Antheridia diameters, for example, are around 1.3 times larger when measured on wet material compared to dry (own data), and a direct comparison of values given in different monographs therefore feels hazardous. Because of these shortcomings, we decided to use our own descriptions and measurements for this difficult group, instead of information given in monographs and other publications. In many cases, the ranges given for our own measurements and descriptions exceed ranges given in the literature, perhaps because a larger number of individuals and populations were included in this study.

All descriptions and measurements were obtained from dry (herbarium) material under a stereomicroscope with 90x magnification, unless specified otherwise. Where possible, at least 20 individuals from a minimum of 10 populations were included. Bract-cells, bracteoles, spine-cells and stipulodes were preferably measured on the 2nd to 3rd internode (from the apex). Axis diameter was measured on the widest internode, which in most cases was situated on the lower parts of the plant. Oogonia length:width ratio was calculated for individual oogonia. In the descriptions, "occasionally" and "rarely" means that the character in question is shown by roughly 20% and $\leq$10%, respectively, of all collections.

Speciation is an ongoing process. We are observing populations at a moment in evolutionary time, and those populations exhibit different degrees of genetic isolation and adaptation to local conditions. In *Characeae* 'species ' have been designated as a matter of convenience and regional historic convention. Schubert et al. (Chap. 5) go into this in more detail. Authors have a high degree of freedom in their decision about taxonomic level. Half a century ago, crossing experiments revealed discrepancies in this phenetic species concept (Croy, 1982; Grant and Proctor, 1972; Proctor, 1971a,

1971b, 1975). Later, clusters in genetic analyses often failed to mirror the "classical" taxonomic entities (for an overview, see Nowak et al., Chap. 6). Still, the designation of species is justified by consistency of characters, historical discrimination, and to a degree, distribution. This is important so that named species can be managed and conserved, and for entities to be recognised in political protection frameworks.

Within the *Hartmania* group, we are maintaining names and taxonomic ranks, as long as description and type material are correct, valid and available, though the taxonomic ranks sometimes feel rather arbitrary. For example, *Chara baltica* var. *liljebladii* is morphologically not less different from to *C. baltica* than *C. horrida*. All three taxa are connected by transitions and could have been combined within one species instead. In contrast we did not accept the name "*Chara baltica*" for specimens collected from the Mediterranean region or "*Chara baltica* var. *affinis*" as a variety, as both taxa differ morphologically from *Chara baltica* (see Chaps. 17 and 80). Further, we did not accept *Chara visianii* as a separate species, as the characters given in the species description all are within the variation exhibited by *C. papillosa* (see Chap. 44). For the same reason, *C. dominii* was included in *C. globata* (see Chap. 33).

The dichotomous determination key (Chap. 10) can also be used for the determination of specimens of the *Hartmania* group, but the high variability within and broad overlap among taxa may hamper identification. Additionally (or alternatively), the following matrix key can be used, which apart from morphological characters also includes habitat and distribution as criteria which may support determination. In the Mediterranean area, the matrix key (Table 15.2) should be used for determination.

Table 15.2 Matrix key for the identification of specimens of the *Hartmania* group

taxon	Sex		axial cortex						spine-cells						stipulodes					bract-cells						habitat			distribu-tion			diameter of antheridia		
			arrange-ment			no. of rows			length			arrange-ment			length			no. tiers		arrange-ment		length of adaxials				coa-stal	inland					measured dry!		
	di	mon	aul	iso	tyl	2	2-3	3	< axis	= axis	> axis	1	2-3	4-6	short	int.	long	2	>2	vert.	not vert.	<500	500-1000	1000-3500	> 3500		brack	fresh	N	C	M	< 500	500-550	> 550
Cluster 1																																		
hellenica	x				x	x	x			x	x	x	x		x			x			x			x	x			x			x			x
Cluster 2																																		
hispida		x	x	x		x			x	x	x	x	x		x	x	x	x		(x)	x	(x)	x	x			(x)	x	x	x	x	x	x	x
subspinosa		x	x			x			x			x	x		x	x		x			x	x	x	x				x	x	x		x	x	x
Cluster 3																																		
horrida		x	x	(x)	(x)	x				x	x	x	x	x		x	x		x	x	x		x	x		x			x	x				x
baltica		x	(x)	x	x	x	x		x	x	(x)	x	x		x	x	x	x	(x)	x	x	x	x	(x)		x			x	x		(x)	x	x
Cluster 4																																		
corfuensis		x		(x)	x	x	(x)			x	x	x	x	(x)		x	x	x		x	(x)	(x)		x	x		x*	?			x	x	x	(x)
aculeolata		x		(x)	x	x				x	x		x	x		x	x	x		x	x		(x)	x				x	x	x	x	x	x	(x)
papillosa		x		(x)	x	x	(x)		x	x		x	x		x	x	x	x			x	x	x	x			x	x	x	x	x	x	x	(x)
Mediterra-nean		x		x	x	x			x	x	x	x	x		x	x	x	x		x	x		x	x			x*	?			x	x	x	(x)
affinis		x		(x)	x		x				x	x	x				x	x		(x)	x		x			x	x	x		x	(x)	x	x	(x)
globata		x		x	x	x	x	(x)	x	(x)		x	(x)			x	x	x		x				x	x		x	x			x	x	x	x

Abbreviations and symbols: Sex: di = dioecious; mon = monoecious. Axial cortex: aul = aulacanthous; iso = isostichous; tyl = tylacanthous. Spine cells, length: < axis = shorter than axis diameter; = axis = as long as axis diameter; > axis = longer than axis diameter. Spine cells, arrangement: 1 = single-spined; 2–3 = spine-cells in clusters of 2–3; 4–6 = spine-cells in clusters of 4–6. Stipulodes, length: int. = intermediate. Stipulodes: no tiers = no. of tiers. Stipulodes, no. tiers: 2 = two regular tiers; >2 = irregular, additional stipulodes among the two tiers. Bract-cells, arrangement: vert. = verticillate (adaxial and abaxial bract-cells of equal length); not vert. = not verticillate (adaxial bract-cells longer than abaxial ones). Bract cells, length of adaxials is given in µm. Habitat: brack = brackish. x* = brackish water sites near the coast but disconnected from the open sea. Distribution: N = Northern Europe; C = Central Europe; M = Mediterranean area. Diameter of antheridia is given in. µm x = common character in the taxon. (x) = occasionally occurring character in the taxon. Diagnostic characters are marked blue

References

Blaženčić, J., & Randjelović, V. (1994). *Chara visianii* J. Blaženčić et V. Randjelović sp. nov. (Characeae) from the river Krka (Croatia). *Cryptogamie, Algologie, 15*, 73–79.

Croy, C. D. (1982). *Chara aspera* (Charophyta) breeding pattern in the northern hemisphere. *Phycologia, 21*, 243–246.

Grant, M. C., & Proctor, V. W. (1972). *Chara vulgaris* and *Chara contraria*: Patterns of reproductive isolation for two cosmopolitan species complexes. *Evolution, 26*, 267–281.

Herbst, A., Henningsen, L., Schubert, H., & Blindow, I. (2018). Encrustations and element composition of charophytes from fresh or brackish water sites – habitat- or species-specific differences? *Aquatic Botany, 148*, 29–34.

Nowak, P., & Schubert, H. (2019). Genetic variability of charophyte algae in the Baltic Sea area. *Botanica Marina, 62*, 75–82.

Proctor, V. W. (1971a). *Chara globularis* Thuillier (= *C. fragilis* Desvaux): Breeding patterns within a cosmopolitan complex. *Limnology and Oceanography, 16*, 422–436.

Proctor, V. W. (1971b). Taxonomic significance of monoecism and dioecism in the genus *Chara*. *Phycologia, 10*, 299–307.

Proctor, V. W. (1975). The nature of charophyte species. *Phycologia, 14*, 97–113.

Winter, U., & Kirst, G. O. (1991). Partial turgor pressure regulation in *Chara canescens* and its implications for a generalized hypothesis of salinity response in charophytes. *Botanica Acta, 104*, 37–46.

Wood, R. D. (1965). *Monograph of the Characeae*. J. Cramer.

Chapter 16
Chara aculeolata

Irmgard Blindow, Roman Romanov, Hendrik Schubert, Nick Stewart, and Klaus van de Weyer

Scientific name: ***Chara aculeolata*** **Kütz. in Rchb., Fl. Germ. Excurs.: 843. 1832**

Lectotype (Wood, 1965: 136): Tennstaedt/[no date]/[no collector] (L, annotated as *Chara aculeolata* in Kützing's script; L 2067797 type fragment from collection van Raam).

= *Chara pedunculata* Kütz. in Flora 17: 706. 1834.
Lectotype (Wood, 1965: 138): Germany, close to Halle/[no date]/[Kützing] (L).

= *Chara polyacantha* A. Braun ex Braun, Rabenhorst & Stizenberger, Charac. Eur. 2/ 48. 1859, n. inv. [The name was published without a valid morphological description, see Abdelahad et al. 2019.]

≡ *Chara hispida* L. var. *hispida* f. *polyacantha* (A. Braun ex A. Braun, Rabenh. & Stizenb.) R. D. Wood, Taxon 11: 9. 1960, n. inv.

I. Blindow (✉)
Biologische Station Hiddensee, Universität Greifswald, 18565 Kloster, Germany
e-mail: blindi@uni-greifswald.de

R. Romanov
Nišice Ujtin Potok b.b., Dobra Voda, 85356 Bar, Montenegro
e-mail: streptophytes@gmail.com

H. Schubert
Universität Rostock Institut für Biowissenschaften, Albert-Einstein-Straße 3, 18059 Rostock, Germany
e-mail: hendrik.schubert@uni-rostock.de

N. Stewart
4 Hexton Road, Glastonbury BA6 8HL, UK
e-mail: nfstewart@freeuk.com

K. van de Weyer
lanaplan, Lobbericher Str. 5, 41334 Nettetal, Germany
e-mail: klaus.vdweyer@lanaplan.de

H. Schubert et al. (eds.), *Charophytes of Europe*,
https://doi.org/10.1007/978-3-031-31898-6_16

= *Chara polyacantha* A. Braun f. *dasyacantha* Mig., Charac. Eur.: 483. 1893.
Type not designated.

Taxonomic notes: Migula (1889–1897) erroneously used *Chara aculeolata* as a synonym of *C. intermedia* (= *C. papillosa*) which was followed by several authors (e.g., Dąmbska, 1964; Hollerbach and Krassavina, 1983; Krause, 1997; Olsen, 1944).

Local names: BELARUSIAN: Хара многакалючкавая; DUTCH: Fijnstekelig kransblad; ENGLISH: Hedgehog Stonewort; GERMAN: Vielstachelige Armleuchteralge; NORWEGIAN: Hårpiggkrans; POLISH: Ramienica wielokolczasta, RUSSIAN: Хара мелкошиповатая; SWEDISH: Törnstäfse; UKRAINIAN: Хара багатоколючкова.

16.1 Morphology

See Figs. 16.1, 16.2, 16.3, 16.4, 16.5.

Habitus: *Chara aculeolata* is greyish green to nearly whitish due to strong encrustation, mid-green if less encrusted (Fig. 16.2). Especially on the upper internodes, the axis is hardly visible due to the extremely dense spine-cells, giving the species

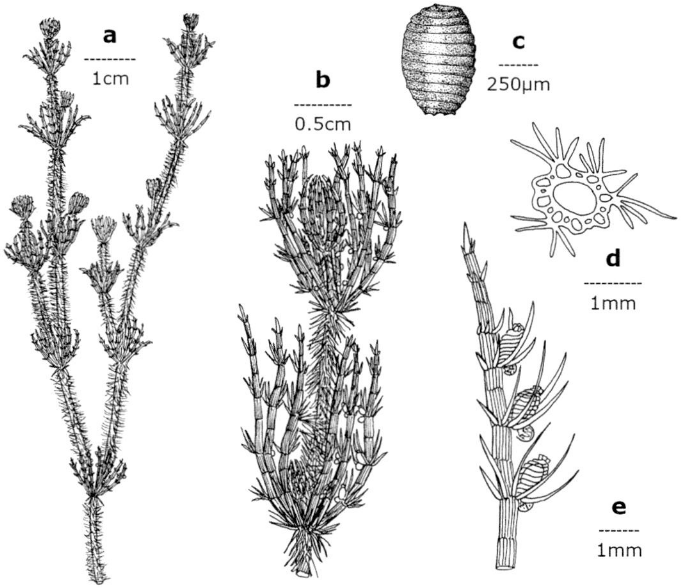

Fig. 16.1 *Chara aculeolata*. **a** Habitus; **b** Part of main axis showing long spine-cells, stipulodes, and verticillate bract-cells; **c** Oospore; **d** Cross-section of main axis; **e** Part of branchlet with gametangia. Drawings after Werner Krause, modified by Dagmar Wassong in Weyer and Schmidt (2018)

Fig. 16.2 Habitus of *Chara aculeolata*. **a** In situ photograph, the spiny appearance is clearly visible; **b** Fresh plants, showing long, dense spine-cells on the upper internodes and sparse spine-cells in the lower internodes. Adaxial bract-cells are distinctly longer than abaxial bract-cells in this collection; **c**, **d** *Chara aculeolata* from Vrana Lake, Croatia, leg. K. van de Weyer. **c** Collection from shallow water with long, verticillate bract-cells and long, clustered spine-cells; **d** Collection from 24 m depth with short spine-cells and bract-cells; **e** Collection from Arrie, southern Sweden, in LE, showing the typical "brush-like" appearance. Photographs: **a**—Egbert Korte; **b**—Klaus van de Weyer; **c**, **d**, **e**—Irmgard Blindow

its typical "brush-like" appearance, while spine-cells are far less dense on the lower internodes. Long side branches are common, the lower internodes are far longer than the branchlets.

Size: The plants are medium-sized to long, 10–75 cm, occasionally up to 150 cm (Fig. 16.5).

Axis: Diameter varies between about 750–2100 μm.

Cortex: *Chara aculeolata* is diplostichous, rarely with a transition to triplostichous cortication, tylacanthous, rarely almost isostichous, but never aulacanthous, only weakly twisted. The size difference between primary and secondary cortex cell rows can be large, in which case the secondary cortex cell rows are almost covered by the primary ones.

Fig. 16.3 Details of *Chara aculeolata*. Left panel: Clustered spine-cells. Altenburg, Thüringen, Germany; middle panel: Long, clustered spine-cells, long stipulodes. Parsteiner See, Germany; right panel: Long, verticillate bract-cells. Parsteiner See, Germany. Photographs: Klaus van de Weyer

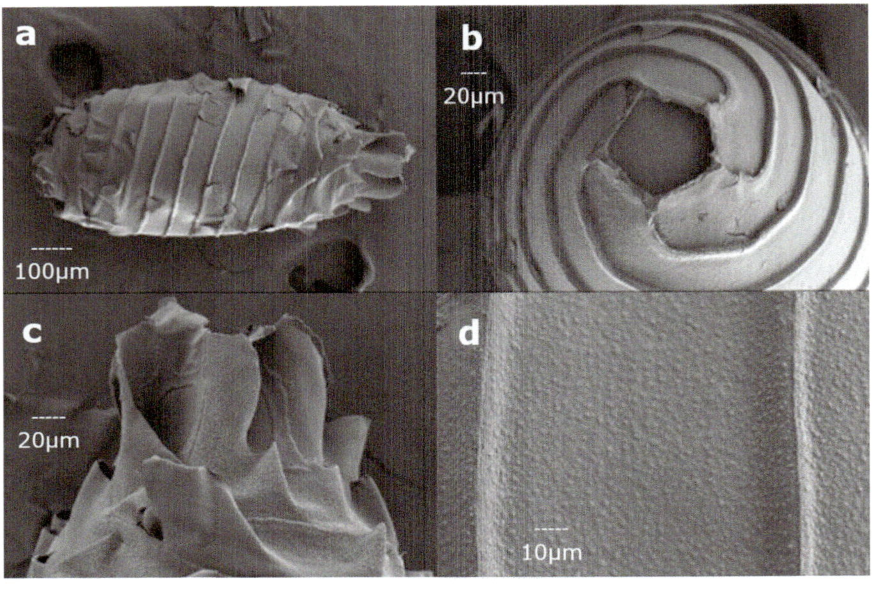

Fig. 16.4 SEM images of oospores of *Chara aculeolata*. **a** Oospore, side view; **b** Basal plate; **c** Apex; **d** Smooth oospore-membrane; pictures taken from a herbarium specimen gathered at Garboldisham Fen, Norfolk, 1897, coll. G. R. Bullock-Webster (BM001216841). Photographs: Michelle T. Casanova

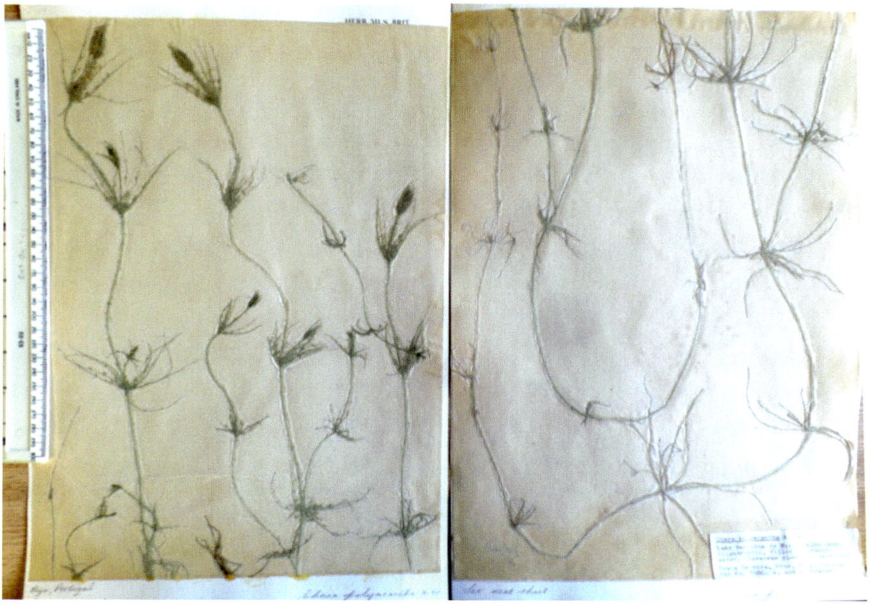

Fig. 16.5 Extremely large individuals of *Chara aculeolata* from Lake Barrhinha de Mira, Portugal, in BM. Photographs: Irmgard Blindow

Spine-cells: On the upper internodes, spine-cells are arranged in bunches, which consist of 4–6 spine-cells (about 80%) of the collections (Fig. 16.3), occasionally of 2–3 spine-cells (about 20% of the collections). Spine-cells are dense, acute and variable in length, but normally as long as, to up to four times as long as, the axis diameter, 900–5600 μm, mean value around 2100 μm. The arrangement of the spine-cells is variable, they can point in various directions forming a "brush-like" pattern along the axis, be arranged perpendicular to the axis, or, as in *Chara affinis*, the spine-cells point downwards on the upper half of the internode, but upwards on the lower part, which can cause a "traffic-accident-like" appearance in the middle of the axis. On older internodes and in deep water forms, the spine-cells can be distinctly reduced (van de Weyer and Schmidt, 2018, own data; Fig. 16.2), and can detach in herbarium collections (Abdelahad et al., 2019).

Stipulodes: *Chara aculeolata* has two tiers of dense, normally regularly arranged, acute stipulodes, of which the upper tier can be somewhat longer than or equally as long as the lower tier, about 1–2x as long as the main axis diameter, occasionally shorter, ranging from 700 to 3700 μm (Abdelahad et al., 2019, own data).

Branchlets: Whorls consist of 8–10 branchlets with 5–9 segments, of which the terminal 1- to 2, rarely 3 are ecorticate. The ultimate cell is far narrower than the penultimate cell and small to very small; occasionally, it can be covered by the bract-cells of the ultimate branchlet node.

Bract-cells: Spreading, well developed bract-cells are present along the whole branchlet. In most individuals, both adaxial and abaxial bract-cells are far longer than the diameter of the branchlets. Adaxial bract-cells range from 500 to 3300 μm and are acute. Abaxial bract-cells are variable in size, but always acute and never rudimentary, and range from 300 to 2000 μm. Adaxial bract-cells are longer than abaxial in about 65% of the collections, but also a verticillate or near-verticillate arrangement is common.

Bracteoles: Long, longer than the adaxial bract-cells, a bit longer to about 4 times longer than the ripe oogonium. They range from 900 to 4300 μm, mean value is around 2800 μm (Wood, 1965, own data).

Sexuality: *Chara aculeolata* is monoecious. Gametangia are conjoined at the lowest 3–4 branchlet nodes (Wood, 1965).

Oogonia: The coronula is often spreading or slightly spreading (Krause, 1997, own data). For morphometry, see Table 16.1.

Oospores: Drawings show an inconspicuous claw with 3–4 teeth (Krause, 1997, Fig. 16.1). Migula (1889–1897) and Wood (1965) mention a basal cage. Figure 16.4 shows SEM images of oospores. For morphometry, see Table 16.2.

Antheridia: Antheridia are yellowish. Though information about antheridia sizes varies considerably, all authors agree with Migula (1889–1897) that they are small. Mean values of antheridia diameter is around 500 μm (own data). Table 16.3 gives information about diameter of antheridia.

Bulbils: Multicellular bulbils are occasionally present around the lower axis nodes.

Table 16.1 Oogonia morphometry

Author	Length	Width	Coronula length	Coronula width	Number of helical stripes
Migula (1889–1897)	up to 1200	700	120	–	12–13
Groves and Bullock-Webster (1924)	850–1100	575–675	150	300	12–15
Wood (1965)	850–1100	575–675	95–180	225–300	11–15
Krause (1997)	850–1100	600–700	100–180	200–300	12–15
Holtz (n.d.)	1200	540	–	–	–
Own data	730–1100	480–730	110–200	210–310	10–15
Total range	**730–1200**	**480–730**	**95–200**	**200–310**	**(10) 11–15**

Length in the first column excludes the coronula, its length and width given separately; all values for length and width in μm

Table 16.2 Oospore morphometry

Author	Length	Width	Number of striae	Colour
Migula (1889–1897)	700–850	500 (550)	12	Dark brown to black
Groves and Bullock-Webster (1924)	650–800	325–475	11–14	Dark brown to black
Wood (1965)	650–800	325–475	11–14	Dark brown to black
Krause (1997)	650–900	350–450	11–14	Black
Holtz (n.d.)	920	480	–	Dark brown
Total range	**650–920**	**325–550**	**11–14**	

All values for length and width in μm

Table 16.3 Antheridia morphometry

Author	Diameter (μm)
Groves and Bullock-Webster (1924)	500
Wood (1965)	480–615
Krause (1997)	250–500
Own data	390–560 (670)
Total range	**250–560 (670)**

Wood (1965) measured rewetted antheridia. Other authors did not give any information on methods to measure antheridia. Own data are from dry antheridia

16.2 Phenology and Reproduction

Phenological data for *Chara aculeolata* are sparse. The species is predominantly perennial and hibernates as green plants, but also reproduces by means of oospores, and possibly bulbils (Langangen, 2007). *Chara aculeolata* is often richly fertile. In Scandinavia, ripe oospores are found from August (Langangen, 2007).

16.3 Habitat Conditions and Ecology

Habitat type: *Chara aculeolata* occurs in natural and artificial lakes, but more commonly in small water bodies such as ditches, fishponds, peat excavations, spring ponds, calcareous mires, dune slacks, gravel pits, running water with low flow velocities, rarely even in abandoned retention ponds for wastewater of a pulp and paper mill (Becker, 2019; Kabus and Mauersberger, 2011; Romanov et al., 2017; Vishnyakov et al., 2021, own data). In Italian volcanic lakes, it forms dense vegetation (Azzella, 2014). In stony marl lakes in Ireland it is an associate in the *Chara curta* zone between the cyanobacterial crust zone in the shallows and the deeper water beds of e.g. *C. subspinosa*, in up to 2 metres (Roden et al., 2020).

Sediment: *Chara aculeolata* grows on sand, gravel, peat, and calcareous gyttja (Langangen, 2007; Pätzold, 2011, own observations).

Nutrient conditions: *Chara aculeolata* occurs at a wide range of nutrient conditions. At its sites, PO_4-concentrations between 1 and 73 $\mu g\ L^{-1}$ and total P-concentrations between 2 and 130 $\mu g\ L^{-1}$ have been recorded (Alegro et al., 2016; Azzella, 2014; Doege et al., 2016; Gąbka, 2009; Romanov et al., 2017; van Raam, 1998). Exceptionally, PO_4-concentrations can be as high as 150 $\mu g\ L^{-1}$ (Gąbka, 2009) and total P-concentrations as high as 1400 $\mu g\ L^{-1}$ (Lambert, 2007).

Water chemistry: *Chara aculeolata* is a typical hard water species, which only occasionally occurs in soft water. According to Moore (1986), the species prefers calcium-rich water rich in humic substances. Ca concentrations of 215 mg L^{-1} have been recorded (Romanov et al., 2017). pH values at its sites range from 6.6 to 9.1 (Alegro et al., 2016; Doege et al., 2016; Romanov et al., 2017; Vishnyakov et al., 2021). The species is restricted to freshwater and only occurs at low salinities (Korsch, 2013; van Raam, 1998), with conductivities ranging from 54 to 3,000 $\mu S\ cm^{-1}$ (Alegro et al., 2016; Azzella, 2014; Doege et al., 2016; Romanov et al., 2017; Vishnyakov et al., 2021, own data). Herbarium records from the Baltic Sea in BILAS, GFW, ROST, Stralsund and WI, some of which were published by Lindner (1978), turned out to be misidentifications (Blindow, 2004; C. Blümel pers. comm., own data).

Depth range: *Chara aculeolata* inhabits a wide depth range from just a few cm depth (own observations) to very deep water (Azzella, 2014; Langangen, 2007). In Germany, it has been recorded down to 10 m depth (Pätzold, 2011). In volcanic lakes in Italy, the species mainly inhabits a middle depth range, but occurs down to 18 m (Azzella, 2014). In Lake Vrana, Croatia, historic records report depths down to 21 m (Golubić, 1961); here, the species has recently been found even down to 28 m (Alegro et al., 2016; Fig. 16.2).

Associated species: *Chara aculeolata* can form dense monospecific stands which have been described as *Charetum polyacanthae* Dąmbska, 1966 and *Charetum aculeolatae* Dąmbska, 1966, respectively. It has been recorded together with other charophytes (*Chara aspera, C. contraria, C. papillosa, C. tomentosa, C.virgata, C. vulgaris, C. intermedia, C. hispida, C. gymnophylla, Nitellopsis obtusa, Nitella hyalina, N. gracilis, Lychnothamnus barbatus*), various vascular plants and mosses (e.g. *Fontinalis antipyretica, Potamogeton coloratus, Stuckenia pectinata*) (Alegro et al., 2016; Azzella, 2014; Korsch, 2013; Romanov et al., 2017; Spiess et al., 2010, own data).

16.4 Variability

Migula (1889–1897) described 7 forms, of which *Chara polyacantha* f. *dasyacantha* is characterized by extremely dense spine-cells. Krause (1997) distinguished 3 forms of "*Chara polyacantha*", one with short branchlets adpressed to the main axis, hardly branched and with dense spine-cells, the second richly branched with long branchlets and dense spine-cells, the third with distinctly elongated internodes and dense spine-cells only on the upper internodes.

Deep water plants can have a deviating morphology with few, single, short spine-cells, short stipulodes, and unusually long internodes (Fig. 16.2).

16.5 Identification and Risk of Confusion

Chara aculeolata belongs to the *Hartmania* group, a complex of coarse, diplostichous *Chara* species that are hard to separate from each other (see Chap. 15).

Chara aculeolata has sometimes been confused with *C. horrida* (Blindow, 2004; Wood, 1965). Both species have a similar appearance with long, dense spine-cells and long, densely arranged stipulodes. In contrast to *C. horrida*, which is restricted to coastal brackish sites and has irregularly arranged stipulodes, the stipulodes of the freshwater species *C. aculeolata* are arranged in two regular tiers. Most collections of *C. aculeolata* are distinctly tylacanthous and can therefore easily be distinguished from *C. horrida* which is aulacanthous to isostichous. Fertile *C. horrida* can be distinguished from *C. aculeolata* by their typical two oospore claws and larger antheridia. *C. aculeolata* is often strongly encrusted in contrast to *C. horrida* which only has weak encrustation. This is, however, not a suitable criterion, as this character is influenced by the site conditions rather than being species-specific (Herbst et al., 2018).

The brush-like appearance with very dense spine-cells separates *C. aculeolata* from other freshwater species. Tylacanthous cortex separates *C. aculeolata* from the aulacanthous (to isostichous) *C. hispida*, which also has long, clustered spine-cells, but not more than 3 spine-cells per cluster. In contrast to *C. aculeolata* which often has verticillate to near-verticillate bract-cells, adaxial bract-cells are distinctly longer than abaxial ones in almost all collections of *C. hispida*.

Also *Chara papillosa* is tylacanthous but has single spine-cells or spine-cells in clusters of 2–3, while *C. aculeolata* never is single-spined (at least not on the upper internodes), and most collections have more than 3 spine-cells per cluster. Most *C. aculeolata* have longer and dense spine-cells, longer stipulodes, and longer bracteoles which are at least twice as long as the ripe oogonium. Also bract-cells are longer than in *C. papillosa*. Abaxial bract-cells are sometimes as long as the adaxial ones, which is only rarely the case in *C. papillosa*, and acute even if they are short, while many *C. papillosa* have papilliform or even rudimentory abaxial bract-cells. Transition forms do occur between both taxa, however (Korsch, 2013, own data).

In the Mediterranean region, additional tylacanthous, "spiny" taxa can be hard to separate from *C. aculeolata*, namely, *C. corfuensis*, *C. globata*, and "Mediterranean Chara" (a taxon with uncertain affiliation). In the West European region, *Chara affinis*, another "spiny" taxon, might be confused with *C. aculeolata*. High number of spine-cells per cluster is a character diagnostic for *C. aculeolata* and separates it from all these taxa. A record from Turkey (Cabi et al., 2019) is treated as doubtful, as the number of spine-cells per cluster is not mentioned. For more details to distinguish these taxa, see the species chapters in question and the matrix key in Chap. 15.

16.6 Distribution

Chara aculeolata occurs across the whole of Europe from Scandinavia in the north to the Mediterranean Sea and from the Iberian peninsula and Ireland in the west to Russia, Belarus and Ukraine in the east and the Balkans to the southeast (Fig. 16.6). The species is lacking from Finland (Langangen et al., 2002). Outside of Europe, the species is known from North Africa (Corillion, 1978) and has been mentioned from southern Africa (Wood, 1965). A record from Turkey (Cabi et al., 2019) is treated as doubtful (see above). It has also been published for a number of Asian countries such as Iran, Kazakhstan, Turkmenistan and Uzbekistan (bibliography available in Romanov et al., 2017), but Central Asian records are mainly based on misidentifications of *C. globata* and *C. papillosa* (R. Romanov unpubl. data).

16.7 Status and Threats

Chara aculeolata has been assessed as RE (Regionally Extinct) in Bulgaria and Czechia, CR (Critically Endangered) in the Netherlands, Denmark and Switzerland, as EN (Endangered) in Germany, Poland and SE Europe, as VU (Vulnerable) in Norway and Belarus and as "Depleting" in Ukraine. For the European part of Russia, the status "Rare" was suggested (Romanov et al., 2017). In Iberia, Ireland, Britain, Sweden, Lithuania and Estonia, the species is too frequent to be included in the Red Lists (see Chap. 9). *Chara aculeolata* is assessed as LC (Least Concern) in the European Red List (see Chap. 9).

The species has probably declined from many sites but has colonised new sites especially in artificial lakes and small water bodies. *C. aculeolata* is a characteristic species of the habitat 3140 of the Habitats Directive (European Commission, 1992). It is threatened by eutrophication and—especially in lakes—introduction of fish. In small water bodies clearances can favour the species and prevent overgrowth by strong competitors.

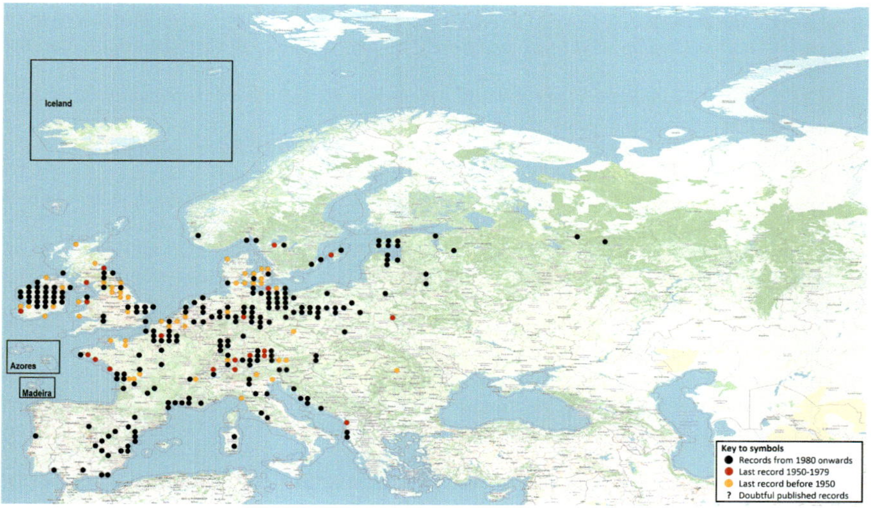

Fig. 16.6 European distribution of *Chara aculeolata*. For worldwide distribution see Korsch (2018). Black dots are records from 1980 onwards, red dots last record from 1950 to 1979 and yellow dots last record before 1950; doubtful published records (if applicable) are marked with "?" superimposed. Map prepared by Heiko Korsch

References

Abdelahad, N., Millozza, A., & Iberite, M. (2019). The '*Chara hispida* species complex' (Charales, Charophyceae) kept in the Herbarium RO of the Sapienza University of Rome. *Webbia, 74*, 121–132.

Alegro, A., Stanković, ĆI., Šegota, V., van de Weyer, K., & Blaženčić, J. (2016). Macrophytic vegetation in the oligotrophic Mediterranean Lake Vrana (Island of Cres, Northern Adriatic). New insight after 50 years. *Flora Serbica, 40*, 137–144.

Azzella, M. M. (2014). Italian volcanic lakes: A diversity hotspot and refuge for European charophytes. *Journal of Limnology, 73*, 502–510.

Becker, R. (2019). The Characeae (Charales, Charophyceae) of Sardinia (Italy): Habitats, distribution and conservation. *Webbia, 74*, 83–101.

Blindow, I. (2004). *Chara polyacantha*. In H. Schubert & I. Blindow (Eds.), *Charophytes of the Baltic Sea* (pp. 129–130). A. R. G. Gantner.

Cabi, E., Şekerciler, F., Çingay, B., & Turan, N. (2019). A new stonewort record for Turkey: *Chara polyacantha* A. Braun. *Acta Biologica Turcica, 32*, 206–210.

Corillion, R. (1978). Les Characées du Nord de l'Afrique: éléments floristiques et distribution. *Bulletin de la Société d'Etudes Scientifiques de L'anjou, Nouvelle Série, 10*, 27–34. [in French].

Dąmbska, I. (1964). *Charophyta—ramienice*. Państwowe Wydawnictwo Naukowe [in Polish].

Dąmbska, I. (1966). Zbiorowiska ramienic Polski. *Prace Komisji Biologicznej, Poznanskie Towarzystwo Przyjaciol Nauk, Wydzial Matematyczno Przyrodniczy, 31*, 1–76. [in Polish].

Doege, A., van de Weyer, K., Becker, R., & Schubert, H. (2016). *Bioindikation mit Characeen*. In Arbeitsgruppe Characeen Deutschlands, Armleuchteralgen: Die Characeen Deutschlands (pp. 97–138). Springer, Berlin Heidelberg [in German].

European Commission (1992). *Council Directive 92/43/EEC of 21 May 1992 on the conservation of natural habitats and of wild fauna and flora*. Official Journal of the European Communities L 206/7.

Gąbka, M. (2009). *Charophytes of the Wielkopolska region (NW Poland): distribution, taxonomy, and autecology*. Bogucki Wydawnictvo.

Golubić, S. (1961). Der Vrana-See an der Insel Cres—ein Chara-See. *Verhandlungen der Internationalen Vereinigung für Limnologie, 14*, 846–849. [in German].

Groves, J., & Bullock-Webster, G. R. (1924). *The British Charophyta. Volume II. Chareæ with plates, concluding articles, geological sketch, bibliography and index*. Ray Society.

Herbst, A., Henningsen, L., Schubert, H., & Blindow, I. (2018). Encrustations and element composition of charophytes from fresh or brackish water sites—habitat- or species-specific differences? *Aquatic Botany, 148*, 29–34.

Hollerbach, M. M., & Krassavina, L. K. [Голлербах, М. М. & Л. К. Красавина]. (1983). Определитель пресноводных водорослей СССР. 14. *Charophyta [Determination key of freshwater algae of USSR]*, 189 pp. Nauka [in Russian].

Holtz, L. (n.d.). *Zeichnungen und Messungen von Characeen Pommerns von Dr. Paul Hauptfleisch, entsprechend meinen Publikationen*, Inv. Nr. 5551. University of Greifswald, Germany [in German].

Kabus, T., & Mauersberger, R. (2011). Liste und Rote Liste der Armleuchteralgen (Characeae) des Landes Brandenburg 2011. *Naturschutz und Landschaftspflege in Brandenburg, 20* (4, Beilage), 1–31 [in German].

Korsch, H. (2013). Die Armleuchteralgen (Charophyceae) Sachsen-Anhalts. *Berichte des Landesamtes für Umweltschutz Sachsen-Anhalt, 1*(2013), 1–85 [in German].

Korsch, H. (2018). The worldwide range of the Charophyte species native to Germany. *Rostocker Meeresbiologische Beiträge, 28*, 45–96.

Krause, W. (1997). Charales (Charophyceae) Süßwasserflora von Mitteleuropa. In H. Ettl, G. Gärtner, H. Heynig, & D. Mollenhauer (Eds.), *Süßwasserflora von Mitteleuropa 18*. Fischer [in German].

Lambert, S. J. (2007). *The environmental range and tolerance limits of British Stoneworts (Charophytes)*. Ph.D. Thesis, University of East Anglia, Norwich, Great Britain, 452 pp.

Langangen, A. (2007). *Charophytes of the Nordic countries* (p. 102). Saeculum ANS.

Langangen, A., Koistinen, M., & Blindow, I. (2002). The charophytes of Finland. *Memoranda Societatis pro Fauna Et Flora Fennica, 78*, 17–48.

Lindner, A. (1978). Soziologisch-ökologische Untersuchungen an der submersen Vegetation in der Boddenkette südlich des Darß und des Zingst (südliche Ostsee). *Limnologica, 11*, 229–305. [in German].

Migula, W. (1889–1897). Fünfter Band: Die Characeen. In A. Fischer, E. Fischer, F. Hauck, G. Limpricht, C. Luerssen, W. Migula, H. Rehm, P. Richter, & G. Winter (Eds.), *Dr. L. Rabenhorst's Kryptogamenflora von Deutschland, Oesterreich und der Schweiz*, 2nd ed. Eduard Kummer, Leipzig [in German].

Moore, J. A. (1986). *Charophytes of Great Britain and Ireland*. BSBI Handbook 5. Botanical Society of the British Isles, London.

Olsen, S. (1944). Danish Charophyta. *Det Kongelige Danske Videnskabernes Selskab, Biologiske Skrifter, 3*(1), 1–240, pl. 1–2.

Pätzold, F. (2011). Beobachtungen zur Verbreitung von Characeen in Baggerseen der Badischen Oberrheinaue. *Berichte der Botanischen Arbeitsgemeinschaft Südwestdeutschland Beiheft, 3*, 37–42. [in German].

Roden, C., Murphy, P., & Ryan, J. (2020). *Benthic vegetation in Irish marl lakes: monitoring habitat 3140. Condition 2011 to 2018*. Irish Wildlife Manuals 124, National Parks and Wildlife Service, Department of Housing, Local Government and Heritage.

Romanov, R., Zhakova, L., Efimova, V., & Leostrin, A. (2017). *Chara aculeolata* (Charophyceae, Charales): First reliable record for Russia. *Turczaninowia, 20*, 70–81. [in Russian].

Spiess, H. J., Bolbrinker, P., Möbius, F., & Waterstraat, A. (2010). Ergebnisse der Untersuchungen submerser Makrophyten in ausgewählten Gewässern Mecklenburg-Vorpommerns. *Botanischer Rundbrief Mecklenburg-Vorpommern, 47*, 4–182. [in German].
van de Weyer, K., & Schmidt, C. (2018). Bestimmungsschlüssel für die aquatischen Makrophyten (Gefäßpflanzen, Armleuchteralgen und Moose) in Deutschland: Band 1: Bestimmungsschlüssel, ed. 2. Landesamt für Umwelt (LfU) Brandenburg. *Potsdam, Fachbeiträge des LfU Brandenburg, 119*, 1–180. [in German].
van Raam, J. C. (1998). *Handboek Kranswieren.* Chara boek [in Dutch].
Vishnyakov, V. S., Romanov, R. E., Komarova, A. S., Belyakov, E. A., Moseev, D. S., Churakova, E., Czhobadze, A. B., & Philippov, D. A. (2021) New charophyte records (Characeae) European Russia. Botanicheskii *Zhurnal*, 106(1), 61–76 [in Russian].
Wood, R. D. (1965). *Monograph of the Characeae.* J. Cramer.

Chapter 17
Chara affinis

Irmgard Blindow, Nick Stewart, and Hendrik Schubert

Scientific name: ***Chara affinis*** **(H. Groves & J. Groves) Blindow, N. Stewart & Hendr. Schub. comb.nov.**

≡ *Chara baltica* (Hartm.) Bruzelius var. *affinis* H. Groves & J. Groves in J. Bot. 18: 354. 1881.
Lectotype (designated here Blindow, N. Stewart & Hendr. Schub.): The Lizard, by cliff/VII-1878/W. Curnow (BM000771974).

Taxonomic note: The taxon, originally described by Groves and Groves (1881), is morphologically clearly distinct from *Chara baltica* (see below) and is therefore separated here from that species.

Local names: ENGLISH: Lizard Stonewort.

I. Blindow (✉)
Biologische Station Hiddensee, Universität Greifswald, 18565 Kloster, Germany
e-mail: blindi@uni-greifswald.de

N. Stewart
4 Hexton Road, Glastonbury BA6 8HL, UK
e-mail: nfstewart@freeuk.com

H. Schubert
Universität Rostock Institut für Biowissenschaften, Albert-Einstein-Straße 3, 18059 Rostock, Germany
e-mail: hendrik.schubert@uni-rostock.de

H. Schubert et al. (eds.), *Charophytes of Europe*,
https://doi.org/10.1007/978-3-031-31898-6_17

17.1 Morphology[1]

See Figs. 17.1, 17.2, and 17.3.

Habitus: Plants without or with (heavy) encrustation, light to fresh to dark green but becoming greyish green with encrustation. Often, the plants are sparsely branched.

Size: Stout to rather delicate, 15–50 cm long.

Main axis: Diameter is about 0.7–2.1 mm; unusually slender plants with axis diameters of only 0.3 mm named as this taxon are doubtful (see section 17.5).

Cortex: Diplostichous and isostichous or slightly tylacanthous, rarely with triplostichous sections.

Spine-cells: Single or arranged in bunches of 2–3, dense, slender and acute, longer than or as long as the main axis diameter, 1.2–2.2 mm long. According to Groves and Groves (1881), Cornish specimens often have clustered spine-cells, while spine-cells are solitary in collections from Guernsey. Collections in BR from Belgium had both solitary and clustered spine-cells. Apart from the youngest, still elongating internodes, the spine-cells are often inclined at a low angle towards the axis, most of them directed towards the centre of the internodes. This often causes a "traffic-accident-like" pattern in the middle of the internode where these long and dense spine-cells meet and entangle (see Figs. 17.1b and 17.2d). Spine-cells are often sparse on lower internodes.

Stipulodes: Long and slender, arranged in two tiers of almost equal length, which often are dense and irregular. Length measurements vary between 1.1 and 2.2 mm.

Branchlets: Whorls consist of 9–10 nearly straight, or slightly incurved branchlets, 4–8 segments, 1–2(3) outermost ecorticate, the corticate part of the branchlet far longer than the ecorticate part.

Bract-cells: Acute, long and spreading. Adaxial bract-cells are longer than the abaxial ones. (Almost) verticillate bract-cells are only rarely found.

Bracteoles: Longer or far longer than the ripe oogonium, reaching up to 4 times its length (own data).

Sexuality: Monoecious. Gametangia are conjoined on the 3–5 lowest branchlet nodes.

Oogonia: Slender with a high length:width ratio (average 1.7, max 2.0). Coronula cells are spreading, often distinctly so (Groves and Groves, 1881; collections in BM). A straight coronula was observed, however, in the French collections. Table 17.1 gives an overview about the morphometry of the oogonia.

---

[1] Morphological details summarized below originate from Groves and Groves (1881) combined with observations of the collections in BM (many of which were deposited by H. Groves & J. Groves) and BR, if not stated otherwise.

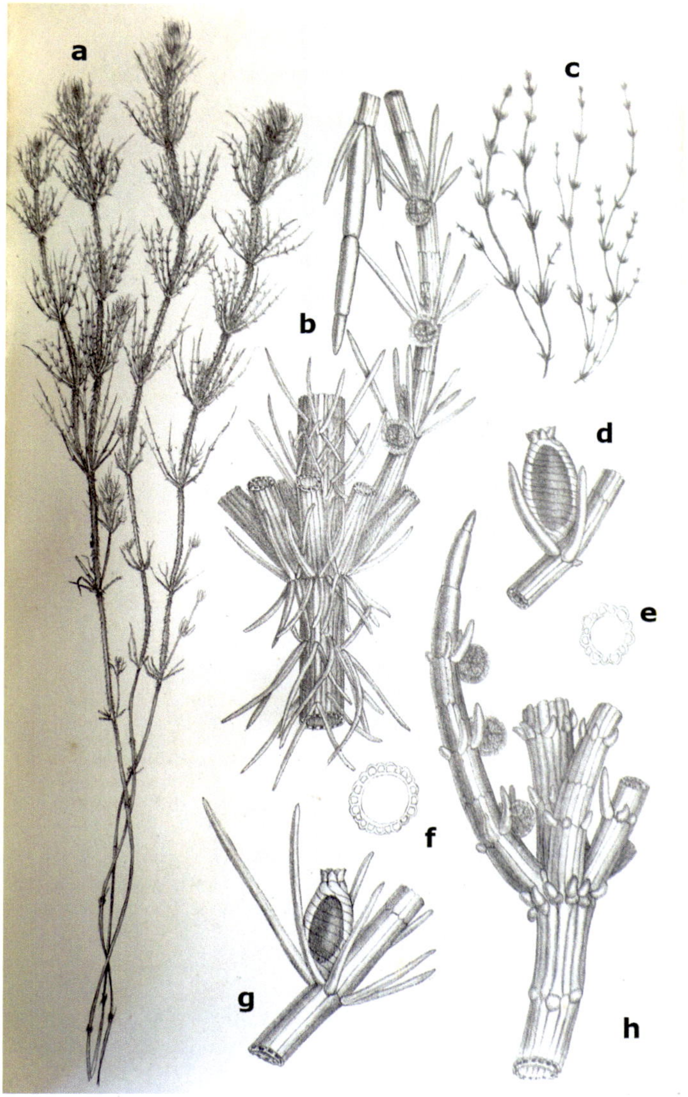

Fig. 17.1 *Chara baltica* var. *affinis*. **a** Habitus; **b** Whorl base and branchlet; **c** Elongated growth form; **d** Oogonium; **e** Tylacanthous main axis section; **f** Isostichous main axis section; **g** Oogonium with verticillate bract-cells and long bracteoles; **h** Whorl base and branchlet with reduced spine-cells, stipulodes and bract-cells. Drawings from Groves and Groves (1881)

Fig. 17.2 Habitat and habitus of *Chara affinis*. **a** Mullion Quarry in Mullion Cove, UK, a site of *Chara affinis*; **b** Habitus of plants from Mullion Quarry; **c** Richly fertile collection of *Chara affinis* from Guersney in BM; **d** Internode section of a collection in BM, showing downwards directed spine-cells in the upper and upwards directed spine-cells in the lower part of the internode, as well as the "traffic-accident-like" pattern in the middle of the internode. Photographs: **a**, **b**—Klaus van de Weyer; **c**—Irmgard Blindow, **d**—Hendrik Schubert

Oospores: Figure 17.3e, f shows SEM images of two oospores, one of them being unusually slender. An overview about morphometry of oospores is given in Table 17.2.

Antheridia: Antheridia are yellowish to orange. Measurements on 21 dry antheridia gave a mean diameter of around 450 μm (range of 340–620 μm).

Bulbils: Multicellular, semi-translucent bulbils are present around the basal main axis nodes in several collections.

Fig. 17.3 Details of *Chara affinis*. **a** Collection with tylacanthous cortex; **b** Fertile collection in BM, oogonia with spreading coronula, long bracteoles are visible; **c** Collection with clustered spine-cells, BM, **d** Shoot cross-sections showing isostichous cortex; **e** SEM image of an oospore of a collection from Grand Mare, Guernsey, UK, coll. by Marquand, E.D. 1889 (BM000771978); **f** SEM image of an oospore of a collection from Mullion Quarry, near Mullion Cove, UK, coll. 2019 by G. S. Hislop, M.T. Casanova, K. van de Weyer, N.F. Stewart. Photographs: **a**—Jo Wilbraham, **b**, **c**—Hendrik Schubert; **d**—Klaus van de Weyer; **e**, **f**—Michelle T. Casanova

Table 17.1 Oogonia morphometry

Author	Length	Width	Coronula length	Coronula width	Number of helical stripes
Groves and Groves (1881)	–	–	–	–	11
Own data and total range	**670–1080**	**410–600**	**120–260**	**240–530**	**10–13**

Length in the first column excludes the coronula; its length and width are given separately, all values for length and width in μm

Table 17.2 Oospore morphometry

Author	Length	Width	Number of striae	Colour
Groves and Groves (1881)	–	–	–	Black
Own data	700–840	410–480	15	Black
Total range	**700–840**	**410–480**	**11–15**	

All values for length and width in μm

17.2 Phenology and Reproduction

Little is known about the phenology of this species. It has been recorded throughout the year and can be perennial. Gametangia are present in the summer, and plants are often richly fertile. However, there is little information on the timing of growth stages.

17.3 Habitat Conditions and Ecology

Habitat type: In its British sites, *Chara affinis* occurs in small quarries excavated in serpentinite (a base-rich igneous rock). These vary in size from small pits a few metres across (quarried for ornamental stone for carving) to a large stone quarry. The species has also previously occurred in small streams, most likely where there were static areas out of the main flow. In Guernsey it occurred in some ditches and a shallow pool. In France it occurs in a former stone quarry (Mireval), a brackish water marsh, and a brackish pond. In Belgium, it has been recorded from an artificial lake and a ditch.

Sediment: Sandy sediments were recorded in France, silt over a stony base in Britain.

Nutrient conditions: Information is lacking. The French sites are probably mesotrophic.

Water chemistry: In France *Chara affinis* grows in brackish water (6–7.9 psu). In Britain it occurs in waters that are around neutral. The British and Guernsey sites are close to the sea (up to 5 km inland) on exposed coasts facing the Atlantic Ocean. They may receive some wind-carried salt but there is no evidence of this in the associated flora. However, the rock-type is rich in magnesium (but low in calcium), and therefore may have a high Mg:Ca ratio similar to sea water.

Depth range: In France, the taxon was found in water 30–80 cm deep (more than knee-deep) and probably so in similar depths in Guernsey. In Britain it can occur in depths from 10 to 300 cm.

Associated species: In Britain, *Chara affinis* is associated with a soft water flora, notably *Chara fragifera, Potamogeton polygonifolius, P. natans, Eleogiton fluitans* and *Myriophyllum alterniflorum.* In shallow water it may be mixed among small emergents such as *Juncus articulatus, Eleocharis palustris, E. quinqueflora, Molinia caerulea, Carex demissa* and *Ranunculus flammula.* Little is known about its associates in Guernsey except that *Chara virgata* and *Potamogeton coloratus* were recorded from the site in the same period.

17.4 Variability

Like other taxa within the *Hartmania* group, *Chara affinis* shows a rather high variability. Axis diameter and arrangement of spine-cells especially (single or in clusters) vary among populations.

17.5 Identification and Risk of Confusion

Chara affinis belongs to the *Hartmania* group, a complex of coarse, diplostichous *Chara* spp. that are hard to separate from each other. Within this group, it was described as a variety of *Chara baltica* by Groves and Groves (1881), but it is morphologically closer to the cluster combining tylacanthous to isostichous monoecious taxa from freshwater and inland brackish water sites (see Chap. 15).

From *C. baltica, C. affinis* is separated by long, dense and acute spine-cells, as well as long stipulodes and bract-cells. Antheridia are smaller than in *C. baltica.*

From *C. papillosa*, the taxon can be distinguished by long and dense spine-cells, long stipulodes, and long bracteoles. Most *C. aculeolata* have spine-cells arranged in groups of more than 3, never single, and often have verticillate bract-cells. *C. corfuensis*, which has been recorded from the Mediterranean region, is similar, but coarser, and has rather round oogonia (mean value for length:width ratio = 1.2). Most collections of *C. corfuensis* have verticillate and often squarrose bract-cells. The distinction from "Mediterranean Chara", a taxon under evaluation (see Chap. 80), needs to be clarified.

The identification of *Chara affinis* is complicated by its often isostichous or only rather weakly tylacanthous cortication. Therefore, the species is hard to separate from isostichous forms of *Chara hispida* (Wood, 1965: 148). In *C. hispida*, however, spine-cell clusters are star-shaped and pointing in different directions, while most spine-cells are directed towards the centre of the internode in *C. affinis*.

Slender collections can be confused with *Chara contraria*. Some of the collections of "*Chara baltica* var. *affinis*" in BM look similar to *Chara contraria* var. *hispidula* reported from Canada, which also has solitary, long spine-cells, directed towards the centre of the internode, and often isostichous cortication. Genetically, *C. contraria* var. *hispidula* is clearly placed in the *C. contraria* clade (Mann et al., 2022).

An unusually high length:width ratio of oogonia and oospores was observed in herbarium and fresh material (see Fig. 17.3f). This character needs verification.

17.6 Distribution

Chara affinis is restricted to Europe where it has been found in Britain, the Channel Islands (Iles Anglo-Normande), France and Belgium (Fig. 17.4). An occurrence in the Netherlands has not been confirmed yet but is probable. Van Raam (1998) reported *C. baltica* (including var. *affinis*) from the country. His description and drawings of *C. baltica* var. *affinis* resemble *C. affinis*, but antheridia diameters of up to 600 μm are given, which is more in the range of *C. baltica* than *C. affinis*. Leiden contains collections of "*Chara baltica*" from the Netherlands and Belgium, some of which resemble *Chara affinis* in appearance.

17.7 Status and Threats

Because this taxon has not previously been treated at species level it has not been assessed for Red Lists. However, it is likely that it would be classed as Threatened in Britain and France and Regionally Extinct in the Channel Islands.

In Britain, because most sites are small water bodies, the main threat is succession. This includes the type locality which has been partly engulfed by scrub and partly infilled. Three other localities (two recent, one historic) have probably been lost due to increased bog vegetation. Another recent site seems to have been lost due to enrichment causing excessive algal growth. However, the two remaining sites seem to be stable. The Guernsey site (last seen in 1894) was in an area of grazing marsh which has since been drained and is now a golf course. Some water bodies remain but seem to be too enriched. Because of these losses it has been assessed as Vulnerable in Europe (and globally) (see Chap. 9)

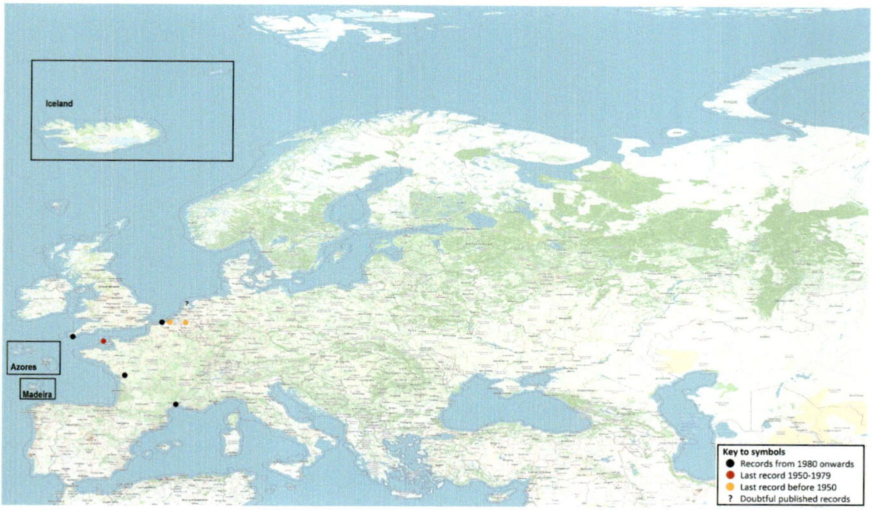

Fig. 17.4 European distribution of *Chara affinis*. Black dots are records from 1980 onwards, red dots last record from 1950 to 1979 and yellow dots last record before 1950; doubtful published records (if applicable) are marked with "?" superimposed. Map prepared by Heiko Korsch

Acknowledgements We are grateful to Jo Wilbraham for checking the material in BM and providing photos. Emile Nat and Luc Denys thoroughly investigated (and provided) herbarium material from L and BR. The curator of BR is acknowledged for the loan of specimens.

References

Groves, H., & Groves, J. (1881). Notes on the British Characeæ. *Journal of Botany, 18*, 353–356.

Mann, H., Hanel, C., Langangen, A., & Nowak, P. (2022). *Chara contraria* var. *hispidula* Braun (Charales) in Newfoundland, Canada a new variety described from North America. *Botany Letters, 169*, 250–258.

van Raam, J. C. (1998). *Handboek Kranswieren*. Chara boek [in Dutch].

Wood, R. D. (1965). *Monograph of the Characeae*. J. Cramer.

Chapter 18
Chara altaica

Roman Romanov and Vasily Vishnyakov

Scientific name: ***Chara altaica*** **A. Braun in A. Braun & Nordst., Abh. Königl. Akad. Wiss. Berlin 1882: 148. 1882**

Lectotype (Wood, 1965: 159): [RUSSIA]: Altai/1826/[ex Herb.] Ledebour 112 (BM).

= *Chara sibirica* Mig. in Trudy Imp. S.-Peterburgsk. Bot. Sada 23: 537. 1904. Lectotype (designated here by R. Romanov): [RUSSIA]: Sibiria, gub. Tomsk, in lacu Czany. Д. Средн. Ярки, на мелком илистом берегу озера Чаны [the village of Srednie Yarki, at shallow silty shore of Lake Chany]/16 Aug. 1890/Korshinsky (LE A0001000).

Taxonomic note: The original material of *Chara altaica* collected by Bunge and labelled as a part of the herbarium by Ledebour consisted of two species: *Chara altaica* and *C. canescens* (LE!). The presence of these two species of *Chara* in the lectotype of *C. altaica* was already noted by Wood (1965). Also, in the original material of *C. sibirica* these two species, *C. altaica* and *C. canescens*, were present (Hollerbach, 1945; LE!).

Local names: RUSSIAN: Хара алтайская; UKRAINIAN: Хара алтайська.

R. Romanov (✉)
Nišice Ujtin Potok b.b, Dobra Voda, Bar 85356, Montenegro
e-mail: streptophytes@gmail.com

V. Vishnyakov
Papanin Institute for Biology of Inland Waters of the Russian Academy of Sciences, Laboratory for Algology, 152742 Yaroslavl Region, Nekouz District, Borok, Russia

H. Schubert et al. (eds.), *Charophytes of Europe*,
https://doi.org/10.1007/978-3-031-31898-6_18

18.1 Morphology

See Figs. 18.1, 18.2, 18.3, and 18.4.

Habitus: The plants are bright green, mostly not encrusted, scarcely branching, having a diffusely spiny appearance of its main axis and shortly spiny outline of the branchlets, usually consisting of several shoots (up to 50) arising from the common base.

Size: *Chara altaica* is a small to medium plant, from 1–3 cm up to 20–30 cm and even 60 cm high (Fig. 18.3).

Main axis: The main axis is 285–750 μm in diameter, slender, flaccid, and completely corticated. Length of internodes is nearly equal to branchlets or 2–4 times longer, usually a few cm long, up to 7.5 cm.

Cortex: Main axis cortex is irregular, tylacanthous, with well developed primary rows and imperfectly appearing narrow secondary rows, mostly between a haplo-diplostichous to diplostichous, rarely diplo-triplostichous arrangement (Fig. 18.2c, d).

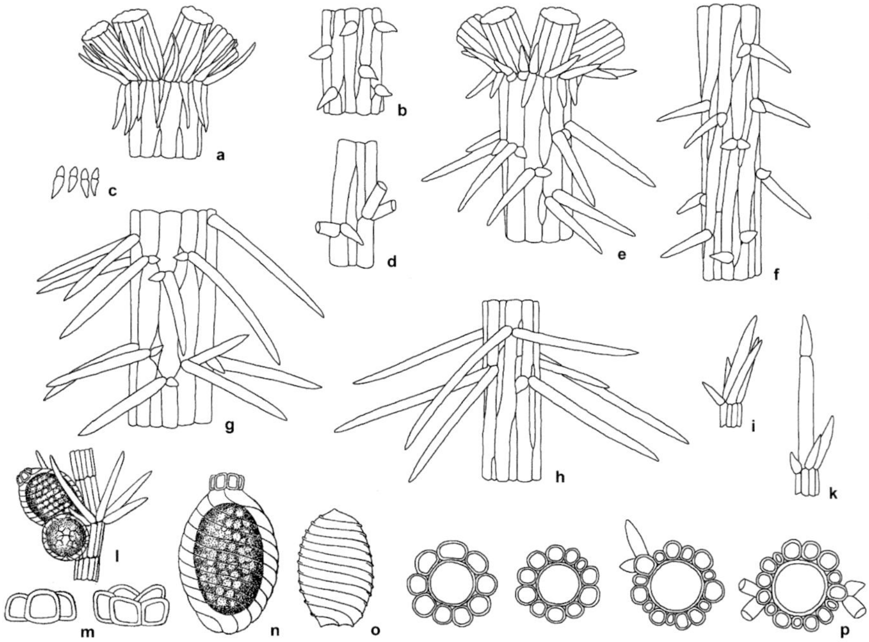

Fig. 18.1 *Chara altaica*. **a–h** Different arrangement of main axis cortex, spine-cells and stipulodes; **i**, **k** Ends of branchlets; **l** Gametangia; **m** Coronulae of oogonia; **n** Oogonium; **o** Oospore; **p** Cross sections of main axis, showing variability of cortex arrangement. Figures **a, b, c, d, g, i, k, l, m, n, o**, last one at **p** were taken from the isotype of *C. altaica*; **e, f, h**, first three ones at **p**—from the lectotype specimen of *C. sibirica*. Drawings: Hollerbach (1945, 1949), Hollerbach and Krassavina (1983)

Fig. 18.2 *Chara altaica,* details. **a** Condensed whorls of somewhat inflated plant from more brackish environment; **b** Whorl of branchlets; **c, d** Main axis cortex. Scale: **a, b**—1 mm, **c**—0.5 mm, **d**—0.2 mm. Photographs: Roman Romanov

Spine-cells: The spine-cells are both long and papillate, nearly equal to main axis diameter or up to 2–2.5 longer, always pointed, 1–3-fasciculate, unequal (one long, the others short), sometimes appearing in unclear rings.

Stipulodes: The stipulodes are well developed, diplostephanous, in pairs next to each branchlet (duplicate), slender, acuminate, uppers are occasionally much longer than lowers (up to 700 μm), lowers are sometimes pointed papillate to blunt.

Branchlets: The branchlets are 9–11 in spreading regular whorls, more or less arcuate at upper parts or usually nearly straight (Figs. 18.2a, b and 18.3), 0.9–2(3) cm length, consisting of (1–3)4–6 corticate segments and a short or somewhat elongated ecorticate part. The ecorticate part is 1–3(5)-celled, always without nodes, slightly longer than adjacent bract-cells, nearly equal in length to the last corticate segment or longer. The end-cell is acuminate. Branchlet cortex varies from haplostichous to irregular, haplo-diplostichous.

Bract-cells: The bract-cells are well developed, up to 1.5 mm length, verticillate, acuminate. Adaxial and abaxial bract-cells are nearly equal to the oogonium or longer, usually the adaxial bract-cells are somewhat longer than abaxial ones.

Bracteoles: The bracteoles are clearly shorter than the oogonium.

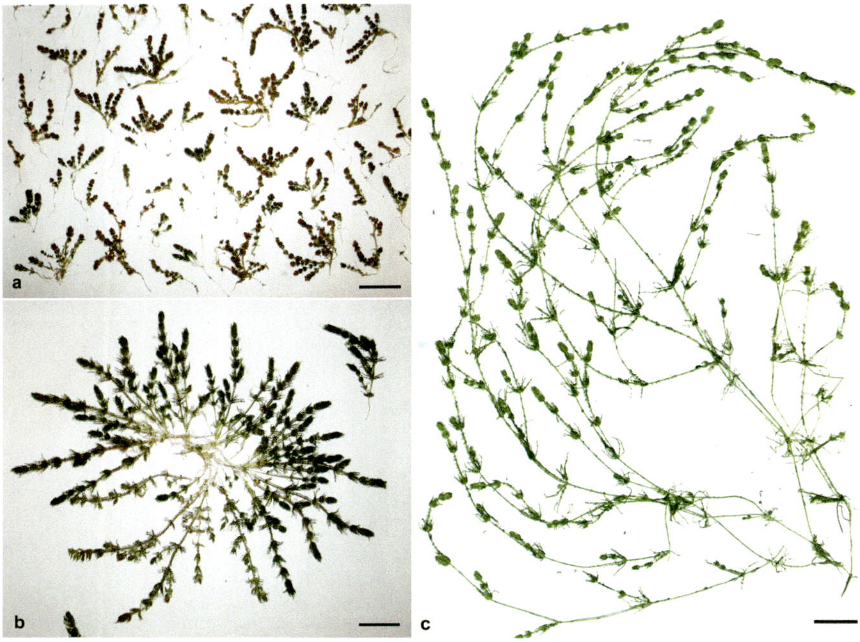

Fig. 18.3 Variability of general habit of *Chara altaica*. **a** Tiny plants from steppe shallow brackish lake, few cm depth; **b** Medium plants from the same lake, 20–30 cm depth; **c** Large plants from brackish drainage channel in steppe depression, 40–50 cm depth. Scale: 2 cm. Photographs: Roman Romanov

Sexuality: *Chara altaica* is monoecious. Gametangia are solitary, always conjoined, at the first 3–4 nodes of the branchlet.

Oogonia: The oogonia are ellipsoid, significantly enlarging after fertilization. The spiral cells are sometimes somewhat inflated below coronula. The coronula is compact, with tight alignment of cells to each other, variable in shape, mostly narrowing to the apex (conical), rarely straight to slightly spreading. Table 18.1 gives an overview about published morphometrical data.

Oospores: Brown to black, ellipsoid, 450–741 μm in length and 243–614 μm in width, with 8–13 low spiral ridges ending with tiny conical claws surrounding basal cell impression (Fig. 18.4a). Oospore surface is nearly smooth to obscurely tuberculate in LM (Hollerbach and Krassavina, 1983), very scarcely granulate in LM and SEM (Kato et al., 2010) or irregularly scarcely granulate up to nearly smooth (Romanov, 2011) or irregularly roughened (Fig. 18.4b). The granules are uniform (Kato et al., 2010) or irregular in shape and size, but smaller ones are most abundant. The surface between granules is smooth. The ribs are irregularly scarcely granulate or nearly smooth (Romanov, 2011). Table 18.2 gives an overview about published morphometric data.

Fig. 18.4 Oospores of *C. altaica*, topotype (LE) scanning electron microscopy. **a** General view, **b** Oospore surface. Scale: **a**—100 μm, **b**—10 μm. Photographs: Roman Romanov

Antheridia: The antheridia are octoscutate, at the left side of the base of oogonium in view from abaxial side of branchlet. Table 18.3 gives an overview about published morphometric data.

Bulbils: Nodal bulbils are known from Northern Kazakhstan (Sviridenko and Sviridenko, 2016).

18.2 Phenology and Reproduction

Chara altaica reproduces by oospores, which is indicated by the typically richly fertile plants. The arrangement of whole plant—usually without signs basal decay and the rare occurence of dense stands, growing mainly in shallow water, suggests that *C. altaica* is mostly annual species.

Table 18.1 Oogonia morphometry

Author	Length	Width	Coronula length	Coronula width	Number of helical stripes
Braun and Nordstedt (1882)	600–840	330–540	50	120	12
Migula (1904), *C. sibirica*	400–670	230–340	80	170	-
Wood and Imahori (1964)	675	480	135	225	14
Wood and Imahori (1964), *C. sibirica*	450	300	75	150	ca. 13
Wood (1965) *C. sibirica*	400–600(670?)	230–340	78–80	150–170	ca. 13
Hollerbach and Krassavina (1983)	435–600(840)	250–420(540)	50–84	117–185	10–13
Han and Li (1994)	700–900	450–480	70	150–160	12–14
Kato et al. (2010)	620–730	350–450	50–80	150–180	12–14
Romanov and Kipriyanova (2010)	405–794	348–589	63–102	154–218	-
Total range	**400–900**	**230–589**	**50–102**	**117–218**	**10–14**

Length in the first column is excluding the coronula; its length and width are given separately, all values for length and width in μm

Table 18.2 Oospore morphometry

Author	Length	Width	Number of striae	Colour
Braun and Nordstedt (1882)	450–600	240–400	10, 11	–
Migula (1904), *C. sibirica*	480	280	–	Opaque black
Wood and Imahori (1964)	585	275	12	Yellow brown (nearly ripe)
Wood (1965), *C. sibirica*	450	280	–	Black (immature)
Hollerbach and Krassavina (1983)	554–660	337–400	9–12	Dark brown, almost black
Han and Li (1994)	560–720	300–370	11, 12	–
Kato et al. (2010)	610–700	330–400	11–13	Black
Romanov and Kipriyanova (2010)	472–614	243–614	–	Black brown to brown
Romanov (2011)	472–741	243–378	8–10	Black
Total range	**450–741**	**240–614**	**8–13**	**Brown to black**

All values for length and width in μm

Table 18.3 Antheridia morphometry

Author	Diameter (μm)
Braun and Nordstedt (1882)	420–480
Migula (1904), *C. sibirica*	250
Wood and Imahori (1964)	330–375
Wood and Imahori, (1964), *C. sibirica*	225–250
Wood (1965), *C. sibirica*	225–250
Hollerbach and Krassavina (1983)	265–420
Han and Li (1994)	400–450
Kato et al. (2010)	350–450
Romanov and Kipriyanova (2010)	269–371
Total range	**225–480**

18.3 Habitat Conditions and Ecology

Habitat type: Brackish lakes and wetlands, rarely small water bodies and channels associated with groundwater discharge in inland semiarid and arid regions, only one occurrence reported from a lake connected directly to the sea (Hollerbach and Krassavina, 1983; Kato et al., 2010; Kipriyanova and Romanov, 2013; Romanov and Kipriyanova, 2010; Sviridenko and Sviridenko, 2016; own data).

Sediment: Sand, silty sand and silt, rarely silt over gravel (Sviridenko and Sviridenko, 2016; own data).

Nutrient conditions and water chemistry: Associated with brackish waters (Table 18.4), having high concentrations of Cl^- and Na^+, able to grow in mesotrophic and eutrophic water bodies with inorganic nitrogen content up to 200–942 mg/m^3 (Kipriyanova and Romanov, 2013; Sviridenko and Sviridenko, 2016; own data).

Depth range: 0.1–2.3 m, but clearly prefers shallow water less than 1 m deep (Kipriyanova and Romanov, 2013; Sviridenko and Sviridenko, 2016; own data).

Table 18.4 Total dissolved solids (TDS), pH and water hardness of some habitats of *Chara altaica*

Author	Localities	pH	Hardness (mequiv/l)	TDS (mg/l)
Kipriyanova and Romanov (2013)	South of West Siberia	8.20–9.12	–	0.61–6.40
Sviridenko and Sviridenko (2016)	South of West Siberia	7.2–8.6	5.3–37.3	0.79–8.01
Sviridenko and Sviridenko (2016)	Kazakh Upland	–	5.5–7.4	0.67–1.43
Total range	**–**	**7.2–9.12**	**5.3–37.3**	**0.61–8.01**

Associated species: *Chara altaica* is able to form its own communities (Kipriyanova, 2005) but in a much narrower range of salinity known for this species, frequently in association with *Chara canescens*, *C. aspera*, *C. contraria*, *C. tomentosa*, *Stuckenia pectinata*, *Cladophora* spp., *Zannichellia pedunculata* (Kipriyanova and Romanov, 2013; own data).

18.4 Variability

The depth and water transparency resulting in different insolation seem to be main causes of morphological variability of *C. altaica*. It can be tiny and condensed or elongated slender plants. The habit of branchlets is variable too, from clearly arcuate to nearly straight (Fig. 18.3). The insolated whorls are always condensed. This species can be somewhat inflated in more saline environments (Fig. 18.2a). A juvenile plant having long ecorticate branchlet parts, growing in the shallows of a large brackish lake in forest-steppe was described as *C. sibirica* (Migula, 1904; Wood, 1965). The main axis and branchlet cortex are commonly variable. The stipulodes can be shortened in both rows or the lower row only. Spine-cells are variable in length.

18.5 Identification and Risk of Confusion

Chara altaica is really similar in general appearance with *C. canescens*. Both species frequently grow together. *Chara altaica* differs by being monoecious, having irregular cortication on the main axis and branchlets, and is usually less densely spiny on the main axis. *Chara altaica* is always fertile, so its monoecious character should allow for correct identification. Because *C. canescens* is more well-known and regarded as being easily recognised by the naked eye, a thorough check of all specimens, at least the ones from Eastern and Mediterranean Europe, is recommended for detecting further occurrences of *C. altaica*.

The species has some similarity in habit with *C. strigosa* but is clearly different with respect to ecology and distribution. The delineation with other monoecious species from the section *Desvauxia*, namely *C. evoluta* Allen (North America, China) and *C. hirsuta* Allen (North America), *C. piniformis* W.Q. Chen & F.S. Han (China), *C. longiarticulata* F.S. Han (China) needs to be clarified. *C. piniformis* differs by having an haplostichous main axis cortex, solitary spine-cells only, bracteoles clearly exceeding oogonium length and short, nearly rudimentary, blunt stipulodes (Han and Li, 1994). Delineation of *C. altaica* and *C. longiarticulata* is possible due to the short corticated segments and long ecorticate segments (longer than the corticate part) in the latter, but the reliability of these traits is questionable.

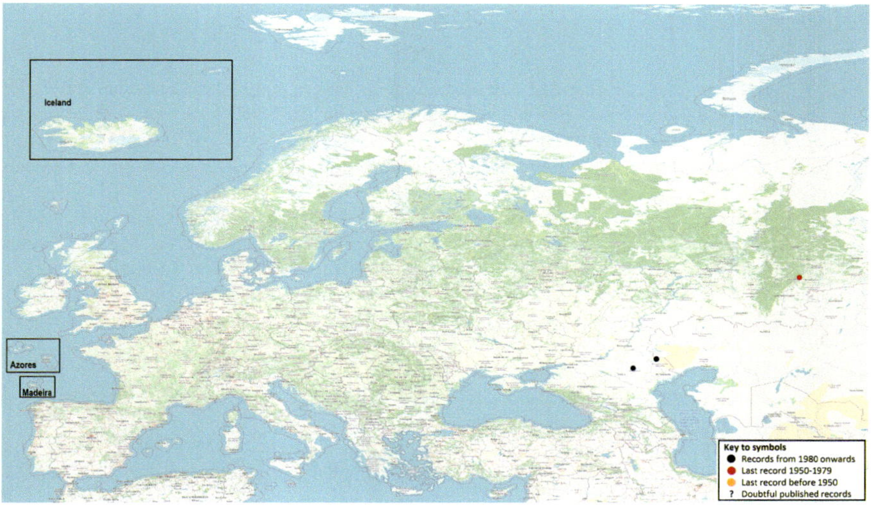

Fig. 18.5 Distribution of *Chara altaica* in Europe. Black dots are records from 1980 on, red dots from 1950 to 1979 and yellow dots before 1950; unconfirmed/doubtful published records (if applicable) are marked with "?" on top. Map prepared by Heiko Korsch

18.6 Distribution

See Fig. 18.5.

South-East Europe: a single locality in steppe between the Azov Sea and Volga River and a single locality in the steppe east of the Volgograd Reservoir (R. Romanov and V. Vishnyakov unpubl.). Outside of Europe: mainly the centre of Asia, subarid and arid regions—Central Asia, Mongolia, South Siberia, inland China with northernmost records in forest-steppe of South Ural and West Siberia and easternmost locality in Japan (Han and Li, 1994; Hollerbach and Krassavina, 1983; Kato et al., 2010; Romanov and Kipriyanova, 2010; Romanov et al., 2014, 2022; Sviridenko and Sviridenko, 2016; LE!, own data).

18.7 Status and Threats

Chara altaica has been assessed as CR (Critically Endangered) in the European Red List (see Chap. 9). The Eutrophication seems to be an important threat, because of its reduced competitive ability in comparison with *Cladophoraceae* in brackish environments. Long-term drought because of climate change could also be important for this species of arid and subarid regions.

References

Braun, A., & Nordstedt, C. F. O. (1882). Fragmente einer Monographie der Characeen. Nach den hinterlassen Manuscripten A. Braun's herausgegeben von Dr. Otto Nordstedt. *Abhandlungen der Königlichen Akademie der Wissenschaften zu Berlin für das Jahr 1882*, 1–211. [in German].

Han, F., & Li, Y. Y. (1994). *Flora algarum sinicarum aquae dulcis. Tomus 3. Charophyta*. Science. [in Chinese].

Hollerbach, M. M. (1945). De corticis structura caulis *Charae sibiricae* Mig. et de positione ejusdem specie in systemate. *Notulae Systematicae e Sectione Cryptogamica Instituti Botanici nomini V.L. Komarovii Academiae Scientiarum URSS, 5*, 138–142. [in Russian and Latin].

Hollerbach, M. M. (1949). De *Chara jubata* A. Br. et *Chara contraria* A.Br. f. *jubataeformis* Vilh. in florae URSS notula. *Notulae Systematicae e Sectione Cryptogamica Instituti Botanici nomini V.L. Komarovii Academiae Scientiarum URSS, 6*, 59–65. [in Russian].

Hollerbach, M. M. & L. K. Krassavina, [Голлербах, М. М. & Красавина, Л. К.]. (1983). *Определитель пресноводных водорослей СССР. 14. Charophyta* [*Determination key of freshwater algae of USSR*]. Nauka. [in Russian].

Kato, S., Sakayama, H., Morishima, H., Sano, S., Oomori, Y., Kato, N., Ito, M., Kasai, F., Watanabe, M., & Nozaki, H. (2010). Morphology and molecular phylogeny of *Chara altaica* (Charales, Charophyceae), a monoecious species of the section *Desvauxia*. *Cytologia, 75*, 211–220.

Kipriyanova, L. M. (2005). The current state of aquatic and shoreline vegetation of the Chany Lake system. *Sibirskiy Ekologicheskiy Zhurnal, 2*, 201–213. [in Russian].

Kipriyanova, L. M., & Romanov, R. E. (2013). Communities of charophytes in waterbodies and water courses in the north of endorheic basin of the Ob-Irtysh Interfluve (Western Siberia). *Inland Water Biology, 6*, 184–193.

Migula, W. (1904). Characeae Rossicae ex Herbarion Horti Petropolitani. *Trudy Imperatorskago S.-Peterburgskago Botaniceskago Sada. Acta Horti Petropolitani, 23*, 533–539. [in German].

Romanov, R. E. (2011). The preliminary data about ultrasculpture of wall external surface of the oospore of *Chara altaica* (Streptophyta: Charales). *Rastitel'nyj Mir Aziatskoj Rossii [Plant Life of Asian Russia], 2*, 15–19. [in Russian].

Romanov, R. E., & Kipriyanova, L. M. (2010). Charophyte species diversity and distribution on the south of the West-Siberian Plain. *Charophytes, 2*, 72–86.

Romanov, R. E., Zhakova, L. V., Bazarova, B. B., & Kipriyanova, L. M. (2014). The charophytes (Charales, Charophyceae) of Mongolia: A checklist and synopsis of localities, including new records. *Nova Hedwigia, 98*, 127–150.

Romanov, R. E., Anissimova, O. V., Anishchenko, L. N., Efimov, D. Y., Kapitonova, O. A., Kipriyanova, L. M., Konechnaya, G. Yu., Kopyrina, L. I., Kotovshchikov, A. V., Kurganov, A. A., Nikolaenko, S. A., Shilov, M. P., Smirnova, E. V., Vodeneeva, E. L., Zarubina, E. Yu., & Zhakova, L. V. (2022). The noteworthy new records of charophytes (Charales, Charophyceae) from Russian regions: improvement of species distribution ranges in Eurasia. *Botanicheskii Zhurnal, 107*, 466–477.

Sviridenko, T. V., & Sviridenko, B. F. [Свириденко, Т. В. & Свириденко, Б. Ф.]. (2016). *Харовые водоросли Западно-Сибирской равнины. Монография* [*Charophyta of the West Siberian Plain. Monograph*]. Amphora. [in Russian].

Wood, R. D. (1965). *Monograph of the Characeae*. J. Cramer.

Wood, R. D., & Imahori, K. (1964). *Iconograph of the Characeae*. J. Cramer.

Chapter 19
Chara aspera

Lars Baastrup-Spohr

Scientific name: ***Chara aspera Willd*****., Mag. Neuesten Entdeck. Gesammten Naturk. Ges. Naturf. Freunde Berlin 3: 298. 1809**

Lectotype (Wood, 1965: 203): in Ducato Megapolitano/[no date]/Detharding (B-W17100-010).

= *Chara intertexta* Desv. in Loisel., Not. Fl. France: 138. 1810.
Lectotype (Wood, 1965: 203): Gallia/[no date]/[Desvaux?] (PC).

≡ *Chara aspera* var. *subinermis* Kütz., Sp. Alg.: 521. 1849.

= *Chara delicatula* Desv. in Loisel., Not. Fl. France: 137. 1810.
Lectotype (Wood, 1965: 203): Gallia/[no date]/[Desvaux?] (PC).

= *Chara canescens* Loisel. subsp. *hoelii* Langangen in Polar Res. 19: 148. 2000.
Holotype (Langangen, 2000: 148): Svalbard, Norway; the Trollkjeldane (Troll Springs) in Bockfiorden, the "cold" spring/28.08.1992/Sissel Aarvik (O).

= *Chara fischeri* Mig. in Trudy Imp. S.-Peterburgsk. Bot. Sada 23: 538. 1904.
Holotype (Wood, 1965: 189): Chiwa, Chanka, in stagnis/17. Aug. 1873/Korolkow & Krause (LE).

Taxonomic note: DNA analyses have shown that *Chara canescens* subsp. *hoelii* from Svalbard is an aberrant form of *Chara aspera* (Langangen et al., 2020). Morphological study of original material of *C. fischeri* has shown it is identical with *C. aspera* (Romanov, 2022).

L. Baastrup-Spohr (✉)
Freshwater Biological Laboratory, University of Copenhagen, Universitetsparken 4, 3rd floor, 2100 Copenhagen Ø, Denmark
e-mail: lbaastrupspohr@bio.ku.dk

H. Schubert et al. (eds.), *Charophytes of Europe*,
https://doi.org/10.1007/978-3-031-31898-6_19

Local names: BELARUSIAN: Хара шурпатая; DANISH: Ru kransnål; DUTCH: ruw kransblad; ENGLISH: Rough stonewort; ESTONIA: Kare mändvetikas; FINNISH: Mukulanäkinparta; FRENCH: Chara rugueuse; GERMAN: Raue Armleuchteralge; LATVIAN: Skarbà hara; LITHUANIAN: Šiurkštusis maurabragis; NORWEGIAN: Bustkrans; POLISH: Ramienica szorstka (rarely used); RUSSIAN: Хара шероховатая; SWEDISH: Borststräfse (Sweden); Knölsträfse (Finland); UKRAINIAN: Хара шорстка.

19.1 Morphology

Habitus: Plants are yellowish green to dark green and found in both freshwater and brackish environments. Plants can be slightly to moderately incrusted. Branchlets are typically shorter than the internodes. The appearance of plants depends on growth site, but commonly they have well developed spine-cells resulting in a pipe cleaner-like look (Figs. 19.1, 19.2, and 19.3).

Size: *Chara aspera* is rather small to medium-sized (5–15 cm), but may reach up to 70 cm if growing in deeper water (Blindow and Schütte, 2007).

Main axis: The main axis diameter is mostly about 0.2–0.4 mm but can be longer (0.5–0.6 mm); there are few branches. Plants in brackish environments are on average shorter than freshwater plants (Blindow and Schütte, 2007).

Fig. 19.1 *Chara aspera,* details. **a** Stipulodes and cortex; **b** Long-spined form; **c** Specimen with short spine-cells; **d** Bulbils; **e** Male individual; **f** Female plants. Photographs: **a–d**—Chris Carter, **e**—Bjarne Moeslund, **f**—Jens Christian Schou

Fig. 19.2 *Chara aspera* in different environments. **a** In a shallow pond on Öland (Sweden); **b** Meadow-forming population in a lake in a former coal mining area near Köckern (Germany); **c** Scattered individuals in brackish water in Blekinge (Sweden). Photographs: **a**—Bjarne Moeslund; **b**—Heiko Korsch; **c**—Peter Holm

Cortex: The cortex is triplostichous tylacanthous. Transitions to diplostichous cortex do appear. Isostichous cortex can be found on older internodes.

Spine-cells: The needle-like spine-cells are as long or longer than the axis diameter. Forms with papilliformous spine-cells do occur.

Stipulodes: Stipulodes are diplostephanous; the upper row is often slightly longer than the lower row. The stipulodes are shorter but otherwise similar to the spine-cells.

Branchlets: Whorls consist of 8–9 short branchlets which are straight to incurved (female plants), or strongly incurved (fertile male plants). Branchlets contain 6–8 corticated segments while the outermost segment is ecorticated.

Bract-cells and bracteoles: Well developed pointed bract-cells are present along the whole branchlet.

Sexuality: The species is dioecious, with males and females co-occuring over the entire range.

Oogonia: Oogonia are solitary, oblong ovate and up to 900 μm long. The coronula short and spreading. Table 19.1 gives an overview about morphometric data on oogonia in the literature.

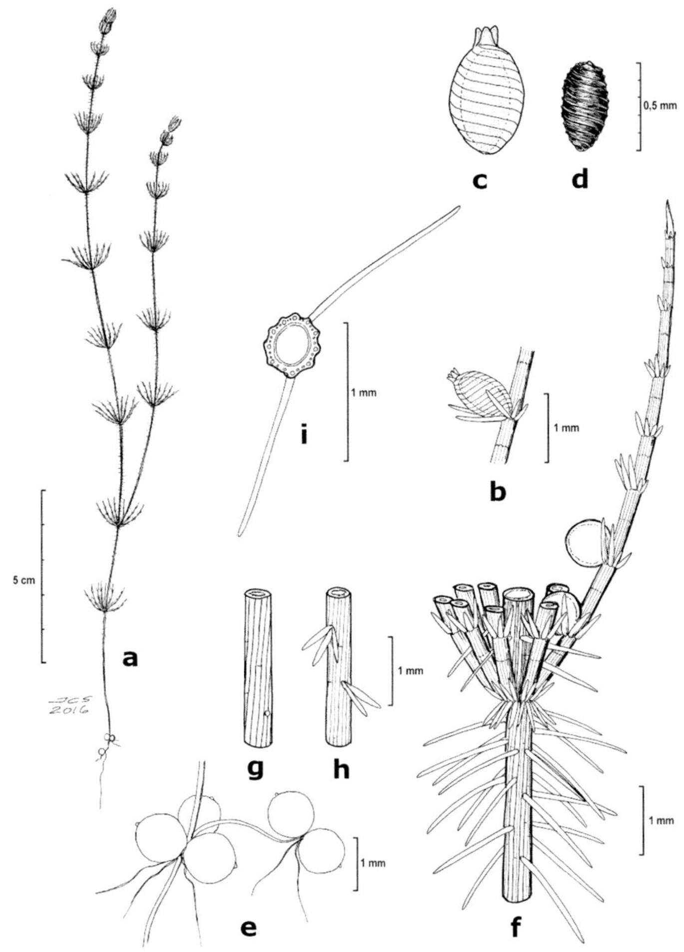

Fig. 19.3 Drawings of *Chara aspera*, details. **a** Freshwater plant habitus; **b**, **c** Oogonium; **d** Oospore; **e** Bulbils; **f** Long-spined form; **g** Short-spined form; **h** *Chara curta ;* **i** Cross section of main axis and cortex. Drawing: Jens C. Schou

Oospores: The oospore is slender with flat ribbons terminating in basal claws (Groves and Bullock-Webster, 1924) and shiny black at maturity from May to September. The outer membrane of the oospore is thick, laminated, brittle, opaque, dull brown and delicately granulated (Groves and Bullock-Webster, 1924). The granules are around 5 per 10 μm and are coarser than those of *Chara curta* (see Chap. 27 *Chara curta*). Table 19.2 gives an overview about morphometric data on oospores in literature.

Table 19.1 Oogonia morphometry

Author	Length	Width	Coronula length	Coronula width	Number of helical stripes
Migula (1889–1897)	750–900	400–450	80	140	
Groves and Bullock-Webster (1924)	650–700	450–525	75–100	200	13–15
Corillion (1957)	600–700	400–450	75–100	120–200	13–15
Wood (1965)	600–800	400–550	75–100	120–200	13–15
Krause (1997)	600–700	400–450	75–120	100–200	13–15
Bailly and Schaefer (2010)		400–500			
Total range	**600–900**	**400–550**	**75–120**	**100–200**	**13–15**

Length in the first column is excluding the coronula; its length and width are given separately, all values for length and width in μm

Table 19.2 Oospore morphometry

Author	Length	Width	Number of striae	Colour
Migula (1889–1897)	500–650	400–450	11–13	Black
Groves and Bullock-Webster (1924)	400–600	255–375	12–14	Black
Corillion (1957)	400–650	250–400	12–14	Black
Wood (1965)	400–650	250–400	12–14	Dark brown-black
Krause (1997)	400–650	250–400	12–14	Shiny black
Vedder (2003)	(325)425–675	250–450	11–15	Shiny black
Bailly and Schaefer (2010)		400–450	12–14	
Total range	**(325)400–675**	**250–450**	**11–15**	

All values for length and width in μm

Antheridia: The antheridia are solitary and found on the inner (1)2(4) nodes on the branchlets. They are spherical, orange to red in colour. Table 19.3 gives an overview about morphometric data on antheridia in literature.

Table 19.3 Antheridia morphometry

Author	Diameter (μm)
Migula (1889–1897)	550–700
Corillion (1957)	400–600
Wood (1965) and Krause (1997)	390–600
Total range	**390–700**

Bulbils: The formation of numerous, white, apparently single-celled bulbils (up to 1 mm) on the rhizoids is common.

19.2 Phenology and Reproduction

Chara aspera can be either annual or perennial. In perennial forms the whole or parts of the plant survive the winter and give rise to new shoots in the spring (Langangen, 2007). Bulbils and oospores normally sprout in spring and experimental tests with bulbils and oospores from the Netherlands show that temperatures of at least 10 °C are needed for plant establishment (Van den Berg et al., 1998, 2001); there was also a positive relationship between temperature and sporeling emergence.

In the laboratory, 100% of bulbils produced plants but only 1–15% of oospores germinated (Van den Berg et al., 2001). Indicating that bulbils serve as propagules for regeneration after winter while oospores are adapted to long term survival. A study on both marine and freshwater populations of *C. aspera* showed that high salinity (20 psu) reduced germination of oospores and establishment from bulbils while establishment at lower salinities depended on site of origin (Blindow et al., 2009).

19.3 Habitat Conditions and Ecology

Habitat type: *Chara aspera* occurs both in freshwater and brackish water habitats.

Sediment: *Chara aspera* has been reported from sheltered as well as wave-exposed shores on a broad range of sediments such as sand, clay, silt, sand, gravel.

Nutrient conditions: *Chara aspera* is found under a very wide range of nutrient conditions from oligotrophic (<1 μg TP l^{-1}) to highly eutrophic (>270 μg TP l^{-1}) (Blindow, 1992; Kragh et al., 2017).

Water chemistry: *Chara aspera* is found at salinity ranges from 0 up to 17 or 18 PSU (Blindow, 2000; Nielsen 2004; Olsen, 1944), which agrees with results from laboratory investigations showing that the plants can grow in water up to 20 PSU (Blindow et al., 2003). However, plants growing at at salinities of 5 and 10 PSU performed better than plants grown at 20 PSU which is probably close to the physiological limit of this species (Winter and Kirst, 1992). In freshwater *Chara aspera* is primarily found in alkaline waters (alkalinity >1 meq l^{-1}) and the corresponding pH values are therefore rather high 7.5–10. In brackish waters in the southern Baltic Sea, pH fluctuates regularly between pH 8 und pH 9 with extremes of pH 10 and pH 7 (Mecklenburg-Vorpommern, 1995).

Depth range: *Chara aspera* has been found in at least 6 m depth with the lower depth limit situated at about 15% of incident surface irradiance (Blindow and Schütte, 2007), but is most frequently found down to 50 cm deep. In more saline environments (7–8 PSU) the depth limit has been shown to be restricted to 40% of incident light (Blindow and Schütte, 2007).

Associated species: Because of the very broad tolerance with respect to salinity as well as nutrient content a list of associated species would be of little relevance and is omitted here. In shallow water the species often grows without other charophytes.

19.4 Variability

For *Chara aspera* a number of subspecies and varieties have been described indicating substantial morphological variation (e.g., Migula, 1889–1897; Nielsen 2004). Specimens with pappiliformous spine-cells and acute stipulodes have been called *Chara aspera* var. *subinermis*. Freshwater populations of *C. aspera* tend to be thinner and to form mats while brackish water populations appear to be stouter and grow in tussocks (Blindow et al., 2009; Olsen, 1944). This is supported by experimental evidence showing a higher growth rate of individuals in brackish water (4–7.5 PSU) compared to freshwater conditions (Flor-Arnau et al., 2006; Rojo et al., 2017). The variations within *C. aspera* populations may favor speciation in relatively new ecosystems such as the Baltic Sea (Blindow et al., 2009).

19.5 Risk of Confusion

Chara aspera is close to *Chara curta* but can be separated mainly by the number of spine-cells in each cluster. While *C. aspera* predominantly has single spine-cells *C. curta* has spine-cells in clusters on mature plants (for further details see *C. curta*).

Chara galioides resembles *Chara aspera*. Both species are dioecious and have a triplostichous cortication. However, *Chara galioides* is larger and stouter than *Chara aspera*. This is particularly true for the size of the antheridia which are 800–1500 μm in diameter in *Chara galioides*. Bulbils are rarely found in *Chara galioides* and if they are found they are transparent and not white as in *Chara aspera*. The differentiation is supported by detailed morphological studies (Flor-Arnau et al., 2006) and partly by molecular studies (Schneider et al., 2016). Beside *Chara galioides*, when *C. aspera* is long spined and fertile it is clearly different from any other *Chara* species because of the triplostichous cortex, the long spine-cells and dioecy. Infertile plants with long spine-cells are difficult to differentiate from the monoecious *Chara tenuispina* (Urbaniak and Gąbka, 2014).

Forms without spine-cells (*Chara aspera* f. *subinermis*) can resemble *Chara connivens* which is also triplostichous and dioecious, or if infertile, *Chara virgata*

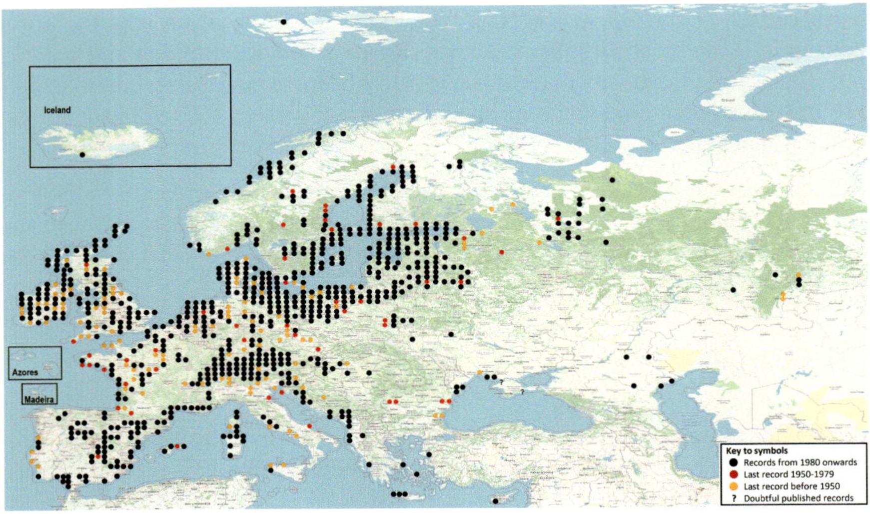

Fig. 19.4 Distribution of *Chara aspera* in Europe. For worldwide distribution see Korsch (2018). Black dots are records from 1980 on, red dots from 1950 to 1979 and yellow dots before 1950; unconfirmed/doubtful published records are marked with red dots . Map prepared by Heiko Korsch

or *Chara globularis*. In all cases features of the gametangia and bulbils are most characteristic (Nielsen, 2004).

In brackish waters, infertile *Chara aspera* can resemble *Chara baltica* as both species can be irregularly corticated with transitions between diplostichous and triplostichous cortex.

19.6 Distribution

See Fig. 19.4.

Chara aspera is widely distributed in the Northern hemisphere (Corillion, 1975; Wood, 1965), in North America (from Canada to Mexico), North Africa, the entire European continent and Central Asia (Korsch, 2018).

19.7 Status and Threats

The species is rather common in northern and western countries and is listed as LC (Least Concern) in all published Red Lists, except in Norway where it categorised as NT (Near Threatened). Further south, it is listed as Threatened in several countries

(Germany, Switzerland, Czechia, Poland and Ukraine) but in the Balkans *C. aspera* was listed as Low risk by Blaženčić et al. (2006).

In brackish water physical disturbances such as coastal engineering have been suggested as the main threat along with eutrophication leading to reduced water clarity and increased competition from filamentous algae (Nielsen, 2004). In freshwater, eutrophication also represents a severe threat to populations of *Chara aspera*, even though it is one of the charophyte species most resistant to eutrophic waters (Blindow, 1992).

References

Bailly, G., & Schaefer, O. (2010). *Guide illustré des Characées du nord-est de la France* (96 pp). Conservatoire Botanique National de Franche-Comté [in French].

Blaženčić, J., Stevanović, B., Blaženčić, Z., & Stevanović, V. (2006). Red data list of charophytes in the Balkans. *Biodiversity and Conservation, 15*(11), 3445–3457. https://doi.org/10.1007/s10531-005-2008-5

Blindow, I. (1992). Decline of charophytes during eutrophication: a comparison to angiosperms. *Freshwater Biology, 28*, 9–14. https://doi.org/10.1111/j.1365-2427.1992.tb00557.x

Blindow, I. (2000). Distribution of charophytes along the Swedish coast in relation to salinity and eutrophication. *International Review of Hydrobiology, 85* (5-6), 707–717. https://doi.org/10.1002/1522-2632(200011)85:5/63.0.CO;2-W

Blindow, I., Dietrich, J., Möllmann, N., & Schubert, H. (2003). Growth, photosynthesis and fertility of *Chara aspera* under different light and salinity conditions. *Aquatic Botany, 76* (3), 213–234. https://doi.org/10.1016/S0304-3770(03)00053-6

Blindow, I., Möllmann, N., Boegle, M. G., & Schütte, M. (2009). Reproductive isolation in *Chara aspera* populations. *Aquatic Botany, 91* (3), 224–230. https://doi.org/10.1016/j.aquabot.2009.06.009

Blindow, I., & Schütte, M. (2007). Elongation and mat formation of *Chara aspera* under different light and salinity conditions. *Hydrobiologia, 584*, 69–76. https://doi.org/10.1007/s10750-007-0578-9

Corillion, R. (1957). Les Charophycées de France et d'Europe Occidentale. *Bulletin de la Société Scientifique de Bretagne, 32*(Hors série I & II), 1–499 [in French].

Corillion, R. (1975). *Flore et végétation du Massif Armoricain. T. IV. Flore des Charophytes (Characées) du Massif Armoricain et des contrées voisines d'Europe occidentale*. Jouve [in French].

Flor-Arnau, N., Reverter, F., Soulié-Märsche, I., & Cambra-Sánchez, J. (2006). Morphological differentiation of *Chara aspera* Detharding *ex* Willdenow and *Chara galioides* De Candolle under different environmental variables. *Cryptogamie, Algologie, 27* (4), 435–449.

Groves, J., & Bullock-Webster, G. R. (1924). *The British Charophyta. Volume II. Chareæ with plates, concluding articles, geological sketch, bibliography and index*. Ray Society.

Korsch, H. (2018). The worldwide range of the Charophyte species native to Germany. *Rostocker Meeresbiologische Beiträge, 28*, 45–96.

Kragh, T., Andersen, M. R., & Sand-Jensen, K. (2017). Profound afternoon depression of ecosystem production and nighttime decline of respiration in a macrophyte-rich, shallow lake. *Oecologia, 185*, 157–170. https://doi.org/10.1007/s00442-017-3931-3

Krause, W. (1997). Charales (Charophyceae). In H. Ettl, G. Gärtner, H. Heynig, & D. Mollenhauer (Eds.), *Süßwasserflora von Mitteleuropa 18*. Fischer [in German].

Langangen, A. (2000). Charophytes from the warm springs of Svalbard. *Polar Research, 19* (2), 143–153. https://doi.org/10.1111/j.1751-8369.2000.tb00339.x

Langangen, A. (2007). *Charophytes of the Nordic countries*. Saeculum ANS.

Langangen, A., Ballot, A., Nowak, P., & Schneider, S. C. (2020). Charophytes in warm springs on Svalbard (Spitsbergen): DNA barcoding identifies *Chara aspera* and *Chara canescens* with unusual morphological traits. *Botany Letters, 167* (2), 179–186. https://doi.org/10.1080/23818107.2019.1672104

Ministerium für Bau, Landesentwicklung und Umwelt Mecklerburg-Vorpommern. (1995). *Gewässergütebericht 1995 Gütezustand der oberirdischen Gewässer, der Küstengewässer und des Grundwassers in Mecklenburg-Vorpommern*. [in German].

Migula, W. (1889–1897). Fünfter Band: Die Characeen. In A. Fischer, E. Fischer, F. Hauck, G. Limpricht, C. Luerssen, W. Migula, H. Rehm, P. Richter, & G. Winter (Eds.), *Dr. L. Rabenhorst's Kryptogamenflora von Deutschland, Oesterreich und der Schweiz* (2 ed.). Eduard Kummer [in German].

Nielsen, R. (2004). *Chara aspera*. In H. Schubert & I. Blindow (Eds.), *Charophytes of the Baltic Sea* (pp. 42–52). A. R. G. Gantner.

Olsen, S. (1944). Danish Charophyta. *Det Kongelige Danske Videnskabernes Selskab, Biologiske Skrifter, 3*(1), 1–240, pl. 1–2 [in Danish].

Rojo, C., Carramiñana, M., Cócera, D., Roberts, G. P., Puche, E., Calero, S., & Rodrigo, M. A. (2017). Different responses of coexisting *Chara* species to foreseeable Mediterranean temperature and salinity increases. *Aquatic Botany, 138*, 53–63. https://doi.org/10.1016/j.aquabot.2017.01.003

Romanov, R. E. (2022). Typification and clarification of species concepts of *Chara fischeri* Migula and *C. gobiana* Vilhelm (*Charales, Charophyceae*). *Notulae Algarum, 233*, 1–4.

Schneider, S. C., Nowak, P., von Ammon, U., & Ballot, A. (2016). Species differentiation in the genus *Chara* (Charophyceae): Considerable phenotypic plasticity occurs within homogenous genetic groups. *European Journal of Phycology, 51*, 282–293. http://dx.doi.org/10.1080/09670262.2016.1147085

Urbaniak, J., & Gąbka, M. (2014). *Polish Charophytes. An illustrated guide to identification*. Uniwersytet Przyrodniczy we Wroclawiu.

Van den Berg, M. S., Coops, H., & Simons, J. (2001). Propagule bank buildup of *Chara aspera* and its significance for colonization of a shallow lake. *Hydrobiologia, 462*, 9–17. https://doi.org/10.1023/A:1013125603555

Van den Berg, M. S., Coops, H., Simons, J., & de Keizer, A. (1998). Competition between *Chara aspera* and *Potamogeton pectinatus* as a function of temperature and light. *Aquatic Botany, 60* (3), 241–250. https://doi.org/10.1016/S0304-3770(97)00099-5

Vedder, F. (2003). *Morphologie und Taxonomie rezenter und subfossiler Characeen-Oosporen aus der Ostsee* (62 pp). Diplomarbeit, Ernst-Moritz-Arndt Universität Greifswald [in German].

Winter, U., & Kirst, G. O. (1992). Turgor pressure regulation in *Chara aspera* (Charophyta): The role of sucrose accumulation in fertile and sterile plants. *Phycologia, 31* (3-4), 240–245. https://doi.org/10.2216/i0031-8884-31-3-4-240.1

Wood, R. D. (1965). *Monograph of the Characeae*. J. Cramer.

Chapter 20
Chara baltica

Irmgard Blindow, Roman Romanov, Hendrik Schubert, and Nick Stewart

Scientific name: *Chara baltica* (Hartm.) Bruzelius, Observ. Charae: 11. 1824

≡ *Chara hispida* L. var. *baltica* Hartman, Handb. Skand. Fl.: 376. 1820.
Lectotype (Wood, 1965: 150): From Böda harbour. Öland/Sweden/1816/Dr. Christ. Stenhammer 10/1816 (S, Artdatabanken 90,308).

Infraspecific taxa

***Chara baltica* (Hartm.) Bruzelius var. b *altica* , Observ. Charae: 11. 1824**

***Chara baltica* (Hartm.) Bruzelius var. *densa* (Mig.) Hendr. Schub., Blindow & Weyer, comb. nov.**

Basionym: *Chara baltica* (Hartm.) Bruzelius f. *densa* Mig., Charac. Deutschl.: 533. 1894.

Neotype (designated here by Blindow): Vitter Bodden, Hiddensee, Germany/22 July 2021/I. Blindow (B 400045359).

I. Blindow (✉)
Biologische Station Hiddensee, Universität Greifswald, 18565 Kloster, Germany
e-mail: blindi@uni-greifswald.de

R. Romanov
Nišice Ujtin Potok b.b., Dobra Voda, 85356 Bar, Montenegro
e-mail: streptophytes@gmail.com

H. Schubert
Universität Rostock Institut für Biowissenschaften, Albert-Einstein-Straße 3, 18059 Rostock, Germany
e-mail: hendrik.schubert@uni-rostock.de

N. Stewart
4 Hexton Road, Glastonbury BA6 8HL, UK
e-mail: nfstewart@freeuk.com

H. Schubert et al. (eds.), *Charophytes of Europe*,
https://doi.org/10.1007/978-3-031-31898-6_20

***Chara baltica* (Hartm.) Bruzelius var. *liljebladii* (Wallman), H. Schub., Blindow & K. Weyer, Armleuchteralgen. Charac. Deutschl.: 237. 2015**

Chara liljebladii Wallman, Förs. Syst. Uppställn. Charac.: 86. 1853.
Lectotype (designated here by Blindow): Ytterhultsmar, Westerums socken, Clm. län, Sweden/1839/Wallman (LD 1573889).

***Chara baltica* (Hartm.) Bruzelius var. *rigida* J. Groves & Bull.-Webst. in J. Bot. 62: 34. 1924**
Lectotype (designated here by Blindow): Hickling Broad, Norfolk/Aug. 1898/G. R. Bullock-Webster (BM000771984).

Taxonomic note: *Chara baltica* var. *affinis* is considered to be separate from *C. baltica* based on morphological grounds and treated as *Chara affinis* (see Chap. 17). Also "*Chara baltica*" collected from the Mediterranean region (e.g. Krause, 1997) is considered to be a separate taxon and treated as a taxon of uncertain affiliation (see "Mediterranean Chara", Chap. 80). Several varieties have been retained in *Chara baltica* (e.g. var. *liljebladii*, var. *densa*) because they have been commonly distinguished in the past.

Local names: DANISH: Baltisk kransnål; DUTCH: Kust kransblad; ENGLISH: Baltic stonewort; ESTONIAN: Balti mändvetikas; FINNISH: Itämerennäkinparta; GERMAN: Baltische Armleuchteralge; LATVIAN: Baltijas chara; LITHUANIAN: Baltijinis maurabragis; NORWEGIAN: Grønnkrans; POLISH: Ramienica bałtycka, RUSSIAN: Хара балтийская; SWEDISH: Grönsträfse; långsträfse (*C. baltica* var. *liljebladii*); UKRAINIAN: Хара балтійська.

20.1 Morphology[1]

See Figs. 20.1, 20.2, 20.3, and 20.4.

Habitus: Plants are light to dark green, hardly encrusted.

Size: *Chara baltica* varies in size depending on habitat and variety. Small forms (*C. baltica* var. *densa*) are only 1–15 cm long (Figs. 20.1 and 20.3). Both *C. baltica* var. *baltica* and *C. baltica* var. *liljebladii* reach around 30–150 cm.

Main axis: Diameter is about 1–1.5 mm but can be smaller (0.5 mm) in *C. baltica* var. *densa*.

Cortex: *Chara baltica* is regularly diplostichous, but transitions between diplo- and triplostichous cortication are common (Blümel, 2004; Olsen, 1944; see Fig. 20.1) and have been observed in around 40% of all collections of var. *baltica* and var. *densa*. All investigated collections of var. *liljebladii* were diplostichous. The species

[1] *Chara baltica* is a highly variable taxon. The morphological description follows Migula (1889–1897) and Krause (1997), as well as own observations, if not stated otherwise.

Fig. 20.1 *Chara baltica.* **a** *Chara baltica*, in situ photograph; **b** Habitus, plant with multicellular bulbil; **c** Details of stipulodes, cortex and spine-cells; **d** Detail of cortex showing diplo- to triplostichous cortication, spine-cells are wide at base and narrow into an acute tip. Photographs: **a**—Volker Krautkrämer; **b**, **c**—Klaus van de Weyer; **d**—Irmgard Blindow

is predominantly tylacanthous, although often rather weakly so, and some plants are isostichous.

Spine-cells: Solitary, occasionally in bunches of 2–3, typically wide at base and abruptly narrowing into an acute tip, not needle-like (Fig. 20.1). The length of spine-cells is highly variable (300–1600 μm), although most collections have short spine-cells (shorter than the axis diameter). Also in var. *liljebladii*, spine-cell lengths are highly variable (400–3000 μm), but often longer than the axis diameter.

Stipulodes: Two rows of regular, equally long stipulodes with two pairs of stipulodes below each branchlet, which reach a length of 400–1100 μm (200–1200 μm in var. *liljebladii*). Like the spine-cells, stipulodes are not needle-like but wide at base and narrowing towards the tip.

Fig. 20.2 Variability of *Chara baltica*. **a** *Chara baltica* var. *liljebladii*; **b** Transition between *Chara baltica* and *C. horrida* ; **c–e** Three different collections of *Chara baltica* from Loch Stenness, Orkney, in BM, the right one **e** resembling var. *liljebladii*. Photographs: Irmgard Blindow

Branchlets: Whorls consist of 6–11 branchlets with 5–9 segments, the last 1–4 of which are ecorticate. Commonly, the penultimate ecorticate segment(s) is/are inflated (see Luther, 1979). Another common pattern, especially in var. *densa*, is that only the distant, very short segment is ecorticate and forms a "tricellular tip" together with the bract-cells of the last branchlet node (see Fig. 20.3). Rarely, all segments can be ecorticate.

Bract-cells: Spreading, regularly positioned along the branchlet. Adaxial bract-cells are 300–1700 μm in length (400–1900 μm in var. *liljebladii*), longer than the abaxial ones. Rarely, the bract-cells are almost verticillate. Near-verticillate bract-cells are found in about 30 % of all collections of var. *liljebladii*.

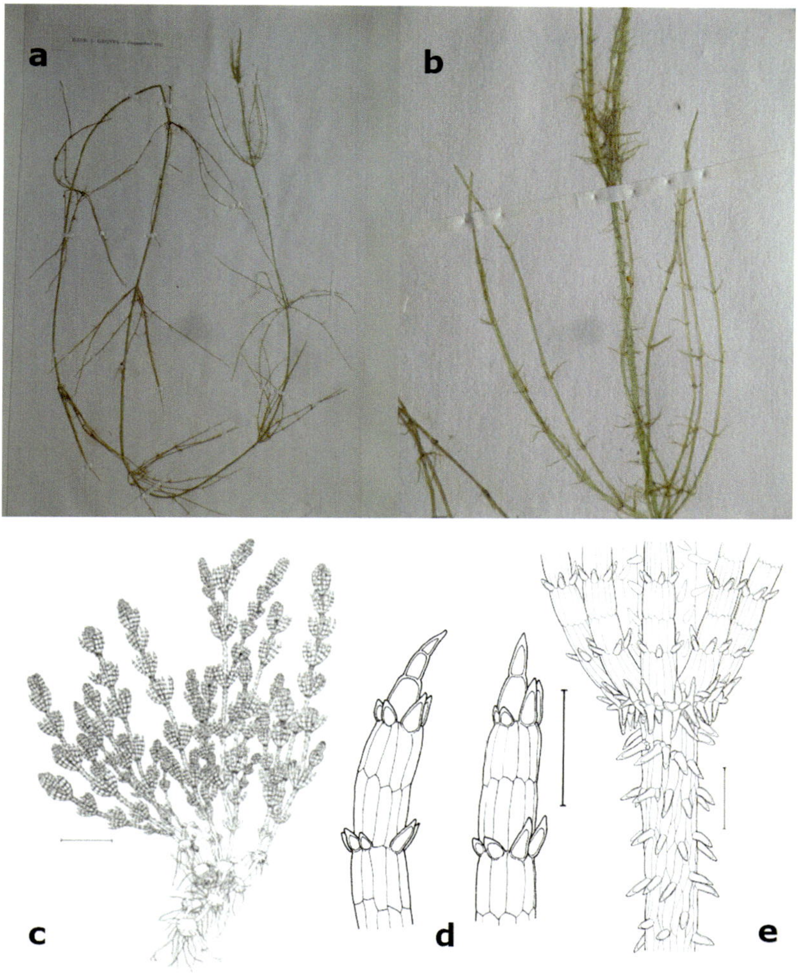

Fig. 20.3 Variability of *Chara baltica*. **a**, **b** *Chara baltica* var. *rigida* from Hickling Broad in BM (**a** Whole plant, **b** Detail showing long bract-cells and long spine-cells); **c**–**e** *Chara baltica* var. *densa* from the Baltic Sea (**c** Habitus, plant with numerous bulbils, **d** Branchlet tips with short ecorticated end segments, **e** Internode section, showing spine-cells and stipulodes. Photographs and drawings: **a**, **b**—Irmgard Blindow; **c**, **d**, **e**—Mindaugas Ryla

Bracteoles: Two, 1–2 times as long as the mature oogonium (Wood, 1965). Their length is 900–1700 μm (own measurements), but extremely long bracteoles (3400 μm) were recorded in a collection from Zingst in Germany (GFW). The variation in bracteole lengths is especially high in var. *liljebladii* (700–4400 μm).

Sexuality: *Chara baltica* is monoecious. Gametangia are conjoined singly and can be found on all corticated branchlet segments.

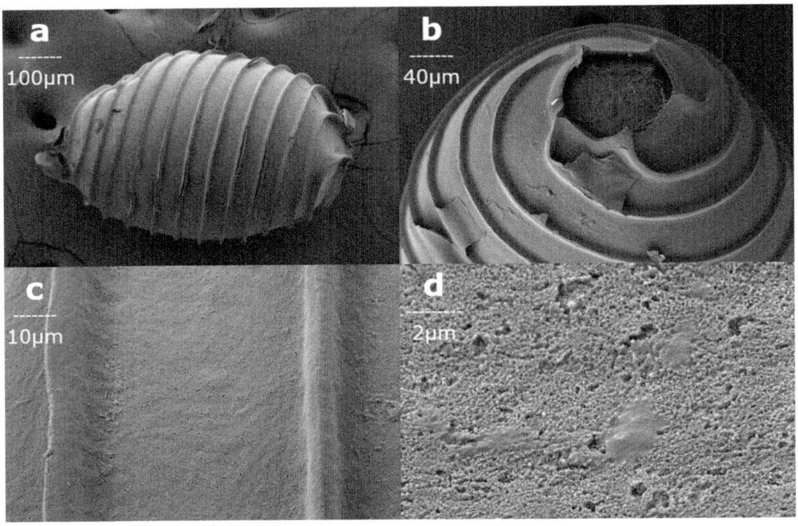

Fig. 20.4 SEM images of an oospore of *Chara baltica*, collection from Sweden, PC0577978, Dan, G., s.d. **a** Side view; **b** Apex; **c**, **d** Oospore-membrane at different magnification. Photographs: Michelle T. Casanova

Oogonia: Information in the literature about oogonia size is highly variable (Table 20.1). This is due partly to the fact that plants collected from near the Mediterranean Sea, here treated as a separate taxon with uncertain affiliation (see "Mediterranean Chara", Chap. 80) were (probably) included by several authors (Corillion, 1957; Krause, 1997). Information given in the tables below is restricted to plants that originate from the Baltic Sea. The coronula is straight in most collections but can occasionally be spreading. Connivent coronulae have been observed in var. *liljebladii*. One Swedish collection of var. *liljebladii* had an unusually high number of helical stripes (16–17).

Oospores: As for oogonia, the literature information about oospore dimensions varies considerably. In Table 20.2, the information is restricted to plants which most probably originate from the Baltic Sea. Oospores of *C. baltica* collected from the Baltic Sea were found to be significantly narrower than oospores of "*Chara baltica*" collected from southern France, and shorter than oospores of *C. horrida*. Holzhausen et al. (2014) could not separate *C. baltica* var. *baltica* and *C. baltica* var. *liljebladii* by means of quantitative oospore parameters, and there was a considerable overlap among both these taxa and *C. papillosa*. SEM images of oospores are shown in Fig. 20.4.

Antheridia: Verdam (1938) mentions antheridia diameter for "*Chara baltica*" collected in the Netherlands as 225–300 µm, which is far outside the range of this species (see Table 20.3). These measurements are unusually small, even if the plants

Table 20.1 Oogonia morphometry

Author	Length	Width	Coronula length	Coronula width	Number of helical stripes
Migula (1889–1897)	Up to 1300	–	–	–	14–16
Wood (1965); var. *baltica*	900–1000	570–650	158–210	350–400	14–15
Wood (1965); var. *liljebladii*	1065	570	165	335	15–16
Hollerbach and Krassavina (1983)	900–1000	570–650	158–250	350–400	14–15
Holtz (n.d.), var. *baltica* & *densa*	1000–1310	660–720	–	–	–
own data, Baltic Sea, var. *baltica* & *densa*	710–1050	520–780	140–280	240–360	(10)11–15
own data, Baltic Sea, var. *liljebladii*	790–1140	540–730	(80)110–225	(135)170–280	11–15(17)
own data, Groves and Bullock-Webster (1924), var. *rigida*	800–920	490–680	190–210	240–350	14
Total range	**710–1310**	**520–780**	**110–280**	**170–400**	**(10)11–16(17)**

Length in the first column excludes the coronula; its length and width are given separately, all values for length and width in μm

Table 20.2 Oospore morphometry

Author	Length	Width	Number of striae	Colour
Migula (1889–1897)	700–800	450–500	11–14	Black
Wood (1965); var. *baltica*	645–705	420–500	13–14	Black
Wood (1965); var. *liljebladii*	950	480	(10–14)15–16	Black
Hollerbach and Krassavina (1983)	650–750	420–525	13–14	Black
Holtz (n.d.), var. *baltica* & *densa*	720–810	370–550	–	Brown to black
Vedder (2003)	650–800	375–525	11–15	Black
Blume et al. (2009)	525–875	350–550	10–15	Black
Total range	**525–950**	**350–550**	**10–16**	

All values for length and width in μm

Table 20.3 Antheridia morphometry

Author	Diameter (μm)
Migula (1889–1897)	500–800
Wood (1965); var. *baltica*	450–600
Wood (1965); var. *liljebladii*	500–700
Hollerbach and Krassavina (1983)	450–600
Own data, Great Britain	430–720
Own data, Baltic Sea, var. *baltica* & *densa*	510–710
Own data, Baltic Sea, var. *liljebladii*	470–730
Own data, Groves and Bullock-Webster (1924), var. *rigida*	530–790
Total range	**430–800**

Own data are measurements of dry material, Wood (1965) measured rewetted antheridia

possibly were confused with *Chara affinis*, which has considerably smaller antheridia than *C. baltica*.

Bulbils: Multicellular bulbils are common and appear from mid summer on the lower nodes (Luther, 1979, see Figs. 20.1 and 20.3).

20.2 Phenology and Reproduction

In the southern Baltic Sea, *Chara baltica* appears in March to April, while first germlings were observed during May to June along the coast of middle Sweden (Blümel, 2004). In mild winters or in deeper water, the species can hibernate as green plants, as the plants do not die back after oospore formation. In shallow water, plants hibernate by means of oospores or bulbils. According to Olsen (1944), both main axis bulbils and multicellular rhizoid bulbils are formed, and perennial forms are more common than annual ones. Fertile plants can be found between June and October, but as early as end of May in the southern Baltic Sea (Blümel, 2004; Langangen, 2007). Ripe oospores have been observed from July (Olsen, 1944). Both richly fertile and sterile specimens are commonly found (Olsen, 1944).

20.3 Habitat Conditions and Ecology

Habitat type: *Chara baltica* can be found in protected or moderately exposed sites of the Baltic Sea. In the North Sea area, it has not been observed in "eulitoral" sites affected by the tides. In Great Britain, Ireland and Norway, the species was found in brackish water bodies along the coast (Bryant and Stewart, 2011; Groves and

Bullock-Webster, 1924; Langangen, 2007; Moore, 1986). A record from Greenland was from a brackish lake (Langangen and Bennike, 1997).

Sediment: *Chara baltica* can be found on sandy to soft, muddy bottoms.

Nutrient conditions: *Chara baltica* occurs in oligotrophic sites in the Bothnian Bay, but has, in the Baltic Sea, also been observed at total phosphorus concentrations between 52 and 400 $\mu g\ L^{-1}$ (Doege et al., 2016). In the Netherlands, it has been observed between 400 and 2000 $\mu g\ L^{-1}$ TP (Nat et al., 1994). Occurrences at high nitrogen concentrations (1200 $\mu g\ L^{-1}$) have been reported from Sweden (Länsstyrelsen Stockholms Län, 1997).

Water chemistry: The species only occurs in brackish water, mainly in a salinity range between about 2–10 psu (Blümel, 2004; Langangen, 2007; Scott et al., 2011), with salinities up to 20 psu in the Black Sea sites (R. Romanov pers. comm.). Rarely, the species has been found in higher (18 psu, Denmark, Olsen, 1944) or lower salinities such as the almost fresh water of the Curonian lagoon (Blümel, 2004). The pH ranges between 7.0 and 8.3 (Doege et al., 2016; Scott et al., 2011).

Depth range: In the Baltic Sea, most records are from shallow water (less than 1 m depth). Maximum depth ranges observed are 3–5 m. Small bushy forms (var. *densa*) are mainly found in shallow water. Possibly, the maximum depth range has decreased during the last decades. The fact that most records are from shallow water cannot exclusively be explained by the investigation methods, as scuba divers also only rarely find specimens occurring at greater depth. *C. baltica* thus prefers shallower water than *Tolypella nidifica*, but deeper water than *C. canescens*. Variety *liljebladii* seems to occur in deeper water (about 0.5–1.5 m) than the other forms (Blümel, 2004; Skuja, 1924; Wallentinus, 1979). In more saline sites in Orkney and the Outer Hebrides (Scotland), the species seems to be restricted to the depth range above the halocline (Scott et al., 2011).

Associated species: *Chara baltica* is often associated with other charophytes, but also vascular plants. In sheltered, shallow sites, it has been found together with *C. tomentosa, C. canescens* and *C. horrida, Ruppia* spp., *Zannichellia palustris* s.l., *Zostera marina*, *Stuckenia pectinata* and *Najas marina*. In more wind-exposed sites it grows together with *C. aspera*, *C. canescens* and *Stuckenia pectinata* (Blindow et al., 2016; Lindner, 1978; Munsterhjelm, 1997, Scott et al., 2011). Also associations with *Tolypella nidifica* have been described. *Chara baltica* var. *liljebladii* can form monospecific stands but is often associated with *C. aspera.* Dense vegetation with a dominance of *C. baltica* has been described as *Charetum balticae* Kornas 1959 (= *Chara baltica*-association sensu A. Lindner, 1978). In the Black Sea, *C. baltica* was found together with *Lamprothamnium papulosum* and *Chara canescens* (Romanov et al., 2020).

20.4 Variability

Chara baltica is one of the most variable charophytes. The taxonomic relevance of many of the described forms and varieties (see Migula, 1889–1897) is doubtful. Most authors assume that they are ecotypes or local adaptations (Blümel, 2004), which, according to Krause (1997), are segregated along a depth gradient. At Hiddensee, Germany, C. *baltica* var. *liljebladii* occurs in deeper water (around 1 m); *C. baltica* var. *densa* is mainly found in shallow water (around 0.5 m), but also in deeper water and may there grow adjacent to var. *liljebladii* (own observations). We distinguish as variants:

1. var. *baltica*: See description of the species above.
2. var. *densa*: Plants are small, compact, bushy, mostly shorter than 15 cm long, dark green, in autumn sometimes almost black, without any visible encrustation, often with dense spine-cells of variable length. Mainly in shallow water on silty sand, sand and gravel (Fig. 20.3).
3. var. *liljebladii*: Plants are large, sometimes encrusted. Cortication is tylacanthous, rarely isostichous, diplostichous. Spine-cells are solitary in most collections, but occasionally occur in groups of 2–3, short to twice as long as the axis diameter. Branchlets are 9–11 per whorl, elongate, with 6–8 segments, of which the last 1–4 segments are ecorticate. Oospores are black, and their membrane is smooth (Wood, 1965, own data) (Fig. 20.2).
4. var. *rigida*: only recorded from Hickling Broad, E. Norfolk. Coarse, isostichous to slightly tylacanthous, diplo- to triplostichous; stipulodes, spine-cells, bract-cells and bracteoles rather long, spine-cells can exceed the axis diameter, bracteoles are about twice as long as the ripe oogonium. Spine-cells single, occasionally fasciculate. Multicellular bulbils are common (Groves and Bullock-Webster, 1924, own data) (Fig. 20.3).

20.5 Identification and Risk of Confusion

Chara baltica belongs to the *Hartmania* group, a complex of coarse, diplostichous *Chara* spp. that are hard to separate from each other. Within the *Hartmania* group, the species belongs, together with *C. horrida,* to a cluster of taxa characterised by large antheridia, all occurring in brackish water coastal sites (see Chap. 15).

The main characters which separate *C. baltica* from *C. horrida* are a tylacanthous to isostichous, not aulacanthous cortex, and stipulodes arranged in two regular rows. Further, spine-cells are mainly single in *C. baltica*, which is only occasionally the case in *C. horrida*. Adaxial bract-cells are distinctly longer than abaxial ones in *C. baltica*, while many collections of *C. horrida* have near-verticillate bract-cells. The lengths of spine-cells, bracteoles and stipulodes are highly variable in both taxa, but often longer in *C. horrida*. A clear separation between both taxa is not possible, however, due to the existence of transition forms such as individuals with irregular/

additional stipulodes, but tylacanthous cortex, which have been recorded from the Baltic Sea, the Black Sea and the Hebrides (Romanov et al., 2020; see Fig. 20.2).

Morphologically, a separation between *Chara baltica* and *C. papillosa* is often not possible. Both taxa are tylacanthous, and single-spined collections are more common than those with clustered spine-cells. Antheridia sizes overlap, but are smaller in *C. papillosa* (Migula, 1889–1897, own data). Typically, *Chara papillosa* is heavily encrusted, while *C. baltica* is only weakly encrusted (Migula, 1889–1897). However, weak encrustation in coastal brackish water and heavy encrustation in freshwater have been shown to occur even within the same taxon, indicating that this character is influenced by water chemistry and not of any taxonomic value (Herbst et al., 2018). A good criterion is colour which typically is mid to dark green in *C. baltica* and often olive to brownish green in *C. papillosa*. Genetically, both species are clearly separated from each other (Nowak and Schubert, 2019).

Small individuals (*C. baltica* var. *densa*) can be hard to distinguish from *C. aspera*, if gametangia and bulbils are lacking, which may be the case during spring. In both species, transitions between diplostichous and triplostichous cortex are common. When collecting charophytes from shallow brackish water in the Baltic Sea during spring, the colour (yellowish green in *Chara aspera*, fresh to dark green in *C. baltica*) turned out to be the best character to distinguish both species (own observations).

Chara affinis (Chap. 17) and "Mediterranean Chara" (Chap. 80) can be distinguished from *C. baltica* by smaller antheridia and less inflated and often longer spine-cells, stipulodes and bract-cells (own data). Oospores of "Mediterranean Chara", a taxon restricted to the Mediterranean area, were morphometrically clearly separated from oospores of both *C. baltica* and *C. horrida* (Blume et al., 2009).

20.6 Distribution

Chara baltica is probably restricted to Europe (see Fig. 20.5) and Greenland (Langangen and Bennike, 1997). The species was reported from Bolivia and Brazil (Braun and Nordstedt, 1882; Hasslow, 1934), but these records have been doubted as no herbarium material seems available (Bicudo, 1977; García, 1993; Wood, 1965; see also Langangen et al., 1996). Confusion with other taxa such as the South American *Chara andina* seems probable (Blindow et al., 2018). Collections from Bolivia (in L and BM) were revised as *C. leptosperma* or *C. andina*, respectively (own observations). *C. baltica* was also described from several sites in Argentina, but plants collected from the Mediterranean area were used as reference (García, 1993, 1994). A collection from Mongolia (in BM) was redetermined as *C. papillosa* (own observations).

The main distribution area of *C. baltica* is the Baltic Sea. Several authors (e.g. Krause, 1997) have described *C. baltica* also from the Mediterranean Sea, but these collections are morphologically separated from C. *baltica* and here treated as "Mediterranean Chara", a taxon with uncertain affiliation (see Chap. 80). Outside of the Baltic Sea, *C. baltica* also has been reported from Scotland (Outer Hebrides,

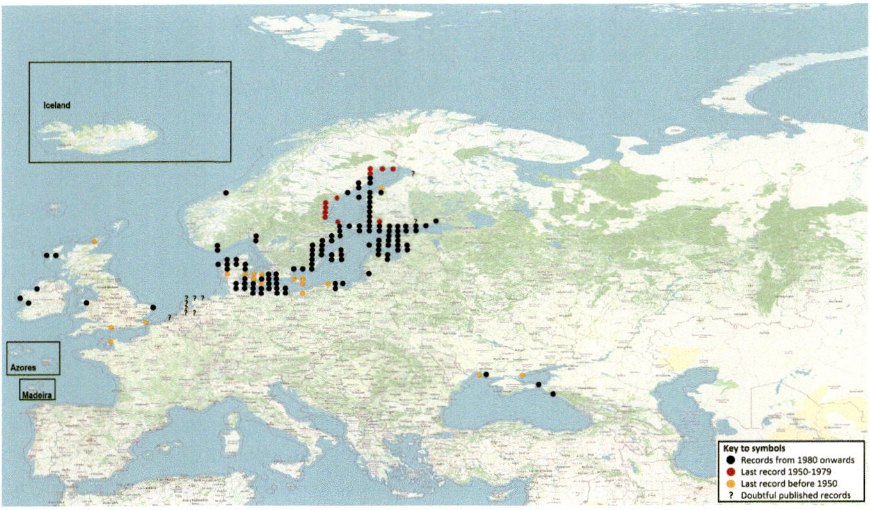

Fig. 20.5 European distribution of *Chara baltica*. For worldwide distribution see Korsch (2018). Black dots are records from 1980 onwards, red dots last record from 1950 to 1979 and yellow dots last record before 1950; doubtful published records (if applicable) are marked with "?" superimposed. Map prepared by Heiko Korsch

Orkney; see Fig. 20.3), Faroe Islands, England (Norfolk, Dorset, Kent), Ireland (Galway), the Danish Belt region and Norway (Corillion, 1957; Langangen, 2007; Stewart and Church, 1992, collections in BM and GFW). Van Raam (1998) reported 18 records of *C. baltica* (including var. *affinis* = *Chara affinis*) from the Netherlands and mentioned the species also from Belgium. His description and drawings correspond to *C. baltica*, but have unusually long stipulodes and long, thin spine-cells, and the plants are light, not dark green. Most records from the Netherlands in L have spine-cells exceeding the axis diameter in length and small antheridia, which does not correspond to *C. baltica* (E. Nat pers. comm.). Also records in BR have long spine-cells and small antheridia and were determined as *C. affinis*. *Chara baltica* has also been found in the Black Sea (records in LE; Romanov et al., 2020).

20.7 Status and Threats

In the national Red Lists, *Chara baltica* is assessed as Endangered in the Netherlands, Ireland, Norway, Germany and Lithuania, as Vulnerable in Britain and Poland and as Near Threatened in Denmark. Because of its frequent occurrence, the species is not included in the Swedish, Finnish and Estonian Red Lists. It is assessed as LC (Least Concern) in the European Red List (see Chap. 9).

In most countries, the data are not sufficient to give information about changes in occurrence. In many regions of the Baltic Sea, there has probably been no major change in occurrence, but decreases were observed in Germany (Blümel, 2004). In Sweden, a decline was mentioned for the southern and western coast, while no change in occurrence was observed along the east coast except for var. *liljebladii* which possibly decreased in this region (Blindow, 2000; Blümel, 2004).

The main threats in the Baltic Sea are eutrophication, coastal engineering and cover by filamentous algae (Blümel, 2004; Langangen, 2007). For Estonia and Germany, Blümel (2004) mentioned tourism, coastal fishery and intense boat traffic as additional potential threat factors; for southern Finland, filamentous algae and in isolated lagoons, mechanical disturbance, e.g. dredging which causes sediment resuspension were assumed to be the main threat factors. In Sweden eutrophication is probably the most important threat factor. Along the west coast, coverage by filamentous algae due to increased eutrophication may be responsible for the decline of *C. baltica*. Along the east coast, light is probably the most important factor limiting the occurrence at higher nutrient concentrations. This is indicated by the fact that a decrease only was observed for var. *liljebladii* which seems to be restricted to deeper water (Blindow, 2000; Blümel, 2004).

In the UK, eutrophication is the main factor threatening *Chara baltica* in its last remaining English sites in the eastern Norfolk Broads and it has caused the loss from several other sites in this lake system. In Kent, the species disappeared due to eutrophication and/or salinity changes. Loss from its sites in Dorset was caused by a reduction in salinity due to natural landscape changes (build up of dunes). Increasing salinity caused its disappearance in Orkney but the site has subsequently been impacted by eutrophication causing increased algal cover, flotsam accumulations and the development of planktonic algal blooms.

Acknowledgements We thank Klaus van de Weyer, Luc Denys and Emile Nat for valuable comments.

References

Bicudo, R. M. T. (1977). *O gênero Chara (Charophyceae) no Brasil. Trabalhos do XXVI Congresso Nacional de Botânica, Rio de Janeiro* (pp. 23–32). Academia Brasieleira de Ciências, Rio de Janeiro [in Portuguese].

Blindow, I. (2000). Distribution of Charophytes along the Swedish coast in relation to salinity and eutrophication. *International Review of Hydrobiology, 85*, 707–717.

Blindow, I., Dahlke, S., Dewart, A., Flügge, S., Hendreschke, M., Kerkow, A., & Meyer, J. (2016). Long-term and interannual changes of submerged macrophytes and their associated diaspore reservoir in a shallow southern Baltic Sea bay: Influence of eutrophication and climate. *Hydrobiologia, 778*, 121–136.

Blindow, I., Marquardt, R., Schories, D., & Schubert, H. (2018). Charophyceae of Chile—taxonomy and distribution. Part 1. *Nova Hedwigia, 107*, 1–47.

Blümel, C. (2004). *Chara baltica*. In H. Schubert & I. Blindow (Eds.), *Charophytes of the Baltic Sea* (pp. 53–63). A. R. G. Gantner.

Blume, M., Blindow, I., Dahlke, S., & Vedder, F. (2009). Oospore variation in closely related *Chara* taxa. *Journal of Phycology, 45*, 995–1002.

Braun, A., & Nordstedt, C. F. O. (1882). Fragmente einer Monographie der Characeen. Nach den hinterlassen Manuscripten A. Braun's herausgegeben von Dr. Otto Nordstedt. *Abhandlungen der Königlichen Akademie der Wissenschaften zu Berlin für das Jahr 1882*, 1–211. [in German].

Bryant, J. A., & Stewart, N. F. (2011). Phylum Chlorophyta. Order Charales. In D. M. John, B. A. Whitton, & A. J. Brook (Eds.), *The freshwater algal flora of the British Isles. An identification guide to freshwater and terrestrial algae* (2nd ed., pp, 742–765). Cambridge University.

Corillion, R. (1957). Les Charophycées de France et d'Europe Occidentale. *Bulletin de la Société Scientifique de Bretagne, 32* (Hors série I & II), 1–499 [in French].

Doege, A., van de Weyer, K., Becker, R., & Schubert, H. (2016). Bioindikation mit Characeen. In *Arbeitsgruppe Characeen Deutschlands, Armleuchteralgen: Die Characeen Deutschlands* (pp. 97–138). Springer [in German].

García, A. (1993). *Chara baltica* Bruz. (Charophyta) en Argentina. *Candollea, 48*, 297–303. [in Spanish].

García, A. (1994). Pleistocene Charophyta from Arroyo Perrucho Verna, Province of Entre Ríos, Argentina. *Journal of Paleolimnology, 10*, 53–58. [in Spanish].

Groves, J., & Bullock-Webster, G. R. (1924). *The British Charophyta. Volume II. Chareæ with plates, concluding articles, geological sketch, bibliography and index*. Ray Society.

Hasslow, O. J. (1934). Några characeer från Amerika. *Botaniska Notiser, 1934*, 346–348. [in Swedish].

Herbst, A., Henningsen, L., & Schubert, H. (2018). Encrustations and element composition of charophytes from fresh or brackish water sites—habitat- or species-specific differences? *Aquatic Botany, 148*, 29–34.

Hollerbach, M. M., & Krassavina, L. K. [Голлербах, М. М. & Л. К. Красавина]. (1983). *Определитель пресноводных водорослей СССР. 14. Charophyta* [*Determination key of freshwater algae of USSR*]. (189 pp). Nauka [in Russian].

Holtz, L.(n.d.). *Zeichnungen und Messungen von Characeen Pommerns von Dr. Paul Hauptfleisch, entsprechend meinen Publikationen. Inv. Nr. 5551*. University of Greifswald. [in German].

Holzhausen, A., Nowak, P., Niedrig, C., Feike, M., & Schubert, H. (2014). Morphometry of *Chara aspera, C. canescens, C. baltica* var. *baltica, C. baltica* var. *liljebladii* and *C. intermedia* oospores: Local variation versus taxonomic differences. *Aquatic Botany, 120*, 60–66.

Korsch, H. (2018). The worldwide range of the Charophyte species native to Germany. *Rostocker Meeresbiologische Beiträge, 28*, 45–96.

Krause, W. (1997). Charales (Charophyceae) Süßwasserflora von Mitteleuropa. In H. Ettl, G. Gärtner, H. Heynig, & D. Mollenhauer (Eds.), *Süßwasserflora von Mitteleuropa 18*. Fischer [in German].

Langangen, A. (2007). *Charophytes of the Nordic countries* (p. 102). Saeculum ANS.

Langangen, A., & Bennike, O. (1997). *Chara baltica* found in Greenland. *Cryptogamie, Algologie, 18*, 375–376.

Langangen, A., Hansen, J. B., & Mann, H. (1996). The charophytes of Greenland. *Cryptogamie, Algologie, 17*, 239–257.

Länsstyrelsen Stockholms Län. (1997). *Vegetation i trösklade havsvikar i Stockholms län*. Rapport nr. 33, 155 pp. [in Swedish]

Lindner, A. (1978). Soziologisch-ökologische Untersuchungen an der submersen Vegetation in der Boddenkette südlich des Darß und des Zingst (südliche Ostsee). *Limnologica, 11*, 229–305. [in German].

Luther, H. (1979). Fynden av *Chara baltica* i Bottniska viken. *Memoranda Societatis pro Fauna et Flora Fennica, 55*, 12–15. [in Swedish].

Migula, W. (1889–1897). Fünfter Band: Die Characeen. In: A. Fischer, E. Fischer, F. Hauck, G. Limpricht, C. Luerssen, W. Migula, H. Rehm, P. Richter, & G. Winter (Eds.), *Dr. L. Rabenhorst's Kryptogamenflora von Deutschland, Oesterreich und der Schweiz* (2 ed.). Eduard Kummer [in German].

Moore, J. A. (1986). *Charophytes of Great Britain and Ireland.* BSBI Handbook 5. Botanical Society of the British Isles.

Munsterhjelm, R. (1997). The aquatic macrophyte vegetation of flads and gloes, S coast of Finland. *Acta Botanica Fennica, 157*, 1–68.

Nat, E., Simons, J., de la Haye, M. A. A., & Coops, H. (1994). *Historisch en actueel verspreidingsbeeld van kranswieren in Nederland in samenhang met waterkwaliteitsfactoren* (77 pp). RIZA werkdocument 94.148X [in Dutch].

Nowak, P., & Schubert, H. (2019). Genetic variability of charophyte algae in the Baltic Sea area. *Botanica Marina, 62*, 75–82.

Olsen, S. (1944). Danish Charophyta. *Det Kongelige Danske Videnskabernes Selskab, Biologiske Skrifter, 3*(1), 1–240, pl. 1–2.

Romanov, R., Korolesova, D., Afanasyev, D., & Zhakova, L. (2020). *Chara baltica* (Charophyceae, Charales) from the Black Sea region and taxonomic implications of extrastipulodes. *Botanica, 26*, 126–139.

Scott, S., Darwell, A., & Stewart, N. (2011). *Bird's nest stonewort Tolypella nidifica in Loch an t-Sruith Mhoir and Loch an Duin, North Uist.* Scottish Natural Heritage Commissioned Report No. 461, 98 pp.

Skuja, H. (1924). Mersraga-Ragaciema piekrastes algas (Beitrag zur Algenflora des Rigaschen Meerbusens). *Acta Universitatis Latviensis, 10*, 337–393. [in Latvian].

Stewart, N. F., & Church, J. M. (1992). *Red data books of Britain and Ireland: Stoneworts.* The Joint Nature Conservation Committee.

van Raam, J. C. (1998). *Handboek Kranswieren.* Chara boek [in Dutch].

Vedder, F. (2003). *Morphologie und Taxonomie rezenter und subfossiler Characeen-Oosporen aus der Ostsee. Diploma thesis* (62 pp). University of Greifswald, Germany [in German].

Verdam, H. D. (1938). The Netherlands Charophyta. *Blumea, 3*, 5–33.

Wallentinus, I. (1979). Environmental influences on benthic macrovegetation in the Trosa-Askö area, northern Baltic proper II. The ecology of macroalgae and submersed phanerogams. *Contributions from the Askö Laboratory, University of Stockholm, 25*, 1–210.

Wood, R. D. (1965). *Monograph of the Characeae.* J. Cramer.

Chapter 21
Chara baueri

Michelle T. Casanova

Scientific name: *Chara baueri* A. Braun, Neue Denkschr. Allg. Schweiz. Ges. Gesammten Naturwiss. 10(3): 13. 1847, n. nov. pro Chara scoparia Bauer ex Rchb.

≡ *Chara scoparia* G.H. Bauer ex Rchb., Handb. Gewächsk., ed. 2, vol. 3: 1596. 1829, n. ill. non Chevalier 1827.
Neotype (Wood, 1965: 319): Weißensee & Lankwitz b. Berlin/[no date]/[no collector] (L 54,658).

Local names: BELARUSIAN: Хара Бауэра; ENGLISH: Bauers Chara; GERMAN: Bauers Armleuchteralge; POLISH: Ramienica Bauera, RUSSIAN: Хара Бауэра; SWEDISH: Bauersträfse; UKRAINIAN: Хара Бауера.

21.1 Morphology

Habitus: *Chara baueri* is an annual species. The branchlets that occur on the middle axis nodes are as long as internodes, but often considerably shorter on the lower nodes. The plant is richly branched from the ground up and often has a bushy habit. It usually occurs as single plants (Figs. 21.1 and 21.2b, c).

Size: *Chara baueri* is usually 8–10(15) cm in height.

Main axis: Axis diameter is about 1–1.2 mm. Internodes are shorter than or up to the length of the branchlets (Fig. 21.2c, e).

M. T. Casanova (✉)
273 Casanova Rd, Westmere, VIC 3351, Australia
e-mail: m.casanova@nhm.ac.uk; mt.casanova@federation.edu.au

H. Schubert et al. (eds.), *Charophytes of Europe*,
https://doi.org/10.1007/978-3-031-31898-6_21

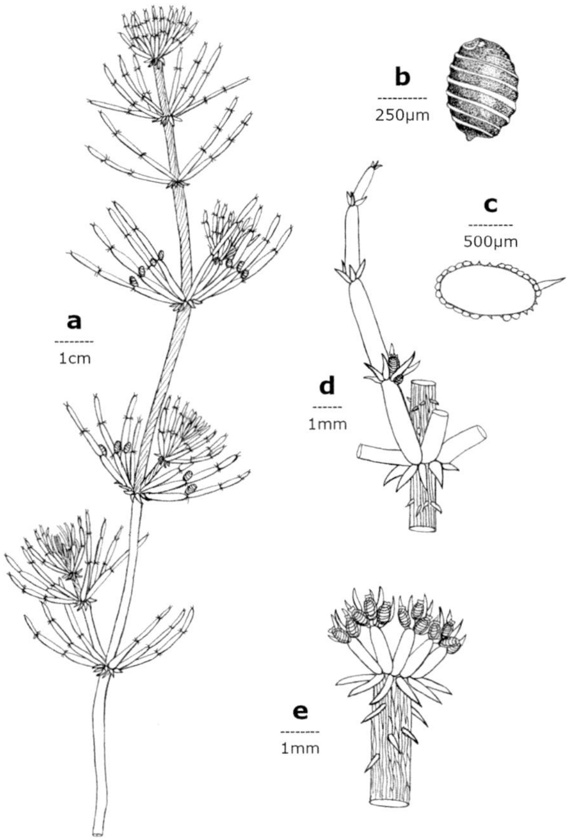

Fig. 21.1 *Chara baueri,* drawings. **a** Habitus; **b** Oospore; **c** Cross section of main axis; **d** Branchlet; **e** Fertile whorl and stipulodes. Drawings after Werner Krause, modified by Dagmar Wassong in van de Weyer and Schmidt (2018)

Cortex: According to Migula (1889–1897) *Chara baueri* is triplostichous and isostichous. The material from Kazakhstan is diplo-triplostichous (Langangen and Sviridenko, 1995).

Spine-cells: The spine-cells are arranged individually, tapered and about half as long as the shoot diameter. especially on the younger parts of the shoots.

Stipulodes: *Chara baueri* has a single tier of long and pointed stipulodes, approximately as long as the main axis diameter (Figs. 21.1 and 21.2d).

Branchlets: There are 8–9 (11) branchlets in a whorl, somewhat spreading, with 3–4 segments, completely ecorticate, the end segment is sometimes swollen and provided with 2–4 bract-cells of equal length forming a structure called a corona (Fig. 21.1).

Bract-cells: There are generally 6–7 bract-cells at the first node, with fewer at the upper branchlet nodes.

Fig. 21.2 Habitat appearance and morphology of *Chara baueri*. **a** Habitat of *Chara baueri* in field pond in a ploughed field in Poland; **b** *Chara baueri* in situ; **c** Overall habit of pressed plant; **d** Diagnostic features on specimen *F. Bauer* 1830, Berolin (Berlin) MEL.: Corticated axis, ecorticate branchlets, acute stipulodes in one row, a corona of cells at the branchlet tips; **e** Apices of branchlets and branchlet tips; **f** Cortex and spine cells. Photographs: **a**, **b**, **d**—Michelle T. Casanova; **c**, **e**, **f**—Klaus van de Weyer

Bracteoles: Two, similar to the anterior bract-cells, slightly exceeding mature oogonium.

Sexuality: *Chara baueri* is monoecious. Gametangia are conjoined singly and in pairs (geminate), on the lowest 3 branchlet nodes.

Oogonia: The oogonia are long and ovoid, 580–590 μm long (Table 21.1). The coronula is large and spreading, apiculate but blunt tipped.

Table 21.1 Oogonia morphometry

Author	Length	Width	Coronula length	Coronula width	Number of helical stripes
Migula (1889–1897)	650	500	150	200	8–10
Braun and Nordstedt (1882)	700–880	380–480	-	-	-
Wood (1959)	650	-	-	-	8–10
Wood (1965)	800	375–500	150–180	200–225	8–10
Urbaniak and Gąbka (2014)	585–950	365–610	-	-	-
Total range	**580–950**	**365–610**	**150–350**	**120–275**	**8–10**

Length in the first column is excluding the coronula; its length and width are given separately, all values for length and width in μm

Table 21.2 Oospore dimensions for *Chara baueri*

Author	Length	Width	Number of striae	Colour
Migula (1889–1897)	500–550	280–340	8	Black
Braun and Nordstedt (1882)	500–560	310–350	-	-
Wood (1965)	465–550	280–340	8–9	Black
Urbaniak and Gąbka (2014)	465–740	280–340	-	Black or dark brown
Pukacz et al. (2012)	420–550	220–300	8–10	Black
Total range	**420–740**	**220–350**	**8–10**	**Black**

All values for length and width in μm. In Krause (1997), values of 650–700 μm are given for the oospore width, which is most probably, erroneous

Oospores: Oospores are black and have 8–9 striae (Table 21.2) of somewhat raised ridges (pachygyra), the fossa is c. 60 μm across, membrane smooth and unornamented (Fig. 21.3).

Antheridia: Antheridia range in size from 250 to 370 μm (Table 21.3).

Bulbils: Bulbils have not been described for this annual species.

21.2 Phenology and Reproduction

Gametangia are formed from July to October. The occurrence of *Chara baueri* is erratic, and often cannot be found in the same location from one year to the next.

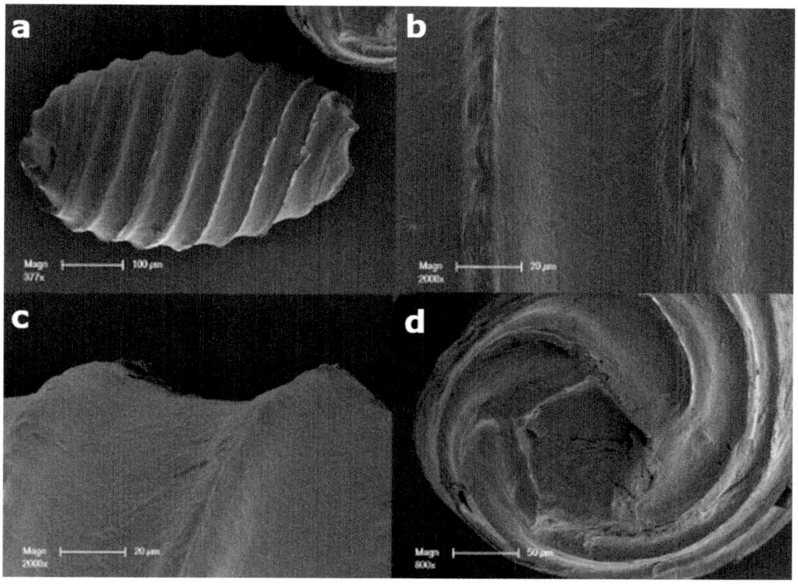

Fig. 21.3 Oospores of *Chara baueri*: Specimen *Jahn,* in Teuchen bei Weipenpe (HBG). **a** Side view of oospore, apex to the left; **b** Detail of oospore wall and striae, showing little ornamentation; **c** View of the raised striae, with distinct pachygyra; **d** View of the basal cell impression at the base of the oospore. Photographs: Michelle T. Casanova

Table 21.3 Antheridia morphometry

Author	Diameter (µm)
Migula (1889–1897)	250–300
Braun and Nordstedt (1882)	280–370
Wood (1965)	250–330
Urbaniak and Gąbka (2014)	250–330
Total range	**250–370**

21.3 Habitat Conditions and Ecology

Habitat type: *Chara baueri* occurs in small, shallow field ponds that dry out in the winter (Fig. 21.2a).

Sediment: The species has mostly been detected on clayey and fine silt sediments.

Nutrient conditions: These parameters have rarely been recorded. On the basis of the currently known habitat type (field ponds and temporary wetlands) there would normally be plenty of nutrients. Pukacz et al. (2012) recorded (among other parameters) total phosphorus of 1.04–1.12 mg L^{-1} and total nitrogen at 3.89–5.15 mg

L^{-1} indicating eutrophic conditions. Trbojević et al. (2020) also recorded eutrophic conditions (total phosphorus of 0.04–0.08 mg L^{-1} and total nitrogen at 1.98–3.33 mg L^{-1}).

Water chemistry: A pH of 7.9–8.7 and conductivity of 350–630 μS/cm have been demonstrated for various *C. baueri* sites (Pukacz et al., 2012; Trbojević et al., 2020). Water hardness of 4.17–13.2 °dH and various calcium (14.9–71.8 mg L^{-1}) and magnesium (7.3–16.3 mg L^{-1}) concentrations have also been recorded (Pukacz et al. 2012; Trbojević et al., 2020).

Depth range: *Chara baueri* is found in shallow water (as little as c. 5 cm). Urbaniak and Gąbka record the deepest depth of c. 1 m.

Associated species: *Chara baueri* is associated with species of ephemeral water bodies such as *Ceratophyllum submersum, Elatine alsinastrum, Nitella mucronata* or *Sphaerochara prolifera,* but also with more common species such as *Chara globularis* or *Bolboschoenus maritimus.*

21.4 Variability

In general, the plant is not very variable.

21.5 Identification and Risk of Confusion

Chara baueri can be distinguished from the similar *C. braunii* on the basis of its more bulky, compact and stiffer growth form. The presence of an axial cortex prevents confusion with *C. braunii.* This distinction was not well-made in very early collections, so the determination of older specimens deposited in herbaria should be checked. *Chara gymnophylla* might also be mistaken for *C. baueri* but should be distinguished on the basis of two rows of short and obtuse stipulodes, rather than a single row of acute stipulods as in *C. baueri.*

21.6 Distribution

Chara baueri was amalgamated with the Australian species *Chara muelleri* (A. Braun) F. Muell. by Wood (1962). However, the Australian material has been shown to be a different species (Casanova, 2014). *Chara baueri* seems to have always been rare. It was collected several times in the nineteenth century in what is now Berlin, the last time in 1869. After that, the plant was presumed extinct. It was surprisingly recorded from Kazakhstan (Langangen and Sviridenko, 1995), and rediscovered

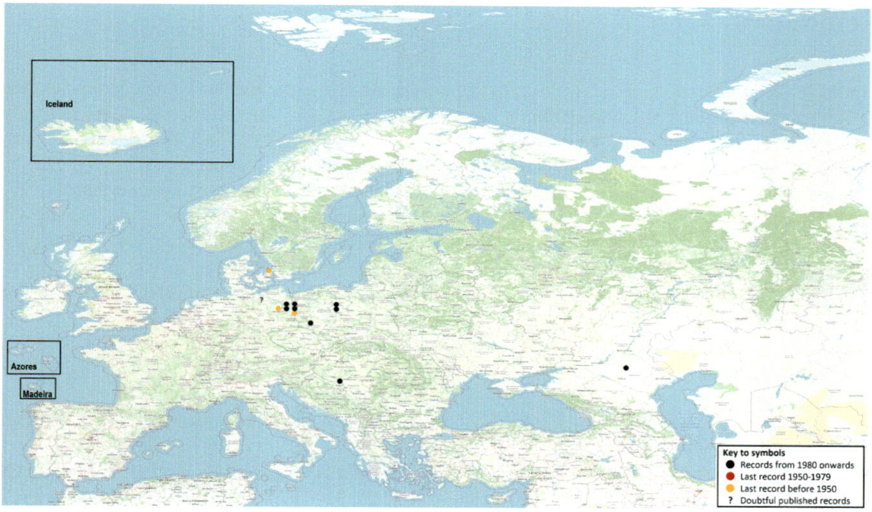

Fig. 21.4 European distribution of *Chara baueri*. For worldwide distribution see Korsch (2018). Black dots are records from 1980 on, red dots from 1950 to 1979 and yellow dots before 1950; unconfirmed/doubtful published records (if applicable) are marked with "?" on top. Map prepared by Heiko Korsch

(Raabe, 2009) in a few small bodies of water in Brandenburg and Poland (Urbaniak and Gąbka, 2014). The data for Sweden, Belarus and Lithuania were considered unlikely by Migula (1889–1897). Most recently it has been found in Serbia in very small and shallow ponds (Trbojević et al., 2020). Trbojević et al. (2020) hypothesise that the species has been overlooked because its habitat is atypical for charophytes (Fig. 21.4).

21.7 Status and Threats

Chara baueri is a rare species throughout its range, perhaps one of the rarest species in Europe. Although it was described early in the nineteenth century, many of its habitats disappeared within a short period of time (Pukacz et al., 2012). After thorough searching it has been rediscovered in both northern Germany (Raabe, 2009) and Poland, and more recent discoveries have been in regions outside its original distribution (Langangen and Sviridenko, 1995; Urbaniak and Gąbka, 2014). It has been listed on the Red List of several countries (e.g. Siemińska et al., 2006; Gärdenfors, 2010), but is often classified as Data Deficient due to a lack of records. It has been assessed as "Endangered" in Europe and globally (see Chap. 9). Conservation of *Chara baueri* will likely depend on conservation of its particular habitat, i.e. temporary wetlands that are shallowly flooded.

References

Braun, A., & Nordstedt, C. F. O. (1882). Fragmente einer Monographie der Characeen. Nach den hinterlassen Manuscripten A. Braun's herausgegeben von Dr. Otto Nordstedt. *Abhandlungen der Königlichen Akademie der Wissenschaften zu Berlin für das Jahr, 1882*, 1–211. [in German].

Casanova, M. T. (2014). A revision of *Chara* sect. *Charopsis* (Characeae, Charophyceae) in Australia, including specimens collected for Bush Blitz. *Australian Systematic Botany, 27*, 403–414.

Gärdenfors, U. (ed). (2010). Rödlistade arter i Sverige 2010. *The 2010 red list of Swedish species*. ArtDatabanken, SLU, Uppsala [in Swedish].

Korsch, H. (2018). The worldwide range of the Charophyte species native to Germany. *Rostocker Meeresbiologische Beiträge, 28*, 45–96.

Krause, W. (1997). Charales (Charophyceae) Süßwasserflora von Mitteleuropa. In H. Ettl, G. Gärtner, H. Heynig, & D. Mollenhauer (Eds.), *Süßwasserflora von Mitteleuropa* 18. Fischer [in German].

Langangen, A., & Sviridenko, B. F. (1995). *Chara baueri* A. Br., a charophyte with a disjunct distribution. *Cryptogamie, Algologie, 16*, 125–132.

Migula, W. (1889–1897). Fünfter Band: Die Characeen. In A. Fischer, E. Fischer, F. Hauck, G. Limpricht, C. Luerssen, W. Migula, H. Rehm, P. Richter, & G. Winter (Eds.), *Dr. L. Rabenhorst's Kryptogamenflora von Deutschland, Oesterreich und der Schweiz* (2nd ed.). Eduard Kummer [in German].

Pukacz, A., Boszke, P., Pełechaty, M. J., & Raabe, U. (2012). Comparative study of the oospore morphology of two populations of a rare species *Chara baueri* A. Braun in Cedynia (Poland) and Batzlow (Germany). *Acta Societatis Botanicorum Poloniae, 81*, 131–136.

Raabe, U. (2009). *Chara baueri* rediscovered in Germany. Plus additional notes on Gustav Heinrich Bauer (1794–1888) and his herbarium. *International Research Group Charophytes News, 20*, 13–15.

Siemińska, J., Bąk, M., Dziedzic, J., Gąbka, M., Gregorowicz, P., Mrozińska, T., Pełechaty, M., Owsianny, P. M., Pliński, M., & Witkowski, A. (2006). Red list of algae in Poland. In Mirek, Z., K. Zarzycki, W. Wojewoda & Z. Szeląg (Eds.), *Red list of plants and fungi in Poland. Szafer Institute of Botany* (pp. 36–52). Kraków: Polish Academy of Sciences.

Trbojević, I, Milovanović, V, Subakov Simić, G. (2020). The Discovery of the Rare *Chara baueri* (Charales, Charophyceae) in Serbia. *Plants, 9*(11), 1606. https://doi.org/10.3390/plants9111606.

Urbaniak, J., & Gąbka, M. (2014). *Polish Charophytes. An illustrated guide to identification* (120 pp.). Uniwersytet Przyrodniczy we Wroclawiu [in Polish].

van de Weyer, K., & Schmidt, C. (2018). Bestimmungsschlüssel für die aquatischen Makrophyten (Gefäßpflanzen, Armleuchteralgen und Moose) in Deutschland: Band 1: Bestimmungsschlüssel, ed. 2. Landesamt für Umwelt (LfU) Brandenburg. *Potsdam, Fachbeiträge Des LfU Brandenburg, 119*, 1–180. [in German].

Wood, R. D. (1959). Gametangial constants of extant charophyta for use in micropaleobotany. *Journal of Paleontology, 33*, 186–194.

Wood, R. D. (1962). New combinations and taxa in the revision of Characeae. *Taxon, 11*, 7–25.

Wood, R. D. (1965). *Monograph of the Characeae*. J. Cramer.

Chapter 22
Chara braunii

Michelle T. Casanova and Angela Doege

Scientific name: ***Chara braunii*** **C.C. Gmelin, Fl. Bad. 4: 646. 1826**

Neotype (Wood, 1965: 262) Prope Oldenico/aetate 1856/Malvinverni (NY 945366).

= *Chara coronata* Ziz ex Bisch. Krypt. Gew. 1: 3. 1828.
Holotype (Bischoff, 1828): Tafel 1, Fig. 7.

Infraspecific taxa

Chara brauni **C.C. Gmelin var.** ***braunii***

Chara brauni **C.C. Gmelin var.** ***soleirolii*** **(A. Braun) Bég. & Formigg. in Boll. Soc. Bot. Italiana 1907: 112. 1907**

≡ *Chara coronata* Ziz ex Bisch. var. *soleirolii* A. Braun 1849: 296.

type not designated.

Taxonomic note: *Chara braunii* has been thought to be cosmopolitan (Wood, 1965). Monoecious ecorticate specimens with a corona of cells at the branchlet tips are widely distributed in temporary water bodies on all continents except Antarctica (Proctor, 1970). However, *C. braunii* consists of a "complex of geographical races or incipient species" that are not genetically compatible (Proctor, 1970). Variation in the structure of the oospore wall coincides with some of the previously designated varieties or forms, and these could well be recognised as separate species. All specimens within Europe (except for historical records from UK) conform to the type

M. T. Casanova (✉)
273 Casanova Rd., Westmere 3351, Australia
e-mail: m.casanova@nhm.ac.uk; mt.casanova@federation.edu.au

A. Doege
Morokkajärventie 150, 99800 IVALO, Finnland
e-mail: ad@morokka.net

H. Schubert et al. (eds.), *Charophytes of Europe*,
https://doi.org/10.1007/978-3-031-31898-6_22

and can confidently be called *Chara braunii*. Records from outside Europe require further study.

Local names: DUTCH: kroontjeskransblad; ENGLISH: Braun's Chara; FINNISH: Dutch: kroontjeskransblad; Silonäkinparta; GERMAN: Brauns Armleuchteralge; POLISH: Ramienica wieńcowa; RUSSIAN: Хара Брауна, SWEDISH: Barklöst sträfse (Sweden), Svedsträfse (Finland); UKRAINIAN: Хара Брауна.

22.1 Morphology

Habitus: *Chara braunii* is usually somewhat rigid, light green, shiny and translucent, with darker axial nodes. It often has a shrubby, branching habit (Fig. 22.1); the whorls of stipulodes occur in a single row alternate to the branchlets, are clearly visible, and outward pointing (Fig. 22.1).

Size: *Chara braunii* is small to medium in size. The length ranges from a few centimeters to about 80 cm.

Fig. 22.1 *Chara braunii*. **a** Habit of shoots with abundant fructification, ecorticate branchlets and shoots; **b** Nodal complex showing single row of alternate stipulodes, ecorticate branchlet and axis; **c** In situ growing in a wetland; **d** Branchlet tip with corona (little crown) of cells. Photographs: **a**, **b**, **c**—Klaus van de Weyer, **d**—Michelle T. Casanova

Main axis: The axis diameter is 1–2 mm; the internodes are shorter in shallow water but longer than the branches in deeper water.

Cortex: No cortex is present on the axis or branchlets.

Spine-cells: No spine-cells are present.

Stipulodes: The stipulodes are in a single row and are up to three times as wide as the main axis diameter arranged alternate with the branchlets.

Branchlets: The 8–12 branchlets are spreading, shorter or longer than the internodes. They have 2–5 ecorticate segments with gametangia and bract-cells, with the uppermost segment with usually 3 small cells. The branchlet cells can be constricted at the nodes.

Bract-cells: The bract-cells are narrow and pointed, usually 2 on the inside of the branchlet and about as long as the oogonium. The uppermost bract-cells form a crown (corona) with the apical cell (Soulié-Märsche and Vautier, 2004, Fig. 3). The bract-cells on the abaxial of the branchlet are thin and inconspicuous.

Bracteoles: There are two bracteoles, usually longer than the oogonium.

Sexuality: *Chara braunii* is monoecious.

Oogonia: The oogonia are relatively large and elongated (Table 22.1), they are usually solitary, but 2–3 can occur together at the lowest branchlet nodes. The coronula is low but wide, its cells are usually narrowed and typically acuminate.

Table 22.1 Oogonia morphometry

Author	Length	Width	Coronula length	Coronula width	Number of helical stripes
Migula (1889–1897)	550	440	200	200–300	11
Corillion (1957)	525–650	400–450	150	200	9–10
Groves and Bullock-Webster (1920, 1924)	525–650	400–450	150	200	9–10
Krause (1997)	550–880	400–500	180	250	9–10
Wood (1965)	630–675	480–570	165–180	210–300	10–11
Caisová and Gąbka (2009)	510–850	375–530	-	-	-
Soulié-Märsche and Vautier (2004)	700		150	200	-
Total range	**510–880**	**375–530**	**150–200**	**200–300**	**9–11**

Only studies that record specimens from Europe are given here. Length in the first column excludes the coronula; its length and width are given separately, all values for length and width in μm

Table 22.2 Oospore morphometry

Author	Length	Width	Number of striae	Colour
Braun (1847)	400–480	-	-	Black
Braun (1867)	420–550	-	9	Black
Migula (1889–1897)	420–550	320	8–9 (10)	Black to dark red brown
Filarszky (1893)	520	320	9–10	Black
Groves and Bullock-Webster (1920, 1924)	450–550	275–300	7–9	Dull brown to black
Wood (1965)	570	375–405	9	Dark brown
Haas (1994)	420–620	270–360	7–10	Dark brown to black
Boszke et al. (2008)	466–600	250–366	8–10	
Caisová and Gąbka (2009)	-	-	-	Dark brown to black
Soulié-Märsche and Vautier (2004)	520	320	8–9	Black
Total range	**400–620**	**250–405**	**7–10**	

Only studies that record specimens from the continent of Europe are given here; all values for length and width in μm

Oospores: The oospores are ellipsoidal to almost cylindrical, with a 'shoulder' (i.e., flattened at the apex) and somewhat truncate at the base, and have well developed striae (Table 22.2). The ornamentation is granular (Urbaniak, 2007), tuberculate (Boszke et al., 2008) or smooth (Fig. 22.2).

Antheridia: The diameter of antheridia ranges from 250 to 330 μm (Table 22.3).

Bulbils: *Chara braunii* does not have bulbils.

22.2 Phenology and Reproduction

Chara braunii grows and becomes fertile in summer and autumn. It is usually an annual species, reproducing solely by oospores. It germinates late in the season (May) but produces gametangia almost immediately and abundantly. The reproduction takes place via oospores, produced before its (usually) temporary habitat dries out over winter. In perennial warm water habitats, it can be longer-lived (Shaw, 1948).

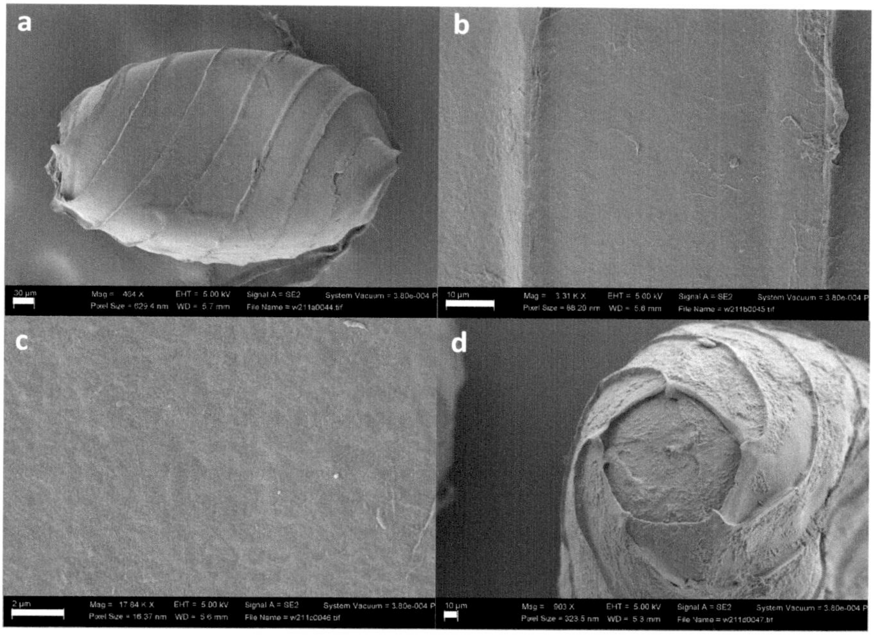

Fig. 22.2 *Chara braunii* oospores from specimen BM013738705 (Indebetou Aug. 1874, Sweden). **a** Side view of ovate oospore with flattened basal area, and 7–8 striae; **b** Detail of fossa with wide ridges (striae) and relatively smooth oospore wall; **c** Fine detail of oospore surface, still looking smooth; **d** View of the basal area with the impression of a single end-cell. Photographs: Michelle T. Casanova

Table 22.3 Antheridia morphometry

Author	Diameter (μm)
Migula (1889–1897)	250–280
Urbaniak and Gąbka (2014)	250–330
Groves and Bullock-Webster (1920, 1924)	250–275
Wood (1965)	275–300
Krause (1997)	275
Caisová and Gąbka (2009)	300
Soulié-Märsche and Vautier (2004)	250–280
Total range	**250–330**

Only studies that record specimens from the continent of Europe are given here.

22.3 Habitat Conditions and Ecology

Habitat type: *Chara braunii* has been described as the "exemplar of a sporadic species" (Schmidt et al., 1996), as it occurs in temporary, winter-dry, summer-wet pools, for example in fishponds, and as a summer-growing pioneer species. More permanent habitats are known (Korsch, 2010; Vesić et al., 2014), such as thermal spring pools and water bodies with artificially elevated temperature (Groves and Bullock-Webster, 1924). The species also occurs in very shallow water in the Bothnian Bay (Baltic Sea) on substrates exposed to mechanical stress from ice (Havs-och Vattenmyndigheten, 2020; Pekkari, 1953). Proctor (1999) showed that in USA it fails to thrive in habitats with large populations of herbivores (particularly amphipods). In South America the species has been found in large lakes, rivers and creeks (Schubert et al., 2015).

Sediment: *Chara braunii* occurs predominantly on low-lime sand as well as loam and clay soils and also tolerates small amounts of fine silt and organic deposits.

Nutrient conditions: *Chara braunii* colonises nutrient-poor to moderately nutrient-rich waters that are poor in organic substances dissolved in the water. Measured on the standard red Total N can be up to 2.9 mg/L, nitrate up to 5.0 mg/L, and Total P up to 0.033 mg/L. Occurrence of *C. braunii* is often not as related to the nutrient content as it is to the lack of competition with established plants.

Water chemistry: *Chara braunii* colonises acidic (pH 5) to neutral (rarely weakly alkaline), low-calcium, electrolyte-poor to moderately electrolyte-rich fresh water. It can also be found in mesohaline locations and fishponds when lime has been added.

Depth range: *Chara braunii* prefers shallow waters; in Europe the species is absent in depths greater than 2 m (but was found down to 33 m depth in Chilean lakes as reported by Blindow et al., 2018).

Associated species: *Chara braunii* is a poor competitor and often forms a single-species population. It can also occur in temporary systems and at the edges of deeper ponds with other charophytes (*C. contraria, C. globularis, C. virgata, C. vulgaris, Nitella confervacea, N. mucronata, Sphaerochara prolifera*), and with angiosperms (*Alisma lanceolatum, Alopecurus aequalis, Carex* spp., *Echinochloa crus-galli, Juncus* spp., *Limosella aquatica, Lysimachia nummularia, Potentilla* spp., *Typha latifolia* and *Veronica anagallis-aquatica*) and mosses (Raabe, 2017). In the Bothnian Bay, *C. braunii* has been found together with angiosperms (*Callitriche* spp., *Elatine* sp., *Eleocharis acicularis, Isoetes* spp., *Potamogeton berchtoldii, P. pusillis, Subularia aquatica* and *Utricularia vulgaris*), and other charophytes (*Chara globularis, Nitella flexilis, N. opaca, N. syncarpa* and *N. wahlbergiana*) (Havs-och Vattenmyndigheten, 2020; Pekkari, 1953; Zhakova, 2004).

22.4 Variability

Chara braunii can be quite variable. Water depth and water transparency can influence the size and overall morphology of the plant. Although in good specimens it has a shrubby habit, it can also be very small and unbranched (when young) or somewhat elongate, depending on the light conditions and competition for space. There is evidence that populations with partial or complete reproductive barriers exist, even in close proximity (Proctor, 1970). Within Europe it is possible that more than one variety will be recognised (e.g., *C. braunii* var. *braunii* in the central part of its range on continental Europe, *C. braunii* var. *soleirolii* in Corsica). More collections and study are required to determine the validity of var. *soleirolii*.

22.5 Identification and Risk of Confusion

The lack of cortication and the distinctive single whorl of stipulodes (approx. 2 mm long), make it difficult to confuse *Chara braunii* with other species (Fig. 22.3). *Chara socotrensis* (reported from Scotland) has similar ecorticated axis and branchlets but does not have the same development of bract-cells, and no corona of bract-cells at the apex of the branchlets. *Chara baueri* is similar in its overall morphology and habitat but has at least a partially corticated axis and spine cells. *Lamprothamnium* species have no cortex, but they are generally confined to higher salinity water bodies, whereas *C. braunii* prefers fresh water, tolerating low salinity conditions only.

22.6 Distribution

Chara braunii has an apparent worldwide distribution (Wood and Imahori 1965), and because of this, and ease of culturing, has been the first charophyte for which the entire genome has been documented (Nishiyama et al., 2018). *Chara braunii* has been considered to be widely distributed (Norway, Sweden, Blindow, 2000), Finland, Germany (Migula, 1889–1897), Belgium (Crepin, 1863), France (Hy, 1913), The Netherlands (Bruinsma, 2010), Corsica, Spain, Italy, Austria, Hungary (Filarszky, 1893), Croatia (Istria), Libya, Syria, East Indies, Canada, USA (Robinson, 1906), Mexico, Hawaii, Australia (Wood, 1972), Chile (Schubert et al., 2015) and New Zealand, with some of the specimens described as different forms or varieties. Additional localities, UK (Groves and Bullock-Webster, 1924), Greece (Koumpli-Sovantzi, 1997), Poland (Urbaniak and Gąbka, 2014), Serbia (Vesić et al., 2016), Czech Republic (Caisová and Gąbka, 2009), have since been recorded. The work of Proctor (1970) demonstrated that many of the specimens within the worldwide distribution of the species cannot be crossed to produce fertile offspring, and yet,

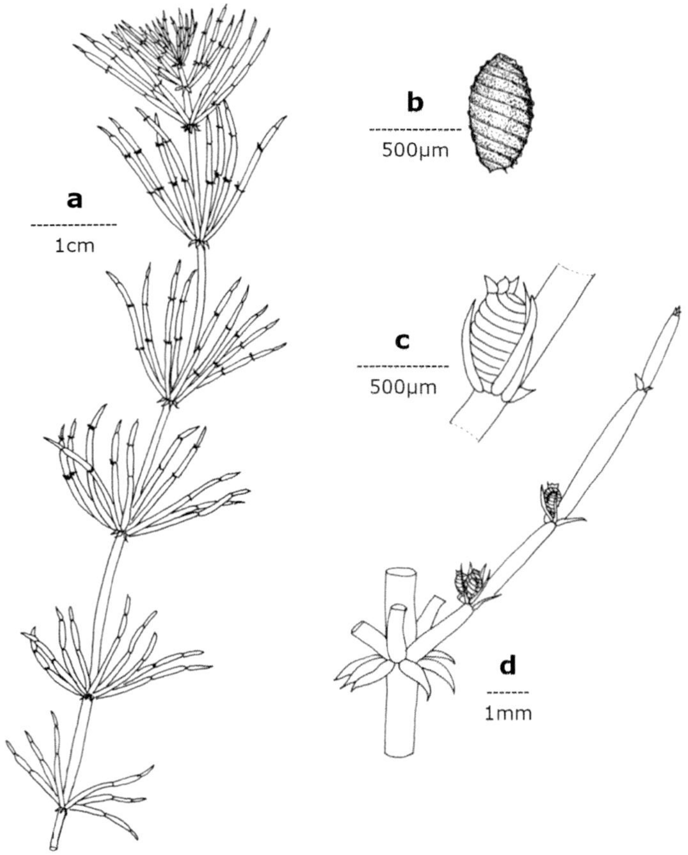

Fig. 22.3 *Chara braunii*. **a** Main axis and branchlets, **b** Oospore, **c** Oogonium, **d** Fertile branchlet and node. Drawings after Werner Krause, modified by Dagmar Wassong in van de Weyer and Schmidt (2018)

because of the similarity of morphology and habitat (mostly temporary wetlands and warm water bodies) they are all retained within the concept of *Chara braunii*.

22.7 Status and Threats

Chara braunii is listed as 'Vulnerable' in Finland and Sweden and 'Data Deficient' in Poland (Urbaniak, 2007). The UK records are thought to be based on an introduction of the species in the vicinity of cotton-processing mills, where it was able to persist in abnormally warm water temperatures (Groves and Bullock-Webster, 1924), and the locality has since been destroyed, resulting in its listing as 'extinct' in the UK (Stewart

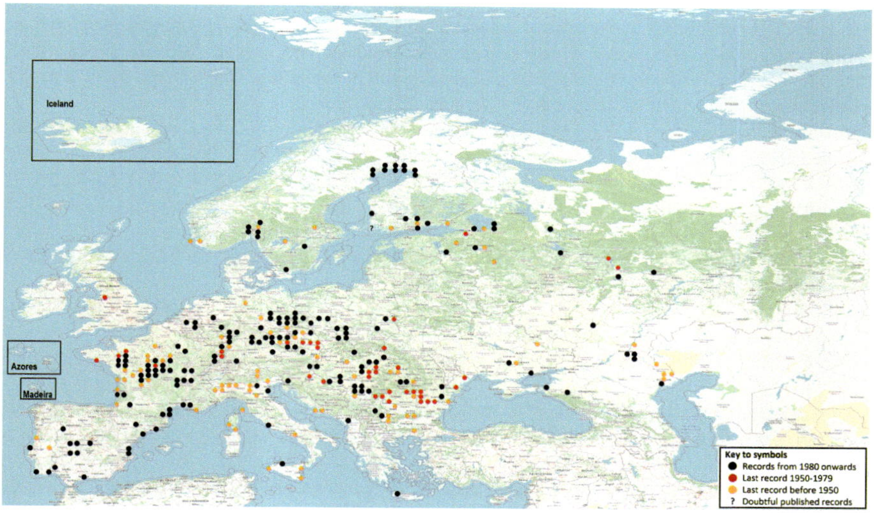

Fig. 22.4 Distribution of *Chara braunii* in Europe. For worldwide distribution see Korsch (2018). Black dots are records from 1980 on, red dots from 1950 to 1979 and yellow dots before 1950; unconfirmed/doubtful published records (if applicable) are marked with "?" on top. Map prepared by Heiko Korsch

and Church, 1992). It has been described as 'on the Red List of many countries in Europe' and was 'close to extinction' in Southern France (Soulié-Märsche and Vautier, 2004). The number of sites recorded in the Camargue have risen over the last few years due to the increasing production of organic rice (Mouronval et al., 2015).

Throughout the rest of Europe *Chara braunii* is considered relatively widespread and abundant and may even be increasing (Doege et al., 2016). Its particular habitat (shallow temporary wetlands and fishponds) is under increasing threat via land-use change and inappropriate cultivation or flooding regimes. Conservation of this species can only be undertaken through conservation and appropriate management of its habitat.

References

Bischoff, G. W. (1828). *Die kryptogamische Gewächse mit besonderer Berücksichtigung der Flora Deutschlands und der Schweiz, organographisch, anatomisch, physiologisch und systematisch bearbeitet. 1: Die Chareen und Equiseteen*. Johann Leonhard Schrag, Nürnberg [in German].

Blindow, I. (2000). Distribution of Charophytes along the Swedish coast in relation to salinity and eutrophication. *International Review of Hydrobiology, 85*, 707–717.

Blindow, I., Marquardt, R., Schories, D., & Schubert, H. (2018). Charophyceae of Chile—taxonomy and distribution. Part 1. *Nova Hedwigia, 107*, 1–47.

Boszke, P., Pełechaty, M., & Pukacz, A. (2008). Oospore dimensions and wall ornamentation in *Chara braunii*. *Biologia, 63*, 457–460.

Braun, A. (1847). Uebersicht der schweizerischen Characeen. *Neue Denkschriften der Allgemeinen Schweizerischen Gesellschaft für die Gesammten Naturwissenschaften, 10*, 1–23. [in German].

Braun, A. (1867). *Conspectus Systematicus Characearum Europaeum* (8 pp.). Dresden [in German; attachment to issue 3 of Rabenhorst's Characeae Exsiccatae.].

Braun, A., & Nordstedt, C. F. O. (1882). Fragmente einer Monographie der Characeen. Nach den hinterlassen Manuscripten A. Braun's herausgegeben von Dr. Otto Nordstedt. *Abhandlungen der Königlichen Akademie der Wissenschaften zu Berlin für das Jahr, 1882*, 1–211. [in German].

Bruinsma, J. (2010). Distribution of charophytes in North Brabant, The Netherlands, 1986–2007. *Charophytes, 2*, 31–37.

Caisová, L., & Gąbka, M. (2009). Charophytes (Characeae, Charophyta) in the Czech Republic: Taxonomy, autoecology and distribution. *Fottea, 9*, 1–43.

Corillion, R. (1957). Les Charophycées de France et d'Europe Occidentale. *Bulletin de la Société Scientifique de Bretagne, 32*(Hors série I & II), 1–499 [in French].

Crepin, F. (1863). Les Characées de Belgique. *Bulletin de la Société royale de Botanique de Belgique, 2*, 115–131. [in French].

Doege, A., van de Weyer, K., Becker, R., & Schubert, H. (2016). Bioindikation mit Characeen. In *Arbeitsgruppe Characeen Deutschlands, Armleuchteralgen: Die Characeen Deutschlands* (pp. 97–138). Springer [in German].

Filarszky, N. (1893). *A charafélék (Characeæ L. Cl. Richard) különös tekintettel a Magyarországi fajokra*. Kiadja a K.M. Természettudományi társul [in Hungarian].

Groves, J., & Bullock-Webster, G. R. (1920). *The British Charophyta* (p. 405). Ray Society.

Groves, J., & Bullock-Webster, G. R. (1924). *The British Charophyta. Volume II. Chareæ with plates, concluding articles, geological sketch, bibliography and index*. Ray Society.

Haas, J. N. (1994). First identification key for charophyte oospores from central Europe. *European Journal of Phycology, 29*, 227–235.

Havs-och Vattenmyndigheten. (2020). *Åtgärdsprogram för sällsynta kransalger längs kusten* (57 pp.) [in Swedish].

Hy, F. (1913). Les Characées de France. *Bulletin de la Société Botanique de France, 60*(sup 1), 1–47 [in French].

Korsch, H. (2010). Ergebnisse der Kartierung der Armleuchteralgen im Süden von Sachsen-Anhalt. 2. *Beitrag Mitteilungen zur Floristischen Kartierung in Sachsen-Anhalt, 15*, 135–139. [in German].

Korsch, H. (2018). The worldwide range of the Charophyte species native to Germany. *Rostocker Meeresbiologische Beiträge, 28*, 45–96.

Koumpli-Sovantzi, L. (1997). The charophyte flora of Greece, I. *Flora Mediterranea, 7*, 173–179.

Krause, W. (1997). Charales (Charophyceae) Süßwasserflora von Mitteleuropa. In H. Ettl, G. Gärtner, H. Heynig, & D. Mollenhauer (Eds.), *Süßwasserflora von Mitteleuropa 18*. Fischer [in German].

Migula, W. (1889–1897). Fünfter Band: Die Characeen. In A. Fischer, E. Fischer, F. Hauck, G. Limpricht, C. Luerssen, W. Migula, H. Rehm, P. Richter, & G. Winter (Eds.), *Dr. L. Rabenhorst's Kryptogamenflora von Deutschland, Oesterreich und der Schweiz* (2nd ed.). Eduard Kummer [in German].

Mouronval, J. B., Baudouin, S., Borel, N., Soulié-Märsche, I., Klesczewski, M., & Grillas, P. (2015). *Guide des Characées de France méditerranéenne* (211 pp.). Office national de la Chasse et de la Faune Sauvage [in French].

Nishiyama, T., Sakayama, H., de Vries, J., Buschmann, H., Saint-Marcoux, D., Ullrich, K. K., Haas, F. B., Vanderstraeten, L., Becker, D., Lang, D., Vosolsobě, S., Rombauts, S., Wilhelmsson, P. K. I., Janitza, P., Kern, R., Heyl, A., Rümpler, F., Calderón Villalobos, L. A. I., Clay, J. M., et al. (2018). The *Chara* Genome: Secondary complexity and implications for plant terrestrialization. *Cell, 174*, 448–464.

Pekkari, S. (1953). Fynd av *Chara braunii* vid Haparanda. *Botaniska Notiser, 1953*, 73–77. [in Swedish].

Proctor, V. W. (1970). Taxonomy of *Chara braunii*: An experimental approach. *Journal of Phycology, 6*, 317–321.
Proctor, V. W. (1999). Charophytivory, playas y papalotes, a local paradigm of global relevance. *Australian Journal of Botany, 47*, 399–406.
Raabe, U. (2017). Zur Verbreitung der Schimmernden Glanzleuchteralge (*Nitella translucens*) in Westfalen. *Rostocker Meeresbiologische Beiträge, 27*, 53–62.
Robinson, C. B. (1906). The Chareae of North America. *Bulletin of the New York Botanical Garden, 4*, 244–308.
Schmidt, D., van de Weyer, K., Krause, W., Kies, L., Garniel, A., Geissler, U., Gutowski, A., Samietz, R., Schütz, W., Vahle, H.-C., Vöge, M., Wolff, P., & Melzer, A. (1996). Rote Liste der Armleuchteralgen (Charophyceae) Deutschlands. 2. Fassung, Stand: Februar 1995. *Schriftenreihe Für Vegetationskunde, 28*, 547–576. [in German].
Schubert, H., Marquardt, R., Schories, D., & Blindow, I. (2015). Biogeography of Chilean charophytes. *Aquatic Botany, 120*, 129–141.
Shaw, S. (1948). *Chara Braunii* Gmelin. *Northwestern Naturalist, 23*, 166–167.
Soulié-Märsche, I., & Vautier, J. (2004). Ecologie et cycle de vie de *Chara braunii* Gmelin (Charales) dans un habitat méditerranéen. *Vie Et Milieu, 54*, 37–45. [in French].
Stewart, N. F., & Church, J. M. (1992). *Red data books of Britain and Ireland: Stoneworts*. The Joint Nature Conservation Committee, Peterborough.
Urbaniak, J. (2007). Distribution of *Chara braunii* Gmelin 1826 (Charophyta) in Poland. *Acta Societatis Botanicorum Poloniae, 76*, 313–320.
Urbaniak, J., & Gąbka, M. (2014). *Polish Charophytes. An illustrated guide to identification* (120 pp.). Uniwersytet Przyrodniczy we Wroclawiu [in Polish].
Vesić, A., Blaženčić, J., & Šinžar-Sekulić, J. (2014). Ecological preferences of charophytes in Serbia in relation to habitat type and other aquatic macrophytes. *Plant Biosystems, 150*, 490–500.
Vesić, A., Blaženčić, J., & Šinžar-Sekulić, J. (2016). Contribution to knowledge of the charophytes (Charales) of Vojvodina (Serbia)–20 years after the first review. *Botanica Serbica, 40*, 237–247.
van de Weyer, K., & Schmidt, C. (2018). *Bestimmungsschlüssel für die aquatischen Makrophyten (Gefäßpflanzen, Armleuchteralgen und Moose) in Deutschland: Band 1: Bestimmungsschlüssel* (2nd ed., pp. 1–180). Landesamt für Umwelt (LfU) Brandenburg, Potsdam, Fachbeiträge des LfU Brandenburg 119 [in German].
Wood, R. D. (1965). *Monograph of the Characeae*. J. Cramer.
Wood, R. D. (1972). Characeae of Australia. *Nova Hedwigia, 22*, 1–120.
Zhakova, L. V. (2004). *Chara braunii*. In H. Schubert & I. Blindow (Eds.), *Charophytes of the Baltic Sea* (pp. 64–69). A. R. G. Gantner.

Chapter 23
Chara canescens

Hendrik Schubert, Irmgard Blindow, Klaus van de Weyer, and Anders Langangen

Scientific name: ***Chara canescens*** **Desv. & Loisel. in Loisel, Not. Fl. France: 139. 1810**

Lectotype (Wood, 1965: 159): in aquis stag. circa Tolonum/[no date]/[no collector] (PC 641358 Herb. Desvaux).

= *Chara crinita* Wallr., Annus Bot.: 190. 1815.
Lectotype (T. Gregor, Hendr. Schub. Blindow & K. Weyer in Schubert and Blindow, 2016: 261): Wallr., Annus Bot.: t. III. 1815.

Epitype (Romanov, 2022: 3): In stagno Koelmensi pone Halam. cop./[no date]/ Wallroth/Hb. Schrader (LE A0001536).

= *Chara karelinii* Less. in Linnaea 9: 213. 1835 "Karelini".
Lectotype (Romanov, 2021: 3): [Kazakhstan, Atyrau Oblast], in aquis ad osti fluvii Ural/no date [1832]/G. Karelin (LE A0000408).

= *Chara sphagnoides* Griff., Not. Pl. Asiat. 4(2): 278, pl. 63. 1849.
Holotype (Griffith 1849): pl. 63.

H. Schubert (✉)
Universität Rostock Institut für Biowissenschaften, Albert-Einstein-Straße 3, 18051 Rostock, Germany
e-mail: hendrik.schubert@uni-rostock.de

I. Blindow
Biologische Station Hiddensee, Universität Greifswald, 18565, Germany
e-mail: blindi@uni-greifswald.de

K. van de Weyer
lanaplan, Lobbericher Str. 5, 41334 Nettetal, Germany
e-mail: klaus.vdweyer@lanaplan.de

A. Langangen
Natural History Museum, University of Oslo, P.O.Box 1172 Blindern, 0318 Oslo, Norway

H. Schubert et al. (eds.), *Charophytes of Europe*,
https://doi.org/10.1007/978-3-031-31898-6_23

= *Chara pusilla* Floerke ex Kütz., Sp. Alg. 526. 1849.
Type not designated.

Taxonomic note: Not included in *Chara canescens* are the monoecious species *Chara altaica* A. Braun in Braun and Nordstedt (1882), Abh. Kgl. Akad. Wiss. Berlin: 148, Fig. 228–231, *Chara evoluta* Allen 1882 in Bull. Torrey Bot. Club 9: 41, pl. 19 and *Chara hirsuta* Allen 1900 in Bull. Torrey Bot. Club 27: 301, pl. 11, 1.

Local names: DANISH: Grå kransnål; DUTCH: Brakwaterkransblad; ENGLISH: Bearded stonewort; ESTONIAN: Kähar mändvetikas; FINNISH: Kalvasnäk-inparta; FRENCH: Chara blanchâtre; GERMAN: Graue Armleuchteralge; LATVIAN: Iesirmā hara; LITHUANIAN: Pilkšvasis menturdumblis; NORWE-GIAN: Hårkrans; POLISH: Ramienica włochata; RUSSIAN: Хара седеющая; SWEDISH: Hårsträfse; UKRAINIAN: Хара сивіюча.

23.1 Morphology

See Figs. 23.1 and 23.2.

Habitus: Plants are light green to greyish, rarely slightly encrusted. Below the apex, branchlets are typically shorter than the internodes. Dense, well-developed spine-cells give them a brush-like appearance.

Size: *Chara canescens* is small to medium-sized (5–15 cm) but may reach up to 30 cm if growing in deeper water. In culture up to 40 cm length was reported by Langangen (2000).

Main axis: Axis diameter is about 0.5 mm; often repeatedly branched, but always with a prominent main axis.

Cortex: *Chara canescens* is strictly haplostichous. At the lower internodes, however, secondary cortical row cells may develop, visible by means of a hand lens. Cortex cells can, especially at inland brackish water sites, become almost colorless.

Spine-cells: Spine-cells are longer than the axis diameter and appear in dense bunches of 2–6 all over the axis.

Stipulodes: *Chara canescens* is diplostephanous; both rows of stipulodes are equal in appearance and similar to the spine-cells.

Branchlets: Whorls consist of 7–10 short branchlets, which are corticated except for the end-cell. Partial to total ecortication of branchlets is described for specimens from warm springs on Svalbard (Langangen, 2000).

Bract-cells: Spreading, well developed, pointed bract-cells are present along the whole branchlet.

Bracteoles: Inconspicuous, hardly visible with a hand lens.

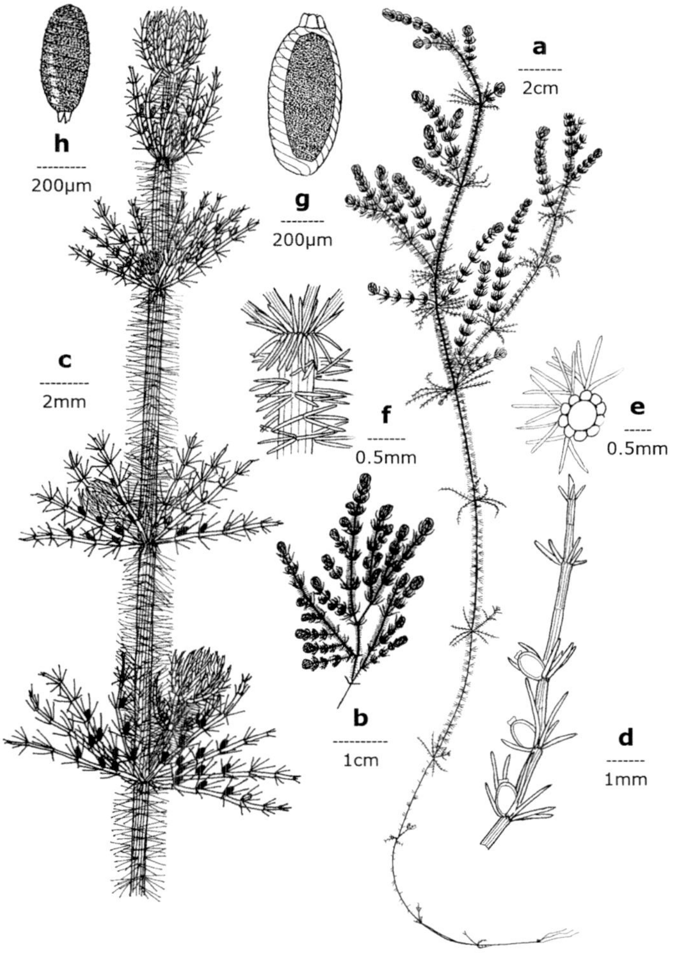

Fig. 23.1 *Chara canescens*, female parthenogenetic plants. **a** Plant collected from 1 m depth; **b** Plant collected from 20 cm depth; **c** Shoot tip; **d** Branchlet; **e** Shoot transection, spine-cells in bunches of 2–4; **f** cortex, spine-cells and stipulodes; **g** Oogonium; **h** Oospore. Drawings by Werner Krause

Bractlets: Inconspicuous, hardly visible with a hand lens.

Sexuality: The species is dioecious, but sexually reproducing populations, consisting of males and females, are rare in Europe where most populations consist of ovoapogameous females. The species is the only known charophyte able to reproduce parthenogenetically. Recent sexual populations are known from Spain (Cirujano et al., 2007), Austria (Schaible et al., 2009) Italy (Becker, 2019; Nowak and Schubert,

Fig. 23.2 *Chara canescens*, details. **a** With dense groups of spine-cells; **b** Less spiny form, both from the Baltic Sea; **c** Specimen from an inland brackish water site; **d** Male individual from Neusiedler See; **e** Antheridium; **f** Oogonium. Photographs: **a**, **b** —Klaus van de Weyer, **c**—Heiko Korsch, **d**—Hendrik Schubert, **e**, **f**—Ralph Schaible

2019; Nowak et al., 2019) and Serbia (Sabovljević et al., 2022) only. A former occurrence in France near Montpellier (Corillion, 1957) is well represented by herbarium material in, e.g., St. Petersburg, Leiden and Paris. However, the exact coordinates of the site are unknown. When searching the area in 2004 and 2005 (own observation) no recent occurrence could be found, so probably the site is lost. Another occurrence on Sicily also seems to be extinct (Troia, 2020). All records of sexual specimens checked so far from Greece (Corillion, 1957) turned out to be misidentified *C. galioides* (e.g., Schaible and Schubert, 2008 and references therein). Migula (1889–1897) also refers to occurrences of male *C. canescens* from Romania (Hermannstadt), Gurjen (probably Guryev, now Atyrau, Kazakhstan) Greece (Piraeus) and France ("Courteison", probably Courthezon, Vaucluse) but no herbarium specimen from these sites could be traced yet except for Gurjen (Atyrau), the *locus classicus* for *C. karelinii*. Hungarian occurrences of male plants, represented by herbarium material stored in, e.g., Jena, geographically continue east of the recent Austrian ones as far as Budapest and seem to represent the same type of habitats temporary ponds); whether they still exist is not known.

Oogonia: Oogonia are elongated ovals, with constricted coronulae. Data about quantitative morphometry differ among authors (Table 23.1), which probably can be at least in part explained by habitat (salinity) dependent development of oogonia.

Oospores (Fig. 23.3): There is a huge range of morphometrical data for oospores of *C. canescens* as can be seen from Table 23.2. For specimens from the Baltic Sea Niedrig (2010) reported length/width ratios between 1.3 and 1.49. Width of striae was found to vary between 58 and 67 μm. Oospores from inland brackish water sites (Angersdorf, DE) differed substantially from these values, exhibiting a median length/width ratio of 1.9 (Holzhausen et al., 2014). Similar differences between oospores from inland brackish water and marine habitats have been mentioned already by Hollerbach and Palamar-Mordwinzewa (1991).

Antheridia: Irrespective of the fact, that only a few sites with male individuals exist, many authors reported morphometrical data about antheridia (Table 23.3). Probably many of these data are taken from dried material or may refer to divergent species concepts (e.g., Wood, 1962, including the monoecious taxa excluded here).

Bulbils: For *Chara canescens*, no bulbils have been found yet.

Table 23.1 Oogonia morphometry

Author	Length	Width	Coronula length	Coronula width	Number of helical stripes
Migula (1889–1897)	550–850	360–550	80	180	11–13
Groves and Bullock-Webster (1924)	600–750	400–500	50–80	100–150	13–15
Corillion (1957)	600–750	400–500	50–80	100–150	13–15
Wood (1965)	550–850	278–550	50–110	100–220	13–15
Hollerbach and Krassavina (1983)	(450)600–750(840)	(340)400–500(570)	50–80(120)	100–150(200)	(11)13–15
Krause (1997)	550–800	300–450	50–80	100–200	11–15
Calero and Rodrigo (2018)	497 ± 67	296 ± 44	–	–	–
Total range	**(450)550–850**	**278–550(570)**	**50–110(120)**	**100–220**	**11–15**

Probably all of the data are from parthenogenic populations, there are no data explicitly traceable to sexual ones. Length in the first column is excluding the coronula; its length and width are given separately, all values for length and width in μm

Fig. 23.3 Oospores of *Chara canescens*. Upper row: Apex view (left), basal shield-cell (middle) and side view (right) of *C. canescens* oospores from a parthenogenetic population of the Baltic Sea. Scale bars 100 μm. Lower row: SEM photographs of an oospore from the lectotype material (Desvaux 1810 In aquis stag. circa Tolonum) left panel—side view, scale bar 100 μm; right panel—oospore-membrane, scale bar 10 μm. Photographs: upper row—Anja Holzhausen; lower row—Michelle T. Casanova

Table 23.2 Oospore morphometry

Author	Length	Width	Number of striae	Colour
Migula (1889–1897)	360–600	200–400	11–12	Black
Groves and Bullock-Webster (1924)	450–625	275–400	10–13	Black
Corillion (1957)	450–650	275–400	10–13	Black
Wood (1965)	(350)450–680	(225)275–400	9–19	Black
Hollerbach and Krassavina (1983)	450–625	275–400	10–13	Black
Krause (1997)	400–700	200–400	10–13	–
Vedder (2003)	425–600	275–425	9–13	Black
Niedrig (2010)	530–590	380–430	9–11	Black
Calero and Rodrigo (2018)	392 ± 34	219 ± 24	–	Black
Total range	**340–700**	**140–430**	**9–19**	**Black**

Probably all of the data are from parthenogenic populations, there are no data explicitly traceable to sexual ones. All values for length and width in μm

Table 23.3 Antheridia morphometry

Author	Diameter (μm)
Migula (1889–1897)	560–700
Groves and Bullock-Webster (1924)	600–650
Corillion (1957)	560–650 (700)
Wood (1965)	(240) 500–650 (700)
Hollerbach and Krassavina (1983)	(480) 600–650 (830)
Krause (1997)	500–700
Becker (pers. comm, n = 120, dried material)	(400) 500–800 (925)
Total range	**(240) 500–800 (925)**

23.2 Phenology and Reproduction

Chara canescens germinates from oospores in spring. Vegetative overwintering is rare, but possible (described for deep-water occurrences by Wahlstedt (1864) and Schubert et al. (2016) as well as for lab cultures and shallow-water populations by Calero and Rodrigo, 2018). Oospore formation and ripening can be very fast in southern Europe, where ripe oospores are found as early as mid-April (e.g., Neusiedler Lacken, Sardinia and Spain); more northern localities start growing in April and ripe oospores appear in late July (Schubert et al., 2016). *Chara canescens* reproduces by means of oospores; after ripening the plants decay. *Chara canescens* is regarded as an opportunist with low competitive ability. Beside ion anomalies and brackish conditions, both reducing the number of potential competitors, ice scraping has been shown to favor the species (Blindow et al., 2016; Idestam-Almquist, 1998).

23.3 Habitat Conditions and Ecology

Habitat type: *Chara canescens* is restricted to brackish water habitats. Apart from brackish inland waters (permanent as well as ephemeral ones), the species is found in sheltered coastal lagoons and permanent as well as temporary beach lakes.

Sediment: *Chara canescens* has been reported from a broad range of sediments from sandy to muddy but was also found on submerged peat. However, in a recent study of British inland sites, Lambert (2007) found for British inland sites *C. canescens* prefers sediments which are characterized by high shear stress values – whether this is consequence of (caused by, e.g., compaction of the sediment by rhizoids) or requirement for *C. canescens* stands is not yet clear. However, noticeable compact sediment texture was also observed at other inland occurrences (e.g., Neusiedler Lacken, remainders of Salziger See, Bodon Blanco).

Nutrient conditions: *Chara canescens* is mainly characterized as a heliophilous brackish-water species, exhibiting no clear preference for nutrient conditions. In the Baltic Sea the species is frequent also under highly eutrophic conditions. On the other hand, *C. canescens* also occurs in oligotrophic, groundwater fed lakes (e.g., Winter et al., 1987). In Austria (Neusiedler Lacken) the species is even able to penetrate through solid microbial mats, covering the sediment surface of highly eutrophic temporary ponds.

Water chemistry: *Chara canescens* is one of the few true brackish water species, which neither occurs in freshwater nor under fully marine conditions. Salinity of *C. canescens* habitats ranges between 1.5[1] and 22.4 psu (Blindow and Schubert, 2004, Becker pers. comm.), which is in good agreement with the results of laboratory investigations, where the species has been shown to manage turgor regulation up to a salinity of 20 (Winter and Kirst, 1991). At inland brackish sites the species withstands extreme ion anomalies. Such anomalies do not allow for conversion of conductivity readings to salinity, so salinity data from inland brackish water sites must be critically reviewed.

The species is able to tolerate a broad range of pH. At the southern Baltic Sea, pH fluctuates regularly between pH 8 und pH 9 with irregular extremes of pH 10 and pH 7 (MBLU-MV 1995). Interestingly, encrustations are rare irrespective of Ca^{2+}-concentrations between 100 and 150 mg L^{-1} (Friess, 2007; Nessim, 1980).

In contrast to all other Charophytes investigated, *C. canescens* was not found in negative interstitial redox conditions during an extensive survey of British inland brackish sites (Lambert, 2007). High interstitial Cu- as well as P- concentrations were shown to be negatively correlated with occurrence of *C. canescens* (Lambert, 2007).

Depth range: As a heliophilous species, *Chara canescens* prefers shallow water between 5 und 50 cm depth (Küster et al., 2004). However, the few diving surveys available reported well developed stands for depths down to at least 3 m (Schubert et al., 2016), which supports former reports of Lakowitz (1907), Stroede (1931) and Olsen (1944) who found the species in dredged material down to 5 m.

Associated species: Large monospecific dense stands of *Chara canescens* (Fig. 23.4) are reported from just a few places in Europe, namely Neusiedler Lacken and Angersdorf. Typically, the species is associated with a number of halotolerant species, in many cases growing just in patches or individuals between them. In shallow bights and lagoons of the Baltic Sea the most common associated species are *C. baltica*, *C. aspera*, *Tolypella nidifica*, *Ruppia cirrhosa*, *R. maritima* and *Najas marina* (Blindow and Schubert, 2004). At higher salinities, *Lamprothamnium papulosum* is commonly found together with *C. canescens* in northern as well as southern Europe; in the Mediterranean area *C. galioides* often co-occurs with the species.

---

[1] Becker (pers. comm.) reported about an exceptional case where a stand was found in a dammed river mouth on Sardinia at 0.2 psu actual salinity but no information is available about salinity variation at this site.

Fig. 23.4 Habitat appearance of *Chara canescens*. Dense, monospecific stand of an inland brackish water site (Angersdorf, left panel) and scattered individuals at a sheltered brackish water site (Orther Reede, right panel). Photographs: left panel—Heiko Korsch, right panel—Klaus van de Weyer

For Sardinia, Becker (pers. comm.) often observed *Tolypella hispanica* as an accompanying species; rarely *T. nidifica*, *T. salina* and *Riella helicophylla* co-occurred as well. For inland brackish water sites, no distinct pattern has been observed yet. *Chara papillosa* has been frequently found growing together with *C. canescens*, as well as *C. vulgaris* and *C. filiformis* (Korte et al., 2010).

23.4 Variability

For *Chara canescens* a number of subspecies, varieties and forms have been described (see, e.g., Filarszky, 1935; Blindow and Schubert, 2004). However, the existence of a parthenogenetic and a sexually reproducing morphotype is the most striking feature of this species. In brackish water, especially in the central parts of the Baltic Sea, plants are usually stout and green, whereas plants from inland brackish water sites as well as high-saline brackish water are pale green to greyish (Fig. 23.2). Cortex cells of such plants are often colorless with contrasting green spine-cells.

23.5 Identification and Risk of Confusion

As the only haplostichous dioecious species in Europe, *Chara canescens* should be easy to recognise. However, irregular cortex has led to misidentifications as *Chara aspera* (which is triplostichous and has round bulbils). Both species are known for irregular cortex, but for *C. canescens* irregular cortex is restricted to the lower internodes, so concentrating on the upper ones should help. An extreme case appeared at Svalbard (Spitsbergen) where an ecorticated specimen of *C. aspera* resulted in some doubts about delineation (Langangen, 1979, 2000). At this site (Troll Springs), *C.*

canescens and *C. aspera* co-exist, and the haplostichous cortex, together with the lack of bulbils allows for reliable discrimination (Langangen et al., 2000). However, there are examples of ecorticated specimens of *C. canescens* as well (e.g., at Pantelleria, Sicily, pers. comm. Mauro Iberite; Baltic Sea, collections in LD), which may be confused with, e.g., *Lamprothamnium papulosum*, which often shares the habitat with *C. canescens*. In such cases the contracted heads of *L. papulosum* may help immediately; taking a magnifying glass, the haplostephanous character and the round bulbils of *L. papulosum* will help to distinguish between them. Another species *C. canescens* might be confused with is *C. altaica*. Being haplostichous (with a tendency to develop secondary rows) as well, the only reliable character for discrimination is sexuality. *C. altaica* is a monoecious species, which should allow for unambiguous determination, and it is recommended to carefully check haplostichous specimens, especially when gathered in South-East Europe. If you find a monoecious, haplostichous specimen, antheridia must be also be examined because *C. altaica* can be distinguished from other monoecious members of *Desvauxia* (*C. evoluta*, *C. hirsuta*—both described from North America) only by having octoscutate antheridia (as it is the case for *C. canescens*); the North American relatives are described to be tetrascutate species (Wood and Imahori, 1964). A special case is *C. pseudocanescens* (Yuanjie et al., 1991), which is distinct from all other members of *Desvauxia* by having gametangia on the basal branchlet node (see also Chap. 6, this volume), a feature usually not noted for by determination of *Chara* spp. Because no material or publication about this taxon, except for its original description, is known to the authors attention should be paid to this feature.

23.6 Distribution

See Fig. 23.5.

Chara canescens is neither found in freshwater nor in fully marine water bodies. In the Baltic Sea it is widespread except for the northernmost part of the Bothnian Bay (Norrbotten). Its global distribution range is hard to estimate because of many misidentifications as well as differences between the taxonomic concepts.

23.7 Status and Threats

Most of the marine-brackish stands are reported to be stable over the last decade, sites lost over the period of heavy eutrophication (1960s to late 1980s) have been reoccupied since the mid-1990s (Blindow and Schubert, 2004; Schubert et al., 2016). A decline in depth range has, however, been reported from Baltic Sea lagoons probably caused by eutrophication (Blindow et al., 2016). However, some inland brackish water sites (e.g., Röblingen/Wansleben) have been lost; several lagoons and beach lakes have been drained or turned to freshwater habitats by diking (e.g., Becker,

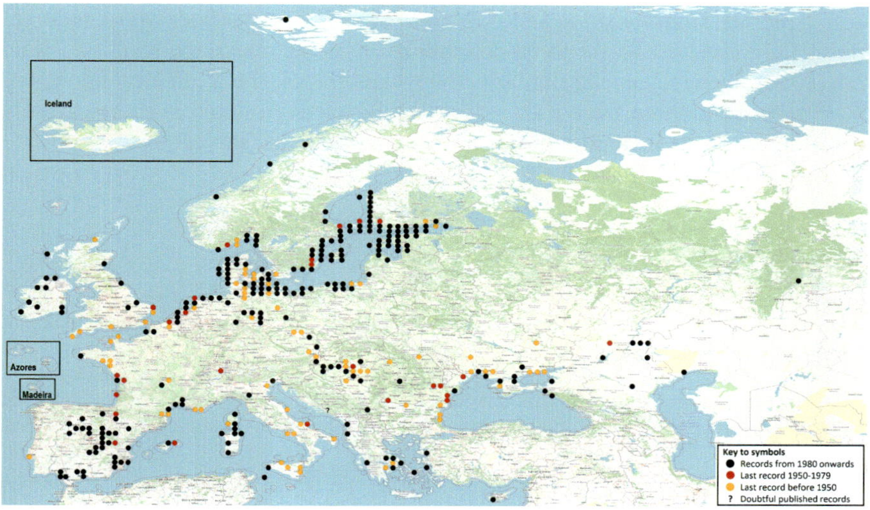

Fig. 23.5 European distribution of *Chara canescens*. For worldwide distribution see Korsch (2018). Black dots are records from 1980 on, red dots from 1950 to 1979 and yellow dots before 1950. Map prepared by Heiko Korsch

2019). On the other hand, intense mapping revealed occurrences not known before, so general trends especially for inland sites are hard to detect.

Habitat destruction, however, seems to be the most important threat. Eutrophication probably is an indirect threat, acting mainly via competition for light with phytoplankton and filamentous algae. Reduced frequency of ice-winters due to global warming is assumed to threaten the species in Baltic Sea coastal lagoons (Blindow et al., 2016).

23.8 Remarks

Wood combined monoecious and dioecious "strains" into one species (Wood, 1962). The authors of this book do not follow this species concept and consider *Chara canescens* to be exclusively dioecious.

Being the only haplostichous species mentioned in standard monographs which apply the microspecies-concept, field botanists might not be aware of other haplostichous species which may appear in Europe. In the macrospecies concept all haplostichous microspecies are lumped to *C. canescens* anyway. Special attention should be paid on the occurrence of bisexual populations because other haplostichous species which may occur in Europe (e.g., *C. hirsuta*, *C.* ***altaica***, *C. canescentiformis*) haven't been recognised yet, except for the East-European countries where they appeared in the determination keys. Two additional haplostichous, monoecious species from

North America, *Chara evoluta* and *C. hirsuta* have been described, whose relationship to European taxa is yet to be determined. In this context, the results of Kato et al. (2010), showing genetic similarity (by means of *rbc*L) between the monoecious *Chara altaica* and some, but not all, samples of dioecious *C. canescens* are extremely interesting and is an invitation for for further studies on the subject.

References

Becker, R. (2019). The Characeae (Charales, Charophyceae) of Sardinia (Italy): Habitats, distribution and conservation. *Webbia, 74*, 83–101.

Blindow, I., Dahlke, S., Dewart, A., Flügge, S., Hendreschke, M., Kerkow, A., & Meyer, J. (2016). Long-term and interannual changes of submerged macrophytes and their associated diaspore reservoir in a shallow southern Baltic Sea bay: Influence of eutrophication and climate. *Hydrobiologia, 778*, 121–136.

Blindow, I., & Schubert, H. (2004). *Chara canescens*. In H. Schubert & I. Blindow (Eds.), *Charophytes of the Baltic Sea* (pp. 70–81). A. R. G. Gantner, Ruggell.

Braun, A., & Nordstedt, C. F. O. (1882). Fragmente einer Monographie der Characeen. Nach den hinterlassen Manuscripten A. Braun's herausgegeben von Dr. Otto Nordstedt. *Abhandlungen der Königlichen Akademie der Wissenschaften zu Berlin für das, 1882*, 1–211. [in German].

Calero, S., & Rodrigo, M. A. (2018). The life cycle of a parthenogenetic population of *Chara canescens* from an interdunal Mediterranean pond. *Botany Letters, 165*, 1–11.

Cirujano, S., García Murillo, P., Meco, A., & Fernández Zamudio, R. (2007). Los carófitos ibéricos. *Anales Del Jardín Botánico De Madrid, 64*, 87–102. [in Spanish].

Corillion, R. (1957). Les Charophycées de France et d'Europe Occidentale. *Bulletin de la Société Scientifique de Bretagne, 32*(Hors série I & II), 1–499 [in French].

Filarszky, N. (1935). Monographia Characearum enumeratione formarum in Hungaria ad huc observatorum. *Matematikai És Természettudományi Értesitö, 52*, 459–469.

Friess, R. (2007). *Wechselwirkung von Osmoregulation und CO_2-Aufnahme bei Characeen.* Diploma thesis, Universität Rostock, Rostock [in German].

Giesenhagen, K. (1898). Untersuchungen über die Characeen. II. der Bau Der Sprossknoten. *Flora, 85*, 19–64 [in German].

Griffith, W. (1849). P*osthumous papers bequeathed to the honorable the East India company in Notulae ad plantas asiaticas*. 4(II).

Groves, J., & Bullock-Webster, G. R. (1924). *The British Charophyta. Volume II. Chareæ with plates, concluding articles, geological sketch, bibliography and index*. Ray Society.

Hollerbach, M. M., & Palamar-Mordvintseva, G. M. [Голлербах, М. М. & Г. М.Паламар-Мордвинцева], 1991. *Визначник прісноводних водоростей України. IX. Харові водорості (Charophyta) [The identification manual of freshwater algae of Ukraine: IX. Charophytes (Charophyta)*]. Nauk. dumka [in Ukrainian].

Hollerbach, M. M., & Krassavina, L. K. [Голлербах, М. М. & Л. К. Красавина]. (1983). *Определитель пресноводных водорослей СССР. 14. Charophyta [Determination key of freshwater algae of USSR]* (189 pp.). Nauka [in Russian].

Holzhausen, A., Nowak, P., Niedrig, C., Feike, M., & Schubert, H. (2014). Morphometry of *Chara aspera, C. canescens, C. baltica* var. *baltica, C. baltica* var. *liljebladii* and *C. intermedia* oospores: Local variation versus taxonomic differences. *Aquatic Botany, 120*, 60–66.

Idestam-Almquist, J. (1998). *Temporal and spatial variation of submersed aquatic plants in the Baltic Sea*. Ph. D. thesis. Department of Botany, Stockholms University, Stockholm.216 pp.

Kato S., Sakayama, H., Morishima, H., Sano, S., Oomori, Y., Kato, N., Ito, M., Kasai, F., Watanabe, M., & Nozaki, H. (2010). Morphology and molecular phylogeny of *Chara altaica* (Charales, Charophyceae), a monoecious species of the section Desvauxia. *Cytologia, 75*, 211–220.

Korsch, H. (2018). The worldwide range of the Charophyte species native to Germany. *Rostocker Meeresbiologische Beiträge, 28*, 45–96.

Korte, E., Krautkrämer, V., Pätzold, F., & van de Weyer, K. (2010). Brackwasser-Armleuchteralge (*Chara canescens*) und Faden-Armleuchteralge (*Chara filiformis*) im Borkener See. Zwei bemerkenswerte Neunachweise für Hessen. *Botanik Und Naturschutz in Hessen, 23*, 55–62. [in German].

Krause, W. (1997). Charales (Charophyceae) Süßwasserflora von Mitteleuropa. In H. Ettl, G. Gärtner, H. Heynig, & D. Mollenhauer (Eds.), *Süßwasserflora von Mitteleuropa 18*. Fischer [in German].

Küster, A., Schaible, R., & Schubert, H. (2004). Light acclimation of photosynthesis in three charophyte species. *Aquatic Botany, 79*, 111–124.

Lakowitz, K. (1907). *Die Algenflora der Danziger Bucht. Ein Beitrag zur Kenntnis der Ostseeflora*. W. Engelmann [in German].

Lambert, S. J. (2007). *The environmental range and tolerance limits of British stoneworts (Charophytes)*. Ph.D. Thesis, University of East Anglia, Norwich, Great Britain.

Langangen, A. (1979). *Chara canescens* reported from Spitzbergen. *Phycologia, 18*, 436–437.

Langangen, A. (2000). Charophytes from the warm springs of Svalbard. *Polar Research, 19*, 143–153.

Migula, W. (1889–1897). Fünfter Band: Die Characeen. In A. Fischer, E. Fischer, F. Hauck, G. Limpricht, C. Luerssen, W. Migula, H. Rehm, P. Richter, & G. Winter (Eds.), *Dr. L. Rabenhorst's Kryptogamenflora von Deutschland, Oesterreich und der Schweiz* (2nd ed.). Eduard Kummer [in German].

Nessim, R. B. (1980). *Untersuchungen zur Verteilung der Hauptkomponenten des Salzgehaltes im Wasser und Sediment der Darss-Zingster Boddengewässer unter besonderer Berücksichtigung der Ionenanomalie sowie erste Erhebungen über den Schwermetallgehalt*. PhD thesis, Universität Rostock [in German].

Niedrig, C. (2010). *Untersuchungen zur Übereinstimmung zwischen rezenter Characeenvegetation und Diasporenbank in Flachgewässern—ein Beitrag zur Gewässergütebewertung*. MSc thesis, Universität Rostock, Rostock [in German].

Nowak, P., & Schubert, H. (2019). Genetic variability of charophyte algae in the Baltic Sea area. *Botanica Marina, 62*, 75–82.

Nowak, P., van de Weyer, K., & Becker, R. (2019). The occurrence of sexual *Chara canescens* in Sardinia. *Webbia, 74*, 103–109.

Olsen, S. (1944). Danish Charophyta. *Det Kongelige Danske Videnskabernes Selskab, Biologiske Skrifter, 3*(1), 1–240, pl. 1–2 [in Danish].

Romanov, R. E. (2021). Typification of *Chara* (Charales, Charophyceae) species described by C.F. Lessing, F.J. Ruprecht and M.M. *Hollerbach. Notulae Algarum, 213*, 1–7.

Romanov, R. E. (2022). Typification of *Chara connivens* Salzmann ex A.Braun f. *brevifolia* Vilhelm, *C. crinita* Wallroth and *C. glomerulifera* Ruprecht and clarification of conflicting species concepts for *C. arcuatifolia* Vilhelm (Charales, Charophyceae). *Notulae Algarum, 223*, 1–4.

Sabovljević, M. S., Tomović, G., Pantović, J. P., Djurović, S. Z., Buzurović, U., Denchev, T. T., Denchev, C. M., Boycheva, P., Dimitrova, T., Marković, A., Sabovljević, A. D., Stefănu, S., Bîrsan, C. C., Šabanović, E., Djordjević, V., Niketić, M., Šovran, S., Mašić, E., Stoykov, D., et al. (2022). New records and noteworthy data of plants, algae and fungi in SE Europe and adjacent regions, *Botanica Serbica ,46*.

Schaible, R., & Schubert, H. (2008). The occurrence of sexual *Chara canescens* populations (Charophyceae) is not related to ecophysiological potentials with respect to salinity and irradiance. *European Journal of Phycology, 43*, 309–316

Schaible, R., Bergmann, I., & Schubert, H. (2009). A survey of sexually reproducing female and male populations of *Chara canescens* (Charophyta) in the National Park Neusiedler See-Seewinkel (Austria). *Cryptogamie, Algologie, 30*, 279–294.

Schubert, H., Blindow, I., & van de Weyer, K. (2016). *Chara canescens*. In Arbeitsgruppe Characeen Deutschlands (Eds.), *Armleuchteralgen: Die Characeen Deutschlands* (pp. 235–262). Springer [in German].

Stroede, W. (1931). *Ökologie der Characeen* (119 pp.). Dissertation Friedrichs-Wilhelm-Universität zu Berlin [in German].

Troia, A. (2020). Homage to Proserpina, or: why did the charophytes of the Pergusa lake vanish? In T. La Mantia, E. Badalamenti, A. Carapezza, P. Lo Cascio, & A. Troia (Eds.), *Life on islands. 1. Biodiversity in Sicily and surrounding islands. Studies dedicated to Bruno Massa* (pp. 47–51). Edizioni Danaus.

Vedder, F. (2003). *Morphologie und Taxonomie rezenter und subfossiler Characeen-Oosporen aus der Ostsee*. Diplomarbeit, Ernst-Moritz-Arndt Universität Greifswald [in German].

Wahlstedt, L. J. (1864). *Om characeernas knoppar och öfvervintring. Botanisk avhandling*. University of Lund [in Swedish].

Winter, U., & Kirst, G. O. (1991). Partial turgor pressure regulation in *Chara canescens* and its implications for a generalized hypothesis of salinity response in charophytes. *Botanica Acta, 104*, 37–46.

Winter, U., Meyer, M. I. B., & Kirst, G. O. (1987). Seasonal changes of ionic concentrations in the vacuolar sap of *Chara vulgaris* L. growing in a brackish water lake. *Oecologia, 74*, 122–127.

Wood, R. D. (1962). New combinations and taxa in the revision of Characeae. *Taxon, 11*, 7–25.

Wood, R. D. (1965). *Monograph of the Characeae*. J. Cramer.

Wood, R. D., & Imahori, K. (1964). *Iconograph of the Characeae*. J. Cramer.

Yuanjie, L., Shulian, X., & Lichuan, Q. (1991). 山西轮藻属新植物 [Some new taxa of Chara from Shanxi]. *Acta Hydrobiologica Sinica, 15*, 350–355. [in Chinese].

Chapter 24
Chara connivens

Kaire Torn and Emile Nat

Scientific name: *Chara connivens* Salzm. ex A. Braun, Flora 18: 73. 1835

Neotype (Wood, 1965: 177): Tanger/[no date]/Salzmann (PC 577722).

= *Chara arcuatifolia* Vilh. in Spisy Prír. Fak. Karlovy Univ. 80: 23. 1928 "arcuatofolia"
Lectotype: (Wood, 1965: 179): Czelkar, in fl. Kauldzhur/[no date]/N. Androssow 77 (LE).

Local names: DANISH: Glat Kransnål; DUTCH: Gebogen kransblad; ENGLISH: Convergent stonewort; ESTONIAN: Sile mändvetikas; FINNISH: Suppunäkinparta; GERMAN: Gebogene Armleuchteralge; LITHUANIAN: Glaustašakis maurabragis; POLISH: Ramienica zagięta, RUSSIAN: Хара сходящаяся; SWEDISH: Tuvsträfse; UKRAINIAN: Хара збігаюча.

24.1 Morphology

Habitus: Plants are bright green, sometimes yellowish or dark green and usually not encrusted (Fig. 24.1). Branchlets of male plants are often curved inward (Fig. 24.2a). Plants can be compact and stiff (shallow water, Fig. 24.3a) or slender with long internodes (deeper water, Fig. 24.3b, c).

K. Torn (✉)
Estonian Marine Institute, University of Tartu, Mäealuse 14, 12618 Tallinn, Estonia
e-mail: kaire.torn@ut.ee

E. Nat
Dutch Information Center on Charophytes, Naturalis Biodiversity Center, P.O. Box 9517, 2300 RA Leiden, The Netherlands

H. Schubert et al. (eds.), *Charophytes of Europe*,
https://doi.org/10.1007/978-3-031-31898-6_24

Fig. 24.1 Habitus appearance of *Chara connivens*. *Chara connivens* in the Baltic Sea. Photograph: Kaire Kaljurand, Kõiguste Bay, 2013

Fig. 24.2 Details of *Chara connivens*, photos. **a** Incurved branchlets of male *C. connivens*, **b** Bulbils, **c** Branchlet with antheridia, **d** Branchlet with oogonia, **e** Cortex without spine-cells (nodal cells of cortex are visible), **f** Stipulode rows. Photographs: **a**—Heiko Korsch; **b**—Klaus van de Weyer; **c**, **e**, **f**—Anastasiia Kovtun-Kante; **d**—Kaire Torn

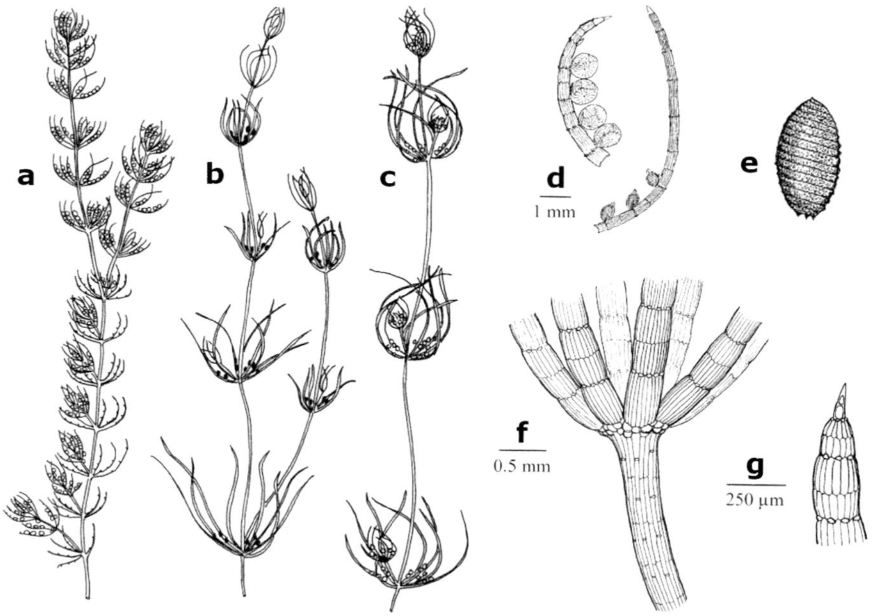

Fig. 24.3 Details of *Chara connivens*, drawings. **a** Habitus of male plant from shallow conditions, **b** Female plant from deep water, **c** Male plant from deep water, **d** Fertile branchlets of male (left) and female (right) plants, **e** Oospore, **f** Cortex with node and stipulodes, **g** Tip of branchlet. Drawings: **a**—Klaus van de Weyer; **b–e**—Werner Krause; **f**, **g**—Mindaugas Ryla

Size: The plants are small to medium-sized (5–20 cm) but may reach up to 50 cm if growing in deeper water.

Main axis: The shoot diameter is 0.3–0.6 (1.2) mm, but based on Urbaniak and Gąbka (2014) can reach up to 1.4 mm. The internodes are generally longer than the branches, 0.3–2 (5.3) cm in length, in the upper whorls mainly shorter or of equal length.

Cortex: The cortex is triplostichous and isostichous, the surface smooth and finely striped. The cortex can be slightly twisted (Figs. 24.2e and 24.3f).

Spine-cells: Spine-cells are lacking or rudimentary. Cortex nodal cells are visible or solitary rudimentary cells are present (Fig. 24.2e).

Stipulodes: Stipulodes are small and globular, arranged in two rows, but often rudimentary and form irregular rows (Figs. 24.2f and 24.3f).

Branchlets: Predominantly 6–9 (up to 10) branchlets are arranged in one whorl. Each branchlet has 8–10 segments of which the upper 1–2 short segments are ecorticated. The end cell is narrow, short and pointed. The branchlets of the male plants are mainly curved inward (Figs. 24.2c–d and 24.3d–f).

Bract-cells: Rudimentary papillous cells are present.

Bracteoles: Elongated and slender. On male plants bracteoles are mainly shorter than ripe antheridia, rarely of the same length. On female plants they are shorter than or as long as the oogonia.

Bractlets: Elongate, shorter than or as long as the oogonia (Fig. 24.2d).

Sexuality: The species is dioecious. Gametangia solitary, only developed on the lowest 1–4 (rarely 5) segments of the branchlets (Figs. 24.2c, d and 24.3d).

Oogonia: Solitary, coronula strongly convergent. Table 24.1 gives an overview about morphometrical data available (see also Fig. 24.2d).

Oospores: Oospores are slender, dark brown to black when mature (Fig. 24.3e). Membrane of the oospore is granular (Corillion, 1957; Groves and Bullock-Webster, 1924; Urbaniak and Blaženčić, 2012). Table 24.2 gives an overview about morphometrical data.

Antheridia: Antheridia are inconspicuous; maximum diameter 1100 μm. Colour is orange red to brownish red, located relatively close to each other (Figs. 24.2c and 24.3d). Table 24.2 gives an overview about morphometrical data available Table 24.3.

Bulbils: According to most authors bulbils are missing. Based on German, Danish and Swedish material multicellular bulbils may be present (Becker, 2016; Blindow, 2008; Schou et al., 2017; see Fig. 24.2b).

24.2 Phenology and Reproduction

Chara connivens is mostly annual, data about germination are limited. Fertile plants were found from June to September (Becker, 2016; Mouronval et al., 2015; Torn and Martin, 2004). The plants in the Mediterranean Basin can begin to fructify in March (R. Becker own data), then die off in temporary waters as they dehydrate in summer and germinate again after the rainfall (Becker, 2016).

24.3 Habitat Conditions and Ecology

Habitat type: *Chara connivens* has been found in lakes, ditches, temporary ponds, swamps, artificial irrigation systems, as well as in the brackish Baltic Sea and its associated lagoons and coastal lakes and in the Mediterranean area up to Israel especially along the coast.

Sediment: Grows on sand, mud, silty sand, muddy sand, muddy clay, loamy conditions, peat bogs and in the Baltic Sea also on mixtures of sand with pebbles.

Table 24.1 Oogonia morphometry

Author	Length	Width	Coronula length	Coronula width	Number of helical stripes
Migula (1889–1897)	850–1150*	320–550	–	–	–
Groves and Bullock-Webster (1924)	750	400	200	185	13–14
Corillion (1957)	650–900	320–550	200	185	13–14
Wood (1965)	675–750	330–400	200–240	150–180	13–15
Moore (1986)	–	< 500	–	–	–
Compère (1992)	600–750	300–400	–	–	–
Krause (1997)	650–750	330–400	200–240	160	14–15
Torn and Martin (2003)	650–1100*	320–550	–	–	13–15
Langangen (2007)	< 1100*	–	–	–	–
Cirujano et al. (2008)	675–800	330–400	150–240	–	–
Ahmadi et al. (2012)	850–930**	450–520	130–200	180–200	14–15
Urbaniak and Gąbka (2014)	605–775	330–410	–	–	–
Mouronval et al. (2015)	–	–	200–240	–	–
Becker (2016)	625–800	375–400	125–200	–	11
Schou et al., (2017)	500–1150*	150–550	–	–	–
Total range	**600–900(1150*)**	**(150)300–550**	**125–240**	**150–200**	**11–15**

* Uncertain whether length is without coronula, ** with coronula
Length in the first column excludes the coronula except for values marked by asterisks; coronula length and width are given separately, all values for length and width in μm.

Nutrient conditions: *Chara connivens* is found from oligotrophic to eutrophic conditions. It is reported to tolerate high concentrations of nutrients (Becker, 2016). It is found at total phosphorus concentrations up to 6.9 mg/l, at total nitrogen concentrations up to 4.7 mg/l and showed positive correlation with orthophosphates (Del Pozo et al., 2011).

Water chemistry: *Chara connivens* can occur in fresh calcareous and brackish water. In the Netherlands, *C. connivens* was detected at chloride concentrations between 0 and 1070 mg/l (Simons and Nat, 1996), the optimum growth of the species is at a chloride concentration of 124 mg/l (Simons et al., 1994). In the Baltic Sea the species is more frequent at salinity 2–6 psu, but can be found in salinity conditions up to

Table 24.2 Oospore morphometry

Author	Length	Width	Number of striae	Colour
Migula (1889–1897), Corillion (1957)	580–700	240–350	12–13	Black
Groves and Bullock-Webster (1924)	500–600	250–300	12–13	Dark brown to black
Wood (1965)	500–700	240–350	11–13	Dark brown to black
Krause (1997)	500–700	250–350	12–14	Dark brown to black
van Raam (1998)	< 650	< 300	–	–
Torn and Martin (2004)	500–700	240–350	11–14	Dark brown to black
Cirujano et al. (2008)	500–650	250–350	–	Dark brown to black
Ahmadi et al. (2012)	630–700	350–430	12–13	Brown to black
Urbaniak and Gąbka (2014)	485–595	205–325	–	Dark brown to black
Becker (2016)	500–550	300–350	–	Black
Schou et al. (2017)	490–700	210–350	12–14	Dark brown
Total range	**485–700**	**205–430**	**11–14**	

All values for length and width in μm

Table 24.3 Antheridia morphometry

Author	Diameter (μm)
Migula (1889–1897)	800–1100
Groves and Bullock-Webster (1924), Compère (1992)	600–700
Corillion (1957)	600–1100
Wood (1965)	500–1100
Moore (1986), van Raam (1998), Langangen (2007)	< 1000
Krause (1997)	600–1000
Torn and Martin (2004)	600–1100
Cirujano et al. (2008)	700–1000
Ahmadi et al. (2012)	720–930
Urbaniak and Gąbka (2014)	650–1000
Becker (2016)	(450)500–1000
Schou et al. (2017)	(450)500–1100
Total range	**(450)500–1100**

9 psu (Langangen, 2007; Mouronval et al., 2015). Based on the overview in Doege et al. (2016), the range of water chemistry is: pH value 7–9, electrical conductivity 34–3070 μS/cm, carbonate hardness 4 °dH, general hardness 4.5 °dH, Ca^{2+} 16.8–180 mg/l, and Cl^- 50–1070 mg/l.

Depth range: The species is more often found below 2 m water depth but may be present up to 4 m depth (Torn and Martin, 2004). In Germany, the Netherlands and in Sardinia *Chara connivens* is often found at depths below 1 m.

Associated species: A large range of associated species is listed in literature. As accompanying species are among others mentioned: *Chara aspera*, *C. globularis*, *C. hispida*, *C. tomentosa*, *C. vulgaris*, *C. virgata*, *Myriophyllum spicatum*, *Stuckenia pectinata*, *Najas marina* (Becker, 2016 and references therein). *Chara connivens* also occurs in monospecific stands.

24.4 Variability

Variability of the species is minor. Plants from deeper water or more dense communities are thin and slender with relatively long internodes and do not hold shape. Branchlets of sterile specimens do not curve inward, also some fertile specimens may have straight branchlets.

24.5 Identification and Risk of Confusion

Chara connivens cannot be differentiated from *C. globularis* when sterile. The differences are also minor with *C. virgata* if it is sterile and when the stipulodes of *C. virgata* are undeveloped. The spine-cells of *C. virgata* are larger compared to missing or rudimentary spine-cells of *C. connivens*. From spineless *C. aspera* or *C. aspera* with papilliformous spine-cells it can be differentiated by acute tips of stipulodes of *C. aspera* (Nielsen, 2004a). Dioecious *C. fragifera* can be distinguished from *C. connivens* by the characteristic multicellular bulbils of *C. fragifera* (Nielsen, 2004b).

24.6 Distribution

See Fig. 24.4.

The main distribution of *Chara connivens* in Europe is in Spain. Outside Europe, it has been detected along the Southern Mediterranean coast (Morocco, Algeria, Tunisia, Egypt, Syria, Israel), and Asia (e.g., Iran, Russia, India) (Ahmadi et al., 2012; Becker, 2016; Krause, 1997; Rybak and Woyda-Ploszczyca, 2019).

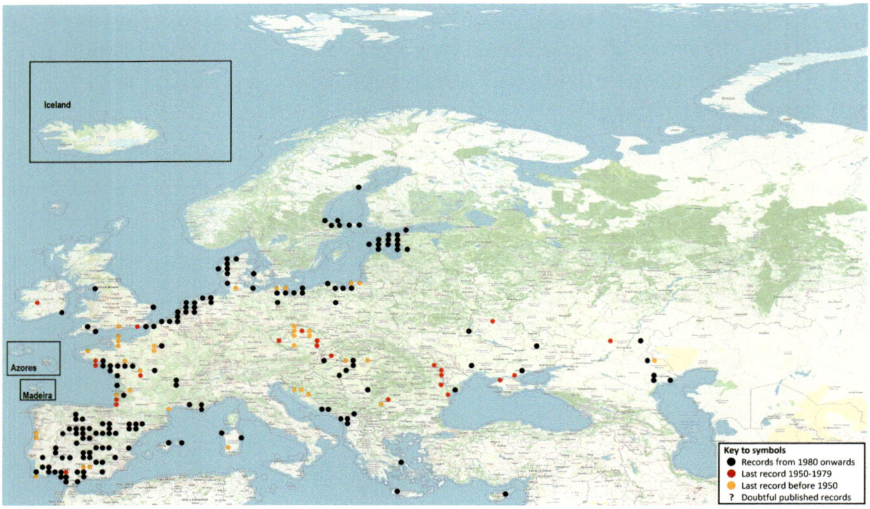

Fig. 24.4 Distribution of *Chara connivens* in Europe. For worldwide distribution see Korsch (2018). Black dots are records from 1980 on, red dots from 1950 to 1979 and yellow dots before 1950; unconfirmed/doubtful published records are marked with "?" on top. Map prepared by Heiko Korsch

24.7 Status and Threats

The species is extinct in several localities (Becker, 2016) where it had been described by Migula (1889–1897). *Chara connivens* was included in the previous HELCOM list of threatened and/or declining species. The species is not considered threatened in the Baltic Sea at the moment. In the Estonian waters *C. connivens* has become more common in recent decades and the species is rather frequent also along the Uppland coast in Sweden (HELCOM, 2013) with probably stable occurrence. One of the *Chara* collections in the Linnaean herbarium is *Chara connivens* (Gregor et al., 2019). In Germany it is a rare species in the past and at present. In the Netherlands it is rare at the Wadden Islands and the dune area of the North Sea and has more or less stable vegetations in peat puddles. *Chara connivens* is very rare in Italy and in other parts of the Mediterranean area (Becker, 2019) except the Iberian Peninsula, where it is regarded as rather common (Cirujano et al., 2008).

24.8 Remarks

The distribution vector of the species into the Baltic Sea is considered controversial. The species is assumed to have been introduced into the Baltic Sea by shipping (Leppäkoski and Olenin, 2000; Luther, 1979; Paavola et al., 2005). However, a natural dispersal by waterfowl is rather probable (Blindow, 2009). In Finland, the species has been considered unsuitable for IUCN Red List assessment as an introduced species (Hyvärinen et al., 2019). Although, HELCOM (HELCOM, 2013), Sweden (SLU Artdatabanken, 2020) and Estonia (eElurikkus, 2021) have assessed the threats to the species. Trbojević et al. (2020) reported *Chara connivens* from Serbia, but it did not cluster with other *C. connivens*, but more closely with *C. globularis*.

References

Ahmadi, A., Riahi, H., Sheidai, M., & van Raam, J. (2012). Some Charophytes (Characeae, Charophyta) from central and western of Iran Including *Chara kohrangiana* species nova. *Cryptogamie, Algologie, 33*, 359–390.

Becker, R. (2016). *Chara connivens*. In Arbeitsgruppe Characeen Deutschlands (Eds.), *Armleuchteralgen: Die Characeen Deutschlands* (pp. 270–282). Springer.

Becker, R. (2019). The Characeae (Charales, Charophyceae) of Sardinia (Italy): Habitats, distribution and conservation. *Webbia, 74*, 83–101.

Blindow, I. (2008). Schwedische Artenschutzprogramme für bedrohte Characeen. *Rostocker Meeresbiologische Beiträge, 19*, 23–28.

Blindow, I. (2009). *Åtgärdsprogram för hotade kransalger: tuvsträfse och barklöst sträfse 2008–2011. Tuvsträfse (Chara connivens). Barklöst sträfse (Chara braunii).* Naturvårdsverket.

Cirujano, S., Cambra, J., Sánchez Castillo, P. M., Meco, A., & Flor Arnau, N. (2008). *Flora ibérica. Algas continentales. Carófitos (Characeae).* Real Jardín Botánico.

Compère, P. (1992). Charophytes. In *Flore pratique des algues d'eau douce de Belgique*. Jardin botanique national de Belgique.

Corillion, R. (1957). Les Charophycées de France et d'Europe Occidentale. *Bulletin de la Société Scientifique de Bretagne, 32*(Hors série I & II), 1–499.

Del Pozo, R., Fernández-Aláez, C., Fernández-Aláez, M., & Santiago, N.-F. (2011). Assessment of eutrophication effects on charophytes in Mediterranean ponds (North-Western Spain). *Fundamental and Applied Limnology, 178*, 257–264. https://doi.org/10.1127/1863-9135/2011/0178-0257

Doege, A., van de Weyer, K., Becker, R., & Schubert, H. (2016). Bioindikationmit Characeen. In Arbeitsgruppe Characeen Deutschlands (Eds.), *Armleuchteralgen: Die Characeen Deutschlands* (pp. 97–138). Springer.

eElurikkus. (2021). *Chara connivens* Salzm.ex A.Braun. Red List assessment. https://elurikkus.ee/iehub/species/3630?lang=en#redlist

Gregor, T., Blindow, I., Casanova, M. T., & Schubert, H. (2019). (2721) Proposal to conserve the name *Chara flexilis* (*Nitella flexilis*) (Characeae) with a conserved type. *Taxon, 68*, 1363–1371. https://doi-org.ezproxy.utlib.ut.ee/10.1002/tax.12165

Groves, J., & Bullock-Webster, G. R. (1924). *The British Charophyta. Volume II. Chareæ with plates, concluding articles, geological sketch, bibliography and index*. Ray Society.

HELCOM. (2013). *Red list of Baltic Sea species in danger of becoming extinct* (No. 140). Baltic Sea Environmental Proceedings. https://helcom.fi/post_type_publ/bsep140/

Hyvärinen, E., Juslén, A., Kemppainen, E., Uddström, A., & Liukko, U.-M. (Eds.) (2019). *Suomen lajien uhanalaisuus. Punainen kirja 2019.* Ympäristöministeriö & Suomen ympäristökeskus. http://hdl.handle.net/10138/299501

Korsch, H. (2018). The worldwide range of the Charophyte species native to Germany. *Rostocker Meeresbiologische Beiträge, 28*, 45–96. https://www.oekologie.uni-rostock.de/storages/uni-rostock/Alle_MNF/Bio_Oekologie/RMB/RMB28/RMB28_04_Korsch.pdf

Krause, W. (1997). Charales (Charophyceae). In H. Ettl, G. Gärtner, H. Heynig & D. Mollenhauer (Eds.), *Süßwasserflora von Mitteleuropa*. Fischer.
Langangen, A. (2007). *Charophytes of the Nordic countries*. Saeculum ANS.
Leppäkoski, E., & Olenin, S. (2000). Non-native species and rates of spread: Lessons from the brackish Baltic Sea. *Biological Invasions, 2*, 151–163.
Luther, H. (1979). *Chara connivens* in the Baltic Sea area. *Annales Botanici Fennici, 16*, 141–150.
Migula, W. (1889–1897). Fünfter band: Die Characeen. In: A. Fischer, E. Fischer, F. Hauck, G. Limpricht, C. Luerssen, W. Migula, H. Rehm, P. Richter & G. Winter (Eds.), *Dr. L. Rabenhorst's Kryptogamenflora von Deutschland, Oesterreich und der Schweiz* (ed. 2). Eduard Kummer.
Moore, J. A. (1986). *Charophytes of Great Britain and Ireland*. BSBI Handbook No. 5. Botanical Society of the British Isles.
Mouronval, J. B., Baudouin, S., Borel, N., Soulié-Märsche, I., Klesczewski, M., & Grillas, P. (2015). *Guide des Characées de France méditerranéenne*. Office national de la Chasse et de la Faune Sauvage.
Nielsen, R. (2004a). *Chara aspera*. In H. Schubert, & I. Blindow (Eds.), *Charophytes of the Baltic Sea* (pp. 42–52). A. R. G. Gantner.
Nielsen, R. (2004b). *Chara fragifera*. In H. Schubert, & I. Blindow (Eds.), *Charophytes of the Baltic Sea* (pp. 95–98). A. R. G. Gantner.
Paavola, M., Olenin, S., & Leppäkoski, E. (2005). Are invasive species most successful in habitats of low native species richness across European brackish water seas? *Estuarine, Coastal and Shelf Science, 64*, 738–750.
Rybak, A., & Woyda-Ploszczyca, A. (2019). Ecology and distribution patterns of *Chara connivens* (Charophyta, Characeae) on the Canary Islands the first record from Fuerteventura. *Oceanological and Hydrological Studies, 48*, 368–380. https://doi.org/10.2478/ohs-2019-0033
Schou, J. C., Moeslund, B., Båstrup-Spohr, L., & Sand-Jensen, K. (2017). *Danmarks vandplanter*. BFN's.
Simons, J., & Nat, E. (1996). Past and present distribution of stoneworts (Characeae) in The Netherlands. *Hydrobiologia, 340*, 127–135.
Simons, J., Ohm, M., Daalder, R., Boers, P., & Rip, W. (1994). Restoration of Botshol (The Netherlands) by reduction of external nutrient load: Recovery of a characean community, dominated by *Chara connivens*. *Hydrobiologia, 275/276*, 243–253. https://doi.org/10.1007/978-94-017-2460-9_21
SLU Artdatabanken. (2020). *Rödlistade arter i Sverige 2020*. SLU. https://www.artdatabanken.se/publikationer/bestall-rodlista-2020/
Torn, K., & Martin, G. (2003). Changes in the distribution of charophyte species in enclosed seabays of western Estonia. *Proceedings of the Estonian Academy of Sciences, Biology and Ecology, 52*, 134–140.
Torn, K., & Martin, G. (2004). *Chara connivens*. In: H. Schubert, & I. Blindow (Eds.), *Charophytes of the Baltic Sea* (pp. 131–141). A. R. G. Gantner.
Trbojević, I., Marković, A., Blaženčić, J., Subakov Simić, G., Nowak, P. Ballot, A., & Schneider, S. (2020). Genetic and morphological variation in *Chara contraria* and a taxon morphologically resembling *Chara connivens*. *Botany Letters*, 167(2), 187–200. https://doi.org/10.1080/23818107.2019.1703808
Urbaniak, J., & Blaženčić, J. (2012). SEM study of oospore characteristics in endemic and endangered balkan charophytes. *Cryptogamie, Algologie, 33*, 277–288.
Urbaniak, J., & Gąbka, M. (2014). *Polish Charophytes. An illustrated guide to identification*. Uniwersytet Przyrodniczy we Wroclawiu.
van Raam, J. C. (1998). *Handboek Kranswieren*. Chara boek.
Wood, R. D. (1965). *Monograph of the Characeae*. J. Cramer.

Chapter 25
Chara contraria

Karl-Georg Bernhardt and Thomas Gregor

Scientific name: *Chara contraria* A. Braun ex Kütz., Phycol. Germ.: 258. 1845

Lectotype (Wood, 1965: 93): Carlsruhe/1839/A. Braun (L 546).

≡ *Chara vulgaris* L. var. *contraria* (A. Braun ex Kütz.) J. A. Moore, Charophytes Great Britain Ireland: 53. 1986, n. inv.

≡ *Chara vulgaris* L. f. *contraria* (A. Braun ex Kütz.) R. D. Wood in Taxon 11: 8. 1962, n. inv.

= *Chara muscosa* J. Groves & Bullock-Webster in J. Bot. 62: 33. 1924.
Lectotype (Wood, 1965: 96): Lough Mullaghony, West Donegal/11.8.1919/G. R. Bullock-Webster (BM).

? = *Chara contraria* A. Braun ex Kütz. var *hispidula* A. Braun, Neue Denkschr. Allg. Schweiz. Ges. Gesammten Naturwiss. 10(3): 16. 1847.
Type not designated.

≡ *Chara vulgaris* L. var. *hispidula* (A. Braun) J. A. Moore, Charophytes Great Britain Ireland: 53. 1986, n. inv.

= *Chara inconnexa* Allen in Bull. Torrey Bot. Club 9: 40. 1882
Lectotype (Wood 1965: 119): Storm Lake, Iowa / 6.8.1872 / C. B. Bessey (NY).

Taxonomic notes: Schneider et al. (2016) mention an undescribed taxon with the tentative name "*Chara arcadiensis*". Its sole specimen was collected 2005 in Greece

K.-G. Bernhardt (✉)
University of Natural Resources and Life Sciences, Gregor Mendel-Straße 33, 1180 Vienna, Austria
e-mail: karl-georg.bernhardt@boku.ac.at

T. Gregor
Department of Botany and Molecular Evolution, Senckenberg Research Institute and Natural History Museum Frankfurt, Senckenberganlage 25, 60325 Frankfurt am Main, Germany
e-mail: thomas.gregor@senckenberg.de

H. Schubert et al. (eds.), *Charophytes of Europe*,
https://doi.org/10.1007/978-3-031-31898-6_25

by U. Raabe. It morphologically resembles *Chara contraria* but differs by dioecy. According to the plastidial matK gene it belongs to the *Chara contraria*-cluster (Schneider et al., 2016).

Local names: BELARUSIAN: Хара процілеглая; DANISH: Buttornet kransnål; DUTCH: Brokkelig kransblad; ENGLISH: Opposite Stonewort; ESTONIAN: Näsa-mändvetikas; FINNISH: Harmaanäkinparta; FRENCH: Chara oposée; GERMAN: Gegensätzliche Armleuchteralge; LITHUANIAN: Mažasis maurabragis; NORWEGIAN: Gråkrans; POLISH: Ramienica przeciwstawa; RUSSIAN: Хара противоположная; SWEDISH: Gråsträfse; UKRAINIAN: Хара протилежна.

25.1 Morphology

See Figs. 25.1 and 25.2.

Habitus: Plants are nearly always incrusted and greyish green and moderately branched. In the upper part, branchlets are longer than the internodes forming "fox-tails". In shallow water plants grow in small groups, in deeper water they form dense underwater lawns.

Size: The length is (5)30(50) cm.

Main axis: The main axis diameter is (0.3)1.0(1.5) mm, the internodes being (2)4(6) cm, they are 2–4 times as long as branchlets. The main axis diameter is smaller in var. *hispidula* (to 0.8 mm).

Cortex: *Chara contraria* is typically diplostichous and tylacanthous, partly triplostichous plants occur. Older parts of plants tend to be isostichous. In the middle of internodes secondary rows may be missing. Individuals with partly lacking main axis cortex are similar to *C. dissoluta*.

Spine-cells: Spine-cells are solitary and generally inconspicuous; they disappear in older parts of the plant. *C. contraria* var. *hispidula* has protruding spine-cells as long as to twice as long as the main axis diameter.

Stipulodes: Stipulodes are inconspicuous in two rows, mostly only the upper row is well developed; in var. *hispidula* both rows are well developed with the upper row slightly longer than the lower.

Branchlets: Whorls consist of 6–8(12) branchlets. They consist of usually 5 (2–7) corticated and usually 3 (2–8) uncorticated terminal segements. which are mostly as long as the last corticated segment but can be up to 6 times as long. The length of the branchlets is (5)8(25) mm. Older whorls can have effuse branchlets.

Bract-cells: Mostly 5, adaxials elongate, longer than the oogonium, abaxials usually papillate.

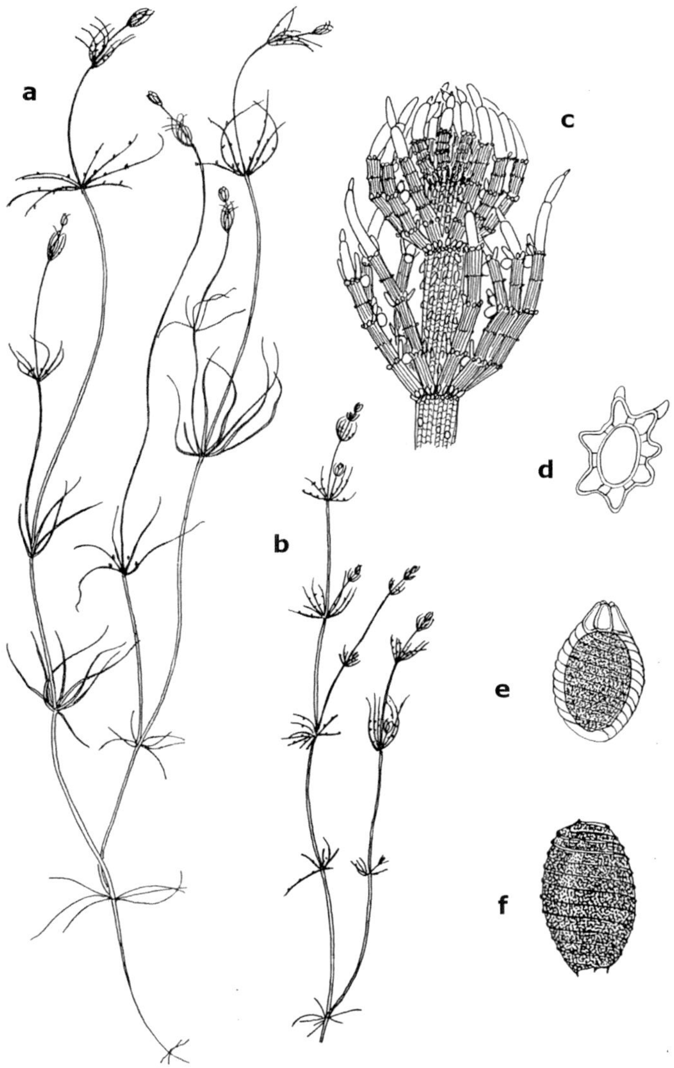

Fig. 25.1 *Chara contraria*. **a** Plants from 1–2 m depth; **b** Plants from 0.5 m depth; **c** Apex of the shoot; **d** Shoot transection; **e** Oogonium; **f** Oospore. Drawings by Werner Krause

Bracteoles: 2, longer than adaxial bract-cells, from shorter to double the length as long (in var. *hispidula*) of the oogonium.

Sexuality: *Chara contraria* is monoecious and conjoined. Gametangia are found on the (1)3(4) basal nodes.

Oogonia: Oogonia are elongated and cylindrical. For morphometry, see Table 25.1.

Fig. 25.2 Habitus and habitat appearance of *Chara contraria*. **a** Protruding shoots in clear water of a quarry pond (Reeser Meer, Rees, Germany); **b** Fertile apex (Boker Heidesee, Paderborn, Germany); **c** Habitus of richly fertile specimen (Wegeleben, Germany); **d** Dense stands in shallow water (Teichwiesengraben, Riedstadt, Germany); **e** Cortication, spine-cells and stipulodes of the uppermost shoot section. Photographs: **a**, **b**, **d**—Klaus van de Weyer; **c**—Heiko Korsch; **e**—Karin Pall

Oospores: Oospores are elliptic to cylindrical and often smaller than the oospores of other *Chara* species. The top is apiculate with a shoulder. Upper ribs are more distinct than the rest of ribs. The membrane is rather more or less smooth at lower magnification, somewhat granulous at higher magnification. For morphometry, see Table 25.2.

Antheridia: Antheridia are spherical reddish yellow to light-red and 250–400 μm in diameter (Table 25.3).

Bulbils: Bulbils are absent.

Table 25.1 Oogonia morphometry

Author	Length	Width	Coronula length	Coronula width	Number of helical stripes
Migula (1889–1897)	750–1000	550–750	160	–	12–16
Groves and Bullock-Webster (1924)	650–975	500–575	125–190	260–360	13–15
Corillion (1957)	650–1000	550–650	120–190	220–360	13–15
Wood (1965)	645–705	375–420			–
Krause (1997)	650–900	400–600			–
Dekere (2004)	600–900	> 600			
Total range	**600–1000**	**375–750**	**120–190**	**220–360**	**13–15**

Length in the first column excludes the coronula; its length and width are given separately, all values for length and width in μm.

Table 25.2 Oospore morphometry

Author	Length	Width	Number of striae	Colour
Migula (1889–1897)	(450)550–670	(220)350–420	11–14	(Light brown) dark brown to black
Groves and Bullock-Webster (1924)	525–725	375–425	11–14	–
Corillion (1957)	525–725	350–490	10–14	–
Wood (1965)	560–615	285–315	13–14	–
Haas (1994)	(300)500–700(830)	(170)340–490(530)	10–14	Brown, dark brown to black
Krause (1997)	550–700	300–450	12–14	Black
Vedder (2003)	450–650	250–425	10–15	Brown, dark brown
Dilger (2010), pers. Comm	(430)520–730(840)	(200)260–390(480)	(9)11–14(17)	Dark brown to black
Dekere (2004)				Black
Total range	**(300)450–730(840)**	**(170)250–490(530)**	**(9)10–14(17)**	

All values for length and width in μm

Table 25.3 Antheridia morphometry

Author(s)	Diameter (μm)	Colour
Migula (1889–1897)	280–350	Light red
Wood (1965)	255–400	
Krause (1997)	250–400	Reddish yellow
Dekere (2004)	Up to 400	Reddish yellow
Total range	**250–400**	

Wood (1965) measured rewetted antheridia. other authors did not give any information on methods to measure antheridia

25.2 Phenology and Reproduction

Autumn and spring populations occur (van Raam, 1998). The former germinate in autumn, fructify in April/May of the next year and produce oospores in June/July, they mostly die in August. The latter germinate in March/April and produce oospores in September to October. Both types occur in the Netherlands, Germany, Belgium and Northern France. However, only spring germinating populations are known from the British Isles, Ireland, Denmark and southern Sweden.

Fructification takes place from beginning of April to end of October, mature oospores are found from June/July on (Corillion, 1957; Holtz, 1903; Langangen, 2007; Olsen, 1944; Schmidt, 1981; van Raam, 1998). In shallow water the species is annual, in deeper water perennial.

25.3 Habitat Conditions and Ecology

Habitat type: *Chara contraria* occurs from lowland to mountain regions (ca. 2000 m in Val di Travers/Switzerland (Migula, 1889–1897), 2150 in Almsee, Gröden/Italy (Hohla et al., 2020) in a wide array of habitats. As a pioneer species, it can colonise natural and artificial small water bodies with changing water levels. The majority of occurrences is in deep waters of lakes with groundwater influence. *C. contraria* avoids running waters. *C. contraria* colonises fine sandy substrates, calcareous gyttja and marl clay but also organogenous horizons (Holtz, 1903; Langangen, 2007; Schmidt, 1981; Schmidt et al., 1996; Stroede, 1933; van Raam, 1998). *Chara contraria* var. *hispidula* is ecologically divergent; in Austria it is only found in oligo-mesotrophic hard-water lakes with rather fine substrate in intermediate depth (K. Pall pers. comm.).

Nutrient conditions: *Chara contraria* tolerates eutrophication (Jäger, 1999, 2000; Melzer, 1988; Nat et al., 1994; Pietsch, 1982). Habitats of *C. contraria* show TN values up to 6.5 mg/l (Simons and Nat, 1996), NO_3-Values up to 10 mg/l, NH_4-values up to 4.5 mg/l (LfULG Sachsen, 2012). TP-values up to 0.5 mg/l (Simons and

Nat, 1996) and orthophosphate-values up to 0.186 mg/l (Blindow, 1992) demonstrate that *C. contraria* can tolerate eutrophic conditions which allows plants to provide areas with clear water in otherwise turbid lakes (Scheffer et al., 1994), and build up considerable biomass in eutrophicated lakes (Schanz, 1978).

Water chemistry: The species tolerates a broad pH range between 3.8 and 9.2 (Corillion, 1957; Humberg and Beck, 2006; Nat et al., 1994; Pietsch, 1982; Simons and Nat, 1996). *C. contraria* grows best in neutral, alkaline to calcium carbonate-rich waters (Doege and van de Weyer, 2016). Humberg and Beck (2006) reported an electrical conductivity in its habitats of 1535 μS/cm, LfULG Sachsen (2012) of 2540 μS/cm in sulphate-rich waters. According to Dekere (2004), *C. contraria* prefers α-mesohaline habitats. It also occurs in slightly brackish waters (van Raam, 1998) and habitats rich or poor in calcium carbonate (Simons and Nat, 1996; van Raam, 1998). *C. contraria* var. *hispidula* tends to occur in harder waters.

Depth range: *Chara contraria* prefers deeper water of up to 30 m (Corillion, 1957; Krause, 1981; Langangen, 2007; Schmidt et al., 1996), but can also grow in shallow water (Bernhardt et al., 2017).

Associated species: In shallow water *Chara contraria* is often associated with *C. vulgaris* and *C. globularis*, sometimes with *C. aspera* or *C. virgata*. In hard-water lakes *C. tomentosa* is often admixed, but also *C. hispida, C. papillosa* or *Tolypella glomerata*.

25.4 Variability

Chara contraria shows a similarly large variability as *C. vulgaris*. Migula (1889–1897) distinguishes 26 forms with the length of the uncorticated branchlet cells as an important character. Low growing plants regarded as *C. muscosa* in the British Isles (Stewart and Church, 1992) are included in *C. contraria* following van de Weyer (2012). *Chara contraria* var. *hispidula* is probably a morphologically and ecologically divergent form. Eleven of the 26 of the forms of Migula (1889–1897) belong to this taxon. However barcoding (with matK) found no difference between *C. contraria* (var. *contraria*) and *C. c.* var. *hispidula* (Mann et al., 2022). Transitions exist from forms with complete cortex to forms with no cortex. This makes the distinction from *C. dissoluta* difficult.

25.5 Identification and Risk of Confusion

C. contraria is similar to *C. vulgaris* and probably often confused with this taxon. Differentiation needs close inspection of the cortication which is in tylacanthous in *C. contraria* and aulacanthous in *C. vulgaris*, best seen in the uppermost whorls.

Fresh plants have a greyish green colour rarely found in *C. vulgaris*. In *C. vulgaris* oospores are less than 550 μm long (more than 550 μm in *C. contraria*). Slender collections of *C. papillosa* with mostly single spine-cells may be hard to distinguish from *Chara contraria*. Plants with reduced branchlets are similar to *C. filiformis*. Individuals with a higher number of ecorticated terminal branchlet segments can be confused with *Chara squamosa* but can be distinguished from this taxon by absence of bract-cells and gametangia at the ecorticated branchlet nodes and their far smaller antheridia. *C. contraria* var. *hispidula* can be distinguished from *C. aculeolata*, which always has fasciculate spine cells, by its solitary spine cells and by a smaller internode diameter (most *C. contraria* <1 mm, most *C. aculeolata* > 1 mm). A distinction from slender collections of *Chara affinis* is difficult.

25.6 Distribution

See Fig. 25.3.

C. contraria is an extratropical cosmopolite, mainly occurring in the northern hemisphere but also in South America (Blindow et al., 2018), South-Africa and Australia (Korsch, 2018). It is common throughout Europe and occurs in America (Langangen, 2002), Asia (Dixit, 1935; Faridi, 1955; Hussain et al., 2003; Naz et al.,

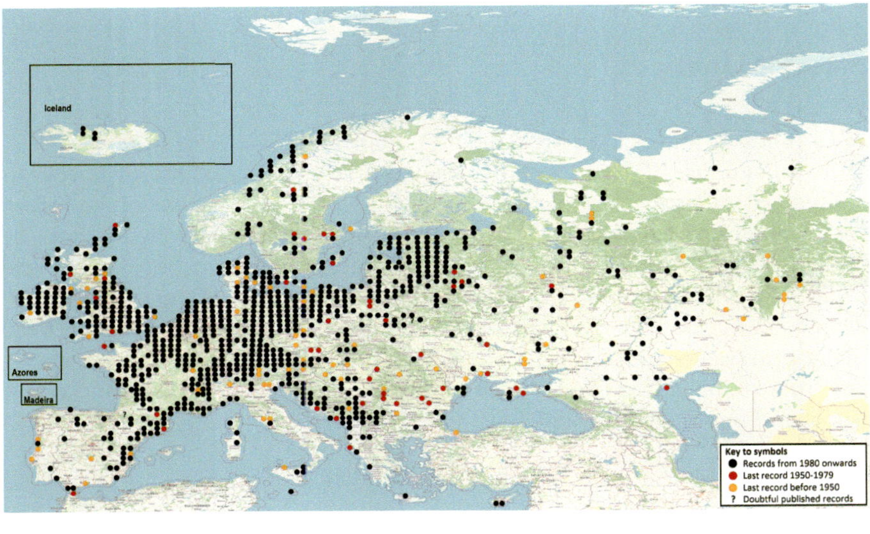

Fig. 25.3 European distribution of *Chara contraria*. For worldwide distribution see Korsch (2018). Black dots are records from 1980 on, red dots from 1950 to 1979 and yellow dots before 1950; unconfirmed/doubtful published records (if applicable) are marked with "?" on top. Map prepared by Heiko Korsch

2011; Pal et al., 1962) and Australia (Naz et al., 2011). *Chara contraria* var. *hispidula* is known from Western and Central Europe.

25.7 Status and Threats

C. contraria is common in Europe and generally not endangered (see Chap. 9).

References

Bernhardt, K.-G., Gregor, T., & Wernisch, M. M. (2017). Ausbreitung von Typha laxmannii entlang der neu ausgebauten Westbahnstrecke im Tullnerfeld (Niederösterreich). *Wissenschaftliche Mitteilungen des Niederösterreichischen Landesmuseums, 27*, 231–242.
Blindow, I. (1992). Decline of Charophyta during eutrophication: A comparison to angiosperms. *Freshwater Biology, 28*, 9–14. https://doi.org/10.1111/j.1365-2427.1992.tb00557.x
Blindow, I., Marquardt, R., Schories, D., & Schubert, H. (2018). Charophyceae of Chile—taxonomy and distribution. Part 1. *Nova Hedwigia, 107*, 1–47. https://doi.org/10.1127/nova_hedwigia/2017/0452
Corillion, R. (1957). Les Charophycées de France et d'Europe Occidentale. *Bulletin de la Société Scientifique de Bretagne, 32* (Hors série I & II), 1–499.
Dekere, Z. (2004). *Chara contraria.* In H. Schubert, & I. Blindow (Eds.), *Charophytes of the Baltic Sea* (pp. 89–94). Gantner.
Dilger, M. (2010). Merkmale zur Erkennung von *Nitella-* und *Tolypella-*Arten der Characeaen [sic] über Oosporen aus Sedimenten oder von Pflanzen. *Rostocker Meeresbiologische Beiträge, 23*, 40–59.
Dixit, S. L. (1935). The Charophytes of the Bombay Presidency. *Journal of the Indian Botanical Society, 16*, 257–263.
Doege, A., & van de Weyer, K. (2016). *Chara contraria.* In: Arbeitsgruppe Characeen Deutschlands, *Armleuchteralgen: Die Characeen Deutschlands* (pp. 282–292). Springer. https://doi.org/10.1007/978-3-662-47797-7
Faridi, M. A. F. (1955). Contribution to the Charales of West Pakistan. *Biologia, 1*, 70–81.
Groves, J., & Bullock-Webster, G. R. (1924). *The British Charophyta. Chareæ with plates, concluding articles, geological sketch, bibliography and index* (Vol. II). Ray Society.
Haas, J. N. (1994). First identification key for charophyte oospores from central Europe. *European Journal of Phycology, 29*, 227–235. https://doi.org/10.1080/09670269400650681
Hohla, M., Wilhalm, T., & Gregor, T. (2020). Die Armleuchteralgen (Characeae) Südtirols. *Gredleriana, 20*, 5–27.
Holtz, L. (1903). *Characeen. Kryptogamenflora der Mark Brandenburg. Band* 4(1). Borntraeger.
Humberg, B., & Beck, M. (2006). Der Makrophytenbestand in ausgewählten Baggerseen der Oberrheinaue. *LUBW, Oberirdische Gewässer, Gewässerökologie, 102*, 1–348.
Hussain, M. I., Reginald, V., & Khoja, M. T. (2003). Charophytes of the sultanate of Oman, Southern Arabia. *Nova Hedwigia, 77*, 429–444. https://doi.org/10.1127/0029-5035/2003/0077-0429
Jäger, D. (1999). *Beiträge zur Characeen-Flora Vorarlbergs.* Diplomarbeit Univ. Innsbruck.
Jäger, D. (2000). Beiträge zur Characeen-Flora Vorarlbergs (Österreichs). *Berichte des naturwissenschaftlich-medizinischen Vereins in Innsbruck, 87*, 67–85.
Korsch, H. (2018). The worldwide range of the Charophyte species native to Germany. *Rostocker Meeresbiologische Beiträge, 28*, 45–96.

Krause, W. (1981). Characeen als Bioindikatoren für den Gewässerzustand. *Limnologica, 1*, 399–418.

Krause, W. (1997). Charales (Charophyceae). In: H. Ettl, G. Gärtner, H. Heynig, & D. Mollenhauer (Eds.), *Süßwasserflora von Mitteleuropa* (Vol. 18). Fischer.

Langangen, A. (2002). Some Charophytes (Chlorophyte, Charales) from Alaska, USA. *Acta Musei Nationales Pragae, Series b, Historia Naturalis, 57*, 51–56.

Langangen, A. (2007). *Charophytes of the Nordic countries*. Saeculum ANS.

LfULG [Landesamt für Umwelt, Landwirtschaft und Geologie] Sachsen. (2012). *Daten der Gewässeruntersuchungen nach WRRL 2005–2012*. Staatliche Umweltbetriebsgesellschaft Sachsen.

Mann, H., Hanel, C., Langangen, A., & Nowak, P. (2022). *Chara contraria* var. *hispidula* Braun (Charales) in Newfoundland, Canada a new variety described from North America. *Botany Letters, 169*, 250–258. https://doi.org/10.1080/23818107.2021.2018041

Melzer, A. (1988). *Der Makrophytenindex – eine biologische Methode zur Ermittlung der Nährstoffbelastung von Seen*. Habilitationsschrift TU München.

Migula, W. (1889–1897). Die Characeen Deutschlandsm Oesterreichs und der Schweiz. In A. Fischer, E. Fischer, F. Hauck, G. Limpricht, C. Luerssen, W. Migula, H. Rehm, P. Richter, & G. Winter (Eds.), *Dr. L. Rabenhorst's Kryptogamenflora von Deutschland, Oesterreich und der Schweiz* (2nd ed., 5. Band). Eduard Kummer.

Nat, E., Simons, J., de la Haye, M. A. A., & Coops, H. (1994). *Historisch en actueel verspreidingsbeeld van kranswieren in Nederland in samenhang met waterkwaliteitsfactoren*. RIZA Werkdocument 94.148.

Naz, S., Diba, N. J., & Schubert, H. (2011). *Monograph on Charophytes of Bangladesh*. VDM Dr. Müller.

Olsen, S. (1944). Danish Charophyta. *Det Kongelige Danske Videnskabernes Selskab, Biologiske Skrifter, 3*(1), 1–240, pl. 1–2.

Pal, B. P., Kundu, B. C., Sundaralingam, V. S., & Venkataraman, G. S. (1962). *Charophyta*. Indian Council of Agricultural Research.

Pietsch, W. (1982). Biologische, mikrobiologische und toxikologische Methoden. Makrophytische Indikatoren für die ökochemische Beschaffenheit der Gewässer. In C. Legler, & G. Breitig (Eds.), *Ausgewählte Methoden der Wasseruntersuchung, Band 2: Biologische , mikrobiologische und toxikologische Methoden* (pp. 67–88). Gustav Fischer.

Schanz, F. (1978). Algal populations in a mountain pond (Swiss National Park). *Internationale Vereinigung für theoretische und angewandte Limnologie, Verhandlungen, 20*, 2188–2193.

Scheffer, M., van den Berg, M. S., van den Breukelaar, A., Breukers, C., Coops, H., Doef, R., & Meijer, M.-L. (1994). Vegetated areas with clear water in turbid shallow lakes. *Aquatic Botany, 49*, 193–196. https://doi.org/10.1016/0304-3770(94)90038-8

Schmidt, D. (1981). Die Characeen – eine im Aussterben begriffene Pflanzengruppe unserer Gewässer. *Gleditschia, 8*, 141–157.

Schmidt, D., van de Weyer, K., Krause, W., Kies, L., Garniel, A., Geissler, U., Gutowski, A., Samietz, R., Schütz, W., Vahle, H.-C., Vöge, M., Wolff, P., & Melzer, A. (1996). Rote Liste der Armleuchteralgen (Charophyceae) Deutschlands. 2. Fassung, Stand: Februar 1995. *Schriftenreihe für Vegetationskunde, 28*, 547–576.

Schneider, S. C., Nowak, P., von Ammon, U., & Ballot, A. (2016). Species differentiation in the genus *Chara* (Charophyceae): Considerable phenotypic plasticity occurs within homogenous genetic groups. *European Journal of Phycology, 51*, 282–293. https://doi.org/10.1080/09670262.2016.1147085

Simons, J., & Nat, E. (1996). Past and present distribution of stoneworts (Characeae) in The Netherlands. *Hydrobiologia, 340*, 127–135. https://doi.org/10.1007/BF00012744

Stewart, N. F., & Church, J. M. (1992). *Red data books of Britain and Ireland: Stoneworts*. The Joint Nature Conservation Committee.

Stroede, W. (1933). Über die Beziehungen der Characeen zu den chemischen Faktoren der Wohngewässer und des Schlammes. *Archiv für Hydrobiologie, 25*, 192–229.

van de Weyer, K. (2012). Anmerkungen zu *Chara muscosa* J. Groves et Bullock-Webster in Deutschland und Österreich. *Rostocker Meeresbiologische Beiträge, 24*, 119–124.
van Raam, J. C. (1998). *Handboek Kranswieren*. Chara boek.
Vedder, F. (2003). *Morphologie und Taxonomie rezenter und subfossiler Characeen-Oosporen aus der Ostsee*. Diplomarbeit Ernst-Moritz-Arndt Universität Greifswald.
Wood, R. D. (1965). *Monograph of the Characeae*. J. Cramer.

Chapter 26
Chara corfuensis

Chrysoula Christia, Eva Papastergiadou, Irmgard Blindow, and Roman Romanov

Scientific name: ***Chara corfuensis*** **J. Groves ex Fil. in Mat. Term. Értes. 55: 477. 1937**

Lectotype (Wood, 1965: 139): Between Moraitika and Braganiotika, Corfu/19. Sep. 1929/T. B. Blow E 101 (BM 840041).

≡ *Chara hispida* L. f. *corfuensis* (J. Groves ex Fil.) R. D. Wood in Taxon 11: 9. 1962, n. inv.

Local names: Unknown.

26.1 Morphology

See Figs. 26.1 and 26.2.

Habitus: The plants are robust, green to greyish with distinct and heavy incrustations and robust, spiny main axis. The long verticillate bract-cells result in branchlet whorls

C. Christia (✉) · E. Papastergiadou
Department of Biology, University of Patras, 265 00 Patras, Greece
e-mail: xchristia@gmail.com

E. Papastergiadou
e-mail: evapap@upatras.gr

I. Blindow
Universität Greifswald, Biologische Station Hiddensee, 18565 Kloster, Germany
e-mail: blindi@uni-greifswald.de

R. Romanov
Nišice Ujtin Potok b.b., Dobra Voda, 85356 Bar, Montenegro
e-mail: streptophytes@gmail.com

H. Schubert et al. (eds.), *Charophytes of Europe*,
https://doi.org/10.1007/978-3-031-31898-6_26

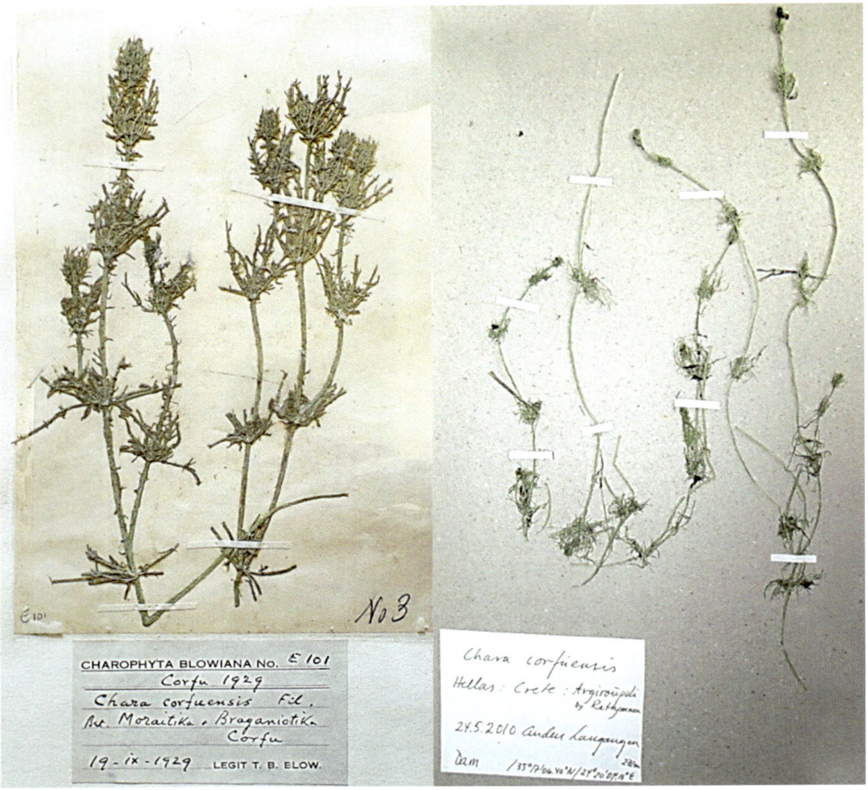

Fig. 26.1 *Chara corfuensis*, herbarium material. Left panel: Lectotype. Legit T. B. Blow, V 124, 17/9/1929, collected from a small stream entering Laguna Kalichiopoulos Cassopitna, Corfu Island, Greece: (Charophyta Blowiana No. E 101), BM00840041); right panel: from Argiroupolis spring (Crete) collected by Langangen (2010). Photographs: Irmgard Blindow

appearing as "filled" inside and spiny outside, being most spiny at lowest nodes of the branchlets.

Size: Ranging from small (7–8 cm) to large (60 cm) as reported by Langangen (2012) in Lake Kournas (Crete). Most collections are of medium size, ranging from 15–30 cm.

Main axis: Stout, firm, fragile; from 900–1600 μm in diameter, mean value 1100 μm (own measurements).

Cortex: Diplostichous, rarely diplo-triplostichous, tylacanthous to strongly tylacanthous, rarely isostichous (Langangen, 2007; own data).

Spine-cells: Mostly single, well developed, long, acute, some in pairs, occasionally in clusters of 3–4, pointing in the same direction. Spine-cells are longer than the

Fig. 26.2 *Chara corfuensis*, details. **a** Upper and central part of individuals gathered in Kaiafas lagoon on May 2019 by Chrysoula Christia and Eva Papastergiadou; **b** Apex section from lectotype (BM 840041); **c** Apex section of a specimen from Prokopos lagoon collected by George Fyttis and Eva Papastergiadou on May 2009; **d** Branchlets and axial node of a specimen from Prokopos lagoon; **e** Shoot tip of a specimen from lectotype (BM 840041); **f** SEM of an oospore from lectotype (BM 840041). Photographs: **a**, **c**, **d**—Chrysoula Christia; **b**, **e**—Irmgard Blindow; **f**—Michelle T. Casanova

diameter of axis in most cases, rarely as long as the axis diameter, from 1000–4000 μm, occasionally up to 6000 μm (mean value 2600 μm; own data).

Stipulodes: In 2 tiers, both well developed, 2 sets per branchlet, as long as or shorter than spine-cells, often irregular in length, 500–2000 μm, occasionally up to 4400 μm long, acuminate to abruptly pointed. The mean values (own measurements) are 1700 μm.

Branchlets: Whorls consist of 9 to 11 stout branchlets, 1.4–3.5 cm long (Langangen, 2007), completely differentiated into nodes and internodes, with 4–7 segments, of which 3–4 are 2 corticated, the distal 2–3 segments ecorticate, occasionally, all or most parts of the branchlet are ecorticate; the ecorticate segments are usually somewhat longer than corticated ones; terminal segments 1–2 celled, end cell long conical, acuminate. The cortication of the branchlet segments is highly variable. In many collections, possibly more common than in other taxa of *Hartmania* group, the ecorticated part of the branchlets is longer than the corticated one.

Bract-cells: Most *Chara corfuensis* specimens have verticillate bract-cells in most (or all) whorls [as originally described by Filarszky (1937) and Wood (1965)], in some collections adaxial bract-cells are only slightly longer than the abaxial ones. Out of the 17 collections examined by us, only one had distinctly longer adaxial than abaxial bract-cells. Adaxial bract-cells range from 450–6740 μm (mean 3310 μm; own measurements). The longest abaxial bract-cells are at the lowest node of the branchlet, their size decreases upwards, becoming significantly shorter than the adaxial bract-cells.

Bracteoles: 2, similar in shape, but longer than bract-cells, far exceeding the oogonia and adjacent segment in length. 2800 μm were measured in one collection (own data).

Sexuality: The species is monoecious. Plants are relatively fertile, but unripe oospores are common.

Gametangia: Conjoined at the 2–3 lowest branchlet nodes.

Oogonia: The quantitative data collected by different authors (Table 26.1) show high variation of oogonium dimensions. In 3 out of 4 collections, coronula cells were not spreading, spreading only in 1 collection (own data).

Oospores: Oospores are commonly immature and difficult to see. The wall ornamentation at different magnifications (130x, 300x) showed that oospores from Lake Desne, Croatia had a smooth surface; the well-developed flanged striae were about 4.0–5.1 μm wide and 10–12 μm high; the internal and external surface of the striae have a smooth ornamentation; a delicate ribbon-like structure arises on the ridge of each stria (Urbaniak and Blaženčić, 2012). Oospore removed and images taken with Scanning Electron Microscope (SEM) by Michelle T. Casanova on 11.7.2019 from the type collection (BM 840041) had 9–11 striae, with a pronounced cage and ribbon. The fossa wall appears smooth, but has irregularly arranged, smooth pustules. The ribbon is easily removed from the striae, it has a nodulose/fibrous surface. The cage of at the base of the oospore can be partial or complete and fenestrate (Table 26.2).

Table 26.1 Oogonia morphometry

Author	Length	Width	Coronula length	Coronula width	Number of helical stripes
T. B. Blow 17.9.1929 (BM)	-	-	-	-	11–13
Wood (1965)	750	650	165	290	8
Langangen (2007)	Up to 1000	-	-	-	-
Langangen (2010)	550	450	-	-	12
Own data	560–910	450–815	145–265	240–410	11–13
Total range	**550–1000**	**450–815**	**145–265**	**240–410**	**11–13**

Length in the first column excludes the coronula; its length and width are given separately, all values for length and width in μm

Table 26.2 Oospore morphometry

Author	Length	Width	Number of striae	Fossa width	Length/ width ratio
Langangen (2012)	800	-	-	-	-
Urbaniak and Blaženčić (2012)	745–871 (829 ± 31.3)	388–462 (430 ± 22.2)	10–12 (11 ± 0.6)	70–100 (80 ± 6.7)	1.65 ± 2.13 (1.95 ± 0.11)
M. T. Casanova pers. comm.	-	-	9–11	-	-

All values for length and width in μm. Min, max values, median ± standard error are given

Antheridia: Antheridia are orange, 8-scutate, solitary. Table 26.3 gives an overview about diameter values known so far.

Table 26.3 Antheridia morphometry

Author	Diameter (μm)
Wood (1965)	525–600 (+)
Langangen (2007)	Up to 500
Own measurements	360–530 (mean 420)
Total range	**360–600**

All own measurements were done on dry material. Wood (1965) measured rewetted antheridia. Own measurements of the type material collected by T. B. Blow (17/9/1929) in BM ranged from 408 to 528 μm

26.2 Habitat Conditions and Ecology

Habitat type: *Chara corfuensis* inhabits mainly coastal brackish water habitats. As well as brackish inland waters, the species is found in temporary ponds and lakes. According to Habitats Directive (European Commission, 1992) the species belongs to the priority for conservation habitat type "1150* Lagoons".

Sediment: The species is usually established in sandy mud to muddy sand or surface sediments rich in sand of biogenic origin composed by shells and their fragments or grayish lime marl. In Kaiafas lagoon the sediment contents are: Sand 69.1%, Silt 14.6%, Clay 16.3%, C org 2.8%, and $CaCO_3$ 40.5% (Bouzos and Kontopoulos, 2004). In Lake Koumoundourou the average $CaCO_3$ content is 69.1%, and C-org ranged from 2 to 7.5%, indicating anthropogenic inputs and/or temporarily anoxic conditions in the lake. In Prokopos lagoon sediments were composed of fine silt to medium sand (Katsaros et al., 2017). In Lake Kournas the bottom was sand or grayish lime marl, soft silty and loamy (Langangen, 2012).

Nutrient conditions: In Kaiafas lagoon the species was found in waters with low ortho-phosphates (~10 μg/L), mean TP values lower than 100 μg/L, nitrate higher than 200 μg/L, and Chl-*a* concentrations from 3.27 to 5.04 μg/L. The species was absent from the southern part of the lagoon where the water quality was poor (ammonium values 2.10 mg/L in 2006) due to sewage discharges (Christia et al., 2014). In Lake Koumoundourou, high values of nitrates were reported (0.22 to 9.13 mg/l, average 4.06 mg/l), while ammonium ranged from 0 to 5.94 mg/l, (average 0.35 mg/l), and ortho-phosphates ranged from 0 to 0.90 mg/l (Mentzafou and Dimitriou, 2019; Dimitriou et al., 2012). In Prokopos lagoon, 21.4 μg/l mean ortho-phosphates, 58.8 μg/l mean nitrates and Chl-*a* mean concentrations 20.4 mg/l, were measured (Fyttis et al., 2019; Katsaros et al., 2017).

Water chemistry: *Chara corfuensis* inhabits shallow to deep mesohaline water environments, with high dissolved oxygen values. It is adapted to oligotrophic and eutrophic waters with high light intensity, but also grows in deep waters with high turbidity. In Kaiafas lagoon, salinity ranged from 6.9 to 14.2‰, D.O. from 4.92 to 12.65 mg/l, and pH from 7.27 to 9.00 (Christia et al., 2011, 2014). Generally, as referred from Lake Koumoundourou and Prokopos lagoon, *Chara corfuensis* prefers slightly acidic to alkaline water (pH 6.4 to 9) with relatively high D.O. which ranges from 4.9 mg/l to 15 mg/l (Fyttis et al., 2019; Mentzafou et al., 2016). In Lake Kournas conductivity was 1680 μS/cm and water pH 8.4 (Langangen, 2012).

Depth range: The species could be found from very shallow waters to over 3 m depth. In Kaiafas lagoon, the species was found at a mean depth of about 3.2 m, and in Koumoundourou and Prokopos at a mean depth of around 1 m.

Associated species: *Chara corfuensis* can form large monospecific dense perennial mats or grow together with *Stuckenia pectinata*, *Ruppia cirrhosa*, *Lamprothamnium papulosum*, *Potamogeton filiformis*, *Zannichellia* sp. and *Najas marina*.

26.3 Variability

Chara corfuensis is not a variable charophyte.

26.4 Identification and Risk of Confusion

Chara corfuensis belongs to the *Hartmania* group, a complex of robust, diplostichous *Chara* spp. difficult to separate from each other (see Chap. 15).

The most important characters to separate *Chara corfuensis* from other species belonging to the *Hartmania* group are its extraordinarily long bract-cells, which are verticillate in most (or all) whorls [as originally descripted by Filarszky (1937) and Wood (1965).

Chara corfuensis can be confused with *C. aculeolata*, *C. papillosa* and especially, *C. globata* and "Mediterranean Chara". *Chara corfuensis* can be distinguished from *C. aculeolata* by the spine-cells which are single in most cases, sometimes in pairs or group of 3, and only rarely in groups of up to 4. More than 70% of all *C. aculeolata* collections examined have spine-cells in clusters of more than 3 cells. Collections with predominantly single spine-cells have not been found in *C. aculeolata* (see Chap. 16). While *C. aculeolata* is almost exclusively found in freshwater, *C. corfuensis* prefers brackish waters.

"Mediterranean Chara" (see Chap. 80) differs from *C. corfuensis* by bract-cell patterns. In most collections, adaxial bract-cells are considerably longer than abaxial ones, and only in less than 30% of the collections, slightly longer (= "near-verticillate"). Also, most collections of *C. papillosa* (Chap. 44) have far longer adaxial than abaxial bract-cells. Spine-cell length is another useful character. They are long in *C. corfuensis* (up to almost 7 mm), evidently exceeding the axis diameter, while "Mediterranean Chara" has spine-cells ranging from 400 to 1700 μm and spine-cells of *C. papillosa* are usually shorter than the axis diameter. *Chara globata* (see Chap. 33) differs from *C. corfuensis* by the short spine-cells, usually obviously shorter than the axis diameter, and frequently globate upper whorls because of arcuate branchlets.

26.5 Distribution

See Fig. 26.3.

Chara corfuensis is found in Greece, Croatia, Albania and Montenegro. In Greece the species was reported from Corfu, Crete, Samos, and Sifnos Islands. The type locality between Moraitika and Braganiotika on Corfu no longer exists (Langangen, 2010, own search). The species was also recorded from Lake Koumoundourou, from the northeastern part of the lake, close to the springs and in different parts around it, as

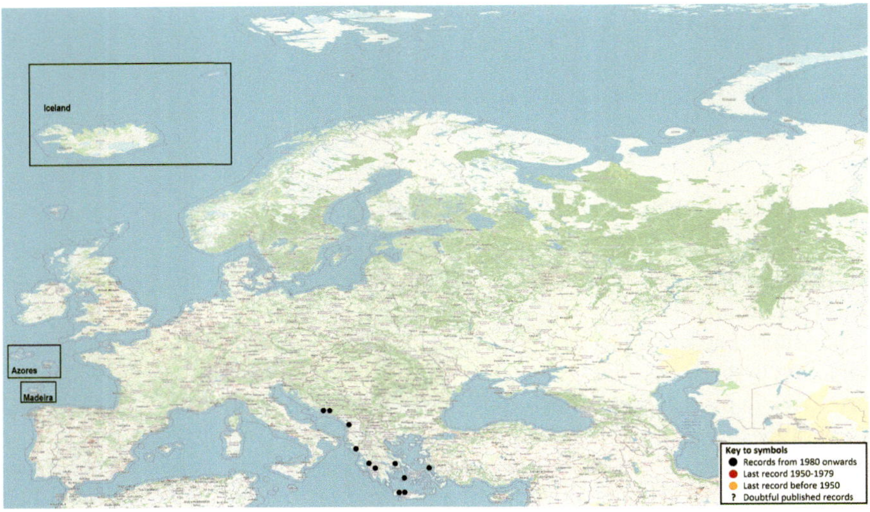

Fig. 26.3 Distribution of *Chara corfuensis* in Europe. Black dots are records from 1980 onwards, red dots last record from 1950 to 1979 and yellow dots last records before 1950; doubtful published records (if applicable) are marked with "?" superimposed. Map prepared by Heiko Korsch

well as from its centre (Kousouris et al., 1994; Pavlidou et al., 2004; Mentzafou et al., 2016), but it seems that today the species occurrence is doubtful due to decreased water quality (Mentzafou and Dimitriou, 2019).

C. corfuensis was also collected in Kaiafas and Prokopos lagoons (Western Greece). Kaiafas lagoon is a very well-preserved location for the species which is still present and in high abundance. In Kaiafas lagoon, sonar data in conjunction with targeted ground truthing sampling revealed that *C. corfuensis* appeared to be the dominant species, (Christia et al., 2014).

The species is also recorded from Aegean Sea islands of Sifnos (Lake Kamares) and Samos (Mesokampos, dam-hidden lake). In Crete it occurs in Lake Kournas (Chania) and Argyroupolis spring (Rethymno). In the Balkan area *C. corfuensis* was recorded in Lake Butrint, Lake Bufi (Sarande, Albania) and in Croatia at the Bacinska lakes complex and especially in Lake Desne (Blaženčić and Blaženčić, 2002, 2003; Urbaniak and Blaženčić, 2012; BEOU, confirmed by R. Romanov and I. Trboević). In Montenegro *C. corfuensis* was found in 2023 in an interdunal coastal pond with a highly variable water level that drops severely during summer (c. 1 m) (Romanov, unpublished data).

Reports of an occurrence in Spain (Pedrola et al., 1990) as *Chara* aff. *corfuensis* at Peniscola (Castellon), are not confirmed by herbarium material.

26.6 Status and threats

Chara corfuensis is an endemic species in the Balkans and is included in the Red List of charophytes for the Balkans as Critically Globally Endangered according to IUCN Threat Categories and Criteria (Blaženčić et al., 2006). It is assessed as EN (Endangered) in the European Red List (see Chap. 9).

The most important threats seem to be pollution, water abstraction, hydrological interventions, anthropogenic activities and habitat destruction. Eutrophication probably is an indirect threat, acting mainly by competition for light with phytoplankton and filamentous algae. Since the species only is known from a few sites, there is a great need for monitoring of its populations and conservation actions.

References

Blaženčić, J., & Blaženčić, Ž. (2002). Rare and threatened species of Charophytes (Charophyta) in Southeast Europe. *Phytologia Balcanica, 8*, 315–326.

Blaženčić, J., & Blaženčić, Ž. (2003). An overview of the existing data on living charophytes (Charales) of the Balkan Peninsula. *Acta Micropaleontologica Sinica, 20*, 103–110.

Blaženčić, J., Stevanović, B., Blaženčić, Ž, & Stevanović, V. (2006). Red data list of charophytes in the Balkans. *Biodiversity and Conservation, 15*, 3445–3457.

Bouzos, D., & Kontopoulos, N. (2004). Oceanographic observations in the lagoon of Kaiafa (W. Peloponnese). *Bulletin of the Geological Society of Greece, 36*, 1026–1035. [in Greek].

Christia, C., Tziortzis, I., Fyttis, G., Kashta, L., & Papastergiadou, E. (2011). A survey of the benthic aquatic flora in transitional water systems of Greece and Cyprus (Mediterranean Sea). *Botanica Marina, 54*, 169–178.

Christia, C., Papastergiadou, E., Papatheodorou, G., Geraga, M., & Papadakis, E. (2014). Seasonal and spatial variations of water quality, substrate and aquatic macrophytes based on side scan sonar, in an eastern Mediterranean lagoon (Kaiafas, Ionian Sea). *Environmental Earth Sciences, 71*, 3543–3558.

Dimitriou, E., Mentzafou, A., Zogaris, S., Koutsikos, N., Colombari, E., Markogianni, V., Karaouzas, I., Konstantinopoulou, A., Stathopoulou, E., Dasenakis, M., Katsiapi, S., & Moustaka, M. (2012). Monitoring of the ecological quality of Koumoundourou Lake and designing of management, restoration and developmental actions. In *Final Technical Report, Hellenic Centre for Marine Research, Anavyssos* [in Greek].

European Commission. (1992). Council directive 92/43/EEC of 21 May 1992 on the conservation of natural habitats and of wild fauna and flora. *Official Journal of the European Communities L, 206*, 7.

Filarszky, N. (1937). Idegenfoldi Charafelek hatarozasa - Determinatio Characearum exoticarum. *Magyar Tudományos Akadémia Budapest, 55*, 476–495. [in Hungarian].

Fyttis, G., Reizopoulou, S., & Papastergiadou, E. (2019). Assessing the ecological quality of coastal lagoons of Western Greece: The response of benthic communities to environmental variability. *International Journal of Environmental Science and Development, 10*, 91–95.

Katsaros, D., Panagiotaras, D., Kontopoulos, N., & Avramidis, P. (2017). Sediments characteristics and heavy metals distribution of a very shallow protected coastal lagoon, Prokopos lagoon, Mediterranean Sea, Western Greece. *Fresenius Environmental Bulletin, 26*, 6093–6103.

Kousouris, T., Diapoulis, A., Bertahas, I., et al. (1994). *Study of environmental upgrading in Koumoundourou Lake*. National Centre for Marine Research. [in Greek].

Langangen, A. (2007). Some charophytes from the Western Cyclades (Greece). *Journal of Biological Research, 7*, 109–113.
Langangen, A. (2010). Some charophytes (Charales) collected on the island of Evia, Greece in 2009. *Flora Mediterranea, 20*, 149–157.
Langangen, A. (2012). Charophytes (Charales) from Crete collected in 2010. *Flora Mediterranea, 22*, 25–32.
Mentzafou, A., Dimitriou, E., & Zogaris, S. (2016). Integrated ecological assessment and restoration planning in a heavily modified peri-urban Mediterranean lagoon. *Environmental Earth Sciences, 75*, 983.
Mentzafou, A., & Dimitriou, E. (2019). Nitrogen loading and natural pressures on the water quality of a shallow Mediterranean lake. *Science of Total Environment, 646*, 134–143.
Pavlidou, A., Dasenakis, E., Diapoulis, A., et al. (2004). *Study of the groundwater, the water of the Koumoundouros Lake and Elefsis Gulf, in relation to the landfill area of the Western part of Attica Region.* Hellenic Centre for Marine Research. [in Greek].
Pedrola, J., Acuña, J. D., & Bastida, J. (1990). Observaciones sobre algunos caracteres de valor taxonómico en el grupo de los carófitos. *Collectanea Botanica, 18*, 21–43. [in Spanish].
Urbaniak, J., & Blaženčić, J. (2012). SEM study of oospore characteristics in endemic and endangered Balkan charophytes. *Cryptogamie, Algologie, 33*, 277–288.
Wood, R. D. (1965). *Monograph of the Characeae.* J. Cramer, Weinheim.

Chapter 27
Chara curta

Nick Stewart

Scientific name: *Chara curta* Nolte ex Kütz., Tab. Phyc. 7: 22, pl. 53. 1857

≡ *Chara aspera* Deth. ex Willd. var. *crassior* Kütz., Sp. Alg. 521. 1849 [basionym]. Lectotype (Wood, 1965: 205–206): Bej Schallsee u. Wittbeck—in Lauenburg [Germany]/[no date]/[collector not stated but likely to be E. F. Nolte] (L).

≡ *Chara globularis* Thuill. var. *aspera* (Deth. ex Willd.) R. D. Wood f. *curta* (Nolte ex Kütz.) R. D. Wood, Taxon 11: 11. 1962.

= *Chara aspera* Deth. ex Willd. subsp. *desmacantha* H. Groves and J. Groves in J. Bot. 36: 410. 1898.
Lectotype (Wood, 1965: 206): Canal running into Lough Owel, Westmeath, Ireland/ 5 VIII 1892/H. C. Levinge, H. Groves and J. Groves (BM).

≡ *Chara desmacantha* (H. Groves and J. Groves) J. Groves and Bullock-Webster in Brit. Charoph. 1: 94. 1920.

Taxonomic note: The publication of the name *Chara curta* by Kützing (1857) is based on an illustration with the brief accompanying text merely explaining what plant parts are shown. However, Kützing refers to an earlier taxon, *Chara aspera* γ *crassior* described by him in Species Algarum (1849) and here "*Chara curta* Nolte" is mentioned as an unpublished name. This is described as "brachyphylla, maxime incrustata" [short-branched, greatly encrusted] and this matches an annotation by Braun in 1840 on the specimen later designated as the type "scheint eine stark incrustirte, kurzblättrige *Chara aspera* zu sein" [seems to be a heavily encrusted short leaved *Chara aspera*]. It is clear from other herbarium specimens that short branches and encrustation were the characters generally used in the nineteenth century for distinguishing this taxon. However, although commonly present in *Chara curta*, as now understood, they can also be found in *Chara aspera* and are not therefore

N. Stewart (✉)
4 Hexton Road, Glastonbury BA6 8HL, United Kingdom
e-mail: nfstewart@freeuk.com

H. Schubert et al. (eds.), *Charophytes of Europe*,
https://doi.org/10.1007/978-3-031-31898-6_27

diagnostic. In Kützing (1857), the close-up of the main axis shows the spine-cells occurring singly, but this is clearly not an accurate representation as the number of cortical rows shown suggest a diplostichous species and the spine-cells are shown emerging from every row. The identity of the name *Chara curta* therefore depends on the type. At some point, part of this has been cleaned with acid and this has shown that the spine-cells are clustered, and *Chara curta* is therefore the correct name for the species described here.

There are also discrepancies around the type localities. The type specimen mentions two localities Schaalsee (written as Schallsee) and Wittbeck, giving both as in Lauenburg (today in Germany but at that time in Denmark), while Kützing (1857) gives the locality as Schleswig. Both are only partly correct; Wittbeck is in Schleswig while Schaalsee is in Lauenberg. However, the protologue in Kützing (1849) gives no locality and these discrepancies are not important.

Local names: ENGLISH: Lesser Bearded Stonewort; NORWEGIAN: Knippe-bustkran; RUSSIAN: Хара пучковатоколючковая; UKRAINIAN: Хара пучкуватоколючкова.

27.1 Morphology

See Figs. 27.1, 27.2, 27.3 and 27.4.

Habitus: Plants light to mid green but normally encrusted and often whitish to greyish due to strong encrustation.

Size: Small to medium-sized, not usually exceeding 30 cm but sometimes to 45 cm.

Main axis: Fairly slender, 0.4–0.7 mm but appearing wider due to the dense covering of spine-cells. Internodes to 4 cm long, usually twice to four times the length of the branchlets.

Cortex: Triplostichous, usually regularly, with the secondary cortical cells usually meeting squarely but sometimes obliquely, tylacanthous.

Spine-cells: Well developed and persistent, in groups of 2–6 but usually 3–4, sometimes solitary in early growth stages or in wave-washed conditions. The spine-cells are often irregular in size and occasionally only one is developed with the others rudimentary or globular. Developed spine-cells mostly long (1–2 mm) up to four times the axis diameter, slender, acuminate and sometimes enlarged at base. Groves and Bullock-Webster (1924) note that they observed extra cells between the node cells of the cortex and the spine cells, but this has not been confirmed by other authors. See below under "possible confusion" for further discussion about spine-cell development.

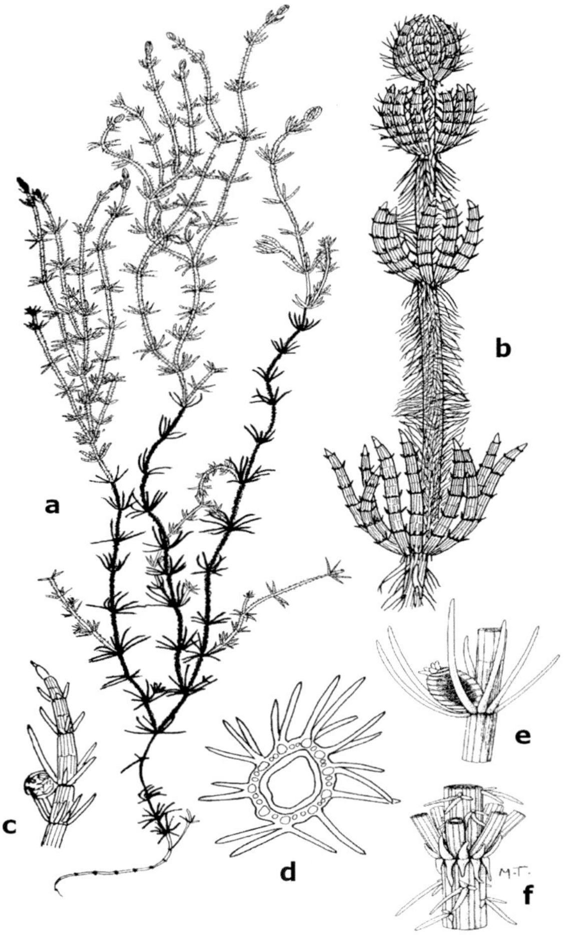

Fig. 27.1 *Chara curta*. **a** Habit; **b** Single axis; **c** Male branchlet; **d** Main axis cross-section; **e** Female branchlet node with oogonium; **f** Detail of axis and node. Drawings: **a–d**—reproduced from Krause (1997), **e–f**—reproduced from Moore (1986) with permission from the Botanical Society of Britain and Ireland

Stipulodes: Diplostephanous, both sets developed more or less equally and usually long (0.25–1.0 times as long as main axis diameter), tapering, acuminate and sometimes enlarged at base.

Branchlets: In whorls of (7)8–11, 0.5–5 cm long, moderately stout, straightish or incurved, of usually 7–8 segments of which the upper 1–2 are reduced and ecorticate.

Bract-cells: 5–7, near-verticillate, usually all long and slender and 1–2 x branchlet diameter but abaxial bract-cells usually a little shorter than the adaxial ones.

Bracteoles: 2, similar to the bract-cells but equalling or slightly exceeding them, to 2.2 mm, longer than the oogonium.

Bractlets: 1, usually as long as or slightly shorter than the bracteoles.

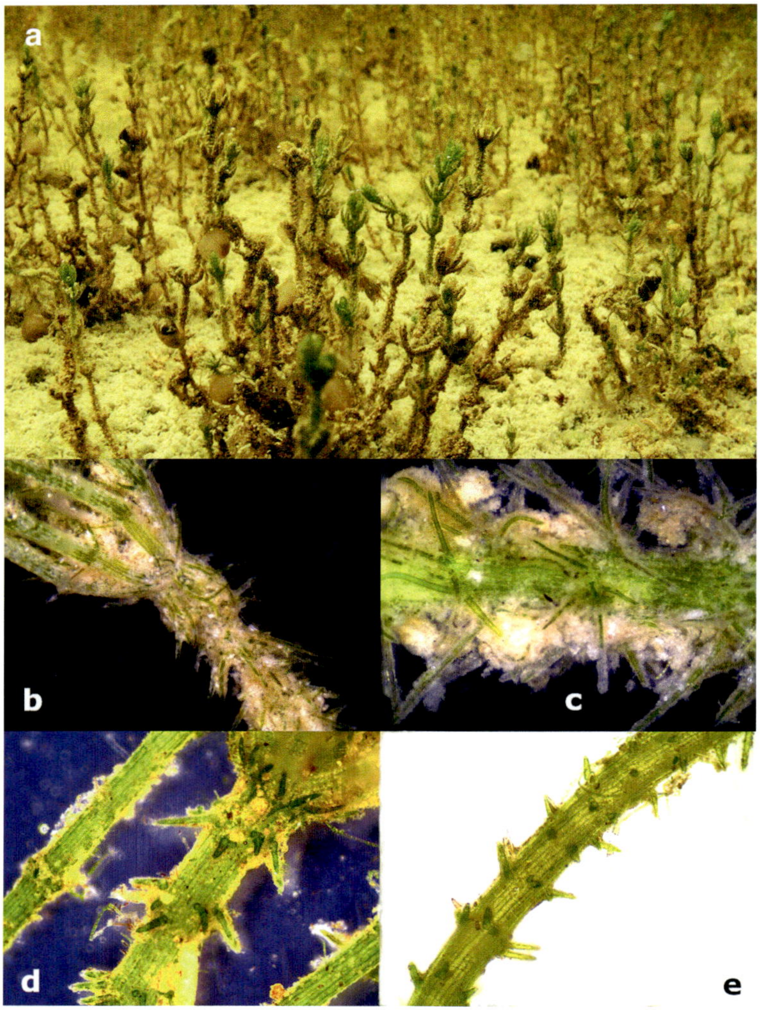

Fig. 27.2 Habitat, appearance and cortication of *Chara curta*. **a** Stand of *Chara curta* showing common Irish form with short branchlets and long internodes; **b, c** Typical form with well-developed spine-cells (from Grand Canal, Edenderry, Offaly, Ireland, coll. Nick Stewart 16/5/2018). As is often the case, the main axis and spine bases are difficult to see due to encrustation. However, note that many of the susbidiary spine-cells in each cluster are more tangentially arranged. In *C. aspera* the spine-cells tend to be more uni-directional and hence more ordered in appearance; **d, e** Less spiny forms (from Baston Fen, coll. Jon Graham 14/8/2019) with the dominant spine-cells developed but subsidiary spine-cells in each cluster much reduced. Photographs: **a**—Klaus van de Weyer, **b–e**—Chris Carter

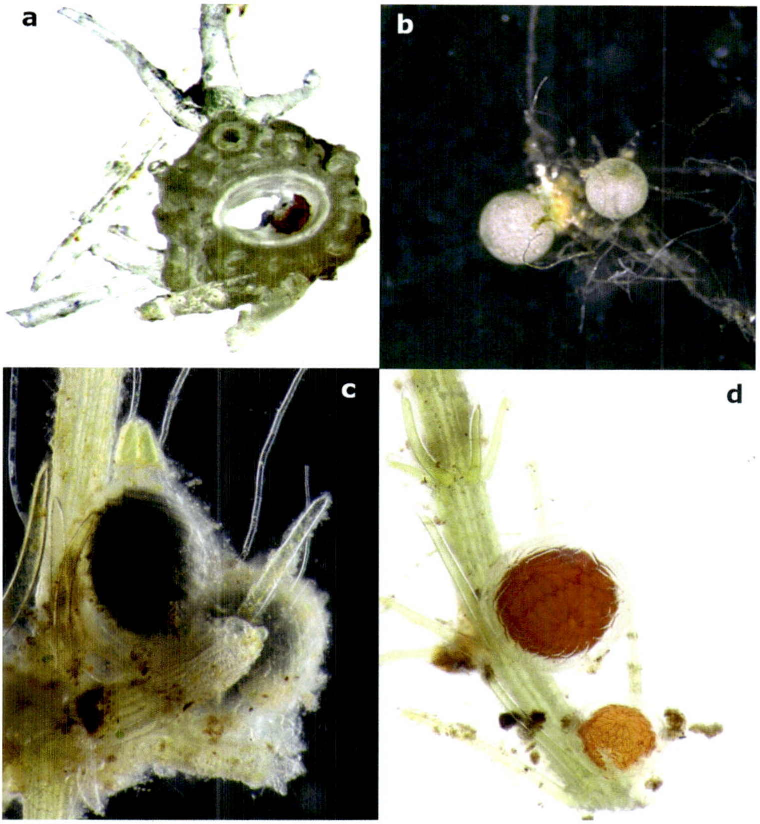

Fig. 27.3 *Chara curta,* details. **a** Main axis cross-section; **b** Rhizoid bulbils; **c** Ripe oogonia; **d** Male branchlet. Photographs: **a**, **c**, **d**—Claudia Ferguson-Smyth (specimen from Royal Canal WSW of Keenagh, Longford, Ireland, coll. Sylvia Reynolds 22/7/2016); **b**—Chris Carter (specimen from Grand Canal, Edenderry, Offaly, Ireland, coll. N. F. Stewart 16/5/2018)

Sexuality: Dioecious. Gametangia solitary, produced at the lowest 1–3 branchlet nodes.

Oogonia: Dimensions see Table 27.1. Coronula spreading.

Oospores: Dimensions see Table 27.2. Ridges weak, occasionally prolonged downwards at base into basal claws (Fig. 27.4a). Outer membrane thick-laminated, semi-rigid, semi-opaque, light brown to reddish-brown, granulated with granules of varying prominence (Fig. 27.4; Groves and Bullock-Webster, 1924). The granules are around 7 per 10 μm and are finer than those of *Chara aspera* (Fig. 27.4c, Groves and Bullock-Webster, 1920).

Antheridia: Dimensions see Table 27.3.

Table 27.1 Oogonia morphometry

Author	Length	Width	Coronula length	Coronula width	Number of helical stripes
Groves and Bullock-Webster (1924) and Corillion (1957)	625–750	375–450	c.125	c. 250	14–17
Wood (1965)	625–750	375–450	c.125	c. 250	14–17
Krause (1997)	600–750	375–450	c.125	c. 250	14–17
Bailly and Schaefer (2010)	-	400–500	-	-	-
Total range	**600–750**	**375–500**	**c. 125**	**c. 250**	**14–17**

Length in the first column is excluding the coronula; its length and width are given separately, all values for length and width in μm

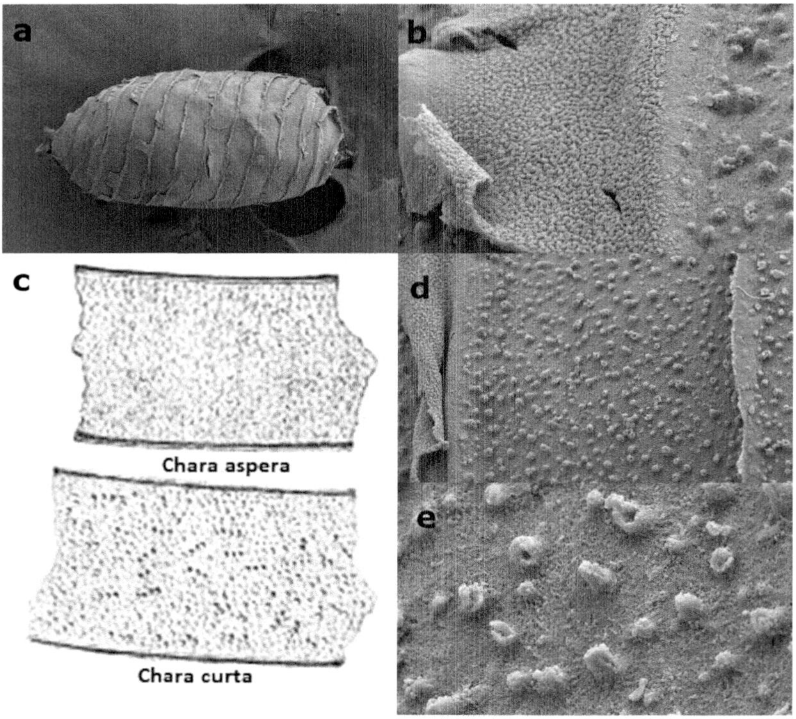

Fig. 27.4 *Chara curta*, oospore. **a** SEM of an oospore from Cotswold Water Park, Lake 29, M. T. Casanova and N. Stewart 2019; **b**, **d**, **e** Surface ornamentation and ornamentation details of the oospore shown in a; **c** Light microscope—comparison of *Chara aspera* and *Chara curta*. Surface ornamentation pattern. Photographs: **a**, **b**, **d**, **e**—M. T. Casanova; **c**—reproduced from Groves and Bullock-Webster (1924)

Table 27.2 Oospore morphometry

Author	Length	Width	Number of striae	Colour
Groves and Bullock-Webster (1920, 1924) and Corillion (1957)	c.525	275–325	13–16	Bright red brown
Wood (1965)	450–525	275–325	13–16	-
Krause (1997)	450–525	250–350	15–16	Brown
Bailly and Schaefer (2010)	-	350–450	12–14	Blackish
Total range	**450–525**	**275–450**	**12–16**	

All values for length and width in μm

Table 27.3 Antheridia morphometry

Author	Diameter (μm)
Groves and Bullock-Webster (1924)	Up to 650
Corillion (1957)	650–750
Wood (1965)	480–650
Krause (1997)	500–650
Bailly and Schaefer (2010)	400–600
Total range	**480–750**

Bulbils: Single-celled, spherical, whitish bulbils, 1–2 mm diameter are often produced on the rhizoids, singly or in groups but these are less frequent than in *Chara aspera*.

27.2 Phenology and Reproduction

Rather little is known about the early growth stages of *Chara curta*. It predominantly grows as a perennial but can be annual in ephemeral pools and in shallow water, particularly in wave-exposed conditions. Here it regrows from bulbils as well as oospores. It can be found fertile throughout the summer. Groves and Bullock-Webster (1924) suggest that it does not fruit freely and that ripe oospores are rare, but it seems no less fertile than other charophytes.

27.3 Habitat Conditions and Ecology

Habitat type: This is a species of hard waters of various types from small ponds to large lakes. Situations include marl lakes, mesotrophic lakes in areas of calcareous geology, gravel and clay pits, quarry lakes, dune slack pools and calcareous fen pools, but also less commonly in ditches and canals. In stony marl lakes, particularly in central Ireland, it is characteristic of a zone between the cyanobacterial crust zone (krustenstein) in the shallows and the *Chara subspinosa* zone in deeper water (Roden et al., 2020). This is optimally developed in the 1–2 metre depth band but varying with exposure to wind/waves and water clarity. In gravel and clay pits it is characteristic of the more sparsely vegetated zone up to 1 metre depth kept open by wave-wash and water fluctuation. Here the depth is limited by the deep-water communities such as *Elodea* species.

Sediment: Peat, clay, marl, sand, gravel or organic mud.

Nutrient conditions: Details of water chemistry from 40 lakes in the United Kingdom are given in the Table 27.4. Although there are some outliers, the majority of sites have TP less than 40 μg/l and TKN less than 1.2 mg/l. However, healthy populations are restricted to levels much lower than this. For example, in a study of marl lakes Roden et al. (2020) gives data for 17 sites with healthy populations with TP ranging from 3–19 μg/l, median 9.5 μg/l.

Water chemistry: See Table 27.4; most poulations are in hard water, but it can occur in softer water especially if magnesium levels are high. The species is exclusively a freshwater one, but some sites are near exposed coasts where some wind-carried salt is likely. However it is largely absent from sites underlain by calcareous ("machair") sand on the exposed coasts on the western fringes of Ireland and Scotland where the related *Chara aspera* is very abundant. Records from two brackish sites mentioned in Stewart (2001) are almost certainly from isolated locations close to freshwater inflows.

Depth range: The species can occur in depths from 0.1–4 metres, rarely up to 5 metres if water clarity is good (Roden et al., 2020). In small water bodies it is mostly restricted to permanent waters, but some sites have significant water fluctuations and it can tolerate being exposed in dry summers.

Associated species: This species often occurs in monospecific stands and often in some density. It typically occurs in sites that are rich in other charophytes, but the most frequent direct associates are *Chara contraria, C. aculeolata* and *C. tomentosa*. It can also occur with *Chara aspera* in some situations.

In more vascular plant dominated lakes, scattered plants of the deeper water communities can occur amongst the *Chara curta*. Many such sites have a diverse mesotrophic aquatic flora including a variety of *Potamogeton* species such as *P. gramineus, P. lucens, P. pusillus, P. crispus, P. praelongus* and *P. perfoliatus*. Other associates include *Stuckenia pectinata, Myriophyllum* species (mainly *M. spicatum* but sometimes also *M. verticillatum* or *M. alterniflorum*), *Persicaria amphibia* and *Elodea* species. In shallow water it may also be associated with *Littorella uniflora, Baldellia ranunculoides* or *Eleocharis acicularis*. In sites affected by eutrophication *Lemna trisulca, Lemna minor* or *Ceratophyllum demersum* may also be present.

In lakes where the alkaline influence is more due to basaltic rocks it can also occur with a suite of softer water species such as *Lobelia dortmanna, Utricularia vulgaris*

Table 27.4 Chemical data from 40 lakes in the United Kingdom

			n	Min	1st Quartile	Median	3rd Quartile	Max
pH	pH	-	40	6.7	7.9	8.1	8.4	9.5
Conductivity	COND	$\mu S\ cm^{-1}$	39	110	258	315	399	951
Alkalinity	ALK	$m\ eql^{-1}$	39	0.56	1.52	2.16	2.56	4.91
Soluble reactive phosphorous	SRP	μgl^{-1}	13	1	2	2	6	224
Total soluble phosphate	TSP	μgl^{-1}	31	2.2	4.9	8.7	14	252
Total phosphate	TP	μgl^{-1}	34	5.1	15.3	19.2	24.9	345
Ammonium	NH4	μgl^{-1}	11	3	9	27	49	354
Nitrates	NO3	μgl^{-1}	11	0.01	0.01	0.01	0.15	0.48
Total Kjeldahl nitrogen	TKN	mgl^{-1}	33	0.07	0.25	0.61	0.99	2.48
Potassium	K	mgl^{-1}	12	0	0.6	1.2	1.8	7.2
Calcium	Ca	mgl^{-1}	12	16.6	22	30.8	54.7	90
Magnesium	Mg	mgl^{-1}	12	2.1	3.4	5.1	13.5	21.3
Silica	SiO2	μgl^{-1}	11	0.4	1.2	1.7	2.6	10.3
Chloride	Cl	mgl^{-1}	12	9.6	12.9	23.3	29	280.8
Sulphate	SO4	mgl^{-1}	12	6.4	8.6	10.4	11.9	44.8
Chlorophyll a	CHLOR	μgl^{-1}	12	2.6	3.7	6.2	21.2	40
Sodium	Na	mgl^{-1}	12	2.6	5.3	10.8	14.1	100.5

Most data are based on single water samples collected in summer. (Source data from Scottish Natural Heritage, Northern Ireland Environment Agency, Cotswold Water Park Society and Freshwater Habitats Trust)

agg. and the aquatic form of *Juncus bulbosus*. In such sites the *Chara curta* is often very little encrusted (C. Roden pers. comm.).

27.4 Variability

Plants vary from forms with stout, rigid main axes, with short branchlets and long internodes, reminiscent of a smaller *Chara canescens*, to more slender forms with more flexuous branchlets with an appearance closer to *Chara aspera*. More rarely there are also small, condensed, stiff forms in shallow water which are akin to *Chara aspera* var. *lacustris* and *Chara virgata* var. *annulata*. There is some variation in the development of the spine-cells, stipulodes and bract-cells, as discussed below (under "Identification").

27.5 Identification and Risk of Confusion

This species is near to *Chara aspera* but differs primarily in the clustered spine-cells. The spine-cells are typically dense and hide the main axis, but some forms are more open and *aspera*-like in appearance. In each cluster the subsidiary spine-cells tend to be angled away from the line of the main axis in a tangential way, and this can be a useful guide when it is difficult to see whether the spine-cells are clustered or just very close together (e.g. due to encrustation, Fig. 27.2). By contrast the spine-cells of *Chara aspera* are more uni-directionally angled towards the centre of the internode.

Chara aspera can rarely produce the occasional double or even triple spine-cell group among predominantly single spine-cells, while on the mature parts of *Chara curta* multiple spine-cells are usually numerous. Difficulties arise on juvenile plants and on plants growing in wave-washed conditions. In such cases, both species tend to have less spine-cell development. In *Chara curta* the earliest axes are often single-spined but become multiple-spined on the upper parts of mature axes. There are also examples where only one spine-cell is well-developed and subsidiary spine-cells are present only as reduced globular cells at the base of the developed spine-cell. Close examination may be needed to see these, particularly if there is significant encrustation. Such clusters can be found on the axes as they develop towards maturity but in some samples, it may be difficult to find clusters with all of the spine-cells well-developed.

The species also differ in the ornamentation of the oospores. While both species have a granular decoration, those of *Chara curta* are finer (c. 7 granules per 10 μm) than those of *Chara aspera* (c. 5 granules per 10 μm) (Groves and Bullock-Webster, 1920).

According to Groves and Bullock-Webster (1920, 1924) other characters that differentiate *Chara curta* from *Chara aspera* are:

- It tends to be stouter and have more rigid main axes and branchlets.
- The bract-cells are more numerous (5–7 compared to around 5 in *C. aspera*), are near-verticillate and usually exceed the oogonia (compared to the outer bract-cells being shorter than the inner ones and these are usually shorter to sometimes longer than the oogonia in *C. aspera*).
- The taller coronula (c. 125 μm in *C. curta,* 75–100 μm in *C. aspera*).
- The larger antheridia (to 650 μm in *C. curta,* 500–600 μm in *C. aspera*).
- The lighter oospore (bright red brown in *C. curta,* very dark to blackish in *C. aspera*).
- The more numerous cortical nodes in each internode (14–18 in *C. curta,* 10–12 in *C.aspera* counting the opposed ascending and descending rows).

However, it has not been possible to test the strength of these characters, and some may be indicative more than diagnostic.

In its well-developed form, *Chara curta* resembles a more slender version of *Chara canescens* in its dense spine-cells and neat short branchlets with relatively long internodes (this is reflected in its name in English). The latter can easily be distinguished by the milky-translucent axes, the lack of male plants (in most of Europe) and also microscopically by the fewer rows of cortex (haplostichous). Slender forms of *Chara aculeolata* may also look similar but this species is diplostichous. However the tendency in the latter species for the secondary cortex to meet very obliquely and overlap and therefore appear partially triplostichous can cause confusion. Also both species are densely spiny and often strongly encrusted making it difficult to view the cortex without (and even with) decalcification with acid. If fertile, then *Chara aculeolata* can be easily distinguished because it is monoecious. The branchlet whorls also tend to have a rougher bristlier appearance compared to the softer, more fur-like appearance of *Chara curta* (Fig. 27.2).

27.6 Distribution

See Fig. 27.5.

The distribution of this species is imperfectly known due to its treatment in some countries as a variety of *Chara aspera* and perhaps sometimes not distinguished at any level.

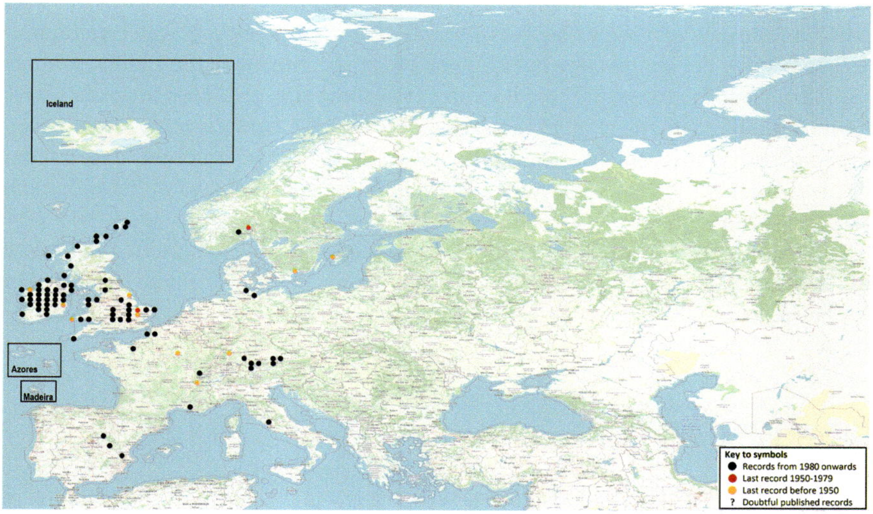

Fig. 27.5 European distribution of *Chara curta*. Note that there are no dots for some countries from which it has been confirmed because in past decades it was not treated as a species. Black dots are records from 1980 onwards, red dots last recorded from 1950 to 1979 and yellow dots last recorded before 1950; unconfirmed published records (if applicable) are marked with "?" superimposed. Map prepared by Heiko Korsch

It is most frequent in Ireland where it is locally frequent in the central limestone area and also scattered but less frequent in Britain (Bryant and Stewart, 2011). It also seems to be frequent in the Jura Mountains on the border of France and Switzerland (Bailly and Schaefer, 2010). Elsewhere there are scattered records from Sweden and Norway (Langangen, 2007), Germany (Arbeitsgruppe Characeen Deutschlands 2016—included under *C. aspera*), Austria (specimens seen), Italy (Bazzichelli and Abdelahad, 2009—illustrated as *C. aspera*), Spain (Cirujano et al., 2007, 2008 as *C. aspera* var. *curta*) and Morocco (Feldmann, 1953 fide Corillion 1957 as *C. desmacantha*).

27.7 Status and Threats

A number of sites have been completely lost due to eutrophication (e.g., Stewart, 2001) and this is the most widespread threat. Because of its preference for shallower water, it is less vulnerable to increased turbidity but the zone where it occurs is more vulnerable to smothering by filamentous algae, increased vascular plant competition and substrate change (i.e. becoming softer and more organic due to algal deposition). See also under "Nutrient conditions" for chemical details.

In the deeper water of vascular plant dominated lakes *Chara curta* is often constrained by the vascular plant communities. This is particularly an issue with spreading introduced species such as *Elodea nuttallii* where deep water communities can be very dense and large detached rafts can accumulate in the shallows where the *Chara curta* is growing. There is evidence at some sites that *Elodea* species have significantly reduced the extent of, and even excluded, *Chara curta* communities.

In gravel pits, this species is particularly associated with younger pits and in older sites it is frequently pushed out by a combination of marginal swamp development and overshading by trees.

In smaller sites, such as ponds, there is often an issue of increased competition from swamp and vascular plants due to successional changes. Periodic clearance or other disturbance may be necessary to maintain open conditions.

27.8 Remarks

There has been some disagreement over whether this taxon deserves specific rank or should be considered at a lower level within *Chara aspera* (compare Krause, 1997; van Raam, 1998). It has long been recognised as a separate species in Britain and Ireland which appears to be its stronghold and in some parts of central Ireland it is much more common than *Chara aspera*. Some populations can be difficult to place because of poor spine-cell development but this is usually because of ecological conditions or early growth stages (see above). A study of the genetics of the two species using AFLP markers failed to find a difference between them (O'Reilly et al., 2007) but further work is needed here.

Nevertheless, in addition to the characters described above under "risk of confusion", particularly the clustered spine-cells and differences in the oospore ornamentation, there are also differences in the ecology and distribution of the two taxa. While *Chara curta* is largely restricted to hard, fresh waters, *Chara aspera* is more broad ranging extending into softer waters as well as brackish waters. This is particularly striking in Ireland where the usual plant in marl lakes on the limestones of central Ireland is *Chara curta* while *Chara aspera* is the usual plant in lakes on the calcareous (machair) sand on the west coast. There is therefore a good case for it to continue to be treated as a separate species.

References

Bailly, G., & Schaefer, O. (2010). *Guide illustré des Characées du nord-est de la France* (96 pp.). Conservatoire Botanique National de Franche-Comté, Besançon [in French].

Bazzichelli, G., & Abdelahad, N. (2009). *Alghe d'acqua dolce d'Italia. Flora analitica delle Caroficee.* Sapienza Università di Roma. Ministero dell'Ambiente e della Tutela del territorio e del mare. Roma, Centro Stampa Università (73 pp.) [in Italian].

Bryant, J. A. & Stewart, N. F. (2011). Order Charales. In D. M. John, B. A. Whitton, & A. J. Brook (Eds.), *The freshwater algal flora of the British Isles. An identification guide to freshwater and terrestrial algae* (2nd ed, pp. 742–765). Cambridge University, Cambridge.

Cirujano, S., Cambra, J., Sánchez Castillo, P. M., Meco, A., & Flor Arnau, N. (2008). *Flora ibérica. Algas continentales. Carófitos (Characeae).* Real Jardín Botánico, Madrid (132 pp.) [in Spanish].

Cirujano, S., García Murillo, P., Meco, A., & Fernández Zamudio, R. (2007). Los carófitos ibéricos. *Anales del Jardín Botánico de Madrid, 64*, 87–102. [in Spanish].

Corillion, R. (1957). Les Charophycées de France et d'Europe Occidentale. *Bulletin de la Société Scientifique de Bretagne, 32* (Hors série I & II), 1–499 [in French].

Feldmann, G. (1953). Les Charophycées marocaines. *Bulletin de la Société des Sciences Naturelles et Physiques du Maroc, 33*, 13–22. [in French].

Groves, J., & Bullock-Webster, G. R. (1920). *The British Charophyta. I Nitelleae.* Ray Society, London (405 pp.).

Groves, J. & Bullock-Webster, G. R. (1924). *The British Charophyta. II Chareæ.* Ray Society, London.

Krause, W. (1997). Charales (Charophyceae) Süßwasserflora von Mitteleuropa. In H. Ettl, G. Gärtner, H. Heynig, & D. Mollenhauer (Eds.), *Süßwasserflora von Mitteleuropa, 18.* Fischer [in German].

Kützing, F. (1849). *Species Algarum.* F.A. Brockhaus, Lipsiae [in Latin].

Kützing, F. T. (1857). *Tabulae phycologicae; oder, Abbildungen der Tange* (Vol. 7). W. Köhne, Nordhausen [in German].

Langangen, A. (2007). *Charophytes of the Nordic countries* (102 pp). Saeculum ANS.

Moore, J. A. (1986). *Charophytes of Great Britain and Ireland.* BSBI Handbook No. 5, Botanical Society of the British Isles, London.

O'Reilly, C. L., Cowan, R. S., & Hawkins, J. A. (2007). Amplified fragment length polymorphism genetic fingerprinting challenges the taxonomic status of the near-endemic species *Chara curta* Nolte ex Kütz. (Characeae). *Botanical Journal of the Linnean Society, 155*, 467–476.

Roden, C., Murphy, P., & Ryan, J. (2020). *Benthic vegetation in Irish marl lakes: monitoring habitat 3140. Condition 2011–2018*. Irish Wildlife Manuals 124. National Parks and Wildlife Service, Department of Housing, Local Government and Heritage, Dublin (36 pp.).
Stewart, N. F. (2001). *Review of the status of Biodiversity Action Plan stonewort species*. Plantlife Report 170. Plantlife International, Salisbury.
van Raam, J. C. (1998). *Handboek Kranswieren*. Chara boek, Hilversum [in Dutch].
Wood, R. D. (1965). *Monograph of the Characeae*. J. Cramer, Weinheim.

Chapter 28
Chara dissoluta

Roman Romanov, Nick Stewart, Klaus van de Weyer, Cilian Roden, Ivana Trbojević, Vera Biberdžić, Sonja Trajanovska, Antun Alegro, and Lefter Kashta

Scientific name: *Chara dissoluta* **A. Braun ex Leonh. in Verh. Naturf. Vereins Brünn Abh. 2: 182. 1864**

Lectotype (designated here by R. E. Romanov): [Switzerland] e lacu Neocomensi [from the Lake Neuchâtel/Neuenburgersee]/1854/Bulnheim (LE A0001500).

≡ *Chara contraria* A. Braun ex Kütz. subsp. *dissoluta* (A. Braun ex Leonh.) A. Braun in A. Braun & Nordst., Abh. Königl. Akad. Wiss. Berlin 1882: 145. 1882.

≡ *Chara dissoluta* A. Braun ex Leonh. f. *helvetica* Mig., Charac. Deutschl. 384. 1891, n. inv.

= *Chara dissoluta* A. Braun ex Leonh. var. *ohridana* Kostić in Acta Bot. Croat. 11: 69. 1937.
Lectotype (designated here by A. Alegro): Ohridsko Jezero: Gumnišće, 7 km daleko od Ohrida, dubina 5 m [Lake Ohrid: Gumnišće, 7 km away from Ohrid, 5 m deep]/ 4 September 1936/Lj. Kostić (ZA 68923).

R. Romanov (✉)
Nišice Ujtin Potok b.b., Dobra Voda, 85356 Bar, Montenegro
e-mail: streptophytes@gmail.com

N. Stewart
4 Hexton Road, Glastonbury BA6 8HL, United Kingdom
e-mail: nfstewart@freeuk.com

K. van de Weyer
lanaplan, Lobbericher Str. 5, D-41334 Nettetal, Germany
e-mail: klaus.vdweyer@lanaplan.de

C. Roden
Heritage Department, Atlantic Technological University, Dublin Road, Galway, Ireland
e-mail: cilianroden@icloud.com

H. Schubert et al. (eds.), *Charophytes of Europe*,
https://doi.org/10.1007/978-3-031-31898-6_28

≡ *Chara ohridana* (Kostić) W. Krause, Süßwasserflora 18: Charales: 123. 1997, n. inv.

Nomenclatural note: Krause (1975) made the combination *Chara ohridana* without fully citing the basionym and it is therefore invalid.

Taxonomic note: The species rank of *Chara dissoluta* has been questioned due to uncertainty over delineation from morphotypes of *C. contraria* with reduced cortication (e.g. Arbeitsgruppe Armleuchteralgen Deutschlands, 2016; Corillion, 1960; Krause, 1997; Krause and Krause, 1984–1986). This is discussed in greater detail under "Variability" and "Risk of confusion" below. For many years this taxon has been equated with South African plants collected originally by Drège (1843) and described as *Chara denudata* A. Braun (Braun, 1847, see Braun, 1868; Groves and Bullock-Webster, 1924 etc.). Although *C. denudata* shares many common morphological traits with European plants, they are distant according to *rbc*L sequences (Romanov et al., 2023) and accepted here as a separate species. They also seem to be ecologically distinct. *Chara denudata* can be recognised as a shallow water species of arid regions; it is usually richly fertile. *C. dissoluta* is (mostly) a deep-water species of deep, stratified, transparent, temperate lakes; it has mostly low fertility. The original material of *Chara dissoluta* var. *ohridana* is stored in ZA (Fig. 28.3), allowing designation of the lectotype. *C. dissoluta* var. *dissoluta* and *C. dissoluta* var. *ohridana* have similar morphology and these taxa are indistinguishable from each other (and from *C. contraria*) according to *mat*K sequences (Schneider et al., 2016). The detailed study of numerous specimens of *C. dissoluta* var. *ohridana* including original material (see below under "Variability") failed to find any evidence for its delineation from *C. dissoluta*. Therefore, it is accepted here as a synonym of *C. dissoluta*. This epithet has been used at species level, "*Chara ohridana*", by various

I. Trbojević
Institute of Botany and Botanical Garden Jevremovac, Faculty of Biology, University of Belgrade, Takovska 43, 11000 Belgrade, Serbia
e-mail: itrbojevic@bio.bg.ac.rs

V. Biberdžić
Natural History Museum of Montenegro, Bećir bega Osmanagića 16, P.O.Box 374, 81000 Podgorica, Crna Gora, Montenegro
e-mail: vera.biberdzic@pmcg.co.me

S. Trajanovska
Hydrobiological Institute Ohrid, Naum Ohridski 50, 6000 Ohrid, Macedonia
e-mail: sonjat@hio.edu.mk

A. Alegro
Division of Botany, Department of Biology, Faculty of Science, University of Zagreb, Marulićev trg, 20/II, 10000 Zagreb, Croatia
e-mail: antun.alegro@biol.pmf.hr

L. Kashta
Research Center for Flora and Fauna, University of Tirana, Place Mother Tereza, P.O. Box 183, Tirana, Albania
e-mail: lefter.kashta@fshn.edu.al

authors (e.g. Blaženčić et al., 2006, 2018; Krause, 1997; Urbaniak and Blaženčić, 2012) but has never been validly published.

Local names: DANISH: Barkløs Kransnål; ENGLISH: Naked Stonewort; GERMAN: Nackte Armleuchteralge; RUSSIAN: Хара обнаженная; UKRAINIAN: Хара оголена.

28.1 Morphology

See Figs. 28.1, 28.2, 28.3, 28.4, 28.5 and 28.6.

Habitus: Plants are light to green and olive green, always with a matt surface, more or less evenly encrusted, small to moderately tall, slender and simple in appearance, rough to touch, scarcely branching. The smooth appearance of the main axis and branchlets gives a resemblance in general appearance to a *Nitella* with simple branchlets (Figs. 28.1, 28.2 and 28.3). A useful field character for separation from the latter is that the main axis becomes whitish when broken. Plants from south-east European populations, especially specimens from shallow water, often have a bushy appearance.

Size: A small to medium and elongated plant, from 5 cm high in shallow water to >60 cm in deep water.

Main axis: Main axis (170)400–750 μm, up to 1500 μm (south-east European populations) in diameter, fragile because of encrustation.

Cortex: Rudimentary or incomplete, consisting of initials only appearing as a knot of cells around the nodes or equally spaced isolated primary tubes (Figs. 28.1c, g, t, z, 28.3d, e, g and 28.5d–g) which are rudimentary or evidently elongated, not contacting at the centre of the internode and not in contact with each other in the case of ones originating from the same node, frequently peeling off the internodal cell and inclined at acute irregular angles to the internode (Fig. 28.5e). Variations with more complete primary cortication are discussed in greater detail under "Variability" below. Secondary tubes are missing or just rudimentary to a little elongated (Fig. 28.5h) but complete tylacanthous, diplostichous main axis cortex is not formed. Imperfect triplostichous axis cortex consisting of narrow, scarcely contacting cells has very rarely been seen on some plants from Lake Ohrid (Fig. 28.5i).

Spine-cells: Spine-cells are always solitary, sparse, globular to rarely papillate, short, hardly recognisable.

Stipulodes: Diplostephanous, short, obtuse to ovoid, the lower set are somewhat shifted and spaced closer to each other relative to the upper set at each node (Fig. 28.4e–g). The lower stipulodes can be inconspicuous (Figs. 28.1s and 28.5b).

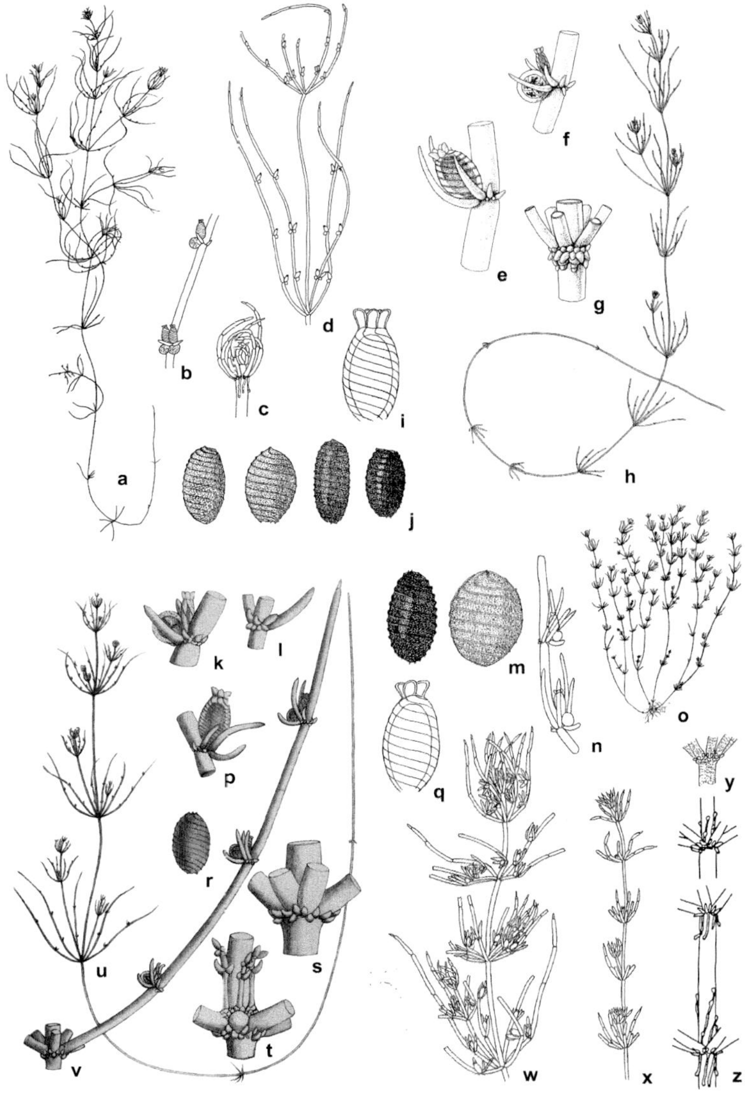

Fig. 28.1 *Chara dissoluta*. **a**, **h**, **u**, **o** General appearance; **b**, **e**, **f**, **k**, **l**, **n**, **p**, **m** Nodes of branchlets; **c**, **d**, **w**, **x** Apical parts; **g**, **s**, **t**, **y**, **z** Bases of branchlet whorls with different expression of cortication and stipulodes; **i**, **q** Oogonia; **j** Gyrogonites (two left) and oospores (two right); **r** Oospore; **m** Oospore (left) and gyrogonite (right); **a–d**, **i**, **j** central European populations**;** **e–h**, **k**, **l**, **p**, **r–v** Irish populations; **m–o**, **q**, **w–z** Balkan populations. More illustrations of cortication variability for central European populations are available elsewhere (Jäger, 2010; Sluiter, 1910). Drawings: **a–d**, **i**, **j**, **m–o**, **q**, **w–y**—Krause (1997); **e–h**—Moore (1986), **k**, **l**, **p**, **r–v**—Groves and Bullock-Webster (1924); **z**—Krause and Krause (1984–1986)

Fig. 28.2 General habit of *Chara dissoluta*. **a** Lake Constance/Bodensee, Germany; **b** Lough Owel, Ireland. Photographs: **a**—Klaus van de Weyer; **b**—Chris Carter

The initials of the primary cortical tubes are situated below and between them, opposite the branchlets, and can easily be erroneously recognised as stipulodes from the lower row.

Branchlets: (5)6–10 branchlets in a whorl, (5)6–8 in the case of south-east European populations, 0.5–2-4(6) cm length, predominantly without cortex, consisting of 1–3(4) segments with one or two basal segments with short cortex initials forming a ring at a node, some of which may elongate resulting in incomplete to complete diplostichous cortex, and a short or long undifferentiated/nodeless (2)3–4-celled end segment.

Bract-cells: The abaxial are invariably short, papillate. The adaxial are shorter, equal to or somewhat longer than the oogonium length (up to 2.5 times in south-east European populations), bluntly pointed to slightly acute in south-east European populations (Figs. 28.4d and 28.5g, m–o).

Bracteoles: Shorter or somewhat longer than the adaxial bract-cells.

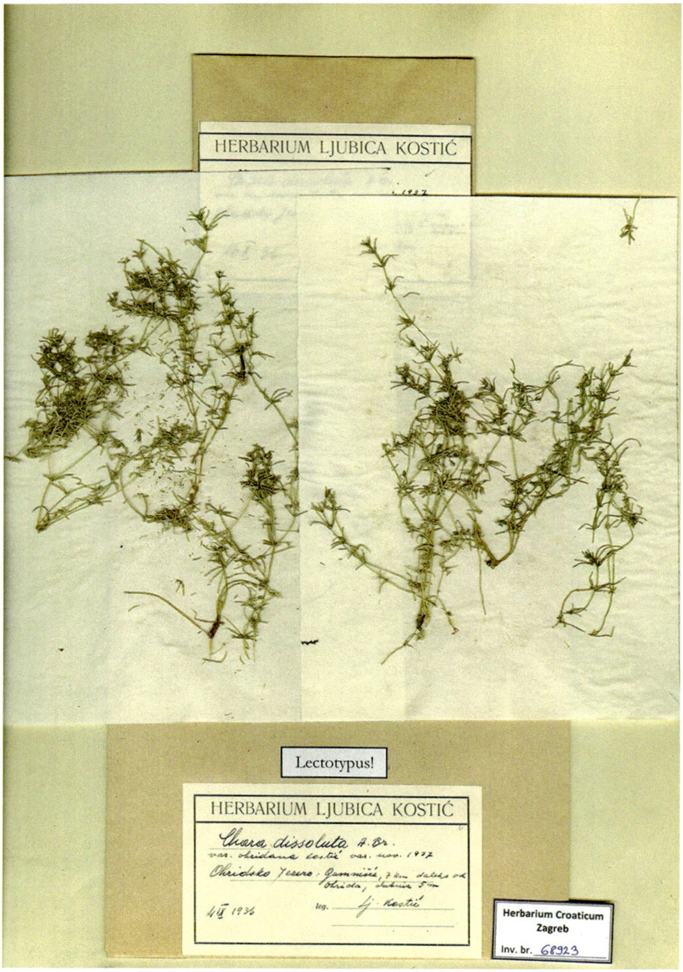

Fig. 28.3 Lectotype of *Chara dissoluta* var. *ohridana* (ZA). Photograph: Antun Alegro

Sexuality: Monoecious. Gametangia are solitary or geminate, conjoined (rarely sejoined (?), cf. Fig. 28.1e, p), at the 1–3 lowest nodes of the branchlet.

Oogonia: The oogonia are ellipsoid. The coronula is wide and slightly spreading. Table 28.1 gives an overview of the morphometric data. The gametangia measurements were taken from fixed material in the case of our own data. Gyrogonites are present, at least in Central European (Krause, 1997; Migula, 1889–1897) and south-east European populations (Krause, 1997; Krause and Krause, 1984–1986; own data), but not described.

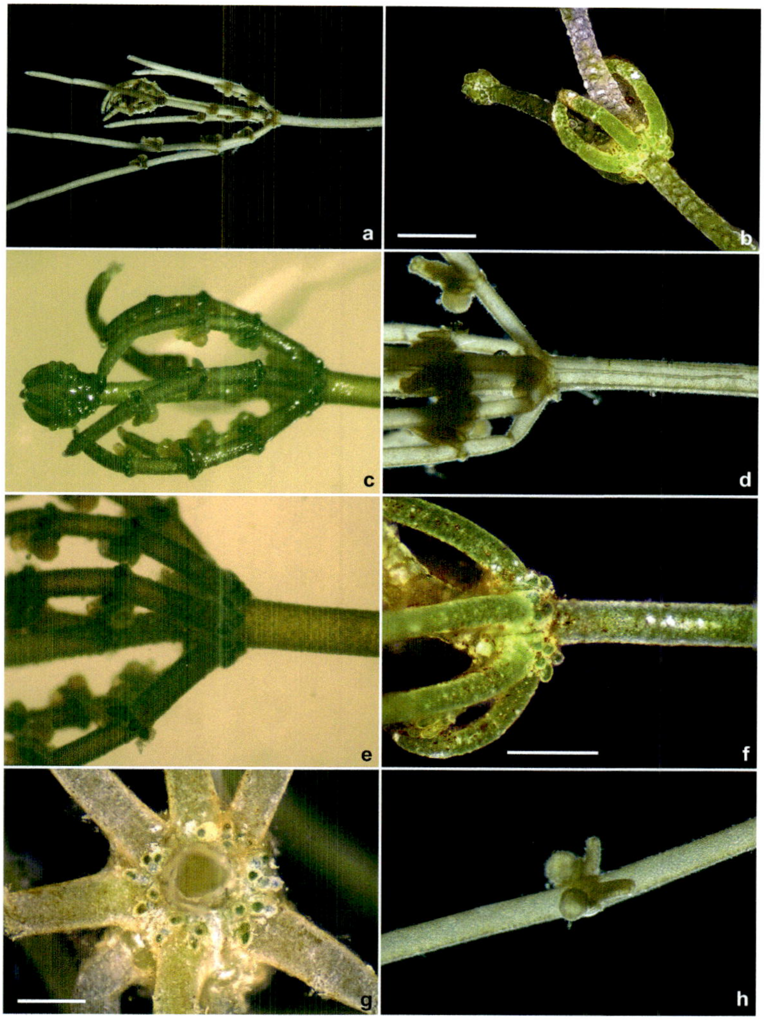

Fig. 28.4 *Chara dissoluta*, details. **a–c** Branchlets with different number of nodes, length of nodeless end part and bract-cells; **d** Base of the whorl and upper part of the internode with narrow elongated primary tubes without secondary ones; **e**, **f** Base of branchlet whorl with recognisable upper stipulodes and initials of primary tubes below them placed between their pairs; **g** Stipulodes and main axis cortex initials viewed from the base of branchlet whorl; **h** Geminate gametangia; **a**, **d**, **h** Lac Leman, Switzerland; **b**, **f**, **g** Lough Owel, Ireland; **c**, **e** Lake Constance, Germany. Scale: **b**—1 mm, **f**—0.5 mm **g**—0.5 mm. Photographs: **a**, **d**, **h**—Aurélie Boissezon; **b**, **f**, **g**—Chris Carter; **c**, **e**—Klaus van de Weyer

Fig. 28.5 *Chara dissoluta* var. *ohridana*. **a** Upper part of plant without cortex; **b** Base of branchlet whorl with short stipulodes (lower row is invisible) and initials of axis cortex below them, initials of branchlet cortex at branchlet nodes; **c** Upper part of plant with better developed axis and branchlet cortex and anterior bract-cells exceeding length of oogonia; **d** Base of branchlet whorl and part of axis internode with few elongated primary tubes of axis cortex; **e** Base of branchlet whorl and part of axis internode with elongated peeling off primary tubes of axis cortex; **f** Elongated primary tubes of axis cortex with globular spine cells (arrowhead) and short secondary tubes; **g** Base of fertile branchlet whorl and elongated primary tubes of axis cortex above and below main axis node; **h**: More developed axis cortex with long primary tubes and elongated secondary tubes (arrowhead); **i** Irregular triplostichous axis cortex (arrowhead); **j–l** Ends of branchets; **m** Conjoined gametangia between ecorticate branchlet segments; **n** Conjoined gametangia between ecorticate branchlet segments with unpaired bract-cells below antheridia (arrowheads); **o** Fertile branchlet node with elongated cortex tubes below; **p** Two basal nodes of branchlet with noncontiguous cortex tubes; **q** Gyrogonites; **r** Starchless nodal bulbil. Photographs: **a–h**, **j–r**—Vera Biberdžić; **i**—Ivana Trbojević

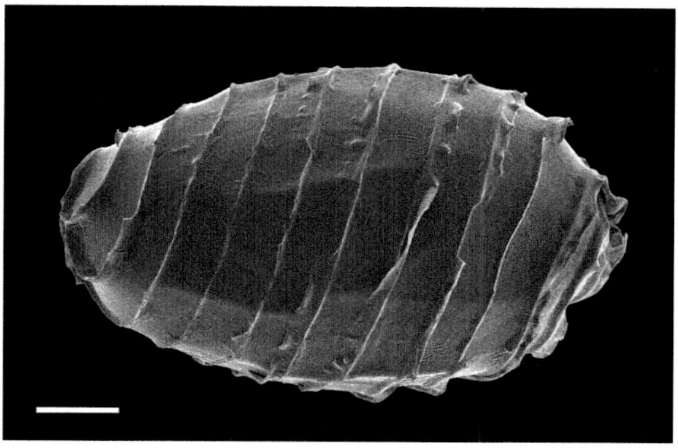

Fig. 28.6 Oospore of *Chara dissoluta* var. *ohridana*. SEM photograph of an oospore taken from a specimen from Lake Ohrid (depth 3.5 m, 17 July 2005, BEOU 1194). Scale: 100 μm. Photograph: Roman Romanov

Oospores: Ellipsoid, variable in outline, from slightly ovoid to ellipsoid or nearly cylindrical, with rather weak thin spiral ridges ending with short claws at the base of oospore. The oospore membrane is finely granulate to papillate under light microscope. Var. *ohridana* has a smooth surface in SEM and flanged ribs (Urbaniak and Blaženčić, 2012; Fig. 28.6). Table 28.2 gives an overview of the morphometric data.

Antheridia: The antheridia are sessile. Table 28.3 gives an overview of the morphometric data.

Bulbils: Not described for many populations but starchless nodal bulbils are common in south-east European populations (Fig. 28.5r).

28.2 Phenology and Reproduction

Low fertility is typical, and plants are not continuously fertile. It seems to be perennial. Oospores are ripe in July and August in northern Europe (Schou et al., 2017). Gametangia are more common in south-east European populations (Kashta et al., 2013; Trajanovska and Blaženčić, 2008; own data) and most abundant during summer and in plants from the shallows (Trajanovska and Blaženčić, 2008). The plants from deeper sites are described as sterile (Kostić, 1937) but this is not always the case (Trajanovska and Blaženčić, 2008).

28.3 Habitat Conditions and Ecology

Habitat types: *Chara dissoluta* is an exclusively freshwater species of strongly alkaline stable standing water bodies, mainly deep transparent lakes on marl and limestone (Stewart and Church, 1992). Here it grows in intermediate to greater depths,

Table 28.1 Oogonia morphometry

Author	Length	Width	Coronula length	Coronula width	Number of helical stripes
Migula (1889–1897)	800–900	500–600	100	250	12–14
Sluiter (1910)	750–1000	400–570	80–130	200–250	12–14
Groves and Bullock-Webster (1924)	ca. 1000–1100	ca. 550–600	ca. 175–200	350–400	13–15
Kostić (1937), var. *ohridana*	896–980	434–462	98–140	182–364	12–16
Corillion (1957)	800–900	500–600	100	250	–
Wood (1965)	800–900	500–600	ca. 100	ca. 250	12–14
Corillion (1975)	800–1100	550–600	(105)175–200	255–400	13–15
Hollerbach and Krassavina (1983)	800–1100	540–600	105–120	255–400	13–15
Moore (1986)	–	>500	–	–	–
Krause (1997)	800–1000	500–700	100–150	200–300	11–14
Krause (1997), var. *ohridana*	950	ca. 700	ca. 150	ca. 300	12–14
Langangen (2007)	Up to 1000	–	–	–	–
Trajanovska and Blaženčić (2008), var. *ohridana*	335–1105	124–790	52–268	130–460	–
Bailly and Schaefer (2010)	500–700	–	–	–	–
Schou et al. (2017)	500–1100	400–700	–	–	–
Own data, var. *ohridana*	460–1200	400–670	75–290	150–400	11–13
Total range	**(335)500–1200**	**(124)400–790**	**(52)75–290**	**130–460**	**11–16**

Length in the first column is excluding the coronula; its length and width is given separately, all values for length and width in μm

where it can form its own zone at the lower end of the euphotic zone often forming dense swards > 50 cm in height (Roden et al., 2020; Schou et al., 2017). In Ireland it is particularly associated with a type of stony marl lake where the shallows are dominated by encrusting cyanobacteria (Roden et al., 2020). It is mostly unknown from shallow and temporary water bodies but var. *ohridana* has also been reported from a river and puddles (Blaženčić et al., 2018). *C. dissoluta* has been recently

Table 28.2 Oospore morphometry

Author	Length	Width	Number of striae	Colour
Migula (1889–1897)	780	530	–	Black
Sluiter (1910)	600–780	300–410	–	Black
Groves and Bullock-Webster (1924)	675–825	400–450	12–14	Reddish-brown to nearly black
Kostić (1937), var. *ohridana*	616–798	322–490	11–14	Dark brown to black
Corillion (1957)	670–700(780)	530	10–14	Black
Wood (1965)	660–700(780)	c. 530	10–14	Black
Corillion (1975)	585–825	375–450	12–14	Dark brown to black
Hollerbach and Krassavina (1983)	585–825	375–450	12–14	Dark reddish brown to almost black
Krause (1997)	600–800	350–600	10–13	Black
Krause (1997), var. *ohridana*	ca. 750	ca. 440	10–12	Black
Langangen (2007)	–	–	–	Black
Trajanovska and Blaženčić (2008), var. *ohridana*	–	–	10	Dark brown to black
Bailly and Schaefer (2010)	350–600	–	–	–
Jäger (2010)	600–900	–	11–13 (average)	Black
Urbaniak and Blaženčić (2012), var. *ohridana*	661–682	472–493	9–11	–
Schou et al. (2017)	350–900	300–600	10–14	–
Own data, var. *ohridana*	600–900	400–750	11–13	Dark brown, brown, black
Total range	**350–900**	**300–750**	**9–14**	

All values for length and width are in μm. Own data were taken from fixed material

found in two sheltered sites, distant from each other, at eastern part of the Gulf of Finland (Romanov et al., 2023; own data).

Sediment: In Lake Constance *Chara dissoluta* grows on sand and silt. Var. *ohridana* is known on sandy, silty and rocky bottoms (Blaženčić and Stevanović, 2015; Blaženčić et al., 2018; Trajanovska and Blaženčić, 2008; own data) but most reports do not record the substrate for *C. dissoluta*. It is likely that sand, marl and silt are more common substrates.

Nutrient conditions: There are only few data on nutrient conditions of *Chara dissoluta.* The total phosphorus range is reported to vary between 9–38 μg/l in Lake Constance (Arbeitsgruppe Characeen Deutschlands, 2016) while in Ireland it has been reported between 5–17 μg/l (see Table 28.4).

Table 28.3 Antheridia morphometry

Author	Diameter (μm)
Migula (1889–1897)	480–500
Sluiter (1910)	300–420
Groves and Bullock-Webster (1924)	400–425
Corillion (1957)	480–500
Wood (1965)	480–500
Corillion (1975)	375–425
Hollerbach and Krassavina (1983)	375–425
Moore (1986)	<500
Krause (1997)	400–450
Krause (1997), var. *ohridana*	ca. 550
Langangen (2007)	up to 400
Trajanovska and Blaženčić (2008), var. *ohridana*	224–511
Bailly and Schaefer (2010)	400–450
Schou et al. (2017)	300–500
Own data, var. *ohridana*	347–560
Total range	**(224)300–560**

Own data were taken from fixed material

Table 28.4 Euphotic depth and water chemistry data from 8 Irish lakes with *Chara dissoluta* populations. Based on annual means from the year when the species was recorded most recently (data provided by the Environmental Protection Agency, Ireland)

	Min	Median	Max
Euphotic depth (m)	5.0	6.3	9.0
Alkalinity (mg/l CaCo3)	94	136	188
Total hardness (mg/l CaCO3)	100	123	193
pH	8.0	8.2	8.4
Colour (hazen units)	2.8	16.0	32.6
TP (μg/l)	5.0	9.8	17.6
TON (mg/l)	0.07	0.24	0.42
Conductivity (μS/cm)	210	279	356

Water chemistry: Data on the water chemistry of *Chara dissoluta* are limited. It is a strictly freshwater species of hard, alkaline freshwater and Olsen (1944) describes *C. dissoluta* (as *C. denudata*) as a halophobic species (chloride up to 42 mg/l). The pH range is reported to vary between (6.2)7.0–8.1 (Arbeitsgruppe Characeen Deutschlands, 2016) and 7.4–8.6 for var. *ohridana* (Blaženčić and Stevanović, 2015; Trajanovska and Blaženčić, 2008; BEOU). Total hardness is reported to vary between 80 and 187 mg/l $CaCO_3$ (Arbeitsgruppe Characeen Deutschlands, 2016; Olsen, 1944). Conductivity range is reported to vary between 210–370 μS/cm (Arbeitsgruppe Characeen Deutschlands, 2016). Similar results have been recorded from

Ireland (Table 28.4) and here water colour resulting from humic content is also an important factor limiting the maximum depth of vegetation and therefore the suitability of sites for this species (Roden et al., 2020).

Depth range: *Chara dissoluta* is predominantly found in deep water (2–20 m) at the lower end of the macrophyte growth (Arbeitsgruppe Characeen Deutschlands, 2016; Auderset Joye and Rey-Boissezon, 2013; Krause and Krause 1984–1986; Moore, 1986; Roden et al., 2020). In Ireland, it grows in greater depth and in a wider band in clearer lakes with greater maximum vegetation depths and has not been found in lakes where the maximum vegetation depth is less than 5 m (Fig. 28.7). Nevertheless, occurrences in shallower water (from beyond 0.5 m) have been reported elsewhere (Auderset Joye and Rey-Boissezon, 2013; Krause, 1997, own data). Var. *ohridana* grows at a broader range of depths, from 0.3 to 17 m, but it is also mostly associated with depths of several metres (Blaženčić and Stevanović, 2015; Blaženčić et al., 2018; Kashta et al., 2013; Kostić, 1937; Krause, 1990; Krause and Krause, 1984–1986; Trajanovska and Blaženčić, 2008; Trajanovski et al., 2016; labels of the specimens studied).

Associated species: *Chara dissoluta* usually grows in monospecific stands (Roden et al., 2020) or can be associated with *C. aspera, C. contraria, C. virgata* or *Nitella syncarpa* (Arbeitsgruppe Characeen Deutschlands, 2016; Corillion, 1960; Jäger, 2010), *Nitella flexilis/opaca* or *Tolypella glomerata* (Roden et al., 2020). Other occasional associates in Ireland are *C. aculeolata, C. aspera, C. curta, C. subspinosa, C.*

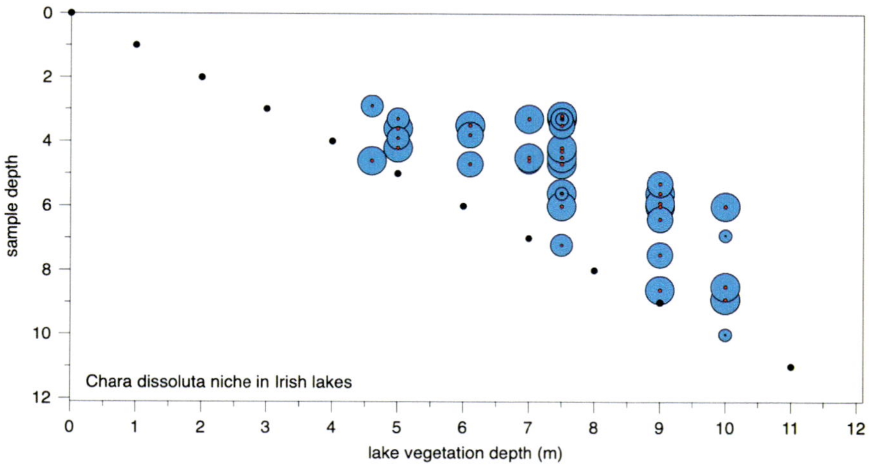

Fig. 28.7 Depth and water transparency (measured as maximum depth of vegetation) preferences of *Chara dissoluta* from about 40 samples in 5 Irish lakes. The blue circles are proportionate in size to the species cover value with the largest circle size representing covers of 75–100%. Each sample is plotted against sample depth (y axis) and maximum vegetation depth along the transect which included the sample (x axis). The dotted black line shows maximum depth of vegetation (Cilian Roden pers. data)

tomentosa, Elodea canadensis, Potamogeton berchtoldii, P. friesii, P. perfoliatus, P. x angustifolius and red cyanophytes but often these are merely stragglers from shallower depths (C. Roden pers. data). *C. dissoluta* var. *ohridana* forms monospecific communities in belts of a few metres to more than 2 km width in Lake Ohrid, or grows in small individual patches or occurs in the mixed community *Scirpo-Phragmitetum*, together with *Chara aspera, C. contraria, C. globularis, C. imperfecta, C. tomentosa, Nitella hyalina, N. opaca, N. syncarpa, Nitellopsis obtusa, Cladophora* sp., *Elodea canadensis, Myriophyllum spicatum, Najas minor, N. marina, Potamogeton acutifolius, P. lucens, P. perfoliatus, P. pusillus, Ranunculus trichophyllus, Stuckenia pectinata, Vallisneria spiralis, Zannichellia palustris*, mostly with species indicating slight eutrophication (Blaženčić et al., 2018; Kashta et al., 2013; Krause and Krause, 1984–1986; Schneider et al., 2020; Trajanovska, 2009; Trajanovska and Blaženčić, 2008; Trajanovski et al., 2016).

28.4 Variability

In its classic form, this taxon has predominantly only rudimentary cortex on the main axis and branchlets. It is markedly different in appearance from *Chara contraria*, to which it is clearly related, resembling a *Nitella* more than a *Chara*. On the other hand, *C. contraria* is more prone than most species to occur with reduced cortication. Usually this is expressed as partial cortication of the branchlets to varying degrees but rarely can also include reductions in the secondary tubes of axis cortex. This has raised questions over the limits of *C. dissoluta* in relation to *C. contraria* or, indeed, whether they are truly distinct (e.g. Arbeitsgruppe Armleuchteralgen Deutschlands, 2016; Corillion, 1960; Krause, 1997; Krause and Krause, 1984–1986). Further research is needed before a conclusion can be reached.

In Ireland, at least, there appears to be a distinct discontinuity in characters between populations where the cortex is predominantly rudimentary or developed for a few millimetres and those where the primary cortex is complete, the latter being placed in *C. contraria*. A similar situation appeared to be the case in the Balkan region where near-ecorticate plants have been considered a separate var. *ohridana* (Kostić, 1937), but checking its specimens failed to confirm this pattern (see below).

Wood (1965) attempted to make a separation on similar grounds. He distinguished between near-ecorticate plants as forma *denudata* and ones where the primary cortex is at least partially complete as forma *dissoluta,* but this distinction is not sharp (Krause, 1997; own observations). Jäger (2010) investigated differences in the oospore proportions in relation to different degrees of primary tube development in one central European population (Lake Constance) testing the assumption that both of Wood's forms were present in the lake. He found no significant difference between them nor with *C. contraria* except in the proportions of the oospore; *C. contraria* tended to have thicker and the form *dissoluta* more elongated oospores, with the form *denudata* somewhere in between, having bulkier oospores than form *dissoluta.*

However, in Central Europe the picture is more complicated; although there are some populations with low levels of cortication, there are others with a variable degree of cortical development from rudimentary to more or less elongate even within the same stand (Arbeitsgruppe Characeen Deutschlands, 2016; Doege, 2012; Jäger, 2010; Krause, 1976; Migula, 1889–1897; Sluiter, 1910). Plants from France and Switzerland are illustrated with plants having rudimentary cortex cells forming knots around the nodes with some cortex tubes being elongated (Bailly and Schaefer, 2010; Corillion, 1960; Krause and Krause, 1984–1986; Info Flora, 2022), with completely or incompletely corticate branchlets (Corillion, 1960) or with branchlets with cortex initials only (Bailly and Schaefer, 2010; Info Flora, 2022). This variation led A. Braun in Leonhardi (1864) to broaden his circumscription of this taxon in his description of *C. dissoluta* and indeed original material of *C. dissoluta* (seen by R. Romanov in LE) has elongate to complete primary cortex with only rarely very short secondary cortex initials. Its branchlet segments are short with short bract-cells and with some diplostichous cortex. Therefore, it represents the most corticated morphotype of this species in a continuum of different cortication expression known from the same population (Jäger, 2010; Sluiter, 1910).

The occurrence of one or two normally corticate branchlet segments in the case of some plants from the mainland of Europe can be recognised as a notable difference from plants from Ireland. These branchlet segments with more or less developed cortication are always shorter than the ecorticate segments. However, Groves and Bullock-Webster (1924) note that in Irish plants, the branchlet cortex, which is typically represented by a ring of cells at each fruiting node, often has one anterior cell lengthening and projecting below the antheridium. The same pattern was found in south-east European plants (Figs. 28.1n and 28.5n). Therefore, the difference in branchlet cortication between plants from different regions is expressed in the length of the branchlet cortex cells rather than in terms of presence/absence.

On the other hand, populations of *Chara contraria* with reduced cortex, particularly on the branchlets, occur occasionally and appear to form a continuum with normal forms. More extreme forms coming within the range of *C. dissoluta* are unusual but there are occurrences which could be named *C. dissoluta* but which ecologically and distributionally make more sense as variants of *C. contraria*. Some examples seen by the authors include plants lacking developed secondary cortex and with only occasional cortex on the lowest branchlet node growing in 1–2 m depth in a disused gravel pit in Cotswold Water Park in southern England. Normal *C. contraria* is common within this large gravel pit system and similar poorly corticated plants have not been refound since. Plants with *C. dissoluta*-like morphology have also been seen in small water bodies in a brackish minerotrophic fen overlying old sea sediments in NW Russia and in a minerotrophic fen over a gypsum layer in northern Central Russia (within European Russia). The partly corticate forms of *C. contraria* from North Norway (Langangen, 2021) could also be easily referred to *C. dissoluta*.

It has been suggested that *Chara dissoluta* could be a deepwater ecotype of *C. contraria* (Krause and Krause, 1984–1986), but this is not supported by its distribution in Ireland. Here there are apparently similar clear water calcareous lakes, some with one, some with the other and some with both growing in the same deep-water conditions. For example, *C. dissoluta* has not been found in Loughrea (euphotic depth 12 m) where *C. contraria* is instead abundant, while in Lough Owel both *C. contraria* and *C. dissoluta* occur at depth.

Chara dissoluta could perhaps be developing as a neospecies or infraspecific taxon, evolving from *C. contraria* in the hypolimnion of deep transparent lakes with cold to cool waters, i.e. in a harsh and unfavorable, but stable environment (Krause and Krause, 1984–1986). Here perhaps the stable combination of low to moderate water temperature with limited insolation has resulted in the formation of the morphology typical of *C. dissoluta*. Dispersal is also difficult for a deepwater plant with negative buoyancy, increasing the genetic isolation. This point of view would imply independent speciation in Irish, north European, central European and south-east European lakes. A review of the genetic affinity between these populations is needed for testing this hypothesis. However, initial studies using *mat*K sequences failed to distinguish between *C. dissoluta* var. *dissoluta, C. dissoluta* var. *ohridana* and *C. contraria* (Schneider et al., 2016).

An additional observation, particularly in relation to the central European populations, is that the more ecorticate populations seem to be restricted to the regions that were glaciated in the last Ice Age while the ones with greater cortical development occur outside the limit of glaciation.

Var. *ohridana* was initially described as always having ecorticate branchlets together with longer anterior bract-cells, up to 1.5–2.5 times the length of oogonium. It also has a more compact and rigid habit (Krause, 1997). The revision of wet and pressed specimens (>25, incl. original material) stored in ZA, BEOU, NHMM, PSI Hydrobiological Institute in Ohrid, TIR, W and IBIW revealed that single cortex cells or cortex initials can be seen in the basal two segments of branchlets in almost all fertile specimens (Fig. 28.5). Main axis cortication pattern (where present) is variable—sometimes only primary tubes are developed in spiral or regular/parallel appearance, and sometimes secondary tubes are also notably developed showing tylacanthous, irregularly diplostichous to rarely imperfect triplostichous cortex. Specimens are variable in size, the small ones sometimes richly branched forming a compact, bushy appearance. Gametangia are situated on the 1–2 lowest nodes of the branchlets. Bracteoles are variable in length, usually equal to or somewhat longer than the oogonium, but also sometimes shorter or longer. Therefore, a clear and consistent delineation of var. *ohridana* with var. *dissoluta* seems to be impossible and var. *ohridana* is accepted as a synonym of *C. dissoluta*.

28.5 Identification and Risk of Confusion

As discussed above, plants with more developed cortex are not far from growth forms of *Chara contraria* with partial to complete loss of developed cortex on the branchlets, but well-developed axis cortex is not a trait of *C. dissoluta*. Care is also needed with juvenile plants, particularly in the shallow margins of pools and in wave-washed conditions. Germlings of *C. contraria* can be slow to develop cortex with 2–3 main axis internodes forming before the development of, first, axis cortex and then branchlet cortex. Other species with reduced cortication are *C. squamosa* which differs from *C. dissoluta* in having complete tylacanthous axis cortex but with branchlet cortex absent and frequently long bract-cells, *C. gymnophylla* which is distinct with its complete aulacanthous to isostichous axis cortex, *C. imperfecta* which has complete primary cortex with secondary cortex absent and is dioecious and *C. socotrensis* which is a glossy plant with minute cortex initials which are visible only at high magnification.

28.6 Distribution

See Fig. 28.8.

Chara dissoluta is known from Ireland (Groves and Bullock-Webster, 1924; Moore, 1986; Roden et al., 2020), France (Bailly and Schaefer, 2010; Corillion 1957, 1960, 1975), Italy (Migula, 1889–1897, later not confirmed, cf. Bazzichelli and Abdelahad, 2009), Switzerland (Auderset Joye and Schwarzer, 2012; Auderset Joye et al., 2002; Migula, 1889–1897), Austria (Jäger, 2010), Germany (Arbeitsgruppe Armleuchteralgen Deutschlands, 2016; Doege, 2012; Krause, 1976), Denmark (Olsen, 1944; Schou et al., 2017), the Balkans (Lakes Ohrid and Skadar) (Talevska et al., 2009), Slovakia and Russia, where it has been found in eastern part

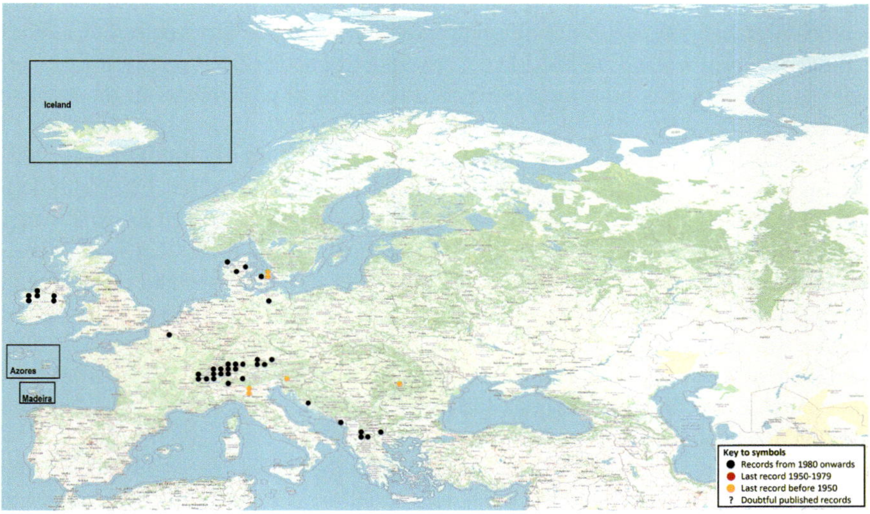

Fig. 28.8 Distribution of *Chara dissoluta* in Europe. For world-wide distribution see Romanov et al. (2023). Black dots are records from 1980 on, red dots last records from 1950 to 1979 and yellow last records dots before 1950; unconfirmed/doubtful published records (if applicable) are marked with "?" superimposed. Map prepared by Heiko Korsch

of the Gulf of Finland (Romanov et al., 2023) and in big limnocrene (lake-forming spring–Lake Geyzernoe) in the Altai Mountains (R. Romanov unpubl. data). The records from Iraq, Kurdistan (Corillion, 1960) and India (Pundhir and Vidyavati, 1994) need confirmation. *Chara dissoluta* var. *ohridana* is reported only from the Balkans (Albania, Greece, Makedonia, Montenegro), mostly from Lake Ohrid, but it is recorded for lakes Prespa and Skadar, the Crni Drim River flowing out of Lake Ohrid and in the puddles around and in the shallows of Lake Dojran (Blaženčić, 2004; Blaženčić and Blaženčić, 2002; Blaženčić et al., 2018; Kashta et al., 2013; Kostić, 1937; Talevska et al., 2009; Trajanovska, 2010; Trajanovska and Blaženčić, 2008; Trajanovska et al., 2014, 2015; ZA, BEOU, NHMM, PSI Hydrobiological Institute in Ohrid, TIR, W, IBIW).

28.7 Status and Threats

Chara dissoluta has been assessed as NT (Near Threatened, European and Globally) in the European Red List (see Chap. 9). Because it is primarily a deep-water species, *C. dissoluta* is particularly vulnerable to the effects of water pollution and particularly increased turbidity (Auderset Joye and Schwarzer, 2012; Lang, 1981; Stewart and Church, 1992). Roden et al. (2020) note that declines in euphotic depth are associated with increases in angiosperm cover with consequent declines in the charophyte flora. Blindow (1992) also observed that at greater depths angiosperms have an advantage in turbid water because their buoyancy and often larger propagules allows them to reach higher in the water column before requiring light. Nevertheless, in Lough Ennell in central Ireland and in Lake Constance it has survived a period of lower water quality and it is now recovering as a result of improvements to the sewage treatment.

Chara dissoluta var. *ohridana* is included in the Red Lists for the Balkans (globally Endangered) and Macedonia (Blaženčić et al., 2006b; IUCN ECARO, 2018). Eutrophication, extensive tourist development, recreational activities, and water abstraction are the main treats for var. *ohridana* (Blaženčić et al., 2018). *C. dissoluta* is currently rated as "Rare" under the old IUCN system in Ireland (Stewart and Church, 1992) but it has not been reassessed since more extensive information has become available (e.g. Roden et al., 2020). Of 11 known sites, it has been lost or is significantly impacted by nutrient enrichment at over half of these. Many Irish marl lakes are close to cut-over peat bogs and increasing peat staining is also a significant threat in at least one site.

Chara dissoluta was not evaluated according to the IUCN criteria in Switzerland because of insufficient data, although a decline in its occurrence has been observed (Auderset Joye and Rey-Boissezon, 2013; Auderset Joye and Schwarzer, 2012), nor in Germany because of its taxonomic uncertainty (Arbeitsgruppe Armleuchteralgen Deutschlands, 2016). In Vorarlberg, a Federal State of Austria, *C. dissoluta* is included in the Red List as Endangered (Jäger, 2012).

Acknowledgements The curators of the herbaria listed above are kindly acknowledged for their encouragement and guidance, Thomas Gregor for valuable suggestions, Ricky Taylor and Matthew Janks for really helpful clarification of environment in type locality of *C. denudata*, Luc Denys for the help with an inaccessible reference.

References

Arbeitsgruppe Characeen Deutschlands. (2016). *Armleuchteralgen: Die Characeen Deutschlands*. Springer. [in German]. https://doi.org/10.1007/978-3-662-47797-7

Auderset Joye, D., Castella, E., & Lachavanne, J.B. (2002). Occurrence of Characeae in Switzerland over the last two centuries (1800–2000). *Aquatic Botany, 72*, 369–385. https://doi.org/10.1016/S0304-3770(01)00211-X

Auderset Joye, D., & Rey-Boissezon, A. (2013). Les Characées de Genève et environs. In *Distribution et Ecologie* (92 pp.). Groupe d'Ecologie aquatique, Institut Forel et Institut des Sciences l'Environnement, Université de Genève. [in French]. https://archive-ouverte.unige.ch/unige:42980

Auderset Joye, D., & Schwarzer, A. (2012). Rote Liste Armleuchteralgen. In *Gefährdete Arten der Schweiz, Stand 2010* (72 pp.). Bundesamt für Umwelt BAFU & Laboratoire d'Ecologie et de Biologie Aquatique der Universität Genf. [in German]. https://www.bafu.admin.ch/dam/bafu/de/dokumente/biodiversitaet/uv-umwelt-vollzug/rote_liste_armleuchteralgen.pdf.download.pdf/rote_liste_armleuchteralgen.pdf

Bailly, G. & Schaefer, O. (2010). *Guide illustré des Characées du nord-est de la France* (96 pp.). Conservatoire Botanique National de Franche-Comté. [in French].

Bazzichelli, G., & Abdelahad, N. (2009). *Alghe d'acqua dolce d'Italia. Flora analitica delle Caroficee*. Centro Stampa Università. [in Italian].

Blaženčić, J., & Blaženčić, Ž. (2002). Rare and threatened species of Charophytes (Charophyta) in Southeast Europe. *Phytologia Balcanica, 8*, 315–326.

Blaženčić, J., & Stevanović, B. (2015). Catalogue of charophytes (Charales) of Montenegro. *Montenegrian Academy of Sciences and Arts. Catalogues 11. The Section of Natural Sciences, 10*, 1–51. [in Serbian].

Blaženčić, J., Stevanović, B., Blaženčić, Ž, & Stevanović, V. (2006). Red data list of charophytes in the Balkans. *Biodiversity and Conservation, 15*, 3445–3457. https://doi.org/10.1007/978-1-4020-5734-2_6

Blaženčić, J., Kashta, L., Vesić, A., Biberdžić, V., & Stevanović, B. (2018). Charophytes (Charales) of Lake Skadar/Shkodra: Ecology and distribution. In V. Pešić, G. Karaman, & A. Kostianoy (Eds.), *The Skadar/Shkodra Lake environment* (pp. 169–202). Springer. https://doi.org/10.1007/698_2018_265

Blaženčić, J. (2004). The Ohrid Lake an important centre of Charophyta diversity. In *Proceedings of the 2nd Congress of Ecologists of the Republic of Macedonia with International Participation, 25–29.10.2003, Ohrid, 6*. Special issues of Macedonian Ecological Society.

Blindow, I. (1992). Decline of charophytes during eutrophication: A comparison to angiosperms. *Freshwater Biology, 28*, 9–14. https://doi.org/10.1111/j.1365-2427.1992.tb00557.x

Braun, A. (1847). Uebersicht der schweizerischen Characeen. *Neue Denkschriften der Allgemeinen Schweizerischen Gesellschaft für die Gesammten Naturwissenschaften, 10*, 1–23. [in German].

Braun, A. (1868). Die Characeen Afrika's. *Monatsberichte der Königlich Preussischen Akademie der Wissenschaften zu Berlin, 1867*, 782–800. [in German].

Corillion, R. (1957). Les Charophycées de France et d'Europe Occidentale. *Bulletin de la Société Scientifique de Bretagne, 32* (Hors série I & II), 1–499. [in French].

Corillion, R. (1960). Sur la présence et la signification de formes à cortication incomplète du *Chara contraria* Kützing dans l'Est armoricain. *Bulletin de la Société Mayenne-Sciences, Laval année, 1960,* 49–60. [in French].
Corillion, R. (1975). Flore et végétation du Massif Armoricain. In *T. IV. Flore des Charophytes (Characées) du Massif Armoricain et des contrées voisines d'Europe occidentale*. Jouve. [in French].
Doege, A. (2012). Zum Vorkommen von *Chara denudata* A. Br. 1847 in Sachsen und zur morphologischen Abgrenzbarkeit einiger unvollständig berindeter Formen der Gattung *Chara. Rostocker Meeresbiologische Beiträge, 24,* 86–99. [in German].
Drège, J. F. (1843). Zwei pflanzengeographische Documente. *Flora, 26*(Besondere Beigabe), 1–230. [in German].
Groves, J., & Bullock-Webster, G. R. (1924). The British Charophyta. In *Chareæ with plates, concluding articles, geological sketch, bibliography and index* (Vol. II). Ray Society.
Hollerbach, M. M., & Krassavina, L. K. [Голлербах, М. М. & Красавина, Л. К.]. (1983) *Определитель пресноводных водорослей СССР. 14. Charophyta* [*Determination key of freshwater algae of USSR, 14, Charophyta*]. Nauka. [in Russian].
IUCN ECARO. (2018). *National Red Lists for species conservation in Macedonia: Setting the stage* (viii + 53 pp.). IUCN Regional Office for Eastern Europe and Central Asia, Belgrade. https://redlist.moepp.gov.mk/wp-content/uploads/2019/11/NationalRedLists_Macedonia_IUCN2018.pdf
Info Flora. (2022). info-Flora. Das nationale Daten- und Informationszentrum der Schweizer Flora. https://www.infoflora.ch
Jäger, D. (2010). Exemplare des formenkreises *Chara denudata* A. Braun 1847 und *Chara dissoluta* A. Braun ex Leonhard 1864 aus dem Bodensee. *Rostocker Meeresbiologische Beiträge, 23,* 29–39. [in German].
Jäger, D. (2012). Rote Liste der Characeen Vorarlbergs (Österreich). *Rostocker Meeresbiologische Beiträge, 24,* 75–85 [in German].
Kashta, L., Perçini, D., & Molla, O. (2013). Të dhëna mbi diversitetin dhe ekologjinë e algave Charophyta në Liqenin e Ohrit. *Buletini i Shkencave të Natyrës, Universiteti i Tiranës, 15,* 142–150. [in Albanian].
Kostić, L. (1937). Prinos poznavanju haraceja Ohridskog Jezera i okoline. Ein Beitrag zur Kenntnis der Characeen des Ochridsees. *Acta Botanica Instituti Botanici Universitatis Zagrebensis, 11,* 63–84. [in Croatian].
Krause, W. (1975). Siedlungen gefährdeter Pflanzen in Baggerseen der Oberrheinebene. *Beiträge Zur Naturkundlichen Forschung in Südwestdeutschland, 34,* 187–199. [in German].
Krause, W. (1976). Characeen aus Bayern. Teil 1. Bestimmungsschlüssel und Abbildungen. *Berichte der Bayerischen Botanischen Gesellschaft zur Erforschung der Heimischen Flora, 47,* 229–257. [in German].
Krause, W. (1990). Zum neufund der *Chara kokeilii* A. Br. im Skutarisee. *Carinthia II, 180,* 705–713. [in German].
Krause, W. (1997). Charales (Charophyceae). In H. Ettl, G. Gärtner, H. Heynig, & D. Mollenhauer (Eds.), *Süßwasserflora von Mitteleuropa, 18.* Fischer. [in German].
Krause, W., & Krause, H. (1984–1986). *Exsikkate Europäischer Characeen* (2nd ed, 5 Bände). [in German].
Lang, G. (1981). Die submersen Makrophyten des Bodensees 1978 im Vergleich mit 1967. *Bericht der Internationalen Gewässerschutzkommission für den Bodensee, 26,* 1–54, Anhang. [in German].
Langangen, A. (2007). *Charophytes of the Nordic countries.* Saeculum ANS.
Langangen, A. (2021). Charophytes from alpine lakes in Láhko National Park, Northern Norway, with comments on *Tolypella canadensis* Sawa and a description of the observed morphological variation of *Chara contraria* A. Br. https://www.researchgate.net/publication/352762645

von Leonhardi, H. (1864). Die bisher bekannten österreichischen Armleuchter-Gewächse besprochen von morphogenetischen Standpuncte. *Verhandlungen des naturforschenden Vereins in Brünn, Abhandlungen, 2*, 122–224. [in German].

Migula, W. (1889–1897). Fünfter Band: Die Characeen. In A. Fischer, E. Fischer, F. Hauck, G. Limpricht, C. Luerssen, W. Migula, H. Rehm, P. Richter, & G. Winter (Eds.), *Dr. L. Rabenhorst's Kryptogamenflora von Deutschland, Oesterreich und der Schweiz* (2nd ed). Eduard Kummer. [in German].

Moore, J. A. (1986). *Charophytes of Great Britain and Ireland*. BSBI Handbook No. 5. Botanical Society of the British Isles.

Olsen, S. (1944). Danish Charophyta. *Det Kongelige Danske Videnskabernes Selskab, Biologiske Skrifter, 3*(1), 1–240, pl. 1–2. [in Danish].

Pundhir, H. S., & Vidyavati. (1994). New karyomorphological observations on charophytes from Uttar Pradesh, India. *Nucleus, 37*, 39–45.

Roden, C., Murphy, P., & Ryan, J. (2020). *Benthic vegetation in Irish marl lakes: Monitoring habitat 3140. Condition 2011–2018*. Irish wildlife manuals, 124, National Parks and Wildlife Service, Department of Housing, Local Government and Heritage. https://www.npws.ie/sites/default/files/publications/pdf/IWM124.pdf

Romanov, R. E., Barinova, S. S., Nikulin, V. Yu., & Gontcharov, A. A. (2023). 'Unfinished' morphogenesis hides different speciation pathways in charophytes: Evidence from 190-year-old original material of *Chara denudata* (Charales, Charophyceae). *Diversity, 15*, 249. https://doi.org/10.3390/d15020249

Schneider, S. C., Nowak, P., von Ammon, U., & Ballot, A. (2016). Species differentiation in the genus *Chara* (Charophyceae): Considerable phenotypic plasticity occurs within homogenous genetic groups. *European Journal of Phycology, 51*, 282–293. https://doi.org/10.1080/09670262.2016.1147085

Schneider, S. C., Trajanovska, S., Biberdžić, V., Marković, A., Talevska, M., Imeri, A., Veljanoska-Sarafiloska, E., Đurašković, P., Jovanović, K., & Cara, M. (2020). The Balkan Macrophyte Index (BMI) for assessment of eutrophication in lakes. *Acta Zoologica Bulgarica, 72*, 439–454.

Schou, J. C., Moeslund, B., Båstrup-Spohr, L., & Sand-Jensen, K. (2017). *Danmarks vandplanter*. BFN's. [in Danish].

Sluiter, C. P. (1910). Beiträge zur Kenntnis von *Chara contraria* A. Braun und *Chara dissoluta* A. *Braun*. *Botanische Zeitung, 68*, 125–167. [in German].

Stewart, N. F., & Church, J. M. (1992). *Red data books of Britain and Ireland: Stoneworts*. The Joint Nature Conservation Committee.

Talevska, M., Petrovic, D., Milosevic, D., Talevski, T., Maric, D., & Talevska, A. (2009). Biodiversity of macrophyte vegetation from Lake Prespa, Lake Ohrid and Lake Skadar. *Biotechnology and Biotechnological Equipment, 23*(supl), 931–935.

Trajanovska, S. (2009). Taxonomy, ecology and status of charophyte flora (Charophyta) in the Ohrid Lake. Doctoral thesis, University "Св. Кирил и Методиј" [of St. Ciril and Metodij], Faculty of Science and Mathematics.

Trajanovska, S. (2010). The anthropogenic pressure and the changes in the Charophyta composition in Lake Ohrid. In *The 39th Annual Conference of the Serbian Water Pollution Control Society "WATER 2010", Conference Proceeding, Divcibare, Serbia*, (pp. 153–158). [in Macedonian].

Trajanovska, S., & Blaženčić, J. (2008). Morphological variability, fructification, distribution and ecology of endemic *Chara ohridana* (Kostic) Krause 1997 from Lake Ohrid. *PSI Hydrobiological Institute Ohrid. Review, 41*, 15–22. [in Macedonian].

Trajanovska, S., Talevska, M., Imeri, A., & Schneider, S. C. (2014). Assessment of littoral eutrophication in Lake Ohrid by submerged macrophytes. *Biologia, 69*, 756–764. https://doi.org/10.2478/s11756-014-0365-9

Trajanovska, S., Talevska, M., Budzakoska-Gjoreska, B., & Trajanovski, S. (2015). Review of the extant researches of the stoneworts (Charophyta) from Lake Ohrid. *PSI Hydrobiological Institute Ohrid. Review, 43*, 96–103. [in Macedonian].

Trajanovski, S., Budzakoska-Gjoreska, B., Trajanovska, S., Talevska, M., & Zdraveski, K. (2016). Macrophyte vegetation as a structuring factor of the macrozoobenthic communities in Lake Ohrid. *Botanica Serbica, 40*, 145–151. https://doi.org/10.5281/zenodo.162212
Urbaniak, J., & Blaženčić, J. (2012). SEM study of oospore characteristics in endemic and endangered Balkan charophytes. *Cryptogamie, Algologie, 33*, 277–288. https://doi.org/10.7872/crya.v33.iss3.2012.277
Wood, R. D. (1965). *Monograph of the Characeae*. J. Cramer

Chapter 29
Chara fibrosa sensu lato

Ingeborg Soulié-Märsche and Nadia Abdelahad

Scientific name: ***Chara fibrosa*** **C. Agardh ex Bruzelius, Observ. Charae: 21. 1824**

Lectotype (Wood, 1965: 291): [no location, probably Mariana Islands]/[no date]/ Gaudichaud 158 (LD).

= *Chara pelosiana* Avetta in Malpighia: 230. 1898.
Lectotype (Langangen, 2000: 249): Valli e risaie di S. Anna presso S. Cesario. Prov. Di Modena/1 Ottobre 1886/E. Ferrari (O).

≡ *Chara baueri* A. Braun f. *pelosiana* (Avetta) R. D. Wood, Rev. Charac.: 319. 1965, n. inv.

= *Chara benthamii* A. Braun in Monatsber. Königl. Preuss. Akad. Wiss. Berlin 1867: 799. 1868 "Benthami".
Lectotype (Wood, 1965: 290): Hong Kong/Feb. 1858/Wilford 238 (K).

≡ *Chara fibrosa* C. Agardh ex Bruzelius subsp. *benthamii* (A. Braun) Zaneveld in Blumea 4: 373. 1940.

Taxonomic notes: *Chara pelosiana* was published 1898 by Carlo Avetta, based on plants from rice fields near Modena. Noticing the special characters of a voucher specimen kept in the herbarium in Oslo, Krause (1997) and Langangen (2000) attributed *C. pelosiana* to *Chara fibrosa*. Complementary studies of topotype materials of *C.*

I. Soulié-Märsche (✉)
Université Montpellier, Institut des Sciences de l'Évolution (ISEM), Case Courrier 061, Place Eugène Bataillon, 34095 Montpellier Cedex 05, France
e-mail: insouma43@gmail.com

N. Abdelahad
Dipartimento di Biologia Ambientale Sapienza, Università di Roma, P.le Aldo Moro 5, 00185 Roma, Italy
e-mail: nadia.abdelahad@uniroma1.it

H. Schubert et al. (eds.), *Charophytes of Europe*,
https://doi.org/10.1007/978-3-031-31898-6_29

pelosiana from herbaria in Italy confirmed the affiliation of *C. pelosiana* to *Chara* sect. *Agardhia* Wood (Millozza and Abdelahad, 2021).

More recent finds from rice fields in South France and Italy (Abdelahad and Piccoli, 2017; Soulié-Märsche et al., 2013) were attributed to *C. fibrosa* subsp. *benthamii* based on Zaneveld (1940) who highlighted the presence of only one stipulode per branchlet for subsp. *benthamii* compared to *C. fibrosa* s.str. This determination is revised here after examination of the types of both *C. fibrosa* (Casanova, 2013) and *C. benthamii*, by (Casanova and Karol 2023), (this vol. Table 14.1).

As a consequence, it appears that neither the plants described as *C. fibrosa* subsp. *benthamii* (with a single stipulode per branchlet) nor *C. pelosiana* with 1–3 stipulodes per branchlet should be attributed to *C. fibrosa* s.str. However, they obviously belong to the *Agardhia*-complex (Casanova, see Chap. 14). The current state of knowledge about the more than twenty, relatively poorly known members of subsection *Agardhia* does presently not allow a precise determination of the European material.

Local names: **Unknown**.

29.1 Morphology

See Figs. 29.1 and 29.2.

Habitus: Plants grow straight, upright with no or very few lateral branches. Internodes are three to five times longer than the whorls. Whorls are regularly spaced with branchlets curved towards the axis. Whorls are of bushy appearance, separated by long internodes. Incrustation is lacking or weak.

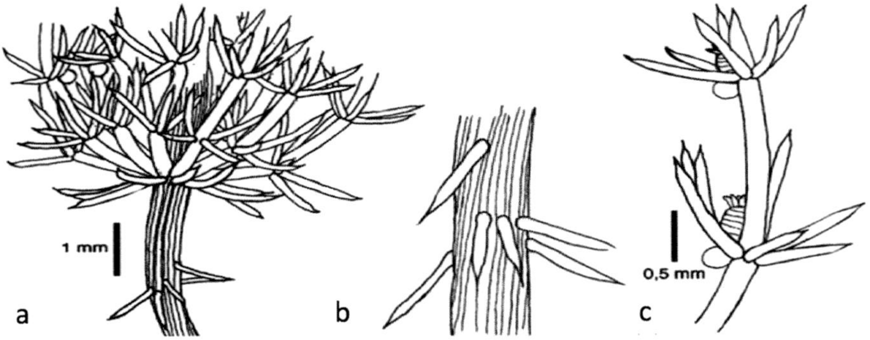

Fig. 29.1 Morphological characteristics of *Chara fibrosa* s. l.. **a** Whorl with one stipulode inserted below each branchlet (haplostephanous); **b** Diplostichous axis with spine cells; **c** Fertile branchlet with crown of bract-cells. Drawings: Aurélie Boissezon, from Soulié-Märsche et al. (2013), reproduction with permission by Botany Letters

Fig. 29.2 Habitus of *Chara fibrosa* s.l.. Shown are plants from KRD pond, Camargue, South France, collected on 28 July 2009. The inlet (a) shows a ripe oospore photographed under natural light. Photographs: from Soulié-Märsche et al. (2013), reproduction with permission by Botany Letters

Size: The plants are 40–50 cm high, the plants from Italy are much smaller (10–15 cm). The herbarium specimen in Oslo is just 3 cm and heavily incrusted (Langangen, 2000).

Main axis: Main axis diameter varies from 0.3 to 0.7 mm, internodes 3–5 times longer than the whorls.

Cortex: Cortex is diplostichous-isostichous tending towards tylacanthous (Fig. 29.1a).

Spine-cells: Spine-cells are sparse, solitary and acuminate, often forming a ring around the axis; spine-cell length is equal to main axis diameter or slightly shorter, sometimes also 1 mm long. Spine cells are denser at the apex.

Stipulodes: The present materials show one ring of well-developed, acute stipulodes, longer than the main axis diameter, up to 1–1.6 to 2 mm long, spreading ± perpendicular to the axis at the base of the whorls. The number of stipulodes is equal to the number of branchlets with one stipulode per branchlet inserted below the branchlets (opposite, not alternate) (Fig. 29.1b). The number of stipulodes is variable, 1–2(3) in the material of *C. pelosiana* from the 19th century with length up to 1350 μm.

Branchlets: Whorls are mostly composed of 8–10(12) branchlets (phylloids) bent towards the axis. Branchlets totally ecorticate, composed of a variable number of 3–4 (6) segments, 2–3 of them being fertile. Frequently they show banded incrustation.

Bract-cells: Both anterior and posterior bract-cells are well developed. They form a crown of mostly 6 acute cells at each branchlet node up to the top of branchlets including the terminal one, which is also verticillate. The bract-cells are as long as or about 2 times longer than the oogonia (Fig. 29.1c).

Bracteoles: 2, as long as the bract-cells.

Sexuality: The species is monoecious with conjoined gametangia, oogonia above the antheridia. Gametangia are formed on two to three branchlet nodes.

Oogonia: Solitary oogonia of regularly ovoidal shape. Oogonia c. 600 μm high, 440 μm wide (measured wet, with swollen spiral cells). The coronula is composed of five acuminate cells, 160 μm high. Table 29.1 gives an overview about morphometric data.

Oospores: Mature oospores are brown, 400–500(550) μm high and 250–290 μm wide (Table 28.2). They display 6–8 short ridges separated by fossae of 55–70 μm width (Fig. 29.3). The oospore wall appears smooth at low magnification. *Chara fibrosa* oospores do not calcify (no gyrogonites).

Antheridia: Antheridia are relatively small, average diameter 300 μm (Table 29.3), solitary, situated below the oogonia.

Bulbils: Not known for *Chara fibrosa.*

Table 29.1 Oogonia morphometry

Author	Length	Width	Coronula height	Coronula width	Number of helical stripes
Krause (1997), as *C. fibrosa*	420–650	270–380	100–150	100–160	–
Langangen (2000), as *C. fibrosa*	420 (unripe?)–650	270–380	100–150	100–160	–
Soulié-Märsche et al. (2013), as *C. fibrosa* subsp. *benthamii*	600	440	160	–	–
Abdelahad and Piccoli (2017), as *C. fibrosa* subsp. *benthamii*	540–600	340–380	60–80	–	7–8
Millozza and Abdelahad (2021), as *C. pelosiana*	425–450	312–340	–	–	–
Total range	**420–650**	**270–440**	**60–150**	**100–160**	**7–8**

Length in the first column excludes the coronula; its length and width are given separately, all values for length and width in μm

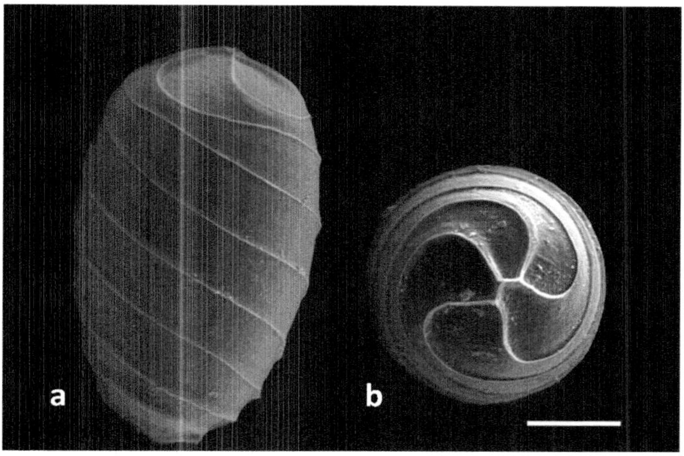

Fig. 29.3 SEM views of oospores of *Chara fibrosa* s.l.. **a** Lateral view; **b** Apical view. Scale bar 100 μm. Photographs: from Soulié-Märsche et al. (2013), reproduction with permission by Botany Letters

Table 29.2 Oospore morphometry

Author	Length	Width	Number of striae	Colour
Avetta (1898), as *C. pelosiana*	416–440	240–256	–	–
Krause (1997), as *C. fibrosa*, dry material	370–400	250	–	Noted unripe
Langangen (2000), as *C. fibrosa*, dry material	370–400 (ripe ?)	250	–	Brown
Soulié-Märsche et al. (2013), as *C. fibrosa* subsp. *benthamii*, wet material	400–500	–	7–9	Brown
Abdelahad and Piccoli (2017), as *C. fibrosa* subsp. b*enthamii*, formaldehyde preserved	350–390	240–316	7–8	Brown
Millozza and Abdelahad (2021), as *C. pelosiana*, dry material	350	235	7–8	Golden-brown
Total range	**350–500**	**240–316**	**7–9**	

All values for length and width in μm

29.2 Phenology and Reproduction

The European representatives of *Chara fibrosa* s.l. are monoecious and richly fertile from July to the end of summer. They develop annually from oospores and disappear with the drying of of the habitat.

Table 29.3 Antheridia morphometry

Author	Diameter (μm)
Krause (1997), as *C. fibrosa*, dry material	240–250
Langangen (2000), as *C. fibrosa*, dry material	240–250
Soulié-Märsche et al. (2013), as *C. fibrosa* subsp. *benthamii*-wet material	c. 300
Abdelahad and Piccoli (2017), as *C. fibrosa* subsp. *benthamii*—formaldehyde preserved material	(280)300(320)
Milloza and Abdelahad (2021), as *C. pelosiana*, dry material	250–270
Total range	**240–320**

29.3 Habitat Conditions and Ecology

Habitat type: The occurrence of *Chara fibrosa* in Europe may seem surprising. However, the species was only found in particular habitats. It was recorded (as *C. pelosiana*) in connection with rice fields in Italy, which are usually flooded in late spring providing warm water for germination. Similarly, the single locality in a highwater detention pond in the Crau plain, Camargue, South France, becomes flooded only by the end of May, at temperatures > 20 °C. Due to a reverse hydrological cycle in that pond, *C. fibrosa* can outcompete the native Characeans (such as *C. vulgaris* and *C. globularis*) (Soulié-Märsche et al., 2013). One locality in Greece refers to a freshwater lake (lake Kournas, Crete) where *Chara fibrosa* was mentioned in 1997 (Bergmeier and Abrahamczyk, 2008) but seems to be locally extinct as reported by Langangen (2012).

Sediment: *Chara fibrosa* grows usually on soft, silty substrate and calcareous mud.

Nutrient conditions: There are no reports about nutrient content in the water. In one case, nitrite and phosphate were below detectable amount (Soulié-Märsche et al., 2013).

Water chemistry: *Chara fibrosa* grows in freshwater. At the single locality in France, pH of 8.5 at 22.5°C was reported. The water was well oxygenated (DO 5.9–6.0 mg L^{-1}; 81.7–82.3 % saturation), with electrical conductivity (C25) of 102.0 μS cm^{-1}. The chlorophyll a concentration was 7.0–7.6 μg L^{-1} (Soulié-Märsche et al., 2013). In a Hungarian rice paddy, the pH increased up to 10.38 while temperature rose to 30–33.5 C. EC there was of 269–393 μS cm^{-1} (J. Csiky pers.comm.).

Depth range: *Chara fibrosa* is restricted to shallow water (<1.5 m)

Associated species: In South France, it was associated with well-developed *Nitella hyalina* and sparse plants of *Chara vulgaris* and *C. globularis* (Soulié-Märsche et al., 2013). Samples from the rice fields in Italy contained individuals of *Chara braunii* and *C. globularis* (Abdelahad and Piccoli, 2017).

29.4 Variability

Two different morphological types corresponding to the macrospecies *Chara fibrosa* sensu Wood (*Chara* sect. *Agardhia*) were recognised in Europe and are treated here as *C. fibrosa* s.l.:

The plants of the original collection of Avetta (1898) described as *C. pelosiana* and later determined as *C. fibrosa* (Krause, 1997, Langangen, 2000). They display a variable number (1–3) rows of stipulodes (Millozza and Abdelahad, 2021) (Fig. 29.4).

The more recent collections previously described as *C. fibrosa* subsp. *benthamii* (Soulié-Märsche et al., 2013; Abdelahad and Piccoli, 2017), which are composed of plants showing only one stipulode per branchlet, in opposite position.

These plants present the fundamental characteristics of *Chara* subsect. *Agardhia* as defined by Casanova (this vol.) i.e. diplostichous cortication, one row of stipulodes, naked branchlets. However, the European plants do not correspond to *C. fibrosa* s.str. The question remains open whether the two morphological types represent two different species or not.

Fig. 29.4 *Chara fibrosa* s.l., morphology of herbarium specimens. **a** Habitus of a plant collected in 1999 in rice fields of Northern Italy; **b–e** Herbarium specimens collected in 1886 in rice fields of Northern Italy; **b** and **c** Samples from the type locality (S. Anna, near S. Cesario, province of Modena), showing variable number of stipulodes (MOD). **d** Dry sample from the province of Pavia (PAV). **e** Decalcified sample from Nonantola (province of Modena) (TO). Photographs: **a**—from Abdelahad and Piccoli (2017, as *C. fibrosa* subsp. *benthamii*); **b–e**—from Millozza and Abdelahad (2021, as *Chara pelosiana*), reproduction with permission by Plant Biosystems and Plants

29.5 Identification and Risk of Confusion

The plants of *Chara fibrosa* are very different from all other European species and easy to identify due to its single row of stipulodes combined with naked branchlets. However, detailed observation of the cortication is necessary to distinguish *Chara fibrosa* (diplostichous) from *Chara baueri*, which has a triplostichous axis (Soulié-Märsche et al., 2009) but was also observed sometimes with a diplo-triplostichous cortication (Langangen and Sviridenko, 1995). According to Avetta (1898: 234), A. Pelosi, the student who first studied *C. pelosiana*, noted already its resemblance with *C. baueri* (Millozza and Abdelahad, 2021).

29.6 Distribution

See Fig. 29.5.

Chara fibrosa s.l. is known from exceptional occurrences in South France and Italy. Bergmeier and Abrahamczyk (2008) mentioned *C. fibrosa* from Crete (Greece), where it was later again seen in lake Kournas (van de Weyer pers. comm.). Plants with the same characters that belong undoubtedly to *Chara* subsect. *Agardhia*, were also identified in rice fields near Szarvas, Hungary (Mesterházy and Csiky pers. comm.).

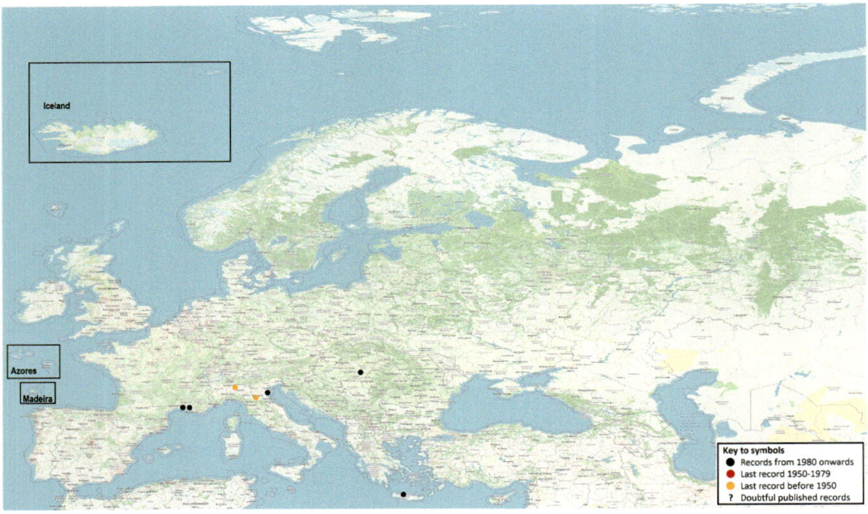

Fig. 29.5 Distribution of *Chara fibrosa* s.l. in Europe. Black dots are records from 1980 on, red dots from 1950 to 1979 and yellow dots before 1950. Map prepared by Heiko Korsch

29.7 Status and Threats

As a non-native, exotic species, *Chara fibrosa* s.l. requires particular habitat conditions that are rarely fulfilled in Europe. There is no risk that the species should become invasive.

29.8 Remarks

The rare attested records of *Chara fibrosa* s.l. in Europe are closely related to rice fields. Presumably, the original oospores were accidentally introduced together with rice seeds imported from South-East Asia. Depending on the country/region of the original rice-seed material, different alien species may have arrived (see *C. oryzae,* this vol.) or may come in to Europe in the future. In most cases, the foreign materials will lead to ephemeral appearance of the exotic species, which are not able to adapt to the climatic context and cannot develop stable populations in Europe. These taxa are included in the present monograph for their witness of the viability of oospores and for their contribution to the record of plant dissemination through human activity.

References

Abdelahad, N., & Piccoli, F. (2017). Report on Charophytes from rice fields in northern Italy including the alien species *Chara fibrosa* ssp. *benthamii*. *Plant Biosystems*, 152, 599–603. https://doi.org/10.1080/11263504.2017.1320310

Avetta, C. (1898). Nuova specie di *Chara* (*Chara pelosiana* mihi). *Malpighia*, 12, 229–235. [in Italian]

Bergmeier, E. & Abrahamczyk, S. (2008). Current and historical diversity and new records of wetland plants in Crete, Greece. *Willdenowia*, 38, 433–453. https://doi.org/10.3372/wi.38.38205

Casanova, M. T. (2013). Review of the species concepts *Chara fibrosa* and *C. flaccida* (Characeae, Charophyceae). *Australian Systematic Botany*, 26, 291–297. https://doi.org/10.1071/SB13020

Casanova, M. T. and Karol, K. G. (2023). Charophytes of Australia's Northern Territory. I Tribe Chareae. Australian Systematic Botany 36: 38–79. https://doi.org/10.1071/SB22023

Comelles, M. (1984). Noves citacions de Caròfits a Espanya. *Butlletí de la Institució Catalana d'història Natural, Secció de Botánica*, 51, 35–39. [in Spanish]

Corillion, R. (1957). Les Charophycées de France et d'Europe Occidentale. *Bulletin de la Société Scientifique de Bretagne*, 32, Hors série. [in French]

Krause, W. (1997). Charales (Charophyceae) In H. Ettl, G. Gärtner, H. Heynig & D. Mollenhauer (Eds.) *Süßwasserflora von Mitteleuropa* vol. 18, Fischer, Jena. [in German]

Langangen, A. (2012). Charophytes (Charales) from Crete collected in 2010. *Flora Mediterranea*, 22, 25–32. https://doi.org/10.7320/FlMedit22.025

Langangen, A., & Sviridenko, B.F. (1995). *Chara baueri* A. Br., a charophyte with a disjunct distribution. *Cryptogamie, Algologie*, 16, 125–132.

Millozza, A. & Abdelahad, N. (2021). The contributions of historical and morphological studies on herbarium specimens to a better definition of *Chara pelosiana* Avetta (Charales, Charophyceae). *Plants*, 10, 2488. https://doi.org/10.3390/plants10112488

Soulié-Märsche, I., Triboit, F., Despréaux, M., Rey-Boissezon, A., Laffont-Schwob, I. & Thiéry, A. (2009). Evidence of an "alien" *Chara* species in South France. In *16th Meeting of the Group of European Charophytologists (GEC),* 12–16 September, Ohrid, Macedonia.

Soulié-Märsche, I., Triboit, F., Despréaux, M., Rey-Boissezon, A., Laffont-Schwob I. & Thiéry, A. (2013). Evidence of *Chara fibrosa* Agardh ex Bruzelius, an alien species in South France. *Acta Botanica Gallica: Botany Letters,* 160, 157–163. https://doi.org/10.1080/12538078.2013.806872

Wood, R. D. (1965). *Monograph of the Characeae.* J. Cramer.

Zaneveld, J. S. (1940). The Charophyta of Malaysia and adjacent countries. *Blumea,* 4, 1–224.

Chapter 30
Chara filiformis

Zofija Sinkevičienė

Scientific name: ***Chara filiformis*** **Hertzsch ex A. Braun in [Hertzsch], Hedwigia 1(12): 81. 1855**

Lectotype (Raam in Armleuchteralgen, Characeen Deuschlands 2016: 292): Angermünde, im Paarstein-See/[no date]/H. Hertzsch (L, left specimen); Rabenhorst, Algen Sachs. Eur. 478.

= *Chara jubata* A. Braun ex Kütz., Tab. Phycol.: 19. 1857, n. ill.
Holotype (Kützing, 1857): Fig. 1 of Table 47.

Local names: BELARUSIAN: Хара ніткападобная; ESTONIAN: Juusmändvetikas, GERMAN: Faden-Armleuchteralge; LATVIAN: Pavedienu mieturīte; LITHUANIAN: Kuoduotasis maurabragis; POLISH: Ramienica grzywiasta; RUSSIAN: Хара нитевидная; SWEDISH: Trådsträfse; UKRAINIAN: Хара ніткоподібна.

Taxonomic notes:

Whether *Chara filiformis* is an independent species or a morphologically extreme form of *C. contraria* is discussed controversially. Many transitional forms have been observed (Dąmbska, 1964; Migula, 1889–1897). No genetic differences were found by Nowak et al. (2016) and Schneider et al. (2016). Olsen (1944) stated that *C. filiformis* cultivated from oospores in an aquarium did not change their characteristic morphology. *Chara contraria* oospores in cultivation developed under extreme light-deficiency conditions plants similar to *C. filiformis*, while *C. filiformis* kept their characteristic morphotype, regardless of the light conditions (Teppke, 2016).

For a long time *Chara filiformis* was associated with *C. hungarica* collected in Lukács thermal pool in Budapest and described by Filarszky (1930). In this monograph *C. hungarica* is treated as conspecific with *C. squamosa* (see Chap. 45).

Z. Sinkevičienė (✉)
Nature Research Centre, Institute of Botany, Žaliųjų Ežerų Str. 47, 12200 Vilnius, Lithuania
e-mail: zofijasin@gmail.com

H. Schubert et al. (eds.), *Charophytes of Europe*,
https://doi.org/10.1007/978-3-031-31898-6_30

Along with *Chara jubata* another heterotypic synonym of *C. filiformis* mentioned in the literature (Hollerbach and Krassavina, 1983; Migula, 1889–1897; Mowszowicz, 1947) is "*Chara tyzenhausii* Gorski, Icon. Potam. Charac. Cyp. Gram. 1849, Tab. XI". We failed to locate this work in libraries and suppose that it was never published.

30.1 Morphology

See Figs. 30.1 and 30.2.

Habitus: *Chara filiformis* is slender, greyish green, usually heavily encrusted. It is characterised by its 'threadlike' appearance due to the elongated internodes and very short, only a few millimetres long, branchlets.

Size: Usually medium-sized, to 40 cm (Krause, 1997; Wood, 1965), but may be small (about 10 cm) when growing in shallow water and can reach up to 90 cm in deep waters.

Main axis: The main axis is often moderately branched, with elongated internodes, slender—diameter range from 0.3 to 0.5 mm (Krause, 1997; Teppke, 2016; Wood, 1965).

Cortex: The cortex is diplostichous and heterostichous, tylacanthous.

Spine-cells: The spine-cells are solitary, papilliform, located on more prominent primary cortex cells; developed on younger internodes.

Stipulodes: The stipulodes are arranged in two rows (diplostephanous); both rows of small ellipsoid stipulodes are equal in appearance.

Branchlets: Branchlets 6–8 in a whorl (Wood, 1965), extremely short and slightly curved. They consist of 1, rarely 2 corticated basal segments and ecorticated end members. The two corticated segments are of equal length (Wood, 1965), or the second ones is prolonged (Hollerbach and Krassavina, 1983). The length of the branchlets ranges from 0.5 mm (Dąmbska, 1964) to 2 mm (Krause, 1997; Olsen, 1944; Teppke, 2016; Wood, 1965).

Bract-cells: Bract-cells 6 unilateral; anteriors 4 times longer than wide; posteriors small, globular.

Bracteoles: Bracteoles 2, similar to anterior bract-cells but slightly larger, not exceeding half of the length of mature oogonia.

Sexuality: Plants are monoecious; gametangia conjoined, usually situated at the lowest (rarely on the second) corticated node of branchlets, rarely two pairs of gametangia (Krause, 1997).

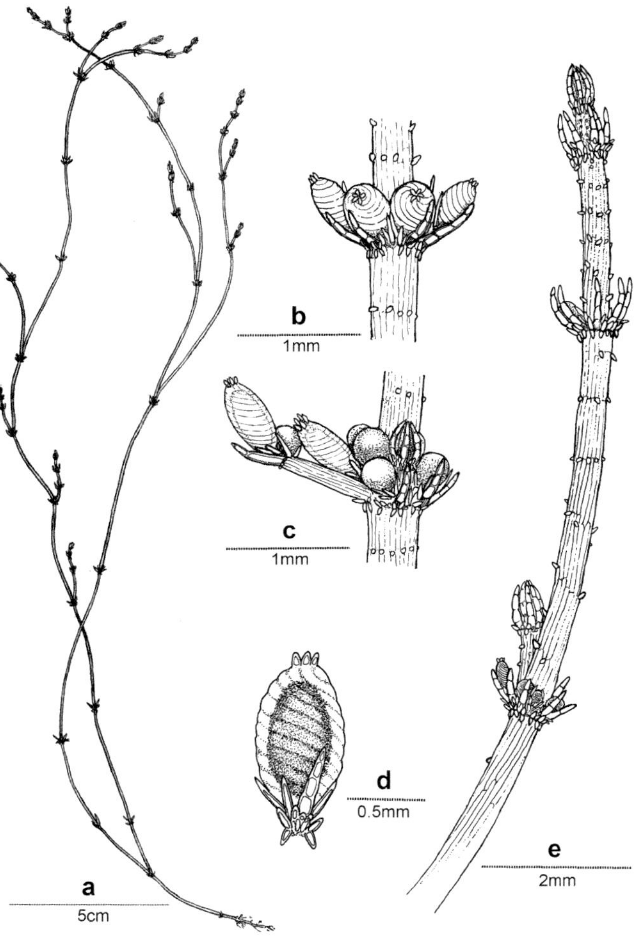

Fig. 30.1 *Chara filiformis,* morphological features. **a** Habitus; **b** Fertile branchlet whorl with immature oogonia; **c** Fertile branchlet whorl with gametangia and one prolonged branchlet; **d** Oogonium; **e** Upper part of the main axis with sterile and fertile branchlet whorls. Drawing: Mindaugas Ryla

Oogonia: Oogonia are solitary (rarely geminate acc. to Krause, 1997), ovoid, subglobose or ellipsoid, heavily encrusted. Variability presented in Table 30.1.

Oospores: Data about quantitative morphometry and colour of oospores differ among between authors (Table 30.2). The latest measurements of oospores revealed quite wide variability in shape and sizes (Sinkevičienė, 2019). The oospores were significantly longer and with a smaller number of ridges than in the cited literature. The colour of oospores was different in wet and dry condition. Calcified oospores become gyrogonites. They were measured (n = 50) from Lake Germantas and varied in shape from quite broad to elongate slender. The length of gyrogonites ranged from 750 to

Fig. 30.2 Habitus, habitat appearance and details of *Chara filiformis*. **a** Dense stands of *C. filiformis* in Lake Borken (Germany/Hesse); **b** Branchlets whorl with crowded gametangia; **c** Habitus from shallow water (height 10 cm); **d** Wet oospores; **e** Dry oospore. Photographs: **a**—Klaus van de Weyer; **b**—Liucija Kamaitytė-Bukelskienė; **c–e**—Zofija Sinkevičienė

920 μm, with an average of 844 ± 39.2 μm, whereas width range was 370–550 μm (average 452 ± 39.5 μm). The number of ridges ranged from 9 to 12, but often it was 10 (Sinkevičienė, 2019).

Antheridia: Antheridia are located below oogonia. Variability in diameter reported by various authors presented in Table 30.3.

Bulbils: Nodular shaped, greenish-yellow bulbils are 1–2 mm in size (Teppke, 2016). First reported by van de Weyer in the Large Stechlin/Mecklenburg Lake district and observed in specimens from Lithuanian lakes also.

Table 30.1 Oogonia morphometry

Author	Length	Width	Coronula length	Coronula width	Number of helical stripes
Migula (1889–1897)	900	600	150	–	12–15
Corillion (1957)	900	600	150	–	–
Wood (1965)	675–750	450–525	120–150	225–255	–
Hollerbach and Krassavina (1983)	675–750(915)	450–525(630)	120–150	225–255	12–15
Krause (1997)	650–850	450–550	120–150	225–250	12–15
Urbaniak and Gąbka (2014)	650–770	440–565	–	–	–
Total range	**650–900(915)**	**440–600(630)**	**120–150**	**225–255**	**12–15**

Length in the first column excludes the coronula; its length and width are given separately, all values for length and width in μm

Table 30.2 Oospore morphometry

Author	Length μm	Width μm	Number of ridges	Colour
Migula (1889–1897)	500–660	350–420	12–14	Dark brown-dark red brown
Corillion (1957)	500–600	350–420	12–14	Black
Wood (1965)	600–660	390–420	11–12	Dark chestnut brown to almost black
Hollerbach and Krassavina (1983)	500–660(680)	350–420(472)	11–14	Dark chestnut brown to almost black
Krause (1997)	500–700	350–450	11–14	Black
Urbaniak and Gąbka (2014)	510–705	305–385	–	Dark brown
Sinkevičienė (2019)	680–810	300–460	9–12	Dark brown—wet, almost black—dry
Total range	**500–810**	**300–460(472)**	**9–14**	

All values for length and width in μm

30.2 Phenology and Reproduction

Very little data on the reproduction and phenology of *Chara filiformis* are available. The focus of its occurrence is from summer to early autumn (June-September). Individual plants can also be observed in late autumn (Teppke, 2016). *Chara filiformis*

Table 30.3 Antheridia morphometry

Author	Diameter (μm)
Migula (1889–1897), Corillion (1957)	300–360
Wood (1965)	350–405
Dąmbska (1964)	200–300
Hollerbach and Krassavina (1983)	300–405
Krause (1997)	300–400
Urbaniak and Gąbka (2014)	255–450
Total range	**200–450**

can hibernate as green plants in deep waters (Olsen, 1944). Fructification takes place from July to August (Dąmbska, 1964; Holtz, 1903; Olsen, 1944). In herbarium material, ripe oospores were mainly observed in plants collected at the end of July and in August. According to Olsen (1944) *C. filiformis* can germinate at irregular times of the year, generally after a rest of one or several years. Vegetative propagation from nodal bulbils is possible also.

30.3 Habitat Conditions and Ecology

Habitat type: *Chara filiformis* is restricted to freshwater habitats. Large clear lakes with calcium-rich water are preferable habitats of this species (Šarkinienė and Trainauskaitė, 1973; Urbaniak and Gąbka, 2014; Zviedre, 2008). Occasionally, it was found in quarries (Latvia) (Zviedre, 2008) and in a waterbody—e.g., Lake Borken (Germany), formed by lignite mining (Korte et al., 2010).

Sediment: *Chara filiformis* has been reported from a broad range of sediments from sand and gley in shallow localities to mud or calcareous gyttja in deeper parts of lakes (Korte et al., 2010; Šarkinienė and Trainauskaitė, 1973; Teppke, 2016).

Nutrient conditions: *Chara filiformis* prefers oligotrophic to mesotrophic waters bodies but can sometimes grow in eutrophic waters as well. Concentrations of nutrients are presented in Table 30.4 (mainly based on data from Lithuanian and Latvian lakes) (Environmental Protection Agency, 2013–2015; Zviedre, 2008).

Water chemistry: *Chara filiformis* is a freshwater species. The species is known mainly from large deep mesotrophic lakes with calcium-rich waters and good light conditions (Urbaniak and Gąbka, 2014). Ranges of physical properties of waters are presented in Table 30.4 (mainly based on data from Lithuanian and Latvian lakes) (Environmental Protection Agency, 2013–2015; Zviedre, 2008)

Depth range: Preferable depth range recorded is between 1(2) m and 5(6) m (Auderset Joye and Schwarzer, 2012; Urbaniak and Gąbka, 2014; Urbaniak et al., 2011; Olsen, 1944, Teppke, 2016); however, it occurs also in the deepest and the

Table 30.4 Physico-chemical characteristics of the lakes with occurring *C. filiformis*

	Lithuania, n = 56			Latvia, n = 8		
	Min	Mean	Max	Min	Mean	Max
pH	8.1	8.5	9.3	7.7	7.97	8.3
Secchi depth m	1.0	4.31	7.48	1.9	3.6	6.5
Alkalinity meq/l	0.5	2.7	4.2	–	–	–
Conductivity μS/cm	129	316	508	–	–	–
Ca^{2+} mg/l	24.8	45.5	79.8	22.5	32.4	43.03
NH_4^+–N mg/l	0.007	0.025	0.135	–	–	–
NO_2–N mg/l	0.000	0.012	0.353	0.004	0.006	0.008
NO_3–N mg/l	0.003	0.075	0.535	0.9	1.037	1.1
TN mg/l	0.245	0.663	1.725	–	–	–
$PO4^{3-}$–P mg/l	0.001	0.006	0.020	0.003	0.022	0.085
TP mg/l	0.006	0.020	0.067	–	–	–

shallowest places. A growing depth of 8 m was recorded by Teppke et al. (2016), 10 m by Korte et al. (2010), 13 m Šarkinienė and Trainauskaitė (1973). The maximal growing depth of 22 m was recorded in 1860 in Lake Levrasjön (Hasslow, 1931).

Associated species: *Chara filiformis* forms dense monodominant stands [as. *Charetum filiformis* (Jeschke, 1959; Krausch, 1964)] or with co-occurring *Characeae* (*Chara contraria, C. aspera, C. globularis, C. tomentosa, C. subspinosa, Nitellopsis obtusa*) or species of angiosperms (*Myriophyllum spicatum, Potamogeton* spp., *Utricularia vulgaris* s.l.) and bryophytes (*Fontinalis antipyretica, Drepanocladus* spp., *Scorpidium scorpioides*) (Dąmbska, 1966; Tomaszewicz, 1979). *Chara filiformis* is a common component of communities formed by other *Chara* species especially *C. tomentosa* or *C. subspinosa*, where it forms a more or less dense lower layer. As accompanying species, it can be found in communities dominated by *Chara aspera, C. globularis* or other submerged macrophytes (*Myriophyllum spicatum, Potamogeton perfoliatus, P. lucens*).

30.4 Variability

Chara filiformis varies mainly in the length of internodes depending on the growing depth. Migula (1889–1897) described 5 forms (under the name *Chara jubata*) varying in the length of internodes and branchlets. However, *C. jubata* f. *subcontraria* and *C. jubata* f. *longifolia* are long-leaved and most probably belong to *C. contraria* rather than *C. filiformis*.

30.5 Identification and Risk of Confusion

Due to its characteristic threadlike appearance typical *Chara filiformis* is an easily recognisable species. The general appearance of *C. filiformis* is very similar to *C. kirghisorum*, however in fertile state monoecious *C. filiformis* clearly differs from dioecious *C. kirghisorum*. Moreover, *C. filiformis* can be confused with forms of *C. contraria*, having a shortened or incomplete branchlets. These two species often grow together. The main difference is in the first corticated basal segment, which although shortened in *C. contraria* is visible by naked eye, while the cortex of *C. filiformis* always is rudimentary. *C. filiformis* only rarely has one or several branchlets per whorl with second, long corticated segment. Extensive review of specimens suggests that plants with elongate branchlets can be attributed to *Chara filiformis* in this case only if the first corticated segments of branchlets are rudimentary. Conjoined gametangia of *C. filiformis* usually are located at the first node of the lowest corticated segment, rarely at the second corticated node. Because the first fertile branchlet segment in *C. filiformis* is extremely short (rudimentary), gametangia and oospores appear to be located directly on the main axis (Fig. 30.2b). Atypical *C. contraria* forms with shortened branchlets often have only one corticated fertile segment or completely ecorticated branchlets. In these cases, a careful examination of specimens is required to find a normally developed (not rudimentary) basal segment, usually occurring on the main axis.

30.6 Distribution

See Fig. 30.3.

Chara filiformis is an endemic European species mainly distributed southeast of the Baltic Sea. The most recent locations are concentrated in north-eastern Germany (Teppke, 2016), north-eastern Poland (Gąbka, 2009; Urbaniak and Gąbka, 2014), Lithuania (Sinkevičienė, 2019; Sinkevičienė and Gudžinskas, 2021), and Latvia (Zviedre, 2008) where the species is still quite widespread and abundant. The species is rare in Denmark (Baastrup-Spohr et al., 2013), Sweden (Blindow, 2008; Westling, 2015), Belarus (Mikheyeva, 2015), Russia (Hollerbach and Krassavina, 1983) and Switzerland (Auderset Joye and Schwarzer, 2012). No specimens were seen from France where it should occur in the Pyrenees (Corillion, 1957). The species has not been found during the inventory in 2011 but taking join into account its occurrence in neighbouring Latvia its presence has been suspected also in Estonia (Kovtun-Kante, 2015).

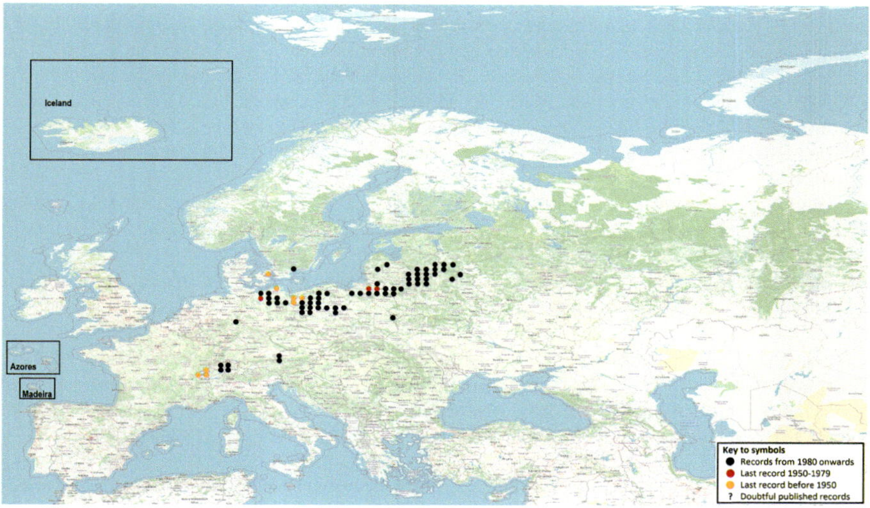

Fig. 30.3 European distribution of *Chara filiformis*. For worldwide distribution see Korsch (2018). Black dots are records from 1980 onwards, red dots from 1950 to 1979 and yellow dots before 1950; unconfirmed/doubtful published records (if applicable) are marked with "?" on top. Map prepared by Heiko Korsch

30.7 Status and Threats

Eutrophication and increasing water turbidity could be the most serious threat for *Chara filiformis* (Langangen, 2007). The species can survive in eutrophic waters for quite a long time if the water remains clear, but the normal development of oospores is usually disturbed. *C. filiformis* is included in the national Red Lists as VU (Vulnerable) in Belarus (Mikheyeva, 2015), RE (Regionally Extinct) in Denmark (Baastrup-Spohr et al., 2013), 2 (CR) Critically Endangered in Germany (Korsch et al., 2013), EN in Poland (Urbaniak and Gąbka, 2014), CR (Critically Endangered) in Sweden (Westling, 2015), DD (Data Deficient) in Switzerland (Auderset Joye and Schwarzer, 2012) and Latvia; it is not assessed as threatened in Lithuania.

References

Auderset Joye, D., & Schwarzer, A. (2012). Rote Liste Armleuchteralgen. Gefährdete Arten der Schweiz, Stand 2010. *Bundesamt für Umwelt BAFU & Laboratoire d'Ecologie et de Biologie Aquatique der Universität Genf, Bern* (72 pp.) [in German].

Baastrup-Spohr, L., Iversen, L. L., Dahl-Nielsen, J., & Sand-Jensen, K. (2013). Seventy years of changes in the abundance of Danish charophytes. *Freshwater Biology, 58*, 1682–1693.

Blindow, I. (2008). Schwedische Artenschutzprogramme für bedrohte Characeen. *Rostocker Meeresbiologische Beiträge, 19*, 23–28. [in German].

Corillion, R. (1957). Les Charophycées de France et d'Europe Occidentale. *Bulletin de la Société Scientifique de Bretagne, 32*(Hors série I & II), 1–499 [in French].
Dąmbska, I. (1964). *Charophyta – ramienice*. Państwowe Wydawnictwo Naukowe [in Polish].
Dąmbska, I. (1966). Zbiorowiska ramienic Polski. *Prace Komisji Biologicznej, Poznanskie Towarzystwo Przyjaciol Nauk, Wydzial MatematycznoPrzyrodniczy, 31*, 1–76. [in Polish].
Environmental Protection Agency. (2013–2015). Results of monitoring of lakes and ponds. Average hydrochemical data [in Lithuanian].
Filarszky, N. (1930). A Budai hőforrások nyílt vizeinek Chara-féléi és néhány más érdekesebb, ritkább algafaja. Matematikai Természettudományi Értesítö A. *M. Tud. Akadémia III. Osztályának Folyóirata, 47*, 652–674. [in Hungarian].
Gąbka, M. (2009). *Charophytes of the Wielkopolska region (NW Poland): Distribution, taxonomy, and autecology*. Bogucki Wydawnictvo.
Hasslow, O. J. (1931). Sveriges characeer. *Botaniska Notiser, 1931*, 63–136. [in Swedish].
Hollerbach, M. M., & Krassavina, L. K., [Голлербах, М. М. & Л. К. Красавина]. (1983). Определитель пресноводных водорослей СССР. 14. *Charophyta [Determination key of freshwater algae of USSR]* (189 pp.). Nauka, [in Russian].
Holtz, L. (1903). Characeen. In *Kryptogamenflora der Mark Brandenburg. Band* (Vol. 4, no. 1). Borntraeger [in German].
Jeschke, L. (1959). Die Pflanzengesellschaften einiger Seen bei Feldberg in Mecklenburg. *Feddes Repertorium Beiheft, 138*, 161–214. [in German].
Korsch, H. (2018). The worldwide range of the Charophyte species native to Germany. *Rostocker Meeresbiologische Beiträge, 28*, 45–96
Korsch, H., Doege, A., Raabe, U., & van de Weyer, K. (2013). Rote Liste der Armleuchteralgen (Charophyceae) Deutschlands. 3. Fassung, Stand: Dezember 2012. *Haussknechtia Beiheft, 17*, 1–32. [in German].
Korte, E., Krautkrämer, V., Pätzold, F., & van de Weyer, K. (2010). Brackwasser-Armleuchteralge (*Chara canescens*) und Faden-Armleuchteralge (*Chara filiformis*) im Borkener See. Zwei bemerkenswerte Neunachweise für Hessen. *Botanik und Naturschutz in Hessen, 23*, 55–62. [in German].
Kovtun-Kante, A. (2015). *Charophytes of Estonian inland and coastal waters: distribution and environmental preferences* (p. 82). Dissertaciones biologicae Universitatis Tartuensis.
Krausch, H. D. (1964). Die Pflanzengesellschaften des Stechlinseegebietes I: Die Gesellschaften des offenen Wassers. *Limnologica, 2*, 145–203 [in German].
Krause, W. (1997). Charales (Charophyceae) Süßwasserflora von Mitteleuropa. In H. Ettl, G. Gärtner, H. Heynig, & D. Mollenhauer (Eds.), *Süßwasserflora von Mitteleuropa* (Vol. 18). Fischer [in German].
Kützing, F. T. (1857). *Tabulae phycologicae; oder, Abbildungen der Tange* (Vol. 7). W. Köhne [in German].
Langangen, A. (2007). *Charophytes of the Nordic countries* (p. 102). Saeculum ANS.
Migula, W. (1889–1897). Fünfter Band: Die Characeen. In A. Fischer, E. Fischer, F. Hauck, G. Limpricht, C. Luerssen, W. Migula, H. Rehm, P. Richter & G. Winter (Eds.), *Dr. L. Rabenhorst's Kryptogamenflora von Deutschland, Oesterreich und der Schweiz* (2nd ed). Eduard Kummer [in German].
Mikheyeva, T. M. (2015). Algae. In: I. M. Kachanovskii, V. I. Nikiforov, & V. I. Parfenov (Eds.), Red data book of Belarus. In *Plants* (4th ed, pp. 297–324). Belaruskia Encyklapedya imia Petrusia Brouki [in Russian].
Mowszowicz, J. (1947). Characeae Wilna i jego okolic. *Acta Societatis Botanicorum Poloniae, 18*, 1–8. [in Polish].
Nowak, P., Schubert, H., & Schaible, R. (2016). Molecular evaluation of the validity of the morphological characters of three Swedish Chara sections: Chara, Grovesia, and Desvauxia (Charales, Charophyceae). *Aquatic Botany, 134*, 113–119.
Olsen, S. (1944). Danish Charophyta. *Det Kongelige Danske Videnskabernes Selskab, Biologiske Skrifter, 3*(1): 1–240, pl. 1-2 [in Danish].

Šarkinienė, I. B., & Trainauskaitė, I. Y., [Шаркинене И. Б. & И. Ю.Трайнаускайте]. (1973). Экология харовых водорослей Литовской ССР [Ecology of charophytes occurring in Lithuania]. In K. Jankevičius (Ed.), Харовые водоросли и их использование в исследованиях биологических процессов клетки *[Charophyta and their use in the investigations of the biological processes of cells]* (pp. 104–118). Vilnius [in Russian].

Schneider, S. C., Nowak, P., von Ammon, U., & Ballot, A. (2016). Species differentiation in the genus *Chara* (Charophyceae): considerable phenotypic plasticity occurs within homogenous genetic groups. *European Journal of Phycology, 51*, 282–293.

Sinkevičienė, Z. (2019). Peculiarities of *Chara filiformis* (Charales, Charophyceae) distribution and oospore sizes in Lithuania. *Webbia, 74*, 133–138.

Sinkevičienė, Z., & Gudžinskas, Z. (2021). Revision of the Characeae (Charales, Charophyceae) species and their distribution in Lithuania. *Botanica, 27*, 95–124.

Teppke, M. (2016). *Chara filiformis*. In *Arbeitsgruppe Characeen Deutschlands, Armleuchteralgen: Die Characeen Deutschlands* (pp. 292–299). Springer [in German].

Teppke, M., Möbius, F., Schubert, H., & Spieß, H.-J. (2016). *Rote Liste der Armleuchteralgen (Charophyceae) Mecklenburg-Vorpommerns. Ministerium für Landwirtschaft* (59 pp.). Umwelt und Verbraucherschutz Mecklenburg-Vorpommern [in German].

Tomaszewicz, H. (1979). *Roślinność wodna i szuwarowa Polski: klasy Lemnetea, Charetea, Potamogetonetea, Phragmitetea w stanu zbadania na rok 1975*. Rozprawy Uniwersytetu Warszawskiego [in Polish].

Urbaniak, J., & Gąbka, M. (2014). *Polish Charophytes. An illustrated guide to identification* (120 pp.). Uniwersytet Przyrodniczy we Wroclawiu [in Polish].

Urbaniak, J., Sugier, P., & Gąbka, M. (2011). Charophytes of the Lubelszczyzna region (Eastern Poland). *Acta Societatis Botanicorum Poloniae, 80*, 159–168.

Westling, A. (ed). (2015). *Rödlistade arter i Sverige 2015*. ArtDatabanken, SLU [in Swedish].

Wood, R. D. (1965). *Monograph of the Characeae*. J. Cramer.

Wood, R. D., & Imahori, K. (1964). *Iconograph of the Characeae*. J. Cramer.

Zviedre, E. (2008). *Latvijas saldûdens mieturalģu (Charophyta) flora un ekoloģija* (100 pp.). Dissertation [in Latvian].

Chapter 31
Chara fragifera

Nick Stewart

Scientific name*: Chara fragifera* Durieu, Bull. Soc. Bot. France 6: 185. 1859

Lectotype (Wood, 1965: 178): Très abundant mais tout à fait detruit, Étangs de Cazau à Cazau, Gironde/8 juillet 1855/M. C. Durieu de Maisonneuve (PC, top left of three specimens on sheet).

≡ *Chara globularis* Thuill. var. *globularis* f. *fragifera* (Durieu) R. D. Wood, Rev. Charac.: 177. 1965, n. inv.

≡ *Chara galioides* DC. subsp. *fragifera* (Durieu) A. Braun in Monatsber. Königl. Preuss. Akad. Wiss. Berlin 1867: Tab. I. 1868.

Local names: BELARUSIAN: Хара клубняносная; ENGLISH: Strawberry Stonewort; FRENCH: Chara fragifère; RUSSIAN: Хара клубненосная; UKRAINIAN: Хара бульбоносна.

31.1 Morphology

See Figs. 31.1 and 31.2.

Habitus: Plants are mid to dark green, unencrusted or rarely a little encrusted, semi-translucent, slender and flexuous.

Size: A small to medium species up to 20 cm tall or sometimes reaching up to 30 cm when growing in deeper water.

Main axis: The main axis is slender, normally to 0.4 mm diameter but rarely up to 0.7 mm diameter, branching from the lower main axis nodes and rhizoid nodes.

N. Stewart (✉)
4 Hexton Road, Glastonbury BA6 8HL, United Kingdom
e-mail: nfstewart@freeuk.com

H. Schubert et al. (eds.), *Charophytes of Europe*,
https://doi.org/10.1007/978-3-031-31898-6_31

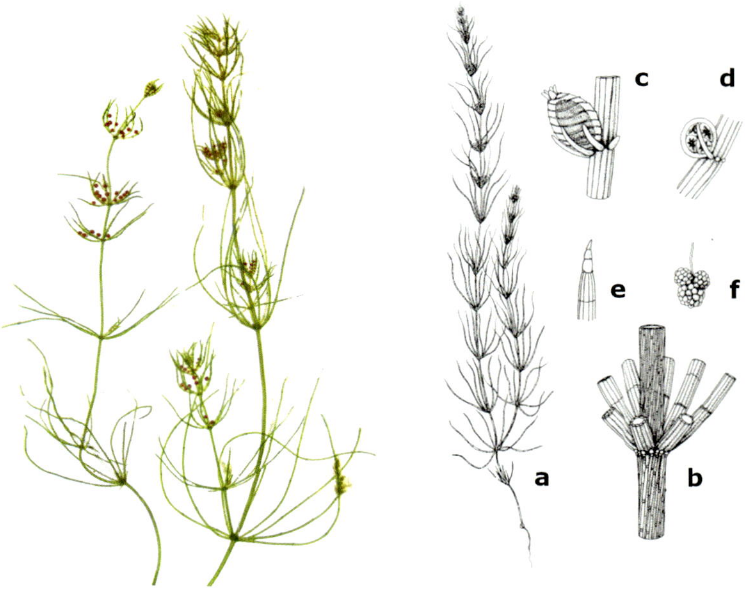

Fig. 31.1 Habitus of *Chara fragifera*. Left panel: Habit of male plants collected by P. A. Gainey from The Lizard, Cornwall, UK. Right panel: Drawing of *C. fragifera*, illustrating **a** General habit; **b** Portion of branchlet whorl; **c** Branchlet node with oogonium; **d** Branchlet node with antheridium; **e** Branchlet apex; **f** Rhizoid bulbil. Photograph (left panel): Chris Carter. Drawings (right panel) by Margaret Tebbs, reproduced from Moore (1986) with permission from the Botanical Society of Britain and Ireland

Internodes 0.5–2 × as long as branchlets, usually approximately equal to shorter than the branchlets.

Cortex: Triplostichous, usually tylacanthous but sometimes nearly isostichous.

Spine-cells: Usually more or less rudimentary, but sometimes weakly papillate.

Stipulodes: Diplostephanous, the lower rudimentary, but the upper are usually globular or slightly elongate, obtuse and up to two times as long as wide (to 70 × 42 μm).

Branchlets: In whorls of 6–9, to 2(3) cm long, slender and flexuous, those of the male plants often shorter and sometimes a little incurved. Segments (7)9–13, with the upper 1–3 segments ecorticate. Sometimes slightly swollen at the nodes and this can become more prominent in dried herbarium specimens.

Bract-cells: 5–6, the outer rudimentary to globular, the inner pair shortly developed and acute, up to 0.5 × length of the gametangia but usually much less than this, particularly in male plants.

Bracteoles: 2, acute, up to 0.5(0.8) × length of the gametangia.

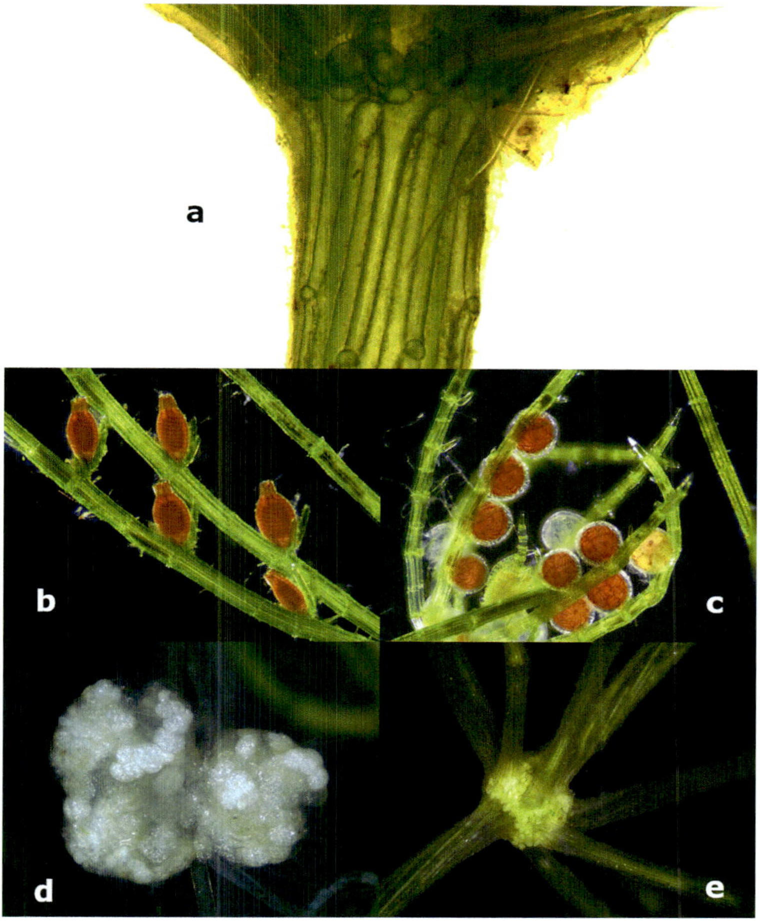

Fig. 31.2 *Chara fragifera* details. **a** Main axis cortex and node; **b** Female branchlets and oogonia; **c** Male branchlets and antheridia; **d** Rhizoid bulbils; **e** Axial node bulbil of specimens collected by P. A. Gainey from The Lizard, Cornwall, UK in 2016 and 2017. Photographs: Chris Carter

Bractlets: 1, acute, up to 0.35(0.5) × length of the gametangia.

Sexuality: Dioecious. Gametangia solitary or the oogonia very rarely geminate, at the lowest (2)3–5 nodes of the branchlets.

Oogonia: Often reddish orange. Coronula usually connivent, but sometimes straight or spreading. Morphometric data are listed in Table 31.1.

Oospores: For dimensions see Table 31.2. Oblong-ellipsoid. Ridges, rather faint (Fig. 31.3a), prolonged downwards at the base into spines or cage (Fig. 31.3a, b). Outer oospore-membrane thick-laminated, semi-rigid, semi-opaque, dark red brown,

Table 31.1 Oogonia morphometry

Author	Length	Width	Coronula length	Coronula width	Number of helical stripes
Migula (1889–1897)	900–1200	450–600	-	-	-
Groves and Bullock-Webster (1924), Corilion (1957)	875–1000	550–675	125–175	175–200	12–14
Wood (1965)	675–1000	525–675	105–175	175–200	12–14
Krause (1997)	750–1000	550–700	100–200	150–200	10–12
Cirujano et al. (2008)	675–1000	525–675	105–175	-	-
Total range	**675–1200**	**450–700**	**100–200**	**150–200**	**10–14**

Length in the first column is excluding the coronula; its length and width are given separately; all values for length and width in μm

finely and evenly granulate under optical microscopy (Fig. 31.3d) (Corillion, 1957; Groves and Bullock-Webster, 1924) but minutely pitted under SEM (Fig. 31.3c).

Antheridia: Morphometric data are listed in Table 31.3.

Bulbils: Produced at the lower axial nodes (axial) and on the rhizoids (lateral), in both cases multicellular, whitish, to 7 mm in diameter but normally 1–3 mm in diameter. The latter are said to resemble miniature strawberries, but larger bulbils are often more irregular with low rounded lobes.

Table 31.2 Oospore morphometry

Author	Length	Width	Number of striae	Colour
Migula (1889–1897)	600–800	300–400	12	Dark red-brown
Groves and Bullock-Webster (1924)	550–675	375–450	9–12	Chestnut-brown to nearly black
Corillion (1957)	550–700	375–450	9–12	Brown to blackish
Wood (1965)	550–700	345–450	9–12	Chestnut-brown to black
Krause (1997)	550–700	350–450	9–11	-
Cirujano et al. (2008)	550–700	350–450	-	Dark brown to black
Total range	**550–800**	**300–450**	**9–12**	

All values for length and width in μm

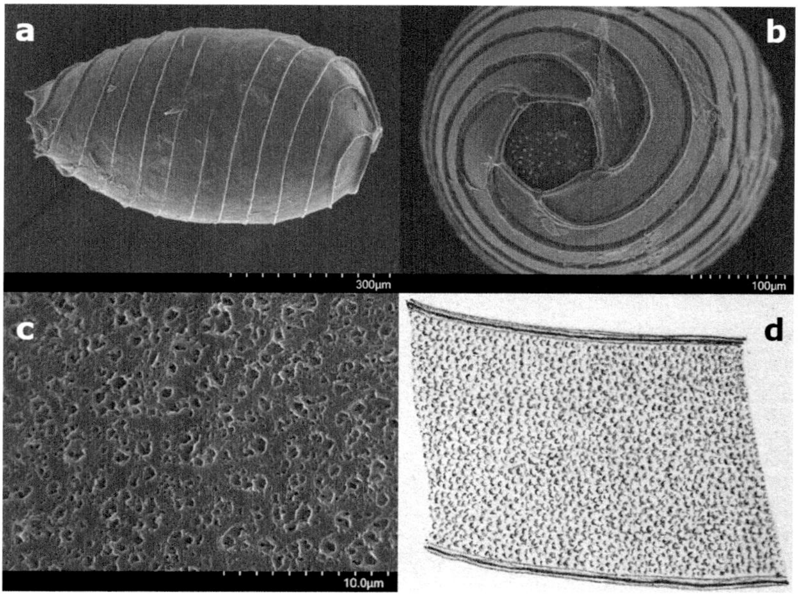

Fig. 31.3 *Chara fragifera* oospore ornamentation. **a–c** Appearance under SEM; **a** Whole spore; **b** Base plate; **c** Detail of oospore-membrane of a specimen from Hayle Kimbro, Cornwall, UK, collected by G. R. Bullock-Webster in 8/1898 (BM); **d** Oopsore-membrane appearance under optical microscope. Photographs: **a–c**—Michelle T. Casanova; **d**—from Groves and Bullock-Webster (1924)

Table 31.3 Antheridia morphometry

Author	Diameter (μm)
Migula (1889–1897)	450–800
Groves and Bullock-Webster (1924)	500–575
Corillion (1957)	500–575(800)
Wood (1965)	450–575(625)
Comelles (1985)	700–800
Krause (1997)	500–600
Cirujano et al. (2008)	450–600
Total range	**450–800**

31.2 Phenology and Reproduction

In its typical French habitat, *Chara fragifera* mainly overwinters as bulbils and green plants begin to develop from these as soon as the water temperature reaches 10°C. The presence of bulbils and the tendency of this species to grow in shallow water that heats up quickly when there is a period of sunshine means that it can produce axes early in the season from April onwards. However, if early growth is set back by conditions becoming unfavourable again, there are sufficient reserves of bulbils to produce new shoots again when favourable conditions return (Corillion, 1953, 1957).

Plants continue to grow and fruit until late in the season (October–November) if conditions remain suitable. In areas of little frost near the coast and sometimes in mild winters inland, plants can over-winter in a green state and these plants can become fertile from around March onwards. However, in sites which dry out over the summer, plants can complete their growth cycle by May or June and die back as the pools dry out. In such situations, oospores may play a more important role; timing of oospore germination has not been reported but seems likely to be in the autumn as water levels rise.

Bulbils tend to be most frequent when the plants age later in the season and can be sparse or absent early in the season when the plants are putting most of their energy into vegetative growth.

31.3 Habitat Conditions and Ecology

Habitat type: *Chara fragifera* is a species of soft water that is low in nutrients (oligotrophic) and around neutral pH where it grows predominantly in shallow water in a few centimetres to a metre depth. It likes open, well-lit situations where the water warms up quickly in periods of sunshine. Here, it is able to grow rapidly from its bulbils early in the season while the deeper water remains more constantly cool (Corillion, 1953, 1957). Such habitats would normally be susceptible to drying out quickly but its preference for regions with higher humidity and rainfall means that the season is prolonged by periodic replenishment by wet weather (Corillion, 1953, 1957).

In its French stronghold it most commonly occurs on the very shallowly shelving sandy edges of lakes, and the flat bottoms of shallow pan-shaped pools. The latter, lacking a reservoir of cold water in the deeper parts, are able to warm up more quickly throughout. Here it can form extensive populations on wind-exposed shores and its slender flexuous form and lack of encrustation means that it is not fragile, and it is able to withstand significant wave-wash (Corillion, 1953, 1957).

In Britain (Stewart, 2003), it is particularly associated with areas of heathland. Here it sometimes occurs in similar large shallow pools but more frequently it occurs in

- small quarry pits excavated in the past for serpentinite. More rarely it occurs in shallow pools on the bottom of larger quarries. It also occurs in a basalt quarry in its only locality on the Mediterranean coast of France (Mouronval et al., 2015),
- rutted trackways where it is confined to the lower areas that remain inundated for longest,

Corillion (1957) and Stewart (2003) also note that it occurs occasionally in a variety of other small water bodies, including rice fields. There are also a few records from flowing water, either in seepages or in areas of slack water in streams. Corillion (1953, 1957) notes that its distribution in France and Britain is restricted to areas with

- annual mean temperature 10°C or above,
- January mean temperature 2–6°C,
- maximum 60 days of frost per year,
- maximum 10 days of snow per year.

On the other hand, its restriction in the Mediterranean region is probably limited by its shallow water habitat drying out too quickly in the hot dry summers (e.g., Cirujano et al., 2008).

The niches in which it grows tend to have considerable fluctuations in temperature, depending on the weather. As discussed above (under phenology) it is able to exploit brief periods of warm weather to begin growth early in the season with a water temperature of 10°C identified as the minimum for growth by Corillion (1953, 1957).

Sediment: The main requirement is a loose sediment in which the bulbils can develop and grow. In its French stronghold this is predominantly sand which is acid to mildly alkaline (Corillion, 1953, 1957). Stiff clays are unsuitable, but it can grow in lighter clays with a loose texture (Corillion, 1953, 1957) and this is its typical substrate in Britain (Stewart, 2003). Here it grows on soft clay which may sometimes have a covering of peat or organic mud and it only more rarely occurs on sand or gravel.

It seems to be particularly associated with areas of igneous geology although this varies from granite (Corillion, 1953, 1957) to basalt (Mouronval et al., 2015). In Britain it previously occurred in areas of granite but it is now restricted to an area of serpentinite, a soft, base-rich rock of igneous origin.

Nutrient conditions: No water chemistry data is available, but nitrate and phosphate levels are likely to be low at all sites. The species has been lost from some sites that have suffered from eutrophication. It seems to require clear water and situations where it is little affected by epiphytes and larger competing vegetation (e.g., Corillion, 1957).

Water chemistry: *Chara fragifera* usually occurs in slightly acid to slightly alkaline water, between pH 6.3–7.5 (Mouronval et al., 2015). Exceptionally it can occur in more acid and more alkaline situations, but in the latter case it remains unencrusted (Corillion, 1953). It seems to be exclusively a freshwater species, but many sites in Britain and France are near exposed coasts where some wind-carried salt is likely. Current British sites are thought to have low Calcium but high Magnesium content.

Depth range: This species predominantly grows in depths from a few centimetres to a metre. However, there are a few records from up to 4.5 metres (Mouronval et al., 2015).

Associated species: This species prefers very open situations with little or no competition from filamentous algae or vascular plants and therefore there are often few associated species. Where other plants are present, in neutral waters the typical associates are *Littorella uniflora, Lobelia dortmanna, Juncus bulbosus, Eleogiton fluitans, Helosciadum inundatum* and *Potamogeton polygonifolius*. It also occurs with other charophytes, *Nitella opaca, N. translucens, N. confervacea, N. mucronata, Chara virgata* and *C. braunii* (Corillion, 1953, 1957; Mouronval et al. 2015; Stewart, 2003). In more alkaline sites it can grow with *Chara aspera, C. vulgaris, Potamogeton gramineus, P. perfoliatus, Utricularia vulgaris* agg. and *Zannichellia palustris* (Corillion, 1953, 1957). In a few locations it occurs with the rare *Pilularia globulifera* and *Chara affinis* (Stewart, 2003).

It can persist in low swamps with *Eleocharis* spp. or *Agrostis* spp. providing these are not too dense or matted with decaying vegetation. However, it is not able to persist in taller swamps or under shade. At the other end of the successional scale it is unable to compete with dense growths of permanent aquatics such as *Myriophyllum alterniflorum* and *Potamogeton natans*, although it does sometimes occur mixed with these when they are less dense.

31.4 Variability

Chara fragifera shows rather little variability. Some forms have been described by Migula (1889–1897) and Braun and Nordstedt (1882), differing from the type in height, slenderness, internode length and fewer branchlet segments. However, these variations are minor and probably largely responses to ecological conditions. Groves and Bullock-Webster (1924) report finding a single branchlet whorl on a female plant bearing antheridia as well as oogonia but at different nodes. This seems to be just an unusual aberration. The size of the antheridia appears to decrease northwards, and this is discussed in more detail in the next section in relation to separation from *Chara connivens*.

31.5 Identification and Risk of Confusion

Chara fragifera belongs to a group of triplostichous species without or with minute papilliform spine-cell development which includes *Chara globularis, Chara virgata* and *Chara connivens*. It differs from all three of these (in most cases—but see below) by the production of multicellular whitish bulbils on the lower nodes and rhizoids. However, bulbils are sometimes difficult to find particularly in the period of maximum vegetative growth in the spring and early summer. At this time it is usually fertile and can be distinguished from all but *Chara connivens* by being dioecious.

The slender flexuous branchlets with numerous segments contrast with the rather stiff, straight or upwardly curved branchlets of *Chara globularis* and *Chara virgata* and the inwardly curved branchlets of male *Chara connivens*. The nodes are frequently slightly swollen, and this can become more prominent in dried specimens when the whorls can be reminiscent of the skeleton bones of a hand.

Chara fragifera resembles *Chara virgata* in size, tylacanthy and sometimes with some development of papillate spine-cells and of the upper stipulodes, in contrast to *Chara globularis* and *Chara connivens* where the spine-cells are more or less invisible and the stipulodes no more than globular. *Chara fragifera's* preference for soft water also means that there is ecological overlap much more with *Chara virgata* than with the other species. It is therefore the distinction from the latter species that is most likely to be an issue. The presence of bulbils and/or dioecious state is normally sufficient to distinguish the two but, in the absence of either feature, the semi-translucence of the plants and the slender, flexuous branchlets are indicative. A further complication is that in a few cases both *Chara virgata* and *Chara globularis* have been recorded with whitish bulbil-like growths and these tend to be formed when the plants are senescing and antheridia are difficult to find. However, in these species the growths are formed by swellings around the lower axis nodes and normally appear cartilagenous with some translucence compared to the more opaquely whitish bulbils of *C. fragifera*. Nevertheless, there are examples of plants that have been incorrectly identified as *C. fragifera* because of opaque axis-node bulbils but on further examination they have been found to be monoecious. Other such examples which are non-fertile but lack the typical semi-translucence and flexuous branchlets of *C. fragifera* need to be viewed with caution. *Chara fragifera* can produce axis-node bulbils which appear cartilagenous with some translucence (like many *Chara* species) as well as ones that are opaquely whitish, but it more usually produces bulbils on the rhizoids to which they are attached laterally (in contrast to the axis-node bulbils which are arranged around the axes).

Between the two dioecious species, *Chara fragifera* has smaller antheridia than *Chara connivens* but the sizes in both species are smaller towards the northern part of the range. Thus Commelle (1985) gives the antheridia diameter as 700–800 μm for *Chara fragifera* and 800–1000 μm for *Chara connivens* in Iberia while Groves and Bullock-Webster (1924) gives 500–575 μm for *Chara fragifera* and 600–700 μm for *Chara connivens* in Britain. Information from France (e.g., Corillion, 1957, Mouronval et al., 2015) suggests more variability but this may be because measurements from non-French sources have been incorporated. *Chara connivens* also differs from *Chara fragifera* in having more cylindrical oospores with a distinct shoulder towards the apex and the coronula is taller (>200 μm). Male *C. connivens* also have fewer branchlet segments (up to 9).

Confusion is possible with spiny species such as *Chara aspera* and *Chara galiodes* with very reduced spine-development (such as in wave-washed conditions). These can usually be distinguished by the stipulodes which are acute and both rows equally developed but also by the single-celled bulbils which are commonly present in such ecological conditions.

31.6 Distribution

See Fig. 31.4.

A primarily Atlantic-Mediterranean species, recorded from western Iberia (e.g., Cirujano et al., 2007), western France (e.g., Corillion, 1957, Mouronval et al., 2015) and reaching the northern edge of its distribution in south-western England (e.g., Bryant and Stewart, 2011). It is most frequent in France and, while elsewhere it is uncommon away from the coast, in France it extends further east to the Paris basin, some 250 km from the coast (Corillion, 1953, 1957). There are also a few records from Belgium (Compère, 1992) where it is now extinct (L. Denys pers. comm.).

Within this range it can be locally frequent in areas with suitable habitat. For example Corillion (1953, 1957) indicates over 100 localities in France, particularly in the north-west and centre, while Cirujano et al. (2007) describe it as frequent in western Iberia. In Britain its range is extremely restricted but around 60 small colonies are known within an area of around 25 square kilometres.

Its range extends very sparsely in the Mediterreanean region with a few records from southern France (Mouronval et al., 2015), Algeria and Tunisia (Feldmann, 1946), Albania (Blaženčić et al., 2006), Cyprus (E. Papastergiadou pers. comm.) and Romania (Blaženčić et al., 2006; Caraus, 2002, 2017). However, a record from the Russian enclave of Kaliningrad is an error for *Chara globularis* (Nielsen, 2004) and records from Belarus (Burdyko, 1973) and Ukraine (Borisova et al., 2016) are considered doubtful (R. Romanov pers. comm.). Elsewhere there are isolated records in

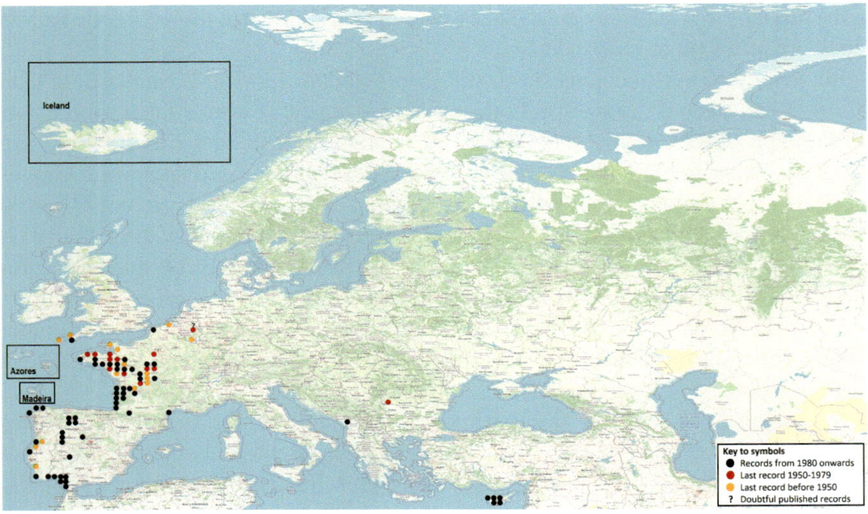

Fig. 31.4 European distribution of *Chara fragifera*. Black dots are records from 1980 onwards, red dots last recorded from 1950 to 1979 and yellow dots last recorded before 1950; unconfirmed published records (if applicable) are marked with "?" superimposed. Map prepared by Heiko Korsch

Africa from Togo (Corillion, 1953) and South Africa (Groves and Bullock-Webster, 1924).

31.7 Status and Threats

Information on recent population trends is limited in much of its range. The greatest threat is thought to be from enrichment pollution, causing increased turbidity, changes to a more organic sediment and greater competition from filamentous algae and vascular plants.

In smaller water bodies, there are greater threats from habitat loss, either through direct infilling or from neglect resulting in succession to swamp and wet woodland. A widespread problem in much of the former British range is that former heathland has been agriculturally improved to improved pasture leaving only a few unmanaged wild relics which have largely been overcome by scrub (Stewart, 2003). Nevertheless, within its remaining area in Britain, improved management has resulted in an increase in the number of colonies. This has included the creation of ponds and the re-activation of some disused trackways.

In Britain, the introduced aquatic plant *Lagarosiphon major* has colonised two ponds and outcompetes the stonewort in the deeper parts of the ponds (Stewart, 2003).

31.8 Remarks

Examples of *Chara fragifera* collected by Durieu from Étang de Lacanau were distributed by Braun et al. (1857–1878: 73a in 1867). Here, Durieu claimed to have found single-celled bulbils very rarely in two localities, occurring on decomposing material in shallow water. However, the packet containing material with single-celled bulbils (at least in BM) clearly contains at least some *Chara aspera* in a decayed state and therefore with spine-cells poorly developed. It is probable that none of the single-celled bulbils belong to *C. fragifera.*

References

Blaženčić, J., Stevanović, B., Blaženčić, Ž, & Stevanović, V. (2006). Red data list of Charophytes in the Balkans. *Biodiversity and Conservation, 15*, 3445–3457.

Borisova, O. V., Palamar-Mordvintseva, G. M., & Tsarenko, P. M., [Борисова, О. В., Паламар-Мордвинцева, Г. М. & Царенко, П. М.] (2016). *Флора водоростей України. Том 12. Харофітові водорості. Вип. 2.* [*Flora algarum Ucrainicae. Charophyta*] (Vol. 12, no. 2, 282 pp.). Kholodny Institute of Botany of the National Academy of Science of Ukraine [in Ukrainian].

Braun, A., & Nordstedt, C. F. O. (1882). Fragmente einer Monographie der Characeen. Nach den hinterlassen Manuscripten A. Braun's herausgegeben von Dr. Otto Nordstedt. *Abhandlungen der Königlichen Akademie der Wissenschaften zu Berlin für das Jahr, 1882*, 1–211. [in German].
Braun, A., Rabenhorst, L., & Stizenberger, E. (eds) (1857–1878) *Die Characeen Europa's in getrockneten Exemplaren. i-iv*. Nr. 1-121. Dresden [in French, German & Latin].
Bryant, J. A., & Stewart, N. F. (2011). Order Charales. In D. M. John, B. A. Whitton, & A. J. Brook (Eds.), *The freshwater algal flora of the British Isles. An identification guide to freshwater and terrestrial algae* (2nd ed, pp. 742–765). Cambridge University, Cambridge.
Burdyko, P. I., [Бурдыко И. И.]. (1973). Харовые водоросли и некоторые особенности их экологии в водоемах Белоруссии. [Charophytes and some peculiarities of their ecology in waterbodies of Belarus]. In *Материалы к Всесоюзному симпозиуму по изучению харовых водорослей*. [*The material for All-Union symposium of the investigation of Charophyta algae*] (pp. 60–66). Vilnius [in Russian].
Caraus, I. (2002). The algae of Romania. *Studii si Cercetari, Universitatea Bacau, Biologie, 7*, 1–694. [in Romanian].
Caraus, I. (2017). Algae of Romania. A distributional checklist of actual algae. Version 2.4. *Studii si Cercetari Biologie, 7*, 1–1002, 1 map.
Cirujano, S., Cambra, J., Sánchez Castillo, P. M., Meco, A., & Flor Arnau, N. (2008). *Flora ibérica. Algas continentales. Carófitos (Characeae)* (132 pp.). Real Jardín Botánico, Madrid [in Spanish].
Cirujano, S., García Murillo, P., Meco, A., & Fernández Zamudio, R. (2007). Los carófitos ibéricos. *Anales del Jardín Botánico de Madrid, 64*, 87–102. [in Spanish].
Comelles, M. (1985). *Clave de identificación de las especies de carófitos de la Península Ibérica*. Asociación Española de Limnologia, Madrid. [in Spanish].
Compère, P. (1992). *Charophytes. Flore pratique des algues d'eau douce de Belgique*. Jardin botanique national de Belgique.
Corillion, R. (1953). *Chara fragifera* Durieu. Notes phytogéographiques, écologiques, phytosociologiques. *Bulletin de la Société Scientifique de Bretagne, 28*, 45–54. [in French].
Corillion, R. (1957). Les Charophycées de France et d'Europe Occidentale. *Bulletin de la Société Scientifique de Bretagne, 32* (Hors série I & II), 1–499 [in French].
Feldmann, G. (1946). Les Charophycées d'Afrique du Nord. *Bulletin de la Société d'Histoire Naturelle de l'Afrique du Nord, 37*, 64–118. [in French].
Groves, J., & Bullock-Webster, G. R. (1924). *The British Charophyta. II Chareæ*. Ray Society, London.
Krause, W. (1997). Charales (Charophyceae) Süßwasserflora von Mitteleuropa. In H. Ettl, G. Gärtner, H. Heynig, & D. Mollenhauer (Eds.), *Süßwasserflora von Mitteleuropa, 18*. Fischer, Jena [in German].
Migula, W. (1889–1897). Fünfter Band: Die Characeen. In A. Fischer, E. Fischer, F. Hauck, G. Limpricht, C. Luerssen, W. Migula, H. Rehm, P. Richter, & G. Winter (Eds.), *Dr. L. Rabenhorst's Kryptogamenflora von Deutschland, Oesterreich und der Schweiz* (2nd ed). Eduard Kummer, Leipzig [in German].
Moore, J. A. (1986). *Charophytes of Great Britain and Ireland*. BSBI Handbook 5, Botanical Society of the British Isles, London.
Mouronval, J. B., Baudouin, S., Borel, N., Soulié-Märsche, I., Klesczewski, M., & Grillas, P. (2015). *Guide des Characées de France méditerranéenne* (211 pp.). Office national de la Chasse et de la Faune Sauvage [in French].
Nielsen, R. (2004). *Chara fragifera*. In H. Schubert, & I. Blindow (Eds.), *Charophytes of the Baltic Sea* (pp. 95–98). A. R. G. Gantner, Ruggell.
Stewart, N. F. (2003). Review of the records for *Chara fragifera* in Britain. *Botanical Cornwall, 12*, 1–8.
Wood, R. D. (1965). *Monograph of the Characeae*. J. Cramer, Weinheim.

Chapter 32
Chara galioides

Elisabeth Lambert and Hendrik Schubert

Scientific name: ***Chara galioides*** **DC., Cat. Pl. Horti Monsp.: 93. 1813**

Lectotype (Wood, 1965: 207): Des mares près de la mer à Montpellier/[no date]/[De Candolle?] (G, Herb. Moricand).

= *Chara aspera* Willd. var. *macrosphaera* A. Braun, Ann. Sci. Nat., Bot., sér. 2, 1: 356. 1834.
Lectotype (designated here by E. Lambert & Hendr. Schub. [proposed by J. van Raam]): pr. Cette [Sète]/[Unio Iteneraria, 1831]/W. Schimper (L, ex herbarium Kützing, ex herbarium Weber-Van Bosse).

Local names: FRENCH: Chara faux-gaillet; GERMAN: Große Raue Armleuchteralge; RUSSIAN: Хара галиоидная; SPANISH: Ova espinosa, Asprella espinosa, Madejas del agua; UKRAINIAN: Хара галіоідна.

32.1 Morphology

See Figs. 32.1 and 32.2.

Habitus: Plants are erect, brownish to greyish green (Cirujano et al., 2008), rough (Langangen, 2004), fairly robust and slightly or not encrusted (Corillion, 1975).

E. Lambert (✉)
Faculté des Sciences, Département Biologie-Environnement, Université Catholique de l'Ouest (UCO), 3, Place André Leroy—BP 10808, 49008 Angers, Cedex 01, France
e-mail: lambert.elisa49@gmail.com

H. Schubert
Institut für Biowissenschaften, Universität Rostock, Albert-Einstein-Straße 3, 18051 Rostock, Germany
e-mail: hendrik.schubert@uni-rostock.de

H. Schubert et al. (eds.), *Charophytes of Europe*,
https://doi.org/10.1007/978-3-031-31898-6_32

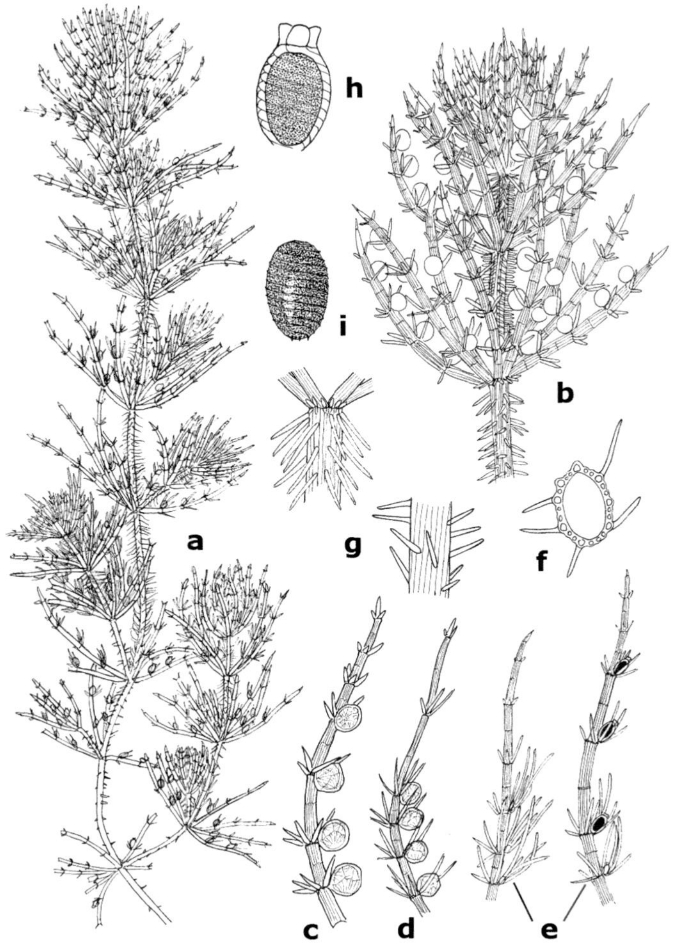

Fig. 32.1 *Chara galioides*. **a** Female plant, habitus; **b** Upper part of a male plant; **c** Male branchlet; **d** Male branchlet of *C. aspera* for comparison; **e** Female branchlets of *C. galioides*; **f** Cortex, spine-cells and stipulodes; **g** Axis cross-section; **h** Oogonium; **i** Oospore. Drawings by Werner Krause

Size: *Chara galioides* is a medium-sized plant of about 15–20 cm length but can grow up to 50 cm long (Bazzichelli and Abdelahad, 2009; Cirujano et al., 2008; Corillion, 1975; Hollerbach and Krassavina, 1983; Krause, 1997; Langangen, 2004).

Main axis: Moderately stout with diameter up to 0.8–1(1.5) mm (Bazzichelli and Abdelahad, 2009; Cirujano et al., 2008; Hollerbach and Krassavina, 1983; Krause,

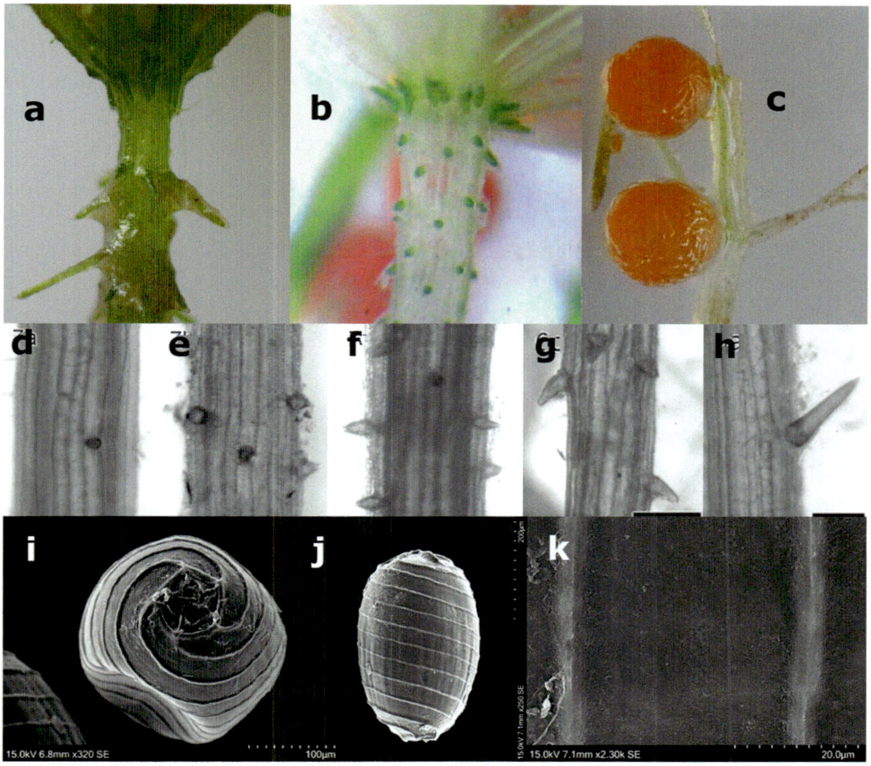

Fig. 32.2 *Chara galioides* details. **a** Specimen with well-developed spine-cells and stipulodes; **b** Regular cortex with small stipulodes and spine-cells, in both examples the typical pairwise arrangement of stipulodes can be seen; **c** Antheridia; **d–h** Gradient of spine-cell appearance from almost invisible, papilliform (**d, e**) to well-developed ones, equalling the main axis diameter (**h**); **i, j** SEM-photographs of ripe oospores, side and apex view; **k** SEM photograph of oospore membrane. Photographs: **a, c**—Klaus van de Weyer; **b**—Elisabeth Lambert; **d–h**—Nuria Flor-Arnau; **i–k**—Michelle T. Casanova

1997; Langangen, 2004; Olsen, 1944; Wood, 1965)[1]. Internodes 1–2(3) times as long as the branchlets (Hollerbach and Krassavina, 1983), up to 2–(3)4 cm long (Corillion, 1975; Wood, 1965).

Cortex: Irrespective of being triplostichous, cortication can be irregular occasionally (Krause, 1997; Wood, 1965), or slightly tylacanthous to isostichous.

Spine-cells: Spine-cells very variable in number, form and size in the same internode, from abundant, especially in the upper part, to sparse, sometimes almost triangular, more often rudimentary or globular; solitary and occasionally fasciculated in groups

[1] According to Krause (1997) values given by another source (Verdam 1938) are probably based on misidentification.

of 2 on the same plant, acute (Cirujano et al., 2008; Corillion, 1957a; Krause, 1997). Spine-cells of variable density and length, usually quite short (75–150 μ long, 50–75 μ wide), simple, swollen, conical (Corillion, 1975); as long as the main axis diameter or shorter (Bazzichelli and Abdelahad, 2009; Krause, 1997); usually solitary when short and blunt, and elongate when fasciculated (Wood, 1965).

Stipulodes: In a double whorl, 2 sets per branchlet, usually well developed (Corillion, 1957a); occasionally rudimentary (Cirujano et al., 2008). Typically in separated pairs (Fig. 32.2). With respect to relative measurments, authors have different observations. Upper rows are described being longer than the lower (Wood, 1965); or shorter than the lower (Hollerbach and Krassavina, 1983). The same applies for relative lengths compared to the spine-cells, where they are described as being similar to spine-cells (Bazzichelli and Abdelahad, 2009), shorter (Krause, 1997) as well as longer than spine-cells (Hollerbach and Krassavina, 1983). Lower row of stipulodes ½–1 times as long as the axis diameter (Hollerbach and Krassavina, 1983; Wood, 1965); up to 275 μm (Corillion, 1975).

Branchlets: Whorls consist of 6–8(10) branchlets often curved or slightly curved towards the axis or straight and extended; 1–2 cm long; with 4–7 corticated segments and an ecorticate end segment of 1–2 reduced cells (Bazzichelli and Abdelahad, 2009; Cirujano et al., 2008; Corillion, 1957a, 1975; Wood, 1965).

Bract-cells: 4–5 unilateral (rarely whorled). Anterior (adaxial) ½–1 times as long as the bracteoles, 250–900 μm long (Corillion, 1975), 2–5 times longer than the oogonium (Bazzichelli and Abdelahad, 2009), shorter or slightly exceeding the antheridia (Cirujano et al., 2008); posterior (abaxial) very short, conical, tiny to as long as the branchlet diameter (Corillion, 1975; Wood, 1965).

Bracteoles: There is a large variability with respect to length of the two bracteoles of *C. galioides*. Wood (1965), Cirujano et al. (2008) and Bazzichelli and Abdelahad (2009) describe them to be much shorter to (slightly) longer than mature oogonia, whereas Corillion (1975) found them to be as long or longer (750–900 μm) than the mature oogonia. Migula (1889–1897) takes the relative length of bracteoles to bract-cells as a distinguishing character for *C. aspera* (having equal length for bract-cells and bracteoles) and *C. galioides* (bracteoles half the length of the bract-cells).

Bractlets: Bractlets are described as being similar to the bracteoles (Corillion, 1975) but Wood (1965) reported that they are often longer than the bracteoles.

Sexuality: Dioecious, very fertile (Langangen, 2004). Gametangia at the 1–4 lowest nodes of the branchlets (Corillion, 1975; Migula, 1889–1897, Wood, 1965). Chromosome number: n = 12 (Ernst 1918 in Corillion, 1975); n = 14 (Guerlesquin, 1967; Krause, 1997).

Oogonia: Oogonia are described to be large (Bazzichelli and Abdelahad, 2009) and solitary (Cirujano et al., 2008). With respect to the appearance of the coronula, authors describe them differently as spreading or fairly straight (Corillion, 1957a, 1975; Hollerbach and Krassavina, 1983; Migula, 1889–1897). Data about the quantitative morphometry also differ considerably between authors (Table 32.1), which can be at least in part explained by habitat (salinity) dependent development of the oogonia.

Table 32.1 Oogonia morphometry

Author	Length	Width	Coronula length	Coronula width	Number of helical stripes
Migula (1889–1897)	*850–1000*	400–550	–	–	–
Reyes Prósper (1910)	*850–1100*	400–550	–	–	–
Wood (1965)	*(670)850–1000*	400–550	c. 120	c. 160	c. 10
Hollerbach and Krassavina (1983)	(670)850–1000	400–550	120–165	160–250	13–15
Corillion (1957a)	850–1000	400–550	90–150	~ 200	12–15
Corillion (1975)	700–850	400–550	120–160	250–270	12–15
Krause (1997)	700–800	550–600	90–150	150–200	12–15
Cirujano et al. (2008)	650–1000	400–550	80–120	–	–
Bazzichelli and Abdelahad (2009)	*800–1100*	–	–	–	–
Total range	**650–1000**	**400–600**	**80–165**	**150–270**	**10–15**

Length in the first column is excluding the coronula, except for values in italics (also not taken into account for total range); length and width of the coronula are given separately, all values for length and width in μm

Oospores: Oospores ellipsoidal or slightly ovoid (Bazzichelli and Abdelahad, 2009; Hollerbach and Krassavina, 1983), with a smooth oospore wall, sometimes slightly nodulate or pitted; striae are slightly protuding (Corillion, 1975). In some specimens there is a claw and a cage around the pentagonal basal cell but there is no evidence of a ribbon on the striae (M. T. Casanova pers. comm.). Table 32.2 gives an overview of the available morphometric data.

Table 32.2 Oospore morphometry

Author	Length	Width	Number of striae	Colour
Migula (1889–1897)	500–600	350–450	11–13	Black opaque
Reyes Prósper (1910)	500–600	350–450	11–13 blunt ridges	Black
Wood (1965)	500–600	350–450	11–13 blunt ridges	Black
Hollerbach and Krassavina (1983)	(460)500–600	(300)350–450	11–13	Black
Krause (1997)	500–600	350–450	11–13	Black
Corillion (1957a, 1975)	500–600	350–450	11–13	Dark brown to black
Cirujano et al. (2008)	500–600	350–450	–	Black
Bazzichelli and Abdelahad (2009)	500–600	350–450	11–13	–
Total range	**(460)500–600**	**(300)350–450**	**11–13**	

All values for length and width in μm

Antheridia: *Chara galioides* is remarkable for its large antheridia. However, this character varies with age; the first antheridia are always larger than those at the ends of the branchlets and in the upper whorls (Hy, 1913). Corillion (1952) mentions the smaller diameter of antheridia (775–850 μm) for *C. galioides* f. *condensata.*

Because antheridia size is often cited as the main criterion for distinction from *C. aspera*, Table 32.3 contains data about both species, if available from the same author, for comparison.

Bulbils: Spherical bulbils are absent according to Corillion (1975). Migula (1889–1897) observed rarely transparent bulbils; this observation was also confirmed by Küster and Nielsen (2003). Migula's text might be the reason for Krause's (1997) statement that bulbils may rarely occur. Wood (1965; 206), referring to Migula (1889–1897: 685 et seq.) mentioned the occurrence of *C. aspera*-like white bulbils ("Bulbils on rhizoids, simple, resembling those of f. *aspera*") but, as a footnote on the same page, refers to Migula (1889–1897): "Migula (1889–1897: 320) considered that *C. galioides* lacks bulbils" (referring to Migula's key, where this character is taken for distinction between *C. aspera* and *C. galioides*), which is confusing to the reader. Indeed, Migula's text is confusing with regard to its comparison with the multicellular bulbils of *C. fragifera* (Migula, 1889–1897: 687), but pretty clear with regard to single-celled bulbils, where he explicitly states that sometimes round, single-celled, transparent bulbils can be found. On p. 690 he refers to Sonder (1890), who made similar observations, and argued that maybe these transparent bulbils later turn white—but at no place did he observe such white bulbils himself. The original

Table 32.3 Antheridia morphometry

Author	*Chara galioides* diameter (μm)	*Chara aspera* diameter (μm)
Migula (1889–1897)	800–1100	550–750
Reyes Prósper (1910)	800–1100 vermilion red	550–750 red 'saturn'!
Hy (1913)	> 700	< 700
Groves and Bullock-Webster (1920)	> 1100	500–600
Feldmann (1946)	800–1100	500–750
Corillion (1957a)	800–1100	400–600
Guerlesquin (1958)	800–1100	400–600
Wood (1965)	800–1100	390–600
Corillion (1975)	800–1100	400–600
Hollerbach and Krassavina (1983)	(650)800–1100	400–600(750)
Comelles (1985)	800–1000	300–750
Comelles (1986)	700–1100	400–700
Bonis et al. (1993)	(676)685–976	475–726(759)
Krause (1997)	800–1100	400–600
Langangen (2004)	750–1000	–
Cirujano et al. (2008)	800–1000 solitary	–
Zeneli and Kashta (2016)	> 800	–
Blindow pers. comm.	706–941	–
Total range	**(650)800–1100**	**(390)400–600(759)**

material of Sonder (1890)[2] were not found at Kiel herbarium (KI) by Küster and Nielsen (2003) but material from the same site sampled 10 years later were regarded as misidentifications (Küster and Nielsen, 2003). Having not seen white bulbils on any fresh or herbarium material of *C. galioides* the authors consider that the lack of white bulbils is a good criterion for distinction between *C. aspera* and *C. galioides*. Bulbil-like transparent structures on the rhizoids have been observed at least on specimens from Spain (Marais de Donana) and France (Camargue).

[2] The herbarium of Sonder was purchased and shipped to Australia in 1880 and is lodged in MEL (pers. comm. M. T. Casanova).

32.2 Phenology and Reproduction

Chara galioides occurs very early in the season, especially on the Mediterranean coast, where plants can be found in December (southern Coast). On the northern Mediterranean coast (France) the species usually starts growing in January, fructifying from April to August (Corillion, 1975). More northerly populations, on the Atlantic Coast of France, appear to grow later (April–May). Development may cease prematurely (June) by the drying of shallow biotopes or continue into July–August at greater depths (2 m), (Corillion, 1957a).

32.3 Habitat Conditions and Ecology

See Fig. 32.3.

Fig. 32.3 *Chara galioides,* habitat and habitus. **a** Fully developed male plant; **b** Early stage with ripe males and inconspicuous unripe female plants; **c** Typical stands with scattered *C. galiodes* in associated vegetation; **d** Almost discoloured ripe females with black oospores, males are already decayed; **e, f** Dense, almost monospecific stands in a shallow brackish lagoon at the Golfo de Cadiz (Spain). Photographs: **a, b, d–f**—Hendrik Schubert; **c**—Elisabeth Lambert

Habitat type: *Chara galioides* is restricted to saline habitats ranging from slightly brackish to hypersaline conditions. As it is rarely found in regions with a continental climate (Cirujano et al., 2008; Guerlesquin, 1974), the species prefers shallow marine and brackish water bodies of all kinds: lagoons, ditches, channels, beach lakes, saltmarshes, etc. irrespective of whether they are permanent or temporary as long as they provide water for a period sufficient for fructification. Salinity often varies during development as is typical for lagoonal sites. *C. galioides* is regarded as a thermophilous species (Flor-Arnau et al., 2006).

Sediment: *Chara galioides* has been observed on a large variety of substrates, ranging from muddy to sandy bottoms (Corillion, 1957a), but there are records from silty substrate (Langangen, 2004; Zeneli and Kashta, 2016) and also gravel (Verhoeven and Van Vierssen, 1978), indicating the species' insensitivity to sediment conditions.

Nutrient conditions: *Chara galioides,* as with most of the brackish and marine species, is best characterized as a heliophilous species without clear preferences for certain nutrient conditions. It covers the full range from eutrophic conditions, as reported by Langangen (2004), to oligotrophic ones as reported by Mouronval et al. (2015). A report containing distinct nutrient concentrations for a *C. galioides* site is provided by Zouaïdia et al. (2015). Based on Becker (2019) from 12 Sardinian waterbodies the following values are known (single measurements): TP 0.024–0.163 mg L^{-1} (mean value 0.065 mg L^{-1}), $PO_4{}^{3}$-P 0.010–0.063 mg L^{-1} (mean value 0.026 mg L^{-1}), TN 0.75–3.97 mg L^{-1} (mean value 1.95 mg L^{-1}), NO_3-N 0.39–1.59 mg L^{-1} (mean value 0.9 mg L^{-1}), and NH_4-N <0.03–1.51 mg L^{-1} (mean value 0.35 mg L^{-1}).

Water chemistry: *Chara galioides* is found from brackish to hypersaline conditions, Comín and Alonso (1988) report a maximum salt content of 300 g L^{-1}! With respect to pH, sources give a range from 6.8 to 9.4 (Zeneli and Kashta, 2016, Becker unpubl. data), the latter demonstrating its ability to withstand strongly alkaline conditions. This is supported by the data of Carretero (1993), reporting an alkalinity of 2.15–5.10 meq L^{-1} and $Ca^{2+} + Mg^{2+}$ concentrations of 32–65 meq L^{-1}. In Sardinia salinity in *C. galioides* habitats ranged mostly from 3.0 to 22.6 psu (with a minimum of 1.2 and a mean value of 9.99 psu). Conductivity varied between 2.2 and 46.2 mS cm^{-1} (mean value 15.7 mS cm^{-1}) and hardness between 17 to 226 °dH with a mean value of 110°dH (R. Becker own data).

Depth range: On the Atlantic coast, *Chara galioides* grows exclusively in shallow waters and is found between 5–20 cm (Beaubert et al., 2015; Corillion, 1957a). Along the Western Mediterranean coast, the depth range can reach 2 m (France: Corsica: Verhoeven and Van Vierssen, 1978, Camargue: Charpentier et al., 2005, Pyrénées-Orientales, Corillion, 1957a; Spain: Carretero, 1993). On the Adriatic coast, *Chara galioides* has been found from 10–50 cm depth (Italy: Corillion and Guerlesquin, 1963; Albania: Zeneli and Kashta, 2016).

Associated species: *Chara galioides* forms dense submerged meadows, predominantly in monospecific populations. However, it may also be associated with other charophytes (*Chara aspera, C. baltica, C. canescens, C. connivens, C. major, C. aculeolata, C. vulgaris, Lamprothamnium papulosum, Tolypella glomerata, T. hispanica, T. nidifica, T. salina*), as well as vascular plants (*Althenia filiformis, Najas marina, Ruppia cirrhosa, R. drepanensis, R. maritima, Stuckenia pectinata, Zannichellia obtusifolia, Z. pedunculata, Z. pedicellata, Zostera noltei*) and the very rare liverwort *Riella helicophylla* (Amat, 2000; Beaubert et al., 2015; Becker, 2019; Becker and Homm, 2018; Bergmeier and Abrahamczyk, 2008; Carretero, 1993; Comín and Alonso, 1988; Corillion, 1975; Corillion and Guerlesquin, 1959; Espel et al., 2019; Margalef Mir, 1981; Martinez-Taberner and Pericàs, 1988; Martínez-Taberner and Moyà, 1993). *C. galioides* is diagnostic for the *Charetalia canescentis* F. Fukarek ex W. Krause 1997 and the *Charion canescentis* F. Fukarek 1961 vegetation types. In these orders and alliances, *C. galioides* is the specific species of the *Charetum galioidis* Corill. 1957. It can also participate in the *Lamprothamnietum papulosi* Corill. 1953 or the *Charetum conniventis* Ionescu-Teculescu 1972 (Felzines, 2012, 2016) vegetation types. *C. galioides* can also grow in a *Charo-Tolypelletum* or be in contact with other phytosociological groups such as the *Chaetomorpho-Ruppietum* Braun-Blanquet 1952 and the *Ranunculetum baudotii* Braun-Blanquet 1952. It can also be present in the understorey of the *Scirpetum maritimi*, at the edge of this (Krause, 1997) or sometimes coexist with the *Arthrocnemo-Juncetum subulati* Brullo and Furnari 1976 (Espinar et al., 2002).

32.4 Variability

See Fig. 32.2.

For *Chara galioides* a number of varieties and formae have been described (e.g. Corillion, 1952; Hy, 1913; Migula, 1889–1897; Reyes Prósper, 1910). However, given the variability with respect to length, appearance of bract-cells, bracteoles, spine-cells and stipulodes demonstrated above (which can be observed sometimes even within a single plant) distinction between defined morphotypes seems to be unwarranted.

32.5 Identification and Risk of Confusion

See Fig. 32.4.

Although stronger, rougher and more rigid, *C. galioides* is, by its morphology, closely related to *C. aspera* and the validity of the taxon has been questioned by many authors. In some stations with fluctuating environments these two taxa appear with intermediate forms (of spine-cells, bract-cell, main axis diameter) next to each other (Bazzichelli and Abdelahad, 2009; Bonis et al., 1993; Cirujano et al., 2008; Corillion, 1952, Flor-Arnau, 2014; Flor-Arnau et al., 2006; Krause, 1997; Zeneli and Kashta, 2016). However, several studies have provided evidence that the two taxa are

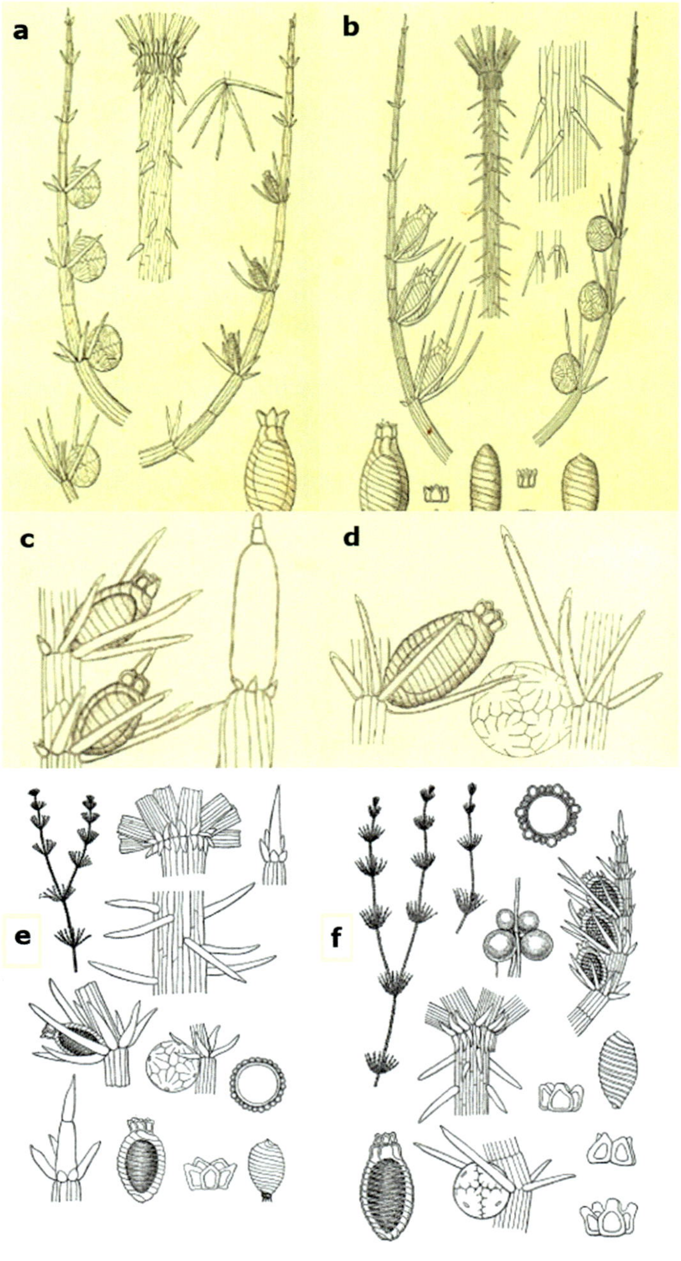

Fig. 32.4 *Chara galioides* vs. *Chara aspera*. Comparison of drawings of *C. galioides* (left column: **a, c, e**) and *C. aspera* (right column, **b, d, f**) by Migula (1889–1897; **a, b**); Feldmann (1946; **c, d**) and Hollerbach and Krassavina (1983; **e, f**)

separate entities by means of morphological, enzymatic as well as genetic characters (Bonis et al., 1993; Flor-Arnau, 2014; Schneider et al., 2011, 2016).

The most frequently used criteria for distinguishing between *C. galioides* and *C. aspera* are the absence of white bulbils and the larger size of the antheridia for *C. galioides*; main axis diameter, which is larger for *C. galioides*, is also often mentioned as a useful criterion. Another character, first introduced by Corillion (1975) are stipulodes. They are always well developed (both rows) in *C. aspera* (relatively small if nearby spine-cells are small, or long if nearby spine-cells are long), whereas in *C. galioides* the upper row is usually longer than the lower one and, in any case, shorter than the spine cells of the adjacent internode. The last criterion is the fertility: generally, the number of gametangia per branchlet of *C. galioides* is greater than for *C. aspera.*

The variable salinity, light and their interaction are the main factors that determine the growth rate of the two taxa (Flor-Arnau et al., 2006). Whereas increasing temperature (15–25 °C) increases growth of *C. galioides*, it hinders the growth of *C. aspera* (Flor-Arnau et al., 2006). In contrast to *C. aspera, C. galioides* has not yet been recorded from freshwater.

Küster and Nielsen (2003) reported that all herbarium material available for records from the Baltic Sea region, turned out to be misidentifications of either *C. aspera* or *C. virgata*, the latter being a monoecious species.

32.6 Distribution

See Fig. 32.5.

Chara galioides is known from many places in Europe and North Africa. Feldmann (1946) as well as Guerlesquin (1974) and Muller et al. (2017) indicate a distribution in Northern Africa between Egypt and Morocco. In Europe this species is mainly present in brackish wetlands along the occidental and oriental Mediterranean mainland coasts (France: Mouronval et al., 2015—Italy: Corillion and Guerlesquin, 1963—Spain: Comelles, 1984, Ortega et al., 2004, Cirujano et al., 2008—Greece: Raabe and Koumpli-Sovantzi, 2000, 2002—Albania: Zeneli and Kashta, 2016) and islands (Becker, 2019; Bergmeier and Abrahamczyk, 2008; Corillion, 1952; Langangen, 2004; Martínez-Taberner and Moyà, 1993, Romanov et al., 2019). It is also identified on the Atlantic coast (Beaubert et al., 2015; Corillion, 1957b, 1975, Lafon and Lambert, 2020) and found along the Sea of Azov (Black Sea) (Borisova, 2016; Palamar-Mordvintseva, 1998). Former reports of occurrences in the Baltic Sea region turned out to be misidentifications (see Blümel and Raabe, 2004; Küster and Nielsen, 2003, Sinkevičienė, 2004).

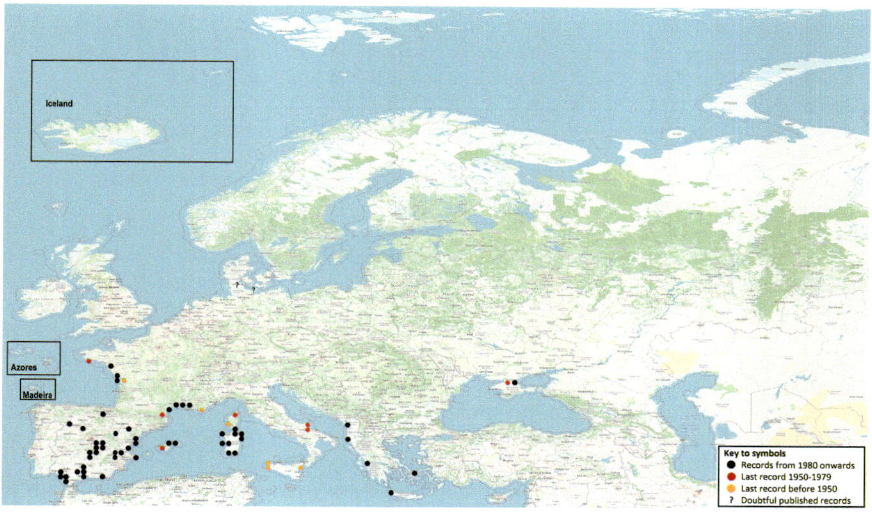

Fig. 32.5 European distribution of *Chara galioides*. Black dots are records from 1980 on, red dots from 1950 to 1979 and yellow dots before 1950; unconfirmed/doubtful published records (if applicable) are marked with "?" on top. Map prepared by Heiko Korsch

32.7 Status and Threats

In Sardinia, protection of *Chara galioides* is considered a highest priority (Becker, 2019), while in the Balkans the species is listed as Critically Endangered (CR), criterion B1 (Blaženčić et al., 2006). Several suitable sites are endangered directly by human activities; from the Cyclades, Langangen (2004) reported about a brackish water locality of unique biological value where dumping of sewage must be stopped; Corillion (1952) labeled *C. galioides* as one of the "rarest species of Charophyceae observable on the whole of the French territory". Rodrigo and Alonso-Guillen (2016) highlighted that *C. galioides* is in low abundance and discontinuous in time, which also implies the need for protection measures.

Acknowledgements We thank Nuria Flor-Arnau for sharing the results of her work, Irmgard Blindow for providing us with data from different herbaria and Michelle T. Casanova for her advice during the proofreading.

References

Amat, J. A. (2000). Courtship feeding, food sharing, or tolerated food theft among paired Red-crested Pochards (*Netta rufina*)? *Journal für Ornithologie, 141*, 327–334.

Bazzichelli, G., & Abdelahad, N. (2009). *Alghe d'acqua dolce d'Italia. Flora analitica delle Caroficee.* (73 pp.). Sapienza Università di Roma. Ministero dell' Ambiente e della Tutela del territorio e del mare. Centro Stampa Università, Roma [in Italian].

Beaubert, R., Desmots, D., Gernigon, J., Lambert, E., Lemesle, J.-C., & Micheau, A. (2015). *Macrophytes patrimoniales des lagunes de l'île de Ré* (147 pp.). Réserve Naturelle « Lilleau des Niges », LPO [in French].

Becker, R. (2019). The Characeae (Charales, Charophyceae) of Sardinia (Italy): Habitats, distribution and conservation. *Webbia, 74*, 83–101.

Becker, R., & Homm, T. (2018). *Riella helicophylla* (Bory & Mont.) Mont. In L. T. Ellis, M. Aleffi, G. Asthana, C. Bhagat, V. A. Bakalin, K. Baráth, R. Becker, H. Bednarek-Ochyra, M. Boiko, M. R. Brito, C. Pimentel, M. Brugués, L. Sáez, P. Dřevojan, J. Enroth, P. Erzberger, V. E. Fedosov, S. Fontinha, E. Fuertes Lasala, R. Gabriel, M. Gallego, S. R. Gradstein, Th. Homm, V. Hugonnot, T. G. Ivchenko, K. G. Klimova, J. Kučera, P. Lamkowski, E. D. Lapshina, M. Lebouvier, A. L. López González, W.-Z. Ma, P. Manolaki, J. Monteiro, C. Vieira, A. P. Portela, M. Sim-Sim, A. I. Maksimov, N. Norhazrina, N. Syazwana, S. Asyifaa, S. Poponessi, R. Venanzoni, D. Gigante, F. Prosser, A. D. Potemkin, V. M. Kotkova, M. S. Sabovljević, A. D. Sabovljević, A. Schäfer-Verwimp, C. Sérgio, C. Garcia, J. R. Shevock, A. Stebel, J. Drobnik, G. Vončina, & Y.-M. Wei, New national and regional bryophyte records. *Journal of Bryology, 57*, 411.

Bergmeier, E., & Abrahamczyk, S. (2008). Current and historical diversity and new records of wetland plants in Crete, Greece. *Willdenowia, 38*, 433–453.

Blaženčić, J., Stevanović, B., Blaženčić, Ž., & Stevanović, V. (2006). Red data list of Charophytes in the Balkans. *Biodiversity and Conservation, 15*, 3445–3457.

Blümel, C., & Raabe, U. (2004). Vorläufige Checkliste des Characeen Deutschlands. *Rostocker Meeresbiologische Beiträge, 13*, 9–26. [in German].

Bonis, A., Gemayel, P., & Grillas, P. (1993). Polymorphisme enzymatique et morphologique comparé chez *Chara aspera* et *C. galioides* (Charophyceae) en Camargue. *Cryptogamie, Algologie, 14*, 115–131. [in French].

Borisova, E. V. (2016). Analyse de la Flore de Charales (Charophyta) d'Ukraine. *Algologia, 26*, 403–417. [in French].

Carretero, J.-L. (1993). Aportaciones a la distribución y ecología de la Carofíceas de la Provincia de Valencia. *Acta Botanica Malacitana, 18*, 31–37. [in Spanish].

Charpentier, A., Grillas, P., Lescuyer, F., Coulet, E., & Auby, I. (2005). Spatio-temporal dynamics of a *Zostera noltii* dominated community over a period of fluctuating salinity in a shallow lagoon, Southern France Estuarine. *Coastal and Shelf Science, 64*, 307–315.

Cirujano, S., Cambra, J., Sánchez Castillo, P. M., Meco, A., & Flor Arnau, N. (2008). *Flora ibérica. Algas continentales. Carófitos (Characeae)* (132 pp.). Real Jardín Botánico, Madrid [in Spanish].

Comelles, M. (1984). Noves Citacions de Caròfits a Espanya. *Butlletí de la Institució Catalana d'Història Natural Secció de Botánica, 51*, 35–39. [in Spanish].

Comelles, M. (1985). *Clave de identificación de las especies de carófitos de la Península Ibérica.* Asociación Española de Limonolgía, Madrid. [in Spanish].

Comelles, M. (1986). *Tolypella salina* Corillion, carofícea nueva para España. *Anales del Jardín Botánico de Madrid, 42*, 293–298. [in Spanish].

Comín, F. A., & Alonso, M. (1988). Spanish salt lakes: Their chemistry and biota. *Hydrobiologia, 158*, 237–245.

Corillion, R. (1952). *Chara galioides* DC. Ses localités françaises et leur répartition géographique. *Bulletin de la Société Botanique de France, 99*, 318–320. [in French].

Corillion, R. (1957a). Les Charophycées de France et d'Europe Occidentale. *Bulletin de la Société Scientifique de Bretagne, 32* (Hors série I & II), 1–499 [in French].

Corillion, R. (1957b). Sur la répartition géographique des Charophycées vivantes in Atlas de biogéographie. *Compte rendu des séances de la Société de Biogéographie, 300*, 122–156. [in French].

Corillion, R. (1975). *Flore et végétation du Massif Armoricain. T. IV. Flore des Charophytes (Characées) du Massif Armoricain et des contrées voisines d'Europe occidentale*. Jouve, Paris [in French].

Corillion, R., & Guerlesquin, M. (1959). Observations charologiques (Ouest, centre et Sud-Est de la France). *Bulletin de la Société Scientifique de Bretagne, 34*, 209–215 [in French].

Corillion, R., & Guerlesquin, M. (1963). Contribution à l'étude des végétations de Charophycées d'Italie péninsulaire (étage méditerranéen). *Bulletin de la Société Scientifique de Bretagne, 37*, 193–211 [in French].

Espel, D., Diepens, N. J., Boutron, O., Buffan-Dubau, E., Chérain, Y., Coulet, E., Grillas, P., Probst, A., Silvestre, J., & Elger, A. (2019). Dynamics of the seagrass *Zostera noltei* in a shallow Mediterranean lagoon exposed to chemical contamination and other stressors. *Estuarine, Coastal and Shelf Science, 222*, 1–12.

Espinar, J. L., Garcia, L. V., Garcia Murillo, P., & Toja, J. (2002). Submerged macrophyte zonation in a Mediterranean salt marsh: A facilitation effect from established helophytes? *Journal of Vegetation Science, 13*, 831–840.

Feldmann, G. (1946). Les Charophycées d'Afrique du Nord. *Bulletin de la Société d'Histoire de l'Afrique du Nord, 37*, 64–118 [in French].

Felzines, J. -C., & Lambert, E. (2012). Contribution au prodrome des végétations de France: Les Charetea fragilis F. Fukarek 1961. *Le Journal de Botanique de la Société Botanique de France, 59*, 133–188 [in French].

Felzines, J. -C., & Lambert, E. (2016). Contribution au prodrome des végétations de France: Modification de la structure syntaxonomique des Charetea et compléments. *Le Journal de Botanique de la Société Botanique de France, 74*, 41–55 [in French].

Flor-Arnau, N. (2014). *Diversitat, ecologia i usos en bioindicació d'algues caràcies i macròfits a la península Ibèrica*. Tesi doctoral (412 pp.). Barcelona [in Catalan].

Flor-Arnau, N., Reverter, F., Soulié-Märsche, I., & Cambra-Sánchez, J. (2006). Morphological differentiation of *Chara aspera* Detharding ex Willdenow and *Chara galioides* De Candolle under different environmental variables. *Cryptogamie, Algologie, 27*, 435–449.

Groves, J., & Bullock-Webster, G. R. (1920). *The British Charophyta, I Nitelleae* (405 pp.). Ray Society, London.

Guerlesquin, M. (1958). *Recherches sur l'anthéridie chez les Charophycées. Etude morphologique, biologique et systématique* (107 pp.). Diplôme d'Etudes Supérieures, Faculté des Sciences de Toulouse [in French].

Guerlesquin, M. (1967). *Recherches caryotypiques et cytotaxinomiques chez les Charophycées d'Europe occidentale et d'Afrique du Nord*. Thèse Université de Toulouse, Ed. Jouve, UCO, Angers (265 pp.) [in French].

Guerlesquin, M. (1974). Nouvelle contribution à l'étude des Charophycées du Maroc Occidental. In CNRS, *Étude de certains milieux du Maroc et de leur évolution récente. Travaux RCP249, Fasc., 2*, 187–207 [in French].

Hollerbach, M. M., & Krassavina, L. K., [Голлербах, М. М. & Красавина Л. К.]. (1983). *Определитель пресноводных водорослей СССР. 14. Charophyta [Determination key of freshwater algae of USSR, 14, Charophyta]* (189 pp.). Nauka, Leningrad [in Russian].

Hy, F. (1913). Les Characées de France. *Bulletin de la Société Botanique de France, 60* (sup 1), 1–47 [in French].

Krause, W. (1997). Charales (Charophyceae) Süßwasserflora von Mitteleuropa. In H. Ettl, G. Gärtner, H. Heynig, & D. Mollenhauer (Eds.), *Süßwasserflora von Mitteleuropa* (Vol. 18). Fischer [in German].

Küster, A., & Nielsen, R. (2003). *Chara galioides*. In H. Schubert, & I. Blindow (Eds.), *Charophytes of the Baltic Sea* (pp. 97–98). A. R. G. Gantner, Ruggell.

Lafon, P., & Lambert, E. (2020). État des connaissances sur les Charophytes d'Aquitaine et de Poitou-Charentes, pour un inventaire actualisé. *Bulletin de la Société Botanique du Centre-Ouest, 51*, 15–44 [in French].

Langangen, A. (2004). Charophytes from four Cyclade Islands (Mykonos, Naxos, Paros and Antiparos) in Greece. *Journal of Biological Research, 1*, 31–38.

Margalef Mir, R. (1981). *Distribución de los macrófitos de las lagunas dulces y salobres del Este y Noreste de España y dependencia de la composición química del medio* (63 pp.). Departamento de Botanica de la Facultad de Biologia de la Universidad de Barcelona e Instituto Botánico, Madrid [in Spanish].

Martínez-Taberner, A., & Moyà, G. (1993). Submerged vascular plants and water chemistry in the coastal marsh Albufera de Mallorca (Balearic Islands). *Hydrobiologia, 271*, 129–139.

Martinez-Taberner, A., & Pericàs, J. (1988). Notes Floristiques: Les Characeae de l'Albufera de Mallorca. *Bolletí de la Societat d'Historia Natural de les Balears, 32*, 145–150 [in Spanish].

Migula, W. (1889–1897). Fünfter Band: Die Characeen. In: A. Fischer, E. Fischer, F. Hauck, G. Limpricht, C. Luerssen, W. Migula, H. Rehm, P. Richter, & G. Winter (Eds.), *Dr. L. Rabenhorst's Kryptogamenflora von Deutschland, Oesterreich und der Schweiz* (2nd ed). Eduard Kummer, Leipzig [in German].

Mouronval, J. B., Baudouin, S., Borel, N., Soulié-Märsche, I., Klesczewski, M., & Grillas, P. (2015). *Guide des Characées de France méditerranéenne* (211 pp.). Office National de la Chasse et de la Faune Sauvage, Paris [in French].

Muller, S. D., Rhazi, L., & Soulié-Märsche, I. (2017). Diversity and distribution of Characeae in the Maghreb (Algeria, Morocco, Tunisia). *Cryptogamie, Algologie, 38*, 201–251.

Olsen, S. (1944). Danish Charophyta. *Det Kongelige Danske Videnskabernes Selskab, Biologiske Skrifter, 3*(1), 1–240, pl. 1–2 [in Danish].

Ortega, F., Paracuellos, M., & Guerrero, F. (2004). Corología de macrófitos acuáticos en Andalucía oriental. *Lazaroa, 25*, 179–185. [in Spanish].

Palamar-Mordvintseva, G. M. [Паламарь-Мордвинцева, Г. М.]. (1998). CHAROPHYTA КРЫМСКОГО ПОЛУОСТРОВА (УКРАИНА). [Charophyta of the Crimean Peninsula (Ukraine)]. *Algologia, 8*, 14–22 [in Russian].

Raabe, U., & Koumpli-Sovantzi, L. (2000). Contribution to the flora of the Strophylia area (NW. Peloponnisos). *Praktika, 8*, 368–371. [in Greek].

Raabe, U., & Koumpli-Sovantzi, L. (2002). Studies on the flora of some endangered wetlands of Argolis (Greece). *Praktika, 9*, 286–289. [in Greek].

Reyes Prósper, E. R. (1910). *Las Carofitas de España. Singularmente las que crecen en sus estepas.* Imprenta Artistica Española, Madrid [in Spanish].

Rodrigo, M. A., & Alonso-Guillén, J. L. (2016). Charophyte flora in a Ramsar Mediterranean wetland (Albufera de València Natural Park, Spain) during the period 2007–2010. *Botanica Serbica, 40*, 205–215.

Romanov, R., Napolitano, T., van de Weyer, K., & Troia, A. (2019). New records and observations to the Characean flora of Sicily (Italy). *Webbia, 74*, 111–119.

Schneider, S., Mannschreck, B., Bögle, M., & Melzer, A. (2011). Genetische Differenzierung von Characeen. *Berichte der Botanischen Arbeitsgemeinschaft Südwestdeutschland Beiheft, 3*, 43–48 [in German].

Schneider, S. C., Nowak, P., von Ammon, U., & Ballot, A. (2016). Species differentiation in the genus *Chara* (Charophyceae): Considerable phenotypic plasticity occurs within homogenous genetic groups. *European Journal of Phycology, 51*, 282–293.

Sinkevičienė, Z. (2004). Charophyta of the Curonian Lagoon. *Botanica Lithuanica, 10*, 33–57.

Sonder, C. (1890). *Die Characeen der Provinz Schleswig-Holstein und Lauenburg nebst eingeschlossenen fremden Gebietstheilen.* Dissertation Universität Rostock (64 pp.) [in German].

Verdam, H. D. (1938). The Netherlands Charophyta. *Blumea, 3*, 5–33.

Verhoeven, J. T. A., & Van Vierssen, W. (1978). Structure of macrophyte dominated communities in two brackish lagoons on the island of Corsica, France. *Aquatic Botany, 5*, 77–86.

Wood, R. D. (1965). *Monograph of the Characeae.* J. Cramer, Weinheim.

Zeneli, V., & Kashta, L. (2016). Some Charophyta (Charales) from coastal temporary ponds in Velipoja area (North Albania). *Journal of Environmental Science and Engineering, B5*, 69–77.

Zouaïdia, H., De Bélair, G., Benslama, M., Soulié-Märsche, I., & Müller, S. D. (2015). Intérêt des Characeae comme bioindicateurs de la qualité des eaux: Le cas des zones humides de Numidie (Nord-Est Algérien). *Revue d'Ecologie (Terre et Vie), 70*, 121–133 [in French].

Chapter 33
Chara globata

Roman Romanov and Irmgard Blindow

Scientific name: *Chara globata* Mig in Trudy Imp. S.-Peterburgsk. Bot. Sada 23: 537. 1904

Lectotype (designated here by R. E. Romanov & Blindow [designated as 'Isotype or Holotype?' by Wood, 1965: 213]): [China] Pekin [Beijing]/1877/Bretschneider 903 (LE A0000174).

= *Chara dominii* Vilh. in Spisy Prír. Fak. Karlovy Univ. 80: 13. 1928 "*domini*". Lectotype (designated here by R. E. Romanov): [Uzbekistan] Turkestan, St[antia] Syr-Dar'inskaja, Ozero Ush-Tepe [Lake Ush-Tepe]/17 August 1911/V. V. Nikolsky (LE A0000192; envelope label: *Chara domini* Vilh./Rossia asiatica, Turkestan, Reg. Syr-Darja, St. Syr-Darjinskaja, lacus Usch-Tepe/V. V. Nicolsky [Nikolsky]/ 17 August 1911/No. Ch. 97.).

Taxonomic note: The search for morphological traits separating *Chara globata* and *C. dominii* was unsuccessful. The presence of only solitary spine-cells for *C. globata* vs. mainly solitary with some geminate and even triplicate for *C. dominii* had been accepted as a working hypothesis (Romanov, 2019; Romanov et al., 2015, 2018) but was not confirmed after detailed study of many specimens across the whole distribution range for both taxa (see below, variability). This outcome is supported by the successful crossing of *C. globata* from Western Asia and *C. dominii* from Uzbekistan (Proctor, 1975) and genetic studies (R. Romanov, V. Nikulin, & A. Gontcharov unpubl. data). Therefore, *C. dominii* is accepted as a heterotypic synonym for *C. globata.*

R. Romanov (✉)
Nišice Ujtin Potok b.b., Dobra Voda, 85356 Bar, Montenegro
e-mail: streptophytes@gmail.com

I. Blindow
Universität Greifswald, Biologische Station Hiddensee, D-18565 Kloster, Germany
e-mail: blindi@uni-greifswald.de

H. Schubert et al. (eds.), *Charophytes of Europe*,
https://doi.org/10.1007/978-3-031-31898-6_33

Local names: RUSSIAN: Хара кругловатая, Хара Домина; UKRAINIAN: Хара Доміна.

33.1 Morphology

See Figs. 33.1, 33.2, 33.3, 33.4 and 33.5.

Habitus: The plants are light green or green, with light brownish yellow to brownish and brownish-red upper parts in insolated spots, usually large with a coarse texture, slightly to mostly moderately encrusted, frequently with arcuate upper branchlets producing the typical rough somewhat moniliform habit of upper parts with a capitate apex, exemplified by the species name. The combination of "dense", "spiny" whorls with "smooth", short-spined axes results in the typical habit of the species.

Size: It is a medium to large plant, from 10 to 50 cm to ~1 m high.

Axis: The axis is stout, from 435–793 (858) μm, in one case 1500 μm in diameter, completely corticate. The internodes in the apical parts are up to 2-times shorter than the branchlets or several times longer in strongly elongated plants, and usually somewhat longer than the branchlets in lower parts.

Cortex: The axis cortex is diplostichous, frequently partly diplo-triplostichous, moderately or slightly, but distinctly tylacanthous, to sometimes isostichous. Rarely, the cortex is regularly triplostichous and slightly tylacanthous to isostichous.

Spine-cells: The spine-cells are solitary (in the type of *Chara globata*) or mainly solitary (in the type of *C. dominii*), squarrose, at right angles to the axis if solitary. In the type of *C. dominii*, some spine-cells are geminate or even in threes, different in length and inclined to the axis. The spine-cells are acute, with distinctly thickened cell walls at their apex, variable in length from short conical-papillose, aculeolate, several times shorter than the axial diameter, to conical and long subulate, exceeding the axial diameter, even within the same population. The spine-cells have a typical extended apex (Fig. 33.4). The geminate and clustered spine-cells are mostly distinctly shorter than the axial diameter, but the former can be elongated and unequal with each other, whereas the latter are invariably short and look star-shaped. The spine-cells can be lost on old internodes.

Stipulodes: The stipulodes are arranged in two tiers of medium length, well-developed to elongated, conical to long aculeolate with acute ends.

Branchlets: The branchlets are usually straight, slightly arcuate to obviously arcuate at the top of a plant, or very rarely reflexed, 9–11 in a whorl, 1.6–5 cm in length, and 300–510 μm in diameter. The branchlets are completely or nearly completely differentiated, i.e. have regularly formed nodes up to the branchlet end or nearly so, which can be easily revealed by the presence of conspicuous bract-cells. They consist of 5–6 segments (excluding the ultimate one), of which the ultimate 0–6

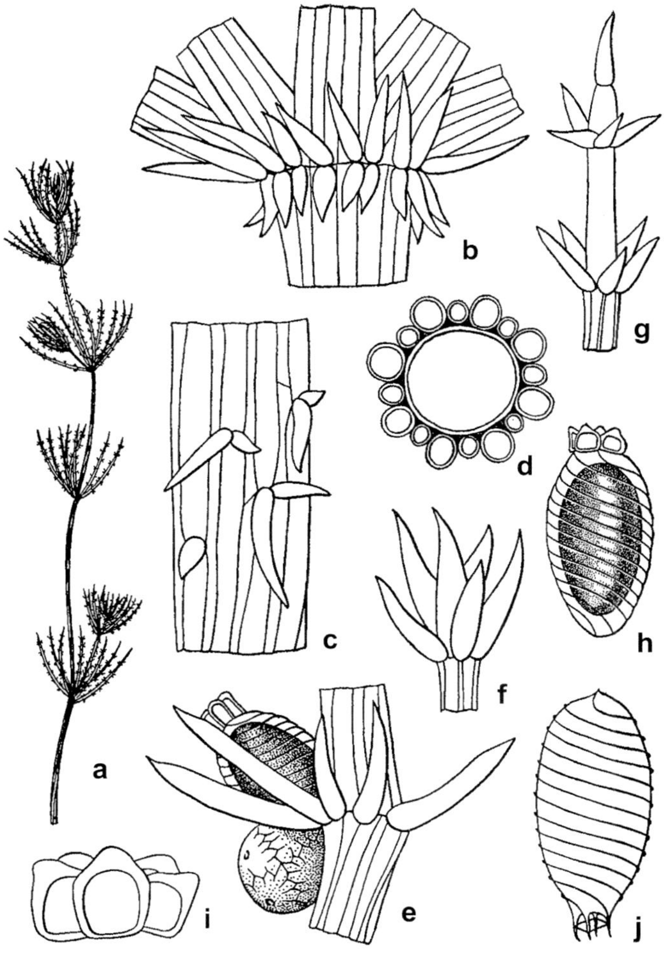

Fig. 33.1 *Chara globata* (as *C. dominii*, based on original material by Nikolsky). **a** Single shoot; **b** Stipulodes; **c** Main axis cortex (infrequent case of abundant duplicate spine-cells); **d** Cross axis section; **e** Branchlet node with conjoined gametangia; **f**, **g** Ends of branchlets; **h** Oogonium; **i** Coronula of oogonium; **j** Oospore. Drawings: Hollerbach and Krassavina (1983), except for (**a**) which was redrawn from Wood and Imahori (1964). The latter figure is a reconstruction based on photo of part of original material from Vilhelm (1928). It is dissimilar with the actual habit of branchlet whorls because the illustrated bract-cells are too short (LE, checked by R. Romanov)

are ecorticate, although mostly and completely ecorticate branchlets are rare. The branchlets end with one short conical cell or two confluent cells surrounded with the bract-cells of the ultimate node. The cortex of corticate branchlet segments is complete, diplostichous.

Fig. 33.2 *Chara globata* from Ekaterinisky Liman in the Krasnodar Krai. Photograph: Vasily Vishnyakov

Bract-cells: The bract-cells are subulate, aculeolate and needle-shaped, never inflated, with distinctly thickened acute ends, with a conspicuous extension at the top, in most cases distinctly verticillate or nearly so, typically squarrose to spreading, up to 4.5 mm in length, 1.5–3.0 times longer than the oogonium, and vary between (0.12–0.2)–(0.5–2) of the length of adjacent corticate segment. The abaxial bract-cells are nearly equal, slightly longer to rarely 1.8–3-times shorter than the adaxial ones at the same plant.

Bracteoles: The bracteoles are similar to adaxial bract-cells both in shape and length.

Sexuality: The species is monoecious. The gametangia are solitary, conjoined, occurring at the 2–4 lowest branchlet nodes between corticate segments, only once observed between ecorticate segments.

Oogonia: The oogonia are ellipsoid, large; their dimensions are variable according to different authors (Table 33.1). The coronula is slightly to distinctly spreading. Gyrogonites are common but have not been studied.

Oospores: Ripe oospores are black in reflective and transmitted light. Striae are of moderate height, ending at the base of the oospores with a basal cage without windows, which surrounds the basal plate. Table 33.2 gives an overview about

Fig. 33.3 *Chara globata* from inland freshwater water reservoir in Crimea. **a** Globate upper whorls; **b** Base of the whorl with stipulodes of moderate length and short spine-cells at the main axis. Photographs: Sergey Svirin

published morphometric data. The oospore ornamentation varies from nearly smooth in probably the unripe state and indistinctly pustular through to pustular or papillate in SEM (Fig. 33.5). The ornamentations of the fossa and striae are identical.

Antheridia: Antheridia are sessile and octoscutate. Table 33.3 gives an overview of published morphometric data. Measurements by Wood and Imahori (1964), Wood (1965), Hollerbach and Krassavina (1983) and Romanov et al. (2015, 2018) are based on rewetted specimens; the method used by other authors is unknown.

Bulbils: Not described.

33.2 Phenology and Reproduction

Chara globata is a perennial species, usually forming stable dense underwater meadows. It reproduces by oospores, but vegetative reproduction also seems to be common, which is indicated by the typically perennial stands and poorly fertile specimens in some collections. Upper globate whorls or short apical fragments are easily

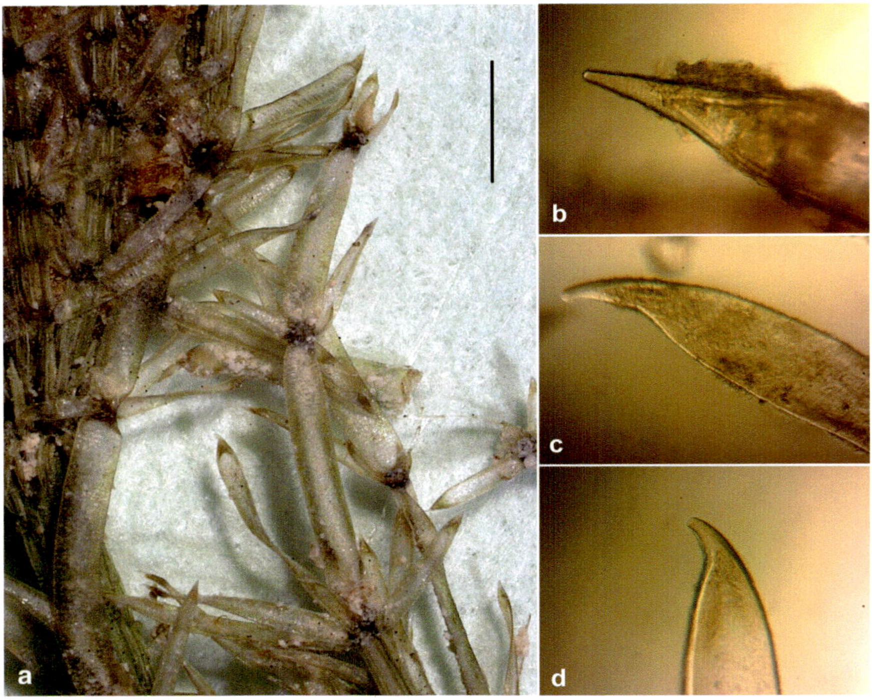

Fig. 33.4 *Chara globata*, details of a plant from Israel. **a** Ends of a branchlet; **b**, **c** Bract-cells; **d** Ultimate cell of a branchlet. Scale: 0.5 mm. Photographs: Roman Romanov

detached by waves from the plants. While charophytes commonly have negative bouyancy and sink when detached, these apical fragments of *C. globata* can float at the water surface and are outwashed by waves at the shore, especially along the water line. This mode of fragmentation could be a mechanism of dispersal within and between water bodies.

33.3 Habitat Conditions and Ecology

Habitat type: *Chara globata* is known from permanent water bodies, lakes of different origins, rarely water reservoirs and inundated quarries, permanent water bodies due to stable groundwater discharge in upper sections of wadis, very rarely from lower sections of highly regulated steppe rivers, drainage channels of irrigation systems and desert ponds. The most extensive stands are known from the most stable environments.

Table 33.1 Oogonia morphometry

Author	Length	Width	Coronula length	Coronula width	Number of helical stripes
Migula (1904)	~ 1000	550–600	120	160	16–18
Wood and Imahori (1964), Wood (1965)	Up to 900	Up to 645	Confusing data	180–195	>10
Shoyakubov (1979), *C. dominii*	1105–1158	600–686	72–129	200–286	13–15
Hollerbach and Krassavina (1983), *C. dominii*	1000–1200	500–700	–	130–205	13–15
Noedoost et al. (2015)	800–1200	450–600	100–200	150–250	≥ 10
Romanov et al. (2015)	(744)800–1041	(418)469–641	–	–	(13)14–15
Romanov et al. (2018), *C. dominii*	920–1000	600–650	–	–	–
Israel, own data	880–990	505–605	90	160	12–13
Total range	**800–1200**	**450–700**	**72–200**	**130–286**	**(10?)12–15(18)**

Length in the first column excludes the coronula; its length and width are given separately, all values for length and width in μm

Sediment: *Chara globata* is most commonly found on muddy, frequently anoxic sediments.

Nutrient conditions and water chemistry: *Chara globata* grows in oligo- to meso-eutrophic water bodies with hard, fresh to brackish (TDS up to several g/L, but not higher than 10 g/L) circumneutral to alkaline waters (pH 7.1–8.5). Brackish water populations are known only from Central Asia, slightly brackish water ones from the Lower Volga Delta and Black Sea Region. In Romania and Tunisia, the species was found in coastal lakes of unknown salinities.

Depth range: *Chara globata* grows at different depths, from <1 to 15 m.

Associated species: The species forms both monospecific associations and associations with *Chara tomentosa* and *Nitellopsis obtusa*, rarely with *C. kirghisorum*, *C. connivens*, *C. aspera*.

Table 33.2 Oospore morphometry

Author	Length	Width	Number of striae	Colour
Migula (1904)	~ 800	400	15	Black
Wood and Imahori (1964), Wood (1965)	c. 800	400	15	Black
Shoyakubov (1979), *C. dominii*	772–815	386–457	11–14	Black
Hollerbach and Krassavina (1983), *C. dominii*	(772)850–920	386–470	13, 14	Black
Noedoost et al. (2015)	850–860	480–520	12–13	Black
Romanov et al. (2015)	652–776 or 745–881 incl. basal cage	(303)343–503	11–14	Black
Romanov et al. (2018), *C. dominii*	800–810	450–480	13	Black
Total range	**652–920**	**(303)343–520**	**11–15**	

All values for length and width in μm

33.4 Variability

Two growth forms are known. The more common variants are medium-sized or mostly large plants, scarcely branched or almost unbranched, growing in dense stands, with the lowest parts decaying, i.e. with the typical habit of perennial large charophytes. The second variant is rare and seems to be restricted to shallow water. These are medium-sized, tufted plants up to 18 cm in length, consisting of numerous main axes emerging from a tangle of rhizoids.

In many populations, only solitary spine-cells were found. The frequency of occurrence of geminate spine-cells is highly variable, from extremely rare (only once on one plant) to scarce. Clusters with three spine-cells are extremely rare. Sometimes plants with solitary spine-cells grow together with plants having some geminate and triplicate spine-cells. Rarely, collections from the same localities have only solitary spine-cells in one year and some grouped spine-cells in another.

The length of the bract-cells is variable among populations. Verticillate, well-developed bract-cells are typical for plants in well-insolated environments. Deep water forms have very short verticillate conical bract-cells.

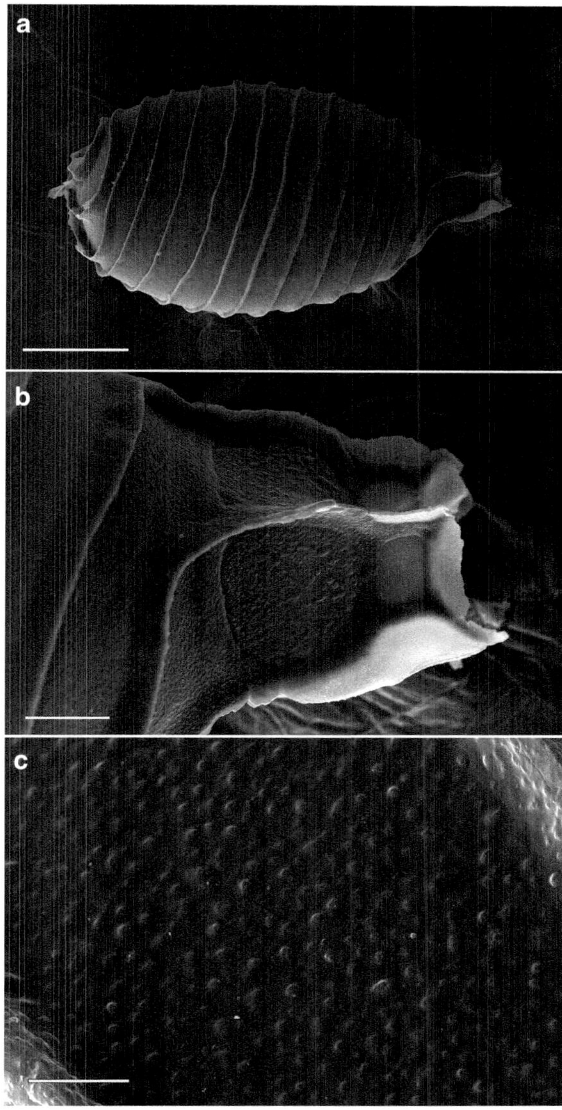

Fig. 33.5 Oospores of *Chara globata* from Uzbekistan, scanning electron microscopy. **a** General view; **b** Basal cage; **c** Oospores surface. Scale: **a**—200 μm, **b**—40 μm, **c**—10 μm. Photographs: Roman Romanov

33.5 Identification and Risk of Confusion

Chara globata belongs to the *Hartmania* group, a complex of coarse diplostichous *Chara* spp. that are hard to separate from each other (see Chap. 15). Wood (1965) placed *C. globata* within the section *Grovesia*, which is not supported by its morphological and genetic affinity to the *Hartmania* group (Romanov et al., 2015; Saber et al., 2021; Romanov et al., 2023; R. Romanov et al. unpubl. data).

Table 33.3 Antheridia morphometry

Author	Diameter (μm)
Migula (1904)	320
Wood and Imahori (1964), Wood (1965)	(320)450–525
Shoyakubov (1979), *C. dominii*	Unripe 372–472
Hollerbach and Krassavina (1983), *C. dominii*	640–760
Noedoost et al. (2015)	600–750
Romanov et al. (2015)	(458)515–583(618)
Total range	**(320)450–760**

Chara globata is placed between several species of this subsection in a multidimensional space of morphological trait variability. Other tylacanthous taxa such as *Chara aculeolata*, *C. corfuensis* and *C. papillosa*, *C. affinis* and "Mediterranean Chara" are most similar. *Chara baltica*, with short to medium-sized bract-cells, together with *C. horrida,* forms a separate 'coastal cluster' with large antheridia and occurs mainly in central and northern Europe. According to our present knowledge, the traits of oospores and molecular genetics seem to be of little value for defining the borders between most of these species, but *C. globata* can be differentiated from all other species with *rbc*L sequences (Romanov et al., 2015; Saber et al., 2021; R. Romanov, V. Nikulin, & A. Gontcharov unpubl. data). Another good criterion is the cortex, which sometimes (but not always) is triplostichous in *C. globata*, but never consistently so in any of the other taxa. The taxonomic value of the extended ends of bract-cells and end cells of branchlet needs investigation.

Maybe the best character to separate *Chara globata* morphologically from all these other taxa, which are either "overall smooth" or "overall spiny", is its unique combination of long, verticillate bract-cells with sparse and mainly single spine-cells, normally (far) shorter than the axis diameter. The combination of "fluffy", "dense", "spiny" whorls consisting of arcuate branchlets and "empty" or "smooth" internode sections, gives the species a moniliform appearance and is diagnostic for *C. globata*. In many cases, it allows the species to be identified at first glance.

Chara papillosa is the "smoothest" among the taxa which easily can be confused with *C. globata*. While it shares the short, often single and sparse spine-cells with *C. globata*, it has mainly short bract-cells, and the adaxial bract-cells are distinctly longer than the abaxial ones. Juvenile or suppressed individuals as well as lower parts of fully developed plants of *C. globata* can easily be confused with *C. papillosa*.

Chara aculeolata, one of the "spiny" taxa, has long and often verticillate bract-cells similar to *C. globata*, but the spine-cells are always at least as long as the axis diameter and always clustered. Also *C. corfuensis*, another "spiny" taxon with long and verticillate bract-cells, has long spine-cells, which often are single.

In Europe, *C. globata* has only been found in the Black Sea region and between the Azov Sea and Volga River. It can also be expected in the European part of the Mediterranean region, which is the most challenging region for identification of taxa belonging to the *Hartmania* group. Besides *C. corfuensis*, *C. hellenica* (the only dioecious taxon in the group) is restricted to the Mediterranean region as well as "Mediterranean Chara", a taxon under evaluation (see Chap. 80) and with an apparent high variability. Also, *Chara affinis* (see Chap. 17), has been recorded from the Mediterranean region. While the characters of both *Chara affinis* and "Mediterranean Chara" must still be clarified, absence of a triplostichous cortex can separate them from *C. globata*. *Chara affinis* always has long spine-cells, and "Mediterranean Chara" seems to be either "overall spiny" or "overall smooth" in contrast to *C. globata*. Still, the identification of *C. globata* is sometimes tricky, especially in plants with scarcely clustered spine-cells, e.g. in collections from Lake Ohrid and Lake Vrana, Cres (R. Romanov unpubl. data).

33.6 Distribution

See Fig. 33.6.

In Europe *Chara globata* is known from Croatia, North Macedonia, Romania, Ukraine, and Russia (Barinova et al., 2014; Hollerbach and Krassavina, 1983; Hollerbach and Palamar-Mordvintseva, 1991; Romanov, 2019; Romanov et al., 2015; Romanov et al. unpubl. data). Approximately 20 European sites were found in the Adriatic Coast, Lake Ohrid, the west and north-east of the Black Sea Region, Crimea, the eastern steppe region of Ukraine, North Caucasus and Lower Volga delta. Its whole distribution range covers semiarid and arid regions of Eurasia and

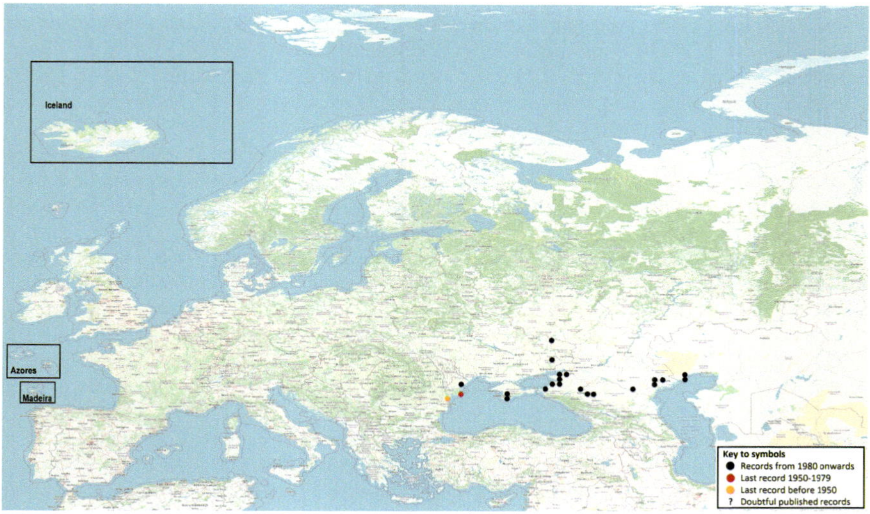

Fig. 33.6 The distribution range of *Chara globata* in Europe. Black dots are records from 1980 on, red dots from 1950 to 1979 and yellow dots before 1950; unconfirmed/doubtful published records (if applicable) are marked with "?" superimposed. Map prepared by Heiko Korsch

North Africa (most records were discovered or confirmed via examination of individual specimens by R. Romanov)—eastern and southern Europe, Middle East (Israel and Iran), Caucasus (Lake Sevan, Lchashenskaya Cove, 07 October 2006, A. A. Bobrov, det. R. Romanov, IBIW: 54128; Romanov et al., 2023), West Siberia, south of Yenisey Siberia, China, Central Asia, with majority of localities from the last region (Hollerbach and Krassavina, 1983; Noedoost et al., 2015; Romanov, 2019, unpubl.; Romanov and Abdullin, 2018; Romanov et al., 2015, 2018, 2022; Saber et al., 2021; Shoyakubov, 1979). The species could be found in Turkey, Afghanistan and Mongolia. *C. globata* was not found in the West Mediterranean region, except for Tunisia, despite a targeted search (own data, collections in B, BCN, COI, H, LE, LISU, MA, W).

33.7 Status and Threats

Chara globata has been assessed as NT (Near Threatened) in the European Red List (see Chap. 9). The main threats are human alteration of hydrological regimes of water bodies and excessive water abstraction.

References

Barinova, S., Fatyukha, A., & Romanov, R. (2014). Macrophytes and charophytes in ecological assessment of the protected lakes in Donetsk Region, Ukraine. *Natural Resources and Conservation, 2*, 71–79. https://doi.org/10.13189/nrc.2014.020501

Hollerbach, M. M., & Krassavina, L. K. [Голлербах, М. М. & Красавина, Л. К.]. (1983). *Определитель пресноводных водорослей СССР. 14. Charophyta* [*Determination key of freshwater algae of USSR, 14, Charophyta*]. Nauka. [in Russian].

Hollerbach, M. M., & Palamar-Mordvintseva, G. M. [Голлербах, М. М. & Паламар-Мордвинцева, Г. М.]. (1991). *Визначник прісноводних водоростей України. IX. Харові водорості (Charophyta)* [*The identification manual of freshwater algae of Ukraine. IX. Charophytes (Charophyta)*]. Naukova Dumka. [in Ukrainian].

Migula, W. (1904). Characeae Rossicae ex herbario horti Petropolitani. *Trudy Imperatorskago S.-Peterburgskago Botanicheskago Sada. Acta Horti Petropolitani, 23*, 533–539. [in Latin].

Noedoost, F., Riahi, H., Sheidai, M., & Ahmadi, A. (2015). Distribution of charophytes from Iran with three new records of Characeae (Charales, Chlorophyta). *Cryptogamie, Algologie, 36*, 389–405. https://doi.org/10.7872/crya/v36.iss4.2015.389

Proctor, V. W. (1975). The nature of charophyte species. *Phycologia, 14*, 97–113. https://doi.org/10.2216/i0031-8884-14-2-97.1

Romanov, R. (2019). New interesting records of charophytes (Charales, Charophyceae) from Eurasia and Africa. *Webbia, 74*, 159–166. https://doi.org/10.1080/00837792.2019.1609337

Romanov, R. E., & Abdullin, S. R. (2018). *Chara kirghisorum* (Charales): Lectotypification, first reliable record in Europe and update of species distribution in Russia. *Phytotaxa, 362*, 220–226. https://doi.org/10.11646/phytotaxa.362.2.8

Romanov, R. E., Gontcharov, A. A., & Barinova, S. S. (2015). *Chara globata* Mig. (Streptophyta: Charales): Rare species revised. *Fottea, 15*, 39–50. https://doi.org/10.5507/fot.2015.004

Romanov, R. E., Kapitonova, O. A., & Zarubina, E. Y. (2018). *Chara dominii* Vilh. (Streptophyta: Charales): A new species in the flora of Russia. *Inland Water Biology, 11*, 1–5. https://doi.org/10.1134/S1995082918010133

Romanov, R. E., Efimov, D. Yu., Makeeva, E. G., Shaulo, D. N., Kipriyanova, L. M., Zotina, T. A., Pospelov, I. N., Ebel, A. L., Zarubina, E. Yu., & Polyanskaya, D. Yu. (2022). The charophytes (Characeae, Charophyceae) from the Yenisey Siberia (Russia). *Turczaninowia, 25(2),* 19–46. https://doi.org/10.14258/turczaninowia.25.2.2

Romanov, R. E., Mallaliev, M. M., Nikulin, V. Yu., Gontcharov A. A., & Barinova, S. S. (2023). The charophytes (Charophyceae, Characeae) from Dagestan aquatic habitats, North Caucasus: biogeographical and barcoding perspectives. *Environments, 10(9),* 153. https://doi.org/10.3390/environments10090153

Saber, A. A., Gontcharov, A. A., Nikulin, A. Y., Nikulin, V. Y., Rayan, W. A., & Cantonati, M. (2021). Integrative taxonomic, ecological and genotyping study of charophyte populations from the Egyptian Western-Desert Oases and Sinai Peninsula. *Plants, 10*, 1157. https://doi.org/10.3390/plants10061157

Shoyakubov, R. S., [Шоякубов, Р. Ш.] (1979). *Харовые водоросли Узбекистана* [*Charophytes of Uzbekistan*]. Fan. [in Russian].

Vilhelm, J. (1928). Characeae Europae orientalis et Asiae ex herbario instituti cryptogamici horti botanici reipublicae rossicae (ante Petropolitani). *Spisy Vydávané Přírodovědeckou Fakultou Karlovy University, 80*, 1–24. [in French].

Wood, R. D. (1965). *Monograph of the Characeae*. J. Cramer.

Wood, R. D., & Imahori, K. (1964). *Iconograph of the Characeae*. J. Cramer.

Chapter 34
Chara globularis

Egita Zviedre and Hendrik Schubert

Scientific name: *Chara globularis* **Thuill., Fl. Env. Paris, ed. 2: 472. 1799**

Lectotype (Zaneveld, 1940: 193): [no location]/[no date]/Thuillier (L 54649). Epitype (Sakayama et al., 2009: 926): Hab. in Gallia/[no date]/[Thuillier?] (PC 74031).

Nomenclatural comment: The taxon was often referred to as *Chara fragilis*. According to the description given, this younger name must be regarded as an illegitimate synonym of *Chara vulgaris*.

Local names: BELARUSIAN: Хара ломкая; DANISH: Skør kransnål; DUTCH: Breekbar kransblad; ENGLISH: Fragile stonewort; ESTONIAN: Rabe mänd-vetikas; FINNISH: Hapranäkinparta; GERMAN: Zerbrechliche Armleuchteralge; LATVIAN: Trauslā mieturīte; LITHUANIAN: Trapusis; maurabragis; POLISH: Ramienica krucha; RUSSIAN: Хара шариковая, Хара ломкая; SWEDISH: Skörsträfse; UKRAINIAN: Хара кулькоподібна, Хара ломка.

34.1 Morphology

See Figs. 34.1 and 34.2.

E. Zviedre (✉)
Faculty of Biology, University of Latvia, Jelgavas Street 1-514, Riga, Latvia LV-1004
e-mail: egita.zviedre@lu.lv

H. Schubert
Institut für Biowissenschaften, Universität Rostock, Albert-Einstein-Straße 3, 18059 Rostock, Germany
e-mail: hendrik.schubert@uni-rostock.de

H. Schubert et al. (eds.), *Charophytes of Europe*,
https://doi.org/10.1007/978-3-031-31898-6_34

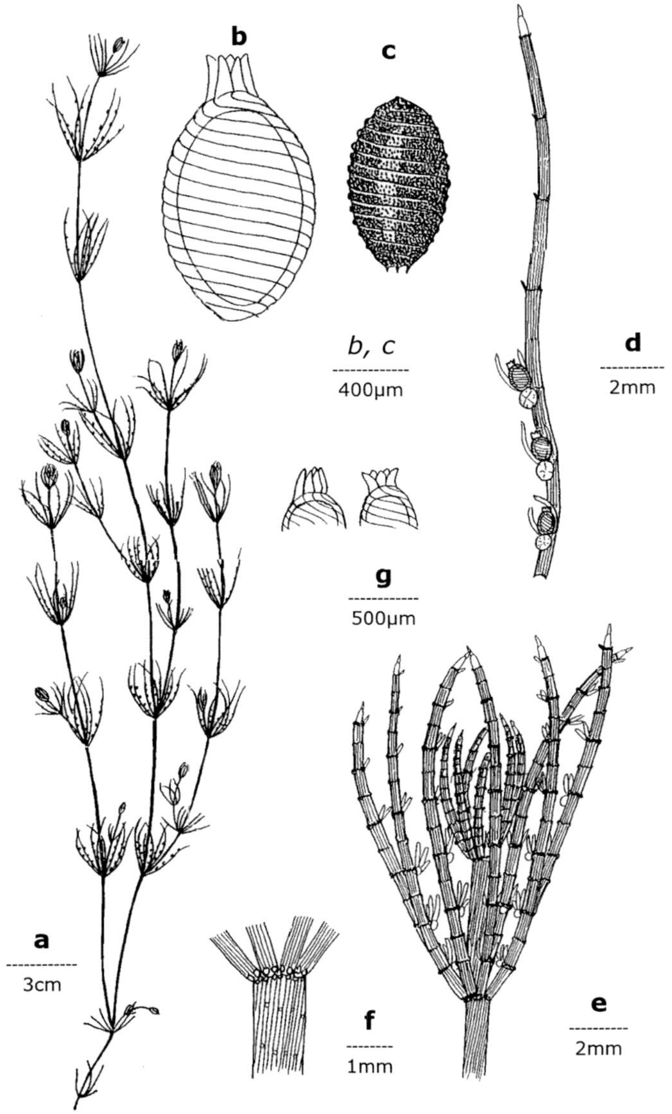

Fig. 34.1 *Chara globularis.* **a** Habitus; **b** Oogonium; **c** Oospore; **d** Fertile branchlet; **e** Apex of shoot; **f** Cortex with rudimentary spine–cells and stipulodes; **g** Types of coronula. Drawings by Werner Krause

Fig. 34.2 *Chara globularis*. **a** Typical appearance of stands; **b**, **c** Apex of individuals exhibiting the typical dark green (nodes)—bright green (internodes) pattern of branchlets; **d** Branchlet base with two equal rows of papilliform stipulodes; **e** Gametangia with well-developed bracteoles and rudimentary bract-cells; **f**–**h** SEM photographs, side view (**f**), basal plate (**g**) and surface (**h**) of *C. globularis* oospores taken from herbarium material (Beleze, M. 1897 Gallia fort de Ramboulet, Loire et Cher.). Photographs: **a**–**e**—Klaus van de Weyer; **f**–**h**—Michelle T. Casanova

Habitus: Typical plants are characterised by long internodes and short branchlets (but see also for "variability", because the species is very variable this trait). Because of the less developed bract-cells, the rudimentary stipulodes and the missing spine-cells, the plant has an "empty" and "smooth" appearance irrespective of being repeatedly branched which may give them a bushy character. The colour is green to olive-brown, only slightly greyish by incrustations on their smooth cortex.

Size: *Chara globularis* usually is small to medium-sized (5–25 cm) but may reach up to 120 cm in deeper water.

Main axis: Main axis diameter is very variable and ranges between 0.3 and 1.4 mm; often repeatedly branched.

Cortex: The species is regularly triplostichous and isostichous.

Spine-cells: Spine-cells, if developed at all, are hardly recognisable or almost invisible.

Stipulodes: *Chara globularis* is diplostephanous; both rows of stipulodes are rudimentary and require a hand-lens at least for being seen.

Branchlets: Whorls consist of (6)7–8(9) short branchlets which are corticated except for the last 2–3 branchlets. Branchlet nodes are much darker than the corticated internodes, giving the branchlet a characteristic pattern of dark spots along a brighter thread.

Bract-cells: Bract-cells are numerous (4–6, rarely 7) but less developed or even rudimentary. On the inner part of the branchlet node they are always papilliform, only on the outside they might be visible to the naked eye.

Bracteoles: Bracteoles are always present and are about as long as the oogonium. However, length of bracteoles may differ even within one plant and might be shorter as well as slightly longer than the oogonium.

Bractlets: absent (monoecious species).

Sexuality: The species is monoecious.

Oogonia: Oogonia are developed at the lower 3–4 branchlet nodes. Table 34.1 gives an overview about published morphometric data.

Oospores: Oospores are large and black; their shape is ellipsoidal to elongated (Fig. 34.2f). Complete oospores are clawed, upper ribs without shoulders. Table 34.2 gives an overview about published morphometrical data. The oospore membrane is opaque and punctuated or granulated (Corillion, 1975; Haas, 1994; Kasaki, 1964, see also Fig. 34.2h).

Antheridia: Antheridia are about 300–550 μm in diameter (Table 34.3).

Bulbils: Bulbils are not reported for *Chara globularis*.

34.2 Phenology and Reproduction

Chara globularis is a perennial species, fructifying already in its first year. However, depending on site conditions stands may disappear in winter. Fructification starts in May and ends in October; ripe oospores are observed from June onwards at least in northern Europe and France (Corillion, 1957; Langangen, 2007; Schmidt, 1981). In addition, vegetative reproduction by fragments as well as adventive shoots is often reported; Sonder (1890) as well as Olsen (1944) reported regeneration of stands from

Table 34.1 Oogonia morphometry

Author	Length	Width	Coronula length	Coronula width	Number of helical stripes
Migula (1889–1897)	800–1100	590–700	60–150	–	15–16
Groves and Bullock-Webster (1924)	750–1000	500–675	„Variable "	250	14–15
Corillion (1957, 1975)	800–950 (1200)	500–675	175–250	200–375	14–15 (17)
Wood (1965)	900–1000	600–750	200–230	200–270 (345)	14–15
Pal et al. (1962)	625–850	500–700	175–250	200–375	13–17
Kasaki (1964)	880–1130	570–680	150–170	190–230	13
Krause (1997)	700–1100	600–700	125–200	175–300	14–16
Total range	**625–1100(1200)**	**500–750**	**(60)125–250**	**175–375**	**13–17**

Length in the first two columns is excluding the coronula; its length and width is given in the next two columns, all values for length and width in μm

Table 34.2 Oospore morphometry

Author	Length	Width	Number of striae	Colour
Migula (1889–1897)	500–700	350–400 (500)	10–12–14 (15)	Black
Groves and Bullock-Webster (1924)	625–700	350–475	12–14	Black
Corillion (1957, 1975)	550–750	350–450	12–15	Almost black
Wood (1965)	675–750	440–510	11–13	Black
Pal et al. (1962)	500–700	350–450	11–14	Black
Kasaki (1964)	550–730	400–550	11–12	Black
Krause (1997)	500–800	350–450	12–14	Black
Total range	**500–800**	**350–550**	**10–15**	

All values for length and width in μm

Table 34.3 Antheridia morphometry

Author	Diameter (μm)
Migula (1889–1897)	300
Groves and Bullock-Webster (1924)	450
Corillion (1957, 1975)	450–550
Pal et al. (1962)	300–500
Wood (1965)	480–525
Krause (1997)	300
Total range	**300–550**

lower segments filled with starch in the precedent autumn and developing new shoots and rhizoids in spring.

34.3 Habitat Conditions and Ecology

Habitat type: *Chara globularis* can be found in largely different habitats, ranging from streams to small ponds. The species is found in backwaters, peat ditches, dune ponds, fishponds, ditches, springs, quarry ponds and lakes and was even found in alpine lakes (Becker, 2010; Corillion, 1957; 1975; Haas, 1994; Krause, 1997; Korsch, 2013; Migula, 1889–1897; Simons & Nat, 1996).

Sediment: *Chara globularis* does not have specific requirements with respect to sediment conditions. Sediments range from sand to mud and gyttja; the species is also found on peat with high organic content and calcareous mud (Langangen, 2007; Schmidt, 1981; Stroede, 1933; van Raam, 1998). Nevertheless, according to Holtz (1903) the species always prefers or even requires an organic cover of the sediment.

Nutrient conditions: *Chara globularis* is often observed in nutrient-rich water bodies. In some investigations its habitats were characterised by very high P and N concentrations (van de Weyer and Doege, 2016). The species is reported consistently by many authors to be rather insensitive to high nutrient concentrations (Blindow, 1992; Doege et al., 2016; Jäger, 1999; Krause, 1981, 1997; Langangen, 2007; Schmidt, 1981; Simons and Nat, 1996; Spiess et al., 2010; Stroede, 1933; van Raam, 1998).

Water chemistry: *Chara globularis* inhabits acidic, neutral and basic waters with a broad range of Ca^{2+}-contents (Krause, 1997; Olsen, 1944; Schmidt, 1981). pH values ranging between (3.8) 6.5 and 8.5, the carbonate hardness values between 0.7 and 15.9 ° dH. Total hardness ranges between 1.0 and 70.0 ° dH (Corillion, 1957; Humberg and Beck, 2006; Pietsch, 1978; Stroede, 1933). Van Raam (1998) describes *Chara globularis* from clear freshwater to slightly brackish waters in the Netherlands. In Scandinavia it also occurs in brackish water (Langangen, 2007; Sinkevicienė, 2004a, 2004b). Also Forsberg (1965) noted a high degree of flexibility with regard to salinity; maximum conductivity of the habitats is given Langangen (1972) with up to 8100 µS/cm.

Depth range: *Chara globularis* is usually found between 0.5–10 m depth (Corillion, 1957; Haas, 1994; Langangen, 2007; Schmidt, 1981), a maximum depth of 12 m is reported by Corillion (1957).

Associated species: *Chara globularis* often forms extensive monospecific stands (described as *Charetum fragilis* Corillion (1949) 1957). The species endures stronger competition with other aquatic plants for rather long periods, still being present in the stands with individual specimens between higher plants.

34.4 Variability

For *Chara globularis*, a large number of varieties and forms are described, reflecting an acclimation to habitat conditions rather than specific genotypes (see, e.g., Jäger, 1999, 2000; Migula, 1889–1897; Wood, 1965). Krause (1997) distinguished three forms that can be linked to habitat conditions, a "normal" form in large, oligosaprobic lakes; an elongated form, found in deep water as well as running water forms, and a short, small and fragile form in shallow water. A morphotype of *C. globularis* with clearly arcuate branchlets was erroneously reported as *C. arcuatifolia* (Romanov, 2022). It occurs mostly in sites with excessive insolation, especially in wave-exposed habitats. The habitus, however, always remains easily recognisable, so discrimination between forms and varieties isn't necessary.

34.5 Identification and Risk of Confusion

Due to its triplostichous-isostichous cortication, *Chara globularis* is easy to recognise when being fertile. When sterile, *C. globularis* cannot be distinguished from *C. connivens*. Also separation of *Chara globularis* and *Chara virgata* is sometimes difficult; both species may even occur together in mixed stands. *Chara virgata* is usually smaller and delicate. It is characterized by elongated (up to 2 mm) upper stipulodes, frequently in combination with tylacanthous main axis cortex and papillose spine cells, whereas both rows of stipulodes are usually papilliform for *C. globularis*, although its upper stipulodes can be infrequently elongated (1.5–2(3) or even longer than their width). The latter cases are source of confusion and doubts (Doege, 2004; Korsch, 2013) which could be reduced if all important morphological traits are taken into consideration. The branchlet length: internode length ratio is higher in case of *C. virgata* (Pukacz et al., 2011). Another helpful and unique character of *C. globularis* is that branchlet nodal cells are dark green, contrasting with the brighter branchlet internodes and resulting in a striped pattern of the branchlets of living plants – however, this character can appear in *Chara virgata* (and *C. uzbekistanica*) as well. Complete oospores are clawed, the upper ribs without shoulders, which may help to distinguish them from oospores of *C. hispida* (Krause, 1986, van de Weyer and Doege, 2016) irrespective of overlapping size (Fig. 34.3).

34.6 Distribution

See Fig. 33.3.

Chara globularis is a cosmopolitan species, reported frequently from all continents, except Antarctica (see Korsch, 2018). Its centre of occurrence, however, is in

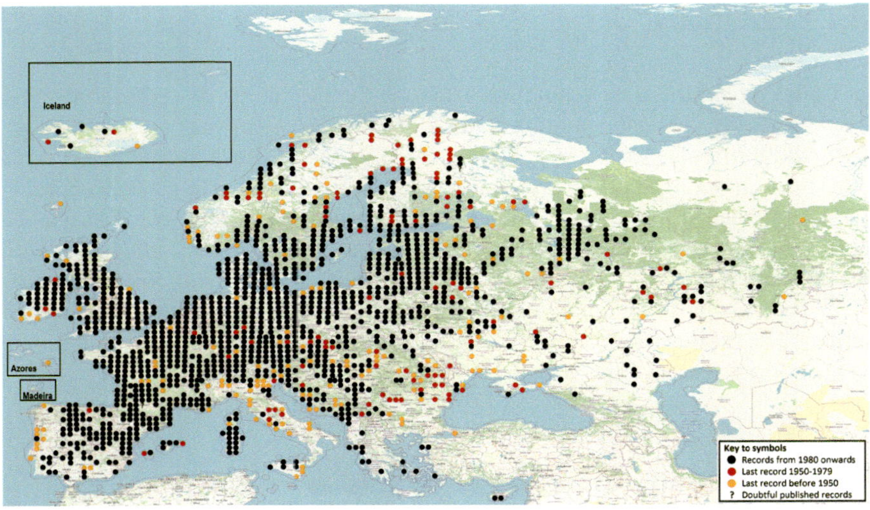

Fig. 34.3 European distribution of *Chara globularis*. For worldwide distribution see Korsch (2018). Black dots are records from 1980 on, red dots from 1950 to 1979 and yellow dots before 1950; unconfirmed/doubtful published records (if applicable) are marked with "?" on top. Map prepared by Heiko Korsch

the temperate and nemoral northern hemisphere with just a few occurrences north of the Polar circle (Krause, 1997, Langangen, 2007).

34.7 Status and Threats

The ecological niche occupied by *Chara globularis* species is very broad, compared to other charophytes; the species can be regarded as a generalist. Moreover, being a widespread species with respect to its geographical distribution and also being eutrophication-tolerant, *C. globularis* is not seen as an endangered species (Chap. 9). Except for a few countries (e.g., Poland and Hungary) where it is regarded as "Vulnerable", the species is "too frequent" to get red-listed at all (see Chap. 9). The loss of primary sites is at least balanced by successful installation in newly formed habitats in Europe.

References

Becker, R. (2010). Bemerkenswerte Characeen-Funde in Nordwest-Deutschland. *Abhandlungen Des Naturwissenschaftlichen Vereins Zu Bremen, 46*, 409–418. [in German].

Blindow, I. (1992). Decline of Charophyta during eutrophication: A comparison to angiosperms. *Freshwater Biology, 28*, 9–14.

Corillion, R. (1957a). Les Charophycées de France et d'Europe Occidentale. *Bulletin de la Société Scientifique de Bretagne, 32*(Hors série I & II), 1–499 [in French].

Corillion, R. (1975). *Flore et végétation du Massif Armoricain. T. IV. Flore des Charophytes (Characées) du Massif Armoricain et des contrées voisines d'Europe occidentale*. Jouve [in French].

Doege, A. (2004). Neue Kenntnisse über die Armleuchteralgen (Charophyceae) Sachsens. *Rostocker Meeresbiologische Beiträge, 13*, 163–172. [in German].

Doege, A., van de Weyer, K., Becker, R., & Schubert, H. (2016). Bioindikation mit Characeen. In Arbeitsgruppe Characeen Deutschlands, Armleuchteralgen: Die Characeen Deutschlands (pp. 97–138). Springer [in German].

Forsberg, C. (1965). Environmental conditions of Swedish Charophytes. *Symbolae Botanicae Upsalensis, 43*, 1–67.

Groves, J., & Bullock-Webster, G. R. (1924). *The British Charophyta. Volume II. Chareæ with plates, concluding articles, geological sketch, bibliography and index*. Ray Society.

Haas, J. N. (1994). First identification key for charophyte oospores from central Europe. *European Journal of Phycology, 29*, 227–235.

Holtz, L. (1903). *Characeen. Kryptogamenflora der Mark Brandenburg. Band* (Vol. 4, no. 1). Borntraeger [in German].

Humberg, B., & Beck, M. (2006). Der Makrophytenbestand in ausgewählten Baggerseen der Oberrheinaue. *LUBW, Oberirdische Gewässer, Gewässerökologie, 102*, 1–348. [in German].

Jäger, D. (1999). *Beiträge zur Characeen-Flora Vorarlbergs* (161 pp.). Diplomarbeit Univ. [in German].

Jäger, D. (2000). Beiträge zur Characeen-Flora Vorarlbergs (Österreichs). *Berichte des naturwissenschaftlich-medizinischen Vereins in Innsbruck, 87*, 67–85 [in German].

Kasaki, H. (1964). The Charophyta from the lakes of Japan. *Journal of the Hattori Botanical Laboratory, 27*, 217–235. [in Japanese].

Korsch, H. (2013). Die Armleuchteralgen (Charophyceae) Sachsen-Anhalts. *Berichte Des Landesamtes Für Umweltschutz Sachsen-Anhalt, 1*(2013), 1–85. [in German].

Korsch, H. (2018). The worldwide range of the Charophyte species native to Germany. *Rostocker Meeresbiologische Beiträge, 28*, 45–96.

Krause, W. (1981). Characeen als Bioindikatoren für den Gewässerzustand. *Limnologica, 1*, 399–418. [in German].

Krause, W. (1986). Zur Bestimmungsmöglichkeit subfossiler Characeen-Oosporen an Beispielen aus Schweizer Seen. *Vierteljahrschrift der Naturforschenden Gesellschaft in Zürich, 131*, 295–313.

Krause, W. (1997). Charales (Charophyceae) Süßwasserflora von Mitteleuropa. In H. Ettl, G. Gärtner, H. Heynig, & D. Mollenhauer (Eds.), *Süßwasserflora von Mitteleuropa* (Vol. 18). Fischer [in German].

Langangen, A. (2007). *Charophytes of the Nordic countries* (p. 102). Saeculum ANS.

Langangen, A. (1972). The Charophytes of Iceland. *Astarte, 5*, 27–31.

Migula, W. (1889–1897). Fünfter Band: Die Characeen. In A. Fischer, E. Fischer, F. Hauck, G. Limpricht, C. Luerssen, W. Migula, H. Rehm, P. Richter, & G. Winter (Eds.), *Dr. L. Rabenhorst's Kryptogamenflora von Deutschland, Oesterreich und der Schweiz* (2nd Ed). Eduard Kummer [in German].

Olsen, S. (1944). Danish Charophyta. *Det Kongelige Danske Videnskabernes Selskab, Biologiske Skrifter, 3*(1), 1–240, pl. 1–2 [in Danish].

Pal, B. P., Kundu, B. C., Sundaralingam, V. S., & Venkataraman, G. S. (1962). *Charophyta* (130 pp.). Indian Council of Agricultural Research.

Pietsch, W. (1978). Zur Soziologie, Ökologie und Bioindikation der *Eleocharis multicaulis*-Bestände der Lausitz. *Gleditschia, 6*, 209–264. [in German].

Pukacz, A., Pełechaty, M., & Petrović, S. (2011). The use of morphometric characteristics in the identification of two morphologically similar charophytes: *Chara globularis* and *Chara virgata. Biologia, 66*, 425–428.
Romanov, R. E. (2022). Typification of *Chara connivens* Salzmann ex A.Braun f. *brevifolia* Vilhelm, *C. crinita* Wallroth and *C. glomerulifera* Ruprecht and clarification of conflicting species concepts for *C. arcuatifolia* Vilhelm (Charales, Charophyceae). *Notulae Algarum, 223*, 1–4.
Sakayama, H., Kasai, F., Nozaki, H., Watanabe, M. M., Kawachi, M., Shigyo, M., Nishihiro, J., Washitani, I., Krienitz, I., & Ito, M. (2009). Taxonomic reexamination of *Chara globularis* (Charales, Charophyceae) from Japan based on oospore morphology and rbcL gene sequences, and the description of *C. leptospora* sp. nov. *Journal of Phycology, 45*, 917–927.
Schmidt, D. (1981). Die Characeen – eine im Aussterben begriffene Pflanzengruppe unserer Gewässer. *Gleditschia, 8*, 141–157. [in German].
Simons, J., & Nat, E. (1996). Past and present distribution of stoneworts (Characeae) in The Netherlands. *Hydrobiologia, 340*, 127–135.
Sinkevicìènė, Z. [Sinkevičienė]. (2004a). *Chara globularis.* In H. Schubert, & I. Blindow (Eds.), *Charophytes of the Baltic sea* (pp. 99–106). A. R. G. Gantner.
Sinkevičienė, Z. (2004). Charophyta of the Curonian Lagoon. *Botanica Lithuanica, 10*, 33–57.
Sonder, C. (1890). *Die Characeen der Provinz Schleswig-Holstein und Lauenburg nebst eingeschlossenen fremden Gebietstheilen* (64 pp.). Dissertation Universität Rostock [in German].
Spiess, H. J., Bolbrinker, P., Möbius, F., & Waterstraat, A. (2010). Ergebnisse der Untersuchungen submerser Makrophyten in ausgewählten Gewässern Mecklenburg-Vorpommerns. *Botanischer Rundbrief Mecklenburg-Vorpommern, 47*, 4–182. [in German].
Stroede, W. (1933). Über die Beziehungen der Characeen zu den chemischen Faktoren der Wohngewässer und des Schlammes. *Archiv Für Hydrobiologie, 25*, 192–229. [in German].
van de Weyer, K., & Doege, A. (2016). *Chara globularis.* In *Arbeitsgruppe Characeen Deutschlands, Armleuchteralgen: Die Characeen Deutschlands* (pp. 235–262). Springer [in German].
van Raam, J. C. (1998). *Handboek Kranswieren.* Chara boek [in Dutch].
Wood, R. D. (1965). *Monograph of the Characeae.* J. Cramer.
Zaneveld, J. S. (1940). The Charophyta of Malaysia and Adjacent Countries. *Blumea, 4*, 1–224.

Chapter 35
Chara gymnophylla

Roman Romanov and Ralf Becker

Scientific name: ***Chara gymnophylla*** **(A. Braun) A. Braun in Flora 18: 62. 1835**

≡ *Chara foetida* A. Braun var. *gymnophylla* A. Braun in Ann. Sci. Nat., Bot. 1: 354. 1834
Lectotype (Wood 1964: 108): Algier/3. 1832/W. Schimper (PC 616867)

≡ *Chara vulgaris* L. var. *gymnophylla* (A. Braun) Nyman, Consp. Fl. Eur.: 875. 1884

=?*Chara rohlenae* Vilh. in Hedwigia 53: 30. 1912
Lectotype (Wood, 1965: 109): [Montenegro] Jugoslavia, flora montenegrina: in rivulo Mratinje sub monte Maglic/no date/J. Rohlena (PRC)

Taxonomic note: For a long time, *Chara squamosa* was accepted to be conspecific with *C. gymnophylla* (e.g. Corillion, 1957; Wood, 1965; Krause, 1997) and, therefore, the legitimate name for this taxon. However, there are results showing that *C. squamosa* is conspecific with *C. conimbrigensis* (for details see Chap. 45) and not a correct name for this taxon.

Local names: POLISH: Ramienica nagonibylistna; RUSSIAN: Хара гололистная, UKRAINIAN: Хара гололиста.

R. Romanov (✉)
Nišice Ujtin Potok b.b., Dobra Voda, 85356 Bar, Montenegro
e-mail: streptophytes@gmail.com

R. Becker
Am Schützenplatz 58, 26121 Oldenburg, Germany
e-mail: becker.r@posteo.de

H. Schubert et al. (eds.), *Charophytes of Europe*,
https://doi.org/10.1007/978-3-031-31898-6_35

35.1 Morphology

See Figs. 35.1, 35.2 and 35.3.

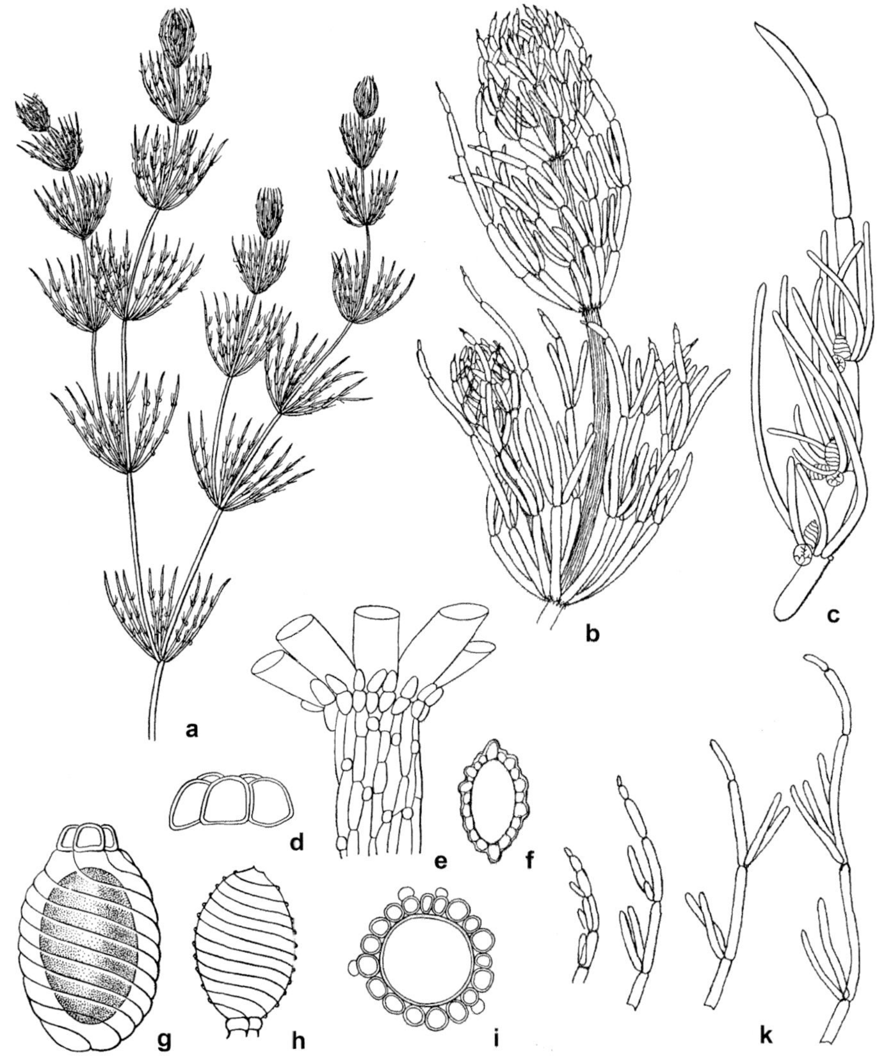

Fig. 35.1 *Chara gymnophylla.* **a** General habit of upper part; **b** Apical part; **c** Fertile branchlet; **d** Coronula; **e** Base of branchlet whorl with stipulodes and main axis cortex; **f**, **i** Cross-sections of main axis; **g** Oogonium; **h** Oospore; **k** Sterile branchlets. Drawings: **a**, **c**, **d**, **e**, **g–i**—Hollerbach and Krassavina (1983); **b**, **f**, **k**—Krause (1997)

Fig. 35.2 *Chara gymnophylla*, habitus and details. **a**, **b** Examples of different general appearances; **c** Unusually elongated upper stipulodes and usually short spine cells; **d** Spine-cells of moderate length and usual appearance of stipulodes; **e** Usual appearance of stipulodes and unusually elongated spine cells (arrowhead); **f** Usual appearance of stipulodes and spine cells; **g** Branchlets with basal node between ecorticate segments (arrowhead), second node between ecorticate segments above single basal corticate segment (double arrowhead), third node between ecorticate segments above two basal corticate segments (triple arrowhead); **h** Vertically geminate oogonia (arrowheads); **a**, **b**, **d**, **e**, **f**, **h**—plants from Spain (MA), **g**—plant from Egypt (TELA). Scale: **g**—2 mm. Photographs: Roman Romanov

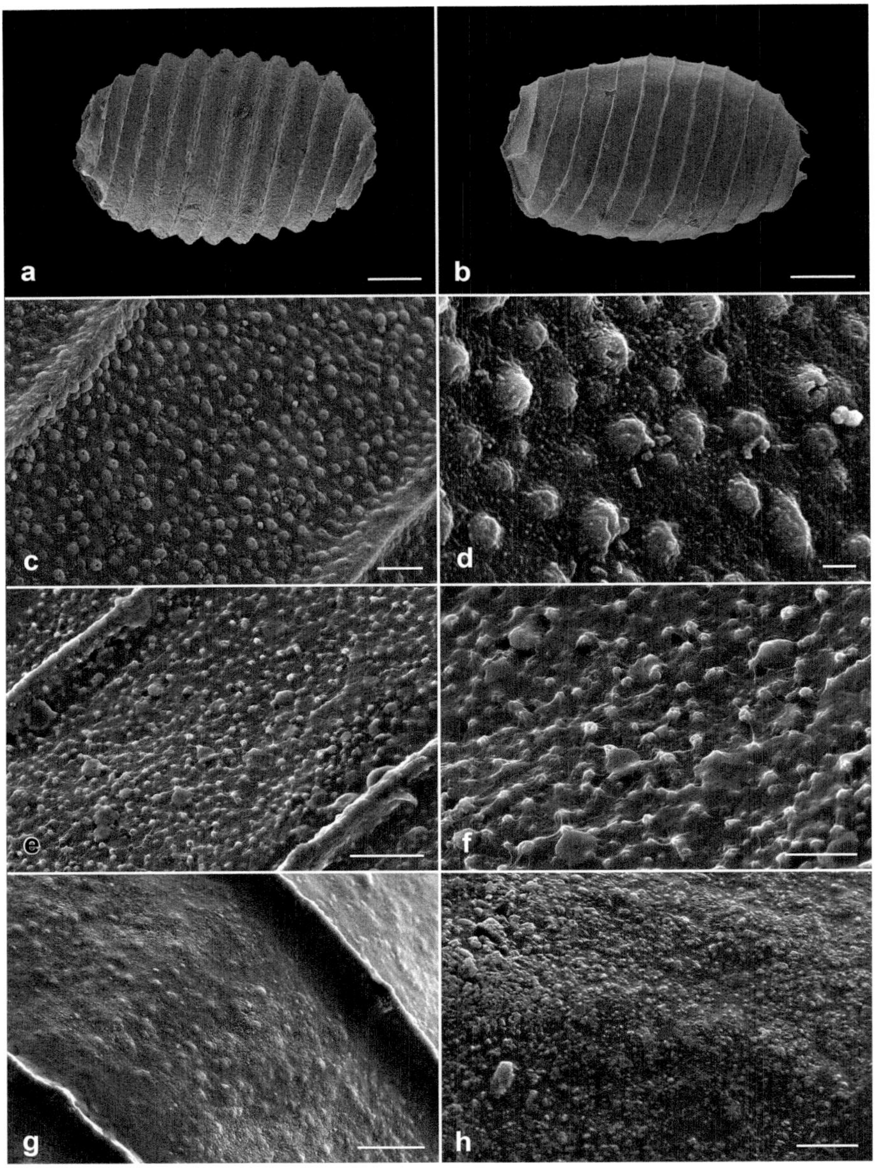

Fig. 35.3 *Chara gymnophylla*, oospores and gyrogonite (LE), scanning electron microscopy. **a** Gyrogonite; **b–h** Oospores; **a–d** Dagestan, North Caucasus, Russia; **e**, **f** Sierra Nevada, Spain; **g**, **h** Travnik, Bosnia and Herzegovina. Scale: **a**, **b**—100 μm, **c**, **f**, **h**—5 μm, **d**—1 μm, **e**, **g**—10 μm. Photographs: Roman Romanov

Habitus: The plants are bright green in life, but greyish green in pressed specimens, heavily or slightly incrusted, small or moderately high, sometimes flaccid (flexible).

Size: It is generally a small to medium plant, to 15 cm high, however it can reach a height up to 50 cm (Migula, 1889–1897; Mouronval et al., 2015; Muller et al., 2017).

Main axis: The main axis is 350–880 μm in diameter, corticated in a similar way to *C. vulgaris*, with cortical rows often dextrally twisted. The length of the internodes varies greatly, from up to 6 times as long as the branchlets to shorter than the branchlets, especially in the upper whorls.

Cortex: The axial cortex is perfectly or irregularly diplostichous, basically aulacanthous, but sometimes it can be nearly isostichous.

Spine-cells: The spine-cells are solitary, somewhat rudimentary and significantly shorter than the axis diameter, adpressed and obtuse, occasionally as long as axis diameter (Wood, 1965; Mouronval et al., 2015) or more rarely up to 3.5 mm long (Tortić-Njegovan, 1965; Fig. 35.2).

Stipulodes: The stipulodes are short, obtuse, in 2 rows, two per branchlets, and not always of similar length, the lower row is sometimes sparsely developed and hardly discernable, the upper can be significantly elongated (Tortić-Njegovan, 1965; Cirujano et al., 2008; Fig. 35.2).

Branchlets: There are from 6 to 12 branchlets in a whorl, constricted between ecorticate cells, typically entirely ecorticate or with 1 or 2 corticate segments at the base, always with nodes (with bract-cells) between ecorticate segments. The nodes between ecorticate cells are without cortex initials. The branchlets terminate with a 1–3 celled end-segment without nodes. Sometimes the terminal whorls have swollen segments. The end cell of the branchlets is obtuse, rounded, sometimes more or less mucronate.

Bract-cells: 4–6, the anterior bract-cells are long, up to 3.5–5 mm, the posterior ones rudimentary or absent or occasionally up to 6 mm long.

Bracteoles: Similar to the anterior bract-cells, 2–3 times longer than the oogonium.

Sexuality: Monoecious. The gametangia are mostly conjoined, rarely sejoined, solitary or geminate, developing at nodes between ecorticate segments and between or at the top of corticate segments if they are present, in contrast to the typical *C. vulgaris* where gametangia occur only on nodes between or adjacent to corticate segments. Fertility is less pronounced than in *C. vulgaris.*

Oogonia: Dimensions vary between 500–800 μm in length and 350–550 μm in width, without coronula. Table 35.1 gives an overview about published morphometric data. Since *C. gymnophylla* and *C. squamosa* were often not delineated in the past or it cannot be traceable in some cases (Bazzichelli and Abdelahad, 2009; Boira and Carretero, 1985; Caisová and Gąbka, 2009; Krause, 1997) these data for gametangia and oospore traits are not listed in Table 35.1. Some populations produce gyrogonites (Fig. 35.3). Their morphology and variability are poorly known.

Table 35.1 Oogonia morphometry

Authors	Length	Width	Coronula length	Coronula width	Number of helical stripes
Migula (1889–1897)	800	450	100	200	13–14
Zaneveld (1940)	~ 800	~ 450	110	200	13–14
Corillion (1957)	800	450	110	200	13–14
Wood (1965)	~ 800	450	c. 100	200	13–14
Hollerbach and Krassavina (1983)	640–800	420–450	67–100	185–200	12–14
Cirujano et al. (2008)	500–800	300–550	90–120	–	–
Muller et al. (2017)	~ 600–800	–	–	–	–
Total range	**500–800**	**300–550**	**67–120**	**185–200**	**12–14**

Length in the first column is excluding the coronula; its length and width given separately, all values for length and width in μm

Oospores: The oospores have short basal claws or rarely with a small basal cage without windows (own data). The membranes are pustulate or roughened in SEM (Fig. 35.3; M. T. Casanova pers. comment; MEL Herbarium; own data). Table 35.2 gives an overview of published morphometric data. Own data were taken from the oospores studied in SEM.

Antheridia: Table 35.3 gives an overview of published morphometric data.

Table 35.2 Oospore morphometry

Authors	Length	Width	Number of striae	Colour
Migula (1889–1897)	500	320	11	Brown, rarely black
Zaneveld (1940)	500–600	300–350	11–12	Brown
Feldmann (1946)	500	320	11	Brown
Corillion (1957)	500–600	300–350	11–12	Brown
Wood (1965)	500–600	320–350	11–12	Brown
Hollerbach and Krassavina (1983)	500–590	320–370	11–13	Brown up to almost black
Cirujano et al. (2008)	500–600	325–350	–	Brown to black
Own data	441–568	294–352	9–13	Brown
Total range	**441–600**	**294–370**	**9–13**	

All values for length and width in μm

Table 35.3 Antheridia morphometry

Authors	Diameter (μm)
Migula (1889–1897)	300
Zaneveld (1940)	300–400
Feldmann (1946)	300
Corillion (1957)	300–400
Wood (1965)	300–400
Hollerbach and Krassavina (1983)	275–450
Cirujano et al. (2008)	300–500
Muller et al. (2017)	400–600
Total range	**275–600**

All values for length and width in μm

Bulbils: Nodal starchless bulbils are common in permanent water bodies (I. Soulié-Märsche pers. comment; R. Romanov & S. Barinova unpubl. data).

35.2 Phenology and Reproduction

Reproduction takes place generally by oospores. The fertility of populations varies. In general, reproduction is higher in the Mediterranean basin and Asia than in Central Europe.

35.3 Habitat Conditions and Ecology

Habitat type: *Chara gymnophylla* grows in shallow fresh waters, lentic and slightly flowing, aquifer springs ("fontanili"), including thermal and mineral springs, temporary ponds, ditches, small troughs in calcareous mountain regions, pools, small, shallow 'ditches' beside gravel roads, lakes, shallow rivers with low water flow, outflows from thermal springs, brooks, fishponds, and rice fields (Corillion, 1957; Krause, 1997; Manolaki et al., 2011; Langangen, 2014; Soulié-Märsche pers. comm.). According to Corillion (1957), *C. gymnophylla* was common in rice fields of the Camargue (France), but Mouronval et al. (2015) did not find it in that region. *C. gymnophylla* is most commonly found in lowland localities but can also be found in mountain areas (Migula, 1889–1897; Corillion, 1957) up to 1800 m a.s.l. (Italy).

Sediment: *Chara gymnophylla* prefers, like *C. vulgaris* soils and substrates with a high carbonate content (Corillion, 1957) but is also found on rocky substrate and gravel downstream of karstic and thermal springs.

Nutrient conditions and water chemistry: *Chara gymnophylla* is strictly limited to freshwater and prefers low concentrations of chlorides (Corillion, 1957). In Mollarino springs (Lazio, Italy), the water is oligomineral, but the substrate is rich in carbonates (M. Iberite pers. comment). The parameters vary a lot according to the various authors, particularly pH and conductivity (see Table 35.4). Data from some references with hydrochemical data (Aboal, 1985; Azzella, 2014; Boira and Carretero, 1985; Carretero, 1993; Manolaki and Papastergiadou, 2016; Zouaïdia et al., 2015) were omitted below because delineation between *C. gymnophylla* and *C. squamosa* is not traceable in these cases.

Depth range: *Chara gymnophylla* prefers shallow waters (see habitat conditions).

Associated species: Both *Chara vulgaris* and *C. gymnophylla* can grow at the same site (Langangen, 2008; R. Romanov & S. Barinova unpubl. data). *Chara gymnophylla* can be associated with many other algae, sometimes with *C. aspera* (Corillion 1957). Some aquatic bryophytes like *Cratoneuron filicinum*, or phanerophytes like *Ranunculus trichophyllus* are able to grow with *C. gymnophylla* (E. Papastergiadou pers. comm.).

35.4 Variability and Risk of Confusion

Chara gymnophylla is very polymorphic and closely resembles *C. vulgaris* (Migula, 1889–1897; Feldmann, 1946; Corillion, 1957; Mouronval et al., 2015; see Chap. 52). Genetic analyses support the close relationship between the two species (Schneider et al., 2016; P. Nowak pers. comm.; R. Romanov, A. Gontcharov, & V. Nikulin unpubl. data). The variability of *C. gymnophylla* is mirrored in the range known for *C. vulgaris* in the different length of internodes, branchlets, bract-cells, spine cells and nodeless end segments of branchlets, resulting from different habitat conditions (depth, insolation, flowing/standing) and density of plants. The number of corticate segments is also variable even on the same plant and in the same whorl. Some morphotypes do no have corticate segments; some, including original material, produce few corticate segments as a stable character in combination with ecorticate segments above them.

Chara gymnophylla differs from *C. vulgaris* and *C. oedophylla* by having only ecorticate branchlets, or having ecorticate segments dominant, and gametangia borne between ecorticate segments. In frequent cases when 1 or 2 basal segments of branchlets are corticate, they always have at least one ecorticate internode segment above the corticate part of branchlet. *Chara gymnophylla* is distinct from *C. oedophylla* in having mostly conjoined gametangia, and no, or fewer inflated bract-cells, 1 or rarely 2 oogonia at single nodes and a much longer terminal branchlet segment (usually over one third of branchlet length), consisting of several cells. *Chara gymnophylla* differs from *C. lipkinii* R. E. Romanov et al. by being monoecious and having an aulacanthous axis cortex (Romanov et al., 2022).

Table 35.4 Environmental variables in some habitats of *Chara gymnophylla*

Author	Localities	pH	Conductivity (µS/cm)	Temperature (°C)	Cl (mg/l)	Ca (mg/l)	Total dissolved solids (mg/l)
Corillion (1957)	Different Mediterranean regions	7.3–7.8	–	Up to more than 30 °C	Low concentrations	High limestone content	–
Langangen (2008)	Khios Island, Greece	–	654	–	–	100	–
Langangen (2014)	Karlovasi River, Greece	–	1010	–	40	152	–
Barinova and Romanov (2015)	Ein Quinia Pool, Israel	6.7–8.1	1900–5300	16–19	–	–	137–385
M. Iberite pers. comment	Mollarino springs, Italy	8.7	339	25.8	–	–	–
Total range		**6.7–8.7**	**339–5300**	**16–25.8 (>30)**	**40**	**100–152**	**137–385**

Chara gymnophylla is described as having an aulacanthous or isostichous cortex (Feldmann, 1946; Corillion, 1957; Wood, 1965) which agrees with the type specimen (Wood and Imahori, 1964; Wood, 1965; isotypes in W, LE checked by R. Romanov), but Krause (1997) included tylacanthous *C. conimbrigensis* and isostichous *C. rohlenae* in this species, an opinion accepted by several authors (e.g. Bazzichelli and Abdelahad, 2009). *C. gymnophylla* differs from *C. squamosa* (= *C. conimbrigensis*) in having an aulacanthous axis cortex in contrast to a tylacanthous cortex in the latter species. The isostichous, or nearly so, axis cortex is not uncommon in both species. The study of fresh or fixed abundant plants reduces the possibility of their misidentification.

Chara imperfecta and *C. dissoluta* differ from *C. gymnophylla* in having a partial or incomplete axis cortex.

35.5 Distribution

See Fig. 35.4.

Chara gymnophylla is well represented in the Mediterranean: Greece, Cyprus, Albania, Italy (from north to south; Sicily), France, Spain and Portugal (Cunha, 1942; Corillion, 1957; Cirujano et al., 2008; Bazzichelli and Abdelahad, 2009; Manolaki et al., 2011; Langangen, 2014; Mouronval et al., 2015; Papastergiadou et al., 2015;

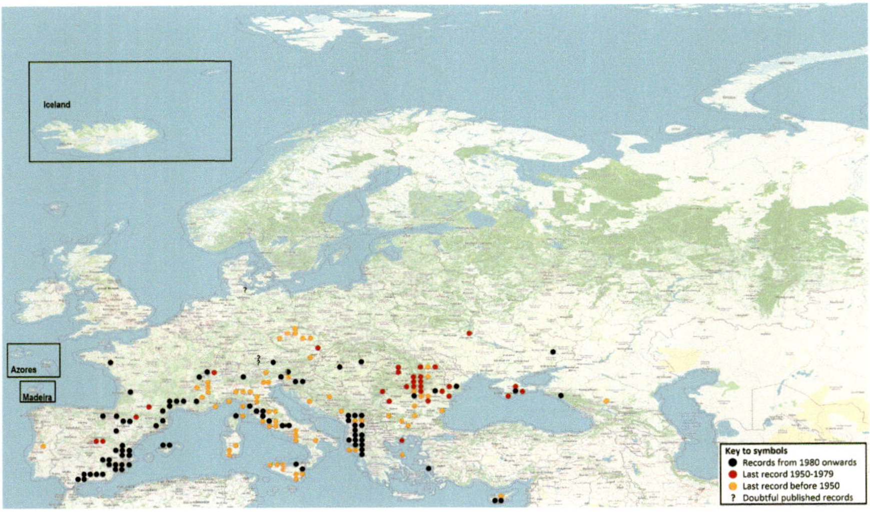

Fig. 35.4 Distribution of *Chara gymnophylla* in Europe. Black dots are records from 1980 on, red dots last records from 1950 to 1979 and yellow dots last records before 1950; unconfirmed/doubtful published records (if applicable) are marked with "?" superimposed. Map prepared by Heiko Korsch

Kashta and Miho, 2016; I. Soulié-Märsche pers. comm.). The record of *C. rohlenae* from Greece (Langangen, 2006) belongs to this species. In some cases, delineation from *C. squamosa* is not undertaken. Thus, voucher specimens have to be studied to confirm the evidence of *C. gymnophylla*.

The species is less frequently reported from Western, Central and Eastern Europe. Records from Great Britain are based on specimens of *C. vulgaris* (Moore, 1986; Bryant et al., 2002; Bryant and Stewart, 2002; N. Stewart pers. comm.). In Germany, Migula (1889–1897) mentioned a few locations, but there is no extant evidence of *C. gymnophylla* (Arbeitsgruppe Characeen Deutschlands, 2016; I. Blindow pers. comm.). In Austria the species is reliably known from historical collections only (Hohla et al., 2020; W!). It is also reported for Czech Republic (Caisová and Gąbka, 2009), Hungary (Filarszky, 1931), Serbia, Montenegro (Blaženčić, 2014; Blaženčić et al., 2018), Romania (Ştefureac and Ionescu-Ţeculescu, 1967; W!), Bulgaria (Petkoff, 1913, 1914; Vilhelm, 1913), Ukraine (Crimea) (Borisova et al., 2016), and Russia (North–East Black Sea region and North Caucasus; Hollerbach, 1950; Romanov et al., 2023; R. Romanov unpubl. data).

Specimens from Portugal, Spain, Montenegro, Serbia, old collections from Greece, Cyprus, Austria, Bosnia and Herzegovina, Croatia, Romania, Russia, and Ukraine were confirmed (R. Romanov unpubl. data; Trbojević et al., 2024). Specimens from Sardinia (Becker, 2019) were checked by R. Romanov and R. Becker and proved to be *C. squamosa*. Numerous records outside Europe are known (e.g. Corillion, 1957; Wood, 1965; Krause, 1997) but they might belong to different species, except reliable ones from North Africa (Muller et al., 2017), West Asia, incl. Caucasus (Ahmadi et al., 2012; Barinova and Romanov, 2015; Romanov et al., 2023; R. Romanov unpubl. data) and China (Han and Li, 1994; W checked by R. Romanov). The species is also known from South Africa and South America (Wood, 1978; Guerlesquin, 1981; Blindow et al., 2018).

35.6 Status and Threats

Chara gymnophylla has been assessed as LC (Least Concern) in the European Red List (see Chap. 9). According to Krause (1997) its dispersal capacity is limited by low fertility. The species is Critically Endangered in Czech Republic (Caisová and Gąbka, 2009), and Vulnerable in Ukraine (Palamar-Mordvintseva and Tsarenko, 2004).

Pollution, water abstraction, hydrological interventions and habitat destruction seem to be the most important threats. Eutrophication probably is an indirect threat, acting mainly via competition for light with phytoplankton and filamentous algae.

Acknowledgements The curators of the herbaria listed above are kindly acknowledged for their encouragement and guidance, Nadia Abdelahad and Luc Denys for the very valuable discussion of *C. gymnophylla*, Michelle T. Casanova for sharing SEM images of oospores.

References

Aboal, M. (1985). Aportación al conocimiento de las algas del SE de España I. Caráceas (Characeae). *Anales de Biologia, 6*, 7–17. [in Spanish].

Ahmadi, A., Riahi, H., Sheidai, M., & van Raam, J. (2012). Some charophytes (Characeae, Charophyta) from Central and Western of Iran including *Chara kohrangiana* species nova. *Cryptogamie, Algologie, 33*, 359–390. https://doi.org/10.7872/crya.v33.iss4.2012.359.

Arbeitsgruppe Characeen Deutschlands (2016). *Armleuchteralgen – Die Characeen Deutschlands.* Springer Spektrum. [in German]. https://doi.org/10.1007/978-3-662-47797-7.

Azzella, M. M. (2014). Italian volcanic lakes: A diversity hotspot and refuge for European charophytes. *Journal of Limnology, 73*, 502–510. https://doi.org/10.4081/jlimnol.2014.950.

Barinova, S., & Romanov, R. (2015). The new high mountain locality Ein Qinia with charophytes in the northern Israel. *Universal Journal of Plant Science, 3*, 109–119. https://doi.org/10.13189/ujps.2015.030503.

Bazzichelli, G., & Abdelahad, N. (2009). *Alghe d'acqua dolce d'Italia. Flora analitica delle Caroficee.* Centro Stampa Università. [in Italian].

Becker, R. (2019). The Characeae (Charales, Charophyceae) of Sardinia (Italy): habitats, distribution and conservation. *Webbia 74(1):* 83–101. https://doi.org/10.1080/00837792.2019.1607497.

Blaženčić, J. (2014). Overview of the stoneworts (Charales) of Serbia with the estimation of the threat status. *Botanica Serbica, 38*, 121–130.

Blaženčić, J., Kashta, L., Vesić, A., Biberdžić, V., & Stevanović, B. (2018). Charophytes (Charales) of Lake Skadar/Shkodra: Ecology and distribution. In V. Pešić, G. Karaman, & A. Kostianoy (Eds.), *The Skadar/Shkodra Lake Environment* (pp. 169–202). Springer. https://doi.org/10.1007/698_2018_265.

Blindow, I., Marquardt, R., Schories, D., & Schubert, H. (2018). Charophyceae of Chile—Taxonomy and distribution. Part 1. *Nova Hedwigia, 107*, 1–47. https://doi.org/10.1127/nova_hedwigia/2017/0452.

Boira, H., & Carretero, J. L. (1985). Las carofíceas de la provincias de Castellón y Valencia. *Collectanea Botanica, 16*, 13–19. [in Spanish].

Borisova, O. V., Palamar-Mordvintseva, G. M., & Tsarenko, P. M. (2016). *Flora algarum Ucrainicae. Charophyta, 12(2).* Kholodny Institute of Botany of the National Academy of Science of Ukraine. [in Ukrainian].

Bryant, J. A., & Stewart, N. F. (2002). Order Charales. In D. M. John, B. A. Whitton, & A. J. Brook (Eds.), *The freshwater algal flora of the British Isles* (pp. 593–612). Cambridge University.

Bryant, J. A., Stewart, N. F., & Stace, C. A. (2002). A checklist of Characeae of the British Isles. *Watsonia, 24*, 203–208.

Caisová, L., & Gąbka, M. (2009). Charophytes (Characeae, Charophyta) in the Czech Republic: Taxonomy, autoecology and distribution. *Fottea, 9*, 1–43. https://doi.org/10.5507/fot.2009.001.

Carretero, J.-L. (1993). Aportaciones a la distribución y ecología de la Carofíceas de la Provincia de Valencia. *Acta Botanica Malacitana, 18*, 31–37. [in Spanish]. https://doi.org/10.24310/abm.v18i.8959.

Cirujano, S., Cambra, J., Sánchez Castillo, P. M., Meco, A., & Flor Arnau, N. (2008). *Flora ibérica. Algas continentales. Carófitos (Characeae).* Real Jardín Botánico. [in Spanish]. https://bibdigital.rjb.csic.es/records/item/15323-flora-iberica-algas-continentales-carofitos-characeae?offset=2.

Corillion, R. (1957). Les Charophycées de France et d'Europe Occidentale. *Bulletin de la Société Scientifique de Bretagne, 32* (Hors série I & II), 1–499. [in French].

Cunha, A. G. (1942). Contribüção para o estudo dos Carofitos portugueses. *Revista da Faculdade de Ciências, 2*, 227–357. [in Portuguese].

Feldmann, G. (1946). Les Charophycées d'Afrique du Nord. *Bulletin de la Société d'histoire naturelle de l'Afrique du Nord, 37*, 64–118. [in French].

Filarszky, N. (1931). A Balaton es Vidénék Charaféléi. *A Magyar Biologiai Kutató Intézet Munkái, 4*, 249–270. [in Hungarian].

Guerlesquin, M. (1981). Contribution a la connaissance des Characees d'Amérique du Sud (Bolivie, Equateur, Guyane française). *Revue d'hydrobiologie tropicale, 14*, 381–404. [in French].

Han, F., & Li, Y. Y. (1994). *Flora algarum sinicarum aquae dulcis. Tomus 3. Charophyta*. Science. [in Chinese].

Hohla, M., Wilhalm, T., & Gregor, T. (2020). Die Armleuchteralgen (Characeae) Südtirols. *Gredleriana, 20*, 5–27. [in German]. https://doi.org/10.5281/zenodo.4245006.

Hollerbach, M. M., & Krassavina, L. K. [Голлербах, М. М. & Красавина, Л. К.] (1983). *Определитель пресноводных водорослей СССР. 14. Charophyta* [*Determination key of freshwater algae of USSR. 14. Charophyta*]. Nauka. [in Russian].

Hollerbach, M. M. [Голлербах М. М.] (1950). Систематический список харовых водорослей, обнаруженных в пределах СССР по 1935 г. включительно [The systematic list of charophytes, found on USSR territory till 1935 inclusively]. *Trudy Botanicheskogo Instituta im. V.L. Komarova AN SSSR. Ser. 2. Sporovye Rasteniya, 5*, 20–94. [in Russian].

Kashta, L., & Miho, A. (2016). The more frequently occurring macroalgae in Albanian running waters. *Buletini i Shkencave të Natyrës, 21*, 31–40.

Krause, W. (1997). Charales (Charophyceae). In: Ettl, H., G. Gärtner, H. Heynig & D. Mollenhauer (Eds.), *Süßwasserflora von Mitteleuropa, 18*. Fischer. [in German].

Langangen, A. (2006). *Chara rohlenae* Vilhelm (Charales) found in Greece. *Allionia, 40*, 31–33.

Langangen, A. (2014). Charophytes (Charales) from Samos and Ikaria (Greece) collected in 2013 and report on some localities in Skiathos (Greece). *Flora Mediterranea, 24*, 139–151. https://doi.org/10.7320/FlMedit24.139.

Langangen, A. (2008). Charophytes from some Aegean Islands (Khios, Lesvos and Limnos) in Greece. *Flora Mediterranea, 18*, 379–384.

Manolaki, P., & Papastergiadou, E. (2016). Environmental factors influencing macrophytes assemblages in a middle-sized Mediterranean stream. *River Research and Applications, 32*, 639–651. https://doi.org/10.1002/rra.2878.

Manolaki, P., Tsakiri, E., & Papastergiadou, E. (2011). Inventory of aquatic and riparian flora of Acheron and Louros river, and Zirou lake in Western Greece. *Fresenius Environmental Bulletin, 20*, 861–874.

Migula, W. (1889–1897). Fünfter band: Die Characeen. In A. Fischer, E. Fischer, F. Hauck, G. Limpricht, C. Luerssen, W. Migula, H. Rehm, P. Richter & G. Winter (Eds.), *Dr. L. Rabenhorst's Kryptogamenflora von Deutschland, Oesterreich und der Schweiz* (ed. 2). Eduard Kummer. [in German].

Moore, J. A. (1986). *Charophytes of Great Britain and Ireland*. BSBI Handbook 5. Botanical Society of the British Isles.

Mouronval, J. B., Baudouin, S., Borel, N., Soulié-Märsche, I., Klesczewski, M., & Grillas, P. (2015). *Guide des Characées de France méditerranéenne*. Office national de la Chasse et de la Faune Sauvage. [in French].

Muller, S. D., Rhazi, L., & Soulié-Märsche, I. (2017). Diversity and distribution of Characeae in the Maghreb (Algeria, Morocco, Tunisia). *Cryptogamie, Algologie, 38*, 201–251. https://doi.org/10.7872/crya/v38.iss3.2017.201.

Palamar-Mordvintseva, G. M., & Tsarenko, P. M. (2004). Red list of Charales of the Ukraine. *International Journal on Algae, 6*, 305–318. https://doi.org/10.1615/InterJAlgae.v6.i4.10.

Papastergiadou, E., Manolaki, P., Dorflinger, G., Giannouris, E., Kostara, K., & Stefanidis, K. (2015). Inventory of aquatic and riparian flora in fluvial corridors of a mediterranean island. *Fresenius Environmental Bulletin, 24*, 2179–2192.

Petkoff, S. (1914). Les Characées de Bulgarie. *Nuova Notarisia, 25*, 35–56. [in Bulgarian].

Petkoff, S. (1913). Les Characées de Bulgarie. *Spisanie na Bulgarskata akademiia na naukite, 7*, 1–44. [in Bulgarian].

Romanov, R., Barinova, S., Nikulin, V. Y., & Gontcharov, A. A. (2022). *Chara lipkinii* (Charales, Charophyceae): A new dioecious Mediterranean species under risk of extinction in the wild and some implications for the taxonomy of the genus *Chara*. *Fottea, 22*, 1–12. https://doi.org/10.5507/fot.2021.012.

Romanov, R. E., Mallaliev, M. M., Nikulin, V. Yu., Gontcharov, A. A., & Barinova, S. S. (2023). The charophytes (Charophyceae, Characeae) from Dagestan aquatic habitats, North Caucasus: biogeographical and barcoding perspectives. *Environments, 10(9),* 153. https://doi.org/10.3390/environments10090153.

Schneider, S. C., Nowak, P., von Ammon, U., & Ballot, A. (2016). Species differentiation in the genus *Chara* (Charophyceae): Considerable phenotypic plasticity occurs within homogenous genetic groups. *European Journal of Phycology, 51*, 282–293. https://doi.org/10.1080/09670262.2016.1147085.

Ştefureac, T. I., & Ionescu-Ţeculescu, V. (1967). Contribuţii la cunoaşterea Characeelor din România (III). *Studii si Cercetări de Biologie, Seria Botanică, 19*, 441–448. [in Romanian].

Trbojević, I., Marković, A., Šinžar Sekulić, J., Subakov Simić, G., & Romanov, R. (2024). *Chara squamosa* (Characeae, Charophyceae) in Serbia—insights from the taxonomic revision of the BEOU charophyte collection and recent field records. *Botanica Serbica, 48* (1), 85–92. https://doi.org/10.2298/BOTSERB2401085T

Tortić-Njegovan, M. (1965). *Chara gymnophylla* i neke druge haraceje Jugoslavije. *Acta Botanica Croatica, 14–15*, 145–162. [in Croatian].

Wood, R. D. (1965). *Monograph of the Characeae*. J. Cramer.

Zaneveld, J. S. (1940). The charophyta of Malaysia and adjacent countries. *Blumea, 4*, 1–224.

Zouaïdia, H., De Bélair, G., Benslama, M., Soulié-Märsche, I., & Müller, S. D. (2015). Intérêt des Characeae comme bioindicateurs de la qualité des eaux: Le cas des zones humides de Numidie (Nord-Est Algérien). *Revue d'écologie – la Terre et la Vie, 70*, 121–133. [in French].

Chapter 36
Chara hellenica

Anders Langangen

Scientific name: *Chara hellenica* Langangen in Studia Bot. Hung. 41: 25. 2010

Holotype (Langangen, 2010: 25): Greece, Corfu: Lake Bertonou/26.05.2008/Anders Langangen (O)

Local names: Unknown.

36.1 Morphology

See Figs. 36.1 and 36.2.

Habitus: Plants are grey, medium to heavily incrusted.

Size: *Chara hellenica* is a medium-sized species, to 20 cm long.

Main axis: Main axis diameter is up to 1.0 mm. Internodes are 1–3 times as long as the branchlets, which are 1.2–1.5 cm long.

Cortex: The cortex is commonly diplostichous (1-corticate on some internodes and 3-corticate on some old internodes), strongly tylacanthous.

Spine-cells: Spine-cells are commonly single or in pairs, or three together, if so, the third one is short, acute. Spine-cells are densely abundant in young internodes, pointing upwards in the lower part and downwards in the upper part of the internodes. They are as long as the axes diameter, or longer, up to 1.5 mm long.

Stipulodes: *Chara hellenica* is diplostephanous. Stipulodes are short and obscure, missing in many specimens.

A. Langangen (✉)
Natural History Museum, University of Oslo, P.O. Box 1172 Blinder, 0318 Oslo, Norway
e-mail: langangen@hotmail.com

H. Schubert et al. (eds.), *Charophytes of Europe*,
https://doi.org/10.1007/978-3-031-31898-6_36

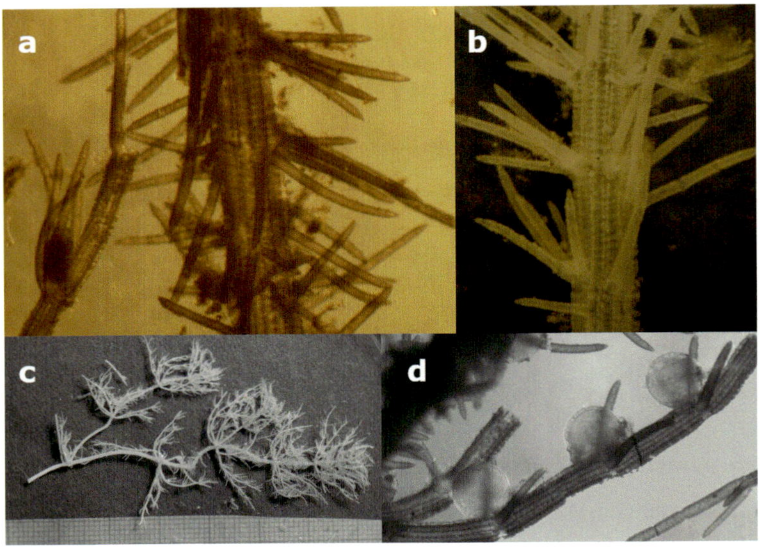

Fig. 36.1 *Chara hellenica*, habitus and details. **a**, **b** Details of main axis cortex and spine-cells (main axis diameter 650 μm); **c** Upper part of a specimen, 7 cm long; **d** Branchlet with antheridia, diameter 600 μm. Photographs: Anders Langangen

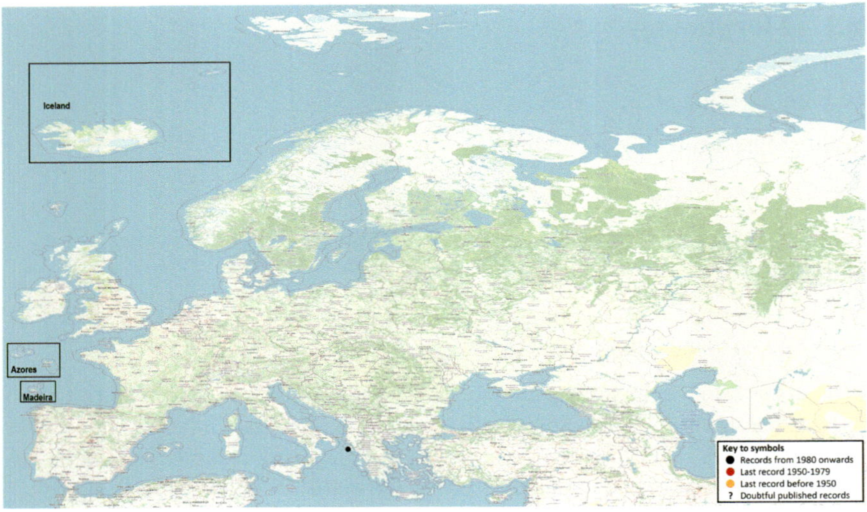

Fig. 36.2 European distribution of *Chara hellenica*. Black dots are records from 1980 on, red dots from 1950 to 1979 and yellow dots before 1950; unconfirmed/doubtful published records (if applicable) are marked with "?" on top. Map prepared by Heiko Korsch

Branchlets: Whorls consist of 7–8 branchlets, 1.2–1.5 cm long, with 5–6 segments of which 4–5 are 3-corticated; end segments are 2–3-celled, naked.

Bractlets: Not described.

Sexuality: The species is dioecious.

Female plants have 2 adaxial bract-cells, 1–3 mm long. Abaxial bract-cells are papillate or short, up to 100 μm. Bracteoles 2, 1–1.5 mm long. The bractlet is as long as the bracteoles.

Oogonia are 800 μm long (ripe?).

Oospores are black, 700 μm long, 500 μm wide, with striae of 10 ridges.

Male plants have 2 bracteoles and 2 adaxial bract-cells, which are not developed in all plants, as long as or shortly longer than the antheridia. Bractlets have not been seen.

Antheridia are 600–750 μm in diameter (measured on fresh material).

36.2 Phenology and Reproduction

Nothing is known about the life cycle of this species.

36.3 Habitat Conditions and Ecology

Chara hellenica is a freshwater species. Lake Bertonou is a highly eutrophic calcareous lake with different water plants and surrounded by trees and shrubs. The species was found in dense stands down to 20 cm depth.

The lake is situated close to Temploni, and runoff from this landfill caused extremely high values of nitrogen and phosphorus, with the disappearance of *Chara hellenica* from Lake Bertonou as result.

36.4 Identification and Risk of Confusion

Chara hellenica has unique characteristics which separate it from other species, especially the dioecious condition in a diplostichous species. The spine-cells also exhibit key characteristics, varying from single to pairs and three together, and commonly longer than the main axis diameter. Also, the features of the bract-cells are special as the adaxial are long and the abaxial are rudimentary.

36.5 Distribution

The species is only known from its type locality in Corfu.

36.6 Status and Threat

The species is assessed as CR (Critically Endangered) in the European Red List (see Chap. 9). Eutrophication has probably caused the disappearance of the species from its the only known locality (Langangen, 2017). A search of the species in the region is greatly needed.

References

Langangen, A. (2010). *Chara hellenica* spec. nov. and other charophytes collected in Corfu and Amorgos islands (Greece) in 2008. *Studia Botanica Hungarica, 41*, 21–30.

Langangen, A. (2017). The disappearance of *Chara hellenica* (Characeae) from the only one known locality, Lake Bertonou in Corfu (Greece). *Flora Mediterranea, 27*, 111–116.

Chapter 37
Chara hispida

Irmgard Blindow, Roman Romanov, Hendrik Schubert, Nick Stewart, and Klaus van de Weyer

Scientific name: ***Chara hispida*** **L., Sp. Pl. 2: 1156. 1753**

Conserved type (Gregor et al., 2014: 933): Germany, Mecklenburg-Western Pomerania, Rügen, 2 km N of Gummanz/30 Apr 2014/H. Schubert (B)

Infraspecific taxa:

Chara hispida **L. f.** ***hispida*** **L., Sp. Pl. 2 : 1156. 1753**

Chara hispida **L. f.** ***crassicaulis*** **A. Braun in Braun, Rabenhorst & Stizenberger, Charac. Eur. 1/3. 1857 "** ***Chara hispida var. micracantha crassicaulis*** **"**

≡ *Chara equisetina* Kütz., Flora 17: 706. 1834
Lectotype (Wood, 1965: 144): Germany, Todtenlache, Schleusingen/[no date]/ Kützing (L)

I. Blindow (✉)
Universität Greifswald, Biologische Station Hiddensee, 18565 Kloster, Germany
e-mail: blindi@uni-greifswald.de

R. Romanov
Nišice Ujtin Potok B.B, Dobra Voda, 85356 Bar, Montenegro
e-mail: streptophytes@gmail.com

H. Schubert
Universität Rostock, Institut für Biowissenschaften, Albert-Einstein-Straße 3, 18059 Rostock, Germany
e-mail: hendrik.schubert@uni-rostock.de

N. Stewart
4 Hexton Road, Glastonbury BA6 8HL, UK
e-mail: nfstewart@freeuk.com

K. van de Weyer
lanaplan, Lobbericher Str. 5, 41334 Nettetal, Germany
e-mail: klaus.vdweyer@lanaplan.de

H. Schubert et al. (eds.), *Charophytes of Europe*,
https://doi.org/10.1007/978-3-031-31898-6_37

≡ *Chara hispida* L. f. *equisetina* (Kütz.) Mig., Charac. Deutschl.: 637. 1895

= *Chara major* Vaill. ex Hy in Mém. Soc. Bot. France 26: 37. 1913, n. ill.

≡ *Chara hispida* L. var. *major* (Vaill. ex Hy) R. D. Wood in Taxon 11: 9. 1962, n. inv.

Nomenclatural note: Linnaeus (1753) created the name *Chara hispida* to cover all the conspicuously spiny *Chara* species then known. As separate species were distinguished in the early nineteenth century, these were carved off leaving the name *C. hispida* for the taxon described here and this usage soon became well-established. However, the collection in the Linnean herbarium, selected as lectotype by Wood, is *Chara aspera*. To avoid an unused application of the name, typification of *Chara hispida* with new type material was proposed (Gregor et al., 2014) and, later, accepted (Prud'homme van Reine, 2017).

Taxonomic note: Although Nordstedt noted that the Linnean type of *Chara hispida* belonged to *C. aspera* (see Groves and Bullock-Webster, 1924: 33), this was not recognised by Wood (1965). As a result, Wood (1965), and subsequent followers of his taxonomy, allocated the nominate variety or form of *Chara hispida* (Wood, 1962) to a tylacanthous, mainly single-spined species. The currently recognised *Chara hispida* was designated as *Chara hispida* var. *major* or *C. h.* f. *major*. This caused confusion and some authors (e.g., Cirujano et al., 2008) amalgamated *C. hispida* with tylacanthous taxa such as *C. aculeolata* or *C. papillosa*, some treated *C. hispida* in a very wide sense (Moore, 1986). Others tried to avoid the confusion by adopting the name *Chara major* for the species described here (van Raam, 1998).

Local names: BELARUSIAN: Хара шчацiнiставаласiстая; DUTCH: Stekelharig kransblad; ENGLISH: Bristly Stonewort; ESTONIAN: Karvane mändvetikas; GERMAN: Steifborstige Armleuchteralge; NORWEGIAN: Breitaggkrans; POLISH: Ramienica kosmata; RUSSIAN: Хара щетинистоволосистая; SWEDISH: Taggsträfse; UKRAINIAN: Хара щетинистоволосиста.

37.1 Morphology

See Figs. 37.1, 37.2, 37.3 and 37.4.

Habitus: *Chara hispida* is green, typically light green, often with a greyish-green appearance due to encrustations (Fig. 37.1).

Size: *Chara hispida* is robust and one of the largest charophyte species worldwide. It is 20 cm–1 m long in most cases, but the plants can reach up to 2 m.

Main axis: Diameter is 1–2.5 mm (2.5–5 mm in f. *crassicaulis*); internodes can reach a length of up to 15 cm; young shoot sections are often dark green, older shoots are often totally encrusted and greyish green.

Fig. 37.1 Habitat appearance of *Chara hispida*. Left panel: Dense vegetation of *Chara hispida*, together with *Potamogeton natans*; right panel: Habitus, the long, protruding spine-cells are clearly visible. Photographs: Klaus van de Weyer

Cortex: Regularly diplostichous, aulacanthous in the upper sections (Fig. 37.2), but isostichous in the older parts. Both primary and secondary cell rows are clearly visible, the primaries are not usually covered by the secondaries. The cortical cell rows are often, but not always, spirally twisted around the main axis and often detach from the internode cell.

Spine-cells: Arranged in bunches of 2–3, rarely 4, always distinctly protruding from the cortex and pointing in different directions, inclined and sticking out from the axis (Figs. 37.2 and 37.3), not in an anti-parallel orientation and not appressed towards the axis. Occasionally, both clustered and single spine-cells are observed on the same individual plant. Rarely, a collection shows only single spine-cells. Spine-cell lengths are highly variable, ranging from about 500–3000 (4000) μm and often exceeding the axis diameter. A collection from southern Sweden (in B) has spine-cells up to 6 mm in length, about 3 times as long as the axis diameter (Fig. 37.3). In deep water, the spine-cells can be reduced or even lacking.

Stipulodes: Stipulodes are well-developed and arranged in two rows (Fig. 37.3), which are almost equal in length and shorter or longer than the axis diameter, ranging from 600 to 2600 μm.

Branchlets: The whorls consist of 8–9 (11) branchlets with 6–9 segments, of which the last 1–2 (3) ones are ecorticate.

Bract-cells: Spreading bract-cells are present along the whole branchlet. Adaxial bract-cells are variable in length (about 400–2200 μm), distinctly longer than the abaxial ones, which can be rudimentary or even lacking (Fig. 37.4). Only rarely, the

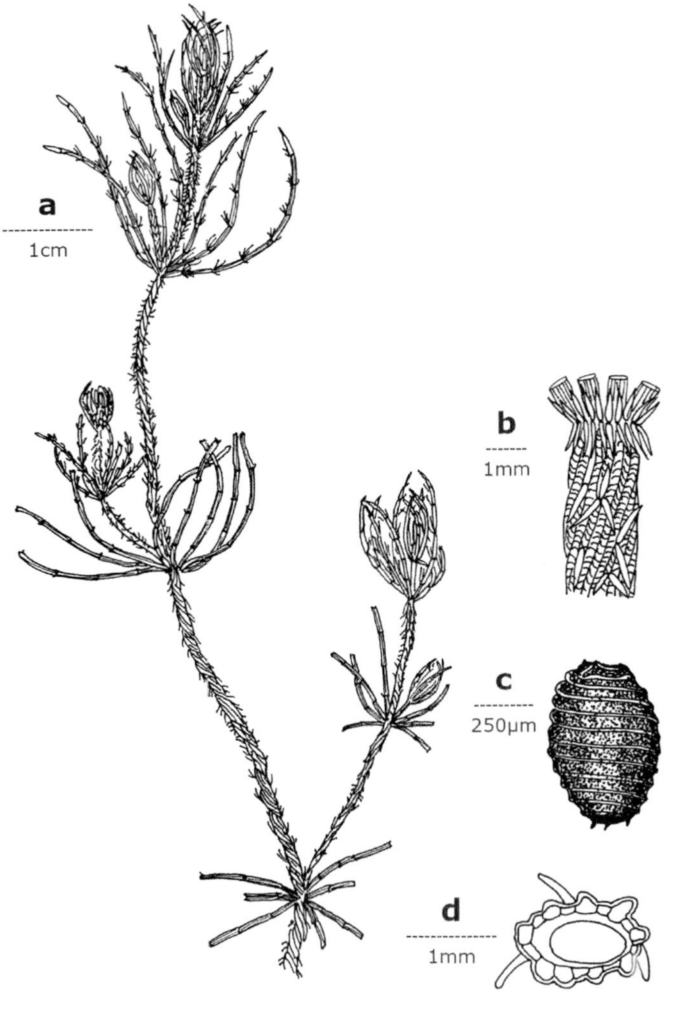

Fig. 37.2 *Chara hispida*. **a** Habitus; **b** Main axis section showing long, inclined spine-cells; **c** Oospore; **d** Internodal cross-section showing the aulacanthous cortication. Drawings after Werner Krause, modified by Dagmar Wassong in van de Weyer and Schmidt (2018)

bract-cells are "near-verticillate", i.e., the adaxial ones only slightly longer than the abaxial bract-cells.

Bracteoles: Similar to the adaxial bract-cells, but usually longer and stouter, variable in length, about 1000–4000 µm, about as long as or (far) longer than the ripe oogonium (Fig. 37.4).

Fig. 37.3 Herbarium specimens of *Chara hispida*. **a** *Chara hispida* f. *crassicaulis*, collections from the locus classicus (Braun et al. 1857–1878: 3 in 1857); **b** *Chara hispida* f. *crassicaulis*, collection in B; **c** Collection with extremely long spine-cells from Alnarp, south Sweden, in B; **d** Main axis section with long, clustered, inclined spine-cells and long stipulodes. Photographs: **a**–**c**—Irmgard Blindow; **d**—Klaus van de Weyer

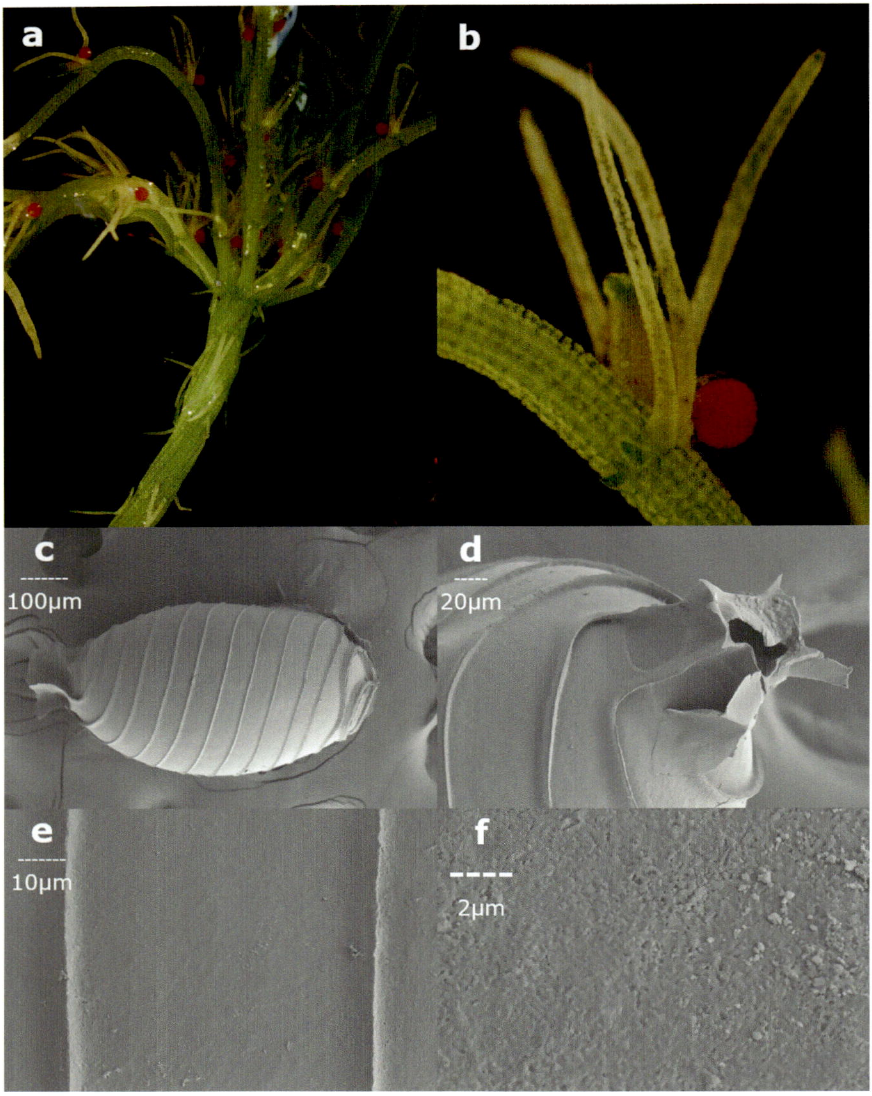

Fig. 37.4 *Chara hispida* details. **a** Axis with cortex, long, inclined spine-cells, long bract-cells and bracteoles (collection from Öland, Sweden); **b** Details of a branchlet node, showing antheridium and young oogonium, bract-cells and bracteoles. Adaxial bract-cells are long and considerably longer than abaxial ones which are very short (collection from Öland, Sweden); **c** Oospore, side view; **d** Oospore apex; **e**, **f** Oospore-membrane in different magnification (panels **c**–**f** are taken from an oospore from Lackalänga, Sweden, in PC (PC0577512), leg. Wahlstedt 1896). Photographs: **a**, **b**—Roland Bengtsson; **c**–**f**—Michelle T. Casanova

Table 37.1 Oogonia morphometry

Author	Length	Width	Coronula length	Coronula width	Number of helical stripes
Migula (1889–1897)	1200–1350	600–700	120	240	13–15
Groves and Bullock-Webster (1924)	1000–1100	650–750	150–250	250–350	13–14
Corillion (1957)	1000–1250(1500)	650- > 750	150–250	250–300	13–14
Wood (1965)*	1000–1200	650–825	135–250	250–350	13–18
Hollerbach and Krassavina (1983)	1000–1100(1350)	600–830	120–250	250–350	13–15
Krause (1997)	1000–1350	650–850	120–250	180–350	13–15
Holtz [no date]	1060	750	–	–	–
Own data	790–1100	550–760	120–250	250–390	10–12
Total range	**790–1350**	**550–850**	**120–250**	**180–390**	**(10) 11–15 (18)**

Length in the first column excludes the coronula; its length and width given separately; all values for length and width in μm
***Chara hispida* var. *major* f. *major*

Sexuality: Monoecious. Gametangia are conjoined singly and can be found on the first 3–5 branchlet nodes, often on all corticated branchlet segments except for the last one.

Oogonia: Data on morphometry are rather similar among authors (Table 37.1). Coronula cells are straight, but can occasionally be slightly spreading, which also was mentioned by Groves and Bullock-Webster (1924).

Oospores: Different authors give rather different information on oospore characters, especially Vedder (2003) gives far smaller oospore dimensions (Table 37.2). Oospores have a claw with three teeth at the base; several authors mention the heavy calcification of the oospores. Figure 37.4 shows an SEM image of the oospore.

Antheridia: Table 37.3 gives information about the diameter of the antheridia.

Bulbils: Giesenhagen (1896) mentions that *C. hispida* has multicellular bulbils.

37.2 Phenology and Reproduction

Chara hispida is often perennial and starts growing in spring from hibernating green plants or nodal fragments, either by continued growth from the shoot tips or by accessory shoots from old nodes. In a pond in southern Sweden, hibernating, violet

Table 37.2 Oospore morphometry

Author	Length	Width	Number of striae	Colour
Migula (1889–1897)	700–800	"Variable"	–	Dark brown to black
Groves and Bullock-Webster (1924)	700–750	500	12–13	Chestnut brown
Corillion (1957)	700–750(800)	–	12–13	–
Wood (1965)*	700–875	500–540	12–14	Black
Hollerbach and Krassavina (1983)	700–800(900)	400–500	12–13	Dark brown to black
Krause (1997)	700–900	450–650	12–13	Black
Vedder (2003)	575–900	325–525	12–13	Brown, rarely black
Holtz [no date]	960	650	–	–
Total range	**(575) 700–900 (960)**	**325–650**	**12–14**	

All values for length and width in μm
** Chara hispida* var. *major*

Table 37.3 Antheridia morphometry

Author	Diameter (μm)
Migula (1889–1897)	500
Groves and Bullock-Webster (1924)	375–500
Corillion (1957)	375–500
Wood (1965)*	375–675
Hollerbach and Krassavina (1983)	350–530
Krause (1997)	400–600
Own data	360–580
Total range	**350–675**

Wood (1965) measured rewetted antheridia; own data are from dry antheridia. No information is given in the other literature sources
** Chara hispida* var. *major* f. *major*

plants of *Chara hispida* were found under snow-covered ice totally covered by purple bacteria (own observations). The species can also hibernate by means of oospores, germinating from May to July (Wahlstedt, 1864; own observations). Fertile material can be found from late April to October (Olsen, 1944).

37.3 Habitat Conditions and Ecology

Habitat type: *Chara hispida* can be found in a large variety of aquatic ecosystems, mainly in standing water. The species grows in small water bodies such as ponds, ditches, peat excavations, calcareous mires, gravel and sand pits, but also in natural and artificial lakes. Occasionally, it occurs in streams and flushes but only the latter where there is near-permanent seepage.

Sediment: *Chara hispida* grows on sand, gravel, peat and calcareous gyttja (Pätzold, 2011).

Nutrient conditions: *Chara hispida* can mainly be found in oligo- to mesotrophic conditions, but also in eutrophic water (Korsch, 2013). Where it grows, PO_4-P concentrations between 1 and 50 $\mu g\ L^{-1}$, total P concentrations between 1 and 480 $\mu g\ L^{-1}$ and total N concentrations between 0.4 and 1300 $\mu g\ L^{-1}$ have been measured (Baastrup-Spohr et al., 2013; Doege et al., 2016; Simons and Nat, 1996). In Britain, *C. hispida* has been recorded at total P concentrations as high as 1400 $\mu g\ L^{-1}$ (Lambert, 2007) but this is very unsual. Laboratory experiments showed that *C. hispida* grows well at PO_4-P concentrations above 1000 $\mu g\ L^{-1}$ (Blindow, 1988). This indicates that, like other charophytes, *Chara hispida* is absent from eutrophic water not by direct physiological mechanisms, but by competition with other primary producers.

Water chemistry: *Chara hispida* is a typical hard water species. It has been found at alkalinities between 0.3 and 2.6 meq L^{-1} (Baastrup-Spohr et al., 2013) and Ca concentrations between around 26 and 180 mg L^{-1} (Stroede, 1931, 1933; Corillion, 1957; Barinova et al., 2014; own data). Simons and Nat (1996) give a wide range of Ca concentrations between 6 and 1130 mg L^{-1}. pH ranges from 6.3 to 9.1 (Doege et al., 2016; Gąbka, 2009; Nat et al., 1994). The species has been found at conductivities between 55 and 3000 $\mu S\ cm^{-1}$ (Forsberg, 1965; Gąbka, 2009; Humberg and Beck, 2006; own data). No records from the Baltic Sea sites with higher salinities could be confirmed, as the records in question turned out to be misidentifications (see Mannschreck, 2004). In an experiment with *C. hispida* from Hickling Broad, UK, biomass decreased at high (2500 mg Cl L^{-1}) salinities (Barker et al., 2008).

Depth range: *Chara hispida* can be found in shallow water of about 50 cm depth in small water bodies. In lakes, it mainly occurs between 1 and 10 m (Olsen, 1944; Pätzold, 2011; Langangen, 2007), but can occur down to 15 m (Krause, 1997). While it can be found in very shallow water in small water bodies, it avoids shallower water in larger lakes, probably because of mechanical stress from waves, and it is therefore excluded from the most eutrophic, turbid lakes (Blindow, 1992).

Associated species: *Chara hispida* forms dense, monospecific stands ("Magnocharetum hispidae Corillion") of up to 2 m height, which can hardly be colonised by other species. It can, however, also grow associated with other species, often together with other charophytes (*C. vulgaris*, *C. aculeolata*, *C. globularis*, *C. virgata*,

C. contraria, *C. aspera*, *C. tomentosa*, *Nitellopsis obtusa*, *Tolypella glomerata*, *Nitella tenuissima* and *N. confervacea*) and vascular plants such as *Stuckenia pectinata*, *Potamogeton berchtoldii*, *P. panormitanus*, *P. natans*, *P. lucens*, *Utricularia vulgaris*, *Myriophyllum spicatum*, *M. verticillatum*, *Najas marina*, *Ceratophyllum demersum*, *Polygonum amphibium* (Corillion, 1957; Krause, 1975; Humberg and Beck, 2006; Barinova et al., 2014; R. Becker pers. comm., H. Korsch pers. comm., own observations). In the former gypsum pit mine Angersdorfer Teiche close to Halle, Germany, it was found together with *Chara canescens* (own observation). In the south Swedish Lake Krankesjön, it grows together with *C. subspinosa* with which it forms transitional forms (own observations).

37.4 Variability

The length of stipulodes and spine-cells is rather variable. Forms without spine-cells have been described by Migula (1889–1897) and can be found in deep water (van de Weyer and Schmidt, 2018). Rarely, bract-cells can be rudimentary (Migula, 1889–1897; Doege, 2006).

Chara hispida f. *crassicaulis* has a distinctly different appearance from the nominate form of *C. hispida*. The plants are very robust with axis diameters between 2.5 and 5 mm (Migula, 1889–1897, own measurements; see Fig. 37.3). Metsch, who collected the type material, mentioned that it was "equally hard to deposit in a herbarium as a tree" (see label of NY 2017518). Krause (1997) mentioned short spine-cells, but lengths of up to 2000 μm and exceeding the axis diameter have been recorded (own data). The taxon has verticillate bract-cells according to Wood (1965), but in all collections included in this study, adaxials are longer than abaxials. Bract-cells in this form (400–1100 μm) and especially stipulodes (300–700 μm) are short, but still within the range of *C. hispida* (own measurements). *Chara hispida* f. *crassicaulis* is rarely fertile and gametangia are sparse, which is the probable explanation for the assumption that it is dioecious (Kützing, 1845, 1849; Wood, 1965). Rabenhorst (label of NY 2017518), however, mentioned conjoined gametangia.

Cultures of C. *hispida* f. *crassicaulis* collected at its locus classicus (Todtenlache, Germany) developed long spine-cells and conjoined gametangia and "turned" into "typical" *C. hispida*, when grown at higher temperatures (Raida, 2020). We assume that this taxon is an ecotype of *C. hispida*, and therefore prefer to keep this taxon at an infraspecific rank.

The taxon has been recorded from rather few sites in Sweden, Germany, France and Spain (Migula, 1889–1897; Krause and Krause, 1984–1986; Krause, 1980, 1985, 1997; Cirujano et al., 2007). Collections are available in GFW, L, LD, MICH, NY, ROST, S and UPS.

37.5 Identification and Risk of Confusion

Chara hispida belongs to the *Hartmania* group, a complex of robust, diplostichous *Chara* spp. that are hard to separate from each other (see Chap. 15).

From *C. papillosa*, *C. hispida* can be distinguished by its aulacanthous cortication and the clustered arrangement of the spine-cells, which are often single in *C. papillosa*. Colour is often brownish in *C. papillosa*, but greyish green in *C. hispida*. The same characters separate *C. hispida* from *C. baltica*, from which it also can be distinguished by smaller antheridia and its occurrence mainly in freshwater. *C. aculeolata* is distinguished by being tylacanthous. Rarely occurring isostichous specimens of *C. aculeolata* can be separated from *C. hispida* by the higher density and higher number of spine-cells (up to 6) in each cluster. The spine-cells are also long-persistent while in *C. hispida* they have often fallen from older parts of the main axis. The abaxial bract-cells are often more than half the length of the adaxial, compared to less than half in *C. hispida* but there is some variation in this. *C. affinis* is typically only weakly tylacanthous to isostichous but can be distinguished from isostichous individuals of *C. hispida* by the fact that spine-cells are mainly directed towards the centre of the internode, not stellate-arranged.

From the aulacanthous *C. horrida, C. hispida* can be distinguished by its regularly arranged stipulodes and smaller antheridia. Oospores have 2 claws in *C. horrida*, but 3 in *C. hispida*. While *C. horrida* is restricted to coastal lagoons with brackish water, *C. hispida* is a freshwater species which is only occasionally found in brackish water with low salinity.

While the above taxa are rarely confused with *C. hispida*, the distinction from *C. subspinosa* is difficult. In contrast to *C. hispida*, *C. subspinosa* is extremely aulacanthous with primary cell rows often almost covered by the secondary ones. In *C. subspinosa*, spine-cells are in clusters of 2–3 cells and arranged in a longitudinal direction within the deep cortex furrows (see Fig. 47.3), or spine-cells are in y-shaped clusters of 3, protruding from but appressed to the cortex (see Fig. 47.3). In *C. hispida*, the stellate-arranged spine-cells are prominent and projecting away from the cortex, not aligned or appressed (see Fig. 37.3). Krause (1997) who did not separate these taxa in his determination key, mentions the "empty" character of *C. subspinosa*, caused by the fact that spine-cells, stipulodes, bract-cells and bracteoles are short compared to the long branchlets, and in contrast to the "spiny" appearance of *C. hispida*. These characters can give an additional help in determination, but do not allow a clear separation between both taxa because of a high overlap and the occasional existence of transitional forms. Genetically, *C. hispida* is not separated from *C. subspinosa* (Nowak et al., 2011; Urbaniak and Combik, 2013; Urbaniak and Sakayama, 2017). Slender individuals of *C. hispida* can be confused with robust *C. vulgaris*.

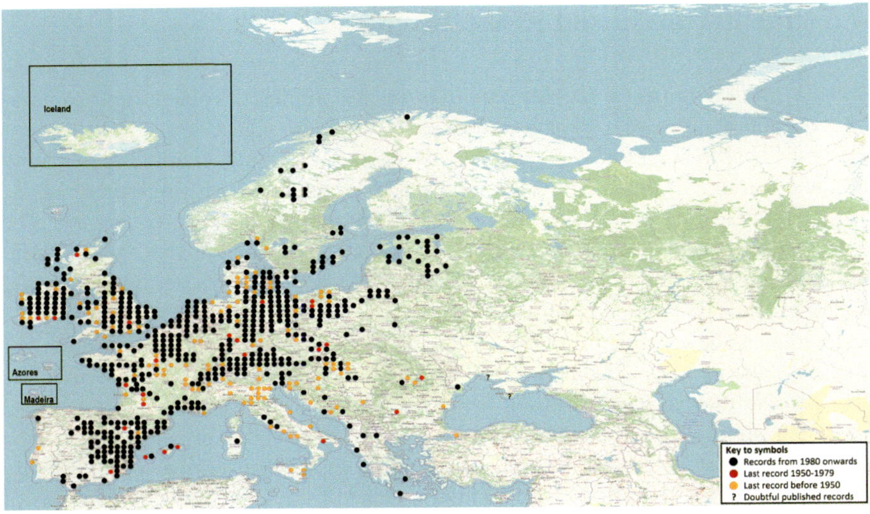

Fig. 37.5 European distribution of *Chara hispida*. For worldwide distribution see Korsch (2018). Black dots are records from 1980 onwards, red dots last record from 1950 to 1979 and yellow dots last record before 1950; doubtful published records (if applicable) are marked with "?" superimposed. Map prepared by Heiko Korsch

37.6 Distribution

See (Fig. 37.5).

Chara hispida has been mainly recorded from Europe. Outside of Europe, the species seems to have a scattered distribution with occurrences in several Asian and North African countries such as Azerbaijan, Turkey, Oman, Jordan, Morocco, Algeria and Tunisia (Barinova et al., 2014, and references therein). Former records from Russia turned out to be misidentifications. Any occurrence in Kazakhstan (see Barinova et al., 2014) or Russia could not be confirmed (Romanov et al., 2018, R. Romanov unpubl. data).

In Europe, the species occurs from the Mediterranean Sea in the south to Scandinavia in the north, Estonia and Belarus in the east to the Iberian Peninsula in the west. The species has not been recorded in Finland (Langangen et al., 2002).

37.7 Status and Threats

Chara hispida is assessed as CR (Critically Endangered) in the national Red List of Montenegro. In Poland and Serbia, the species is classified as Endangered, as Vulnerable in Switzerland and Hungary, as ES (equivalent to Vulnerable) in Czechia, and as DD (Data Deficient) in Bulgaria and Ukraine. Because of its frequent occurrence, the species is not included in the Red Lists of the Netherlands, Ireland, Britain, Sweden and Estonia. It is assessed as LC (Least Concern) in the European Red List (see Chap. 9).

A decline in many older sites has been observed; however, the species is also able to colonise newly created water bodies (Korsch et al., 2013). *Chara hispida* is threatened by eutrophication, especially in lakes where it is absent from deeper water due to low light availability (Blindow, 1992). Like other charophyte species, it is damaged by dense carp populations. In small water bodies, clearance management can help *C. hispida* to resist competition from other plants.

37.8 Remarks

Krause (1997) did not distinguish *Chara hispida* and *C. subspinosa* in his determination key. As this monograph has been used widely in Europe as a handbook for determination of charophytes, both taxa may have been misidentified and confused in many countries and collections. If in doubt, collections belonging to these taxa should be checked.

Acknowledgements We thank Anders Langangen for valuable comments.

References

Barinova, S., Romanov, R., & Solak, C. N. (2014). New record of *Chara hispida* (L.) Hartm (Streptophyta: Charophyceae, Charales) from the Işıklı Lake (Turkey) and critical checklist of Turkish charophytes. *Natural Resources and Conservation, 2*, 33–42.

Baastrup-Spohr, L., Iversen, L. L., Dahl-Nielsen, J., & Sand-Jensen, K. (2013). Seventy years of changes in the abundance of Danish charophytes. *Freshwater Biology, 58*, 1682–1693.

Barker, T., Hatton, K., O'Connor, M., Connor, L., Bagnell, L., & Moss, B. (2008). Control of ecosystem state in a shallow, brackish lake: Implications for the conservation of stonewort communities. *Aquatic Conservation: Marine and Freshwater Ecosystems, 18*, 221–240.

Blindow, I. (1988). Phosphorus toxicity in *Chara*. *Aquatic Botany, 32*, 393–395.

Blindow, I. (1992). Decline of Charophyta during eutrophication: A comparison to angiosperms. *Freshwater Biology, 28*, 9–14.

Braun, A., Rabenhorst, L., & Stizenberger, E. (Eds.) (1857–1878). *Die Characeen Europa's in getrockneten Exemplaren* i–iv, Nr. 1–121. Dresden [in French, German & Latin].

Cirujano, S., Cambra, J., Sánchez Castillo, P. M., Meco, A., & Flor Arnau, N. (2008). *Flora ibérica. Algas continentales. Carófitos (Characeae)*. Real Jardín Botánico, Madrid, 132 pp. [in Spanish].

Cirujano, S., García Murillo, P., Meco, A., & Fernández Zamudio, R. (2007). Los carófitos ibéricos. *Anales del Jardín Botánico de Madrid, 64*, 87–102 [in Spanish].

Corillion, R. (1957). Les Charophycées de France et d'Europe Occidentale. *Bulletin de la Société Scientifique de Bretagne, 32* (Hors série I & II), 1–499 [in French].

Doege, A., van de Weyer, K., Becker, R., & Schubert, H. (2016). Bioindikation mit Characeen. In: Arbeitsgruppe Characeen Deutschlands, *Armleuchteralgen: Die Characeen Deutschlands*. Springer, pp. 97–138 [in German].

Doege, A. (2006). Die Charales des Kulkwitzer Sees bei Leipzig (Sachsen). *Lauterbornia, 58*, 109–122. [in German].
Forsberg, C. (1965). Environmental conditions of Swedish charophytes. *Symbolae Botanicae Upsalensis, 43*, 1–67.
Gąbka, M. (2009). *Charophytes of the Wielkopolska region (NW Poland): Distribution, taxonomy, and autecology*. Bogucki Wydawnictvo, Poznañ.
Gavrilović, B., Ćirić, M., Vesić, A., Vidaković, D., Novaković, B., & Živanović, M. (2018). Biodiversity overview of soda pans in the Vojvodina region (Serbia). *Journal of the Geographical Institute Jovan Cvijic SASA, 68*, 195–214.
Giesenhagen, K. (1896). Untersuchungen über die Characeen. *Flora, 82*, 381–433. [in German].
Gregor, T., Blindow, I., Raabe, U., Schubert, H., & Stewart, N. (2014). (2303) Proposal to conserve the name *Chara hispida* (Characeae) with a conserved type. *Taxon, 63*, 933–934.
Groves, J., & Bullock-Webster, G. R. (1924). *The British Charophyta. Volume II. Chareæ with plates, concluding articles, geological sketch, bibliography and index*. Ray Society, London.
Hollerbach, M. M., & Krassavina, L. K. [Голлербах, М. М. & Л. К. Красавина] (1983). Определитель пресноводных водорослей СССР. 14. *Charophyta [Determination key of freshwater algae of USSR*]. Nauka, Leningrad, 189 pp. [in Russian].
Holtz, L. [no date]. *Zeichnungen und Messungen von Characeen Pommerns von Dr. Paul Hauptfleisch, entsprechend meinen Publikationen*. Inv. Nr. 5551. University of Greifswald, Germany [in German].
Humberg, B., & Beck, M. (2006). Der Makrophytenbestand in ausgewählten Baggerseen der Oberrheinaue. *LUBW, Oberirdische Gewässer, Gewässerökologie, 102*, 1–348. [in German].
Korsch, H. (2013). Die Armleuchteralgen (Charophyceae) Sachsen-Anhalts. *Berichte des Landesamtes für Umweltschutz Sachsen-Anhalt, 1*(2013), 1–85. [in German].
Korsch, H. (2018). The worldwide range of the charophyte species native to Germany. *Rostocker Meeresbiologische Beiträge, 28*, 45–96.
Korsch, H., Doege, A., Raabe, U., & van de Weyer, K. (2013). Rote Liste der Armleuchteralgen (Charophyceae) Deutschlands. 3. Fassung, Stand: Dezember 2012. *Haussknechtia Beiheft, 17*, 1–32. [in German].
Krause, W. (1997). Charales (Charophyceae) Süßwasserflora von Mitteleuropa. In H. Ettl, G. Gärtner, H. Heynig & D. Mollenhauer (Eds.), *Süßwasserflora von Mitteleuropa 18*. Fischer, Jena [in German].
Krause, W., & Krause, H. (1984–1986). *Exsikkate Europäischer Characeen* (ed. 2). 5 Bände. Aulendorf [in German].
Krause, W. (1975). Siedlungen gefährdeter Pflanzen in Baggerseen der Oberrheinebene. *Beiträge zur Naturkundlichen Forschung in Südwestdeutschland, 34*, 187–199. [in German].
Krause, W. (1980). Zur Gesellschaftsbildung der Characeen in der Oberrheinebene. *Phytocoenologia, 7*, 305–317. [in German].
Krause, W. (1985). Über die Standortansprüche und das Ausbreitungsverhalten der Stern-Armleuchteralge *Nitellopsis obtusa* (Desvaux). J. Groves. *Carolinea, 42*, 31–42. [in German].
Kützing, F. (1845). *Phycologia germania*. Köhne, Nordhausen [in Latin].
Kützing, F. (1849). *Species Algarum*. F.A. Brockhaus, Lipsiae [in Latin].
Lambert, S. J. (2007). *The environmental range and tolerance limits of British Stoneworts (Charophytes)*. Ph.D. Thesis, University of East Anglia, Norwich, Great Britain, 452 pp.
Langangen, A. (2007). *Charophytes of the Nordic countries*. Saeculum ANS, Oslo, 102 pp.
Langangen, A., Koistinen, M., & Blindow, I. (2002). The charophytes of Finland. *Memoranda Societatis pro Fauna Et Flora Fennica, 78*, 17–48.
Mannschreck, B. (2004). *Chara hispida*. In H. Schubert & I. Blindow (Eds.), *Charophytes of the Baltic Sea*. A. R. G. Gantner, Ruggell, pp. 107–112.
Migula, W. (1889–1897). Fünfter Band: Die Characeen. In A. Fischer, E. Fischer, F. Hauck, G. Limpricht, C. Luerssen, W. Migula, H. Rehm, P. Richter, & G. Winter (Eds.), *Dr. L. Rabenhorst's Kryptogamenflora von Deutschland, Oesterreich und der Schweiz* (ed. 2). Eduard Kummer, Leipzig [in German].

Moore, J. A. (1986). *Charophytes of Great Britain and Ireland.* BSBI Handbook 5. Botanical Society of the British Isles, London.

Nat, E., Simons, J., de la Haye, M. A. A., & Coops, H. (1994). *Historisch en actueel verspreidingsbeeld van kranswieren in Nederland in samenhang met waterkwaliteitsfactoren.* RIZA werkdocument 94.148X. Lelystad, 77pp. [in Dutch].

Nowak, P., Schubert, H., & Schaible, R. (2011). *Genetic analysis of the Characeae.* Report for the year 2010/2011. University of Rostock, 21 pp.

Olsen, S. (1944). Danish Charophyta. *Det Kongelige Danske Videnskabernes Selskab, Biologiske Skrifter, 3*(1), 1–240, 1–2.

Pätzold, F. (2011). Beobachtungen zur Verbreitung von Characeen in Baggerseen der Badischen Oberrheinaue. *Berichte der Botanischen Arbeitsgemeinschaft Südwestdeutschland Beiheft, 3,* 37–42. [in German].

Prud'homme van Reine, W. F. (2017). Report of the nomenclature committee for Algae: 15. *Taxon, 66,* 191–192.

Raida, R. M. (2020). *Morphologische Veränderungen von Chara hispida und Chara equisetina an die Standortfaktoren unterschiedlicher chemischer Quellen.* Bachelor Thesis, Universität Rostock, Germany, 42 pp. + IX [in German].

Romanov, R. E., Patova, E. N., Teteryuk, B. Y., & Chemeris, E. V. (2018). Charophytes (Charales, Charophyceae) on the northeastern edge of Europe: Is it something different across Northern Europe in their diversity and biogeography? *Nova Hedwigia, Beiheft, 147,* 161–182.

van Raam, J. C. (1998). *Handboek Kranswieren.* Chara boek, Hilversum [in Dutch].

Simons, J., & Nat, E. (1996). Past and present distribution of stoneworts (Characeae) in The Netherlands. *Hydrobiologia, 340,* 127–135.

Stroede, W. (1931). *Ökologie der Characeen.* Dissertation Friedrichs-Wilhelm-Universität zu Berlin, Berlin, p. 119 [in German].

Stroede, W. (1933). Über die Beziehungen der Characeen zu den chemischen Faktoren der Wohngewässer und des Schlammes. *Archiv Für Hydrobiologie, 25,* 192–229. [in German].

Urbaniak, J., & Combik, M. (2013). Genetic and morphological data fail to differentiate *Chara intermedia* from *C. baltica*, or *C. polyacantha* and *C. rudis* from *C. hispida. European Journal of Phycology, 48,* 253–259.

Urbaniak, J., & Sakayama, H. (2017). Taxonomical analysis of closely related species of *Chara* L. section *Hartmania* (Streptophyta: Charales) based on morphological and molecular data. *Fottea, 17,* 222–239.

van de Weyer, K., & Schmidt, C. (2018). Bestimmungsschlüssel für die aquatischen Makrophyten (Gefäßpflanzen, Armleuchteralgen und Moose) in Deutschland: Band 1: Bestimmungsschlüssel, ed. 2. Landesamt für Umwelt (LfU) Brandenburg. *Potsdam, Fachbeiträge des LfU Brandenburg, 119,* 1–180. [in German].

Vedder, F. (2003). *Morphologie und Taxonomie rezenter und subfossiler Characeen-Oosporen aus der Ostsee.* Diploma thesis, Universität Greifswald, Germany, 62 pp. [in German].

Wahlstedt, L. J. (1864). *Om characeernas knoppar och öfvervintring.* Botanisk avhandling. University of Lund, Schweden, 48 pp. [in Swedish].

Wood, R. D. (1962). New combinations and taxa in the revision of Characeae. *Taxon, 11,* 7–25.

Wood, R. D. (1965). *Monograph of the Characeae.* J. Cramer.

Chapter 38
Chara horrida

Irmgard Blindow, Roman Romanov, and Hendrik Schubert

Scientific name: *Chara horrida* Wallman ex Wahlst., Bidrag Skand. Charac.: 24. 1862

Lectotype (designated here by Blindow & Ljungstrand): Westervik, Sweden [Småland], Baltic Sea/Aug. [probably 1860]/J. Scheutz (GB-0029741); Areschough, Algæ Scand. Exs. Fasc. 44 (128), 1861

= *Chara baltica* var. *fastigiata* Wallman in Kongl. Vetensk. Akad. Handl. 1852: 314. 1853
Type non designated

≡ *Chara hispida* L. var. *baltica* Hartm. f. *fastigiata* (Wallmann) R.D.Wood, Taxon 11: 9. 1962, n. inv.

Local names: ESTONIAN: Liht-mändvetikas; FINNISH: Piikkinäkinparta; GERMAN: Struppige Armleuchteralge; POLISH: Ramienica gęstokolcowa; RUSSIAN: Хара ощетиненная; SWEDISH: Raggsträfse; UKRAINIAN: Хара ощетинена.

I. Blindow (✉)
Universität Greifswald, Biologische Station Hiddensee, 18565 Kloster, Germany
e-mail: blindi@uni-greifswald.de

R. Romanov
Nišice Ujtin Potok B.B, Dobra Voda, 85356 Bar, Montenegro
e-mail: streptophytes@gmail.com

H. Schubert
Universität Rostock, Institut für Biowissenschaften, Albert-Einstein-Straße 3, 18059 Rostock, Germany
e-mail: hendrik.schubert@uni-rostock.de

H. Schubert et al. (eds.), *Charophytes of Europe*,
https://doi.org/10.1007/978-3-031-31898-6_38

38.1 Morphology

See Figs. 38.1, 38.2 and Fig. 38.3.

Habitus: Plants are fresh to dark green, rarely greyish green due to (weak!) encrustation (Fig. 38.1). The long and dense spine-cells, long stipulodes, bract-cells and

Fig. 38.1 *Chara horrida,* appearance. **a** Field appearance at Grieben Bay, Hiddensee, Germany; **b** *Chara horrida* from Valje, Skåne, Sweden, in LD; **c** A "spiny" form of *Chara baltica* from Gråen, Skåne, in LD. The main determination criterion was the aulacanthous cortication in *C. horrida* and the tylacanthous cortication in *C. baltica*. Photographs: **a**—Sven Dahlke; **b, c**—Irmgard Blindow

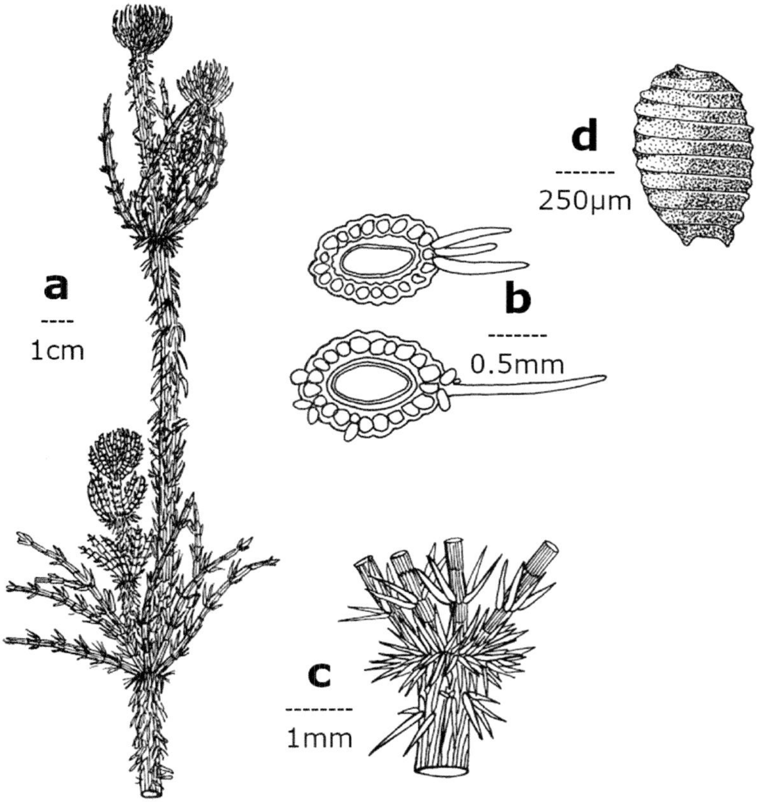

Fig. 38.2 *Chara horrida.* **a** Habitus; **b** Internodal cross-section showing the aulacanthous cortication and clustered spine-cells; **c** Main axis section showing long, clustered spine-cells, dense, irregular stipulodes and verticillate bract-cells; **d** Oospore. Drawings after Werner Krause, modified by Dagmar Wassong in van de Weyer and Schmidt (2018)

bracteoles give a rough appearance (Fig. 38.2). Sparsely branched; branches short, not much longer than the branchlets.

Size: *Chara horrida* is large and robust, 15–70 cm long, up to 100 cm.

Main axis: Diameter is about 0.9–1.5 mm; internodes are about 2 cm long.

Cortex: *Chara horrida* is regularly diplostichous. The plants are aulacanthous, but about 20% of the collections checked are (almost) isostichous.

Spine-cells: Spine-cells are arranged in bunches of 3–7 (Fig. 38.3), rarely single, the longer ones are at least as long as the main axis diameter, in most cases (far) longer, about 900–3000 µm long.

Fig. 38.3 *Chara horrida*, details. **a** Long stipulodes in more than two irregular tiers and long, clustered spine-cells as well as almost verticillate bract-cells (note that the cortex is more or less isostichous!), **b** Whorl and internode appearance of *Chara horrida* from Hiddensee/Germany, **c** Branchlet with gametangia, long bracteoles and non-verticillate bract-cells, **d** SEM image of an oospore of *Chara horrida* from Gotland/Sweden, 1859, leg. E. Westöö. Photographs: **a**, **c**—Gustav Johansson; **b**—Sven Dahlke, **d**—Roman Romanov

Stipulodes: Stipulodes are often unequally long and about as long as or longer than the main axis diameter, about 700–2400 µm. They are very dense, and contrary to the description by Wahlstedt (1862), typically appear in more than two irregular, not regular, rows with numerous extra stipulodes (Romanov et al., 2020), giving a "crowded" appearance (Krause, 1997; Figs. 38.2 and 38.3). According to Migula (1889–1897), the stipulodes are arranged in two regular rows; his drawing shows a plant with regular stipulodes, sparse spine-cells and isostichous cortex (i.e. possibly an example of *C. baltica*).

Branchlets: Whorls consist of 7–10 branchlets, which according to Wahlstedt (1862), Krause (1997) and our own results are corticate except for the end cell and sometimes 1–2 penultimate cells. The corticate part of the branchlet is far longer than the ecorticate part. Sometimes the ecorticate end cell is very short, forming a "tricellular tip"/"multicellular tip" together with the bract-cells attached to the distant branchlet node.

Bract-cells: Spreading, well developed and long bract-cells are present along the whole branchlet. Adaxial bract-cells are from c. 700–2400 μm long. According to Migula (1889–1897), bract-cells are verticillate (adaxial and abaxial bract-cells of equal length; Fig. 38.2), a character treated as diagnostic and also mentioned by Krause (1997). According to Wood (1965), the adaxial bract-cells are distinctly longer than the abaxial ones. Our own investigations showed that in 60% of the collections, adaxial bract-cells were distinctly longer than the abaxial ones (Fig. 38.3), while 40% of the collections were "near-verticillate".

Bracteoles: Two, similar to the adaxial bract-cells (Wood, 1965) and long, 1700–2400 μm long (own data).

Sexuality: *Chara horrida* is monoecious. Gametangia are conjoined and can be found on the first 4 branchlet nodes.

Oogonia: Oogonia are large, the tips of the coronula are (widely) spreading (Krause, 1997; Blümel, 2004, own data), but oogonia with straight coronula cells have also been observed (own data). For data on morphometry, see Table 38.1.

Table 38.1 Oogonia morphometry

Author	Length	Width	Coronula length	Coronula width	Number of helical stripes
Migula (1889–1897)	1200–1300	600–700	120	180	13–15
Corillion (1957)	1000–1050	600–700	120	180	12–13
Wood (1965)	1200–1300	600–700	120	180	13–15
Hollerbach and Krassavina (1983)	1200–1300	600–700	120–170	180–280	13–15
Krause (1997)	1000–1300	600–700	–	120	12–15
Holtz [without date]	1000–1250	690–770	–	–	–
Own data	800–1200	550–750	170–250	200–450	11–15
Total range	**800–1300**	**550–770**	**170–250**	**(120)180–450**	**11–15**

Length in the first column excludes the coronula; its length and width given separately; all values for length and width in μm

Table 38.2 Oospore morphometry

Author	Length	Width	Number of striae	Colour
Migula (1889–1897)	700–800	450–650	11–12	Dark brown
Corillion (1957)	700–800	450–650	11–12	–
Wood (1965)	700–800	450–650	11–12	Dark brown (never black)
Hollerbach and Krassavina (1983)	635–800	370–650	10–11	Dark brown
Krause (1997)	700–800	450–650	10–12	Dark brown to black
Vedder (2003)	750–925	425–500	12–15	Dark brown to black
Blume et al. (2009)	600–925	375–525	12–16	–
Holtz [without date]	770–910	470–700	–	–
Total range	**600–925**	**370–700**	**12–16**	

All values for length and width in μm

Table 38.3 Antheridia morphometry

Author	Diameter (μm)
Migula (1889–1897)	500
Corillion (1957)	500
Wood (1965)	500
Hollerbach and Krassavina (1983)	500–800
Krause (1997)	500
Own data	550–750
Total range	**500–800**

Own measurements are from dry antheridia, Wood (1965) measured rewetted antheridia. No information about methods is available for the other literature sources

Oospores: Different authors give similar information about oospore characters (Table 38.2), as usually dark brown to black with 11–12 (16) striae. Oospores have two long claws at the base. A basal cage may be present (R. Romanov own obs.).

Antheridia: *Chara horrida* has large antheridia (Table 38.3).

Bulbils: Giesenhagen (1896) and Krause (1997) mention that *C. horrida* has multicellular main axis node bulbils.

38.2 Phenology and Reproduction

Phenological data of *Chara horrida* are sparse. According to Havs- och Vattenmyndigheten (2020), the species hibernates by means of bulbils or as green plants, which have been observed together with *Chara baltica* under ice cover. In southern

Sweden, juvenile plants have been found from the 20th of June. No data for the onset of germination are available for any other country. Swedish herbaria contain plants that were collected as late as November (Blümel, 2004). Senescent plants with numerous oospores were found in Germany as early as September (Blindow, 2009), which suggests that these plants hibernate by means of oospores.

In Finland and Sweden, fertile plants have been found from June to September, and ripe oospores can be found as early as June (Olsen, 1944). Many Finnish and Swedish herbaria specimens are sterile, and the same was observed in Denmark (Olsen, 1944). According to Migula (1889–1897), fructification is rather sparse, but Wahlstedt (1875) mentioned that shallow-water plants have numerous gametangia, while deep-water plants have few.

Vegetative reproduction by means of fragmentation has been observed in Sweden (Havs- och Vattenmyndigheten, 2020).

38.3 Habitat Conditions and Ecology

Habitat type: *Chara horrida* prefers protected inner coastal lagoons of the Baltic Sea (Olsen, 1944).

Sediment: *Chara horrida* can be found on sandy, but mainly on soft, muddy bottoms with moderately to rather high organic contents (Stroede, 1933; Olsen, 1944; Krause, 1997; Blümel, 2004; Blindow, 2009).

Nutrient conditions: Only few data are available on nutrient concentrations at sites of *Chara horrida*. Total nitrogen concentrations between 300 and 1500 μg/l and total P concentrations between 15 and 50 μg/l have been recorded at sites in Germany and Sweden (Meyer et al., 2019; G. Johansson pers. comm.).

Water chemistry: The species has been found in a salinity range between about 3.5–9.5 psu (Blümel, 2004; Båstrup-Spohr, 2013, G. Johansson pers. comm.). Salinity values of 15 psu, given by Olsen (1944), are doubtful. These values, originally given by Sonder (1890), are not based on direct measurements, but derived from the prevailing salinities in the coastal regions in question, whereas the actual sites may have been more or less disconnected from the open Baltic Sea with possibly lower salinities. The species has been observed in a pH range between 7.6 and 8.7 (Stroede, 1931, 1933; own data).

Depth range: *Chara horrida* has been found between 0.3 and 3 m (Blümel, 2004; Pettersson, 2007; Havs- och Vattenmyndigheten, 2020).

Associated species: *Chara horrida* has rarely been found in monospecific stands. In most cases, solitary plants grow together with other charophytes (*C. tomentosa, C. aspera, C. canescens, C. baltica*), higher plants (*Myriophyllum spicatum, Stuckenia pectinata, Ruppia cirrhosa, Najas marina, Ceratophyllum demersum, C. submersum*) and macroalgae (*Fucus vesiculosus*) (Blümel, 2004; Blindow, 2009; G. Johansson and Å. Widgren pers. comm.; Havs- och Vattenmyndigheten, 2020).

38.4 Variability

Migula (1889–1897) describes different forms that can be distinguished by size, the length of stipulodes and spine-cells, length of internodes and branchlet characters. Generally, *C. horrida* seems to be rather uniform (but see below under "risk of confusion").

38.5 Identification and Risk of Confusion

Chara horrida belongs to the *Hartmania* group, a complex of coarse, diplostichous *Chara* spp. that are hard to separate from each other. Within the *Hartmania* group, *C. horrida* belongs together with *C. baltica* and *C. baltica* var. *liljebladii* to a cluster of taxa which are restricted to coastal brackish sites and characterised by large antheridia. (see Chap. 15).

These two characters and the irregularly arranged, dense stipulodes separate *C. horrida* from all freshwater and inland brackish water taxa belonging to the *Hartmania* complex, including the aulacanthous species *C. hispida* and *C. subspinosa*.

From *Chara baltica* var. *baltica* and *C. baltica.* var. *liljebladii*, *C. horrida* is separated by the (mainly) aulacanthous cortication and the dense, irregularly arranged stipulodes. In contrast to these two taxa, most collections of *C. horrida* have clustered spine-cells. All modern literature sources use these stipulode, cortication and spine-cell characters as the main criteria to separate *C. horrida* from *C. baltica.* Further differences are the distinctly longer adaxial than abaxial bract-cells in *C. baltica*, whereas many collections of *C. horrida* have near-verticillate bract-cells. The lengths of spine-cells, bracteoles and stipulodes are highly variable in *C. baltica*, but they are long in almost all collections of *C. horrida.* (Partially) triplostichous cortication, which is sometimes observed in *C. baltica*, has not been found in *C. horrida.* In agreement with Wahlstedt (1862) and Migula (1889–1897), Olsen (1944) mentions that the ecorticate end segment of the branchlets is always short, whereas it can be long, even swollen, in *C. baltica.* Also ripe oospores may serve to distinguish *C. horrida* from other taxa, as several authors refer to the two long claws at the base of the oospores.

A clear separation between *Chara baltica* and *C. horrida* is not possible, however, because of numerous transitions and the fact that isostichous cortication is not uncommon in both taxa. "Spiny" plants with additional stipulodes, but tylacanthous cortex have been found both in the Baltic Sea (see Fig. 38.1), the Black Sea and the Outer Hebrides, and are here determined as *C. baltica*, as cortication is treated as the primary character (see Chap. 10, determination key). The genetic differences between both taxa are smaller than "normal" interspecific differences, and it was assumed that the taxa are either in an ongoing speciation process, or that there is a high plasticity in morphological characters (Nowak et al., 2011). Ripe oospores are formed in crossings between both taxa (Blindow and Erichsen, 2010).

38.6 Distribution

See Fig. 38.4.

Chara horrida only occurs in Europe and is here probably restricted to the coastal regions of the southern and central Baltic Sea. It has been found in Denmark, Germany, Poland, Estonia, Finland and Sweden (Blümel, 2004) with the majority of its recent sites in Sweden (Blindow, 2009).

All records outside of this region are doubtful. The species is not reported from Norway (Langangen, 2007). Hollerbach and Krassavina (1983) doubt its occurrence in the Black Sea, and the vouchers seem to be lost. Recent collections from the Black Sea turned out to be tylacanthous and therefore, were determined as *C. baltica* (see Chap. 20). Therefore the occurrence of *C. horrida* in the Black Sea thus is doubtful, and records published from the region should be checked (Romanov et al., 2020). Also, recent records from Scotland (Loch Obasaraigh, North Uist, Outer

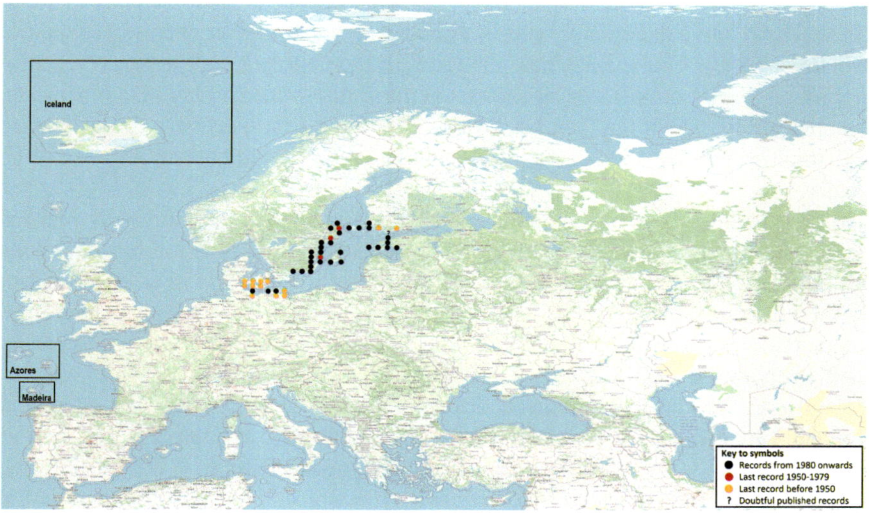

Fig. 38.4 European distribution of *Chara horrida*. For worldwide distribution see Korsch (2018). Black dots are records from 1980 onwards, red dots last record from 1950 to 1979 and yellow dots last record before 1950; doubtful published records (if applicable) are marked with "?" superimposed. Map prepared by Heiko Korsch

Hebrides) have irregular stipulode tiers with additional stipulodes but are isostichous to tylacanthous and therefore not included in *C. horrida* (own data). Hy (1913) reports the species from Giseux (now Gizeux; Etang du Mur, France), an inland site; however no herbarium collection from this site is known. The same is the case for a record from Finistere (France) which is mentioned by Corillion (1975) and included in his distribution map.

38.7 Status and Threats

Chara horrida is assessed as RE (Regionally Extinct) in the national Red Lists of Denmark and Poland, as 1 (threatened with extinction) in Germany, as Endangered in Finland, as NT (Near-Threatened) in the Baltic Sea Red List and the Red Lists of Sweden and Estonia, and as DD (Data Deficient) in Ukraine. It is assessed as NT (Near Threatened) in the European Red List (see Chap. 9).

Blümel (2004) reports a severe decline for the southern Baltic Sea. In Germany, the species is now known only from a limited area in Vorpommern and has disappeared from a number of its former sites. (Fig. 38.4). In Denmark, it has disappeared from all of its former sites (Baastrup-Spohr et al., 2013). While Blümel (2004) and Blindow (2009) also assumed a decline in Sweden, recent investigations have shown that *C. horrida* is rather common. The number of recorded locations (38 before 1930; 10 between 1930 and 1990; 57 between 1980 and 2006; more than 70 between 2010 and 2021) reflects investigation intensity, but also a probable recent increase in abundance. The relative increase of records is far higher than in other charophyte species (Blindow, 2009). The species has increased in the Östergötland province recently (J. Edlund pers. comm.) and is rather common in Sweden today (G. Johansson, presentation in Linköping October 2017). Also the recent discovery of the species in Estonia, where the species was not found before, may indicate an increase of *C. horrida*.

Eutrophication and coastal engineering are given as the most important reasons for decline in Germany (Blümel 2004b). In Sweden, eutrophication has also been assumed as the main threat for *C. horrida*. Boat traffic and digging activities might additionally threaten the species in shallow, protected lagoons which are its preferred habitat (Borgiel, 2007; Blindow, 2009; Johansson and Gustavsson, 2013). Pettersson (2014) reported a decline of *C. horrida* in two coastal lagoons around Gotland, probably due to eutrophication; experiments to re-establish *C. horrida* in several plots are obviously the first of their kind. *Chara horrida* is included in a former and ongoing Swedish action plan for threatened species (Blindow, 2009, Havs-och Vattenmyndigheten, 2020), as Sweden houses the majority of the global population of this species.

Acknowledgements We thank Gustav Johansson for information on water chemistry. Nick Stewart and Klaus van de Weyer gave valuable comments.

References

Blindow, I. (2009). *Åtgärdsprogram för bevarande av hotade kransalger. I. Arter i brackvatten och hav.* Naturvårdsverket (Swedish Environmental Protection Agency). [in Swedish].

Blindow, I., & Erichsen, I. (2010). Cross-fertilisation within the *Chara hispida* species complex. *Charophytes, 2*, 59–65.

Blume, M., Blindow, I., Dahlke, S., & Vedder, F. (2009). Oospore variation in closely related *Chara* taxa. *Journal of Phycology, 45*, 995–1002.

Blümel, C. (2004). *Chara horrida*. In H. Schubert & I. Blindow (Eds.), *Charophytes of the Baltic Sea*. A. R. G. Gantner, Ruggell, pp. 113–121.

Borgiel, M. (2007). *Rassa vikar. Mariobiologisk kartläggning och naturvärdesbedömning*. Länsstyrelsen i Stockholms Län, Rapport, 2, 68 pp. [in Swedish].

Corillion, R. (1957). Les Charophycées de France et d'Europe Occidentale. *Bulletin de la Société Scientifique de Bretagne 32 (Hors série I & II)*, 1–499 [in French].

Corillion, R. (1975). *Flore et végétation du Massif Armoricain. T. IV. Flore des Charophytes (Characées) du Massif Armoricain et des contrées voisines d'Europe occidentale*. Jouve, Paris [in French].

Giesenhagen, K. (1896). Untersuchungen über die Characeen. *Flora, 82*, 381–433. [in German].

Havs- och Vattenmyndigheten (2020). *Åtgärdsprogram för sällsynta kransalger längs kusten*. 57 pp. [in Swedish].

Hollerbach, M. M., & L. K. Krassavina, [Голлербах, М. М. & Л. К. Красавина]. (1983). Определитель пресноводных водорослей СССР. *14. Charophyta [Determination key of freshwater algae of USSR]*. Nauka, Leningrad, 189 pp. [in Russian].

Holtz, L. [no date]. Zeichnungen und Messungen von Characeen Pommerns von Dr. Paul Hauptfleisch, entsprechend meinen Publikationen. Inv. Nr. 5551. University of Greifswald, Germany [in German].

Hy, F. (1913). Les Characées de France. *Bulletin de la Société Botanique de France, 60*(sup 1), 1–47 [in French].

Johansson, G., & Gustavsson, A. (2013). *Raggsträfse i Norra Kalmar Län*. Naturvatten i Roslagen AB, Hydrophyta Ekologikonsult, Report 2013. 62 pp. [in Swedish].

Korsch, H. (2018). The worldwide range of the Charophyte species native to Germany. *Rostocker Meeresbiologische Beiträge, 28*, 45–96.

Krause, W. (1997). Charales (Charophyceae) Süßwasserflora von Mitteleuropa. In H. Ettl, G. Gärtner, H. Heynig & D. Mollenhauer (Eds.), *Süßwasserflora von Mitteleuropa 18*. Fischer, Jena [in German].

Langangen, A. (2007). *Charophytes of the Nordic countries* (p. 102). Saeculum ANS.

Meyer, J., Dahlke, S., Kafka, M., Kerkow, A., Lindner, C., Kube, S., Nawka, B. N., Schubert, H., Schumann, R., & Blindow, I. (2019). Submerged vegetation in a shallow brackish lagoon does not enhance water clarity but offers substantial refuge for zooplankton. *Aquatic Botany, 154*, 1–10.

Migula, W. (1889–1897). Fünfter Band: Die Characeen. In A. Fischer, E. Fischer, F. Hauck, G. Limpricht, C. Luerssen, W. Migula, H. Rehm, Richter P. & G. Winter (Eds.), *Dr. L. Rabenhorst's Kryptogamenflora von Deutschland, Oesterreich und der Schweiz* (ed. 2). Eduard Kummer, Leipzig [in German].

Nowak, P., Schubert, H., & Schaible, R. (2011). Genetic analysis of the Characeae. Report for the year 2010/2011. University of Rostock, 21 pp.

Olsen, S. (1944). Danish Charophyta. *Det Kongelige Danske Videnskabernes Selskab, Biologiske Skrifter, 3*(1), 1–240, pl. 1–2.

Pettersson, M. (2007). *Eftersök av kransalgen Chara horrida i tre gotländska vikar*. Castor & Pollux, Report, 64 pp. [in Swedish].

Pettersson, M. (2014). *Störningsförsök raggsträfse, Chara horrida* 2012–2014. Castor & Pollux, Report, 10 pp. [in Swedish].

Romanov, R., Korolesova, D., Afanasyev, D., & Zhakova, L. (2020). *Chara baltica* (Charophyceae, Charales) from the Black Sea region and taxonomic implications of extrastipulodes. *Botanica, 26*, 126–139.

Sonder, C. (1890). *Die Characeen der Provinz Schleswig-Holstein und Lauenburg nebst eingeschlossenen fremden Gebietstheilen*. Dissertation Universität Rostock, 64 pp. [in German].

Stroede, W. (1931). Ökologie der Characeen. Dissertation Friedrichs-Wilhelm-Universität zu Berlin, Berlin, 119 pp. [in German].

Stroede, W. (1933). Über die Beziehungen der Characeen zu den chemischen Faktoren der Wohngewässer und des Schlammes. *Archiv Für Hydrobiologie, 25*, 192–229. [in German].

van de Weyer, K., & Schmidt, C. (2018). Bestimmungsschlüssel für die aquatischen Makrophyten (Gefäßpflanzen, Armleuchteralgen und Moose) in Deutschland: Band 1: Bestimmungsschlüssel, ed. 2. Landesamt für Umwelt (LfU) Brandenburg. *Potsdam, Fachbeiträge des LfU Brandenburg, 119*, 1–180. [in German].

Vedder, F. (2003). *Morphologie und Taxonomie rezenter und subfossiler Characeen-Oosporen aus der Ostsee*. Diploma thesis, University of Greifswald, Germany, p. 62 [in German].

Wahlstedt, L. J. (1862). *Bidrag till kännedomen om de Skandinaviska Arterna af Växtfamiljen Characeæ*. Lund, i–viii, 1–43 [in Swedish].

Wahlstedt, L. J. (1875). *Monografi öfver Sveriges och Norges characeer*. Bolagets, Christianstad [in Swedish].

Wood, R. D. (1965). *Monograph of the Characeae*. J. Cramer.

Chapter 39
Chara imperfecta

Ingeborg Soulié-Märsche and Serge D. Muller

Scientific name: ***Chara imperfecta*** **A. Braun in Durieu, Expl. Sci. Algérie 1: Pl. 39. 1849(?)**

Holotype (Durieu 1849[?]): pl. 39, Fig. 1

≡ *Chara vulgaris* L. var. *imperfecta* (A. Braun) R. D. Wood in Rev. Charac: 126. 1965, n. inv.

Nomenclatural note: The publication of *Chara imperfecta* was given effect by drawings without accompanying text. It was certainly planned to publish a text but Durieu's publication of the "Exploration scientifique de l'Algérie pendant les années 1840, 1841, 1842" remained fragmentary. Also, we do not know the exact date of the publication of plate 39. Stafleu and Cowan (1976) give "1848/9" as publication date.

Local names: FRENCH: Chara imparfaite.

39.1 Morphology

See Fig. 39.1.

Habitus: Small plants, with internodes three to five times longer than the whorls, with only a few, tiny lateral branches. Whorls form dense heads towards the apex; Incrustation is lacking or weak.

I. Soulié-Märsche (✉) · S. D. Muller
Université Montpellier, Institut des Sciences de l'Évolution (ISEM), Cc 061, Place Eugène Bataillon, 34095 Montpellier Cedex 05, France
e-mail: insouma43@gmail.com

S. D. Muller
e-mail: serge.muller@umontpellier.fr

H. Schubert et al. (eds.), *Charophytes of Europe*,
https://doi.org/10.1007/978-3-031-31898-6_39

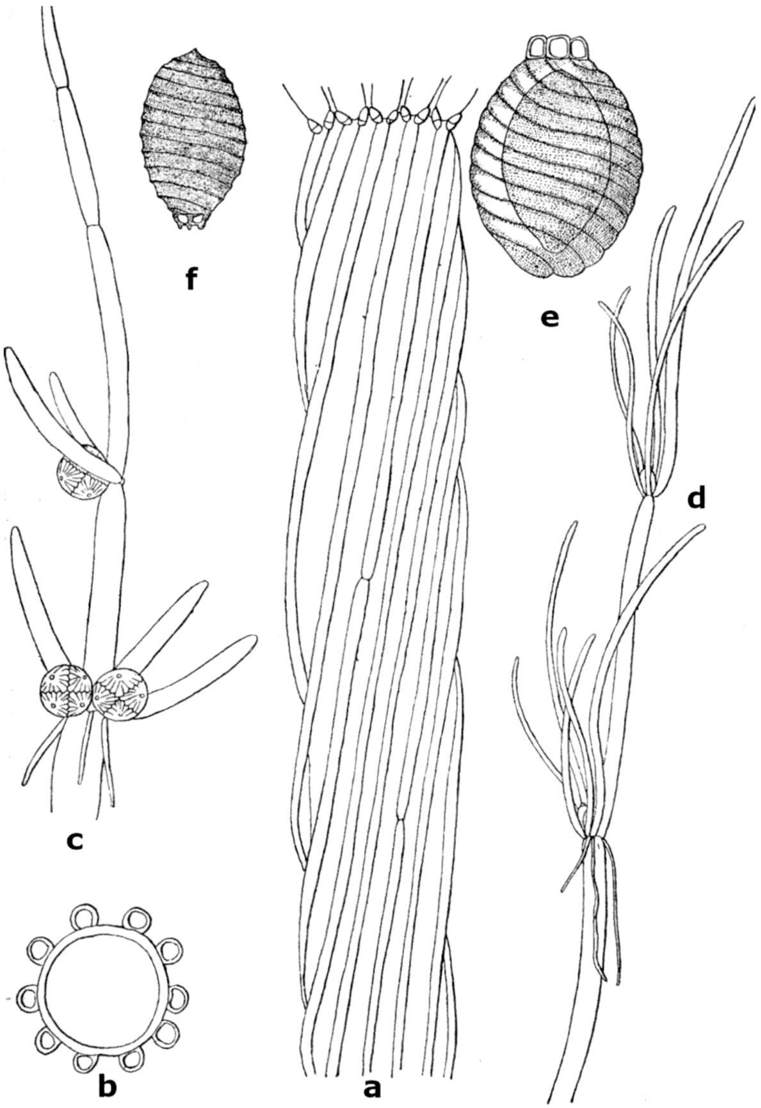

Fig. 39.1 *Chara imperfecta*, morphology. **a** Part of an internode with haplostichous cortication and reduced stipulodes; **b** Cross section of axis; **c** Branchlet of male plant with rudimentary cortication; **d** Branchlet of a female plant showing irregular, incomplete cortication tubes detached from the first branchlet segment; **e** Oospore; **f** Oogonium. Drawings from Migula (1989–1897: Fig. 86)

Size: The plants are about 10–20 (40) cm in length.

Main axis: Main axis diameter is up to 0.6–0.9 mm; internodes are 3–5 times longer than the whorls.

Cortex: The cortex is haplostichous disjoint, meaning the cortical cells do not touch each other. Only primary cortical tubes, directly below the branchlets are present. They leave a space of regular width, without cortication, between the primary cortical cells (Fig. 39.2a, b). Thus, on fresh material, the internodes of the axis appear like a succession of light green and grey green striae, the latter corresponding to the incrusted cortical cells. The cortical cells may be slightly twisted (Migula, 1889–1897; Guerlesquin, 1962). Occasionally, some very short secondary cortical cells appear below the stipulodes, but they do not adhere to the internodal cell.

Spine-cells: Spine cells absent, as the primary cortical cells do not have cortical nodes.

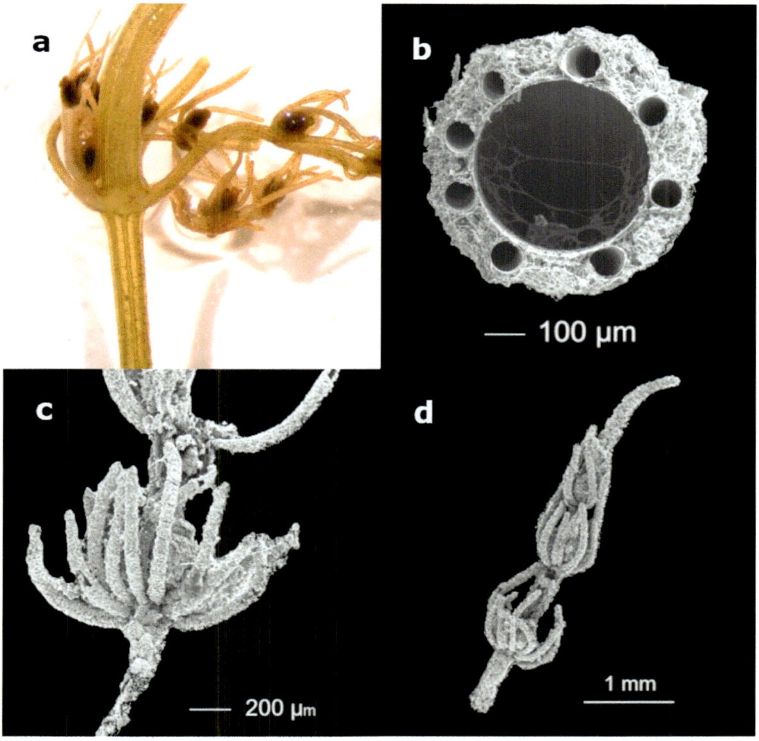

Fig. 39.2 *Chara imperfecta* from Bonne Cougne lake, France. **a** Haplostichous, disjoint cortication (natural light); **b–d** SEM photographs. **b** Cross section of axis showing the haplostichous cortication with lime incrustation in between the cortical cells; **c** Female branchlet node with a bunch of bract-cells hiding two oogonia placed side by side; **d** Branchlet of a female plant with three fertile nodes. Photographs: Ingeborg Soulié-Märsche

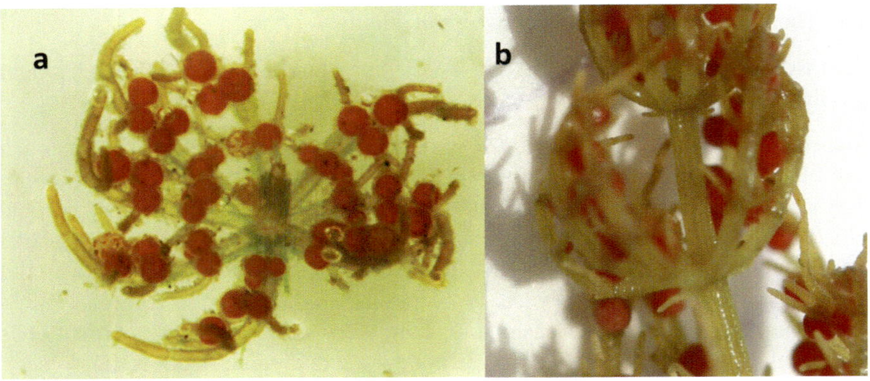

Fig. 39.3 *Chara imperfecta*, male plants. **a** Fertile whorl with triplate antheridia on the first branchlet node from Bonne Cougne lake, France; **b** Male plant showing irregular stipulodes from Hammare pond, Morroco. Photographs: **a**—from Soulié-Märsche (2003); **b**—Serge D. Muller

Stipulodes: Stipulodes are principally in a double row, but poorly and irregularly developed, mostly rudimentary as small knots at the base of the whorls (Fig. 39.1a); exceptionally the upper row shows irregular development of rather long, acute stipulodes (Fig. 39.3b).

Branchlets: The whorls are mostly composed of 8–10 branchlets (phylloids). Branchlets are totally ecorticate; exceptionally, traces of rudimentary cortication were mentioned on the lowermost segment (Migula, 1889–1897; Guerlesquin, 1962). The sterile branchlet segments are mostly longer on the female plants than on the male plants. Branchlets are composed of 5–7 segments. Usually, three branchlet nodes are fertile, followed by 2 or 3 cells without bract-cells. Branchlets end with a rounded, obtuse cell.

Bract-cells: The bract-cells are different on the male and female plants (sexual dimorphism). 2 relatively short bract-cells are present on the branchlet nodes of male plants beside the big antheridia. The branchlet nodes of the female plants are exceptional insofar as they display a bunch of 6–8(10) long bract-cells positioned in front and laterally to the oogonia. These are usually at least twice as long and sometimes up to 3(5) times longer than the oogonia (Fig. 39.2c, d).

Bracteoles: 2, as long as the bract-cells.

Bractlets: Not described.

Sexuality: The species is dioecious and both male and female plants are very fertile. Gametangia are formed on the two to three lower branchlet nodes. Antheridia and oogonia are often geminate or clustered (three placed side by side).

Oogonia: The female plants are more difficult to detect in the field because the small oogonia are hidden behind a bundle of bract-cells (Fig. 39.2c, d). Oogonia are mostly geminate or three together (clustered), exceptionally 6 female fructifications at the same node were seen (Mouronval et al., 2015). Multiple oogonia are arranged horizontally (side by side) at the same node. They are relatively small and of regularly ovoid shape. The coronula is about 250 μm wide and 80 μm high (Table 39.1).

Oospores: Mature oospores are black (Table 39.2). John et al. (1990) found that the oospores of *C. imperfecta* were different from those of *vulgaris*. *Chara imperfecta* forms calcified oospores in the form of small, ovoidal gyrogonites, very similar to *C. vulgaris*.

Antheridia: The male gametangia occur solitary, geminate or up to 3 per branchlet node (Fig. 39.3a, b). As for most dioecious species, the antheridia are large, up to 600–750(800) μm (Table 39.3).

Table 39.1 Oogonia morphometry

Author	Length	Width	Coronula height	Coronula width	Number of helical stripes
Migula (1889–1897)	650–700	450–500	80	250	11–12
Corillion (1957)	650–700	450–500	80	250	11–12
Guerlesquin (1962)	650–700	450–500	–	–	–
Wood (1965)	650–750	450–500	80 (?)	150–250	11–13
Krause (1997)	600–750	450–500	60–80	200–250	11–14
Cirujano et al. (2008)	480–650	280–450	75–85	–	–
Total range	**480–750**	**280–500**	**75–85**	**150–250**	**11–14**

Length in the first column excludes the coronula, its length and width given separately; all values for length and width in μm

Table 39.2 Oospore morphometry

Author	Length	Width	Number of striae	Colour
Braun (1867)	660	–	10	–
Migula (1889–1897)	500–700	350–500	10–12	–
Corillion (1957)	500–700	350–500	9–11	Dark
Guerlesquin (1962)	500–700	350–500	9–11	–
Wood (1965)	500–700	350–500	10–12	Dark brown
Krause (1997)	500–700	–	10–13	Black
Cirujano et al. (2008)	370–650	250–450	–	Dark brown
Total range	**370–660**	**250–500**	**9–13**	

All values for length and width in μm

Table 39.3 Antheridia morphometry

Author	Diameter (μm)
Migula (1889–1897)	**650–800**
Corillion (1957), Guerlesquin (1962) and Muller et al. (2017)	600–750
Wood (1965)	(400)600–800
Krause (1997)	650–800
Cirujano et al. (2008)	500–800
Mouronval et al. (2015)	400–600(800)
Total range	**400–800**

Bulbils: Not known for *Chara imperfecta*

39.2 Phenology and Reproduction

Chara imperfecta is an annual, vernal species. In South France, development starts in February and is already finished in March. The plants decay with rising temperature or through drying out of the habitat. The density of the population depends on the amount of winter precipitation and thus is very irregular from one year to another.

39.3 Habitat Conditions and Ecology

Habitat type: *Chara imperfecta* was found in natural and artificial temporary ponds, and in mountain lakes (Cirujano and Medina, 1994; Muller et al., 2017) but also in an artificial basin and on the shallow shore of a semi-temporary lake (Krause, 1997; Soulié-Märsche, 2003; Mouronval et al., 2015). Its altitudinal distribution spans from lowland to >1300 m a.s.l. The species is also known from permanently flowing water in mountain areas in South Spain (Sierra Nevada) and in irrigation canals with cold freshwater (Krause, 1997). The record from a saline inland lake (Reyes Prósper, 1910) appears doubtful (Cirujano and Medina, 1994; Cirujano et al. (2008).

Sediment: *Chara imperfecta* grows usually on calcareous mud and clay.

Nutrient conditions: There are no reports about nutrient content in the water. Most sites correspond to cold oligotrophic freshwater.

Water chemistry: *Chara imperfecta,* which prefers calcareous freshwater with pH of 7–7.1 close to neutral (Cirujano and Medina, 1994), was categorised as "basiphilous" (Krause, 1997). Several sites in South France had karstic inflow (Emblanc et al., 2002). From three inland ponds in the province of Cuenca, which were observed at different time slots between 1991 and 1993, Cirujano and Medina (1994) reported slightly variable but always low values of CO_3^{-2}, Cl^-, SO^4, Na^+, K^+, Ca^{2+}, Mg^{2+} with a total salt content of maximum 262 mg L^{-1}.

Depth range: *Chara imperfecta* occupies only very shallow parts of its habitats, from 0.2 to 1 m.

Associated species: *Chara imperfecta* forms mostly short-lived monospecific stands. *C. imperfecta* can co-exist with *C. vulgaris* (Corillion, 1957). Its early development coincides also with *Tolypella glomerata*.

39.4 Variability

No special forms or varieties of *Chara imperfecta* have ever been proposed. Guerlesquin (1962) studied two populations in detail: one from a small lowland river (Fuente de Piedra), the second from a small river with running water in the Sierra Nevada (1650 m a.s.l.) but did not find significant differences. These populations, known since Reyes Prósper (1910) and again visited by Krause (1997) proved to be extremely stable.

39.5 Identification and Risk of Confusion

The characters of *Chara imperfecta* are sufficiently distinct to prevent confusion. However, some doubt can appear with respect to species with poorly developed, short rudimentary cortical cells as for instance *C. dissoluta* described by Corillion (1960). The recently described dioecious *Chara lipkinii* from the eastern Mediterranean shows similar habitus but differs by a complete diplostichous tylacanthous cortication (Romanov et al., 2022).

39.6 Distribution

Chara imperfecta was long considered a cryptic species, known only from its type locality in Algeria (Durieu, 1846–1869), and from a few sites in western France (Rochebrune, 1862). Braun and Nordstedt (1882) reported the species from South Spain (Province of Malaga and Granada). At the beginning of the twentieth century it was found near the Atlantic coast in West France (Corillion, 1962; Hy, 1913) where it could not be found recently. Rediscovered in Spain (Comelles, 1984), *Chara imperfecta* was recently identified also in South France (Soulié-Märsche, 2003; Mouronval et al., 2015) as well as in a new locality in Algeria (Muller et al., 2017). It is known from Morocco and Spain (Guerlesquin, 1962; Cirujano et al., 2008, Rodrigo and Alonso-Guillén, 2016), South France (Mouronval et al., 2015) and from one locality

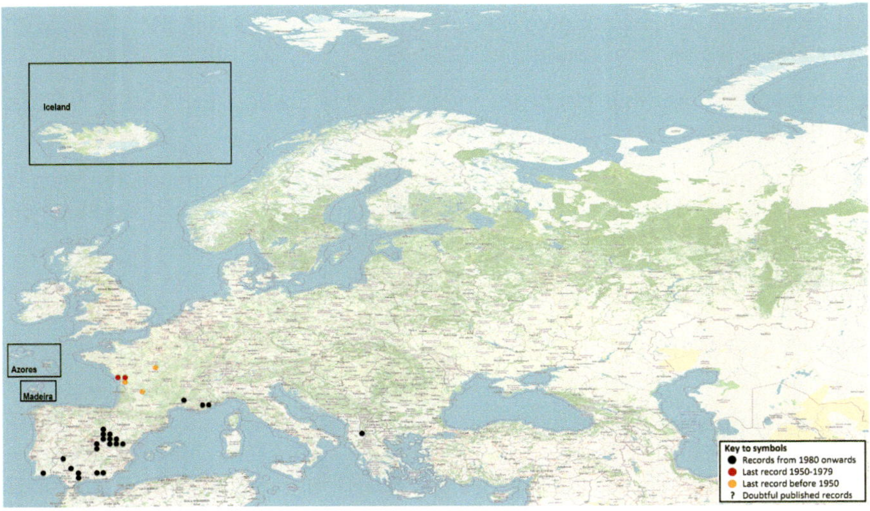

Fig. 39.4 European distribution of *Chara imperfecta*. Black dots are records from 1980 onwards, red dots last record from 1950 to 1979 and yellow dots last record before 1950; doubtful published records (if applicable) are marked with "?"superimposed. Map prepared by Heiko Korsch

in Portugal (Algarve) (Mendes, 1947). One exceptional occurrence was mentioned by Blaženčić et al. (2006) in Macedonia. At present, *Chara imperfecta* appears restricted to the western Mediterranean. The map (Fig. 39.4) summarizes all known localities of *Chara imperfecta* in Europe.

39.7 Status and Threats

Chara imperfecta is a very rare species in Europe. Because of its need for oligotrophic water the species is vulnerable. As a vernal species, its germination strongly depends on sufficient winter precipitation. It has been assessed as Near Threatened in Europe (see Chap. 9) and it should be included on national Red Lists in the future.

Remarks *Chara imperfecta* is one of the rare taxa for which R. D. Wood didn't see good material. Based on the examination of a fragmentary specimen of 3–4 cm (Wood, 1965: 126), he combined *C. imperfecta* as a variety to *C. vulgaris*. His description, measurements and figures were redrawn from Migula (1889–1897). Similar to his considerations for other taxa, Wood did not take into account the fact that *C. imperfecta* is exclusively dioecious. Further, the characteristic features of the position of its gametangia and bract-cells as well as the constantly haplostichous disjoint cortication are arguments that justify its status as a separate species.

References

Blaženčić, J., Stevanović, B., Blaženčić, Z., & Stevanović, V. (2006). Distribution and ecology of charophytes recorded in the west and central Balkans. *Cryptogamie, Algologie, 27*, 311–322.

Braun, A., & Nordstedt, C. F. O. (1882). Fragmente einer Monographie der Characeen. Nach den hinterlassen Manuscripten A. Braun's herausgegeben von Dr. Otto Nordstedt. *Abhandlungen der Königlichen Akademie der Wissenschaften zu Berlin für das Jahr 1882*. [in German].

Braun, A. (1867). *Conspectus Systematicus Characearum Europaeum*. Dresden. [in German; attachment to issue 3 of Rabenhorst's Characeae Exsiccatae.].

Cirujano, S., Cambra, J., Sánchez Castillo, P. M., Meco A., & Flor Arnau, N. (2008). *Flora ibérica. Algas continentales*. Carófitos (Characeae). Real Jardín Botánico, Madrid. [in Spanish].

Cirujano, S., & Medina, L. (1994). Data about the ecology and distribution of *Chara imperfecta* A. Braun in Spain. Anales *Jardín Botánico de Madrid, 52*, 95–98.

Comelles, M. (1984). Noves Citacions de Caròfits a Espanya. *Butletí de la Institució Catalana d'història Natural Secció de Botánica, 51*, 35–39. [in Spanish].

Corillion, R. (1957). Les Charophycées de France et d'Europe Occidentale. *Bulletin de la Société Scientifique de Bretagne 32, Hors série*. [in French].

Corillion, R. (1960). Sur la présence et la signification de formes à cortication incomplète du *Chara contraria* Kützing dans l'Est armoricain. *Bulletin de la Société Mayenne-Sciences, Laval Année 1960*, 49–60. [in French].

Corillion, R. (1962). Nouvelle contribution à l'étude des Charophycées de la péninsule ibérique et du Maroc occidental. *Bulletin de la Société Scientifique de Bretagne, 37*, 65–80. [in French].

Durieu de Maisonneuve, M. C. (1846–1869). *Exploration scientifique de l'Algérie*. Botanique vol. 1. Cryptogamie. Paris [in French].

Emblanc, C., Soulié-Märsche, I., Simler, R., & Thiéry, A. (2002). A karstic inflow versus functioning and biota of a temporary pond in Provence: The case study of the Bonne-Cougne temporary pond (Var, southern France). In F. Carrasco, J.J. Duran & B. Andreo (Eds.), *Karst and Environment*, (pp. 139–143).

Guerlesquin, M. (1962). Variations de la cortication et gymnophyllie chez *Chara imperfecta* A. Br. de la péninsule ibérique. *Bulletin de la Société Scientifique de Bretagne, 37*, 85–90. [in French].

Hy, F. (1913). Les Characées de France. *Bulletin de la Société Botanique de France, 60*, (suppl.1), 1–47. [in French].

John, D. M., Moore, J. A., & Green, D. R. (1990). Preliminary observations on the structure and ornamentation of the oosporangial wall in *Chara* (Charales, Chlorophyta). *British Phycological Journal, 25*, 1–24. https://doi.org/10.1080/00071619000650011

Krause, W. (1997). Charales (Charophyceae). In Ettl, H., G. Gärtner, H. Heynig & Mollenhauer D. (Eds.), *Süßwasserflora von Mitteleuropa* 18. Fischer, Jena. [in German].

Mendes, E. J. (1947). Uma espécie nova para o gen. *Chara* L. *Portugaliae Acta Biologicae*, Séries B, *2*, 286–288. [in Portuguese].

Migula, W. (1889–1897). Fünfter Band: Die Characeen. In A. Fischer, E. Fischer, F. Hauck, G., Limpricht, C., Luerssen, W., W. Migula, H. Rehm, P. Richter & G. Winter (Eds.), *Dr. L. Rabenhorst's Kryptogamenflora von Deutschland, Oesterreich und der Schweiz* (Ed. 2). Eduard Kummer, Leipzig. [in German].

Mouronval, J. B., Baudouin, S., Borel, N., Soulié-Märsche, I., Klesczewski, M., & Grillas, P. (2015). *Guide des Characées de France méditerranéenne*. Office national de la Chasse et de la Faune Sauvage, Paris. [in French].

Muller, S. D., Rhazi, L., & Soulié-Märsche, I. (2017). Diversity and distribution of Characeae in the Maghreb (Algeria, Morocco, Tunisia). *Cryptogamie, Algologie, 38*, 201–251. https://doi.org/10.7872/crya/v38.iss3.2017.201

Reyes Prósper, E. R. (1910). *Las Carofitas de España. Singularmente las que crecen en sus estepas*. Imprenta Artistica Española, Madrid. [in Spanish].

Rochebrune, A. de. (1862). Note sur le *Chara imperfecta* A. Braun. *Bulletin de la Société Botanique de France, 1862*, 336–338. [in French].

Rodrigo, M. A., & Alonso-Guillén, J. L. (2016). Charophyte flora in a Ramsar Mediterranean wetland (Albufera de València Natural Park, Spain) during the period 2007–2010. *Botanica Serbica, 40*, 205–215.

Romanov, R., Barinova, S., Nikulin V. Y. & Gontcharov, A. A. (2022). *Chara lipkinii* (Charales, Charophyceae): A new dioecious Mediterranean species under risk of extinction in the wild and some implications for the taxonomy of the genus *Chara*. *Fottea, 22*, 1–12. https://doi.org/10.5507/fot.2021.012

Soulié-Märsche, I. (2003). *Les Characées de trois lacs du Centre Var (Bonne Cougne—Redon—Gavoti)*. Rapport d'expertise pour le site FR 9301621 du programme NATURA 2000. [in French]

Stafleu, F. A., & Cowan, R. S. (1976). Taxonomic literature. In *A selective guide to botanical publications and collections with dates, commentaries and types*, (Vol. I.) A-G. Bohn, Scheltema & Holkema, Utrecht.

Wood, R. D. (1965). *Monograph of the Characeae*. J. Cramer.

Chapter 40
Chara kirghisorum

Roman Romanov

Scientific name: ***Chara kirghisorum*** **Less. in Linnaea 9: 212. 1835**

Lectotype (Romanov and Abdullin, 2018: 221): Russia. Ural: in riculo [rivulo] Bolshe Mandlibay/[no date]/Lessing (LE).

= *Chara contraria* A. Braun ex Kütz. f. *jubataeformis* Vilh., Izv. Glavn. Bot. Sada S.S.S.R. 29: 589. 1930
Lectotype (Romanov and Abdullin, 2018: 221): Kazakhstan. Ural Governorate [Ural Oblast], Dzhambeyty Uezd: on River Barbastau, in water/27.VII.1927/R. Ju. Roshevitz, N. G. Zhezhel (LE).

Local names: RUSSIAN: Хара киргизская; UKRAINIAN: Хара киргизька.

40.1 Morphology

See Figs. 40.1, 40.2 and 40.3.

Habitus: Plants are light or brownish green, heavily or slightly incrusted, small or moderately high, rough to touch, with (unusual for charophytes) a whip-like appearance with regularly distributed small nodules on the main axis formed by whorls of short and nearly rudimentary branchlets, multiaxial or mostly uniaxial from the base, scarcely branching.

Size: Small to medium plant, from few cm to 20 cm high.

Main axis: Main axis diameter is 285–530 μm, completely corticate. The internodes are several times longer than the branchlets.

R. Romanov (✉)
Nišice Ujtin Potok b.b., Dobra Voda, 85356 Bar, Montenegro
e-mail: streptophytes@gmail.com

H. Schubert et al. (eds.), *Charophytes of Europe*,
https://doi.org/10.1007/978-3-031-31898-6_40

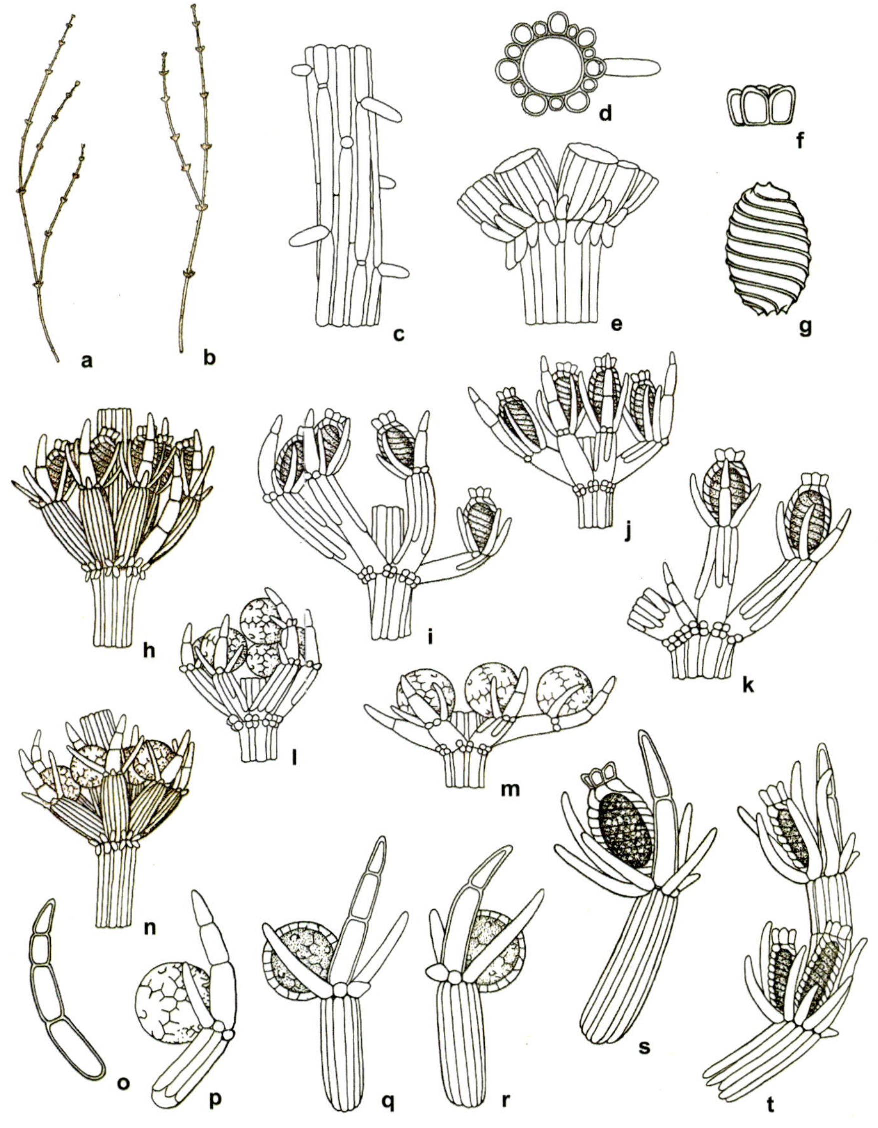

Fig. 40.1 *Chara kirghisorum*. **a** Upper part of male plant; **b** Upper part of female plant; **c** Main axis cortex; **d** Cross-section of main axis; **e** Base of branchlet whorl with stipulodes; **f** Coronula; **g** Oospore; **h–j** Branchlet whorls of female plants with different cortication of branchlet segments; **k, s, t** Branchlet of female plants; **l, m, n** Branchlet whorls of female plants with different cortication of branchlet segments; **o** Nodeless ecorticate branchlet; **p–r** Branchlets of male plants. Drawings: **a–h, n, o, q, r, s, t**—Hollerbach (1941); **i–m, p**—Hollerbach (1949)

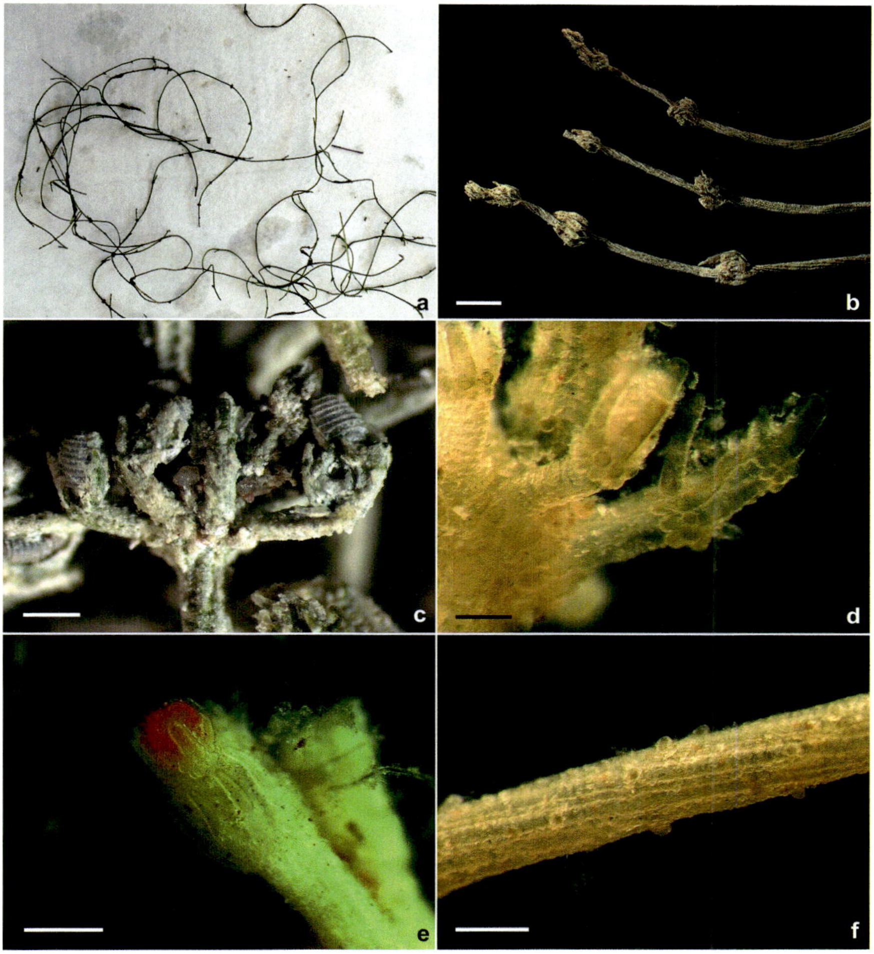

Fig. 40.2 *Chara kirghisorum*, details. **a** General habits; **b** Upper parts; **c** Fertile whorl of female plant; **d** Branchlets with cortex initials; **e** Abbreviated branchlets of male plants; **f** Main axis cortex. Origin of plants: **a**—Taman Peninsula, **b**, **c**—South Kazakhstan, **d**–**f**—South Ural. Scale: **b**—2 mm, **c**, **e**, **f**—0.4 mm, **d**—0.2 mm. Photographs: **a**—Vasily Vishnyakov; **b**–**f**—Roman Romanov

Cortex: Diplostichous, tylacanthous.

Spine-cells: Spine-cells are solitary, as an unusual exception geminate, papillose to elongate with rounded or slightly narrowing apex, significantly shorter than main axis diameter or sometimes approaching it, squarrose.

Stipulodes: Short or slightly elongate, obtuse, similar in appearance to spine cells, in 2 rows, usually uniform in appearance.

Fig. 40.3 Oospores of *Chara kirghisorum*, Aktobe Region, Kazakhstan, scanning electron microscopy. **a** General view; **b** Smooth fossa surface. Scale: **a**—100 μm, **b**—10 μm. Photographs: Roman Romanov

Branchlets: 7–8 in a whorl, slightly or distinctly arcuate, somewhat adpressed to the main axis (especially in the sterile state), 0.8–1.5(2.9) mm length, consisting of one or two nodes and four cells, resulting in a branchlet arrangement of one or a two short basal segments and a two or three-celled undifferentiated ecorticate part. If nodes are missing whole branchlet is ecorticate and four-celled. Sometimes nodes have only cortex initials developed to different degrees resulting in formation of subgymnophyllous branchlets with incomplete cortication of their segments. The end cell is short and conical, not mucronate.

Bract-cells: Abaxial ones are invariably short, adaxial bract-cells are somewhat shorter or nearly equal to gametangia, rarely somewhat longer of antheridium.

Bracteoles: Similar with to bract-cells.

Bractlets: Somewhat shorter or nearly equal to oogonium.

Sexuality: Dioecious. Gametangia are solitary or rarely geminate at one or two basal nodes of the branchlet. Sterile plants occur frequently.

Oogonia: Ellipsoid, (410)540–640(675) μm in length without coronula and (265)330–385(470) μm in width, with 10–13 convolutions of spiral cells, with low but sharp ridges. The coronula is 84–150 μm in length and 165–235 μm in width, thick-set, slightly or evidently spreading. The species produces gyrogonites; their lime shell is thin, wide and ellipsoid with 10–11 spiral crests and sharp ridges, 500–542 μm in length, 413–457 μm in width, without basal cage or claws.

Oospores: Dark brown to black, ellipsoid, (370)420–470(485) μm in length and (200)220–285(335) μm in width, with 9–11 low sharp spiral ridges, obviously tuberculate with big irregular elevations appearing as indistinct parallel rows in LM (Hollerbach and Krassavina, 1983), smooth to pustulate in SEM (R. Romanov own obs.).

Antheridia: (370)405–590 μm in diameter (rewetted pressed specimens, Hollerbach and Krassavina, 1983), 370–420 μm in northern populations (wet specimens, R. Romanov own obs.).

Bulbils: Starchless multicellular nodal bulbils are known for some populations.

40.2 Phenology and Reproduction

The species mainly reproduces by oospores, when bulbils are present vegetative reproduction could be expected. Northern populations are frequently sterile; therefore, their reproduction by fragmentation seems to be possible.

40.3 Habitat Conditions and Ecology

Habitat type: The ecology of *Chara kirghisorum* is poorly known (Hollerbach, 1941, 1949; Dobrokhotova, 1952, 1953; Kostin, 1987; Sviridenko and Sviridenko, 2014; LE!; R. Romanov own obs.); it grows mainly in lakes, rarely in rivers with low discharge that are slow-flowing or nearly stagnant during the summer-autumn low level periods, sometimes in freshwater and brackish inundated excavations and artificial water bodies. Northern populations are restricted to mesotrophic freshwater or, as an exception, brackish lakes with circumneutral waters and enough transparency. In Europe it is known from few karst and single tectonic lakes, floodplain water bodies in delta and brackish seaside delta lakes only (Romanov and Abdullin, 2018;

R. Romanov & V. Vishnyakov unpubl. data). It seems to be unable to grow in truly temporary water bodies. It frequently grows at low light intensity and probably could withstand direct wave action in lakes.

Sediment: *Chara kirghisorum* occurs on sandy and oozy bottom sediments with low, moderate or high content of fine organic particles.

Nutrient conditions and water chemistry: It seems to be absent in highly eutrophic water bodies. In eutrophic water bodies it can grow in shallow water only (R. Romanov own obs.). Therefore, eutrophication may be the main threat for the species at least in its northern distribution. The species was found in alkaline fresh or sometimes brackish water with total dissolved solids 0.4–8.2 g per littre, pH (6.1)7.9–9.45 at different combinations of prevailing ions, i.e., sulfates, chlorides and hydrocarbonates depending on the salinity values, but usually with sodium plus potassium as major ions (Kostin and Shoyakubov, 1974; R. Romanov unpubl. data).

Depth range: From few centimeters to 2–4 m.

Associated species: *Chara kirghisorum* mostly grows in charophyte-rich lakes in monodominant stands or most commonly in association with other species of *Characeae*: *Chara altaica, C. aspera, C. canescens, C. contraria, C. globata, C. globularis, C. papillosa, C. tomentosa, Nitellopsis obtusa, Nitella hyalina*. The composition of individual associations evidently depends on salinity values. Further investigations using the presence of *C. kirghisorum* and its vitality as an indicator for the sustainable existence of other charophyte species are required.

40.4 Variability and Risk of Confusion

Chara kirghisorum is easily recognisable at the field by its distinct habit, which is really unusual for charophytes due to very short branchlets adpressed to the main axis resulting in a knot-like appearance of the whorls. These are widely distributed along the thin main axis due to long internodes. Therefore, it can be easily overlooked. The length of the plant is the most variable trait of the species; it could be few centimeters only or sometimes exceeds 20 cm. The cortication of branchlets is variable from nearly gymnophyllous to completely corticate. The shallow water form is usually multiaxial from the base, whereas plants from deeper sites are uniaxial.

Chara kirghisorum is identical in general appearance with *C. filiformis* but can be easily differentiated in fertile state, being dioecious in contrast with monoecious *C. filiformis*. Moreover, there is a signifcant gap in their distribution. *Chara filiformis* is mainly a Central-European species unknown east of 29 °E in Pskov Oblast of Russia and Belarus and south of 46 °N in France and Switzerland (see Chap. 29), whereas *C. kirghisorum* is mainly a Central-Asian species whose westernmost distribution ranges from at 37°E in the North Black Sea region to a northernmost occurrence at 55°N in South Ural (Romanov and Abdullin, 2018; V. Vishnyakov & R. Romanov unpubl. data).

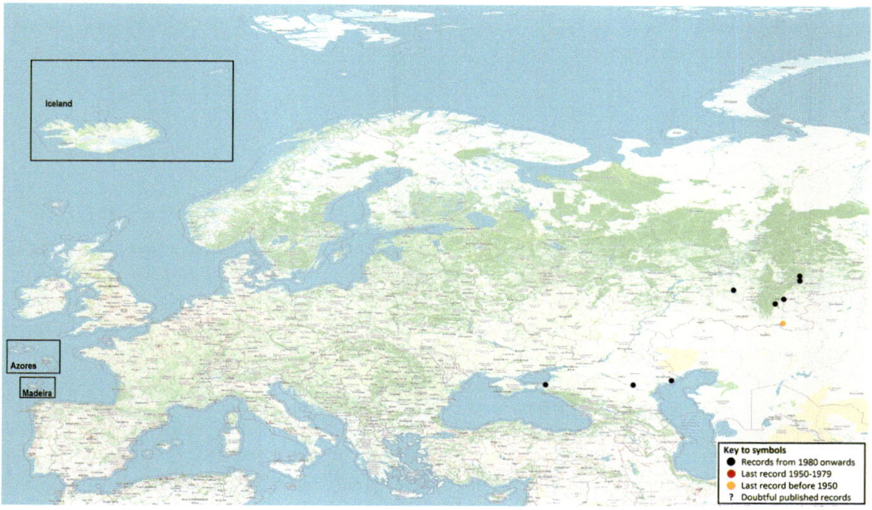

Fig. 40.4 Distribution of *Chara kirghisorum* in Europe. Black dots are records from 1980 on, red dots from 1950 to 1979 and yellow dots before 1950; unconfirmed/doubtful published records (if applicable) are marked with "?" superimposed. Map prepared by Heiko Korsch

40.5 Distribution

See Fig. 40.4.

South-Eastern Europe: North-Eastern Black Sea region, Kumo-Manych Depression, Volga Delta, Southern Ural (Russia), Northern Asia: south-west of West Siberia (Russia), Central Asia: Kazakhstan, Uzbekistan, Turkmenistan, Western Asia: Iran (Hollerbach, 1941, 1949; Shoyakubov, 1979; Hollerbach and Krassavina, 1983; Ahmadi et al., 2012; Romanov and Abdullin, 2018, 2023; Sviridenko et al., 2018; Veisberg and Isakova, 2018; V. Vishnyakov & R. Romanov unpubl. data, ALTB!, LE!, UFA!, W!). The total number of its localities in the world does not exceed 20, including a few recent records from Iran, Kazakhstan, and Russia (Ahmadi et al. 2012; Sviridenko and Sviridenko, 2014; Romanov and Zhamangara, 2017; Kosterin and Ahmadi, 2018; Romanov and Abdullin, 2018, 2023; Sviridenko et al., 2018; Veisberg et al., 2018; R. Romanov & V. Vishnyakov unpubl. data; LE!).

40.6 Status and Threats

Chara kirghisorum has been assessed as VU (Vulnerable) in the European Red List (see Chap. 9). It has been suggested for inclusion in the Red Data Book of Russian Federation as a rare stenobiont species at the border of its distribution range (Romanov and Abdullin, 2018).

Acknowledgements I am grateful to Vasily Vishnyakov for sharing his distributional records, images, specimens of charophytes.

References

Ahmadi, A., Riahi, H., Sheidai, M., & van Raam, J. (2012). Some charophytes (Characeae, Charophyta) from central and Western of Iran including *Chara kohrangiana* species nova. *Cryptogamie, Algologie, 33*, 359–390. https://doi.org/10.7872/crya.v33.iss4.2012.359

Dobrokhotova, K. V. [Доброхотова, К. В.]. (1952). Ботаническая характеристика водоемов нижней части дельты реки Или [Botanical characteristic of water bodies in low part of delta of Ili River]. *Труды Всесоюзного гидробиологического общества [Proceedings of the All-Union Hydrobiological Society], 4,* 260–275. [in Russian].

Dobrokhotova, K. V. [Доброхотова, К. В.]. (1953). Харовые водоросли в ценозах гидромакрофитов [The charophytes in coenoses of hydromacrophytes]. *Труды Всесоюзного гидробиологического общества [Proceedings of All-Union Hydrobiological Society], 5,* 258–263. [in Russian].

Hollerbach, M. M., & Krassavina, L. K. [Голлербах, М. М. & Красавина, Л. К.]. (1983). *Определитель пресноводных водорослей СССР. 14. Charophyta* [*Determination key of freshwater algae of USSR. 14. Charophyta*]. Nauka. [in Russian].

Hollerbach, M. M. (1941). Res nova de specie rara *Chara kirghisorum* Lessing. *Notulae Systematicae e Sectione Cryptogamica Instituti Botanici nomini V.L. Komarovii Academiae Scientiarum URSS, 5*, 37–46. [in Russian and Latin].

Hollerbach, M. M. (1949). De *Chara jubata* A. Br. et *Chara contraria* A.Br. f. *jubataeformis* Vilh. in florae URSS notula. *Notulae Systematicae e Sectione Cryptogamica Instituti Botanici nomini V.L. Komarovii Academiae Scientiarum URSS, 6*, 59–65. [in Russian].

Kosterin, O. E., & Ahmadi, A. (2018). Odonata observed in Central Zagros, Iran, in late May 2017. *International Dragonfly Fund Report, 117*, 1–65.

Kostin, V. A. & Shoyakubov, R. S. [Костин, В. А. & Шоякубов, Р. Ш.]. (1974). Распределение и биомасса харовых водорослей в озере Балхаш [The distribution and biomass of charophytes in Lake Balkhash]. In *Пятая конференция по споровым растениям Средней Азии и Казахстана, 26–27 апреля 1974 г., 1 [The fifth conference for cryptogamous plants of Central Asia and Kazakhstan, 26–27 April 1974. Abstracts, 1].* (pp. 182–183). [in Russian].

Kostin, V. A. [Костин, В. А.] (1987). Материалы к изучению экологии харовых водорослей водоемов Или-Балхашского бассейна [Materials for the study of the ecology of charophytes in reservoirs of the Ili–Balkhash basin]. *Ботанические материалы гербария Института ботаники Академии Наук Казахской ССР [Botanicheskie Materialy Gerbariya Instituta Botaniki Akademii Nauk Kazakhskoy SSR], 15*, 128–133 [in Russian].

Romanov, R. E., & Abdullin, S. R. (2018). *Chara kirghisorum* (Charales): Lectotypification, first reliable record in Europe and update of species distribution in Russia. *Phytotaxa, 362*, 220–226. https://doi.org/10.11646/phytotaxa.362.2.8

Romanov, R. E., & Abdullin, Sh. R. (2023). *Charophytes* (Charales, Characeae) of the republic of Bashkortostan (South Ural). *Botanicheskii Zhurnal, 108*(7), 628–640. [in Russian]. https://doi.org/10.31857/S0006813623070050

Romanov, R. E., & Zhamangara, A. (2017). Pre-Symposium field excursion report: Results in context of regional charophyte knowledge. *IRGC News, 28*, 11–13.

Shoyakubov, R. S. [Шоякубов, Р. Ш.] (1979). *Харовые водоросли Узбекистана [Charophytes of Uzbekistan].* Fan. [in Russian].

Sviridenko, T. V., & Sviridenko, B. F. (2014). Distribution, ecology and phytoindication profile of *Chara kirghisorum* (Charophyta) in the West Siberian Plain and the Kazakh Uplands. In

Proceedings of 13th International Scientific and Practical Conference on Problems of Botany of South Siberia and Mongolia. (pp. 187–192). [in Russian].

Sviridenko, B. F., Sviridenko, T. V., Murashko, Y. A., Efremov, A. N. & Tokar, O. E. (2018). New data on locations and ecology of Charophyta in the West Siberian Plain. In *Proceedings of the IVth All-Russian conference on Algae: Problems of taxonomy, ecology and use in monitoring.* (pp. 387–392). [in Russian].

Veisberg, E. I., & Isakova, N. A. (2018). Addition to the flora of charophytes and xanthophytes (Charales, Vaucheriales) in Chelyabinsk Region (South Ural, Russia). *Turczaninowia, 21*, 47–54. [in Russian]. https://doi.org/10.14258/turczaninowia.21.2.6

Chapter 41
Chara neglecta

Roman Romanov

Scientific name: ***Chara neglecta*** **Hollerb. in Novosti Sist. Nizsh. Rast. 18: 3. 1981**

Holotype (Hollerbach, 1981: 3): USSR: insula Kujuk-Tup in Sivasch Occidentali, laculus subsalsus in parte insulae boreali, 0.25 m prof./19 Aug. 1936/A. I. Proschkina-Lavrenko 36 (17) (LE A0001553).

Local names: RUSSIAN: Хара незамеченная, UKRAINIAN: Хара непомічена.

41.1 Morphology

See Figs. 41.1, 41.2, 41.3 and 41.4.

Habitus: The plants are light green to grey green, more or less evenly slightly encrusted, scarcely branching with smooth to clearly spiny appearance due to different length of stipulodes, spine-cells and bract-cells, with nearly completely corticate branchlets. Male plants are thicker, more robust and rougher in comparison with female ones (Fig. 41.3).

Size: It is a medium plant, up to 20–30 cm high.

Main axis: The main axis is 350–670 μm in diameter in the case of female plants, up to 850 μm in case of male plants, flaccid, completely corticate. The internodes are more or less equal in length to the branchlets, shortened in the upper parts and elongated in the lowermost ones, usually 20–25 mm length.

R. Romanov (✉)
Nišice Ujtin Potok b.b., Dobra Voda, 85356 Bar, Montenegro
e-mail: streptophytes@gmail.com

H. Schubert et al. (eds.), *Charophytes of Europe*,
https://doi.org/10.1007/978-3-031-31898-6_41

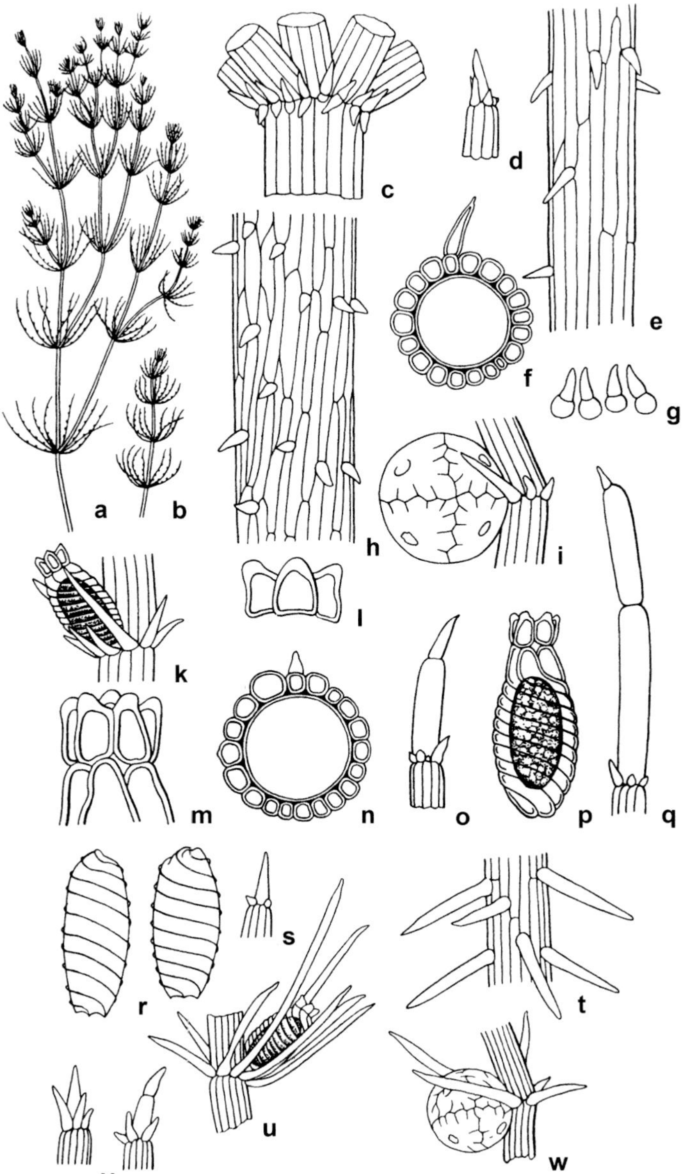

Fig. 41.1 *Chara neglecta*. **a** Part of female plant; **b** Apical part of male plant; **c** Base of branchlet whorl; **d**, **o**, **q**, **s**, **v** Ends of branchlets; **e**, **t** Diplostichous main axis cortex; **f, n** Transverse sections of main axis; **g** Stipulodes; **i**, **w** Fertile nodes of male plant; **k**, **u** Fertile nodes of female plant; **l** Coronula; **m** Upper part of oogonium with coronula; **p** Oogonium; **r** Oospores. Drawings: **a**–**s** from holotype, **t**–**w** from paratype, collected in West Kazakhstan. Drawings: Hollerbach (1981)

Fig. 41.2 Holotype of *Chara neglecta* (LE: A0001553). **a** Male plants; **b** Female plants. Photograph: Roman Romanov

Cortex: The axial cortex is irregular, mostly diplostichous, up to diplo-triplostichous, but never perfectly triplostichous; isostichous or nearly so, sometimes indistinctly aulacanthous and more rarely even indistinctly tylacanthous.

Spine-cells: The spine-cells are always solitary, sparse, bluntly pointed, variable in length, from papillate to elongated, exceeding main axis diameter.

Stipulodes: The stipulodes are diplostephanous, two pairs next to each branchlet, pointed, mostly of moderate length, rarely elongated, the lower row sometimes papillate.

Fig. 41.3 General appearance of *Chara neglecta:* spiny form from the steppe between Azov and Caspian seas. **a** Whole plants; **b**, **c** Apical parts. More delicate and smaller female plants nearly inconspicuous in comparison with male ones are indicated with arrowheads. Photographs: Roman Romanov

Branchlets: The branchlets are 7–8 in moderately spreading regular whorls, more or less arcuate, or nearly straight, more or less clustering at the apex because of shortened upper internodes, 0.5–1 cm length, consisting of (2)3–5(6) segments with a diplostichous cortex (the number of cortical rows is twice the number of bract-cells) and a short ecorticate part. The segments between fertile nodes of female plants are

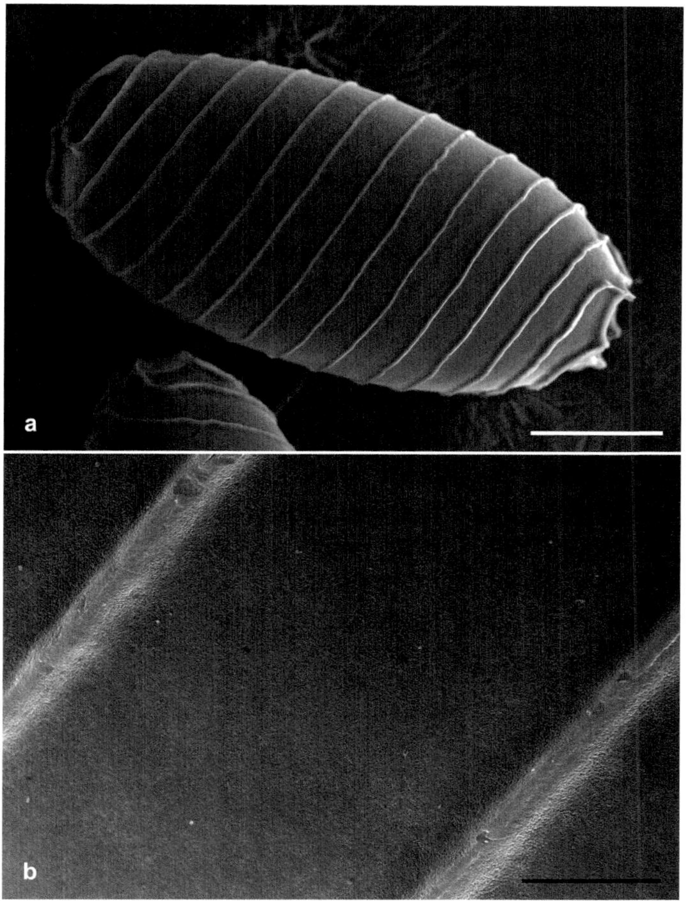

Fig. 41.4 Oospores of *Chara neglecta:* paratype, West Kazakhstan, scanning electron microscopy. **a** General view; **b** Oospore surface. Scale: **a**—100 μm, **b**—10 μm. Photographs: Roman Romanov

usually abbreviated in comparison with upper ones, which results in densely placed oogonia inside the whorl having a filled appearance at the base. The ecorticated part is usually equal to the neighboring corticate segment, two-celled, rarely short, single-celled or greatly elongated and three-celled. Its end-cell is always pointed, conical.

Bract-cells: The bract-cells are well developed, conical, pointed, nearly equal to the oogonium, abaxial ones and adaxial ones 1.5–2 times shorter or all of them 2 times longer on female plants; in male plants usually the 2 adaxial ones are shorter than the antheridium but still exceeding the abaxial ones are in length or, alternatively (rarely) adaxial ones and abaxial ones are 2 times longer than the antheridium.

Bracteoles: The bracteoles are usually 1.5–2 times shorter than, sometimes equal to the oogonium.

Bractlets: Evidently shorter or longer than the oogonium.

Sexuality: This is a dioecious species. Gametangia are always solitary at the 3–5 lowermost nodes of the branchlet.

Oogonia: The oogonia are ellipsoid, sometimes elongated into a neck at the apex. The coronula is straight or weakly crown-like loosely placed at the apex of the spiral cells. Coronula cells become narrow toward the apex, with thickened cell walls at the apices. Table 41.1 gives an overview about published morphometric data.

Oospores: The oospores are narrowly ellipsoid, with low spiral ridges. Oospore surface is brownish-olive, finely punctate in LM, roughened in SEM (Fig. 41.4). Table 41.2 gives an overview of the morphometric data available.

Antheridia: The antheridia are octoscutate. Table 41.3 gives an overview of the published morphometric data.

Bulbils: Bulbils are unknown.

Table 41.1 Oogonia morphometry

Author	Length	Width	Coronula length	Coronula width	Number of helical stripes
Hollerbach (1981), rewetted pressed specimens	(570)600–700(750)	(285)320–350(450)	(100)115–150	185–200(230)	11–12
Hollerbach and Krassavina (1983), rewetted pressed specimens	(570)600–700(750)	(285)320–350(450)	(100)115–150	185–200(230)	11–12
Own data, rewetted pressed specimens	630–722	282–377	76–172	158–241	13(14)
Total range	**570–750**	**282–450**	**76–150**	**158–241**	**11–14**

Length in the first column is excluding the coronula, its length and width given separately; all values for length and width in μm

Table 41.2 Oospore morphometry

Author	Length	Width	Number of striae	Colour
Hollerbach (1981), rewetted pressed specimens	435–500(600)	(185)200–240(350)	10–11(12–13)	Brownish black
Hollerbach and Krassavina (1983), rewetted pressed specimens	435–500(600)	(185)200–230(350)	10–11(12–13)	Brownish black
Own data, rewetted pressed specimens	568–654	254–310	12–15	Mostly black
Total range	**435–654**	**185–350**	**10–15**	

All values for length and width in μm

Table 41.3 Antheridia morphometry

Author	Diameter (μm)
Hollerbach (1981), rewetted pressed specimens	(550)670–840(870)
Hollerbach and Krassavina (1983), rewetted pressed specimens	(550)670–840(870)
Total range	**550–870**

41.2 Phenology and Reproduction

The phenology of the species is poorly known. All plants collected during summer are fertile (LE!; own data).

41.3 Habitat Conditions and Ecology

Habitat type: Desalinated sea bays, brackish continental shallow lakes and ponds, bays and nearshore of inland hyposaline seas (Black, Azov and Caspian seas) in semi-arid and arid continental environment are the main habitats of *C. neglecta* (Raspopov, 1959; Hollerbach and Krassavina, 1983; Sadogurskiy and Belich, 2005; Gromov, 2012; Romanov et al., 2014; Sviridenko and Sviridenko, 2016; own data).

Sediment: The species was collected from silty, clayish and sandy sediments (Raspopov, 1959; Sadogurskiy and Belich, 2005; Gromov 2012; own data).

Nutrient conditions and water chemistry: The environmental variables are poorly known. *C. neglecta* is associated with brackish waters, having a salinity of several g/l and high content of chlorides. It is able to grow in mesotrophic and eutrophic water bodies.

Depth range: *Chara neglecta* is a shallow water species, growing not deeper than 1 m (Raspopov, 1959; Sadogurskiy and Belich, 2005; Gromov, 2012; own data).

Associated species: *Chara neglecta* is able to grow in association with *Lamprothamnium papulosum, Myriophyllum spicatum, Ruppia maritima, Stuckenia pectinata, Vaucheria* sp., *Zannichellia pedunculata, Zostera noltii*, sometimes forming a dense bottom layer or in pure dense stands (Raspopov, 1959; Sadogurskiy and Belich, 2005; Gromov, 2012; LE!; own data).

41.4 Variability

The high variability of morphological traits noted in description of *C. neglecta* (Hollerbach, 1981) is expressed mainly in the arrangement of the axial cortex, length of spine-cells and bract-cells, and the shape of oospores, from almost cylindrical to broadly ellipsoid.

41.5 Identification and Risk of Confusion

Forms of *Chara neglecta* having a smooth appearance are similar in habit to *C. connivens*. The bigger antheridia, sexual dimorphism, i.e. more robust male plants (in comparison with female ones and with *C. aspera*), and absence of white rhizoidal bulbils allow it to be differentiated from *C. aspera*. It differs from *C. connivens* with more or less, and put a comma after developed to well-developed pointed spine-cells, from both *C. connivens* and *C. galioides* by having mostly diplostichous main axis cortex, sometimes weakly aulacanthous and never perfectly triplostichous, and smaller antheridia (although the values are slightly overlapping), from *C. uzbekistanica* by having diplostephanous stipulodes and usually longer spine-cells.

41.6 Distribution

See Fig. 41.5.

Chara neglecta a scattered scattered but not really wide distribution range. It is known from South-Eastern Europe: Sivash Bay, northern shore of the Black Sea, east shore of the Azov Sea, steppes between the Azov Sea and the Caspian Sea, eastern of the Volgograd Reservoir, between Volga and Ural rivers, and the west coast of the Caspian Sea (Raspopov, 1959, 1961; Hollerbach, 1981; Hollerbach and Krassavina, 1983, Palamar-Mordvintseva, 1998, Sadogurskiy and Belich, 2005; Gromov, 2012; Hollerbach and Palamar-Mordvintseva, 1991; Borisova et al., 2016; LE!; own data). Several localities are known outside of Europe: south-western shore of Caspian

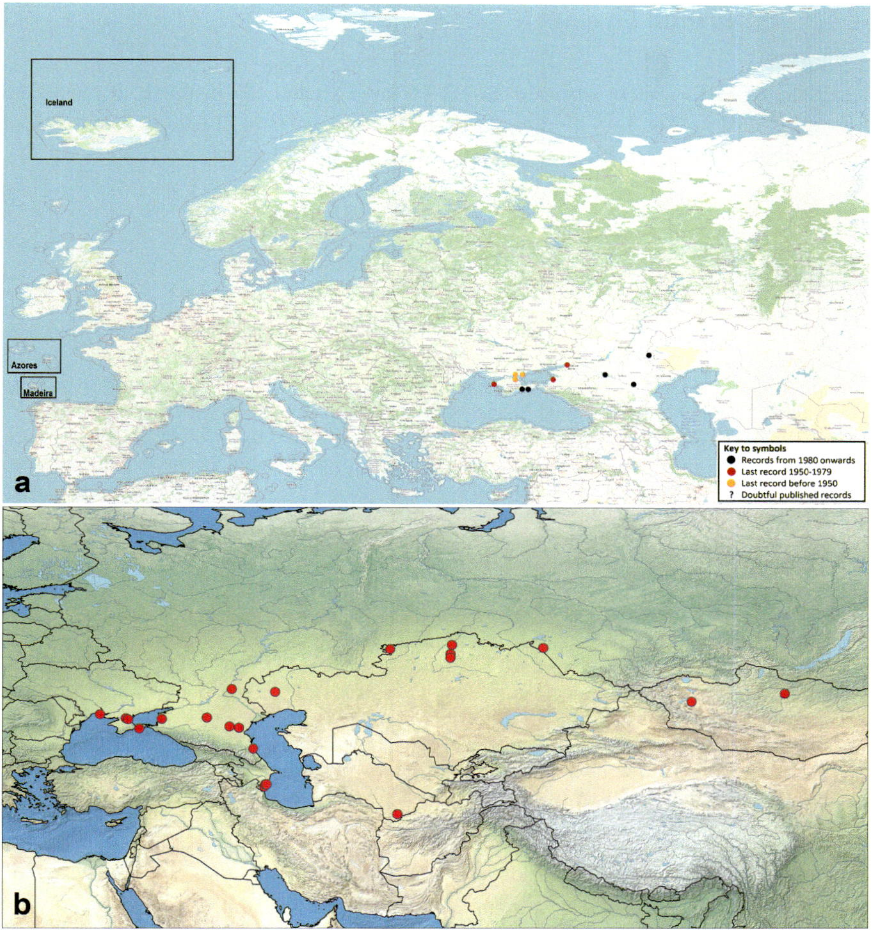

Fig. 41.5 Distribution of *Chara neglecta*. **a** European distribution, **b** Total distribution range. Symbols map a: Black dots are records from 1980 on, red dots from 1950 to 1979 and yellow dots before 1950; unconfirmed/doubtful published records (if applicable) are marked with "?" superimposed. Maps: **a**—prepared by Heiko Korsch, **b**—prepared by Roman Romanov

Sea (Azerbaijan, Dagestan) (Hollerbach, 1981; Romanov et al., 2023; LE!), North Kazakhstan (Sviridenko and Sviridenko, 2016), Mongolia (Romanov et al., 2014), Turkmenistan and south of West Siberia (own data). The total number of localities in the world for *C. neglecta* does not exceed 25, including 6 recent sites from Russia (own data).

41.7 Status and Threats

Chara neglecta has been assessed as NT (Near Threatened) in the European Red List (see Chap. 9). The loss of habitats because of long-lasting drought as a result of climate change seems to be an important threat for this species.

References

Borisova, O. V., Palamar-Mordvintseva, G. M., & Tsarenko, P. M. [Борисова, О. В., Паламар-Мордвинцева, Г. М., & Царенко, П. М.]. (2016). *Флора водоростей України. Том 12. Харофітові водорості. Вип. 2.* [*Flora algarum Ucrainicae. Charophyta, 12(2)*]. Kholodny Institute of Botany of the National Academy of Science of Ukraine. [in Ukrainian].

Gromov, V. V. (2012). Aquatic and coastal vegetation of the northern and western coast of the Azov Sea. *Journal of Siberian Federal University, Biology, 2*, 121–137. [in Russian]. https://doi.org/10.17516/1997-1389-0140

Hollerbach, M. M. & Palamar-Mordvintseva, G. M. [Голлербах, М. М. & Паламар-Мордвинцева, Г. М.]. (1991). *Визначник прісноводних водорослей України. IX. Харові водорості (Charophyta)* [*The identification manual of freshwater algae of Ukraine: IX. Charophytes (Charophyta)*]. Naukova Dumka. [in Ukrainian].

Hollerbach, M. M., & Krassavina, L. K. [Голлербах, М. М. & Красавина, Л. К.]. (1983). *Определитель пресноводных водорослей СССР. 14. Charophyta* [*Determination key of freshwater algae of USSR. 14. Charophyta*]. Nauka. [in Russian].

Hollerbach, M. M. (1981). De specie nova *Chara neglecta* Hollerb. (Charophyta). *Systematicae Plantarum non Vascularium, 18*, 3–9. [in Russian].

Palamar-Mordvintseva, G. M. [Паламарь-Мордвинцева, Г. М.]. (1998). Charophyta Крымского полуострова (Украина). [Charophyta of the Crimean Peninsula (Ukraine)]. *Algologia, 8*, 14–22. [in Russian].

Raspopov, I. M. [Распопов, И. М.]. (1959). Зарастание озёр Карташево и Степаново из группы Камыш-Самарских озёр [The vegetation cover of lakes Kartashevo and Stepanovo from the group of Kamysh-Samarskie lakes]. *Botanicheskii Zhurnal (Moscow & Leningrad), 44*, 860–865. [in Russian].

Raspopov, I. M. [Распопов, И. М.]. (1961). Геоботанический очерк озёр Карташево и Большое Степаново из группы Камыш-Самарских озёр [The geobotanical overview of lake Kartashevo and Bolshoe Stepanovo of the group of Kamysh-Samarskie lakes]. In *Озёра Нижнего Поволжья и Арало-Каспийской низменности. Труды Лаборатории озероведения АН СССР. Т. XIV.* [*The lakes of Lower Volga Region and Aral-Caspian lowland. The proceedings of the laboratory of limnology AS USSR. Vol. XIV*]. (pp. 206–217). [in Russian].

Romanov, R. E., Zhakova, L. V., Bazarova, B. B., & Kipriyanova, L. M. (2014). The charophytes (Charales, Charophyceae) of Mongolia: a checklist and synopsis of localities, including new records. *Nova Hedwigia, 98*, 127–150.

Romanov, R. E., Mallaliev, M. M., Nikulin, V. Yu., Gontcharov, A. A., & Barinova, S. S. (2023). The charophytes (Charophyceae, Characeae) from Dagestan aquatic habitats, North Caucasus: biogeographical and barcoding perspectives. *Environments, 10(9)*, 153. https://doi.org/10.3390/environments10090153

Sadogurskiy, S. Ye., & Belich, T. V. (2005). Macrophytobenthos in the shore aquatic complex at the Cape Chauda (Black Sea, Crimea). *Algologia, 15*, 181–194. [in Russian].
Sviridenko, T. V., & Sviridenko, B. F. (2016). *Харовые водоросли Западно-Сибирской равины. Монография [Charophyta of the West Siberian Plain: Monograph]*. Amphora. [in Russian].

Chapter 42
Chara oedophylla

Ingeborg Soulié-Märsche, Amina Daoud-Bouattour, and Serge D. Muller

Scientific name: ***Chara oedophylla*** **Feldm.-Maz. [as Feldmann] in Bull. Soc. Hist. Nat. Afrique N. 36, 168. 1945.**

Holotype (Feldm.-Maz., 1945: 168): dans une mare pres de l'oued Tindja, en Tunisie/ 1926/Mme L. Gauthier (P).

= *Chara vulgaris* L. var. *oedophylla* (Feldm.-Maz.) R. D. Wood, Rev. Charac.: 114. 1965, n. inv.

Taxonomic notes: *Chara oedophylla* was established as a new species by Feldmann (1945) based on fragmentary herbarium material (at PC) collected at Oued Tinja in north-eastern Tunisia in 1926 by Gauthier-Lièvre. Feldmann described and showed the main characters of this taxon i.e. the disjoint (sejoined) gametangia with big solitary antheridia and the swollen bract-cells reminiscent of an oedema, a feature that inspired the species name to Feldmann. Wood (1965) designated this material as the lectotype of a variety of *C. vulgaris* and mentioned that he could examine only inadequate immature specimens.

The same type of plants was found nearly 60 years later at several localities in Spain by Comelles (1981) who followed Wood (1965) regarding the taxon as a variety of *C. vulgaris,* adopted also by Cirujano et al. (2008). New finds in Tunisia, close to the type locality, revealed dense and homogenous populations of *C. oedophylla.*

I. Soulié-Märsche (✉) · S. D. Muller
Université Montpellier, Institut des Sciences de l'Évolution (ISEM), Case courrier 061, Place Eugène Bataillon, 34095 Montpellier Cedex 05, France
e-mail: insouma43@gmail.com

S. D. Muller
e-mail: serge.muller@umontpellier.fr

A. Daoud-Bouattour
Département de Biologie, Faculté Des Sciences de Tunis, Université de Tunis El Manar, Campus Le Belvédère, 2092 Tunis, Tunisia
e-mail: daoudamina200@gmail.com

H. Schubert et al. (eds.), *Charophytes of Europe*,
https://doi.org/10.1007/978-3-031-31898-6_42

These collections allowed a more precise definition of the differences with respect to *C. vulgaris* and led to the reconsideration of *C. oedophylla* as a separate species (Soulié-Märsche and Muller, 2016).

Local names: ENGLISH: Swollen-leaved stonewort; FRENCH: (not usual) *Chara* à rameaux renflés.

42.1 Morphology

See Fig. 42.1.

Habitus: Plants small to medium sized, with spreading whorls and usually many branches; sometimes heavily incrusted.

Size: 10–40(60) cm

Main axis: Main axis diameter 0.5–1 mm; internodes 2–3 times longer than the whorls are wide, about 2 cm long in the middle part of the main axis.

Cortex: The cortex is diplostichous, strongly aulacanthous.

Fig. 42.1 Characteristic features of *Chara oedophylla*. **a** Apex of plant showing dense short branches and bract-cells; **b** Antheridia and oogonia on different nodes (sejoined gametangia) accompanied by short, inflated lateral bract-cells; **c** Antheridia solitary, oogonia geminate or clustered with a fan of swollen bract-cells; **d** Aulacanthous cortication; **e** Aulacanthous cortication and branchlet ending with a short ecorticate cell. Plants from Garâa Sejenane, Tunisia. Photographs and drawings: Serge D. Muller

Spine-cells: Spine-cells sparse and solitary, 75–125 μm long, obtuse.

Stipulodes: Stipulodes in two rows, 2 per branchlet, short and obtuse.

Branchlets: The whorls are mostly composed of 6–8 branchlets. Branchlets are short (0.5 cm long), generally composed of 4–5 corticate segments bearing gametangia, followed by 1–2 inflated ecorticate cells and a very short conical, obtuse distal tip. In contrast to *C. vulgaris*, the ecorticate part of the branchlets of *C. oedophylla* seems to be always short in Europe.

Bract-cells: Bract-cells are typically, but not always, short and large, inflated, swollen, also described as "hinchadas" (Cirujano et al., 2008), ending with a tiny acuminate tip. Length of the bract-cells is variable. The development of the bract-cells is different on the male and female nodes. Nodes with antheridia show only one pair of bract-cells. The female nodes are accompanied by 3–4 pairs of lateral bract-cells forming a fan on each side of the oogonium. The abaxial bract-cells are rudimentary.

Bracteoles: 2, as long as and similar to the bract-cells, inflated, acuminate similar to the anterior bract-cells.

Sexuality: *Chara oedophylla* is a monoecious species, characterised by predominantly sejoined gametangia. The branchlet nodes bear either antheridia or oogonia: only 3–5% of all fertile nodes present conjoined gametangia with one antheridium below the oogonium. One to 4 nodes of the branchlets are fertile. The same branchlet can bear any combination of oogonia and antheridia such as for instance one solitary antheridium on the two lower nodes and 2–3 oogonia on the third and fourth node. Up to 4 oogonia at the same node is not unusual. No geminate antheridia were seen.

Oogonia: Oogonia are mostly geminate or clustered; sometimes up to 4 female fructifications occur at the same node. In contrast to the geminate oogonia of other *Chara* species, these oogonia are vertically arranged. Even when only two oogonia are present, they are not side by side but one below the other. Average size is about 700 × 500 μm (Table 42.1).

Oospores: Mature oospores are dark brown to black (Cirujano et al., 2008, see also Table 42.2), without basal claws. *Chara oedophylla* forms calcified oospores in the form of ellipsoidal gyrogonites with concave spiral cells, similar to *C. vulgaris*. Measurements of 50 gyrogonites indicate a size range of 640–760 μm for length, and 420–540 μm for width (Soulié-Märsche and Muller, 2016) (Fig. 42.2).

Antheridia: The antheridia are solitary in 97% of the fertile male nodes. They occur mostly on the lower branchlet nodes. Their large size (600–800 μm) corresponds to that of a dioecious species (Table 42.3). In the rare cases where they are coupled with an oogonium, they are significantly smaller (390 μm; Guerlesquin, 1960). Measurements of the antheridia from 25 plants collected at the French locality Bonne Cougne in May 2018 covered the range of 640–900 μm (Soulié-Märsche unpubl.)

Table 42.1 Oogonia morphometry

Author	Length	Width	Coronula length	Coronula width	Number of helical stripes
Feldmann (1945)	800	500	150	100	10–12
Guerlesquin (1960), immature	450–545	235–325	65–90	135–165	8–10
Comelles (1981)	700–750	350–510	–	110–115(200)	–
Cirujano et al. (2008)	600–800	350–500	–	100–160	–
Soulié-Märsche and Muller (2016)	600(700)800	440(500)560	100	225	–
Romanov et al. (2019)	< 800	–	–	–	–
Total range	**600–800**	**350–560**	**100–150**	**100–225**	**10–12**

Length in the first column excludes the coronula, its length and width given separately; all values for length and width in μm

Table 42.2 Oospore morphometry

Author	Length	Width	Number of striae	Colour
Feldmann (1945)	450	300	7–9	
Guerlesquin (1960), immature	270–350	195–255	7–8	
Comelles (1981)	450–575	400–450	7–9	Brown
Cirujano et al. (2008)	400–500	300–350	–	Dark brown
Total range	**400–575**	**300–450**	**7–9**	

All values for length and width in μm

Fig. 42.2 SEM photographs of *Chara oedophylla*. **a** Gyrogonite and oospore; **b** Incrusted terminal node of branchlet. Plants from Bonne Cougne lake, France. Photographs; Ingeborg Soulié-Märsche

Table 42.3 Antheridia morphometry

Author	Diameter (μm)
Feldmann (1945)	600–650
Guerlesquin (1960)	(390)560–665
Comelles (1981)	600–800
Cirujano et al. (2008)	600–800
Soulié-Märsche and Muller (2016)	600–800
Muller et al. (2017)	600–800
Romanov et al. (2019)	742–830
Total range	**560–830**

Bulbils: Not known for *Chara oedophylla.* Romanov et al. (2019) mention the presence of nodal bulbils on the lower parts of the plants.

42.2 Phenology and Reproduction

Chara oedophylla is an annual species forming abundant oospores and gyrogonites in spring. The plants continue growing as long as the pond holds water.

42.3 Habitat Conditions and Ecology

Habitat type: *Chara oedophylla* grows in shallow temporary ponds and at the edge of semi-permanent lakes.

Sediment: *Chara oedophylla* grows usually on calcareous mud and clay.

Nutrient conditions: No data are available about nutrient content in the water.

Water chemistry: *Chara oedophylla* grows in freshwater and in slightly saline ponds (Cirujano et al., 2008; Comelles, 1982). Conductivity reported from two Spanish sites was between 2.69 and 7.5 mS/cm (Espinar et al., 1997). A single water analysis from a Sicilian pond showed NaCl = 6.3%, conductivity = 3.38 mS/cm, TDS = 1.69 g/l (Romanov et al., 2019).

Depth range: 0.5–2 m; the plants decline with diminishing water depth; still green in 12 cm (Espinar et al., 1997).

Associated species: In Spain, *Chara oedophylla* was reported growing in the same ponds as *C. aspera*, *C. galioides* and *Tolypella glomerata* (Espinar et al., 1997). In South France, the species was found together with *C. contraria*, *C. globularis*, *Nitella tenuissima* and *Tolypella glomerata* (Soulié-Märsche, 2003). Finally, in Tunisia, *C.*

oedophylla forms often monospecific populations, but mixed populations with *C. aspera*, *C. connivens*, *C. galioides*, *C. vulgaris* and *Tolypella hispanica* were also found (Hammami, 2018).

42.4 Variability

The species is clearly defined by its sejoined gametangia. Variation concerns mainly the development of the bract-cells. On male nodes, the length of the bract-cells can vary from as long as the antheridium diameter to three times longer. The bract-cells on female nodes vary in length and number, from 2 to 4 pairs of bract-cells and can be more or less swollen.

42.5 Identification and Risk of Confusion

The diplostichous aulacanthous cortication creates links from *Chara oedophylla* to *C. vulgaris*. Due to the large solitary antheridia, specimens could be wrongly considered as dioecious. However, the general habitus with the spread-out whorls and the large conspicuous antheridia allow the distinction in the field. In contrast to *C. vulgaris*, the ecorticate part of the branchlets is generally very short, with an obtuse end cell. Detailed comparison of *C. oedophylla* and *C. vulgaris* through hierarchical cluster analysis, based on numerous (> 100) countings and measurements of both species, showed highly significant statistical differences of the number and length of the bract-cells and the diameter of the antheridia (Hammami, 2018).

42.6 Distribution

Chara oedophylla appears to be a typically Mediterranean taxon (Fig. 42.3). The species occurs frequently in Tunisia where it was first described (Feldmann, 1945). Recently, it was found in 10 sites in north-eastern Tunisia, in Garâa Sejenane (Mogods, northwest), on the Cap Bon peninsula, as well as in the eastern central region and the south up to Ghomrassen near Tataouine (Hammami, 2018, unpubl. data). The species was also recorded from one locality in Morocco (Guerlesquin, 1960). In Europe, *C. oedophylla* was first known from five dispersed localities in Spain (Cirujano et al., 2008; Comelles, 1981). It was later described from two temporary lakes in South France (Soulié-Märsche, 2003; Mouronval et al., 2015) and more recently from a locality in Sicily (Romanov et al., 2019). The species was also identified on an herbarium specimen collected in 1939 in Portugal (M. Rodrigo pers. comm.). A disjoint occurrence was identified on recent collection from the Kalmykia region, southern Russia (R. Romanov pers. comm.).

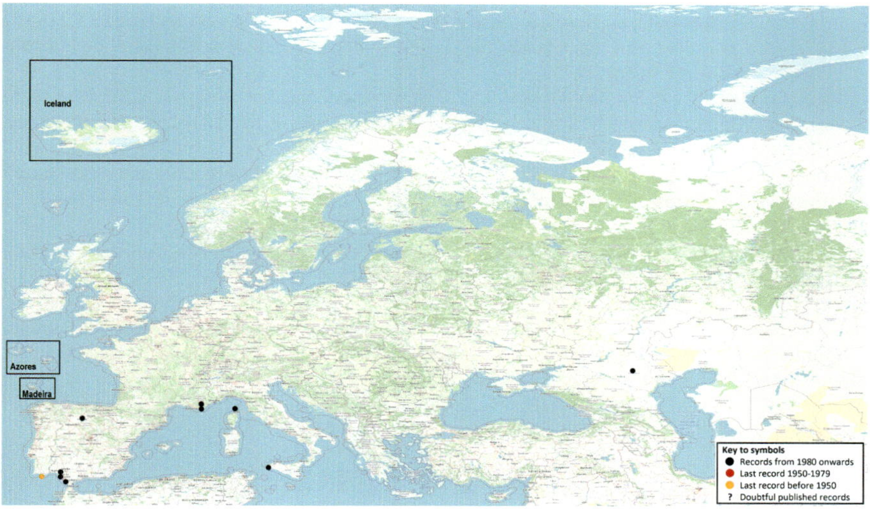

Fig. 42.3 Distribution of *Chara oedophylla* in Europe. Black dots are records from 1980 onwards, red dots last record from 1950 to 1979 and yellow dots last record before 1950; doubtful published records (if applicable) are marked with "?"superimposed. Map prepared by Heiko Korsch

42.7 Status and Threats

In Europe, *Chara oedophylla* appears to be a very rare species and for this reason it warrants particular conservation interest. The French site is defined as a special area of conservation, included in the "Natura 2000" Network.

References

Cirujano, S., Cambra, J., Sánchez Castillo, P. M., Meco, A., & Flor Arnau, N. (2008). *Flora ibérica. Algas continentales. Carófitos (Characeae).* Real Jardín Botánico, Madrid. [in Spanish].

Comelles, M. (1981). Contribució al coneixement de les carofícies d'Espanya. *Collectanea Botanica, 12*, 97–103. [in Spanish].

Comelles, M. (1982). Noves localitats i revisió de la distribució de les espècies de caròfits a Espanya. *Facultad de Biología, Universidad Central de Barcelona* [in Spanish].

Espinar, J. L., Garcia, L. V., Garcia Murillo, P., & Toja, J. (1997). Submerged macrophyte zonation in a Mediterranean salt marsh: a facilitation effect from established helophytes? *Journal of Vegetation Science, 13*, 831–840.

Feldmann, G. (1945). Deux nouvelles espèces de *Chara* de l'Afrique du Nord. *Bulletin de la Société d'Histoire Naturelle de l'Afrique du Nord, 36*, 168–173. [in French].

Guerlesquin, M. (1960). Contribution à l'étude des variations de la répartition des organes reproducteurs chez les Charophycées. *Bulletin de la Société Scientifique de Bretagne, 35*, 273–284. [in French].

Hammami, I. (2018). Contribution à l'étude des Characées de Tunisie. *Mémoire de Mastère, Spécialité Écologie évolutive, Faculté des Sciences de Tunis.* [in French].

Mouronval, J. B., Baudouin, S., Borel, N., Soulié-Märsche, I., Klesczewski, M., & Grillas, P. (2015). *Guide des Characées de France méditerranéenne.* Office national de la Chasse et de la Faune Sauvage, Paris. [in French]

Muller, S. D., Rhazi L., & Soulié-Märsche, I. (2017). Diversity and distribution of Characeae in the Maghreb (Algeria, Morocco, Tunisia). *Cryptogamie, Algologie, 38*, 201–251. https://doi.org/10.7872/crya/v38.iss3.2017.201

Romanov, R., Napolitano, T., van de Weyer K., & Troia, A. (2019). New records and observations to the Characean flora of Sicily (Italy). *Webbia, 74*, 111–119. https://doi.org/10.1080/00837792.2019.1609258

Soulié-Märsche, I. (2003). *Les Characées de trois lacs du Centre Var (Bonne Cougne – Redon – Gavoti).* Rapport d'expertise pour le site FR 9301621 du programme NATURA 2000, [in French].

Soulié-Märsche, I., & Muller, S. D. (2016). New finds of *Chara oedophylla* Feldmann in Tunisia: Significance of sejoined gametangia. *Botanica Serbica, 40*, 229–236.

Wood, R. D. (1965). *Monograph of the Characeae.* J. Cramer.

Chapter 43
Chara oryzae

Roman Romanov

Scientific name: ***Chara oryzae*** **R. E. Romanov, V. Y. Nikulin & A. A. Gontch. in R. E. Romanov et al. in Eur. J. Phycol. 57: 329. 2022**

Holotype (Romanov et al., 2022): 3): SPAIN: autonomous community of Valencia, Natural Park Albufera de Valencia, municipality (town) of Sueca (Valencia), near El Palmar village, rice fields, 39° 18′ 25.2 ″N 00° 19′ 16.4″ W, collected as spreading plants from wet soil after water drainage and in thin water film/20 September 2017/ R. E. Romanov (LE A0000351).

Local names: Unknown.

43.1 Morphology

See Figs. 43.1 and 43.2.

Habitus: Plants are green, slightly lime incrusted, moderately branched. No rhizoid formation was found.

Size: *Chara oryzae* is a medium-sized plant, up to 27.5 cm high.

Main axis: Main axis diameter is (320)353–630 μm; it is thin, flexible. The internodes at most parts of the plant are longer than branchlets; they are only significantly shorter than the internodes in the apical parts (Fig. 43.1a).

Cortex: Regularly isostichous triplostichous, sometimes slightly tylacanthous at higher magnification, completely covering the main axis (Fig. 43.1b, c). The secondary tubes joined mostly slightly obliquely, infrequently overriding each other.

R. Romanov (✉)
Nišice Ujtin Potok b.b., Dobra Voda, Bar 85356, Montenegro
e-mail: streptophytes@gmail.com

H. Schubert et al. (eds.), *Charophytes of Europe*,
https://doi.org/10.1007/978-3-031-31898-6_43

Fig. 43.1 *Chara oryzae*, isotype, Spain. **a** Upper part of a plant with typical arcuate curved branchlets; **b** Triplostichous slightly tylacanthous main axis cortex with solitary papillate spine cells (arrowhead); **c** Triplostichous slightly tylacanthous main axis cortex with strip of diplostichous cortex (arrowhead); **d** Base of branchlet whorl with gametangia appearing hidden inside whorl; **e–h** Base of branchlet whorl with haplostephanous bistipulate stipulodes and different length of branchlet segments and their cortication pattern, main axis cortex with invisible spine cells (single short acute spine cell is visible at **h**); **i–k** Incomplete cortication of branchlet segments resulting in oblique ecorticate part (arrowheads), wide (**i**) or narrow (**j**); **k** Rings of main axis cortex initials around branchlet nodes (arrowheads); **l** Basal part of branchlet with incomplete diplostichous branchlet cortex; **m** Base of branchlet with long internodes, cortex initials and conjoined gametangia; **n** Ends of branchlets; **o**: Triangular plates of antheridium. Scale: **a**—2 mm, **b**—100 μm, **c**, **j**, **l**, **m**, **o**—200 μm, **d**, **n**—1 mm, **e–i**, **k**—400 μm. Photographs: Roman Romanov

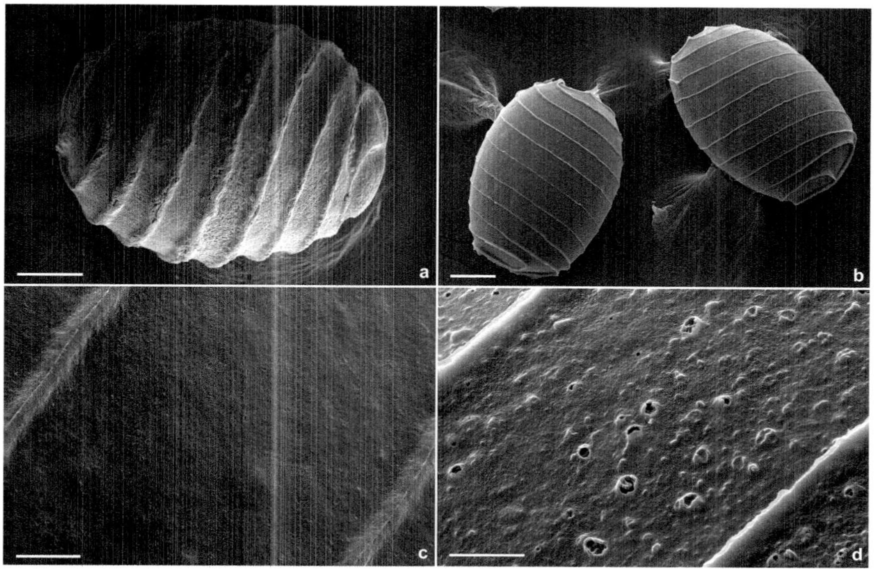

Fig. 43.2 SEM of gyrogonite and oospores of *Chara oryzae*, isotype. **a** Gyrogonite; **b** Two oospores, left one with roughened surface (magnified shown in **c**), right one with scarcely pustulate surface (magnified shown in **d**). Scale: **a**, **b**—100 μm, **c**, **d**—10 μm. Photographs: Roman Romanov

Spine-cells: Spine-cells are always solitary, short, papillose, nearly rudimentary, rarely with a pointed apex, and are usually inconspicuous at low magnification, sometimes up to 70 μm in length.

Stipulodes: Always haplostephanous, 2 per branchlet (bistipulate), short (not rudimentary) or of moderate length, not exceeding the basal segment of the branchlets, usually clearly shorter than the basal branchlet segment, gradually pointed to acute apex, commonly incurved upright, and adpressed to the basal segment of the branchlet, 150–373 × 60–84 μm (Fig. 43.1d–h).

Branchlets: 10 per whorl, 1.4–4.63 cm in length, 202–438 μm in diameter. Branchlets at upper whorls are arcuately curved (Fig. 43.1a), gametangia appear hidden inside the whorls, which are all upswept, hiding the uppermost whorls. The uppermost branchlet whorls are in-curved over the apex where plants are growing in very shallow places or persisting in wet soil after water drawdown from the habitat. The corticate part of the branchlets is less or significantly less than 20–30% of its total length (Fig. 43.1a). Each branchlet has two to four, usually three nodes. The two lowermost segments are commonly abbreviated and short.

The cortex of the branchlet segments is frequently incomplete (i.e. covering only part of the segment), which is usually most conspicuous in the basal segment. The cortex is mostly clearly recognisable at low magnification in the two lowest ones branchlet segments, rare in the third one (Fig. 43.1). The branchlet cortex is diplostichous. The lowest row of branchlet cortex initials can be haplo-diplostichous due to the rare formation of solitary cortical filaments instead of the common pattern of filament formation between bract-cells. The cortical filaments extend from the nodes but commonly do not join together and do not cover the segment completely. The branchlet cortical initials are present at all branchlet nodes, at least as a short band composed of upper and lower rows of short cells, 59–90 × 55–72 μm, equal to each other and forming a complete ring at the branchlet node. The degree of individual cortex tube development is different if they are elongated (Fig. 43.1i). Usually, the downward growing filaments are longer or significantly longer than the upward growing ones within the same segment. The cortication of the basal segment consists of only the filaments growing downward from the first node, which agrees with the only known pattern for the genus *Chara*. The filaments originating from the lowermost node sometimes reach or almost reach the base of a branchlet. The terminal pluricellular segment of the branchlet (without nodes) is ecorticate. It constitutes the majority of the branchlet length, always being many times longer than the other segments (Fig. 43.1a). It consists of 2–4 long cells and short end cell. The end cell is short and conical but neither tiny nor mucronate, 113–420 × 78–150 μm (Fig. 43.1n).

Bract-cells: Unilateral. The anterior ones are commonly from one third as long to nearly equal, or slightly longer than oogonium length (if the oogonia are unripe) or significantly shorter, gradually pointed, 221–540 μm in length. The posterior and lateral are always rudimentary but frequently pointed, ca. 66–79 μm in length (Fig. 43.1d, m). They are commonly nearly half the length of ripe oogonia.

Bracteoles: Similar in appearance and length to anterior bract-cells.

Sexuality: Monoecious. Gametangia are always conjoined, solitary, at first, second, third, and fourth nodes (Fig. 43.1d–f). Sometimes they are not formed at the third node, which can have cortex initials and bract-cells.

Oogonia: Elliptical, without a distinct neck below the coronula, (386)426–690 μm in length without coronula and (285)372–550 μm in width. The coronula is 109–236 μm in length and 161–260(290) μm in width. Gyrogonites are ellipsoidal, 405–545 μm in length and 310–442 μm in width, isopolarity index (100 × L/w)—120–138, with 8–10 convolutions visible in lateral view separated with low spiral crests (Fig. 43.2). Spiral cell calcification is nearly always concave. The apex is indistinctly pointed or rounded; the base is rounded, without a basal cage. The gyrogonite is thin and sometimes incomplete (i.e. absent at the apical part).

Oospores: Black or dark brown, ellipsoidal, 311–490 μm in length and 204–375 μm in width, with 9–10 convolutions separated by low spiral ridges, without a basal cage or claws (Fig. 43.2). Oospore surfaces appear smooth and are very sparsely and irregularly punctulate when observed via light microscopy. In SEM, the oospore wall is roughened and finely, irregularly pitted (Fig. 43.2c), except for a single oospore having scarce and irregular pustules, in addition to a roughened surface (Fig. 43.2d).

Antheridia: Octoscutate, 270–378 μm in diameter.

Bulbils: Unknown.

43.2 Phenology and Reproduction

Chara oryzae seems to be annual, richly fertile at least before rice harvest in September (own data).

43.3 Habitat Conditions and Ecology

Habitat type: Water bodies of rice fields.

Sediment: Inundated soils of rice fields.

Nutrient conditions and water chemistry: Poorly known, associated with warm fresh waters, evidently able to grow in eutrophic water bodies.

Depth range: Shallow, ca. 0.5–10 cm.

Associated species: In association with *Oryza sativa*, *Chara braunii* and *C. vulgaris* (own data).

43.4 Variability, Identification and Risk of Confusion

Chara oryzae is clearly distinct from all other species on the basis of its morphology and genetic affinity (Romanov et al., 2022), including all European species having haplostephanous stipulodes. It has corticate branchlets in contrast with *C. braunii, C. benthamii, C. "fibrosa* sensu lato" and *C. pelosiana*, having ecorticate branchlets. Its partially corticate short branchlet segments allow its differentiation from *C. uzbekistanica* and the haplostephanous morphotype of *C. strigosa* (*C. locuples*) with completely corticate branchlets.

43.5 Distribution

See Fig. 43.3.

South-Western Europe: rice fields in the Natural Park Albufera de Valencia, Spain (own data), and Mediterranean France (as *C.* cf. *chrysospora*; Mouronval et al., 2015). The total number of its localities in the world does not exceed three, but this species should be expected at rice fields and perhaps in other temporary water bodies in other Mediterranean regions. Targeted searches in rice fields, distant from each other, in the south of European Russia, i.e. outside of Mediterranean, during summer 2021 were unsuccessful (own data).

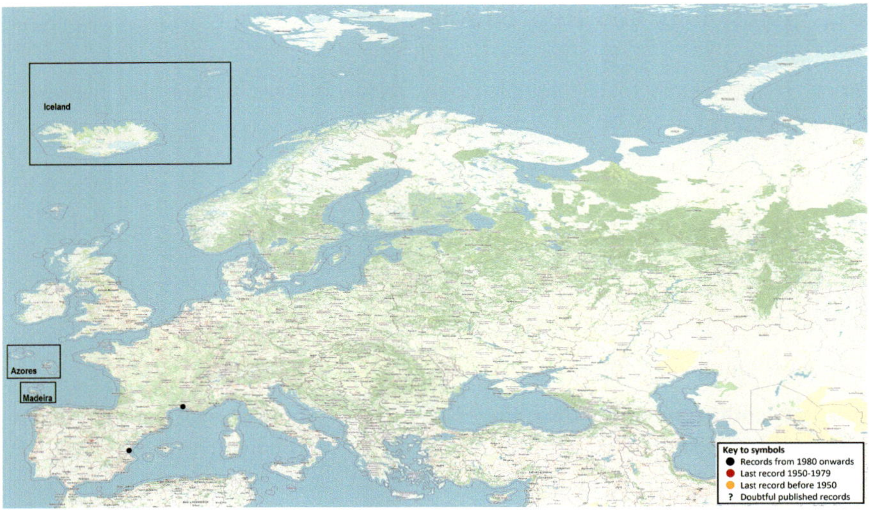

Fig. 43.3 Distribution area of *Chara oryzae*. Black dots are records from 1980 on, red dots from 1950 to 1979 and yellow dots before 1950; unconfirmed/doubtful published records (if applicable) are marked with "?" superimposed. Map prepared by Heiko Korsch

43.6 Status and Threats

Chara oryzae has been assessed as VU (Vulnerable) in the European Red List (see Chap. 9). Although it is known only from a few European rice fields and is assumed to be a non-native species in this region, it is not known elsewhere in the world and has been given a status until its native origins are discovered.

Acknowledgements Michelle T. Casanova is kindly acknowledged for valuable suggestions and English improvement.

References

Mouronval, J. B., Baudouin, S., Borel, N., Soulié-Märsche, I., Klesczewski, M., & Grillas, P. (2015). *Guide des Characées de France méditerranéenne*. Office national de la Chasse et de la Faune Sauvage, Paris, p. 1–211 [in French].

Romanov, R. E., Nikulin, A. Y., Nikulin, V. Y., & Gontcharov, A. A. (2022). New species *Chara oryzae* and a new section *Corillionia* of *Chara* (Charales, Charophyceae) from European Mediterranean rice fields. *European Journal of Phycology* 57(3): 328–342. https://doi.org/10.1080/09670262.2021.1987527

Chapter 44
Chara papillosa

Irmgard Blindow, Roman Romanov, Hendrik Schubert, Nick Stewart, and Klaus van de Weyer

Scientific name: *Chara papillosa* Kütz., Flora 17: 707. 1834

Lectotype (Wood, 1965: 100): in lacu salso Mansfeldensi/[no date]/[Kützing] (L 54646)

≡ *Chara intermedia* A. Braun ex Lange, Haandb. Danske Fl., ed. 2: 703. 1856–1859 [probably 1859], n. ill.

≡ *Chara intermedia* A. Braun in Braun, Rabenhorst and Stizenberger, Charac. Eur. 2/45. 1859, n. ill.

= *Chara visianii* Blaženčić and V. Randjel. in Cryptol. Allgol. 15: 74. 1994
Holotype (Blaženčić and Randjelović, 1994: 74): In aquis fluminis Krka, in stagnis Visovac, prope Cascata Skradinski Buk (Croatia)/17.7.1990/J. Blaženčić (BEOU)

I. Blindow (✉)
Universität Greifswald, Biologische Station Hiddensee, 18565 Kloster, Germany
e-mail: blindi@uni-greifswald.de

R. Romanov
Nišice Ujtin Potok B.B, Dobra Voda, 85356 Bar, Montenegro
e-mail: streptophytes@gmail.com

H. Schubert
Universität Rostock, Institut für Biowissenschaften, Albert-Einstein-Straße 3, 18059 Rostock, Germany
e-mail: hendrik.schubert@uni-rostock.de

N. Stewart
4 Hexton Road, Glastonbury BA6 8HL, UK
e-mail: nfstewart@freeuk.com

K. van de Weyer
lanaplan, Lobbericher Str. 5, 41334 Nettetal, Germany
e-mail: klaus.vdweyer@lanaplan.de

H. Schubert et al. (eds.), *Charophytes of Europe*,
https://doi.org/10.1007/978-3-031-31898-6_44

Nomenclatural note: The often-used name *Chara intermedia* is not available. Between 1856 and (probably) 1859, J. Lange published *Chara intermedia* in his second edition of "Handbog i den danske Flora". In this he cites the earlier name *Chara papillosa* as a synonym making *Chara intermedia* an illegitimate name. Also, in 1859 *Chara intermedia* was published in Charac. Eur. as an invalid name without a description as numbers 2/45, 2/46 and 2/47; number 2/45 also cites *Chara papillosa* as a synonym.

Taxonomic notes: Several authors (e.g., Olsen, 1944; Krause, 1997; Hollerbach and Krassavina, 1983) regarded *C. aculeolata* as a synonym of *C. papillosa*, and distinguished *C. papillosa* (*C. intermedia*) from *C. polyacantha*, which should correctly be given the name *C. aculeolata*. Corillion (1957) merged *C. aculeolata* and *C. papillosa* but later separated the taxa (as *C. polyacantha* and *C. intermedia*) (Corrillion, 1975). Krause (1997), however, clearly separated *Chara intermedia* from *C. polyacantha*.

Chara visianii was described by Blaženčić and Randjelović (1994) and characterised as a species occurring in freshwater, with a connivent coronula, with smaller coronula width, lower number of oospore striae and a lower diameter of antheridia than *Chara intermedia* (= *C. papillosa*). However, *C. papillosa* occurs in both freshwater and brackish water, and the coronula of the oogonia varies from non-spreading to connivent. The diameter of antheridia and coronula width are highly variable, and the measures of both *C. visianii* and *C. intermedia* given by Blaženčić and Randjelović (1994) are within the ranges observed. We do not regard *Chara visianii* as a taxon separate from *C. papillosa*.

Local names: BELARUSIAN: Хара дробнашыпаватая; DUTCH: Grijs kransblad; ENGLISH: Intermediate Stonewort; ESTONIAN: Keskmine mändvetikas; FINNISH: Kalkkinäkinparta; GERMAN: Kurzstachelige Armleuchteralge; LITHUANIAN: Bukadyglis maurabragis; NORWEGIAN: Piggkrans; POLISH: Ramienica kolczasta; RUSSIAN: Хара сосочковая, Хара мелкошиповатая; SWEDISH: Mellansträfse; UKRAINIAN: Хара перехідна, хара дрібношипувата.

44.1 Morphology

See Figs. 44.1, 44.2, 44.3, 44.4 and 44.5.

Habitus: Fresh plants are light, reddish, olive or brownish green, herbarium material often appears greyish green due to the often-abundant encrustation. The plants are sparsely branched.

Size: *Chara papillosa* ranges from medium size to large and robust, usually 8–80 cm long and rarely to 105 cm (Eaton et al., 1999).

Fig. 44.1 Details of *Chara papillosa*. **a** Upper shoot section showing sparse, single spine-cells and reddish colour; **b**, **c** Main axis sections showing stipulodes, tylacanthous cortex and short, sparse, single and papilliform spine-cells (note the reddish colour!); **d** Main axis section showing tylacanthous cortex and clustered spine-cells; **e** Multicellular axial bulbil. Photographs: **a**–**c**—Chris Carter; **d**, **e**—Klaus van de Weyer

Fig. 44.2 Photographs of *C. papillosa*. Left: Dense vegetation. Right: Habitus of *C. papillosa*. Photographs: Klaus van de Weyer

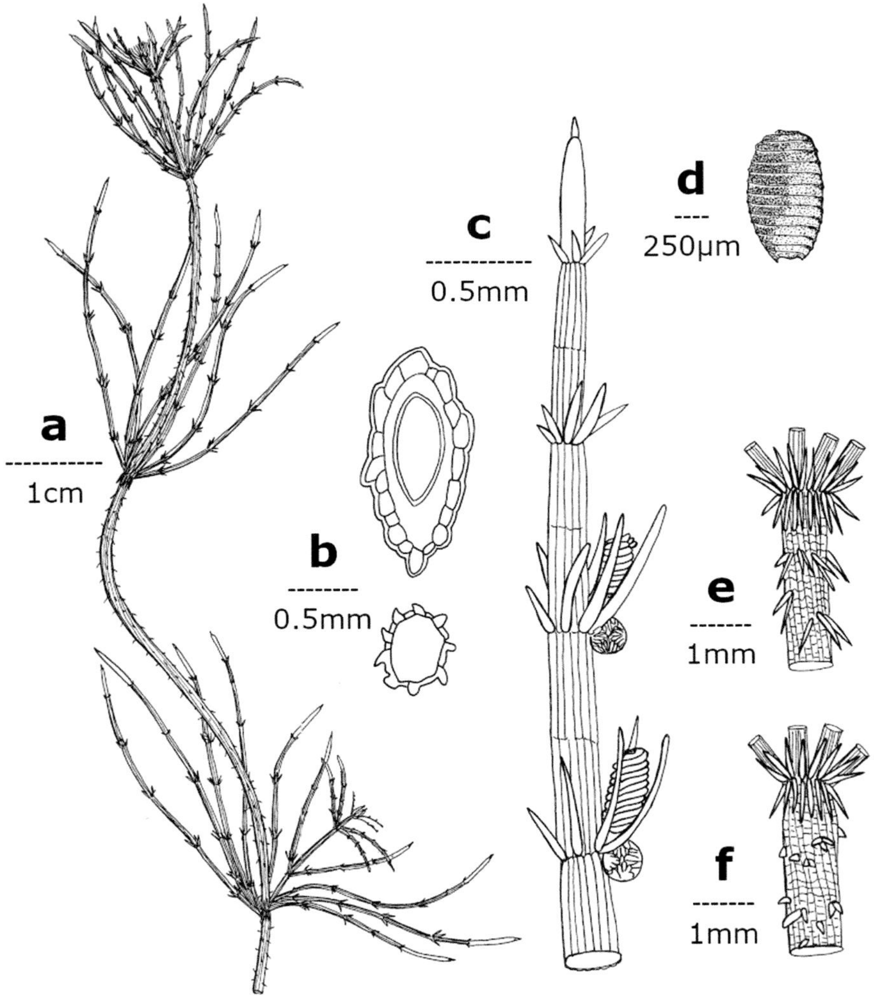

Fig. 44.3 *Chara papillosa*. **a** Habitus; **b** Main axis cross-sections; **c** Part of branchlet with gametangia; **d** Oospores; **e** and **f** Internode sections with stipulodes and spine-cells. Drawings after Werner Krause, modified by Dagmar Wassong in van de Weyer and Schmidt (2018)

Main axis: The axis diameter is highly variable, about 0.6–2.5 mm; branchlets are short compared to the length of the internodes.

Cortex: *Chara papillosa* is diplostichous and heterostichous, tylacanthous (often strongly so; see Figs. 44.1 and 44.3), occasionally isostichous. Spiralling of the cortex has been observed.

Fig. 44.4 Unusual habitus of *Chara papillosa*. Left: Collection from Heigham Sound in BM with unusually long branchlets; right: Unusual branchlet cortication patterns, several ecorticate distant branchlet segments and ecorticate basal branchlet segment (scale length = 1 mm). Photographs: left—Irmgard Blindow; right—Roman Romanov

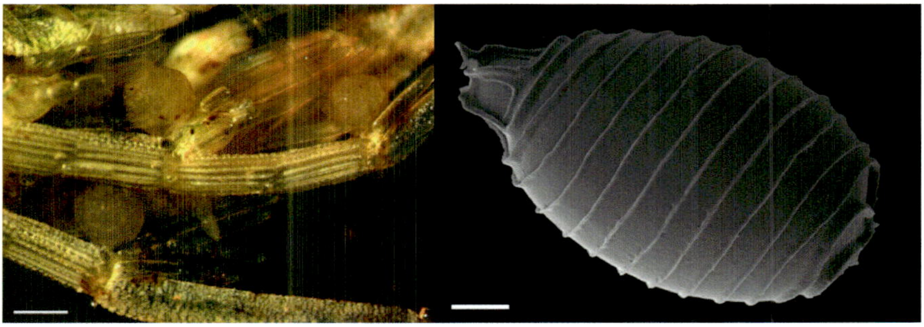

Fig. 44.5 Gametangia and oospore of *Chara papillosa*. Left: Part of branchlet with gametangia, bract-cells and bracteoles; right: Oospore from a karst lake plant, collected in Vladimir Oblast (European Russia). Scale bars = 100 μm. Photographs: Roman Romanov

Spine-cells: Single or arranged in bunches of 2–3, very rarely in clusters of 4 (Fig. 44.1), papilliform to acute, shorter than or as long as the main axis diameter, occasionally longer. Most spine-cells range from 0.6 to 1.3 mm long, but values between 0.3 and 2.6 mm have been measured.

Stipulodes: Arranged in two regular tiers of almost equal length (Fig. 44.1). The length varies between less than half the main axis diameter to almost as long as the main axis diameter. Length measurements vary between 210 and 1350 μm.

Branchlets: Whorls consist of 8–12 branchlets, which are corticate except for the end cell and sometimes penultimate cell, in which case the end cell is narrower than the apex of the penultimate cell. Rarely, more branchlet segments are ecorticate (Fig. 44.4).

Bract-cells: Regularly arranged along the whole branchlet and spreading. Adaxial bract-cells are longer or far longer than abaxial ones, which can be rudimentary (Figs. 44.1 and 44.3); (almost) verticillate bract-cells are rarely found.

Bracteoles: Two, as long as or longer than the ripe oogonium (Wood, 1965, own observations; Fig. 44.3).

Sexuality: Monoecious. Gametangia are conjoined on the 3–5 lowest branchlet nodes (Wood, 1965; own observations).

For the description of oogonia, oospores and antheridia, the information given in Migula (1889–1897), Olsen (1944), Corillion (1957) and Hollerbach and Krassavina (1983) cannot be used, as *C. papillosa* was not (clearly) separated from *C. aculeolata*.

Oogonia: Coronula cells are not spreading in most collections, appearing connivent in some cases, but may also be spreading (own data). One Swedish collection (S080P) has an unusually low number (8–11) of helical stripes. For data about quantitative morphometry, see Table 44.1).

Oospores: Table 44.2 gives an overview about morphometric data. Figure 44.5 shows an SEM image of an oospore.

Table 44.1 Oogonia morphometry

Author	Length	Width	Coronula length	Coronula width	Number of helical stripes
Wood (1965)	950–1200	675–850	185–225	310–390	12–14
Krause (1997)	950–1200	700–800	180–225	300–400	12–14
Holtz [no date]	1000–1480	630–780	–	–	–
Own data	700–1060	430–820	100–260	180–380	(8)10–14
Total range	**700–1480**	**430–850**	**100–260**	**180–400**	**(8)10–14**

Length in the first column excludes the coronula, its length and width given separately; all values for length and width in μm

Table 44.2 Oospore morphometry

Author	Length	Width	Number of striae	Colour
Wood (1965)	660–820	435–500	11–13	Dark brown to black
Krause (1997)	700–850	450–500	11–13	Brown to black
Holtz [no date]	740–850	460–540	–	Dark brown to black
Own data	430	340	11	Black
Total range	**430–850**	**340–540**	**11–13**	

All values for length and width in μm.

Table 44.3 Antheridia morphometry: own data are from dry antheridia, Wood (1965) measured rewetted antheridia

Author	Diameter (μm)
Wood (1965)	500–550
Krause (1997)	400–500
Own data	(290)320–600(710)
Total range	**(290)320–600(710)**

Krause (1997) does not give any information on the method

Antheridia: Antheridia are orange. Table 44.3 gives an overview of the morphometric data.

Bulbils: Multicellular bulbils derived from lower main axis nodes can occasionally be found (Giesenhagen, 1896, own data; Fig. 44.1).

44.2 Phenology and Reproduction

Chara papillosa is perennial and hibernates as a green plant (Olsen, 1944; Langangen, 2007) or starts growing in spring by means of its multicellular bulbils (Giesenhagen, 1896). In northern Europe, gametangia were found between June and September, ripe oospores from July (Langangen, 2007).

A detailed study by J. Harris in Hickling Broad (UK) when growing conditions were good in the late 1990s (Eaton et al., 1999) indicate that, while remaining green over the winter, the main axis senesces and collapses with nutrients being stored in swellings at the lower nodes. By February new shoots were already beginning to form at the nodes on the old main axis and by April the beds were increasing in height. In early summer the plants were firmly anchored by rhizoids in the substrate but as the season progressed the older, basal parts became intensely shaded by the extension of the newer upper parts. Senescence started in September and was a prelude to the general collapse of the plants in the autumn.

J. Harris also found that vegetative reproduction was by far the dominant method of spread within the ecosystem; this was either by layering out (main axis falling over and rooting from the nodes) or by fragmentation, with fragments remaining dormant over the winter and regrowth beginning by May. Despite very good oospore production, she failed to find any evidence of germinating sporelings, but this may in part be due to the soft mobile sediment which would be difficult for sporelings to anchor into.

44.3 Habitat Conditions and Ecology

Habitat type: *Chara papillosa* can be found in a variety of habitats such as lakes, streams, ditches, reservoirs, smaller natural and artificial ponds and peat holes (Krause, 1997; Auderset Joye and Rey-Boissezon, 2013; Kabus and Mauersberger, 2011).

Sediment: *Chara papillosa* grows on sand, gravel, peat and gyttja (Eaton et al., 1999; Mauersberger, 2004; Pätzold, 2011; Doege et al., 2016).

Nutrient conditions: *Chara papillosa* occurs mainly in oligo- to mesotrophic water but has also been found in nutrient-rich habitats (Spiess et al., 2010; Korsch, 2013). In Germany, the species has been found in broad nutrient ranges with total-P-concentrations between 5 and 430 $\mu g\ L^{-1}$ and PO_4-P-concentrations between 1 and 61 $\mu g\ L^{-1}$ (Doege et al., 2016). Nutrient concentrations of British sites are within these ranges (Eaton et al., 1999).

Water chemistry: *Chara papillosa* occurs mainly in calcium-rich habitats. In Germany, it has been found at pH values between 6.4 and 9 (Doege et al., 2016). *Chara papillosa* is a freshwater species but can also be found in inland brackish water sites (Krause, 1997; van Raam, 1998; Schneider, 2004; Korsch, 2013). Concentrations of 1750–2500 mg Cl L^{-1} were measured in British brackish water sites (mean values, Eaton et al., 1999), corresponding to around 3.1–4.8 PSU. In experiments, *C. papillosa* originating from brackish-water Hickling Broad, was favoured by a moderate increase of salinity (Lambert, 2007).

Depth range: *Chara papillosa* has been found between 0.5 and 9 m (Blaženčić and Randjelović, 1994).

Associated species: *Chara papillosa* often forms monospecific stands, but has also been found together with other characeans (*C. filiformis*, *C. contraria*, *C. hispida* and *C. tomentosa, C. baltica* var. *rigida, C. connivens, C. globularis, C. hispida, C. aspera, C. vulgaris, Nitellopsis obtusa)* as well as vascular plants (*Potamogeton coloratus, P. crispus, P. pusillus, Stuckenia pectinata, Najas marina, Hippuris vulgaris, Myriophylum spicatum*) and *Batrachospermum ectocarpum* (Krause, 1997; Eaton et al., 1999; Spiess et al., 2010; Korsch, 2013).

44.4 Variability

The length and arrangement of spine-cells, stipulodes, bract-cells, number of spine-cells in a cluster and difference between primary and secondary cortex cell rows are most variable. The variable length of branchlets and internodes as well as proportions between them and the rough/slender general texture results in a slightly or markedly different habit of plants. Also the number of ecorticate distal branchlet segments is variable (Fig. 44.5). Migula (1889–1897) described as many as 21 different forms of *C. papillosa*, which are separated by these morphological characters.

AFLP analyses indicate genetic differences between a population of *C. papillosa* originating from freshwater and a population of *C. papillosa* originating from brackish water (Boegle et al., 2007). A detailed analysis, however, showed no detectable morphological differences between freshwater and brackish water populations (own data).

44.5 Identification and Risk of Confusion

Chara papillosa belongs to the *Hartmania* group, a complex of coarse, diplostichous *Chara* spp. that are hard to separate from each other (see Chap. 15).

Distinctly tylacanthous cortication separates *Chara papillosa* from *C. hispida* and *C. subspinosa*. Short, mainly single spine-cells and short stipulodes separate the species from *C. aculeolata*, *C. corfuensis* and *C. horrida* (the latter also has aulacanthous cortex). From *C. affinis*, the species can be distinguished by shorter and sparser spine-cells. *Chara baltica* (including var. *liljebladii* and var. *rigida*) and *C. horrida* have larger antheridia, although there is overlap in this feature; those species are restricted to coastal brackish water and are fresh to dark green, while *C. papillosa* is often olive to brownish green. A lack of encrustation in *C. baltica* is explained by the ion composition in the Baltic Sea and therefore is not a useful taxonomic criterion. The distinction from "Mediterranean Chara" (see Chap. 80) is under investigation and further details can be found under that taxon description. *Chara globata* was frequently misidentified as *C. papillosa* in South-East Europe and Central Asia, irrespective of the obviously verticillate bract-cells and the typically arcuate upper branchlets of the former. Such misidentifications may also happen in the Mediterranean area, and collections from that area resembling *Chara papillosa* should be treated carefully.

Slender collections of *C. papillosa* with single spine-cells might be hard to distinguish from *Chara contraria*. LE, for example, contains a collection of delicate plants from the locus classicus close to Halle determined by A. Braun: "*Chara intermedia* A. Br. *gracilescens* ad *Ch. contrariam* accendens".

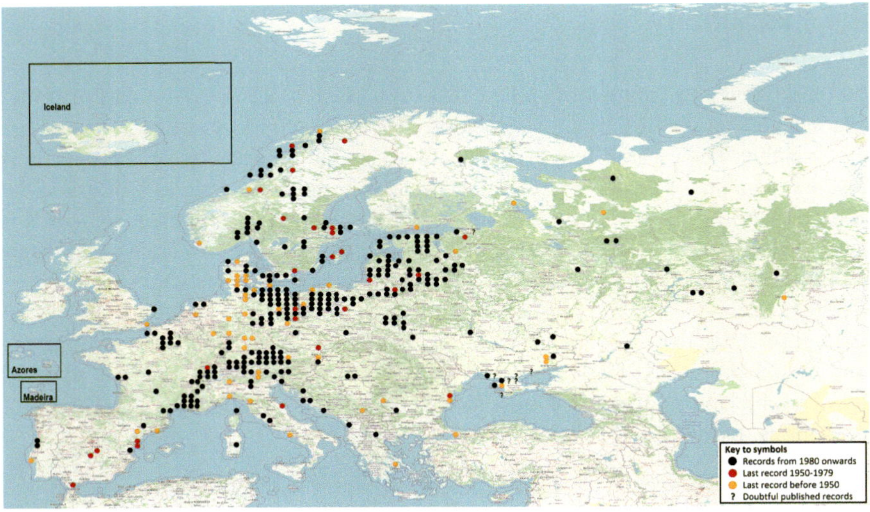

Fig. 44.6 European distribution of *Chara papillosa*. For worldwide distribution see Korsch (2018). Black dots are records from 1980 cnwards, red dots last record from 1950 to 1979 and yellow dots last record before 1950; doubtful published records (if applicable) are marked with "?" superimposed. Map prepared by Heiko Korsch

44.6 Distribution

Chara papillosa occurs in Europe, Asia and Africa. In Europe, it has been recorded sparsely from northern Scandinavia to the Mediterranean Sea in the south, to Russia, the Balkans and Turkey in the east and to France and Great Britain in the west (Fig. 44.6).

44.7 Status and Threats

Chara papillosa was assessed as Regionally Extinct in Bulgaria, as Critically Endangered in Montenegro and Serbia, as Endangered in Britain, the Netherlands, Denmark, Switzerland, Poland and SE Europe, as VS ("vanishing species") in the Czech Republic 3 (equivalent to Vulnerable) in Germany and Ukraine and as Near Threatened in Norway and Finland. In Sweden, Estonia and Lithuania, the species is not included in the Red List as it is too frequent. *Chara visianii* is assessed as Critically Endangered in SE Europe (Blaženčić et al., 2006). *C. papillosa* is assessed as LC (Least Concern) in the European Red List (see Chap. 9).

Due to problems of identification, there is little reliable information on trends and threats. *C. papillosa* is a rare species, and a marked decline is assumed from the available information. As a result it is listed as threatened on many country Red

Lists. However there are also examples of colonisation of new habitats. The species is threatened by eutrophication which causes increasing competition with other submerged plants, overgrowth by filamentous algae or shading and de-oxygenation connected to algal blooms. Conflict with boat users (sailing and motorboats) is also an issue at some of the British sites. Introduction of carp fish is another severe threat factor and should be avoided in sites for *C. papillosa*. In small water bodies and ditches, periodic clearance can help to protect this species, especially if other plants are forming a thick cover.

Acknowledgements Jelena Blaženčić is gratefully acknowledged for a fruitful and positive discussion about *Chara visianii*.

References

Auderset Joye, D., & Rey-Boissezon, A. (2013). *Les Characées de Genève et environs. Distribution et Ecologie.* Groupe d'Ecologie aquatique, Institut Forel et Institut des Sciences l'Environnement, Université de Genève, 92 pp. [in French].

Blaženčić, J., Stevanović, B., Blaženčić, Ž., & Stevanović, V. (2006). Red data list of charophytes in the Balkans. *Biodiversity and Conservation, 15*, 3445–3457.

Blaženčić, J., & Randjelović, V. (1994). *Chara visianii* J. Blaženčić et V. Randjelović sp. nov. (Characeae) from the river Krka (Croatia). *Cryptogamie, Algologie, 15*, 73–79.

Boegle, M. G., Schneider, S., Mannschreck, B., & Melzer, A. (2007). Differentiation of *Chara intermedia* and *C. baltica* compared to *C. hispida* based on morphology and amplified length polymorphism. *Hydrobiologia, 586*, 155–166.

Corillion, R. (1957). Les Charophycées de France et d'Europe Occidentale. *Bulletin de la Société Scientifique de Bretagne, 32* (Hors série I & II), 1–499 [in French].

Corillion, R. (1975). *Flore et végétation du Massif Armoricain. T. IV. Flore des Charophytes (Characées) du Massif Armoricain et des contrées voisines d'Europe occidentale.* Jouve, Paris [in French].

Doege, A., van de Weyer, K., Becker, R., & Schubert, H. (2016). Bioindikation mit Characeen. In *Arbeitsgruppe Characeen Deutschlands, Armleuchteralgen: Die Characeen Deutschlands* (pp. 97–138). Springer [in German].

Eaton, J. W, Harris, J., Moss, B., Sharpe, J., Smith, K., Stewart, N. F., & Walthern, P. (1999). *Assessment of the proposed cutting of aquatic plants in Hickling Broad, final report of the Assessment Team, October 1999.* Broads Authority, Norwich, 276 pp.

Giesenhagen, K. (1896). Untersuchungen über die Characeen. *Flora, 82*, 381–433. [in German].

Hollerbach, M. M., & Krassavina, L. K. [Голлербах, М. М. & Л. К. Красавина]. (1983). Определитель пресноводных водорослей СССР. 14. *Charophyta [Determination key of freshwater algae of USSR].* Nauka, Leningrad, 189 pp. [in Russian].

Holtz, L. [no date]. Zeichnungen und Messungen von Characeen Pommerns von Dr. Paul Hauptfleisch, entsprechend meinen Publikationen. Inv. Nr. 5551. University of Greifswald, Germany [in German].

Kabus, T., & Mauersberger, R. (2011). Liste und Rote Liste der Armleuchteralgen (Characeae) des Landes Brandenburg 2011. *Naturschutz und Landschaftspflege in Brandenburg, 20*(4, Beilage), 1–31 [in German].

Korsch, H. (2013). Die Armleuchteralgen (Charophyceae) Sachsen-Anhalts. *Berichte des Landesamtes für Umweltschutz Sachsen-Anhalt, 1*(2013), 1–85. [in German].

Korsch, H. (2018). The worldwide range of the Charophyte species native to Germany. *Rostocker Meeresbiologische Beiträge, 28*, 45–96.
Krause, W. (1997). Charales (Charophyceae) Süßwasserflora von Mitteleuropa. In: H. Ettl, G. Gärtner, H. Heynig & D. Mollenhauer (Eds.), *Süßwasserflora von Mitteleuropa 18*. Fischer, Jena [in German].
Lambert, S. J. (2007). *The environmental range and tolerance limits of British Stoneworts (Charophytes)* (452 pp.). Ph.D. Thesis, University of East Anglia, Norwich, Great Britain.
Langangen, A. (2007). *Charophytes of the Nordic countries*(102 pp.). Saeculum ANS, Oslo.
Mauersberger, R. (2004). Zum Vorkommen von Armleuchteralgengewächsen (Characeae) im Norden Brandenburgs. *Rostocker Meeresbiologische Beiträge, 13*, 85–104. [in German].
Migula, W. (1889–1897). Fünfter Band: Die Characeen. In A. Fischer, E. Fischer, F. Hauck, G. Limpricht, C. Luerssen, W. Migula, H. Rehm, P. Richter & G. Winter (Eds.), *Dr. L. Rabenhorst's Kryptogamenflora von Deutschland, Oesterreich und der Schweiz* (ed. 2). Eduard Kummer, Leipzig [in German].
Olsen, S. (1944). Danish Charophyta. *Det Kongelige Danske Videnskabernes Selskab, Biologiske Skrifter, 3*(1), 1–240, 1–2.
Pätzold, F. (2011). Beobachtungen zur Verbreitung von Characeen in Baggerseen der Badischen Oberrheinaue. *Berichte der Botanischen Arbeitsgemeinschaft Südwestdeutschland Beiheft, 3*, 37–42. [in German].
van Raam, J. C. (1998). *Handboek Kranswieren*. Chara boek, Hilversum [in Dutch].
Schneider, S. (2004). *Chara intermedia*. In H. Schubert & I. Blindow (Eds.), *Charophytes of the Baltic Sea* (pp. 122–128). A. R. G. Gantner, Ruggell.
Spiess, H. J., Bolbrinker, P., Möbius, F., & Waterstraat, A. (2010). Ergebnisse der Untersuchungen submerser Makrophyten in ausgewählten Gewässern Mecklenburg-Vorpommerns. *Botanischer Rundbrief Mecklenburg-Vorpommern, 47*, 4–182. [in German].
van de Weyer, K., & Schmidt, C. (2018). Bestimmungsschlüssel für die aquatischen Makrophyten (Gefäßpflanzen, Armleuchteralgen und Moose) in Deutschland: Band 1: Bestimmungsschlüssel, ed. 2. Landesamt für Umwelt (LfU) Brandenburg. *Potsdam, Fachbeiträge des LfU Brandenburg, 119*, 1–180. [in German].
Wood, R. D. (1965). *Monograph of the Characeae*. Weinheim: J. Cramer.

Chapter 45
Chara squamosa

Roman Romanov, Ralf Becker, Mauro Iberite, Sophia Barinova, Thomas Gregor, and Attila Mesterházy

Scientific name: *Chara squamosa* Desf., Flor. Atlant. 2 : 331. 1799

Lectotype (Wood, 1965: 108): Barbarie/[no date]/[no collector] (PC 610826, ex herb. Maire)

= *Chara conimbrigensis* A. G. Cunha in Bull. Soc. Portug. Sci. Nat. 12(8): 48. 1935
Lectotype (Wood, 1965: 110): [Portugal] Maldormes, Santa Clara et Conraria (Coimbra)/15.6.1933/M. L. Gonçalves Sobrinho (LISU P 45315).

=? *Chara rohlenae* Vilh. in Hedwigia 53: 30. 1912

R. Romanov (✉)
Nišice Ujtin Potok b.b., Dobra Voda, 85356 Bar, Montenegro
e-mail: streptophytes@gmail.com

R. Becker
Am Schützenplatz 58, 26121 Oldenburg, Germany
e-mail: becker.r@posteo.de

M. Iberite
Dipartimento di Biologia Ambientale, Università di Roma "La Sapienza", Piazzale Aldo Moro 5, 00185 Roma, Italia
e-mail: mauro.iberite@uniroma1.it

S. Barinova
Institute of Evolution, University of Haifa, Mount Carmel, 199 Abba Khoushi Ave, 3498838 Haifa, Israel
e-mail: sophia@evo.haifa.ac.il

T. Gregor
Department of Botany and Molecular Evolution, Senckenberg Research Institute and Natural History Museum Frankfurt, Senckenberganlage 25, 60325 Frankfurt am Main, Germany
e-mail: thomas.gregor@senckenberg.de

A. Mesterházy
Hunyadi Utca 55, 9500 Celldömölk, Hungary
e-mail: amesterhazy@gmail.com

H. Schubert et al. (eds.), *Charophytes of Europe*,
https://doi.org/10.1007/978-3-031-31898-6_45

Lectotype (Wood, 1965: 109): [Montenegro] Jugoslavia, flora montenegrina: in rivulo Mratinje sub monte Maglic/no date/J. Rohlena (PRC).

= *Chara hungarica* Fil. in Mat. Term. Értes. 47: 655. 1930
Lectotype (designated here by R. E. Romanov and Mesterházy): Figs. 1–8 at page 658 in Filarszky (1930), Mat. Term. Értes 47: 652–676.

Epitype (designated here by R. E. Romanov and Mesterházy): [Hungary, Budapest], a Lukács fürdör tó legmélyebb helyén [at the deepest point of Lukács Bath Lake]/[no date]/N. Filarszky (BP).

Taxonomic note: For a long time *Chara squamosa* was accepted to be conspecific with aulacanthous *C. gymnophylla* and at that time the legitimate name for this taxon (e.g. Migula, 1889–1897). However, Wood (1965) described the cortex of the lectotype as "nearly aequistriate". Material by Desfontaines stored in Madrid (MA) has isostichous to tylacanthous main axis cortex. Moreover, a plant depicted in Kützing (1857: p. 72) from the herbarium Willdenow (B) which originated from "Barbarie" has an isostichous to weakly tylacanthous main axis cortex. Therefore, *C. squamosa* is the legitimate name for the taxon called formerly *C. conimbrigensis*. But this should be settled by an investigation of the lectotype in Paris. Since its description *C. hungarica* was associated with *C. filiformis* mostly because similarities in general habit, although checking of the specimens revealed more similarity with *C. squamosa* (see details below in "variability and risk of confusion"). *C. hungarica* is accepted here conspecific with *C. squamosa*. A specimen being in a good agreement with protologue, descriptions and illustrations of *C. hungarica* (Filarszky, 1930; Wood and Imahori, 1964; Wood, 1965) collected from the type locality but with missing label data was selected as an epitype, because none of the specimens cited by Wood was found in BP.

Local names: Unknown.

45.1 Morphology

See Figs. 45.1, 45.2 and 45.3.

Habitus: The plants are light green to greyish-green and green, slightly or heavily incrusted, or unincrusted, small or moderately high, condensed habit if small and with abbreviated upper whorls—condensed apical part, sometimes with moss-like habitus, not abbreviated to abbreviated upper whorls if medium-sized, rarely elongated with largely spaced whorls of short branchlets (*C. hungarica*). The plants are weakly to moderately incrusted in shallow water or unincrusted in deeper water, condensed and short in groundwater seepage or uncondensed and long in deeper waters, growing as one shoot or more commonly as several shoots arising from the common base with rhizoids, sparsely branched or unbranched.

Size: *Chara squamosa* is a small to medium plant, from a few to 30 cm high.

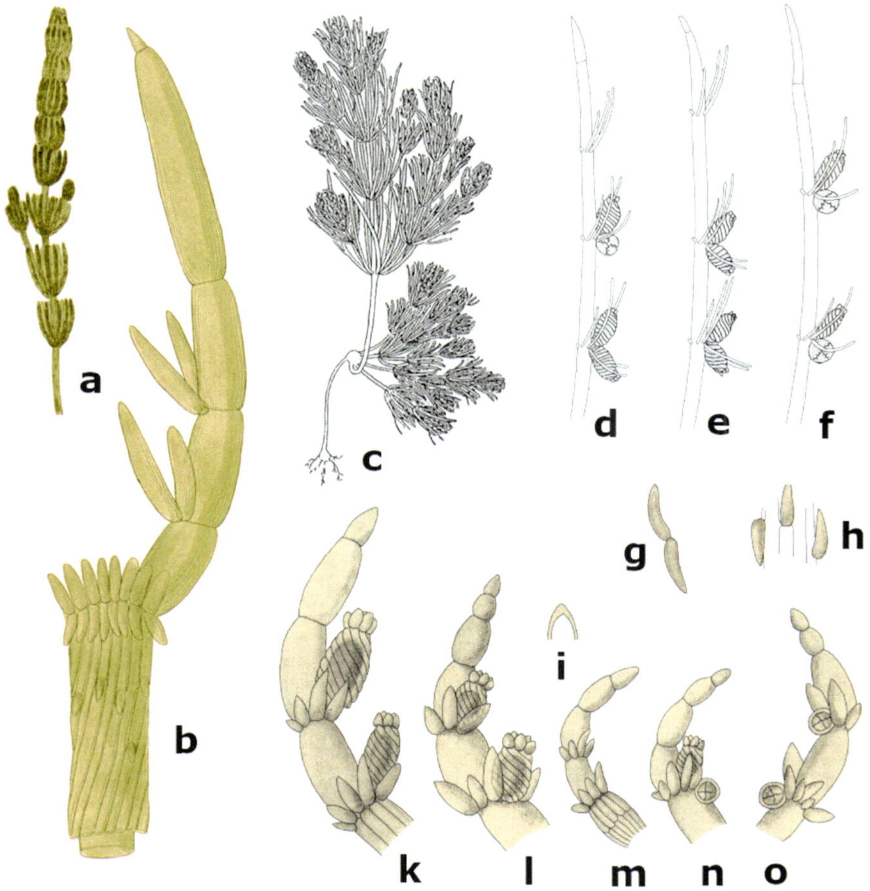

Fig. 45.1 *Chara squamosa*. **a**, **c** General habit; **b** Sterile branchlet, stipulodes and main axis cortex; **d–f**, **k–o** Branchlets with different node numbers, gametangia and cortex arrangement; **g** Stipulodes; **h** Part of main axis cortex; **i** Tip of branchlet end cell. The illustration of branchlet basal segment cortication (**m**) needs detailed study, being dissimilar with most common pattern known for *Chara*, when cortex cell at basal branchlet segment arose downwards from basal node of branchlet only. Drawings: **a**, **b**—Kützing (1857); **c–f**—Cunha (1935) as *C. conimbrigensis*; **g–o**—Filarszky (1930) as *C. hungarica*

Main axis: The main axis is rather thin, 390–740 µm in diameter, completely corticate. Rarely some lowest internodes are ecorticate. Internodes are predominantly shorter, sometimes much longer than the branchlets, mainly very short near the apex.

Cortex: Axis cortex is complete, diplostichous, especially on young internodes rarely irregular triplostichous. It is mostly slightly tylacanthous (this trait is often barely recognisable in pressed specimens) and isostichous, sometimes clearly tylacanthous

Fig. 45.2 *Chara squamosa*, morphological features. **a**, **b**, **c**, **e** Upper parts of plants; **d** Whorl of short branchlets and upper part of plant; **f** Sejoined and conjoined (arrowhead) gametangia; **g** Axial cortex; **h** Axial nodal bulbil. Origin of plant material: **a**, **b**—nearly terrestrial compact plants with variable length of anterior bract-cells from groundwater seepage in Nebrodi Mountains, Sicily; **c**—"deep water" lax plant from a pond in Nebrodi Mountains; **d**—topotype of *C. hungarica*; **e**—topotype of *C. conimbrigensis*; **f–h**—plant from a pond in Nebrodi Mountains. Scale: **a–c**, **e**—2 mm, **d**—1 mm, **f**—0.5 mm, **g**—0.2 mm, **h**—1 mm. Photographs: Roman Romanov

Fig. 45.3 Oospores of *Chara squamosa*, collected by Mauro Iberite at Monte Pollino, Italy, scanning electron microscopy. **a** General view; **b** Oospore surface near the apex. Scale: **a**—100 μm, **b**—10 μm. Photographs: Roman Romanov

in apical parts. Infrequently few short weakly aulacanthous parts can exist in addition to the dominating tylacanthous cortex. Main axis cortex is absent or peeling off in lowermost internodes. Secondary tubes are commonly slightly overriding each other at their ends.

Spine-cells: Spine-cells are solitary, mainly papillate, papilliform or elongated, thick, obtuse, predominantly shorter than main axis diameter, up to four-fifths of its diameter, sometimes slim and longer, at least in running waters, up to 3x as long as main axis diameter, emerging at acute angle or almost adpressed, adpressed if short, up to 240–1600 μm in length.

Stipulodes: The stipulodes are in double rows, obtuse to narrow rounded, short or rarely irregularly elongated, sometimes as long as the main axis diameter, upper ones are sometimes more elongated than lower ones. Upper stipulodes are 170–280 μm long, rarely up to 500 μm, lower ones are 170–260 μm long or rudimentary at lowermost nodes, or rarely somewhat longer than upper ones (*C. hungarica*).

Branchlets: The branchlets are 8–10 in a whorl, arcuate and nearly hiding the upper internodes and apex in groundwater seepage and nearly straight or slightly arcuate and spreading in deeper habitats, (3)8.7–11(25) mm in length. The internodes are mainly shorter or much shorter than the branchlets at least in apical parts, more rarely somewhat longer or several times longer. The branchlets are entirely ecorticate, sometimes the lowest (rarely the lowest two) segments corticate, in most cases completely corticate, but very rarely the cortex can be incomplete (*C. hungarica*). In all cases of developed branchlets, at least one node is always present between ecorticate cells of the branchlet. The branchlets consist of one or mainly two, rarely three ecorticate unabbreviated segments and an ultimate undifferentiated part formed by three or four cells. The end cell is short but not mucronate, conical or conical-cylindrical, with an acute or obtuse apex. Segments of the branchlets are sometimes inflated and constricted at the nodes, which is more expressed in pressed specimens.

Bract-cells: The bract-cells are clearly unilateral, posterior bract-cells are papilliform or very rarely slightly elongated. Anterior bract-cells are long, up to 7.5 mm, usually slightly longer than the adjacent branchlet segment and from several to many times longer than the oogonium (in these cases they can have clearly uneven length), sometimes short, several times shorter than the adjacent branchlet segment, from conical, i.e. much shorter than oogonium (extremely rare), to slightly longer than the oogonium.

Bracteoles: Bracteoles are usually several times shorter than anterior bract-cells, sometimes their difference is not so drastic (own data), or they are 0.5–1.5-times of oogonium length (Sánchez, 1984).

Bractlet: In case of sejoined gametangia the bractlet is shorter than the oogonium.

Sexuality: *Chara squamosa* is monoecious. The gametangia are conjoined on one, two, rarely on the three lowest branchlet nodes. Sometimes gametangia are sejoined, but this is more frequent for antheridia, and it is extremely rare for oogonia. This is not usual but characteristic for the plants from groundwater seepage and found in thermal springs, too (*C. hungarica*).

Oogonia: The oogonia are solitary (one oogonium in a node, irrespective of the presence of an antheridium of antheridium) or very rarely vertically geminate (own data) or even, more rarely, triplicate (Sánchez, 1984; R. Becker pers. obs.) with an antheridium below. The oogonium rarely forms a neck below the coronula. The coronula is slightly spreading (Cunha, 1942). Table 45.1 gives an overview about published morphometric data. Gyrogonites are not described; the population described as *C. hungarica* formed gyrogonites.

Table 45.1 Oogonia morphometry

Author	Length	Width	Coronula length	Coronula width	Number of helical stripes
Cunha (1942)	500–570	350–470	15(?)	50(?)	–
Wood and Imahori (1964); Wood (1965), *C. hungarica*	675–750	375–450	87–120	225–255	12–13
Sánchez (1984)	660–704	488–528	105–132	195–202	–
Own data	568–740	444–460	–	–	–
Total range	**500–750**	**350–528**	**87–132**	**195–255**	**12–13**

Length in the first column is excluding the coronula, its length and width given separately; all values for length and width in μm

Oospores: The oospores are black to dark brown (*C. hungarica*), ellipsoid, with low sharp spiral ridges. Oospore surface is pustulate with varying density of pustulae among different oospores and at the same oospore, tending to be denser towards apices, sometimes in combination with a finely granulated pattern even at pustulae in SEM (Fig. 45.3). Table 45.2 gives an overview about morphometric data available.

Antheridia: The antheridia are octoscutate. Table 45.3 gives an overview of the published morphometric data. The population described as *C. hungarica* has smaller antheridia.

Table 45.2 Oospore morphometry

Author	Length	Width	Number of striae	Colour
Cunha (1942)	360–430	310–370	8	Black
Wood and Imahori (1964); Wood (1965), *C. hungarica*	495–570	270–300	10–12	Dark brown
Sánchez (1984)	545–598	369–396	–	–
Own data	475–600	288–350	10–12	Black
Total range	**360–600**	**270–396**	**(8)10–12**	

All values for length and width in μm

Table 45.3 Antheridia morphometry

Author	Diameter (μm)
Cunha (1942)	450–610
Wood and Imahori (1964), *C. hungarica*	270–350
Sánchez (1984)	475–551
Own data	360–575
Total range	**270–610**

Bulbils: Multicellular multicellular bulbils up to 1.15 mm in diameter are sometimes formed at lowest main axis nodes.

45.2 Phenology and Reproduction

Chara squamosa reproduces by oospores, vegetative reproduction could be expected because of bulbil presence and less fertility. Plants collected in Sardinia (Italy) in May had ripe oospores and antheridia (Becker, 2019). In comparison with other charophyte species *C. squamosa* often seems to produce a smaller number of oospores.

45.3 Habitat Conditions and Ecology

Habitat type: The ecology of *Chara squamosa* is poorly known; it grows mainly in small water bodies associated with ground water seepage, and in small or moderate-sized artificial water bodies—ponds, drainage and irrigation channels, troughs and pits. Moreover, it thrives in small shallow rivers and brooks with at least temporary high flow rates, and in water bodies at thermal springs. It seems to be associated with sunlight exposed habitats. *Chara squamosa* grows both in lowland and mountainous areas. In the type locality of *C. hungarica* it was found in thermal springs with water temperature of 26 °C.

Sediment: *Chara squamosa* grows oozy or sandy bottom sediments, sometimes gravelly.

Nutrient conditions and water chemistry: Hydrochemical data are lacking because they are most probably hidden dissolved in datasets for *Chara gymnophylla*. *C. squamosa* seems to be a freshwater species, growing in mesotrophic water bodies with fresh circumneutral and mild alkaline waters. In Sardinia it occurs mainly in calcareous regions (Becker, 2019).

Depth range: *Chara squamosa* occupies a depth from a few cm to ca. 0.5 m.

Associated species: The association with other species is poorly known. In Sardinia it grows together with *Chara contraria, C. globularis, C. vulgaris, Nitella tenuissima* or *Tolypella glomerata*. In the type locality of *C. hungarica* it was associated with *C. papillosa*.

45.4 Variability and Risk of Confusion

Chara squamosa is a variable species, mainly because of different length of branchlets, internodes and bract-cells resulting in a compact abbreviated to more or less lax general appearance. The plants from shallow water and nearly terrestrial habitats are short and condensed, growing in compact turfs. The plants from the water body associated with a thermal spring (described as *C. hungarica*) have compact whorls separated with largely elongated internodes, in general appearance looking as extraordinary robust and coarse *C. filiformis* (this rough textured morphotype is unknown for *C. filiformis*).

In general appearance *Chara squamosa* mostly resembles morphotypes of *C. vulgaris* and *C. gymnophylla* with long bract-cells. Frequently plants in populations from very shallow habitats have mostly isostichous main axis cortex varying from slightly tylacanthous to slightly aulacanthous. Nonetheless, delineation between primarily aulacanthous and tylacanthous diplostichous forms is widely accepted. Therefore, at present we can tentatively accept *C. squamosa* as a monoecious species having slightly tylacanthous to isostichous main axis cortex and mostly ecorticate branchlets, always having bract-cells and gametangia adjacent to ecorticate segments. This combination of traits allows general delineation between this species and the three taxa mentioned above as well as with *C. contraria*. However, even detailed examination of pressed specimens can be unsatisfactory for their identification as *C. squamosa* or *C. gymnophylla*, and study of living plants is essential. *C. squamosa* differs from *C. contraria* and *C. vulgaris* by having the stable presence of nodes with bract-cells and gametangia between ecorticate segments of branchlets. In frequent cases when basal segments have a cortex the branchlets always bear one or few nodes between ecorticate segments above them.

Chara squamosa can be delineated from *C. gymnophylla* by its tylacanthous main axis cortex. However populations of both species frequently have an isostichous or nearly isostichous main axis cortex.

The colour of ripe oospores is also helpful. *Chara squamosa* and *C. contraria* have black ripe oospores (except *C. hungarica,* having which has dark brown oospores) in contrast with the typically golden or yellow or truly brown oospores of *C. vulgaris* and *C. gymnophylla*. The fact that *C. squamosa* and *C. contraria* sometimes occur in the same waterbody and that *C. contraria* exhibits polymorphism can complicate their delineation. The rarely sejoined gametangia and vertically geminate oogonia of *C. squamosa* can be useful too, but they seem to occur mostly on plants from very shallow habitats and do not allow general delineation from *C. gymnophylla* which sometimes has these traits, too (R. Romanov unpubl. data). *C. squamosa* differs from the similar *C. gymnophylla* mainly by its generally tylacanthous main axis cortex (although aulacanthous main axis cortex is known for *C. squamosa*; see above) and for the occasional presence of sejoined gametangia. However the last trait seems to be typical mainly for plants growing in very shallow water, or nearly as terrestrial plants, as, it was known for the type specimen (Wood and Imahori, 1964; Wood, 1965; LISU, checked by R. Romanov). Therefore, delineation of these species is still unsatisfactory, although they form their own clusters within the big cluster of *C. vulgaris* in *rbc*L sequences (R. Romanov, A. Gontcharov, & V. Nikulin unpubl. data). This can be recognised as evidence of ongoing diversification and speciation within a group of taxa closely related to *C. vulgaris*. This case evidently disagrees with results by Schneider et al. (2016), where aulacanthous and tylacanthous "*C. gymnophylla*" belong to distant clusters of *mat*K sequences. However tylacanthous "*C. gymnophylla*" sequenced by Schneider et al. (2016) originated from Nepal and might belong to another species, possibly *C. grovesii* Pal.

Dioecious *Chara lipkinii* R. E. Romanov et al., known only from a small area in the eastern Mediterranean (Romanov et al., 2022), is similar in its vegetative traits to monoecious *C. squamosa*, but delineation is easy when plants are fertile. A morphotype described as *C. hungarica* and associated with *C. filiformis* by Wood (1965) differs from *C. filiformis* with robust general appearance, common presence of 2 and rarely even 3 branchlet nodes, mostly ecorticate branchlets, always having bract-cells between ecorticate segments, not abbreviated basal branchlet segments and commonly sejoined gametangia (Filarszky, 1930; Wood and Imahori, 1964; Wood, 1965; BP, checked by R. Romanov & A. Mesterházy). The differences with most similar species are outlined elsewhere (Romanov et al., 2019). The identification of sterile juvenile or suppressed plants from this species group, especially when they have undifferentiated ecorticate branchlets, seems to be nearly impossible without cultivation or genetic studies.

45.5 Distribution

See Fig. 45.4.

Mediterranean: Portugal, Spain (mainland and Baleares), Italy (mainland, Sardinia, Sicily), Hungary, Serbia, and outside of Europe from North Africa—Tunisia, Morocco, Middle East—Israel, and Caucasus (as *C. hungarica*: Filarszky, 1930; as *C. conimbrigensis*: Cunha, 1935, 1942, 1943; Mendes, 1948; Wood, 1965; Corillion, 1977; Guerlesquin, 1985; Sánchez, 1984; Romanov et al., 2019; as *C. rohlenae*: Blaženčić and Stevanović, 2018; Romanov et al., 2023; R. Romanov & S. Barinova unpubl. data; COI, LE, LISU, MA, TELA, checked by R. Romanov; BP, checked by A. Mesterházy & R. Romanov). Targeted search for specimens of *C. squamosa* in different collections undertaken by R. Romanov was not as successful as expected. *C. squamosa* seems to be rare in comparison to *C. gymnophylla* in the Mediterranean. The total number of its localities in the world did not exceed 35, including a new record for Israel, a few recent records from Italy (M. Iberite pers. comment) and Serbia until recent update (Trbojević et al., 2024). Current records of *C. gymnophylla* sensu Krause (1997) from Sardinia (Becker, 2019) with predominantly isostichous or tylacanthous main axis cortex and only very rarely additional aulacanthous parts were verified by R. Becker and R. Romanov as *C. squamosa*. Moreover, many former Italian records of *C. gymnophylla* mentioned in Bazzichelli

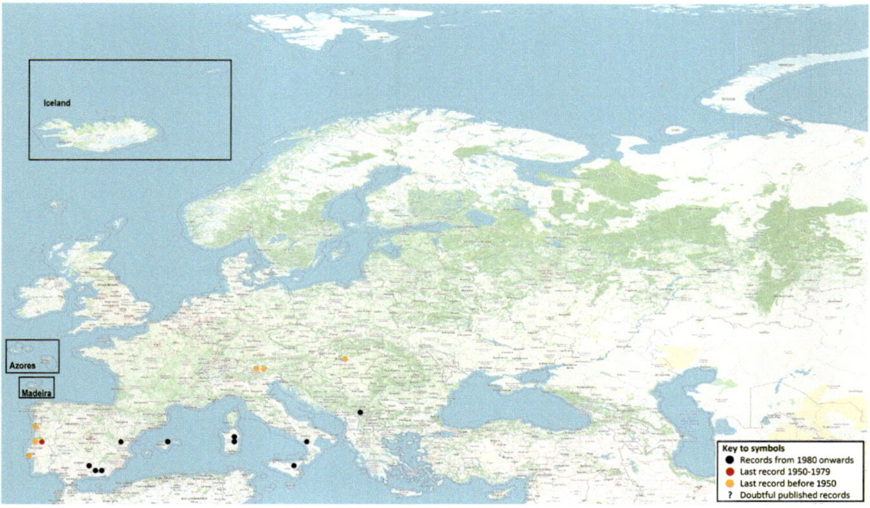

Fig. 45.4 Distribution of *Chara squamosa* in Europe. Black dots are records from 1980 on, red dots from 1950 to 1979 and yellow dots before 1950; unconfirmed/doubtful published records (if applicable) are marked with "?" superimposed. Map prepared by Heiko Korsch

and Abdelahad (2009) were tylacanthous (N. Abdelahad pers. comm.). Thus, verification of herbarium specimens should be done in order to clarify whether they are to be treated as *C. squamosa*. The records from Croatia and Macedonia (Blaženčić and Blaženčić, 1997; Blaženčić et al., 2006) were based on juvenile plants of *C. papillosa* (BEOU, revised by J. Blaženčić & R. Romanov). A recent record of *C. rohlenae* in Serbia (Blaženčić and Stevanović, 2018) seems to belong to *C. squamosa* because the plants were described as having irregular main axis cortex varying from tylacanthous to isostichous, rarely even aulacanthous. The merging of tylacanthous and aulacanthous morphotypes in *C. gymnophylla* (Krause, 1997) further complicated this case, resulting in few records of *C. conimbrigensis* published for Portugal and Spain only. The problem can be overcome by restudy of all available specimens and further field studies. The comprehensive study updated the distribution of *C. squamosa* in Serbia, where it is a rather common species (Trbojević et al., 2024).

45.6 Status and Threats

Chara squamosa has been assessed as DD (Data Deficient) in the European Red List (see Chap. 9). The population from a thermal spring in Budapest was lost at the beginning of 1900s. At that time natural thermal springs were transformed to concrete pools. Boros (1937) visited this place many times between 1915 and 1926 and he did not find *C. hungarica* again.

References

Bazzichelli, G., & Abdelahad, N. (2009). Alghe d'acqua dolce d'Italia. *Flora analitica delle Caroficee*. Centro Stampa Università. [in Italian].

Becker, R. (2019). The Characeae (Charales, Charophyceae) of Sardinia (Italy): habitats, distribution and conservation. *Webbia, 74*, 83–101. https://doi.org/10.1080/00837792.2019.1607497.

Blaženčić, J., & Blaženčić, Ž. (1997). Florističke karakteristike vodene i močvarne vegetacije Dojranskog Jezera. *Glasnik Instituta za botaniku i Botaničke baste Univerziteta u Beogradu, 31*, 43–54. [in Serbian].

Blaženčić, J., & Stevanović, B. (2018). Rediscovery of the endemic species *Chara rohlenae* Vilh. 1912 (Characeae)—believed extinct—on the Balkan Peninsula. *Botanica Serbica, 42*, 109–115. https://doi.org/10.5281/zenodo.1173564.

Blaženčić, J., Stevanović, B., Blaženčić, Ž., & Stevanović, V. (2006). Red Data List of charophytes in the Balkans. *Biodiversity and Conservation, 15*, 3445–3457. https://doi.org/10.1007/978-1-4020-5734-2_6.

Boros, Á. (1937). A magyarországi hévizek felsőbbrendű növényzete. *Botanikai Közlemények, 34*, 85–118. [in Hungarian].

Corillion, R. (1977). Contribution à l'étude des Characées de Tunisie et bilan actuel de la flore charologique tunisienne. *Bulletin de la Société Phycologique de France, 22*, 47–59. [in French].

Cunha, A. G. (1934). *Liste des Characées du Portugal*. [in French].

Cunha, A. G. (1935). Quelques espèces nouvelles de Characées du Portugal. *Bulletin de la Société Portugaise des Sciences Naturelles, 12*, 41–58. [in French].

Cunha, A. G. (1942). Contribüção para o estudo dos Carofitos portugueses. *Revista da Faculdade de Ciências, 2*, 227–357. [in Portuguese].

Cunha, A. G. (1943). Additions à la flore charologique du Portugal. III. *Bulletin de la Société Portugaise des Sciences Naturelles, 14*, 139–143. [in French].

Filarszky, N. (1930). A Budai hőforrások nyílt vizeinek Chara-féléi és néhány más érdekesebb, ritkább algafaja. Matematikai Természettudományi Értesítö A. M. Tud. *Akadémia III. Osztályának Folyóirata, 47*, 652–674. [in Hungarian].

Guerlesquin, M. (1985). Les Characées du Maroc: bilan et répartition de la flore. *Internationale Vereinigung für theoretische und angewandte Limnologie, Verhandlungen, 22*, 2879–2884. [in French].

Krause, W. (1997). Charales (Charophyceae). In H. Ettl, G. Gärtner, H. Heynig & D. Mollenhauer (Eds.), *Süßwasserflora von Mitteleuropa, 18*. Fischer. [in German].

Kützing, F. T. (1857). *Tabulae phycologicae; oder, Abbildungen der Tange 7*. W. Köhne. [in German].

Mendes, E. J. (1948). Notas criptogámicas. III. Carófitos I. *Portugaliae Acta Biologica. Séries B, 2*, 432–435. [in Portuguese].

Migula, W. (1889–1897). Fünfter Band: Die Characeen. In A. Fischer, E. Fischer, F. Hauck, G. Limpricht, C. Luerssen, W. Migula, H. Rehm, P. Richter & G. Winter (Eds.), *Dr. L. Rabenhorst's Kryptogamenflora von Deutschland, Oesterreich und der Schweiz*, 2. Eduard Kummer. [in German].

Romanov, R., Barinova, S., Nikulin, V. Y., & Gontcharov, A. A. (2022). *Chara lipkinii* (Charales, Charophyceae): A new dioecious Mediterranean species under risk of extinction in the wild and some implications for the taxonomy of the genus *Chara*. *Fottea, 22*, 1–12. https://doi.org/10.5507/fot.2021.012.

Romanov, R., Napolitano, T., van de Weyer, K., & Troia, A. (2019). New records and observations to the Characean flora of Sicily (Italy). *Webbia, 74*, 111–119. https://doi.org/10.1080/00837792.2019.1609258.

Romanov, R. E., Mallaliev, M. M., Nikulin, V. Yu., Gontcharov, A. A., & Barinova, S. S. (2023). The charophytes (Charophyceae, Characeae) from Dagestan aquatic habitats, North Caucasus: biogeographical and barcoding perspective. *Environments, 10(9)*, 153. https://doi.org/10.3390/environments10090153.

Sánchez, P. M. (1984). Contribución al conocimiento del género *Chara* en Andalucía. *Acta Botanica Malacitana, 9*, 79–84. [in Spanish].

Schneider, S. C., Nowak, P., von Ammon, U., & Ballot, A. (2016). Species differentiation in the genus *Chara* (Charophyceae): Considerable phenotypic plasticity occurs within homogenous genetic groups. *European Journal of Phycology, 51*, 282–293. https://doi.org/10.1080/09670262.2016.1147085.

Trbojević, I., Marković, A., Šinžar Sekulić, J., Subakov Simić, G., & Romanov, R. (2024). *Chara squamosa* (Characeae, Charophyceae) in Serbia – insights from the taxonomic revision of the BEOU charophyte collection and recent field records. *Botanica Serbica, 48(1)*, 85–92. https://doi.org/10.2298/BOTSERB2401085T

Wood, R. D. (1965). *Monograph of the Characeae*. J. Cramer.

Wood, R. D., & Imahori, K. (1964). *Iconograph of the Characeae*. J. Cramer.

Chapter 46
Chara strigosa

Anders Langangen

Scientific name: *Chara strigosa* A. Braun in Neue Denkschr. Allg. Schweiz. Ges. Gesamten Naturwiss. 10(3): 16. 1847

Neotype (Wood, 1965: 208): Thumsee bei Reichenhall/gegen Ende August und Anfang September 1869/A. Braun (NY 945627); Braun, Rabenhorst & Stizenberger, Characeen Eur. 92.

= *Chara locuples* Hollerb., Bot. Mater. Otd. Sporov. Rast. Bot. Inst. Komarova Akad. Nauk S.S.S.R. 13: 105. 1960.
Lectotype (Romanov, 2021: 3): [Kazakhstan, Akmola Oblast] Заповедник «Боровое», оз. Большое Чебачье, против «Холодного» ключа под г. Синюха, гл. 3 м [Reserve 'Borovoe', Lake Bol'shoe Chebach'e, in front of'Kholodny' spring at the foot of Sinyukha Mountain, 3 m depth]/07.VIII.1948/ Л.Демченко [L.Demczenko] (LE A0000409).

Local names: BELARUSIAN: Хара шчацінiстая; ESTONIAN: Okas-mändvetikas; FINNISH: pohjannäkinparta; GERMAN: Striemen-Armleuchteralge; NORWEGIAN: Stivkrans; POLISH: Ramienica szczeciniasta, RUSSIAN: Хара щетинистая; SWEDISH: Skäggträfse (Sweden); Alpsträfse (Finland); UKRAINIAN: Хара щетиниста.

A. Langangen (✉)
Natural History Museum, PO-Box 1172 Blindern, 0318 Oslo, Norway
e-mail: langangen@hotmail.com

H. Schubert et al. (eds.), *Charophytes of Europe*,
https://doi.org/10.1007/978-3-031-31898-6_46

46.1 Morphology

See Figs. 46.1 and 46.2.

Habitus: Plants are green to greyish, slightly to heavily incrusted, often tufted and richly branched from the base. Spine-cells are normally well developed which gives the plants a brush-like appearance.

Size: *Chara strigosa* is a small to medium-sized (4–25 cm) species.

Main axis: The axis is slender and up to 650 μm in diameter (Wood, 1965: to 1 mm).

Cortex: *Chara strigosa* is regularly triplostichous, sometimes diplo- to triplostichous, rarely diplostichous, normally strongly tylacanthous. The tylacanthous character can in some cases be so dominant that the cortex at a first sight looks haplostichous.

Spine-cells: Spine-cells are well developed, most commonly single, but often in pairs or fascicles of 3, frequently obscuring the cortex view. In some cases, there are 1–2 long cells with 2 short to papillate cells at the basis. Spine-cells are 1–3 times as long as axis diameter, often abundant on younger internodes, pointing downwards on the upper part and upwards on the lower part of the internode (Fig. 46.2).

Fig. 46.1 Habitus of *Chara strigosa*. Drawings: left panel—from Migula (1889–1897); right panel—from Urbaniak and Gąbka (2014)

Fig. 46.2 Details of *Chara strigosa*. **a** Top of specimen showing long spine-cells and long stipulodes (scale bar 1 mm); **b** SEM images of an oospores of *Chara strigosa* from Altai Mountains, side view (scale bar 100 μm); **c** SEM image of the oospore-mebrane of a specimen from North Kazakstan, lectotype of *C. locuples* (scale bar 10 μm). Photographs: **a**—Andrzej Hutorowicz; **b**, **c**—Roman Romanov

Stipulodes: Stipulodes in (1)2 rows, upper row always well developed, to 800 μm long, lower row often irregular (Fig. 46.2) or just partly developed; papilliform to more or less normal, to 500 μm long. *Chara locuples* is haplostephanous, only the upper row of stipulodes is developed. This taxon has been found in Norway, Sweden, Switzerland, Russia and Kazakhstan (Nonweiler, 1907, Langangen 2000; R. Romanov pers. comm.).

Branchlets: Branchlets are 7–9 in a whorl, 3–10 mm long, average 6 mm. Segments 6–9, with 1–3 celled end-segment, 450–750 μm and acute end cell, 250–600 μm.

Bract-cells: The adaxial bract-cells are 250–1100 μm long, the abaxial ones 150–750 μm long.

Bracteoles: Bracteoles are 0.75–1.75 X times as long as the oogonium, most commonly exceeding the length of the oogonium.

Sexuality: *Chara strigosa* is monoecious. Gametangia are onjoined singly, rarely geminate at 3–5 lowest branchlet nodes.

Oogonia: Table 46.1 gives an overview of the morphometric data published. Gyrogonites are produced in some populations (R. Romanov pers. comm.), but not described.

Ooopores: Oospores are black to dark brown, ovoid, with low ridges (Fig. 46.2). Their surface was described as minutely to heavily granulate in LM (Haas, 1994), very rough and irregular due to the presence of numerous depressions and variously sized pits and pores in SEM (John et al., 1990; see Fig. 46.2), with granular and elongated projections in SEM (Mandal et al., 2002) and even smooth or slightly pitted in SEM (Urbaniak, 2011). Fossae are 50–55 μm. Table 46.2 gives an overview of the published morphometrical data.

Antheridia: Antheridia are octoscutate. Table 46.3 gives an overview of the published morphometrical data.

Table 46.1 Oogonia morphometry

Authors	Length	Width	Coronula length	Coronula width	Number of helical stripes
Migula (1889–1897)	800–1000	600–700	150	350	–
Corillion (1957)	800–1000	600–700	150	350	12–14
Wood and Imahori (1964), Wood (1965)	800–920(1000)	480–700	150–270	135–350	12–13
Hollerbach and Krassavina (1983)	740–1000	480–700	150–270	135–350	12–13
Hollerbach and Krassavina (1983), *C. locuples*	815–940	595–625	168–215	252–268	13–14
Krause (1997)	800–900	500–700	150–200	135–200	12–14
Langangen (2007b)	Up to 1100	–	–	–	–
Bailly and Schaefer (2010)	500–700	500–700	–	–	–
Romanov et al. (2014)	(670)800–850	540–580	(170)200–230	190–210	–
Urbaniak and Gąbka (2014)	670–830	386–515	–	–	–
Total range	**(500)670–1000(1100)**	**390–700**	**150–270**	**135–350**	**12–14**

Length in the first column excludes the coronula, its length and width given separately; all values for length and width in μm

Table 46.2 Oospore morphometry

Authors	Length	Width	Number of striae	colour
Migula (1889–1897)	500–700	200–460	8–10	Black
Corillion (1957)	500–700	200–460	10–12	Black
Wood and Imahori (1964), Wood (1965)	500–720	200–460	8–12	Dark brown to black
Hollerbach and Krassavina (1983)	500–720	200–460	8–12	Black
Hollerbach and Krassavina (1983), *C. locuples*	685–725	420–455	11–12	Black
John et al. (1990)	–	–	8–12	–
Haas (1994)	500–700	200–460	8–10(12)	Black
Krause (1997)	500–700	200–450	(8)10–12	Black
Langangen (2007b)	–	–	–	Black
Bailly and Schaefer (2010)	200–500	–	10–15	Black
Urbaniak (2011)	–	–	12–13	–
Romanov et al. (2014)	600–675	325–450	–	Black in reflected light, black and dark brown black in transmitted light
Urbaniak and Gąbka (2014)	460–620	270–360	–	Dark brown
Total range	**(200)460–725**	**200–460**	**(8)10–15**	

All values for length and width in μm

Table 46.3 Antheridia morphometry

Authors	Diameter (μm)
Migula (1889–1897)	350
Corillion (1957)	350
Wood and Imahori (1964), Wood (1965)	(320)450–525
Hollerbach and Krassavina (1983)	300–350
Hollerbach and Krassavina (1983), *C. locuples*	235–250
Krause (1997)	ca. 300
Langangen (2007b)	up to 450
Bailly and Schaefer (2010)	300
Romanov et al. (2014)	300–450
Urbaniak and Gąbka (2014)	310–370
Total range	**(235)300–525**

Bulbils: Bulbils of the *Chara aspera*-type are reported from Norway and Sweden (Langangen, 2000). Bulbils are 350–1400 μm in diameter. Nodal starchless bulbils are found in some populations from Russia (Romanov et al., 2014).

46.2 Phenology and Reproduction

In Fennoscandia, Estonia and western Russia, *Chara strigosa* is found fertile from June to September. Young plants are fertile from July to October, and again in May the coming year. This means that *Chara strigosa* is perennial with overwintering oospores. The many shoots from nodes of old plants are also proof that *Chara strigosa* is perennial. In Estonia fertile plants are found in July, when brown (unripe) and black (ripe) oospores are also found. Specimens from Bosnia collected in August were richly fertile with black oospores. In Fennoscandia *Chara strigosa* seems to have its optimal fructification in *Chara*-lakes and humic *Chara*-lakes. In such localities the species is often found richly fertile. In *Potamogeton*-lakes and more dystrophic lakes, the plants are often sterile or have a low fertility. In the canton of Geneva, Switzerland, plants are sterile from January till May at the depth 1.5–5 m, start producing gametangia from May, bear ripe gametangia from June to August, die off from end of August to beginning of October, and are present as propagule bank of oospores and nodal bulbils between October and January. *Chara strigosa* also reproduces asexually by fragmentation and bulbils.

46.3 Habitat Conditions and Ecology

Habitat type, sediment and temperature: Brown soil or black silt are the bottom sediments preferred by *Chara strigosa*. Clearwater-lakes with peat sediments, as found in the province of Jämtland / Sweden (so-called humic *Chara* lakes) are often covered with a dense vegetation of *C. strigosa*. The species also occurs in *Lobelia-Chara* (LC) lakes and in shallow parts of the *Chara* lakes (Blindow and Langangen, 1995). Small lakes in minerotrophic fens with negligible seasonal variability of water level and soft sediments are one of the main habitat type in northern Russia (Vishnyakov et al., 2021).

As *Chara strigosa* is found in alpine and arctic areas. Temperature is presumed to be a factor of importance for this species. In Jura, France, the temperature in waters with *Chara strigosa* is between 15 and 18 °C (Corillion, 1957; Vaarama, 1954), in Switzerland 0–20 °C (Auderset Joye, 1993) and in Scandinavia 9.0–25.6 °C.

In Scandinavia the vegetation period lasts from May/June to October/November, often with relatively high-water temperatures. For parts of Central Europe the situation differs, as Nonweiler (1907) commented: "Da sie hauptsächlich in den kalten Gebirgsseen vorkommt, so verbringt sie eine sehr lange Zeit des Jahres unter Eis; nach dem Auftauen des Eises verbleiben ihr zu Vegetation nur wenige Monate mit günstigen Entwicklungsbedingungen". [Since it is mainly found in cold mountain lakes, it spends a very long time of the year under ice; after the ice has thawed, it only has a few months with favorable development conditions for vegetation.]

Nutrient conditions: Few data are available for nutrient condition of the sites of *C. strigosa.* According to Jorda and Becker (2016), the species prefers nutrient-poor, unpolluted habitats. Most localities are oligotrophic lakes. For Germany, they report nitrate concentrations of 0.475–1.13 mg/L, ammonia concentrations of up to 0.152 mg/L, and PO_4-P-concentrations of 3–50 μg/L.

Water chemistry: In all known localities with *Chara strigosa* the lime content is moderate to high. *Chara strigosa* is found in soft to hard water with a conductivity of 78–435 μS/cm, alkaline pH (7.2–8.2), relatively high content of lime (Ca^{2+}) (10–90 mg/L) and low values of chloride (1.0–20.0 mg/L), pH 7.2–8.4 (Haas, 1994).

Depth range: *Chara strigosa* is commonly found in shallow water but can reach down to 10 m depth (Corillion, 1957).

Associated species: *Chara strigosa* is most often found in association with other charophytes as *C. papillosa, C. globularis, C. aspera, C. hispida, C. tomentosa, C. virgata, Nitellopsis obtusa, Nitella flexilis* and *N. opaca.*

46.4 Variability

Chara strigosa shows variation in many characters.

The stipulodes are often irregular, poorly developed or completely missing in the lower row. All variants can be observed within the same plant from Switzerland, Russia and Kazakhstan, or plants can be purely haplostephanous (R. Romanov pers. comm.). In both Norway and Sweden haplostephanous specimens with only one row of stipulodes are found. Such forms were described as *Chara locuples* (see above).

The bracteoles are often longer than the oogonium, but in some localities shorter. Both in Norway and Sweden specimens with bract-cells on the lowest branchlet nodes, developed into short additional horizontal branchlets, are found.

On extremely rich fertile specimens the gametangia have so little place on the branchlets that antheridia are laterally positioned. Such specimens also have a bursting (corymbose) top seen from above. Also specimens with geminate oogonia and lateral antheridia can be found.

46.5 Identification and Risk of Confusion

In the group *Triplostichae* or subsection *Grovesia*, to which *Chara strigosa* belongs, *C. tenuispina* is most similar, but differs by having only solitary spine-cells. *C. aspera* sometimes looks very similar (Krause, 1983) but can be easily delineated by its dioecy and solitary spine cells.

46.6 Distribution

See Fig. 46.3.

The species has a disjunct distribution, with southern and northern centres. The southern centre includes Morocco, Central Europe (including Bosnia), Caucasus (Azerbaijan) and Tajikistan, while the northern centre includes Fennoscandia (including the Murmansk area in Russia; Blinova and Koistinen, 2013), Poland, Belarus, Russia, Latvia, Lithuania, Estonia and North Kazakhstan. A record from Uzbekistan is erroneous (LE, checked by R. Romanov), records from South Kazakhstan (Kostin, 1987) are doubtful because of clear differences in environment of reported sites (R. Romanov pers. comm.). The southern finds are located in alpine areas at high altitudes (Morocco 2090 m, Switzerland c. 1500 m, Bosnia 1400 m, France 569–1059 m, Azerbaijan 1553 m, Tajikistan 2195 m). Haas (1994) reports altitudes of 1000–2000 m for Central Europe. Northern records are from boreal to arctic areas and lower altitudes (20–485 m). The occurrence of *Chara strigosa* in Central Europe, Caucasus, Tajikistan and North Africa can be explained by dispersal by birds. It has been demonstrated both by experiments and by observations that such long distant dispersal is possible. Also eastwords different birds are possible vectors for dispersal of *Chara strigosa*, for example greylag goose (*Anser anser*) (Madge and Burn, 1992).

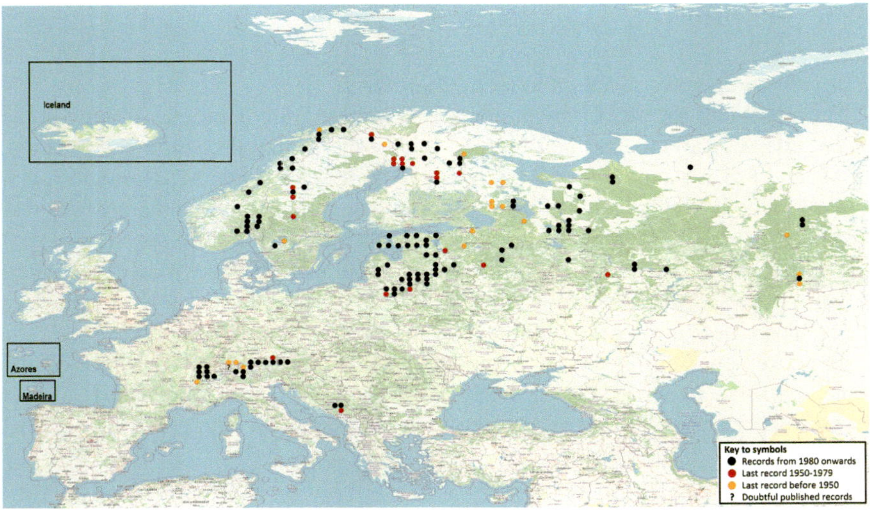

Fig. 46.3 European distribution of *Chara strigosa*. Black dots are records from 1980 on, red dots from 1950 to 1979 and yellow dots before 1950. Map prepared by Heiko Korsch

46.7 Status and Threats

The species is assessed as LC (Least Concern) in the European Red List (see Chap. 9). *Chara strigosa* is threatened by eutrophication (Hutorowicz and Langangen, 1998; Langangen, 2007a) and climate change (Auderset Joye and Schwarzer, 2012). While its occurrence in the German alpine lakes seems to be more or less stable (Jorda and Becker, 2016), declines have been observed in Switzerland (Auderset Joye et al., 2002) and Sweden (Langangen, 2007a).

46.8 Remarks

There is a difference in the fertility of the species in its northern and southern area, with a presumed much higher fertility in Fennoscandia. This fact is important when one suggests in which area the species can have evolved. High fertility and relatively common occurrence in Fennoscandia and adjacent areas can be taken as circumstantial evidence for an evolution of the species in this area.

References

Auderset Joye, D. (1993). *Contribution à l'Écologie des Characées de Suisse*. Thesis, Université de Genève [in French].

Auderset Joye, D., & Schwarzer, A. (2012). *Rote Liste Armleuchteralgen. Gefährdete Arten der Schweiz, Stand 2010* (72 pp.). Bundesamt für Umwelt BAFU & Laboratoire d'Ecologie et de Biologie Aquatique der Universität Genf [in German].

Auderset Joye, D., Castella, E., & Lachavanne, J.-B. (2002). Occurrence of Characeae in Switzerland over the last two centuries (1800–2000). *Aquatic Botany, 72*, 369–385.

Bailly, G., & Schaefer, O. (2010). *Guide illustré des Characées du nord-est de la France* (96 pp.). Conservatoire Botanique National de Franche-Comté [in French].

Blindow, I., & Langangen, A. (1995). Kransalger i Jämtland. *Svensk Botanisk Tidskrift, 89*, 111–117. [in Swedish].

Blinova, I. V., & Koistinen, M. (2013). A review of Characeae (Charophyta) in Murmansk Region (Russia) with notes on a new record of *Chara virgata*. *Memoranda Societatis pro Fauna Et Flora Fennica, 89*, 57–64.

Corillion, R. (1957a). Les Charophycées de France et d'Europe Occidentale. *Bulletin de la Société Scientifique de Bretagne, 32*(Hors série I & II), 1–499 [in French].

Haas, J. N. (1994). First identification key for charophyte oospores from central Europe. *European Journal of Phycology, 29*, 227–235.

Hollerbach, M. M., & Krassavina, L. K. [Голлербах, М. М. & Л. К. Красавина] (1983). Определитель пресноводных водорослей СССР. 14. *Charophyta* [Determination key of freshwater algae of USSR]. Nauka, Leningrad [in Russian].

Hutorowicz, A., & Langangen, A. (1998). *Chara strigosa* A. Braun (Characeae) in Poland. *Acta Societatis Botanicorum Poloniae, 67*, 287–290.

John, D. M., Moore, J. A., & Green, D. R. (1990). Preliminary observations on the structure and ornamentation of the oosporangial wall in *Chara* (Charales, Chlorophyta). *British Phycological Journal, 25*, 1–24.

Jorda, C., & Becker, R. (2016). *Chara strigosa*. In: Arbeitsgruppe Characeen Deutschlands (Ed.), *Armleuchteralgen: Die Characeen Deutschlands* (pp. 336–345). Springer [in German].

Kostin, V. A. [Костин, В.А.] (1987). Материалы к изучению экологии харовых водорослей водоемов Или-Балхашского бассейна [Materials for the study of the ecology of charophytes in reservoirs of the Ili–Balkhash basin]. *Ботанические материалы гербария Института ботаники Академии Наук Казахской ССР [Botanicheskie Materialy Gerbariya Instituta Botaniki Akademii Nauk Kazakhskoy SSR], 15*, 128–133 [in Russian].

Krause, W. (1997). Charales (Charophyceae) Süßwasserflora von Mitteleuropa. In: Ettl, H., G. Gärtner, H. Heynig, & D. Mollenhauer (Eds.), *Süßwasserflora von Mitteleuropa* (Vol. 18). Fischer, Jena [in German].

Krause, W. (1983). Zum Formenkreis der *Chara aspera* Will. in Bayern. *Berichte der Bayerischen Botanischen Gesellschaft, 54*, 151–160. [in German].

Langangen, A. (2000). On the morphology and ecology of *Chara strigosa* A. Braun (Charophyceae) in Scandinavia, with notes on its total distribution. *Aquilo Serie Botanica, 38*, 1–12.

Langangen, A. (2007a). *Charophytes of the Nordic countries*, Oslo: Saeculum ANS.

Langangen, A. (2007b). *Kransalger og deres forekomst i Norge: med korte beskrivelser av viktige norske lokaliteter* (76 pp.). Saeculum, Oslo [in Norvegian].

Madge, S., & Burn, H. (1992). *Wildfowl. An identification guide to the ducks, geese and swans of the world*. Helm, London.

Mandal, D. K., Blaženčić, J. M., & Ray, S. (2002). SEM study of compound oospore wall ornamentation of some members of Charales from Yugoslavia, Croatia and Slovenia. *Archives of Biological Sciences, 54*, 28–34.

Migula, W (1889–1897). Fünfter Band: Die Characeen. In: A. Fischer, E. Fischer, F. Hauck, G. Limpricht, C. Luerssen, W. Migula, H. Rehm, P. Richter, & G. Winter (Eds.), *Dr. L. Rabenhorst's Kryptogamenflora von Deutschland, Oesterreich und der Schweiz* (2nd ed.). Eduard Kummer [in German].

Nonweiler, G. (1907). *Morphologische und Physiologische Untersuchungen an Chara strigosa Braun* (48 pp.) (Dissertation Universität Zürich) [in German].

Romanov, R. E. (2021). Typification of *Chara* (Charales, Charophyceae) species described by C.F. Lessing, F.J. Ruprecht and M.M Hollerbach. *Notulae Algarum, 213*, 1–7.

Romanov, R. E., Chemeris, E. V., Vishnyakov, V. S., Chepinoga, V. V., Azovskii, M. G., Kuklin, A. P., & Timofeeva, V. V. (2014). *Chara strigosa* (Streptophyta) in Russia. *Botanicheskii Zhurnal, 99*, 1148–1161. [in Russian].

Urbaniak, J., & Gąbka, M. (2014). *Polish Charophytes. An illustrated guide to identification* (120 pp.). Uniwersytet Przyrodniczy we Wroclawiu [in Polish].

Urbaniak, J. (2011). A SEM and light microscopy study of the oospore wall ornamentation in Polish charophytes (Charales, Charophyceae)—genus *Chara*. *Nova Hedwigia, 93*, 1–28.

Vaarama, A. (1954). On the ecology and Fennoscandian area of *Chara strigosa* A. Br. *Vegetatio, 1954*, 177–184.

Vishnyakov, V. S., Romanov, R. E., Komarova, A. S., Belyakov, E. A., Moseev, D. S., Churakova, E. Yu., Czhobadze, A. B., & Philippov, D. A. (2021). New charophyte records (Characeae) in European Russia. *Botanicheskii Zhurnal, 106*, 61–76 [in Russian].

Wood, R. D. (1965). *Monograph of the Characeae*. Weinheim: J. Cramer.

Wood, R. D., & Imahori, K. (1964). *Iconograph of the Characeae*. Weinheim: J. Cramer.

Chapter 47
Chara subspinosa

Irmgard Blindow, Roman Romanov, Hendrik Schubert, Nick Stewart, and Klaus van de Weyer

Scientific name: *Chara subspinosa* Rupr., Symb. Pl. Ross.: 225. 1846

Lectotype (T. Gregor et al. in Arbeitsgruppe Characeen Deutschlands 2016: 345): in lacu horti Imp. Gatschinae/3. Aug. 1846/R[uprecht] (LE A0000421).

= *Chara hispida* var. *rudis* A. Braun in Braun, Rabenhorst & Stizenberger, Charac. Eur. 1/4. 1857.
Lectotype (Wood, 1965: 145): In einem Teiche bei Constanz/1856/Stizenberger (NY 9945641); Braun, Rabenhorst & Stizenberger, Characeen Eur. 4a. 1857.

≡ *Chara rudis* (A. Braun) Leonh., Verh. Naturf. Vereins Brünn 2: 66. 1864.

≡ *Chara hispida* L. f. *rudis* (A. Braun) R. D. Wood, Taxon 11: 9. 1962, n. inv.

= *Chara spinosa* var. vaillantii Rupr., Distr. Crypt. Vasc. Ross. 15. 1845, n. ill.

I. Blindow (✉)
Universität Greifswald, Biologische Station Hiddensee, 18565 Kloster, Germany
e-mail: blindi@uni-greifswald.de

R. Romanov
Nišice Ujtin Potok B.B, Dobra Voda, 85356 Bar, Montenegro
e-mail: streptophytes@gmail.com

H. Schubert
Institut für Biowissenschaften, Universität Rostock, Albert-Einstein-Straße 3, 18059 Rostock, Germany
e-mail: hendrik.schubert@uni-rostock.de

N. Stewart
4 Hexton Road, Glastonbury BA6 8HL, UK
e-mail: nfstewart@freeuk.com

K. van de Weyer
lanaplan Lobbericher Str. 5, 41334 Nettetal, Germany
e-mail: klaus.vdweyer@lanaplan.de

H. Schubert et al. (eds.), *Charophytes of Europe*,
https://doi.org/10.1007/978-3-031-31898-6_47

Lectotype: (Romanov, 2021: 4): [Latvia], Curonia, Durben [Durbe]/[no date]/ Fleischner (LE A0000411).

Nomenclatural note: Romanov (2021) interpreted the name *Chara spinosa* Vaillantii* [the asterisk indicating a new species] described by Ruprecht (1845) as *Chara spinosa-vaillantii* Rupr. and lectotypified it with a specimen from Latvia. But it seems also possible to regard this name as *Chara spinosa* var. *vaillantii* Rupr.

Local names: BELARUSIAN: Хара грубая; ENGLISH: Rugged stonewort; ESTONIAN: Krobe mändvetikas; GERMAN: Furchenstachelige Armleuchteralge; LITHUANIAN: Stambusis maurabragis; NORWEGIAN: Smaltaggkrans; POLISH: Ramienica zwyczajna; RUSSIAN: Хара почти-колючая, Хара грубая; SWEDISH: Spretsträfse; UKRAINIAN: Хара груба.

47.1 Morphology

See Figs. 47.1, 47.2 , 47.3 and 47.4.

Habitus: Plants are dark to greyish green due to encrustation. Although they are robust and regularly branched, the plants have an "empty" appearance due to their long branchlets, but relatively short spine-cells, stipulodes, and bract-cells (Figs. 47.1 and 47.2).

Size: *Chara subspinosa* is of medium to large size and robust, 20-90 cm long.

Main axis: Axis diameter is about 0.9 to 2.5 mm. Internodes are usually long; they can reach a length of up to 10 cm, far longer than the branchlets.

Cortex: *Chara subspinosa* is regularly diplostichous and always distinctly aulacanthous (Figs. 47.1 and 47.3). The primary cortex cell rows are often almost covered by the secondary cortex cell rows.

Spine-cells: Spine-cells are acute or blunt, (far) shorter to about as long as or somewhat longer than the axis diameter, 300 to 2200 μm long. In most individuals, they are macroscopically obscure at least in some internodes, with two spine-cells arranged in opposite directions within the deep cortex furrows and hardly protruding. Spine-cells can also occur in bunches of 3, also in this case arranged in the longitudinal direction and/or protruding from the cortex, forming y-shaped clusters, appressed to the cortex and often deciduous on older main axis parts. Single spine-cells occur occasionally (Figs. 47.1 and 47.3).

Stipulodes: *Chara subspinosa* has two stipulode tiers of equal length. Stipulodes are acute to obtuse and range from 300 to 2000 μm, in most individuals shorter than the axis diameter (Fig. 47.3).

Branchlets: Whorls consist of 7–10 long branchlets with 6–8 segments, of which 4–7 are corticate and 1–2(4) ecorticate. The corticate part of the branchlet is far

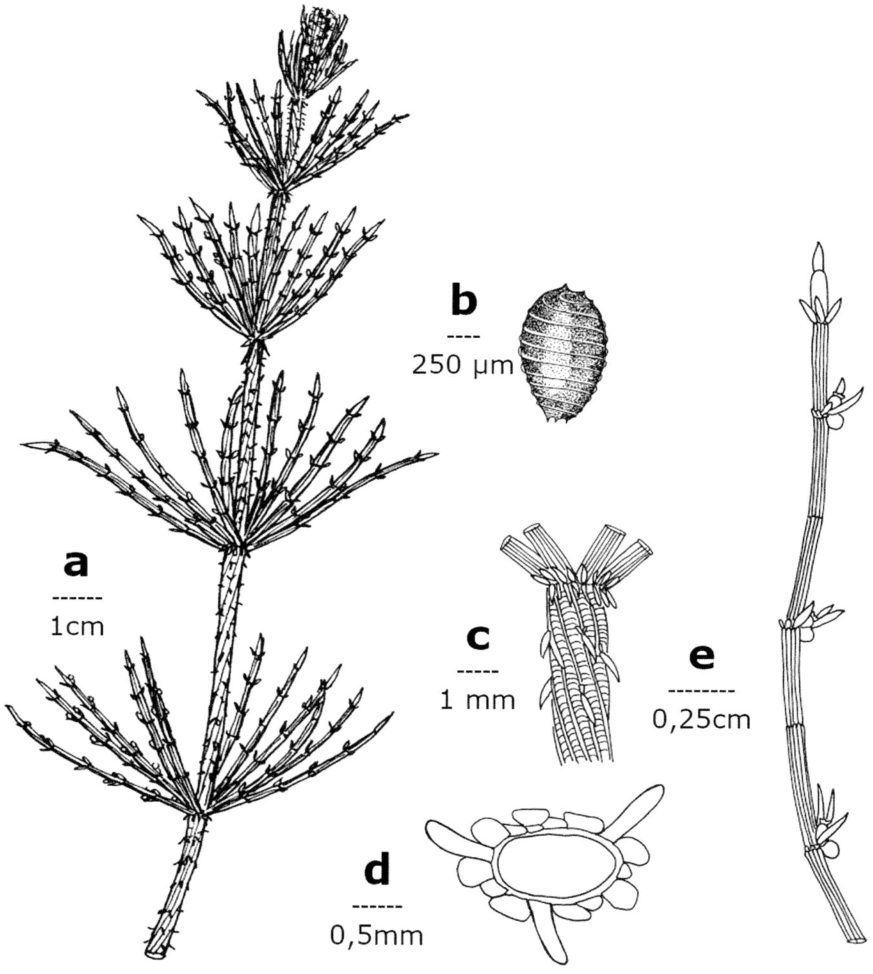

Fig. 47.1 *Chara subspinosa.* **a** Habitus; **b** Oospore; **c** Main axis section showing the strongly aulacanthous cortication; **d** Main axis cross-section; **e** Part of branchlet with gametangia. Drawings after Werner Krause, modified by Dagmar Wassong in van de Weyer and Schmidt (2018)

longer than the ecorticate part, only exceptionally have several ecorticate, elongate segments been observed.

Bract-cells: The adaxial bract-cells range from 300 to 2300 μm and are longer than, in most collections more than twice the length of the abaxial ones, but short compared to the length of, the branchlets (Fig. 47.1).

Bracteoles: Bracteoles are (almost) as long as the ripe oogonium or considerably longer and range from 400 to 3000 μm.

Fig. 47.2 In situ photographs of *Chara subspinosa*. Left: Dense vegetation. Right: Habitus of *C. subspinosa*, showing long branchlets and short spine-cells and bract-cells, which gives the plant its typical "empty" appearance. Photographs: Klaus van de Weyer

Sexuality: *Chara subspinosa* is monoecious. Gametangia are conjoined and can be found on the first 3–5 branchlet segments. The species is often sterile (Blindow et al., 2021; Bociąg and Rekowska, 2012).

Oogonia: According to Wood (1965), the coronula is spreading, but only oogonia with straight coronula were found in our own material. For data about quantitative morphometry, see Table 47.1.

Oospores: Groves and Bullock-Webster (1924) mention a basal cage, which in their drawing is replaced by claws. This cage is also mentioned by Wood (1965). A cage and four claws are visible in the drawing by Hollerbach and Krassavina (1983), but not mentioned in their text. Boszke and Bociąg (2008) found a high intrapopulation variability in oospore sizes. Oospores of plants growing in deeper water were smaller than oospores of plants growing in shallow water, and oospores were larger in plants growing in dense aggregations than in scattered individuals. Table 47.2 gives information about the quantitative characters and colour of the oospores. Fig. 47.4 shows an SEM image of an oospore.

Antheridia: Table 47.3 gives information about the diameter of the antheridia.

Bulbils: Bulbils have not been described.

47.2 Phenology and Reproduction

Chara subspinosa hibernates as a green plant (Blindow, 2009; Langangen, 2007; Olsen, 1944; Pereyra-Ramos, 1981; Skurzyński and Bociąg, 2011; Wahlstedt, 1875). Young shoots can be found even during the winter (Pereyra-Ramos, 1981) but are more common in the spring (Olsen, 1944). The species fruits between May and September. Ripe oospores are formed from July onwards, but oospores from the

Fig. 47.3 Details of *Chara subspinosa*. **a** Shoot section, spine-cells are protruding from the cortex; **b** Stipulodes arranged in two regular tiers; **c** Distinctly heterostichous cortex; spine-cells are arranged in longitudinal direction in the cortex furrows; **d** *C. subspinosa* from Lake Flämsjön in Sweden. The cortex is distinctly aulacanthous, spine-cells are y-shaped and protruding, but appressed to the cortex. Photographs: **a**, **b**, **c**—Klaus van de Weyer, **d**—Tina Kyrkander

year before can still be found on the plants in the following spring (Langangen, 2007; Olsen, 1944). Sterile plants are common and dominate in Sweden (Blindow, 2009).

Chara subspinosa can form dense meadows in calcium-rich lakes (see Fig. 47.1). Biomasses of up to 7.9 kg FW m^{-2} have been measured in a Polish lake (Pereyra-Ramos, 1981). In the shallow Lake Krankesjön, Sweden, meadows of *C. subspinosa*

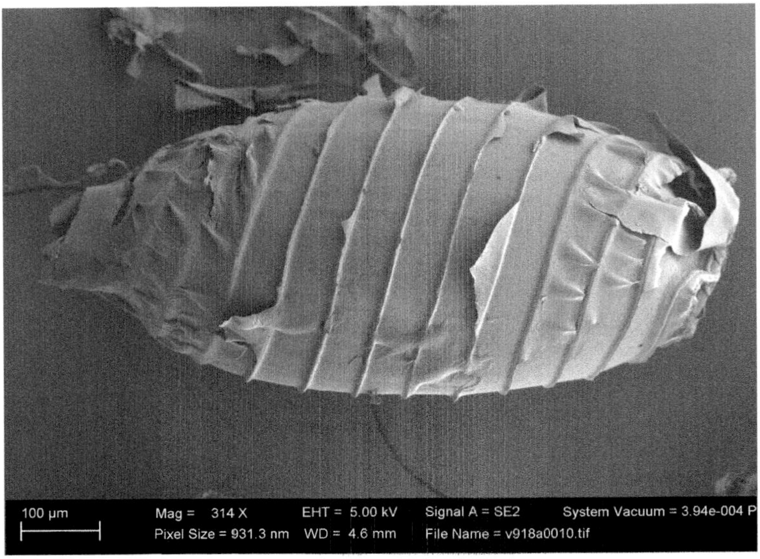

Fig. 47.4 Oospore details. SEM image of an oospore from Olga's Pond, St. Petersburg, Photograph: Michelle T. Casanova

Table 47.1 Oogonia morphometry

Authors	Length	Width	Coronula length	Coronula width	Number of helical stripes
Migula (1889–1897)	1000–1100	–	–	–	–
Groves and Bullock-Webster (1924)	900–1000	625–700	180	350	13–14
Corillion (1957)	900–1000	625–700	–	–	–
Wood (1965)	900–1050	600–630	180–225	350–375	12–14
Hollerbach and Krassavina (1983)	900–1150	600–720	165–225	285–375	12–14
Krause (1997)	1000–1100	600–700	200	350	12–14
Holtz [no date]	1210	630–660	–	–	–
Own data	760–940	570–690	135–190	215–350	11–15
Total range	**760–1210**	**570–720**	**135–225**	**215–375**	**11–15**

Length in the first column excludes the coronula; its length and width are given separately, all values for length and width in μm

Table 47.2 Oospore morphometry

Authors	Length	Width	Number of striae	Colour
Migula (1889–1897)	600–700	–	12	Dark brown/black
Groves and Bullock-Webster (1924)	675–750	425–450	12–13	Dark brown
Wood (1965)	645–750	425–495	10–13	Dark brown
Hollerbach and Krassavina (1983)	645–840	425–495	10–13	Dark brown
Krause (1997)	650–700	–*	11–13	Dark reddish brown to black
Boszke and Bociąg (2008)	600–817	350–516	8–13	–
Holtz [no date]	700–850	460–590	–	Dark brown
Total range	**600–850**	**350–590**	**8–13**	

All values for length and width in μm
*Krause (1997) gives values of 650–700 μm for oospore width, which most probably is erroneous.

Table 47.3 Antheridia morphometry

Authors	Diameter (μm)
Groves and Bullock-Webster (1924)	400–500
Wood (1965)	380–500
Hollerbach and Krassavina (1983)	400–500
Krause (1997)	400
Own data	390–620(730)
Total range	**380–620**

Own data are measurements of dry material, Wood (1965) measured rewetted antheridia. Krause (1997) does not give any information on the method

have increased during the last decade, while other charophytes such as *C. tomentosa* have decreased (Blindow et al., 2021), which indicates that once established, the species is highly competitive. Thallus fragments form a propagule bank which is assumed to be more important in population maintenance than oospores, which have been found to have only low germination rates (Skurzyński and Bociąg, 2011).

47.3 Habitat Conditions and Ecology

Habitat type: *Chara subspinosa* is restricted to freshwater and prefers standing water. Its most important habitats are mesotrophic natural and artificial lakes (Korsch et al., 2011; Krause, 1997; Langangen, 2007; Olsen, 1944; Stewart and Church,

1992; Stroede, 1933). In Sweden, *C. subspinosa* is restricted to the Natura 2000 habitat 3140 (calcium-rich lakes with dominance of charophytes; Blindow et al., 2021; European Commission, 1992). Also in other countries, for example Britain and Ireland, marl lakes are the most important habitat. Only occasionally, the species has been found in running water (Groves and Bullock-Webster, 1924; Hamann, 1999; Moore, 1986), but it can be dominating even in this ecotype, such as in some streams in southern Germany (K. van de Weyer, own observations). It also occurs in small water bodies (Hamann, 1999; Olsen, 1944; Stewart and Church, 1992), but (unlike *C. hispida*) never in habitats that dry out.

Sediment: *Chara subspinosa* occurs on soft bottoms, mainly on calcium-rich gyttja (Blindow, 2009; Kabus and Mauersberger, 2011), in some sites covered with organic material such as leaves and detritus from reeds (Schubert et al., 2016).

Nutrient conditions: *Chara subspinosa* occurs in oligo- and mesotrophic water. In its sites, concentrations of 2–10 $\mu g\ L^{-1}$ SRP and 11–132 $\mu g\ L^{-1}$ total phosphorus have been recorded (Boszke and Bociąg, 2008; Pełechaty et al., 2010; Périllon, 2018; Schubert et al., 2016).

Water chemistry: *Chara subspinosa* is restricted to freshwater and prefers hard water (Blaženčić et al., 2006; Blindow, 2009; Krause, 1997; Langangen, 2007). pH-values vary between 7.0 and 8.5. Ca-concentrations of 34–65 mg L^{-1} were recorded (Boszke and Bociąg, 2008; Pełechaty et al., 2010; Périllon, 2018; Schubert et al., 2016, own data).

Depth range: *Chara subspinosa* occurs between 0.5 and around 9 m depth (Olsen, 1944; Pereyra-Ramos, 1981; Roden et al., 2020; Langangen, 2007). A maximum depth of 9.5 m has been recorded (Spiess et al., 2010).

Associated species: *Chara subspinosa* forms dense monospecific stands with biomasses of >2 kg m^{-2} fresh weight (Pereyra-Ramos, 1981). Accompanying species are other characeans (*Chara tomentosa, C. globularis, C. aspera, C. contraria, C. filiformis, C. papillosa, Nitellopsis obtusa, Nitella flexilis, N. opaca*) and various vascular plants (e.g. *Myriophyllum spicatum, Stuckenia pectinata, Potamogeton lucens, P. perfoliatus, P. acutifolius, Utricularia vulgaris, Hippuris vulgaris, Elodea canadensis, Stratiotes aloides, Ranunculus circinatus, Ceratophyllum demersum, Fontinalis antipyretica*) (Bailly and Schaefer, 2010; Blindow, 2009; Kufel and Kufel, 2002; Kyrkander, 2007; Pereyra-Ramos, 1981, own observations). Sometimes the species is associated with *Chara hispida,* in which case the distinction of both species can be hampered by the presence of transition forms (Blindow, 2009).

47.4 Variability

Lengths of spine-cells, stipulodes, and bract-cells are highly variable, as is the arrangement of the spine-cells (see above).

47.5 Identification and Risk of Confusion

Chara subspinosa belongs to the *Hartmania* group, a complex of coarse, diplostichous *Chara* spp. that are hard to separate from each other. Within the *Hartmania* group, *C. subspinosa* belongs together with *C. hispida* f. *hispida* and *C. hispida* f. *crassicaulis* to a cluster of aulacanthous freshwater taxa (see Chap. 15). Outside of this cluster, only *C. horrida* is aulacanthous. In contrast to *C. subspinosa*, *C. horrida* is restricted to coastal brackish water, has a very "spiny" appearance, irregularly arranged stipulode tiers and larger antheridia. Oospore characters may further distinguish both species: Krause (1997) mentioned the presence of two claws in oospores of *C. horrida*, while oospores of *C. subspinosa* are characterised by a basal cage and the presence of 3–4 claws. Because of it's strongly aulacanthous cortex, *C. subspinosa* is unlikely to be confused with the tylacanthous species of the *Hartmania* group.

The distinction between *C. subspinosa* and *C. hispida* is difficult even for experts and hampered by the fact that the species-specific characters were not previously agreed upon. Consequently, different authors used different criteria to identify *C. subspinosa*, in most cases (with the exception of Wood, 1965) without discussing these different concepts. Due to problems in distinguishing both taxa, several authors gave *C. subspinosa* a taxonomic rank below the species level. Corillion (1957) treated *C. subspinosa* as a subspecies, Moore (1986) as a variety of *C. hispida*, while Migula (1889–1897), Groves and Bullock-Webster (1924), Hollerbach and Krassavina (1983) and Stewart and Church (1992) attribute the taxon to species rank. Also Krause (1997) treated *C. subspinosa* as a separate species, but included it within *C. hispida* in his determination key. Collectors who followed his key may therefore have misdetermined *C. subspinosa* as *C. hispida*. A thorough check of herbarium collections is therefore advised.

The extremely heterostichous, aulacanthous cortex (Fig. 47.3) is the only character all authors agree upon, also visible in the drawing by Migula (1889–1897: 620) and mentioned by him. In *C. subspinosa*, spine-cells are "antiparallel" and rather "hidden" within the furrows of the cortex or arranged in y-shaped clusters and appressed to the cortex (Fig. 47.3), while *C. hispida* always has spine-cells protruding from the cortex (see Fig. 37.3). This character was also suggested by Groves and Bullock-Webster (1924) to distinguish both taxa. Extremely heterostichous cortex and spine-cells appressed to the cortex are also visible in the type material (in LE) and by us treated as the main criteria to separate the two taxa. Other characters are regarded as secondary criteria to distinguish the taxa. The combination of long branchlets with short bract-cells and short stipulodes gives the plants a rather "empty" and “smooth” appearance in contrast to the “spiny” *C. hispida* (Krause, 1997). Mean values of stipulodes, bract-cells, and spine-cells are in fact lower in *C. subspinosa* than in *C. hispida*, while mean values of branchlet lengths are longer, but a high overlap between these taxa hampers a separation (Urbaniak, 2010; own results). *C. subspinosa* has blunter spine-cells, obtuse to broadly acute and never slender and needle-like as is common in *C. hispida*.

A clear distinction from *C. hispida* is not always possible, as transition forms also occur (Blindow, 2009; Urbaniak and Combik, 2013). However, in Britain and Ireland the distinction seems to be more marked than in central Europe. *C. subspinosa* and *C. hispida* have not been distinguished genetically, while *C. subspinosa* and *C. papillosa* are clearly separated (Nowak et al., 2011; Urbaniak and Combik, 2013; Urbaniak and Kwiatkowski, 2019).

Chara subspinosa can be confused with *Chara vulgaris* var. *papillata*, which is robust and strongly aulacanthous, but can be separated from this taxon by the presence of clustered spine-cells.

47.6 Distribution

Chara subspinosa occurs in central, eastern and northern Europe including the Balkans (Blaženčić et al., 2006; Corillion, 1957; Wood, 1965), but is lacking in Finland (Langangen et al., 2002). While the species in Sweden occurs as far north as Jämtland (Blindow and Langangen, 1995), in Norway it is restricted to the southern part of the country (Langangen, 2007). The taxon is lacking from the brackish water of the Baltic Sea (Schubert and Blindow, 2004). In contrast to *C. hispida*, it is restricted to the area of last glaciation (Fig. 47.5).

In Asia, *Chara subspinosa* has a scattered distribution with records from low-mountain landscapes of Southern Ural, Altai and Sayan Mountains, Baikal Siberia, Georgia and Mongolia (Hollerbach and Krassavina, 1983; Romanov et al., 2022a).

47.7 Status and Threats

Chara subspinosa is assessed as Critically Endangered in the national Red List of Denmark, as Endangered in Germany, Poland and SE Europe, as Vulnerable in Norway, Sweden and Montenegro and as Near Threatened in Britain. Because of its frequent occurrence, the species is not included in the Red Lists of Ireland, Estonia and Lithuania. Due to determination problems, the species has not been assessed in several countries. In the European Red List, it is assessed as LC (Least Concern) (see Chap. 9).

As *C. subspinosa* has often been mistaken for *C. hispida*, it is not possible to give reliable information on changes in population sizes. In Germany, declines were

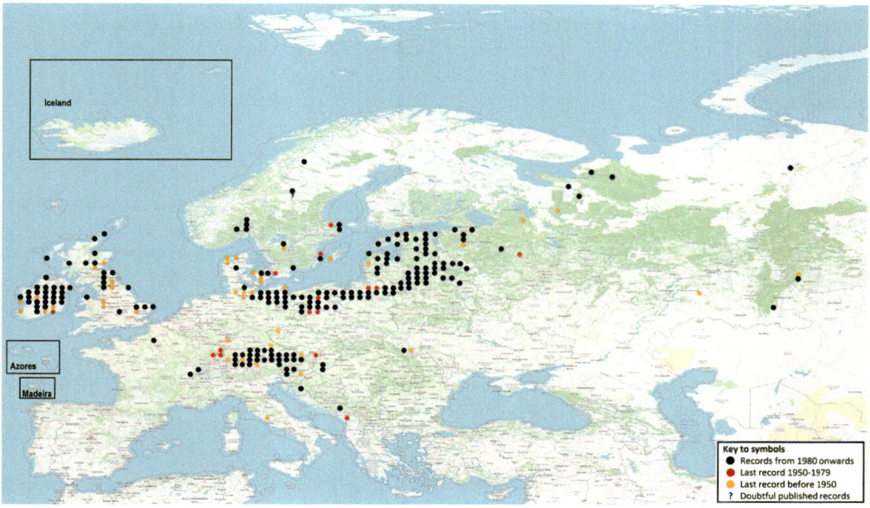

Fig. 47.5 European distribution of *Chara subspinosa*. For worldwide distribution see Korsch (2018). Black dots are records from 1980 onwards, red dots last record from 1950 to 1979 and yellow dots last record before 1950; doubtful published records (if applicable) are marked with "?" superimposed. Map prepared by Heiko Korsch

mentioned for both Schleswig–Holstein and Brandenburg (Hamann, 1999; Kabus and Mauersberger, 2011), while in Bavaria, establishment in several artificial water bodies has been observed (H. Korsch pers. comm.). Eutrophication seems to be a major reason for decline causing increased turbidity and encouraging competition from vascular plants and filamentous algae. In Ireland and Scotland an additional problem is the leaching of humic material from cut-over boglands. This colours the water, reducing light penetration and causing the reduction and eventual break-up of the deeper water *Chara* beds (Roden et al., 2020). Dense populations of carp species and grass carp are also a problem in some countries (Hutorowicz and Dziedzic, 2008; Kłosowski et al., 2006). In Sweden, *C. subspinosa* has disappeared from several eutrophicated lakes such as Lakes Ringsjön and Häljasjön in Scania and Lake Ändsjön in Jämtland (Blindow, 2009; Blindow et al., 2021; Forsberg and Forsberg, 1961), while some new sites have been colonised (Blindow, 2009).

References

Bailly, G., & Schaefer, O. (2010). *Guide illustré des Characées du nord-est de la France* (96 pp.). Conservatoire Botanique National de Franche-Comté, Besançon [in French].

Blaženčić, J., Stevanović, B., Blaženčić, Z., & Stevanović, V. (2006). Distribution and ecology of charophytes recorded in the west and central Balkans. *Cryptogamie, Algologie, 27*, 311–322.

Blindow, I. (2009). *Åtgärdsprogram för hotade kransalger: Arter i kalkrika sjöar 2008–2011. Trådsträfse (Chara filiformis), spretsträfse (Chara rudis)*, stjärnslinke (*Nitellopsis obtusa*) (76 pp.). Naturvårdsverket. Rapport 5848 [in Swedish].
Blindow, I., & Langangen, A. (1995). Kransalger i Jämtland. *Svensk Botanisk Tidskrift, 89*, 111–117. [in Swedish].
Blindow, I., Carlsson, M., & van de Weyer, K. (2021). Re-establishment techniques and transplantations of Charophytes to support threatened species. *Plants, 2021*(10), 1830.
Bociąg, K., & Rekowska, E. (2012). Are stoneworts (Characeae) clonal plants? *Aquatic Botany, 100*, 25–34.
Boszke, P., & Bociąg, K. (2008). Morphological variation of oospores in the population of *Chara rudis* A. Braun in a mesotrophic lake. *Polish Journal of Ecology, 56*, 139–147.
Corillion, R. (1957a). Les Charophycées de France et d'Europe Occidentale. *Bulletin de la Société Scientifique de Bretagne, 32* (Hors série I & II), 1–499 [in French].
European Commission. (1992). *Council Directive 92/43/EEC of 21 May 1992 on the conservation of natural habitats and of wild fauna and flora*. Official Journal of the European Communities L 206/7.
Forsberg, C., & Forsberg, B. (1961). *Chara rudis* ny for Uppland. *Botaniska Notiser, 114*, 108–110 [in Swedish].
Groves, J., & Bullock-Webster, G. R. (1924). *The British Charophyta. Volume II. Chareæ with plates, concluding articles, geological sketch, bibliography and index*. Ray Society, London.
Hamann, U. (1999). *Situationsbericht über die Armleuchteralgen (Charophyceae) Schleswig-Holsteins, Rote Liste der Armleuchteralgen Schleswig-Holsteins* (81 pp.). Landesamt für Natur und Umwelt, Flintbeck [in German].
Hollerbach, M. M., & Krassavina, L. K. [Голлербах, M. M. & Л. К. Красавина] (1983). Определитель пресноводных водорослей СССР. 14. Charophyta [*Determination key of freshwater algae of USSR* (189 pp.)]. Nauka, Leningrad [in Russian].
Holtz, L. [no date]. Zeichnungen und Messungen von Characeen Pommerns von Dr. Paul Hauptfleisch, entsprechend meinen Publikationen. Inv. Nr. 5551. University of Greifswald, Germany [in German].
Hutorowicz, A., & Dziedzic, J. (2008). Long-term changes in macrophyte vegetation after reduction of fish stock in a shallow lake. *Aquatic Botany, 88*, 265–272.
Kabus, T., & Mauersberger, R. (2011). Liste und Rote Liste der Armleuchteralgen (Characeae) des Landes Brandenburg 2011. *Naturschutz und Landschaftspflege in Brandenburg, 20*(4, Beilage), 1–31 [in German].
Kłosowski, S., Tomaszewicz, G. H., & Tomaszewicz, H. (2006). The expansion and decline of charophyte communities in lakes within the Sejny Lake District (north-eastern Poland) and changes in water chemistry. *Limnologica, 36*, 234–240.
Korsch, H. (2018). The worldwide range of the Charophyte species native to Germany. *Rostocker Meeresbiologische Beiträge, 28*, 45–96.
Korsch, H., Raabe, U., Diewald, W., Scheuerer, M., & Göding, H. (2011). Zur Verbreitung der Characeae in Südostbayern. *Hoppea, 72*, 119–138.
Krause, W. (1997). Charales (Charophyceae) Süßwasserflora von Mitteleuropa. In H. Ettl, G. Gärtner, H. Heynig, & D. Mollenhauer (Eds.), *Süßwasserflora von Mitteleuropa* (Vol. 18). Fischer, Jena [in German].
Kufel, L., & Kufel, I. (2002). *Chara* beds acting as nutrient sinks in shallow lakes a review. *Aquatic Botany, 72*, 249–260.
Kyrkander, T. (2007). *Inventering av kransalger i sötvatten 2007* (91 pp.). Länsstyrelsen Västra Götalands Län, Rapport 2007 [in Swedish].
Langangen, A. (2007). *Charophytes of the Nordic countries*. Saeculum ANS.
Langangen, A., Koistinen, M., & Blindow, I. (2002). The charophytes of Finland. *Memoranda Societatis pro Fauna Et Flora Fennica, 78*, 17–48.
Migula, W. (1889–1897). Fünfter Band: Die Characeen. In A. Fischer, E. Fischer, F. Hauck, G. Limpricht, C. Luerssen, W. Migula, H. Rehm, P. Richter, & G. Winter (Eds.), *Dr. L. Rabenhorst's*

Kryptogamenflora von Deutschland, Oesterreich und der Schweiz (2th ed.). Eduard Kummer, Leipzig [in German].
Moore, J. A. (1986). *Charophytes of Great Britain and Ireland*. BSBI Handbook (No. 5). Botanical Society of the British Isles, London.
Nowak, P., Schubert, H., & Schaible, R. (2011). *Genetic analysis of the Characeae. Report for the year 2010/2011* (21 pp.). University of Rostock.
Olsen, S. (1944). Danish Charophyta. *Det Kongelige Danske Videnskabernes Selskab, Biologiske Skrifter,* 3(1), 1–240, pl. 1–2.
Pełechaty, M., Apolinarska, K., Pukacz, A., Krupska, J., Siepak, M., Boszke, P., & Sinkowski, M. (2010). Stable isotope composition of *Chara rudis* incrustation in Lake Jasne, Poland. *Hydrobiologia, 656*, 29–42.
Pereyra-Ramos, E. (1981). The ecological role of Characeae in the lake littoral. *Ekologia Polska, 29*, 167–209.
Périllon, C., van de Weyer, K., Päzolt, J., Kasprzak, P., & Hilt, S. (2018). Changes in submerged macrophyte colonization in shallow areas of an oligomesotrophic lake and the potential role of groundwater. *Limnologica, 68*, 168–176.
Roden, C., Murphy, P., & Ryan, J. (2020). *Benthic vegetation in Irish marl lakes: monitoring habitat 3140. Condition 2011 to 2018* (36 pp.). Irish Wildlife Manuals 124. National Parks and Wildlife Service, Department of Housing, Local Government and Heritage, Dublin.
Romanov, R. E., Anisimova, O. V., Anishchenko, L. N., Efimov, D. Y., Kapitonova, O. A., Kipriyanova, L. M., Konechnaya, G. Yu., Kopyrina, L. I., Kotovshchikov, A. V., Kurganov, A. A., Nikolaenko, S. A., Shilov, M. P., Smirnova, E. V., Vodeneeva, E. L., Zarubina, E. Y., & Zhakova, L. V. (2022a). The noteworthy new records of charophytes (Charales, Charophyceae) from Russian regions: improvement of species distribution ranges in Eurasia. *Botanicheskii Zhurnal, 107*, 466–477.
Romanov, R. E., Efimov, D. Yu., Makeeva, E. G., Shaulo, D. N., Kipriyanova, L. M., Zotina, T. A., Pospelov, I. N., Ebel, A. L., Zarubina, E. Yu., & Polyanskaya, D. Yu. (2022b). The charophytes (Characeae, Charophyceae) from the Yenisey Siberia (Russia). *Turczaninowia, 25*, 19–46. [in Russian].
Romanov, R. E. (2021). Typification of *Chara* (Charales, Charophyceae) species described by C.F. Lessing, F.J. Ruprecht and M.M. Hollerbach. *Notulae Algarum, 213*, 1–7.
Ruprecht, F. J. (1845). *Distributio cryptogamarum vascularium in Imperio Rossico*. Petropoli, Buchdruckerei der Kaiserlichen Akademie der Wissenschaften [in Latin].
Schubert, H., Blindow, I., & van de Weyer, K. (2016). *Chara subspinosa*. In A. C. Deutschlands (Ed.), *Armleuchteralgen: Die Characeen Deutschlands* (pp. 345–354). Springer [in German].
Schubert, H., & Blindow, I. (Eds.). (2004). *Charophytes of the Baltic Sea*. A. R. G. Gantner.
Skurzyński, P., & Bociąg, K. (2011). Vegetative propagation of *Chara rudis* (Characeae, Chlorophyta). *Phycologia, 50*, 194–201.
Spiess, H. J., Bolbrinker, P., Möbius, F., & Waterstraat, A. (2010). Ergebnisse der Untersuchungen submerser Makrophyten in ausgewählten Gewässern Mecklenburg-Vorpommerns. *Botanischer Rundbrief Mecklenburg-Vorpommern, 47*, 4–182. [in German].
Stewart, N. F., & Church, J. M. (1992). *Red data books of Britain and Ireland: Stoneworts*. The Joint Nature Conservation Committee, Peterborough.
Stroede, W. (1933). Über die Beziehungen der Characeen zu den chemischen Faktoren der Wohngewässer und des Schlammes. *Archiv für Hydrobiologie, 25*, 192–229. [in German].
Urbaniak, J. (2010). Analysis of morphological characters of *Chara baltica, C. hispida, C. horrida,* and *C. rudis* from Europe. *Plant Systematics and Evolution, 286*, 209–221.
Urbaniak, J., & Combik, M. (2013). Genetic and morphological data fail to differentiate *Chara intermedia* from *C. baltica*, or *C. polyacantha* and *C. rudis* from *C. hispida*. *European Journal of Phycology, 48*, 253–259.
Urbaniak, J., & Kwiatkowski, P. (2019). Taxonomic studies on the *Chara* section *Hartmania* in Poland based on morphological and molecular data. *PhytoKeys, 135*, 71–90.

van de Weyer, K., & Schmidt, C. (2018). Bestimmungsschlüssel für die aquatischen Makrophyten (Gefäßpflanzen, Armleuchteralgen und Moose) in Deutschland: Band 1: Bestimmungsschlüssel, ed. 2. Landesamt für Umwelt (LfU) Brandenburg. *Potsdam, Fachbeiträge des LfU Brandenburg, 119*, 1–180. [in German].

Wahlstedt, L. J. (1862). *Bidrag till kännedomen om de Skandinaviska Arterna af Växtfamiljen Characeæ*. Lund, i–viii, 1–43 [in Swedish].

Wahlstedt, L. J. (1875). *Monografi öfver Sveriges och Norges characeer*. Bolagets, Christianstad [in Swedish].

Wood, R. D. (1965). *Monograph of the Characeae*. J. Cramer.

Chapter 48
Chara tenuispina

Maciej Gąbka and Jacek Urbaniak

Scientific name: *Chara tenuispina* A. Braun in Flora 18: 68. 1835.

Neotype (Wood, 1965: 209): In Torflöchern auf Wiesen bei Spandau mit *Chara fragilis*, *Utricularia vulgaris* und *minor*, *Sparganium minimum*/Ende Juni bis Anfang August/A. Braun (NY); Braun, Rabenhorst & Stizenberger Char. Exs. No. 74a.

≡ *Chara globularis* var. *tenuispina* (A.Braun) R. D. Wood in Taxon 11: 11. 1962.

? = *Chara kokeilii* A. Braun in Flora 30: 27. 1847.
Lectotype (Wood, 1965: 212): In Gräben am Wörthsee bei Klagenfurth/1844/Kokeil (LD).

Taxonomic note: It is unclear if *Chara kokeilii* should be treated as a synonym of *C. tenuispina*. Plants with ecorticated branches but otherwise similar to *C. tenuispina* are known from scattered sites in South-East Europe (Koletić et al., 2021) but were originally described from Austria (Braun, 1847).

Local names: GERMAN: Dünnstachlige Armleuchteralge; POLISH: Ramienica cienkokolczasta; RUSSIAN: Хара тонкоколючковая; UKRAINIAN: Хара тонкоколючкова.

M. Gąbka (✉)
Department of Hydrobiology, Faculty of Biology, Adam Mickiewicz University Poznań, Uniwersytetu Poznańskiego 6 Str., 61-614 Poznań, Poland
e-mail: gmaciej@amu.edu.pl

J. Urbaniak
Department of Botany and Plant Ecology, Wrocław University of Life and Environmental Sciences, Pl. Grunwaldzki 24a, 50-363 Wrocław, Poland
e-mail: jacek.urbaniak@upwr.edu.pl

H. Schubert et al. (eds.), *Charophytes of Europe*,
https://doi.org/10.1007/978-3-031-31898-6_48

48.1 Morphology

See Figs. 48.1, 48.2 and 48.3.

Habitus: *Chara tenuispina* is a small charophyte, which grows solitary or as several plants together. The plants are delicate and slender, light to fresh green and incrusted (without incrustation to heavily incrusted). *C. tenuispina* is highly branched and spiny (sometimes very spiny) plant.

Size: Plants are usually 10–25 cm high, but 1.5–3 cm high specimens are often found in peatlands. The tallest *C. tenuispina* (45 cm) was found in salt marshes at the Balkans (Kostić, 1937; M. Gąbka unpubl. data).

Main axis: Long and straight (mostly), with 2–5 branches. The main axis is slender to moderately stout, usually 0.3–0.7(1.5) mm in diameter. Internodes are as long as branches or 1–3 times longer than branchlets.

Cortex: Cortification clearly visible, finely ribbed. The cortex is triplostichous, sometimes irregularly, tylacanthous. On older parts of the plant it is isostichous.

Spine-cells: *Chara tenuispina* has very long and very thin spine-cells, usually up to 2–3(4) times longer than the axis diameter. Most spine-cells occur only in the upper parts of main axis; solitary and slender. Spine cells are longest on upper parts of plants, below stipulodes.

Stipulodes: In two rows, elongate and acute; longer than the axis diameter. Upper stipulodes are longer than lower in general.

Branchlets: 7–8(10) in a whorl, straight and short, about 1–2 cm long, occasionally sterile branches up to 7 cm long (Kostić, 1937). Single branchlets are divided into 7–11 segments of which 5–6 are corticated, the end segments are naked, elongate and

Fig. 48.1 Habitus of *Chara tenuispina*. **a** Fresh material; **b** Herbarium specimen; **c** Detail of shoot apex. Photographs: **a**, **c**—Jacek Urbaniak; **b**—Maciej Gąbka

Fig. 48.2 Detail of axis cortication, stipulodes and spine-cells in *Chara tenuispina*. **a** Lower internode with short spine-cells, illustrating triplostichous cortication; **b** Upper internode with very long spine-cells and triplostichous cortex; **c** Upper internode with long stipulodes and spine-cells; **d** Several fertile corticated branchlets with antheridia and oogonia, bract-cells and bracteoles; **e**: a Fertile branchlet in the corticated part, showing long, narrow bract-cells and gametangia. Photographs: Jacek Urbaniak

acute. Individuals with only 1–2 corticated segments of branchlets (or completely uncorticated) were observed.

Bract-cells: Slender and acute, well developed; anteriors very long, 3–4 times longer than oogonia and usually longer than posteriors.

Bracteoles: Bracteoles similar in length to the bract-cells.

Sexuality: *Chara tenuispina* is a monoecious plant, with gametangia conjoined singly at lowest branchlet nodes.

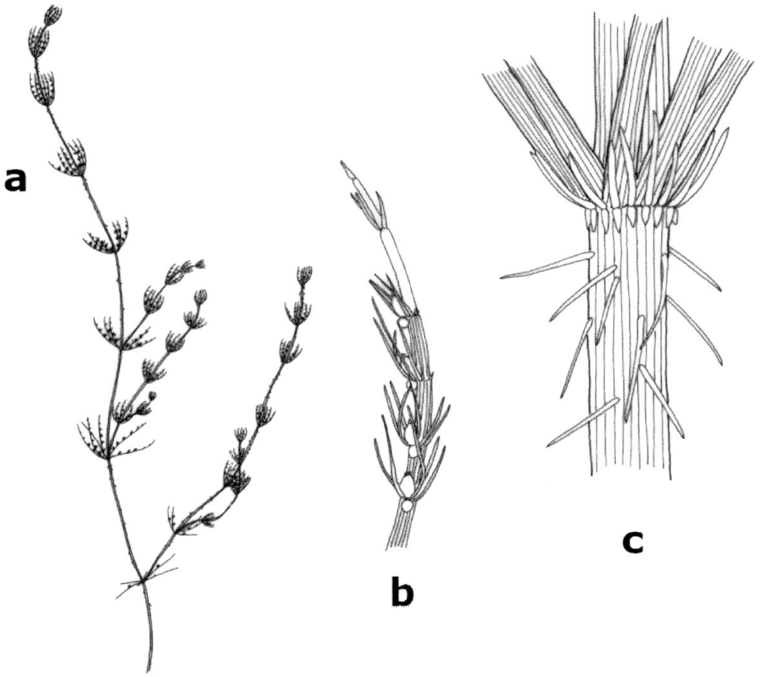

Fig. 48.3 Drawing of *Chara tenuispina*. **a** Whole plant; **b** Fertile branchlet; **c** Whorl base and cortication. Drawings: Kinga Gawrońska, Maciej Gąbka

Oogonia: Elliptical, yellow or yellow-green, 500–800 μm long, 320–560 μm wide (variability of oogonia are presented in Table 48.1). *Chara tenuispina* is a very fertile species, sometimes 2–3 oogonia in 1 whorl of branchlets.

Oospores: Elliptical, mainly dark brown or almost black; variability presented in Table 48.2.

Antheridia: Up to three times smaller than oogonia; orange or reddish. The variability in antheridia diameter is given in Table 48.3.

Bulbils: Not reported on *C. tenuispina* rhizoids.

48.2 Phenology and Reproduction

Phenology and reproduction of *Chara tenuispina* seem to be strictly dependent on water level fluctuations especially in relation to gradual drying of habitats. In Central Europe, *C. tenuispina* is an annual species, and there is no data available for overwintering. The species starts vegetative growth in spring and finishes in autumn.

Table 48.1 Oogonia morphometry of *Chara tenuispina*

Authors	Length	Width	Coronula length	Coronula width	Number of helical stripes
Migula (1889–1897)	650–800	400–650	–	150	12–15
Kostić (1937)	700–798	364–504	–	–	13–14
Dąmbska and Karpiński (1954), Dąmbska (1964)	650–800	400–650	–	150	12–15
Wood (1965)	525–610(800)	375–450	105–150	150–210	12–15
Shoyakubov (1979)	620–1200	320–560	105–120	180–230	12–15
Krause (1997)	500–600(800)	400–500	100–150	150–200	12–15
Urbaniak and Gąbka (2014)	650–765	450–555	–	–	–
Total range	**500–800(1200)**	**320–560**	**100–150**	**150–230**	**12–15**

Length in the first column excludes the coronula, its length and width given separately; all values for length and width in μm

Table 48.2 Oospore morphometry of *Chara tenuispina*

Authors	Length (μm)	Width (μm)	Number of striae	Colour
Migula (1889–1897)	450–520	280–340	11–14	Dark brown
Kostić (1937)	504–546	308–350	11–14	–
Dąmbska and Karpiński (1954), Dąmbska (1964)	450–520	280–340	11–14	Dark brown
Wood (1965)	450–720	280–435	11–14	Dark brown to black
Shoyakubov (1979)	500–800	230–300	11–13	–
Haas (1994)	450–520	280–340	11–14	Brown
Krause (1997)	450–650	300–450	11–14	Black
Urbaniak (2011)	441–522	315–351	9–12	Dark brown to black
Urbaniak and Gąbka (2014)	440–520	315–350	–	Brown to dark brown
Total range	**440–720(800)**	**230–450**	**11–14**	

Table 48.3 Antheridia morphometry of *Chara tenuispina*

Authors	Diameter (μm)
Migula (1889–1897)	250–300
Kostić (1937)	252–308
Dąmbska and Karpiński (1954), Dąmbska (1964)	250–300
Wood (1965)	250–300
Shoyakubov (1979)	300–520
Krause (1997)	250–350
Urbaniak and Gąbka (2014)	252–308
Total range	**250–350(520)**

In summer it can be found in very shallow water depths in small water reservoirs. It also grows in the shade between *Phragmites australis*, but then plants are more elongate. *C. tenuispina* is an ephemeral species that is sometimes not recorded for several years in the same place. Under the influence of competition, like shade from rush vegetation, the abundance of *C. tenuispina* changes. The species may not be observed for several years and after competing vegetation is removed, the abundance rapidly increases. *Chara tenuispina* forms oospores from June to October. In favorable conditions, when the water level is high plants can produce oospores twice a year. Specimens of C. *tenuispina* can also be present in wet places, where water is barely present, producing reproductive organs and fertilised oospores.

48.3 Habitat Conditions and Ecology

Habitat type: *Chara tenuispina* occurs generally in peatlands or in marshes or soda pans. In Central Europe it is a very rare species that can be found in calcareous fens overgrown with *Cladium mariscus,* or in alkaline fens, and connected with peatland habitats like hollow peats, peatland exploitation ponds or ditches (Gąbka, 2007, 2009; Krause and Grüttner, 1990; Urbaniak and Gąbka, 2014). *Chara tenuispina* is not found in lakes (especially in deep and large lakes), however it can occur in the littoral zone overgrown by marsh vegatation characteristic of shallow and vegetated lakes. On the other hand, the occurrence of *C. tenuispina* in shallow and salty, very often ephemeral, aquatic water bodies, is also known. In Southern Europe (e.g., in Hungary, Serbia or Montenegro) it grows in seasonal water bodies, inland polders and in shallow shoreline tracts of lakes (e.g., Blaženčić, 2014; Gavrilović et al., 2018; Vesić et al., 2016). This species was also found in the shallow part of Skadar lake (Blaženčić, 1984) and in gravel pits (Krause and Grüttner, 1990).

Sediment: Organic peat sediment or soft organic substratum rich in calcium. In Skadar Lake *C. tenuispina* is found growing on a silty or silty-sandy substrate (Blaženčić, 1984).

Nutrient conditions: No detailed data on trophic requirements from the entire range of *Chara tenuispina* in Europe are available. In Poland, the species is usually found in fens in very fertile water. Nutrient content depends strictly on water depth. In shallow water (hollow peats) high levels of ammonium nitrogen and phosphate have been measured (M. Gąbka unpubl. data). The ranges of mineral forms of phosphate and ammonium in Polish waters on fens are presented in Table 48.4. Due to the absence of *C. tenuispina* in lakes, it is impossible to determine trophic requirements of the species typical for the classification of lakes. In the fen types and the poor to rich fertility gradient classification, it can be said that this species inhabits moderately to extremely rich fens.

Water chemistry: *Chara tenuispina* is mostly known from freshwater habitats, but also from salty waters as well (it is a moderately halophylic species). The detailed physico-chemical properties of waters are presented in Table 48.4 (M. Gąbka unpubl. data; the data were collected mostly in rich fens in Poland). The content of Ca^{2+}, SO_4^{2-}, NO_3^- and: water color are particularly important in the ecology of *C. tenuispina*. In general, this species inhabits neutral water (pH 7.0), rich in calcium and sulphates, usually with high mineralization and wide range of conductivity (270 up to above 1000 μS cm^{-1}) (M. Gąbka unpubl. data). Waters inhabited by *C. tenuispina* are usually highly coloured.To date, there are no detailed data on the conditions of *C. tenuispina* occurring in saline waters in southern Europe. However, the water habitats are probably rich in soda and sulphate with salinity below 1000 mg dm^{-1} (Vesić et al., 2016). In the continental climate, the occurrence of *C. tenuispina* is

Table 48.4 Water chemistry from Polish fertile fen localities of *Chara tenuispina* (M. Gąbka unpubl. data)

Parameter	Unit	Minimum	Maximum
Colour	[mg Pt l^{-1}]	48	173
pH	–	6.2	7.6
O_2	[mg l^{-1}]	0.4	6.8
Conductivity	[μS cm^{-1}]	272	1010
N-NH_4	[mg l^{-1}]	0.440	5.350
N-NO_3	[mg l^{-1}]	0.000	0.180
P-PO_4	[mg l^{-1}]	0.000	0.153
Ca	[mg l^{-1}]	32.8	203.0
Mg	[mg l^{-1}]	1.8	15.2
Na	[mg l^{-1}]	1.0	37.8
K	[mg l^{-1}]	0.4	17.6
Fe	[mg l^{-1}]	0.01	3.25
SO_4	[mg l^{-1}]	10	189

associated with strong fluctuations in water levels and ion concentrations in shallow waters, which can be very important for the ecology and distribution of the species.

Depth range: *Chara tenuispina* has usually been found at very shallow water depths usually up to 0.3 (0.5) m. The deepest depth for this species was only 0.7 m (Blaženčić, 1984; Blaženčić and Blaženčić, 1983). It often grows at a minimum depth of several centimeters or just on moist ground. It should be noted that water level fluctuations are very important for the ecology of the species.

Associated species: Extremely fertile fens and calcareous fens with *Chara tenuispina* correspond to the *Caricion davallianae* type vegetation. In rich fens *C. tenuispina* co-occurs with brown mosses (e.g., *Drepanocladus polycarpos*, *Calliergon giganteum*, *Calliergonella cuspidata*, *Scorpidium scorpioides* and *Limprichtia* spp.), other charophytes as *C. vulgaris* and *C. virgata* and species from the genus *Utricularia* (especially *U. australis* and *U. minor*). *Chara tenuispina* shows great affinity to places with occurrences of *Schoenoplectus tabernaemontana*, *Thelypteris palustris*, *Phragmites australis* or *Cladium mariscus*. The species forms its own association: *Charetum tenuispinae* Dąmbska 1966 ex Tomaszewicz 1979 inhabiting small water bodies (Gąbka, 2007). In southern Europe, *C. tenuispina* often occurs with *Phragmites australis*. In the Pannonian biogeographic region, *C. tenuispina* coexists together with *C. canescens*, *C. aspera*, *C. tomentosa* and also with *Scirpus maritimus* (Filarszky, 1931; Kusel-Fetzmann, 1979). In a shallow part of Skadar lake *C. tenuispina* also occurs in the community *Potametum perfoliati* (together with the other charophyte species as: *C. globularis*, *C. vulgaris*, *C. kokeilii*, *Nitellopsis obtusa*, *Nitella gracilis*, *N. confervacea*, *Sphaerochara prolifera*, *Tolypella glomerata*, and vascular plants as: *Najas marina* and *N. minor* (Blaženčić, 1984; Blaženčić and Blaženčić, 1983).

48.4 Variability

Migula (1889–1897) distinguished 4 forms of *Chara tenuispina* and Kostić (1937) a further two. However the variability of *C. tenuispina* appears to depend on the particular locality and, most importantly, on the depth of the water.

48.5 Identification and Risk of Confusion

Chara tenuispina can easily be distinguished from *C. globularis* (and it was sometimes treated as a form of *C. globularis;* Wood, 1965) and from *C. virgata* by the presence of long spine-cells, below the whorl of branchlet, which are solitary and longer than the axis diameter. Stipulodes are longer than the axis diameter. Delicate and longer forms of *C. virgata* can be confused with *C. tenuispina*. Sterile *C. aspera* is also similar to *C. tenuispina.* An important feature that distinguishes *C. aspera* from *C. tenuispina* are the single-celled bulbils present on rhizoids of *C. aspera.*

48.6 Distribution

See Fig. 48.4.

Chara tenuispina is found in Southeastern Europe and as yet is unknown from Western Europe. The species was found in 44 localities from 11 countries (Poland—11, Germany—9, North Macedonia—5, Serbia—4, Greece—3, Hungary—3, Austria—2, Montenegro—2, Cyprus—1, Romania—1, Ukraine—1). The species is currently listed from 11 localities in Poland, and from Balkan Peninsula (2 in Montenegro, 3 in Serbia and 3 in Croatia). In the past, it was recorded for 8 localities in Germany, but recently only one occurrence could be confirmed (Krause, 1997; Krause and Grüttner, 1990). Three locations are known from Greece (Langangen and Pavlides, 1999), a few from soda pans from Hungary as well as from Ukrainian steppe (Iakushenko and Borysova, 2012; Palamar-Mordvintseva and Tsarenko, 2004). A common feature of *C. tenuispina*, is that it exists in shallow water sites that are rich in electrolyte content (Table 48.4). In Central Europe *C. tenuispina* is known mainly from grasslands fens, but in Southern Europe from shallow, very often ephemeral, salty aquatic waterbodies.

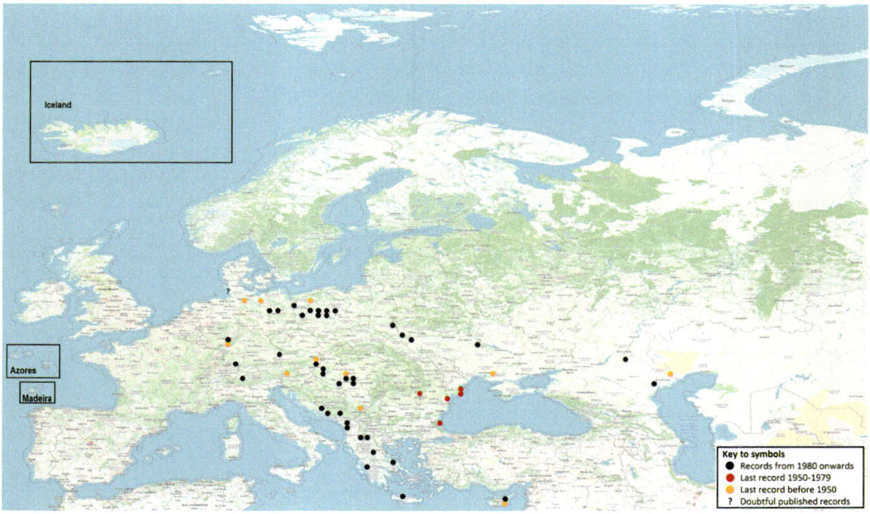

Fig. 48.4 European distribution of *Chara tenuispina*. For worldwide distribution see Korsch (2018). Black dots are records from 1980 on, red dots from 1950 to 1979 and yellow dots before 1950. Map prepared by Heiko Korsch

48.7 Status and Threats

In general, *Chara tenuispina* is one of the rarest and most threatened charophyte species in Europe. The main threats to the occurrence of *C. tenuispina* are anthropogenic draining of peatlands and vegetation succession in these localities.

References

Blaženčić, J. (1984). Rasprostranjenje i ekologija vrste *Nitella gracilis* (Smith) Ag. (Nitellaceae) u Jugoslaviji. *Glasnik Instituta Za Botaniku i Botanicke Bašte Univerziteta u Beogradu, 18*, 31–36. [in Serbian].

Blaženčić, J. (2004). The Ohrid Lake an important centre of Charophyta diversity. In *Proceedings of the 2nd Congress of Ecologists of the Republic of Macedonia with International Participation* (Vol. 6), Oct 25–29, 2003, Ohrid. Special issues of Macedonian Ecological Society, Skopje.

Blaženčić, J. (2014). Overview of the stoneworts (Charales) of Serbia with the estimation of the threat status. *Botanica Serbica, 38*, 121–130.

Blaženčić, J., & Blaženčić, Ž. (1983). Fitocenološka studija zajednica *Charetum fragilis* Corillion 1957 i *Chareto-Nitellopsidetum obtusae* J. Blaž. ass. nova kod Plavnice, na Skadarskom jezeru. Glasnik Republičkog zavoda za zaštitu prirode. *Prirodnjačkog Muzeja Titograd, 16*, 7–13.

Braun, A. (1847). *Chara Kokeilii*, eine neue deutsche Art. *Flora, 30*, 17–29.

Dąmbska, I. (1964). *Charophyta—ramienice.* Państwowe Wydawnictwo Naukowe, Warsawa [in Polish].

Dąmbska, I., & Karpiński, J. (1954). *Ramienice. Klucz do oznaczania gatunków krajowych* (p. 128). Państwowe Wydawnictwo Naukowe, Warszawa [in Polish].

Filarszky, N. (1931). Adatok Horvat-Szavonorszag es Balkan nehany mas orszaganak Chara-vegetaciojahoz. *Magyar Botanikai Lapok. Budapest, 30*, 81–99. [in Hungarian].

Gąbka, M. (2007). Distribution of *Chara tenuispina* A. Braun 1835 (Characeae) in Poland. *Oceanological and Hydrobiological Studies, 36*(Suppl. 1), 241–248.

Gąbka, M. (2009). *Charophytes of the Wielkopolska region (NW Poland): distribution, taxonomy, and autecology*. Bogucki Wydawnictwo Naukowe, Poznań.

Gavrilović, B., Ćirić, M., Vesić, A., Vidaković, D., Novaković, B., & Živanović, M. (2018). Biodiversity overview of soda pans in the Vojvodina region (Serbia). *Journal of the Geographical Institute Jovan Cvijic SASA, 68*, 195–214.

Haas, J. N. (1994). First identification key for charophyte oospores from central Europe. *European Journal of Phycology, 29*, 227–235.

Iakushenko, D., & Borysova, O. (2012). Plant communities of the class Charetea Fukarek ex Krausch 1964 in Ukraine: An overview. *Biodiversity Research and Conservation, 25*, 75–82.

Koletić, N., Rimac, A., Šegota, V., Vuković, N., Blaženčić, J., & Alegro, A. (2021). Updated distribution of rare stonewort *Chara kokeilii* A. Braun (Charophyta) in Southeast Europe with environmental notes. *Cryptogamie, Algologie, 42*, 59–66.

Korsch, H. (2018). The worldwide range of the Charophyte species native to Germany. *Rostocker Meeresbiologische Beiträge, 28*, 45–96.

Kostić, L. (1937). Prinos poznavanju haraceja Ohridskog Jezera i okoline. Ein Beitrag zur Kenntnis der Characeen des Ochridsees. *Acta Botanica Instituti Botanici Universitatis Zagrebensis, 11*, 63–84. [in Croatian].

Krause, W. (1997). Charales (Charophyceae) Süßwasserflora von Mitteleuropa. In: H. Ettl, G. Gärtner, H. Heynig, & D. Mollenhauer (Eds.), *Süßwasserflora von Mitteleuropa* (Vol. 18). Fischer, Jena [in German].

Krause, W., & Grüttner, A. (1990). Über einen Fund der *Chara tenuispina* im Bodenseegebiet mit Blick auf die Gesamtverbreitung der Pflanze. *Carolinea, 48*, 31–46. [in German].

Kusel-Fetzmann, E. (1979). The algal vegetation of Neusiedlersee. In H. Löffler (Ed.), *Neusiedlersee: The limnology of a Shallow Lake in Central Europe* (pp. 171–202). Springer.

Langangen A., & Pavlides, G. (1999). *Chara tenuispina* A. Braun (Charales) found in Greece. *Allionia, 36*, 93.

Migula, W. (1889–1897). Fünfter Band: Die Characeen. In: A. Fischer, E. Fischer, F. Hauck, G. Limpricht, C. Luerssen, W. Migula, H. Rehm, P. Richter, & G. Winter (Eds.), *Dr. L. Rabenhorst's Kryptogamenflora von Deutschland, Oesterreich und der Schweiz* (2 ed.). Eduard Kummer, Leipzig [in German].

Palamar-Mordvintseva, G. M., & Tsarenko, P. M. (2004). Red list of Charales of the Ukraine. *International Journal on Algae, 6*, 305–318.

Shoyakubov, R. Sh. [Шоякубов, Р. Ш.] (1979). Харовые водоросли Узбекистана [*Charophytes of Uzbekistan*]. Fan, Tashkent [in Russian].

Urbaniak, J., & Gąbka, M. (2014). *Polish Charophytes. An illustrated guide to identification* (p. 120). Uniwersytet Przyrodniczy we Wrocławiu, Wrocław.

Urbaniak, J. (2011). A SEM and light microscopy study of the oospore wall ornamentation in Polish charophytes (Charales, Charophyceae)—genus *Chara*. *Nova Hedwigia, 93*, 1–28.

Vesić, A., Blaženčić, J., & Šinžar-Sekulić, J. (2016). Contribution to knowledge of the charophytes (Charales) of Vojvodina (Serbia)—20 years after the first review. *Botanica Serbica, 40*, 237–247.

Wood, R. D. (1965). *Monograph of the Characeae*. J. Cramer.

Chapter 49
Chara tomentosa

Kaire Torn

Scientific name: ***Chara tomentosa*** **L., Sp. Pl. 2: 1156. 1753**

Lectotype (Wood, 1960: 220). [no location]/[no date]/[no collector] (LINN 1088.1).

= *Chara ceratophylla* Wallr., Annus Bot.: 192. 1815
Lectotype (van de Weyer & Jorda in Arbeitsgruppe Characeen Deutschlands 2016: 362): Wallr., Annus Bot.: t. V. 1815.

Local names: BELARUSIAN: Хара лямцавая; DANISH: Tyk Kransnål; ENGLISH: Coral stonewort; ESTONIAN: Ruuge mändvetikas; FINNISH: Punanäkinparta; FRENCH: Chara tomenteuse; GERMAN: Gehörnte Armleuchteralge; LITHUANIAN: Kietasis maurabragis; NORWEGIAN: Rødkrans; POLISH: Ramienica omszona; RUSSIAN: Хара войлочная; SWEDISH: Rödsträfse; UKRAINIAN: Хара повстяна.

49.1 Morphology

Habitus: The plants are robust with thick swollen spine-cells and a rope like main axis (Fig. 49.1). Plants are reddish to brownish (growing in shallow water or below surface), green (growing in deeper water) or grayish (due to incrustation) (Fig. 49.2a, b). Freshwater forms are normally encrusted, brackish water forms are rarely encrusted. Plants are very rigid and break easily even when not encrusted.

Size: The plants are generally medium to large-sized. Height of the upward-facing part is 25–60 cm. Total length of the plant may reach up to 2 m, but the majority of

K. Torn (✉)
Estonian Marine Institute, University of Tartu, Mäealuse 14, 12618 Tallinn, Estonia
e-mail: kaire.torn@ut.ee

H. Schubert et al. (eds.), *Charophytes of Europe*,
https://doi.org/10.1007/978-3-031-31898-6_49

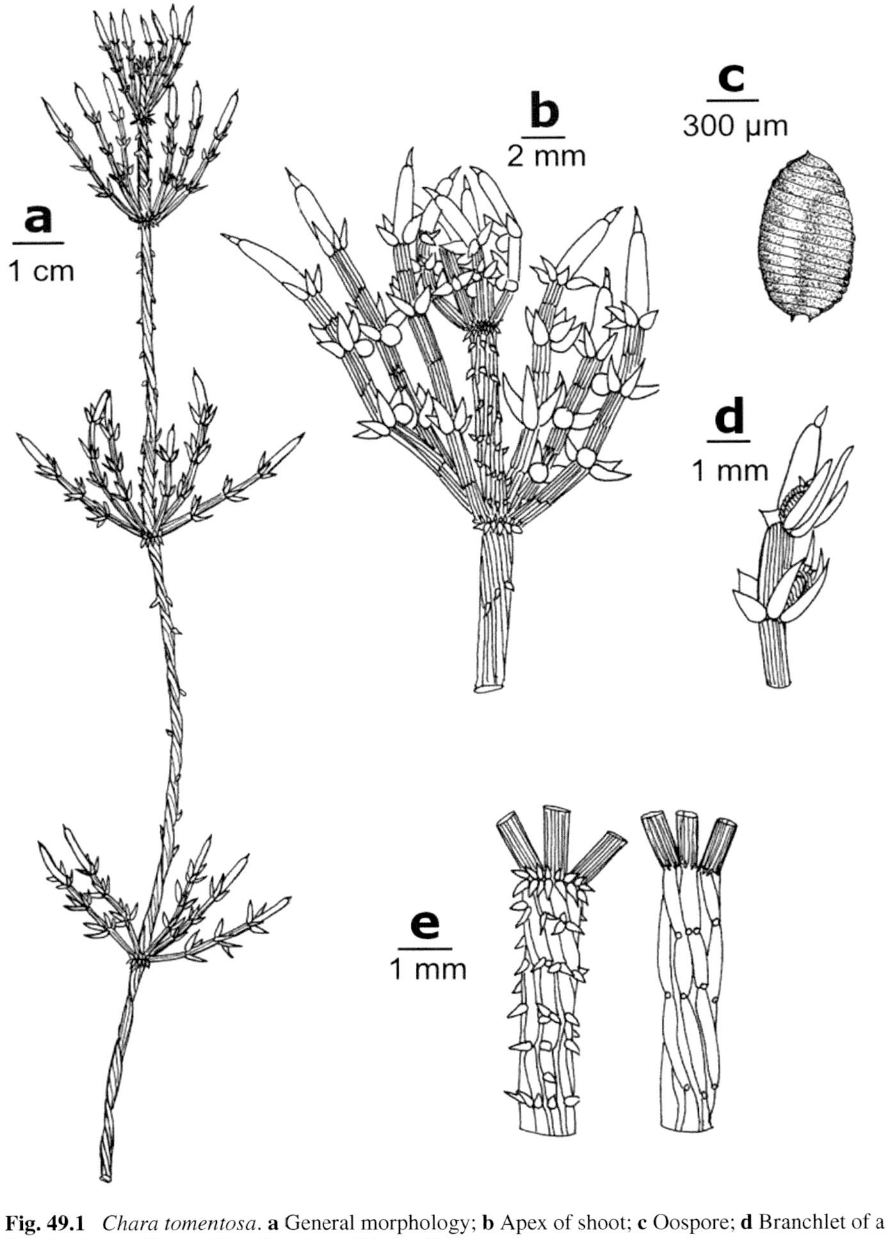

Fig. 49.1 *Chara tomentosa*. **a** General morphology; **b** Apex of shoot; **c** Oospore; **d** Branchlet of a female plant; **e** Cortication. Drawings: Werner Krause

Fig. 49.2 *Chara tomentosa*, appearance. **a** Reddish specimen; **b** Green specimen with swollen tips of branchlets; **c** Dense stands of *Chara tomentosa*. Photographs: **a**, **c**—Heiko Korsch; **b** Klaus van de Weyer

the plant is laid down on the sediment. Tough specimens with a height of only a few centimeters have been found in shallow waters.

Main axis: The axis is thick to very thick, (0.7)1–2(3) mm in diameter. The internodes may be shorter or longer than the branchlets. The cortex cells are often twisted like a rope (Fig. 49.3b–d).

Cortex: The cortex is diplostichous and prominently tylacanthous. Cortical abnormalities are common: the cortex may be irregularly triplostichous (Fig. 49.3b, c) and according to Mouronval et al. (2015) the cortex may be locally aulacanthous in the lower parts.

Fig. 49.3 *Chara tomentosa*, details. **a**, **b** Different length and shape of spine-cells and stipulodes; **b–d** Twisted cortex cells of the main axis; **b** A unique character of the species is the two-celled cortication at the lowermost branchlet section (white circle); **b**, **c** Irregularly triplostichous cortex; **e** Branchlet with ecorticated and swollen tip. Photographs: **a**, **d**, **e**—Anastasiia Kovtun-Kante, **b**—Aurélie Boissezon; c—Gustav Johansson

Spine-cells: Spine-cells are swollen oblong or ovoid shaped, with an acuminate tip. Length of spine-cells varies from very short to longer than the axis diameter (Fig. 49.3a–c). Spine-cells are solitary, geminate or in bundles of three.

Stipulodes: There are two rows of stipulodes, occasionally even three rows, and their, appearance is similar to the spine-cells (Fig. 49.3a, b). Stipulodes in different rows may vary somewhat in size.

Branchlets: The branchlets are shorter to longer than the internodes, up to 12 cm long. Each whorl has 6–7 branchlets, formed of 4–5 segments which are corticate except for the 2–3 end-cells. Lowest ecorticate cell is conspicuously large and swollen (Fig. 49.3e).

Bract-cells: 5–6 equally well-developed bract-cells, inflated appearance with acuminate tip, up to 2 mm long, are present along the whole branchlet.

Bracteoles: Similar or more slender than bract-cells, these can be equal or exceed the length or diameter of gametangia. Slenderness and length comparison with bract-cells is more pronounced on female plants.

Bractlets: Similar or more slender and longer than bract-cells.

Sexuality: The species is dioecious. The gametangia are placed singly at the first 2 or 3 nodes of the branchlets (Mouronval et al., 2015).

Oogonia: Oogonia are elongated ovoidal, with constricted coronulae. Table 49.1 gives an overview about morphometrical data available.

Oospores: There is scanty formation of oospores (Olsen, 1944). The oospore-membrane is granulate. The granules vary in size (Groves and Bullock-Webster, 1924). Table 49.2 gives an overview about morphometrical data available.

Antheridia: Very large antheridia, largest among European *Characeae*, orange or bright red. Table 49.3 gives an overview about morphometrical data available.

Bulbils: Rarely small bulbils produced (Mouronval et al., 2015; Schou et al., 2017).

Table 49.1 Oogonia morphometry

Author	Length	Width	Coronula length	Coronula width	Number of helical stripes
Migula (1889–1897)	<1300	<800	160	400	16–18
Groves and Bullock-Webster (1924)	750–1050	675–700	225–300	405–420	14–15
Pork (1954)	<1300	<800	–	–	18
Corillion (1957)	<1000	<800	200	500	15
Wood (1965)	750–1050	675–700	225–300	405–420	14–15
Krause (1997)	900–1100	700–800	250	400	15–18
Urbaniak and Gąbka (2014)	950–1150	675–800	–	–	–
Total range	**750–1300**	**675–800**	**160–300**	**400–500**	**14–18**

Length in the first column excludes the coronula, its length and width given separately; all values for length and width in μm

Table 49.2 Oospore morphometry

Author	Length μm	Width μm	Number of striae	Colour
Migula (1889–1897)	900–1050	600–700	12–13	Light yellowish-brown
Groves and Bullock-Webster (1924)	800–875	525–600	14	Light yellowish, fine granulate
Pork (1954)	900–1050	600–700	14–16	Light yellowish-brown
Corillion (1957)	800–875	525–600	14	–
Wood (1965)	765–900	465–645	13–14	Light yellowish-brown
Krause (1997)	850–1050	600–700	14–16	Yellowish-brown, black brown
Bailly and Schaefer (2010)	600–700	–	12–16	Yellowish-brown
Korsch (2013)	700–800	400–450	14–16	Light brown
Urbaniak and Gąbka (2014)	765–900	505–645	–	Brown to dark brown or black
van de Weyer and Jorda (2016)	(765)800–1050	(465)525–700	(13)14–16	–
Total range	**765–1050**	**400–700**	**12–16**	

Table 49.3 Antheridia morphometry

Authors	Diameter (μm)
Migula (1889–1897)	800–1100
Groves and Bullock-Webster (1924)	1325–1425
Pork (1954)	800–1000
Corillion (1957)	1300–1400
Wood (1965)	900–1000
Moore (1986)	<1500
Krause (1997)	900–1200
Bailly and Schaefer (2010)	700–1500
Korsch (2013)	750–1000
Urbaniak and Gąbka (2014)	740–1100
Becker et al. (2016)	800–1425
Total range	**700–1500**

49.2 Phenology and Reproduction

Vegetative reproduction is recorded, sexual reproduction is uncommon (Moore, 1986). Male plants have been found more often than female plants. *Chara tomentosa* is normally perennial. In autumn the specimens break into a great number of fragments or hibernate as fully-grown plants. In spring fresh shoots emerge from the nodes or continue to grow from the apices. The formation of antheridia and oogonia occurs in spring (Korsch, 2013), ripe oospores were observed from the beginning of August (Torn et al., 2004).

49.3 Habitat Conditions and Ecology

Habitat type: *Chara tomentosa* can be found in different water bodies (lakes, pits, peat cuts, small ponds and diches, and coastal lagoons). It cannot stand running water or wave action and is present in sheltered to very sheltered areas.

Sediment: Plants exhibit no preferences for certain types of substrate but grow mainly on organic mud. Plants have been also found on sand, gravel, and limestone gyttja.

Nutrient conditions: *Chara tomentosa* tolerates a wide spectrum of trophic conditions (Korsch, 2013), The species is found at total phosphorus concentrations 0.006–0.44 mg/l and total nitrogen 0.2–23.8 mg/l (Pełechaty et al., 2015; van de Weyer and Jorda, 2016). It is noted in Poland that higher concentrations of phosphorus (0.25 mg/l) positively influence development of specimens (Hirsz-Siwicka, 2013). Based on experiments, the growth of the species is not inhibited even at phosphorus concentration 1 mg/l (Blindow, 1988). Deviating from this, Melzer (1994) describes *C. tomentosa* as an indicator of very low nutrient loads.

Water chemistry: *Chara tomentosa* occurs both in fresh and brackish water. In the Baltic Sea, the majority of records originates where the salinity ranges between 4 and 7 psu, the species is observed at salinities up to 7.5 psu. In freshwater it is found in calcareous waters. Based on overview in Doege et al. (2016) the range of water chemistry where the species occurs is: pH value 6.6–8.7, electrical conductivity 80–10,000 μS/cm, carbonate hardness 0–2.5 °dH, total hardness 0.1–60.0 °dH, Ca^{2+} 31–155.5 mg/l, and Cl^{-} 4.5–2595 mg/l.

Depth range: In lakes mainly between 2 and 6 m, occasionally up to 30 m depth (Krause, 1997). In the Baltic sea this species is found in a depth of 0.5–4 m, most records are from shallow water of up to 1.5 m depth.

Associated species: Large dense monospecific stands of *C. tomentosa* are common both from fresh and brackish water (Fig. 49.2c). In mixed stands the species occurs together with other charophytes or some higher plants. Species lists of associated species vary based on different sources. Examples from freshwaters *Chara hispida*, *C. papillosa*, *C. subspinosa*, *C. filiformis*, *C. contraria*, *C. aculeolata*, and *N. obtusa*

(Krause, 1997; Mouronval et al., 2015; Urbaniak and Gąbka, 2014; van de Weyer and Jorda, 2016). In the Baltic Sea this species occurs together with *C. aspera*, *C. baltica*, *C. canescens*, *Ceratophyllum demersum*, *Stuckenia pectinata*, *Najas marina* and *Myriophyllum spicatum*.

49.4 Variability

Chara tomentosa is variable in regard to spine-cells, forms with few isolated spine-cells to very dense and long spine-cells have been observed. Very small forms have been described in shallow water, green specimens with elongated internodes are known from deeper water.

49.5 Identification and Risk of Confusion

In general, *Chara tomentosa* is easy to identify by its robust appearance, reddish colour, rope-like main axis, thick spine-cells, inflated end-cells of the branchlets, and large gametangia. Less spiny forms can be confused with *Chara hispida*, *C. baltica*, *C. papillosa* or *C. subspinosa* (van de Weyer and Jorda, 2016). The unique difference compared to other charophytes is the presence of two following rows of cortical cells in the lowest section of branchlets (Fig. 49.3b). In the first half of the nineteenth century, what is currently known as *Chara hispida* was published by a number of authors under the name *C. tomentosa*. Unpubl. references are therefore often uncertain (van de Weyer and Jorda, 2016).

49.6 Distribution

Chara tomentosa is sub cosmopolitan and occurs in Europe, Asia, America, and Northern Africa. In Europe this species is found throughout the continent, but mostly in Central Europe and it is especially common along the Baltic coast (Fig. 49.4, Torn et al., 2004).

49.7 Status and Threats

Chara tomentosa is listed on several Red Lists (e.g., Norway, Switzerland, and the Balkans) (Mouronval et al., 2015). In the Baltic Sea this species exhibited a dramatic decline during the second half of the twentieth century in several locations. Therefore *C. tomentosa* was included in the previous HELCOM list of threatened and/or

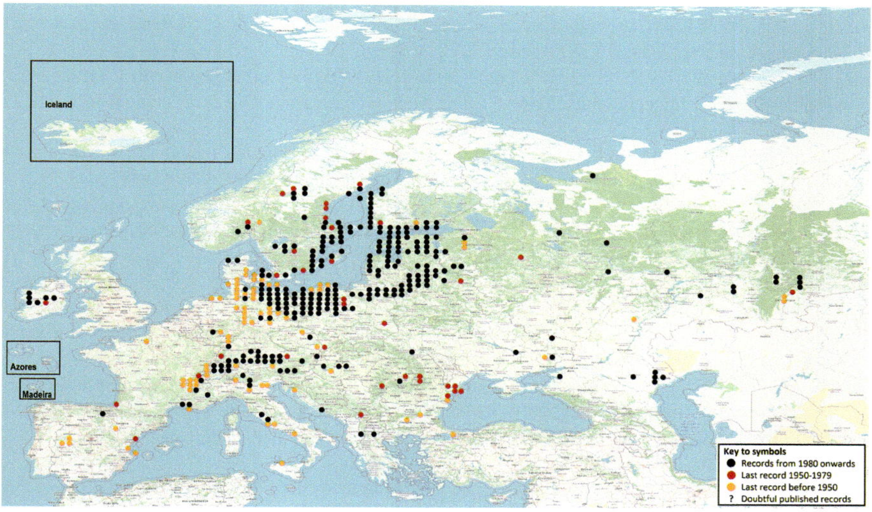

Fig. 49.4 Distribution of *Chara tomentosa* in Europe. For worldwide distribution see Korsch (2018). Black dots are records from 1980 on, red dots from 1950 to 1979 and yellow dots before 1950. Map prepared by Heiko Korsch

declining species (HELCOM, 2007). However, this species is currently not endangered, in northern regions of the Baltic Sea it is widespread and a rather common species (HELCOM, 2013; Torn et al., 2004).

49.8 Remark

Chara tomentosa is the type species of the genus *Chara* and, therefore, of the *Characeae.*

References

Bailly, G., & Schaefer, O. (2010). *Guide illustré des Characées du nord-est de la France.* Conservatoire Botanique National de Franche-Comté.

Becker, R., Blindow, I., Doege, A., Franke, T., Gregor, T., Hamann, U., Jäger, D., Jorda, C., Kabus, T., Korsch, H., Korte, E., Kusber, W. H., Pätzold, F., Raabe, U., Schubert, H., Teppke, M., van de Weyer, K., & Wolff, P. (2016). Beschreibung der Characeen-Arten Deutschlands. In Arbeitsgruppe Characeen Deutschlands (Ed.), *Armleuchteralgen: Die Characeen Deutschlands* (pp. 209–572). Springer.

Blindow, I. (1988). Phosphorus toxicity in *Chara. Aquatic Botany, 32*, 393–395. https://doi.org/10.1016/0304-3770(88)90110-6

Corillion, R. (1957). Les Charophycées de France et d'Europe Occidentale. *Bulletin de la Société Scientifique de Bretagne, 32*(Hors série I & II), 1–499.

Doege, A., van de Weyer, K., Becker, R., & Schubert, H. (2016). Bioindikation mit Characeen. In Arbeitsgruppe Characeen Deutschlands (Eds.), *Armleuchteralgen: Die Characeen Deutschlands* (pp. 97–138). Springer.

Groves, J., & Bullock-Webster, G. R. (1924). *The British Charophyta. Volume II. Chareæ with plates, concluding articles, geological sketch, bibliography and index*. Ray Society.

HELCOM. (2007). *HELCOM lists of threatened and/or declining species and biotopes/habitats in the Baltic Sea area* (No. 113). Baltic Sea Environmental Proceedings. https://helcom.fi/wp-content/uploads/2019/08/BSEP113.pdf

HELCOM. (2013). *Red List of Baltic Sea species in danger of becoming extinct* (No. 113). Baltic Sea Environmental Proceedings. https://helcom.fi/wp-content/uploads/2019/08/BSEP140.pdf

Hirsz-Siwicka, E. (2013). Structure variability of *Chara tomentosa* L. specimen on stands of diverse habitat conditions. *Polish Journal of Natural Sciences, 28*, 31–42.

Korsch, H. (2013). *Die Armleuchteralgen (Charophyceae) Sachsen-Anhalts*. Berichte des Landesamtes für Umweltschutz Sachsen-Anhalt. https://lau.sachsen-anhalt.de/fileadmin/Bibliothek/Politik_und_Verwaltung/MLU/LAU/Wir_ueber_uns/Publikationen/Berichte_des_LAU/Dateien/Berichte_LAU_2013_1.pdf

Korsch, H. (2018). The worldwide range of the Charophyte species native to Germany. *Rostocker Meeresbiologische Beiträge, 28*, 45–96. https://www.oekologie.uni-rostock.de/storages/uni-rostock/Alle_MNF/Bio_Oekologie/RMB/RMB28/RMB28_04_Korsch.pdf

Krause, W. (1997). Charales (Charophyceae) Süßwasserflora von Mitteleuropa. In H. Ettl, G. Gärtner, H. Heynig, & D. Mollenhauer (Eds.), *Süßwasserflora von Mitteleuropa* (Vol. 18). Fischer.

Melzer, A. (1994). Möglichkeiten einer Bioindikation durch submerse Makrophyten—Beispiele aus Bayern. *Beiträge zur angewandten Gewässerökologie Norddeutschlands, 1*, 92–102.

Migula, W. (1889–1897). Fünfter Band: Die Characeen. In A. Fischer, E. Fischer, F. Hauck, G. Limpricht, C. Luerssen, W. Migula, H. Rehm, P. Richter, & G. Winter (Eds.), *Dr. L. Rabenhorst's Kryptogamenflora von Deutschland, Oesterreich und der Schweiz* (2nd ed.). Eduard Kummer.

Moore, J. A. (1986). *Charophytes of Great Britain and Ireland*. BSBI Handbook No. 5, Botanical Society of the British Isles.

Mouronval, J. B., Baudouin, S., Borel, N., Soulié-Märsche, I., Klesczewski, M., & Grillas, P. (2015). *Guide des Characées de France méditerranéenne*. Office national de la Chasse et de la Faune Sauvage.

Olsen, S. (1944). Danish Charophyta. *Det Kongelige Danske Videnskabernes Selskab, Biologiske Skrifter, 3*(1), 1–240.

Pełechaty, M., Ossowska, J., Pukacz, A., Apolinarska, K., & Siepak, M. (2015). Site-dependent species composition, structure and environmental conditions of Chara tomentosa L. meadows, western Poland. *Aquatic Botany, 120*, 92–100.

Pork, M. (1954). Eesti NSV Mändvetiktaimed (*Charophyta*). *Abiks Loodusvaatlejale, 16*, 1–31.

Schou, J. C., Moeslund, B., Båstrup-Spohr, L., & Sand-Jensen, K. (2017). *Danmarks vandplanter*. BFN's, Klitmøller.

Torn, K., Martin, G., & Munsterhjelm, R. (2004). *Chara tomentosa*. In H. Schubert, & I. Blindow (Eds.), *Charophytes of the Baltic Sea* (pp. 131–141). A. R. G. Gantner.

Urbaniak, J., & Gąbka, M. (2014). *Polish Charophytes. An illustrated guide to identification*. Uniwersytet Przyrodniczy we Wroclawiu.

van de Weyer, K., & Jorda, C. (2016). *Chara tomentosa*. In Arbeitsgruppe Characeen Deutschlands (Eds.), *Armleuchteralgen: Die Characeen Deutschlands* (pp. 361–369). Springer.

Wood, R. D. (1960). Linnaean types for Characeae. *Transactions of the American Microscopical Society, 79*, 219–226.

Wood, R. D. (1965). *Monograph of the Characeae*. Weinheim: J. Cramer.

Chapter 50
Chara uzbekistanica

Roman Romanov and Liubov Zhakova

Scientific name: ***Chara uzbekistanica*** **Hollerb. in Bot. Mater. Otd. Sporov. Rast. Bot. Inst. Komarova Akad. Nauk S.S.S.R. 13: 102. 1960**

Holotype (Hollerbach, 1960: 102): [Uzbekistan] Uzbekistania, delta fl. Amu-Darja, lacus Sudoczje (in parte lacus centrali)/24 July 1951/R. S. Denjgina (LE).

Local names: RUSSIAN: Хара узбекистанская, UKRAINIAN: Хара узбецька.

50.1 Morphology

See Figs 50.1 and 50.2.

Habitus: Plants are green to grey-brownish-green, more or less evenly, slightly incrusted, small or moderately high, rough to touch, scarcely branching with a smooth appearance due to short stipulodes, spine cells and bract-cells, completely-corticate branchlets and obvious branchlet nodes, mostly easily seen with the naked eye because of visible bract-cells.

Size: Moderately small to medium plant, from 10 to 30 cm high, up to 120 cm (Tkachenko, 2005).

Main axis: The main axis is (280)400–700 μm in diameter, flaccid, completely-corticate. The internodes are two or more times longer than the branchlets at the

R. Romanov (✉)
Nišice Ujtin Potok b.b., Dobra Voda, 85356 Bar, Montenegro
e-mail: streptophytes@gmail.com

L. Zhakova
Laboratory of Brackish Water Biology, Zoological Institute of the Russian Academy of Science, University Embankment, 1, 199034 St. Petersburg, Russia
e-mail: luba_zhakova@mail.ru

H. Schubert et al. (eds.), *Charophytes of Europe*,
https://doi.org/10.1007/978-3-031-31898-6_50

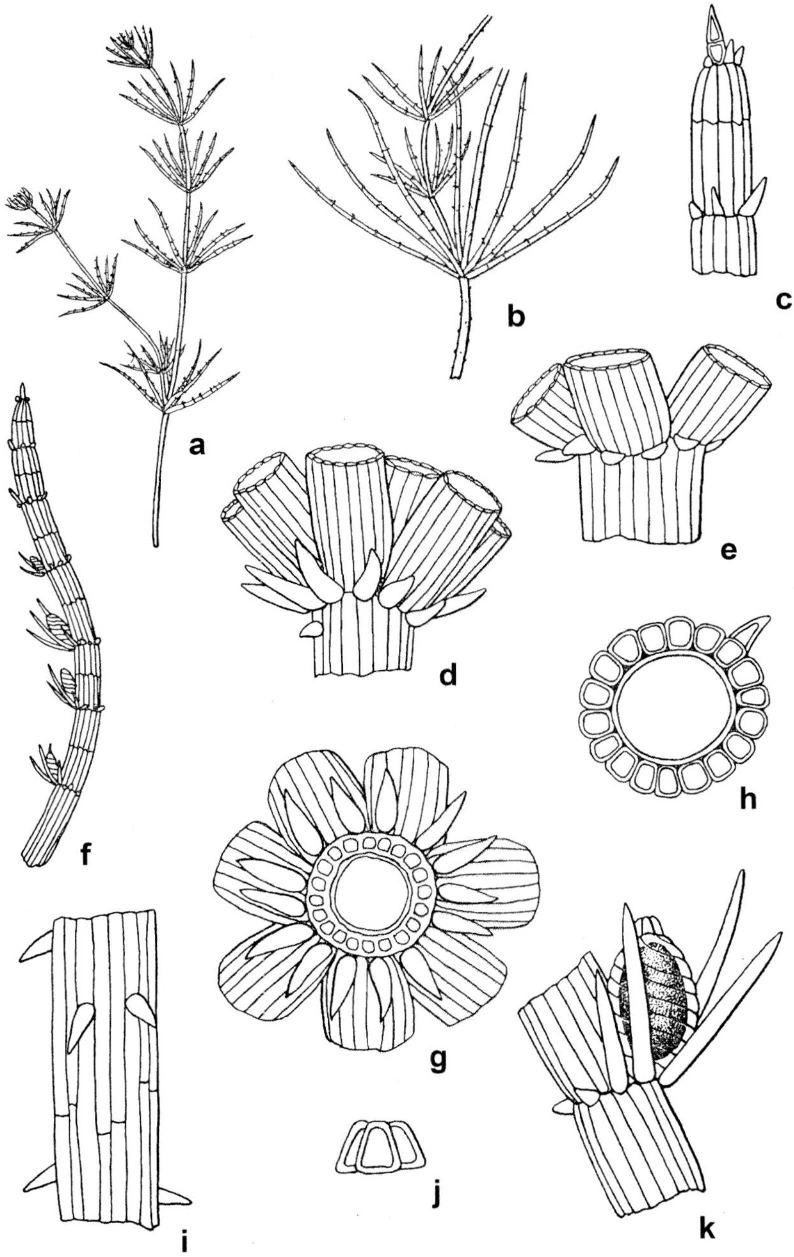

Fig. 50.1 *Chara uzbekistanica*. **a** Upper part of plant; **b** Lower part of plant; **c** End of branchlet; **d** Typical arrangement of the base of branchlet whorl; **e** The base of branchlet whorl at lower part of plant; **f** Female branchlet; **g** Stipulodes from below; **h** Cross-section of the main axis; **i** Main axis cortex; **j** Coronula; **k** Fertile node of a female branchlet. Drawings: Hollerbach and Krassavina (1983)

Fig. 50.2 General appearance of male plant of *Chara uzbekistanica* from Lower Don Region. Photograph: Liubov Zhakova

lower and middle sections of the plants, up to 40 mm at lowest one, 15–30 mm in middle, equal or somewhat shorter than branchlets near the apex.

Cortex: Regularly triplostichous and isostichous.

Spine-cells: Spine-cells are always solitary, sparse, papillose and pointed to subulate, usually not exceeding 30% (50%) of the main axis diameter.

Stipulodes: Diplostephanous but appearing haplostephanous because the upper row of cells is mostly well developed, subulate or pointed papillose, elongate, slightly curved upward or adpressed to branchlets and rudimentary cells in the lower row (type specimen, checked by R. Romanov).

Branchlets: 6(7)–8 in a widely spreading whorl, slightly sigmoid or somewhat arcuate, nearly 1 cm long in the middle part of the plant, up to 2 cm in the lowest parts, consisting of 5–8, (mostly 6–7) corticate segments and a very short, narrow and pointed, 1 or 2-celled terminal ecorticate segment.

Bract-cells: Short at sterile nodes, well developed at fertile nodes only, 8–9 (incl. bracteoles), posterior ones invariably short, papillose and pointed, anterior bract-cells are somewhat shorter than the oogonium. All bract-cells are rudimentary at nodes with antheridia (Tkachenko, 2005).

Bracteoles: Longer than oogonium, up to almost 2-times.

Bractlets: Nearly equal to bracteoles.

Sexuality: Dioecious. Gametangia are always solitary, at the lowest 3 or 4 nodes of the branchlet.

Oogonia: Ellipsoid, 455–575(750) μm in length without coronula and 315–400(500) μm in width (holotype; Hollerbach, 1960) or 480–960 × 180–660 μm (Tkachenko, 2005), with (8)9–10 convolutions of spiral cells. The coronula is 72–85(140, rarely up to 200) μm in length and 145–155(185, rarely up to 270) μm in width (holotype, Hollerbach, 1960) or 200–300 × 144–300 μm (Tkachenko, 2005), regular, thick-set, cone-like and narrowing toward the apex. Gyrogonites are unknown.

Oospores: Ellipsoid, 514–600 μm in length and 285–340 μm in width, with 11–12 spiral ridges.

Antheridia: 700–800 μm in diameter (holotype; Hollerbach, 1960) or 324–960 μm (Tkachenko, 2005).

Bulbils: Unknown.

50.2 Phenology and Reproduction

Unknown.

50.3 Habitat Conditions and Ecology

Habitat type: *Chara uzbekistanica* is known from fresh and brackish shallow standing waters, including marginal lakes, generally in deltas of large rivers (Volga, Amu-Darya, Ili), and in water bodies in arid and semiarid environments (Kostin, 1982; Kostin and Shoyakubov, 1974; Tkachenko, 2005; Zhivoglyad and Krivonosov, 1982). This species is frequent in shallow parts of the open subaqueous parts of the delta of the Volga River (Zhivoglyad and Krivonosov, 1982) and in the northern part of

Volga-Akhtuba floodplain at intermediate stages of aquatic plant community succession (Klinkova and Zhakova, 2014; Klinkova et al., 2012), rare in the water bodies of the western and eastern margins of delta (Zhivoglyad and Krivonosov, 1982). In addition, a few localities are known from lakes and swamps in Iran (Dogadina et al., 2007), fishponds and lakes near the delta of the Danube River in Ukraine (Hollerbach and Palamar-Mordvitseva, 1991; Tkachenko, 2005) and floodplain lakes of the Ili River in Kazakhstan (Kostin, 1982), as well as in the cooling reservoir (Tsimlyanskoe Reservoir) at the Don River (L. Zhakova unpubl. data). This species is able to withstand periodic desiccation of the at (habitat) (Zhivoglyad and Krivonosov, 1982).

Sediment: *Chara uzbekistanica* grows in silty (Denjgina, 1957), silty-sandy (Tkachenko, 2005) to sandy bottom sediments with gravel and stones (L. Zhakova unpubl. data).

Nutrient conditions and water chemistry: Few data are available. This species is known at salinities from 0.3 g per L^{-1} (Klinkova et al., 2012), 0.385–1.6 g per L^{-1} (Kostin and Shoyakubov, 1974), 1 g per L^{-1} (Zhakova unpubl. data) and up to 5–7 g per L^{-1} (Kostin, 1987); the habitat pH is usually slightly above 8, Cl^- has been measured at 4.6 g per L^{-1} in the type locality during sampling of original material (Denjgina, 1957); as well as in waters with sulphates, hydrocarbonates and sodium as main ions in Tsimlyanskoe Reservoir (L. Zhakova unpubl. data).

Depth range: *Chara uzbekistanica* is known from shallow water, up to 0.45 m, but mainly 1–2.5 m, and up to 2 m in the lower Ili region and neighboring Lake Balkhasch (Kostin, 1982). High water level in seaside lakes in combination with less heating of the water favours mass development of the species possibly due to suppression of flowering plants (Tkachenko, 2005).

Associated species: *Chara uzbekistanica* grows in association with big species of *Characeae*: *Chara globata*, *C. tomentosa*, *Nitellopsis obtusa* or with different species of flowering plants (Denjgina, 1957; Tkachenko, 2005; R. Romanov unpubl. data). It occurs frequently in monodominant stands in the Volga Delta (Zhivoglyad and Krivonosov, 1982).

50.4 Variability and Risk of Confusion

Chara uzbekistanica is very similar in general appearance with *C. connivens* and *C. virgata* but can be easily differentiated in the fertile state from the latter, being dioecious in contrast with monoecious *C. virgata*. The elongated upper stipulodes that are usually not rudimentary allow it to be distinguished from *C. connivens*, which has short diplostephanous stipulodes. This difference seems to be not so crucial. The pointed elongated spine cells of *C. uzbekistanica* vs. short papillose ones of *C. connivens* can also be useful for separation, but they are only formed sometimes. *Chara uzbekistanica* is mainly a Central-Asian species but can be locally abundant in the Volga delta. In addition, three distant populations are also known in Europe, i.e., two in Ukraine and single one in Russia.

50.5 Distribution

The European distribution of *Chara uzbekistanica* is limited to South-Eastern Europe (Fig. 50.3): seaside lakes at vicinity of the Danube Delta, fishponds in vicinity of Kiev (Ukraine), the Lower Don Region (Rostov Oblast), the Lower Volga Region—almost entirely restricted to the Volga floodplain, the Volga Delta (Russia); in Central Asia: Kazakhstan—the Ili Delta, Uzbekistan—the Amu-Darya Delta and in Western Asia: Iran (Borisova et al., 2016; Dogadina et al., 2007; Hollerbach, 1960; Hollerbach and Krassavina, 1983; Hollerbach and Palamar-Mordvintseva, 1991; Klinkova and Zhakova, 2014; Kostin, 1987; Tkachenko, 2005; LE!). The total number of its localities in the world does not exceed 30, including a few recent records from Iran, Ukraine, and Russia (Dogadina et al., 2007; Tkachenko, 2005; LE!; L. Zhakova unpubl. data).

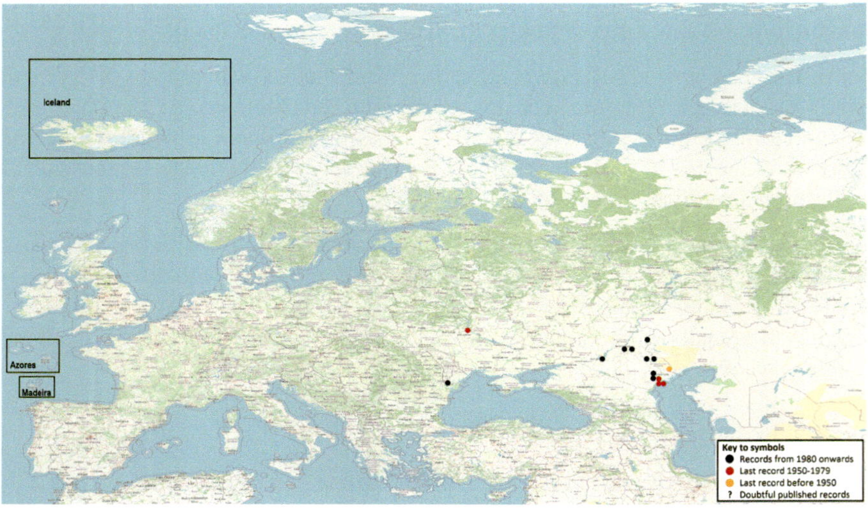

Fig. 50.3 Distribution of *Chara uzbekistanica* in Europe. Black dots are records from 1980 on, red dots from 1950 to 1979 and yellow dots before 1950; unconfirmed/doubtful published records (if applicable) are marked with "?" superimposed. Map prepared by Heiko Korsch

50.6 Status and Threats

Chara uzbekistanica has been assessed as DD (Data Deficient) in the European Red List (see Chap. 9). It has been suggested for the inclusion in the Red Data Book of the Russian Federation as a rare stenobiont species at the edge of its distribution range (Romanov, 2015).

Acknowledgements Michelle T. Casanova is kindly acknowledged for English improvement.

References

Borisova, E. V. (2016). Analysis of the flora of Charales (Charophyta) of Ukraine. *Algologia, 26*(4), 403–417. [in Russian]. https://doi.org/10.15407/alg26.04.403

Denjgina, R. S. [Деньгина, Р. С.]. (1957). Материалы гидробиологических исследований озер северо-западной части дельты р. Аму-Дарьи [The materials of hydrobiological study of the lakes of north-western part of Amu-Darya Delta]. *Труды Лаборатории озероведения АН СССР* [*Proceedings of the Laboratory of Limnology of the Academy of Sciences of USSR*], *4*, 269–305. [in Russian].

Dogadina, T. V., Zarei Darki, B., & Gorbulin, O. S. (2007). *Algal flora of Iran*. V.N. Karazin Kharkiv National University.

Hollerbach, M. M., & Palamar-Mordvintseva, G. M. [Голлербах, М. М. & Паламар-Мордвинцева, Г. М.]. (1991). *Визначник прісноводних водоростей України. IX. Харові водорості (Charophyta) [The identification manual of freshwater algae of Ukraine: IX. Charophytes (Charophyta)]*. Naukova dumka. [in Ukrainian].

Hollerbach, M. M., & Krassavina, L. K. [Голлербах, М. М. & Красавина, Л. К.]. (1983). *Определитель пресноводных водорослей СССР. 14. Charophyta* [*Determination key of freshwater algae of USSR. 14. Charophyta*]. Nauka. [in Russian].

Hollerbach, M. M. (1960). De speciebus novis generis *Chara* e sectione *Haplostephanae Triplostichae. Notulae Systematicae e Sectione Cryptogamica Instituti Botanici nomine V. L. Komarovii Academiae Scientarum URSS – Botanicheskie Materialy Otdela Sporovykh Rasteniy, 13*, 101–107. (in Russian).

Klinkova, G. Yu., Zhakova, L. V., Gorsky, K., & Gorelov, V. P. (2012). The algae of Charophyta within the territory of "The Volga-Akhtuba Floodplain". *Byulleten Glavnogo Botanicheskogo Sada, 3*, 52–58. [in Russian].

Klinkova, G. Yu., & Zhakova, L. V. (2014). New and rare species of Charales in the flora of the Lower Volga Region. *Byulleten MOIP. Otdel biologicheskiy [Bulletin of Moscow Society of Naturalists. Biological series], 119*, 61–66. [in Russian].

Kostin, V. A., & Shoyakubov, R. Sh. [Костин, В. А. & Шоякубов, Р. Ш.]. (1974). Распределение и биомасса харовых водорослей в озере Балхаш [The distribution and biomass of charophytes in Lake Balkhash]. In *Пятая конференция по споровым растениям Средней Азии и Казахстана, 26–27 апреля 1974 г., ч. 1 [The fifth conference for cryptogamous plants of Central Asia and Kazakhstan, Apr 26–27, 1974. Abstracts, 1]*. (pp. 182–183). [in Russian].

Kostin, V. A. [Костин В. А.] (1982). Редкие и исчезающие виды харовых водорослей водоемов реки Или и озера Балхаш [The rare and endangered species of charophytes from water bodies of Ili River and Lake Balkhash]. *Ботанические материалы гербария Института ботаники Академии Наук Казахской ССР [Botanicheskie Materialy Gerbariya Instituta Botaniki Akademii Nauk Kazakhsky SSR], 12*, 114–118. [in Russian].

Kostin, V. A. [Костин, В. А.] (1987). Материалы к изучению экологии харовых водорослей водоемов Или-Балхашского бассейна [Materials for the study of the ecology of charophytes in reservoirs of the Ili–Balkhash basin]. *Ботанические материалы гербария Института ботаники Академии Наук Казахской ССР [Botanicheskie Materialy Gerbariya Instituta Botaniki Akademii Nauk Kazakhskoy SSR], 15*, 128–133. [in Russian].
Romanov, R. E. (2015). Charophytes in the Red Data Book of Russian Federation: new species proposed to inclusion. In *Problems of taxonomy and geography of aquatic plants: proceedings of International conference, Borok, Russia, Oct 21–24, 2015.* (pp. 65–66). Filigran. [in Russian].
Tkachenko, F. P. (2005). The new location of rare species of Ukraine flora *Chara uzbekistanica* Hollerb. *Algologija, 15*(2), 230–235. [in Russian].
Zhivoglyad, A. F., & Krivonosov, G. A. (1982). On the specific composition and the productivity of the Charophyta at the lower reaches of the Volga Delta and Northern Caspy. *Botanicheskii Zhurnal, 67*, 672–674. [in Russian].

Chapter 51
Chara virgata

Emile Nat

Scientific name: ***Chara virgata*** **Kütz. in Flora 17: 705. 1834**

Lectotype (Wood, 1965: 184): Güderott-Moor near Boren/1833 Jun. 17/[no collector] (L 820940); see Fig. 51.1.

≡ *Chara globularis* Thuill. var. *virgata* (Kütz.) R. D. Wood in Taxon 11: 10. 1962, n. inv.

= *Chara diffusa* Lilj. var. *annulata* Wallman in Lilj., Utkast Sv. Fl., ed. 3, 684. 1816
Type not designated.

≡ *Chara virgata* Kütz. var. *annulata* (Wallman in Lilj.) N. F. Stewart & J. A. Bryant in J. A. Bryant, N. F. Stewart & C. A. Stace in Watsonia 24: 206. 2002.

= *Chara delicatula* C. Agardh f. *bulbillifera* A. Braun ex Mig., Charac. Deutschl.: 755. 1897
Type not designated.

Nomenclatural note: Migula (1889–1897), Krause (1997) and many others named this taxon *Chara delicatula*. But as Agardh (1824) cited "*Chara delicatula* Desv. in Loisel., Not. Fl. France: 137. 1810" as a synonym of *Chara aspera*, this name belongs to *Chara aspera* (Wood, 1965).

Local names: BELARUSIAN: Хара вытанчаная; DANISH: Busket kransnål; DUTCH: Teer kransblad; ENGLISH: Delicate stonewort; ESTONIAN: Õrn mändvetikas; FINNISH: Sironäkinparta; FRENCH: Chara délicate; GERMAN: Feine Armleuchteralge; LITHUANIAN: Gležnasis maurabragis; NORWEGIAN: Skjørkrans; POLISH: Ramienica delikatna, RUSSIAN: Хара изящная; SWEDISH: Papillsträfse (Sweden); Spädsträfse (Finnland); UKRAINIAN: Хара витончена.

E. Nat (✉)
Dutch Information Center On Charophytes, Naturalis Biodiversity Center, P.O. Box 9517, 2300, RA Leiden, The Netherlands
e-mail: e.nat@kranswieren.nl

H. Schubert et al. (eds.), *Charophytes of Europe*,
https://doi.org/10.1007/978-3-031-31898-6_51

Fig. 51.1 *Chara virgata*, lectotype at University of Rhode Island. A syntype can be found at the Leiden collection. Photographs: Joop van Raam

51.1 Morphology

See Figs. 51.2 and 51.3.

Habitus: The plants are small to medium size, fresh green to greyish green, unencrusted to slightly encrusted (Figs. 51.2 and 51.3a).

Size: The height of *Chara virgata* is usually between 5 and 15 cm but can be up to 30 cm. In deeper water the size can be up to 50 cm (van Raam, 1998).

Main axis: The main axis diameter ranges between 0.3 and 0.6 mm. The internodes are usually a bit longer than the branchlets, but occasionally twice as long and noticeably rough and encrusted with lime.

Cortex: *Chara virgata* is triplostichous and slightly heterostichous-tylacanthous (Migula, 1889–1897). Also true isostichous (Corillion, 1975) and definite tylacanthous (Langangen, 2007a) appearances have been described. The cells of the secondary rows are little more than half the diameter of the primary row cells (Fig. 51.3b,c).

Spine-cells: *Chara virgata* has small spine-cells which are solitary, papillate to spherical, rarely elongated and sometimes even missing (Corillion, 1975). See Figs. 51.2c–e and 51.3b, c.

Stipulodes: *Chara virgata* is diplostephanous, in the lower row the stipulodes are small and papillate, in the upper row they are mostly larger (up to 2 mm) and acute.

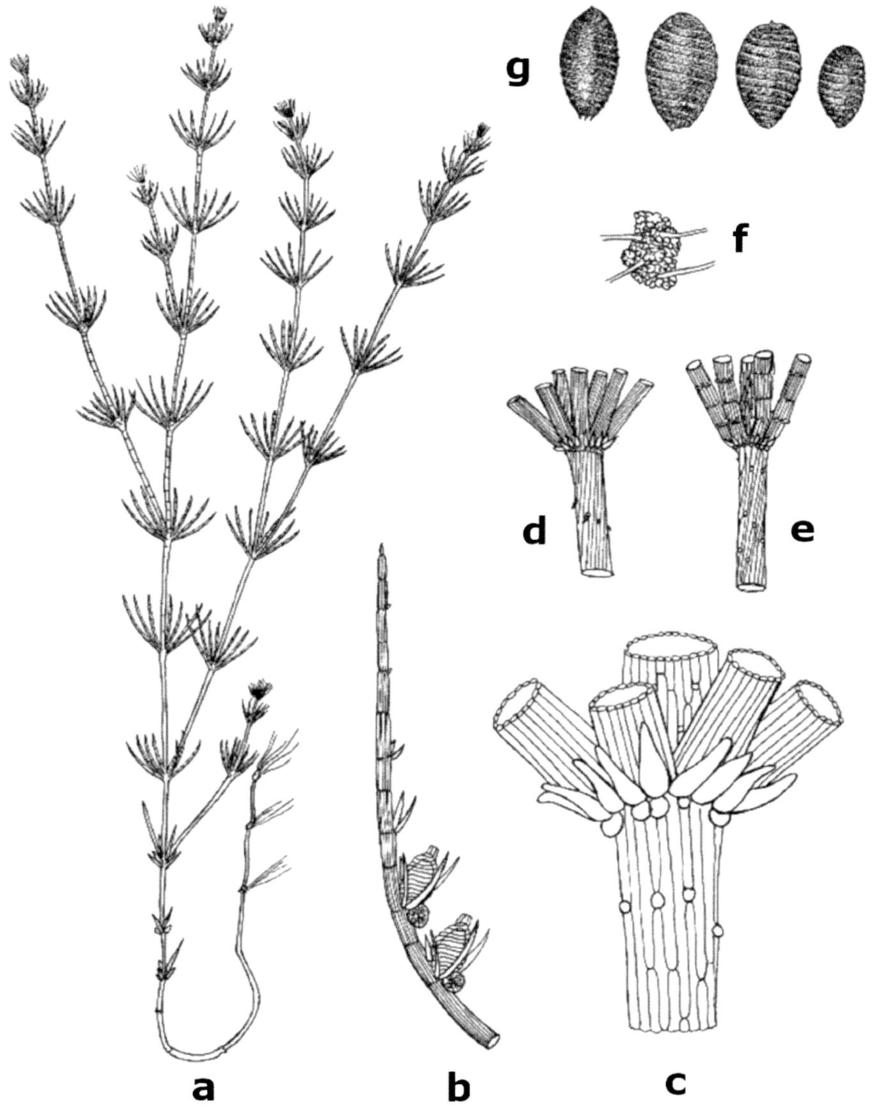

Fig. 51.2 *Chara virgata*. **a** Habitus; **b** Branchlet with two fertile nodes; **c**–**e** Axis with branchlet whorl; **f** Bulbils; **g** Oospores. Drawings after Werner Krause, modified by Dagmar Wassong in Weyer and Schmidt (2018)

Fig. 51.3 *Chara virgata*, habitat appearance and details. **a** Typical appaerance in shallow water; **b** Stipulodes; **c** Fertile branchlets; **d** Branchlet detail with ripe antheridia and unripe oogonia; **e** Bulbils. Photographs: **a**—Klaus van de Weyer; **b**,**c**, **d**—Chris Carter; **e**—Emile Nat

Sometimes the stipulodes of plants in the same stand differ considerably. Both rows of stipulodes can be rudimentary or the stipulodes of the upper row can have different dimensions from papillate to elongated. Compare the nodes shown in Figs. 51.2c–e and 51.3b, c.

Branchlets: Whorls consist of 7–8(9) short branchlets, usually curved but they also can be stiff and standing out, with 5–9(11) segments, the upper 1–3 are ecorticated (Figs. 51.2a and 51.3b, c).

Bract-cells: *Chara virgata* develops 5–7 bract-cells per branchlet node. They are hardly formed on sterile plants, on fertile plants the adaxials are acuminate to about as long as the oogonium; the abaxials are usually rudimentary but occasionally developed. See Figs. 51.2b and 51.3d.

Bracteoles: 1–2 times as long as the oogonium. The length may differ even in one plant. See Figs. 51.2b and 51.3d.

Bractlets: As a monoecious species with conjoined gametangia, bractlets are missing.

Sexuality: *Chara virgata* is monoecious.

Oogonia: The oogonia are usually conjoined with antheridia at 1–3 lowest branchlet nodes. The oogonia are solitary. Data about quantitative morphometry are very different among between authors (see Table 51.1 and Fig. 51.2b).

Table 51.1 Oogonia morphometry

Authors	Length	Width	Coronula length	Coronula width	Number of helical stripes
Migula (1889–1897)	950	600	–	–	13–14
Groves and Bullock-Webster (1924)	750–1000	500–675	Variable	250	14–15
Corillion (1975, 1957)	750–1000	500–675	100–240	180–260	14–15
Wood 1(965)	700–900	550–600	150–210(225)	210–225(230)	12–14
Pal et al., (1962)	650–760	500–700	100–240	180–260	14–15
Hollerbach and Krassavina (1983)	750–1000	500–675	165–205	218–257	14–15(16)
Krause (1997)	750–1000	500–700	100–200	200	13–15
Schou et al., (2017)	650–1000	500–700	–	–	–
Total range	**650–1000**	**500–700**	**100–240**	**180–260**	**12–15(16)**

Length in the first column excludes the coronula, its length and width given separately; all values for length and width in μm

Oospores: The ripe oospores are mostly black or sometimes dark brown. Unripe oospores have a strong orange colour (sometimes yellow). Their shape is ellipsoidal with a narrower apical and basal area (Becker et al. 2016). The oospore-membrane is almost black, thick, semi-rigid, semi-opaque and dotted with well-defined granules of different size (Groves and Bullock-Webster, 1924). Table 51.2 gives an overview about morphometrical data available; see also Fig. 51.4.

Antheridia: The antheridia are solitary and are about 300–560 μm in diameter; morphometrical data about antheridia are presented in Table 51.3 (see also Fig. 51.3d).

Bulbils: Bulbils can be formed with a diameter up to 2 mm. Such plants have also been described as a form: *Chara delicatula* f. *bulbillifera*. In some countries, like the UK, bulbils can sometimes be developed (Moore, 1986) and in others, like the Netherlands, bulbils are hardly formed (van Raam, 1998). See also Fig. 51.2f and 51.3e. Bulbils can be seen as cartilage-like swellings of the lower nodes, typically greenish or half-translucent, present late in the season when the main axis is senescing.

Table 51.2 Oospore morphometry

Author	Length	Width	Number of striae	Colour
Migula (1889–1897)	500–600	–	11–12	Black
Groves and Bullock-Webster (1924)	625–700	350–475	12–14	Black
Corillion (1975, 1957)	625–720	340–550	12–14	Black
Wood (1965)	615–735(750)	420–450(465)	11–12	Dark brown-black
Pal et al. (1962)	625–720	340–550	12–14	Black
Hollerbach and Krassavina (1983)	(500)600–700(720)	350–475	11–12(14)	Black
Krause (1997)	600–800	400–600	11–13	Black
Dilger (2010)	(490)570–740(850)	(300)350–460(490)	(9)10–14(15)	Black, predominantly matt black
Schou et al. (2017)	500–800(850)	(300)340–600	(9)11–14(15)	Dark brown-black
Total range	**(490)500–800(850)**	**(300)340–600**	**(9)11–14(15)**	

All values for length and width in μm

Fig. 51.4 *Chara virgata*, oospore. **a** Lateral view, **b** Basal view, **c** Oospore-membrane with well-defined granules. All oospores are taken from BM013734921, UK (w294). Photographs: Michelle T. Casanova (photo **a** is edited by Wil Leurs)

Table 51.3 Antheridia morphometry

Authors	Diameter (μm)
Migula (1889–1897)	300
Groves and Bullock-Webster (1924)	450
Corillion (1975, 1957)	350–560
Hollerbach and Krassavina (1983)	375–470
Krause (1997)	)350–550
Schou et al. (2017)	300–560
Total range	**300–560**

51.2 Phenology and Reproduction

Chara virgata is mostly an annual species but in deeper water it can be perennial. This species can be found in all months of the year. Jäger (1999, 2000) observed this species under ice during wintertime. *Chara virgata* can overwinter and in that case, plants can grow fast in Spring and can dominate large habitat areas. Oospores germinate in early Spring (March–April). Fructifying plants can be seen in May and are optimally developed during June-July but can survive well into late Autumn. When the circumstances are good the abundant oospores can germinate in Autumn (van Raam, 1998).

51.3 Habitat Conditions and Ecology

Habitat type: *Chara virgata* is common in freshwater bodies and occurs only occasionally in brackish water (Sinkevicìênė et al., 2004). *Chara virgata* populates low electrolyte to calcareous lakes, small ponds and rivers, shallow water areas of larger lakes, especially in boggy areas, often acidic sites (e.g., *Isoetes*-Lakes) on silicate and limestone sediments (Doege and van de Weyer, 2016). It occurs mainly in habitats with a highly variable environment (van Raam, 1998). In the Netherlands *Chara virgata* is found in dune pools low in calcium, hardwater shallow lakes, fens and ditches (Simons and Nat, 1996). This species is able to tolerate lower pH levels than most other species of *Chara* (Moore, 1986). Wood (1965) states that it extends to higher altitudes and further into polar latitudes than most other charophytes.

Sediment: *Chara virgata* grows best on sandy, silty and organic sediments, but also on lime mud, marl and peat (Langangen, 2007a; Schmidt, 1981; van Raam, 1998).

Nutrient conditions: *Chara virgata* populates oligotrophic to mesotrophic habitats and disappears when water features deteriorate. This species is less tolerant to eutrophication than *Chara globularis* (Jäger, 2000; Melzer, 1988; Pietsch, 1982; Schmidt, 1981; Stroede, 1933; van Raam, 1998). Total nitrogen (TN) values 0.2–6.5 mg/l and total phosphorus (TP) values 0.001–0.303 mg/l are reported (Doege et al., 2016).

Water chemistry: *Chara virgata* inhabits both calcium-poor as well as calcium-rich waters (Doege and van de Weyer 2016). PH values ranging between 4.7 and 9.6, the carbonate hardness values between 0 and 12.5° dH. Total hardness ranges between 0.1 and 60.0° dH (Doege et al., 2016). The conductivity is reported from 10 up to 3000 μS/cm. According to Stroede (1933) *Chara virgata* can cope with less Ca^{2+} than other species. Sinkevicìênė et al. (2004) and Langangen (2007a) reported a high degree of flexibility with regard to salinity.

Depth range: *Chara virgata* is mostly found in shallow water up to 4–6 m deep. Nevertheless some deepwater records show this species can be found at considerable depths. Frantz and Cordone (1967) report *Chara virgata* to be found in Lake Tahoe (California and Nevada) at a depth of 200 ft (about 61 m). For Europe Krause (1997) claims 10 m to be the maximum depth.

Associated species: *Chara virgata* can form extensive monospecific vegetations described as *Charetum virgatae* Doll, 1989 corr. Täuscher (= *Charetum delicatulae* Doll, 1989, = *Chara delicatula*-society sensu Pietsch, 1987) (Felzines and Lambert, 2012). Because of its broad ecological niche, *Chara virgata* can be found in vegetations with many other (charophyte) species.

51.4 Variability

Bushy growth forms with multicellular or unicellular bulbils are common in the UK (Moore, 1986), Russia (Hollerbach and Krassavina, 1983) and the Baltic area (Sinkevicienė et al., 2004). Delicate, slightly encrusted forms with lots of bulbils were described as *Chara delicatula* f. *bulbillifera*. In deep waters *Chara virgata* can be perennial and then form a large slender plants resembling *Chara globularis* but can still be distinguished from this species by having the upper whorl of stipulodes well developed (van Raam, 1998).

51.5 Identification and Risk of Confusion

Chara virgata is closely related to *C. globularis*. It is often regarded as a variety or subspecies of *C. globularis*. These two species may occur together (Pełechaty et al., 2004) and then it can be difficult to distinguish them. The papillate spine-cells and the more or less equally tylacanthous cortex of *Chara virgata* distinguishes this taxon from *C. globularis*. The latter taxon has no developed spine-cells and stipulodes and the cortex is regularly isostichous. Distinction from *C. globularis* is possible by means of the stipulodes but it is much less reliable than the spine-cells and tylacanthy characters. In contrast to *C. globularis*, *C. virgata* has, mostly a well-developed upper row of short and pointed stipulodes. The problem is that all possible intermediate forms exist. Where to draw the line? *Chara virgata* often has a stronger smell than *C. globularis* and is always slender. *C. globularis* can be quite robust. However these are subjective attributes. Genetic research has shown *Chara globularis* and *C. virgata* are two different species (Krienitz and Nowak, 2016). Even so it is sometimes not possible to distinguish both species on a morphologic basis. As well, the ecological range is generally overlapping. *Chara globularis* is found in oligotrophic-eutrophic habitats and the range of *C. virgata* is oligotrophic-mesotrophic.

Small, less developed plants can sometimes be confused with *Chara contraria* and *C. vulgaris*. In *C. contraria* and *C. vulgaris* the upper and lower whorl of stipulodes are well developed, whereas *C. virgata* has the upper whorl of stipulodes well developed and the lower row is rudimentary (papillate to spherical). Apart from that *C. contraria* and *C. vulgaris* are diplostichous and *C. virgata* is triplostichous.

Chara aspera var. *subinermis* always has symmetric, closely appressed stipulodes which are narrowly acute whereas the stipulodes of *C. virgata* are often more obtusely tipped to broadly acute. Also, the spine-cells in *C. aspera* var. *subinermis* are usually triangular acute rather than rounded to obtuse tipped as in *C. virgata*. Also, *C. aspera* is clearly dioecious when fertile.

Chara fragifera can also be confused with C. *virgata*. For the differences see under that species.

There is another triplostichous species with which *C. virgata* may be confused: *C. connivens*. *C. virgata* is monoecious but because of protandry it may look like the dioecious species *C. connivens*. Both species have a strong smell but *C. connivens* has two rows of rudimentary stipulodes and has no or rudimentary spine-cells. The branchlets of *C. connivens* can be strongly connivent especially in the top whorls of fertile plants whereas the branchlets of *C. virgata* mostly are slightly curved or stiff standing out.

51.6 Distribution

Chara virgata is a cosmopolitan species (Corillion, 1957; Groves and Bullock-Webster, 1924; Migula, 1889–1897; van Raam, 1998). It occurs in North America, Asia, Africa, Australia, New Zealand and Europe. In Fig. 51.5 the distribution of *Chara virgata* in Europe is presented.

51.7 Status and Threats

In comparison with other species of the genus, *Chara virgata* is one of the most common species (Doege and van de Weyer, 2016; van Raam, 1998). Like *Chara globularis* it is one of the few charophyte species not to be regarded as endangered in most countries of Europe.

51.8 Remarks

Most charophyte species have a specific smell. *Chara virgata*, like *C. connivens*, has a very distinguished (harsh) smell. Possibly the smell of *Chara virgata* makes people say charophytes are 'skunkweeds'. Allelopathic effects have never been conclusively demonstrated for this species (Blindow and Hootsmans, 1991; Mulderij, 2006).

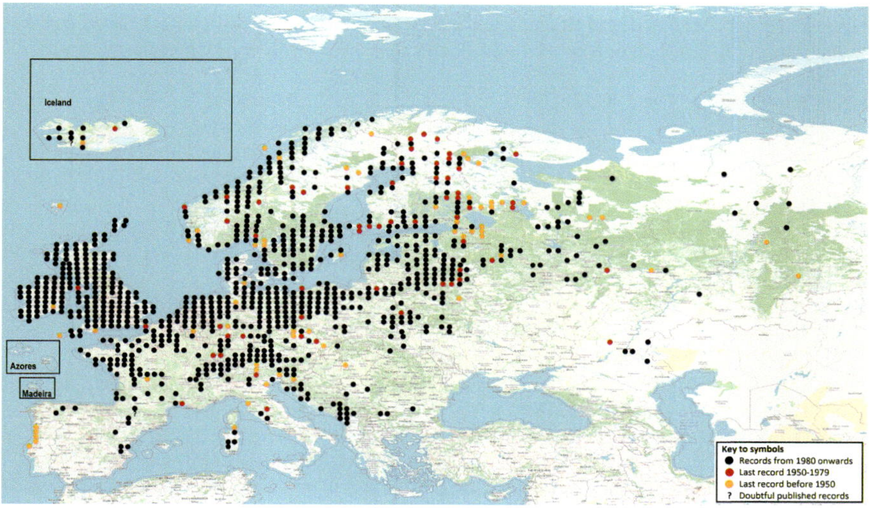

Fig. 51.5 Distribution of *Chara virgata* in Europe. For worldwide distribution see Korsch (2018). Black dots are records from 1980 on, red dots from 1950 to 1979 and yellow dots before 1950; unconfirmed/doubtful published records (if applicable) are marked with "?" on top. Map prepared by Heiko Korsch

For a long period in Europe *Chara virgata* was not distinguished from *Chara globularis*. To research its history one has to examine the literature of *C. globularis* as well.

References

Agardh, C. A. (1824). *Systema Algarum*. Berling, Lundae [in Latin].

Becker, R., Blindow, I., Doege, A., Franke, T., Gregor, T., Hamann, U., Jäger, D., Jorda, C., Kabus, T., Korsch, H., Korte, E., Kusber, W. H., Pätzold, F., Raabe, U., Schubert, H., Teppke, M., van de Weyer, K., & Wolff, P. (2016). Beschreibung der Characeen-Arten Deutschlands. In A. C. Deutschlands (Ed.), *Armleuchteralgen: Die Characeen Deutschlands* (pp. 209–572). Springer [in German].

Blindow, I., & Hootsmans, M. J. M. (1991). Allelopathic effects from *Chara* spp. on two species of unicellular green algae. In M. J. M. Hootsmans, & J. E. Vermaat (Eds.), *Macrophytes, a key to understanding changes caused by eutrophication in shallow freshwater ecosystems* (pp. 139–144).

Corillion, R. (1957). Les Charophycées de France et d'Europe Occidentale. *Bulletin de la Société Scientifique de Bretagne, 32*(Hors série I & II), 1–499 [in French].

Corillion, R. (1975). *Flore et végétation du Massif Armoricain. T. IV. Flore des Charophytes (Characées) du Massif Armoricain et des contrées voisines d'Europe occidentale*. Jouve, Paris [in French].

Dilger, M. (2010). Merkmale zur Erkennung von *Nitella-* und *Tolypella-*Arten der Characeaen [sic] über Oosporen aus Sedimenten oder von Pflanzen. *Rostocker Meeresbiologische Beiträge, 23*, 40–59. [in German].

Doege, A., & van de Weyer, K. (2016). *Chara virgata*. In A. C. Deutschlands (Ed.), *Armleuchteralgen: Die Characeen Deutschlands* (pp. 369–379). Springer [in German].

Doege, A., van de Weyer, K., Becker, & Schubert, H. (2016). Bioindikation mit Characeen. In A. C. Deutschlands (Ed.), *Armleuchteralgen: Die Characeen Deutschlands* (pp. 97–138). Springer [in German].

Doll, R. (1989). Die Pflanzengesellschaften der stehenden Gewässer der DDR. Teil 1: Die Gesellschaften des offenen Wassers (Characeen-Gesellschaften). *Feddes Repertorium, 100*, 281–324. [in German].

Felzines, J.-C., & Lambert, E. (2012). Contribution au prodrome des végétations de France: les *Charetea fragilis* F. Fukarek 1961. *Le Journal de Botanique de la Société Botanique de France, 59*, 133–188. [in French].

Frantz, T. C., & Cordone, A. J. (1967). Observations on deepwater plants in Lake Tahoe, California and Nevada. *Ecology, 48*, 709–714.

Groves, J., & Bullock-Webster, G. R. (1924). *The British Charophyta. Volume II. Chareæ with plates, concluding articles, geological sketch, bibliography and index*. Ray Society.

Hollerbach, M. M., & Krassavina, L. K. [Голлербах, М. М. & Л. К. Красавина]. (1983). Определитель пресноводных водорослей СССР. 14. Charophyta [*Determination key of freshwater algae of USSR* (189 pp.)]. Nauka, Leningrad [in Russian].

Jäger, D. (1999). *Beiträge zur Characeen-Flora Vorarlbergs* (161 pp.). Diplomarbeit Univ. Innsbruck [in German].

Jäger, D. (2000). Beiträge zur Characeen-Flora Vorarlbergs (Österreichs). *Berichte des naturwissenschaftlich-medizinischen Vereins in Innsbruck, 87*, 67–85 [in German].

Korsch, H. (2018). The worldwide range of the Charophyte species native to Germany. *Rostocker Meeresbiologische Beiträge, 28*, 45–96.

Krause, W. (1997). Charales (Charophyceae) Süßwasserflora von Mitteleuropa. In H.Ettl, G. Gärtner, H. Heynig, & D. Mollenhauer (Eds.), *Süßwasserflora von Mitteleuropa* (Vol. 18). Fischer, Jena [in German].

Krienitz, L., & Nowak, P. (2016). Systematik der Charophyceae. In Arbeitsgruppe Characeen Deutschlands (Ed.), *Armleuchteralgen: Die Characeen Deutschlands*. (pp. 17–21). Springer [in German].

Langangen, A. (2007). *Charophytes of the Nordic countries* (p. 102). Saeculum ANS.

Melzer, A. (1988). *Der Makrophytenindex—eine biologische Methode zur Ermittlung der Nährstoffbelastung von Seen* (249 pp.). Habilitationsschrift, TU München [in German].

Migula, W. (1889–1897). Fünfter Band: Die Characeen. In A. Fischer, E. Fischer, F. Hauck, G. Limpricht, C. Luerssen, W. Migula, H. Rehm, P. Richter, & G. Winter (Eds.), *Dr. L. Rabenhorst's Kryptogamenflora von Deutschland, Oesterreich und der Schweiz* (2nd ed.). Eduard Kummer, Leipzig [in German].

Moore, J. A. (1986). *Charophytes of Great Britain and Ireland*. BSBI Handbook No. 5, Botanical Society of the British Isles.

Mulderij, G. (2006). *Chemical warfare in freshwater—allelopathic effects of macrophytes on phytoplankton*. Thesis Radboud University Nijmegen.

Pal, B. P., Kundu, B. C., Sundaralingam, V. S., & Venkataraman, G. S. (1962). *Charophyta* (130 pp.). Indian Council of Agricultural research.

Pełechaty, M., Pukacz, A., & Pełechata, A. (2004). Co-occurrence of two stoneworts, of reverse ecological spectra in the same lake ecosystem. Habitat requirements of *Chara delicatula* Agardh and *Chara globularis* Thuillier in the context of bioindication. *Polish Journal of Environmental Studies, 13*, 551–556.

Pietsch, W. (1982). Biologische, mikrobiologische und toxikologische Methoden. Makrophytische Indikatoren für die ökochemische Beschaffenheit der Gewässer. In C. Legler, & G. Breitig (Eds.),

Ausgewählte Methoden der Wasseruntersuchung, Band 2: Biologische, mikrobiologische und toxikologische Methoden (pp. 67–88). Gustav Fischer [in German].
Pietsch, W. (1987). Zur Vegetation der *Charetea*-Gesellschaften der Mitteleuropäische Tiefebene. *tudia Phytologica Nova Pecsi Akademiai Bizottsag*, 69–86.
Schmidt, D. (1981). Die Characeen—eine im Aussterben begriffene Pflanzengruppe unserer Gewässer. *Gleditschia, 8*, 141–157. [in German].
Schou, J. C., Moeslund, B., Båstrup-Spohr, L., & Sand-Jensen, K. (2017). *Danmarks vandplanter*. BFN's, Klitmøller [in Danish].
Simons, J., & Nat, E. (1996). Past and present distribution of stoneworts (Characeae) in The Netherlands. *Hydrobiologia, 340*, 127–135.
Sinkeviciènė, Z. [Sinkevičienė], I. Blindow, M. Koistinen & J. Urbaniak, 2004. *Chara virgata*. In H. Schubert, & Blindow, I. (Eds.), *Charophytes of the Baltic sea* (pp. 142–148). A. R. G. Gantner.
Stroede, W. (1933). Über die Beziehungen der Characeen zu den chemischen Faktoren der Wohngewässer und des Schlammes. *Archiv Für Hydrobiologie, 25*, 192–229. [in German].
van Raam, J. C. (1998). *Handboek Kranswieren*. Chara boek, Hilversum [in Dutch].
van de Weyer, K., & Schmidt, C. (2018). Bestimmungsschlüssel für die aquatischen Makrophyten (Gefäßpflanzen, Armleuchteralgen und Moose) in Deutschland: Band 1: Bestimmungsschlüssel, ed. 2. Landesamt für Umwelt (LfU) Brandenburg. *Potsdam, Fachbeiträge Des LfU Brandenburg, 119*, 1–180. [in German].
Wood, R. D. (1965). *Monograph of the Characeae*. J. Cramer.

Chapter 52
Chara vulgaris

Luc Denys, Chrysoula Christia, and Eva Papastergiadou

Scientific name: ***Chara vulgaris*** **L., Sp. Pl. 2: 1156. 1753**

Lectotype (Wood, 1960: 220): [no place]/[no date]/[no collector] (LINN 1088.3 as "2. *vulgaris*").

= *Chara batrachosperma* Thuill. 1799, Fl. Env. Paris, ed. 2: 473, n. ill.
Lectotype (Gregor et al., 2012): [no place]/[no date]/[no collector] (Herbarium Persoon in L, specimen annotated by J. L. Thuillier as "*Chara batrachosperma* Thuillier" and by J. van Raam as "*C. vulgaris*" and "Type *Chara batrachosperma* Thuillier. 7/1996.").

≡ *Nitella batrachosperma* (Thuill.) C. Agardh, 1824, Syst. Alg.: 126.

= *Chara fragilis* Desv. in Loisel., Not. Fl. France: 137. 1810, n. ill.
? = *Chara decipiens* Desv. in Loisel., Not. Fl. France: 138. 1810.
Type not designated.

= *Chara vulgaris* L. var. *papillata* Wallr., Annus Bot.: 183. 1815.
Type not designated.

= *Chara longibracteata* Kütz. in Rchb., Fl. Germ. Excurs. 2: 843. 1832.
Type not designated.

L. Denys (✉)
Research Institute for Nature and Forest (INBO), VAC Herman Teirlinck, Havenlaan 88 Bus 73, 1000 Brussels, Belgium
e-mail: luc.denys@inbo.be

C. Christia · E. Papastergiadou
Department of Biology, University of Patras, University Campus Rio, 26500 Patras, Greece
e-mail: xchristia@gmail.com

E. Papastergiadou
e-mail: evapap@upatras.gr

H. Schubert et al. (eds.), *Charophytes of Europe*,
https://doi.org/10.1007/978-3-031-31898-6_52

≡ *Chara vulgaris* L. var. *longibracteata* (Kütz.) Kütz., Sp. Alg.: 523. 1849.

≡ *Chara vulgaris* L. f. *longibracteata* (Kütz.) H. Groves & J. Groves in J. Bot. 18: 133. 1880.

= *Chara foetida* A. Braun in Ann. Sci. Nat., Bot. Sér. 2, 1: 354. 1834.
Type not designated.

= *Chara foetida* A. Braun var. *crassicaulis* Schleich. ex A. Braun 1847 in Neue Denkschr. Allg. Schweiz. Ges. gesammten Naturwiss. 10(3): 15.
Type not designated.

≡ *Chara vulgaris* L. var. *crassicaulis* (Schleich. ex A. Braun) Kütz., Sp. Alg.: 523. 1849.

≡ *Chara crassicaulis* (Schleich. ex A. Braun) Kütz., Tab. Phycol.: 25. 1857.

? = *Chara rabenhorstii* A. Braun in Rabenhorst in Flora 33: 524. 1850.
Type not designated.

≡ *Chara vulgaris* L. var. *vulgaris* f. *rabenhorstii* (A. Braun) R. D. Wood, Rev. Charac.: 86. 1965, n. inv.

= *Chara subhispida* A. Braun in Cohn, Krypt.-Fl. Schlesien: 407. 1876.
Type not designated.

≡ *Chara foetida* A. Braun subsp. *subhispida* A. Braun in Braun & Nordstedt in Abh. Königl. Akad. Wiss. Berlin 1882: 167. 1882.

≡ *Chara foetida* A. Braun f. *subhispida* G. Bailly & O. Schaefer, Guide Illustré Characées Nord-Est France: 33. 2010, n. inv.

? = *Chara pistianensis* Vilh. in Bull. Int. Acad. Tchéque Sci., Cl. Sci. Math. Nat. Méd. 1923: 9. 1924.
Lectotype (Wood, 1965: 92): In aqua calida 32 °C. ad balneas Irma (hotel Thermia), Pieštany, Slovakia occidentalis: opp/Aug. 1922/Dr. Jan Vilhelm (PRC).

Nomenclatural remark: *Chara rabenhorstii* is only documented by material from the locus typicus, Lago San Egidio (Promontorio del Gargano, San Giovanni Rotondi, Apulia), where it occurred with *C. vulgaris*. According to Wood (1965), original material is present in BM (*C. Rabenhorstii,* Lago St. Egidio, Gargano/June 1847/ Rabenhorst/ex Museo Botanico Berolinensi).

Local names: BELARUSIAN: Хара звычайная; DANISH: Almindelig kransnål; DUTCH: Gewoon kransblad; ENGLISH: Common stonewort,—chara; ESTONIAN: Haisev mändvetikas; FINNISH: Haisunäkinparta; FRENCH: Chara commune (chara à axe épais,—à bractées longues,—subhispide), charagne commune; GERMAN: Gemeine Armleuchteralge, Gewöhnliche Armleuchteralge, Stinkende Armleuchteralge; GREEK: Χάρα η κοινή; LATVIAN: Parastā hara; LITHUANIAN: Paprastasis maurabragis; NORWEGIAN: Stinkkrans; POLISH: Ramienica pospolita, RUSSIAN: Хара обыкновенная; SWEDISH: Busksträfse; UKRAINIAN: Хара звичайна.

52.1 Morphology

Habitus: Plants bright to medium dark green or greyish, rigid to flexible, often with flexed main axis. Incrustation absent to heavy. Overall habitus very variable, ranging from fairly robust and dense to very slender and lush (Figs. 52.1, 52.2a and b). Odor conspicuous.

Size: Small to fairly large; 1 to c. 70(80) cm.

Main axis: Diameter usually about 0.5–1 mm, sometimes less ($\geq$0.2 mm) or more (up to 1.5–2 mm for the *C. crassicaulis* morph; Bailly and Schaeffer, 2010; Bazzichelli and Abdelahad, 2009). Main axis often repeatedly branched, but the primary axis remains the most prominent one. Lower internodes can be 4-6 times longer than the branchlets ($\leq$c. 8 cm), with length decreasing upward. Internodes usually shorter or of almost equal length along the main axis in shallow water and etiolated with reduced light (Wang et al., 2008).

Cortex: Diplostichous and, essentially, aulacanthous (see 'Variability'); the lower, older part of the main axis often $\pm$ isostichous (Figs. 52.2c and 52.3). Primary and secondary cell rows usually well differentiated in the upper part of the main axis, primary rows being narrower. Cortical cells straight or torsioned. The cortex of the

Fig. 52.1 *Chara vulgaris*. General aspect upper part, *longibracteata* morphotype. Photograph: Klaus van de Weyer.

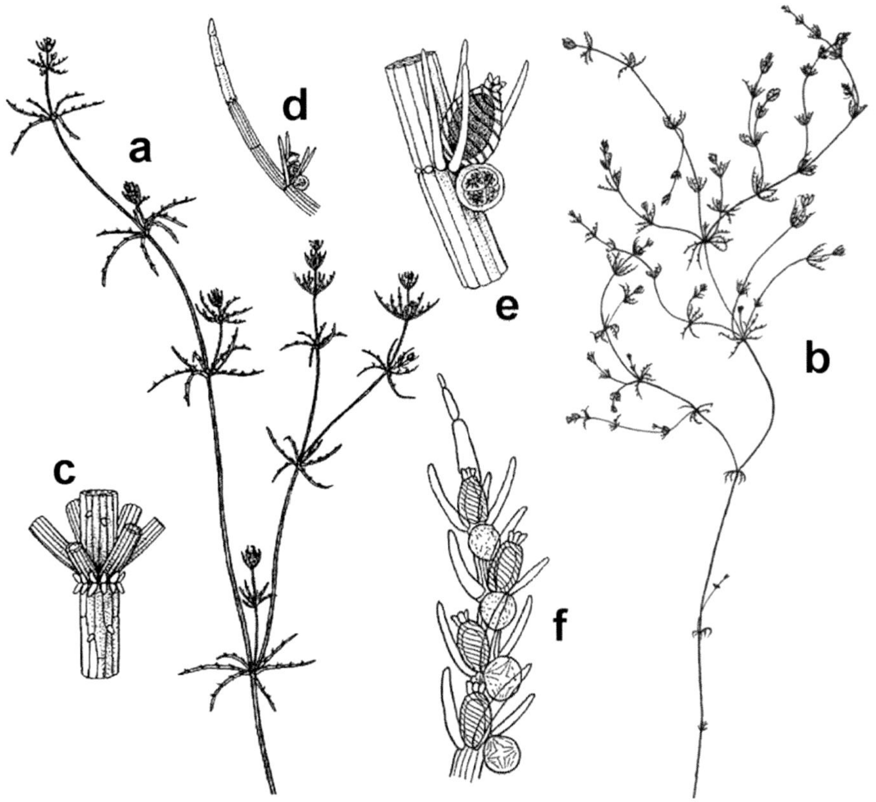

Fig. 52.2 *Chara vulgaris*. **a** General aspect, *vulgaris* morphotype; **b** General aspect, *crassicaulis* morphotype; **c** Detail of primary axis with node; **d** Tip of branchlet; **e** Gametangia, bracteolae and bract-cells; **f** Tip of branchlet where gametangia 'overlap' due to very short internodes, commonly observed in the *crassicaulis* morphotype. Drawings: **a**, **c–e**—Margaret Tebbs, reproduced by courtesy of BSBI; **b, f**—Werner Krause

main axis or branchlets is reduced in some populations, but normally not lacking entirely (see 'Variability'). Very young axes, however, are often ecorticate.

Spine-cells: Abundant to very scarce and varying in length from rudimentary up to, and sometimes exceeding, the diameter of the axis ($\leq$1.8 mm; Figs. 52.3 and 52.4a). Deciduous and commonly better developed and denser on younger internodia and ± lacking on the lower part of the plant. Normally solitary, but in plants with long spine-cells, a few may be geminate, in which case the second spine is much reduced. Pairs of rudimentary spine-cells were reported for *C. vulgaris* f. *pistianensis* (Wood, 1965). The shape is conical, cylindrical or subcylindrical, thinning somewhat towards the usually blunt tip. Longer spine-cells are often more or less adpressed against the axis (<30°) and may remain partly embedded within the cortical furrows. Usually, longer spine-cells drooping downwards on the upper half of the internodes, those

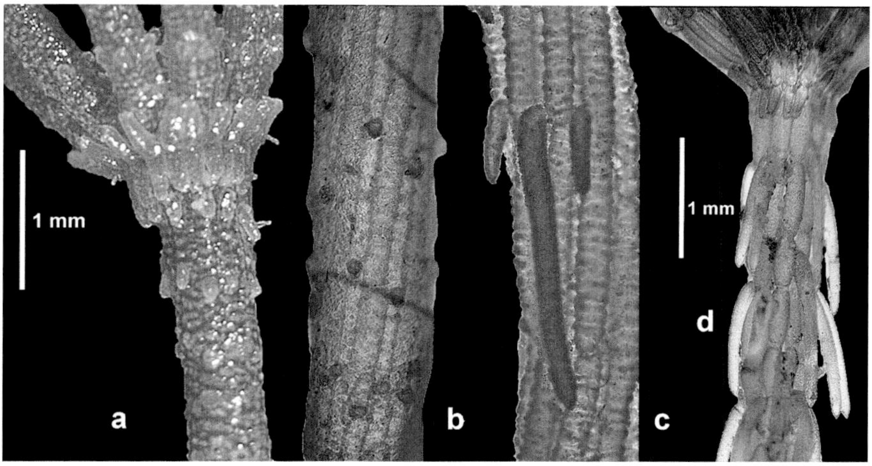

Fig. 52.3 *Chara vulgaris*, details I. **a** Primary axis with slight carbonate encrustation, showing stipulae, spine-cells and the first branchlet nodes with abaxial bract-cells; **b** Detail of primary axis showing diplostichous, aulacanthous cortex with rudimentary spine-cells; **c** As b but with well-developed spine-cells of variable length; **d** As a but with length of spine-cells exceeding the diameter of the axis as in the *papillata* morphotype. Scale bars: **a** left, **b–d** right. Photographs: **a**—Luc Denys; **b–d**—Chris Carter

on the lower part tending more to point upwards. Plants with spine-cells equaling at least the diameter of the main axis may be attributed to particular morphotypes (see 'Variability').

Stipulodes: *Chara vulgaris* is diplostephanous with isomorphic, oblong-ovate, obtusely ending stipulodes; both rows are well-developed (Figs. 52.2c, 52.3a, d and 52.4a). Twice as many stipulodes as branchlets in each tier. Often somewhat irregular in size. Length usually attaining only about 1/5 to 1/2 of the axis diameter, occasionally also more (up to 150 μm according to Langangen, 2015 and as much as 170 ± 68 μm according to Hu et al., 2009).

Branchlets: Whorls of 5–10(11) diplostichous branchlets of similar length. Length varying from 0.4 to 6 cm. Each branchlet with (0)2–6(7) corticate segments and up to 3 terminal ecorticate cells (Fig. 52.2d). The nude part may be longer than the corticated part (see 'Variability') but is usually less than half as long. Terminal cell c. 0.1-0.34 mm long, conical and acute or blunt (Fig. 52.4b); commonly slightly curved in the *C. crassicaulis* morph (Urbaniak, 2011). Branchlets spreading, reflexed or, especially in the upper whorls, incurved. Long, slender branchlets often somewhat torsioned.

Bract-cells: At least 3 and usually 4 or 5 (up to 6 according to Wood and Mason, 1977) persistent bract-cells develop before the gametangia and also on sterile nodes (Muller et al., 2017). Adaxial ones well developed (also on sterile branchlets) and more or less acute. Length up to 6(7) mm (see Variability; Fig. 52.1). Abaxial bract-cells

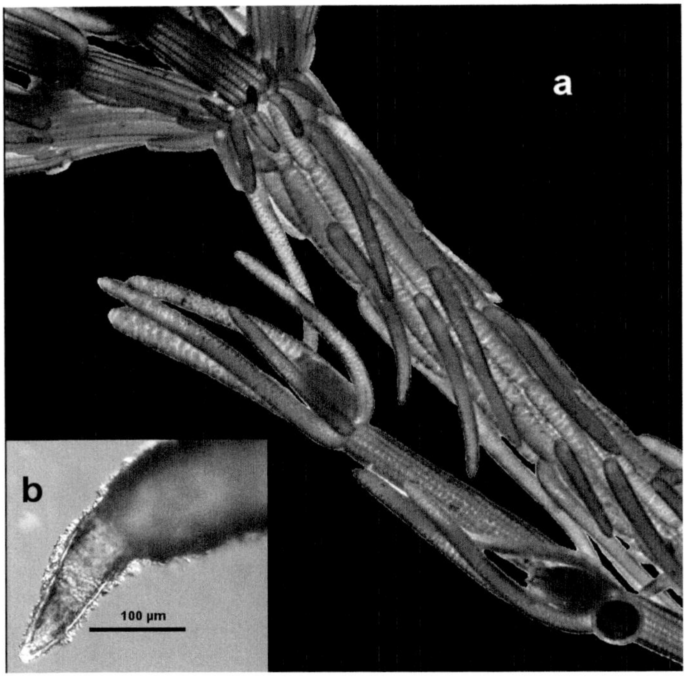

Fig. 52.4 *Chara vulgaris*, details II. **a** Detail of primary axis with long spine-cells and fertile branchlet with long bracteoles and adaxial bract-cells typical of the *subhispida* morphotype; **b** Terminal branchlet cell. Photographs: **a**—Chris Carter; **b**—Luc Denys

usually rudimentary or globose, bluntly tipped and sometimes absent (Figs. 52.2e and 52.3a).

Bracteoles: Four well developed bracteoles. Length ranging from somewhat longer than the oogonia, to up to 12 times longer (Figs. 52.2d–f and 52.4a), i.e. c. 1 to 7 mm, rarely even exceeding 1 cm (Tortić-Njegovan, 1956). Variation in length results in phenotypic differentiation (see 'Variability').

Bractlets: Absent.

Sexuality: Monoecious; conjoined gametangia abundantly formed at the corticate branchlet nodes (Fig. 52.2d–f). Oogonia and antheridia usually solitary and occasionally geminate, although laterally conjoined gametangia occur more frequently in some regions, e.g. southern France (I. Soulié-Märsche pers. comm.). Oogonia placed adaxially, antheridia abaxially. Gametangia formation is slightly protrandric and both types usually occur together on the plant. Protandry might explain the often sejoined gametangia in the *C. crassicaulis* morph (Bailly and Schaefer, 2010).

Oogonia: Oogonia are ellipsoid to cylindrical, symmetrical or somewhat narrowing in the upper half and with constricted coronulae (Fig. 52.2e). Usually, 500-800 µm long and 350–550 μm wide (excl. coronula; Table 52.1), occasionally also larger (up to 1100 μm long and 680 μm wide) with (11)12–16 helical stripes. Coronula very variable in size and shape, 75–125(150) μm high with cells usually more or less spreading and varying from spherical to twice as long as wide (Groves and Bullock-Webster, 1924).

Oospores: Terete in apical view and oval, ellipsoid to ellipsoid-cylindrical in side view, narrowing slightly near the base (Fig. 52.5a–c). No distinct shoulders and 11–15 narrow, low but distinct ridges with delicate flanges. Length (387)420–810 μm (often <700 μm), width 225–540 μm (mostly <500 μm), length/width ratio c. 1.7 (1.4–2) (Table 52.2). Striae 2–3 μm wide, midpoint of striae height is <15 μm. Fossa width variable: 30–81.7 μm (Urbaniak, 2011). Claw small, basal plate pentagonal and almost circular (Fig. 52.5e). Colour yellow, golden to dark brown or dull black at maturity. Membrane thin; irregularly granulate or verrucate to nearly smooth under the light microscope. Figures 52.6a–c show spaced or short series of aligned granules of irregular size (a, b) and sometimes low wart-like protuberances (c).

Different ornamentation patterns were reported for acetolysed spores in SEM (see also Ahmadi et al. 2012b): minutely granulate (Haas, 1994), irregularly granular (De Winton et al., 2007), granulate, papillate and with occasionally conical projections, with or without pores (John et al., 1990), as well as pust(ic)ular with apical pores (Mandal et al., 2002; Ray et al., 2001). Ray et al. (2001) observed pustular elevations of 0.43–0.85 μm (c. 8–10/10 μm) with pores at the tip, small granulae and some simple pores on the fossa wall in one population and pustulae of 0.64–0.85 μm with some pores in another. Urbaniak (2011) reports granules of 1.5–2.8 μm with pores or fissures at the tip and slightly granulate projections with round or elongated pores on both sides of the striae. Oospores from a range of specimens in BRM were examined by Casanova and Wibraham (2019; see also Holzhausen et al., 2024). Their membrane varies from finely granulate to more or less perforate verrucate (Fig. 52.6d–f). In part, the reported variation may be due to differences in treatment and developmental stage, with verrucae and perforations becoming more pronounced with ageing.

The species develops gyrogonites, which are ellipsoid with concave (almost flat, slightly concave to slightly convex according to Harris, 1939) spiral cells with mostly 10–13 striae (Fig. 52.5d). Both ends often somewhat drawn out, more so at the apex, to rounded. Size variable: height 400–760 μm, width 270–545 μm (Table 52.3). Isopolarity index 1.18–2.13 (Sanjuan et al., 2017). The lime crust is thin and fragile (Krause, 1986, 1997).

Antheridia: Spherical and octoscutate, (213)250 to 500(519) μm in diameter (Table 52.4). Gillet (1960) and Kwiatkowska et al. (1997) report considerably larger antheridia from cultured material. Orange to red in colour, becoming paler with age.

Bulbils: According to most authors without rhizoidal bulbils, but Giesenhagen (1896) mentions small unlobed starch rich 'Wurzelknöllchen' (rhizoidal bulbils) for *Chara foetida longibracteata* and Lambert (2007) and Wang et al. (2015) also report very

Table 52.1 Oogonia morphometry

Author	Length	Width	Coronula length	Coronula width	Number of helical stripes
Ahmadi et al., (2010)	500–900	(280)250–550(660)	–	–	–
Ahmadi et al. (2012a)	500–650	350–450	75–120	150–300	c. 14
Boissezon et al. (2017)	500–800	350–500	–	–	–
Caisová and Gąbka (2009)	500–1100	(280)335–525	–	–	–
Bailly and Schaeffer (2010)[a]	–	350–500	–	–	–
Bailly and Schaeffer (2010)[b]	–	300–500	–	–	–
Cirujano et al. (2008)[c]	500–800	350–475	75–125	–	–
Cirujano et al. (2008)[d]	600–750	375–500	90–105	–	–
Cirujano et al. (2008)[e]	500–800	350–500	60–100	–	–
Cirujano et al. (2008)[f]	520–750	380–510	80–140	–	–
Compère (1992)	500–800	350–500	–	–	–
Corillion (1975)[g]	500–800	350–475	75–125	200–325	14–16
Corillion (1975)[h]	c. 600	335–425	c. 105	c. 170	12–15
Gillet (1960)[i]	560 ± 75–575 (±80)	385 ± 75–480 ± 75	–	–	–

(continued)

Table 52.1 (continued)

Author	Length	Width	Coronula length	Coronula width	Number of helical stripes
Hollerbach and Krassavina (1983)	750–800	450–550	75–125	160–325	14–16
Corillion (1957), Groves and Bullock-Webster (1924)	525–800	350–475	75–125	200–325	14–16
Hu et al. (2009)[j]	535 ± 81.8–642.3 ± 121.5	329.3 ± 53.4–380 ± 84.7	–	–	9.4 ± 1.7–12.3 ± 3.2
Korsch (2013)	–	–	–	–	12–16
Krause (1997)	500–800	350–500	100	200–300	12–15
Kwiatkowska et al. (1997)	(438)475–588	300–450	–	–	–
Langangen (2007)	Up to 900	–	–	–	–
Maier (1972)	600–925	–	–	–	–
Migula (1889–1897)	750–800	450–550	90	160	12–15
Moore (1986)	–	Up to 500	–	–	–
Muller et al. (2017)	500–750	–	–	–	–
Noedoost et al. (2015)[k]	598.5–762	383.7–457	105–153.9	177.3–236.6	–

(continued)

Table 52.1 (continued)

Author	Length	Width	Coronula length	Coronula width	Number of helical stripes
Saber et al. (2018)	366–590	355–437	83–118	–	–
Urbaniak and Gąbka (2014)[l]	600–700	335–425	–	–	–
Urbaniak and Gąbka (2014)[m]	575–990	360–680	–	–	–
van Raam (1998)	Up to 850	Up to 550	–	–	–
Wood (1959)	525–800	–	–	–	13–16
Wood (1959)[n]	600–610?	–	–	–	12–13
Wood and Imahori (1964)[o]	675–750	450–510	135–160	225–300	14–15
Wood and Imahori (1964)[p]	525–750	315–480	60–120	120–210	11–13
Wood (1965)	500–800	350–475	75–125	200–325	14–16

(continued)

Table 52.1 (continued)

Author	Length	Width	Coronula length	Coronula width	Number of helical stripes
Wood and Muenscher (1956)	1000	618	105	160	c. 16
Total range	**366–1100**	**280–550(680)**	**60–154**	**(120?)150–325**	**(11)12–16**

Length in the first column is excluding the coronula (also not taken into account for total range); length and width of the coronula are given separately, all values for length and width in μm

[a] *Chara vulgaris* var. *vulgaris*, *C. vulgaris* f. *longibracteata*, *C. vulgaris* f. *subhispida*, *C. vulgaris* var. *papillata*
[b] *Chara vulgaris* var. *crassicaulis*
[c] *Chara vulgaris* var. *vulgaris*
[d] *Chara vulgaris* var. *crassicaulis*
[e] *Chara vulgaris* var. *longibracteata*
[f] *Chara vulgaris* var. *papillata*
[g] *Chara vulgaris* var. *vulgaris*
[h] *Chara vulgaris* var. *crassicaulis*
[i] Minimum and maximum in populations presenting smallest, resp. largest size, with 2 SD
[j] Smallest and largest average with SD for population in four seasons
[k] Range of population mean values; according to bract-cell length inferred from Noedoost et al. (2015; population mean ranging from 2.4 to 6.2 mm) attributable to the *longibracteata* morph
[l] *Chara crassicaulis*
[m] *Chara vulgaris*
[n, o] *Chara pistianensis*, *Chara vulgaris* f. *pistianensis*
[p] *Chara vulgaris* emend., *C. vulgaris* f. *crispa*

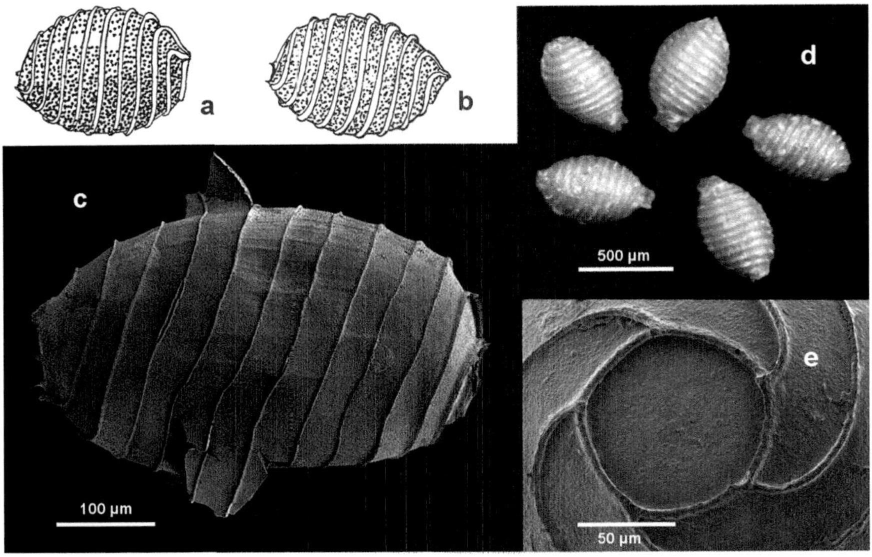

Fig. 52.5 *Chara vulgaris*, oospores. **a** Lateral view of oospore, *vulgaris* morphotype; **b** Lateral view of oospore, *crassicaulis* morphotype; **c** Lateral view of oospore (SEM), *vulgaris* morphotype; **d** Lateral view of fresh gyrogonites, *longibracteata* morphotype; **e** Detail of oospore basal plate (SEM). Photographs and drawings: **a**, **b**—Werner Krause; **c**, **e**—Michelle T. Casanova; **d**—Luc Denys

small bulbils (≤0.1 mm). Occasionally, nodes may thicken to bulbil-like structures (Boissezon et al., 2017, own observations).

52.2 Phenology and Reproduction

Chara vulgaris is a perennial hemicryptophyte or annual therophyte. Productivity is highest in early summer (Hough and Putt, 1988; Winter et al., 1987). Reproduction usually from mid spring to early autumn but may start in February in warmer regions (I. Soulié-Märsche pers. comm.). Oospores are formed even in winter if temperatures permit. In Europe, ripe oospores usually occur from April to October. Vegetative reproduction by prostrate main axis or main axis fragments forming rhizoids from the lower nodes. Overwintering by means of thickened nodes as well as wintergreen main axis or apical parts; the latter unlikely to survive complete freezing (Olsen, 1944).

Table 52.2 Oospore morphometry

Authors	Length	Width	Number of striae	Colour
Ahmadi et al. (2010)	–	–	–	Dark brown or black
Ahmadi et al. (2012a)	400–550[a]	300–375[b]	c. 10[c]	Dark brown or black[d]
Ahmadi et al. (2012b)[e]	386.9 ± 3	266.8 ± 4	11.6 ± 0.5	Dark brown to black
Ahmadi et al. (2012b)[f]	458 ± 4	310.2 ± 3	11.4 ± 0.5	Dark brown
Ahmadi et al. (2012b)[g]	394.8 ± 1	257.5 ± 3	12.4 ± 0.5	Dark brown to black
Auderset Joye and Rey-Boissezon (2013)	420–650	225–450	12–15	Light brown to dark brown
Bailly and Schaefer (2010)[h]	–	250–350	12–15	Light brown
Bailly and Schaefer (2010)[i]	–	400	10–12	Dark brown to black
Bazzichelli and Abdelahad (2009)	420–650	250–350	10–15	–
Blindow et al. (2018)	–	–	–	Black
Boissezon et al. (2017)	400–700	200–450	12–15	Dark brown to black
Caisová and Gąbka (2009)	–	–	–	Dark brown to black
Cirujano et al. (2008)[j]	425–675	225–400		Dark brown to black
Cirujano et al. (2008)[k]	540–640	300–480	–	Dark brown
Cirujano et al. (2008)[l]	420–650	250–350	–	Black
Cirujano et al. (2008)[m]	470–575	320–425	–	Black

(continued)

Table 52.2 (continued)

Authors	Length	Width	Number of striae	Colour
Corillion (1957)	425–675	225–400	12–15	Light brown to dark brown
Corillion (1975)[n]	425–675	225–400	12–15	Light brown to dark brown, rarely black
Corillion (1975)[o]	540–620	300–480	10–14	Brown
Daily (1944)	530–740	–	10–14	Brown to black
De Winton et al. (2007)	579–677	344–418	13–14	Gold brown
Gillet (1960)	–	–	–	Brown or black
Hollerbach and Krassavina (1983)	420–550	280–350	12–15	Light yellow–dark brown–black
Groves and Bullock-Webster (1924)	425–675	7225–400	12–15	Pale golden brown–dark brown
Haas (1994)	420–550(700)	(220)250–350	11–15	Light to dark–dull brown
Hu et al., (2009)[p]	493.6 ± 32.4–552 ± 63.2	254.2 ± 29.8–294.3 ± 52.6	8.4 ± 0.7–11.8 ± 3.5	–
Korsch (2013)	525–600	300–400	11–15	Yellowish–(dark)brown
Krause (1986)	–	–	11–13	Light to dark brown

(continued)

Table 52.2 (continued)

Authors	Length	Width	Number of striae	Colour
Krause (1997)	420–650	250–350[q]	11–14	Brown–dark brown
Krause (1997), Anhang	500–700	300–500	12–14	Brown, rarely black
Leistner (1978)	425–675	–	12–15	Opaque
Langangen (2007)	–	–	–	Normally brown
Maier (1972)	425–675	–	–	–
Migula (1889–1897), (1898)	420–550	280–350	11–14	Light brown–chestnut brown (rarely black)
Noedoost et al. (2015)[r]	440.3–738.1	278.1–383.9	10–12	–
Ray et al. (2001)[s]	478	312.5	12–15	–
Rodrigo et al. (2010)[t]	495–627(559)	231–380 (297)	–	–
Saber et al. (2018)	510–590	270–350	11–13	Brown–dark brown to black
Urbaniak (2011)[u]	–	–	9–10	–
Urbaniak (2011)[v]	423–810; 699 ± 63.4; 469 ± 19.1	279–540; 421 ± 52.1; 298 ± 14.1	10–14; 11.2 ± 0.7; 12.2 ± 0.7	–
Urbaniak and Gąbka (2014)[w]	540–620	300–480	–	Reddish brown–dark brown

(continued)

Table 52.2 (continued)

Authors	Length	Width	Number of striae	Colour
Urbaniak and Gąbka (2014)[x]	420–810	280–540	–	Brown–dark brown
van Raam (1998)	Up to 750	Up to 450	–	–
Wood (1959)	425–675	–	12–15	–
Wood (1959)[y]	450–480	–	–	–
Wood and Imahori (1964)[z]	510–600	315–420	11–14	Pale yellowish brown–golden brown
Wood and Imahori (1964)[aa]	450–525	255–300	10–11	Black, opaque

(continued)

Table 52.2 (continued)

Authors	Length	Width	Number of striae	Colour
Wood (1965)	425–675	225–400	12–15	Dark brown–(black)
Wood and Muenscher (1956)	770	450	c. 13	Brown (becoming black?)
Total range	**(387)420–810**	**(220)225–540**	**10–15**	

All values for length and width in μm
[a] *Chara crassicaulis* up to 620 μm
[b] *Chara crassicaulis* up to 400 μm
[c] *Chara crassicaulis* 11–13
[d] *Chara crassicaulis* light brown
[e] *Chara vulgaris* f. *vulgaris*, mean and standard deviation
[f] *Chara crassicaulis*, mean and standard deviation
[g] *Chara vulgaris* f. *longibracteata*, mean and standard deviation
[h] *Chara vulgaris* var. *vulgaris*, *C. vulgaris* f. *longibracteata*, *C. vulgaris* f. *subhispida*, *C. vulgaris* var. *papillata*
[i] *Chara vulgaris* var. *crassicaulis*
[j] *Chara vulgaris* var. *vulgaris*
[k] *Chara vulgaris* var. *crassicaulis*
[l] *Chara vulgaris* var. *longibracteata*
[m] *Chara vulgaris* var. *papillata*
[n] *Chara vulgaris* var. *vulgaris*
[o] *Chara vulgaris* var. *crassicaulis*
[p] Smallest and largest average with SD for population in four seasons
[q] *Chara crassicaulis* c. 400 μm
[r] Range of population mean values; see remark Table 52.1
[s] Mean values, derived from graph
[t] In sediment, average between brackets
[u] *Chara crassicaulis*
[v] *Chara vulgaris*, range and means ± SD for two populations
[w] *Chara crassicaulis*
[x] *Chara vulgaris*
[y,z] *Chara pistianensis*, *Chara vulgaris* f. *pistianensis*
[aa] *Chara vulgaris* emend., *C. vulgaris* f. *crispa*

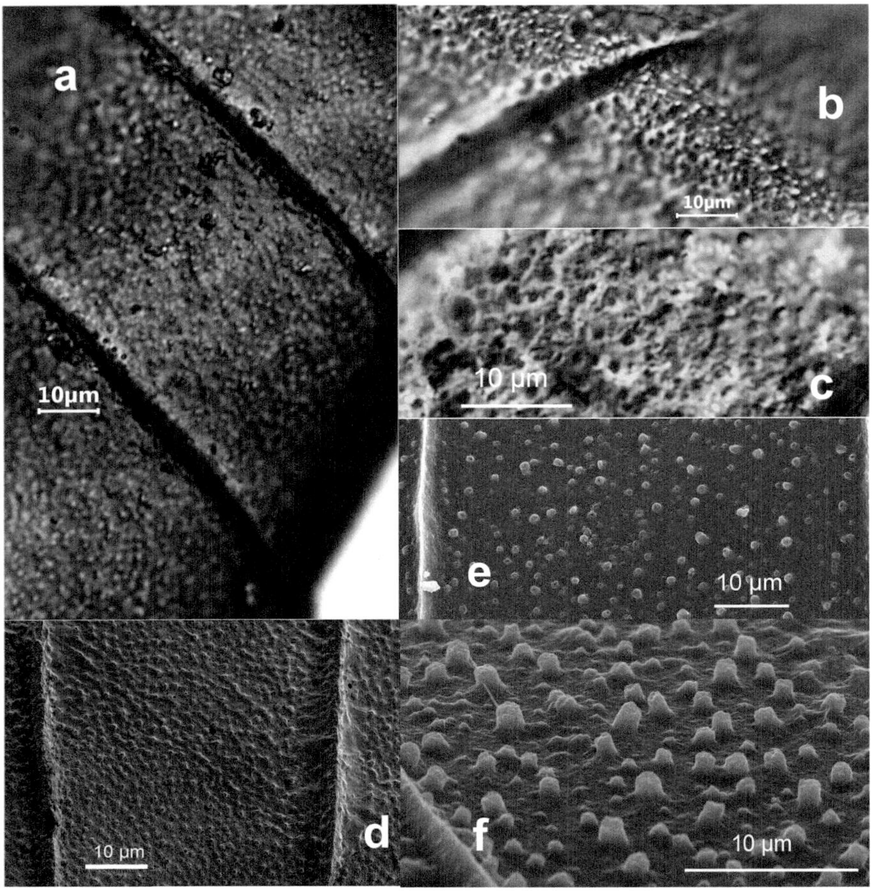

Fig. 52.6 *Chara vulgaris*, oospore membrane. **a–c** Details of oospore membrane (LM) showing granulate to verrucate microstructure; **d–f** Oospore surface microstructure (SEM), note the low ridges and perforation of some protuberances. Photographs: **a–c**—Luc Denys; **d–f**—Michelle T. Casanova

52.3 Habitat Conditions and Ecology

Habitat type: *Chara vulgaris* grows in all kinds of fresh to brackish stagnant waters, permanent, periodic and ephemeral, as well as in marshes and seeps with only a thin film of water. At least in some regions, temporary waters appear to be the predominant habitat, but occurrence is not restricted to shallow conditions (see below). *Chara vulgaris* also inhabits running waters, usually with mild or intermittent flow. It develops rapidly after spates inside channels of larger rivers but is less frequently found in their main channel (Combroux et al., 2001). More occasionally, isolated

Table 52.3 Gyrogonite morphometry

Authors	Length (μm)	Width (μm)	Length/width	Striae
Djamali et al. (2006)[a]	400–610(495–528)	320–480(384–394)	1.07–1.69 (1.29–1.36)	–
Harris (1939)[b]	(480)490–730(760)	–	1.2–2	(9)11 ± 1(13)
Rodrigo et al. (2010)	512–759	363–545	–	11 ± 1
Sanjuan et al. (2017)[c]	430–728	271–519	–	(6)10–15
Soulié-Märsche (1989)[d]	511–631 (516–628)	343–400 (349–398)	1.35–1.75; 1.87–1.98; 1.44–1.47	–
Soulié-Märsche (1989)[e]	659–673; 528–538	409–420; 331–337	1.58–1.63; 1.57–1.61	–
Soulié-Märsche (1989)[f]	537–560	369–384	1.45	–
Total range	**(400)490–730(760)**	**(271)320–520(545)**	**(1.07)–1.69(2)**	**(6?)10–13(15)**

[a] Range of average values in brackets
[b] Single population with mean length c. 610 μm and width c. 420 μm; note mixed use of the terms oogonia, oospores and gyrogonites in this publication
[c] *Chara vulgaris* f. *longibracteata*; values from cultures, the reported minimum of 6 convolutions is exceptional
[d] *Chara vulgaris* var. *vulgaris*; range of 95% confidence interval for five populations from France with range of average values in brackets; length/width for a single plant from France, a single cultured plant from Morocco and its original population
[e] *Chara vulgaris* f. *calveraënsis*; culture in tap water; original station
[f] *Chara vulgaris* f. *longibracteata*

tussocks of sturdy plants occur in quite swiftly flowing water (Wolff, 2011; pers. obs. L. Denys). In Poland it grows together with mosses in lowland streams with a mean flow of 0.29 m.s^{-1} (Jakubas and Gąbka, 2015). Dispersing profusely, it is often one of the first macrophytes to appear in newly created habitat (Crawford, 1979; Jeffries, 2008). Coastal bays, lagoons and salt lakes are also inhabited, as well as inland sites influenced by salt-water intrusion or evaporitic concentration of minerals, desert springs and wells of oases. *Chara vulgaris* is a eurythermal species, growing also in thermal waters with average water temperature up to 31–34 °C (Saber et al., 2018), or even 40 °C (Krause, 1997). With respect to altitude, the species is reported in Europe from sites beneath sea-level (in polders) to well above the tree line. In the European Alps it occurs e.g. at 2070 m.a.s.l. in Austria (Jäger, 2012) and above 2400 m.a.s.l. in Switzerland (Auderset Joye and Rey-Boissezon, 2013; Boissezon, 2008).

Substrate : *Chara vulgaris* most commonly occurs on a more mineral substrate. Its texture ranges from clay to gravel, but most often sand or silt are the main components. Less often on organic mud, gyttja or peat. Usually, the soil is calcareous or with a high content of base cations (Gillet, 1960; Wolff, 2011); the species is uncommon on an acid, siliceous substrate.

Table 52.4 Antheridia morphometry

Authors	Diameter (μm)
Ahmadi et al. (2010), (2012a)	Up to 480
Auderset Joye and Rey-Boissezon (2013)	325–500
Bailly and Schaefer (2010)[1], Bazzichelli and Abdelahad (2009)	250–450
Bailly and Schaefer (2010)[2]	400–500
Blindow et al. (2018)	325–500/250–540
Boissezon et al. (2017)	300–500
Caisová and Gąbka (2009)	Up to 500
Cirujano et al. (2008)[3]	325–500
Cirujano et al. (2008)[4]	420–480
Cirujano et al. (2008)[5]	500
Cirujano et al. (2008)[6]	c. 500
Compère (1992)	350–500
Corillion (1957a)	325–500
Corillion (1975)[7]	325–500
Corillion (1975)[8]	420–480
Daily (1944)	Up to 580
Gillet (1960)[9]	380 ± 55–575 ± 80
Hollerbach and Krassavina (1983)	325–500
Groves and Bullock-Webster (1924)	325–500
Hu et al. (2009)	255.6 ± 116.2–323.6 ± 71.0
Korsch (2013)	275–375
Krause (1997)	250–450
Kwiatkowska et al. (1997)	213–519
Langangen (2007)	Up to 500
Maier (1972)	325–500
Moore (1986)	Up to 500
Migula (1889–1897), (1898)	250–300
Muller et al. (2017)	250–450
Noedoost et al. (2015)[10]	370–471
Saber et al. (2018)	366–375
Urbaniak and Gąbka (2014)[11]	c. 420–500
Urbaniak and Gąbka (2014)[12]	400–510
van Raam (1998)	Up to 500
Wood and Imahori (1964)[13]	300–495
Wood and Imahori (1964)[14]	300–345
Wood (1965)	325–500

(continued)

Table 52.4 (continued)

Authors	Diameter (μm)
Wood and Muenscher (1956)	535
Total range	**250–500 (>575)**

[1] *Chara vulgaris* var. *vulgaris*, *C. vulgaris* f. *longibracteata*, *C. vulgaris* f. *subhispida*, *C. vulgaris* var. *papillata*
[2] *Chara vulgaris* var. *crassicaulis*
[3] *Chara vulgaris* var. *vulgaris*
[4] *Chara vulgaris* var. *crassicaulis*
[5] *Chara vulgaris* var. *longibracteata*
[6] *Chara vulgaris* var. *papillata*
[7] *Chara vulgaris* var. *vulgaris*
[8] *Chara vulgaris* var. *crassicaulis*
[9] Minimum and maximum ± 2 s.d. in populations presenting resp. smallest and largest size
[10] Range of population mean values; see remark Table 52.1
[11] *Chara crassicaulis*
[12] *Chara vulgaris*; lower limit erroneous
[13] *Chara vulgaris* emend., *C. vulgaris* f. *crispa*
[14] *Chara vulgaris* f. *pistianensis*

Nutrient conditions: *Chara vulgaris* spans the entire trophic gradient, from oligo- to eutrophic (Doege et al., 2016) and even up to hypertrophic conditions (Zouaïdia et al., 2015); it is not characteristic of nutrient-poor water. Optimum development often occurs in to somewhat more productive situations, i.e. in the range eutrophic to meso-eutrophic (Bailly and Schaefer, 2010; Boissezon, 2008; Iakushenko and Borysova, 2012; Lambert-Servien et al., 2006; Nat et al., 1994). Total phosphorus concentrations of several mg L^{-1} and reactive phosphorus levels exceeding 1 mg L^{-1} have been reported (Doege, 2016; Nat et al., 1994). Barinova and Romanov (2015) recorded healthy populations of *C. vulgaris* f. *longibracteata* at NO_3-N concentrations of 5.3–10.2 mg L^{-1}. Puche et al. (2018) found similar NO_3-N-levels in a Spanish spring (7.4 ± 2.9 mg L^{-1}, 4.5–11.1 mg L^{-1}). In extreme cases, NH_4-N can exceed 20 mg L^{-1} (Nat et al., 1994).

Water chemistry: *Chara vulgaris* is a freshwater species but tolerates some salinity due to its capacity to regulate turgor to a certain extent (Winter et al., 1987). In much of Europe, it grows well at salinities from 0 up to 5% but its salinity range is more extensive. Tziortzis et al. (2014) report it in low abundance at an average salinity of 16.9% from Kotychi lagoon (Greece). It usually occurs in well-buffered, neutral to alkaline water up to pH 9.5 (Iakushenko and Borysova, 2012; Mouronval et al., 2015; Wolff, 2011), but its full pH range is broader on both sides of the scale, excluding only very acid water (pH < 5). Calcium concentrations may exceed 500 mg L^{-1} (Nat et al., 1994) and sodium ranges from a few up to more than 900 mg L^{-1} (Lambert, 2007; Nat et al., 1994). With respect to pore-water chemistry, *Chara vulgaris* presents a very wide range for many variables (Nat et al., 1994). Lambert (2007) reports Eh values from −488 to 154 mV (median -138 mV), a pH range of 6.2 to 7.7 and conductivities as low as 19 μS cm^{-1} for interstitial pore water.

Depth range: *Chara vulgaris* can exist from subaerial conditions to a depth of 19 m (Korsch, 2013).

Associated species: Besides often being the only macrophyte in more extreme or pioneer conditions (e.g. Panzeca et al., 2021), a wide-spread occurrence and broad environmental range allow *C. vulgaris* to occur with almost any other vascular macrophyte from non-acid waters (e.g. Bensetitti et al., 2002; Doll, 1989; Felzines and Lambert, 2012; Guerlesquin and Mériaux, 1981; Iakushenko and Borysova, 2012; Guarino et al., 2019; Šumberová et al., 2011; van Raam, 1998).

52.4 Variability

Chara vulgaris is notorious for its morphological variability, even within the same population or on the same plant, which led to numerous epithets from formae to species. So far there is no full consensus whether some of these merit formal separation or not (viz. e.g. Korsch, 2016 vs. Mouronval et al., 2015). Nevertheless, the persistence of different morphologies next to each other at certain sites and apparently subject to identical conditions remains enigmatic. Due to the qualitative or inconsistent quantitative criteria suggested in the literature for their discrimination, with some schools lumping them entirely, linking morphological, genetic and environmental data is difficult for the *C. vulgaris* complex. It remains unknown whether potential morphological plasticity differs between populations and to what extent reported variation depends on sampling bias. To avoid possible loss of information and allow more thorough comparisons at a wider geographical scale, we propose to distinguish the following five morphologies until matters are further clarified, but refrain from attributing them any formal status beyond the nominal epithet:

crassicaulis morph (Fig. 52.2b,f): main axis thick (c. 0.9–2 mm in diameter, although as low as 0.65 mm according to Ahmadi et al., 2012a), distinctly heterostichous and with stout rigid branchlets. The latter connivent in the upper part of the main axis and, except for the uppermost whorls, much shorter than the internodes. Spine-cells shorter than the axis diameter and plump. Cortical cells walls thick. Bracteoles equaling anterior bract-cells and up to three times as long as the oogonia. On richly fruiting branchlets short internodes may result in partial overlap of consecutive gametangia. Antheridia in the upper size range, c. 450 μm in diameter;

longibracteata morph (Fig. 52.1): bracteoles considerably longer than the anterior bract-cells and at least some of them more than c. three times as long as the oogonia to even more than ten times longer. Their length ranges from c. 2 to 7 mm and can vary strongly within the same individual but at least some will exceed 3 mm. Cortex regular, branchlets not strongly incurved, main axis up to 1.5 mm thick, spine-cells shorter than the diameter of the main axis. Some branchlets may be entirely ecorticate;

papillata morph (Fig. 52.3d): many spine-cells as long as the diameter of the main axis or even longer (up to c. 1.5 mm), but bracteoles not longer than twice the length of ripe oogonia (not more than c. 2 mm long). Cortex irregular. Branchlets not strongly incurved and main axis <1.5 mm thick;

subhispida morph (52.4 a): with both long spine-cells (as in *papillata* morph) and bracteoles (as in *longibracteata* morph), branchlets not strongly incurved, main axis <1.5 mm thick. The distinction from '*papillata*' follows Mouronval et al. (2015). Note that Braun (1835, 1847), Migula (1889–1897), etc. considered both identical, whilst Bailly and Schaeffer (2010) restrict plants with long bracteoles to *C. vulgaris* f. *longibracteata* and separate *C. vulgaris* f. *subhispida* from *C. vulgaris* var. *papillata* by protruding instead of adpressed spine-cells;

vulgaris morph (Fig. 52.2 a, c–e): lacking the particular character combinations of all above.

The delineation of the *longibracteata* morph is especially arbitrary. Although morphometric analyses indicate that separation should be possible (Ahmadi et al., 2012c), obvious discontinuities in the length of bract-cells and bracteoles are lacking.

Noedoost et al. (2015) found no agreement between the considerable genetic diversity among *C. vulgaris* populations in Iran and their morphology, suggesting that morphological variability results from environmental triggers or reflects development. However, according to the accompanying morphometric data, it cannot be excluded that all their populations belonged to the *longibracteata* morph. Analysis of more strongly differentiated European material (including *crassicaulis* as well as *longibracteata* and *papillata* morphs) also suggests that genetic distinction from classic *C. vulgaris* is impossible (Schneider et al., 2016; P. Nowak pers. comm.). Culture experiments and field observations to some extent support the relationship between phenotypic plasticity and environmental conditions (Gillet, 1960; Henwood, 2014; Proctor, 1975; Saber et al., 2018; pers. obs. E. Nat & L. Denys). Small plants (ca. 1 cm high) with a tuft-like growth form and naked branchlets, or even no cortex at all have been reported as *C. vulgaris* from subaerial habitats (de Zwart, 2000; Romanov and Bulionkova, 2016).

52.5 Identification and Risk of Confusion

In most cases, identification of *C. vulgaris* poses little risk of confusion. Isostichous, spineless material may be similar to *C. contraria*, an essentially tylacanthous species with less dense and shorter bracteoles, a lower row of smaller stipulodes and often less corticated branchlets. Partly ecorticate material can be more difficult to identify and vegetative material might also be confused with a few aulacanthous or certain haplostichous taxa, especially in the Mediterranean region (see 'Remarks'). Although exceptions may occur in extreme environments, plants exclusively presenting completely ecorticate branchlets should receive special attention, in particular with respect to *C. gymnophylla* (see Romanov and Becker, 2024; Romanov et al., 2024). Culture may be necessary for correct identification if cortication type and position of gametangia cannot be determined. Material from coastal waters also warrants particular scrutiny (Blümel, 2004). Occasionally, similarity to *Chara globularis* is mentioned (Langangen, 2016).

52.6 Distribution

Chara vulgaris is cosmopolitan, lacking only in Antarctica (Proctor, 1980). The species is present in every European country (Fig. 52.7), except Finland (Langangen et al., 2002), and up to 67° N (Russia), but its natural range only extends to c. 60° N (Romanov et al., 2018). Very common throughout most of the continent, except in regions with only dilute low-alkalinity waters. It occurs in many island localities but remains absent from Iceland (Hrafnsdottir et al., 2019) and the Faeroes.

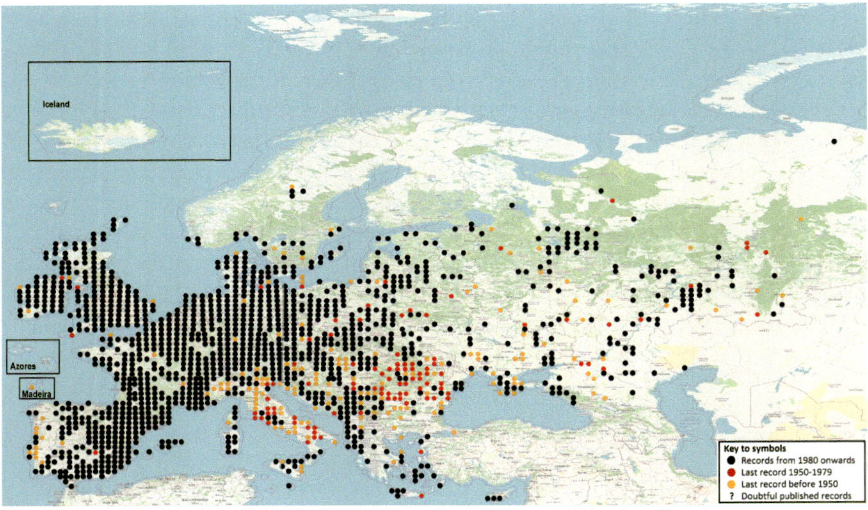

Fig. 52.7 Distribution of *Chara vulgaris* in Europe. For worldwide distribution see Korsch (2018). Black dots are records from 1980 onwards, red dots last record from 1950 to 1979 and yellow dots last record before 1950; doubtful published records (if applicable) are marked with "?" on top. Map prepared by Heiko Korsch

52.7 Status and Threats

Chara vulgaris s.l. is one of the commonest charophytes in a large part of Europe and not considered endangered or a Red List species in any Member State of the European Union. Locally, *C. vulgaris* may suffer from eutrophication or habitat destruction, but its eurytopic/euryecious character and ability to colonise secondary water bodies allow it to persist at a regional scale even where land use is quite intensive.

52.8 Remarks

For the moment, only plants where formation of gametangia remains restricted to nodes between corticate segments are retained in *C. vulgaris*. Those also developing gametangia on the nodes of ecorticate segments are referred to *C. gymnophylla* (see R. Romanov and R. Becker, 2024). Isostichous, tylacanthous as well as aulacanthous specimens have been attributed to *C. gymnophylla*, but presumably only those with aulacanthous cortication belong within the *vulgaris* cluster.

Chara oedophylla, a Mediterranean taxon previously considered as a variety or forma of *C. vulgaris* by some (e.g. Cirujano et al., 2008), presents a character set that is sufficiently different to consider it as a separate species (see Soulié-Märsche et al., 2024). Equally, *C. imperfecta*, although merged with *C. vulgaris* by Wood (1962), is distinct due to its dioecious reproduction and morphology (Soulié-Märsche and Muller, 2024).

Chara pistianensis, referring to plants with partly corticated as well as completely ecorticate branchlets, nearly isostichous to aulacanthous cortex and black oospores from a Slovakian thermal bath (Vilhelm, 1924) and China (Xing-Qiu et al., 1988) requires further examination. It is most similar to the *longibracteata* morph.

C. rabenhorstii, reported from Italy (Rabenhorst, 1850) and Nebraska (Andersen and Walker, 1920), was referred to *C. vulgaris* by J. van Raam (unpubl.). It is only documented by material from the locus typicus, Lago San Egidio (Promontorio del Gargano, San Giovanni Rotondi, Apulia), where it occurred with *C. vulgaris*. It is assumed to differ by the formation of most oogonia and antheridia at different nodes of the same branchlet, whilst its antheridia are larger than for typical *C. vulgaris*: diameter 480–540 μm (Braun and Nordstedt, 1882), >480 μm for solitary ones, as low as 440 μm for conjoined ones (Migula, 1889–1897), c. 550 μm (Groves and Bullock-Webster, 1924; Corillion, 1975). So far, its status remains unclear.

The South American *Chara calveraënsis* (R. D. Wood) R. D. Wood (= *C. vulgaris* var. *vulgaris* f. *calveraënsis* R. D. Wood) was reported from a basin fed by a thermal well in Southern France (Geyser de Lodève at Le Puech) by Soulié-Märsche (1989) and also from Tunisia (Corillion, 1977). After culturing in tap water with and without addition of 2 g $CaCO_3$, the French plants acquired the normal aspect and stipulodes of *C. vulgaris* (Soulié-Märsche, 1989). Blindow et al. (2018) confirm that the material from France and Tunisia identified as *C. calveraënsis* by R. Corillion and kept in

ANGUC, shows the exceptionally long, irregular stipulodes (upper ones ≤3 mm in S-America, Blindow et al., 2018) differentiating this taxon from *C. vulgaris* var. *vulgaris*, but argue that its particular morphology might follow from extreme growth conditions. Other morphologically aberrant strains of *C. vulgaris* were found in thermal waters and it appears that certain combinations of high temperature and hydrochemistry can induce exceptional stipulode development as well. Plants from a ditch in Guangxi, China, with very long stipulodes (upper ones ≤2800, lower ones ≤1640 μm) and bracteoles up to 4200 μm have also been aligned with *C. calveraënsis* (Han and Li, 1994; Ling et al., 1989). According to Schneider et al. (2015), *C. calveraënsis* from Argentina clustered with species in the subsection *Hartmania* and not *Chara*, contrary to a specimen of *C.* cf. *vulgaris*, supporting the distinctiveness of this taxon, for which the occurrence in Europe, and presumably N-Africa, cannot be upheld.

Acknowledgements We are particularly grateful to Ingeborg Soulié-Märsche for providing valuable information and many helpful comments on previous drafts of this chapter, as well as to Nadia Abdelahad for constructive discussions and additions. Emile Nat meticulously reviewed the first and final versions. We would also like to thank Thomas Gregor for streamlining the nomenclature and comments on the text, Irmgard Blindow for details on *Chara calveraënsis* and Michelle T. Casanova for sharing her extensive collection of SEM oospore photographs.

References

Ahmadi, A., Riahi, H., Sheidai, M., & van Raam, J. (2010). *Some charophytes (Characeae, Charophyta) in Iran: taxonomy and distribution*. Unpubl. manuscript.

Ahmadi, A., Riahi, H., Sheidai, M., & van Raam, J. (2012a). Some charophytes (Characeae, Charophyta) from Central and Western of Iran including *Chara kohrangiana* species nova. *Cryptogamie, Algologie, 33*, 359–390.

Ahmadi, A., Riahi, H., Sheidai, M., & van Raam, J. (2012b). A study of the oospore characteristics in some Charophytes (Characeae) of Iran. *Nova Hedwigia, 94*, 487–504.

Ahmadi, A., Riahi, H., Sheidai, M., & van Raam, J. (2012c). Numerical taxonomy of some charophytes in Iran. *Nordic Journal of Botany, 30*, 206–214.

Andersen, E. N., & Walker, E. R. (1920). An ecological study of the algae of some Sandhill lakes. *Transactions of the American Microscopical Society, 39*, 51–85.

Auderset Joye, D., & Rey-Boissezon, A. (2013). *Les Characées de Genève et environs. Distribution et Ecologie*. Institut Forel et Institut des Sciences de l'Environnement, Université de Genève.

Bailly, G., & Schaefer, O. (2010). *Guide illustré des Characées du nord-est de la France*. Conservatoire Botanique National de Franche-Comté.

Barinova, S., & Romanov, R. (2015). The Ein El Balad charophyte locality in the Mount Carmel Biosphere Reserve, Israel. *International Journal of Advanced Research in Botany, 1*, 1–12.

Bazzichelli, G., & Abdelahad, N. (2009). *Alghe d'acqua dolce d'Italia. Flora analitica delle Caroficee*. Sapienza Università di Roma, Ministero dell'Ambiente e della Tutela del territorio e del mare.

Bensettitti, F., Gaudillat, V., & Haury, J. (2002). *Cahiers d'habitats Natura 2000. Connaissance et gestion des habitats et des espèces d'intérêt communautaire. Tome 3. Habitats humides*. La Documentation française.

Blindow, I., Marquardt, R., Schories, D., & Schubert, H. (2018). Charophyceae of Chile—taxonomy and distribution. Part 1. *Nova Hedwigia, 107*, 1–47.
Blümel, C. (2004). Die Characeen in Mecklenburg-Vorpommern. *Rostocker Meeresbiologische Beiträge, 13*, 55–72.
Boissezon, A. (2008). Préférences écologiques de quelques espèces de Charophytes [Master's thesis, Université de Genève]. Archive ouverte UNIGE. https://archive-ouverte.unige.ch/unige:38271.
Boissezon, A., Auderset Joye, D., Schwarzer, A., & Prunier, P. (2017). *Système d'information pour les espèces menacées et prioritaires: characées, macrophytes aquatiques*. https://www.infoflora.ch/fr/flore/characeae/identification.html
Braun, A. (1835). Übersicht der genauer bekannten *Chara*-Arten. *Flora, 18*, 49–73.
Braun, A. (1847). Übersicht der schweizerischen Characeen. *Neue Denkschriften der allgemeine Schweizerischen Gesellschaft für die Gesamten Naturwissenschaften, 10*, 1–23.
Braun, A., & Nordstedt, C. F. O. (1882). Fragmente einer Monographie der Characeen. Nach den hinterlassen Manuscripten A. Braun's herausgegeben von Dr. Otto Nordstedt. *Abhandlungen der Königlichen Akademie der Wissenschaften zu Berlin für das Jahr, 1882*, 1–211.
Caisová, L., & Gąbka, M. (2009). Charophytes (Characeae, Charophyta) in the Czech Republic: taxonomy, autoecology and distribution. *Fottea, 9*, 1–43.
Casanova, M., & Wilbraham, J. (2019). *Dataset: Charophytes of Britain and Ireland*. Natural History Museum Data Portal. https://data.nhm.ac.uk/dataset/charophytes-of-britain-and-ireland.
Cirujano, S., Cambra, J., Sánchez Castillo, P. M., Meco, A., & Flor Arnau, N. (2008). *Flora ibérica. Algas continentales. Carófitos (Characeae)*. Real Jardín Botánico.
Combroux, I., Bornette, G., Willby, N. J., & Amoros, C. (2001). Regenerative strategies of aquatic plants in disturbed habitats: the role of the propagule bank. *Archiv für Hydrobiologie, 152*, 215–235.
Compère, P. (1992). *Charophytes. Flore pratique des algues d'eau douce de Belgique*. Jardin botanique national de Belgique.
Corillion, R. (1957). Les Charophycées de France et d'Europe Occidentale. *Bulletin de la Société Scientifique de Bretagne, 32* (Hors série I & II), 1–499.
Corillion, R. (1975). *Flore et végétation du Massif Armoricain. T. IV. Flore des Charophytes (Characées) du Massif Armoricain et des contrées voisines d'Europe occidentale*. Jouve.
Corillion, R. (1977). Contribution à l'étude des Characées de Tunisie et bilan actuel de la flore charologique tunisienne. *Bulletin de la Société phycologique de France, 22*, 47–59.
Crawford, S. A. (1979). Farm pond restoration using *Chara vulgaris* vegetation. *Hydrobiologia, 62*, 17–32.
Daily, F. K. (1944). The Characeae of Nebraska. *Butler University Botanical Studies, 6*, 149–171.
De Winton, M. D., Dugdale, T. M., & Clayton, J. S. (2007). An identification key for oospores of the extant charophytes of New Zealand. *New Zealand Journal of Botany, 45*, 463–476.
de Zwart, A. (2000). Kranswier op het droge. *Nieuwsbrief Kranswieren, 4*(8), 3–4.
Djamali, M., Soulié-Märsche, I., Esu, D., Gliozzi, E., & Okhravi, R. (2006). Palaeoenvironment of a Late Quaternary lacustrine–palustrine carbonate complex: Zarand Basin, Saveh, central Iran. *Palaeogeography, Palaeoclimatology, Palaeoecology, 237*, 315–334.
Doege, A., van de Weyer, K., Becker, R., & Schubert, H. (2016). Bioindikation mit Characeen. In A. C. Deutschlands (Ed.), *Armleuchteralgen: Die Characeen Deutschlands* (pp. 97–138). Springer.
Doll, R. (1989). Die Pflanzengesellschaften der stehenden Gewässer der DDR. Teil 1. Die Gesellschaften des offenen Wassers (Characeen-Gesellschaften). *Feddes Repertorium, 100*, 281–324.
Felzines, J. -C., & Lambert, E. (2012). Contribution au prodrome des végétations de France: les *Charetea fragilis* F. Fukarek 1961. *Journal de Botanique de la Société botanique de France, 59*, 133–188.
Giesenhagen, K. (1896). Untersuchungen über die Characeen. *Flora, 82*, 381–433.
Gillet, C. (1960). Les charophycées de l'Ardenne et des régions voisines. *Bulletin de la Société Royale de Botanique de Belgique, 92*, 197–228.

Gregor, T., Raabe, U., & van Raam, J. (2012). Typification of *Nitella gracilis* (Sm.) C. Agardh var. *confervacea* Brébisson (= *N. confervacea* (Brébisson) A. Braun ex Leonhardi) with remarks on its nomenclatural history. *Cryptogamie, Algologie, 33*, 391–398.
Groves, J., & Bullock-Webster, G. R. (1924). *The British Charophyta. Volume II.* Ray Society.
Guarino, R., Marcenò, C., Ilardi, V., Mannino, A. M., & Troia, A. (2019). One *Chara* does not make *Charetea* in the Mediterranean aquatic vegetation. *Webbia, 74*, 139–147.
Guerlesquin, M., & Mériaux, J.-L. (1981). Characées et végétations associées des milieux aquatiques du Nord de la France. *Colloques Phytosociologiques, 10*, 415–444.
Haas, J. N. (1994). First identification key for charophyte oospores from central Europe. *European Journal of Phycology, 29*, 227–235.
Han, F., & Li, Y. Y. (1994). *Charophyta. Flora algarum sinicarum aquae dulcis* (Vol. 3). Science Press.
Harris, T. M. (1939). *British Purbeck Charophyta.* British Museum.
Henwood, J. (2014). *A first record of Chara vulgaris var. papillata Wallroth ex A. Braun (Charales) in the Maltese Islands (Central Mediterranean).* Xjenza Online. https://doi.org/10.7423/XJENZA.2014.1.07
Hollerbach, M. M., & Krassavina, L. K. [Голлербах, М. М. & Л. К. Красавина]. (1983). Определитель пресноводных водорослей СССР. *14. Charophyta [Determination key of freshwater algae of USSR].* Nauka [in Russian].
Holzhausen, A., Casanova, M. T., Stewart, N., Sayer, C. D., & Goldsmith, B. (2024). Extant oospores. In H. Schubert, I. Blindow, M. T. Casanova, L. Denys, T. Gregor, H. Korsch, E. Nat, R. Romanov, N. Stewart & K. van de Weyer (Eds.), *Charophytes of Europe* (pp. 35–63), Springer.
Hough, R. A., & Putt, D. A. (1988). Factors influencing photosynthetic productivity of *Chara vulgaris* L. in a moderately productive hardwater lake. *Journal of Freshwater Ecology, 4*, 411–418.
Hrafnsdottir, T., Ingimarsson, F., Ingvason, H. R., Stefánsson, S. M., Þorvaldsdóttir, E. G., Malmquist, H. J., & Langangen, A. (2019). New finds of charophytes in Iceland with an update on the distribution of the charophyte flora. *Nordic Journal of Botany, 2019*, e02111.
Hu, B., Xie, S., Feng, J., & Zhang, M. (2009). Study on the phenology of *Chara vulgaris* in Xin'an Spring, north China. *Frontiers of Biology in China, 4*, 207–213.
Iakushenko, D., & Borysova, O. (2012). Plant communities of the class *Charetea* Fukarek ex Krausch 1964 in Ukraine: an overview. *Biodiversity Research and Conservation, 25*, 75–82.
Jäger, D. (2012). Rote Liste der Characeen Vorarlbergs (Österreich). *Rostocker Meeresbiologische Beiträge, 24*, 75–85.
Jakubas, E., & Gąbka, M. (2015). Significance of current velocity gradients for distribution patterns of charophytes versus mosses and vascular plant communities in a lowland stream. *Oceanological and Hydrobiological Studies, 44*, 139–150.
Jeffries, M. (2008). The spatial and temporal heterogeneity of macrophyte communities in thirty small, temporary ponds over a period of ten years. *Ecography, 31*, 765–775.
John, D. M., Moore, J. A., & Green, D. R. (1990). Preliminary observations on the structure and ornamentation of the oosporangial wall in *Chara* (Charales, Chlorophyta). *British Phycological Journal, 25*, 1–24.
Korsch, H. (2013). Die Armleuchteralgen (Charophyceae) Sachsen-Anhalts. *Berichte des Landesamtes für Umweltschutz Sachsen-Anhalt, 1*(2013), 1–85.
Korsch, H. (2016). *Chara vulgaris.* In Arbeitsgruppe Characeen Deutschlands (Eds.), *Armleuchteralgen: Die Characeen Deutschlands* (pp. 379–387). Springer.
Korsch, H. (2018). The worldwide range of the Charophyte species native to Germany. *Rostocker Meeresbiologische Beiträge, 28*, 45–96.
Krause, W. (1997). Charales (Charophyceae) Süßwasserflora von Mitteleuropa. In H. Ettl, G. Gärtner, H. Heynig & D. Mollenhauer (Eds.), *Süßwasserflora von Mitteleuropa* (Vol. 18). Fischer. [in German].

Krause, W. (1986). Zur Bestimmungsmöglichkeit subfossiler Characeen-Oosporen an Beispielen aus Schweizer Seen. *Vierteljahrschrift Der Naturforschenden Gesellschaft in Zürich, 131*, 295–313.

Kwiatkowska, M., Pala, A., & Maszewski, J. (1997). Size-variation in the antheridia and oogonia of *Chara vulgaris* under different experimental conditions. *Acta Societatis Botanicorum Poloniae, 66*, 29–32.

Lambert, S. J. (2007). *The environmental range and tolerance limits of British Stoneworts (Charophytes).* [Unpublished doctoral dissertation]. University of East Anglia.

Lambert-Servien, E., Clemenceau, G., Gabory, O., Douillard, E., & Haury, J. (2006). Stoneworts (Characeae) and associated macrophyte species as indicators of water quality and human activities in the Pays-de-la-Loire region, France. *Hydrobiologia, 570*, 107–115.

Langangen, A. (2007). *Charophytes of the Nordic countries*. Saeculum ANS.

Langangen, A. (2015). Some finds of charophytes from East-Africa (Zambia, Tanzania, Kenya and Somalia). *Acta Musei Nationalis Pragae B. Historia Naturalis, 71*, 239–248.

Langangen, A. (2016). *Chara vulgaris* L. (Charales) in springs on Amorgos Island (Greece). *Botanica Serbica, 40*, 195–203.

Langangen, A., Koistinen, M., & Blindow, I. (2002). The charophytes of Finland. *Memoranda Societatis pro Fauna Et Flora Fennica, 78*, 17–48.

Leistner, O. A. (1978). *Flora of Southern Africa. Cryptogams* (Vol. 9). Department of Agricultural Technical Services.

Ling, Y. J., Deng, G. Z., & Li, Z. (1989). A preliminary report on the Characeae from Bose of Guangxi in China. *Journal of Shanxi University (natural Science Edition), 12*, 225–232.

Maier, E. X. (1972). De kranswieren (Charophyta) van Nederland. In *Wetenschappelijke Mededelingen Koninklijke Nederlandse Natuurhistorische Vereniging, 93*, 1–44.

Mandal, D. K., Blaženčić, J. M., & Ray, S. (2002). SEM study of compound oospore wall ornamentation of some members of Charales from Yugoslavia, Croatia and Slovenia. *Archives of Biological Sciences, 54*, 28–34.

Migula, W. (1889–1897). Fünfter Band: Die Characeen. In A. Fischer, E. Fischer, F. Hauck, G. Limpricht, C. Luerssen, W. Migula, H. Rehm, P. Richter & G. Winter (Eds.), *Dr. L. Rabenhorst's Kryptogamenflora von Deutschland, Oesterreich und der Schweiz* (2nd ed.). Eduard Kummer.

Migula, W. (1898). *Synopsis Characearum europaearum. Illustrierte Beschreibung der Characeen Europas mit Berücksichtigung der übrigen Welttheile*. Eduard Kummer.

Moore, J. A. (1986). *Charophytes of Great Britain and Ireland*. BSBI Handbook No. 5, Botanical Society of the British Isles.

Mouronval, J. B., Baudouin, S. N. Borel, Soulié-Märsche, I., Klesczewski, M., & Grillas, P. (2015). *Guide des Characées de France méditerranéenne*. Office national de la Chasse et de la Faune Sauvage.

Muller, S. D., Rhazi, L., & Soulié-Märsche, I. (2017). Diversity and distribution of Characeae in the Maghreb (Algeria, Morocco, Tunisia). *Cryptogamie, Algologie, 38*, 201–251.

Nat, E., Simons, J., de la Haye, M. A. A., & Coops, H. (1994). *Historisch en actueel verspreidingsbeeld van kranswieren in Nederland in samenhang met waterkwaliteitsfactoren* (RIZA werkdocument 94.148X). Rijksinstituut voor Integraal Zoetwaterbeheer en Afvalwaterbehandeling.

Noedoost, F., Sheidai, M., Riahi, H., & Ahmadi, A. (2015). Genetic and morphological diversity in *Chara vulgaris* L. (Characeae). *Acta Biologica Szegediensis, 59*, 127–137.

Olsen, S. (1944). Danish Charophyta. *Det Kongelige Danske Videnskabernes Selskab, Biologiske Skrifter, 3*(1), 1–240, pl. 1–2.

Panzeca, P., Troia, A., & Madonia, P. (2021). Aquatic macrophytes occurrence in Mediterranean farm ponds: preliminary investigations in North-Western Sicily (Italy). *Plants, 10*, 1292.

Proctor, V. W. (1975). The nature of Charophyte species. *Phycologia, 14*, 97–113.

Proctor, V. W. (1980). Historical biogeography of *Chara* (Charophyta): an appraisal of the Braun-Wood classification plus a falsifiable alternative for future consideration. *Journal of Phycology, 16*, 218–238.

Puche, E., Sánchez-Carrillo, S., Álvarez-Cobelas, M., Pukacz, A., Rodrigo, M. A., & Rojo, C. (2018). Effects of overabundant nitrate and warmer temperatures on charophytes: the roles of plasticity and local adaptation. *Aquatic Botany, 146*, 15–22.

van Raam, J. C. (1998). *Handboek Kranswieren.* Hilversum: Charaboek.

Rabenhorst, L. (1850). Systematische Übersicht der auf meiner italienischen Reise beobachteten Kryptogamen. *Flora, 33*, 513–525.

Ray, S., Pekkari, S., & Snoeijs, P. (2001). Oospore dimensions and wall ornamentation patterns in Swedish charophytes. *Nordic Journal of Botany, 21*, 207–224.

Romanov, R., & Becker, R. (2024). *Chara gymnophylla.* In H. Schubert, I. Blindow, M. T. Casanova, L. Denys, T. Gregor, H. Korsch, E. Nat, R. Romanov, N. Stewart & K. van de Weyer (Eds.), *Charophytes of Europe* (pp. 509–522), Springer.

Rodrigo, M. A., Alonso-Guillén, J. L., & Soulié-Märsche, I. (2010). Reconstruction of the former charophyte community out of the fructifications identified in Albufera de València lagoon sediments. *Aquatic Botany, 92*, 14–22.

Romanov, R., & Bulionkova, T. (2016). Observations on a terrestrial charophyte in a temperate environment. *Phycological Research, 64*, 118–120.

Romanov, R. E., Patova, E. N., Teteryuk, B. Y., & Chemeris, E. V. (2018). Charophytes (Charales, Charophyceae) on the northeastern edge of Europe: Is it something different across Northern Europe in their diversity and biogeography? *Nova Hedwigia, Beiheft, 147*, 161–182.

Romanov, R., Becker, R., Iberite, M., Barinova, S., Gregor, T., & Mesterházy, A. (2024). *Chara squamosa.* In: H. Schubert, I. Blindow, M. T. Casanova, L. Denys, T. Gregor, H. Korsch, E. Nat, R. Romanov, N. Stewart & K. van de Weyer (Eds.), *Charophytes of Europe* (pp. 613–625), Springer.

Saber, A. A., Ballot, A., Schneider, S. C., & Cantonati, M. (2018). Morphological and molecular features of a *Chara vulgaris* population from desert springs on the Sinai Peninsula (Springs of Moses, Egypt). *Botany Letters, 165*, 77–89.

Sanjuan, J., Vincente, A., Flor-Arnau, N., Monleón, T., Cambra, J., & Martín-Closas, C. (2017). Effects of light and temperature on *Chara vulgaris* (Charophyceae) gyrogonite productivity and polymorphism – palaeoenvironmental implications. *Phycologia, 56*, 204–212.

Schneider, S. C., Nowak, P., von Ammon, U., & Ballot, A. (2016). Species differentiation in the genus *Chara* (Charophyceae): considerable phenotypic plasticity occurs within homogenous genetic groups. *European Journal of Phycology, 51*, 282–293.

Schneider, S. C., Rodrigues, A., Moe, T. F., & Ballot, A. (2015). DNA barcoding the genus *Chara*: Molecular evidence recovers fewer taxa than the classical morphological approach. *Journal of Phycology, 51*, 367–380.

Soulié-Märsche, I. (1989). *Etude comparée de gyrogonites de charophytes actuelles et fossiles et phylogénie des genres actuels.* Imprimerie des Tilleuls.

Soulié-Märsche, I., & Muller, S. D. (2024). *Chara imperfecta.* In H. Schubert, I. Blindow, M. T. Casanova, L. Denys, T. Gregor, H. Korsch, E. Nat, R. Romanov, N. Stewart & K. van de Weyer (Eds.), *Charophytes of Europe* (pp. 555–564), Springer.

Šumberová, K., Hrivnák, R., Rydlo, J., & Otáhelóvá, H. (2011). Vegetace parožnatek (Charetea). Vegetation of stoneworts. In M. Chytrý (Ed.), *Vegetace České republiky 3. Vodní a mokřadní vegetace* (pp. 248–267). Academia.

Soulié-Märsche, I., Daoud-Bouattour, A., & Muller, S. D. (2024). *Chara oedophyllla.* In H. Schubert, I. Blindow, M. T. Casanova, L. Denys, T. Gregor, H. Korsch, E. Nat, R. Romanov, N. Stewart & K. van de Weyer (Eds.), *Charophytes of Europe* (pp. 587–594), Springer.

Tortić-Njegovan, M. (1956). *Chara gymnophylla* i neke druge haraceje Jugoslavije. *Acta Botanica Croatica, 14–15*, 145–162.

Tziortzis, I., Kadis, K., & Papastergiadou, E. (2014). Use of macrophyte assemblages for the ecological evaluation of two coastal lagoons of Greece according to WFD 2000/60/EC. *Journal of Wetlands Biodiversity, 4*, 97–111.

Urbaniak, J., & Gąbka, M. (2014). *Polish Charophytes. An illustrated guide to identification.* Uniwersytet Przyrodniczy we Wroclawiu.

Urbaniak, J. (2011). A SEM and light microscopy study of the oospore wall ornamentation in Polish charophytes (Charales, Charophyceae)—genus *Chara*. *Nova Hedwigia, 93*, 1–28.
Vilhelm, J. (1924). Nouveau *Chara* d'eau thermale et radioactive des bains de Píšťany en Slovaquie. *Bulletin International de l'Académie Tchéque, Classe des Sciences Mathématiques, Naturelles et de la Médecine, 1923*, 8–10.
Wang, H., Liu, C., & Yu, D. (2015). Morphological and reproductive differences among three charophyte species in response to variation in water depth. *Aquatic Biology, 24*, 91–100.
Wang, H., Yu, D., & Xiao, K. (2008). The interactive effects of irradiance and photoperiod on *Chara vulgaris* L.: concerted responses in morphology, physiology and reproduction. *Hydrobiologia, 61*, 33–41.
Winter, U., Meyer, M. I., & Kirst, G. O. (1987). Seasonal changes of ionic concentrations in the vacuolar sap of *Chara vulgaris* L. growing in a brackish water lake. *Oecologia, 74*, 122–127.
Wolff, P. (2011). Die Armleuchteralgen (Charophyceae) im Saarland und im grenznahen Frankreich: Verbreitung und Ökologie. *Abhandlungen Delattinia, 37*, 5–30.
Wood, R. D. (1959). Gametangial constants of extant charophyta for use in micropaleobotany. *Journal of Paleontology, 33*, 186–194.
Wood, R. D. (1960). Linnaean types for Characeae. *Transactions of the American Microscopical Society, 79*, 219–226.
Wood, R. D. (1962). New combinations and taxa in the revision of Characeae. *Taxon, 11*, 7–25.
Wood, R. D. (1965). *Monograph of the Characeae*. J. Cramer, Weinheim.
Wood, R. D., & Imahori, K. (1964). *Iconograph of the Characeae*. J. Cramer, Weinheim.
Wood, R. D., & Mason, R. (1977). Characeae of New Zealand, New Zealand. *Journal of Botany, 15*, 87–180.
Wood, R. D., & Muenscher, W. C. (1956). The Characeae of the state of New York. *Cornell Experimental Station Memoirs, 338*, 1–77.
Xing-Qiu, C., Lian, L., Hua, Z., & Hua-Shan, G. (1988). Some new taxa and records of Characeae in Hunan, China. *Acta Phytotaxonomica Sinica, 26*, 467–482.
Zouaïdia, H., De Bélair, G., Benslama, M., Soulié-Märsche, I., & Müller, S. D. (2015). Intérêt des Characeae comme bioindicateurs de la qualité des eaux: le cas des zones humides de Numidie (Nord-Est Algérien). *Revue d'Ecologie (la Terre et la Vie), 70*, 121–133.

Chapter 53
Chara zeylanica

Ralf Becker, Petra Nowak, and Hendrik Schubert

Scientific name: *Chara zeylanica* Willd. in Mém. Acad. Roy. Sci. Hist. (Berlin) Ser. 2 1803: 86. 1805

Lectotype (Wood, 1965: 230): Ceylon/May 26, 1798/J.G. Klein (B-Willd. 508).

Nomenclatural note: Wood (1965: 230) selected as lectotype "Coll? 508, May 26, 1798. Ceylon (B, in cover 17099 of herb. Willdenow) as "*C. zeylanica* Nob[is]". This is somewhat in conflict with the protologue who states "On rencontre cette nouvelle espèce á la côte de Malabar". In the Willdenow herbarium two collections have been mixed on one sheet. One collected 1798 in Ceylon (recently Sri Lanka) and one collected 1798 in Tranquebar (South India). It is impossible to know which of the specimens belong to which collection. An epitype should be chosen.

Local names: No local names are existing in Europe.

R. Becker (✉)
Am Schützenplatz 58, 26121 Oldenburg, Germany
e-mail: becker.r@posteo.de

P. Nowak · H. Schubert
Universität Rostock, Institut für Biowissenschaften, Albert-Einstein-Straße 3, 18051 Rostock, Germany
e-mail: petra.nowak@uni-rostock.de

H. Schubert
e-mail: hendrik.schubert@uni-rostock.de

H. Schubert et al. (eds.), *Charophytes of Europe*,
https://doi.org/10.1007/978-3-031-31898-6_53

53.1 Morphology

See Figs. 53.1 and 53.2.

Habitus: *Chara zeylanica* is a small to medium sized species. The plants are erect and straight, stout, fresh to greyish green and slightly incrusted. *C. zeylanica* is slightly branched and has single, long spine-cells. The lowest segment of the branchlets is ecorticate.

Size: *Chara zeylanica* is medium sized. The Sardinian specimens are 30–60 cm high.

Main axis: The main axis is erect, 589–1076 μm in diameter with a mean value of 844 μm. Straight with few (0–4) short branches. Internodes are 4.4–8.0 cm long and usually much longer (up to 4×) than the branchlets. The uppermost 1–2 internodes are only 0.5–2.2 cm long, and generally shorter than the adjacent branchlets.

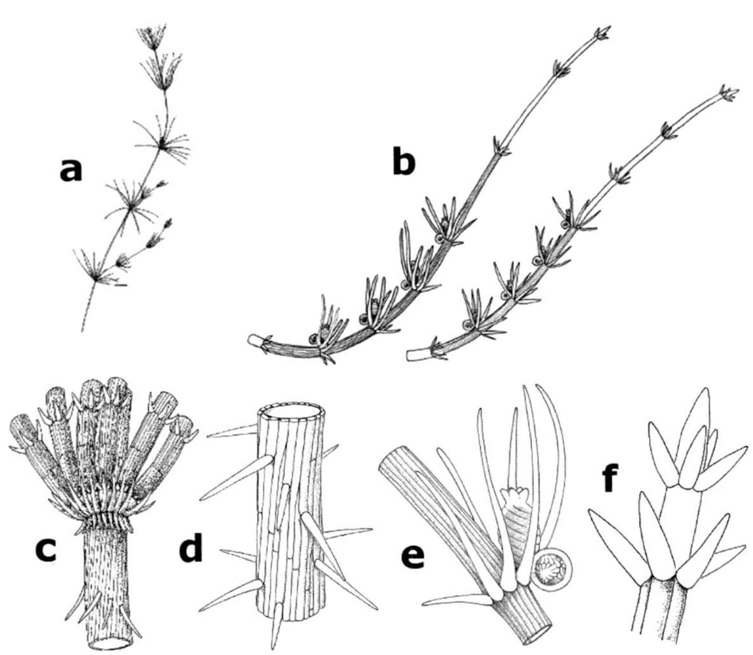

Fig. 53.1 *Chara zeylanica*. Drawings are made from specimens collected at Cala Fuili, Sardinia. **a** Habitus; **b** Branchlets with bract-cells; **c** Diplostephanous stipulodes with lower part of branchlets; **d** Triplostichous main axis cortication with single spine-cells; **e** Conjoined gametangia with tetrascutate antheridia and bract-cells; **f** Branchlet tip with bract-cells. Drawings: Dietmar Jäger

Fig. 53.2 Habitus of *Chara zeylanica*. **a** Whorl of branchlets with ecorticate basal branchlet segments; **b** Branchlet with oogonia, antheridia, bract-cells and bracteoles; **c** Branchlet tip; **d** Stipulodes; **e** Spine cells and cortex. Photographs: **a**, **c**, **d**—Petra Nowak; **b**, **e**—Ralf Becker

Cortex: The cortex of *Chara zeylanica* is mainly triplostichous, rarely (partly) diplostichous and tylacanthous to isostichous.

Spine-cells: Single, acute, thin and needle-like spine-cells exist mainly on young internodes and are rare on older internodes. The spine-cells vary in length even on the same plant. They are 182–1468 μm long with a mean value of 567 μm and are usually 0.5–1× (up to 2×) as long as the axis diameter (Fig. 53.1d and 53.2e). The spine-cells are mainly pointing downwards.

Stipulodes: *Chara zeylanica* is a diplostephanous species. The stipulodes are acute, elongate, in two regular tiers with 2 pairs per branchlet, and well developed. The upper stipulodes are longer than the lower ones, variable in length (Fig. 53.2). The upper stipulodes are 515–1045 μm long (with a mean value of 760 μm), they are mainly longer than the diameter of the main axis and much longer than the lowermost branchlet-segment. The lower stipulodes are sometimes of unequal length, at 161–475 μm long, with a mean value of 293 μm.

Branchlets: The branchlets are 9–12 in a whorl and mainly much shorter than the internodes. They are usually pointing upwards, often at a 45-degree angle. The branchlets are mostly 2.0–4.3 cm long. The branchlets of the uppermost 1–2 youngest whorls are even shorter, at just 0.2–2.0 cm long. The lowermost basal segments of the branchlets are ecorticate and hidden behind the upper stipulodes (Fig. 53.2a). They are very short, at 208–479 μm long (mean value 343 μm) and 189–470 μm wide (mean value 318 μm). The branchlets consist of 7–10 segments, with the lowermost segments always being ecorticate, followed by 4–6 corticate segments and 2–5 ecorticate distal segments with a tiny acute end cell on top. The end cell is ecorticate, at 618–1863 μm long, and is surrounded by a ring of bract-cells. The corticate part of the branchlets is mostly longer than the distal ecorticate part, but sometimes of equal length or even shorter.

Bract-cells: The bract-cells (5–8) are well developed (220–843 μm long), slender, and acute, and are shorter than the bracteoles.

Bracteoles: The two bracteoles are very long (990–1948 μm), at 1–2.5× longer than the oogonia and oospores.

Sexuality: *Chara zeylanica* is a monoecious species with high fecundity. All fertile specimens have conjoined gametangia. Gametangia usually occur only at the nodes of corticate segments, and are rarely observed at the lowest nodes just above the ecorticate segment. The gametangia are mainly solitary, and very rarely geminate.

Oogonia: The oogonia are elliptical to elongated oval in shape, are yellow or greenish in colour, and generally have constricted coronulae. For the Sardinian specimens (see also Chap. 53.8), the length of the oogonia (without coronula) is (600)700–850 (900) μm, and the width of the oogonia is 417–575(600) μm. The length of the coronula is 69–125(150) μm, and the width of the coronula is (127)160–200(250) μm (Table 53.1). The oogonia are usually solitary and conjoined with antheridia on the branchlets; geminate oogonia occur very rarely.

Oospores: The oospores of *Chara zeylanica* are cylindrical to narrow elliptical in shape and black in colour. The Sardinian specimens have a length of (539)600–685 μm, a width of 375–475(500) μm, and 10–13 striae. Table 53.2 presents additional data on oospore morphometry. The structure of the oospore wall of *C. zeylanica* is described from finely granulate under light microscope to pustulate in SEM (Guerlesquin, 1971) as well as smooth, obscurely dotted or granulate (Wood, 1965). The gyrogonites of *C. zeylanica,* measured from 10 American populations, were medium-sized: (550)600–725(800) μm in length and (250)325–400(500) μm in width (Soulié-Märsche, 1999). From Europe no data are available.

Antheridia: The variability of the antheridium diameter of *C. zeylanica* is given in Table 53.3. The antheridia of the Sardinian specimens (see also Chap. 53.8) are

Table 53.1 Oogonia morphometry of *Chara zeylanica*

Authors	Length	Width	Coronula length	Coronula width	Number of helical stripes
Wood (1965)[a]	700–1000(1400?)	400–750	70–143	140–180	11–15(17)
Guerlesquin (1971)	834–1013	417–610	90–145	178–298	11–14
Corillion and Guerlesquin (1971)[a]	693–715	354–417	123	150–230	13–14
Langangen and Leghari (2001)	1000	500	75–150	150–200	12
Leghari and Langangen (2001)	750–800	280–350	120–150	170–200	12–14
Casanova (2008)	< 500	–	–	–	–
Langangen (2015)	< 850	< 500	200	250	11
Muller et al. (2017)	900–1000	–	–	–	–
Ribeiro et al. (2018)	568–1036(1182)	209–657	55–258	83–294	10–13
Becker et al. (2021)[a]	(600)700–850(900)	417–575(600)	69–125(150)	(127)160–200(250)	10–15
Total range	**(<500)568–1036(1400?)**	**(209)280–657(750)**	**(55)69–258**	**(83)140–298**	**10–15(17)**

Length in the first column is excluding the coronula, its length and width are given separately; all values for length and width in μm. Oogonia of own material measured in both fresh and in dry state

[a] Data related to *C. zeylanica* var. *zeylanica*

tetrascutate with a diameter of (300)350–400(450) µm. They are mainly conjoined with oogonia on the branchlets. Table 53.3 presents additional data on antheridia morphometry.

Bulbils: Bulbils are not reported for *C. zeylanica.*

53.2 Phenology and Reproduction

There are few data concerning the phenology and reproduction of *C. zeylanica* in Europe. At the Sardinian site the specimens were healthy and fructifying in the beginning of September with plenty of ripe oospores besides oogonia and antheridia. No data is available for overwintering. During investigations on the same waterbody in May 2015 and 2016 *C. zeylanica* was not found there. Given the fact of the mainly tropical and subtropical distribution of *C. zeylanica,* the species needs high water temperatures of approximately 25 °C for fructification (e.g. Muller et al., 2017).

Table 53.2 Oospore morphometry of *Chara zeylanica*

Authors	Length	Width	Number of striae	Colour
Wood (1965)[a]	(450)600–900	(285)350–440(560)	(8)11–13(16)	Black, occasionally dark brown
Guerlesquin (1971)	506–670(800)	252–417(490)	9–12	–
Corillion and Guerlesquin (1971)[a]	600	360	11–12	Black
Langangen and Leghari (2001)	700	450	–	Black
Leghari and Langangen (2001)	450–500	250–300	–	–
Casanova (2008)	<400	<250	11	–
Langangen (2015)	650	350	11	Dark brown to brown
Ribeiro et al. 2018)	661–700	387–545	10–11	–
Becker et al. (2021)[a]	(539)600–685	375–475(500)	10–13	Black
Total range	**(<400)450–900**	**< 250–545(560)**	**(8)9–13(16)**	

All values for length and width in µm. Oospores of own material were measured in dry state
[a] Data relate to *Chara zeylanica* var. *zeylanica*

Table 53.3 Antheridia morphometry of *Chara zeylanica*

Authors	Diameter (μm)	Antheridial shields
Wood (1965)[a]	(240)350–470(690)	*
Guerlesquin (1971)	239–415	4
Corillion and Guerlesquin (1971)[a]	335–357	4
Langangen and Leghari (2001)	250–350	–
Leghari and Langangen (2001)	350–400	–
Casanova (2008)	<200	4
Langangen (2015)	350	–
Muller et al. (2017)	<500	4
Ribeiro et al. (2018)	252–548	4
Becker et al. (2021)[a]	(300)350–400(450)	–
Total range	**(<200)239–548(690)**	**4**

Antheridia of own material measured in both fresh and dry state
[a]Data relate to *Chara zeylanica* var. *zeylanica*
* Because Wood (1965)—in contrast to the Sardinian specimens and to the results of other authors—combined monoecious and dioecious forms within *Chara zeylanica* var. *zeylanica*, his description of 4(8) antheridial shields in this taxon must be treated with caution (compare Chap. 53.5)

53.3 Habitat Conditions and Ecology

Habitat type: At Cala Fuili, located north of Orosei on the east coast of Sardinia (coordinates: 40°25′03″ N, 9°46′13″ E), *C. zeylanica* was found in a shallow, probably permanent small stream close to the beach, only 110 m from the Mediterranean Sea (Fig. 53.3). Due to lack of rainfall during summer the flow velocity was very low in September 2019 and 2022. The habitat is disturbed by a small bridge, where the small stream is piped through a culvert. The Sardinian waterbody must be considered to be slightly brackish small stream. Throughout tropical and subtropical zones of the world *C. zeylanica* is able to colonise both brackish and freshwater habitats, such as lakes, ponds, pools, ditches, canals, temporarily flooded wetlands, rice fields and retention ponds (e.g. Guerlesquin, 1983; Langangen, 2015; Leghari and Langangen, 2001; Mjelde et al., 2020; Muller et al., 2017; Zaneveld, 1940).

Sediment: The substrate at the Sardinian site was mainly sandy to silty, partly even stony.

Nutrient conditions: The results of a single measurement at Cala Fuili, Sardinia, in September 2019 were as follows: NH_4-N 0.108 mg L^{-1}, NO_3-N 0.279 mg L^{-1}, total N 1.143 mg L^{-1}, PO_4-P 0.073 mg L^{-1}, and total P 0.137 mg L^{-1} (Becker et al., 2021). Few additional hydrochemical datasets originating from tropical and subtropical freshwater habitats range from low-impacted waterbodies (total P below 0.02 mg L^{-1} and total N 0.425 mg L^{-1}, Mjelde et al., 2020) to eutrophic habitats (total P up to 0.093 mg L^{-1}, total N up to 5.2 mg L^{-1}, Duarte et al., 2020).

Fig. 53.3 Habitus and habitat of *Chara zeylanica*. Left panel: Herbarium sheet, showing the general appearance of the plants collected at Cala Fuili, Sardinia. Right panels: Habitat photographs of specimens collected at Cala Fuili, Sardinia. Plants were collected just in front of the bridge. Photographs: left panel—Petra Nowak; right panels—Ralf Becker

Water chemistry: In Sardinia, pH at the sampling date was 8.3 and salinity 1.9. Moreover, the following concentrations were measured: Cl 1290 mg L^{-1}, Ca 62.2 mg L^{-1}, Mg 76.8 mg L^{-1}, water hardness 26.4 °dH (Becker et al., 2021). Because the Sardinian site of *C. zeylanica* is located directly beside the sea a seasonal variation of salinity levels and nutrient concentrations must be assumed. Particularly during wintertime a stronger input of seawater is likely. Hence, at the same site in May 2016 a salinity level of 4.4 and a chloride concentration of 2819 mg L^{-1} was recorded.

Depth range: In Sardinia *C. zeylanica* was found in shallow water at a depth of about 1.0 m.

Associated species: At the Sardinian site *C. zeylanica* was accompanied by "Mediterranean Chara", *Stuckenia pectinata, Ruppia maritima, Najas marina, Phragmites australis* and *Bolboschoenus maritimus.*

53.4 Variability

Chara zeylanica is a very polymorphic taxon (e.g. Becker et al., 2021; Guerlesquin, 1983; Proctor et al., 1971).

53.5 Identification and Risk of Confusion

The Sardinian specimens of *Chara zeylanica* can be confused with *C. tenuispina* because the ecorticate lowermost segments of the branchlets are very short and hidden behind the upper stipulodes and can be easily overlooked.

From a global point of view a risk of confusion with other gymnopodial taxa especially of the genus *Chara* subsect. *Willdenowia* is possible (Becker et al., 2021). Unfortunately concerning the subsect. *Willdenowia* diverging taxonomic treatments exist. According to Wood (1965) *C. zeylanica* is the only species of this subsection and includes a series of varieties and forms with different morphological features. Thus, Wood (1965) combined (i) monoecious and dioecious taxa, (ii) monoecious taxa with sejoined and conjoined gametangia, and, (iii) taxa with tetra- and octoscutate antheridia. In contrast, several authors (e.g. McCracken et al., 1966; van Raam, 2009, 2010) presented alternative taxonomic concepts, dividing *C. zeylanica* sensu Wood (1965) into several species by means of sexuality, gametangia position and antheridia morphology (Becker et al., 2021). This separation is supported by breeding studies of Proctor et al. (1971).

53.6 Distribution

See Fig. 53.4.

Chara zeylanica sensu Wood (1965) shows a worldwide distribution in tropical and subtropical zones of Asia, Australia, Africa, South and Central America, North America, Hawaii and the Caribbean Islands (e.g. Casanova 2005; Guerlesquin, 1983; Langangen, 2015; Langangen and Leghari, 2001; Leghari and Langangen, 2001; Mjelde et al., 2020; Muller et al., 2017; Wood, 1965). Throughout Europe the species has been recorded only at Cala Fuili, Sardinia (Italy) in 2019 and 2022. According to Muller et al. (2017) the closest former population to the current Sardinian habitat was known from a temporary pond in Algeria in the beginning of the twentieth century.

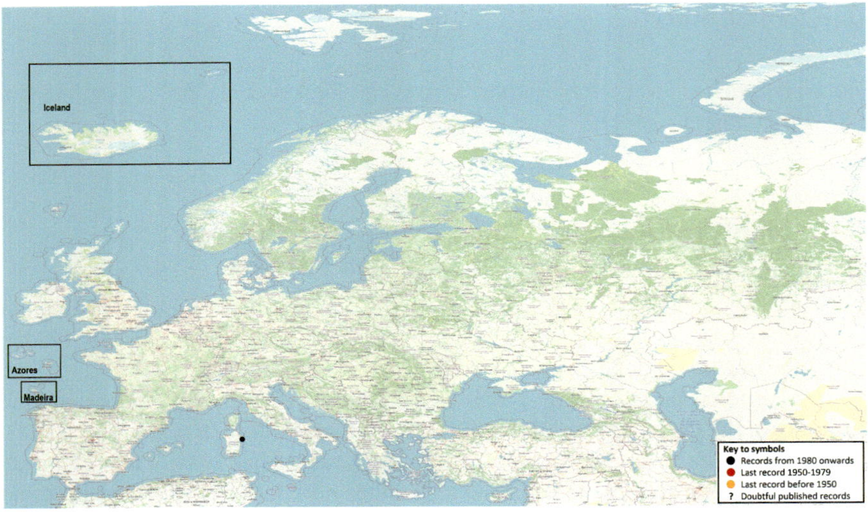

Fig. 53.4 European distribution of *Chara zeylanica*. Map prepared by Heiko Korsch

53.7 Status and Threats

Chara zeylanica is not listed in any European Red List because the record presented here is the first and only one in Europe. Thus, the small Sardinian population has to be considered as threatened. The Sardinian site is potentially jeopardised by habitat destruction due to infrastructural developments and creation of holiday resorts, input of nutrients and pollutants, regulation of running waters as well as by climate change and sea-level rise. Becker (2019) proposed the implementation of action plans to protect threatened charophyte species in Sardinia. *Chara zeylanica* should be added to the group of brackish water charophytes which was given the highest priority for conservation measures.

Currently *C. zeylanica* cannot be considered an invasive species in Europe. The very small Sardinian population is limited to a single and relatively isolated site.

53.8 Remarks

The data presented here mainly focus on specimens collected in September 2019 in Sardinia (Italy), the first European record of *Chara zeylanica* (Becker et al., 2021). In addition, some references from literature data concerning morphological features and ecological data of sites of *C. zeylanica* collected in tropical and subtropical regions are taken into account. Because the Sardinian specimens belong to *C. zeylanica* var. *zeylanica* sensu Wood (1965), the additional data presented focus mainly on *C.*

zeylanica var. *zeylanica* instead of *C. zeylanica* s.l. or other varieties and forms of this taxon.

References

Becker, R. (2019). The Characeae (Charales, Charophyceae) of Sardinia (Italy): habitats, distribution and conservation. *Webbia, 74*, 83–101. https://doi.org/10.1080/00837792.2019.1607497

Becker, R., Schubert, H., & Nowak, P. (2021). *Chara zeylanica* J.G. Klein ex Willd. (Charophyceae, Charales, Characeae): First European record from the island of Sardinia, Italy. *Plants, 10*, 2069. https://doi.org/10.3390/plants10102069

Casanova, M. T. (2005). An overview of *Chara* L. in Australia (Characeae, Charophyta). *Australian Systematic Botany, 18*, 25–39.

Casanova, M. T. (2008). Taxonomy of Daly River Macroalgae. In J. Schult, S. Townsend, M. Douglas, I. Webster, S. Skinner, & M. T. Casanova (Eds.), *Monitoring and managing nutrients in the Daly River* (pp. 34–62). Department of Natural Resources, Environment and the Arts.

Corillion, R., & Guerlesquin, M. (1971). Notes phytogéographiques sur les Charophycées d'Egypte. *Revue Algologique, 2*, 177–191. [in French].

Duarte, T., Almeida, T. S., Albertoni, E. F., & Palma-Silva, C. (2020). Role of the propagule bank in reestablishing submerged macrophytes after removal of free-floating plants for recovery of a shallow lake in Southern Brazil. *Ecología Austral, 30*, 239–250.

Guerlesquin, M. (1971). Recherches sur *Chara zeylanica* Klein ex Willd. (Charophycées) d'Afrique occidentale. *Revue Algologique, 10*, 231–247. [in French].

Guerlesquin, M. (1983). Contribution à l'étude des Characées des Petites Antilles. *Revue d'Hydrobiologie Tropicale, 16*, 213–233. [in French].

Langangen, A. (2015). Some finds of charophytes from East-Africa (Zambia, Tanzania, Kenya and Somalia). *Acta Musei Nationalis Pragae B. Historia Naturalis, 71*, 239–248.

Langangen, A., & Leghari, S. M. (2001). Some charophytes (Charales) from Pakistan. *Studia Botanica Hungarica, 32*, 63–85.

Leghari, S. M., & Langangen, A. (2001). Fresh water algae of Sindh, VI. Charales (Charophyta) from fresh and brackish water of Sindh. Pakistan. *Journal of Biological Sciences, 1*, 461–465.

McCracken, M. D., Proctor, V. W., & Hotchkiss, A. T. (1966). Attempted hybridization between monoecious and dioecious clones of *Chara*. *American Journal of Botany, 53*, 937–940.

Mjelde, M., Swe, T., Langangen, A., & Ballot, A. (2020). A contribution to the knowledge of charophytes in Myanmar; morphological and genetic identification and ecology notes. *Botany Letters, 168*, 102–109. https://doi.org/10.1080/23818107.2020.1847189

Muller, S. D., Rhazi, L., & Soulié-Märsche, I. (2017). Diversity and distribution of Characeae in the Maghreb (Algeria, Morocco, Tunisia). *Cryptogamie, Algologie, 38*, 201–251.

Proctor, V. W., Griffin, D. G., & Hotchkiss, A. T. (1971). A synopsis of the Genus *Chara*, Series *Gymnobasalia* (Subsection *Willdenowia* RDW). *American Journal of Botany, 58*, 894–901.

Ribeiro, C. A., Peixoto Ramos, G. J., Bueno, N. C., Prado, J. F., & Moura, C. W. N. (2018). O genero *Chara* (Charophyceae, Characeae) das Regioes Metropolitanas de Salvador e de Feira de Santana, Bahia, Brasil. *Rodriguesia, 69*, 1987–2017. [in Portuguese]. https://doi.org/10.1590/2175-7860201869432

Soulié-Märsche, I. (1999). Extant gyrogonite populations of *Chara zeylanica* and *Chara haitensis*, implications for taxonomy and palaeoecology. *Australian Journal of Botany, 47*, 371–382.

van Raam, J. (2009). A matrix key for the determination of Characeae. *Rostocker Meeresbiologische Beiträge, 22*, 53–55.

van Raam, J. C. (2010). *Annotated Bibliography of the Characeae* (2nd ed.). Charaboek, Hilversum.

Wood, R. D. (1965). *Monograph of the Characeae*. J. Cramer.

Zaneveld, J. S. (1940). The Charophyta of Malaysia and adjacent countries. *Blumea, 4*, 1–224.

Chapter 54
Lamprothamnium papulosum

Maria A. Rodrigo, Santos Cirujano, Ingeborg Soulié-Märsche, Chrysoula Christia, and Eva Papastergiadou

Scientific name: ***Lamprothamnium papulosum*** **(Wallr.) J. Groves in J. Bot. 54: 337. 1916**

≡ *Chara papulosa* Wallr., Comp. Fl. Germ. 4: 107. 1833
Neotype (Wood, 1965: 354): In den Buchten des Großen Belt, bei Glonö in Seeland/ Sept. 8, 1869/P. Nielsen (NY; Isoneotypes in L etc.); Braun, Rabenhorst and Stizenberger, Characeen Eur. 81 as *Chara alopecuroides* var. *wallrothii.*

= *Chara pouzolsii* J. Gay ex A. Braun in Flora 18: 58. 1835
Lectotype (Wood, 1965): 388): Corse, Bonifacio/[no date]/Gay (PC).

= *Lamprothamnus alopecuroides* Delile ex A. Braun and Nordstedt in Abh. Königl. Akad. Wiss. Berlin 1882: 100. 1882, n. ill.
Type not designated

M. A. Rodrigo (✉)
Universitat de València, Institut Cavanilles de Biodiversitat i Biologia Evolutiva, C/ Catedràtic José Beltran 2, 46980 Paterna (València), Spain
e-mail: maria.a.rodrigo@uv.es

S. Cirujano
C/Ayala 83 3° C, 28006 Madrid, Spain
e-mail: santoscirujano@gmail.com

I. Soulié-Märsche
Université Montpellier, Institut des Sciences de l'Évolution (ISEM), Cc 061 Place Eugène Bataillon, 34095 Montpellier Cedex 05, France
e-mail: insouma43@gmail.com

C. Christia · E. Papastergiadou
University of Patras, Department of Biology, 26500 Patras, Greece
e-mail: xchristia@gmail.com

E. Papastergiadou
e-mail: evapap@upatras.gr

H. Schubert et al. (eds.), *Charophytes of Europe*,
https://doi.org/10.1007/978-3-031-31898-6_54

= *Lamprothamnus hansenii* Sond., Charac. Schlesw.-Holst: 25. 1890
Lectotype (Wood, 1965: 341): Im Drei bei Flensburg [Germany]/[no date]/C. Sonder PC 143204).

≡ *Lamprothamnium hansenii* (Sond.) Corill., Charoph. France Europe Occid.: 144. 1957.

= *Lamprothamnus aragonensis* Reyes y Prosper, Carof. España: 91. 1910
Lectotype (Cirujano and Rodrigo, 2019: 2): [España] Chiprana (Zaragoza)/ 10.09.1905/E. Reyes Prósper (MA-ALGAE 3032/1).

= *Lamprothamnus toletanus* Reyes y Prosper, Carof. España: 95. 1910
Lectotype (Cirujano and Rodrigo, 2019: 2): [España] Toly Quero a Vill/[no date]/E. Reyes Prósper (MA-ALGAE 11230).

= *Lamprothamnium carrissoi* A. G. Cunha in Bull. Soc. Portug. Sci. Nat. 12: 45. 1935
Lectotype (Wood, 1965: 336): Ria de Aveiro, no Braço de Ilhavo, entre a ponte de Ilhavo e a Vista Alegre [Portugal]/[no date]/Taborda de Morais 28 (COI).

= *Lamprothamnium sonderi* Garniel in H. Schubert & Blindow, Charoph. Baltic Sea: 163. 2004
Holotype (Garniel in Schubert and Blindow, 2004: 163): Fastensee (island of Fehmarn, Germany)/01.07.2003/H. Schubert (KIEL).

Local names: DANISH: Rævehaletråd; ENGLISH: Foxtail stonewort; FRENCH: Characée en queue de renard; GERMAN: Fuchsschwanz-Armleuchteralge; GREEK: Χαρόφυτα; RUSSIAN: Лампротамниум пупырчатый; SPANISH: Colas de agua, copetes de agua; SWEDISH: Axsträfse; UKRAINIAN: Лампротамніум пухирчастий.

54.1 Morphology

See Figs. 54.1 and 54.2.

Habitus: Plants are erect and fresh specimens have a pale green or yellowish green colour while dried specimens sometimes become darkened; encrustation is lacking or weak. The upper internodes are short with condensed whorls, so that the upper part of the shoots has a characteristic foxtail-like appearance. In deeper water (>1 m), this appearance might be not so marked.

Size: Variable. Plants can be from small to medium size (4–15 cm), but may reach up to 40 cm. In permanent water in Sweden, these plants are longer as shown by the herbarium sheets of the 19th century.

Main axis: Axis diameter is up to 0.75 mm.

Cortex: Cortex is absent.

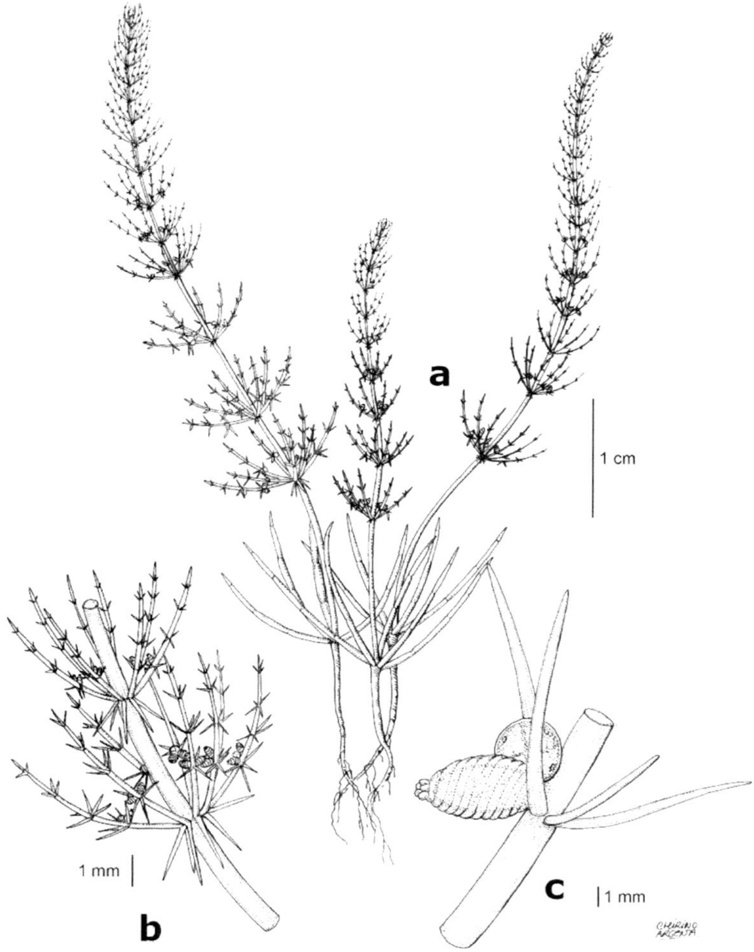

Fig. 54.1 *Lamprothamnium papulosum.* **a** Habitus; **b** Upper foxtail-like part of the plant, relatively long bract-cells and a single row of downward-pointing stipulodes; **c** Fertile branchlet node with oogonium below/besides antheridium. Drawings by Marta Chirino, in Cirujano et al. (2008)

Spine-cells: Spine-cells are absent.

Stipulodes: *Lamprothamnium papulosum* has well-developed and acuminate stipulodes, about twice as long as the axis diameter or even longer, up to 3 mm. The stipulodes occur in a single row pointing downwards. In contrast to the alternate position observed in several *Chara* species, the position of the stipulodes of *L. papulosum* is opposite the bases of the branchlets (Fig. 54.1b).

Branchlets: 6–8(10) undivided branchlets, with 4–5 segments, the last one of which can have 1–2 cells; the terminal cell is of variable length, acute, sometimes inflated.

Fig. 54.2 *Lamprothamnium papulosum*, fructified plants from "les Aresquiers", South France, collected May 22. 2009. Photographs: Ingeborg Soulié-Märsche

The sterile branchlets can reach 7 cm and usually lack bract-cells. The fertile branchlets are shorter (usually less than 2 cm), rigid and curved inward, pointing to the main axis. The elongate protonema can remain visible in some plants. On some occasions the branchlets can appear inflated.

Bract-cells: Conspicuous, spreading, 4–8 bract-cells on each node of the branchlets. The bract-cells are at least 2 times longer than the oogonia.

Bracteoles: 2, similar to bract-cells, sometimes thinner and shorter.

Sexuality: The species is monoecious, with the oogonia below or beside the antheridia (always conjoined) on the branchlets, never at the base of the whorls (Fig. 54.1c). Very exceptionally, the "normal" position with oogonium above antheridium, as in *Chara* is observed, here and there on a single node. Geminate gametangia (two oogonia or two antheridia together at the same node) are very rare.

Oogonia: Mostly solitary oogonia, elongated ovoidal. The coronula is large and consists of 5 oval cells. Data about quantitative morphometry varies greatly among authors (see Table 54.1). The length can range between 600 and 1020 μm, the width between 400 and 660 μm, the coronula length and width between 90–140 and 185–250 μm, respectively and the oogonia have between 10 and 15 convolutions.

Oospores: Mature oospores are black and usually up to 600–660 μm long. They can be distinguished from other charophyte oospores by the existence of small projections close to the poles and by undulating striae (Soulié-Märsche, 1989: pl. V, Figs. 2–4) (Fig. 54.3a, b). The wall is often tuberculate. There is a huge range of morphometrical data for oospores of *L. papulosum* as can be seen from Table 54.2. The length varies between 535 and 792 μm, the width between 149 and 500 μm and they have between 8 and 12 striae. The gyrogonites have a very typical morphology: a cylindrical outline with a flat apex (Fig. 54.3c–f). The apex is only slightly or not calcified (Groves

Table 54.1 Oogonia morphometry

Authors	Length	Width	Coronula length	Coronula width	Number of helical stripes
Migula (1889–1897)	650–800(700)	410–500(600)	–	–	10–12
Groves and Bullock-Webster (1924)	750–950	400–575	90	200–250	12–15
Cunha (1935) (*L. carrissoi*)*	430–490	200–260	180–240	85–100	–
Corillion (1957)	750–950	400–575	90	200–250	12–15
Wood (1965)	870–1020	570–660	112–120	187–213	12
Hollerbach and Palamar-Mordwinzeva (1991)	640–900	400–600	100–140	185–200	10–12
Krause (1997)	750–950	400–600	90–130	200	10–14
Cirujano et al. (2008) (n = 25)	600–1000	400–600	–	–	–
R, Romanov, pers. comm., lake Donuzlav (Crimea)	521–959	405–729	98–137	197–237	–
Total range	**(430)521–1020**	**(200)400–729**	**90–140(240)**	**(85)185–250**	**10–16**

Length in the first column excludes the coronula, its length and width given separately; all values for length and width in μm

* The oogonia dimensions of *Lamprothamnium carrissoi* are markedly smaller

and Bullock-Webster, 1924); the calcification in transverse-section shows a fan-like structure (Soulié-Märsche, 1989). The degree of calcification and elongation is very variable and can vary depending on the hydrochemical environment where the gyrogonites are formed. For example, the gyrogonites studied from Gallocanta lake (Zaragoza, Spain; Soulié-Märsche et al., 2018) growing in waters with high sulfate content display a greater length compared to populations from brackish water sites on the Mediterranean coast (Soulié-Märsche, 1973, 1989; Figs. 29 and 30) (Table 54.3).

Antheridia: The antheridia are mostly solitary, rarely geminate. The most common diameter is between 400 and 600 μm (Table 54.4) with the exception of the measurements made by Cunha (1935) for *L. carrissoi*.

Bulbils: White globular bulbils, up to 1 mm in diameter, in clusters of 3–5 cells, develop in the sediment on the rhizoids early in the year. They represent a means of storing energy (starch-sucrose) and diminish in size and starch content as the thallus grows (Soulié-Märsche pers. observ.).

Table 54.2 Oospore morphometry

Authors	Length	Width	Number of striae	Colour
Migula (1889–1897)	535–620	320–400	(7)9–15	black
Groves and Bullock-Webster (1924)	600–750	275–375	11	black
Cunha (1935) (*L. carrissoi*)	400–460	180–240	–	–
Olsen (1944)	600–660	–	–	black
Corillion (1957)	600–750	275–375	11	black
Horn af Rantzien (1959) (n = 10)	590(639)–730	330(370)–460	9–10	black
Wood (1965)	690–754	390–405	11	black, opaque
Hollerbach and Palamar-Mordwinzeva (1991)	535–750	270–500	8–10(11)	black
Krause (1997)	600–700	275–450	10–12	black
Cirujano et al. (2008)	578–604	334–353	–	black
Rodrigo et al. (2009) (n = 20)	479–792	149–495	9–11	black
R, Romanov, pers. comm., lake Donuzlav (Crimea)	391–757	269–490	–	–
Total range	**391–792**	**149–500**	**8–12**	Black*

All values for length and width in μm

* The oospores of *Lamprothamnium sonderi* have been described as being of brown colour

Table 54.3 Gyrogonite morphometry

Authors	Length	Width	Site
Horn af Rantzien (1959) (n = 10)	660(717)–820	410(482)–550	From herbarium plant, Bohuslän, Sweden
Soulié-Märsche (1973), (1989) (n = 60)	575–775	350–525	From recent plants, Les Onglous, South France
Rodrigo et al. (2009) (n = 16)	347–644	479–842	From Albufera de Valencia lagoon sediments, Spain
Soulié-Märsche et al. (2018) (n = 100)	625–850	300–500	From Gallocanta lake sediments, Spain
Total range	**347–850**	**300–842**	

All values for length and width in μm

Fig. 54.3 *Lamprothamnium papulosum*, oospore and gyrogonites. **a** Oospore, showing waved ridges; **b** Oospore, basal view; **c** Apical view of a gyrogonite; **d–f** Morphological variation of gyrogonites at different calcification stages. Photographs: Ingeborg Soulié-Märsche

Table 54.4 Antheridia morphometry

Authors	Diameter (μm)
Groves and Bullock-Webster (1924)	400–500
Cunha (1935) (*L. carrissoi*)	310–340
Corillion (1957)	400–500
Wood (1965)	420–465
Hollerbach and Palamar-Mordwinzeva (1991)	420–580
Krause (1997)	400–600
Cirujano et al. (2008)	400–500
R. Romanov, pers. comm., lake Donuzlav (Crimea)	324–491
Total range	**(310)324–600**

54.2 Phenology and Reproduction

Lamprothamnium papulosum usually has an annual life history and hibernates by means of oospores and bulbils (Wahlstedt, 1864) in temporary environments, but it can behave as perennial in permanent waters (Corillion, 1957) where it is less fertile. The species has been shown to have early germination and vegetative development. In the French Mediterranean littoral, specimens can be collected with mature oospores in April and, in the Atlantic localities, in May. Fructifications continue to appear until the end of June (Corillion, 1957). *L. papulosum* needs flooded conditions for at least three months to complete its life cycle from germination to the production of ripe reproductive organs (Soulié-Märsche, 1993, Mouronval et al., 2015). After that, due to strong evaporation, the temporary habitats are usually dry for several months in the Mediterranean region. Soulié-Märsche (1998, Fig. 3) described the life cycle of *L. papulosum* as it occurred at the historical location of "Les Onglous" (South France) already sampled by Hy (1913) and later by Corillion (1957). *Lamprothamnium papulosum* germinates there following autumn rains and develops remarkably long protonemas that can hardly be recognised as a charophyte. These germlings remain in quiescence (sort of stand-by, different from the dormancy of the oospores) and develop the typical bulbils during winter. As soon as irradiance and temperature increase, they grow to their typical foxtail habitus. Correlatively, the bulbils become empty and develop additional shots (Soulié-Märsche pers. observ.). Depending on water depth, specimens can be observed until July prior to the formation of a salt crust due to evaporation. As for *Characeae* in general, the stages of this life cycle depend on climate and the time of flooding (Gauthier et al., 2004: Fig. 14). In Sardinia, specimens of *Lamprothamnium* develop later in the year (end of April, May; R. Becker pers. comm.) even in temporary water bodies. In Schleswig–Holstein (Germany), germination usually takes place during the first 20 days of June (A. Garniel pers. obs.). The species is normally richly fertile with fructification from July to September (Olsen, 1944), but this can be extended until October as documented in Swedish herbarium material. Apart from oospores, the very resistent gyrogonites play a considerable role in the reproduction and dispersal of *L. papulosum* in space and time.

54.3 Habitat Conditions and Ecology

See Fig. 54.4.

Habitat type: According to the EU Habitats Directive (European Commission, 1992), *Lamprothamnium papulosum* figures in the priority for conservation habitat type "3130* Lagoons". It occurs in coastal and inland saline or hypersaline lakes like Laguna de Gallocanta, Spain (Cirujano et al., 2008; Soulié-Märsche et al., 2018 and refs therein). The typical habitats are brackish waterbodies that experience salinity fluctuations and dry out for several months. The species can also occur in permanent

Fig. 54.4 *Lamprothamnium papulosum* from diverse lagoons in Sardinia, Italy. **a** Spiaggia Salinedda, 17.03.2012; **b** San Giovanni Di Sinis, 17.04.2014 (specimens with large protonemas); **c** Cabras, 16.04.2014 (specimens with inflated branchlets), **d** Cabras, 16.04.2014. Photographs: Klaus van de Weyer

water in parts of lagoons with low salinity (Feldmann, 1953). In the Mediterranean area, it also colonises abandoned salt pans where the hydrological cycle is marked by seasonal variation in salinity. In Sardinia, *L. papulosum* occurs in lagoons, brackish water pools in dune valleys along the coastline, in small ponds and puddles in temporary flooded salt marshes and in a few deeper and bigger water bodies used as fish farms (R. Becker pers. comm.). In Greece, it inhabits choked or restricted lagoon types with meso- to polyhaline water chemistry, temporary ponds and lakes, and in salt lakes (Christia et al., 2011). The species also inhabits small, sheltered bays in the Baltic Sea (Andersson et al., 2004; Hamann, 2016).

Sediment : *Lamprothamnium papulosum* has been reported from a broad range of sediments, from hard bottom (sand, silty-sand, silty-shells or gravel with pebbles) to soft substrates like clay, although soft bottoms are only recorded for a few specimens. In ponds near beaches it grows on sand mixed with marine shells. In Sardinia, sediments are often very muddy, sometimes sandy with small shells and only a thin, muddy layer on top. In Greece, the species shows a preference for sandy silt substrate or clay-silt.

Nutrient conditions: *Lamprothamnium papulosum* lives mainly in oligo-mesotrophic waters, in most cases with chlorophyll *a* value of below 5 μg/l in the water.

Water chemistry: *Lamprothamnium papulosum* is one of the few truly brackish water species, which occurs neither in freshwater nor under fully marine conditions. It is the most halotolerant species among the extant brackish charophytes (Soulié-Märsche, 1998), but with a clear euryhaline character. *L. papulosum* is often subjected to changes in water volume over daily, seasonal and annual timescales, either as part of regular cycles, or periodic, irregular disturbance events. Hence, salinity can vary widely (Martin, 2001). The species has produced strategies to cope with extremely high variations of salinity. However, there appears to be some contention concerning the salinity tolerance range for the physiological activity of *L. papulosum* on one hand, and the range allowing for its reproduction.

Different field populations of *L. papulosum* may have very different salinity optima. In Britain and Ireland, the species is considered to occur within the salinity range of 8 to 30 g/l (Daniel, 1974; Moore, 1986) (annual mean values: 6–31 g/l, optimum 22–23 g/l for growth). Other authors reported even wider salinities (5–35 g/l, Martin, 2001; 1–95 g/l, Wichmann and Kirst, 1989; 9–59 g/l, Guerlesquin, 1992; 14–46 g/l, Comín et al., 1983; the thallus of the plants can continue growing in salinities up to 125 g/l, Cirujano and Medina, 2002; 14–43 g/l, Christia et al., 2018; up to 85 g/l in lab cultures, M. A, Rodrigo unpubl.). Regarding salinities for oospore production, the salinity values are more restricted (24–28 g/l, Daniel, 1974; 20–40 g/l, Guerlesquin, 1992; in lab experiments, no oospore production at 6 g/l and no ripe oospore at 48 g/l; M. A. Rodrigo unpubl.; up to 40 g/l, Soulié-Märsche, 1998). *L. papulosum* needs low salinity (<20‰) for germination (Dubois, 1968).

With regard to water pH, the species is associated with alkaline waters in the Mediterranean (pH around 9; Mouronval et al., 2015). In Sardinia, it has been found living in water with a pH between 7.2 and 9.4 (R. Becker pers. comm.)

Depth range: *Lamprothamnium papulosum* is restricted to shallow waters of 0.05–2.0 (to 3.2) m depth, therefore is regarded as a heliophilous species.

Associated species: In many places where *L. papulosum* develops well, it builds small monospecific patches of about 30 to 40 cm diameter, but it is frequently associated with other species (with *Althenia filiformis* and *Tolypella glomerata* -but it develops then earlier in the year, at lower salinity-, with *Chara galioides*, *C. canescens*, *C. baltica*, *Tolypella hispanica* and *Tolypella salina* in South France, Mouronval et al., 2015; with *C. aspera*, *C. galioides* and *Ruppia drepanensis* in Gallocanta lake, Comín et al., 1993, M. A. Rodrigo pers. obs.; with *C. galioides*, *T. hispanica*, *R. drepanensis* and *Zannichellia pedunculata* in Pétrola lake, Cirujano and Medina, 2002; with *Tolypella nidifica*, *R. maritima* and *R. cirrhosa*, sometimes also with *Althenia filiformis* in Sardinia, R. Becker pers. comm.; with *Zostera noltii* -in low-salinity, high-transparency, high-nitrogen lagoons-, with *Gracilaria bursa-pastoris* and *Cladophora glomerata* -in mesohaline, deep, high transparent waters—, with *R. cirrhosa* and *Acetabularia acetabulum*, *Gracilariopsis longissima*, *Ulva rigida* and

Chaetomorpha linum, in Greece, Christia et al., 2018; with *Ruppia* spp., *Tolypella nidifica, Chara aspera, Chara canescens, Stuckenia pectinata, Zannichellia palustris* s.l., *Zostera marina*, *Cladophora* spp. and other filamentous algae in the Baltic Sea area, Andersson et al., 2004; with *Ruppia cirrhosa* or *Ruppia maritima* and *Zostera marina* in Ireland, C. Roden pers. comm.

54.4 Variability

The variation of *Lamprothamnium papulosum* mainly concerns the size of the plants. Length and width of the stipulodes are also variable. In small temporary ponds in the Mediterranean region, so-called dwarf forms of 3–5 cm (formerly distinguished as *Chara pouzolsii*) are frequent at the edges of the brackish water body. Towards the deeper parts of the pond, the size of the plants increases progressively forming a continuous population. *L. papulosum* can reach a length of 40 cm for which the characteristic foxtail-like appearance is less pronounced because of relatively long upper internodes. Reyes Prósper (1910) distinguished two other species of *Lamprothamnium* (= *Lamprothamnus; L. aragonensis* and *L. toletanus*), and Cunha (1935) established *L. carrisoi.* Wood (1965) reduced the status of these species to varieties or forms of *L. papulosum.* They are at present considered synonyms of *L. papulosum* (Cirujano and Rodrigo, 2019; Cirujano et al., 2008). Similarly, *L. hansenii* and *L. sonderi* are not sufficiently different to merit species status. The common character of these taxa is the position of oogonia below the antheridia and the absence of gametangia at the base of the whorl.

54.5 Identification and Risk of Confusion

Due to the absence of a cortex and because of the single row of stipulodes and the well-developed bract-cells, the sterile individuals of *Lamprothamnium papulosum* could be confused with *Chara braunii*. However, in *C. braunii* the stipulodes are arranged alternately to the branchlets, not opposite and *C. braunii* displays mostly a crown of tiny bract-cells on the terminal branchlet segment whereas the branchlets in *L. papulosum* end with an acute tip. As well, the bract-cells in *Lamprothamnium* usually spread out at angles of 45° to 90° to the branchlet node. Fertile specimens clearly differ by the position of the antheridia, located above the oogonia in *Lamprothamnium*, not below. In addition, *C. braunii* is restricted to freshwater and brackish water of low salinity.

There could also be confusion with partly ecorticate forms of *C. canescens,* which often shares the habitat with *L. papulosum* (Andersson et al., 2004), but in this case the contracted heads of *L. papulosum* can help to immediately distinguish between them. *Lamprothamnium succinctum* from Africa, which has not yet found in Europe, has a few additional oogonia external to the base of the whorls (Muller et al., 2017; Wood,

1962). *Lamprothamnium sardoum* (see Chap. 55), shows abundant gametangia inside and outside the base of the whorls and is easily distinguished on that basis.

54.6 Distribution

See Fig. 54.5.

Lamprothamnium papulosum, as recognised here, has a restricted distribution in Europe limited to a certain type of waterbodies. It is also recorded from Africa and doubtfully from parts of Asia (Casanova pers. comm.). In Europe, it occurs along the Atlantic coast of Europe, from northernmost UK to France and to the Iberian Peninsula. Numerous sites are known in the western Baltic Sea area. In the Mediterranean domain it extends east to Cyprus and west to Spain. There are a few localities in North Black Sea region (coastal lakes or lagoons of Crimea as well as in a few other bays in north-western parts of Ukraine, Corillion, 1957); the northernmost continental locality in Europe is from a brackish inland lake far north of Caucasus (R. Romanov pers. comm.).

In the Baltic Sea, the main distribution of *L. papulosum* is in the western part. All recent and former records of the species are restricted to Denmark, Germany and the west coast of Sweden. In Sweden, there are recently only two sites left in the province of Bohuslän (Havs- och Vattenmyndigheten, 2020). In Denmark,

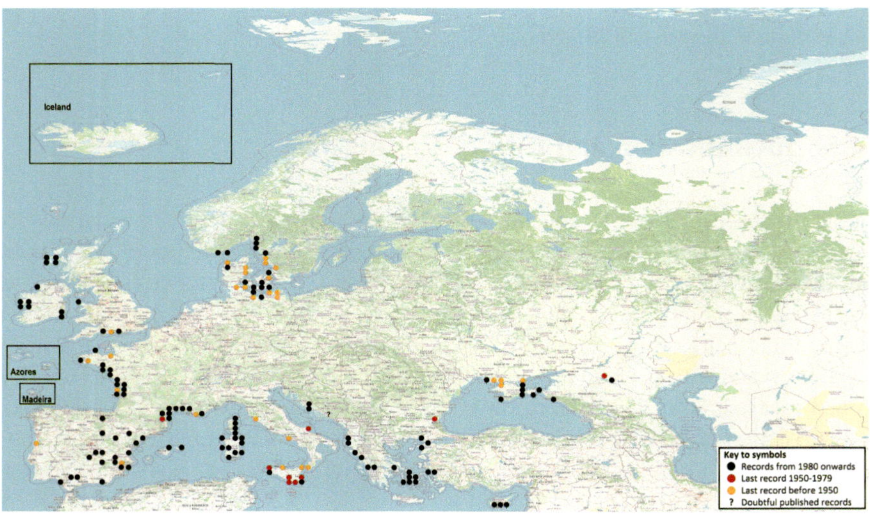

Fig. 54.5 Distribution of *Lamprothamnium papulosum* in Europe. For worldwide distribution see Korsch (2018). Black dots are records from 1980 on, red dots from 1950 to 1979 and yellow dots before 1950; unconfirmed/doubtful published records (if applicable) are marked with "?" on top. Map prepared by Heiko Korsch

there are few records after 1995 (northern Sjælland, Fyn), but none from the last 10-year period. Six records exist for the time-period of 1988–1991 around Lolland and southern Sjælland. No specific locality information could be obtained about recent occurrences in Denmark, but the species was known to occur at 15 different locations historically. In Germany, 15 recent records (time period 2004–2011) exist in bays and lagoons (Orther Bucht, Fastensee, Kirchsee, Salzhaff). It is known to have disappeared from seven locations and two new localities have been found recently. The species is known to be present but rare mainly along the south coast of England from Sussex to Dorset, with 12 stable populations in the Western Isles of Scotland. In Ireland it is present in the southeast, north and west of the island. In France, it has drastically regressed along the Atlantic coast, especially in southern Britany (as already noted by Corillion, 1957). Its main area of occurrence is now along the Mediterranean coast including Corsica. However, the number of localities has also decreased since the 1970s. Most of the recent localities focus on the Rhone river delta (Camargue). In the Iberian Peninsula *L. papulosum* is present along the Mediterranean coast and the Balearic Islands and it is particularly abundant in inland saline waterbodies, but with lesser representation in the western half of Spain. In Portugal, it was recorded only once under the name *L. carrissoi*. In Italy, it is present in Tuscany, Campania, Apulia, Veneto, the Po delta and in the islands of Sicily and Sardinia, the latter being extremely rich with 68 individual sites found recently (Becker, 2019). In Greece, *L. papulosum* is in Amvrakikos gulf lagoons and in Kotychi and Alyki Aigiou lagoons and so far in coastal lagoons of Northern Greece (Orfanidis et al., 2011). Previously, it was reported from the Cyclades islands. *L. papulosum* is known from Cyprus (Akrotiri saline lakes Christia et al., 2011).

54.7 Status and Threats

The populations of *L. papulosum* are severely fragmented due to their occurrence in very few disconnected, specialised habitats. The past and current threats for *L. papulosum* have been mainly: eutrophication, construction, water traffic and tourism (codes H01.05, D03, J02.01.02, J02.12, G01.01.01, G05 according to the Habitats Directive article num. 17). Other threats such as marine aquaculture and climate change (F01, M01, M02) have to be considered as future threats. To conserve the species, several actions should be implemented, such as combating local nutrient inputs (mainly from agriculture) that cause eutrophication, retaining the natural hydrological regime, applying conservation measures, such as restriction of coastal construction and dredging, and regulation of beach tourism in shallow coastal lagoons, bays and fjords.

The species is declining troughout nearly all of Europe and is listed as Threatened or Near Threatened in all countries with published Red Lists. Iberia (VU, Cirujano et al. 2009); Ireland (VU, Moore 1991; Stewart and Church 1992); Britain (NT, Joint Nature Conservation Committee 2004); Norway (VU, Artsdatabanken 2021); Sweden (EN, Artdatabanken 2020); Denmark (EN, Bastrup-Spohr et al. 2013); Baltic

Sea (EN, HELCOM 2013); Germany (CR, Korsch et al. 2013); The Balkans (EN, Blazencic et al. 2006); Bulgaria (DD, Temniskova et al. 2008); Ukraine (VU, Didukh 2009); Crimea (Rare, Yena and Fateryga 2015). It is also under threat in other countries without a Red List. Nevertheless, despite the steady attrition of sites it is still widely but sparsely distributed and it is assessed as Near Threatened in Europe as a whole (see Chap. 9).

54.8 Remarks

Lamprothamnium specimens from Gallocanta lake (Zaragoza), from Santa Pola saltcrops (Alicante) (collector M. A. Rodrigo), from the vicinity of Salinas de la Algaida (Espacio Natural Doñana, Cádiz; collector: A. Holzhausen), in Spain, from Alyki Aigiou (Aigio, Greece) (collector C. Christia), from Sardinia, Italy (collector: R. Becker) and from the Baltic Sea, Germany (collector: A. Holzhausen and T. Steinhardt) as well as a specimen of *Lamprothamnium sonderi* from the collection of H. Schubert sampled in 2003 in the Fastensee (island of Fehmarn), were genetically analysed for *rbc*L and *mat*K sequences by P. Nowak, Univ. Rostock, Germany. The *rbc*L and *mat*K sequences were identical in all these samples suggesting a single common haplotype for the European *Lamprothamnium* (P. Nowak pers. comm.) including the specimen of *L. sardoum* from Sardinia that bore gametangia at the base of the whorls. Thus, the genetic analyses tend to confirm the homogeneity of European *Lamprothamnium* populations (including the new one from Sardinia) for these markers. This result is in opposition to the morphological species concept as proposed for the Australian *Lamprothamnium* species, using diverse arrangements of gametangia at the base of whorls for establishing new species (Casanova, 2013; Casanova et al., 2011). Further comparative investigations may clarify the significance of the genetic results and provide a solution in the future.

References

Andersson, J. A., Garniel, R., & Nielsen, R. (2004). *Lamprothamnium papulosum*. In H. Schubert & I. Blindow (Eds.), *Charophytes of the Baltic Sea* (pp. 156–162). A. R. G. Gantner.

Artsdatabanken (2021). *Norwegian Red List for Species*. https://www.biodiversity.no/Pages/135380/Norwegian_Red_List_for_Species

Båstrup-Spohr, L., Dahl-Nielsen, J., & Sand-Jensen, K. (2013). *Kransnålalger Rummer Mange Truede Arter. Urt, 37*, 66–70. [in Danish].

Becker, R. (2019). The Characeae (Charales, Charophyceae) of Sardinia (Italy): habitats, distribution and conservation. *Webbia, 74*, 83–101.

Blaženčić, J., & Blaženčić, Ž. (2002). Rare and threatened species of Charophytes (Charophyta) in Southeast Europe. *Phytologia Balcanica, 8*, 315–326.

Blaženčić, J., Stevanović, B., Blaženčić, Ž, & Stevanović, V. (2006). Red data list of Charophytes in the Balkans. *Biodiversity and Conservation, 15*, 3445–3457. https://doi.org/10.1007/s10531-005-2008-5

Casanova, M. (2013). *Lamprothamnium* in Australia (Characeae, Charophyceae). *Australian Systematic Botany, 26*, 268–290. https://doi.org/10.1071/SB13026
Casanova, M., Bradbury, S., & Ough, K. (2011). Morphological variation in an Australian species of *Lamprothamnium* (Characeae, Charophyceae) in response to different salinities. *Charophytes, 2*, 87–92.
Christia, C., Giordani, G., & Papastergiadou, E. (2018). Environmental variability and macrophyte assemblages in coastal lagoon types of Western Greece (Mediterranean Sea). *Water, 10*, 151. https://doi.org/10.3390/w10020151
Christia, C., Tziortzis, I., Fyttis, G., Kashta, L., & Papastergiadou, E. (2011). A survey of the benthic aquatic flora in transitional water systems of Greece and Cyprus (Mediterranean Sea). *Botanica Marina, 54*, 169–178. https://doi.org/10.1515/BOT.2011.016
Cirujano, S., & Medina, L. (2002). *Plantas acuáticas de las lagunas y humedales de Castilla-La Mancha.* Junta de Comunidades de Castilla-La Mancha [in Spanish].
Cirujano, S., Cambra, J., Sánchez Castillo, P. M., Meco, A., & Flor Arnau, N. (2008). *Flora ibérica. Algas continentales. Carófitos (Characeae).* Real Jardín Botánico [in Spanish].
Cirujano, S., & Rodrigo, M. A. (2019). Revisión del herbario de Eduardo Reyes Prósper y tipificación de tres especies de carófitos (Charophyceae, Characeae) presentes en el mismo. *Anales del Jardín Botánico de Madrid, 76*, e081. [in Spanish]. https://doi.org/10.3989/ajbm.2525
Comín, F. A., Alonso, M., Lopez, P., & Comelles, M. (1983). Limnology of Gallocanta Lake, Aragon, Northeastern Spain. *Hydrobiologia, 105*, 207–221. https://doi.org/10.1007/BF00025189
Comín, F. A., Rodó, X., & Menéndez, M. (1993). Spatial Heterogeneity of Macrophytes in Lake Gallocanta (Aragón, NE Spain). *Hydrobiologia, 267*, 169–178. https://doi.org/10.1007/BF00018799
Corillion, R. (1957). Les Charophycées de France et d'Europe Occidentale. *Bulletin de la Société Scientifique de Bretagne, 32*(Hors série I & II), 1–499 [in French].
Cunha, A. G. (1935). Quelques espèces nouvelles de Characées du Portugal. *Bulletin De La Société Portugaise Des Sciences Naturelles, 12*, 41–58. [in French].
Daniel, G. F. (1974). The ecology of the charophyte *Lamprothamnium papulosum*, J. Groves. Undergraduate Thesis, University of London.
Didukh, Ya. P. (Ed.). (2009). *The Red Data Book of Ukraine. The Vegetable World.* Globalconsulting [in Ukrainian].
Dubois, A. (1968). Observations sur la morphologie et la biologie des formes naines de *Lamprothamnium papulosum* J. Gr. (Characées). *Naturalia Monspeliensa, Série Botanique, 19*, 37–41. [in French].
European Commission. (1992). *Council Directive 92/43/EEC of 21 May 1992 on the Conservation of Natural Habitats and of Wild Fauna and Flora.* Official Journal of the European Communities L 206/7.
Feldmann, G. (1953). Contribution à l'étude biologique des étangs méditerranéens. La végétation de l'étang de Salses (rive sud). *Vie et Milieu, 4*, 685–700 [in French].
Gauthier, P., Grillas, P., Hugonnot, V., & Hébrard, J. P. (2004). Vegetation. In P. Grillas, P. Gauthier, N. Yavercovski, & C. Perennou (Eds.), *Mediterranean Temporary Ponds* (Vol. 1, pp. 40–45). La Tour du Valat.
Groves, J., & Bullock-Webster, G. R. (1924). *The British Charophyta. Volume II. Chareae with Plates, Concluding Articles, Geological Sketch, Bibliography and Index.* Ray Society.
Guerlesquin, M. (1992). Systématique et biogéographie du genre *Lamprothamnium* (Characées) Caractéristique des biotopes aquatiques saumâtres. *Revue des Sciences de l'Eau, 5*, 313–461. [in French].
Hamann, U. (2016). *Lamprothamnium papulosum.* In *Arbeitsgruppe Characeen Deutschlands, Armleuchteralgen: Die Characeen Deutschlands* (pp. 529–538). Springer [in German].
Havs- och Vattenmyndigheten. (2020). *Åtgärdsprogram för sällsynta kransalger längs kusten* (57 pp.) [in Swedish].

HELCOM. (2013). Red List of Baltic Sea species in danger of becoming extinct. *Baltic Sea Environmental Proceedings, 140*, 1–106.
Hollerbach, M. M., & Palamar-Mordvintseva, G. M. [Голлербах, М. М. & Г. М. Паламар-Мордвинцева] (1991). Визначник прісноводних водорослей України. IX. Харові водорості (Charophyta) [*The identification manual of freshwater algae of Ukraine: IX. Charophytes (Charophyta)*]. Nauk. dumka [in Ukrainian].
Horn af Rantzien, H. (1959). Recent charophyte fructifications and their relations to fossil charophyte gyrogonites. *Arkiv för Botanik, 4*, 167–332.
Hy, F. (1913). Les Characées de France. *Bulletin de la Société Botanique de France, 60*(sup 1), 1–47 [in French].
JNCC [Joint Nature Conservation Committee]. (2004). *Conservation Designations for UK Taxa*. http://jncc.defra.gov.uk/page-3408
Korsch, H. (2018). The worldwide range of the Charophyte species native to Germany. *Rostocker Meeresbiologische Beiträge, 28*, 45–96.
Korsch, H., Doege, A., Raabe, U., & van de Weyer, K. (2013). Rote Liste der Armleuchteralgen (Charophyceae) Deutschlands. 3. Fassung, Stand: Dezember 2012. *Haussknechtia Beiheft, 17*, 1–32. [in German].
Krause, W. (1997). Charales (Charophyceae) Süßwasserflora von Mitteleuropa. In H. Ettl, G. Gärtner, H. Heynig, & D. Mollenhauer (Eds.), *Süßwasserflora von Mitteleuropa* (Vol. 18). Fischer [in German].
Martin, A. (2001). *The Ecology and Palaeoecology of the Charophyte Lamprothamnium papulosum in U.K. Coastal Lagoons*. Theses University College.
Migula, W. (1889–1897). Fünfter Band: Die Characeen. In A. Fischer, E. Fischer, F. Hauck, G. Limpricht, C. Luerssen, W. Migula, H. Rehm, P. Richter, & G. Winter (Eds.), *Dr. L. Rabenhorst's Kryptogamenflora von Deutschland, Oesterreich und der Schweiz* (2nd edn). Eduard Kummer [in German].
Moore, J. A. (1986). *Charophytes of Great Britain and Ireland*. BSBI Handbook No. 5, Botanical Society of the British Isles.
Moore, J. A. (1991). *Lamprothamnium papulosum*, a pioneer in the conservation of Characeae and their habitats. *Bulletin de la Societe Botanique de France. Actualites Botaniques, 138*, 73–74. https://doi.org/10.1080/01811789.1991.10827043
Mouronval, J. B., Baudouin, S., Borel, N., Soulié-Märsche, I., Klesczewski, M., & Grillas, P. (2015). *Guide des Characées de France méditerranéenne*. Office National de la Chasse et de la Faune Sauvage [in French].
Muller, S. D., Rhazi, L., & Soulié-Märsche, I. (2017). Diversity and distribution of Characeae in the Maghreb (Algeria, Morocco, Tunisia). *Cryptogamie, Algologie, 38*, 201–251. https://doi.org/10.7872/crya/v38.iss3.2017.201
Olsen, S. (1944). Danish Charophyta. *Det Kongelige Danske Videnskabernes Selskab, Biologiske Skrifter,* 3(1), 1–240, pl. 1–2.
Orfanidis, S., Panayotidis, P., & Ugland, K. I. (2011). Ecological Evaluation Index continuous formula (EEI-c) application: A step forward for functional groups, the formula and reference condition values. *Mediterranean Marine Science, 12*, 199–231. https://doi.org/10.12681/mms.60
Reyes Prósper, E. R. (1910). *Las Carofitas de España. Singularmente las que crecen en sus estepas*. Imprenta Artistica Española, Madrid [in Spanish].
Rodrigo, M. A., Alonso-Guillén, J. L., Cirujano, S., & Soulié-Märsche, I. (2009). Aproximación a las comunidades de carófitos que existieron en la Albufera de Valencia a partir del estudio de las oósporas del sedimento. *Anales del Jardín Botánico de Madrid, 66*, 51–64. [in Spanish]. https://doi.org/10.3989/ajbm.2214
Schubert, H., & Blindow, I. (Eds.). (2004). *Charophytes of the Baltic Sea*. A. R. G. Gantner.
SLU Artdatabanken (2020) Rödlistade arter i Sverige 2020. SLU, Uppsala. http://www.artdatabanken.se/globalassets/ew/subw/artd/2.-var-verksamhet/publikationer/31.-rodlista-2020/rodlista-2020

Soulié-Märsche, I. (1989). *Etude comparée de gyrogonites de charophytes actuelles et fossiles et phylogénie des genres actuels*. Imprimerie des Tilleuls [in French].

Soulié-Märsche, I. (1973). Données générales sur les Charophytes du Plio-Pléistocène. *Bulletin de L'Association Française pour l'étude du Quaternaire, 2*, 69–77. [in French].

Soulié-Märsche, I. (1993). Apport des Charophytes fossiles à la recherche de phénomènes climatiques abrupts. *Bulletin de la Société Géologique de France, 164*, 123–130. [in French].

Soulié-Märsche, I. (1998). Fossil *Lamprothamnium papulosum* (Charophytes), a Biomarker for Seasonal Rainfall in Northern Mauritania. *Paleoecology of Africa, 25*, 65–76.

Soulié-Märsche, I., Baumhauer, R., & Schütt, B. (2018). Palaeoenvironmental significance of charophyte gyrogonites from Laguna de Gallocanta (Aragón, Spain). *Botany Letters*, 165, 35–44. https://doi.org/10.1080/23818107.2017.1370611

Stewart, N. F., & Church, J. M. (1992). *Red data books of Britain and Ireland: Stoneworts*. The Joint Nature Conservation Committee.

Temniskova, D., Stoyneva, M. P., & Kirjakov, I. K. (2008). Red List of the Bulgarian algae I. Macroalgae. *Phytologia Balcanica, 14*, 193–206.

Wahlstedt, L. J. (1864). *Om characeernas knoppar och öfvervintring. Botanisk avhandling* (48 pp.). University of Lund [in Swedish].

Wichmann, F., & Kirst, G. O. (1989). Adaptation of the euryhaline charophyte *Lamprothamnium papulosum* to brackish and freshwater: Turgor pressure and vacuolar solute concentrations during steady-state culture and after hypo-osmotic treatment. *Journal of Experimental Botany, 40*, 135–141. https://doi.org/10.1093/jxb/40.1.135

Wood, R. D. (1962). New combinations and taxa in the revision of Characeae. *Taxon, 11*, 7–25.

Wood, R. D. (1965). *Monograph of the Characeae*. J. Cramer.

Yena, A. V., & Fateryga, A. V. (Eds.). (2015). *Red Data Book of the Republic of Crimea: Plants, algae and fungi*, Arial, Simferopol [in Russian].

Chapter 55
Lamprothamnium sardoum

Ralf Becker and Michelle T. Casanova

Scientific name: ***Lamprothamnium sardoum*** **Ralf Becker in Casanova and Ralf Becker, Phytotaxa 567(3): 271. 2022**

Holotype (Casanova and Becker, 2022: 271): ITALY. Sardini, circa San Giovanni Sinis, in aquis stag., 39°53′13 N, 8°26′22 E, WGS 84/05 April 2018/Becker 2018 (ROST).

Local names: unknown.

55.1 Morphology

See Fig. 55.1.

Habitus: The monoecious plants are light green, opaque and not encrusted. The protonemal axis is much longer than the height of the plant. Gametangia are clustered inside and outside the base of the whorl and clustered and geminate on the lowest branchlets. There are well developed bract-cells and long stipulodes where they are not replaced by gametangia.

Size: *Lamprothamnium sardoum* is small to medium-sized (9–17 cm including the protonema).

R. Becker (✉)
Am Schützenplatz 58, 26121 Oldenburg, Germany
e-mail: becker.r@posteo.de

M. T. Casanova
273 Casanova Rd, 3351 Westmere, Victoria, Australia
e-mail: mt.casanova@federation.edu.au

H. Schubert et al. (eds.), *Charophytes of Europe*,
https://doi.org/10.1007/978-3-031-31898-6_55

Fig. 55.1 Collected material of *Lamprothamnium sardoum*. **a** Condensed head with developing oogonia and ripe antheridia; **b** Detail with base of the whorl; **c** Dried material showing basal oogonia and antheridia and stipulodes; **d** Oogonium with oval coronula cells and bract-cells at a branchlet node; **e** Branchlet tips with end segments, the end cell mucronate. Photographs: Ralf Becker

Main axis: Main axis diameter is 350 μm up to 600 μm (flattened), mainly 375–550 μm. Ecorticate. The protonemal axis exceeds the terminal fertile heads by up to 7 cm. The internodes are mainly 1–3.2 cm long with one exception up to 6.5 cm. Internodes are much shorter in the apices.

Cortex and spine-cells: Absent.

Stipulodes: Up to one per branchlet, but frequently replaced by gametangia. The stipulodes are up to 1.75 mm long, acute, opposite the branchlets and pointing downward and outward.

Branchlets: Whorls consist of 5–8 ecorticate branchlets, up to 1.8 cm long. The basal branchlet cells are short on fertile branchlets, up to 2.0 mm long. The branchlet end segments are usually 1 cell long. The apical cell of the branchlets is mucronate and up to 750 μm long. All branchlets are fertile on the material seen so far.

Bract-cells: Bract-cells verticillate, 4 or 5 around on the lowest two or three branchlet nodes in fertile whorls, up to 1.7 mm long.

Bracteoles: Not distinguished from the bract-cells.

Sexuality: *Lamprothamnium sardoum* is monoecious. Gametangia are clustered inside and outside the base of the whorl and conjoined, geminate or geminate-conjoined on the lowest 2–3 branchlet nodes.

Oogonia: Oogonia are apparently sessile, clustered inside and outside the base of the whorl. When occurring on the branchlets, they are usually conjoined with antheridia, below or beside the antheridium. Oogonia are oval, up to 750 mm long, with 10–14 convolutions, the coronula cells are oval-shaped (Table 55.1, morphometric data based on dried herbarium material).

Table 55.1 Oogonia morphometry

Authors	Length	Width	Coronula length	Coronula width	Number of helical stripes
Casanova and Becker (2022)	500–750	400–475(600)	75–110	130–150	10–14

Length in the first column is excluding the coronula, length and width of the coronula is given separately, all values for length and width in μm

Oospores: Oospores are up to 510–650 μm long, 250–475 μm wide at the widest point, with 9–11 striae of low ridges. The wall ornamentation is smooth to slightly roughened with a sparsely verrucate fossa wall (Fig. 55.2). Table 55.2 gives the morphometric data of oospores based on dried herbarium material. The only mature oospores available were the ones from which plants germinated, so measurements are approximate.

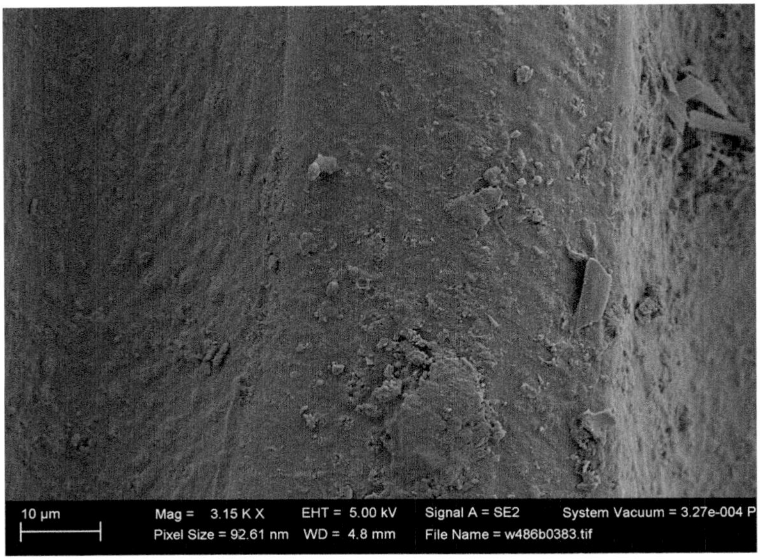

Fig. 55.2 Detail of oospore wall on a germinated oospore. Photograph: Michelle T. Casanova

Table 55.2 Oospore morphometry

Authors	Length	Width	Number of striae	Colour
Casanova and Becker (2022)	510–650	250–475	9–11	black

All values for length and width in µm

Antheridia: Antheridia are up to 450 µm in diameter, solitary or conjoined with oogonia on the branchlets, as well as inside and outside the base of the whorl (Table 55.3, morphometric data based on dried herbarium material).

Table 55.3 Antheridia morphometry

Authors	Diameter (µm)
Casanova and Becker (2022)	325–450

Bulbils: Have not been found.

55.2 Phenology and Reproduction

Lamprothamnium sardoum is an annual species. The plants germinate from oospores in early spring or late winter. Oospore formation and ripening occurs in March and April. Ripe antheridia and immature oogonia were found in early April (Sardinia,

Casanova and Becker, 2022). Thus, ripe oospores can be expected between mid-April and May. Vegetative reproduction is not known. The species shows high fecundity.

55.3 Habitat Conditions and Ecology

Habitat type: *Lamprothamnium sardoum* is restricted to brackish water habitats. The only known European site exists in ephemeral shallow pools between halophilous shrubs in a Mediterranean salt marsh (Fig. 55.3).

Sediment: The species has been reported from muddy substrates.

Nutrient conditions: According to a single measurement *Lamprothamnium sardoum* is characterized as a brackish water species, being able to colonise eutrophic temporary ponds rich in nutrients. Total nitrogen measured at the Sardinian site reached 1.94 mg/l TN, NO_3 4.79 mg/l and TP 0.035 mg/l. Additional data are presented in Table 55.4. It can be expected that *Lamprothamnium sardoum*, similar to *L. papulosum,* is able to grow under even more mesotrophic conditions.

Water chemistry: The salinity of the Sardinian site in April 2018 was 12.3 psu, the conductivity 19.2 mS/cm (own field data, Casanova and Becker, 2022). The pH-value of the Sardinian habitat was recorded as 9.1 and the chloride concentration 5.55 mg/l. Moreover, concentrations of 123.4 mg/l calcium, 272 mg/l magnesium and a total hardness of 79.8°dH were determined (see Table 55.4). These values indicate the species' preference to grow in alkaline, calcium- and magnesium-rich water. But it can be expected that *Lamprothamnium sardoum* is able to tolerate a much broader range of psu-values and ion concentrations.

Depth range: *Lamprothamnium sardoum* was recorded in shallow water. The depth ranged at the time of sampling between 15 and 25 cm.

Associated species: A small population of *Lamprothamnium sardoum* is reported from just one locality in Europe in temporary brackish pools with *Juncus subulatus* and halophilous shrubs in a Mediterranean salt marsh. *Lamprothamnium sardoum* is associated with the halotolerant species *Tolypella nidifica* and *T. hispanica.* In closely neighbouring shallow brackish pools without halophilous shrubs *Lamprothamnium papulosum, Ruppia cirrhosa* and *R. maritima* were recorded (Casanova and Becker, 2022).

55.4 Variability

All specimens have shown a similar morphology and a high fertility. Therefore, other morphotypes are not known.

Fig. 55.3 Habitat of *Lamprothamnium sardoum*. **a** Overview of site; **b** Sparse stands in shallow, brackish water among other vegetation; **c** View of plants through the water; all photographs taken at Sardinia, 2018-04-05. Photographs: Ralf Becker

55.5 Identification and Risk of Confusion

Lamprothamnium sardoum may be confused with the subtropical and tropical species *L. succinctum* and with *L. papulosum*. *Lamprothamnium sardoum* differs from *L. succinctum* by having bigger antheridia both inside and outside the base of the whorl, well developed bract-cells on shorter branchlet segments, and larger oospores with smooth to slightly roughened, somewhat verrucate fossa wall ornamentation (Fig. 55.2). The axis and branchlet diameters are smaller, and the plants are not calcified. *Lamprothamnium papulosum* can be distinguished from *L. sardoum* by the

Table 55.4 Hydrochemical values of the Sardinian site of *Lamprothamnium sardoum*, sampled at 05.04.2018 by R. Becker

Parameter	Value
Salinity (psu)	12.3*
Conductivity (mS/cm)	19.2*
pH	9.1
Chlorid (mg/l)	5.55
TP (mg/l)	0.035
PO_4 (mg/l)	0.044
PO_4-P (mg/l)	0.015
TN (mg/l)	1.945
NH_4 (mg/l)	0.088
NO_3 (mg/l)	4.79
NO_2 (mg/l)	<0.016
Ca (mg/l)	123.4
Mg (mg/l)	272
Total hardness (°dH)	79.8
Carbonate hardness (°dH)	15
Carbonate hardness (mg/l $CaCO_3$)	267

* = field data

absence of gametangia at the base of the whorl, bigger antheridia, longer oogonia and bulbils. Sterile plants of *L. sardoum* could be confused with *L. papulosum* and *L. succinctum*.

55.6 Distribution

See Fig. 55.4.

Lamprothamnium sardoum has been reported from only one European temporary brackish waterbody in Sardinia in a salt marsh east of San Giovanni di Sinis, Province of Oristano (39°53′13" N, 8°26′22" E).

55.7 Status and Threats

Currently only one European site of *Lamprothamnium sardoum* is known, exhibiting a small population. At first the species of the Sardinian locality was erroneously cited as *Lamprothamnium succinctum* (Becker, 2019). A comparison with the type material of *Lamprothamnium succinctum* along with additional material from Libya deposited in BM as well as SEM studies revealed the existence of the new species

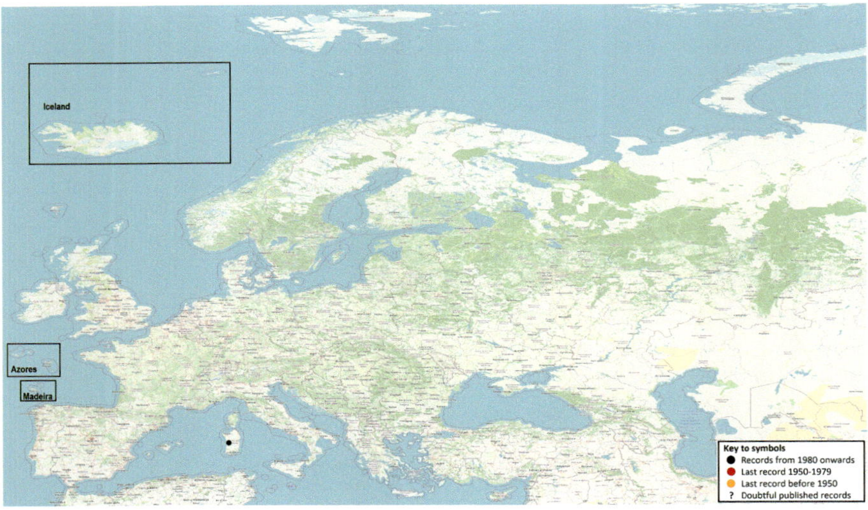

Fig. 55.4 European distribution of *Lamprothamnium sardoum*. Map prepared by Heiko Korsch

Lamprothamnium sardoum. Consequently, *L. sardoum* is very rare throughout the world.

In Sardinia many coastal wetlands have been destroyed or heavily modified during the last few decades (Becker, 2019). Existing wetlands are seriously endangered through sea level rise and climate change (Čížková et al., 2013; European Commission, 2017). Moreover, the Sardinian habitat of *L. sardoum* is severely hampered by drainage, salt marsh succession with an increase in shrubs and reeds, unstable hydrochemical conditions and water levels, limited size of the population and intensification of agricultural land use. By contrast, current extensive grazing by sheep is required to maintain the habitat and to create open spaces for charophytes. A monitoring program for the population of *L. sardoum* should be implemented immediately. Furthermore, a periodical removal of shrub vegetation might be required.

Becker (2019) called for the implementation of valid Action plans for conservation of Sardinian charophytes focusing on *Characeae* in brackish habitats. He mentioned *Lamprothamnium* species as key species and the wetlands of the Sinis Peninsula and around Stagno di Cabras as one of the most important Sardinian hotspots for charophytes. The Sardinian site of *Lamprothamnium sardoum* is part of the Habitats Directive Site of Community Interest "Stagno di Mistras di Oristano" (site code: ITB030034) mentioned in Natura 2000 lists. Unfortunately, precise data on *Characeae* and concrete proposals for conservation measures linked to charophytes are missing in the management plan concerning "Stagno di Mistras di Oristano" (Regione Autonoma della Sardegna, 2006).

55.8 Remarks

Further investigations on other water bodies inhabited by *Lamprothamnium sardoum* should be conducted especially on the Sinis Peninsula (Sardinia). Moreover, special attention should be paid on the potential occurence of *L. sardoum* in other brackish habitats all over Europe and other continents. Former sites and herbarium material of *Lamprothamnium succinctum* should be reviewed critically.

Phylogenetic analyses of the barcode markers *rbc*L, *psb*C and *matK* of Sardinian specimens of *Lamprothamnium sardoum* revealed identical gene sequences to individuals of *L. papulosum* from Sardinia and Spain, indicating a close evolutionary relationship of both species (P. Nowak pers. comm.). In contrast, specimens of *L. sardoum* and *L. succinctum* from GenBank can be phylogenetically clearly distinguished from each other based on the *rbc*L (LC536055-LC536061, KX431014, AB440262) and *psb*C (KX430972) gene sequences (P. Nowak pers. comm.).

Acknowledgements We are very grateful to Petra Nowak, who has conducted several phylogenetic analyses and to Melanie Willen for the hydrochemical analyses.

References

Becker, R. (2019). The Characeae (Charales, Charophyceae) of Sardinia (Italy): Habitats, distribution and conservation. *Webbia, 74*, 83–101. https://doi.org/10.1080/00837792.2019.1607497

Casanova, M. T., & Becker, R. (2022). *Lamprothamnium sardoum* sp. nov. (Characeae, Streptophyta): A new species of *Lamprothamnium* for Europe. *Phytotaxa, 567*(3), 269–277. https://doi.org/10.11646/phytotaxa.567.3.6

Čížková, H., Květ, J., Cumin, F. A., Laiho, R., Pokorný, J., & Pithart, D. (2013). Actual state of European wetlands and their possible future in the context of global climate change. *Aquatic Science, 75*, 3–26.

European Commission. (2017). *LIFE and coastal habitats*. http://ec.europa.eu/environment/life/publications/lifepublications/lifefocus/documents/coastal_habitats.pdf.

Regione Autonoma della Sardegna (Eds.). (2006). *Piano di Gestione Stagno di Mistras di Oristano and San Giovanni di Sinis* [in Italian].

Chapter 56
Lychnothamnus barbatus

Mariusz Pełechaty, Michał Brzozowski, Aleksandra Marković, Andrzej Pukacz, Maciej Gąbka, and Zofija Sinkevičienė

Scientific name: *Lychnothamnus barbatus* (Meyen) Leonh. in Lotos 13: 57. 1863

≡ *Chara barbata* Meyen in Linnaea 2: 75. 1827[Jan.]
Neotype (Wood, 1965: 347): In lacu "Plötzensee" vicinia Berol./aestate 1828 / Bauer (NY 01089212)

Andrzej Pukacz: Deceased

M. Pełechaty (✉)
Adam Mickiewicz University, Poznań, Faculty of Biology, Department of Hydrobiology, Uniwersytetu Poznańskiego Str. 6, 61-614 Poznań, Poland
e-mail: marpel@amu.edu.pl

M. Brzozowski
Poznań University of Life Sciences, Faculty of Environmental and Mechanical Engineering, Department of Ecology and Environmental Protection, Piątkowska Str. 94C, 60-649 Poznań, Poland
e-mail: michal.brzozowski@up.poznan.pl

A. Marković
University of Belgrade, Institute of Chemistry, Technology and Metallurgy, Department of Chemistry, Njegoševa Str. 12, 11000 Belgrade, Serbia
e-mail: aleksandra.z.markovic@gmail.com

A. Pukacz
Adam Mickiewicz University, Poznań, Collegium Polonicum, Kościuszki Str. 1, 69-100 Słubice, Poland

M. Gąbka
Adam Mickiewicz University, Poznań, Faculty of Biology, Department of Hydrobiology, Uniwersytetu Poznańskiego Str. 6, 61-614 Poznań, Poland
e-mail: gmaciej@amu.edu.pl

Z. Sinkevičienė
Nature Research Centre, Institute of Botany, Žaliųjų Ežerų Str. 47, 12200 Vilnius, Lithuania
e-mail: zofijasin@gmail.com

H. Schubert et al. (eds.), *Charophytes of Europe*,
https://doi.org/10.1007/978-3-031-31898-6_56

= *Chara spinosa* Amici in Mem. Accad. Sci. Lett. Arti Modena 1: 215. 1827
Type not designated

≡ *Lychnothamnus barbatus* (Meyen) Leonh. f. *spinosus* (Amici) R. D. Wood, Rev. Charac. 348. 1965

? = *Lychnothamnus longibracteatus* F. S. Han in Acta Phytotax. Sin. 7(2): 174. 1958, n. inv.

= *Lychnothamnus barbatus* (Meyen) Leonh. f. *fasciculatus* Imahori & R. D. Wood in R. D. Wood in Taxon 11: 15. 1962

= *Lychnothamnus barbatus* (Meyen) Leonh. f. *hyalina* Karczm. in Acta Soc. Bot. Poloniae 36: 436. 1967

= *Lychnothamnus barbatus* (Meyen) Leonh. var. *giganteus* Bharati & M. S. Chennaveeraiah in Cytologia 48: 843. 1983 as *gigantii*, n. inv.

Nomenclatural notes: Both, *Chara barbata* and *C. spinosa* were published in 1827. The exact publication date for *C. spinosa* is unknown, *C. barbata* was published in January. But, as the description of *C. spinosa* was read at the session on 25 Jan. 1827 of "Reale Accademia de Scienze, Lettere ed arti in Modena" it was probably published later. Han (1974) designated a "typus" but did not specify a herbarium thus indicating all specimens of that gathering as syntypes. The name is invalid as from 1.1.1958 the indication of a type is mandatory. Bharati and Chennaveeraiah (1983) published a "typus lectus" with three numbers thus failing to designate a type.

Taxonomic note: Wood (1965) considered *C. spinosa* to be on the rank of form different from *L. barbatus*.

Local names: GERMAN: Bart-Glanzleuchteralge, KAZAKHSTANIAN: Лихнотамнус сақалды, LITHUANIAN: Šiurpinis žvakidumblis, POLISH: Lychnotamnus brodaty; RUSSIAN: Лихнотамнус бородатый; UKRAINIAN: Ліхнотамнус бородатий.

56.1 Morphology

See Figs. 56.1 and 56.2.

Habitus: *Lychnothamnus barbatus* belongs to the group of large charophytes. It is light green to dark green, mostly heavily or occasionally incrusted. It is characterized by a very specific, symmetric habitus: long internodes, straight branchlets, evenly distributed in whorls with long bract-cells as well as very long, straight and single-row stipulodes. Hardly to be confused with any other species.

Size: It belongs to the group of large charophytes, usually 20–80 cm long (up to 140 cm). Maximally, up to 189 cm in Lake Kuźnickie (Brzozowski and Pełechaty, 2020).

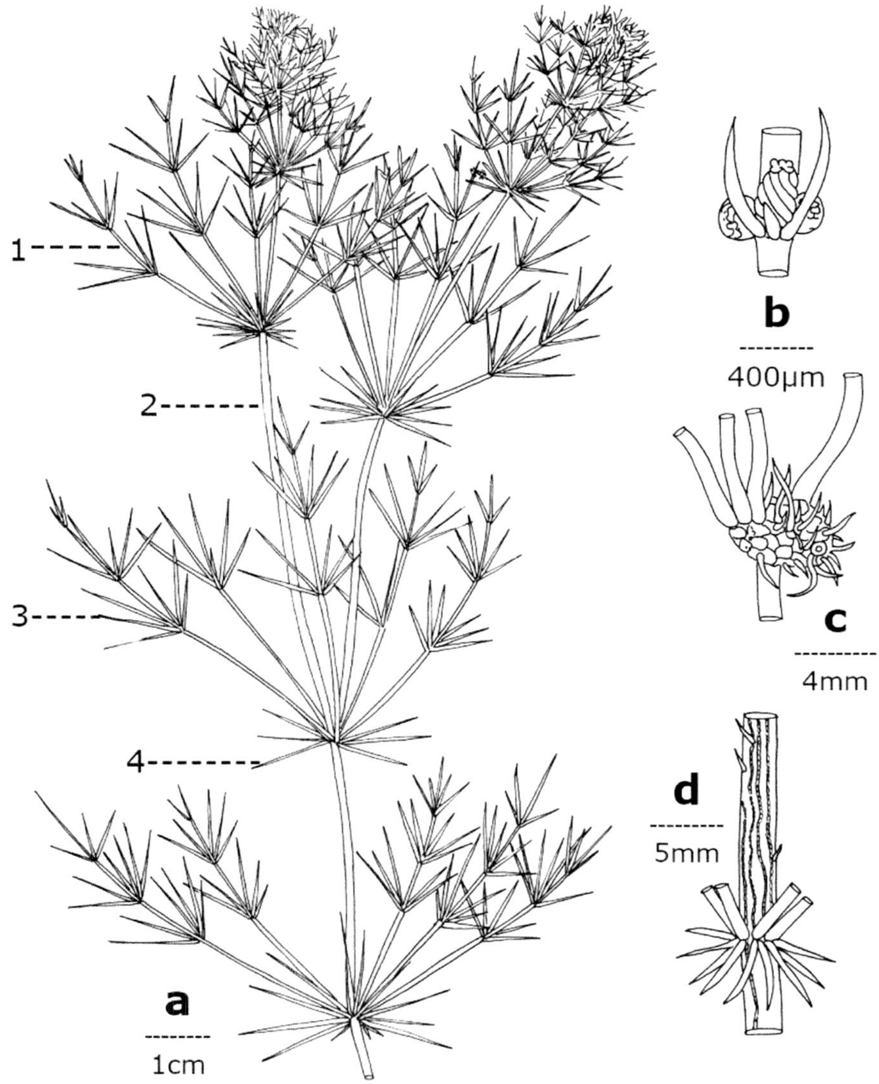

Fig. 56.1 *Lychnothamnus barbatus*, morphological features. **a** Habitus, 1—branchlet, 2—main axis, 3—bract-cells equally developed around the branchlet node, 4—one row of needle-like and acute stipulodes; **b** Branchlet node with centrally located oogonium accompanied by bracteoles and two antheridia; **c** Node with bulbils; **d** Axial node with a row of stipulodes, branchlet arrangement and main axis fragment with rudimentary cortex in a form of detached primary cells. Drawings after Werner Krause, modified by Dagmar Wassong in van de Weyer and Schmidt (2018)

Fig. 56.2 *Lychnothamnus barbatus*, details. **a** Deep water thalli collected from one of lakes of western Poland; **b** Top nodes of the main axis and primary cortex cells detached from the axial internode; **c** Two nodes of main axis with whorls of branchlets and long needle-shaped stipulodes, and the axial internode with cortex and well-developed single spine-cells. Photographs: Mariusz Pełechaty

Main axis: Long and mostly straight, with no branches or 2–4 branches, maximally 7 branches in Lake Kuźnickie (Brzozowski and Pełechaty, 2020), divided into long (up to 10 cm) stout internodes, with a diameter of about 1 mm (up to 2.2 mm), internodes 2–4 × of branchlet length.

Cortex: The rudimentary cortex appears only on upper parts of internodes, growing from the nodes in the form of detached primary cells.

Spine-cells: Lacking or (if cortex is present) only rudimentary.

Stipulodes: Very long (about 0.5 cm, up to 1.2 cm), needle-shaped and acuminated, arranged in one row below the axial node, slightly inclined against the main axis, 2 per branchlet.

Branchlets: 7–10 in a whorl, each of 3–5 ecorticate segments. The first segment is much longer, than the subsequent ones (>½ of the entire branchlet length), end segment is 1-celled and acute.

Bract-cells: 4–7 in each node, all well developed, very long (even up to 21 mm) and acute. The longest are at the first node, the shortest at the last node.

Bracteoles: Two, below gametangia, very often much shorter than bract-cells but may be also the same size.

Sexuality: Monoecious. Gametangia (usually 1 oogonium between 2, rarely 3, antheridia) are conjoined at the lowest 1-3 branchlet nodes.

Oogonia: Elliptical, yellow, yellow-green or light green, variability presented in Table 56.1.

Table 56.1 Oogonia morphometry of *Lychnothamnus barbatus*

Authors	Length	Width	Coronula length	Coronula width	Number of helical stripes
Migula (1889–1897)	1100–1300	700–800	–	–	9–11
Corillion (1957)	1100–1300	700–800	–	–	9–11
Wood (1965)	(800)1050–1140	(665)720–725	70–160	120–160	7–8(11)
Dąmbska (1964)	1100–1300	700–800	–	–	9–11
Karczmarz (1967)	1100–1200	700–770	–	–	8–9(10)
Krause (1997)	1000–1350	600–800	45	130	7–9
Borisova et al., (2016)	1000–1300	700–800	50–60	110–130	9–12
Total range	**(800)1000–1350**	**600–800**	**45–160**	**110–160**	**7–12**

Length in the first column excludes the coronula, its length and width given separately; all values for length and width in μm

Oospores: Oval, mainly brown and black, variability presented in Table 56.2. Oospores become calcified, forming oval gyrogonites. Variability for European gyrogonites is presented in Table 56.3 and Fig. 56.3. Asian and Australian gyrogonites are longer than European (800–1075 μm) and have more convolutions (up to 10 or 11) (Brzozowski et al., 2019).

Table 56.2 Oospore morphometry of *Lychnothamnus barbatus*

Authors	Length	Width	Number of striae	Colour
Migula (1889–1897)	660–720	460–500	7–8	Black
Corillion (1957)	660–720	460–500	7–8	Brown to black
Wood (1965)	660–735	465–495	8–10	Orange brown to black
Dąmbska (1964)	660–720	460–500	7–8	Reddish brown to black
Haas (1994)	580–720	460–500	7–10	Reddish brown to dark brown
Krause (1997)	600–800	500–700	7–8	Black
Boszke et al., (2007)	584–734	383–683	8–9	Dark brown to black
Urbaniak and Gąbka (2014)	650–850	410–490	–	Brown to dark brown
Borisova et al., (2016)	660–720	460–515	7–9	Dark orange
Total range	**580–850**	**383–700**	**7–10**	

All values for length and width in μm

Table 56.3 Gyrogonite morphometry of *Lychnothamnus barbatus*

Authors	Type	Length (μm)	Width (μm)	Convolutions
Horn af Rantzien (1959)	Extant	770–860	590–680	8–9
Soulié-Märsche (1989)	Extant	700–975	550–775	–
Brzozowski et al., (2019)	Extant and subfossil	712–1021	604–836	8–9
Total range	–	**700–1021**	**550–836**	**8–9**

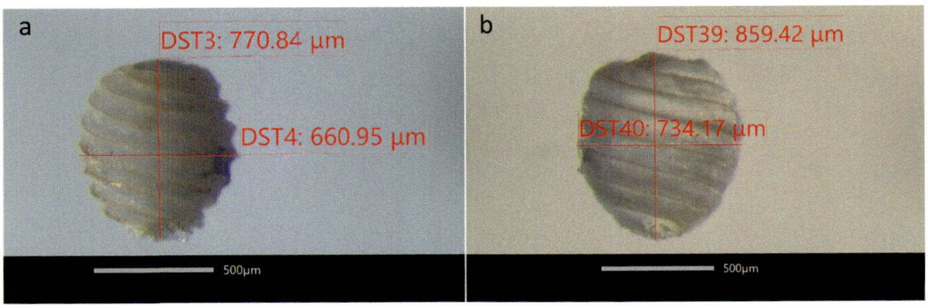

Fig. 56.3 *Lychnothamnus barbatus*, gyrogonites. **a** Slightly and **b** Fully calcified oospore (gyrogonite). Photographs: Marcel Palomares Cabanilles

Table 56.4 Antheridia morphometry of *Lychnothamnus barbatus*

Authors	Diameter (μm)
Migula (1889–1897)	200–250
Corillion (1957)	200–250
Wood (1965)	(200)300–380
Dąmbska (1964)	200–250
Karczmarz (1967)	300–330
Krause (1997)	350
Borisova et al. (2016)	200–315
Total range	**200–380**

Antheridia: Threefold smaller than oogonia, orange or reddish. Variability in diameter is presented in Table 56.4.

Bulbils: Small and irregular bulbils are formed on rhizoids.

56.2 Phenology and Reproduction

Lychnothamnus barbatus starts vegetative growth in spring and finishes in autumn. Data on the phenology of *L. barbatus* related to the species overwintering are scarce. Wintering of single individuals only or parts of thalli, whose nodes produced young sprouts in spring while the rest of thalli decayed, were reported (Meyen, 1827; Migula, 1889–1897; Sinkevičienė, 2010). Pełechaty et al., (2017) provided evidence that the species overwintered in Lake Kuźnickie, Western Poland, in the form of extensive meadows along a wide depth gradient. Importantly, in addition to vegetative propagation, *L. barbatus* reproduces generatively and produces fully developed gyrogonites in this lake, although the lake is characterized by the nutrient concentration typical of slightly eutrophic waters. The number of gyrogonites increased with an increase in depth but their size decreased (Brzozowski et al., 2019). Furthermore, germinating gyrogonites were also observed in Lake Kuźnickie. To a certain extent it is inconsistent with Karczmarz's (1967) opinion that only in nutrient-poor clear waters *L. barbatus* accomplishes the full biological cycle, while in eutrophic waters the species produces only immature oospores. On this basis, fossil gyrogonites were applied in palaeo reconstructions to indicate past oligotrophic conditions (Martín-Closas et al., 2006; Mazzini et al., 2015; Soulié-Märsche and Martín-Closas, 2003). To sum up, the species undertakes vegetative propagation by nodes and node outgrowths and sexual reproduction leading to oospores and, after calcification, gyrogonite production.

56.3 Habitat Conditions and Ecology

Habitat type: In Europe, *L. barbatus* occurs mainly in lakes (Fig. 56.4), but single locations are known from a peat excavation site, old riverbed and rice field (Brzozowski et al., 2022 and references therein).

Sediment: Mineral to soft organic sediment and calcareous gyttja, more often on calcareous organic substrate (Blaženčić and Blaženčić, 1983) than on the mineral one (Bhatia, 2006; Casanova et al., 2003). Shallow sites tend to be mineral, while at deeper sites soft Ca-rich organic bottom dominate where the species contributes to increased carbonate content in sediment (M. Brzozowski & M. Pełechaty own obs.).

Nutrient conditions: The ecological plasticity of *L. barbatus* in relation to trophy of lakes is broader than previously thought, from oligotrophic to eutrophic waters, as evidenced in a variety of European lakes (e.g. Azzella and Abdelahad, 2011; Blaženčić and Blaženčić, 1983; Gąbka et al., 2010; Pełechaty et al., 2009, 2017; Raabe et al., 2012; Sinkevičienė, 2010; Sinkevičienė and Urbaitė-Maževič, 2012; van de Weyer, 2016); Karczmarz (1967) claimed that low-nutrient lakes are optimal for *L. barbatus*. In oligotrophic conditions of Lake Martignano, Italy (Azzella and Abdelahad, 2011) *L. barbatus* grows in low nutrient concentrations reaching 0.01 mgl^{-1} for TP and 0.038–0.048 mgl^{-1} for TN. For other European locations, TP and TN show higher values and broader ranges of 0.003–0.056 mgl^{-1} and 0.26–5.00 mgl^{-1}, respectively (Brzozowski and Pełechaty, 2022; Brzozowski et al., 2018, 2019, 2021a; Raabe et al., 2012; Pełechaty et al., 2009, 2017). The ranges of mineral forms of P and N in European waters are presented in Table 56.5. However, the data on nutrient concentrations and other chemical compounds of waters are rather scarce. The study in Lake Kuźnickie provided evidence of nutrient levels indicating slightly eutrophic conditions but water clarity and chlorophyll-*a* concentrations suggested a lower, mesotrophic state of the lake's water and process of water purification in the last years (Brzozowski and Pełechaty, 2022; Brzozowski et al., 2018; Pełechaty et al., 2017). Thus, *L. barbatus* requires clear waters rather than low nutrient content, and light availability seems more crucial than other indices of water fertility (Pełechaty et al., 2009). High Secchi disk visibility, ranging mainly between 1.5 and 5.0 m, was reported from different *L. barbatus* locations.

Water chemistry: *Lychnothamnus barbatus* is a freshwater species. Ranges of physicochemical properties of waters are presented in Table 56.5 based on the review of data available in published reports (Azzella and Abdelahad, 2011; Blaženčić and Blaženčić, 1983; Blaženčić et al., 1998; Brzozowski and Pełechaty, 2022; Brzozowski et al., 2018; 2019; Gąbka et al., 2010; Sinkevičienė, 2010; Pełechaty et al., 2009 2017; Raabe et al., 2012; Sinkevičienė and Urbaitė-Maževič, 2012; Urbaniak and Gąbka, 2014; van de Weyer, 2016). Wide ranges of solute content, water colour and pH values reflect the different habitats from which *L. barbatus* has been reported. Nevertheless, the species prefers clear, mainly alkaline and calcium-rich hard waters

Fig. 56.4 Habitat appearance of *Lychnothamnus barbatus*. Monospecific stand in a moderately nutrient-rich shallow lake in Poland. Photograph: Mariusz Pełechaty

Table 56.5 Ranges of mineral nutrient forms and other physicochemical water indices from European locations of *L. barbatus*

Parameter	Unit	Minimum	Maximum
Colour	[$mgPtl^{-1}$]	2	10
pH	–	6.5	8.7
O_2	[mgl^{-1}]	5.9	13.3
Conductivity	[μScm^{-1}]	196	568
N-NH_4	[mgl^{-1}]	0.000	0.260(0.464)
N-NO_3	[mgl^{-1}]	0.000	0.78
P-PO_4	[mgl^{-1}]	0.000	0.230
Ca	[mgl^{-1}]	17.5	74.4
Mg	[mgl^{-1}]	5.0	28.1
Na	[mgl^{-1}]	9.1	19.5
K	[mgl^{-1}]	3.1	18.9
Fe	[mgl^{-1}]	0.03	0.25
SO_4	[mgl^{-1}]	20	80

Outstanding maximum values are given in brackets

(Brzozowski and Pełechaty, 2022; Pełechaty et al., 2009; Sugier et al., 2010; van de Weyer, 2016).

Depth range: *Lychnothamnus barbatus* occurs along a wide depth range. Karczmarz (1967) described the species at depths of 3–5(7) m. Pełechaty et al., (2017) observed the species in Lake Kuźnickie at depths between 1.5 and 6.5 m in autumn and 2–6.5 m in spring, with the highest frequencies and coverages of the species between 3 and 5 m. Similar depth range was reported by Sinkevičienė (2010). Single specimens or even dense patches of *L. barbatus* were also observed at shallow littoral locations of less than 0.5 m deep (Brzozowski and Pełechaty, 2020; Brzozowski et al., 2018). In Balkan waters and in the Ukrainian Lake Svitiaz *L. barbatus* was observed at greater depths, between 5 and 10 m (Blaženčić et al., 1998; Iakushenko and Borysova, 2012).

Associated species: In a limited number of European lakes, *L. barbatus* forms dense communities in which it is the only species or frequently co-occurs with *Nitellopsis obtusa*. The species forms its own association *Lychnothamnetum barbati* (Gołdyn, 1984) Brzeg et Wojterska 2001. *Lychnotamnus barbatus* occurs also in the form of small clusters in communities formed by other submerged macrophytes, particularly in patches of *N. obtusa*. From amongst the angiosperm group of submerged macrophytes *L. barbatus* frequently co-occurrs with *Ceratophyllum demersum* and *Myriophyllum spicatum*; *L. barbatus* tends to be a better competitor at deeper sites (Brzozowski and Pełechaty, 2022; Brzozowski et al., 2018). At shallower locations the species can be outcompeted by *M. spicatum* and *C. demersum* (Brzozowski and Pełechaty, 2022; Brzozowski et al., 2019; Gąbka et al., 2010).

56.4 Variability

Wood (1965) listed 3 forms: *L. barbatus* f. *barbatus*, f. *fasciculatus* (with partial 1-corticate cortex and long spine-cells in fascicles of 1–3 spine cells) and additionally f. *spinosus* (similar to f. *barbatus* but with swollen structures). Karczmarz (1967) distinguished 3 morphological forms Wood: *L. barbatus* f. *barbatus* (the typical one), f. *hyalina* (with rudimental cortex and very long internodes) and f. *fasciculatus* (small and partly corticated). A new variety, *L. barbatus f. giganteus*, characterized by significantly thick incrusted main axis (1500–2200 μm) was described and published from India (Bharati and Chennaveeraiah, 1983). Out of the above forms, *L. barbatus* f. *barbatus* and f. *spinosus* were described from material collected from Europe. The other two forms come from India. However, the variability of *L. barbatus* forms seems to be dependent on the character of the site of occurrence, primarily depth. In deep sites the species forms elongated internodes and specimens are slender and less compact, while at shallow sites the thalli are shorter and more compact. Both forms can occur at different depths in the same water body, such as in Lake Balsys (Sinkevičienė, 2010) and Lake Kuźnickie (Brzozowski and Pełechaty, 2020).

56.5 Identification and Risk of Confusion

As the only representative of its genus with a very characteristic habitus *L. barbatus* is easy to distinguish from species of other genera. Since it occurs very often with *Nitellopsis obtusa* it may be confused with its individuals, especially when young. Certain similarities in the overall impression of the shape of the thallus may occur between *L. barbatus*, *Chara braunii* and *Lamprothamnium papulosum*. Closer analysis of morphological features minimizes the risk of confusion between these species.

56.6 Distribution

Thirty three sites in 9 countries can be listed for *L. barbatus* as its current (observed after 1980) locations in European freshwaters (Fig. 56.5). Except for a small water body in the Volga River mouth in Russia, an oxbow lake in Ukraine and two located near each other rice fields in Italy, the species occurs in lakes. One locality was reported from peat excavation site in Poland, but this site does not exist anymore (P. Sugier pers. comm.). Most of the locations can be classified as oligotrophic, oligo-mesotrophic and mesotrophic (Brzozowski et al., 2022). It is of note, the species occurs also in more fertile waters, meso-eutrophic, eutrophic or even highly eutrophic. What is more, the species forms extensive meadows not only in nutrient-poor lakes but also in meso-eutrophic and highly eutrophic lakes. The highest numbers of current *L. barbatus* sites are listed from Poland (13), Lithuania (6) and

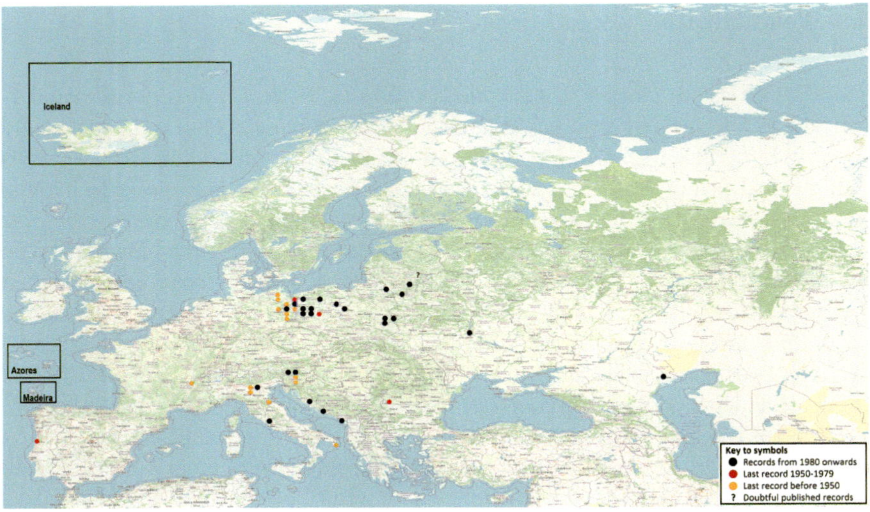

Fig. 56.5 European distribution of *Lychnothamnus barbatus*. For worldwide distribution see Korsch (2018). Black dots are records from 1980 on, red dots from 1950 to 1979 and yellow dots before 1950; unconfirmed/doubtful published records (if applicable) are marked with "?" on top. Map prepared by Heiko Korsch

Croatia (5). Until recently, findings in Croatia (Blaženčić et al., 1998) were the only current findings in the Balkans (Blaženčić et al., 2006). In spite of considerable efforts made in the charophyte distribution studies in the Balkans over the last few years (Schneider et al., 2020; Vesić et al., 2016), only the latest resulted in a new discovery of *L. barbatus* in Montenegro (Biberdžić et al., 2023). Three locations are known from Italy, namely one lake and two above-mentioned rice fields (it is rather doubtful if they are permanent sites considering the fact that they were observed in 1994 in a cultivated area), 2 from Ukraine and 2 from Austria (including a new one known since 2007, K. Pall pers. comm.). Single locations have been reported from Russia and Germany.

56.7 Status and Threats

From the end of the nineteenth century, 63 *L. barbatus* sites have disappeared, which is more than the number of currently known sites of this species worldwide (Pełechaty et al., 2017). Therefore, not only the species but the entire lineage is threatened by extinction, and thus, *L. barbatus* is red-listed as Endangered or Critically Endangered in Poland (Siemińska et al., 2006), Lithuania (Balevičius, 2000; Sinkevičienė and Gudžinskas, 2021), Germany (Korsch et al., 2013) and the Balkans (Blaženčić et al., 2006). Anthropogenic eutrophication of water continues to be one of the most serious threats to the species, as are the related negative impacts among which the most important are improper fish management, recreation and land use, locally, and climate warming, globally (Brzozowski et al., 2021a, b, 2022).

56.8 Remarks

In recent years the colonisation and recolonisation of sites by *L. barbatus* have been reported, including new sites in America from which the species had not been recorded earlier (Karol et al., 2017). It seems that the ability to overwinter in the conditions of warming winters gives the species a competitive advantage over other macrophytes (Brzozowski and Pełechaty, 2022; Pełechaty et al., 2017). Despite this, however, the fact that *L. barbatus* can occur in more fertile waters than previously thought, oligo-mesotrophic water bodies with high water transparency are still habitats preferred by this species and light availability seems more crucial than other indices of water fertility (Brzozowski and Pełechaty, 2020, 2022; Brzozowski et al., 2018).

Acknowledgements The authors would like to express their gratitude to the following persons for all information that helped to recognise the current distribution of *L. barbatus*: Mattia Azzella, Victoriia Berezovska, Vera Biberdžić, Lisa Brancaleoni, Caraus Ioan, Agnieszka Kolada, Karin Pall, Karol Pietruczuk, Uwe Raabe, Roman Romanov and Piotr Sugier.

References

Azzella, M. M., & Abdelahad, N. (2011). Une station actuelle de *Lychnothamnus barbatus* (Charophyceae) en Italie Centrale: le Lac de Martignano (Latium). *Cryptogamie, Algologie, 32*, 301–307. [in French].

Balevičius, A. (2000). *Lychnothamnus barbatus* Leonh. In *Lietuvos raudonoji knyga* (pp. 54–55). Augalu bendrijos [in Lithuanian].

Bharati, S. G., & Chennaveeraiah, M. S. (1983). *Lychnothamnus barbatus* var. *gigantii* var. nov. *Cytologia, 48*, 843–846.

Bhatia, S. B. (2006). Ecological parameters and dispersal routes of *Lychnothamnus barbatus* (Characeae) in the Early Middle Holocene from the Ganga plain, India. *Cryptogamie, Algologie, 27*, 341–356.

Biberdžić, V., Marković, A., Brzozowski, M., & Pełechaty, M. (2023). *Lychnothamnus barbatus* (Meyen) Leonhardi 1863, a new species to the flora of Montenegro. *Aquatic Botany, 184*, 103601.

Blaženčić, J., & Blaženčić, Ž. (1983). *Lychnothamnus* (Rupr.) v. Leonh. (Characeae) a new genus to the flora of Yugoslavia. *Acta Botanica Croatica, 42*, 95–101.

Blaženčić, J., Blaženčić, Ž, Lakušić, D., & Randjelović, V. (1998). Phytogeographical and ecological characteristics of the charophytes of the East Adriatic coast. *Archives of Biological Sciences, 50*, 119–128.

Blaženčić, J., Stevanović, B., Blaženčić, Z., & Stevanović, V. (2006). Distribution and ecology of charophytes recorded in the west and central Balkans. *Cryptogamie, Algologie, 27*, 311–322.

Borisova, O. V., Palamar-Mordvintseva, G. M., & Tsarenko, P. M. (2016). [Борисова, О. В., Г. М. Паламар-Мордвинцева & П. М. Царенко], 2016. Флора водоростей України. Том 12. Харофітові водорості. Вип. 2. [Flora algarum Ucrainicae. *Charophyta, 12*(2)]. Kholodny Institute of Botany of the National Academy of Science of Ukraine, Kyiv [in Ukrainian].

Boszke, P., Pełechaty, M., & Pukacz, A. (2007). Cechy biometryczne oospor *Lychnothamnus barbatus* z Jeziora Łagowskiego. *Ekologia i Technika, 3*, 79–84. [in Polish].

Brzozowski, M., Kowalewski, G., Szczuciński, W., Kaczmarek, L., & Pełechaty, M. (2021a). Preliminary evidence of an endangered species benefiting from moderate climate warming: A palaeolimnological study of the charophyte *Lychnothamnus barbatus. Aquatic Conservation. Marine and Freshwater Ecosystems, 31*, 2673–2689.

Brzozowski, M., Palomares Cabanilles, M., Kowalewski, G., & Pełechaty, M. (2019). Environmental factors responsible for the gyrogonite formation by an endangered macroalga, *Lychnothamnus barbatus*, a fertility indicator of past and present lacustrine ecosystems. *Limnologica, 77*, 125686.

Brzozowski, M., Pełechata, A., Kaczmarek, L., & Pełechaty, M. (2021b). Transformation and simplification of aquatic vegetation structure and reoligotrophication of a lake during the last 40 years. *Acta Societatis Botanicorum Poloniae, 90*, 905.

Brzozowski, M., & Pełechaty, M. (2020). Broad morphological and reproductive variability of the endangered macroalga *Lychnothamnus barbatus* in the depth gradient. *Aquatic Botany, 165*, 103239.

Brzozowski, M., & Pełechaty, M. (2022). Overwintering of an endangered charophyte during milder winters in Central Europe enhances lake water quality. *Limnologica, 92*, 125944.

Brzozowski, M., Pełechaty, M., & Bogawski, P. (2022). A winner or a loser in climate change? Modelling the past, current, and future potential distributions of a rare charophyte species. *Global Ecology and Conservation, 34*, e02038.

Brzozowski, M., Pełechaty, M., & Pietruczuk, K. (2018). Co-occurrence of the charophyte *Lychnothamnus barbatus* with higher trophy submerged macrophyte indicators. *Aquatic Botany, 151*, 51–55.

Casanova, M. T., García, A., & Feist, M. (2003). The ecology and conservation of *Lychnothamnus barbatus* (Characeae). *Acta Microbiologica Sinica, 20*, 118–128.

Corillion, R. (1957). Les Charophycées de France et d'Europe Occidentale. *Bulletin de la Société Scientifique de Bretagne, 32*(Hors série I & II), 1–499 [in French].

Dąmbska, I. (1964). *Charophyta—ramienice*. Państwowe Wydawnictwo Naukowe, Warsawa [in Polish].

Gąbka, M., Owsianny, P. M., & Burchardt, L. (2010). The influence of co-occuring vegetation and habitat variables on distribution of rare charophyte species *Lychnothamnus barbatus* (Meyen) in lakes of western Poland. *Polish Journal of Ecology, 58*, 13–25.

Haas, J. N. (1994). First identification key for charophyte oospores from central Europe. *European Journal of Phycology, 29*, 227–235.

Han, F. H. (1974). A preliminary account of the Characeae from the Western Szechuan plain. *Acta Phytotaxonomica Sinica, 7*, 155–190.

Horn af Rantzien, H. (1959). Recent charophyte fructifications and their relations to fossil charophyte gyrogonites. *Arkiv för Botanik, 4*, 167–332.

Iakushenko, D., & Borysova, O. (2012). Plant communities of the class *Charetea* Fukarek ex Krausch 1964 in Ukraine: An overview. *Biodiversity Research and Conservation, 25*, 75–82.

Karczmarz, K. (1967). Variabilité et distribution géographique de *Lychnothamnus barbatus* (Meyen) Leonh. *Acta Societatis Botanicorum Poloniae, 36*, 431–439. [in French].

Karol, K. G., Skawinski, P. M., McCourt, R. M., Nault, E., Evans, R., Barton, M. E., Berg, M. S., Perleberg, D. J., & Hall, J. D. (2017). First discovery of the charophycean green alga *Lychnothamnus barbatus* (Charophyceae) extant in the New World. *American Journal of Botany, 104*, 1108–1116.

Korsch, H. (2018). The worldwide range of the Charophyte species native to Germany. *Rostocker Meeresbiologische Beiträge, 28*, 45–96.

Korsch, H., Doege, A., Raabe, U. & van de Weyer, K. (2013). Rote Liste der Armleuchteralgen (Charophyceae) Deutschlands, 3. Fassung, Stand: Dezember 2012, *Hauss knechtia Beiheft, 17*, 1–32.

Krause, W. (1997). Charales (Charophyceae) Süßwasserflora von Mitteleuropa. In H. Ettl, G. Gärtner, H. Heynig, & D. Mollenhauer (Eds.) *Süßwasserflora von Mitteleuropa*. Fischer, Jena [in German].

Martín-Closas, C., Wójcicki, J. J., & Fonollá, L. (2006). Fossil charophytes and hydrophytic angiosperms as indicators of lacustrine trophic change. A case study in the Miocene of Catalonia (Spain). *Cryptogamie, Algologie, 27*, 357–379.

Mazzini, I., Gliozzi, E., Koci, R., Souli-Märsche, I., Zanchetta, G., Baneschi, I., Sadori, L., Giardini, M., Van Welden, A., & Bushati, S. (2015). Historical evolution and Middle to Late Holocene environmental changes in Lake Shkodra (Albania): New evidences from micropaleontological analysis. *Palaeogeography, Palaeoclimatology, Palaeoecology, 419*, 47–59.

Meyen, F. (1827). Beobachtungen und Bemerkungen über die Gattung *Chara*. *Linnea, 2*, 55–81.

Migula, W. (1889–1897). Fünfter Band: Die Characeen. In A. Fischer, E. Fischer, F. Hauck, G. Limpricht, C. Luerssen, W. Migula, H. Rehm, P. Richter, & G. Winter (Eds.), *Dr L. Rabenhorst's Kryptogamenflora von Deutschland, Oesterreich und der Schweiz*, 2nd ed. Eduard Kummer, Leipzig [in German].

Pełechaty, M., Brzozowski, M., & Pietruczuk, K. (2017). Overwintering and gyrogonite formation by the rare and endangered indicative macroalga *Lychnothamnus barbatus* (Meyen) Leonh. in eutrophic conditions. *Aquatic Botany, 139*, 19–24.

Pełechaty, M., Gąbka, M., Sugier, P., Pukacz, A., Chmiel, S., Ciecierska, H., Kolada, A., & Owsianny, P. M. (2009). *Lychnothamnus barbatus* in Poland: Habitats and associations. *Charophytes, 2*, 13–18.

Raabe, U., Pukacz, A., Peschel, T., & Müller, R. (2012). Die Bart-Glantzleuchteralge. *Lychnothamnus barbatus* (Meyen) Leonh. in Deutschland wieder aufgefunden. *Verhandlungen des Botanischen Vereins von Berlin und Brandenburg, 145*, 235–248.

Schneider, S. C., Trajanovska, S., Biberdžić, V., Marković, A., Talevska, M., Imeri, A., Veljanoska-Sarafiloska, E., Đurašković, P., Jovanović, K., & Cara, M. (2020). The Balkan Macrophyte Index (BMI) for Assessment of Eutrophication in Lakes. *Acta Zoologica Bulgarica, 72*, 439–454.

Siemińska, J., Bąk, M., Dziedzic, J., Gąbka, M., Gregorowicz, P., Mrozińska, T., Pełechaty, M., Owsianny, P. M., Pliński, M., & Witkowski, A. (2006). Red list of algae in Poland. In Z. Mirek, K. Zarzycki, W. Wojewoda, & Z. Szeląg (Eds.), *Red list of plants and fungi in Poland* (pp. 36–52). Szafer Institute of Botany, Polish Academy of Sciences.

Sinkevičienė, Z. (2010). Morphological variation in *Lychnothamnus barbatus* (Meyen) Leonh. in Lake Balsys (Lithuania). *Charophytes, 2*, 25–30.

Sinkevičienė, Z. & Urbaitė-Maževič, N. (2012). *Lychnothamnus barbatus* (Meyen) Leonh. rediscovered in shallow Lake Šventininkai (Lithuania) after 50 years. *Biodiversity: Research and Conservation, 25*, 91–96.

Sinkevičienė, Z. (2010). Morphological variation in *Lychnothamnus barbatus* (Meyen) Leonh. in Lake Balsys (Lithuania). *Charophytes, 2*, 25–30.

Sinkevičienė, Z. & Gudžinskas, Z. (2021). Revision of the Characeae (Charales, Charophyceae) species and their distribution in Lithuania. *Botanica, 27(2)*, 95–124.

Soulié-Märsche, I., (1989). *Etude comparée de gyrogonites de charophytes actuelles et fossiles et phylogénie des genres actuels*. Imprimerie des Tilleuls [in French].

Soulié-Märsche, I., & Martín-Closas, C. (2003). *Lychnothamnus barbatus* (Charophytes) from the Upper Miocene of La Cerdanya (Catalonia, Spain): Taxonomic and palaeoecological implications. *Acta Micropaleontologica Sinica, 20*, 156–165.

Sugier, P., Pełechaty, M., Gąbka, M., Owsianny, P. M., Pukacz, A., Ciecierska, H., & Kolada, A. (2010). *Lychnothamnus barbatus*: Global history and distribution in Poland. *Charophytes, 2*, 19–24.

Urbaniak, J., & Gąbka, M. (2014). *Polish Charophytes. An illustrated guide to identification*. Uniwersytet Przyrodniczy we Wroclawiu.

van de Weyer, K., & Schmidt, C. (2018). *Bestimmungsschlüssel für die aquatischen Makrophyten (Gefäßpflanzen, Armleuchteralgen und Moose) in Deutschland: Band 1: Bestimmungsschlüssel* (2nd ed., Vol. 119, pp. 1–180). Landesamt für Umwelt (LfU) Brandenburg, Potsdam, Fachbeiträge des LfU Brandenburg [in German].

van de Weyer, K. (2016). *Lychnothamnus barbatus*. In: Arbeitsgruppe Characeen Deutschlands Hrsg., *Armleuchteralgen. Die Characeen Deutschlands*. Springer, Berlin Heidelberg: 397–404 [in German].

Vesić, A., Blaženčić, J., & Šinžar-Sekulić, J. (2016). Contribution to knowledge of the charophytes (Charales) of Vojvodina (Serbia)—20 years after the first review. *Botanica Serbica, 40*, 237–247.

Wood, R. D. (1962). New combinations and taxa in the revision of Characeae. *Taxon, 11*, 7–25.

Wood, R. D. (1965). *Monograph of the Characeae*. J. Cramer.

Chapter 57
Nitella capillaris

Emile Nat

Scientific name: ***Nitella capillaris*** **(Krock.) J. Groves & Bull.-Webst., Brit. Charoph. 1: 96. 1920**

Neotype (Wood, 1965: 419): In an old fen ditch, The washes, near Sutton, Cambridgeshire/25th, Mai. 1885/A. Fryer and H. & J. G.[roves]/(NY 01090196); Groves and Groves, Charac. Brit. Exs. 30.

= *Chara capitata* Nees, Denkschr. Königl.-Baier. Bot. Ges. Regensburg 1(2): 80. 1818
Neotype (Wood, 1965: 419): In an old fen ditch, The washes, near Sutton, Cambridgeshire/25th, Mai. 1885/A. Fryer & H. and J. G.[roves] (NY 01090196); Groves & Groves, Charac. Brit. Exs. 30.

≡ *Nitella capitata* (Nees) C. Agardh, Syst. Alg.: 125. 1824.

≡ *Nitella syncarpa* (Thuill.) Chevall. var. *capitata* (Nees) Kütz., Phycol. Germ.: 256. 1845.

= *Tolypella coutinhoi* A. G. Cunha in Bull. Soc. Phyl. Fr. 12(8): 41–58. 1935
Neotype (Wood, 1965: 418): Palhota, Saboia, Baixo, Alentejo/Jan. 1934/Florival Lino Marnede (LISU).

Nomenclatural note: The priority for *Nitella capillaris* was recognised at the beginning of the twentieth century; before that *Chara capitata*, *Nitella capitata* or *Nitella syncarpa* var. *capitata* were used.

Taxonomic note: R. D. Wood and S. Cirujano regarded *Tolypella coutinhoi* as a synonym of *Nitella syncarpa,* but this has been rejected by R. Romanov after the study of type specimen of *T. coutinhoi.*

E. Nat (✉)
Dutch Information Center On Charophytes, Naturalis Biodiversity Center, P.O. Box 9517, 2300 RA Leiden, The Netherlands
e-mail: e.nat@kranswieren.nl

H. Schubert et al. (eds.), *Charophytes of Europe*,
https://doi.org/10.1007/978-3-031-31898-6_57

Local names: BELARUSIAN: Нітэла воласападобная; DANISH: Vår-Glanstråd; DUTCH: Kleinhoofdig glanswier; ENGLISH: Slimy-fruited stonewort; GERMAN: Haarfeine Glanzleuchteralge; ITALIAN: Pitera vermicolare; POLISH: Krynicznik włosowaty; RUSSIAN: Нителла волосовидная; SWEDISH: Vårslinke; UKRAINIAN: Нітела волосоподібна.

57.1 Morphology

Habitus: *Nitella capillaris* forms erect light to dark green plants and is slender, delicate and often lightly and annularly encrusted. The plants produce many small heads with gametangia in a mucus layer. Sterile branches are long, mostly 1 × divided with three dactyls (Figs. 57.1 and 57.2).

Size: Plants are up to 35 cm but mostly less than 10 cm high. Krause (1997) reported the exceptional growing height of 60 cm. Elongated and stocky plants with small or big heads were described as forms, depending on the water depth and organic contamination.

Main axis: The main axis is rather slender, the diameter is less than 0.9 mm. The internodes are 1–2 times as long as the branchlets.

Cortex, spine-cells, stipulodes: Absent.

Branchlets: Sterile whorls consist of 6–9 branchlets; the primary branchlet segments of sterile whorls are longer than the secondaries. Fertile whorls mostly have short branchlets and form dense small heads. The primary and secondary branchlet segments of fertile whorls are usually of equal length (Figs. 57.1b–e and 57.2b, d). Sterile branchlets are divided once. The primary and secondary branchlet segments (dactyls) are usually of nearly equal length. There are 2–4, one-celled, secondary branchlet segments. Gametangia are usually geminate or aggregated at fertile branchlet nodes, rarely at the base of whorls, in numerous small heads. Dactyls are single celled; dactyls of sterile whorls are elongate, and dactyls of fertile whorls are very short. The shape of the dactyls can be extremely variable: sometimes quite blunt but usually abruptly or gradually narrowing into a sharp point (Fig. 57.1f).

Bract-cells, bracteoles, bractlets: Absent/not observed for *Nitella capillaris*.

Sexuality: The species is dioecious. The gametangia are often, but not always, covered with a (thick) mucus layer.

Oogonia: The gametangia are usually geminate or aggregated at fertile branchlet nodes, enveloped in mucus, in numerous small heads. Oogonia 2–3, sometimes solitary (Figs. 57.1c, e and 57.2c). Table 57.1 gives an overview of the quantitative morphometry data. The coronula of the oogonium falls off quickly (Arbeitsgruppe Characeen Deutschlands 2016; Fig. 57.2c).

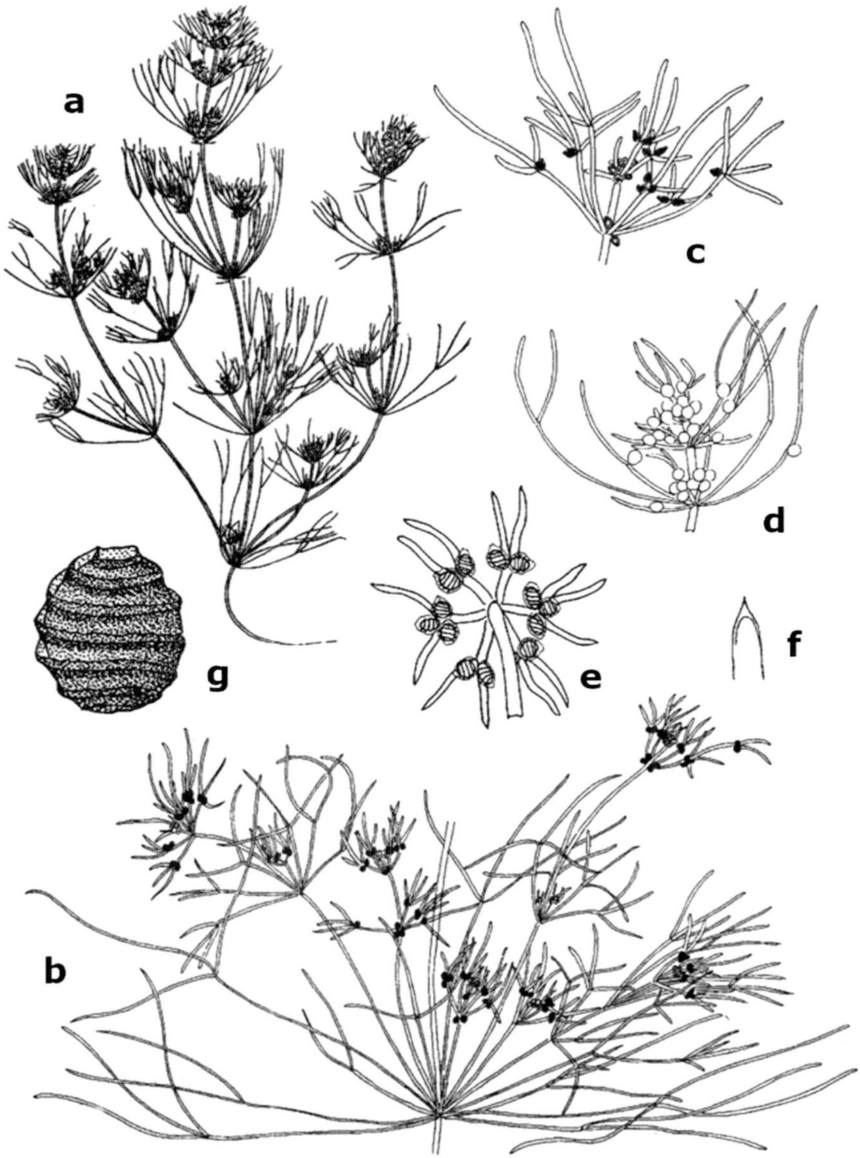

Fig. 57.1 *Nitella capillaris.* **a** Compact appearance; **b** Slender appearance; **c** Top of female plant; **d** Top of male plant; **e** Fertile whorl, female; **f** Apex of branchlet; **g** Oospore. Drawings: Werner Krause

Fig. 57.2 *Nitella capillaris,* habitus and details. **a** Habitus appearance; **b** Male fructifying whorl; **c** Oogonia; **d** Female fructifying whorl; **e** Oospore, SEM, side view; **f** Granulate oospore membrane, SEM. Notice the white stripes on the thallus in panels b and c; this is a calcium deposit as a product of photosynthesis. Panels e and f depicting an oospore of a specimen collected by Juliet Hawkins (JKH), 20–5–2019, Bramfield Park, Suffolk, England, w068. Photographs: **a**, **c**—Emile Nat; **b**, **d**—Klaus van de Weyer; **e**, **f**—Michelle T. Casanova

Oospores: The oospores of *Nitella capillaris* have 5–8 prominent, sharply flanged ridges (see Table 57.2 and Fig. 57.2e). For the colour of ripe oospores the authors give a range from light brown to black brown. For width (270–490 µm) and the length is mostly 280–620 µm, however the values given by different authors vary greatly. It is not clear whether the measurements are made on fresh or dried material. The oospore membrane is figured with linear grains (Fig. 57.2f).

Antheridia: The antheridia are relatively large and often solitary, reaching a diameter up to 1000 µm. The colour is yellow–orange and the mucus layer, if present, is mostly

Table 57.1 Oogonia morphometry

Author	Length	Width	Coronula length	Coronula width	Number of helical stripes
Migula (1889–1897)	420–480	–	26–80	–	6
Groves and Bullock-Webster (1920)	550–575	425–450	25–35	–	7
Corillion (1957a)	550–576	400–500	25–36	–	–
Compère (1992)	450–550	400–450	–	–	5–6
Krause (1997)	450–600	400–500	40–50	60–70	7–8
van Raam (1998)	up to 550	up to 450	–	–	–
Cirujano et al., (2008)	425–525	425–450	42–49	–	–
Caisová and Gąbka (2009)	465–575	425–450	–	–	–
Bailly and Schaefer (2010)	400–500	–	–	–	5–6
Urbaniak and Gąbka (2014)	460–580	425–490	–	–	–
Schou et al., (2017)	400–600	400–500	–	–	5–8
Total range	**400–600**	**400–500**	**25–50**	**60–70**	**5–8**

Length in the first column excludes the coronula, its length and width given separately; all values for length and width in μm

is thicker than the oogonia envelope. See Table 57.3 for an overview of the published diameter ranges.

Bulbils: It is not known if *Nitella capillaris* forms bulbils.

57.2 Phenology and Reproduction

The oospores germinate in Autumn. Depending on the region, mature plants can be found from March–May. *Nitella capillaris* produces gametangia in Spring (March–May) and ripe oospores in June–July after which the plant usually dies and other species take over. In the Mediterranean area ripe gametangia can be found in March–April (Mouronval et al., 2015). Sometimes the oospores germinate in June–July and create a second generation. These plants can be found till September with ripe gametangia.

Table 57.2 Oospore morphometry

Author	Length	Width	Number of striae	Colour
Migula (1889–1897)	280–360	270–350	6	Maroon-black
Groves and Bullock-Webster (1920)	375–400	325–375	6	–
Corillion (1957a)	375–401	325–375	8	–
Compère (1992)	–	–	5–6	Dark brown-blackish
Cirujano et al., (2008)	325–400	285–375	–	–
Caisová and Gąbka (2009)	300–400	–	5–6	Dark brown
Krause (1997)	300–400	280–380	5–6	Black brown
van Raam (1998)	up to 400	up to 300	–	–
Dilger (2010)	350–620	310–490	6–7	Brown-black brown
Bailly and Schaefer (2010)	300–400	–	5–6	–
Urbaniak and Gąbka (2014)	315–410	285–375	–	(light)brown
Schou et al., (2017)	280–620	280–490	5–8	Brown
Total range	**280–620**	**270–490**	**5–8**	

All values for length and width in μm

Table 57.3 Antheridia morphometry

Author	Diameter (μm)
Migula (1889–1897)	Up to 600
Groves and Bullock-Webster (1920)	600
Corillion (1957a)	Up to 600
Compère (1992)	600–700
Cirujano et al., (2008)	500–1000
Caisová and Gąbka (2009)	Up to 600
Krause (1997)	600
van Raam (1998)	Up to 600
Bailly and Schaefer (2010)	600–700
Urbaniak and Gąbka (2014)	600–735
Schou et al., (2017)	500–1000
Total range	**500–1000**

57.3 Habitat Conditions and Ecology

Habitat type: *Nitella capillaris* is a typical freshwater species and grows in sunny peat, sand and clay pits, ditches, pools and fishponds. The water in which *N. capillaris* grows, can be low in calcium, particularly places that are recently excavated (Krause, 1997). This species is also found in waters with a wider Ca-range (Doege et al., 2016). *Nitella capillaris* has not been recorded from brackish water. It is rarely found in deep or running waters and large lakes. *Nitella capillaris* often occurs in monospecific stands or grows in habitats with few other submerged species. In freshly excavated waterbodies *Nitella capillaris* can appear rapidly and mostly it is not found the year after, but reappears when the waterbody is cleared again. Populations of this species in deeper lakes can be more stable.

Sediment: *Nitella capillaris* grows on sand, peat and clay substrates (van Raam, 1998).

Nutrient conditions: *Nitella capillaris* prefers oligotrophic-mesotrophic waters but is found in the trophic range oligotrophic-eutrophic waters. Total phosphorus measured at sites containing *Nitella capillaris* reaches up to 0.11 mg/l (Doege et al., 2016). The concentration of Total Nitrogen has a known maximum level of 2263 mg/l.

Water chemistry: Some authors (Krause, 1997; Migula, 1889–1897) describe *Nitella capillaris* as growing in a small habitat niche, but this species can occur in a great number of water types and have a wide ecological tolerance. With respect to conductivity, *N. capillaris* is a freshwater species. Minimum conductivity of *N. capillaris* habitats were found at 171 μS/cm and maximum values measured were 2600 μS/cm (Doege et al., 2016). The range for Ca^{2+} is 7.9–42.1 mg/l; Cl^- is 8–200 mg/l and total hardness ranges between 9 and 25° dH (Doege et al., 2016).

Depth range: *Nitella capillaris* is found in shallow ditches, pools, fens and other small waterbodies at up to 1.5 m depth. In natural lakes and excavation waters this species can be found in the zone 4–8 m deep water. In Lake Weilerhofer, Germany, *N. capillaris* is found at a depth of 11 m (Korte et al., 2016a).

Associated species: *Nitella capillaris* mostly grows in monospecific stands especially in freshly excavated small waterbodies. It is the character species of the *Nitelletum capillaris* and the *Lemno-Nitelletum capillaris* and associate species of the *Tolypelletum proliferae*. Apart from that it can occur with vascular plants and the following charophyte species: *Chara contraria, C. globularis, C. hispida, C. virgata, C. vulgaris, Nitella flexilis, N. mucronata, N. opaca, N. syncarpa, N. tenuissima, N. translucens, Nitellopsis obtusa, Tolypella glomerata, Sphaerochara intricata* and *S. prolifera.*

57.4 Variability

Nitella capillaris is a (moderately) variable species in which the different growth forms more or less merge (Migula, 1889–1897). The presence or absence of mucus is a variable character of this species, which should not be used for determination. Elongated and stocky plants, with small and bushy heads were described as forms (e.g., Migula, 1889–1897). The variation depends on differences in water depth and the trophic state of the water.

57.5 Identification and Risk of Confusion

Nitella capillaris and *N. syncarpa* have similar characteristics. Both species can form mucus layers over the gametangia. Some authors mention the differences in thickness of the mucus layer: *N. syncarpa* has a thicker layer than *N. capillaris*. This is regarded as a subjective characteristic. Collections in alcohol don't show this feature because the mucus will dissolve in alcohol.

The oogonia of *Nitella capillaris* can be found at bifurcate internodes of the branchlets. The oogonia of *Nitella syncarpa* are situated in clusters of 2–4 at the nodes of unbranched branchlet segments.

The species may also be distinguished by their oospores. *Nitella capillaris* have oospores with six prominent sharply flanged ridges, whereas the oospores of *Nitella syncarpa* bears only faint, low striae.

The membrane of the ripe oospores looks different for these species: the oospores of *N. syncarpa* show no structure (smooth) or are slightly granulated and the oospores of *N. capillaris* are decorated with minute non-contiguous rounded tubercles (see Fig. 57.2f).

What can also help is in which period of the year the species is collected. *Nitella capillaris* can be found in spring (March–May) and falls apart during the summer. *N. syncarpa* develops in summer and can be found in autumn with ripe gametangia. There is an overlap period during the summer when both species can occur. In Lake Silbersee, Ratingen, Germany, *Nitella capillaris* and *N. syncarpa* can grow together at the same time during the summer: *N. capillaris* beneath the thermocline and *N. syncarpa* in shallow water (K. van de Weyer pers. comm.).

57.6 Distribution

Nitella capillaris is not rare in Europe. It can be found from the south of Sweden into north Africa and from Portugal into Ukraine. Figure 57.3 with the distribution of *Nitella capillaris* in Europe is made by Heiko Korsch. There are many locations in

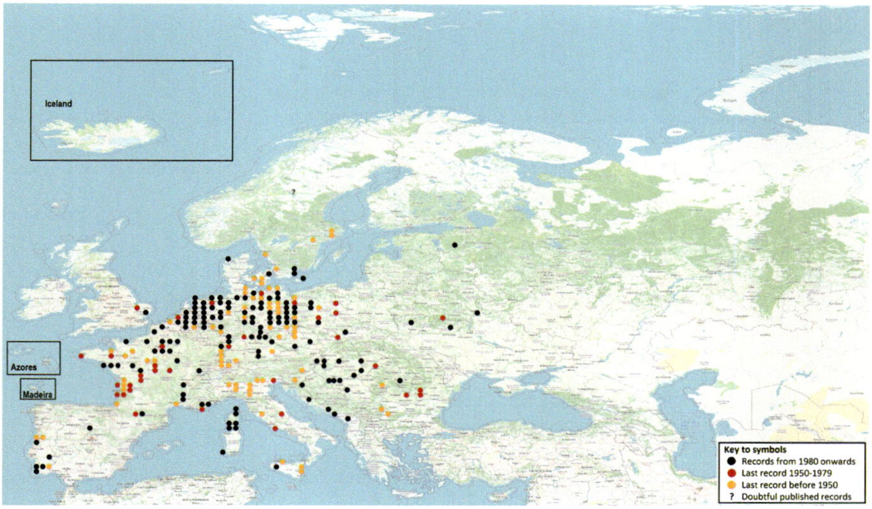

Fig. 57.3 European distribution of *Nitella capillaris*. For worldwide distribution see Korsch (2018). Black dots are records from 1980 on, red dots from 1950 to 1979 and yellow dots before 1950; unconfirmed/doubtful published records (if applicable) are marked with "?" on top. Map prepared by Heiko Korsch

Sweden, Germany, Poland and northern Italy where this species was last seen before 1950. It recently reappeared in Britain.

57.7 Status and Threats

The status of *Nitella capillaris* is assessed as 'Vulnerable' or 'Endangered' in most national Red Lists in Europe. It is threatened especially in small waterbodies that are not cleaned/excavated regularly. Populations in deeper water seem relatively stable. Deferred maintenance of ditches, pools and fens can be seen as a threat to the occurrence of *Nitella capillaris* in its entire area. During early spring this species can be found in small waterbodies. In that period only a few people are looking for charophytes. For this reason the distribution status of *N. capillaris* might be unrealistic.

57.8 Remarks

Nitella capillaris is a species which develops in the period March–May especially at freshly excavated sites. It is possible these sites are over-looked or researched later in the year resulting in missing this species.

References

Agardh, C. A. (1824) *Systema Algarum Berling* [in Latin]
Arbeitsgruppe Characeen Deutschlands (2016). *Armleuchteralgen: Die Characeen Deutschlands*. Springer [in German].
Bailly, G., & Schaefer, O. (2010). *Guide illustré des Characées du nord-est de la France*. Conservatoire Botanique National de Franche-Comté. [in French].
Caisová, L., & Gąbka, M. (2009). Charophytes (Characeae, Charophyta) in the Czech Republic: Taxonomy, autoecology and distribution. *Fottea, 9* 1–43.
Cirujano, S., Cambra, J., Sánchez Castillo, P. M., Meco, A., & Flor Arnau, N. (2008). *Flora ibérica. Algas continentales. Carófitos (Characeae)*. Real Jardín Botánico [in Spanish].
Compère, P. (1992). *Charophytes. Flore pratique des algues d'eau douce de Belgique*. Jardin botanique national de Belgique. [in French].
Corillion, R. (1957a). Les Charophycées de France et d'Europe Occidentale. *Bulletin de la Société Scientifique de Bretagne* 32, (Hors série I & II), 1–499 [in French].
Cunha, A. G. (1935). Quelques espèces nouvelles de Characées du Portugal. *Bulletin de la Société Portugaise des Sciences Naturelles 12*, 41–58 [in French.
Dilger, M. (2010). Merkmale zur Erkennung von *Nitella-* und *Tolypella-*Arten der Characeaen [sic] über Oosporen aus Sedimenten oder von Pflanzen. *Rostocker, Meeresbiologische Beiträge, 23*, 40–59 [in German].
Doege, A., van de Weyer, K., Becker, R., & Schubert, H. (2016). Bioindikation mit Characeen. In Arbeitsgruppe Characeen Deutschlands, *Armleuchteralgen: Die Characeen Deutschlands:* (pp. 97–138). Springer [in German].
Groves, J., & Bullock-Webster, G. R. (1920). *The British Charophyta. Volume I. Nitelleae*. Ray Society.
Korsch, H. (2018). The worldwide range of the Charophyte species native to Germany. *Rostocker Meeresbiologische Beiträge 28*, 45–96.
Korte, E., Pätzold, F., & Becker, R. (2016a). *Nitella capillaris*. In Arbeitsgruppe Characeen Deutschlands, *Armleuchteralgen: Die Characeen Deutschlands:* (pp. 405–412). Springer [in German].
Krause, W. (1997). Charales (Charophyceae) Süßwasserflora von Mitteleuropa. In H. Ettl, G. Gärtner, H. Heynig, & D. Mollenhauer (Eds.), *Süßwasserflora von Mitteleuropa 18*. Fischer, Jena [in German].
Migula, W. (1889–1897). Fünfter band: Die Characeen. In A. Fischer, E. Fischer, F. Hauck, G. Limpricht, C. Luerssen, W. Migula, H. Rehm, P. Richter, & G. Winter (Eds.), *Dr. L. Rabenhorst's Kryptogamenflora von Deutschland, Oesterreich und der Schweiz*, ed. 2. Eduard Kummer, Leipzig [in German].
Mouronval, J. B., Baudouin, S., Borel, N., Soulié-Märsche, I., Klesczewski, M., & Grillas, P. (2015). *Guide des Characées de France méditerranéenne*. Office national de la Chasse et de la Faune Sauvage. [in French].
van Raam, J. C. (1998). *Handboek Kranswieren*. Chara boek, Hilversum [in Dutch].
Schou, J. C., Moeslund, B., Båstrup-Spohr, L., & Sand-Jensen, K. (2017). *Danmarks vandplanter*. BFN's, Klitmøller [in Danish].
Urbaniak, J., & Gąbka, M. (2014). *Polish Charophytes. An illustrated guide to identification*. Uniwersytet Przyrodniczy we Wroclawiu, Wrocław.
Wood, R. D. (1965). *Monograph of the Characeae*. J. Cramer.

Chapter 58
Nitella confervacea

Klaus van de Weyer and Nick Stewart

Scientific name: ***Nitella confervacea*** **(Bréb.) A. Braun ex Leonh. in Lotos 13: 146. 1863**

≡ *Nitella gracilis* (Sm.) C. Agardh var. *confervacea* Bréb., Fl. Normandie, ed. 2: 338. 1849.
Neotype (Gregor et al., 2012): Argentan / Jt. 1837 / Brébisson (PC 719925)

= *Nitella nordstedtiana* H. Groves and J. Groves in J. Bot. 28: 66. (1890), n. nov. pro *Chara batrachosperma* Rchb. 1830

= *Chara batrachosperma* Rchb., Iconogr. Bot. Pl. Crit. 8: 794. 1830, n. ill. non Thuill. 1799.
Lectotype (Gregor et al., 2012: 394): Reichenbach 1830, plate DCCXCIV, fig. 1070.

Nomenclatural notes: The epithet *batrachosperma* has been used frequently for this species but the reasons for its illegitimacy are set out in detail by Groves and Groves (1890), Wood (1965) and Gregor et al. (2012). *Chara batrachosperma* was first used by Weis (1770) for a taxon that is clearly not this species. It was attributed to *Batrachospermum moniliforme* by Groves and Groves (1890) and, if not the latter species, it certainly belongs to the genus *Batrachospermum.* In any case the name is illegitimate as Weis (1770) equates his taxon with the pre-existing *Conferva gelatinosa* L.

The name was used again by Thuillier (1799) based on material from near Paris. This was subsequently identified by Braun et al. (1857–1878) as *Chara vulgaris,* a conclusion confirmed by Gregor et al. (2012). Agardh (1824) made the combination

K. van de Weyer (✉)
lanaplan, Lobbericher Str. 5, Nettetal, Germany
e-mail: klaus.vdweyer@lanaplan.de

N. Stewart
4 Hexton Road, Glastonbury BA6 8HL, United Kingdom

H. Schubert et al. (eds.), *Charophytes of Europe*,
https://doi.org/10.1007/978-3-031-31898-6_58

Nitella batrachosperma based on Thuillier's name, therefore making the combination unavailable for the species under discussion here.

The first use of the epithet *batrachosperma* (as a species of *Chara*) for the current species was by Reichenbach (1830). Reichenbach's *Chara batrachosperma* was then combined as *Nitella batrachosperma* by Braun (1847). However, when it became obvious that *Nitella batrachosperma* was not available due to its prior usage; Groves and Groves (1890) coined the name *Nitella nordstedtiana*.

Nitella confervacea was described in 1863 from material collected by Brébisson from north-western France and was treated as distinct from *N. batrachosperma*/*N. nordstedtiana* by Braun (1867), Braun and Nordstedt (1882) and Migula (1889–1897). However, after having studied populations in north-western France, Hy (1905) concluded that the two were synonymous although he maintained them as subspecies (see discussion under variability below). Subsequent authors have treated them as conspecific, although Corillion (1946, 1957a) followed Hy by treating them as taxa below specific rank. The synonymy of the two taxa was also confirmed by Gregor et al. (2012). Nevertheless, the name *N. batrachosperma* has continued in usage into recent times in some treatments, e.g., Krause (1997) and Bailly and Schaefer (2010).

Local names: DUTCH: Dwergglanswier; ENGLISH: Least Stonewort; ESTONIAN: Väike nitell; FINNISH: Mörkslinke; FRENCH: Nitelle conferve; GERMAN: Zwerg-Glanzleuchteralge; LITHUANIAN: Kurklinis menturdumblis; NORWEGIAN: Dvergglattkrans; POLISH: Krynicznik najmniejszy; RUSSIAN: Нителла конферная; SWEDISH: Dvärgslinke; UKRAINIAN: Нітела конферна.

58.1 Morphology

See Figs. 58.1 and 58.2.

Habitus: *Nitella confervacea* is the smallest of the charophytes. It is mid to dark green and in deeper water it is typically elongate with the whorls forming ball-like clusters separated by filamentous internode sections (moniliform). In shallower water the internodes shorten towards the tip of the main axis so that the upper 2–3 branchlet clusters become semi-continuous and head-like. Occasionally whorls are differentiated into lax sterile whorls and dense heads (in the true sense) of fertile whorls (e.g. see illustration in Bailly and Schaefer, 2010).

This species normally has little or no encrustation and when present this is often restricted to the dactyl tips.

Size: *Nitella confervacea* is a tiny plant, normally 2–6 cm tall. Occasionally larger plants reach 10 cm. One specimen seen was 60 cm tall (F. Pätzold pers. comm.).

Main axis: Main axis diameter is about 0.15–0.3 mm. The lower internodes are 0.5–1.5 cm long, 1–4 times the length of the branchlets. The upper internodes are shorter and at the axis tip may become equal to the whorls, so that branchlet clusters become more or less continuous. In some cases, the fertile whorls are contracted into heads.

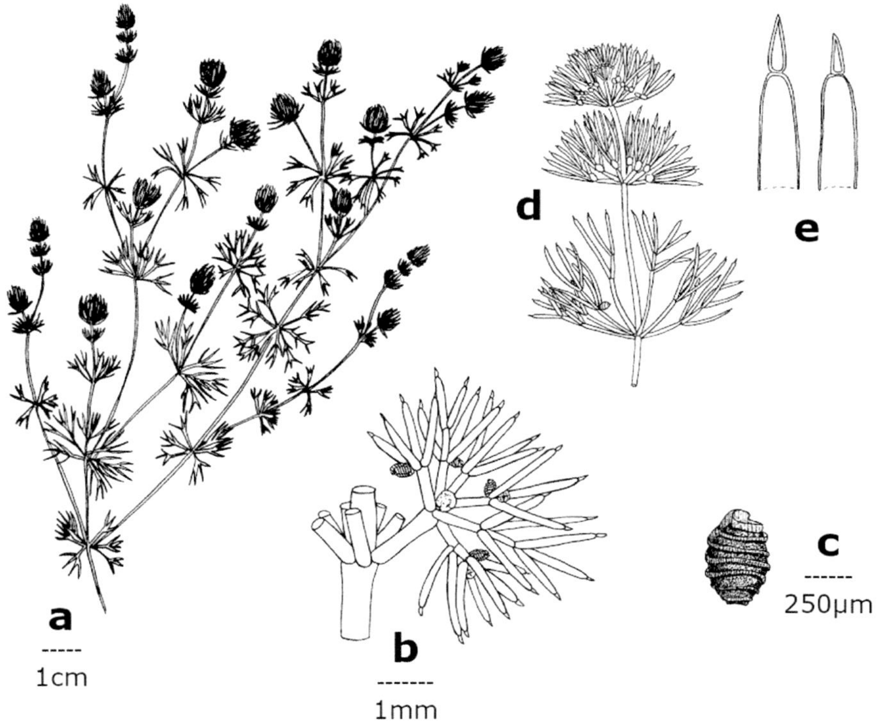

Fig. 58.1 *Nitella confervacea*. **a** Habitus; **b** Branchlet with gametangia; **c** Oospore; **d** Axis tip; **e** Terminal cells of a branchlet. Drawings after Werner Krause, modified by Dagmar Wassong in van de Weyer and Schmidt (2018)

Cortex: *Nitella confervacea* is ecorticate.

Spine-cells: Absent.

Stipulodes: Absent.

Branchlets: *Nitella confervacea* has 6–8 branchlets per whorl. The fertile branchlets are 0.2–0.5 cm long, 1–2 (3) times divided and composed of 3–5 cells. The sterile branchlets are normally similar to the fertile whorls but in some plants with heads they may be up to 2 cm long and only once divided. The dactyls are 2-celled with the end cell narrower and much shorter than the penultimate cell, forming an apiculate tip. The end cell is normally 50–80 μm long and sharply acute but very rarely more elongate (e.g. see Hy, 1905). Two examples with some 3-celled dactyls are described by Hy (1905, 1914) but these seem to be merely aberrant forms.

Bract-cells, bracteoles, bractlets: Absent.

Sexuality: The species is monoecious.

Fig. 58.2 *Nitella confervacea*, details. **a** From a gravel pit near Frankfurt, Germany; **b** Habitus, from a gravel pit near Roppenheim, France; **c** *S*hoot, from a gravel pit near Roppenheim, France; **d** Habitus; **e** Dactyls; **f** Oogonium. Photographs: **a**, **b**, **c**—Klaus van de Weyer; **d**—Frank Pätzold, **e**, **f**—Chris Carter

Gametangia: The gametangia are regularly at the first branchlet furcation or rarely also at the second branchlet node of branchlets that are 3 times divided. In some cases, the gametangia are enveloped in mucus but more usually mucus is lacking (despite the implication in the old epithet *batrachosperma*).

Oogonia: The oogonia are 300–450 μm long (without coronula) and 220–350 μm wide (without coronula). The oogonia have (5)6–10 convolutions. The coronula is 15–50 μm long and 40–67 μm wide (Table 58.1).

Oospores: The oospores are grey, reddish- or yellowish-brown or black. The oospores are 200–350 μm long, 200–300 μm wide and have 6–9 striae (Table 58.2 and Fig. 58.3). There are differences in the description of the oospore membrane between different sources, but this seems to be due to the level of magnification. The membrane appears granulate under light microscopy (Fig. 58.4a, b). However, under electron microscopy the surface is spongy with ropey fibres but with smooth flanges (Fig. 58.4c, d, M. T. Casanova pers. comm.).

Antheridia: The diameter of the antheridia is 120–250 μm (Table 58.3).

Bulbils: Absent.

Table 58.1 Oogonia morphometry

Author	Length	Width	Coronula length	Coronula width	Number of helical stripes
Migula (1889–1897)	290–330	220–270	–	–	6–7
Groves and Bullock-Webster (1920)	375–450	300–350	15–25	–	8–9
Corillion (1957a)	375–450	300–350	15–25	–	6–8
Krause (1997)	375–450	275–350	25–50	40–50	7–9
van Raam (1998)	< 440	< 350	–	–	6–9
Cirujano et al., (2008)	325–400	285–350	42–49	–	–
Caisová and Gąbka (2009)	345–450	285–350	–	–	–
Bailly and Schaefer (2010)	300–350	–	–	–	6–8
Urbaniak and Gąbka (2014)	350–450	285–350	–	–	–
Langangen (2007a)	< 450	–	–	–	–
Wood (1965)	345–450	285–350	15–49	40–67	(5)8–10
Total range	**300–450**	**220–350**	**15–50**	**40–50**	**6–10**

Length in the first column excludes the coronula, its length and width are given separately; all values for length and width in μm

Table 58.2 Oospore morphometry

Author	Length	Width	Number of striae	Colour	Membrane
Bailly and Schaefer (2010)	200–250	–	6–8	–	Granulate and spongy
Caisová and Gąbka (2009)	225–300	–	–	Yellowish- or reddish-brown to dark brown	Granulate
Cirujano et al., (2008)	220–275	–	–	–	Reticulate-rugulate
Corillion (1957a)	–	200–250	6–8	–	–
Dilger (2010)	270–350	220–300	6–7	Grey, brown to black	–
Groves and Bullock-Webster (1920)	225–300	200–250	6–8	–	–
Haas (1994)	220–300	200–250	6–8(9)	Dark reddish-brown to black	Minutely granulate
Koistinen (2004a)	–	–	–	Yellowish-brown or reddish-brown to black	Granulate
Krause (1997)	250–350	200–250	6–8	Greyish-brown to black	Minutely granulate
Langangen (2007a)	–	–	–	Yellowish-brown to black	Granulate
Migula (1889–1897)	200–280	–	6–7	Dark reddish-brown to black	–
Mouronval et al. (2015)	–	–	6–8	–	Granulate and spongy
Urbaniak and Gąbka (2014)	220–335	210–260	–	Light brown to brown	–
van Raam (1998)	< 350	< 250	6–9	–	Granulate
Wood (1965)	–	–	6–8	Dull yellowish- or reddish-brown	Granulate
Total range	**200–350**	**200–300**	**6–9**		

Length and width in μm

58.2 Phenology and Reproduction

Nitella confervacea is normally an annual species. It germinates from early spring until autumn and in many situations, fructification takes place in summer. However, in west France (Hy, 1905; Corillion, 1957), Britain and Ireland fructification is usually

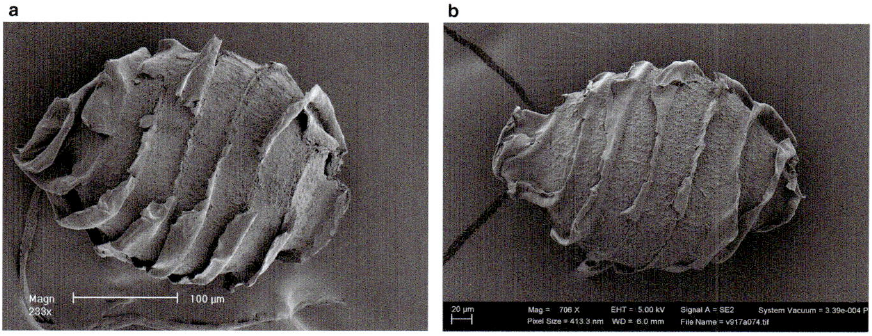

Fig. 58.3 Oospores of *Nitella confervacea*. **a** Near L'Etang de ???, France, 24/6/1909, G. R. Bullock-Webster (MEL); **b** Obbe, Isle of Harris, Hebrides, Scotland, 8/1887, C. F. King (BM 013,828,598). Photographs: Michelle T. Casanova

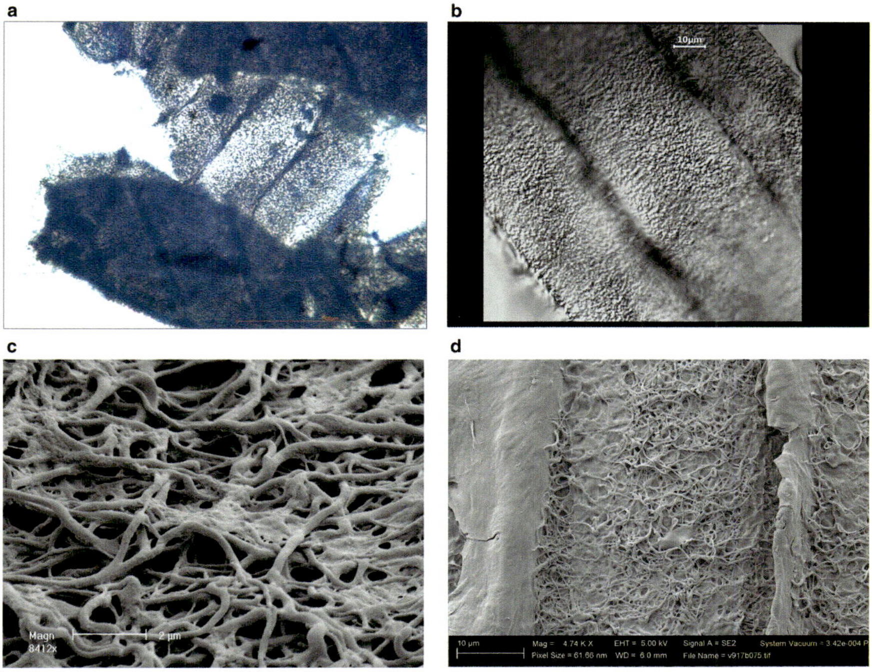

Fig. 58.4 Ornamentation of the oospore membrane of *Nitella confervacea*. **a** Light miscrosope, Upper Rhine valley, France; **b** Light miscrosope, Postel, Belgium; **c** SEM, Near L'Etang de ???, France, 24/6/1909, G. R. Bullock-Webster (MEL); **d** Obbe, Isle of Harris, Hebrides, Scotland, 8/1887, C. F. King (BM 013,828,598). Photographs: **a**—Klaus van de Weyer; **b**—Luc Denys; **c**, **d**—Michelle T. Casanova

Table 58.3 Diameter of the antheridia

Author	Diameter (μm)
Corillion (1957)	175–200
Koistinen (2004)	<250
Cirujano et al., (2008)	120–220
Caisová and Gąbka (2009)	<250
Bailly and Schaefer (2010)	120–220
Wood and Imahori (1964)	120–220
Krause (1997)	120–220
Langangen (2007)	<250
van Raam (1998)	<200
Total range	**120–250**

later, from late August to November and sometimes even December. This may be due to it commonly occurring in deeper water in these areas. However, the complete life cycle can also be very short, i.e. within two months (Denys and Packet, 2008). In certain circumstances, *Nitella confervacea* is able to survive in winter until March. It seems that apart from sexual reproduction there are plants that can develop from vegetative parts of plants that survive the winter.

58.3 Habitat Conditions and Ecology

Habitat type: *Nitella confervacea* is found in ditches, pools, temporary ponds, fens, oxbow lakes, artificial and natural lakes. There appear to be two contrasting habitat types: In many areas it is predominantly a species of soft waters, growing in sandy or peaty lakes at a variety of depths. *N. confervacea* also occurs in ditches through peat bogs and marshy pastures. However, in its stronghold in the Rhine valley it is a species of clear hard water lakes, particularly in gravel and sand pit lakes.

It is a pioneer species and a weak competitor. Being a small annual species, it is important that there is limited competition with other water plants, particularly for germination. In shallow water it probably requires regular disturbance, such as from livestock, wave-wash or boat traffic. It can also survive short periods of drought when embedded among detritus or other small plants (Koistinen, 2004). Indeed, it has been suggested (Hutchinson, 1975) that it is inhibited by bright light and benefits from a covering of silt.

In deeper water the populations seem to be more stable especially where there is an incomplete cover of larger vegetation in the lower part of the photic zone. Alternatively, it can also occur in 1–2 m depth in wind-exposed locations benefitting from the wave turbulence during storms keeping the vegetation more open, while conditions tend to be quieter during the summer growing period.

Sediment: *Nitella confervacea* has been reported from many different sediments including sand, gravel, silt, clay, mud, peat and combinations of these. The main requirement seems to be a stable substrate to anchor into and that is not easily stirred up. However, it can frequently be hidden in a covering of sediment or organic debris, resulting in the observation "Migula pouvait déjà affirmer, avec raison, que *N.* [*confervacea*] devait être recherchée davantage avec les mains qu'avec les yeux!" (Corillion, 1957) [Migula has already claimed, with reason, that it is better to look for *N. confervacea* with the hands than with the eyes!].

Nutrient conditions: There is limited data on the nutrient conditions of *Nitella confervacea.* It has been reported from oligotrophic and mesotrophic waters and probably requires clear, though sometimes peaty, water. The total phosphorus range is reported to vary between 5 and 60 μg/l. Single water chemistry measurement from 3 sites in Northern Ireland gave TP measurements 12–49 μg/l and TKN 0.2–0.9 mg/l (Northern Ireland Environment Agency data).

Water chemistry: There is rather little data on the water chemistry of *Nitella confervacea,* but it is a species of both soft and alkaline water. The pH range is reported to vary between 5.7 (Northern Ireland Environment Agency data) and 8.3 (Doege et al., 2016; Humberg and Beck, 2006) reflecting the different ecologies discussed above. The conductivity range in freshwater is reported to be normally between 270 μS/cm and 504 μS/cm (Denys and Packet, 2008; Doege et al., 2016; Humberg and Beck, 2006). There are also records in Germany from water bodies with conductivity >1000 μS/cm in waters influenced by potash mining. It is not often reported from brackish water but in the Baltic Sea salinity ranges are reported from 0 to 2.25 psu (Koistinen, 2004). Alkalinities measured from nine soft water sites in Scotland and Northern Ireland ranged from 50–600 μeq/l, (mean 270 μeq/l) (Northern Ireland Environment Agency and NatureScot data).

Depth range: *Nitella confervacea* is often reported from shallow water between 5 and 50 cm depth. In Britain and Ireland it sometimes occurs in similar depths (Stewart and Preston, 2000), but it is most commonly found between 1 and 2 m depth. In Germany it has been reported in depths up to 15 m (Pätzold et al., 2016). It seems mainly to be an opportunist for situations with open conditions, irrespective of depth.

Associated species: As *Nitella confervacea* occurs both in soft and hard waters there are different associated species. In soft waters *Nitella confervacea* occurs with *Lobelia dortmanna, Nitella opaca, N. translucens, Juncus bulbosus, Littorella uniflora, Chara virgata, C. fragifera, Myriophyllum alterniflorum, Potamogeton polygonifolius, Eriocaulon aquaticum, Hypericum elodes,* etc. In Britain and Ireland it frequently occurs in close association with the rare *Najas flexilis*, which, being one of the few aquatic vascular plants that are annual, shares its need for open substrate with limited competition.

In hard waters *Nitella confervacea* is often associated with other Charophytes like *Chara aspera, C. hispida, Nitella syncarpa, N. tenuissima, Nitellopsis obtusa, Sphaerochara intricata* and *Tolypella glomerata* (Humberg and Beck, 2006). From the Baltic Sea *Nitella confervacea* has been described from *Eleocharis acicularis* mats (Koistinen, 2004).

58.4 Variability

The distinction between *Nitella confervacea* from *N. nordstedtiana/N. batrachosperma* was based primarily on the presence of mucus around the gametangia of the latter and the colour and ridge prominence of the oospores. However, the presence of mucus has been found to be rather uncommon and perhaps present in both taxa. The oospores of *N. confervacea* are supposed to be pale, with weak ridges, rarely prolonged into delicate and colourless wings while those of *N. nordstedtiana/ batrachosperma* are brown or dark red, with more or less prominent ridges prolonged into rigid and often tinted flanges (Hy, 1905). However, Hy linked this to the ecology of the *N. confervacea* form, growing in deep water and therefore maturing later. He found material of the latter collected in November that had a range of oospores forming a gradation between the two taxa.

Migula (1889–1897) described three varieties of this species while Hy (1905, 1913, 1914) used seven names for different morphotypes. Characters used to separate these variations, in addition to those described in the previous paragraph, include size, axis diameter, presence of branchlets dividing 3 times, presence of some 3-celled dactyls, elongated end-cells of the dactyl and tendency to form heads. In Hy (1905) he presents his named variations as subspecies but in Hy (1913) he notes "comprenant plusieurs forms considérées souvent comme types distincts, mais rattachées les unes aux autres par de nombreux passages" [including many forms, often considered as distinct types but attached one to another by numerous intermediates.] Gregor et al. (2012) traced original specimens and/or illustrations of all of Migula's and some of Hy's named morphotypes and considered all of these to be part of the continuum of variation within *N. confervacea.*

Nevertheless, there does seem to be a marked distinction based on size and ecology which needs further investigation. The plants of soft water seem to be smaller, to a maximum of 5 cm tall and with whorl clusters to 0.3 mm diameter, while plants of hard waters can grow to 10 cm (rarely more) with fertile whorls to 0.5 mm diameter. Also the tendency to form heads of fertile whorls that are distinct in size from the sterile whorls may be restricted to plants in hard water.

58.5 Identification and Risk of Confusion

Nitella confervacea is normally recognised by its diminutive size but could be confused with small versions of other *Nitella* species. A summary of the key differentiating characters is presented in Table 58.4. This species is most likely to be confused with *Nitella tenuissima* which is nearest in size, also has 2-celled dactyls and often

Table 58.4 Key features of small *Nitella* species

	N. confervacea	*N. tenuissima*	*N. hyalina*	*N. gracilis*	*N. wahlenbergiana*	*N. mucronata*
Dactyls	2-celled	2-celled	2-celled	Frequent 3-celled	2-celled	2–3-celled
Gametangia location on branchlets	1st (2nd) node	2nd (3rd) node	All nodes	All nodes	(1st) 2nd node	All nodes
Oospore membrane ornamentation	Granulate	Reticulate	Granulate	Granulate	Granulate	Reticulate

shares the moniliform (beaded) habit. However, the latter bears its gametangia at the second or third furcation of the branchlets and the oospore membrane has reticulated ornamentation (under optical microscopy). It is also restricted to hard waters.

Nitella confervacea has branchlets with uniform length at each node, while N. *hyalina* has primary whorls and additional shorter accessory whorls. *Nitella wahlbergiana* is less slender and most gametangia are located on the second furcation. *Nitella gracilis* has longer, laxer branchlets which do not form heads and has many 3-celled dactyls. *Nitella mucronata* has an oospore membrane which is reticulate (under optical microscopy) compared to the granulate oospore membrane of *N. confervacea*. *Nitella mucronata* is also usually larger (main axis diameter >1 mm) and usually has some 3-celled fertile branchlets.

58.6 Distribution

See Fig. 58.5.

In Europe *Nitella confervacea* has been recorded from many countries but it appears to be sparsely distributed in most of these. This may be partly due to it being overlooked because of its small size but it does appear to be genuinely rare in much of its range. The exception is in Germany where it is locally common in gravel pits in the Rhine valley between Basel and Mainz (Korsch et al., 2008; Pätzold, 2011) and in field ponds in Brandenburg (Gregor et al., 2012; Pätzold et al., 2016). Its near-European range also extends to Algeria and Morocco (Muller et al., 2017).

Beyond this it appears to have a wide but disjunct distribution in eastern North America, southern Africa and south-eastern Asia (see Korsch, 2018). According to Corillion (1957) there are also records from Australia.

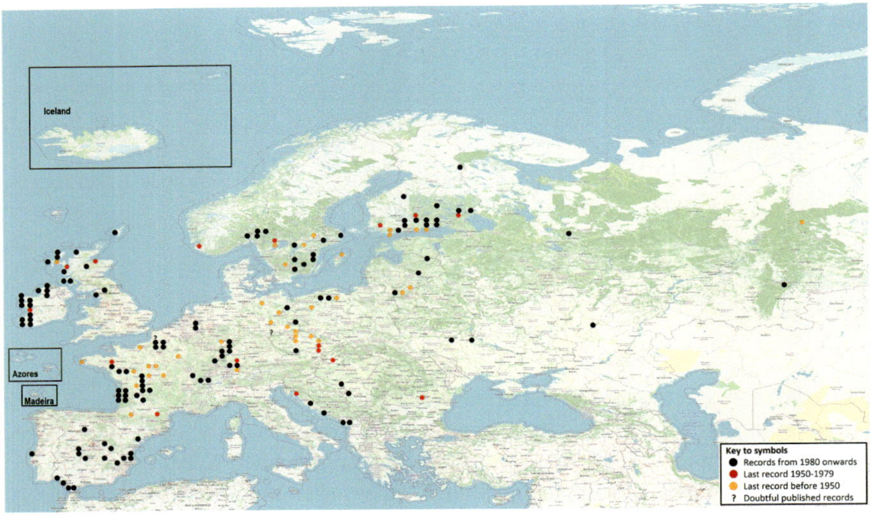

Fig. 58.5 European distribution of *Nitella confervacea*. For worldwide distribution see Korsch (2018). Black dots are records from 1980 on, red dots from 1950 to 1979 and yellow dots before 1950; unconfirmed/doubtful published records (if applicable) are marked with "?" on top. Map prepared by Heiko Korsch

58.7 Status and Threats

Because of its small size, this species requires open conditions. This means that it often requires regular disturbance, particularly when growing in shallow water. For example, cessation of grazing and the subsequent expansion of reed swamp is identified as a cause of its loss from sites on the Gulf of Finland (Koistinen, 2004) and successional changes were the cause of its disappearance from a site in Britain. In the absence of other disturbance, periodic clearance of vegetation may be necessary at some sites.

The other main threat is from nutrient enrichment, both through the encouragement of competition from filamentous algae and vascular plants but also through alteration of the sediment to softer, more mobile organic silt that is difficult to anchor into. Increases in turbidity are also a problem for populations in deeper water. In general, all measures to reduce/prevent eutrophication are recommended. This also includes maintenance of fish communities which are typical for the specific water body type.

Increased peatiness is another threat which has been observed in Ireland.

The current status of this species is difficult to assess due to its small size and it has almost certainly been overlooked at some sites. There have been lost sites documented in several countries. For example, Gregor et al. (2012) report that none of the eight sites reported in Corillion (1946) now had any suitable habitat. But there have also been new sites discovered across its range. Nevertheless, it seems genuinely to be a rare species through most of its range with the exception of the Rhine valley in Germany where it is locally frequent.

Nitella confervacea is listed in several Red Data Books: the Balkans (EN), Finland (VU), Germany (VU), Lithuania (CR), Poland (CR), Norway (VU), Sweden (VU) and Czech Republic (CR). *Nitella confervacea* is extinct in Switzerland. In both Britain and Ireland, it is regarded as Near Threatened (NT) as many of its sites are on the western and northern fringes where human pressures are less intense.

References

Agardh, C. A. (1824). *Systema Algarum*. Literis Berlingianus, Lundae [in Latin].

Bailly, G., & Schaefer, O. (2010). *Guide illustré des Characées du nord-est de la France*. Besançon: Conservatoire Botanique National de Franche-Comté [in French].

Braun, A. (1847). Uebersicht der schweizerischen Characeen. *Neue Denkschriften der Allgemeine Schweizerischen Gesellschaft für die Gesamten Naturwissenschaften, 10*, 1–23 [in German].

Braun, A. (1867). *Conspectus Systematicus Characearum Europaearum*. Dresden: 1–8) [in German; attachment to issue 3 of Rabenhorst's Characeae Exsiccatae].

Braun, A., & Nordstedt, C. F. O. (1882). Fragmente einer Monographie der Characeen. Nach den hinterlassen Manuscripten A. Braun's herausgegeben von Dr. O. Nordstedt *Abhandlungen der Königlichen Akademie der Wissenschaften Zu Berlin 1882*, 1–211 [in German].

Braun, A., Rabenhorst, L., & Stizenberger, E. (1857–1878). *Die Characeen Europa's in getrockneten Exemplaren*. Fasc. I-V: 1–121. Dresden [in French, German & Latin].

Caisová, L., & Gąbka, M. (2009). Charophytes (Characeae, Charophyta) in the Czech Republic: Taxonomy, autoecology and distribution. *Fottea, 9*, 1–43.

Cirujano, S., Cambra, J., Sánchez Castillo, P. M., Meco, A., & Flor Arnau, N. (2008). *Flora ibérica. Algas continentales. Carófitos (Characeae)*. Real Jardín Botánico. [in Spanish].

Corillion, R. (1946). Contribution à l'étude des Charophycées du Massif Armoricain *Bulletin de la Société Botanique de Bretagne, 21*, 77–96 [in French].

Corillion, R. (1957). Les Charophycées de France et d'Europe Occidentale. *Bulletin de la Société Scientifique de Bretagne, 32* (Hors série I & II), 1–499 [in French].

Denys, L., & Packet, J. (2008). *Nitella confervacea* en andere kranswieren (Charophyta) in de Langdonken en het Goor-Asbroek (Aarschot, Herselt, Hulshout; prov. Antwerpen). *Dumorteria, 95*, 18–27 [in Dutch].

Dilger, M. (2010). Merkmale zur Erkennung von *Nitella*- und *Tolypella*-Arten der Characeaen über Oosporen aus Sedimenten oder von Pflanzen. *Rostocker Meeresbiologische Beiträge, 23*, 40–59 [in German].

Doege, A., van de Weyer, K., Becker, R., & Schubert, H. (2016). Bioindikation mit Characeen. In Arbeitsgruppe Characeen Deutschlands, *Armleuchteralgen: Die Characeen Deutschlands*: 97–138. Springer, Berlin [in German].

Gregor, T., Raabe, U., & van Raam, J. (2012). Typification of *Nitella gracilis* (Sm.) C. Agardh var. *confervacea* Brébisson (= *N. confervacea* (Brébisson) A. Braun ex Leonhardi) with remarks on its nomenclatural history. *Cryptogamie Algologie, 33*, 391–398.

Groves, H., & Groves, J. (1890). Notes on the British Characeae for 1897–9. *Journal of Botany British and Foreign, 28*, 65–69, plate 296.

Groves, J., & Bullock-Webster, G. R. (1920). *The British Charophyta. Volume I. Nitelleae*. London: Ray Society.

Haas, J. N. (1994). First identification key for charophyte oospores from central Europe. *European Journal of Phycology, 29*, 227–235.

Humberg, B., & Beck, M. (2006). Der Makrophytenbestand in ausgewählten Baggerseen der Oberrheinaue. *LUBW, Oberirdische Gewässer, Gewässerökologie, 102*, 1–348 [in German].
Hutchinson, G. E. (1975). *A treatise on limnology Vol. III:. Limnological botany*. Wiley.
Hy, F. (1905). Sur le *Nitella confervacea* Braun. *Bulletin de la Société Botanique de France, 52*, 88–94 [in French].
Hy, F. (1913). Les Characées de France. *Bulletin de la Société Botanique de France, 60*, (sup 1): 1–47 [in French].
Hy, F. (1914). Les Characées de France. Note additionnelle. *Bulletin de la Société Botanique de France, 61*, 235–241 [in French].
Koistinen, M. (2004). *Nitella confervacea*. In H. Schubert, & I. Blindow (Eds.), *Charophytes of the Baltic Sea* (pp. 168–173). Gantner.
Korsch, H. (2018). The worldwide range of the Charophyte species native to Germany. *Rostocker Meeresbiologische Beiträge, 28*, 45–96.
Korsch, H., Raabe, U., & van de Weyer, K. (2008). Verbreitungskarten der Characeen Deutschlands. *Rostocker Meeresbiologische Beiträge, 19*, 57–108.
Krause, W. (1997). Charales (Charophyceae) Süßwasserflora von Mitteleuropa. In H. Ettl, G. Gärtner, H. Heynig & D. Mollenhauer (Eds.), *Süßwasserflora von Mitteleuropa 18*. Fischer, Jena [in German].
Langangen, A. (2007). *Charophytes of the Nordic countries*. Oslo: Saeculum.
Migula, W. (1889–1897). Die Characeen Deutschlands, Oesterreichs und der Schweiz: Unter Berücksichtigung aller Arten EuropasFünfter band: In L. Rabenhorst,'*Kryptogamenflora von Deutschland, Oesterreich und der Schweiz* (2nd ed.). Eduard Kummer, Leipzig [in German].
Mouronval, J. B., Baudouin, S., Borel, N., Soulié-Märsche, I., Klesczewski, M., & Grillas, P. (2015). *Guide des Characées de France méditerranéenne*. Office National de la Chasse et de la Faune Sauvage, Paris (211 pp.) [in French].
Muller, S. D., Rhazi, L., & Soulié-Märsche, I. (2017). Diversity and distribution of Characeae in the Maghreb (Algeria, Morocco, Tunisia). *Cryptogamie, Algologie, 38*, 201–251.
Pätzold, F., Korte, E., & Blindow, I. (2016). *Nitella confervacea*. In: Arbeitsgruppe Characeen Deutschlands, *Armleuchteralgen: Die Characeen Deutschlands*, 413–420. Springer, Berlin [in German].
Pätzold, F. (2011). Beobachtungen zur Verbreitung von Characeen in Baggerseen der Badischen Oberrheinaue. *Berichte der Botanischen Arbeitsgemeinschaft Südwestdeutschland, Beiheft, 3*, 37–42 [in German].
Reichenbach, H. G. L. (1830). *Iconographia Botanica, seu plantae criticae*. Lipsiae: Fridericus Hofmeister [in Latin].
Stewart, N. F., & Preston, C. D. (2000). Stoneworts of Coll and Tiree. *Botanical Journal of Scotland, 52*, 91–100.
Thuillier, J. L. (1799). *La flore des environs de Paris, ou distribution méthodique des plantes qui y croissent naturellement, faite après le systême Linnée, ...*, (2nd ed.). Auteur et H.L. Perronneau, Paris.
Urbaniak, J., & Gąbka, M. (2014). *Polish Charophytes. An illustrated guide to identification* (120 pp.). Wrocław: Uniwersytet Przyrodniczy we Wroclawiu.
van Raam, J. C. (1998). *Handboek Kranswieren*. Hilversum: Charaboek [in Dutch].
van de Weyer, K., & Schmidt, C. (2018). Bestimmungsschlüssel für die aquatischen Makrophyten (Gefäßpflanzen, Armleuchteralgen und Moose) in Deutschland: Band 1: Bestimmungsschlüssel, ed. 2. Landesamt für Umwelt (LfU) Brandenburg. *Potsdam, Fachbeiträge des LfU Brandenburg , 119*, 1–180 [in German].
Weis [Weiss], F. G. (1770). *Plantae cryptogamae florae Gottingensis*. Gottingae: Vandenhoeck [in Latin].
Wood, R. D. (1965). *Monograph of the Characeae*. Weinheim: Cramer.
Wood, R. D., & Imahori, K. (1964). *Iconograph of the Characeae*. Weinheim: Cramer.

Chapter 59
Nitella dixonii

Roman Romanov

Scientific name: ***Nitella dixonii*** **H. Groves & J. Groves in J. Bot. 53: 41. 1915**

Lectotype (Wood, 1965: 560): Ditch by roadside, Caldas de Monquiche, Algarve, Portugal/May 1911/H. W. Dixon (BM).

Local names: No local names known.

59.1 Morphology

See Figs. 59.1 and 59.2.

Habitus: Plants are green to brownish, unencrusted, small to medium-sized, thin and flexible, ecorticate, having lax appearance resembling *Nitella gracilis*, without heads and mucilage. The secondary and subsequent branchlet segments can be deciduous at the lowermost branchlets resulting in them appearing simple (Krause, 1997).

Size: Small to medium sized plant, up to 20 cm high.

Main axis: Moderately slender, 400–525 μm in diameter. The internodes are slightly shorter, to 2 times longer than the branchlets, up to 3.5 cm long.

Cortex, spine-cells, stipulodes: Absent.

Branchlets: 5–8 in a whorl, sterile and fertile are similar, up to 2.4 cm in length, 1–3 furcate. The primary segment is 0.4–0.5 x total branchlet length, secondary segments 4–7. They are simpler (mostly) furcate into 3–5 tertiary segments, sometimes further furcate into 2–4 quaternaries. The presence of quaternary branchlet segments was not confirmed by examination of the type (Wood, 1965). Branchlet segments are

R. Romanov (✉)
Nišice Ujtin Potok b.b., Dobra Voda, 85356 Bar, Montenegro
e-mail: streptophytes@gmail.com

H. Schubert et al. (eds.), *Charophytes of Europe*,
https://doi.org/10.1007/978-3-031-31898-6_59

Fig. 59.1 *Nitella dixonii.* **a** Upper part of female plant; **b** Upper part of tufted female plant; **c** Upper part of male plant; **d** Upper part of tufted female plant; **e** Fertile female branchlet with central secondary branchlet segment; **f** Four- and five-celled dactyls; **g, j** Ends of branchlets; **h, n, p** Branchlets; **i** Oospore **k** Oogonium; **l** Branchlet of female plant; **m** Geminate antheridia; **o** Elongated older male plant; **q** Older female plant with mostly first branchlet segments of branchlets remaining. Drawings: **a, b, c, g, i, j, k, l, m**—Groves and Groves (1915); **d, e, f, h, n, o, p, q**—Krause (1997)

unequal in length. The secondary branchlet segments are about 0.6 x the length of primary branchlet segments and more elongated if they are dactyls. The branchlets of female plants have a short central secondary branchlet segments which is absent in male plants. Dactyls can be present at each furcation but mostly at the ultimate and penultimate furcations. There are 2–5 dactyls, which are unequal, elongated, up to the length of primary branchlet segments, (2)4(5)-celled, appearing constricted between cells when pressed. The basal cell of a dactyl is the longest one, end cells are short, conical and acute, confluent to mucronate, evidently deciduous on lower branchlets.

Bract-cells, bracteoles: Absent.

Bractlets: Unknown.

Sexuality: Dioecious, without evident sexual dimorphism, but male plants do not have central secondary segments in their branchlets, mucus is absent.

Fig. 59.2 *Nitella dixonii*, details. **a–e** Whorls; **f** Antheridia at first branchlet furcations; **g, h** Nodal bulbils. Scale: 1 mm. The arrows indicate central secondary branchlet segments (**b**, **e**), solitary oogonium (**c**) and antheridium at first furcations of the branchlet (**f**). Scale: 1 mm. Photographs: Roman Romanov

Oogonia: Always solitary at second and third branchlet nodes, occasionally at the first node, terminal or lateral. The coronula is persistent, cells rows of equal length. Table 59.1 gives an overview of the published morphometric data. Few measurements of gametangia and oospores are actually available (Table 59.1).

Oospores: Light yellowish brown or black, widely ellipsoid, with prominent spiral ridges; oospore surface seems to be roughly anastomosing spongy elements in SEM (Fig. 59.3). Table 59.2 gives an overview of the available morphometric data.

Table 59.1 Oogonia morphometry

Author	Length	Width	Coronula length	Coronula width	Number of helical stripes
Groves and Groves (1915)	475–550	360–410	–	–	–
Cunha (1942)	475–550	360–410	20–25	–	10–11
Corillion (1957)	475–550	360–410	–	–	10–11
Wood and Imahori (1964)	(360)	(330)	25	49	10
Wood (1965)	(360)475–550	(330)360–410	~ 25	~ 49	~ 10
Krause (1997)	~500	~400	25	–	10
Total range	**360–550**	**330–410**	**20–25**	**~ 49**	**10–11**

Length in the first column is excluding the coronula, its length and width are given separately; all values for length and width in μm

Antheridia: Sessile, solitary or geminate at first and second branchlet nodes, the second antheridium is placed below the first one at the branchlet furcations (Groves and Groves, 1915; Fig. 59.1m). Table 59.3 gives an overview of the published morphometric data.

Bulbils: Nodal bulbils can be produced (Krause, 1997; Fig. 59.2g, h).

59.2 Phenology and Reproduction

The species reproduces by oospores as well as nodal bulbils.

59.3 Habitat Conditions and Ecology

Habitat type: The ecology of *Nitella dixonii* is poorly known; it grows in shallow water of small water bodies like ditches by roadside, ponds, water tanks and swamps. It was collected from a lagoon in a single case only.

Sediment: Unknown.

Nutrient conditions and water chemistry: *Nitella dixonii* seems to be a species that grows mainly in freshwater.

Depth range: *Nitella dixonii* seems to be shallow water species (Corillion, 1957).

Associated species: Unknown.

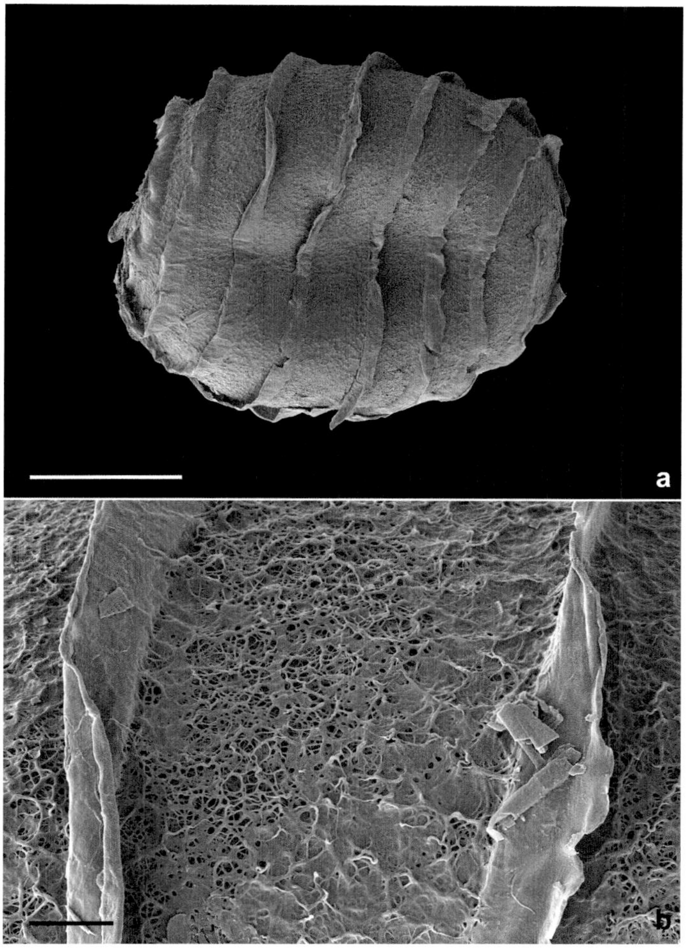

Fig. 59.3 Oospores of *Nitella dixonii*, type specimen, scanning electron microscopy. **a** General view; **b** Fossa surface. Scale: **a**—100 μm, **b**—10 μm. Photographs: Michelle T. Casanova

59.4 Variability

Nitella dixonii can have a lax appearance resembling *N. gracilis* or somewhat tufted because of condensed branchlets, more resembling *N. ornithopoda* (Groves and Groves, 1915). The branchlets can be lax or abbreviated resulting in different general habits (Fig. 59.1a, b, c, d, o).

Table 59.2 Oospore morphometry

Author	Length	Width	Number of striae	Colour
Groves and Groves (1915)	~320	~270(~230)	8–9	–
Cunha (1942)	320	230–270	8–9	Black
Corillion (1957)	320	270	8–9	Blackish
Wood and Imahori (1964)	(330)	(285)	7–8	Light yellowish brown, not fully ripe
Wood (1965)	320–330	230–285	7–9	Black
Krause (1997)	~325	250	7–9	Black
Total range	**320–330**	**230–285**	**7–9**	

All values for length and width in μm

Table 59.3 Antheridia morphometry

Author	Diameter (μm)
Groves and Groves (1915)	325–410
Cunha (1942)	325–410
Corillion (1957)	325–410
Wood and Imahori (1964)	285–315
Wood (1965)	285–410
Krause (1997)	285–400
Total	**285–410**

59.5 Identification and Risk of Confusion

Since its initial description *Nitella dixonii* has been described as similar to monoecious *N. ornithopoda*, differing in being dioecious (Groves and Groves, 1915). Cirujano et al. (2007, 2008) amalgamated these species names later under the name *N. ornithipoda.* However, the branchlets of *N. ornithopoda* frequently have a central secondary branchlet segment, as do female plants of *N. dixonii* (Fig. 59.11, e), but the frequency of this trait needs to be studied. It seems to be always absent in male plants, because it is replaced with an antheridium. The end cells of dactyls in *N. dixonii* are shorter than in *N. ornithipoda*, i.e. not three times longer than its width, usually much less (Groves and Groves, 1915), and they are acute to mucronate in contrast with frequently allantoid end cells of *N. ornithopoda*. The antheridia placement is useful for their delineation too, because they are present at the first furcation of the branchlet in the case of *N. dixonii*. Moreover, they are bigger than antheridia of *N. ornithopoda*, and their diameter is close to the diameter of the main axis. The gametangia of *N. ornithopoda* seem to rarely form at the first furcation of the branchlet because of central secondary branchlet segment presence. This trait was illustrated by Corillion (1957: Fig. 12a) and confirmed by specimen study (COI, checked by R. Romanov).

The oospore surface of *N. ornithopoda* is finely spongy in SEM, most similar to that found in *N. gracilis* (R. Romanov pers. observ.) and looks quite different to the roughly anastomosing spongy pattern known for *N. dixonii*. New studies of undoubtedly ripe oospores of this species are desirable. The statement about monoecy in *N. dixonii* in Krause (1997: 156) is evidently a misprint.

59.6 Distribution

See Fig. 59.4.

To date this species is known only from western and southern Portugal (Groves and Groves, 1915; Cunha, 1934, 1941, 1942, 1951; LISU, MA, checked by R. Romanov) and from a single recent locality in Spain (Huelva: Rivetehilos (Almonte). En charka de turbera dentro de brezal/tojal de *Erico ciliaris-Ulicetum minoris*, 29SPQB1210/08–05-2006/Leg. E. Sánchez Gullón & P. Weickert., det. R. Romanov, MA-Algae 11021; Cirujano et al., 2008: Fig. 104, male plant as *N. ornithopoda*). The total number of its localities in the world does not exceed 10 and some of them are lost (Krause, 1997).

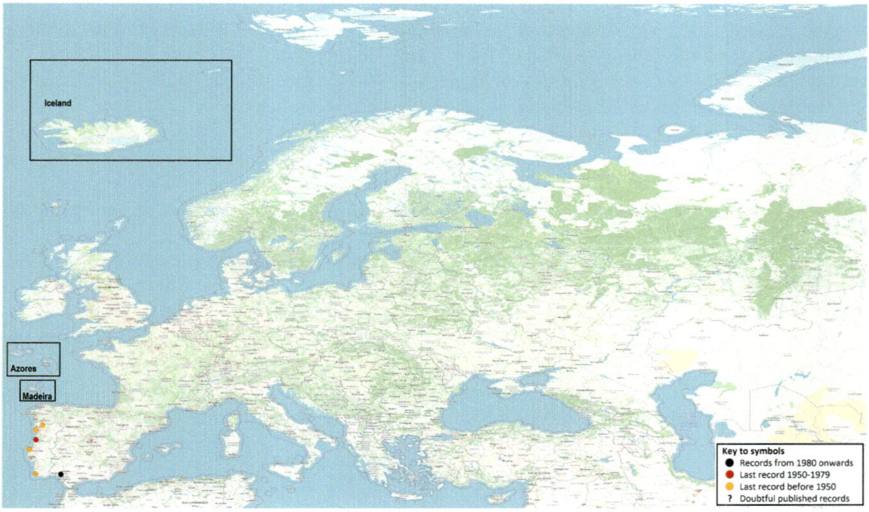

Fig. 59.4 Distribution area of *Nitella dixonii*. Black dots are records from 1980 on, red dots from 1950 to 1979 and yellow dots before 1950; unconfirmed/doubtful published records (if applicable) are marked with "?" superimposed. Map prepared by Heiko Korsch

59.7 Status and Threats

Nitella dixonii has been assessed as CR (Critically Endangered, European and Globally) in the European Red List (see Chap. 9). The evaluation according IUCN categories is problematic due to the absence of contemporary surveys of its localities. The main threat to this species could be eutrophication which is so for many charophytes. *N. dixonii* formally can be recognised as "Extinct" in Portugal because no records are available since 1938 (Cunha, 1942; LISU, checked by R. Romanov), "Critically Endangered" in Spain where it is known from single recent locality and "Critically Endangered" for the whole of its distribution.

Acknowledgements The curators of the herbaria listed above are kindly acknowledged for their encouragement and guidance, Michelle T. Casanova for SEM images of oospores, valuable suggestions and English improvement.

References

Cirujano, S., Cambra, J., Sánchez Castillo, P. M., Meco, A., & Flor Arnau, N. (2008). *Flora ibérica. Algas continentales. Carófitos (Characeae)*. Real Jardín Botánico. [in Spanish].
Cirujano, S., García Murillo, P., Meco, A., & Fernández Zamudio, R. (2007). Los carófitos ibéricos. *Anales del Jardín Botánico de Madrid, 64*, 87–102. [in Spanish].
Corillion, R. (1957). Les Charophycées de France et d'Europe Occidentale. *Bulletin de la Société Scientifique de Bretagne, 32* (Hors série I & II), 1–499. [in French].
Cunha, A. G. (1934). *Liste des Characées du Portugal*. [in French].
Cunha, A. G. (1941). Additions à la flore charologique du Portugal. *Bulletin de la Société Portugaise des Sciences Naturelles, 13*, 153–156. [in French].
Cunha, A. G. (1942). Contribuïção para o estudo dos carofitos portugueses. *Revista da Faculdade de Ciências, 2*, 227–357. [in Portuguese].
Cunha, A. G. (1951). Additions à la flore charologique du Portugal. *IV. Boletim da Sociedade Broteriana, 25*, 25–30. [in French].
Groves, H., & Groves, J. (1915). A new *Nitella*. *Journal of Botany 53*, 41–43, pl. 536.
Krause, W. (1997). Charales (Charophyceae). In H. Ettl, G. Gärtner, H. Heynig, & D. Mollenhauer (Eds.), *Süßwasserflora von Mitteleuropa, 18*. Fischer. [in German].
Wood, R. D. (1965). *Monograph of the Characeae*. J. Cramer.
Wood, R. D., & Imahori, K. (1964). *Iconograph of the Characeae*. J. Cramer.

Chapter 60
Nitella flexilis

Karin Pall, Ralf Becker, and Hendrik Schubert

Scientific name: ***Nitella flexilis*** **(L.) C. Agardh, Syst. Alg.: 124. 1824**

≡ *Chara flexilis* L., Sp. Pl. 2: 1157. 1753
Neotype (Wood, 1960: 224): England. Suffolk, Henly, near Ipswich / [no date] / Buddle (BM-SL) Proposed conserved type (Gregor et al., 2019): [United Kingdom, England], E. Sussex, Catsfield / [Mai 1895]/Salmon (BM 13735408)

= *Chara furcata* Roxb. ex Bruzelius, Observ. Charae: 22. 1822
Neotype (Wood, 1965: 481): [India] Shallow pond, Saharanpur District, United Provinces/Sept.-Oct. 1926/G. O. Allen [NY]; G. O. Allen, Char. Ind. Exs. No. 7

= *Chara commutata* Rupr., Beitr. Pflanzenk. Russ. Reiches: 9. 1845
Lectotype (Romanov, 2021: 1): [Russian Federation, Saint Petersburg] ad viam Peterhofianam, pr. Ulianka/17 Juni 1839/Grewenitz (LE A0000405)

= *Nitella flexilis* (L.) C. Agardh f. *longifolia* A. Braun ex Mig., Charac. Deutschl.: 138. 1890.
Type not designated

= *Nitella spanioclema* J. Groves & Bull.-Webst. ex Bull.-Webst. in J. Bot. 55: 1. 1919
Lectotype (Wood, 1964: 390): [Lough Kindrum, in the Fanad Peninsula, West Donegal]/[Aug.] 1919 / [G. R. Bullock-Webster] (BM)

K. Pall (✉)
SYSTEMA Bio- und Management Consulting GmbH, Bensasteig 8, 1140 Wien, Austria
e-mail: karin.pall@systema.at

R. Becker
Am Schützenplatz 58, 26121 Oldenburg, Germany
e-mail: becker.r@posteo.de

H. Schubert
Universität Rostock, Institut für Biowissenschaften, Albert-Einstein-Straße 3, 18059 Rostock, Germany
e-mail: hendrik.schubert@uni-rostock.de

H. Schubert et al. (eds.), *Charophytes of Europe*,
https://doi.org/10.1007/978-3-031-31898-6_60

Taxonomical note: Wood (1960) selected as neotype ("lectotype") a female dioecious plant which would be assigned to *Nitella opaca*. To save the use of *Nitella flexilis* for monoecious plants in the traditional way a proposal for the botanical congress in Madrid in 2024 was published (Gregor et al., 2019).

Local names: BELARUSIAN: Нітэла гнуткая; DANISH: Almindelig kransnål (also used for *N. opaca*); DUTCH: Buigzaam glanswier; ENGLISH: Smooth stonewort; ESTONIAN: Lookjas nitell; FINNISH: Järvisiloparta; GERMAN: Biegsame Glanzleuchteralge; LITHUANIAN: Lankstusis menturdumblis; NORWEGIAN: Glansglattkrans; POLISH: Krynicznik giętki; RUSSIAN: Нителла гибкая; SWEDISH: Glansslinke (Sweden), sjöslinke (Finland); UKRAINIAN: Нітела гнучка.

60.1 Morphology

See Figs. 60.1 and 60.2.

Habitus: Plants are shining dark green and slender, but rather robust for a *Nitella* species. Encrustations, which are rarely found, exhibit a banding pattern.

Size: *Nitella flexilis* is of small to medium size, most specimens are between 10 cm and 40 cm length; Krause (1997) and van Raam (1998) report maximum lengths up to 1 m. This applies especially to deep-water-plants in lakes. Even in running water the plants are often elongated.

Main axis: Main axis diameter is between 0.5 and 1.2 mm, rarely up to 1.5 mm (Migula, 1889–1897, Urbaniak and Gąbka, 2014). Internodes are between 4 and 9 cm length, about 1–2 times longer than the respective branchlets (Caisová and Gąbka, 2009; Becker and Wolff, 2016 report extremes between 0.5 and 4 cm). Internode length increases with depth and in running water.

Cortex, spine-cells, stipulodes: Absent.

Branchlets: Whorls regularly consist of 6 branchlets which are divided once (often undivided). Lower (down to 3) or higher (up to 10) number of branchlets per whorl may occur exceptionally (Migula, 1889–1897; Groves and Bullock-Webster, 1920; Corillion, 1957; Caisová and Gąbka, 2009; Bailly and Schaefer, 2010; Urbaniak and Gąbka, 2014; Becker and Wolff, 2016). The apical cell of the branchlets forms a small, mainly obtuse tip with a thickened cell wall. Upper branchlet whorls are often funnel-shaped and spreading, rarely directed to one side. Fertile branchlets are infrequently constricted to heads (Groves and Bullock-Webster, 1920; Corillion, 1957; Krause, 1997) and can be rarely divided 2 times (Becker and Wolff, 2016).

Bract-cells, bracteoles, bractlets: Absent.

Sexuality: The species is monoecious; populations in running water are often sterile; plants tend to be protandrous.

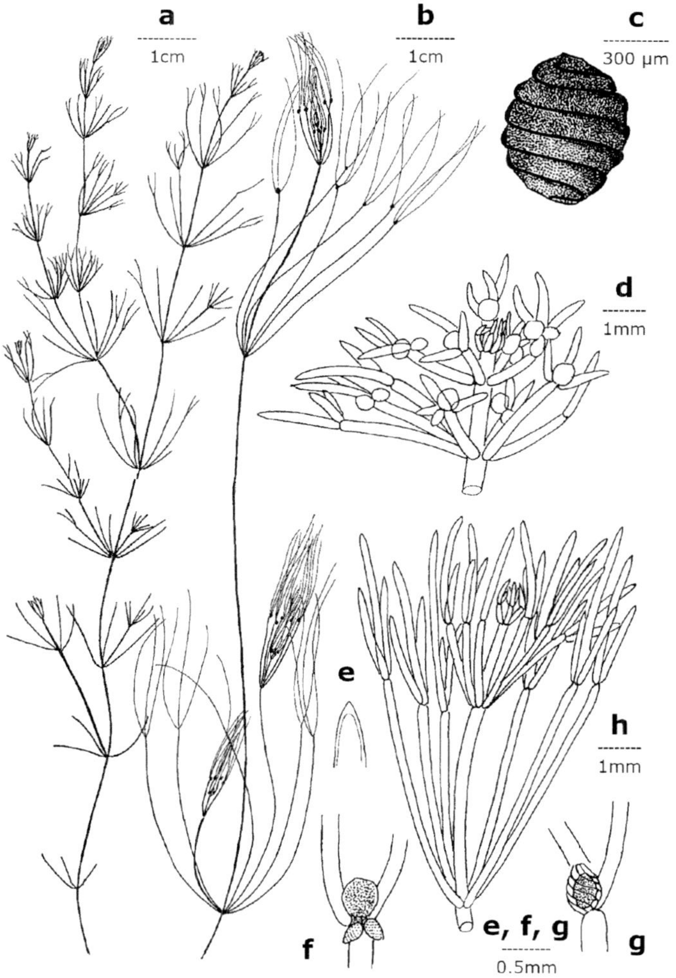

Fig. 60.1 *Nitella flexilis*. **a** Normal type, habitus; **b** Plant collected from deep water; **c** Oospore; **d** Shoot tip, f. *subcapitata* Braun; **e** Branchlet tip; **f** Young gametangia; **g** Mature oogonium; **h** Shoot tip. Drawings: Werner Krause

Oogonia: Oogonia are without mucus and placed above the antheridia in groups of 2–3 (rarely solitary or in groups of 4, see also Fig. 60.3). Table 60.1 gives an overview of the quantitative morphometry data.

Oospores: Oospores of *Nitella flexilis* are bigger than most of the other *Nitella*-species; most of the authors give a range of 400–600 μm length and 400–500 μm width (Table 60.2). The colour of ripe oospores is black to dark red-brown, striae are well developed and thick (Fig. 60.3). Oospore membrane is described as smooth

Fig. 60.2 Appearance of *Nitella flexilis*. **a**, **b** Habitat appearance of *Nitella flexilis*, stout specimen on gravel (**a**) and fragile specimen on muddy sediment at a sheltered site (**b**); **c**, **d** Typical spreading appearance without (**c**) and with (**d**) "banding pattern" of encrustation. Photographs: **a**, **b**—Klaus van de Weyer; **c**, **d**—Heiko Korsch

by several authors (Bailly and Schaefer, 2010; Groves and Bullock-Webster, 1920; Krause, 1997).

Antheridia: Antheridia are relatively large, reaching a diameter of up to 1000 μm. Ripe antheridia are greenish-brown to orange-brown and always without mucus (Figure 60.3). *Nitella flexilis* is protandrous, sometimes even appearing as a sequential hermaphrodite. Table 60.3 gives an overview of the published diameter ranges.

Bulbils: Formation of bulbils is not observed for *N. flexilis*.

Fig. 60.3 Gametangia and details of *Nitella flexilis*. **a** Conjoined stand with a pair of oogonia; **b** Fertile elongated branchlets; **c** Richly fertile condensed branchlets; **d** Apical cell of the branchlets, forming a small obtuse tip with thickened cell wall; **e, f, g** Side view (**e**), oospore membrance (**f**) and shield cell impression (basal plate, **g**) of an oospore sampled by E. S. Salmon in 1895 (Catsfield, near Battle, Sussex) with the typical smooth appearance of the membrane and the apical beak (extension of the apical striae). Fossa walls are often covered by a layer of loose grains, giving them a leathery or shagreened appearance (pers. comm. M. T. Casanova). Photographs: **a**, **b**, **c**—Heiko Korsch; **d**—Chris Carter; **e**, **f**, **g**—Michelle T. Casanova

Table 60.1 Oogonia morphometry

Author	Length	Width	Coronula length	Coronula width	Number of helical stripes
Migula (1889–1897)	600–900	500–790	–	–	7–9
Groves and Bullock-Webster (1920)	625–900	550–750	40	–	8–9
Corillion (1957a)	600–900	550–750	40	–	8–9
Compère (1992)	600–900	500–800	–	–	–
Krause (1997)	600–800	550–700	40–60	70–110	8–9
van Raam (1998)	Up to 750	Up to 700	–	–	–
Caisová and Gąbka (2009)	600–900	550–750	–	–	–
Bailly and Schaefer (2010)	500–700	500–700	–	–	–
Korsch (2013)	–	–	–	–	7–9
Urbaniak and Gąbka (2014)	600–950	550–750	25–63	–	–
Total range	**500–950**	**500–800**	**25–63**	**70–110**	**7–9**

Length in the first column is excluding the coronula, its length and width are given separately; all values for length and width in μm

60.2 Phenology and Reproduction

Nitella flexilis is a perennial species throughout its distribution range (Wahlstedt, 1864; Luther, 1951b; Olsen, 1944, Blindow, 2004). Data about phenology are confusing and hard to interpret. Gametangia ripen between June and September in lakes in Germany (Becker and Wolff, 2016); in small ponds, ditches and shallow areas of lakes sometimes as early as in April (Becker, 2010; Groves and Bullock-Webster, 1920; Olsen, 1944). In France, they may appear as early as in March (Mouronval et al. (2015). Migula (1889–1897) described germination from oospores in autumn, pointing to a lack of dormancy. Notably in shallow ponds the species can complete its life cycle fast enough to occur as an annual component (Schmidt et al., 1996). In low salinity brackish water the plant is known to hibernate green; plants in brackish water rarely become fertile, making it difficult to link the data of Luther (1951a, 1951b) to either *N. flexilis* or *N. opaca* (see "Identification and risk of confusion" below).

Table 60.2 Oospore morphometry

Author	Length	Width	Number of striae	Colour
Migula (1889–1897)	380–500	280–400	6–8	Dark red-brown to black
Groves and Bullock-Webster (1920)	500–575	425–500	5–7	Dark red-brown to black
Corillion (1957a)	500–575	425–500	5–7	Red-brown to black
Krause (1997)	500–600	450–500	5–7	Dark brown to black
van Raam (1998)	Up to 550	Up to 500	–	–
Dilger (2010)	(420)460–550(610)	(360)400–480(520)	5–7 (8)	Black
Bailly and Schaefer (2010)	400–500	400–500	5–7	Red-brown to black
Korsch (2013)	450–500	400–475	6–8	Dark brown to black
Urbaniak and Gąbka (2014)	350–580	370–525	–	Brown to dark brown
Becker and Wolff (2016)	475–525	450–500	6–7 (8)	Dark brown to black
Total range	**350–600(610)**	**280–525**	**5–8**	

All values for length and width in μm

Table 60.3 Antheridia morphometry

Author	Diameter (μm)
Groves and Bullock-Webster (1920)	500–625
Corillion (1957)	500–625
Compère (1992)	500–650
Krause (1997)	500–1000
van Raam (1998)	Up to 600
Caisová and Gąbka (2009)	500–800
Bailly and Schaefer (2010)	500–600
Korsch (2013)	500–800
Urbaniak and Gąbka (2014)	480–900
Total range	**480–1000**

60.3 Habitat Conditions and Ecology

Habitat type: *Nitella flexilis* prefers permanent soft water habitats, thriving in running as well as in stagnant water bodies. Habitat types range from artificial lakes and ponds, puddles, mire ponds, ephemeral ponds and ditches to large lakes and rivers (Becker and Wolff, 2016); the species has also been found in beach lakes (Corillion, 1957; Krause, 1983) and it is frequently found in sheltered bays in the low-salinity regions of the Finnish and Bothnian Bays of the Baltic Sea (Blindow, 2000b; Luther, 1951b). In Germany *N. flexilis* is the most frequent charophyte species of running waters.

Sediment: Sediment types beneath *Nitella flexilis* vary considerably. Stands on sandy as well as muddy sediment are described by e.g., Blindow (2004b) and Becker (2010). Schmidt et al. (1996) found the species on lime mud as well as organogenic substrate. However there are also reports of *Nitella flexilis* growing on clay (Blindow, 2004); peat (van Raam, 1998) and gravel (K. van de Weyer acc. to Becker and Wolff, 2016).

Nutrient conditions: Irrespective of being found most frequently under oligo- to mesotrophic conditions, *Nitella flexilis* is able to tolerate eutrophic conditions as well. In Northern Europe, the species is reported to prefer oligotrophic conditions (Langangen, 2007). Further south, most occurrences are in the oligo-mesotrophic range of conditions (Spiess et al., 2010; Wiegleb, 1981) but including some highly eutrophicated sites as well, as reported by Simons and Nat (1996). The conclusion of several authors (e.g., Krause, 1997; Schmidt et al., 1996) therefore is, that this species is rather eutrophication-tolerant and its occurrence is not directly dependent on nutrient conditions. This is supported by the maximum main nutrient concentrations measured at *N. flexilis* sites, reaching 1.4 mg/l TP, 23.5 mg/l TN and 210 mg/l sulphate in the Netherlands, where the highest values in Europe have been measured so far (Simons and Nat, 1996). However, the data of Simons and Nat (1996) are exceptional when compared to other authors (see Doege et al., 2016), so more data are necessary to describe the nutrient conditions of *N. flexilis* properly. In any case, the reason for the above-mentioned geographic differences in preference for oligo- or meso-/eutrophic conditions might be related to the number of water bodies in the respective range of trophic conditions.

Water chemistry: As with most of *Nitella* spp., *Nitella flexilis* is most frequently found at sub-neutral pH, being able to stand acidic conditions down to pH 4.7; however, maximum values of pH 9.6 indicate the species' ability to withstand alkaline conditions, too (Doege et al., 2016). Given the fact, that in poorly buffered soft-water conditions large diurnal changes in pH can occur, especially under eutrophic conditions, point measurements of pH are of limited use for niche description. With respect to conductivity, *N. flexilis* can be characterised as a freshwater species with limited ability to thrive under oligohaline conditions. Tazawa et al. (1987) found that *N. flexilis* is unable to regulate turgor, relying on transcellular osmosis only for maintaining its inner osmotic potential. Minimum conductivity of *N. flexilis* habitats were found at 10 μS/cm, maximum values measured so far 4700 μS/cm (Forsberg, 1965), most of the occurrences are well below 2000 μS/cm (Becker and Wolff, 2016). The maximum values with respect to salinity were reported from Finland, where Luther (1951b) reported occurrences at salinities up to 3.5 psu in the Tvärminne area. With respect to ion composition, potassium is especially crucial for this species, minimum potassium concentration is as low as 4 mg/l (van Raam, 1998, reporting also maximum values of up to 93 mg/l). For calcium, the species can grow under extreme soft water conditions with <10 mg/l calcium and is found down to 0.1 °dH (Doege et al., 2016, maximum values are 52.5 °dH).

Depth range: In Europe, occurrence of *N. flexilis* is reported down to a maximum depth of 38.5 m in Poland (Krause, 1981). In Lake Constance the species reached 15 m before severe eutrophication occurred (Baumann, 1911). Recent depth maxima are mostly at around 10 m (e.g., Trapp, 2000; Spiess et al., 2010, Pall et al. 2010). The species often indicates the lower limit of macrophytes vegetation (Mauersberger, 2004) but, just as often, can cover the full depth-range of macrophyte vegetation in lakes (e.g., Pall et al., 2010). Thus, the species is even able to grow close to the surface. *Nitella flexilis* is reported from shallow areas as well as fully exposed ponds, indicating a broad range of light acclimation (Doll, 1989; Spiess, 2004; Urbaniak and Gąbka, 2014).

Associated species: Irrespective of being the character species of the *Nitelletum flexilis* (Corillion, 1957), *Nitella flexilis* rather rarely is found in large monospecific stands, except for deep-water occurrences. However, E. Nat (unpubl.) describes a finding of monospecific vegetation of this species in a shallow (90 cm depth) sheltered pond, too (in the Distelvinkplas which is part of the Molenpolder, north of Utrecht, the Netherlands). For most of the occurrences reported, the species is a component of mixed stands without gaining dominance, occurring also in the *Nitello-Vaucherietum dichotomae* (Krausch, 1964), the *Nitellopsidetum obtusae* and the *Charetum hispidae* communities (van Raam, 1998). Given the broad range of habitat conditions described above, a full list of associated species reported would reflect the water flora of Europe, except for the marine and brackish species. For charophytes, the species has been found in Europe together with *Chara aspera, C. braunii, C. connivens, C. contraria, C. filiformis, C. fragifera, C. globularis, C. papillosa, C. subspinosa, C. tomentosa, C. virgata, Lychnothamnus barbatus, Nitella capillaris, N. confervacea, N. gracilis, N. hyalina, N. opaca, N. mucronata, N. syncarpa, N. translucens, N. wahlbergiana, Nitellopsis obtusa* and *Tolypella nidifica* (Becker and Wolff, 2016).

60.4 Variability

Variability of *Nitella flexilis* is mostly due to environmental conditions, notably current and irradiance (depth). Growing in shallow stagnant water, the species is rather stout and thrifty, branchlets rarely condense into to heads (such specimens can be treated as f. *crassa* sensu Migula, 1889–1897), becoming gradually less stout, but still thrifty in appearance and without formation of heads (f. *brevifolia* sensu Migula, 1889–1897). Specimens growing at the lower edge of the macrophyte belts are delicate and fragile with elongated internodes and long, always divided branchlets; this form has not been treated by Migula or Braun, but is mentioned and illustrated by later authors (e.g., Krause, 1997). In running water, the species tends to elongate in general, with branchlets being more affected compared to internodes, resulting in relatively large, but in contrast to deep-water forms, sturdy individuals with branchlets longer than internodes (which might be treated as f. *longifolia*). It is noteworthy that even an author like Migula (1889–1897), who usually listed many formae for all species, highlighted the limited variability of this species (Migula, 1889–1897: 34:"...überhaupt wenig ausgesprochene Formen bildend"). Probably the only morphotype which could be ranked as a variety is *Nitella flexilis* f. *subcapitata* sensu Migula (1889–1897) which, in contrast to the typical appearance of *N. flexilis*, has the uppermost fertile whorls to heads. Until now, no clear link to environmental conditions could be identified and taxonomic status of this morphotype should be investigated in more detail (see also Braun and Nordstedt, 1882).

60.5 Identification and Risk of Confusion

The main problem with *Nitella flexilis* is the distinction from *Nitella opaca*, which is nearly impossible for sterile specimen by means of morphological characters. In contrast to *N. opaca* the thickened cell wall of the end-cell of branchlets of *N. flexilis* (see Migula, 1889–1897) can be a potential discrimination feature (Fig. 60.3d) but this should be verified. Apart from that, the only character available for distinguishing sterile specimen is number of chromosomes, requiring karyological analysis or, alternatively, genetic analysis. For fertile specimens protandry of *N. flexilis* provides another risk of confusion, consequently only clearly monoecious specimens can be identified as *N. flexilis* for sure without molecular genetic tools. Phenological proxies have been shown to be uncertain, *N. flexilis* has been observed to fructify in spring (and autumn), too, as reported by several authors (e.g., Becker, 2010; Groves and Bullock-Webster, 1920; Olsen, 1944). The same applies of morphological proxies such as one-sided orientation of branchlets of *Nitella opaca*, a large and empty space between branchlets, indicating the former presence of an already lost antheridium, and antheridium diameter (Becker and Wolff, 2016). An important fact is that both species occasionally co-occur, so extreme caution is strongly recommended. Distinction from *Nitella mucronata* can be done by means of the end cell of the branchlet, being obtuse or acuminate for *N. flexilis*, but never mucronate. Additionally, branchlets of *N. flexilis* are 1-furcate, whereas *N. mucronata* has 2–3 furcate branchlets. Delicate sterile specimen may resemble *Nitella capillaris* or *Nitella syncarpa*; both specimens, in contrast to *N. flexilis*, form heads and never have obtuse end-cells; *N. capillaris* in addition has extremely fine branchlets, collapsing completely when taken out of the water. *Nitella acuminata,* with single celled dactyls and monoecy, could be confused with *Nitella flexilis*, or infertile plants of *N. opaca*. It is distinguished from those by the length of the dactyls and distinctive acuminate tips of the dactyls. *Nitella acuminata* was initially collected on the island of Reunion in the Indian Ocean. Varieties of the species (sensu Wood, 1965) occur in USA and Mexico, as well as in Africa, India, the Philippines and Australia. As far as is known there are no sites of *N. acuminata* in Europe.

60.6 Distribution

See Fig. 60.4.

Nitella flexilis is a cosmopolitan species. It occurs almost everywhere (except arctic regions) in the northern hemisphere. In the southern hemisphere the species is widespread in the southern part of South America and New Zealand. Beside this, some single findings in the central part of Africa, the southern part of India and the eastern part of Australia are documented (Korsch, 2018). In Europe the species can be found in almost all countries; larger distribution gaps are most likely due to insufficient sampling.

For most countries, *N. flexilis* has been recorded only in freshwater. Reliable records from brackish-water habitats are only available for Finland and Sweden. Eichwald (1849) mentions an occurrence of this species in Haapsalu and Kuressaare

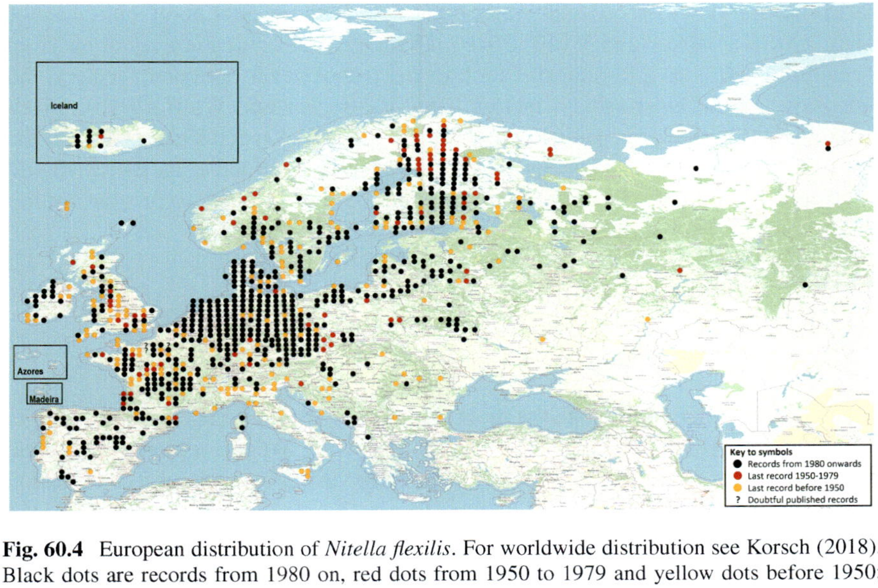

Fig. 60.4 European distribution of *Nitella flexilis*. For worldwide distribution see Korsch (2018). Black dots are records from 1980 on, red dots from 1950 to 1979 and yellow dots before 1950; unconfirmed/doubtful published records (if applicable) are marked with "?" on top. Map prepared by Heiko Korsch

Bay (western Estonia), but recently, it could not be found there. Herbarium specimens to document these records are lacking. There are no recent, but there are three former records from brackish water sites in Finland (Langangen et al., 2002; Luther, 1951b). Additionally, there are 18 recent and 15 former collections with sterile specimens of *N. flexilis* or *N. opaca* from brackish water (Langangen et al., 2002). For Sweden, two records from brackish water exist from the northern part of the Bothnian Bay (Västerbotten, Norrbotten) (Blindow, 2000a). According to Hasslow (1931), the species is occasionally found in brackish water. He mentions one location in Västerbotten and another in the coastal sea of the archipelago of Stockholm. The plants collected from Stockholm, however, are sterile (LD, C) and thus cannot be distinguished from *N. opaca*. Data from the Mediterranean and Maghreb can be found in Blaženčić et al. 2006; Bazzichelli and Abdelahad, 2009; Mouronval et al. 2015; Muller et al., 2017; Becker, 2019).

60.7 Status and Threats

Nitella flexilis is a cosmopolitan species. It is frequent in all regions with soft-water habitats and occurs also in moderate hard-water all over Europe. In the Mediterranean region it seems to be less frequent than in central and northern Europe (Blaženčić et al., 2006; Mouronval et al., 2015; Muller et al., 2017; Becker, 2019). Nevertheless, it is treated as an endangered species in several regional as well as national Red Lists because of being victim to habitat loss and habitat destruction. Especially the European-wide loss of small water bodies is causing a decline of suitable sites for *N. flexilis* (Becker and Wolff, 2016). Drainage, eutrophication resulting from agricultural activities and intensification of the use of water bodies for fishery and recreation are the most critical factors listed by Becker and Wolff (2016). Given the large distribution area and broad ecological niche, the species is probably at the moment not endangered on a European level. However, effective measures should be taken to stop this trend in decline of number of sites in order to prevent further fragmentation of populations.

60.8 Remarks

Nitella flexilis is combined with the dioecious *N. opaca* by several European authors (e.g., Cirujano et al., 2008; Moore, 1986), following Wood's opinion about the insignificance of this character (Wood, 1962). However, his opinion was criticised immediately after being published in more detail (Wood, 1965) by Sarma and Khan (1967), targeting on the consistent difference in chromosome number between *N. flexilis* ($n = 12$) and *N. opaca* ($n = 6$). The taxa are treated as different at the species level here, following not only a long European tradition, but also respecting differences in distribution and phenology (see above). Still some problems need to be mentioned in this context. Most important, there is a genetic overlap between the species, because 6 of the 12 chromosomes of *N. flexilis* have been reported to be identical with the ones of *N. opaca* (Sarma and Khan, 1967). Another problem is that sterile specimens are almost impossible to determine except by genetic analysis and even fertile specimen are only identifiable for sure when they are clearly monoecious (see above, "Identification and risk of confusion").

References

Bailly, G., & Schaefer, O. (2010). *Guide illustré des Characées du nord-est de la France*. Conservatoire Botanique National de Franche-Comté. [in French].

Baumann, E. (1911). Die Vegetation des Untersees (Bodensee). *Archiv für Hydrobiologie, Supplement 1*(I-V), 1–554, Tafeln I–XV. [in German].

Bazzichelli, G., & Abdelahad, N. (2009). *Alghe d'acqua dolce d'Italia. Flora analitica delle Caroficee*. Sapienza Università di Roma. Ministero dell'Ambiente e della Tutela del territorio e del mare. Roma, Centro Stampa Università. [in Italian].

Becker, R., & Wolff, P. (2016). *Nitella flexilis*. In: Arbeitsgruppe Characeen Deutschlands (Ed.), *Armleuchteralgen: Die Characeen Deutschlands* (pp. 413–426). Springer. [in German]

Becker, R. (2010). Bemerkenswerte Characeen-Funde in Nordwest-Deutschland. *Abhandlungen Des Naturwissenschaftlichen Vereins Zu Bremen, 46*, 409–418. [in German].

Becker, R. (2019). The Characeae (Charales, Charophyceae) of Sardinia (Italy): Habitats, distribution and conservation. *Webbia, 74*, 83–101.

Blaženčić, J., Stevanović, B., Blaženčić, Z., & Stevanović, V. (2006). Distribution and ecology of charophytes recorded in the west and central Balkans. *Cryptogamie, Algologie, 27*, 311–322.

Blindow, I. (2000a). Kransalger i Skåne. *Länsstyrelsen i Skåne Län, Rapportserien Skåne i Utvekling, 99*, 1–34. [in Swedish].

Blindow, I. (2000b). Distribution of Charophytes along the Swedish coast in relation to salinity and eutrophication. *International Review of Hydrobiology, 85*, 707–717.

Blindow, I. (2004). *Nitella flexilis*. In H. Schubert, & I. Blindow (Eds.), *Charophytes of the Baltic Sea* (pp. 174–180). A. R. G. Gantner.

Braun, A., & Nordstedt, C. F. O. (1882). Fragmente einer Monographie der Characeen. Nach den hinterlassen Manuscripten A. Braun's herausgegeben von Dr. Otto Nordstedt. *Abhandlungen der Königlichen Akademie der Wissenschaften zu Berlin für das Jahr 1882*, 1–211. [in German].

Caisová, L., & Gąbka, M. (2009). Charophytes (Characeae, Charophyta) in the Czech Republic: Taxonomy, autoecology and distribution. *Fottea, 9*, 1–43.

Cirujano, S., Cambra, J., Sánchez Castillo, P. M., Meco, A., & Flor Arnau, N. (2008). *Flora ibérica. Algas continentales. Carófitos (Characeae)*. Real Jardín Botánico. [in Spanish].

Compère, P. (1992). *Charophytes. Flore pratique des algues d'eau douce de Belgique*. Jardin botanique national de Belgique. [in French].

Corillion, R. (1957). Les Charophycées de France et d'Europe Occidentale. *Bulletin de la Société Scientifique de Bretagne, 32*(Hors série I & II), 1–499. [in French].

Dilger, M. (2010). Merkmale zur Erkennung von *Nitella*- und *Tolypella*-Arten der Characeaen [sic] über Oosporen aus Sedimenten oder von Pflanzen. *Rostocker Meeresbiologische Beiträge, 23*, 40–59. [in German].

Doege, A., van de Weyer, K., Becker, R., & Schubert, H. (2016). Bioindikation mit Characeen. In: Arbeitsgruppe Characeen Deutschlands (Ed.), *Armleuchteralgen: Die Characeen Deutschlands* (pp. 97–138). Springer. [in German].

Doll, R. (1989). Die Pflanzengesellschaften der stehenden Gewässer der DDR. Teil 1: Die Gesellschaften des offenen Wassers (Characeen-Gesellschaften). *Feddes Repertorium, 100*, 281–324. [in German].
Eichwald, E. (1849). Zweiter Nachtrag zur Infusorienkunde Russlands. *Bulletin de la Société des Naturalistes de Moscou, 22*, 400–548. [in German].
Forsberg, C. (1965). Environmental conditions of Swedish Charophytes. *Symbolae Botanicae Upsalensis, 43*, 1–67.
Gregor, T., Blindow, I., Casanova, M. T., & Schubert, H. (2019). (2721) Proposal to conserve the name *Chara flexilis* (*Nitella flexilis*) (Characeae) with a conserved type. *Taxon, 68*, 1363–1371.
Groves, J., & Bullock-Webster, G. R. (1920). *The British Charophyta*. Ray Society.
Hasslow, O. J. (1931). Sveriges characeer. *Botaniska Notiser, 1931*, 63–136. [in Swedish].
Korsch, H. (2013). Die Armleuchteralgen (Charophyceae) Sachsen-Anhalts. *Berichte Des Landesamtes Für Umweltschutz Sachsen-Anhalt, 1*(2013), 1–85. [in German].
Korsch, H. (2018). The worldwide range of the Charophyte species native to Germany. *Rostocker Meeresbiologische Beiträge 28*, 45–96.
Krausch, H. D. (1964). Die Pflanzengesellschaften des Stechlinseegebietes I: Die Gesellschaften des offenen Wassers. *Limnologica, 2*, 145–203. [in German].
Krause, W. (1981). Characeen als Bioindikatoren für den Gewässerzustand. *Limnologica, 1*, 399–418 [in German].
Krause, W. (1983). Characeen-Standorte in Portugal mit besonderer Rücksicht auf den Einfluß des Menschen. *Tuexenia, 3*, 289–296. [in German].
Krause, W. (1997). Charales (Charophyceae) Süßwasserflora von Mitteleuropa. In H. Ettl, G. Gärtner, H. Heynig, & D. Mollenhauer (Eds.), *Süßwasserflora von Mitteleuropa 18*, Fischer. [in German].
Langangen, A. (2007). *Charophytes of the Nordic countries*. Saeculum ANS.
Langangen, A., Koistinen, M., & Blindow, I. (2002). The charophytes of Finland. *Memoranda Societatis pro Fauna Et Flora Fennica, 78*, 17–48.
Luther, H. (1951a). Verbreitung und Ökologie der höheren Wasserpflanzen im Brackwasser der Ekenäs-Gegend in Südfinnland. I Allgemeiner Teil. *Acta Botanica Fennica, 49*, 1–232 [in German].
Luther, H. (1951b). Verbreitung und Ökologie der höheren Wasserpflanzen im Brackwasser der Ekenäs-Gegend in Südfinnland. II Spezieller Teil. *Acta Botanica Fennica, 50*, 1–370. [in German].
Mauersberger, R. (2004). Zum Vorkommen von Armleuchteralgengewächsen (Characeae) im Norden Brandenburgs. *Rostocker Meeresbiologische Beiträge, 13*, 85–104. [in German].
Migula, W. (1889–1897). Fünfter Band: Die Characeen. In A. Fischer, E. Fischer, F. Hauck, G. Limpricht, C. Luerssen, W. Migula, H. Rehm, P. Richter & G. Winter (Eds.), *Dr. L. Rabenhorst's Kryptogamenflora von Deutschland, Oesterreich und der Schweiz* (2nd ed.). Eduard Kummer. [in German].
Moore, J. A. (1986). *Charophytes of Great Britain and Ireland*. BSBI Handbook 5. Botanical Society of the British Isles.
Mouronval, J. B., Baudouin, S., Borel, N., Soulié-Märsche, I., Klesczewski, M., & Grillas, P. (2015). *Guide des Characées de France méditerranéenne*. Office national de la Chasse et de la Faune Sauvage. [in French].
Muller, S. D., Rhazi, L., & Soulié-Märsche, I. (2017). Diversity and distribution of Characeae in the Maghreb (Algeria, Morocco, Tunisia). *Cryptogamie, Algologie, 38*, 201–251.
Olsen, S. (1944). Danish Charophyta. Det Kongelige Danske Videnskabernes Selskab. *Biologiske Skrifter 3*(1), 1–240. [in Danish].
Pall, K., Mayerhofer, V., & Mayerhofer, S. (2010). Makrophytenkartierung am Vorarlberger Bodenseeufer, Bericht und Bewertung nach WRRL. *Schriftenreihe Lebensraum Vorarlberg, 60*, 1–65. [in German].
van Raam, J. C. (1998). *Handboek Kranswieren*. Chara boek, Hilversum. [in Dutch].

Romanov, R. E. (2021). Typification of *Chara* (Charales, Charophyceae) species described by C.F. Lessing, F.J. Ruprecht and M.M. Hollerbach. *Notulae Algarum, 213*, 1–7.

Sarma, Y. S. R. K., & Khan, M. (1967). Dioecism and monoecism as taxonomic criteria in Charophyta. *Current Science, 36*, 245–247.

Schmidt, D., van de Weyer, K., Krause, W., Kies, L., Garniel, A., Geissler, U., Gutowski, A., Samietz, R., Schütz, W., Vahle, H.-C., Vöge, M., Wolff, P., & Melzer, A. (1996). Rote Liste der Armleuchteralgen (Charophyceae) Deutschlands. 2. Fassung, Stand: Februar 1995. *Schriftenreihe Für Vegetationskunde, 28*, 547–576. [in German].

Simons, J., & Nat, E. (1996). Past and present distribution of stoneworts (Characeae) in The Netherlands. *Hydrobiologia, 340*, 127–135.

Spiess, H. J. (2004). Ergebnisse der Untersuchungen submerser Makrophyten in mesotroph-eutrophen Seen Mecklenburg-Vorpommerns. *Rostocker Meeresbiologische Beiträge, 13*, 73–84. [in German].

Spiess, H. J., Bolbrinker, P., Möbius, F., & Waterstraat, A. (2010). Ergebnisse der Untersuchungen submerser Makrophyten in ausgewählten Gewässern Mecklenburg-Vorpommerns. *Botanischer Rundbrief Mecklenburg-Vorpommern, 47*, 4–182. [in German].

Tazawa, M., Shimmen, T., & Mimura, T. (1987). Membrane control in Characeae. *Annual Revue of Plant Physiology, 38*, 95–117.

Trapp, S. (2000). *Zur Vegetationsdynamik anthropogener Stillgewässer Norddeutschlands und ihrer Bedeutung für die Bioindikation*. Dissertation Universität Bremen. [in German].

Urbaniak, J., & Gąbka, M. (2014). *Polish Charophytes. An illustrated guide to identification*. Uniwersytet Przyrodniczy we Wroclawiu. [in Polish].

Wahlstedt, L. J. (1864). *Om characeernas knoppar och öfvervintring. Botanisk avhandling*. University of Lund. [in Swedish].

Wiegleb, G. (1981). Struktur, Verbreitung und Bewertung von Makrophytengesellschaften niedersächsischer Fließgewässer. *Limnologica, 13*, 427–448. [in German].

Wood, R. D. (1960). Linnaean types for Characeae. *Transactions of the American Microscopical Society, 79*, 219–226.

Wood, R. D. (1962). New combinations and taxa in the revision of Characeae. *Taxon, 11*, 7–25.

Wood, R. D. (1965). *Monograph of the Characeae*. J. Cramer.

Chapter 61
Nitella gracilis

Maciej Gąbka and Jacek Urbaniak

Scientific name: ***Nitella gracilis*** **(Smith) C. Agardh Syst. Alg.: 125. Agardh, Systema Algarum, Berling, 1824**

≡ *Chara gracilis* Smith in Engl. Bot. 30: 2140. 1810
Lectotype (Korsch in Arbeitsgruppe Characeen 2015: 435): St. Leonhard's forest / [no date] / W. Borrer (SM)

= *Chara flexilis* L. var. *flabellata* Rchb., Handb. Gewächsk., ed. 2.: 1594. 1829

≡ *Chara flabellata* (Rchb.) Rchb., Handb. Gewächsk., ed. 3.: 1869. 1834
Type not designated.

Taxonomic note: In the synonymy of *Chara flexilis* var. *flabellata* with *Nitella gracilis* we follow Braun (1847a).

Local names: BELARUSIAN: Нітэла грацыёзная, DUTCH: Sierlijk glanswier, ENGLISH: Slender stonewort, ESTONIAN: Peen nitell, FINNISH: Hentosiloparta, GERMAN: Zierliche Glanzleuchteralge, LITHUANIAN: Lieknasis menturdumblis, NORWEGIAN: Skjørglattkrans, POLISH: Krynicznik smukły, RUSSIAN: Нителла стройная, SWEDISH: Spädslinke; UKRAINIAN: Нітела струнка.

M. Gąbka (✉)
Faculty of Biology, Department of Hydrobiology, Adam Mickiewicz University, Poznań, Uniwersytetu Poznańskiego Str. 6, 61-614 Poznań, Poland
e-mail: gmaciej@amu.edu.pl

J. Urbaniak
Department of Botany and Plant Ecology, Wrocław University of Life and Environmental Sciences, pl. Grunwaldzki 24a, 50-363 Wrocław, Poland
e-mail: jacek.urbaniak@upwr.edu.pl

H. Schubert et al. (eds.), *Charophytes of Europe*,
https://doi.org/10.1007/978-3-031-31898-6_61

61.1 Morphology

See Figs. 61.1 and 61.2.

Fig. 61.1 *Nitella gracilis,* details. **a**, **b** General habitus; **c** Acute end-cell; **d** Mucronate apical cell with oogonium and antheridium below; **e** Sterile branchlet 1–2 divided with acute end-cell; **f** Fertile whorl and end-cell; **g** Fertile branches in a whorl; **h** Visible encrustation on the main axis. Photographs and drawings: **a**, **d**, **e**, **g**, **h**—Jacek Urbaniak; **b**, **c**, **f**—Kinga Gawrońska, Maciej Gąbka

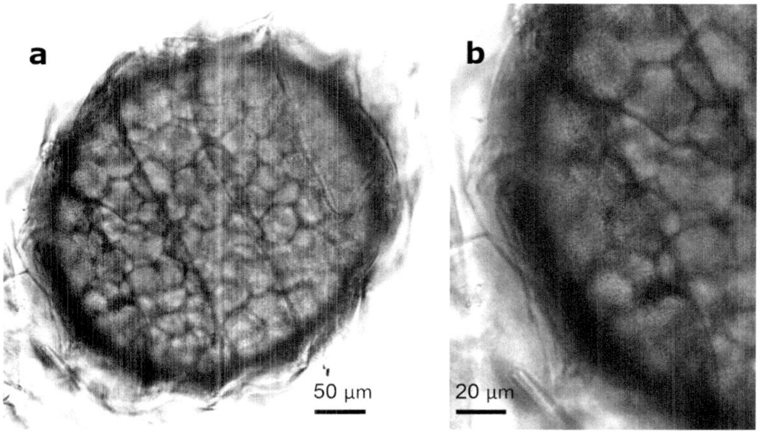

Fig. 61.2 *Nitella gracilis*, oospore. **a** Side view; **b** Membrane, showing the spongy ornamentation typical for this species. Photographs: Jacek Urbaniak

Habitus: *Nitella gracilis* is a small plant that grows in groups of several individuals together. The plants are light yellow green to green, sometimes transparent, slender and with delicate branches, resembling small *N. mucronata* in their habitus. *N. gracilis* is occasionally slightly encrusted. Herbaria specimens are brownish green in colour.

Size: *Nitella gracilis* is a small to medium size plant (3)10–20 cm high.

Main axis: The main axis is slender to moderately stout, 3–5 mm in diameter. Inter-nodes 1–3 times as long as the branchlets and rarely shorter; plants are richly branchlets and bushy in appearance.

Cortex: Absent, does not form cortex.

Spine-cells: Absent, does not form spine-cells.

Stipulodes: Absent, does not form stipulodes.

Branchlets: Fertile branchlets 5–7 in a whorl, occasionally compacted, 1–4 divided. Primary branchlet segment about 1/2–3/4 of entire branchlet length, secondaries 4–6, tertiaries 3–5, quaternaries (when present) 3–4. End segment 2–3 celled, short or elongate, tapered sharply at apex. End-cell acute, at the base about 2/3 as broad as the subapical cell, occasionally mucronate. Sterile branchlets of similar length, 1–2 divided with acute end-cells. Fertile whorls form heads.

Bract-cells: Absent, does not form bract-cells.

Bracteoles: Absent, does not form bracteoles.

Sexuality: *Nitella gracilis* is a monoecious plant. Gametangia present at all branchlet nodes.

Table 61.1 Oogonia morphometry of *Nitella gracilis*

Author	Length	Width	Coronula length	Coronula width	Number of helical stripes
Migula (1889–1897)	290–350	230–270	–	–	8
Corillion (1957a)	525	350	25–30	–	8–9
Wood (1965)	(390)450–525	315–350	25–45	52–63	8–10
Dąmbska and Karpiński (1954), Dąmbska (1964)	290–430	230–270	–	–	8
Krause (1997)	450–525	300–350	30–50	50–60	8–9
Urbaniak and Gąbka (2014)	340–435	230–380	–	–	
Total range	**(800)1050–1350**	**600–800**	**45–160**	**110–160**	**7–12**

[1]Oogonia were measured including coronula
Length in the first column is, except for Wood (1965), excluding the coronula; all values for length and width in μm

Oogonia: Oogonia generally solitary, (rarely more than one per node), 290–525 μm long, 230–350 μm wide (variability presented in Table 61.1).

Oospores: Yellow brown to light brown, punctuate; 220–350 μm long, 150–290 μm wide, variability presented in Table 61.2. More detailed data about oospore features were presented by Urbaniak (2009).

Antheridia: Antheridia are about 150–300 μm in diameter. The variability of antheridia diameter is given in Table 61.3.

Bulbils: Not reported on the rhizoids.

61.2 Phenology and Reproduction

Nitella gracilis can be annual or perennial depending on the habitat conditions. The plant prefers to occur in shallow water that usually freezes in wintertime. Here the species is often lost, and then germinates from oospores in the spring. Similarly, *N. gracilis* is also lost in shallow water places, in shallow littoral or lake margins in late summer. *Nitella gracilis* is an ephemeral species that is sometimes not recorded for several years in the same places. Usually mature and fertile plants are found from June onwards, and ripe oospores can be found between August and October. According to Migula (1889–1897), *N. gracilis* can hibernate and thus occur in one location for several years, but detailed data on wintering and phenology *N. gracilis* are rare. Results of research on the ecology and life cycle of *N. gracilis* in shallow ponds in

Table 61.2 Oospore morphometry of *Nitella gracilis*

Author	Length	Width	Number of striae	Colour
Migula (1889–1897)	225–270	200–250	6–7	Black
Corillion (1957a)	225–350	175–200	6	Brown
Wood (1965)	250–320	225–250	6	Light to dark brown
Dąmbska and Karpiński (1954), Dąmbska (1964)	225–270	200–250	6–7	Black
Haas (1994)	220–300	460–500	5–7	Light (yellowish) brown
Krause (1997)	250–300	–	6–8	Dark brown
Urbaniak (2009)	249–307	151–292	4–7	–
Urbaniak (2011b)	220–310	160–270	5–6	–
Urbaniak and Gąbka (2014)	230–320	150–290	–	Yellow brown to light brown
Total range	**222–350**	**150–290(500)**	**4–8**	

All values for length and width in μm

Table 61.3 Antheridia morphometry of *Nitella gracilis*

Author	Diameter (μm)
Migula (1889–1897)	220–250
Corillion (1957a)	230
Wood (1965)	(130)225–250
Dąmbska and Karpiński (1954), Dąmbska (1964)	220–250
Krause (1997)	250–350
Urbaniak and Gąbka (2014)	150–290
Total range	**150–350**

France clearly indicates that *N. gracilis* may be a perennial species (Auderset Joye and Boissezon, 2017). *N. gracilis* was noticed in late spring and after reproduction, the plants started a second reproduction cycle in late summer. In this second reproduction period oospore germination occurred as well. This plant can therefore be considered as perennial and can reproduce all year round (Auderset Joye and Boissezon, 2017).

61.3 Habitat Conditions and Ecology

Habitat type: *Nitella gracilis* is a typically freshwater species. The species has most commonly been found in shallow water. In Europe, *N. gracilis* occurs mainly in lakes (especially *Lobelia*-lakes, mesotrophic lakes and peat bog lakes). It is common in ponds (also fishponds), pools, ditches and clay pits. Single locations are known from sand pits in an old riverbed.

Sediment: Most specimens were recorded on mineral bottom clay or on peaty silt, without calcium content.

Nutrient conditions: In general, *Nitella gracilis* occurs in water poor in nutrients. However, there are no detailed data on the trophic requirements of the species from the entire range in Europe. Most locations can be classified as oligo-mesotrophic or mesotrophic, e.g., *Lobelia*-lakes and mesotrophic lakes. However the species occurs also in more fertile waters, meso-eutrophic to eutrophic (e.g., Krause, 1981; Langangen 2007a; Caisová and Gąbka, 2009; Gąbka, 2009). *Nitella gracilis* grows in wide ranges of nutrient concentrations between 0.002–0.014 mgL^{-1} for total phosphorus and 0.05–10.00 mgL^{-1} for total nitrogen (Korsch, 2016b).The ranges of P and N in European waters are presented in Table 61.4. *Nitella gracilis* is considered a pioneer species, favored by clear water and lack of competing species.

Water chemistry: *Nitella gracilis* is known only from freshwater habitats. The species is recorded from acidic to neutral to lightly alkaline water (pH 5.0–9.0), usually with a low content of calcium and sulphates. Numerous data indicate a very low calcium content in the water inhabited by *N. gracilis*: France 27–43 mg L^{-1} (Auderset Joye and Boissezon, 2017), Switzerland 4.5–6.0 mg L^{-1} (Rey-Boissezon and Auderset Joye, 2015) and in Germany 15–30 mg L^{-1} (Doege et al., 2016). Such calcium content is probably of great ecological importance for the species. However, scarce data show the possibility of *N. gracilis* occuring in waters rich in calcium (Gąbka, 2009; Pełechaty et al., 2013). Details of physico–chemical properties of waters are presented in Table 61.4 based on the published data (Auderset Joye and Boissezon, 2017; Doege et al., 2016; Gąbka, 2009). However, the species definitely prefers clean, low productivity, low electrolyte content and usually low in calcium water.

Depth range: *Nitella gracilis* can be found in depth ranges from 0.05 to 3 m however, the water depth usually does not exceed 0.5 m. Migula (1889–1897) reports *N. gracilis* also at 4 m depth.

Associated species: *Nitella gracilis* is considered a pioneer species which very rarely forms its own associations. A community with a dominance of *N. gracilis* was described by Corillion (association of *Nitelletum gracilis* Corillion, 1957) but is very rarely found (and documented) in Europe (e.g., Vahle, 1982; Blaženčić, 1984; Doll, 1989; Dąmbska, 1966; Iakushenko and Borysova, 2012). *Nitella gracilis* occurs in different water habitats and can form loose clumps or can be found in between other

Table 61.4 Water chemistry from European locations of *N. gracilis*

Parameter	Unit	Minimum	Maximum
Colour	mg Pt l^{-1}	1	72
pH	–	5.0	9.0
O_2	mg l^{-1}	5.9	10.5
Conductivity	$\mu S\ cm^{-1}$	629	701
N-NH_4	mg l^{-1}	0.00	0.35
N-NO_3	mg l^{-1}	0.00	0.04
P-PO_4	mg l^{-1}	0.00	0.19
Ca	mg l^{-1}	4.5	30.0(93.0)
Mg	mg l^{-1}	2.3	12.1
Na	mg l^{-1}	7.2	18.3
K	mg l^{-1}	3.2	11.8
Fe	mg l^{-1}	0.04	0.06
SO_4	mg l^{-1}	37	58

Outlier values are given in brackets

charophytes or vascular plants. It co-occurs with many species of vascular plants (*Potamogeton polygonifolius, P. berchtoldii, P. gramineus, P. obtusifolius, Stuckenia pectinata, Myriophyllum spicatum, M. verticillatum, Callitriche* sp . and *Utricularia australis*) and charophytes (e.g., *Chara globularis, C. virgata* and *Nitella flexilis*). It can occur with species characteristic of high water trophy, e.g., *Ceratophyllum demersum* (Pełechaty et al., 2013). It is also found growing beneath emergent plants such as: *Eleocharis palustris, Phragmites australis* and *Typha latifolia*.

61.4 Variability

Nitella gracilis is a very variable species, rich in forms and varieties. The main differences relate to the size of different parts of the plant. Migula (1889–1897) distinguished 13 forms, and the variability concerns, among other features, the number of branchlet divisions (and their length) or the presence of heads (bushy or loose). Migula (1889–1897) distinguish three main groups of forms: (1) *genuinae*—heads absent, branchlets divided three times, (2) *heteromorphae*—branchlets divided one to two times with loose heads and (3) *simpliciores* with indistinctly divided branchlets and heads absent. The indicated differences remain within the intraspecific variation that generally occurs within genus *Nitella*.

61.5 Identification and Risk of Confusion

Nitella gracilis is a similar species to *N. mucronata*, which also forms sparse heads. *Nitella mucronata* has branches with the last cell of the end segment short, conical and mucronate (about 1/3 as broad as the subapical cell), whereas the last cell of *N. gracilis* is acute, at the base about 2/3 as broad as the subapical cell. *Nitella gracilis* has a more slender and delicate habit and punctate (granulate) oospore wall ornamentation, whereas the oospore ornamentation of *N. mucronata* is reticulate (light microscope) (Urbaniak, 2009). Very small and delicate specimens of *N. gracilis* can be confused with *N. confervacea* but *N. gracilis* has upper whorls not contracted (sparse appearance) and the branchlet end segments are mostly 3-celled (*N. confervacea*—end segments always 2-celled).

61.6 Distribution

See Fig. 61.3.

Nitella gracilis is a cosmopolitan species recorded from North and South America, Australia, Africa, Asia and Europe (Migula, 1889–1897; Corillion, 1957a; Korsch, 2018). It is widespread in Europe, but not especially common. It's European distribution has a northern border in Fennoscandia (Langangen, 2007a). It probably does

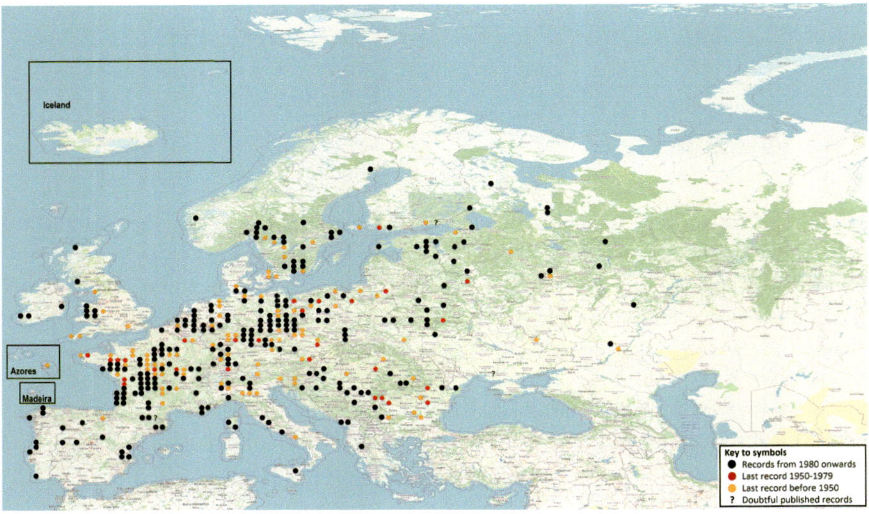

Fig. 61.3 European distribution of *Nitella gracilis*. For worldwide distribution see Korsch (2018). Black dots are records from 1980 on, red dots from 1950 to 1979 and yellow dots before 1950. Map prepared by Heiko Korsch

not occur in Iceland and Luxembourg. No distribution data are available from some regions of the Balkans (Blaženčić, 2014). In Europe it has been reported to be quite common in Finland, Norway, Hungary, Germany, Poland, Spain, France, Portugal and Italy (e.g., Krause, 1997; Langangen, 2007a, Urbaniak and Gąbka, 2014). Currently (observed after 1980) the species has disappeared from many localities in Europe, and it is threatened in much of its distribution range. *N. gracilis* is more frequent in France, Germany and Poland (over 50 currently known localities). The most probable threats are eutrophication and acidification and humification of the lakes.

61.7 Status and Threats

Nitella gracilis is assessed as RE (Regionally Extinct) in Denmark, EN (Endangered) in Switzerland, Netherlands and in southeastern Europe. In other countries as Ireland, Great Britain, Norway, Finland and Poland, the species is classified as Vulnerable.

A decline of *N. gracilis* in previously known sites has been observed. New locations are also being discovered, although the species requires monitoring throughout much of its range. *N. gracilis* is threatened by various threats including eutrophication, which can lead to decreasing light availability in lakes and other water bodies.

References

Agardh, C. A. (1824). *Systema Algarum*. Berling. [in Latin].

Auderset Joye, D., & Boissezon, A. (2017). New insights into the ecology and phenology of two Characeae: *N. opaca* (Bruzelius) C. Agardh and *N. gracilis* (Sm.) C. Agardh. *Botany Letters, 165*, 91–102.

Blaženčić, J. (1984). Rasprostranjenje i ekologija vrste *Nitella gracil*is (Smith) Ag. (Nitellaceae) u Jugoslaviji. *Glasnik Instituta Za Botaniku i Botanicke Bašte Univerziteta u Beogradu, 18*, 31–36. [in Serbian].

Blaženčić, J. (2014). Overview of the stoneworts (Charales) of Serbia with the estimation of the threat status. *Botanica Serbica, 38*, 121–130.

Braun, A. (1847). Uebersicht der schweizerischen Characeen. *Neue Denkschriften der Allgemeinen Schweizerischen Gesellschaft für die Gesammten Naturwissenschaften, 10*, 1–23. [in German].

Caisová, L., & Gąbka, M. (2009). Charophytes (Characeae, Charophyta) in the Czech Republic: Taxonomy, autoecology and distribution. *Fottea, 9*, 1–43.

Corillion, R. (1957a). Les Charophycées de France et d'Europe Occidentale. *Bulletin de la Société Scientifique de Bretagne* 32 (Hors série I & II): 1–499 [in French].

Dąmbska, I. & J. Karpiński, 1954. *Ramienice. Klucz do oznaczania gatunków krajowych*. Państwowe Wydawnictwo Naukowe [in Polish].

Dąmbska, I. (1964). *Charophyta—ramienice*. Państwowe Wydawnictwo Naukowe [in Polish].

Dąmbska, I. (1966). Zbiorowiska ramienic Polski. *Prace Komisji Biologicznej, Poznańskie Towarzystwo Przyjaciół Nauk, Wydział Matematyczno Przyrodniczy, 31*, 1–76. [in Polish].

Doege, A., van de Weyer, K., Becker, R., & Schubert, H. (2016). Bioindikation mit Characeen. In: Arbeitsgruppe Characeen Deutschlands, *Armleuchteralgen: Die Characeen Deutschlands* (pp. 97–138). Springer [in German].

Doll, R. (1989). Die Pflanzengesellschaften der stehenden Gewässer der DDR. Teil 1: Die Gesellschaften des offenen Wassers (Characeen-Gesellschaften). *Feddes Repertorium, 100*, 281–324. [in German].

Gąbka, M. (2009). *Charophytes of the Wielkopolska region (NW Poland): Distribution, taxonomy, and autecology*. Bogucki Wydawnictwo Naukowe.

Haas, J. N. (1994). First identification key for charophyte oospores from central Europe. *European Journal of Phycology, 29*, 227–235.

Iakushenko, D., & Borysova, O. (2012). Plant communities of the class Charetea Fukarek ex Krausch 1964 in Ukraine: An overview. *Biodiversity Research and Conservation, 25*, 75–82.

Korsch, H. (2016b). *Nitella gracilis* In: Arbeitsgruppe Characeen Deutschlands, *Armleuchteralgen: Die Characeen Deutschlands* (pp. 435–443). Springer [in German].

Korsch, H. (2018). The worldwide range of the Charophyte species native to Germany. *Rostocker Meeresbiologische Beiträge, 28*, 45–96.

Krause, W. (1997). Charales (Charophyceae) Süßwasserflora von Mitteleuropa. In H. Ettl, G. Gärtner, H. Heynig, & D. Mollenhauer (Eds.), *Süßwasserflora von Mitteleuropa 18*. Fischer [in German].

Krause, W. (1981). Characeen als Bioindikatoren für den Gewässerzustand. *Limnologica, 1*, 399–418. [in German].

Langangen, A. (2007). *Charophytes of the Nordic countries* (p. 102). Saeculum ANS.

Migula, W. (1889–1897). Fünfter Band: Die Characeen. In A. Fischer, E. Fischer, F. Hauck, G. Limpricht, C. Luerssen, W. Migula, H. Rehm, P. Richter, & G. Winter (Eds.), *Dr. L. Rabenhorst's Kryptogamenflora von Deutschland, Oesterreich und der Schweiz*, Ed. 2. Eduard Kummer [in German].

Pełechaty, M., Pronin, E., & Pukacz, A. (2013). Charophyte occurrence in *Ceratophyllum demersum* stands. *Hydrobiologia, 737*, 111–120.

Rey-Boissezon, A., & Auderset Joye, D. (2015). Habitat requirement of Charophytes—evidence of species discrimination through distribution analysis. *Aquatic Botany, 120*, 84–91.

Urbaniak, J. & Gąbka, M. (2014). *Polish Charophytes. An illustrated guide to identification* (p. 120). Uniwersytet Przyrodniczy we Wroclawiu [in Polish].

Urbaniak, J. (2009). Oospore variation in *Nitella gracilis* and *Nitella mucronata* (Charales, Charophyceae) from Poland. *Biologia, 64*, 252–260.

Urbaniak, J. (2011b). A SEM study of the oospore wall ornamentation in Polish charophytes (Charales, Charophyceae) genera *Lychnothamnus*, *Nitella* and *Nitellopsis*. *Nova Hedwigia, 93*, 537–549.

Vahle, H. C. (1982). Ein Fundort von *Nitella gracilis* (Smith) Ag. in der Lüneburger Heide. *Göttinger Floristische Rundbriefe, 16*, 36–38. [in German].

Wood, R. D. (1965). *Monograph of the Characeae*. J. Cramer.

Chapter 62
Nitella hyalina

Emile Nat and Ralf Becker

Scientific name: ***Nitella hyalina*** **(DC.) C. Agardh, Syst. Alg: 126. 1824**

≡ *Chara hyalina* DC. in Lam. & DC. Fl. Franç. (Lamarck) 5 & 6, ed. 3: 247. 1815
Lectotype (Wood, 1965: 660): Lac de grand lieu/1805/Hectot (G-DC)

= *Nitella hyalina* (DC. in Lam. & DC.) C. Agardh f. *maxima* A. Braun in Migula, Charac. Deutschl.: 196. 1890
Lectotype (Wood, 1965: 664): Dans les eaux de l'étang de Soustons près Bayonne/ 15. Mai 1877/L. Motelay (NY); Braun, Rabenhorst & Stizenberger, Characeen Eur. 107

= *Chara condensata* Rupr., Beitr. Pflanzenk. Russ. Reiches: 10. 1845

≡ *Nitella condensata* (Rupr.) Kütz., Sp. Alg.: 515. 1849
Lectotype (Romanov, 2021: 2): [Kazakhstan, East Kazakhstan Region, formerly Dzhungaria] [near Narym River] на берегу реки [at river bank] [illegible words] тучной земли [rich soil]/Августа [August] 18[illegible characters] [18.VIII.1838]/no collector (LE A0000403)

= *Chara interrupta* Rupr., Beitr. Pflanzenk. Russ. Reiches: 11. 1845

≡ *Nitella interrupta* (Rupr.) Kütz., Sp. Alg.: 516. 1849
Lectotype (Romanov, 2021: 2): [Kazakhstan, Karagandy Oblast], Dshilantschik [River]/25 Jul[i] [18]42/Schrenk (LE A0000413)

E. Nat (✉)
Dutch Information Center On Charophytes Naturalis Biodiversity Center, P.O. Box 9517, 2300 RA Leiden, The Netherlands
e-mail: e.nat@kranswieren.nl

R. Becker
Am Schützenplatz 58, 26121 Oldenburg, Germany
e-mail: becker.r@posteo.de

H. Schubert et al. (eds.), *Charophytes of Europe*,
https://doi.org/10.1007/978-3-031-31898-6_62

Local names: DUTCH: Klein glanswier; ENGLISH: Many-branched stonewort; FINNISH: Kalvassiloparta; FRENCH: Nitelle hyaline; GERMAN: Vielästige Glanzleuchteralge; LITHUANIAN: Žalsvasis menturdumblis; RUSSIAN: Нителла гиалиновая; SWEDISH: Glas-slinke (Sweden), blekslinke (Finland); UKRAINIAN: Нітела гіалінова.

62.1 Morphology

Habitus: Plants of *Nitella hyalina* are rather delicate, consisting of strings of isolated heads with a thick mucus layer around the gametangia. Branchlets are arranged, at each axial node, in two distinctly different series (heteroclemous), comprising a primary whorl of branchlets and accessory whorls (see Figs. 62.1, 62.2c and 62.3a, b), in two rows, one above and one below each primary branchlet. The habitus has the appearance of a string of beads (moniliform), see Figs. 62.1a and 62.2a. The species is relatively resistant to deformation out of the water (Fig. 62.2b). *Nitella hyalina* is bright- to dark or brownish green and shiny. The plants can be encrusted with lime, according to Baumann (1911) even heavily encrusted. Under these circumstances the species has a grey-green appearance.

Size: *Nitella hyalina* is small to medium-sized (5–20[40] cm) but usually below 15 cm. In the Bodensee (Germany), France: Landes, Soustons près Bayonne and some deep pits in Portugal a form exists that can be 1 m in size (van Raam, 1998). In depths between 2 and 4 m in clear quarry ponds in northern Germany specimens are up to 35 cm long (R. Becker unpubl.).

Main axis: The main axis diameter is up to 0.8 mm but is mostly around 0.3 mm. The internodes are 2–4(5) times as long as the branchlets and have a size between 0.3 and 4.5 cm, but rarely exceed 1.5 cm (Becker, 2016b).

Cortex, spine-cells, stipulodes: Absent (but see Chap. 3 for stipulodes).

Branchlets: Whorls consist of 6–8(9) primary and 6–12(16) secondary (accessory) branchlets. The secondary branchlets are often curved downwards. The primary branchlets are between 0.3 and 1.1 cm in size, 2–3(4) times furcate, the accessories are much shorter and less furcate ([0]1–2 times). The dactyls are 2-celled (bicellulate), the end-cell is short and pointed/acute, see Figs. 62.1d, e and 62.2a. Young fertile branchlets are enveloped in mucus which is sometimes difficult to see. These plants attach to herbarium paper, and they glitter in direct sunlight.

Bract-cells, bracteoles, bractlets: Absent.

Sexuality: The species is monoecious and richly fertile, see Fig. 62.3a, c. Gametangia are conjoined at the second and the third (occasionally at the first) nodes of the primary branchlets and sometimes on the accessory branchlets. Usually the gametangia are enveloped in mucus (Fig. 62.3b).

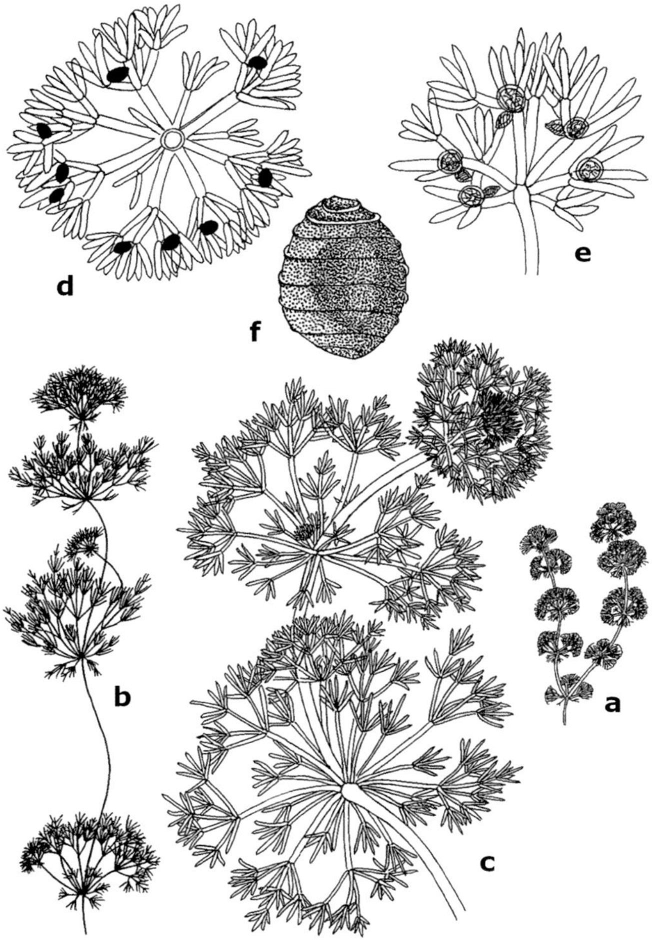

Fig. 62.1 *Nitella hyalina*. **a** Habitus appearance of young plant; **b** Habitus appearance of mature plant; **c** Top of plant; **d** Whorl with primary and secondary branchlets; **e** Fertile top; **f** Oospore. Drawings: Werner Krause

Oogonia: The length of the oogonia are very different among authors. Morphometric data of the oogonia are listed in Table 62.1.

Oospores: The oospore ornamentation is mostly described as fibrous, with fibrous flanges or finely granulate (see Fig. 62.2d, e). The striae are inconspicuous. Table 62.2 gives an overview of the morphometry and colour of the oospores.

Antheridia: The antheridia of *Nitella hyalina* are relatively small and the colour is orange to red. Table 62.3 gives an overview of the literature data on antheridia diameter.

Fig. 62.2 *Nitella hyalina*, details. **a** Habitus appearance; **b** Shape retention when taken out of the water; **c** Node with branchlets (red arrow) and accessory branchlets (blue arrow); **d** SEM of oospore; **e** SEM of oospore-membrane showing fibrous oospore-membrane ornamentation. Photographs: **a**—Klaus van de Weyer; **b**—Emile Nat, **c**—János Csiky; **d**, **e**—Michelle T. Casanova

Bulbils: There are no reports about bulbil formation of *Nitella hyalina*.

62.2 Phenology and Reproduction

Nitella hyalina is predominantly an annual species both in north-western Europe and the Mediterranean region (M. Azzella pers. comm.). The oospores germinate in late Spring (April–May) and fructification is in the summer (August–September). Plants of *Nitella hyalina* can be found in north-west Europe until December and they disappear quickly in winter (van Raam, 1998). In Germany it is seems to be a perennial species in deeper lakes (Becker, 2016b) and can survive a frost period of several weeks (Becker, 2011).

Fig. 62.3 *Nitella hyalina*, gametangia. **a** Branchlets with gametangia and end-cells; **b** Fertile heads enveloped in mucus; **c** Fertile whorls. Photographs: **a**—Dario J. Cavaliere; **b**—Wim Langbroek; **c**—Wil Leurs

62.3 Habitat Conditions and Ecology

Habitat type: *Nitella hyalina* can be found in ponds, pools, brooks, swamps, pits, lakes, oxbow lakes and in Germany and the Netherlands often in quarry ponds. Moreover, a large population was recently recorded in a shallow stormwater pond in north-west Germany (Becker et al., 2022). It can grow in freshwater as well as in brackish waterbodies, estuaries, groundwater wells and in the Finnish part of the Baltic (Corillion, 1957; Krause, 1997; Espinar et al., 2002; Koistinen, 2004; Cirujano et al., 2008; Herbarium specimens Naturalis, Leiden). This species can grow on bare substrates in sheltered openings of reed and bulrush stands, along the shorelines and even in exposed areas of lakes.

Table 62.1 Oogonia morphometry

Author	Length	Width	Coronula length	Coronula width	Number of helical stripes
Migula (1889–1897)	–	–	50–60	–	–
Groves and Bullock-Webster (1920)	500–625	375–400	45	–	9–10
Corillion (1957)	500–625	375–400	45	–	9–10
Wood (1965)	380–625*	260–400	(20)45–70	60–70	8–10
Moore (1986)	up to 400	up to 400	–	–	–
Krause (1997)	500–650	375–400	45–60	60	8–10
van Raam (1998)	up to 500	up to 400	–	–	–
Cirujano et al., (2008)	340–550	260–400	30–70	–	–
Bailly and Schaefer (2010)	350–400	350–400	–	–	–
Becker unpubl	(350)475–625*	(275)350–425	40–75	(45)60–80	8–10
Total range	**340–650**	**260–400(425)**	**(20)45–75**	**(45)60–70(80)**	**8–10**

*Including coronula
Length in the first column excludes the coronula except for values marked by asterisks; coronula length and width are given separately, all values for length and width in μm

Sediment: *Nitella hyalina* occurs on sandy or gritty substrates as well as on mud and sand with a (thin) peat layer (van Raam and Maier, 1995). Other substrates: lime mud (Krause, 1997) and clay, gyttja, silt and coarse sand (Koistinen, 2004).

Nutrient conditions: *Nitella hyalina* has a preference for mesotrophic to moderately nutrient-rich conditions (Schmidt et al., 1996). van Raam (1998) and van Raam and Maier (1992) consider this species as occurring in waterbodies low in nitrogen and phosphorus. There are only few data available for nutrient concentrations of water-bodies with *Nitella hyalina*. The records of total phosphorus range from 0.011 to 0.014 mg/l whereas values for ortho-phosphate are <0.015 mg/l in northern Germany (Doege et al., 2016). Data from Versener Heidesee near Meppen (with a huge population of *N. hyalina*) 2016/2017–2020 (4 samplings/year): TN 0.54–0.98 mg/l; NH4 0.03–0.27 mg/l; TP predominantly 0.016–0.031 mg/l but two single samples 0.5 respectively 0.67 mg/l (Böckermann et al., 2022). Data from six water bodies with *N. hyalina*-populations in north-western Germany have shown the following results between 2014 and 2020: TP 0.012–0.067 mg/l, TN 0.048–1.23 mg/l. (Becker et al., 2022). In the Netherlands the measured range for total phosphorus is 0.02–0.2 mg/l (Simons and Nat, 1996) and Azzella (2014) recorded in five Italian lakes the range 0.011–0.022 mg/l. The measured range for total nitrogen is 1.23–1.4 mg/l in Lower Saxonian waters (Becker, 2016b), 0.6–3.2 mg/l in the Netherlands (Nat et al., 1994;

Table 62.2 Oospore morphometry

Author	Length	Width	Number of striae	Colour
Migula (1889–1897)	300–350	280–320	6–7(8)	Reddish to black brown
Groves and Bullock-Webster (1920)	350	300	7–8	Reddish-brown
Corillion (1957)	350	300	7–8	Maroon to black brown
Wood (1965)	(250)300–350	(210) 280–320	6–7(8)	Light- to dark- to reddish-brown
Krause (1997)	300–350	300	6–8	Brown to black brown
van Raam (1998)	Up to 300	Up to 300	–	–
Cirujano et al., (2008)	300–350	260–320	–	Reddish to black brown
Bailly and Schaefer (2010)	Up to 300	Up to 300	7–9	Reddish-brown
Dilger (2010)	(330)338–380(390)	(250)270–320(330)	6–8(9)	Light brown to black
Becker (2016b), unpubl	325–375(400)	(225)275–325(350)	(6)7–8	Brown black
Total range	**(250)300–380(400)**	**(210)260–325(350)**	**6–8(9)**	

All values for length and width in μm

Table 62.3 Antheridia morphometry

Author	Diameter (μm)
Migula (1889–1897)	Up to 300
Groves and Bullock-Webster (1920)	350–425
Corillion (1957)	350–420
Wood (1965)	300–425
Moore (1986)	up to 400
Krause (1997)	350–400
van Raam (1998)	up to 400
Cirujano et al., (2008)	350–425
Bailly and Schaefer (2010)	Up to 400
R.Becker unpubl	(175)325–450(500)
Total range	**(<300)300–450(500)**

Simons and Nat, 1996) and in Italy 0.2–0.4 mg/l (Azzella, 2014). For ammonia the range is 0.067–0.077 mg/l and for nitrate 0.85–0.98 mg/l in northern Germany (Becker, 2016b). The range for ammonia in the Netherlands is 0.1–0.3 mg/l (Nat et al., 1994; Simons and Nat, 1996). In France and Portugal *Nitella hyalina* only grows in nutrient poor conditions (van Raam and Maier, 1995).

Water chemistry: *Nitella hyalina* prefers water with circumneutral to basic acidity. The recorded pH values range between 6.3 and 9.3. The species can be found in freshwater and also in brackish waterbodies (Moore, 1986; van Raam, 1998; Koistinen, 2004; Becker, 2019). The recorded conductivity values range between 28 and 1100 μS/cm. In north-western Germany conductivity ranges are 128–255 μS/cm, pH range from 7.1 to 9.3 and total hardness from 2,4 to 2,9 (Becker et al., 2022). In the Netherlands *Nitella hyalina* prefers calcareous conditions (van Raam, 1998) but it can be found in waters low in calcium as well (Doege et al., 2016). The species can be encrusted, total hardness ranges in between 2.0° dH and 3.2° dH, carbonate hardness varies between 1.5° dH and 3.0°dH. The maximum calcium concentration reported is 95.8 mg/l and maximum chloride concentration is 226 mg/l.

Depth range: *Nitella hyalina* prefers shallow water habitats and is usually found in locations with a water depth between 0.3 and 4 m. In lakes this species can be found at a depth of 8–11 m (Becker, 2016b). According to Corillion (1957) *Nitella hyalina* can withstand some dehydration when water levels drop in the summer. The thick mucus layer protects the gametangia in these conditions.

Associated species: *Nitella hyalina* is the character species of the *Nitellopsidetum obtusae* Sauer ex Dąmbska 1961. In most of the Dutch vegetation recordings *Nitella hyalina* is accompanied by *Nitellopsis obtusa* (van Raam, 1998). Other accompanying species are: *Nitella flexilis, Chara aspera, C. virgata, Eleocharis acicularis, Potamogeton berchtoldii, Utricularia vulgaris, U. minor, Elodea nuttallii* and *Ranunculus circinatus*. In NW Germany currently it occurs with: *Nitella opaca, N. translucens, Baldellia ranunculoides, Pilularia globulifera, Potamogeton alpinus* and *Myriophyllum alterniflorum*. In France and Portugal *Nitella hyalina* is accompanied by species of the *Littorelletea, Potametea or Phragmitetea*, e.g. *Chara braunii, C. fragifera, C. aspera, Nitella gracilis, N. confervacea* and *N. translucens* (Corillion, 1957; Krause, 1983b). Espinar et al. (2002) report for salt marshes in South-Spain the accompanying species *Chara connivens, C. canescens* and *C. galioides*. In Spain *Nitella hyalina* often grows in rice fields. This species is also the character species of the *Nitelletum hyalinae* Corillion, 1957 (Corillion, 1957; Csiky et al., 2014) and is characteristic as part of the *Nitelletum syncarpo-tenuissimae* W. Krause, 1969 (Krause, 1969). *Nitella hyalina* can also grow in monospecific stands of a few hundred square metres.

62.4 Variability

Nitella hyalina is hardly varying with respect to its habitus, except for its size. In lakes in France, the German Bodensee and deep pits in Portugal a large form is known: *Nitella hyalina* f. *maxima*. In these waters the plants can reach a height of 1 m.

62.5 Identification and Risk of Confusion

Nitella hyalina can be quite small and develop dense whorls. With this appearance it will resemble the following species: *Nitella tenuissima, N. confervacea* and *N. wahlbergiana*. The arrangement of branchlets in two distinctly different (heteroclemous) whorls consisting of a bigger primary whorl of branchlets and two smaller secondary (accessory) whorls of branchlets is a characteristic of *Nitella hyalina* that separates this species from *N. tenuissima, N. confervacea* and *N. wahlbergiana*. *Nitella congesta* is the dioecious counterpart of *Nitella hyalina* but it is rare and endemic in Australia.

62.6 Distribution

Nitella hyalina has a cosmopolitan distribution (Korsch, 2018). It roughly can be found between 65° north latitude and 50° south latitude and a number of oceanic islands like New Caledonia, New Zealand, Madagascar, Japan and Gran Canarias. The species is missing around the equator. In Europe the centre of distribution of *Nitella hyalina* is located in the western Mediterranean area and in France (Fig. 62.4). In southern Finland in the Baltic area there are some isolated habitats. Although the distribution area is large, *Nitella hyalina* is (very) rare at most sites.

62.7 Status and Threats

The disappearance of *Nitella hyalina* at many sites in Europe is quite attributable to the destruction of its habitats. Especially the loss of small waterbodies is a major threat for the species. Apart from this, for the area of Zürich (Switzerland) Schwarzer (2019) mentioned eutrophication, increase in water turbidity, anthropogenic interference, strong predation by non-domestic herbivorous fish and waterfowl and competition from vascular plants as the main risk factors. *Nitella hyalina* can disappear and reappear quite suddenly for no apparent reason. That makes water management and the carrying out of species protection programs difficult. The current example of

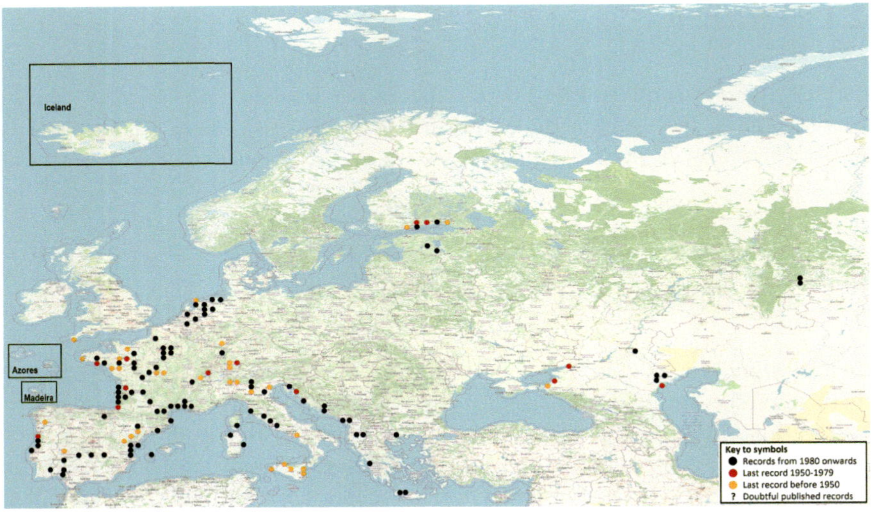

Fig. 62.4 Distribution of *Nitella hyalina* in Europe. For worldwide distribution see Korsch (2018). Black dots are records from 1980 on, red dots from 1950 to 1979 and yellow dots before 1950; unconfirmed/doubtful published records (if applicable) are marked with "?" on top. Map prepared by Heiko Korsch

a mesotrophic lake in north-western Germany with stable and huge stands of *N. hyalina* for more than 15 years, demonstrates, that such a population can almost be destroyed within less than two years by changes in fishery exploitation.

On the other hand this species can tolerate some disturbance: In the Croatian site (on the island of Pag), sheep use the small pond as drinking water, which they use every day in the summer. Therefore the turbidity and eutrophication in the water body is relatively high. Despite these conditions, this species has been found in the pond several times from 2007 to 2017 (July–August). In some cases, the stands were covered by a thin layer of cyanobacteria/filamentous algae (as a result of eutrophication), but they survived. This phenomenon (thin cyanobacteria layer on the thalli) has been reported in other cases with other *Nitella* species, in shallow ponds, puddles (with muddy bottom) in Hungary (J. Csiky pers. comm.).

Nitella hyalina is threatened according to several national European Red Lists (Becker, 2016a). The species is Critically Endangered in the Balkan countries and in Germany (Korsch et al., 2013), Endangered in the Netherlands (van Raam, 2002) and Vulnerable in Finland (Koistinen, 2010) and Spain (Cirujano et al., 2008). *N. hyalina* is Extinct in Great Britain (JNCC 2004) and Regionally Extinct in Switzerland (Auderset Joye and Schwarzer, 2012).

Throughout Europe *N. hyalina* occurs in several habitat types of Community Interest encoded in Annex I of the Habitats Directive 92/43EC (European Commission, 1992). This applies particularly to the oligo- and mesotrophic freshwater habitat types 3110, 3130 and 3140. Moreover, especially in the Mediterranean area, coastal lagoons, encoded as priority habitat type *1150 are in some regions the habitat of *N. hyalina* (e.g., Becker, 2019). From the national point of view various waterbodies are protected as nature reserves or have a related status of conservation.

N. hyalina is not a pioneer species and requires particularly stable oligo- and mesotrophic freshwater habitats.

With regard to the conservation of *Nitella hyalina* appropriate actions are needed. The main actions for protection and conservation are:

- prevention from damage of habitats,
- prevention/reduction of input of nutrient and pollutants,
- prevention/reduction of use-related impairments,
- preservation of the hydrological regimes,
- removal of cyprinids, predominantly carp and bream
- establishment of protected areas
- implementation of valid Action Plans.

In Switzerland an Action Plan focusing on *Nitella hyalina* is already in the process of implementation (Schwarzer, 2019), for Germany and Sardinia national or regional Action Plans are postulated (Becker, 2016b, 2019).

References

Agardh, C. A. (1824). *Systema Algarum*. Berling. [in Latin].

Auderset Joye, D., & Schwarzer, A. (2012). *Rote Liste Armleuchteralgen. Gefährdete Arten der Schweiz, Stand 2010*. Bundesamt für Umwelt BAFU & Laboratoire d'Ecologie et de Biologie Aquatique der Universität Genf (p. 72) [in German].

Azzella, M. M. (2014). Italian Volcanic lakes: A diversity hotspot and refuge for European charophytes. *Journal of Limnology, 73*, 502–510.

Bailly, G., & Schaefer, O. (2010). *Guide illustré des Characées du nord-est de la France* (p. 96). Conservatoire Botanique National de Franche-Comté [in French].

Baumann, E. (1911). Die Vegetation des Untersees (Bodensee). *Archiv für Hydrobiologie, Supplement 1*, I-V, 1–554, Tafeln I-XV [in German].

Becker, R. (2016a). Gefährdung und Schutz von Characeen. In: Arbeitsgruppe Characeen Deutschlands, *Armleuchteralgen: Die Characeen Deutschlands* (pp. 149–191). Springer [in German].

Becker, R. (2016b). *Nitella hyalina*. In: Arbeitsgruppe Characeen Deutschlands, *Armleuchteralgen: Die Characeen Deutschlands* (pp. 443–455). Springer [in German].

Becker, R. (2011). Erstfund der Vielästigen Glanzleuchteralge *Nitella hyalina* (DC.) Agardh in Niedersachsen. *Berichte der Botanischen Arbeitsgemeinschaft Südwestdeutschlands Beiheft, 3*, 5–10. [in German].

Becker, R. (2019). The Characeae (Charales, Charophyceae) of Sardinia (Italy): Habitats, distribution and conservation. *Webbia, 74*, 83–101.

Becker, R., Fuhrmann, K., Meyer-Spethmann, U., Sinkowski, M., & Willen, M. (2022). Verbreitung, Populationsentwicklung und Schutz besonders gefährdeter Arten der Armleuchteralgen (Charales, Characeae) in Nordwestdeutschland. *Drosera, 40*, 81–100. [in German].

Böckermann, T., Becker, R., & Willen, M. (2022). Der Versener Heidesee – vom Baggersee zum fischfreien Hotspot für gefährdete Pflanzenarten in Nordwestdeutschland. *Drosera, 40*, 37–54.

Cirujano, S., Cambra, J., Sánchez Castillo, P. M., Meco, A., & Flor Arnau, N. (2008). *Flora ibérica. Algas continentales. Carófitos (Characeae)* (p. 132). Real Jardín Botánico [in Spanish].
Corillion, R. (1957). Les Charophycées de France et d'Europe Occidentale. *Bulletin de la Société Scientifique de Bretagne, 32* (Hors série I & II), 1–499 [in French].
Csiky, J., Purger, D., & Blaženčić, J. (2014). New occurrence and distribution of *Nitella hyalina* (DC.) Agardh (Characeae) and the first report on *Nitelletum hyalinae Corillion* 1957, in Croatia. *Archiv of Biological Science, 66*, 203–208.
Dilger, M. (2010). Merkmale zur Erkennung von *Nitella-* und *Tolypella-*Arten der Characeaen [sic] über Oosporen aus Sedimenten oder von Pflanzen. *Rostocker Meeresbiologische Beiträge, 23*, 40–59. [in German].
Doege, A., van de Weyer, K., Becker, R., & Schubert, H. (2016). Bioindikation mit Characeen. In Arbeitsgruppe Characeen Deutschlands, *Armleuchteralgen: Die Characeen Deutschlands* (pp. 97–138). Springer [in German].
Espinar, J. L., Garcia, L. V., Garcia Murillo, P., & Toja, J. (2002). Submerged macrophyte zonation in a Mediterranean salt marsh: A facilitation effect from established helophytes? *Journal of Vegetation Science, 13*, 831–840.
European Commission. (1992). Council Directive 92/43/EEC of 21 May 1992 on the conservation of natural habitats and of wild fauna and flora. *Official Journal of the European Communities L 206*, 7.
Groves, J., & Bullock-Webster, G. R. (1920). *The British Charophyta* (p. 405). Ray Society.
JNCC [Joint Nature Conservation Committee]. (2004). *Conservation Designations for UK Taxa*. http://jncc.defra.gov.uk/page-3408.
Koistinen, M. (2004). *Nitella hyalina*. In H. Schubert, & I. Blindow (Eds.), *Charophytes of the Baltic Sea* (pp. 186–192). A. R. G. Gantner.
Koistinen, M. (2010). *Näkinpartaislevät*. In P. Rassi, E. Hyvärinen, A. Juslén, & I. Mannerkoski (Eds.), *Suomen lajien uhanalaisuus: Punainen kirja 2010* (pp. 204–207). Ympäristöministeriö & Suomen ympäristökeskus.
Korsch, H. (2018). The worldwide range of the Charophyte species native to Germany. *Rostocker Meeresbiologische Beiträge, 28*, 45–96.
Korsch, H., Doege, A., Raabe, U., & van de Weyer, K. (2013). Rote Liste der Armleuchteralgen (Charophyceae) Deutschlands. 3. Fassung, Stand: Dezember 2012. *Haussknechtia Beiheft, 17*, 1–32. [in German].
Krause, W. (1969). Zur Characeenvegetation der Oberrheinebene (Characeae in the upper plain of the Rhine). *Archiv Für Hydrobiologie, Supplement, 35*, 202–253. [in German].
Krause, W. (1983). Characeen-Standorte in Portugal mit besonderer Rücksicht auf den Einfluß des Menschen. *Tuexenia, 3*, 289–296. [in German].
Krause, W. (1997). Charales (Charophyceae) Süßwasserflora von Mitteleuropa. In H. Ettl, G. Gärtner, H. Heynig, & D. Mollenhauer (Eds.), *Süßwasserflora von Mitteleuropa BD 18*. Fischer [in German].
Migula, W. (1889–1897). Fünfter Band: Die Characeen. In A. Fischer, E. Fischer, F. Hauck, G. Limpricht, C. Luerssen, W. Migula, H. Rehm, P. Richter, & G. Winter (Eds.) *Dr. L. Rabenhorst's Kryptogamenflora von Deutschland, Oesterreich und der Schweiz* (2nd ed.). Eduard Kummer [in German].
Moore, J. A. (1986). *Charophytes of Great Britain and Ireland*. BSBI Handbook 5. Botanical Society of the British Isles.
Nat, E., Simons, J., de la Haye, M. A. A., & Coops, H. (1994). *Historisch en actueel verspreidingsbeeld van kranswieren in Nederland in samenhang met waterkwaliteitsfactoren*. RIZA werkdocument 94.148X (p. 77). Lelystad [in Dutch].
van Raam, J. C. (1998). *Handboek Kranswieren*. Chara boek [in Dutch].
Romanov, R. E. (2021). Typification of *Chara* (Charales, Charophyceae) species described by C.F. Lessing, F.J. Ruprecht and M.M. Hollerbach. *Notulae Algarum, 213*, 1–7.
Schmidt, D., van de Weyer, K., Krause, W., Kies, L., Garniel, A., Geissler, U., Gutowski, A., Samietz, R., Schütz, W., Vahle, H.-C., Vöge, M., Wolff, P., & Melzer, A. (1996). Rote

Liste der Armleuchteralgen (Charophyceae) Deutschlands. 2. Fassung, Stand: Februar 1995. *Schriftenreihe Für Vegetationskunde, 28*, 547–576. [in German].
Schwarzer, A. (2019). Aktionsplan Vielästige Glanzleuchteralge (*Nitella hyalina* (DC.) C. Agardh). AP ZH 1–53. *Artenschutzmassnahmen für gefährdete Farn- und Blütenpflanzen im Kanton Zürich* (p. 30). Fachstelle Naturschutz Zürich [in German].
Simons, J., & Nat, E. (1996). Past and present distribution of stoneworts (Characeae) in The Netherlands. *Hydrobiologia, 340*, 127–135.
van Raam, J. C. (2002). Rode lijst 2000 voor de Nederlandse kranswieren. *Nieuwsbrief Kranswieren, 6*(11), 7–8.
van Raam, J. C., & Maier, E. X. (1992). Overzicht van de Nederlandse kranswieren. *Gorteria, 18*, 111–116. [in Dutch].
van Raam, J. C., & Maier, E. X. (1995). Nederlandse kranswieren. 5. Klein glanswier [*Nitella hyalina* (DC.) Agardh]. *Gorteria, 21*, 101–106. [in Dutch].
Wood, R. D. (1965). *Monograph of the Characeae*. J. Cramer.

Chapter 63
Nitella mucronata

Maciej Gąbka and Jacek Urbaniak

Scientific name: ***Nitella mucronata*** **(A. Braun) Miq., Fl. Belg. Sept. 2(2): 428. 1840**

≡ *Chara mucronata* A. Braun in Anna. Sci. Nat. Bot. 1: 351. 1834
Neotype (Wood, 1965: 504): Salem/Oktober 1858/Jack (NY); Braun, Rabenhorst and Stizenberger, Characeen Eur. 30

= *Nitella flabellata* Kütz., Phycol. General.: 317. 1843
Lectotype (designated here): Eilenburg/[no date]/[no collector] (L 4096591; upper row, left specimen)

= *Nitella mucronata* (A. Braun) Miq. var. *heteromorpha* A. Braun in Flora 18: 52. 1835 Lectotype (designated here by Gąbka & Urbaniak): Reichenbach, H. G. L. (1830): Iconographia Botanica, seu plantae criticae 8: t. 796/1072 "*Chara flexilis* var. *nidifica* R."

Local names: BELARUSIAN: Нітэла спічастая, DANISH: Brodspidset glanstråd, DUTCH: Puntdragend glanswier, ENGLISH: Pointed Stonewort, ESTONIAN: Teravlevine nitell, GERMAN: Stachelspitzige Glanzleuchteralge, LITHUANIAN: Smailiaviršūnis menturdumblis, NORWEGIAN: Broddglattkrans, POLISH: krynicznik kolczasty, RUSSIAN: Нителла остроконечная, SWEDISH: Uddslinke, UKRAINIAN: Нітела гострокінцева.

M. Gąbka (✉)
Department of Hydrobiology, Faculty of Biology, Adam Mickiewicz University, Poznań, Uniwersytetu Poznańskiego Str. 6, 61-614 Poznań, Poland
e-mail: gmaciej@amu.edu.pl

J. Urbaniak
Department of Botany and Plant Ecology, Wrocław University of Life and Environmental Sciences, pl. Grunwaldzki 24a, 50-363 Wrocław, Poland

H. Schubert et al. (eds.), *Charophytes of Europe*,
https://doi.org/10.1007/978-3-031-31898-6_63

63.1 Morphology

See Figs. 63.1 and 63.2.

Habitus: *Nitella mucronata* is a small to medium size plant, 10-35 cm high with a moderately robust plant axis, 0.8–1.5 mm in diameter. The colour of the plant is green to dark green, sometimes brownish when found on peat. A richly branched and bushy species which occasionally forms heads. Often encrusted with visible characteristic narrow stripes (streaks). *Nitella mucronata* can grow as solitary plants as well as developing dense underwater charophyte meadows (carpets).

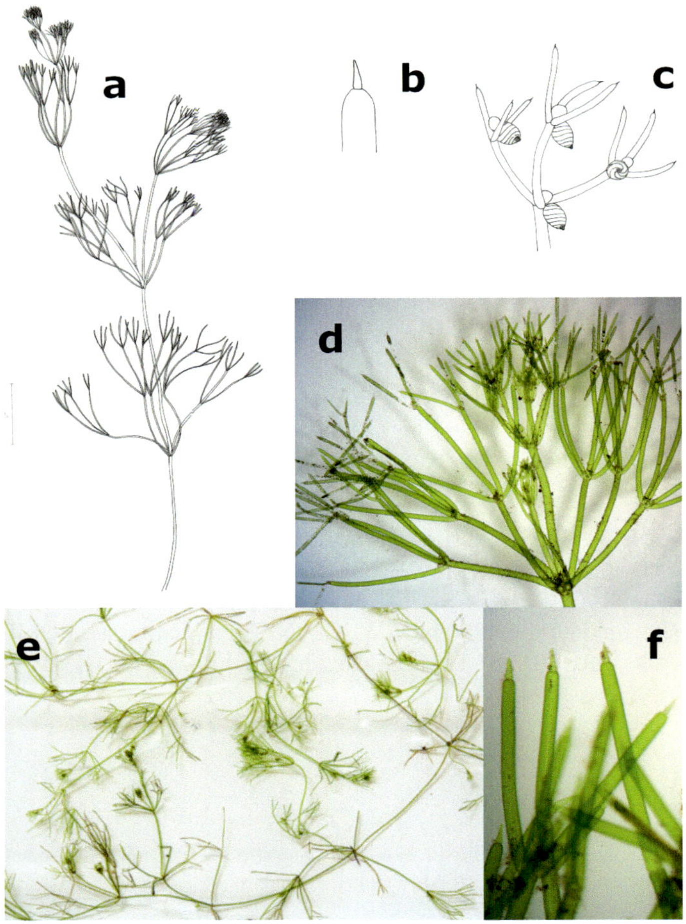

Fig. 63.1 *Nitella mucronata,* details. **a**, **e** General appearance of N. mucronata *;* **b**, **f** Conical, acute and mucronate apical cell—about 1/3 as broad as subapical cell; **d**: Sterile branches in a whorl. Drawings and photographs: **a**, **b**, **c**—Kinga Gawrońska, Maciej Gąbka; **d**, **e**, **f**—Jacek Urbaniak

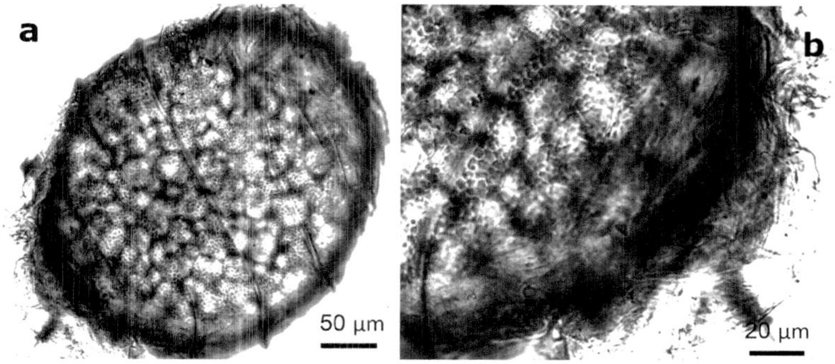

Fig. 63.2 Reticulate oospore ornamentation of *Nitella mucronata* at low (**a**) and high (**b**) magnification. Photographs: Jacek Urbaniak

Size: Medium sized to large plant, 10–35 cm high (up to 120 cm).

Main axis: The axis is slender to moderately stout, 0.8–1.5 mm in diameter, internodes 1–2 times as long as the branchlets.

Cortex: Absent, does not form cortex.

Spine-cells: Absent, does not form spine-cells.

Stipulodes: Absent, does not form stipulodes.

Branchlets: Fertile branchlets 6–7 in a whorl, divided twice or three times, sterile divided once or twice. Fertile branchlet primary branchlet segment 1/2 – 2/3 of total branchlet length, secondary segments 3–5, tertiaries 2–4. Sterile branchlets 4–6 in a whorl. End segment 2–3 celled, variable in length. Last cell is short, conical, acute and mucronate—about 1/3 as broad as the subapical cell.

Bract-cells: Absent, does not form bract-cells.

Bracteoles: Absent, does not form bracteoles.

Sexuality: *Nitella mucronata* is a monoecious species with gametangia at the branchlet nodes, situated on the furcations.

Oogonia: Oogonia are solitary or geminate, often absent from the lowest branchlet nodes (variability presented in Table 63.1).

Oospores: Oospore yellow brown to brown, reticulate. Variability of oospores is presented in Table 63.2. More detailed data about oospore features were presented by Urbaniak (2009).

Antheridia: Variability in diameter is presented in Table 63.3.

Bulbils: Not reported on the rhizoids.

Table 63.1 Oogonia morphometry of *Nitella mucronata*

Author	Length	Width	Coronula length	Coronula width	Number of helical stripes
Migula (1889–1897)	*430–625* [1]	320–400	30–55	90	7–9
Corillion (1957a)	430–625	320–400	30–55	–	8–9
Wood (1965)	*430–575*[1]	352–450	35–55	68–80	7–9
Dąmbska and Karpiński (1954), Dąmbska (1964)	*430–625*[1]	320–400	30–55	90	7–9
Krause (1997)	450–550	320–400	–	–	7–9
Cirujano et al., (2008)	370–575	260–350	35–55	–	–
Urbaniak and Gąbka (2014)	350–550	310–520	–	–	–
Total range	**350–575**	**260–540**	**30–55**	**68–90**	**7–9**

[1]Oogonia were measured including coronula
Length in the first column is, except for values kept in italics, excluding the coronula; all values for length and width in μm

Table 63.2 Oospore morphometry of *Nitella mucronata*

Author	Length	Width	Number of striae	Colour
Migula (1889–1897)	250–350	235–300	7	Dark brown to black
Corillion (1957a)	250–350	235–300	6–7	Brown
Wood (1965)	256–350	234–300	6–7(8)	Dark brown
Dąmbska and Karpiński (1954), Dąmbska (1964)	250–350	235–300	7	Dark brown to black
Haas (1994)	220–300	460–500	5–7	Light (yellowish) brown
Krause (1997)	350–450	235–300	6–8	Dark brown to black
Urbaniak (2009)	275–421	206–430	4–6	–
Cirujano et al., (2008)	250–370	220–300	–	Dark brown to black
Urbaniak and Gąbka (2014)	275–420	205–430	–	Yellow brown to brown
Total range	**220–450**	**205–430(500)**	**4–8**	

All values for length and width in μm

Table 63.3 Antheridia morphometry of *Nitella mucronata*

Author	Diameter (μm)
Migula (1889–1897)	200–310
Corillion (1957a)	240–310
Wood (1965)	240–310
Dąmbska and Karpiński (1954), Dąmbska (1964)	200–310
Krause (1997)	240–300
Cirujano et al., (2008)	240–310
Urbaniak and Gąbka (2014)	160–320
Total range	**150–320**

63.2 Phenology and Reproduction

Nitella mucronata is a perennial species and, especially in the deeper parts of the lakes, green during the winter. The species is richly fertile producing oospores from early summer to autumn (from July to September). Ripe oospores are not commonly observed, probably falling off early.

63.3 Habitat Conditions and Ecology

Habitat type: In Europe, *Nitella mucronata* grows in stagnant or slowly flowing water. It occurs mainly in hardwater oligotrophic and mesotrophic lakes but can be found in dystrophic lakes, containing high amounts of humic substances and organic acids too (humic lakes). It is common in ponds (also fishponds), pools, floodplains (periodically flooded meadows or fields) and in peatlands. *Nitella mucronata* also occurs in artificial water reservoirs such as clay pits, gravel pits and ditches. The species has also been observed in deepened channels and in polder ditches. The species is also often found in river oxbow lakes, and less often in slow flowing water in streams and channels.

Sediment: *Nitella mucronata* has been found on soft organic substrates e.g., detritus gyttia or peat; less frequently on sandy substrates or clay.

Nutrient conditions: *Nitella mucronata* occurs in lakes in a wide trophic spectrum from mesotrophic to moderately eutrophic waters; it is a eurytopic species. This species is relatively eutrophication resistant (e.g., Caisová et al., 2008; Krause, 1981; Lambert-Servien et al., 2006; Langangen, 2007a; Moore, 1986; Romanov and Barinova, 2016). *N. mucronata* grows in wide ranges of nutrient concentrations between 0.02–1.3 mg L^{-1} for total phosphorus and 0.06–17.60 mg L^{-1} for total nitrogen (e.g., Simons and Nat, 1996; Langangen, 2007a; Gąbka, 2009; Urbaniak and Gąbka,

2014). In shallow, dystrophic (humic lakes) it may be the dominant species, growing densely (Gąbka, 2009; Urbaniak and Gąbka, 2014).

Water chemistry: *Nitella mucronata* is mostly known from freshwater habitats and is recorded in from neutral to lightly alkaline water (pH 6.0–9.0), rich in calcium and sulphates. Numerous data indicate a wide spectrum of calcium content in the water where *N. mucronata* grows, however, this species is very often noted in hard waters rich in calcium (Krause, 1981; Moore, 1986; Romanov and Barinova, 2016). It may also occur in acidic and calcium-poor waters in dystrophic (humic) lakes (Gąbka, 2009; Mróz and Gąbka, 2013). Details of physico-chemical properties of waters are presented in Table 63.4 based on the published data (e.g., Bornette and Arens, 2002; Gąbka, 2009; Lambert-Servien et al., 2006; Mróz and Gąbka, 2013; Romanov and Barinova, 2016; Simons and Nat, 1996). *Nitella mucronata* can also be found in brackish water (Langangen, 2007a). Detailed information on the physico-chemical properties of water determined by Simons and Nat from the Netherlands, seems to confirm the possibility of *N. mucronata* occuring in waters with high conductivity (1080 μS cm^{-1}) and high concentrations of Cl^- (226 mg l^{-1}) and $SO_4{}^{2-}$ (137 mg l^{-1}).

Depth range: *Nitella mucronata* is found in shallow water (usually up to 2.5 m), but occasionally plants are found at down to 4.0 m (9 m) depth. Krause (1997) reports *N. mucronata* also at 25 m depth.

Associated species: *Nitella mucronata* often forms dense underwater monospecific meadows especially in lakes. The community with the dominance of *N. mucronata* was described as *Nitelletum mucronatae* Tomaszewicz 1979 ex Hrivnák et al., (2001) and it is widely known in Europe (Dąmbska, 1966; Borhidi and Santa, 1999; Schratt, 1993; Doll, 1989; Hrivnák et al., 2001). *Nitella mucronata* can co-occur with many

Table 63.4 Water chemistry from European locations of *N. mucronata*

Parameter	Unit	Minimum	Maximum
Colour	mg Pt l^{-1}	2	160
pH	–	6.0	9.0
O_2	mg l^{-1}	4.5	9.8
Conductivity	μS cm^{-1}	37	477(1080)
N-NH_4	mg l^{-1}	0.00	0.19
N-NO_3	mg l^{-1}	0.00	0.40(3.76)
P-PO_4	mg l^{-1}	0.01	0.08
Ca	mg l^{-1}	5.7	75.6
Mg	mg l^{-1}	1.2	12.1
Na	mg l^{-1}	1.3	10.5
K	mg l^{-1}	1.12	8.2
Fe	mg l^{-1}	0.04	2.2
SO_4	mg l^{-1}	0	65(137)

Outlier values are given in brackets

species of vascular plants e.g., *Potamogeton natans, Elodea canadensis, Ceratophyllum demersum, Fontinalis antipyretica, Chara globularis, C. virgata, Nitella flexilis* and *Nitellopsis obtusa*.

63.4 Variability

Nitella mucronata is a very variable species, rich in forms and varieties, but the taxonomy of them is complicated. Some authors emphasise the pronounced polymorphic character of *N. mucronata* (Allen, 1928; Migula, 1889–1897). An overview of the variability of this species was given by Wood and Imahori (1964). The variability mainly concerns the size, branchlets and the number of cells in the branchlets. *Nitella mucronata* varies largely in size; the smallest plants are about 5 cm high, but giants of more than 120 cm can be found as well. Another aspect of variability concerns the apical part, which can be condensed, forming a head or, as it is the case for *N. mucronata* var. *heteromorpha*, exhibiting a diffuse and indistinct terminal part.

An interesting variety of *N. mucronata* is var. *gracillima* with a 2–3 celled end segment where the penultimate and apical cell are of similar size at their junction but tapering to the apex (Groves and Bullock-Webster, 1920). *N. mucronata* var. *gracillima* can therefore be confused with *Nitella gracilis*.

63.5 Identification and Risk of Confusion

Nitella mucronata can be confused with *N. gracilis*, which however is more slender and delicate and has different last cells of the end segments. *Nitella mucronata* has branches with the last cell of the end segment short, conical and mucronate (about 1/3 as broad as the subapical cell), whereas the last cell of *N. gracilis* is acute, at the base about 2/3 as broad as the subapical cell. *Nitella gracilis* has a more slender and delicate habit and punctate oospore wall ornamentation, whereas the oospore of *N. mucronata* is reticulate (light microscope). *Nitella mucronata* can also be confused with *Nitella wahlbergiana*, which is known from Scandinavia, but is in general much smaller. It is up to 10 cm high, with the main axis less than 0.5 mm in diameter. The branchlets of *N. wahlbergiana* are divided two or three times, the secondary branchlet segments are much shorter than other branchlet segments, and this gives the whorls a characteristic, condensed appearance, like heads.

63.6 Distribution

See Fig. 63.3.

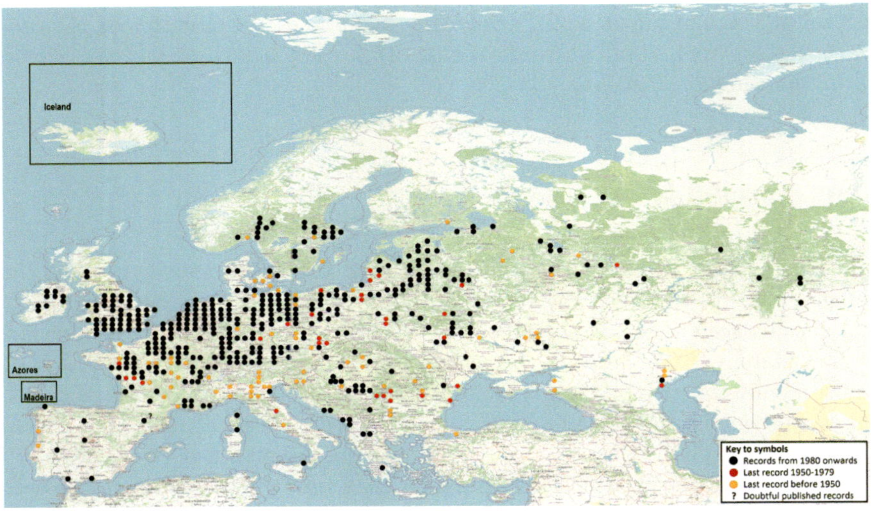

Fig. 63.3 European distribution of *Nitella mucronata*. For worldwide distribution see Korsch (2018). Black dots are records from 1980 on, red dots from 1950 to 1979 and yellow dots before 1950; unconfirmed/doubtful published records (if applicable) are marked with "?" on top. Map prepared by Heiko Korsch

Nitella mucronata is a cosmopolitan species recorded from almost all of the World. It is reported from Central and South America, Africa, Asia and Europe (Corillion, 1957a, Korsch, 2018, Migula, 1889–1897). In Europe it is common from Finland, Hungary, Spain, Germany, Poland and France, locally in Ireland, Norway and Turkey (Krause, 1997). It is especially common in Britain, France, Germany and Poland (over 100 currently known localities). Currently the species is disappearing from many localities in Europe, and it is a vulnerable species in some parts of its distribution range. The most probable threats are eutrophication and acidification and humification of the lakes.

63.7 Status and Threats

Nitella mucronata is assessed as CR (Critically Endangered) in the Iberian Peninsula, EN (Endangered) in Denmark and Switzerland. In other countries such as Norway and Poland, the species is classified as Vulnerable or R (Rare) in Ireland.

A decline of *N. mucronata* in previously known sites has been observed, but new locations are also being found, although the species requires detailed monitoring throughout parts of its range. *Nitella mucronata* is threatened by various threats such as eutrophication, that can lead to decreasing light penetration in lakes and other water bodies where it can be found.

References

Allen, G. O. (1928). Charophyte notes from Saharanpur, UP. *Journal of Indian Botanical Society, 7*, 49–69.

Borhidi, A., & Sánta, A. (Eds.). (1999). *Vörös könyv Magyarország növénytársulásairól*. Természet búvár Alapítvány Kiadó [in Hungarian].

Bornette, G., & Arens, M. F. (2002). Charophyte communities in cut-off river channels the role of connectivity. *Aquatic Botany, 73*, 149–162.

Caisová, L., Husak, S., & Komárek, J. (2008). *Nitella mucronata* (Br.) Miquel (Charophyta) in the Czech Republic. *Fottea, 8*, 105–107.

Cirujano, S., Cambra, J., Sánchez Castillo, P. M., Meco, A., & Flor Arnau, N. (2008). *Flora ibérica. Algas continentales. Carófitos (Characeae)*. Real Jardín Botánico [in Spanish].

Corillion, R. (1957a). Les Charophycées de France et d'Europe Occidentale. *Bulletin de la Société Scientifique de Bretagne* 32 (Hors série I & II), 1–499. [in French].

Dąmbska, I. (1966). Zbiorowiska ramienic Polski. *Prace Komisji Biologicznej, Poznańskie Towarzystwo Przyjaciół Nauk, Wydział Matematyczno Przyrodniczy, 31*, 1–76. [in Polish].

Dąmbska, I., & Karpiński, J. (1954). *Ramienice. Klucz do oznaczania gatunków krajowych*. Państwowe Wydawnictwo Naukowe [in Polish].

Dąmbska, I. (1964). *Charophyta - ramienice*. Państwowe Wydawnictwo Naukowe [in Polish].

Doll, R. (1989). Die Pflanzengesellschaften der stehenden Gewässer der DDR. Teil 1: Die Gesellschaften des offenen Wassers (Characeen-Gesellschaften). *Feddes Repertorium, 100*, 281–324. [in German].

Gąbka, M. (2009). *Charophytes of the Wielkopolska region (NW Poland): distribution, taxonomy, and autecology*. Bogucki Wydawnictwo Naukowe.

Groves, J., & Bullock-Webster, G. R. (1920). *The British Charophyta* (p. 405). Ray Society.

Haas, J. N. (1994). First identification key for charophyte oospores from central Europe. *European Journal of Phycology, 29*, 227–235.

Hrivnák, R., Ot'ahel'ová, H., & Husák, S. (2001). *Nitella mucronata* and *N. translucens*—contribution to occurrence and ecology in Slovakia. *Biologia, 56*, 13–15.

Korsch, H. (2018). The worldwide range of the Charophyte species native to Germany. *Rostocker Meeresbiologische Beiträge, 28*, 45–96.

Krause, W. (1981). Characeen als Bioindikatoren für den Gewässerzustand. *Limnologica, 1*, 399–418. [in German].

Krause, W. (1997). Charales (Charophyceae) Süßwasserflora von Mitteleuropa. In H. Ettl, G. Gärtner, H. Heynig, & D. Mollenhauer (Eds.), *Süßwasserflora von Mitteleuropa 18*. Fischer [in German].

Lambert-Servien, E., Clemenceau, G., Gabory, O., Douillard, E., & Haury, J. (2006). Stoneworts (Characeae) and associated macrophyte species as indicators of water quality and human activities in the Pays-de-la-Loire region, France. *Hydrobiologia, 570*, 107–115.

Langangen, A. (2007). *Charophytes of the Nordic countries*. Saeculum ANS.

Migula, W. (1889–1897). Fünfter Band: Die Characeen. In A. Fischer, E. Fischer, F. Hauck, G. Limpricht, C. Luerssen, W. Migula, H. Rehm, P. Richter, & G. Winter (Eds.), *Dr. L. Rabenhorst's Kryptogamenflora von Deutschland, Oesterreich und der Schweiz* (2nd edn). Eduard Kummer [in German].

Moore, J. A. (1986). *Charophytes of Great Britain and Ireland*. BSBI Handbook 5. Botanical Society of the British Isles.

Mróz, P., & Gąbka, M. (2013). Charophytes (Characeae, Charophyta) of peatland habitats in the vicinity of Drawsko and Miały (Noteć forest), NW Poland. *Botanica Steciana, 17*, 71–75.

Romanov, R. E., & Barinova, S. S. (2016). Species of *Nitella* (Charophyceae, Charales) from Israel: Low species richness and rare occurrence. *Botanica Serbica, 40*, 217–227.

Schratt, L. (1993). *Charetea fragilis*. In G. Grabherr, & L. Mucina (Eds.), *Die Pflanzengesellschaften Österreichs. Teil II. Natürliche waldfreie Vegetation* (pp. 45–54). Gustav Fischer [in German].

Simons, J., & Nat, E. (1996). Past and present distribution of stoneworts (Characeae) in The Netherlands. *Hydrobiologia, 340*, 127–135.
Urbaniak, J. (2009). Oospore variation in *Nitella gracilis* and *Nitella mucronata* (Charales, Charophyceae) from Poland. *Biologia, 64*, 252–260.
Urbaniak, J., & Gąbka, M. (2014). *Polish Charophytes. An illustrated guide to identification.* Uniwersytet Przyrodniczy we Wrocławiu, Wrocław.
Wood, R. D. (1965). *Monograph of the Characeae*. J. Cramer.
Wood, R. D., & Imahori, K. (1964). *Iconograph of the Characeae*. J. Cramer.

Chapter 64
Nitella opaca

Karin Pall, Ralf Becker, and Hendrik Schubert

Scientific name : ***Nitella opaca*** **(C. Agardh ex Bruzelius) C. Agardh, Syst. Alg.: 124. 1824**

≡ *Chara opaca* C. Agardh ex Bruzelius, Observ. Charae: 16: 1824
Lectotype (designated here by Pall, Ralf Becker & Hendr. Schub.): Smith, 1802, Engl. Bot 15: 1070.

Nomenclatural note: According to taxonomic literature (Stafleu and Cowan, 1976) Bruzelius' "Observationes in genus Charae" was published in February 1824, Agardh's "Systema Algarum" between May and September, probably in September, 1824. – Bruzelius (1824) cited the drawing of Smith (1802) which is selected as lectotype and is clearly dioecious.

Local names: BELARUSIAN: Нітела цьмяная, DANISH: Almindelig kransnål (also used for *N. flexilis*); DUTCH: Donker glanswier; ENGLISH: Dark stonewort; ESTONIAN: Tume nitell, FINNISH: Himmeäsiloparta; GERMAN: Dunkle Glanzleuchteralge; LITHUANIAN: Tamsusis menturdumblis; POLISH: Krynicznik ćmy; RUSSIAN: Нителла тусклая; SWEDISH: Mattslinke; UKRAINIAN: Нітела тьмяна.

K. Pall (✉)
SYSTEMA Bio- und Management Consulting GmbH, Bensasteig 8, 1140 Wien, Austria
e-mail: karin.pall@systema.at

R. Becker
Am Schützenplatz 58, 26121 Oldenburg, Germany
e-mail: becker.r@posteo.de

H. Schubert
Universität Rostock, Institut für Biowissenschaften, Albert-Einstein-Straße 3, 18059 Rostock, Germany
e-mail: hendrik.schubert@uni-rostock.de

H. Schubert et al. (eds.), *Charophytes of Europe*,
https://doi.org/10.1007/978-3-031-31898-6_64

64.1 Morphology

Figs. 64.1, 64.2 and 64.3.

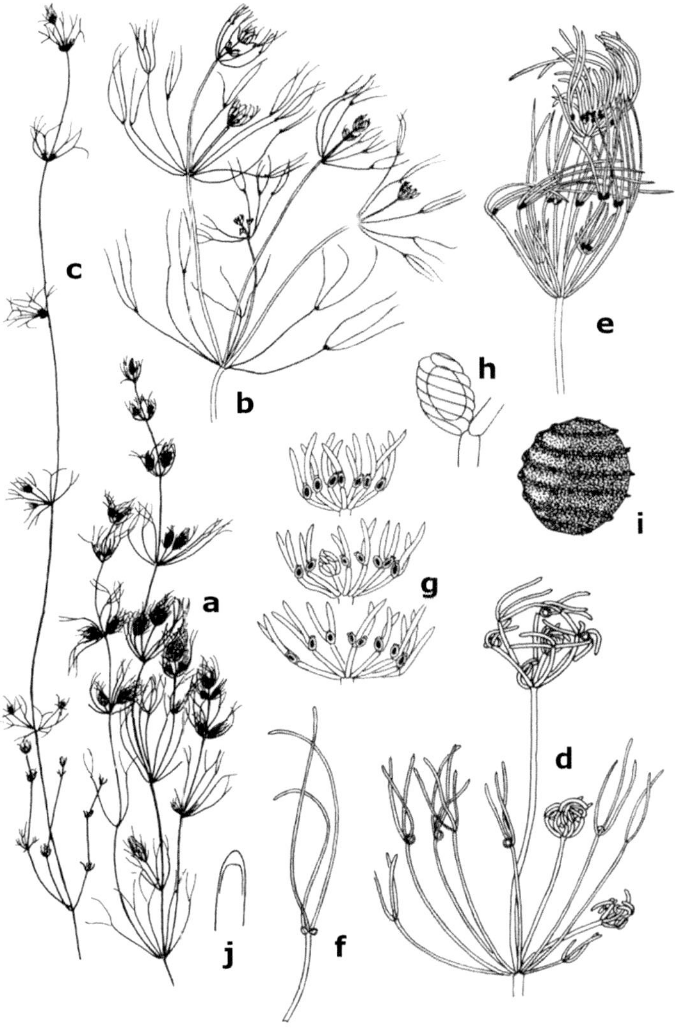

Fig. 64.1 *Nitella opaca*. **a** Shallow water form; **b** Shoot tip; **c** Deep water form; **d**, **e** Shoot tip, branchlet whorls one side combed; **f** Branchlet of female plant; **g** Isolated fertile whorls of female plant; **h** Oogonium; **i** Oospore; **j** Apex of branchlet. Drawings by Werner Krause

Fig. 64.2 Habitus and habitat appearance of *Nitella opaca*. **a** Habitus, female plant; **b** Sparse stands with caddisfly larvae cases; **c** Dense stands in shallow, stagnant water; **d** Luxuring individual. Photographs: **a**—Wil Leurs, Natuurfotografie and Hydrobiologie Oldeberkoop, the Netherlands, **b**, **c**, **d**—Klaus van de Weyer

Fig. 64.3 Habitus details. **a** Male specimen, condensed head with developing and ripe antheridia; **b** Male specimen, long-stalked antheridium; **c** Female specimen, pair of oogonia; **d** Female specimen, condensed head with developing oogonia; **e–g** SEM photographs of an oospore taken from BM013828849, a plant sampled by L. Agassiz at Valorbey, Switzerland, side view (**e**), basal plate view (**f**) and showing the smooth membrane and fossa walls (**g**) typical for *N. opaca*. Photographs: **a**, **b**, **c**, **d**—Heiko Korsch; **e**, **f**, **g**—Michelle T. Casanova

Habitus: Plants are slender and collapsing when taken out of the water. The plant mostly isn't incrusted but can be, depending on habitat. The plant's appearance is shining dark green and somewhat opaque.

Size: *Nitella opaca* is of small to medium size, most specimens are between 10 and 30 cm length, in temporary ponds sometimes as small as 5 cm (Becker and Wolff, 2016); in deeper lakes lengths between 50 to 80 cm commonly are reached. Krause (1997) and van Raam (1998) report maximum lengths up to 1 m.

Main axis: Main axis diameter is between 0.5 mm and 0.8 mm, rarely up to 1.5 mm (Migula, 1889–1897). Internodes are between 1 and 4 cm length, about 1–4 times longer than the respective branchlets, rarely shorter (Becker and Wolff, 2016). Each whorl typically develops 2 (sometimes 3) adventitious axes.

Cortex, spine-cells, stipulodes: Absent.

Branchlets: Whorls consist regularly of 6–8 branchlets which are divided once at the tips (often undivided). The apical cell of the branchlets is pointed, but can also form an obtuse tip. In contrast to *N. flexilis* thickened cell walls of the end cell have never been reported. The short fertile branchlet whorls are often constricted to dense heads and commonly one-side combed (Krause, 1997).

Bract-cells, bracteoles: Absent.

Bractlets: Bractlets are not observed for *Nitella opaca.*

Sexuality: The species is dioecious.

Oogonia: Oogonia are without mucus and placed solitary or in pairs (rarely in groups of three) at the furcation(s) of the branchlets. Table 64.1 gives an overview of the quantitative morphometry data.

Oospores: Oospores of *Nitella opaca* are in the typical range for *Nitella*-species, most of the authors give a range of 350–530 μm length and 300–450 μm width (Table 64.2). However, size data differ largely between the authors. For length, minimum values are 290 and 300 μm reported by Dilger (2010) and Migula (1889–1897), respectively, a maximum length of 610 μm is reported by Dilger (2010). With respect to width Migula (1889–1897) as well as Urbaniak and Gąbka (2014) reported minimum values of 230 and 240 μm. Trajanovska et al. (2012) give a maximum value of 450 μm. The colour of ripe oospores is dark brown to black. Striae are well developed and thick. Oospore membrane is described as smooth or scabrous by several authors (Groves and Bullock-Webster, 1920; Corillion, 1957; Krause, 1997; Becker and Wolff, 2016).

Antheridia: Antheridia are relatively large, reaching a diameter of up to 800 μm. Ripe antheridia are greenish-brown to orange-brown and consistently without mucus. Variability in antheridia diameter within a population can be extreme as between 220 and 815 μm as reported by Trajanovska et al. (2012) for Lake Ohrid (Macedonia). Antheridia are placed solitary or in pairs at the branchlet furcation(s), ripening simultaneous with the oogonia of the female plants. Table 64.3 gives an overview about published diameter ranges.

Bulbils: No recent references to bulbils of *N. opaca* are known and observations on fresh material or herbarium records haven't shown the presence of such. However, presence of bulbils is mentioned by Groves and Bullock-Webster (1920), but without

Table 64.1 Oogonia morphometry

Author	Length	Width	Coronula length	Coronula width	Number of helical stripes
Migula (1889–1897)	450–510	400–460	Up to 30	–	–
Groves and Bullock–Webster (1920)	650–700	500–565	25–40	–	7–9
Corillion (1957a)	580–700	500–570	25–40	–	7–9
Compère (1992)	540–600	500–570	–	–	–
Krause (1997)	600–700	500–600	40	70	7–9
van Raam (1998)	Up to 650	Up to 600	–	–	–
Langangen (2007a)	Up to 700	Up to 600	–	–	–
Caisová and Gąbka (2009)	450–500	400–460	–	–	–
Bailly and Schaefer (2010)	500–600	500–600	–	–	–
Trajanovska et al., (2012)*	*232–648*	*166–579*	*27–99*	*32–165*	*7–9*
Korsch (2013)	–	–	–	–	7–9
Urbaniak and Gąbka (2014)	450–585	410–490	–	–	–
Grillas et al., (2004)	600–700	500–600	-	-	-
Total range	**450–700**	**400–600**	**25–40**	**70**	**7–9**

*The very diverging values of this source were not included into the total range
Length in the first column is excluding the coronula, its length and width are given separately; all values for length and width in μm

giving a reference. Olsen (1944) refers to Wahlstedt (1864), who himselves refers to an observation of Nordstedt, presented the same year (published 1865) at the 9th meeting of the Scandinavian Naturalists—however, it needs to be clarified whether the text published by Wahlstedt (1864) really refers to bulbils: "*Nitella opaca* Ag. är åtminstone understundom flerårig och skjuter då tidigt på våren adventivknoppar och knoppgrenar från de öfvervintrande stjelkarnas leder. Knoppars uppkomst på rötterna har Nordstedt redan visat. Denna art har en seg och fast consistens, hvilket gör öfvervintringen lätt. Frukterna mogna temligen tidigt på sommaren, och groningen tyckes med anledning häraf infalla om efterhösten" [*Nitella opaca* Ag. is at least occasionally perennial and then produces adventitious buds and bud branches in early spring at the joints of the overwintering stems. The emergence of buds on the roots has already been shown by Nordstedt. This species has a tough and firm consistency, which makes overwintering easy. The fruits ripen fairly early in the summer, and it seems reasonable that they fall and germinate in the autumn].

Table 64.2 Oospore morphometry. All values for length and width in μm

Author	Length	Width	Number of striae	Colour
Migula (1889–1897)	300–360	240–300	6–7	Dark brown to black
Groves and Bullock-Webster (1920)	375–425	350–400	7	Chestnut-brown to black
Corillion (1957a)	375–425	350–400	7	Dark brown to black
Compère (1992)	–	–	5–7	Dark brown to black
Krause (1997)	350–450	300–400	6–7	Dark brown to black brown
van Raam (1998)	Up to 400	Up to 400	5	–
Dilger (2010)	(290)350–530(610)	(250)300–450(500)	(5)6–7(8)	Red brown to black
Bailly and Schaefer (2010)	350–400	350–400	5–7	Dark brown to black
Trajanovska et al., (2012)	320–465	310–450	6–8	Dark brown
Korsch (2013)	450–575	350–425	6–7	Dark brown
Urbaniak and Gąbka (2014)	350–470	230–310	–	Brown to dark brown
Becker unpubl.	425–500	300–425	–	–
Grillas et al., (2004)	350–500	350–400	6–7	Brown to black
Total range	**(290)300–575(610)**	**(230)240–450(500)**	**5–8**	

64.2 Phenology and Reproduction

Nitella opaca is often regarded as a spring species, germinating in autumn and developing large numbers of ripe gametangia as early as in April (Germany). In the Mediterranean region fructification starts in February/March (Becker and Wolff, 2016; Grillas et al., 2004; Mouronval et al., 2015) and ends in May (Grillas et al., 2004). There are even reports of fructification as early as in December (Westermann and Westermann, 1998) in springs in southern Germany. But in literature, data about fructification give a more diverse picture. Stands below the thermocline develop gametangia in full summer (van de Weyer and Krautkrämer, 2009) or, as reported from Lake Ohrid, even later in August (Trajanovska et al., 2012). This applies, according to observations of the authors, also for lakes of the Alpine region. There, in high altitudes fructification takes place later than in low altitudes. As well at higher latitudes, fructification is delayed to April-August in Denmark (Olsen, 1944)

Table 64.3 Antheridia morphometry

Author	Diameter (μm)
Migula (1889–1897)	Up to > 600
Groves and Bullock-Webster (1920)	650–775
Corillion (1957)	650–775
Compère (1992)	600–800(900)
Krause (1997)	650–700
van Raam (1998)	Up to 800
Langangen (2007)	Up to 800
Caisová and Gąbka (2009)	600–700
Bailly and Schaefer (2010)	650–800
Trajanovska et al., (2012)*	*220–815*
Korsch (2013)	400–525
Urbaniak and Gąbka (2014)	650–770
Becker and Wolff (2016)	450–800
Grillas et al., (2004)	650–775
Muller et al., (2017)	600–800
Total range	**400–800(900)**

*The very diverging values of this source were not included into the total range

and August–September in Sweden (Blindow, 2004a; Langangen, 2007). The species in principle can be perennial (Wahlstedt, 1864; van de Weyer and Krautkrämer, 2009). This applies specially to stands in deeper lakes. However, in small ponds and ditches and in the Mediterranean *N. opaca* is an annual species and also from larger lakes decay of the stands after fructification is often recorded (Becker and Wolff, 2016; Mouronval et al., 2015; Muller et al., 2017; Pätzold, 2011; van Raam, 1998).

64.3 Habitat Conditions and Ecology

See Fig. 64.2.

Habitat type: *Nitella opaca* is found in almost all kinds of habitats, stagnant as well as running water, from ephemeral small ponds to large and deep lakes, from crystal-clear springs and bog puddles to eutrophic fishponds and mining lakes (for a detailed list see Becker and Wolff, 2016). However, the species is restricted to freshwater conditions, its upper salinity limit of about 3 PSU is well within the oligohaline range (Blindow, 2004b; Grillas et al., 2004). Accordingly, for most countries *N. opaca* has been recorded only in freshwater. *N. opaca* is a type-specific charophyte species of lakes of the Alpine and peri-Alpine region (Pall et al., 2005). *Nitella opaca* is the most frequent charophyte species in running waters in Sardinia (Becker and Ziesmer,

1989). Reliable records from brackish-water habitats are only available for Finland and Sweden. For Denmark, Olsen (1944) describes an occurrence from Nissum Fjord on the Danish west coast, but does not provide information of its exact position. A herbarium specimen documents this record with the following label: "*Nitella capitata* N. E. f. *major*. Nissum Fjord, April 1885. Jeppesen" This specimen was checked by J. Urbaniak in 2002 who confirmed its identity as a male *N. opaca*. The salinity at this site may have been very low (Olsen, 1944). For Finland two former records from slightly brackish water exist: Nylandia (in LD), a location where sterile plants of *Nitella sp.* were found recently which may be specimens of *N. opaca* (Blindow, 2004b) and Karelia australis, Hamina, Lake Kirkkojärvi (in JYV), collected 1952 by M. Raatikainen, determined by M. Koistinen. For Sweden twelve recent (2000–2022) brackish-water locations, all in the northern parts of the Bothnian Bay (Västerbotten, Norrbotten) are reported by Blindow (pers. comm.); Hasslow (1931) mentions one record from brackish water in Medelpad (fertile female plants). In Sweden, sterile plants of *N. flexilis* vel *opaca* were found after 1980 in 24 locations of the northern part of the Bothnian Bay. Three sites have also were detected along the coast of Medelpad (Blindow, 2004b).

Sediment: *Nitella opaca* is found on sandy, muddy and organogenic substrates including lime mud; sometimes also on gravel (Becker, 2010; Blindow, 2004b; Pätzold, 2011; Schmidt et al., 1996; Urbaniak and Gąbka, 2014).

Nutrient conditions: Irrespective of being found most frequently under oligo- to mesotrophic conditions, *Nitella opaca* is able to tolerate eutrophic conditions as well (Becker and Wolff, 2016). The species can be regarded as rather eutrophication-tolerant and its occurrence not being directly dependent from nutrient conditions as can be seen by the maximum main nutrient concentrations reported. Overall, there are only few published data available. Total phosphorus measured at *N. opaca* sites has been recorded up to 0.303 mg/l (Doege et al., 2016), TN can reach 2.8 mg/l, and up to 134 mg/l sulphate has been reported from European sites so far (Simons and Nat, 1996). Based on recent data from six waterbodies in Northwestern Germany with populations of *N. opaca* TP ranged between 0.012–0.041 mg/l and PO_4 between < 0.015–0.071 mg/l (Becker et al., 2022). Moreover, the following hydrochemical data were measured: PO_4-P < 0.005 -0.023 mg/l, TN 0.51–1.89 mg/l and NH_4-N < 0.03–0.277 mg/l.

Water chemistry: *Nitella opaca* is often cited to prefer slightly subneutral conditions, mostly referring to Stroede (1933). In fact, the species is shown to be able to stand pH as low as 3.8 (Doege et al., 2016), but there are also sites with pH-values at or even above 10 listed, indicating the species' ability to withstand alkaline conditions, too (Doege et al., 2016). Recent investigations at six *N. opaca* sites in Northwestern Germany revealed pH-values between 4.7 and 8.9 (Becker et al., 2022). Concerning habitats in Southern France, Italy and the Maghreb, pH-values from 6–8.1 are mentioned (Azella, 2014; Mouronval et al., 2015; Muller et al., 2017). Given the fact that in poorly buffered soft-water conditions large diurnal changes in pH can occur especially under eutrophic conditions, point measurements of pH are of limited use for niche description. With respect to conductivity, *N. opaca* is a freshwater species with very limited ability to thrive under oligohaline conditions. Minimum conductivity of *N. opaca* habitats were found at 16 $\mu S\ cm^{-1}$, maximum values measured so far 2130 $\mu S\ cm^{-1}$, the latter value being already far above the mean (Doege et al., 2016), which is in good agreement with Blindow (2004b) estimating the upper limit at a salinity below 3 PSU for the Baltic Sea. However, laboratory studies could demonstrate survival at slightly higher salinities, Daniel (1980) put the upper limit to salinities above 5–8 PSU. *Nitella opaca* on the one hand is a type-specific species of calcareous Alpine and peri-Alpine lakes in Austria and Germany (Pall et al., 2005). On the other hand, it is often cited as a species growing in extreme soft water (e.g., Stroede, 1933), the data available give a range between 10.3 and 91 mg/l calcium, total hardness ranges between 1.0 and 62.7°dH (Becker et al., 2022; Doege et al., 2016). These values, again, indicate a rather broad range of occurrences and clearly show the species ability to grow in calcium-rich water, but do not prove special adaptations with respect to extremely low calcium concentrations.

Depth range: *Nitella opaca* is grows along the full depth range of the phytolittoral; it occurs in shallow, fully exposed ephemeral ponds as well as at the lowermost edge of the macrophyte belt of large lakes (Baumann, 1911, van de Weyer and Krautkrämer, 2009, Becker, 2010, Pall et al. 2010, Trajanovska et al., 2012, Azzella, 2014, Urbaniak and Gąbka, 2014) indicating the species' broad range of irradiance acclimation. In the Mediterranean region it is a typical charophyte species of shallow temporary water bodies and of running waters (e.g., Mouronval et al., 2015; Becker, 2019). Corillion (1957) refers to *N. opaca* as a typical deep water species, well adapted to low irradiance conditions. In Alpine and peri-Alpine lakes *N. opaca* usually can be found forming the lower limit of the macrophyte vegetation (Pall et al., 2005). Maximum depths reported are 60 m (van Raam, 1998), several sources report stands well below 30 m, as in the German Messinghausen quarry (33.4 m, Becker and Wolff, 2016), the Swedish Lake Vättern (40 m, Corillion, 1957) and the Croatian Lake Vrana (40 m, Golubic, 1961).

Associated species: Similar to its monoecious sister species (*N. flexilis*) *Nitella opaca* also rarely grows in large monospecific stands, except for deep-water meadows. For most of the occurrences reported, the species is a component of mixed stands without gaining dominancy. It is the character species of the *Nitelletum opacae* s.l. Corillion, 1957 (see Preising et al., 1990; Rennwald, 2000). As for *N. flexilis*, the broad range of habitat conditions the species inhabits, a full list of associated species would reflect the aquatic flora of Europe, except for marine and brackish species. For charophytes, the species has been found in Europe together with *Chara aculeolata, C. aspera, C. baltica, C. braunii, C. connivens, C. contraria, C. filiformis, C. fragifera, C. globularis, C. hispida, C. papillosa, C. subspinosa, C. tomentosa, C. virgata, C. vulgaris, Lychnothamnus barbatus, Nitella capillaris, N. confervacea, N. gracilis, N. hyalina, N. flexilis, N. mucronata, N. syncarpa, N. tenuissima, N. translucens, N. wahlbergiana, Nitellopsis obtusa, Sphaerochara canadensis, S. intricata, S. prolifera, Tolypella glomerata* and *T. nidifica,* a list even longer than its monoecious sister species (Becker and Wolff, 2016).

64.4 Variability

Nitella opaca is highlighted as a very variable species by several authors (e.g., Migula, 1889–1897; Langangen, 2007) acclimating to the environmental conditions morphologically to a large extent (Trajanovska et al., 2012). Variability is observed with respect to size, length of branchlets and internodes as well as the extent of formation of heads. Migula (1889–1897) distinguished between a number of formae, which describe the extremes of morphological acclimation rather than distinct morphotypes.

64.5 Identification and Risk of Confusion

Once-furcate branchlets and single-celled dactyls separate *N. opaca* from all other *Nitella* spp. apart from *N. capillaris*, *N. syncarpa* and *N. flexilis*. According to Krause (1997) several morphological differences allow a separation from *N. capillaris* and *N. syncarpa*. The main problem is distinction from *N. flexilis*, which is nearly impossible for sterile specimen by means of morphological characters. The only character available for distinguishing sterile specimen is number of chromosomes, requiring karyological analysis or, alternatively, genetic analysis. For fertile specimen protandry of *N. flexilis* creates another risk of confusion, consequently only clearly monoecious specimens can be identified as *N. flexilis* for sure without molecular genetic tools.

Several authors found that both species can be distinguished by the size of their oospores (far larger in *flexilis*). According to Krause (1997) and Stewart (1998), they do not even overlap. Also Sawa (1965) and Blindow et al. (2018) gave ranges without overlap for Canada and Chile, respectively. In contrast, antheridia are described by some authors to be far smaller in *N. flexilis* (Krause, 1997; Stewart, 1998; Sawa, 1965; Blindow et al., 2018). However, as can be seen from Table 64.2 and Fig. 64.2 as well as Table 64.3 and Fig. 64.3, oospore size and antheridia diameter are very variable and overlapping between the species to an extent making it, overall, an unreliable character.

The same applies for other morphological proxies such as one-side combed orientation of branchlets of *N. opaca*, which is at the very best, but not only, is a character for individuals from running water habitats. However, there is no clear indication that *N. flexilis* does not grow like this. Antheridia diameter is, as can be seen from Table 64.3 and Fig. 64.3 very variable and overlapping between the species to an extent making it an unreliable character too. Phenological proxies have been shown to be unreliable, *N. opaca* has been observed to fructify in summer and autumn, too, as reported by several authors (see above). This leaves just the thickened cell wall of the end cell of branchlets of *N. flexilis* (see Migula, 1889–1897) as a potential discriminative feature, which needs to be verified by further observations.

Both species often co-occur, so extreme caution is strongly recommended. In the Mediterranean region *N. opaca* is more frequent whereas *N. flexilis* is rare in many regions (Blaženčić et al., 2006; Becker, 2019; Mouronval et al., 2015; Muller et al.,

2017). The same applies for the Alpine and peri-Alpine region (Pall et al., 2005). Distinction from *N. mucronata* can be done by means of the end cell of the branchlet, being never mucronate. Additionally, branchlets of *N. opaca* are 1-furcate, whereas *N. mucronata* has 2–3 furcate branchlets.

64.6 Distribution

See Fig. 64.4.

Nitella opaca is a cosmopolitan species. Whereas it occurs almost everywhere in the northern hemisphere, findings in the southern hemisphere are documented only for the south-eastern and the south-western part of South America (Korsch, 2018; Blindow et al., 2018, respectively). In Europe the species can be found in nearly every country, records are missing only for Luxembourg, Kosovo and Moldova. The species may occur in Estonia as well but was not separated from *N. flexilis* in the recent monitoring therefore data from Estonia were not included in Fig. 64.4. Data from the Mediterranean and Maghreb can be found in Blaženčic et al. (2006), Bazzichelli and Abdelahad (2009), Daoud-Bouattour et al. (2011), Mouronval et al. (2015), Muller et al., (2017), and Becker (2019).

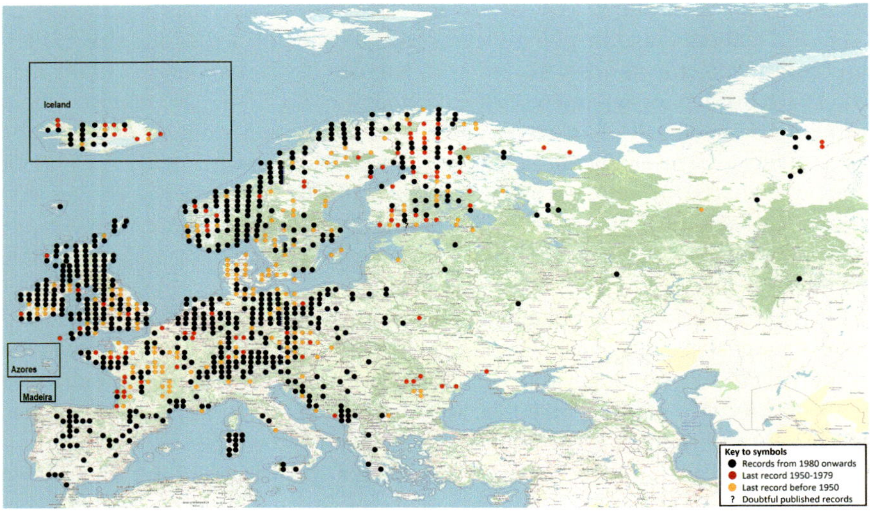

Fig. 64.4 European distribution of *Nitella opaca*. For worldwide distribution see Korsch (2018). Estonian data are not included because the species is not distinguished from *N. flexilis* in the national monitoring. Black dots are records from 1980 on, red dots from 1950 to 1979 and yellow dots before 1950; unconfirmed / doubtful published records (if applicable) are marked with "?" on top. Map prepared by Heiko Korsch

64.7 Status and Threats

Nitella opaca is frequently found all over Europe and, given its very broad niche, it might be surprising that the species is red-listed so often regionally as well as nationally. The reason for this is the declining trend. Being at least one of the most frequent, if not the most frequent species in many European countries, habitat loss and habitat destruction took their share the past 70 years. As well the European-wide loss of small water bodies is responsible for the decline of suitable sites for *N. opaca* (Becker and Wolff, 2016). Drainage, eutrophication resulting from agricultural activities and intensification of the use of water bodies for fishery and recreation are also threatening the stands (Becker and Wolff, 2016). On the other hand, newly created quarry ponds and waterbodies for conservation measures have been colonised by *N. opaca* during the last decade. Given the large distribution area and broad ecological niche, the species itself probably is not at acute risk on a European level. However, effective measures should be taken to stop this trend of decline in number and sites in order to prevent further fragmentation of populations.

64.8 Remarks

Nitella opaca is combined with the monoecious *N. flexilis* by several European authors (e.g., Cirujano et al., 2008; Moore, 1986), following Wood's opinion about the significance of this character for charophytes in general (Wood, 1962). However, his opinion was criticised immediately after being published in more detail (Wood, 1965) by Sarma and Khan (1967), targeting on the consistent difference in chromosome number between *N. flexilis* (n = 12) and *N. opaca* (n = 6). The taxa are treated as different at the species level here, following not only a long European tradition, but also respecting for differences in distribution and phenology (see above). Still some problems need to be mentioned in this context. Most important, there is a genetic overlap between the species, because 6 of the 12 chromosomes of *N. flexilis* have been reported to be identical with the ones of *N. opaca* (Sarma and Khan, 1967). Another one is, that sterile specimens are almost impossible to determine except for genetic analysis and even fertile specimen must be taken with caution because of protandry of *N. flexilis* (see above, "risk of confusion").

References

Agardh, C. A. (1824). *Systema Algarum*. Berling, Lundae. [in Latin].

Azzella, M. M. (2014). Italian Volcanic lakes: A diversity hotspot and refuge for European charophytes. *Journal of Limnology, 73*, 502–510.

Bailly, G., & Schaefer, O. (2010). *Guide illustré des Characées du nord-est de la France*. Conservatoire Botanique National de Franche-Comté. [in French].

Baumann, E. (1911). Die Vegetation des Untersees (Bodensee). *Archiv für Hydrobiologie, Supplement 1*(I–V), 1–554, Tafeln I-XV. [in German].
Bazzichelli, G., & Abdelahad, N. (2009). *Alghe d'acqua dolce d'Italia. Flora analitica delle Caroficee.* Sapienza Università di Roma. Ministero dell'Ambiente e della Tutela del territorio e del mare. Roma, Centro Stampa Università. [in Italian].
Becker, R. & Wolff, P. (2016). *Nitella flexilis.* In Arbeitsgruppe Characeen Deutschlands (Ed.), *Armleuchteralgen: Die Characeen Deutschlands* (pp. 413–426). Springer. [in German].
Becker, R. (2010). Bemerkenswerte Characeen-Funde in Nordwest-Deutschland. *Abhandlungen Des Naturwissenschaftlichen Vereins Zu Bremen, 46*, 409–418. [in German].
Becker, R. (2019). The Characeae (Charales, Charophyceae) of Sardinia (Italy): Habitats, distribution and conservation. *Webbia, 74*, 83–101.
Becker, R., Fuhrmann, K., Meyer-Spethmann, U., Sinkowski, M., & Willen, M. (2022). Verbreitung, Populationsentwicklung und Schutz besonders gefährdeter Arten der Armleuchteralgen (Charales, Characeae) in Nordwestdeutschland. *Drosera, 40*, 81–100. [in German].
Becker, R., & Ziesmer, B. (1989). *Floristische und vegetationskundliche Untersuchungen der Makrophyten in den Fliessgewässern Sardiniens.* Universität Oldenburg. [in German].
Blaženčić, J., Stevanović, B., Blaženčić, Z., & Stevanović, V. (2006). Distribution and ecology of charophytes recorded in the west and central Balkans. *Cryptogamie, Algologie, 27*, 311–322.
Blindow, I. (2004a). *Nitella flexilis.* In H. Schubert, & I. Blindow (Eds.), *Charophytes of the Baltic Sea* (pp. 174–180). A. R. G. Gantner.
Blindow, I. (2004b). *Nitella opaca.* In H. Schubert, & I. Blindow (Eds.), *Charophytes of the Baltic Sea* (pp. 198–203). A. R. G. Gantner.
Blindow, I., Marquardt, R., Schories, D., & Schubert, H. (2018). Charophyceae of Chile—taxonomy and distribution. Part 1. *Nova Hedwigia 107*, 1–47.
Cirujano, S., Cambra, J., Sánchez Castillo, P. M., Meco, A., & Flor Arnau, N. (2008). *Flora ibérica. Algas continentales. Carófitos (Characeae).* Real Jardín Botánico. [in Spanish].
Compère, P. (1992). *Charophytes. Flore pratique des algues d'eau douce de Belgique.* Jardin botanique national de Belgique. [in French].
Corillion, R. (1957). Les Charophycées de France et d'Europe Occidentale. *Bulletin de la Société Scientifique de Bretagne*, 32(Hors série I & II) 1–499. [in French].
Daniel, D. J. (1980). *H+ and K+ transport in Nitella.* Thesis, State University of New York College Brockport.
Daoud-Bouattour, A., Muller, S. D., Ferchichi-Ben Jamaa, H., Ben Saad-Limam, S., Rhazi, L., Soulié-Märsche, I., Rouissi, M., Touati, B., Ben Haj Jilani, I., Gammar, A. M., & Ghrabi-Gammar, Z. (2011). Conservation of Mediterranean wetlands: Interest of historical approach. *C. R. Biologies 334*, 742–756.
Dilger, M. (2010). Merkmale zur Erkennung von *Nitella-* und *Tolypella-*Arten der Characeaen [sic] über Oosporen aus Sedimenten oder von Pflanzen. *Rostocker Meeresbiologische Beiträge, 23*, 40–59. [in German].
Doege, A., van de Weyer, K., Becker, R., & Schubert, H. (2016). Bioindikation mit Characeen. In Arbeitsgruppe Characeen Deutschlands (Ed.), *Armleuchteralgen: Die Characeen Deutschlands* (pp. 97–138). Springer. [in German].
Gąbka, M. (2009). *Charophytes of the Wielkopolska region (NW Poland): distribution, taxonomy, and autecology.* Bogucki Wydawnictvo.
Golubic, S. (1961). Der Vrana-See an der Insel Cres – ein *Chara*-See. *Verhandlungen des Internationalen Vereins Limnologie, 14*, 846–849. [in German].
Grillas, P., Gauthier, P., Yavercovski, N., Perennou, C. (2004). *Mediterranean Temporary Pools, Volume 2. Species information sheets.* Station biologique de la Tour du Valat.
Groves, J., & Bullock-Webster, G. R. (1920). *The British Charophyta.* Ray Society.
Hasslow, O. J. (1931). Sveriges characeer. *Botaniska Notiser, 1931*, 63–136. [in Swedish].
Korsch, H. (2013). Die Armleuchteralgen (Charophyceae) Sachsen-Anhalts. *Berichte Des Landesamtes Für Umweltschutz Sachsen-Anhalt, 1*(2013), 1–85. [in German].

Korsch, H. (2018). The worldwide range of the Charophyte species native to Germany. *Rostocker Meeresbiologische Beiträge, 28*, 45–96.

Krause, W. (1997). Charales (Charophyceae) Süßwasserflora von Mitteleuropa. In H. Ettl, G. Gärtner, H. Heynig, & D. Mollenhauer (Eds.), *Süßwasserflora von Mitteleuropa 18*. Fischer. [in German].

Langangen, A. (2007). *Charophytes of the Nordic countries*. Saeculum ANS.

Migula, W. (1889–1897). Fünfter Band: Die Characeen. In A. Fischer, E. Fischer, F. Hauck, G. Limpricht, C. Luerssen, W. Migula, H. Rehm, P. Richter, & G. Winter (Eds.), *Dr. L. Rabenhorst's Kryptogamenflora von Deutschland, Oesterreich und der Schweiz* (2nd ed.). Eduard Kummer. [in German].

Moore, J. A. (1986). *Charophytes of Great Britain and Ireland*. BSBI Handbook 5. Botanical Society of the British Isles.

Mouronval, J. B., Baudouin, S., Borel, N., Soulié-Märsche, I., Klesczewski, M., & Grillas, P. (2015). *Guide des Characées de France méditerranéenne*. Office national de la Chasse et de la Faune Sauvage. [in French].

Muller, S. D., Rhazi, L., & Soulié-Märsche, I. (2017). Diversity and distribution of Characeae in the Maghreb (Algeria, Morocco, Tunisia). *Cryptogamie, Algologie, 38*, 201–251.

Olsen, S. (1944). Danish Charophyta. Det Kongelige Danske Videnskabernes Selskab. *Biologiske Skrifter 3*(1), 1–240. [in Danish].

Pall, K., Moser, V., Mayerhofer, S., & Till, R. (2005). Makrophyten-basierte Typisierung der Seen Österreichs. *Studie im Auftrag des Bundesministeriums für Land- und Forstwirtschaft, Umwelt und Wasserwirtschaft*. 1–62. [in German].

Pall, K., Hippeli, S., Mayerhofer, V., & Mayerhofer, S. (2010). Makrophytenkartierung Attersee. *Untersuchung im Auftrag der Landesregierung Oberösterreich und des Bundesministeriums für Land- und Forstwirtschaft, Umwelt und Wasserwirtschaft* 1–124. [in German].

Pätzold, F. (2011). Beobachtungen zur Verbreitung von Characeen in Baggerseen der Badischen Oberrheinaue. *Berichte der Botanischen Arbeitsgemeinschaft Südwestdeutschland Beiheft, 3*, 37–42. [in German].

Preising, E., Vahle, H.-C., Brandes, D., Hofmeister, H., Tüxen, J., & Weber, H. E. (1990). Die Pflanzengesellschaften Niedersachsens. Wasser- und Sumpfpflanzengesellschaften des Süßwassers. *Naturschutz Landschaftspflege Niedersachsen 208*, 47–161. [in German].

van Raam, J. C. (1998). *Handboek Kranswieren*. Chara boek. Hilversum. [in Dutch].

Rennwald, E. (2000). Verzeichnis und Rote Liste der Pflanzengesellschaften Deutschlands. *Schriftenreihe Für Vegetationskunde, 35*, 1–820.

Sarma, Y. S. R. K., & Khan, M. (1967). Dioecism and monoecism as taxonomic criteria in Charophyta. *Current Science, 36*, 245–247.

Schmidt, D., van de Weyer, K., Krause, W., Kies, L., Garniel, A., Geissler, U., Gutowski, A., Samietz, R., Schütz, W., Vahle, H.-C., Vöge, M., Wolff, P., & Melzer, A. (1996). Rote Liste der Armleuchteralgen (Charophyceae) Deutschlands. 2. Fassung, Stand: Februar 1995. *Schriftenreihe Für Vegetationskunde, 28*, 547–576. [in German].

Simons, J., & Nat, E. (1996). Past and present distribution of stoneworts (Characeae) in The Netherlands. *Hydrobiologia, 340*, 127–135.

Smith, J. E. (1802). *English botany; or, coloured figures of British plants, with their essential characters, synonyms, and places of growth: To which will be added, occasional remarks*. XV. Taylor and Co.

Stafleu, F. A. & Cowan, R. S. (1976). *Taxonomic literature. A selective guide to botanical publications and collections with dates, commentaries and types (Vol. I)*. A-G. Bohn, Scheltema & Holkema.

Stewart, N. F. (1998). *Nitella flexilis/N. opaca*. In T. C. G. Rich, & A. C. Jermy (Eds.), *Plant Crib 1998* (pp. 1–2). Botanical Society of the British Isles.

Stroede, W. (1933). Über die Beziehungen der Characeen zu den chemischen Faktoren der Wohngewässer und des Schlammes. *Archiv Für Hydrobiologie, 25*, 192–229. [in German].

Trajanovska, S., Blaženčić, J., Trajanovski, S., & Budzakoska-Gjoreska, B. (2012). Distribution, morphological variability, ecology and the present state of *Nitella* from Lake Ohrid and its surroundings. *Archives of Biological Sciences Belgrade, 64*, 549–556.
Urbaniak, J., & Gąbka, M. (2014). *Polish Charophytes. An illustrated guide to identification*. Uniwersytet Przyrodniczy we Wroclawiu. [in Polish].
van de Weyer, K., & Krautkrämer, V. (2009). *Nitella opaca* (Bruzelius) Agardh im Steinbruch Messinghausen (Sauerland) – mit einer Übersicht der maximalen Makrophyten-Tiefengrenzen in Deutschland. *Rostocker Meeresbiologische Beiträge, 22*, 57–64.
Wahlstedt, L. J. (1864). *Om characeernas knoppar och öfvervintring. Botanisk avhandling*. University of Lund. [in Swedish].
Westermann, K., & Westermann, S. (1998). Die Quellgewässer und ihre Vegetation in der südbadischen Oberrheinniederung. *Naturschutz Südlicher Oberrhein, 2*, 1–93. [in German].
Wood, R. D. (1962). New combinations and taxa in the revision of Characeae. *Taxon, 11*, 7–25.
Wood, R. D. (1965). *Monograph of the Characeae*. J. Cramer.

Chapter 65
Nitella ornithopoda

Roman Romanov

Scientific name: ***Nitella ornithopoda*** **A. Braun in Leonh. in Lotos 13: 146. 1863**

Neotype (Wood, 1965: 559): [France, Nouvelle-Aquitaine] Tourbière d'Heurtebise pres Angoulême, Charente/June 1867/A. de Rochebrune (PC 737601).

≡ *Nitella tenuissima* (Desv.) Kütz. subsp. *ornithopoda* (A. Braun in Leonh.) R. D. Wood Rev. Charac.: 557. 1965

Local names: RUSSIAN: Нителла птиценожковая, UKRAINIAN: Нітела пташконіжкова.

65.1 Morphology

See Figs. 65.1, 65.2 and 65.3.

Habitus: The plant is green or usually light green, unencrusted to frequently weakly encrusted and then, mainly at the last segments of the branchlets. It is small to medium-sized, thin and flexible, usually moniliform in the upper part or nearly the whole plant due to condensed and clearly spaced whorls of fertile branchlets without any heads or clasping of the upper whorls on the rarely branched axis. Upper whorls may gradually reduce in size distally. The species resembles *Nitella tenuissima* by habit in these cases but with less tidy and more disheveled whorls. The lowermost part or whole plant has spreading diffuse branchlets resembling lax forms of *N. gracilis*.

Size: Small to medium plant, up to 20 cm high.

R. Romanov (✉)
Nišice Ujtin Potok b.b, Dobra Voda, Bar 85356, Montenegro
e-mail: streptophytes@gmail.com

H. Schubert et al. (eds.), *Charophytes of Europe*,
https://doi.org/10.1007/978-3-031-31898-6_65

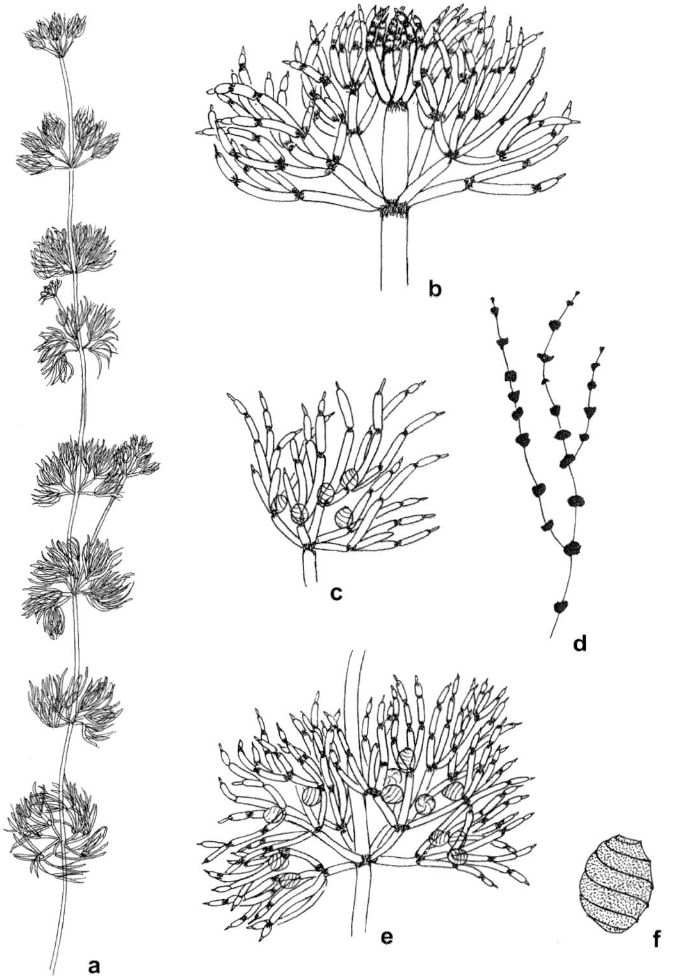

Fig. 65.1 *Nitella ornithopoda*. **a**, **d** Upper part of a plant showing moniliform *Nitella tenuissima*-like appearance; **b** Upper part of a plant; **c** Branchlets; **e** Whorl of branchlets; **f** Oospore. Drawings: Werner Krause

Main axis: Slender to moderately slender, 300–500 μm in diameter. The internodes are always clearly, at least two times, longer than the branchlets.

Cortex, spine-cells, stipulodes: Absent.

Branchlets: 6–7 in a whorl, up to 0.8 cm in length, sterile are 1–2 furcate and mostly diffuse, fertile 2(3) furcate, less diffuse and mostly condensed. Branchlet cells

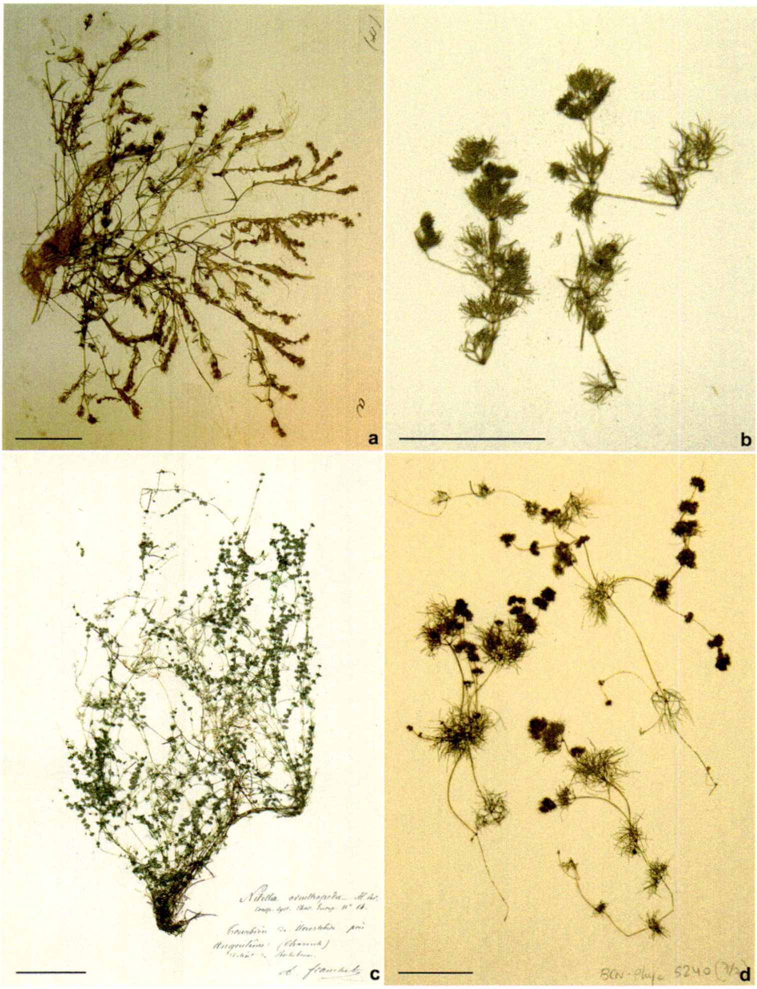

Fig. 65.2 Variability of general appearance of *Nitella ornithopoda*. **a** Plant with neither lax nor compact thin branchlet whorls (LISU); **b** Apical fragments of plants with lax "fluffy" branchlet whorls clustering at axis apex (MA-Algae 5430); **c** Long clearly uniform moniliform habit of upper parts of plants because of compact branchlet whorls (LE); **d** Plants with lax "fluffy" low branchlet whorls and abbreviated compact apical whorls (BCN 5240). Scale: 2 cm. Photographs: Roman Romanov

seem to be somewhat allantoid especially in the dactyls because pressed branchlets are constricted at the nodes and the transverse walls. The primary branchlet segment is not longer, mostly shorter than half of the branchlet length. There are 5–6 secondary branchlet segments, including a central secondary one, 3–4 tertiary

Fig. 65.3 Oospores of *Nitella ornithopoda* (LISU), scanning electron microscopy. **a** General appearance; **b** Basal plate view; **c** Fossa sculpture. Scale: **a**, **b**—100 μm, **c**—10 μm. Photographs: Roman Romanov

branchlet segments, one of them occasionally short and furcating into (2)3–4 quaternary branchlet segments. If the secondary and tertiary branchlet segments are forked, they are the shortest in the branchlets. Dactyls are 3 to 4, 3- to 5-celled, longer than the penultimate branchlet segment, end cell variable in shape, from somewhat mucronate/acute and conical to obtuse and allantoid, 300–400 μm in length (Cunha, 1942) or 60–90 × 20–45 μm (Wood and Imahori, 1964), sometimes easily lost.

Bract-cells, bracteoles, bractlets: Absent.

Sexuality: Monoecious, with mucus cover, gametangia are solitary or conjoined at the second and third furcations only.

Oogonia: The coronula is persistent, its cells are equal in both rows. Table 65.1 gives an overview of the published morphometric data.

Oospores: With flanged ribs, bright yellow brown (R. Romanov own obs.; Fig. 65.3). Oospore surface is fibrous-arachnoid (Cirujano et al., 2008: figs. 106, 107, but these photographs can belong to *N. dixonii* as well) or in other words spongy in SEM (Fig. 65.3). It was described and illustrated as strongly reticulate with ca. 6 meshes across the fossa in LM (Wood and Imahori, 1964: icon 305: 7) and described as punctate (Corillion, 1957a; Cunha, 1942), reticulate-punctate to strongly reticulate (Wood, 1965), or finely dotted in LM (Krause, 1997). Table 65.2 gives an overview of the published morphometric data.

Antheridia: Sessile; Table 65.3 gives an overview of the published morphometric data.

Bulbils: Unknown, but seems to be possible.

Table 65.1 Oogonia morphometry

Author	Length	Width	Coronula length	Coronula width	Number of helical stripes
Cunha (1942)	400–450	300–350	30–35	–	8–9
Corillion (1957a)	400–450	300–350	30–35	–	8–9
Wood and Imahori (1964)	375–390	315–325	~35	~52	~9
Wood (1965)	375–425	315–350	~35	~52	9–10
Krause (1997)	400–450	300–350	~35	–	7–9
Total range	**375–450**	**300–350**	**30–35**	**~ 52**	**7–10**

Length in the first column is excluding the coronula; its length and width is given separately, all values for length and width in μm

Table 65.2 Oospore morphometry

Author	Length	Width	Number of striae	Colour
Cunha (1942)	300–325	210–250	6–7	Black
Corillion (1957a)	300–325	210–250	6–7	Black
Wood and Imahori (1964)	240–250	185–200	6–7	Chestnut brown
Wood (1965)	250–325	185–250	6–7	Yellowish, bright or chestnut brown
Krause (1997)	~300	200–250	6–7	Dark brown to black
Total range	**240–325**	**185–250**	**6–7**	

All values for length in width in μm

Table 65.3 Antheridia morphometry

Author	Diameter (μm)
Cunha (1942)	250–350
Corillion (1957a)	250–350
Wood and Imahori (1964)	195–220
Wood (1965)	195–220(350)
Krause (1997)	250–350
Total range	**195–350**

65.2 Phenology and Reproduction

The species reproduces by oospores. It seems to be a spring—early summer species (Corillion, 1957a; Krause, 1997).

65.3 Habitat Conditions and Ecology

Habitat type: The ecology of *Nitella ornithopoda* is poorly known; it was collected from peat bogs, lakes, ponds, spring waters and in one case from a river probably from shallow water.

Sediment: Unknown.

Nutrient conditions and water chemistry: It seems to be freshwater species. Its encrustation can be recognised as evidence of somewhat alkaline waters (Corillion, 1957a).

Depth range: It seems to be a shallow water species.

Associated species: Unknown.

65.4 Variability, Identification and Risk of Confusion

The whorls can be compact or diffuse on the same or different plants (Fig. 65.2), which could be recognised as modifications under different insolation (Krause, 1997). These extreme forms were described as formae by Migula (1889–1897). The compact dense whorls with fluffy outline are common in the upper parts of plants appearing like moniliform spikes above lax and diffuse fluffy moniliform lower whorls (Fig. 65.2c, d) or the general habit of plants can be more uniform (Fig. 65.2a, b). Some plants have thin neither lax, nor fluffy, nor clearly moniliform whorls (Fig. 65.2a).

Nitella palhinhae (as *Tolypella palhinhae*) (see Chap. 82) was synonymised by Wood (1962, 1965) with *N. ornithopoda* but stands apart from this species because it has diffuse single-furcated (!) branchlets without (!) a central secondary branchlet segment, 5–6 in a whorl, is fertile at the first node and has bigger gametangia and shortly stipitate antheridia. Up to date it is known only from the type locality (LISU!) and new specimens are required to confirm this taxonomic decision.

Nitella ornithopoda is similar to *N. tenuissima* in general appearance, but can be differentiated by the multicellular, 3–5-celled dactyls and the spongy surface of oospore versus strictly 2-celled dactyls and roughly beaded reticulate ornamentation of the oospores of the latter species. The multicellular dactyls of *N. ornithopoda* are a trait common with the dioecious *N. dixonii*, but fertile plants of both species are easily distinguishable (for other traits see Chap. 59). In one single case *N. ornithopoda* was illustrated with a photograph of *N. dixonii* because they were assumed to be conspecific (Cirujano et al., 2008: fig. 104; MA!). The difference from a form of *N. gracilis* described by Hy, (1913) having 3–4-celled dactyls (Corillion, 1957a; Hy, 1913) is not evident and needs to be evaluated. Possibly the central secondary segment of the branchlet typical for *N. ornithopoda* could be a useful trait. The slightly allantoid dactyl cells of *N. ornithopoda* resulting from evident constrictions between them in the pressed state in contrast with the non-allantoid cells and the fewer but overlapping number of cells in the branchlet of the latter form of *N. gracilis* (Hy, 1913: figs. 23, 24) evidently are not enough for their delineation.

65.5 Distribution

See Fig. 65.4.

To date known only from a small area of Charente in France, a few localities in Spain and Portugal and a single site in the Azores (Cunha, 1934, 1941, 1942, 1951; Corillion, 1957a; Pedrola, 1988; Cirujano et al., 2007; BCN!, L!, LE! LISU!, MA!). The total number of its localities in the world does not exceed 15 but most of them were found in the 19th and the first half of the 20th century. The record of 2006 (Cirujano et al., 2007) is based on a specimen of *N. dixonii* (MA!; see Chap. 59). The

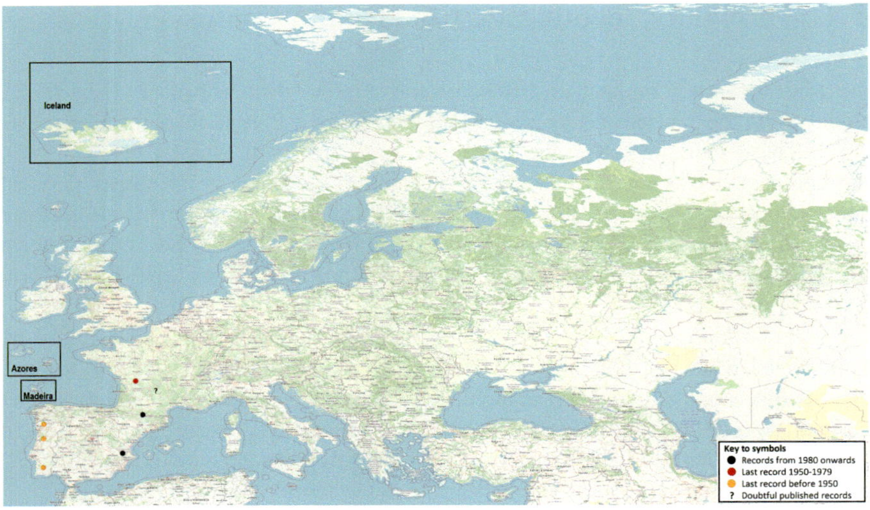

Fig. 65.4 Distribution of *Nitella ornithopoda* in Europe. Black dots are records from 1980 on, red dots from 1950 to 1979 and yellow dots before 1950; unconfirmed/doubtful published records (if applicable) are marked with "?" superimposed. Map prepared by Heiko Korsch

most recent collections of 1987 and 2009 are known from Spain (Flor-Arnau, 2014; MA!). The record from Cambridgeshire, Great Britain (Wood, 1965) seems to be extraordinary from the species distribution range perspective and probably belongs to *N. tenuissima* (N. Stewart pers. comm.).

65.6 Status and Threats

Nitella ornithopoda has been assessed as CR (Critically Endangered, European and Globally) in the European Red List (see Chap. 9). Recent surveys of most of its localities are missing. It is unknown in France since the 19th century, and in Portugal since 1947. Therefore, it can be tentatively referred to as "Extinct" in these countries. In Spain it was evaluated as "Critically Endangered" (Cirujano et al., 2008; Flor-Arnau, 2014) and this assessment seems to fit well to the whole distribution range.

Acknowledgements The curators of the herbaria listed above are kindly acknowledged for their encouragement and guidance, Ana Isabel Correia for the permission to sample oospores for SEM study.

References

Cirujano, S., Cambra, J., Sánchez Castillo, P. M., Meco, A., & Flor Arnau, N. (2008). *Flora ibérica. Algas continentales. Carófitos (Characeae).* Real Jardín Botánico. [in Spanish].

Cirujano, S., García Murillo, P., Meco, A., & Fernández Zamudio, R. (2007). Los carófitos ibéricos. *Anales del Jardín Botánico de Madrid, 64,* 87–102. [in Spanish].

Corillion, R. (1957a). Les Charophycées de France et d'Europe Occidentale. *Bulletin de la Société Scientifique de Bretagne 32 (Hors série I & II),* 1–499. [in French].

Cunha, A. G. (1934). *Liste des Characées du Portugal.* [in French].

Cunha, A. G. (1941). Additions à la flore charologique du Portugal. *Bulletin de la Société Portugaise des Sciences Naturelles, 13,* 153–156. [in French].

Cunha, A. G. (1942). Contribüção para o estudo dos Carofitos portugueses. *Revista da Faculdade de Ciências, 2,* 227–357. [in Portuguese].

Cunha, A. G. (1951). Additions à la flore charologique du Portugal IV. *Boletim da Sociedade Broteriana, 25,* 25–30. [in French].

Flor-Arnau, N. (2014). *Diversitat, ecologia i usos en bioindicació d'algues caràcies i macròfits a la península Ibèrica.* Tesi doctoral. [in Catalan].

Hy, F. (1913). Les Characées de France. *Bulletin de la Société Botanique de France 60* (sup 1), 1–47. [in French].

Krause, W. (1997). Charales (Charophyceae) Süßwasserflora von Mitteleuropa. In H. Ettl, G. Gärtner, H. Heynig, & D. Mollenhauer (Eds.) *Süßwasserflora von Mitteleuropa* (Vol. 18). Fischer. [in German].

Migula, W. (1889–1897). Fünfter Band: Die Characeen. In A. Fischer, E. Fischer, F. Hauck, G. Limpricht, C. Luerssen, W. Migula, H. Rehm, P. Richter, & G. Winter (Eds.), *Dr. L. Rabenhorst's Kryptogamenflora von Deutschland, Oesterreich und der Schweiz (2nd ed).* Eduard Kummer. [in German].

Pedrola, J. (1988). Fragmenta chorologica occidentalia, Algae, 1849–1851. *Anales del Jardín Botánico de Madrid, 45,* 521. [in Spanish].

Wood, R. D. (1962) New combinations and taxa in the revision of Characeae. *Taxon, 11,* 7–25. https://doi.org/10.2307/1216853

Wood, R. D. (1965). *Monograph of the Characeae.* J. Cramer.

Wood, R. D., & Imahori, K. (1964). *Iconograph of the Characeae.* J. Cramer.

Chapter 66
Nitella syncarpa

John Bruinsma and Klaus van de Weyer

Scientific name: ***Nitella syncarpa*** **(Thuill.) Chev., Fl. Gén. Env. Paris 2: 125. 1827**

≡ *Chara syncarpa* Thuill., Fl. Env. Paris, ed. 2: 473. 1799
Neotype (Wood, 1965: 417): In Nebenwassern des Rheines am Isteiner Klotz auf felsigem Grunde/20. August 1891/F. Förster & K. Kopp (NY); Migula, Sydow & Wahlstedt, Characeae Exs. 1 [female plant of *Nitella syncarpa* f. *longifolia*]

Nomenclatural note: The lectotype belongs to "*Nitella syncarpa* (Thuill.) Kütz. f. *longifolia* A. Br. Migula Characeen p. 105". Migula's form is illegitimate because of the older name *Nitella syncarpa* f. *elongata* Wahlst. 1875. Migula (1889–1897) refers to Braun (1847). Here the rank of "*Nitella syncarpa* (a) *laxa* (α) *longifolia*" cannot be interpreted with certainty, but as "*Nitella syncarpa a) laxa α) longifolia*" is without a description, but with "*Ch. syncarpa* Thuill.! [! = seen by A. Braun]" as a synonym, it is a nominate rank.

Local names: BELARUSIAN: Нітэла зрослаплодная; DANISH: Høstslinke; DUTCH: Vruchtrijk glanswier; ESTONIAN: Kobarnitell; FINNISH: Elosiloparta, Höstlinke; FRENCH: Nitelle à fruits groupés; GERMAN: Verwachsenfrüchtige Glanzleuchteralge; LITHUANIAN: Blizgantysis mentudumblis; NORWEGIAN: Höstlinke; POLISH: Krynicznik obskubany; RUSSIAN: Нителла сростноплодная; SWEDISH: Höstslinke; UKRAINIAN: Нітела зростноплідна

J. Bruinsma (✉)
Thorbeckelaan 24, 5694 CR Breugel, The Netherlands
e-mail: jhpbruinsma@gmail.com

K. van de Weyer
lanaplan, Lobbericher Str. 5, 41334 Nettetal, Germany

H. Schubert et al. (eds.), *Charophytes of Europe*,
https://doi.org/10.1007/978-3-031-31898-6_66

66.1 Morphology

See Figs. 66.1 and 66.2.

Habitus: The habit and size of *Nitella syncarpa* are quite variable. The plants are thin and bright green (van Raam, 1998; Mouronval et al., 2015; Korte et al., 2016), green to dark green (Urbaniak and Gąbka, 2014), or bright green to brownish (Cirujano et al., 2008). According to Migula (1889–1897) the plants are bright green but may appear darker to brownish-green due to the presence of epiphytes. Plants can have loose, remote whorls, or the whorls can be quite connivent and dense (usually in male plants).

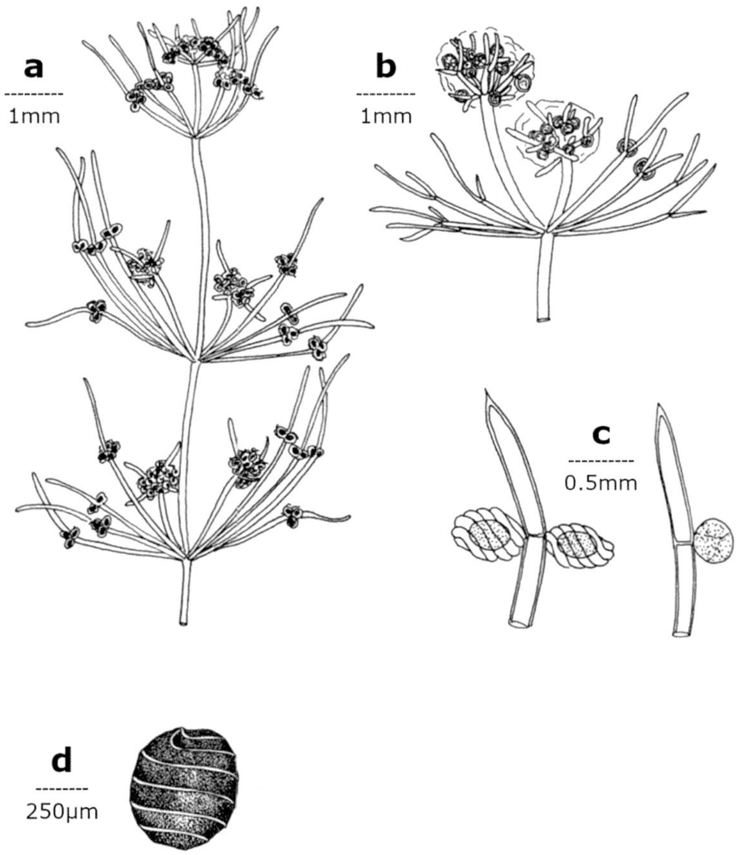

Fig. 66.1 *Nitella syncarpa*, drawings. **a** Female plant; **b** Male plant; **c** Position of oogonia and antheridium; **d** Oospore. Drawings after Werner Krause, modified by Dagmar Wassong in van de Weyer and Schmidt (2018)

Fig. 66.2 *Nitella syncarpa*, details. **a** Female plant with oogonia and mucus; **b** Male plant with antheridia and mucus from a gravel pit near Karlsruhe, Germany; **c** Male plant with pronounced banding pattern from a gravel pit near Düsseldorf; **d** Population of female plants from a mining lake near Dessau, Germany; **e** Female plant from a mining lake near Dessau, Germany; **f** Oogonia. Photographs: **a**, **b**, **c**, **d**, **e**—Klaus van de Weyer; **f**—Heiko Korsch

Size: *Nitella syncarpa* is middle sized and grows from 10 to 40 cm high, exceptionally to 70 cm (Krause, 1997).

Main axis: The diameter of the axis is c. 0.7 mm. The internodes are 2–12 cm long (Krause, 1997) and usually longer than the branchlets.

Cortex, spine-cells, stipulodes: Absent.

Branchlets: Sterile whorls consist of (4)6(8) branchlets. Sterile branchlets may be simple (i.e. without furcations), especially in the lower part of the plant. In higher parts of the plant they can have one furcation with (2–3)4 dactyls. The primary branchlet segment is about 1/2–2/3 of the total length of the branchlet. Fertile whorls are generally shorter than sterile ones. They differ from sterile ones especially in their degree of furcation. Most fertile female branchlets have one dactyl, which is sometimes very long. Sometimes this dactyl is apparently absent or very short. In fertile male branchlets the segments are usually very short, so that the antheridia are clustered in dense heads. Dactyls are single celled, acuminate to acute, van Raam (1998) describes the tips of female branches as long acuminate. Occasionally the tips are rounded (Migula, 1889–1897, photo in Urbaniak and Gąbka, 2014) and the top of the dactyl is thickened.

Bract-cells, bracteoles: Absent.

Bractlets: Absent.

Sexuality: *Nitella syncarpa* is dioecious. The gametangia are surrounded by mucus which sticks to the herbarium paper. This is an important determination character. Gametangia are geminate or clustered. They can be borne on branchlets that are similar to the sterile ones, however the majority are on branchlets near the top of the main axes and the lateral branches.

Oogonia: The oogonia of *Nitella syncarpa* are situated in clusters of 2–4 at the nodes of unbranched branchlets. Often this is so striking and characteristic that is an important indication for the identification.

Table 66.1 presents an overview of the literature about the morphometry of the oogonia. Some care has to be taken in interpreting these figures, as the circumstances of the measurements (dry, fresh, treated) are usually unknown. In summary, the oogonia without coronula are (280)400–700(760) μm long and (400)420–525 μm wide. There are (6)7–9 spirals. The coronula is up to 20–30(59) μm high and 80-90 μm wide. According to Krause (1997) the secondary cells of the coronula are yellow and inflated at the top.

Oospores: The oospore-membrane of *Nitella syncarpa* is usually described as being smooth or (slightly) granulate (Table 66.2, Fig. 66.3). It is not quite clear whether these descriptions are based on observations with Light Microscope and/or with Scanning Electronic Microscope. Under light microscopy the detail of the oospore membrane appears smooth, with at the most very few, very small granules (e.g., Mouronval et al., 2015). John and Moore (1987) describe their result of SEM-study: “The wall is smooth although low irregularities are seen at very high magnification.” Figure 66.3d shows, at very high magnification with the aid of SEM, a uniquely felted and granulate structure.

Most commonly the colour is reported as dark brown or black. Also chestnut brown and orange brown are mentioned. The oospore is 300–450 μm long and 270–380(400) μm wide and has (5)6–7(8) inconspicuous wings.

Antheridia: the antheridia are situated singly in the branchlet axils. The range of their diameter can be summarised as 250–705 μm. The differences between the authors are quite large. Table 66.3 gives an overview of the literature about the morphometry of the antheridia.

Bulbils: *Nitella syncarpa* does not form bulbils.

Table 66.1 Oogonia morphometry

Author	Length	Width	Coronula length	Coronula width	Number of helical stripes
Bailly and Schaefer (2010)	400–500	400–500	...	...	...
Caisová and Gąbka (2009)	460–690	420–525	...	...	...
Cirujano et al., (2008)	450–670	420–425	20–25	...	...
Compère (1992)	450–600	420–500	...	...	...
Corillion (1957)	500–580	...	20–30	...	6
Corillion (1975)	500–650	420–520(580)	(20)30–50	80–90	7–9
Dąmbska (1964)	500–580	420–480	20–30	...	7–8
Hollerbach and Krassavina (1983)	430–450	380–500	50–67	ca 50–67	7–8
Krause (1997)	500–700	400–525	To 59	To 85	9
Langangen (2007)	To 700	To 700	...	...	...
Maier (1972)	c. 280–580	...	...	...	...
Migula (1889–1897)	500–580	...	20–30	...	7–8
Mouronval et al., (2015)	...	...	...	...	6–7
Urbaniak and Gąbka (2014)	450–760	420–525	...	...	...
van Raam (1998)	To 550	To 500	...	...	...
Verdam (1938)	280–500 (average 350)	...	...	...	...
Wood and Imahori (1964)	570–670	480–525	c. 52	82–88	8–9
Wood (1965)	(465)500–690	420–525	c. 52	82–88	7–9
Total range	**(280)400–700(760)**	**(400)420–525**	**20–30(59)**	**80–90**	**(6)7–9**

Length and width of oogonia is without coronula; all values for length and width in µm

66.2 Phenology and Reproduction

Depending on the region *Nitella syncarpa* is a summer or autumn plant. It is seldom seen before May. Occurrence and fructification are mainly from July until September. It is an annual, which persists by means of oospores.

Table 66.2 Oospore morphometry, ornamentation and colour

Author	Length	Width	Number of striae	Ornamentation	Colour
Bailly and Schaefer (2010)	300–400	...	6–7	Smooth or granulate	Light to dark brown
Bazzichelli and Abdelahad (2009)	300–450	270–400	...	Smooth or granulate	Black
Caisová and Gąbka (2009)	...	...	...	Smooth or with marked stripes	Dark brown to black
Cirujano et al., (2008)	300–425	270–380	...	Smooth	Chestnut brown to black
Compère (1992)	...	...	...	...	Blackish brown
Corillion (1957)	300–350		6	...	Black
Corillion (1975)	300–350(440)	270–320	6–7(8)	...	Blackish brown
Dąmbska (1964)	300–350	270–320	...	...	Black
Dilger (2010)	320–400	290–370	5–7	...	Black
Hollerbach and Krassavina (1983)	330–375	280–335	6–7	Smooth or nearly smooth	Brown to almost black
Korte et al. (2016)	...	...	...	Smooth	...
Krause (1997)	300–450	270–400	6–8	Smooth	Brown
Langangen (2007)	300–450	...	...	...	Black
Maier (1972)	300–350	...	6	...	...
Migula (1889–1897)	300–348	270–320	...	...	Chestnut brown to black
Mouronval et al., (2015)	...	...	6–7	Smooth	...
Olsen (1944)	...	...	...	Smooth	...
Ray et al., (2001)	...	...	...	Smooth	...
Urbaniak and Gąbka (2014)	...	...	...	...	Brown to dark brown – almost black
van Raam (1998)	To 350	To 300	...	...	...
Wood and Imahori (1964)	420–435	c. 380	6–7	Smooth	Orange brown
Wood (1965)	300–435	270–380	6(8)	Smooth	Chestnut brown to dark brown or nearly black

(continued)

Table 66.2 (continued)

Author	Length	Width	Number of striae	Ornamentation	Colour
Total range	**300–450**	**270–400**	**(5)6–7(8)**		

All values for length and width in μm

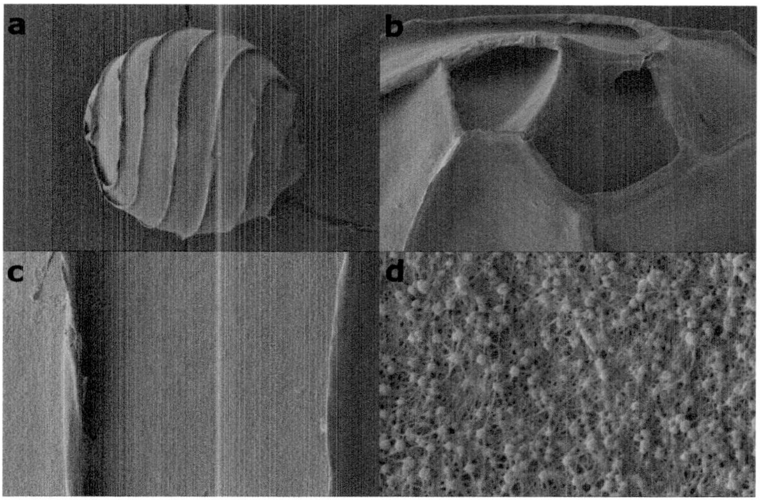

Fig. 66.3 Oospores of *Nitella syncarpa*. **a** Side view of oospore; **b** Basal view showing two end-cell impressions; **c** Detail of the fossa wall, appearing smooth; **d** Fine detail of the oospore wall with a felted-granulate appearance. Oospores collected by Garin and Belyakov 8 Jul. 2015, Nizhny Novgorod Oblast, Russia, *IBIW 56894*. Photographs: Michelle T. Casanova

66.3 Habitat Conditions and Ecology

Habitat types: *Nitella syncarpa* grows in the Baltic Sea, in natural lakes as well as in ponds, pits, dams, oxbow lakes, ditches, channels, rivers and marshes. According to Mouronval et al. (2015) *Nitella syncarpa* also occurs in temporary waters. Krause (1997) mentions the influence of ground water.

Sediment: *Nitella syncarpa* can grow on sand, mud, gravel, peat or clay. It prefers soft substrates, and can tolerate sediment cover.

Nutrient conditions: Mostly *Nitella syncarpa* occurs in mesotrophic water, but it certainly tolerates slight eutrophication. In this case competition is also critical. In more nutrient rich waters *Nitella syncarpa* may be outcompeted by larger *Chara* species or by *Nitellopsis obtusa*. Only few measurements of nutrient level are available.

Table 66.3 Antheridia morphometry

Author(s)	Diameter (μm)
Bailly and Schaefer (2010)	To 400
Caisová and Gąbka (2009)	To 700
Cirujano et al., (2008)	400–500
Compère (1992)	400–500
Corillion (1957)	400–450
Corillion (1975)	400–450
Dąmbska (1964)	400–450
Hollerbach and Krassavina (1983)	375–670
Krause (1997)	400–700
Maier (1972)	c. 250–450
Migula (1889–1897)	408–450
Urbaniak and Gąbka (2014)	360–510
van Raam (1998)	To 450
Verdam (1938)	250–325 (average 290)
Wood and Imahori (1964)	480–705
Wood (1965)	(365)408–705
Total range	**250–705**

Water chemistry: *Nitella syncarpa* thrives in moderately acid as well as in evidently alkaline water bodies. In the latter, the plants can be partially calcified, with bands of calcium carbonate deposited on the thallus. The mean pH values are between 6.2 and 8.2. The few measurements of EC are between 40 and 732 μS/cm. *Nitella syncarpa* grows in soft and hard water (5.7–90 mg/l Ca, Korte et al. 2016). A salinity range from 0.15 to 0.47 psu has been reported (Zhakova, 2004).

Depth range: *Nitella syncarpa* occurs mostly in shallow water bodies or shallow parts of deeper water bodies. However, in lakes it can also occur at greater depths. In sand and gravel pits in the Upper Rhine valley it grows down to 8 m (Humberg and Beck, 2006; Pätzold, 2011). According to Krause (1997) there is a record of *Nitella syncarpa* in Lake Constance at 30 m depth.

Associated species: *Nitella syncarpa* grows in monospecific beds (Fig. 66.2). *Nitella syncarpa* is considered to be a characteristic species of the association *Nitelletum syncarpae* (1957; Corillion, 1957) Dąmbska, 1966 of the alliance *Nitellion syncarpo-tenuissimae* Krause, 1969). *Nitella syncarpa* may also be associated with other *Characeae*. In moderately acid waters it may occur with *Nitella gracilis* and *N. translucens*, in more alkaline water is may co-occur with *N. tenuissima*, *N. confervacea*, *Chara globularis*, *C. papillosa* and *Nitellopsis obtusa*. In extensively used fish ponds *Nitella syncarpa* is associated with *Chara braunii*, *C. globularis* and *C. virgata*. *Nitella syncarpa* also grows together with vascular plants and bryophytes. *Nitella syncarpa* has also been recorded from different reed communities which are dominated by *Phragmites australis* or other helophytes. In more open vegetation *Eleocharis palustris*, *Elatine triandra*, *E. hydropiper*, *Luronium natans*, *Potamogeton* and *Utricularia* species may be associated.

66.4 Variability

Nitella syncarpa is quite variable. Presumably these are environment induced variations. Migula (1889–1897) distinguishes 10 environmentally induced growth forms, based on differences in plant colour, the amount of encrustation and cover with small organisms, plant size, plant vigour, axis diameter, number of internodes at the main axis, presence or absence of accessory branchlets, presence or absence of branchlets at the lower nodes, number of branchlets, branchlets at the same node of equal or unequal length, number of lateral branches, length of lateral branches, length of lateral branches relative to the internode length, size of the dactyls, diameter of dactyls, position of gametangia, the number of gametangia at each node, total number of gametangia, size of gametangia, striae size, presence or absence of mucus and amount of mucus.

66.5 Identification and Risk of Confusion

It is possible to confuse *Nitella syncarpa* with *Nitella capillaris* because both species have mucus around the gametangia. However, the oogonia of *Nitella capillaris* are situated on the branchlet furcations. The species may also be distinguished by their oospores. *Nitella capillaris* bears wide striae, whereas *Nitella syncarpa* bears only faint, low striae. The oospore-membrane of *Nitella syncarpa* is usually described as smooth, by some as (slightly) granulate or with marked stripes (see Table 66.2) whereas *Nitella capillaris* has a granulate oospore-membrane. Also the phenology of the species differ: *Nitella syncarpa* develops later in the season and is still present in autumn, whereas *Nitella capillaris* is a spring plant, often disappearing during summer. It is questionable whether the often-mentioned discriminating character—the amount of mucus, 'sticks less to the herbarium paper'—applies. It certainly seems that in both species the amount of mucus varies, e.g. in relation with age.

66.6 Distribution

The majority of the records are from Central and Western Europe (Fig. 66.4). In other regions *Nitella syncarpa* is rare or even absent: i.e. in Eastern Europe, on the Balkan peninsula, in Southern Europe, on the Iberian peninsula, in the British Isles

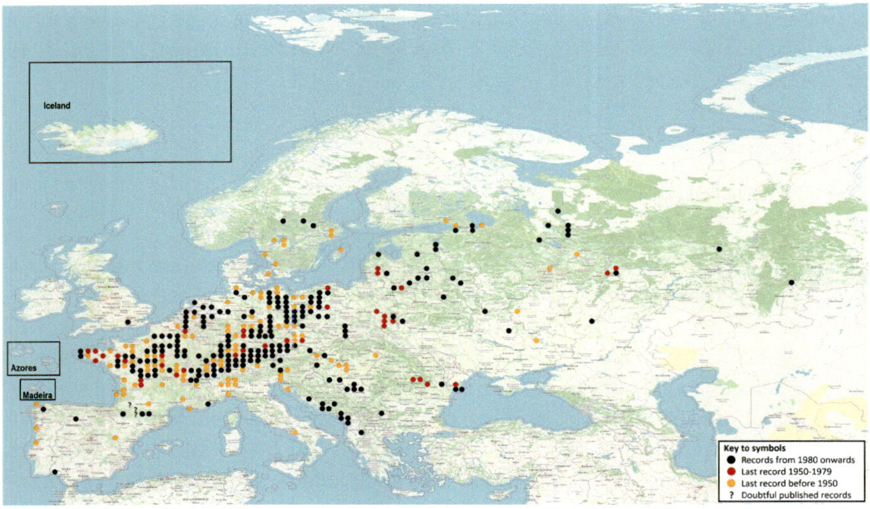

Fig. 66.4 European distribution of *Nitella syncarpa*. For worldwide distribution see Korsch (2018). Black dots are records from 1980 on, red dots from 1950 to 1979 and yellow dots before 1950; unconfirmed/doubtful published records (if applicable) are marked with "?" on top. Map prepared by Heiko Korsch

and in Scandinavia and Northern parts of Europe. In most countries for which there are relevant data, *Nitella syncarpa* is reported as declining.

66.7 Status and Threats

Nitella syncarpa is a pioneer species. For the germination of this annual species it is important that there is little or no competition with other water plants. In deeper lakes the populations of *Nitella syncarpa* seem to be more stable, especially in deeper parts where it is in complete vegetation cover. In the Baltic Sea *Nitella syncarpa* is probably limited by turbidity and wave exposure (Zhakova, 2004). *Nitella syncarpa* is listed in several Red Data Lists see (Chap. 9).

Within ditches and pools it is necessary to maintain areas that are not covered with other macrophytes. It can be useful to remove other macrophytes to facilitate the germination of *Nitella syncarpa*. In general, all measures to reduce/prevent eutrophication are recommended. This also includes the management of fish communities which should be typical for the specific water body type. *Nitella syncarpa* is part of the Swedish action plan for Charophytes in lakes and small water bodies (Blindow et al., 2021).

References

Bailly, G., & Schaefer, O. (2010). *Guide illustré des Characées du nord-est de la France.* Conservatoire Botanique National de Franche-Comté, Besancon [in French].

Bazzichelli, G., & Abdelahad, N. (2009). *Alghe d'acqua dolce d'Italia. Flora analitica delle Caroficee.* Sapienza Università di Roma. Ministero dell'Ambiente e della Tutela del territorio e del mare. Roma, Centro Stampa Università [in Italian].

Blindow, I., M. Carlsson, & K. van de Weyer (2021). Re-Establishment Techniques and Transplantations of Charophytes to Support Threatened Species. *Plants 2021: 10*, 1830.

Braun, A. (1847). Uebersicht der schweizerischen Characeen. *Neue Denkschriften der Allgemeinen Schweizerischen Gesellschaft für die Gesammten Naturwissenschaften, 10*, 1–23. [in German].

Caisová, L., & Gąbka, M. (2009). Charophytes (Characeae, Charophyta) in the Czech Republic: Taxonomy, autoecology and distribution. *Fottea, 9*, 1–43.

Cirujano, S., Cambra, J., Sánchez Castillo, P. M., Meco, A., & Flor Arnau, N. (2008). *Flora ibérica. Algas continentales. Carófitos (Characeae).* Real Jardín Botánico, Madrid [in Spanish].

Compère, P. (1992). *Charophytes. Flore pratique des algues d'eau douce de Belgique.* Jardin botanique national de Belgique [in French].

Corillion, R. (1957). *Les Charophycées de France et d'Europe Occidentale.* Bulletin de la Société Scientifique de Bretagne 32 (Hors série I & II), 1–499 [in French].

Corillion, R. (1975). *Flore des Charophytes (Characées) du Massif Armoricain et des contrées voisines d'Europe occidentale.* Flore et végétation du Massif Armoricain. T. IV. Jouve, Paris [in French].

Dąmbska, I. (1964). *Charophyta – ramienice.* Państwowe Wydawnictwo Naukowe, Warzawa [in Polish].

Dąmbska, I. (1966). Zbiorowiska ramienic Polski. *Prace Komisji Biologicznej, Poznanskie Towarzystwo Przyjaciol Nauk, Wydzial MatematycznoPrzyrodniczy, 31*, 1–76. [in Polish].

Dilger, M. (2010). Merkmale zur Erkennung von *Nitella-* und *Tolypella-*Arten der Characeaen [sic] über Oosporen aus Sedimenten oder von Pflanzen. *Rostocker Meeresbiologische Beiträge, 23*, 40–59. [in German].

Hollerbach, M. M., & Krassavina, L. K. (1983). [Голлербах, М. М. & Л. К. Красавина], *Определитель пресноводных водорослей СССР.* 14. Charophyta [Determination key of freshwater algae of USSR]. Nauka [in Russian].

Humberg, B., & Beck, M. (2006). Der Makrophytenbestand in ausgewählten Baggerseen der Oberrheinaue. *LUBW, Oberirdische Gewässer, Gewässerökologie, 102*, 1–348. [in German].

John, D. M., & Moore, J. A. (1987). An SEM study of the oospore of some Nitella species (Charales, Chlorophyta) with descriptions of wall ornamentation and an assessment of its taxonomic importance. *Phycologia, 26*, 334–355.

Korsch, H. (2018). The worldwide range of the Charophyte species native to Germany. *Rostocker Meeresbiologische Beiträge 28*, 45–96.

Korte, E., Pätzold, F., & Doege, A. (2016). *Nitella syncarpa.* In *Arbeitsgruppe Characeen Deutschlands: Die Characeen Deutschlands.* (pp. 477–485) Springer [in German].

Krause, W. (1997). *Charales (Charophyceae) Süßwasserflora von Mitteleuropa.* H. Ettl, G. Gärtner, H. Heynig, & D. Mollenhauer (Eds.) Süßwasserflora von Mitteleuropa, Teil 18. Fischer, Jena. [in German].

Krause, W. (1969). Zur Characeenvegetation der Oberrheinebene. *Archiv Für Hydrobiologie, Supplement, 35*, 202–253. [in German].

Langangen, A. (2007). *Charophytes of the Nordic countries.* Saeculum ANS.

Maier, E. X. (1972). *De kranswieren (Charophyta) van Nederland.* Wetenschappelijke Mededelingen Koninklijke Nederlandse Natuurhistorische Vereniging. [in Dutch].

Migula, W. (1889–1897). Fünfter Band: Die Characeen. In A. Fischer, E. Fischer, F. Hauck, G. Limpricht, C. Luerssen, W. Migula, H. Rehm, P. Richter, & G. Winter (Eds.), *Dr. L. Rabenhorst's Kryptogamenflora von Deutschland, Oesterreich und der Schweiz* (2nd edn). Eduard Kummer [in German].

Mouronval, J. B., Baudouin, S., Borel, N., Soulié-Märsche, I., Klesczewski, M., & Grillas, P. (2015). *Guide des Characées de France méditerranéenne*. Office national de la Chasse et de la Faune Sauvage [in French].

Olsen, S. (1944). Danish Charophyta. Det Kongelige Danske Videnskabernes Selskab, *Biologiske Skrifter 3*(1), 1–240, pl. 1–2 [in Danish].

Pätzold, F. (2011). Beobachtungen zur Verbreitung von Characeen in Baggerseen der Badischen Oberrheinaue. *Berichte der Botanischen Arbeitsgemeinschaft Südwestdeutschland Beiheft 3*, 37–42 [in German].

van Raam, J. C. (1998). *Handboek Kranswieren*. Chara boek, Hilversum [in Dutch].

Ray, S., Pekkari, S., & Snoeijs, P. (2001). Oospore dimensions and wall ornamentation patterns in Swedish charophytes. *Nordic Journal of Botany, 21*, 207–224.

Urbaniak, J., & Gąbka, M. (2014). *Polish Charophytes. An illustrated guide to identification*. Uniwersytet Przyrodniczy we Wroclawiu [in Polish].

van de Weyer, K., & Schmidt, C. (2018). Bestimmungsschlüssel für die aquatischen Makrophyten (Gefäßpflanzen, Armleuchteralgen und Moose) in Deutschland: Band 1: Bestimmungsschlüssel, ed. 2. *Landesamt für Umwelt (LfU) Brandenburg. Potsdam, Fachbeiträge des LfU Brandenburg, 119*, 1–180. [in German].

Verdam, H. D. (1938). The Netherlands Charophyta. *Blumea, 3*, 5–33.

Wood, R. D. (1965). *Monograph of the Characeae*. J. Cramer.

Wood, R. D., & Imahori, K. (1964). *Iconograph of the Characeae*. J. Cramer.

Zhakova, L. V. (2004). *Nitella syncarpa*. In H. Schubert, & I. Blindow (Eds.), *Charophytes of the Baltic Sea* (pp. 204–208). A. R. G. Gantner.

Chapter 67
Nitella tenuissima

Hendrik Schubert, Egbert Korte, and Frank Pätzold

Scientific name: ***Nitella tenuissima*** **(Desv.) Kütz., Phycol. General.: 319. 1843**

≡ *Chara tenuissima* Desv. in J. Bot. (Desvaux): 313. 1809
Neotype (Gregor & Raabe in Armleuchteralgen Characeen Deutschl. 2016: 485): [no location]/[no date]/Desvaux (B-W 17115-010 [sphal. B-W 17115-015])

= *Nitella tenuissima* (Desv.) Kütz. f. *major* Migula, Charac. Deutschl.: 180. 1890
Type not designated

Nomenclatural note: Wood (1965: 552) designated as lectotype "Coll.? [Desvaux?] (PC)". The sheet PC 143200, with 4 specimens, exists in the herbarium of Desvaux (PC). One plant was collected by Desvaux but is not *N. tenuissima* (Gregor et al., 2012). There exists another sheet in PC (PC 143218) which contains, among other taxa, a specimen annotated as "*Chara tenuissima* Desvaux". This was seen by Wood who noted that this was not *N. tenuissima* but *N. gracilis*. As the lectotypification of Wood seems to be erroneous T. Gregor and U. Raabe selected another specimen as "lectotype" from the Willdenow herbarium in Berlin which was sent from Desvaux to A. Braun. As this specimen is undated it should be better regarded as a neotype.

Local names: DUTCH: Kraaltjesglanswier; ENGLISH: Dwarf Stonewort; FRENCH: Nitelle très grêle; GERMAN: Schirmförmige Glanzleuchteralge; POLISH: Krynicznik parasolowaty; RUSSIAN: Нителла тончайшая; SWEDISH: Pärlslinke; UKRAINIAN: Нітела найтонша.

H. Schubert (✉)
Universität Rostock Institut für Biowissenschaften, Albert-Einstein-Straße 3, 18051 Rostock, Germany
e-mail: hendrik.schubert@uni-rostock.de

E. Korte
INGA, Wiesenstraße 6, 64347 Griesheim, Germany

F. Pätzold
Pätzold-Gewässerökologie, Winzerstrasse 50, 76532 Baden-Baden, Germany

H. Schubert et al. (eds.), *Charophytes of Europe*,
https://doi.org/10.1007/978-3-031-31898-6_67

67.1 Morphology

See Figs. 67.1 and 67.2.

Habitus: *Nitella tenuissima* is a small, yellowish-green species. It develops, almost exclusively, a single branch (or side-shoot) per node, whereas the typical *Nitella*-pattern of two-branches per node is very rarely observed. The species appears as tufts of solitary plants, consisting of axes with globose whorls separated by long internodes (string-of-beads appearance), so the species is easy to recognise. The tufts tend to be horizontal on the substrate surface and individuals growing upright are rare.

Size: *Nitella tenuissima* is small to medium-sized (5–20 cm), but usually well below 10 cm high (Migula, 1889–1897). In contrast to this, Krause (1997) reports a specimen of up to 60 cm length, and similar observations of 40–50 cm long plants were

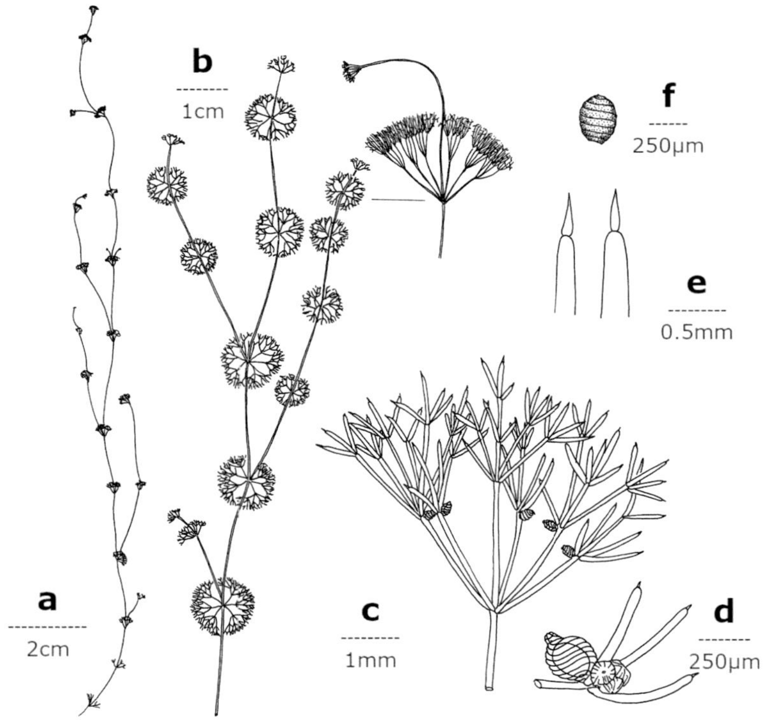

Fig. 67.1. *Nitella tenuissima*. **a** Habitus of deep-water specimen (described by Migula, 1889–1897 as f. *major*, characterized by the triangular appearance of whorls); **b** Typical habitus; **c** Head with solitary oogonia at the first branchlet furcation; **d** Uppermost (3rd) branchlet segment with oogonium and antheridium; **e** Branchlet tips; **f** Oospore. Drawings after W. Krause, modified by Dagmar Wassong in van de Weyer and Schmidt (2018)

Fig. 67.2 *Nitella tenuissima,* stands and heads. **a** Dense patch; **b** Richly branched individual; **c** Spreading head; **d** Condensed heads; **e** Typical form. Photographs: **a**, **b**, **d** - Klaus van de Weyer; **c**, **e** - Frank Pätzold

made by F. Pätzold in the Oberrhein-region, where such specimens seem to occur only in autumn.

Main axis: Axis diameter is between 0.18 and 0.25 mm. The 5–7 internodes of each axis are always longer (4–5 times) than the corresponding branchlets.

Branchlets: Whorls consist of 6 short branchlets, which are 2–3 times divided. The first furcation has 3–5 secondary segments of which 1–4 may be dactyls, and one of which can be central. Second furcations have 2–5 tertiary segments of which 2–5 may be dactyls, and the central one may be further divided. The third furcation, where present, has usually 3 quaternary segments (all of which are dactyls). The whorl diameter is between 3 and 10 mm.

Dactyls: The dactyls are two-rarely three-celled. Where they appear to be three-celled it is likely that it is a suppressed furcation with one segment. The end cell is short, narrowly pointed and falls off easily (Krause, 1997).

Sexuality: The species is monoecious. Whereas Krause (1997) describes plants as often being sterile, Migula (1889–1897) reported the species as richly fructifying. Overwintering plants reproduce sexually from May to November, new plants develop ripe oospores from August onwards. Gametangia develop at the second and third furcation, never at the base of the whorl, or the first furcation. At the second furcation often only oogonia are found.

Table 67.1 Oogonia morphometry

Author	Length	Width	Coronula length	Coronula width	Number of helical stripes
Migula (1889–1897)	240–300	219–260	–	–	8–9
Groves and Bullock-Webster (1920)	Up to 400	up to 260	30–40	–	7–8
Corillion (1957a)	360–370	up to 260	30–40	–	7–8
Krause (1997)	375–425	275–370	30–50	40–55	7–8
van Raam (1998)	Up to 400	up to 350	–	–	–
Cirujano et al., (2007)	330–400	225–270	–	–	–
Caisová and Gąbka (2009)	270–550	225–410	–	–	–
Bailly and Schaefer (2010)	200–300	–	–	–	7–8
Total range	**200–550**	**219–410**	**30–50**	**40–55**	**7–9**

Length in the first column is excluding the coronula, its length and width given separately; all values for length and width in μm

Oogonia: Oogonia are covered by a very thin mucus layer which is sometimes hard to see. Morphometric data of oogonia are listed in Table 67.1. Please note that the data on quantitative morphometry varies among authors especially with respect to the lower length limit.

Oospores: The oospore surface is consistently reticulate and often beaded; striae are inconspicuous (Fig. 67.3). Table 67.2 gives an overview about morphometry and colour of oospores.

Antheridia: Antheridia of *Nitella tenuissima* are slightly compressed and, because of their small size, hard to observe with the naked eye in the dense whorls. Table 67.3 gives an overview of the literature data on antheridia.

Bulbils: There are no reports of bulbil formation.

67.2 Phenology and Reproduction

Nitella tenuissima is a perennial species, overwintering individuals become fertile as early as early spring (e.g., May in Germany) Sexual reproduction continues until September. Newly germinated plants start to become fertile in late summer (August in Germany). However, the species is known for its phenological plasticity, and its capacity to adapt its life-cycle to local conditions (Mouronval et al., 2015).

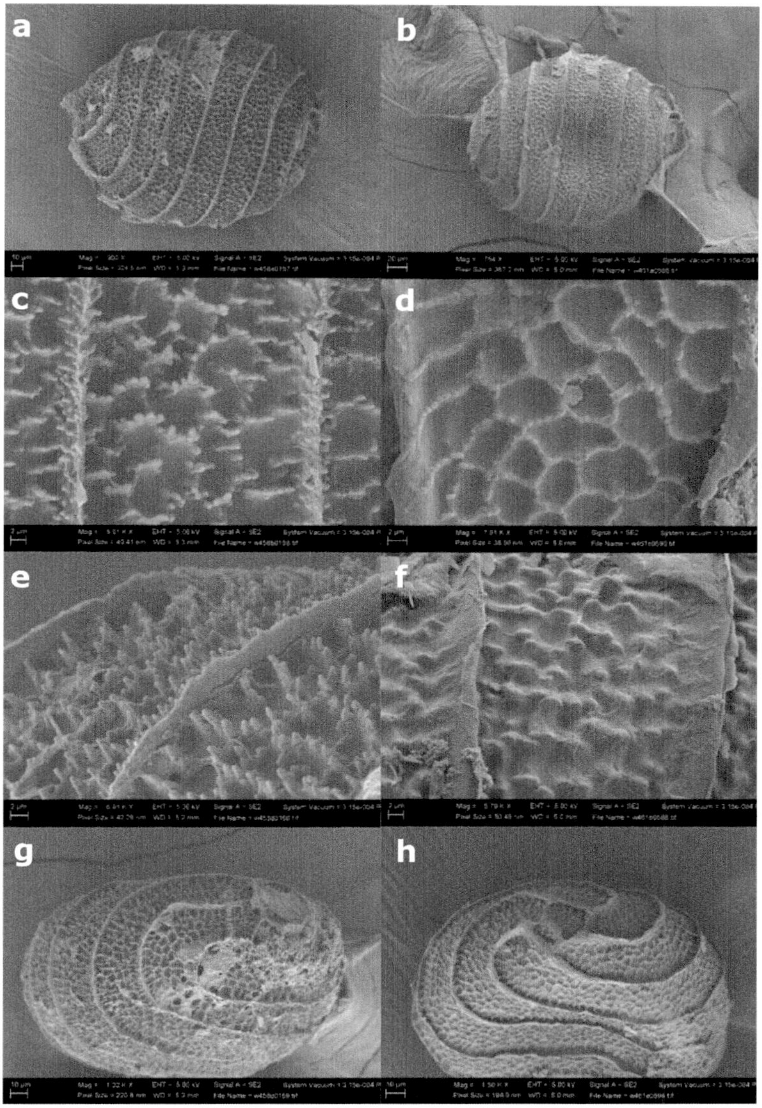

Fig. 67.3 Images of *Nitella tenuissima* oospores. Left row (**a**, **c**, **e**, **g**) w458, G.W. Bullock-Webster, 27/9/1897, Lopham Great Fen, in turf pool, Norfolk East UK (BM000610232); right row (**b**, **d**, **f**, **h**) w461, A. Braun, Aug. 1839, Carlesruhe, Germany, (BM013828851). **a**, **b** View of oospore, **c**, **d**, **e**, **f** Details of oospore wall and striae, **g**, **h** Basal cell impressions. Photographs: Michelle T. Casanova

Table 67.2 Oospore morphometry

Author	Length	Width	Number of striae	Colour
Migula (1889–1897)	180–250	155–200	–	Light brown
Groves and Bullock-Webster (1920)	200–250	175–200	–	–
Corillion (1957a)	200–250	175–201	7–8	–
Krause (1997)	200–300	175–220	7–8	Reddish
van Raam (1998)	Up to 300	Up to 200	–	–
Cirujano et al., (2007)	200–270	175–200	–	–
Caisová and Gąbka (2009)	–	–	–	Yellowish dark-brown to black
Dilger (2010)	210–320	170–280	5–8	Light brown to brown
Total range	**180–320**	**170–280**	**5–8**	

All values for length and width in μm

Table 67.3 Antheridia morphometry

Author	Diameter (μm)
Corillion (1957a)	Up to 175
Krause (1997)	130–200
van Raam (1998)	Up to 200
Cirujano et al., (2007)	90–175
Caisová and Gąbka (2009)	130–200
Bailly and Schaefer (2010)	100–200
Total range	**90–200**

67.3 Habitat Conditions and Ecology

Habitat type: *Nitella tenuissima* grows typically in small water bodies, often ephemeral ones. It is also reported from beach lakes and small, slow-flowing ditches especially in the Mediterranean region. In addition, it is found in mining borrow-pits and quarries, especially in gravel and clay pits.

Sediment: *Nitella tenuissima* prefers muddy sediment. When growing on sandy sediment or peat, a thin layer of mud, at least, is always present.

Nutrient conditions: *Nitella tenuissima* has a clear preference for oligo- to mesotrophic conditions, occurrences in eutrophic conditions are rare. The species is sensitive to competition with fast-growing, large charophytes, like *Chara globularis* or *Nitellopsis obtusa*. There are only few data about nutrient concentrations, maxima recorded so far are 1.13 mg/l nitrate, 0.09 mg/l ammonia, 0.022 mg/l total phosphorous and 0.005 mg/l orthophosphate.

Water chemistry: *Nitella tenuissima* prefers alkaline water with a pH of up to 8.5 but has been found to occur in water with a pH as low as 6.5. The species is frequently encrusted, and the total hardness of its habitat ranges between 4.1 °dH and 19.5 °dH, and carbonate hardness varies between 0 °dH and 9.8 °dH. The species tolerates some salinity and grows at conductivities of up to 732 μS/cm. The maximum chloride concentration reported was 44 mg/l.

Depth range: *Nitella tenuissima* prefers shallow water habits and is rarely found at depths >8 m. Maximum depth recorded is 12 m. However, it is most frequently found between 2 and 6 m depth (Korte et al., 2016).

Associated species: *Nitella tenuissima* is the character species of the *Nitelletum syncarpo-tenuissimae* association. Especially in artificial ponds and mining lakes *Nitella tenuissima* is often accompanied by *Nitella syncarpa, N. confervacea, Tolypella glomerata* and *Sphaerochara intricata*, sometimes also by *Chara aspera, C. aculeolata* und *C. hispida* (Humberg and Beck, 2006; Korte and Gregor, 2008; Pätzold, 2011).

67.4 Variability

Nitella tenuissima varies little with respect to its habit, except for size. The forms listed by Migula (1889–1897) are mainly size-related, and the only notable one is Migula's f. *major*, which differs from the typical habit by having brush-like, triangular whorls instead of the typical globose whorls, resembling slightly *N. confervacea*.

67.5 Identification and Risk of Confusion

Being very small and because of the conspicuous globose whorls and heads, *Nitella tenuissima* is easy to recognise. *Nitella hyalina* can also be quite small and develop dense whorls; however, *N. hyalina* is a heteroclemous species, a character which allows for easy and reliable distinction, leaving *Nitella confervacea* as the only species *N. tenuissima* might be confused with. To distinguish between these two taxa, location of gametangia provides the best character: *N. tenuissima* never develops gametangia at the first branchlet node, whereas this is always the case for *N. confervacea*. In addition, internodes of *N. confervacea* are much shorter (in relation to the branchlet length) and the uppermost whorls are condensed to hedgehog-like heads, whereas *N. tenuissima* has whorls separated by long internodes throughout the plant, giving it its typical 'string-of-beads' appearance.

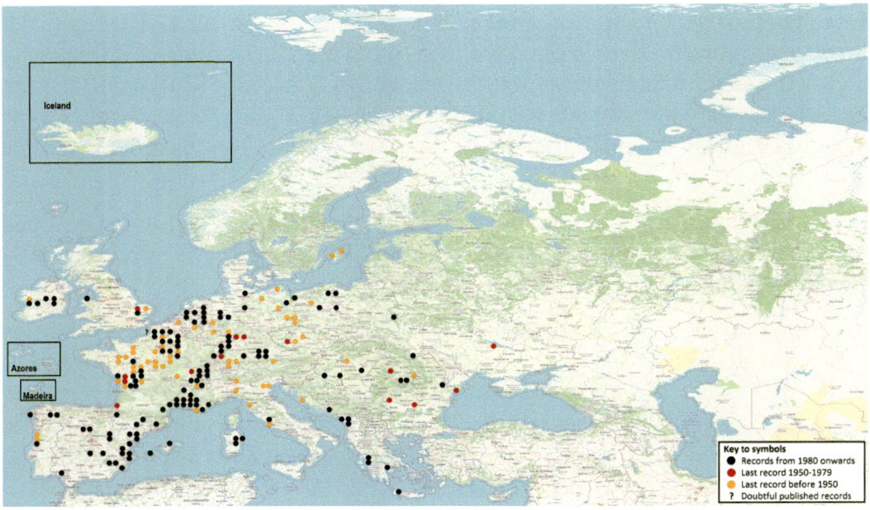

Fig. 67.4 European distribution of *Nitella tenuissima*. For worldwide distribution see Korsch (2018). Black dots are records from 1980 on, red dots from 1950 to 1979 and yellow dots before 1950. Map prepared by Heiko Korsch

67.6 Distribution

See Fig. 67.4.

Nitella tenuissima is found almost all over Europe, except for Iceland, Denmark, Norway, Estonia and Finland. Whereas this part of the distribution pattern may point to a climatic limitation, the absence in other countries (e.g., Belarus) may be due to other reasons as, e.g., lack of data or lack of suitable habitat. *Nitella tenuissima* is a rare species irrespective of its apparent wide geographical distribution (for worldwide distribution see Korsch, 2018). Occurrences are concentrated in restricted regions even in countries where it can be found at many sites. It is reported to have a worldwide distribution, but is rare all over its distribution range. However, there are serious doubts about those reports because it never has been found in Australia and at least some specimens from North America are probably misidentifications as revealed by their oospore ornamentation (M. T. Casanova pers. comm.).

67.7 Status and Threats

The species is rare throughout its range and several sites have been lost during the last few decades. Most, if not all of these lost sites are ponds and ditches, whereas occurrences in large lakes are relatively stable. However, the species most frequently is found in small water bodies, so the Europe-wide loss of these habitats is a major threat for the species. *Nitella tenuissma* is sensitive to both eutrophication and stocking with inappropriate fish species (especially benthic-foraging species) it is one of the species which should be included in species protection programs. Creation of small ponds in its regions of occurrence might help in at least maintaining its European genetic diversity until knowledge about worldwide relationships with other entities identified as the same species are better known.

References

Bailly, G., & Schaefer, O. (2010). *Guide illustré des Characées du nord-est de la France*. Conservatoire Botanique National de Franche-Comté [in French].

Caisová, L., & Gąbka, M. (2009). Charophytes (Characeae, Charophyta) in the Czech Republic: Taxonomy, autoecology and distribution. *Fottea, 9*, 1–43.

Cirujano, S., García Murillo, P., Meco, A., & Fernández Zamudio, R. (2007). Los carófitos ibéricos. *Anales Del Jardín Botánico De Madrid, 64*, 87–102. [in Spanish].

Corillion, R. (1957a). Les Charophycées de France et d'Europe Occidentale. *Bulletin de la Société Scientifique de Bretagne, 32*, 1–499. [in French].

Dilger, M. (2010). Merkmale zur Erkennung von *Nitella*- und *Tolypella*-Arten der Characeen über Oosporen aus Sedimenten oder von Pflanzen. *Rostocker Meeresbiologische Beiträge, 23*, 40–59. [in German].

Gregor, T., Raabe, U., & van Raam, J. (2012). Typification of *Nitella gracilis* (Sm.) C. Agardh var. *confervacea* Brébisson (= *N. confervacea* (Brébisson) A. Braun ex Leonhardi) with remarks on its nomenclatural history. *Cryptogamie Algologie, 33*, 391–398.

Groves, J., & Bullock-Webster, G. R. (1920). *The British Charophyta Volume I. Nitelleae*. Ray Society.

Humberg, B., & Beck, M. (2006). *Der Makrophytenbestand in ausgewählten Baggerseen der Oberrheinaue*. LUBW, Oberirdische Gewässer, Gewässerökologie, 102. [in German].

Korsch, H. (2018). The worldwide range of the Charophyte species native to Germany. *Rostocker Meeresbiologische Beiträge, 28*, 45–96.

Korte, E., Pätzold, F., & Schubert, H. (2016). *Nitella tenuissima*. In Arbeitsgruppe Characeen Deutschlands, *Armleuchteralgen—Die Characeen Deutschlands* (pp. 485–493). Springer [in German].

Korte, E., & Gregor, T. (2008). Neue Characeenfunde aus Hessen. *Rostocker Meeresbiologische Beiträge, 19*, 7–12. [in German].

Krause, W. (1997). Charales (Charophyceae) Süßwasserflora von Mitteleuropa. In H. Ettl, G. Gärtner, H. Heynig, & D. Mollenhauer (Eds.) Süßwasserflora von Mitteleuropa (Vol. 18). Fischer [in German].

Migula, W. (1889–1897). Fünfter Band: Die Characeen. In A. Fischer, E. Fischer, F. Hauck, G. Limpricht, C. Luerssen, W. Migula, H. Rehm, P. Richter, & G. Winter (Eds.) *Dr. L. Rabenhorst's Kryptogamenflora von Deutschland, Oesterreich und der Schweiz* (2nd edn). Eduard Kummer [in German].

Mouronval, J. B., Baudouin, S., Borel, N., Soulié-Märsche, I., Klesczewski, M., & Grillas, P. (2015). *Guide des Characées de France méditerranéenne*. Office national de la Chasse et de la Faune Sauvage [in French].

Pätzold, F. (2011). Beobachtungen zur Verbreitung von Characeen in Baggerseen der Badischen Oberrheinaue. *Berichte Der Botanischen Arbeitsgemeinschaft Südwestdeutschland Beiheft, 3*, 37–42. [in German].

Raam van, J. C. (1998). *Handboek Kranswieren*. Chara boek [in Dutch].

Weyer van de, K. & Schmidt, C. (2018). *Bestimmungsschlüssel für die aquatischen Makrophyten (Gefäßpflanzen, Armleuchteralgen und Moose) in Deutschland: Band 1: Bestimmungsschlüssel* (2nd ed.). Landesamt für Umwelt (LfU) Brandenburg, Fachbeiträge des LfU Brandenburg [in German].

Wood, R. D. (1965). *Monograph of the Characeae*. J. Cramer.

Chapter 68
Nitella translucens

Nadia Abdelahad and Ralf Becker

Scientific name: ***Nitella translucens*** **(Pers.) C. Agardh, Syst. Alg.: 124. 1824**

≡ *Chara translucens* Pers. 1807; Syn. Pl. [Persoon]: 531
Neotype (designated here; formerly designated as a holotype by Wood 1965: 682): Eaton Pool, near Shrewsbury/1800/Rev. E. Williams (LINN, "Smithian Herb." 1432.6)

= *Chara flexilis* Thuill. Fl. Env. Paris, ed. 2: 472. 1799, n. ill. non L. 1753
Type not designated

= *Nitella translucens* (Pers.) C. Agardh f. *confervoides* Thuill. ex Mig. (1890: 146).
Type not designated

Nomenclatural notes: Persoon (1806–1807: 531) did not indicate any specimen, gathering, drawing or reference for *Chara translucens*. Therefore, no holotype or lectotype can be indicated. Wood (1965) indicated a specimen as a "holotype" (see above), which was designated here as a neotype.

The London Natural History Museum preserves in Gay's Herbarium a specimen of probably not published *Chara confervoides* (ex *Chara flexilis* var. *translucens*, Wallr. in litt. ann. 1828, according to the label) (BM13821335), assigned by Braun to *Chara translucens* (label on the specimen; Braun, 1835), and checked by Nick Stewart.

Herbarium Haussknecht (JE) preserves in Migula's Herbarium a specimen which has the features of *N. translucens* f. *confervoides* Thuill. but it is labelled by Migula as *N. translucens* Pers.

N. Abdelahad (✉)
Dipartimento di Biologia Ambientale, Università di Roma "La Sapienza", Piazzale Aldo Moro, 5, Rome 00185, Italy
e-mail: abdelahadnadia3@gmail.com

R. Becker
Am Schützenplatz 58, 26121 Oldenburg, Germany

H. Schubert et al. (eds.), *Charophytes of Europe*,
https://doi.org/10.1007/978-3-031-31898-6_68

Fig. 68.1 Pre-Linnean drawing and recent specimen of *Nitella translucens*. Left panel: Lectotype—Vaillant's drawing of *Chara translucens major flexilis* (reproduced from Vaillant, 1719: Fig. 8, Pl. III looking as a sterile plant of *Chara*, e.g. *C. globularis*, as well), right panel: Ralf Becker's herbarium sheet in Herbarium RO (RO.HG Algae 7-3-17.5-600). Photographs: Left panel—reproduction by BHL (Biodiversity Heritage Library). Right panel—scanned for herbarium RO by Mauro Iberite

Local names: DUTCH: Doorschijnend glanswier; GERMAN: Schimmernde Glanzleuchteralge; NORWEGIAN: Blankglattkrans; POLISH: Krynicznik połyskujący; RUSSIAN: Нителла просвечивающаяся; SWEDISH: Grovslinke; UKRAINAIN: Нітела просвічуюча.

68.1 Morphology

See Figs. 68.1, 68.2, 68.3 and 68.4.

Habitus: *Nitella translucens* is bright to dark green, translucent (Corillion, 1957; Wood, 1965; Krause, 1997). Branchlets appear macroscopically undivided. Dimorphic branchlets, with gametangia on the top of short branchlets, gathered in compact heads. Shape retention out of the water fairly good but when damaged immediately collapsing.

Size: *Nitella translucens* is a medium to large plant, usually 15–80 cm with a maximum of 125 cm. (Mouronval et al., 2015; Becker and Doege, 2016). It is generally a robust, large plant.

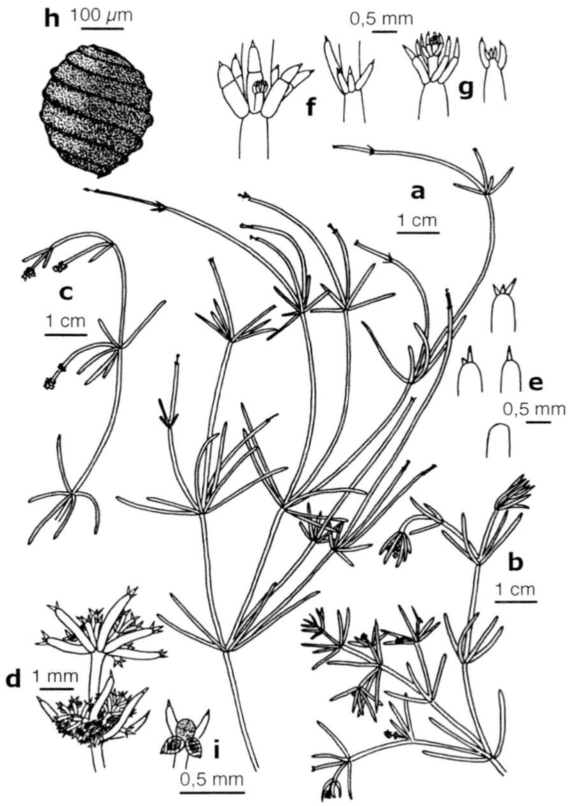

Fig. 68.2 *Nitella translucens*. **a** Plant with long branches; **b** Compact plant; **c** Fertile sprouts; **d** Fertile whorls; **e** End segments of the branches; **f** Whorl of long plant; **g** Summit of a long plant; **h** Oospore; **i** Branch tip with gametangia. Drawings by Werner Krause

Main axis: The axis is ecorticate, moderately slender to stout. The diameter range is 370–2000 μm, and from 750 to 1000 μm in f. *confervoides* (Migula, 1889–1897: 146). The internodes are 1–5 times as long as the sterile branchlets, predominantly 3–8.5 cm long. Lowest internodes up to 20 cm long, in deep water plants are even longer.

Cortex, spine-cells and stipulodes: Absent.

Branchlets: Whorls dimorphic, composed of (2)3–7(8) long sterile branchlets and several short fertile branchlets (Fig. 68.2). Sterile branchlets are 0.5–8(10) cm long, seemingly simple, but are divided at the apex in a rudimentary whorl of (0)3–4(6) dactyls forming a tiny apical crown. Mostly 4–6 short fertile branchlets, 3–40 mm

Fig. 68.3 *Nitella translucens*, details. **a** Habitus with branchlets and shoot apex (plant from a newly created heathland pond, Bad Bentheim, Germany); **b** Head of fertile branchlets with gametangia, fresh material; **c** Habitus in a pond near Breugel, The Netherlands; **d** Young scattered individual in Lake Versen, Meppen, Germany. Photographs: **a**-Ralf Becker; **b**, **c**-Klaus van de Weyer; **d**-Tobias Böckermann

long, 1–4 furcate, the first and the second order branchlets gathered usually in compact heads at the base of the long sterile branchlets (Figs. 68.2d and 68.3a, b). Fertile heads frequently curved downwards (Fig. 68.2c). Diameter of fertile heads up to 10 mm. Dactyls extremely reduced: those of the sterile branchlets of unequal length, 100–250 μm long; those of the fertile branchlets often masked by the gametangia. Basal cell tapering progressively to the conical, acute or acuminate end-cell. Dactyls can break off easily and seem to be missing.

Bract-cells, bracteoles and bractlets: Absent.

Sexuality: Monoecious. Sometimes the plants are permanently sterile or have only few gametangia.

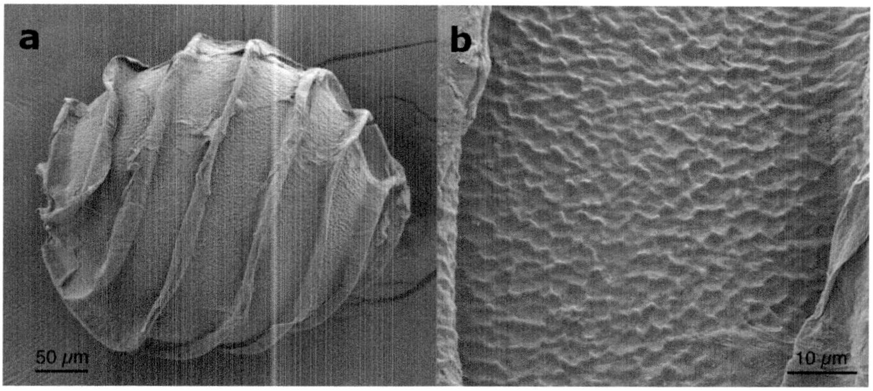

Fig. 68.4 SEM images of the oospore of *N. translucens* from a specimen collected by Pelvet at Vise, Normandy (BM013828852). **a** Side view of oospore; **b** Detail of oospore-membrane showing a finely reticulate ornamentation. Photographs: Michelle T. Casanova

Oogonia: Oogonia in groups of 2–4 or solitary. Without mucus. The dimensions of the oogonia and coronula vary according to different authors (Table 68.1).

Oospores: Oospores of *Nitella translucens* are relatively small. Recent data on fresh plants from Northern Germany show that length and width of oospores can vary significantly even in the same population (R. Becker own data) (Table 68.2). Migula (1889–1897: 144) has carried out checks on oospores in plants collected in Germany

Table 68.1 Oogonia morphometry

Author	Length	Width	Coronulalength	Coronula width	Number of helical stripes
Migula (1889–1897)	–	–	–	–	8–9
Groves and Bullock-Webster (1920)	475–525	400–425	40–45	–	7–8
Corillion (1957)	475–525	400–525	40–50	–	7–8
Wood (1965)	420–525[a]	270–425	40–46	70–80	7–9
Compère (1992)	400–500	300–450	–	–	–
Krause (1997)	450–525	300–425	40–50	70–80	7–8
van Raam (1998)	To 300	To 400	–	–	–
Langangen (2007)	To 500	–	–	–	–

(continued)

Table 68.1 (continued)

Author	Length	Width	Coronulalength	Coronula width	Number of helical stripes
Bailly and Schaefer (2010)	300–500 in diameter	300–500 in diameter	–	–	–
Cirujano et al. (2008)	380–480	270–425	40–46	–	–
Urbaniak and Gąbka (2014)	320–545	260–460	–	–	–
R. Becker, own data	(400)500–675[a]	(325)400–500(550)	25–75	50–90	(6)7–9
Total range	**< 300–545**	**260–550**	**25–75**	**50–90**	**6–9**

Length in the first column, is excluding the coronula, its length and width is given separately; all values for length and width in μm

[a] Including coronula

and Northern France (Neuenkirchen near Osnabrück, Emden, Odenthal, Falaise) and gave the following extreme lengths and widths: 250–296 × 238–270 μm. The average dimensions of the oospores measured by Corillion (1957) on Armorican specimens are nearly the same: average lengths: 265–280 μm; average widths: 240–255 μm. The oospore-membrane is finely reticulate (Groves and Bullock-Webster, 1920; John and Moore, 1987; Haas, 1994; Fig. 68.4).

Table 68.2 Oospore morphometry

Author	Length	Width	Number of striae	Colour
Migula (1889–1897)	250–296	238–274	6–7	Light yellow–brown
Groves and Bullock-Webster (1920)	325–350	250–300	6	Yellow–brown
Corillion (1957)	260–350	240–300	6	Shiny golden-brown
Wood (1965)	300–350	250–300	5–6	Light golden or yellowish brown

(continued)

Table 68.2 (continued)

Author	Length	Width	Number of striae	Colour
Compère (1992)	–	–	5–6	Golden-brown, blackish on drying
Krause (1997)	300–450	250–350	5–6	Brown
van Raam (1998)	To 400	To 300	–	–
Langangen (2007)	–	–	–	Brown
Cirujano et al. (2008)	350–410	250–360	–	Dark brown
Bailly and Schaefer (2010)	300–400 in diameter	300–400 in diameter	5–6	Golden-brown
Urbaniak and Gąbka (2014)	230–350	250–310	–	Yellow brown to dark brown
R. Becker own data	(290)325–425(475)	(225)275–375(400)	(5)6–7	Light brown
Total range	**230–475**	**225–400**	**5–7**	

All values for length and width in μm

Antheridia: Antheridia are small and not enveloped in mucus (Table 68.3). Own measurements, performed on fresh specimens from northern Germany, revealed a relatively broad range of antheridia diameters even in the same population (R. Becker own data).

Table 68.3 Antheridia morphometry

Author	Diameter (μm)
Migula (1889–1897)	Up to 380
Groves and Bullock-Webster (1920)	250–375
Corillion (1957)	250–375
Wood (1965)	225–375
Compère (1992)	250–400
Krause (1997)	250–375
van Raam (1998)	To 400
Langangen (2007)	To 400
Cirujano et al. (2008)	240–310
Bailly and Schaefer (2010)	250–400
Urbaniak and Gąbka (2014)	180–380
R. Becker own data	(200)275–425(475)
Total range	**180–475**

Bulbils: Groves and Bullock-Webster (1920) indicated the presence of composite root bulbils on specimens collected in France (Étang de Cazau (Gironde, France). Corillion (1957) has often observed them both in nature (on Armorican specimens) and in culture (on sterile specimens).

68.2 Phenology and Reproduction

Nitella translucens is annual to perennial. The species germinates generally from oospores from late summer to December, and in mild winters in the Netherlands up to January (van Raam and Maier, 1993). In Western France plants frequently overwinter in perennial state (Corillion, 1957).

68.3 Habitat Conditions and Ecology

Habitat type: *Nitella translucens* is a species of low altitude in Europe and prefers nutrient poor shallow waters, acidic to neutral. The taxon colonises ponds, fish ponds, heathland ponds, oxbow lakes, small ephemeral pools, ditches (Becker and Doege, 2016), peat swamps (Corillion, 1957; Mouronval et al., 2015), and coastal ponds in France, Portugal and Sardinia (Corillion, 1957; Krause, 1983; Becker, 2019). It is also present in lakes and deep sand pits in Germany (Becker and Doege, 2016; Becker et al., 2022), United Kingdom (Groves and Bullock-Webster, 1920) and Nordic countries (Langangen, 2007). The species was even recorded in anthropogenic well shafts in Portugal (Krause, 1983) and stormwater ponds in Germany (Becker et al., 2022). *Nitella translucens* was occasionally found in running waters in Great Britain (Groves and Bullock-Webster, 1920).

Sediment: The species develops mainly on soft, sandy or silty-muddy substrates, sometimes on peaty sediments or substrates rich in humic matter (Corillion, 1957; Krause, 1997; Bailly and Schaefer, 2010; Mouronval et al., 2015).

Nutrient conditions: *Nitella translucens* prefers oligotrophic and mesotrophic freshwater habitats, although nutrient conditions vary a lot according to results of different authors (Becker and Doege, 2016; Doege et al., 2016; Becker et al., 2022). Overall, there are only few published data available. Table 68.4 is based on recent data from 13 ponds and lakes in north-west Germany (Becker et al., 2022). Although TN of *N. translucens*-habitats does mainly not exceed 2.19 mg/l, data, recorded 2014 and 2020, from a single waterbody near Herbrum (north-west Germany) with a permanent, large and vital population of *N. translucens* without losses in vitality reveal, that the species seems to be able to tolerate at least temporary higher TN-concentrations up to 7.918 mg/l without losses in vitality (Becker and Doege, 2016; Doege et al., 2016; Becker et al., 2022; see Tables 68.4 and 68.5).

Table 68.4 Values for selected hydrochemical parameters of 13 *N. translucens*-habitats in Northwestern Germany (Becker et al., 2022)

Parameter	Mean	Minimum	Maximum
TP	0.039	0.015	0.107
PO_4–P	0.015	< 0.005	0.038
TN	1.60	0.48	2.185(7.918[a])
NO_3–N	1.005	0.152	1.745(7.443[a])
NH_4–N	0.047	< 0.03	0.21
NO_2–N	0.008	< 0.005	0.037

All values in mg/l
[a] Single maximum values in one eutrophic waterbody

The data sets of Tables 68.4 and 68.5 as well as Table 68.6 comprise several limitations and problems concerning their interpretation (Doege et al., 2016: 101). Thus, the significance of the presented data varies a lot (e.g., single values versus greater amounts of data). Additional data were presented in Doege et al. (2016).

Table 68.5 Values for selected hydrochemical parameters of *N. translucens*-habitats in the Netherlands (van Raam and Maier, 1993; Nat et al., 1994; Schaminee et al., 1995; Simons and Nat, 1996, van Raam, 1998), and in France (Sellier and Lambert, 2012)

Country	Parameter	Minimum	Maximum
Netherlands	TP	0.013	0.189
Netherlands	NO_3–N	0.0113	0.025
Netherlands	NH_4–N	0.012	0.09
France	TN	0.46	0.71

All values in mg/l

Water chemistry: *Nitella translucens* is a real freshwater species which prefers habitats poor in calcium and in chlorides with low conductivity. Water chemistry varies a lot according to results of various authors (Table 68.6). The example of a deep lake (a former sand-pit) in north-west Germany shows a relatively broad range of pH-values with a strong seasonal variation of between 7.13 and a maximum of 9.0, based on measurements every three month between 2017 and 2020 (Böckermann et al., 2022). In contrast, conductivity varied in the same waterbody during the same period only to a limited extent between 163 and 182 μS/cm. In total, conductivity of *N. translucens*-habitats ranges mainly between 130 and 300 μS/cm. The mean value of 13 waterbodies in north-west Germany, for example, is 176.7 μS/cm (Becker et al., 2022). Only oligotrophic waterbodies with sandy sediments reach lower values.

Depth range: *Nitella translucens* prefers shallow waters (Corillion, 1957; Bazzichelli and Abdelahad, 2009; Mouronval et al., 2015; Becker and Doege, 2016) but can reach depths of 8–10 m in German lakes (Becker and Doege, 2016; Böckermann et al., 2022). *N. translucens* grows across the full depth range; it occurs in shallow,

Table 68.6 pH, conductivity and salt values of *N. translucens*-habitats

Author	pH	Conductivity (μS/cm)	Cl (mg/l)	Ca (mg/l)
Olsen (1944)	5.0–9.4	–	19	0.18–8.4
Corillion (1957)	5.8–7.1	–	–	–
Pietsch (1982)	–	80–600	2.6–10	31–80
Nat et al. (1994)	–	56–230	–	–
Krause (1997)	5.8–7.1	–	–	–
van Raam (1998)	5.8–7.4	56–688	12–40	4–32
Trapp (2000)	7.6–8.2	600–980[a]	140[a]	–
Meyer-Spethmann and Linders (2010)	> 4.5	46–366	–	–
Sellier and Lambert (2012)	5.5–7.86	68–693	–	–
Doege et al. (2016)	4.5–9.4	5.6–980	2.4–140[a]	0.2–80
Becker et al. (2022)	4.7–9.0	36–366	2.4–34.3	2.3–37
Total range	**4.5–9.4**	**5.6–693(980[a])**	**2.4–40(140[a])**	**0.18–80**

[a] Influence of a subterranean salt stock

fully exposed ephemeral ponds not deeper than 0.5 m as well as at the ldeepest parts of deep lakes (Becker and Doege, 2016; Becker et al., 2022).

Associated species: Plants can grow abundantly forming monospecific (*Magnonitelletum translucentis*) or polyspecific associations in Atlantic regions of Western Europe (Corillion, 1957). The taxon is present in the acidic communities of the *Nitellion flexilis* Segal ex Krause (1969) together with *Nitella flexilis*, *N. mucronata*, *N. opaca*, *N. syncarpa*, *Chara braunii*, *C. virgata* and *C. globularis* (Bailly and Schaefer, 2010; Becker and Doege, 2016) and with numerous aquatic plants of the *Littorelletea, Nymphaeetalia* and *Potametalia* (Corillion, 1957; Becker and Doege, 2016). Associated species in Mediterranean France are *Nitella capillaris*, *N. gracilis* and *Chara globularis* (Mouronval et al., 2015). Moreover, in France *Nitellopsis obtusa, Nitella mucronata*, *N. syncarpa*, *Chara braunii*, *C. aspera*, *C. fragifera* (Corillion, 1957). In Portugal in coastal ponds *Nitella hyalina*, *N. gracilis*, *N. confervacea*, *Chara braunii* and *C. fragifera* (Krause, 1983). In Sardinia (Italy) in one temporary coastal pond: *Chara aspera*, *C. galioides*, *C. connivens* and *Nitella opaca* (Becker, 2019). In Denmark it was found together with *Nitella flexilis* and *Chara globularis* (Olsen, 1944). In Germany *N. translucens* occurs predominantly together with *Chara virgata*, *Nitella flexilis* and *N. opaca*. Moreover, in Northern Germany and Saxony with *Chara globularis*, *Nitella gracilis*, *N. capillaris*, *N. syncarpa* and *N. hyalina* (Becker and Doege, 2016; Becker and Böckermann, 2017; Raabe, 2017; Becker et al., 2022).

68.4 Variability

Nitella translucens is not variable. Only one form with a slender axis is known for this species, *Nitella translucens* f. *confervoides* Thuill. ex Migula (Migula, 1889–1897: 146).

68.5 Identification and Risk of Confusion

Nitella translucens is close to *N. brachyteles*, but the two species differ by the ornamentation of the oospore-membrane (see Chap. 86, this volume). *N. brachyteles* has fibrous oospore-membrane ornamentation with SEM (Chap. 86, this volume) while *N. translucens* has a finely reticulate oospore-membrane (John and Moore, 1987 and Fig. 68.4).

N. translucens is closely related to the non-European species *Nitella axillaris* A. Braun, according to Braun (1859: 357), Zaneveld (1940: 70), Corillion (1957: 69) and Wood (1965: 683). SEM micrographs of ripe oospores of a *Nitella* species with features of *N. axillaris*, found in a rice-field of Northern Italy and called previously *N. translucens* (Abdelahad and Piccoli, 2018), were done recently. Effectively, they do not correspond to the oospore ornamentation of *N. translucens*. The identity of these plants is currently under study.

Typical forms of *N. translucens* can be easily distinguished from other species of *Nitella* by the long sterile branchlets ending with a crown of 2–4 very small dactyls (Fig. 68.3). The sterile plants may be confused with *Nitellopsis obtusa*, but this taxon is easily recognisable because of its widespread encrustation and its limestone biotope (Corillion, 1957), and due to the possible presence of star-shaped bulbils (Bazzichelli and Abdelahad, 2009; Bailly and Schaefer, 2010). In Germany, encrustation of *Nitellopsis obtusa* is sometimes missing. The end-cell of *Nitellopsis obtusa* is long and acuminate but a crown of 2–4 dactyls is always missing (Becker and Doege, 2016). Moreover, oogonia and antheridia of *Nitella translucens* are much smaller than those of *Nitellopsis obtusa* (Becker and Doege, 2016). However, *Nitella translucens* and *Nitellopsis obtusa* can rarely occur together in the same waterbody (Corillion, 1957; Becker and Doege, 2016). Attention must also be paid to old records in literature and to old herbarium sheets of *Nitellopsis obtusa*, which were sometimes treated or labelled as *Chara translucens* (Becker and Doege, 2016).

68.6 Distribution

See Fig. 68.5.

Nitella translucens is restricted to Europe and North Africa and was recorded in most European countries. The main distribution area is the Atlantic region of Western and

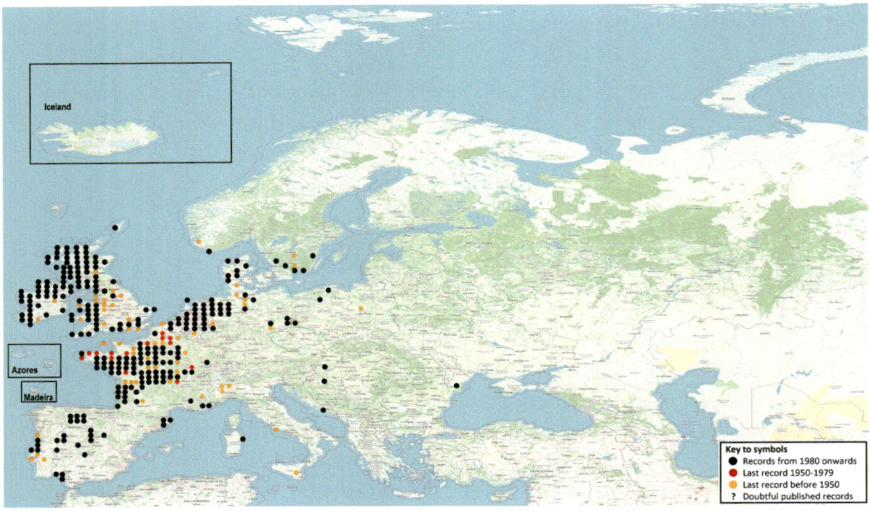

Fig. 68.5 European distribution of *Nitella translucens*. For worldwide distribution see Korsch (2018). Black dots are records from 1980 on, red dots from 1950 to 1979 and yellow dots before 1950; unconfirmed/doubtful published records (if applicable) are marked with "?" on top. Map prepared by Heiko Korsch

Central Europe. In contrast, *N. translucens* is missing mainly in south-east Europe. *Nitella translucens* is well represented in Great Britain, France, Belgium, the Netherlands, Germany and to a lesser account in Ireland, Scandinavia, the most Baltic countries, Poland, Spain, Portugal, few Balkan regions, Ukraine and Russia.

In Germany *Nitella translucens* is restricted to the lowland areas of north-west Germany and to Saxony (Becker and Doege, 2016). Currently nine water bodies with occurrence of *N. translucens* are known from Saxony, 28 from Lower Saxony, 15 from Northrhine-Westfalia and two from Schleswig–Holstein (Becker and Doege, 2016; Raabe, 2017; Becker et al., 2022). In addition, a new record has been made for Rhineland-Palatinate (E. Korte pers. comm.). A recent population of *Nitella translucens* f. *confervoides* was recorded in 2005 in a temporary shallow pond in north-west Germany south-east of Barßel (Becker et al., 2022). Unfortunately, the population became extinct caused by a silting up of the waterbody but can probably be reactivated by desludging.

In the British Isles and Ireland, *N. translucens* is widespread, occasional or locally frequent especially in northern and western parts, rare and restricted to acid habitats elsewhere (Moore, 1986; Bryant and Stewart, 2002, 2011).

The species is widespread along the Atlantic coasts and in the central regions of France, Belgium and the Netherlands (Corillion, 1957; Bailly and Schaefer, 2010; Bruinsma et al., 2018).

In Italy the species was recorded in the nineteenth century in ponds, ditches, swamps and slow-flowing waters. Sardinian specimens of *N. translucens* were recently collected in one ephemeral coastal pool in Sardinia (Becker, 2019) and were verified genetically (P. Nowak pers. comm.).

Nitella translucens f. *confervoides* was recorded in 1860 in rice-fields of Piedmont (Erb. Critt. Ital. Serie I, N. 451, leg. Malinverni [Formiggini, 1908; Cuccuini, 1997]).

Findings outside of Europe are documented only for the Maghreb countries in Northern Africa (Corillion, 1957; Cirujano et al., 2008; Muller et al., 2017).

68.7 Status and Threats

Nitella translucens is threatened according to several national European Red Lists (overview in Becker, 2016: 162–163). The species is Critically Endangered in Norway and in the Balkans (Blaženčić et al., 2006; Sjøtun et al., 2010), Endangered in Germany (Korsch et al., 2013) Vulnerable in Sweden, and Near Threatened in Denmark (Båstrup-Spohr et al., 2013). *Nitella translucens* is Extinct in Poland (Urbaniak and Gąbka, 2014). The species is probably Threatened in Italy where a Red List is not yet available.

Throughout Europe *N. translucens* occurs in several habitat types which are of Community Interest and are encoded in Annex I of the Habitats Directive 92/43EC (European Commission, 1992). This applies to the oligo- and mesotrophic freshwater habitat types 3110 and 3130. In addition, in Central Europe *N. translucens* grows rarely in habitat type 3160. Moreover, especially in the Mediterranean area, coastal lagoons, encoded as priority habitat type *1150 are in some regions habitat of *N. translucens* (e.g. Krause, 1983). Unfortunately, only few habitats of *N. translucens* are part of the European Natura 2000 network and the Water Framework Directive (European Commission, 2000). From the national point of view various waterbodies are protected as nature reserves or related status of conservation.

A thorough analysis focused on Germany has revealed that all former populations of *N. translucens* mentioned by Migula (1889–1897) are extinct and that several localities have been destroyed during the last decades through construction works or urbanisation (Becker and Doege, 2016). On the other hand, at least in Germany several new waterbodies with current occurrence of *N. translucens* have been created during the last 25 years within the scope of species conservation and compensating measures (Becker, 2010; Becker and Doege, 2016; Raabe, 2017; Becker et al., 2022). *N. translucens* is not a pioneer species and requires particularly stable oligo- or mesotrophic freshwater habitats.

A detailed description of threats and pressures as well as appropriate actions for the conservation of *N. translucens* and their habitats has already been worked out by Becker and Doege (2016) with a focus on Germany. Most of the factors and results can be transferred to similar habitats in other European countries. A special aspect might be the threat of some restricted regional populations through sterility (e.g. in Saxony, Germany). However, it is undeniable that many European habitats of *N. translucens* are in need of targeted management actions. Hence, Action Plans focussing on the conservation of *N. translucens* have been elaborated or postulated in countries and regions such as Sweden (Blindow, 2008; Zinko, 2017; Blindow et al., 2021; Törnberg and Widgren, 2021), Germany (Becker, 2016) or Sardinia (Becker, 2019).

Acknowledgements We are very grateful to Jo Wilbraham, Senior Curator, Algae, London Natural History Museum, to Jörn Hentschel, Curator, Cryptogams, Herbarium Haussknecht (JE), to Nick Stewart who checked the specimen BM13821335, to Michelle T. Casanova for introducing us to Jo Wilbraham and for providing SEM images of the oospores of *N. translucens,* to Klaus van de Weyer and Tobias Böckermann for their photographs, and to Melanie Willen, University of Oldenburg, for conducting numerous hydrochemical analyses.

References

Abdelahad, N. & Piccoli, F. (2018). Report on Charophytes from rice-fields in northern Italy including the alien species. Chara fibrosa ssp. benthamii. *Plant Biosystems 152*(4), 599–603. https://doi.org/10.1080/11263504.2017.1320310.

Bailly, G., & Schaefer, O. (2010). *Guide illustré des Characées du nord-est de la France* (96 pp.). Conservatoire Botanique National de Franche-Comté. [in French].

Båstrup-Spohr, L. (2013). *Distribution and functional traits of charophytes and vascular plants. Biodiversity in aquatic systems and along flooding gradients* (132 p.). Ph.D. Thesis, University of Copenhagen, Copenhagen.

Bazzichelli, G., & Abdelahad, N. (2009). *Alghe d'acqua dolce d'Italia. Flora analitica delle Caroficee* (73 pp.). Sapienza Università di Roma. Ministero dell'Ambiente e della Tutela del territorio e del mare. Roma, Centro Stampa Università. [in Italian].

Becker, R. (2010). Bemerkenswerte Characeen-Funde in Nordwest-Deutschland. *Abhandlungen des Naturwissenschaftlichen Vereins zu Bremen, 46*, 409–418. [in German].

Becker, R. (2016). Gefährdung und Schutz von Characeen. In: Arbeitsgruppe Characeen Deutschlands, Armleuchteralgen: *Die Characeen Deutschlands* (pp. 149–191). Springer. [in German].

Becker, R., & Böckermann, T. (2017). Neufund von *Nitella hyalina* (DC.) Agardh (Vielästige Glanzleuchteralge) in Deutschland. *Rostocker Meeresbiologische Beiträge, 27*, 19–26. [in German]

Becker, R., & Doege, A. (2016). *Nitella translucens*. In: Arbeitsgruppe Characeen Deutschlands, Armleuchteralgen: *Die Characeen Deutschlands* (pp. 493–505). Springer. [in German].

Becker, R. (2019). The Characeae (Charales, Charophyceae) of Sardinia (Italy): Habitats, distribution and conservation. *Webbia, 74*, 83–101.

Becker, R., Fuhrmann, K., Meyer-Spethmann, U., Sinkowski, M., & Willen, M. (2022). Verbreitung, Populationsentwicklung und Schutz besonders gefährdeter Arten der Armleuchteralgen (Charales, Characeae) in Nordwestdeutschland. *Drosera, 40*, 81–100. [in German].

Blaženčić, J., Stevanović, B., Blaženčić, Ž., & Stevanović, V. (2006). Red data list of Charophytes in the Balkans. *Biodiversity and Conservation, 15*, 3445–3457.

Blindow, I. (2008). Schwedische Artenschutzprogramme für bedrohte Characeen. *Rostocker Meeresbiologische Beiträge, 19*, 23–28. [in German].

Blindow, I., Carlsson, M., & van de Weyer, K. (2021). Re-Establishment techniques and transplantations of Charophytes to support threatened species (33 pp.). *Plants, 2021*(10), 1830.

Böckermann, T., Becker, R., & Willen, M. (2022). Der Versener Heidesee – vom Baggersee zum fischfreien Hotspot für gefährdete Pflanzenarten in Nordwestdeutschland. *Drosera, 40*, 37–54. [in German]

Braun, A. (1835). Übersicht der genauer bekannten Chara-Arten. *Flora, 18*, 49–73. [in German].

Braun, A. (1859). Characeen aus Columbien, Gyuana und Mittelamerika. *Monatsbericht der Königlichen Akademie der Wissenschaften zu Berlin, 1858*, 349–368. [in German]

Bruinsma, J., Denys, L., Krause, W., Mes, R., & Nat, E., & van Raam, J. (2018). *Determinatietabel van kranswieren in de Benelux* (2nd ed., 112 pp.). Stichting Jeugdbondsuitgeverij. [in Dutch].

Bryant, J. A., & Stewart, N. F. (2002). Order Charales. In D. M. John, B. A. Whitton, & A. J. Brook (Eds.), *The freshwater algal flora of the British Isles* (pp. 593–612). Cambridge University.

Bryant, J. A., & Stewart, N. F. (2011). Phylum Chlorophyta. Order charales. In D. M. John, B. A. Whitton, & A. J. Brook (Eds.), *The freshwater algal flora of the British Isles. An identification guide to freshwater and terrestrial algae* (2nd ed.). Cambridge University, Cambridge: 742–765.

Cirujano, S., Cambra, J., Sánchez Castillo, P. M., Meco, A., & Flor Arnau, N. (2008). *Flora ibérica. Algas continentales. Carófitos (Characeae)* (132 pp.). Real Jardín Botánico. [in Spanish].

Compère, P. (1992). *Charophytes. Flore pratique des algues d'eau douce de Belgique.* Jardin botanique national de Belgique, Meise, 77 pp. [in French].

Corillion, R. (1957). Les Charophycées de France et d'Europe Occidentale. *Bulletin de la Société Scientifique de Bretagne, 32*(Hors série I & II), 1–499. [in French].

Cuccuini, P., (1997). *L'Erbario Crittogamico Italiano. Storia e struttura di una collezione* (165 pp.). Museo Botanico dell'Università di Firenze, Sezione del Museo di Storia Naturale. [in Italian].

Doege, A., van de Weyer, K., Becker, R., & Schubert, H. (2016). Bioindikation mit Characeen. In: Arbeitsgruppe Characeen Deutschlands, Armleuchteralgen: *Die Characeen Deutschlands* (pp. 97–138). Springer. [in German].

European Commission. (1992). *Council Directive 92/43/EEC of 21 May 1992 on the conservation of natural habitats and of wild fauna and flora.* Official Journal of the European Communities L 206/7.

European Commission. (2000). *Richtlinie 2000/60/EG des europäischen Parlaments und des Rates vom 23. Oktober 2000 zur Schaffung eines Ordnungsrahmens für Maßnahmen der Gemeinschaft im Bereich der Wasserpolitik,* Abl. L 327. [in German].

Formiggini, L. (1908). Revisione critica delle Caracee della Flora Veneta compreso il Mantovano. *Atti Dell'accademia Veneta-Trentina-Istriana (classe Di Scienze Naturali, Fisiche e Matematiche) I Ser., 3*, 110–143. [in Italian].

Groves, J., & Bullock-Webster, G. R. (1920). *The British Charophyta* (p. 405). Ray Society.

Haas, J. N. (1994). First identification key for charophyte oospores from central Europe. *European Journal of Phycology, 29*, 227–235.

John, D. M., & Moore, J. A. (1987). An SEM study of the oospore of some *Nitella* species (Charales, Chlorophyta) with descriptions of wall ornamentation and an assessment of its taxonomic importance. *Phycologia, 26*, 334–355.

Korsch, H. (2013). Die Armleuchteralgen (Charophyceae) Sachsen-Anhalts. *Berichte des Landesamtes für Umweltschutz Sachsen-Anhalt, 1*(2013), 1–85. [in German].

Korsch, H. (2018). The worldwide range of the Charophyte species native to Germany. *Rostocker Meeresbiologische Beiträge, 28*, 45–96.

Krause, W. (1983). Characeen-Standorte in Portugal mit besonderer Rücksicht auf den Einfluß des Menschen. *Tuexenia 3*, 289–296. [in German].

Krause, W. (1997). Charales (Charophyceae) Süßwasserflora von Mitteleuropa. In: H. Ettl, G. Gärtner, H. Heynig, & D. Mollenhauer (Eds.), *Süßwasserflora von Mitteleuropa 18*. Fischer. [in German].

Krause, W. (1969). Zur Characeenvegetation der Oberrheinebene (Characeae in the upper plain of the Rhine). *Archiv Für Hydrobiologie, 35*, 202–253. [in German].

Langangen, A. (2007). *Charophytes of the Nordic countries* (p. 102). Saeculum ANS.

Meyer-Spethmann, U., & Linders, H. W. (2010). *Monitoring der niedersächsischen Vorkommen der gemäß FFH-Richtlinie prioritären Art Luronium natans (Linnaeus) Rafinesque-Schmaltz im Jahr 2009* (36 pp.). Im Auftrag des NLWKN. [in German].

Migula, W. (1889–1897). Fünfter Band: Die Characeen. In: A. Fischer, E. Fischer, F. Hauck, G. Limpricht, C. Luerssen, W. Migula, H. Rehm, P. Richter, & G. Winter (Eds.), *Dr. L. Rabenhorst's*

Kryptogamenflora von Deutschland, Oesterreich und der Schweiz (ed. 2). Eduard Kummer. [in German].
Moore, J. A., (1986). *Charophytes of Great Britain and Ireland.* BSBI Handbook 5. Botanical Society of the British Isles.
Mouronval, J. B., Baudouin, S., Borel, N., Soulié-Märsche, I., Klesczewski, M., & Grillas, P. (2015). *Guide des Characées de France méditerranéenne* (211 pp.). Office national de la Chasse et de la Faune Sauvage. [in French].
Muller, S. D., Rhazi, L., & Soulié-Märsche, I. (2017). Diversity and distribution of Characeae in the Maghreb (Algeria, Morocco, Tunisia). *Cryptogamie, Algologie, 38*, 201–251.
Nat, E., Simons, J., de la Haye, M. A. A., & Coops, H. (1994). *Historisch en actueel verspreidingsbeeld van kranswieren in Nederland in samenhang met waterkwaliteitsfactoren.* RIZA werkdocument 94.148X. Lelystad, 77 pp. [in Dutch].
Olsen, S. (1944). Danish Charophyta. *Det Kongelige Danske Videnskabernes Selskab, Biologiske Skrifter, 3*(1): 1–240, pl. 1–2. [in Danish].
Persoon, C. H. (1806–1807). *Synopsis Plantarum seu Enchiridium botanicum, complectens enumerationem systematicam specierum hucusque cognitarum.* Pars secunda. p. 657. (1807). Parisiis Lutetiorum : Apud Bibliopolas Treuttel et Würtz. ; Et Tubingae : Apud J. G. Cottam (Parisiis, Ex Typografia J.-L. Scherff) [in Latin]
Pietsch, W. (1982). Biologische, mikrobiologische und toxikologische Methoden. Makrophytische Indikatoren für die ökochemische Beschaffenheit der Gewässer. In: C. Legler & G. Breitig (Eds.), *Ausgewählte Methoden der Wasseruntersuchung, Band 2: Biologische, mikrobiologische und toxikologische Methoden* (pp. 67–88). Gustav Fischer. [in German].
Raabe, U. (2017). Zur Verbreitung der Schimmernden Glanzleuchteralge (*Nitella translucens*) in Westfalen. *Rostocker Meeresbiologische Beiträge, 27*, 53–62.
Schaminee, J. H. J., Maier, E. X., & van Raam, J. (1995). *De vegetatie van Nederland. Deel 2. Plantengemeenschappen van wateren, moerassen en natte heiden.* Opulus, Uppsala & Leiden.
Sellier, Y., & Lambert, E. (2012). *Initiation à l'étude des characées* (72 pp). Exemple appliqué à la Réserve Naturelle Nationale du Pinail, France. [in French].
Simons, J., & Nat, E. (1996). Past and present distribution of stoneworts (Characeae) in the Netherlands. *Hydrobiologia, 340*, 127–135.
Sjøtun, K., Fredriksen, S., Heggøy, E., Husa, V., Langangen, A., Lindstrøm, E. A., Moy, F., Rueness, J., & Åsen, P. A. (2010). Rødliste over alger. Red list of Cyanophyta, Rhodophyta, Chlorophyta, Ochrophyta. In: J. A. Kålås, Å. Viken, S. Henriksen, & S. Skjelseth (Eds.), *Norsk Rødliste for Arter 2010. The 2010 Norwegian Red List for Species* (pp. 79–86). Norwegian Biodiversity Information Centre.
Thuillier, J. L. (1799). *La flore des environs de Paris, ou distribution méthodique des plantes qui y croissent naturellement, faite après le systême Linnée, ...*, ed. 2. Auteur et H.L. Perronneau.
Törnberg, M., & Widgren, Å. (2021). *Flytt av Grovslinke i Älmtasjön* (11 pp.). Länsstyrelsen Blekinge. [in Swedish].
Trapp, S. (2000). *Zur Vegetationsdynamik anthropogener Stillgewässer Norddeutschlands und ihrer Bedeutung für die Bioindikation* (253 pp.). Dissertation Universität Bremen, Bremen. [in German].
Urbaniak, J., & Gąbka, M. (2014). *Polish Charophytes. An illustrated guide to identification* (120 pp.). Uniwersytet Przyrodniczy we Wroclawiu. [in Polish].
Vaillant, S. (1719). Caractères de quatorze genres de plantes: le dénombrement de leurs espèces; les descriptions de quelques-unes & les figures de plusieurs. *Mémoires de l'Académie royale des sciences (Paris), 1719*, 9–47, 5 pl. [in French].
van Raam, J. C. (1998). *Handboek Kranswieren.* Chara boek. [in Dutch].
van Raam, J. C., & Maier, E. X. (1993). Nederlandse Kranswieren 4. Doorschijnend glanswier (*Nitella translucens* Agardh). *Gorteria, 19*, 88–94. [in Dutch].

Wood, R. D. & Imahory, K. (1965). *A revision of the Characeae. Vol. I. Monograph of the Characeae* (904 pp.). J. Cramer.
Zaneveld, J. S. (1940). The Charophyta of Malaysia and adjacent countries. *Blumea, 4*, 1–224.
Zinko, U. (2017). *Kunskapsuppbyggande program – 15 hotade makrofytarter i permanenta vatten*. Havs- och Vattenmyndighetens Rapport 2017:6, 93 pp. [in Swedish].

Chapter 69
Nitella wahlbergiana

Roman Romanov, Anders Langangen, and Irmgard Blindow

Scientific name: ***Nitella wahlbergiana*** **Wallman, Förs. Syst. Uppställn. Charac.: 26. 1853**

Lectotype (designated here by R. E. Romanov, Langangen & Blindow): [Sweden, Östergöthland, Vist] Stångån/[no date]/Wallman (UPS); ArtDatabanken 92, 447.

= *Nitella gracilis* (Smith) C. Agardh var. *borealis* A. Braun in Braun and Nordstedt (1882: 60)
Type not designated.

Local names: FINNISH: Tupsusiloparta; NORWEGIAN: Hodebroddglattkrans; RUSSIAN: Нителла Вальберга; SWEDISH: Nordslinke (Sweden), Tofsslinke (Finland).

69.1 Morphology

See Figs. 69.1, 69.2 and 69.3.

Habitus: The plants are fresh green, unencrusted, flexible, easily collapsing during withdrawal from the water. They have a tolypelloid appearance because of scarcely

R. Romanov (✉)
Nišice Ujtin Potok b.b., Dobra Voda, 85356 Bar, Montenegro
e-mail: streptophytes@gmail.com

A. Langangen
Natural History Museum, University of Oslo, P.O.Box 1172, Blindern, 0318 Oslo, Norway
e-mail: langangen@hotmail.com

I. Blindow
Universität Greifswald, Biologische Station Hiddensee, 18565 Kloster, Germany
e-mail: blindi@uni-greifswald.de

H. Schubert et al. (eds.), *Charophytes of Europe*,
https://doi.org/10.1007/978-3-031-31898-6_69

Fig. 69.1 *Nitella wahlbergiana*, habitus and details. **a** General habit; **b** Fresh material, showing the condensed heads; **c** Uppermost whorls, note the abbreviated second segments of the branchlets and the dactyls with mucronate end-cell. Photographs: Roland Bengtsson

branching shoots and the combination of widely spreading lower branchlets and condensed apical heads consisting of short fertile branchlets (Figs. 69.1a, b and 69.2a, b). Usually several shoots appearing from the common base result in a "bushy" habit.

Size: The plants are small to medium, from several cm high, up to 10–15 cm, rarely up to 30 cm.

Main axis: The main axis is less than 500 μm in diameter, flaccid. The lower internodes are long, but uppermost are shortened.

Cortex, spine-cells, stipulodes: Absent.

Branchlets: The branchlets are dimorphic. The lower branchlets are long and spreading, 1-furcate (Figs. 69.1a, b and 69.2a, b). The upper branchlets are short and condensed, forming evident apical heads mainly because of abbreviated second segments (Fig. 69.1b, c). The branchlets are mainly 2-, rarely 3-furcate. The dactyls are 2-celled with usually mucronate end-cells.

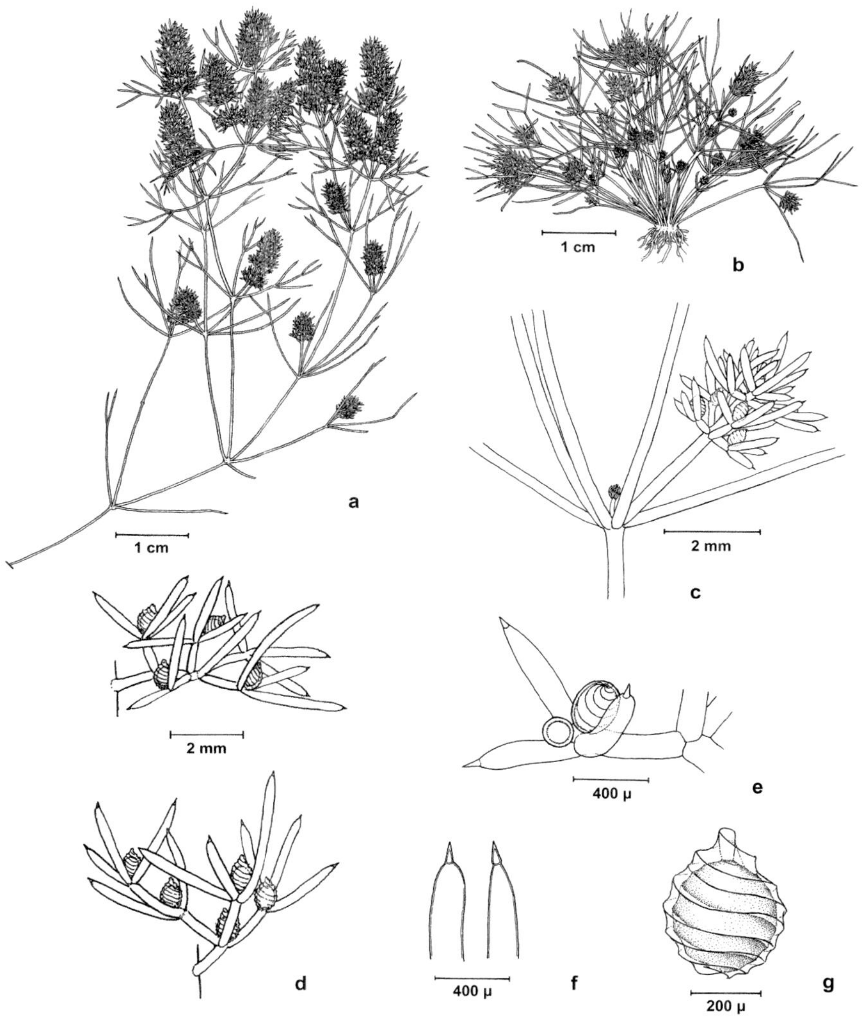

Fig. 69.2 *Nitella wahlbergiana*. **a**, **b** General habit; **c** Head; **d** Branchlets; **e** End segments of branchlet, **f** Ends of dactyls; **g** Oogonium. Drawings: Mindaugas Ryla

Bract-cells, bracteoles, bractlets: Absent.

Sexuality: The plants are monoecious, richly fertile. Gametangia are without mucus, conjoined, mainly at the second, but also at the first furcation of the branchlet, clumped in apical heads.

Oogonia: The oogonia are ellipsoid, 400–470 (500) μm long without coronula, 365–420 μm wide in rewetted state, with convolutions of spiral cells not inflated or elongated below coronula (own data). The coronula is persistent, ca. 55 μm in length and 65 μm in width, cells of upper row are slightly elongated.

Oospores: The oospores are broadly ellipsoid to almost circular, black (?) or light brown in reflected light, light yellow-brownish in transmitted light, 350–450 μm long and 310–390 μm wide in rewetted state, with 6 or 7 spiralled, strongly flanged ridges, nearly horizontal in the middle part, forming a crest at the top up to a little shorter than half of oospore (own data), and roughly spongy surface looking as somewhat "agglutinate" in scanning electron microscopy (Ray et al., 2001, Romanov pers. obs., Fig. 69.3) appearing as densely irregularly reticulate or mostly described as densely granulate in light microscopy.

Antheridia: The diameter of antheridia is (160–170)190–220 μm in rewetted state (own data).

Bulbils: Bulbils are sometimes produced, irregular in shape because of enlargement of basal axis node.

69.2 Phenology and Reproduction

Nitella wahlbergiana is an annual species, producing a lot of oospores during the vegetation period. The possible overwintering of nodal bulbils needs to be checked. This species is normally found from July to October in Sweden and Finland (Blindow and Koistinen, 2004; Langangen, 2007). Ripe oospores are found from July and onwards.

69.3 Habitat Conditions and Ecology

Habitat type: *Nitella wahlbergiana* is mainly a freshwater species, occurring generally in stable oligotrophic to mesotrophic lakes and lowland rivers with soft and somewhat humic water, often in *Lobelia*-lakes, rarely in temporary water bodies (Langangen, 2007; Langangen et al., 2002; Romanov et al., 2018; R. Romanov unpubl. data). It sometimes occurs in slightly brackish waters, i.e. in sheltered bays of the Bothnian Bay, and prefers sites with sparse vegetation, particularly after ice erosion, which indicates a low competitive ability (Blindow and Koistinen, 2004).

Sediment: The species grows on clay, clay-gyttja and sand.

Nutrient conditions and water chemistry: Sparse data are available only, unsuitable for a general characterization (Penning et al., 2008; Romanov et al., 2018). The species occurs in circumneutral and subneutral, but not in distinctly acidic conditions, and tolerates salinities of up to around 2 PSU.

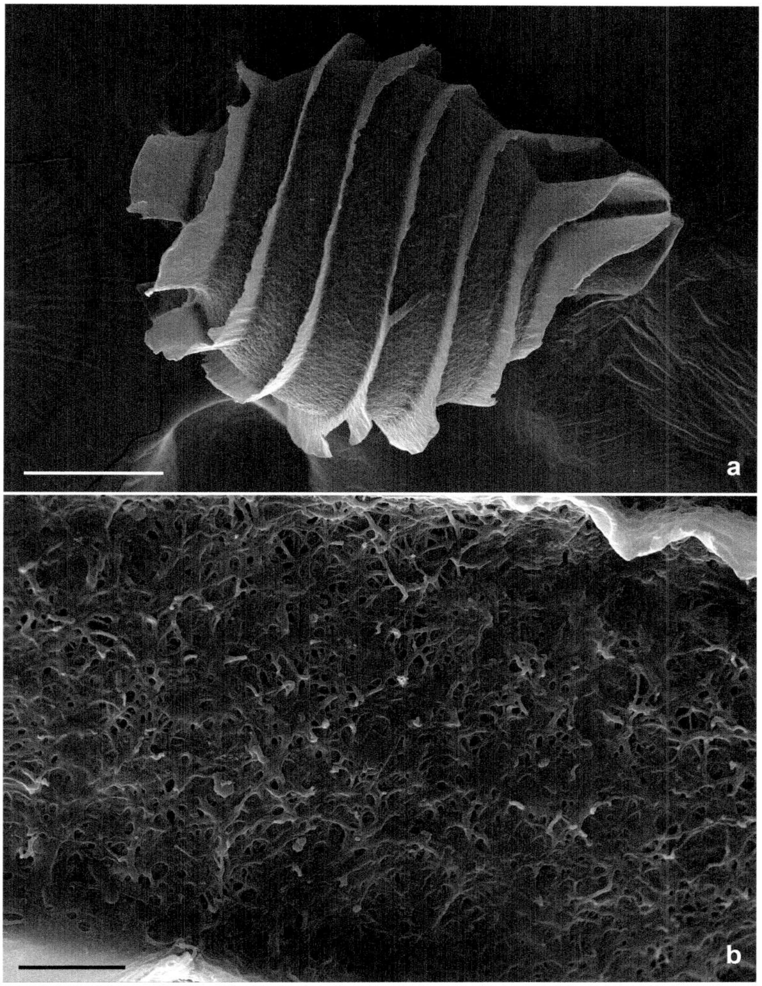

Fig. 69.3 Oospores of *Nitella wahlbergiana*, scanning electron microscopy. **a** General view, showing the strongly flanged ridges; **b** Oospore surface with roughly spongy structure. Scale: **a**—100 μ, **b**—10 μ. Photographs: Roman Romanov

Depth range: *Nitella wahlbergiana* is mainly known from shallow water down to 0.5 m depth, occasionally down to 2 m depth (Blindow and Koistinen, 2004; Langangen, 2007); it was once recorded from 3 m depth (Sandsten, 2009).

Associated species: *Lobelia dortmanna*, *Isoëtes echinospora*, small- and middle-sized species of *Characeae*. *Nitella flexilis* and *N. opaca* are the most common associated species. *Nitella wahlbergiana* is usually not associated with particular species because it mainly grows in sites with very sparse vegetation.

69.4 Variability

The height of plants, number of shoots, originating from the common base, number of heads per shoot and whorls of uncondensed branchlets, as wells as length of internodes are the main expression of species variability which is resulting in different general habit of the plants.

69.5 Identification and Risk of Confusion

Nitella wahlbergiana is in general appearance similar to *N. mucronata* f. *heteromorpha*, but more slender, usually smaller and with a spongy oospore surface appearing as irregularly granulate in light microscopy in contrast to obviously reticulate pattern of the latter species. The abbreviated second segments of the branchlet are also helpful for its delineation with *N. mucronata*. Condensed heads in combination with mucronate end-cells and two-celled dactyls allow to distinguish the species from *N. gracilis*, which has a similar oospore surface. The apical heads and presence of gametangia mostly at the second furcation differentiate it from the more delicate and thinner *N. confervacea*.

69.6 Distribution

See Fig. 69.4.

Nitella wahlbergiana is a North Eurasian species distributed from Fennoscandia, where most records (Langangen, 2007) to the north of Far East (Chemeris et al., 2020). In Russia, *N. wahlbergiana* has a scattered distribution from the Republic of Karelia to Vladimir Oblast in the European part and from the centre of Krasnoyarsk Krai (Yenisseisk) to Southern Siberia, Magadan Oblast and Chukotka in Asian Russia (Blindow and Koistinen, 2004; Romanov, 2017; Romanov et al., 2018; Chemeris et al., 2020, Vishnyakov et al., 2021; Romanov et al., 2022; R. Romanov unpubl. data, H!, KEM!, LE!, IBIW!). The few localities known for Russia seem to be the result of undercollecting. Its presence in North America should be checked.

69.7 Status and Threats

Nitella wahlbergiana is assessed as LC (Least Concern) in the European Red List (see Chap. 9). There are probably few threats to this species, except for eutrophication (Langangen, 2007). The possible negative effects of acidification and brownification (Increased influx of humic substances to lakes) are not evaluated.

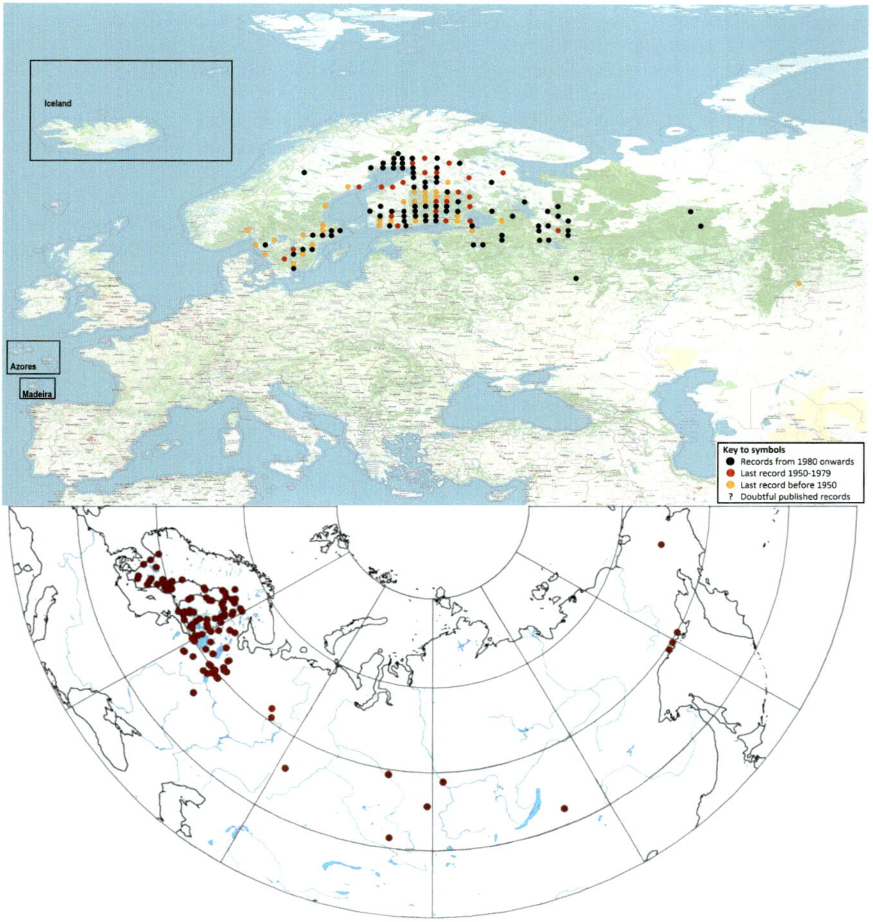

Fig. 69.4 Distribution of *Nitella wahlbergiana*. Upper panel: European distribution; lower panel: World-wide distribution (basing on data from Langangen et al., 2002; Blindow and Koistinen, 2004; Langangen, 2007; Romanov, 2017; Romanov et al., 2018; Chemeris et al., 2020; Vishnyakov et al., 2021; Romanov et al., 2022 and unpubl. data of R. Romanov: H!, KEM!, LE!, IBIW!). Black dots in the upper panel are records from 1980 on, yellow dots last record from 1950 to 1979 and orange dots last record before 1950; doubtful published records (if applicable) are marked with "?" superimposed. Maps prepared by Heiko Korsch (upper panel) and Roman Romanov (lower panel)

References

Blindow, I., & Koistinen, M. (2004). *Nitella wahlbergiana*. In H. Schubert, & I. Blindow (Eds.), *Charophytes of the Baltic Sea* (pp. 209–215). Ruggel.

Braun, A., & Nordstedt, C. F. O. (1882). Fragmente einer Monographie der Characeen. Nach den hinterlassen Manuscripten A. Braun's herausgegeben von Dr. Otto Nordstedt. *Abhandlungen der Königlichen Akademie der Wissenschaften zu Berlin für das Jahr 1882*, 1–211. [in German].

Chemeris, E., Romanov, R., Kopyrina, L., Filippova, V., Mochalova, O., Efimov, D., & Bobrov, A. (2020). How charophytes (Streptophyta, Charales) survive in severe conditions of the permafrost area in Far North-East Asia. *Limnologica, 83*, 125784. https://doi.org/10.1016/j.limno.2020.125784

Langangen, A. (2007). *Charophytes of the Nordic countries*. Saeculum ANS.

Langangen, A., Koistinen, M., & Blindow, I. (2002). The charophytes of Finland. *Memoranda Societatis pro Fauna et Flora Fennica, 78*, 17–48.

Penning, W. E., Mjelde, M., Dudley, B., Hellsten, S., Hanganu, J., Kolada, A., van den Berg, M., Poikane, S., Phillips, G., Willby, N., & Ecke, F. (2008). Classifying aquatic macrophytes as indicators of eutrophication in European lakes. *Aquatic Ecology, 42*, 237–251. https://doi.org/10.1007/s10452-008-9182-y

Ray, S., Pekkari, S., & Snoeijs, P. (2001). Oospore dimensions and wall ornamentation patterns in Swedish charophytes. *Nordic Journal of Botany, 21*, 207–224. https://doi.org/10.1111/j.1756-1051.2001.tb01359.x

Romanov, R. E. (2017). Assessing conservation status of charophytes (Streptophyta, Charales) in southern regions of West Siberia. In *Problems of Botany of South Siberia and Mongolia. Proceedings of the 16h International Scientific and Practical Conference* (Barnaul, 5–8 June 2017), 58–61. [in Russian].

Romanov, R. E., Anisimova, O. V., Anishchenko, L. N., Efimov, D. Y., Kapitonova, O. A., Kipriyanova, L. M., Konechnaya, G. Yu., Kopyrina, L. I., Kotovshchikov, A. V., Kurganov, A. A., Nikolaenko, S. A., Shilov, M. P., Smirnova, E. V., Vodeneeva, E. L., Zarubina, E. Y., & Zhakova, L. V. (2022). The noteworthy new records of charophytes (Charales, Charophyceae) from Russian regions: Improvement of species distribution ranges in Eurasia. *Botanicheskii Zhurnal, 107*, 466–477. https://doi.org/10.31857/S0006813622050076

Romanov, R. E., Patova, E. N., Teteryuk, B. Y., & Chemeris, E. V. (2018). Charophytes (Charales, Charophyceae) on the northeastern edge of Europe: Is it something different across Northern Europe in their diversity and biogeography? *Nova Hedwigia, Beihefte, 147*, 161–182. https://doi.org/10.1127/nova-suppl/2018/016

Sandsten, H. (2009). Undervattensväxter i Södra Vixen och Norra Vixen. Basinventering och eftersök enligt åtgärdsprogram för hotade arter. Calluna A.B. Länsstyrelsen Jönköpings län, Meddelande 2009: 03: 1–15, 3 bilagor. [in Swedish].

Vishnyakov, V. S., Romanov, R. E., Komarova, A. S., Belyakov, E. A., Moseev, D. S., Churakova, E. Yu., Czhobadze, A. B., & Philippov, D. A. (2021). New charophyte records (Characeae) in European Russia. *Botanicheskii Zhurnal, 106*, 61–76. [in Russian]. https://doi.org/10.31857/S0006813621010117

Chapter 70
Nitellopsis obtusa

Karin Pall, Timm Kabus, and Hendrik Schubert

Scientific name: ***Nitellopsis obtusa*** **(Desv.) J. Groves in J. Bot. 57: 127. 1919**

≡ *Chara obtusa* Desv., Not. Fl. France: 136. 1810

≡ *Tolypellopsis obtusa* (Desv.) Bég. & Formigg. in Boll. Soc. Bot. Italiana 1907: 102. 1907
Lectotype (Wood 1965: 353): Gallia/[no date]/[Desvaux] (P)

= *Chara stelligera* Bauer ex Rchb., Handb. Gewächsk., ed. 2.: 1595. 1829

≡ *Nitellopsis stelligera* (Bauer ex Rchb.) Hy in Rev. Bot. Bull. Mens. 8: 46. 1890

≡ *Tolypellopsis stelligera* (Bauer ex Rchb.) Mig., Charac. Deutschl.: 255. 1890
Lectotype (designated here by Casanova & T. Gregor): In lacu "Plötzensee" prope Berolinum / 1828 / Bauer (MEL)

Local names: BELARUSIAN: Нітэлопсіс прытуплены; DANISH: Stjernetråd; DUTCH: Sterkranswier; ENGLISH; Starry stonewort; FINNISH: Tähtimukulaparta, stjärnslinke; GERMAN: Stern-Glanzleuchteralge; LITHUANIAN: Žvaigždėtasis maurašakis; POLISH: Krynicznica tępa; RUSSIAN: Нителлопсис притуплённый; SWEDISH: Stjärnslinke; UKRAINIAN: Нітелопсіс притуплений.

K. Pall (✉)
SYSTEMA Bio- und Management Consulting GmbH, Bensasteig 8, 1140 Wien, Austria
e-mail: karin.pall@systema.at

T. Kabus
Institute of Applied Freshwater Ecology, Schlunkendorfer Straße 2e, 14554 Seddin, Germany
e-mail: kabus@gmx.net

H. Schubert
Universität Rostock, Institut für Biowissenschaften, Albert-Einstein-Straße 3, 18059 Rostock, Germany
e-mail: hendrik.schubert@uni-rostock.de

H. Schubert et al. (eds.), *Charophytes of Europe*,
https://doi.org/10.1007/978-3-031-31898-6_70

70.1 Morphology

See Figs. 70.1, 70.2 and 70.3.

Habitus: Plants are pale green to greyish, often heavily encrusted. Specimen are very large, strong and stout, with undivided branchlets typically bearing very long bract-cells.

Size: *Nitellopsis obtusa* is a huge charophyte, typically between 20 and 100 cm, but may reach up to 200 cm if growing in deeper water.

Main axis: Main axis diameter is about 1–2 mm (Urbaniak and Gąbka, 2014 give 0.5–3 mm, Migula, 1889–1897 even 4 mm); main axes are rarely branched and often with extremely long (up to 20 cm) internodes.

Cortex, spine-cells, stipulodes: Absent.

Branchlets: The number of branchlets per whorl varies. Usually whorls consist of 5–6 branchlets, but several authors reported higher (up to 8: Bailly and Schaefer, 2010; Urbaniak and Gąbka, 2014) as well as lower numbers (4: Cirujano et al., 2007; Korsch, 2013). Branchlets are undivided, consisting of one to three segments, all together between 5 and 12 cm in length. The terminal segment is one-celled, well developed and pointed.

Bract-cells: Spreading, well develop pairs of bract-cells develop from 1–2 branchlet nodes. Rarely three bract-cells are formed per node. These bract-cells resemble branchlets and fall off easily. Consequently, bract-cells exist only on young branchlets at the top of the plants.

Bracteoles, bractlets: Absent.

Sexuality: The species is dioecious. Female and male individuals are often reported to grow apart from each other, occupying different areas of a habitat. The species has been repeatedly reported to be almost sterile, reproducing vegetatively mainly (Luther, 1951; Migula, 1889–1897; Olsen, 1944). According to our own studies, vegetative reproduction occurs mostly via the bulbils. However, in recent decades richly fructifying stands have been observed even from sites where plants were reported as being sterile in the past.

Oogonia: Oogonia are short ovoid and slightly swollen below the coronula (Fig. 70.3). The coronula consists of 5 cells, falling off early. Sub-fossil gyrogonites are heavily calcified. Data about quantitative morphometry are listed in Table 70.1. The gyrogonites of *Nitellopsis obtusa* are the biggest among the living charophytes and they prove extremely variable (Soulié-Märsche et al., 2002; see also Table 70.2).

Oospores: Oospores are elongated with 4–8 fimbriate ridges (Fig. 70.3). In most of the cases they are heavily calcified and their gold-brownish to black colour is hard to observe. Table 70.3 gives an overview of the available morphometrical data.

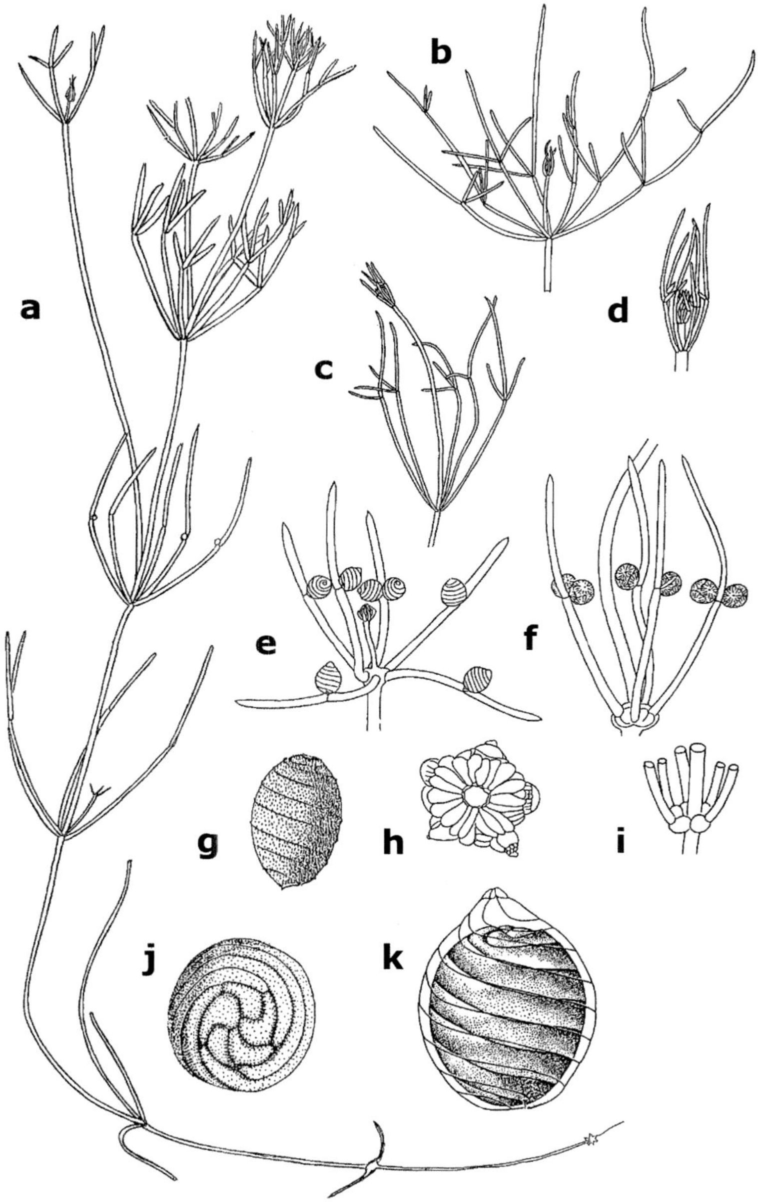

Fig. 70.1 *Nitellopsis obtusa*. **a** Habitus; **b**, **c**, **d** Sterile whorls; **e** Whorl of female plant with oogonia; **f** Whorl of male plant with antheridia; **g** Oospore; **h** Star shaped bulbil; **i** Node; **j** Oospore in apex view; **k** Oogonium with coronula. Drawings: Werner Krause

Fig. 70.2 *Nitellopsis obtusa*, habitus. **a** Sterile plant; **b** *Nitellopsis obtusa* with *Chara globularis* and *Elodea nuttallii*; **c** Typical dense, monospecific stand in lakes; **d** Winter appearance. Photographs: **a**, **b**-Karin Pall; **c**-Klaus van de Weyer; **d**-János Csiky

Antheridia: Antheridia are often reported to be "dark green brown" which is, as can be seen from Fig. 70.3, only the case for young, unripe ones. Ripe antheridia are orange red as typical for charophytes in general. Probably many former authors did not see living antheridia because they were rare, so they had to rely on dried material or written reports. Table 70.4 gives an overview of the available morphometrical data.

Bulbils: The large white bulbils with their very characteristic star shaped form are an important diagnostic feature of this species (Fig. 70.3). They are formed very frequently, usually each individual has at least one of them, making it easy to recognise the occurrence of the species even when growing subdominant in mixed stands.

Fig. 70.3 *Nitellopsis obtusa*, details. **a** Upper whorl with ripe antheridia; **b** Typical star-shaped bulbil; **c** Oospores; **d** Oogonium; **e** Antheridia. Photographs: **a**-Klaus van de Weyer; **b**, **c**, **d**, **e**-Heiko Korsch

70.2 Phenology and Reproduction

Nitellopsis obtusa is a perennial species and green overwintering stands are reported by, e.g., Luther (1951) and Olsen (1944). However, often plants decay in cold winters, regenerating from the bulbils in spring. Gametangia are observed from early summer to autumn, according to Krause (1985). Gametangia are mainly formed in shallow water, whereas deep-water stands stay sterile and reproduce vegetatively.

70.3 Habitat Conditions and Ecology

See Fig. 70.2.

Habitat type: *Nitellopsis obtusa* prefers still water conditions, growing in large as well as small lakes, all sorts of mining lakes but also abandoned or extensively used fish ponds. It can also be found in backwaters, even with slightly flowing water. Typical habitats are Natura 2000 habitat types 3140 and 3150. The species

Table 70.1 Oogonia morphometry

Author	Length	Width	Coronula length	Coronula width	Number of helical stripes
Migula (1889–1897)	*1100–1400*[a]	–	–	–	8–9
Groves and Bullock-Webster (1924)	1200–1400	1000–1200	65	150	9
Corillion (1957)	1200–1400	1000–1200	65	150	9
Wood (1965)	1035–1125(1400)	825–990(1200)	65–105	120–150	9
Krause (1997)	1100–1400	800–1000	90	120	7–9
Cirujano et al. (2008)	1050–1300	850–1100	65–100	–	–
Bailly and Schaefer (2010)	800–1200	800–1200	–	–	–
Urbaniak and Gąbka (2014)	1050–1500	850–1100	–	–	–
Total range	**800–1500**	**800–1200**	**65–105**	**120–150**	**7–9**

Length in the first column, except for values in italics, is excluding the coronula, its length and width is given separately; all values for length and width in μm

[a] Including coronula

Table 70.2 Gyrogonite morphometry of *Nitellopsis obtusa*

Author	Length (μm)	Width (μm)	Material
Soulié-Märsche (1989)	825–1275	625–1100	60 gyrogonites (extant and herbarium)
Soulié-Märsche et al., (2002)	830–1350	896–1156	130 gyrogonites (extant)
Total range	**825–1350**	**625–1156**	

prefers clear water or only slightly turbid conditions. As reported from Hungary (J. Csiky pers comm.), the turbidity value is usually less than 30 NTU.

Sediment: *Nitellopsis obtusa* is most frequently found on lime mud, probably at least in part produced by the stands themselves. In shallow water it can be found on sandy substrate, sometimes even on slightly organic mud.

Nutrient conditions: Irrespective of tolerating a wide range of nutrient conditions, large extensive stands dominating the littoral down to several metres depth, are found under oligo-mesotrophic to mesotrophic clear water conditions only. Increasing nutrient availability results in an upward trend of the stands, which is also observed geographically going north (Krause, 1985). In more detail, maximum TN measured so far was 15 mg/l, maximum nitrate 6.6 mg/l, maximum ammonia 4.94 mg/l and maximum TP 0.43 mg/l (Doege et al., 2016), indicating the competitiveness and

Table 70.3 Oospore morphometry

Author	Length	Width	Number of striae	Colour
Groves and Bullock-Webster (1924)	775	600	7	Gold-brown
Corillion (1957)	775	600	7	Brown
Wood (1965)	775	600	7	Gold-brown
Krause (1997)	700	600	7–8	Dark brown to black
Cirujano et al. (2008)	750–800	500–700	–	Gold-brown
Bailly and Schaefer (2010)	600	600	7	Brown
Korsch (2013)	800–1025	675–825	4–7	Light grey-brown
Urbaniak and Gąbka (2014)	900–1280	720–860	–	Brown to black
Total range	**700–1280**	**500–860**	**4–8**	

All values for length and width in μm

Table 70.4 Antheridia morphometry

Author	Diameter (μm)
Migula (1889–1897)	750–1050
Groves and Bullock-Webster (1924)	1000
Corillion (1957)	1000
Wood (1965)	750–1000
Krause (1997)	1000
Cirujano et al. (2008)	800–1050
Bailly and Schaefer (2010)	800–1000
Korsch (2013)	1050–1350
Urbaniak and Gąbka (2014)	750–1000
Total range	**750–1350**

nutrient tolerance of the species. However, most of the large, dominating stands are found at much lower nutrient concentrations. In a typical pond habitat in Hungary (South Transdanubia) NO_3^- 1.1 mg/l, NH_4^+ 0.13 mg/l, TN 14.38 mg/l and TP 0.18 mg/l were measured (J. Csiky pers. comm.).

Water chemistry: *Nitellopsis obtusa* prefers calcareous, well buffered water bodies and is rarely found at highly alkaline as well as subneutral conditions. However, this preference seems not to reflect the physiological limits, as the species has been found in a pH-range from 5.0–9.8 (with extremes even down to 3.8, see Doege et al., 2016). Total hardness ranged between 1.0 and 60.3 °dH; carbonate hardness between 2.3 and 11.4 °dH. Calcium concentrations ranged between 5.7 and 172 mg/l (Doege et al., 2016). *N. obtusa* is tolerant of brackish conditions to some extent, growing, e.g., at the Finnish Baltic Sea coast at salinities of up to 2.5 (Luther, 1951). Krause

(1985, 1997) reported chloride concentrations of up to 1350 mg/l—however, such conditions are probably beyond its acclimation limit, which is given by Winter et al. (1999) at 500 mg/l Cl^-. According to a measurement in the above mentioned pond in Hungary, the following data were observed: pH 6.25, conductivity 203 μS/cm, DO 0.8 mg/l, Ca^{2+} 33.9 mg/l, Sum HW 63.2 CaO mg/l, Cl^- 10.7 mg/l, HCO_3^- 101.2 mg/l, Sum DM 199 mg/l (J. Csiky pers. comm.).

Depth range: *Nitellopsis obtusa* can successfully compete with higher plants in shallow, mesotrophic to slightly eutrophic lakes. However, its preferred habitat is clear-water lakes of >5 m depth where the species can dominate littoral vegetation, preferably in the medium depth zone, but even down to the lower macrophyte limit. Maximum depth reported for Europe is 30 m in Plöner Lake (Sonder, 1890). More recently, the species was found down to 20 m by Jeschke (1959) in Brandenburg—both values should be critically evaluated, maximum depth confirmed by direct measurement (Scuba diving) is 17.5 m in Lake Attersee, Austria (Pall et al., 2010). In other Austrian (or multinational) lakes stands of the species reach depths of around 10 m, as 12.4 m in Lake Bodensee, 11.0 m in Lake Zellersee and 10.3 m in Lake Wörthersee (Pall et al., 2014).

Associated species: *Nitellopsis obtusa* often forms large monospecific dense stands as long as the conditions are oligo- to mesotrophic (Doll, 1989). *Chara globularis* is probably the most frequent species growing together with *N. obtusa*. However, none of the accompanying charophyte species becomes more than that—*N. obtusa* is dominating in such mixed stands. In shallow water *Chara contraria* can frequently be found together with *Nitellopsis obtusa*. In cases where the stands of *Nitellopsis obtusa* are only sparse, recently *Najas marina* subsp. *intermedia* frequently co-occurs in small to medium water-depth. At the lower depth limit, however, some other species may immigrate into the *Nitellopsis* stands, such as *Nitella opaca*, *Tolypella glomerata* or *Vaucheria dichotoma* and *Fontinalis antipyretica*. Under eutrophic conditions, the presence of higher plants, especially *Potamogeton* spp., *Myriophyllum* spp. and even *Ceratophyllum* spp. indicate reduced viability of the stands, with *Nitellopsis* often being the last charophyte species resisting in eutrophied lakes. In Sweden, *Chara tomentosa*, *C. subspinosa*, *C. hispida*, *Stuckenia pectinata* and *Myriophyllum spicatum* frequently co-occur with *N. obtusa* (Blindow et al., 2021). J. Csiky (pers. comm.) describes a differing situation in Hungary, where associated species in the ponds are *Callitriche cophocarpa*, *Hottonia palustris*, *Lemna trisulca*, *Ricciocarpos natans*, *Riccia fluitans*, *R. rhenana*, *Salvinia natans* and *Utricularia vulgaris*. Due to the frequent droughts these stands are not stable, but rather temporary.

70.4 Variability

The species is almost invariable, except for its size. However, even then appearance and general proportions are similar for some decimetre long shallow water specimen and metre-long deep-water ones. Even Migula (1889–1897), who usually described

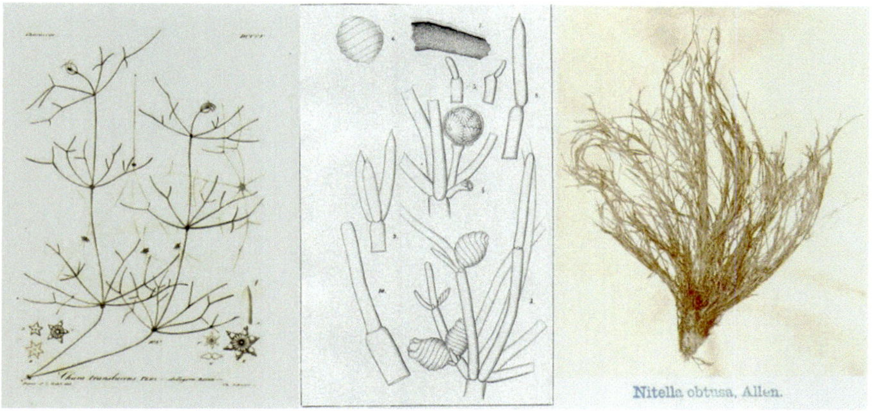

Fig. 70.4 Misleading names. Left: *Chara translucens*, illustration from Reichenbach (Iconographia botanica seu plantae criticae, t. 805, Fig. 1087 (1823–1832) [Bauer & H.G.L. Reichenbach]). This picture clearly refers to *Nitellopsis obtusa* but might be confused with *Nitella translucens*, as is the case with *Nitella obtusa* T. F. Allen (middle and right), which belongs to the genus *Nitella* (Allen, 1892)

long lists of varieties and forms, highlights its observation of continuous forms. He mentions just one divided and weak form, found growing together with *N. mucronata* and therefore probably reflecting the subneutral edge of its niche. The only variety mentioned by Migula (1889–1897), var. *ulvoides* A. Braun seems to be the extreme to the other side: with a main axis diameter of up to 4 mm it is something worth being looked for in old herbaria, these specimens were found at Lago di Bientino in Tuscany, near Mantua (Migula, 1889–1897, p. 272).

70.5 Identification and Risk of Confusion

It seems almost impossible to confuse this species. However, there are examples of confusing badly prepared, small specimens with *Nitella translucens* and, especially when analysing old records of *Chara translucens* (= *Nitella translucens*), at least some of them represent *Nitellopsis obtusa* (Fig. 70.4).

70.6 Distribution

Nitellopsis obtusa seems to be restricted to the northern hemisphere (Korsch, 2018). It is found almost all over Europe and Asia between 18° N and 65° N (Fig. 70.5). In Asia, records exist from Central, North, West, East, South Asia. Its preference for calcareous lakes is reflected by its distribution pattern, mainly lakes with a catchment

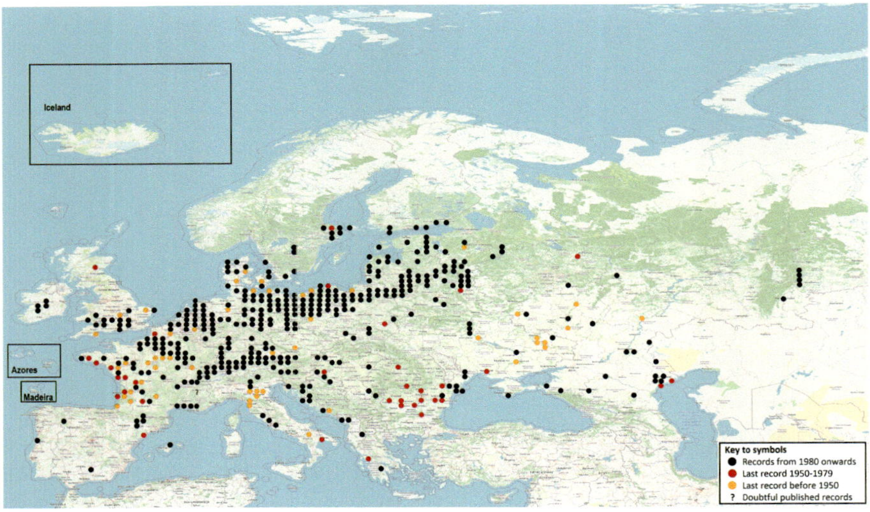

Fig. 70.5 European distribution of *Nitellopsis obtusa*. For worldwide distribution see Korsch (2018). Black dots are records from 1980 on, red dots from 1950 to 1979 and yellow dots before 1950; unconfirmed/doubtful published records (if applicable) are marked with "?" on top. Map prepared by Heiko Korsch

in sedimentary rocks and areas covered by glacial deposits harbour the species. It's not that clear therefore what factor is responsible for the lack of *N. obtusa* in Iceland and Norway. As in Sweden the species northwards to the middle of the country, harsh climatic conditions should not restrict it from these areas. Lack of calcareous bedrock and prevailing acidic bedrock in Iceland and Norway is a probable explanation. The species is able to thrive in low-salinity brackish water, it was found in the Baltic Sea at coastal sites in Denmark and Finland (Willén, 1957) as well as in Lithuania and Russia (Urbaniak, 2004).

The species is found frequently outside Europe and has successfully colonised North American lakes, being regarded an invader since its first appearance in the early 1980s (Alix et al., 2017).

70.7 Status and Threats

Preferring clear water, eutrophication is a threat for *N. obtusa* as for most of the charophytes. Especially increasing turbidity and competition by higher plants reduces depth distribution as well as area inhabited by this species. On the other hand this species is probably one of the least eutrophication-sensitive charophyte species, still growing in dense patches in slightly eutrophic shallow lakes, surrounded by dense meadows of *Potamogeton* spp. and *Myriophyllum* spp. Probably the most threatening factor is ground-foraging fish species, acting both ways, by destroying the stands and by increasing turbidity. It is considered "Threatened" or "Near Threatened" in many northern, eastern and southern European countries but it is fairly widespread in central and western Europe and is increasing in Britain and Ireland (N. Stewart pers. comm). However, it's preferred habitat, mesotrophic calcareous lakes, as well as the typical pond habitats in Hungary are protected all over the European Union by Annex I of the Flora-Fauna-Habitat Directive (European Commission, 1992).

70.8 Remarks

Strong allelopathic activity of *Nitellopsis obtusa*, efficiently supressing phytoplankton growth, has been reported by Berger and Schagerl (2004).

References

Alix, M., Scribailo, R., & Weliczko, C. (2017). *Nitellopsis obtusa* (Desv.) J. Groves, 1919 (Charophyta: Characeae): New records from southern Michigan, USA with notes on environmental parameters known to influence its distribution. *BioInvasions Records, 6*, 311–319.

Allen, T. F. (1892). Notes on some Characeae. *Bulletin of the Torrey Botanical Club, 19*(7), 230.

Bailly, G., & Schaefer, O. (2010). *Guide illustré des Characées du nord-est de la France.* Conservatoire Botanique National de Franche-Comté, Besançon. [in French].

Berger, J., & Schagerl, M. (2004). Allelopathic activity of Characeae. *Biologia, Bratislava, 59*, 9–15.

Blindow, I., Carlsson, M., & van de Weyer, K. (2021). Re-establishment techniques and transplantations of Charophytes to support threatened species. *Plants, 2021*(10), 1830.

Cirujano, S., Cambra, J., Sánchez Castillo, P. M., Meco, A., & Flor Arnau, N. (2008). *Flora ibérica. Algas continentales. Carófitos (Characeae).* Real Jardín Botánico. [in Spanish].

Cirujano, S., García Murillo, P., Meco, A., & Fernández Zamudio, R. (2007). Los carófitos ibéricos. *Anales Del Jardín Botánico De Madrid,* 64, 87–102. [in Spanish].

Doege, A., van de Weyer, K., Becker, R., & Schubert, H. (2016). Bioindikation mit Characeen. In Arbeitsgruppe Characeen Deutschlands (Ed.), *Armleuchteralgen: Die Characeen Deutschlands* (pp. 97–138). Springer. [in German].

Corillion, R. (1957). Les Charophycées de France et d'Europe Occidentale. *Bulletin de la Société Scientifique de Bretagne, 32*(Hors série I & II), 1–499. [in French].

Doll, R. (1989). Die Pflanzengesellschaften der stehenden Gewässer der DDR. Teil 1: Die Gesellschaften des offenen Wassers (Characeen-Gesellschaften). *Feddes Repertorium, 100*, 281–324. [in German].

European Commission. (1992). Council Directive 92/43/EEC of 21 May 1992 on the conservation of natural habitats and of wild fauna and flora. *Official Journal of the European Communities, L 206*(7).

Groves, J., & Bullock-Webster, G. R. (1924). *The British Charophyta (Volume II). Chareæ with plates, concluding articles, geological sketch, bibliography and index*. Ray Society.

Jeschke, L. (1959). Die Pflanzengesellschaften einiger Seen bei Feldberg in Mecklenburg. *Feddes Repertorium Beiheft, 138*, 161–214. [in German].

Korsch, H. (2013). Die Armleuchteralgen (Charophyceae) Sachsen-Anhalts. *Berichte Des Landesamtes Für Umweltschutz Sachsen-Anhalt, 1*(2013), 1–85. [in German].

Korsch, H. (2018). The worldwide range of the Charophyte species native to Germany. *Rostocker Meeresbiologische Beiträge, 28*, 45–96.
Krause, W. (1997). Charales (Charophyceae) Süßwasserflora von Mitteleuropa. In H. Ettl, G. Gärtner, H. Heynig & D. Mollenhauer (Eds.), *Süßwasserflora von Mitteleuropa 18*. Fischer. [in German].
Krause, W. (1985). Über die Standortansprüche und das Ausbreitungsverhalten der Stern-Armleuchteralge *Nitellopsis obtusa* (Desvaux). J. Groves. *Carolinea, 42*, 31–42. [in German].
Luther, H. (1951). Verbreitung und Ökologie der höheren Wasserpflanzen im Brackwasser der Ekenäs-Gegend in Südfinnland II Spezieller Teil. *Acta Botanica Fennica, 50*, 1–370. [in German].
Migula, W. (1889–1897). Fünfter Band: Die Characeen. In A. Fischer, E. Fischer, F. Hauck, G. Limpricht, C. Luerssen, W. Migula, H. Rehm, P. Richter, & G. Winter (Eds.), *Dr. L. Rabenhorst's Kryptogamenflora von Deutschland, Oesterreich und der Schweiz* (2nd ed.). Eduard Kummer. [in German].
Olsen, S. (1944). Danish Charophyta. *Det Kongelige Danske Videnskabernes Selskab, Biologiske Skrifter, 3*(1), 1–240, [in Danish].
Pall K., Hippeli, S., Mayerhofer, V., & Mayerhofer, S. (2010). Makrophytenkartierung Attersee. *Untersuchung im Auftrag der Landesregierung Oberösterreich und des Bundesministeriums für Land- und Forstwirtschaft, Umwelt und Wasserwirtschaft*. 1-124. [in German].
Pall, K., Bertrin, V., Buzzi, F., Boutry, S., Dutartre, A., Germ, M., Oggioni, A., Schaumburg, J., & Urbanic, G. (2014). *Alpine Lake Macrophyte ecological assessment methods – Water Framework Directive Intercalibration Technical Report. JRC Technical Report EU 26507 EN*. Publications Office of the European Union, Luxembourg.
Sonder, C. (1890). *Die Characeen der Provinz Schleswig-Holstein und Lauenburg nebst eingeschlossenen fremden Gebietstheilen*. Dissertation Universität Rostock. [in German].
Soulié-Märsche, I. (1989). *Etude comparée de gyrogonites de charophytes actuelles et fossiles et phylogénie des genres actuels*. Imprimerie des Tilleuls. [in French].
Soulié-Märsche, I., Benammi, M., & Gemayel, P. (2002). Biogeography of living and fossil *Nitellopsis* (Charophyta) in relationship to new finds from Morocco. *Journal of Biogeography, 29*, 1–9.
Urbaniak, J. (2004). *Nitellopsis obtusa*. In H. Schubert & I. Blindow (Eds.), *Charophytes of the Baltic Sea* (pp. 216–222). A. R. G. Gantner.
Urbaniak, J., & Gąbka, M. (2014). *Polish Charophytes. An illustrated guide to identification*. Uniwersytet Przyrodniczy we Wroclawiu. [in Polish].
Willén, T. (1957). On the distribution of *Nitellopsis obtusa* (Desvaux) Groves around the Baltic. *Botanika Notiser, 110*, 313–320.
Winter, U., Kirst, G. O., Grabowski, V., Heinemann, U., Plettner, I., & Wiese, S. (1999). Salinity tolerance in *Nitellopsis obtusa*. *Australian Journal of Botany, 47*, 337–346.
Wood, R. D. (1965). *Monograph of the Characeae*. J. Cramer.

Chapter 71
Sphaerochara canadensis

Anders Langangen

Scientific name: ***Sphaerochara canadensis*** **(Sawa) Soulié-Märsche, comb. nov.**
Basionym: *Tolypella canadensis* Sawa in J. Phycol. 9: 476. 1973

Holotype (Sawa, 1973: 184): Ontario, Canada: Thunder Bay District: the shore of Lake Superior at Rossport Provincial Picnic Park, along Trans-Canada Hwy. (= Ont. Hwy. 176) 0.2 mi east of Rossport/20 July 1970/Sawa-Hotchkiss-Thorpe 70-7-20-3 (TRT).

Nomenclatural comment: Type material was originally deposited at TRTC, the University of Toronto Cryptogamic Herbarium. In 2000, that collection, was transferred to the Royal Ontario Museum (ROM), where it is kept separate as TRT.

Local names: FINNISH: Tunturisykeröparta; NORWEGIAN: Polarglattkrans; SWEDISH: Fjällrufse.

71.1 Morphology

See Fig. 71.1.

Habitus: The plants are green, unencrusted, *Nitella*-like, but unlike this genus, they are stiff, with small terminal and axillary heads. *Sphaerochara canadensis* is a perennial species. The fertile branchlets have commonly the gametangia in dense heads.

Size: *Sphaerochara canadensis*, when found in shallow places, is small to medium-sized (4–14 cm) and in deeper water it can reach up to 27 cm length.

A. Langangen (✉)
Natural History Museum, P.O. Box 1172 Blindern, 0318 Oslo, Norway
e-mail: langangen@hotmail.com

H. Schubert et al. (eds.), *Charophytes of Europe*,
https://doi.org/10.1007/978-3-031-31898-6_71

Fig. 71.1 *Sphaerochara canadensis*. **a**, **b** Habitus; **c** Specimen from Lake Glomdalsvannet in Nordland, Norway; **d** Upper whorls with gametangia, specimen from Lake Lagovatnet, Gildeskål in Nordland, Norway; **e** Whorl with oogonia, some stalked. Photographs: Anders Langangen

Main axis: The diameter of the main axis varies between 320 and 400 μm.

Branchlets: The lowest whorl (protonematal whorl) has sterile branchlets, 6–8 in a whorl and with 3–6 cells, to 2 cm long. Accessory branchlets are simple, sterile. From this whorl (protonema) arise 1–8 mature shoots, with progressively shorter internodes between whorls. Branchlets 6–8 in a whorl, 3–4 celled with a mucro.

This is an important taxonomic character. The branchlets can be sterile or fertile. Accessory branchlets are common.

Sexuality: *Sphaerochara canadensis* is monoecious. Chromosome number: n = 8.

Oogonia: The oogonia are 1–4 aggregated, to 700 μm long (including coronula) and often stalked (stalk cells 250–350 μm), 400–500 wide, with 8–9 convolutions. Dwarf oogonia without content are often found.

Oospores: The oospore is brown, subglobose, to 400 μm long, 300–350 μm wide, with 6–8 prominent ridges, oospore-membrane is smooth (Urbaniak et al., 2012). Basal impression is single.

Antheridia: The antheridia are up to 400 μm in diameter and stalked, 2–5 aggregate, octoscutate.

71.2 Phenology and Reproduction

The species is perennial. In all collections in Scandinavia there were both sterile and fertile specimens. The small tufted plants were often luxuriantly fruiting. Antheridia and oogonia are found from July to October. Antheridia are most common in young plants or shoots, which indicate protandrous development. On older plants and shoots oogonia are often dominant. Ripe oospores are not common, but are nevertheless found in several localities (Karesuando, Sundfjord).

71.3 Habitat Conditions and Ecology

Habitat type: *Sphaerochara canadensis* is found in oligotrophic lakes, often of the lagoon type. It is also found in rivers with relatively low water temperatures. It is presumably a cold-stenotherm species.

Sediment: Preferred sediment is a fine sandy bottom.

Nutrient conditions: *Sphaerochara canadensis* is found in oligotrophic waters. Artificial inorganic enrichment of a locality with *S. canadensis* has shown that the species is heavily affected, and nearly disappeared in the actual lake (Lake Skittenfjordvatn, Finmark, Norway).

Water chemistry: An important parameter is the content of calcium in lakes where it is found, from 4 to 20 mg Ca^{2+}/L. Lakes with high content of lime are found in Sundfjord, Norway and here the following parameters are (Langangen and Breivik, 2010): pH: 7.6–8.7; specific conductivity: 60–150 μS/cm; Ca^{2+}: 10–20 mg/L.

In Lake Glomdalsvatn, Norway the following parameters were measured (Langangen, 1993): pH: 6.8–7.0; specific conductivity: 32–105 μS/cm; Ca^{2+}: 4.0–6.0 mg/L.

Depth range: *Sphaerochara canadensis* is reported from shallow places down to 17 m (Langangen and Breivik, 2010).

Associated species: The species is often found in mixed populations with *Nitella opaca* or *N. flexilis*. Sawa (1973) found it associated with several other charophytes: *Nitella obtusa, N. spanioclema* var. *nidifica* , *Tolypella comosa, Chara globularis* and *C. aspera*. In the mountain area of Sundfjord in Norway, the biggest marble-karst area in Scandinavia, the species was found together with *N. opaca* and *Chara contraria* (Langangen and Breivik, 2010).

71.4 Variability

The phenotypic variation due to depth of growth is large, from small tufted colonies in shallow water to long, thin specimens in deeper water.

71.5 Identification and Risk of Confusion

The mucro and the stiff appearance of *S. canadensis* are two diagnostic characters for this species. As it often is found in association with *Nitella opaca* or *N. flexilis*, it has often been determined to one of these. Hasslow (1939) found the species in Karesuando, Sweden and described it as *Nitella mucronata* f. *haplophylla*.

71.6 Distribution

Sphaerochara canadensis has a circumpolar distribuion. It was found in the following countries: Canada, Alaska, Russia, Finland, Sweden, Norway, Iceland and Greenland (Fig. 71.2).

Three new, important localities:

Russia: Oymyakonsky District of the Sakha Republic at 1014 m above sea level, at the plateau Sordonnokh in the River Indigirka drainage basin, 16 March 2014 (Romanov and Kopyrina, 2016).
Iceland: Högnavatn, Fjallabaki, 20. August 2012, leg. Þóra Katrin Hrafnsdóttir (thora@natkop.is).

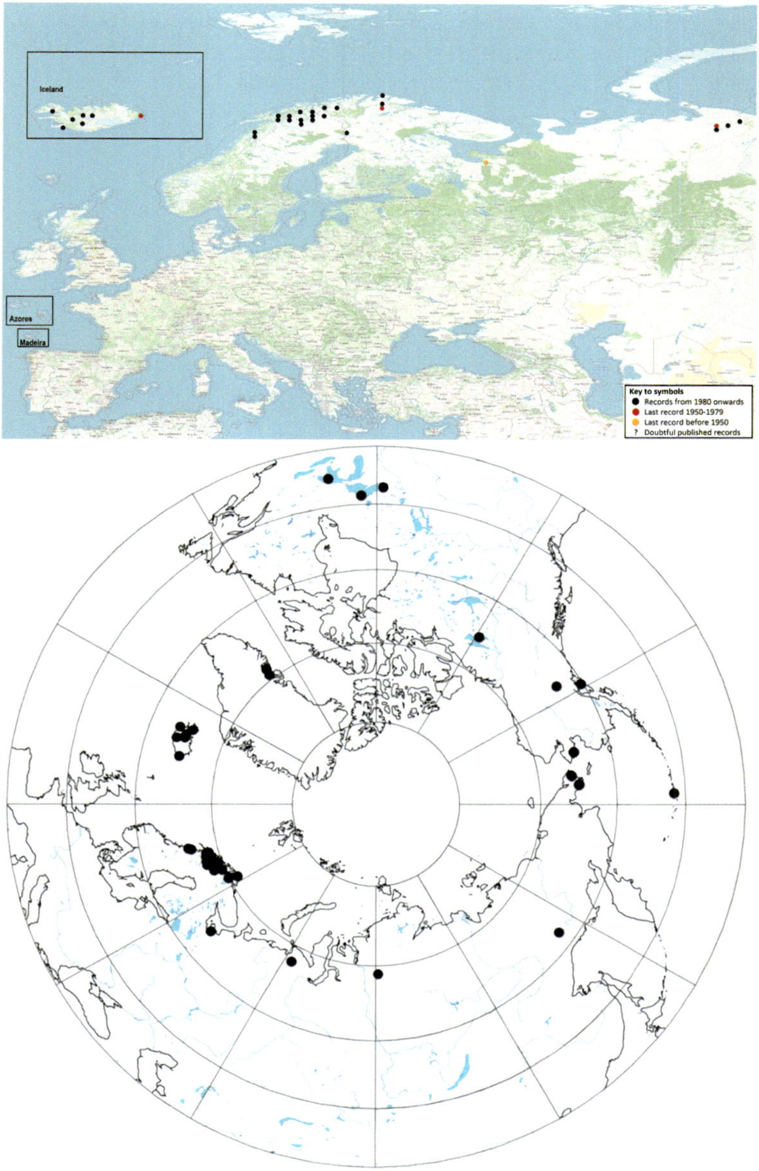

Fig. 71.2 Upper panel: European distribution of *Sphaerochara canadensis*, Lower panel: World-wide distribution of *Sphaerochara canadensis*. Black dots in the upper panel are records from 1980 on, red dots from 1950 to 1979 and yellow dots before 1950: Maps: Upper panel—prepared by Heiko Korsch; lower panel—prepared by Roman Romanov

Iceland: Frostaðavatn, Fjallbaki, 21. July 2012, leg. Idem.; Kleifarvatn, near the southern lake shore. About 30 km south of Reykjavik. 63.911447N, 21.998808W, 15. July 2013, leg. Theo Bakker, det. John Bruinsma (L).

71.7 Status and Threat

Regulation of mountain lakes and rivers can be a threat. The species is redlisted as NT in Norway, Sweden and Finland. In the European Red List, it is assessed as LC (least concern) because of little indication of decline (see Chap. 9).

References

Hasslow, O. J. (1939). Einige Characeenbestimmungen. *Botaniska Notiser, 1939*, 295–301. [in German].

Langangen, A. (1993). *Tolypella canadensis*, a charophyte new to the European flora. *Cryptogamie, Algologie, 14*, 221–231.

Langangen, A., & Breivik, Å. B. (2010). Kalksjøene på Sundsfjordfjellet i Gildeskål kommune (Nordland) og vegetasjonen i dem, en undersøkelse gjort i tiden 2007–2009. *Blyttia, 68*, 201–207. [in Norwegian].

Romanov, R. E., & Kopyrina, L. I. (2016). *Tolypella canadensis* (Charales, Charophyceae) in Asia: Final evidence of its circumpolar distribution. *Nova Hedwigia, 102*, 423–427.

Sawa, T. (1973). Two new species of *Tolypella* (Characeae) from North America. *Journal of Phycology, 9*, 472–482.

Urbaniak, J., Langangen, A., & van Raam, J. (2012). Oospore wall ornamentation in the genus *Tolypella* (Charales, Charophyceae). *Journal of Phycology, 48*, 1538–1545.

Chapter 72
Sphaerochara intricata

Klaus van de Weyer

Scientific name: ***Sphaerochara intricata*** **(Trentep. ex Roth) Feist-Castel & N. Grambart in Riding, Calcareous Algae Stromatolites: 201. 1991.**
Basionym: *Chara intricata* Trentep. ex Roth, Catal. Bot.: 125. 1800

Neotype (Wood, 1965: 741): Hungaria: Budapest-Rákosfalva, in fossis ad "Sárga-Csiko"/8. 5. 1897/F. Filarszky (NY); Migula, Sydow & Wahlstedt, Characeae Exs. 77

≡ *Tolypella intricata* (Trentep. ex Roth) Leonh. in Lotos 13: 57. 1863

Local names: DANISH: Brodspdset Redetråd; DUTCH: Vertakt boomglanswier; ENGLISH: Tassel stonewort; FRENCH: Tolypelle enchevêtrée; GERMAN: Verworrene Baumleuchteralge; POLISH: Rozsocha zawiła; RUSSIAN: Толипелла запутанная; SWEDISH: Uddrufse; UKRAINIAN: Толіпела заплутана.

72.1 Morphology

See Figs. 72.1 and 72.2

Habitus: Typical for *Sphaerochara intricata* are the non regular heads. The sterile branchlets are long and divided.

Size: *Sphaerochara intricata* is small to medium-sized, but may reach up to 50 cm if growing in deeper water.

Main axis: Main axis diameter is about 1–2 mm. The lower internodes may be up to 12 cm long while the upper internodes are short.

K. van de Weyer (✉)
lanaplan, Lobbericher Str. 5, 41334 Nettetal, Germany
e-mail: klaus.vdweyer@lanaplan.de

H. Schubert et al. (eds.), *Charophytes of Europe*,
https://doi.org/10.1007/978-3-031-31898-6_72

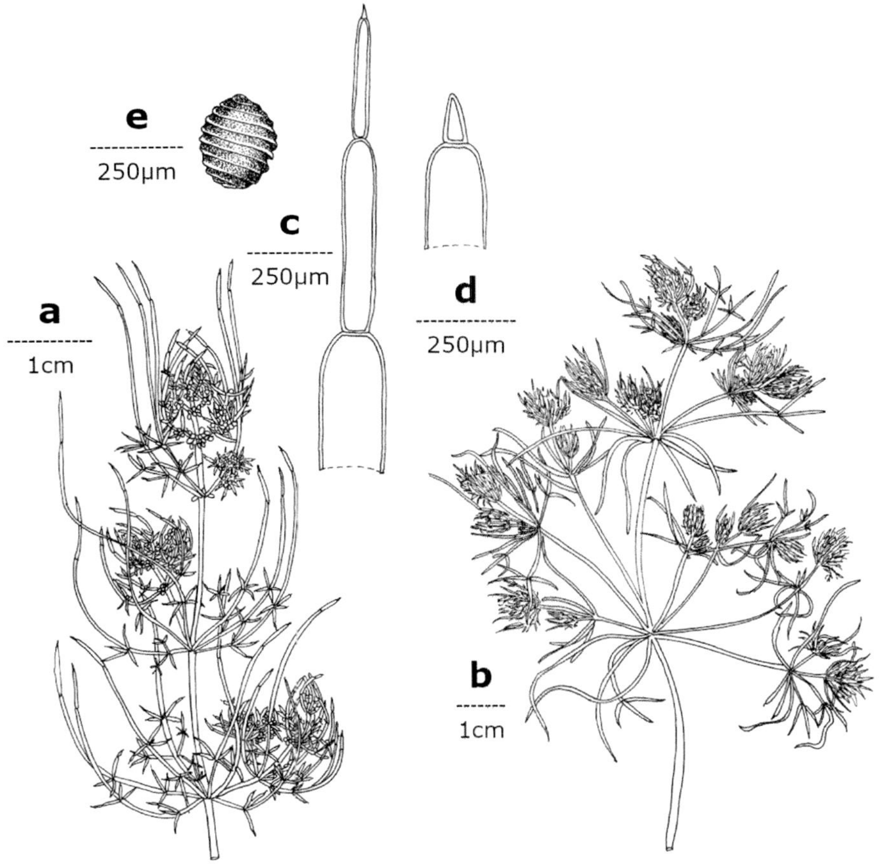

Fig. 72.1 *Sphaerocarpa intricata*, details. **a** Habitus; **b** Shoot tip; **c** Branchlet tip; **d** Terminal cells of a branchlet; **e** Oospore. Drawings after Werner Krause, modified by Dagmar Wassong in van de Weyer and Schmidt (2018)

Cortex: *Sphaerochara intricata* is ecorticate.

Spine-cells: Absent.

Stipulodes: Absent.

Branchlets: *Sphaerochara intricata* has 6–8 branchlets which are divided. The sterile branchlets are 2–8 cm long, the fertile branchlets are 0.4–0.8 cm long. The branchlets are composed of 3–5 cells. The terminal cell is 50–80 µm long, acuminate and shorter than the other cells of the branchlets.

Bract-cells, bracteoles, bractlets: Absent.

Sexuality: The species is monoecious.

Fig. 72.2 *Sphaerochara intricata*, details. **a** Habitus, from a gravel pit near Speyer, Germany; **b** Habitus, from a gravel pit near Frankfurt/Rhine, Germany; **c**, **d** Shoot tips with fertile heads, from a gravel pit near Wesel, Germany; **e** Gametangia; **f** Fertile head, from a gravel pit near Frankfurt/Main, Germany; **g** Oogonium; **h** Fertile head from a gravel pit near Speyer, Germany. Photographs: **a–d**, **f**, **h**-Klaus van de Weyer, **e**, **g**-Chris Carter

Table 72.1 Oogonia morphometry

Author	Length	Width	Coronula length	Coronula width	Number of helical stripes
Migula (1889–1897)	430–450	–	–	–	10–13
Groves and Bullock-Webster (1920)	500	400	30	60	11–12
Corillion (1957)	500	400	30	60	11–12
Wood (1965)	350–560	300–450	30	80	10–13
Krause (1997)	450–800	350–500	30–80	50–80	10–13
Total range	**350–560(800)**	**300–500**	**30–80**	**50–80**	**10–13**

Length in the first column excludes the coronula, its length and width given separately; all values for length and width in μm

Oogonia: The oogonia are 350–560(800) μm long (without coronula) and 300–550 μm wide (without coronula). The oogonia have 10–13 helical stripes. The coronula is 30–80 μm long and 50–80 μm wide (Table 72.1: Corillion, 1957; Groves and Bullock-Webster, 1920; Krause, 1997; Migula, 1889–1897; Wood, 1965).

Oospores: The oospores are brown (Krause, 1997; Wood, 1965) or light golden-yellow (Corillion, 1957; Groves and Bullock-Webster, 1920). The oospores are 300–400(600) μm long, 270–500 μm wide and have 7–11 striae (Table 72.2: Corillion, 1957; Groves and Bullock-Webster, 1920; Krause, 1997; Migula, 1889–1897; Wood, 1965). A reliable character is the oospore-membrane of the oospore which is granulate (Fig. 72.3). The oospores of *Sphaerochara intricata* do calcify and form spherical to slightly ellipsoidal gyrogonites with a length of 325–425 μm and width of 275–325 μm (Thiéry et al., 2020). They are characterised by a single basal plate (Soulié-Märsche and Garcia, 2015) (Fig. 72.5).

Antheridia: The diameter of the antheridia is 200–400 μm (Table 72.3).

Bulbils: Absent.

72.2 Phenology and Reproduction

Sphaerochara intricata is an annual species. For germination it is important that there is no competition with other water plants. According to van Raam (1998) *Sphaerochara intricata* germinates in autumn or winter. In early spring there is strong development with fructification. It is possible that there is a second germination in late spring. In shallow waters *S. intricata* can develop large stands for a short period and than disappear but the oospores can survive for many years. In deeper lakes *Sphaerochara intricata* can be found in the same sites for many years.

Table 72.2 Oospore morphometry

Author	Length	Width	Number of striae	Colour	Oospore-membrane
Migula (1889–1897)	320–400	–	9–11	Light brown	Granulate
Groves and Bullock-Webster (1920)	350	275	9–10	Light gold-yellow	Granulate
Corillion (1957)	350	275	9–10	Light gold-yellow	Granulate
Wood (1965)	300	500	7–10	Light brown or wine-red	–
Krause (1997)	300–600	270–400	7–9	Brown	–
Urbaniak et al. (2012)	–	–	–	–	Smooth/sligthly reticulate
Haas (1994)	320–500	270–400	9–11	Light to dark brown	Finely granulate
Total range	**300–400(600)**	**270–500**	**7–11**		**Granulate/ reticulate**

Length and width in μm

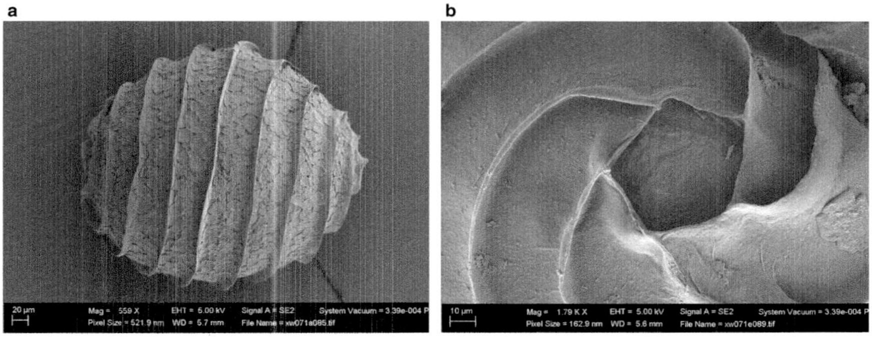

Fig. 72.3 Oospores of *Sphaerochara intricata*. **a**, **b** Earlsway Farm, Bramfield Park, Suffolk, Hawkins, Juliet (JKH), BM013828095. Photographs: Michelle T. Casanova

72.3 Habitat Conditions and Ecology

Habitat type: *Sphaerochara intricata* is found in ditches, streams, pools, temporary waters, oxbow lakes, artificial and natural lakes.

Sediment: *Sphaerochara intricata* has been reported from different sediments including sand, gravel, clay and organic substrate.

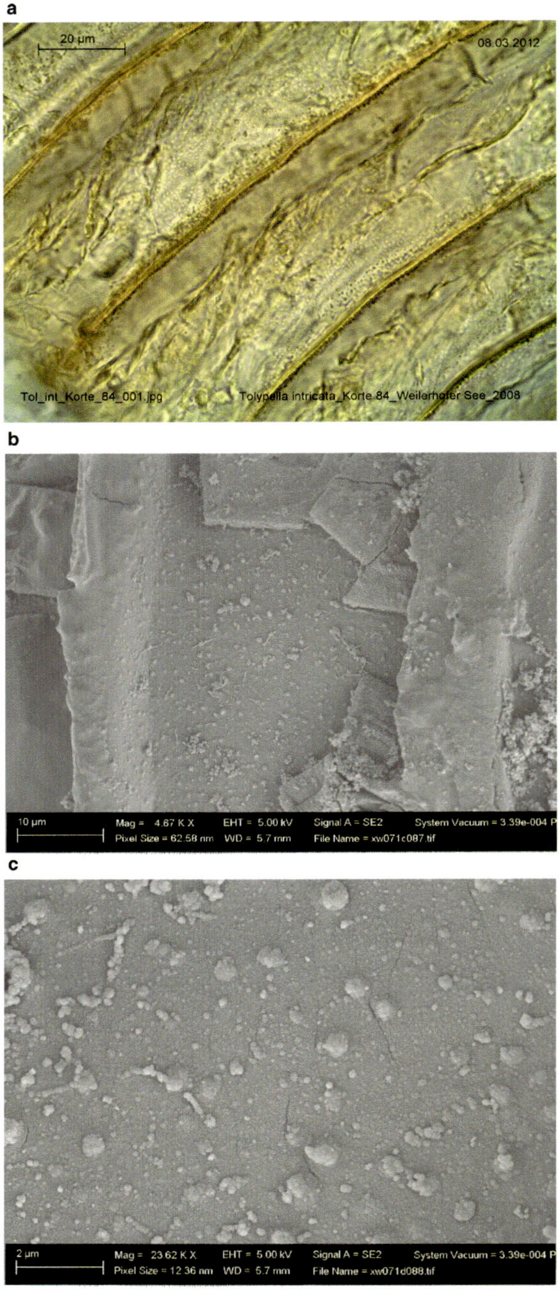

Fig. 72.4 Ornamentation of the oospore-membrane of *Sphaerochara intricata*. **a** Light miscrosope, Weilerhofer See, Germany; **b**, **c** SEM, Earlsway Farm, Bramfield Park, Suffolk, Juliet Hawkins (JKH), BM013828095. Photographs: **a**-Thomas Gregor, **b**, **c**-Michelle T. Casanova

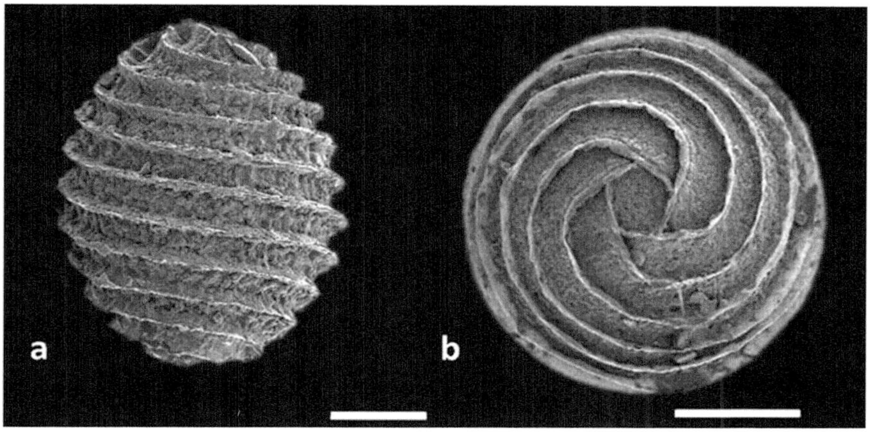

Fig. 72.5 Gyrogonites of *Sphaerochara intricata*. **a** Lateral view; **b** Basal view with pentagonal basal plate visible from the outside. Locality: Lanau pond (South France). Scale bar 100 μm. Photographs: Ingeborg Soulié-Märsche

Table 72.3 Diameter of the antheridia

Author	Diameter (μm)
Corillion (1957)	400
Wood (1965)	(230)255–375
Krause (1997)	200–400
Total range	**200–400**

Nutrient conditions: There are rather little data on the nutrient conditions of *Sphaerochara intricata*. It has been reported from oligotrophic, mesotrophic and eutrophic waters. In lakes *Sphaerochara intricata* has been described from groundwater influenced sites. The total phosphorus range is reported to vary between 0.007 and 0.3 mg/l (Humberg and Beck, 2006; Nat et al., 1994).

Water chemistry: There are rather little data on the water chemistry of *Sphaerochara intricata*. It is a species of alkaline water and there are no descriptions from brackish water. The pH range is reported to vary between 6 and 8.5 (Humberg and Beck, 2006; Nat et al., 1994). Conductivity range is reported to vary between 462 and 918 μS/cm (Humberg and Beck, 2006).

Depth range: *Sphaerochara intricata* occurs in shallow waters and in lakes up to a depth of 15 m.

Associated species: *Sphaerochara intricata* occurs in ditches and pools together with other charophytes such as *Chara vulgaris* or *Chara globularis*. These stands have been described as *Charo-Tolypelletum intricatae* Corillion, 1957 = *Charo-Tolypelletum intricatae* W. Krause 1969). Typically, the species is associated with other charophytes like *Nitella capillaris, Nitella opaca, Nitella mucronata, Chara contraria, C. hispida* and *Vaucheria* spp.. *Sphaerochara intricata* has also been reported in association with a number of vascular plants.

72.4 Variability

According to the season and the habitat it is possible to find small plants with a small number of heads but also large plants with numerous heads.

72.5 Identification and Risk of Confusion

Sphaerochara intricata can be confused with other *Sphaerochara* and *Tolypella* species. *Sphaerochara intricata, S. prolifera* and *S. canadensis* have conical-acute end-cells of the branchlets, while species belonging to the genus *Tolypella* have obtuse end-cells. *Sphaerochara intricata* can be confused with *Sphaerochara prolifera* but the sterile branchlets are divided while those of *S. prolifera*—and *S. canadensis*—are simple. Also, the oospore-membrane of *S. intricata* is granulate while the oospores of *S. prolifera* have a smooth oospore-membrane. *Sphaerochara intricata* can be confused with *Nitella mucronata*, but this has regular heads while the heads of *S. intricata* are not regular. The oospore-membrane of *Nitella mucronata* is reticulate. *Nitella mucronata* has oogonia with 7–9 helical stripes and oospores with 6–8 striae. *Sphaerochara intricata* has oogonia with 10–13 helical stripes and oospores with 7–11 striae.

72.6 Distribution

According to Corillion (1957) and Korsch (2018) *Sphaerochara intricata* is restricted to Europe, Asia and northern Africa. *Sphaerochara intricata* has been recorded from many countries in Europe (Fig. 72.6), but there are no records from most countries in northern and eastern Europe.

72.7 Status and Threats

Sphaerochara intricata is a pioneer in ditches and pools. For the germination of this annual species it is important that there is no or little competition with other water plants. In deeper lakes the populations of *Sphaerochara intricata* seem to be more stable especially in the deeper parts where the cover of vegetation is not complete.

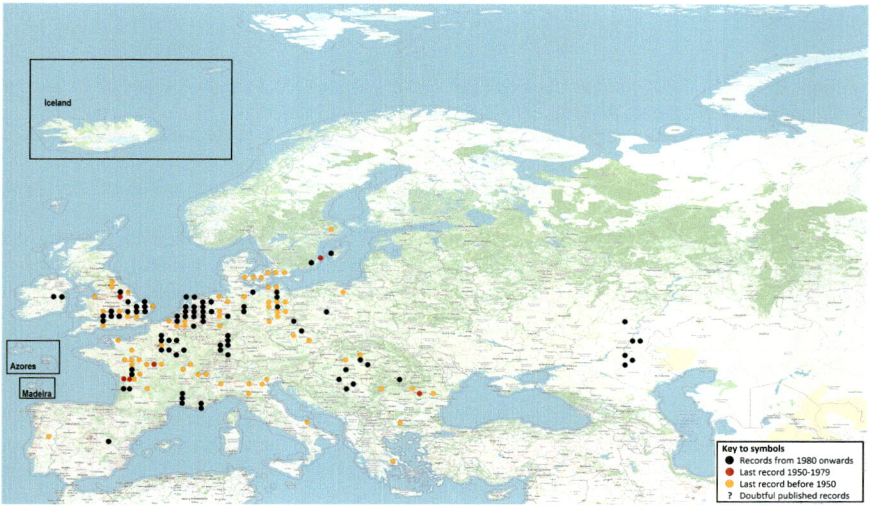

Fig. 72.6 European distribution of *Sphaerochara intricata*. For worldwide distribution see Korsch (2018). Black dots are records from 1980 on, red dots from 1950 to 1979 and yellow dots before 1950; unconfirmed/doubtful published records (if applicable) are marked with "?" on top. Map prepared by Heiko Korsch

Sphaerochara intricata is listed in different Red Data Books: the Balkans (EN), Bulgaria (CR), Germany (EN), Poland (CR), Sweden (VU), Czech Republic (VS), UK and Ireland (EN). *Sphaerochara intricata* is extinct in Denmark and Switzerland.

Within ditches and pools it is necessary to maintain areas which are not covered with other macrophytes. It can be useful to remove other macrophytes to facilitate the germination of *Sphaerochara intricata.* In general, all measures to reduce/prevent eutrophication are recommended. This also includes maintenance of fish communities which are typical for the specific water body type.

72.8 Remarks

Clearly, *Sphaerochara intricata* has been confused with *S. prolifera* as both species have been recorded together from some sites.

References

Corillion, R. (1957). Les Charophycées de France et d'Europe Occidentale. *Bulletin de la Société Scientifique de Bretagne, 32*(Hors série I & II), 1–499. [in French].
Groves, J., & Bullock-Webster, G. R. (1920). *The British Charophyta. Volume I. Nitelleae.* Ray Society.
Haas, J. N. (1994). First identification key for charophyte oospores from central Europe. *European Journal of Phycology, 29*, 227–235.
Humberg, B., & Beck, M. (2006). Der Makrophytenbestand in ausgewählten Baggerseen der Oberrheinaue. *LUBW, Oberirdische Gewässer, Gewässerökologie, 102*, 1–348. [in German].
Korsch, H. (2018). The worldwide range of the Charophyte species native to Germany. *Rostocker Meeresbiologische Beiträge, 28*, 45–96.
Krause, W. (1997). Charales (Charophyceae) Süßwasserflora von Mitteleuropa. In H. Ettl, G. Gärtner, H. Heynig & D. Mollenhauer (Eds.), *Süßwasserflora von Mitteleuropa 18*. Fischer. [in German].
Migula, W. (1889–1897). Fünfter Band: Die Characeen. In A. Fischer, E. Fischer, F. Hauck, G. Limpricht, C. Luerssen, W. Migula, H. Rehm, P. Richter & G. Winter (Eds.), *Dr. L. Rabenhorst's Kryptogamenflora von Deutschland, Oesterreich und der Schweiz* (ed. 2). Eduard Kummer. [in German].
Nat, E., Simons, J., de la Haye, M. A. A., & Coops, H. (1994). *Historisch en actueel verspreidingsbeeld van kranswieren in Nederland in samenhang met waterkwaliteitsfactoren.* RIZA werkdocument 94.148X. Lelystad. [in Dutch].
Soulié-Märsche, I., & Garcia, A. (2015). Gyrogonites and oospores, complementary viewpoints to improve the study of the charophytes (Charales). *Aquatic Botany, 120*, 7–17.
Thiéry, A., Soulié-Märsche, I., Meisch, C., & Simler, R. (2020). Une biocénose remarquable dans la mare temporaire méditerranéenne de Lanau en Crau (Bouches-du-Rhône, sud-est de la France. *Société Linnéenne De Provence, 71*, 45–66.
Urbaniak, J., Langangen, A., & van Raam, J. (2012). Oospore wall ornamentation in the genus *Tolypella* (Charales, Charophyceae). *Journal of Phycology, 48*, 1538–1545.
van de Weyer, K., & Schmidt, C. (2018). Bestimmungsschlüssel für die aquatischen Makrophyten (Gefäßpflanzen, Armleuchteralgen und Moose) in Deutschland: Band 1: Bestimmungsschlüssel, ed. 2. *Landesamt für Umwelt (LfU) Brandenburg, Potsdam, Fachbeiträge des LfU Brandenburg 119*: 1–180. [in German].
van Raam, J. C. (1998). *Handboek Kranswieren.* Chara boek. [in Dutch].
Wood, R. D. (1965). *Monograph of the Characeae.* J. Cramer.

Chapter 73
Sphaerochara prolifera

Klaus van de Weyer

Scientific name: ***Sphaerochara prolifera*** **(Ziz ex A. Braun) Soulié-Märsche, comb. nov.**

Basionym: *Chara prolifera* Ziz ex A. Braun, Ann. Sci. Nat., Bot. 1: 352. 1834
Neotype (Wood, 1965: 744): [England] Ramsey Huntingdonshire/July 12, 1894/A. Fryer and H. G[roves] (NY); Groves and Groves, Charac. Brit. Exs. 50.

≡ *Tolypella prolifera* (Ziz ex A. Braun) Leonh. in Lotos 13: 57. 1863.

Local names: DUTCH: Groot boomglanswier; ENGLISH: Great tassel stonewort; FRENCH: Tolypelle prolifère; GERMAN: Sprossende Baumleuchteralge; POLISH: Rozsocha przerastająca; RUSSIAN: Толипелла пролиферирующая; UKRAINIAN: Толіпела проліферируюча.

73.1 Morphology

See Figs. 73.1 and 73.2

Habitus: Typical for *Sphaerochara prolifera* are the heads with irregular shape. The sterile branchlets are very long and not divided.

Size: *Sphaerochara prolifera* is a large species with a maximum length of about 70 cm. There are also medium-sized forms in shallow waters.

Main axis: Main axis diameter is about 1.5–3.5 mm. The lower internodes are 15–25(45) cm long.

K. van de Weyer (✉)
lanaplan, Lobbericher Str. 5, 41334 Nettetal, Germany
e-mail: klaus.vdweyer@lanaplan.de

H. Schubert et al. (eds.), *Charophytes of Europe*,
https://doi.org/10.1007/978-3-031-31898-6_73

Cortex: *Sphaerochara prolifera* is ecorticate.

Spine-cells, stipulodes: Absent.

Branchlets: *Sphaerochara prolifera* has 6–12 branchlets. The sterile branchlets are 2–8 cm long and usually undivided, the fertile branchlets are 0.4–0.8 cm long and divided. The primary rays are composed of 3–5 cells. The terminal cell is 50–80 μm long, acuminate and shorter than the other cells of the branchlets (Figs. 73.1 and 73.2).

Bract-cells, bracteoles, bractlets: Absent.

Sexuality: The species is monoecious.

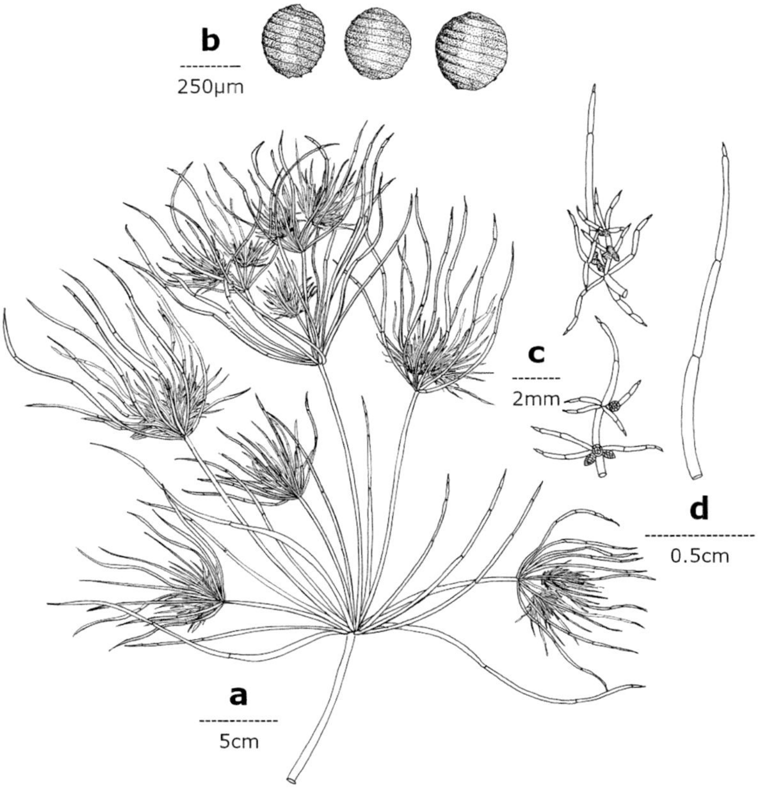

Fig. 73.1 *Sphaerochara prolifera*. **a** Habitus; **b** Oospores; **c** Fertile branchlets, **d** Sterile branchlet. Drawings after Werner Krause, modified by Dagmar Wassong in van de Weyer and Schmidt (2018)

Fig. 73.2 *Sphaerochara prolifera*, details. **a, b** Habitus, from a ditch near Waalwijk, NL; **c, d** Fertile heads, from Lake Vinkenveen, NL; **e** Antheridium; **f** Oogonium. Photographs: **a**, **b**, **c**, **d**-Klaus van de Weyer, **e**, **f**-Chris Carter

Oogonia: The oogonia are 450–560 μm long (without coronula) and 360–450 μm wide (without coronula). The oogonia have 10–12 helical stripes. The coronula is 50–70 μm long and 50–80 μm wide (Table 73.1: Corillion, 1957; Groves and Bullock-Webster, 1920; Krause, 1997; Migula, 1889–1897; Wood, 1965).

Oospores: The oospores are yellow-brown (Corillion, 1957; Groves and Bullock-Webster, 1920; Krause, 1997; Migula, 1889–1897; Wood, 1965). The oospores are

Table 73.1 Oogonia morphometry

Author	Length	Width	Coronula length	Coronula width	Number of helical stripes
Migula (1889–1897)	–	–	–	–	10–12
Groves and Bullock-Webster (1920)	500–560	400–450	50	65	11–12
Corillion (1957)	500–560	400–450	50	65	11–12
Wood (1965)	500–560	360–450	50–70	57–68	11–12
Cirujano et al. (2008)	410–550	300–500	50–70	–	–
Total range	**410–560**	**300–450**	**50–70**	**57–68**	**10–12**

Length in the first column excludes the coronula, its length and width are given separately; all values for length and width in μm

300–400(600) μm long, 250–500 μm wide and have 7–11 striae (Table 73.2: Corillion, 1957; Groves and Bullock-Webster, 1920; Krause, 1997; Migula, 1889–1897; Wood, 1965). A reliable character to separate this species from *S. intricata* is the membrane of the oospore which is smooth (Figs. 73.3 and 73.4).

Antheridia: The diameter of the antheridia is 200–400 μm (Table 73.3).

Bulbils: Absent.

Table. 73.2 Oospore morphometry

Author	Length	Width	Number of striae	Colour	Membrane
Migula (1889–1897)	320–400	–	9–11	Light brown	Smooth
Groves and Bullock-Webster (1920)	350	275	9–10	Light golden-yellow	Smooth
Corillion (1957)	350	275	9–10	Light golden-yellow	Smooth
Wood (1965)	300	500	7–10	Light brown or wine-red	–
Krause (1997)	300–600	270–400	7–9	Brown	–
Urbaniak et al. (2012)	–	–	–	–	Smooth
Haas (1994)	330–400	250–330	–	Light (yellowish) brown to opaque brown black	Smooth
Total range	**300–400(600)**	**250–500**	**7–11**		**Smooth**

Length and width in μm

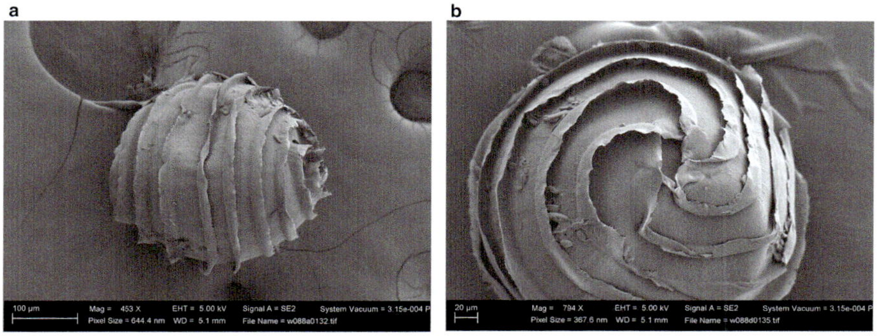

Fig. 73.3 Oospores of *Sphaerochara prolifera*. **a**, **b** Near Barrow Mump, in Burrow Hill Drove Rhyne, Southlake Moor, Bridgwater Somerset, UK. Michelle Casanova & Pat Wolseley. Photographs: Michelle T. Casanova

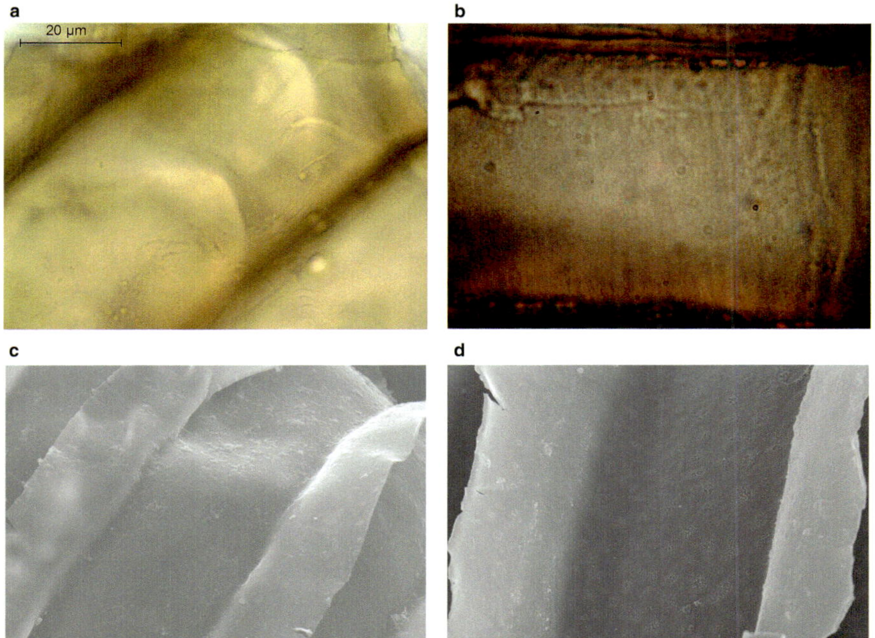

Fig. 73.4 Ornamentation of the oospore membrane of *Sphaerochara prolifera*. **a** Light miscrosope, Vinkenveen, NL; **b** Light miscrosope, ditch near Waalwijk, NL; **c** SEM. from Urbaniak et al. (2012); **d** SEM. Photographs: **a**-Thomas Gregor, **b**-Klaus van de Weyer, **c**-Jacek Urbaniak, **d**-Santos Cirujano Bracamonte

Table. 73.3 Diameter of the antheridia

Author	Diameter (μm)
Corillion (1957)	400
Wood (1965)	(230)255–375
Krause (1997)	200–400
Total range	**200–400**

73.2 Phenology and Reproduction

Sphaerochara prolifera is an annual species. For germination it is important that there is no or little competition with other water plants. Mouronval et al. (2015) describe the development starting in April, fruiting in May and disappearing at the beginning of June. Urbaniak and Gąbka (2014) call *Sphaerochara prolifera* a species of late spring or early summer. In Britain it is a summer species which is difficult to find before June and it continues through till October. According to van Raam (1998) it germinates in autumn or winter. In early spring there is a strong development with fructification of *Sphaerochara prolifera*. It is possible that there is a second germination in late spring. In shallow waters S. *prolifera* can develop large stands for a short period and then disappear for many years, surviving as long-viable oospores. In deeper lakes *Sphaerochara prolifera* can be found in the same sites for many years.

73.3 Habitat Conditions and Ecology

Habitat type: *Sphaerochara prolifera* is found in ditches, streams, pools, temporary waters, oxbow lakes, artificial and natural lakes.

Sediment: *Sphaerochara prolifera* has been reported from different sediments including sand, gravel, clay and organic substrate.

Nutrient conditions: There are only limited data on the nutrient requirements of this species. It has been reported from oligotrophic, mesotrophic and eutrophic waters. The total phosphorus range is reported to vary between 0.008 and 0.142 mg/l (Humberg and Beck, 2006; Nat et al., 1994).

Water chemistry: There are only limited data on the water chemistry of *Sphaerochara prolifera*. It is a species of alkaline water and there are no descriptions from brackish water. The pH range is reported to vary between 7.1 and 8.6 (Doege et al., 2016; Humberg and Beck, 2006; Nat et al., 1994). Conductivity range is reported to vary between 483 and 918 μS/cm (Humberg and Beck, 2006).

Depth range: *Sphaerochara prolifera* occurs in shallow waters and in lakes up to a depth of 9 m (Humberg and Beck, 2006; F. Pätzold & K. van de Weyer own obs.).

Associated species: *Sphaerochara prolifera* occurs in ditches and pools together with other charophytes. These stands have been described as *Tolypelletum proliferate* Guerlesquin 1961. According to Raabe (2008) it can occur in inundated fields, associated with helophytes and *Chara baueri*. *Sphaerochara prolifera* also grows in lakes together with other charophytes (*Nitella opaca, N. mucronata, N. syncarpa, Chara aspera, C. contraria, C. hispida*) (Humberg and Beck, 2006). There are also records from stands of *Elodea nuttallii* (K. van de Weyer own obs.).

73.4 Variability

There is morphological variability according to the season and the habitat. In spring and early summer it is most usual to find large plants with undivided sterile branchlets. In summer there may also be tiny plants with divided sterile branchlets. These plants are hard to separate from *Sphaerochara intricata*.

73.5 Identification and Risk of Confusion

With its dense fertile heads surrounded by long sterile branchlets *Sphaerochara prolifera* is only likely to be confused with *Tolypella* and other *Sphaerochara* species. Like *Sphaerochara intricata* and *S. canadensis* it has conical-acute end-cells of the branchlets compared to *Tolypella* species which all have obtuse/rounded end-cells. *S. prolifera* is most likely to be confused with *S. intricata*; in many situations there is a clear distinction in size but small forms of *S. prolifera* can be more difficult. The sterile branchlets of *S. prolifera*— and *S. canadensis*—are usually simple, while in S. *intricata* there are usually at least some, and often many, divided sterile branchlets. However, as discussed above under "variability", *S. prolifera* can sometimes have divided sterile branchlets, while *S. intricata* may sometimes have undivided sterile branchlets. The tendency for the ends of sterile branchlets to break can also complicate things. Difficult examples can be resolved by examining the membrane of the oospores which is granulate in *S. intricata* and smooth in *S. prolifera*. *S. canadensis* is a circumpolar species and there is no overlap in the distribution areas of *S. prolifera* and *S. canadensis*. Other differences from *S. canadensis* are discussed in more detail under that species.

Sphaerochara prolifera can also be confused with *Nitella translucens* because of their similar large stature. *Nitella translucens* has branchlets which appear superficially to be undivided and single-celled but are tipped by 2–4 very short, 2-celled dactyls. In *Sphaerochara prolifera* the fertile branchlets are monopodially branched and the simple sterile branchlets are multi-celled with normally elongate end-cells. There is also a marked contrast in habitat with *Sphaerochara prolifera* being a species of hard water while *Nitella translucens* occurs in soft water.

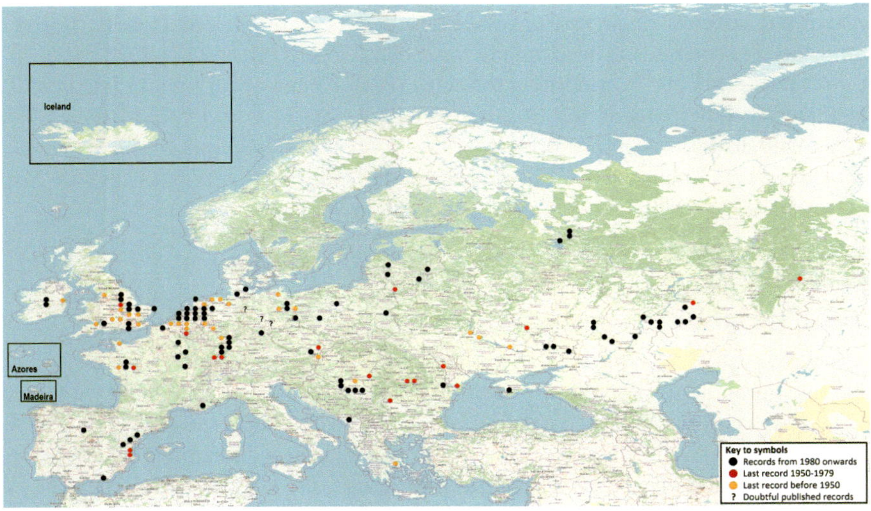

Fig. 73.5 European distribution map of *Sphaerochara prolifera*. For worldwide distribution see Korsch (2018). Black dots are records from 1980 on, red dots from 1950 to 1979 and yellow dots before 1950; unconfirmed/doubtful published records (if applicable) are marked with "?" on top. Map prepared by Heiko Korsch

73.6 Distribution

See Fig. 73.5.

According to Corillion (1957) and Korsch (2018), *Sphaerochara prolifera* is known from Asia, Europe, and both North and South America. *Sphaerochara prolifera* has been recorded from many countries in Europe, but there are no records from most countries in Northern Europe.

73.7 Status and Threats

Sphaerochara prolifera is a pioneer species of open habitats in ditches and pools. For the germination of this annual species it is important that there is little or no competition with other water plants. In deeper lakes the populations of this species seem to be more stable especially in the deeper parts where there is an incomplete cover of vegetation.

Sphaerochara prolifera is listed in many different Red Lists/Red Data Books: the Balkans (EN), Czech Republic (VS), Germany (EN), Poland (CR), Ukraine (VU), Lithuania (EN), Netherlands (VU), Iberia (CR), Ireland (EN) and Britain (VU). It is Extinct in Upper Austria. It also disappeared in Belgium.

Within ditches and pools it is necessary to maintain areas which are not covered with other macrophytes. It can be useful to remove other macrophytes to facilitate the germination of *S. prolifera*. In general, all measures to reduce/prevent eutrophication are recommended. This also includes maintaining fish communities which are typical for the specific water body type.

73.8 Remarks

Obviously *Sphaerochara prolifera* has been confused with *Sphaerochara intricata*. Both species have been recorded together from some sites.

References

Cirujano, S., Cambra, J., Sánchez Castillo, P. M., Meco, A., & Flor Arnau, N. (2008). *Flora ibérica. Algas continentales. Carófitos (Characeae)*. Real Jardín Botánico, Madrid. [in Spanish].

Corillion, R. (1957). Les Charophycées de France et d'Europe Occidentale. *Bulletin de la Société Scientifique de Bretagne, 32*(Hors série I & II), 1–499. [in French].

Doege, A., van de Weyer, K., Becker, R., & Schubert, H. (2016). Bioindikation mit Characeen. In Arbeitsgruppe Characeen Deutschlands, *Armleuchteralgen: Die Characeen Deutschlands* (pp. 97–138). Springer. [in German].

Groves, J., & Bullock-Webster, G. R. (1920). *The British Charophyta. Volume I. Nitelleae.* Ray Society.

Haas, J. N. (1994). First identification key for charophyte oospores from central Europe. *European Journal of Phycology, 29*, 227–235.

Humberg, B., & Beck, M. (2006). Der Makrophytenbestand in ausgewählten Baggerseen der Oberrheinaue. *LUBW, Oberirdische Gewässer, Gewässerökologie, 102*, 1–348. [in German].

Korsch, H. (2018). The worldwide range of the Charophyte species native to Germany. *Rostocker Meeresbiologische Beiträge, 28*, 45–96.

Krause, W. (1997). Charales (Charophyceae) Süßwasserflora von Mitteleuropa. In H. Ettl, G. Gärtner, H. Heynig, & D. Mollenhauer (Eds.), *Süßwasserflora von Mitteleuropa 18*. Fischer. [in German].

Migula, W. (1889–1897). Fünfter Band: Die Characeen. In A. Fischer, E. Fischer, F. Hauck, G. Limpricht, C. Luerssen, W. Migula, H. Rehm, P. Richter & G. Winter (Eds.), *Dr. L. Rabenhorst's Kryptogamenflora von Deutschland, Oesterreich und der Schweiz* (ed. 2). Eduard Kummer. [in German].

Mouronval, J. B., Baudouin, S., Borel, N., Soulié-Märsche, I., Klesczewski, M., & Grillas, P. (2015). *Guide des Characées de France méditerranéenne*. Office national de la Chasse et de la Faune Sauvage. [in French].

Nat, E., Simons, J., de la Haye, M. A. A., & Coops, H. (1994). *Historisch en actueel verspreidingsbeeld van kranswieren in Nederland in samenhang met waterkwaliteitsfactoren*. RIZA werkdocument 94.148X. Lelystad. [in Dutch].

Raabe, U. (2008). Die Sprossende Baumleuchteralge, *Tolypella prolifera* (Ziz ex A. Braun) Leonh., in Brandenburg wieder aufgefunden. *Verhandl. Bot. Ver. Berlin Brandenburg, 141*, 145–152. [in German].

Urbaniak, J., & Gąbka, M. (2014). *Polish Charophytes. An illustrated guide to identification.* Uniwersytet Przyrodniczy we Wroclawiu.

Urbaniak, J., Langangen, A., & van Raam, J. (2012). Oospore wall ornamentation in the genus *Tolypella* (Charales, Charophyceae). *Journal of Phycology, 48*, 1538–1545.
van Raam, J. C. (1998). *Handboek Kranswieren.* Chara boek. [in Dutch].
van de Weyer, K., & Schmidt, C. (2018). Bestimmungsschlüssel für die aquatischen Makrophyten (Gefäßpflanzen, Armleuchteralgen und Moose) in Deutschland: Band 1: Bestimmungsschlüssel, ed. 2. *Landesamt für Umwelt (LfU) Brandenburg, Potsdam, Fachbeiträge des LfU Brandenburg* 119: 1–180. [in German].
Wood, R. D. (1965). *Monograph of the Characeae.* J. Cramer.

Chapter 74
Tolypella glomerata

Angelo Troia and Klaus van de Weyer

Scientific name: *Tolypella glomerata* (Desv. in Loisel) Leonh. in Lotos 13: 129. 1863

≡ *Chara glomerata* Desv. in Loisel in Loisel., Not. Fl. France: 135. 1810
Lectotype (Wood, 1965: 733): in rivulis/[no date]/Desvaux (PC0143220)

= *Tolypella glomerata* (Desv. in Loisel) Leonh. f. *tenuior* A. Braun ex Mig., Charac. Deutschl.: 232. 1890
Type not designated

= *Tolypella glomerata* (Desv. in Loisel) Leonh. f. *littorea* Hy in Mém. Soc. Bot. France: 20. 1913
Type not designated.

Local names: DANISH: Nøgle-Redetråd; DUTCH: Klein boomglanswier; ENGLISH: Clustered Stonewort; FRENCH: Tolypelle agglomérée; GERMAN: Knäuel-Baumleuchteralge; POLISH: Rozsocha skupiona; RUSSIAN: Толипелла клубочковая; SWEDISH: Trubbrufse; UKRAINIAN: Толіпела клубочкова.

74.1 Morphology

See Figs. 74.1 and 74.2.

A. Troia (✉)
Dept. STEBICEF, University of Palermo, Via Archirafi 38, 90123 Palermo, Italy
e-mail: angelo.troia@unipa.it

K. van de Weyer
lanaplan, Lobbericher Str. 5, 41334 Nettetal, Germany
e-mail: klaus.vdweyer@lanaplan.de

H. Schubert et al. (eds.), *Charophytes of Europe*,
https://doi.org/10.1007/978-3-031-31898-6_74

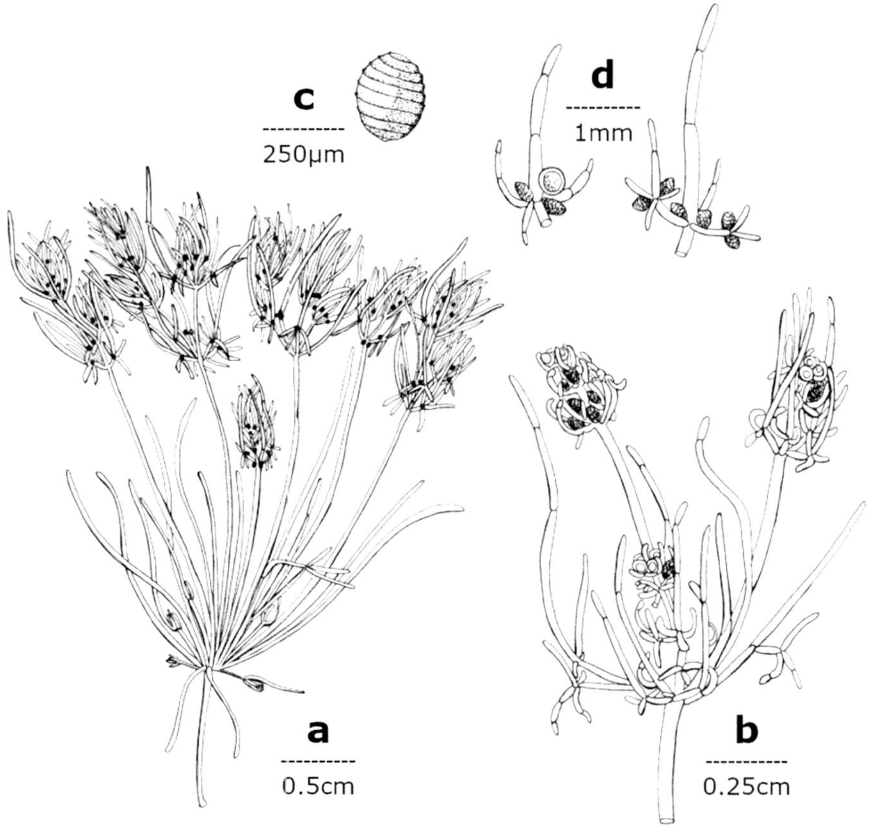

Fig. 74.1 *Tolypella glomerata*. **a** Habitus; **b** Shoot tip; **c** Oospore, **d** Fertile branchlets. Drawings after Werner Krause, modified by Dagmar Wassong in van de Weyer and Schmidt (2018)

Habitus: Small, highly branched, green or grey-green plants, which often are (heavily) encrusted and always appear dull (not shiny). The very dense, irregular interwoven heads are typical. Young plants have very long shooty protonemata ("Proembryones"), whose terminal filaments fall off in old plants. Internodes are 1–2 times the length of the branchlets.

Size: *Tolypella glomerata* is small to medium-sized, from 2 to 50 cm high.

Main axis: Variable in diameter, from 0.3 to 1 mm, with lower internodes up to 17 cm long, the upper mostly short.

Cortex, spine-cells, stipulodes: Absent.

Branchlets: Sterile and fertile whorls are different (dimorphism). The first node of the main axis produces 6 sterile branchlets, and 2–6 secondary axes; the sterile branchlets (6–8 cm long) are undivided and consist of a series of 3–5 elongated cells.

Fig. 74.2 *Tolypella glomerata*, details. **a** Habitus, **b**, **c** Details of the fertile shoots; **d** Detail of an antheridium and an oogonium on the same node; **e** Stalked antheridium; **f** Oogonium. Photographs: **a**-Klaus van de Weyer, **b**-Santos Cirujano, **c**, **d**-Chris Carter, **e**, **f**-Dominique Auderset

The fertile whorls are produced by the secondary axes. They are short and grouped into fertile heads. The fertile branchlets consist in a central row of cells (the "rachis") that is a succession of internodes and nodes. These nodes produce the gametangia as well as 3 rays of 2–3 cells. All terminal cells are rounded, obtuse.

Bract-cells, bracteoles, bractlets: Absent.

Sexuality: The species is monoecious. Gametangia conjoined at the first branchlet nodes.

Oogonia: 1–3(6) at a node. Typically, at a node you find either two oogonia and an antheridium, or 3 oogonia. Table 74.1 summarises references to the morphometry of the oogonia.

Oospores (Fig. 74.3): Mature oospores are golden brown to dark brown, with marked striae. The oospore-membrane is regularly/irregularly reticulate; it was described as "vermiculate" by van Raam (1998), "foveolate/reticulate" by Mouronval et al. (2015), whereas Corillion (1975) described it as having "granulations linéaires". Table 74.2 gives an overview about the available data.

Antheridia: 1–2(3) at a node, small, sessile or stalked. Table 74.3 gives an overview of the published morphometric data.

Table 74.1 Oogonia morphometry

Author	Length	Width	Coronula length	Coronula width	Number of helical stripes
Wood (1965)	(330)425–700[a]	(285)310–545	35–50	65–105	9–10(11)
Groves and Bullock-Webster (1920)	425–500	310–375	35–50	65–90	9–10
Corillion (1957)	325–580	250–375	35–60	65–90	9–10
Krause (1997)	350–600	300–500	35–50	65–100	9–10
Urbaniak and Gąbka (2014)	330–420	280–390	–	–	–
Total range	**325–600**	**250–500**	**35–60**	**65–105**	**9–10(11)**

Length in the first column, is excluding the coronula, whose length and width are given separately; all values for length and width in μm

[a] Including coronula

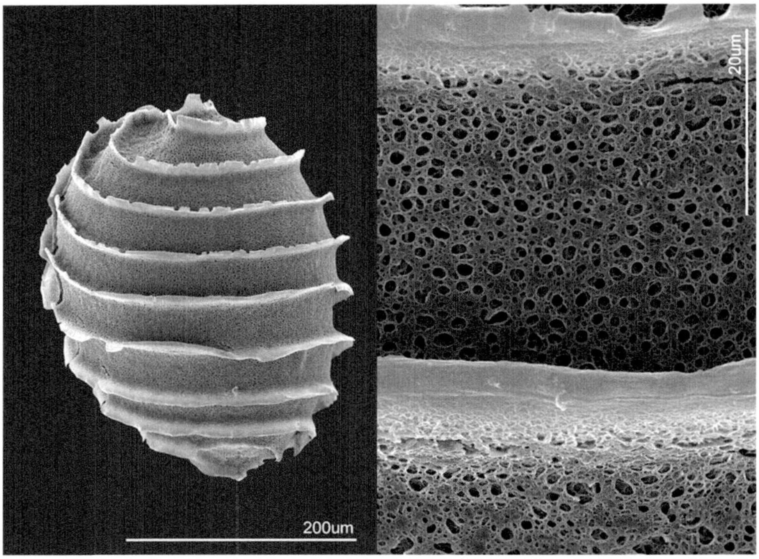

Fig. 74.3 Oospore of *Tolypella glomerata*. Side view (left) and detail of the membrane (right). Scale bars 200 μm (left), 20 μm (right). Photographs (SEM): Santos Cirujano

Bulbils: Absent.

Table 74.2 Oospore morphometry

Author	Length	Width	Number of striae	Colour	Oospore membrane
Migula (1889–1897)	280–350	–	7–9	–	–
Groves and Bullock-Webster (1920)	300–375	250–300	7–9	Yellow to orange-brown or cold dull brown	With linear granulations
Corillion (1957)	200–380	175–325	7–9	Brown to orange-brown	With linear granulations
Wood (1965)	225–475	220–320	7–8(9)	Yellow or golden brown to dull or dark brown to blackish	With dots
Krause (1997)	225–400	200–350	–	(Dark) brown	–
Bazzichelli and Abdelahad (2009)	225–400	200–350	–	Dark brown	–
Urbaniak et al. (2012, Table S2)	275–377	231–310	(6)7–8(9)	–	Fovate
Urbaniak and Gąbka (2014)	275–365	240–310	–	Light brown to brown	–
Mouronval et al. (2015)	(220)300–380	180–350	7–9	Yellowish brown	Foveolate-reticulate
Total range	**200–475**	**175–350**	**7–9**		

All values for length and width in μm

Table 74.3 Antheridia morphometry

Author	Diameter (μm)
Migula (1889–1897)	300–500
Groves and Bullock-Webster (1920)	325–375
Corillion (1957)	220–450
Wood (1965)	325–425
Krause (1997)	325–435
Bazzichelli and Abdelahad (2009)	325–435
Urbaniak and Gąbka (2014)	320–460
Total range	**220–500**

74.2 Phenology and Reproduction

According to Olsen (1944, but see also van Raam, 1998; Korsch, 2013), the species is both winter-annual and summer-annual: the winter-annual germinates in autumn and has ripe oospores in March–June (own observations), and the summer-annual germinates in spring has ripe oospores in August and September. Around Geneve, Switzerland it is "la plus précoce des characées" (ripe oospores in May; in southern France ripe oospores in March) (Auderset Joye and Rey-Boissezon, 2013). During the summer, *T. glomerata* often disappears in shallow waters, whereas in deeper lakes it can be found under the thermocline.

74.3 Habitat Conditions and Ecology

Habitat type: *Tolypella glomerata* shows a wide ecological amplitude. A freshwater species, tolerating moderate salinity (15–8282 mg/l CL^-, Cirujano et al., 2013), it is found in shallow alkaline waters, in ditches, ponds, lakes and temporary pools.

Sediment: *Tolypella glomerata* is found mainly on sandy substrates, but it is also observed on gravelly and loamy substrates.

Nutrient conditions: *Tolypella glomerata* inhabits oligo- to eutrophic waters.

Water chemistry: *Tolypella glomerata* usually occurs in hard water. Ca^{2+} ranges from 31 to 520 mg/l, Mg^{2+} from 14 to 413 mg/l. The electrical conductivity range is 41 to 16,820 μS/cm but usually is between 330 and 2,080 μS/cm. The pH range extends from 3.8 to 10.5, but usually is between 7.0 and 8.5. *T. glomerata* is also found in brackish inland sites (Na^+ range from 6 to 5100 mg/l).

Depth range: *Tolypella glomerata* occurs in shallow waters as well as in lakes up to 21 m depth (van de Weyer, 2016).

Associated species: *Tolypella glomerata* is one of the characteristic species of the association *Charo-Tolypelletum glomeratae* Corillion, 1957. It is found also in the *Charetum canescentis* (in brackish water). In addition, *Tolypella glomerata* can be found in other charophyte communities and communities dominated by aquatic vascular plants.

74.4 Variability

Tolypella glomerata is morphologically very variable depending on location and season. There are both small and stocky (e.g., *T. glomerata* f. *tenuior*) as well as larger plants with prolonged internodes and "dissolved" heads from depths > 10 m.

74.5 Identification and risk of confusion

Tolypella glomerata can be confused with *Nitella opaca* and *N. mucronata:* the latter has regular heads while the heads of the former are not regular. *Tolypella glomerata* can be confused with other *Tolypella* and *Sphaerochara* species. *Sphaerochara intricata, S. prolifera and S. canadensis* have conical-acute end-cells of the branchlets. The other *Tolypella* species have obtuse end-cells of the branchlets. *Tolypella normanniana* has no sterile branchlets while *T. glomerata, T. nidifica* and *T. salina* have fertile and sterile branchlets.

Tolypella glomerata can be confused in brackish waters with *T. nidifica* and *T. salina* . *T. glomerata* is dull (not shiny) while *T. nidifica* and *T. salina* are shiny. These three species can also be separated by the size and colour of the oospores (>0.40 mm long in *T. nidifica*, <0.40 mm and chestnut-brown in *T. salina*, <0.40 mm and yellowish-brown in *T. glomerata*, according to Mouronval et al., 2015) and also by the ornamentation of the oospore-membrane, smooth in *T. salina*, with irregular pits in *T. nidifica*, foveolate/reticulate in *T. glomerata* (Mouronval et al., 2015).

74.6 Distribution

See Fig. 74.4.

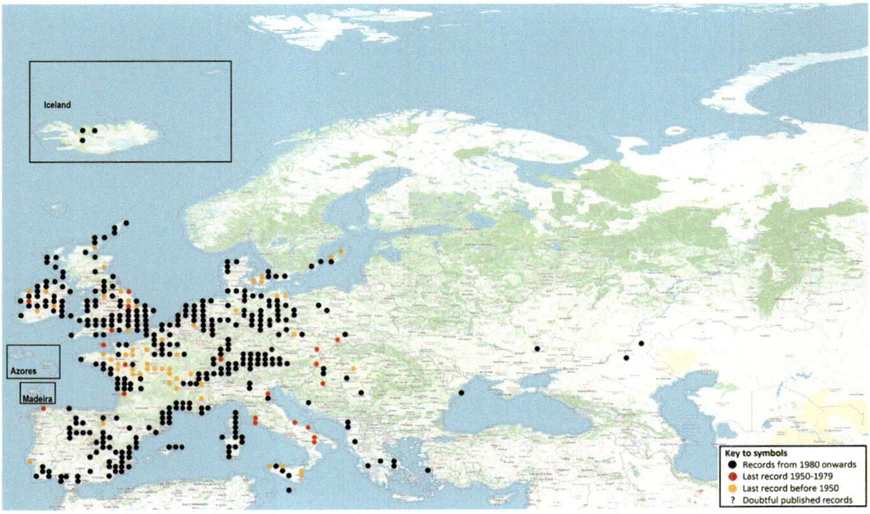

Fig. 74.4 European distribution of *Tolypella glomerata*. For worldwide distribution see Korsch (2018). Black dots are records from 1980 on, red dots from 1950 to 1979 and yellow dots before 1950; unconfirmed/doubtful published records (if applicable) are marked with "?" on top. Map prepared by Heiko Korsch

A subcosmopolitan species, unequally distributed, mainly in western Europe and in the Mediterranean region (Muller et al., 2017). According to Corillion (1957) *Tolypella glomerata* occurs (in addition to most of Europe) in Africa, Asia, America, Australia.

74.7 Status and Threats

It is listed as CR in Poland (Urbaniak and Gąbka, 2014) and in Czech Republic (Caisová and Gąbka, 2009). It has been assessed as EN in Switzerland (Auderset Joye and Rey-Boissezon, 2013), in the Balkans (Blaženčić et al., 2006) and in Sweden (Becker, 2016). It is listed as VU in Denmark and Netherlands (Becker, 2016).

74.8 Remarks

Problematic is *Tolypella glomerata* f. *littorea*, found in low-saline brackish waters in western Europe and described as an intermediate form between *T. glomerata* and *T. nidifica* by Hy (1913). However, irrespective of the debates outlined in Chap. 76 (*Tolypella nidifica*), this form seems to belong rather to *T. glomerata* as described above by all the main characters as being dull and having a reticulate oospore membrane ("finement ponctuée et non vermiculée, presque lisse", Hy, 1913).

Acknowledgements We thank Uwe Raabe and Elisabeth Lambert for valuable comments.

References

Auderset Joye, D., & Rey-Boissezon, A. (2013). *Les Characées de Genève et environs. Distribution et Ecologie.* Groupe d'Ecologie aquatique, Institut Forel et Institut des Sciences l'Environnement, Université de Genève. [in French].

Bazzichelli, G., & Abdelahad, N. (2009). *Alghe d'acqua dolce d'Italia. Flora analitica delle Caroficee.* Sapienza Università di Roma. Ministero dell'Ambiente e della Tutela del territorio e del mare. [in Italian].

Becker, R. (2016). Gefährdung und Schutz von Characeen. In Arbeitsgruppe Characeen Deutschlands (Ed.), *Armleuchteralgen: Die Characeen Deutschlands* (pp. 149–191). Springer. [in German].

Blaženčić, J., Stevanović, B., Blaženčić, Ž., & Stevanović, V. (2006). Red data list of Charophytes in the Balkans. *Biodiversity and Conservation, 15*, 3445–3457. https://doi.org/10.1007/s10531-005-2008-5

Caisová, L., & Gąbka, M. (2009). Charophytes (Characeae, Charophyta) in the Czech Republic: Taxonomy, autoecology and distribution. *Fottea, 9*, 1–43. https://doi.org/10.5507/fot.2009.001

Cirujano B. S., Guerrero M. N., & García Murillo, P. (2013). The genus *Tolypella* (A. Braun) A. Braun in the Iberian Peninsula. *Botany Letters, 160*, 121–129. https://doi.org/10.1080/12538078.2013.801321
Corillion, R. (1957). Les Charophycées de France et d'Europe Occidentale. *Bulletin de la Société Scientifique de Bretagne, 32*(Hors série I & II), 1–499. [in French].
Corillion, R. (1975). *Flore et végétation du Massif Armoricain. T. IV. Flore des Charophytes (Characées) du Massif Armoricain et des contrées voisines d'Europe occidental*e. Jouve. [in French].
Groves, J., & Bullock-Webster, G. R. (1920). *The British Charophyta. Volume I. Nitelleae.* Ray Society.
Hy, F. (1913). Les Characées de France. *Bulletin de la Société Botanique de France, 60*(sup 1), 1–47. [in French].
Korsch, H. (2013). Die Armleuchteralgen (Charophyceae) Sachsen-Anhalts. *Berichte Des Landesamtes Für Umweltschutz Sachsen-Anhalt, 1*(2013), 1–85. [in German].
Korsch, H. (2018). The worldwide range of the Charophyte species native to Germany. *Rostocker Meeresbiologische Beiträge, 28*, 45–96.
Krause, W. (1997). Charales (Charophyceae) Süßwasserflora von Mitteleuropa. In H. Ettl, G. Gärtner, H. Heynig, & D. Mollenhauer (Eds.), *Süßwasserflora von Mitteleuropa 18*. Fischer. [in German].
Migula, W. (1889–1897). Fünfter Band: Die Characeen. In A. Fischer, E. Fischer, F. Hauck, G. Limpricht, C. Luerssen, W. Migula, H. Rehm, P. Richter, & G. Winter (Eds.), *Dr. L. Rabenhorst's Kryptogamenflora von Deutschland, Oesterreich und der Schweiz* (2nd ed.). Eduard Kummer. [in German].
Mouronval, J. B., Baudouin, S., Borel, N., Soulié-Märsche, I., Klesczewski, M., & Grillas, P. (2015). *Guide des Characées de France méditerranéenne* (211 pp.). Office national de la Chasse et de la Faune Sauvage. [in French].
Muller, S. D., Rhazi, L., & Soulié-Märsche, I. (2017). Diversity and distribution of Characeae in the Maghreb (Algeria, Morocco, Tunisia). *Cryptogamie, Algologie, 38*, 201–251. https://doi.org/10.7872/crya/v38.iss3.2017.201
Olsen, S. (1944). Danish Charophyta. *Det Kongelige Danske Videnskabernes Selskab, Biologiske Skrifter, 3*(1), 1–240, pl. 1–2. [in Danish].
van Raam, J. C. (1998). *Handboek Kranswieren.* Chara boek. [in Dutch].
Urbaniak, J., Langangen, A., & van Raam, J. (2012). Oospore wall ornamentation in the genus *Tolypella* (Charales, Charophyceae). *Journal of Phycology, 48*, 1538–1545. https://doi.org/10.1111/jpy.12007
Urbaniak, J., & Gąbka, M. (2014). *Polish Charophytes. An illustrated guide to identification.* Uniwersytet Przyrodniczy we Wroclawiu. [in Polish].
van de Weyer, K. (2016). *Tolypella glomerata.* In Arbeitsgruppe Characeen Deutschlands (Ed.), *Armleuchteralgen: Die Characeen Deutschlands* (pp. 514–522). Springer. [in German].
van de Weyer, K., & Schmidt, C. (2018). Bestimmungsschlüssel für die aquatischen Makrophyten (Gefäßpflanzen, Armleuchteralgen und Moose) in Deutschland: Band 1: Bestimmungsschlüssel, 2nd ed. *Landesamt für Umwelt (LfU) Brandenburg, Potsdam, Fachbeiträge des LfU Brandenburg* 119: 1–180. [in German].
Wood, R. D. (1965). *Monograph of the Characeae.* J. Cramer.

Chapter 75
Tolypella hispanica

Angelo Troia, Ralf Becker, and Maria A. Rodrigo

Scientific name: ***Tolypella hispanica*** **Nordst. ex Allen, Charac. America 1: 51. 1888**

Lectotype (Wood, 1965: 733): [Spain] Guadalhorce à Pizarra/May 30, 1883/N. Hjalmar Nilsson 27 (LD).

= *Tolypella giennensis* Reyes y Prósper, Carofit. España: 83. 1910
Lectotype (Cirujano & Rodrigo, 2019: 2): [Spain] Mancha Real (Jaén)/17 Julio 1906/ E. Reyes Prósper (MA-ALGAE 3031).

Nomenclatural comment: Daily (1954: 145) cited only a specimen collected in Spain and Nilsson's collecting number ("n. 27"), but did not specify a herbarium, thus indicating all specimens of that gathering as syntypes. Wood (1965: 733) specified that the specimen designated as "holotype" by Daily was in LD.

Local names: FRENCH: Tolypelle d'Espagne; RUSSIAN: Толипелла испанская; SPANISH: Corales de rana; UKRAINAIN: Толіпела іспаньска.

A. Troia (✉)
Dept. STEBICEF, University of Palermo, Via Archirafi 38, 90123 Palermo, Italy
e-mail: angelo.troia@unipa.it

R. Becker
Am Schützenplatz 58, 26121 Oldenburg, Germany
e-mail: becker.r@posteo.de

M. A. Rodrigo
Institut Cavanilles de Biodiversitat i Biologia Evolutiva, Universitat de València, C/ Catedràtic José Beltrán, 2, 46980 Paterna, València, Spain
e-mail: maria.a.rodrigo@uv.es

H. Schubert et al. (eds.), *Charophytes of Europe*,
https://doi.org/10.1007/978-3-031-31898-6_75

75.1 Morphology

See Figs. 75.1 and 75.2.

Habitus: *Tolypella hispanica* is easy to recognise with its abundant big orange/reddish antheridia visible with the naked eye. Plants are light green to brownish green, sometimes heavily encrusted. Branching of the axis starts at the first node, which may show up to 30 secondary axes. Sterile whorls are spread, with a few undivided branchlets, which may be short or long but are particularly elongated (5 cm) on male plants. Small plants in very shallow waterbodies can show a compact bushy growth. It differs from all other European *Tolypella* in being dioecious. Female plants are often smaller than male plants.

Size: Plant 3–15(30) cm long. In Greece, plants larger than 30 cm long have been observed.

Main axis: The axis (0.7–1 mm in diameter) shows a first long internode followed by numerous others, short and close to each other.

Cortex: Absent.

Spine-cells: Absent.

Stipulodes: Absent.

Branchlets: Whorls of aspect relatively dimorphic, composed by 5–7(10) branchlets. Mainly or entirely non-fertile whorls are unusual. Fertile branchlets are formed in

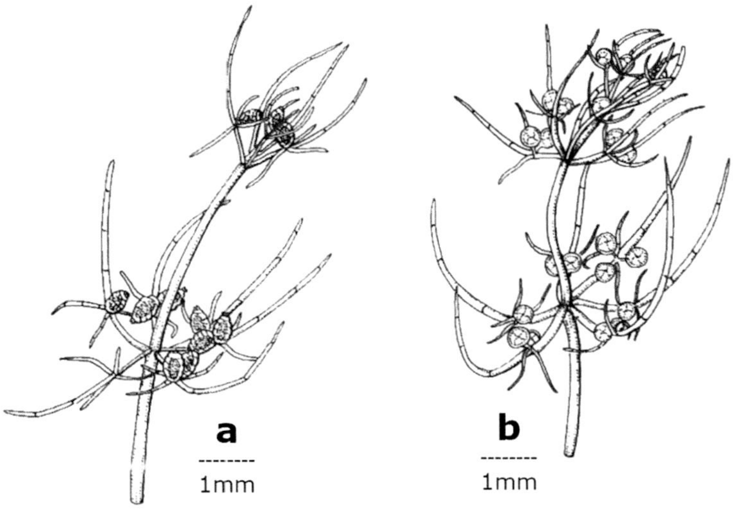

Fig. 75.1 *Tolypella hispanica*. **a** Shoot tip of a female plant with oogonia; **b** Shoot tip of a male plant with antheridia. Drawings by Marta Chirino

Fig. 75.2 *Tolypella hispanica*, details. **a** Shoot tip of a female plant with oogonia; **b** Shoot tip of a male plant with antheridia; **c** Male individuals with showy and abundant antheridia. Photographs: **a**, **b**-Santos Cirujano, **c**-Ralf Becker

great number (8–10 per node) on secondary axes. They are very short, forming dense fertile heads. All terminal cells present are obtuse, rounded, and shorter than the preceding cells. Female plants have thinner terminal cells.

Bract-cells, Bracteoles: Absent.

Bractlets: Not observed.

Sexuality: The species is dioecious.

Oogonia: Small (about 400 μm), visible as black clusters when ripe; clustered in numbers of 4–5(10). Table 75.1 gives an overview of the published data on the morphometry of oogonia.

Oospores (Fig. 75.3): Mature oospores are dark brown to black (yellowish/light brown in young oospores), with thick and prominent (20–23 μm high) flanges and reticulate oospore-membrane. Table 75.2 gives an overview of the available morphometrical data.

Table 75.1 Oogonia morphometry

Author	Length	Width	Coronula length	Coronula width	Number of helical stripes
Cirujano et al. (2008)	300–450	250–350	30	60	–
Corillion (1957)	430	300	30	60	8–9
Krause (1997)	450	300–400	30	60	7–9
Total range	**300–450**	**250–400**	**30**	**60**	**7–9**

Length in the first column is including the coronula; its length and width are given separately in the 3rd and 4th columns, all values for length and width in μm

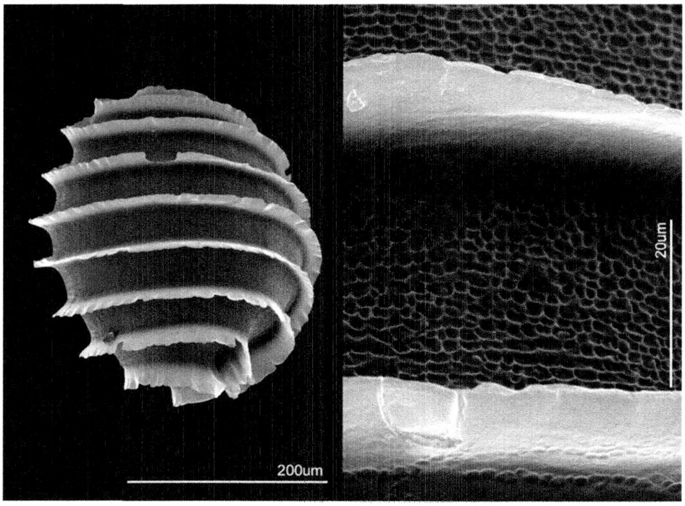

Fig. 75.3 Oospore of *Tolypella hispanica*. Side view (left) and detail of the membrane (right). Scale bars 200 μm (left) and 20 μm (right). Photographs (SEM): Santos Cirujano

Table 75.2 Oospore morphometry

Author	Length	Width	Number of striae	Colour	Oospore-membrane
Migula (1889–1897)	225–250(300)	220–225	7–8	Brown	–
Corillion (1957)	240–315	155–275	6–7	Brown to blackish	"Finement ponctuée"
Krause (1997)	250–300	150–300	6–7	Brown	–
Cirujano et al. (2008)	240–315	150–285	–	Yellow to dark brown to black	Reticulate/foveolate
Bazzichelli and Abdelahad (2009)	250–300	150–300	6–7	Brown	–
Urbaniak et al. (2012, Table S2)	252–317	180–208	5–7	–	Reticulate
Total range	**225–317**	**150–300**	**5–7(8)**		

All values for length and width in μm

Antheridia: 600–1200 μm in diameter, visible with the naked eye, often stalked. Colour from yellow orange to red orange. They can be found solitary, geminated or in groups of 3. Table 75.3 gives an overview of the published morphometrical data.

Bulbils: Not found.

Table 75.3 Antheridia morphometry

Author	Diameter (μm)
Migula (1889–1897)	700–750
Reyes Prósper (1910)	700–750
Corillion (1957)	700–1040
Krause (1997)	700–1000
Cirujano et al. (2008)	650–1200
Bazzichelli and Abdelahad (2009)	700–1000
Mouronval et al. (2015)	650–1200
R. Becker (unpubl.)	600–1000
Total range	**600–1200**

75.2 Phenology and Reproduction

It is an early-spring to spring species which develops from February to April in temporary waterbodies, and decays afterwards due to increasing temperature or drying out. The fructification process has also been described to take place between March and April (until May). Phenology and reproduction is clearly influenced by the precipitation regime and water temperature.

75.3 Habitat Conditions and Ecology

Habitat type: Typically a brackish-water species, *Tolypella hispanica* colonises the edges of lakes, salt marshes, lagoons, ponds, puddles and other shallow endorheic, seasonal and saline wetlands, including small vernal pools in the coastal sand dune systems.

Sediment: *Tolypella hispanica* is found mainly on sandy-silty substrates.

Nutrient conditions: *Tolypella hispanica* occurs in oligotrophic to eutrophic waters. Based on Becker (2019) from nine Sardinian waterbodies the following values are known (single measurements): TP 0.027–0.169 mg/l (mean value 0.066 mg/l), PO_4^{3}–P 0.01–0.041 mg/l, TN 1.16–3.37 mg/l (mean value 1.83 mg/l), NO_3–N 0.39–1.39 mg/l and NH_4–N <0.03–0.29 mg/l.

Water chemistry: *Tolypella hispanica* grows in alkaline waters with moderate/high pH values (7.1–10.5). From oligo-mesohaline waters (0.5–18‰) to hyperhaline conditions (45‰). In Sardinia (Italy) salinity in *T. hispanica*-habitats ranged mostly from 2.1 to 13.2 psu with a minimum of 1.0 and a maximum of 29.2 psu (mean value 7.0 psu); conductivity from 2.2 to 46.2 mS/cm and hardness from 13.5 to 224.2 °dH (R. Becker own data). The conductivity range in Spanish habitats is similar: 1.4–48.9 mS/cm (Cirujano et al., 2013). When growing in calcareous waters the specimens are heavily encrusted.

Depth range: Shallow water (< 50 cm).

Associated species: *Tolypella hispanica* can occur in monospecific stands or in association with other *Characeae* (Fig. 75.4). In oligo- and mesohaline conditions it can be found with *Tolypella salina, T. nidifica, Lamprothamniumpapulosum* , *Chara galioides, C. canescens* and *C. aspera;* at the lowest salinities with *T. glomerata.* It can be found, rarely, with some angiosperms such as *Ruppia* spp., *Stuckenia pectinata* or *Zannichellia pedunculata*, or with the bryophyte *Riella helicophylla*.

Fig. 75.4 Habitat appearance of *Tolypella hispanica*. Dense stand with showy male plants covered with antheridia, together with *T. nidifica* (Sardinia, former lagoon in Pittulongu, left panel) and a submerged view of a male plant in the same place (right panel). Photographs: Ralf Becker (left panel), Frank Pätzold (right panel)

75.4 Variability

Variability in this species relates mainly to the height of the plants and the development of their branchlets.

75.5 Identification and Risk of Confusion

Tolypella hispanica can be confused with other *Tolypella* species, but when fertile the dioecious features prevent any confusion, except with the central-Eurasian *T. mongolica* which has complex protruding elements in the fossae of the oospores and has a different phylogenetic affinity (Romanov et al., 2024). Moreover, the antheridia are approximately twice the size of the other *Tolypella* species. Older female plants could be confused with *Nitella* spp. (i.e. *N. tenuissima*), but the main axes are much thicker than in *N. tenuissima* (U. Raabe pers. obs.). At first glance, male plants could be confused with the freshwater species *Nitella opaca*, whose antheridia can reach diameters of at least 800 μm: compared with *N. opaca* the oogonia and oospores of the oligo- to mesohaline *T. hispanica* are much smaller; but both species only very rarely grow together in coastal pools (R. Becker pers. obs.).

75.6 Distribution

See Fig. 75.5.

According to the current available information, it is a mainly Mediterranean species, present in the south of Europe [from the Iberian Peninsula (Cirujano Bracamonte

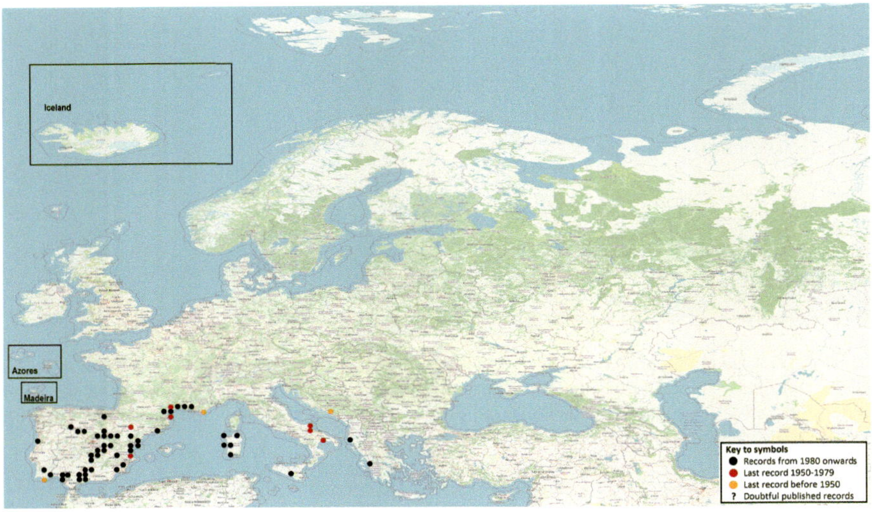

Fig. 75.5 European distribution of *Tolypella hispanica*. Black dots are records from 1980 on, red dots from 1950 to 1979 and yellow dots before 1950; unconfirmed/doubtful published records (if applicable) are marked with "?" on top. Map prepared by Heiko Korsch

et al., 2013), southern France (Corillion, 1961) and Italy (Becker, 2019; Corillion and Guerlesquin, 1963) to Greece], and in North Africa, where it is relatively frequent in the three Maghreb countries (29 localities from 2 to 2090 m a.s.l.) (Muller et al., 2017). Disjunct areas are reported in Asia (Iran, Pakistan, West Siberia) (Cirujano et al., 2008; Corillion, 1957; Romanov et al., 2019).

75.7 Status and Threats

Critically endangered (CR) in the Balkans (Blaženčić et al., 2006). Endangered in Greece (U. Raabe pers. comm.). Rare in Italy (Bazzichelli and Abdelahad, 2009), but common in Sardinia (Becker, 2019). In Spain, its status is Least Concern.

The main threats are habitat destruction, with desiccation of wetlands and eutrophication.

Acknowledgements We thank Uwe Raabe for valuable comments.

References

Bazzichelli, G., & Abdelahad, N. (2009). *Alghe d'acqua dolce d'Italia. Flora analitica delle Caroficee*. Sapienza Università di Roma. Ministero dell'Ambiente e della Tutela del territorio e del mare. Roma, Centro Stampa Università. [in Italian].

Becker, R. (2019). The Characeae (Charales, Charophyceae) of Sardinia (Italy): Habitats, distribution and conservation. *Webbia, 74*, 83–101. https://doi.org/10.1080/00837792.2019.1607497

Blaženčić, J., Stevanović, B., Blaženčić, Ž, & Stevanović, V. (2006). Red data list of Charophytes in the Balkans. *Biodiversity and Conservation, 15*, 3445–3457. https://doi.org/10.1007/s10531-005-2008-5

Cirujano, S., Cambra, J., Sánchez Castillo, P. M., Meco, A., & Flor Arnau, N. (2008). *Flora ibérica. Algas continentales. Carófitos (Characeae)*. Real Jardín Botánico. [in Spanish].

Cirujano, S., & Rodrigo, M. A. (2019). Revisión del herbario de Eduardo Reyes Prósper y tipificación de tres especies de carófitos (Charophyceae, Characeae) presentes en el mismo. *Anales del Jardín Botánico de Madrid, 76*, e081. https://doi.org/10.3989/ajbm.2525 [in Spanish].

Cirujano B. S., Guerrero M. N., & García Murillo, P. (2013). The genus *Tolypella* (A. Braun) A. Braun in the Iberian Peninsula. *Botany Letters, 160*, 121–129. https://doi.org/10.1080/12538078.2013.801321

Corillion, R. (1957). Les Charophycées de France et d'Europe Occidentale. *Bulletin de la Société Scientifique de Bretagne, 32*(Hors série I & II), 1–499. [in French].

Corillion, R. (1961). Les stations à *Tolypella hispanica* Nordstedt (Charophycée) du territoire français. *Revue Algologique. Nouvelle Série, 6*, 33–37. [in French].

Corillion, R., & Guerlesquin, M. (1963). Contribution à l'étude des végétations de Charophycées d'Italie péninsulaire (étage méditerranéen). *Bulletin de la Société Scientifique de Bretagne, 37*, 193–211. [in French].

Daily, F. K. (1954). A rare *Tolypella* new to the United States of America. *Butler University Botanical Studies, 11*, 144–148.

Krause, W. (1997). Charales (Charophyceae) Süßwasserflora von Mitteleuropa. In H. Ettl, G. Gärtner, H. Heynig, & D. Mollenhauer (Eds.), *Süßwasserflora von Mitteleuropa 18*. Fischer. [in German].

Migula, W. (1889–1897). Fünfter Band: Die Characeen. In A. Fischer, E. Fischer, F. Hauck, G. Limpricht, C. Luerssen, W. Migula, H. Rehm, P. Richter, & G. Winter (Eds.), *Dr. L. Rabenhorst's Kryptogamenflora von Deutschland, Oesterreich und der Schweiz* (ed. 2). Eduard Kummer. [in German].

Mouronval, J. B., Baudouin, S., Borel, N., Soulié-Märsche, I., Klesczewski, M., & Grillas, P. (2015). *Guide des Characées de France méditerranéenne*. Office national de la Chasse et de la Faune Sauvage. [in French].

Muller, S. D., Rhazi, L., & Soulié-Märsche, I. (2017). Diversity and distribution of Characeae in the Maghreb (Algeria, Morocco, Tunisia). *Cryptogamie, Algologie, 38*, 201–251. https://doi.org/10.7872/crya/v38.iss3.2017.201

Reyes Prósper, E. (1910). *Las Carofitas de España. Singularmente las que crecen en sus estepas*. Imprenta Artistica Española. [in Spanish].

Romanov, R. E., Nikulin, V. Yu., Vishnyakov, V. S., Tsegmid, B., Efremov, A. N., Zhakova, L. V., & Gontcharov, A. A. (2024). *Tolypella mongolica* (Charophyceae, Characeae): A new dioecious charophyte species from the center of Eurasia and implications for charophyte taxonomy, including the validation of T. porteri and the new combination for *T. jwalae*. *Phytotaxa*, 659 (1), 24–40. https://doi.org/10.11646/phytotaxa.659.1.2

Romanov, R., Napolitano, T., van de Weyer, K., & Troia, A. (2019). New records and observations to the Characean flora of Sicily (Italy). *Webbia, 74*, 111–119. https://doi.org/10.1080/00837792.2019.1609258

Urbaniak, J., Langangen, A., & van Raam, J. (2012). Oospore wall ornamentation in the genus *Tolypella* (Charales, Charophyceae). *Journal of Phycology, 48*, 1538–1545. https://doi.org/10.1111/jpy.12007

Wood, R. D. (1965). *Monograph of the Characeae*. J. Cramer.

Chapter 76
Tolypella nidifica

Elisabeth Lambert and Hendrik Schubert

Scientific name: ***Tolypella nidifica*** **(O. F. Müll.) A. Braun, Abh. Königl. Akad. Wiss. Berlin 1856: 344. 1857**

≡ *Conferva nidifica* O. F. Müll., Fl. Dan. 5(13): t. 761. 1777
Holotype (Müller, 1777): Fl. Dan. 5(13): t. 761

= *Chara stenhammariana* Wallman in Lilj., Utkast Sv. Fl., ed. 3: 686. 1816, "Stenhamariana"
Type not designated.

= *Tolypella nidifica* (O. F. Müll.) A. Braun subsp. *occidentalis* Corill. in Corill. & Guerl. in Bull. Soc. Sci. Bretagne 38: 202. 1963, n. inv.
Type not designated.

Nomenclatural notes: In several publications, the younger *Tolypella nidifica* (O. F. Müll.) Leonh. 1864: 176, footnote, is given as the scientific name.

Local names: DANISH: Redetråd; ENGLISH: Bird's nest stonewort; ESTONIAN: Pesajas tolüpell; FRENCH: Tolypelle en nid d'oiseau; FINNISH: Merisykeröparta; GERMAN: Nestarm-Baumleuchteralge, Ostsee-Baumglanzleuchteralge; LITHUANIAN: Lizdiškasis dmblabragis; NORWEGIAN: Sjøglattkrans; POLISH: Rozsocha morska; RUSSIAN: Толипелла гнездообразующая; SWEDISH: Havsrufse, Havsslinke; UKRAINIAN: Толіпела гніздоутворююча.

E. Lambert (✉)
Faculté des Sciences, Département Biologie-Environnement, Université Catholique de l'Ouest (UCO), 3, Place André Leroy – BP 10808, 49008 Angers Cedex 01, France
e-mail: lambert.elisa49@gmail.com

H. Schubert
Institut für Biowissenschaften, Universität Rostock, Albert-Einstein-Straße 3, 18051 Rostock, Germany
e-mail: hendrik.schubert@uni-rostock.de

H. Schubert et al. (eds.), *Charophytes of Europe*,
https://doi.org/10.1007/978-3-031-31898-6_76

76.1 Morphology

See Fig. 76.1.

Habitus: *Tolypella nidifica* is a highly branched plant which very rarely exhibits encrustation (Corillion, 1975; Moore, 1986). In the Baltic Sea and Scandinavian centre of distribution *T. nidifica* is not encrusted (Olsen, 1944), varying from light green to olive or even dark brownish green in colour (Fig. 76.2). As is typical for *Tolypella* spp., the persisting proembryonic filament towers above the plant (Fig. 76.2c; Krause, 1997; Langangen, 2007a).

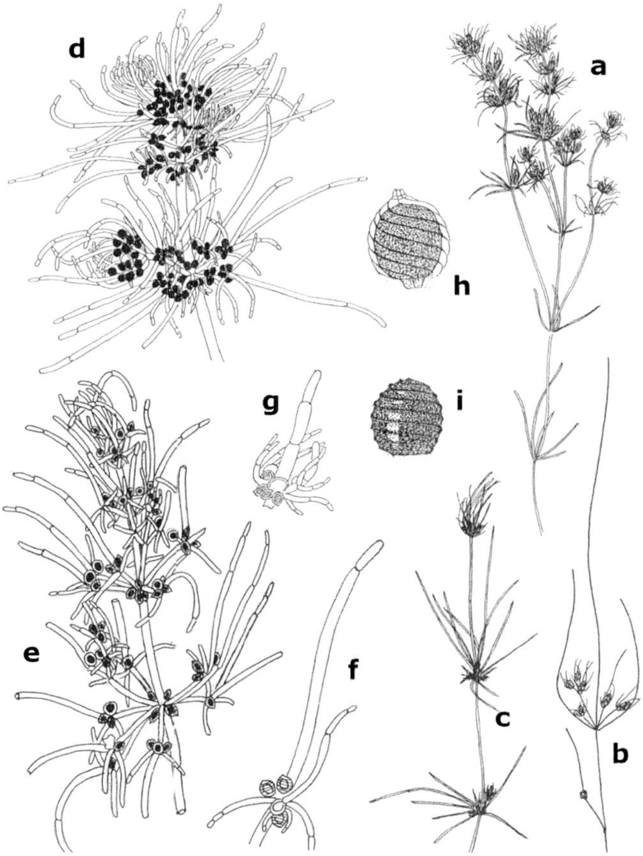

Fig. 76.1 *Tolypella nidifica*. **a** Habitus of a fully developed specimen; **b** Young plant; **c** Elongated deep water plant; **d** Fertile condensed head; **e** Elongated fertile head; **f** Branchlet; **g** Branchlet of *T. nidifica* subsp. *occidentalis*; **h** Oogonium; **i** Oospore. Drawings by Werner Krause

Fig. 76.2 *Tolypella nidifica* habitus. **a** Head of a fertile plant; **b** Field appearance of a stout plant with dense heads; **c** Field appearance of an elongated plant. Photographs: **a**, **b**-Klaus van de Weyer; **c**-Volker Krautkrämer

Size: *Tolypella nidifica* is a small to medium sized plant, typically between 10 and 20 cm long. However, in extreme cases it can reach up to 30 cm in length (Langangen et al., 2002) or, at the other extreme, fully developed plants can be less than 5 cm in length (Langangen et al., 2002).

Main axis: Main axis diameter is 0.3–0.6/0.8 mm (Urbaniak and Gąbka, 2014), occasionally up to 1 mm (e.g. Krause, 1997; Sinkevičienė, 2004). The length of the lower internodes is 3–10 cm, a value summarising the ranges given by various authors (c.f. Wood, 1965; Urbaniak and Gąbka, 2014). The length of the internodes decreases towards the top; lower and middle internodes are 2–3 times as long as the branchlets (Urbaniak, 2004).

Cortex; spine-cells; stipulodes: Absent.

Branchlets: Branchlet tips are rounded to obtuse (sect. *Obtusifolia*); *Tolypella nidifica* develops three kinds of branchlets: (A) short fertile, (B) elongated sterile and (C) tiny accessory ones. Fertile branchlets (length 0.3–0.7cm), usually 6–8(9) per whorl, form dense heads (3–10 per plant), around 0.6–1.2 cm diameter, at the top of the plant. The basal cell of fertile branchlets produces one, rarely two, rays which themselves give rise to 2–4 secondary rays at their nodes. Secondary rays consist of 3–4 cells; the end segment of the primary ray being usually 4-celled. Elongated sterile branchlets are always undivided and 2–4 cm long. They consist of 4–5 cells; 6–7 of these branchlets can be found in each whorl. Tiny accessory branchlets, c. 0.3–0.7 cm long, are often inserted at the base of a fertile whorl (Urbaniak and Gąbka, 2014).

Bract-cells; bracteoles; bractlets: None.

Sexuality: *Tolypella nidifica* is monoecious; on fertile branchlets a central adaxial antheridium is accompanied by 2–4 lateral oogonia at each branchlet node, including the basal node.

Oogonia: Oogonia are more or less globular with a short pedicel (Dąmbska and Karpiński, 1954). Spiral-cells (usually 8–10 helical stripes), slightly swollen-elongated at the top, generally swell at the apex at maturity. The coronula is deciduous and convergent (Langangen, 2007a, 2007b). Data about quantitative morphometry are significantly different between authors, mainly in relation to the maximum length and width (Table 76.1).

Table 76.1 Oogonia morphometry of *Tolypella nidifica*

Author	Length	Width	Coronula length	Coronula width	Number of helical stripes
Migula (1889–1897)	500–550	450	–	–	–
Groves and Bullock-Webster (1920), Corillion (1957)	500–575	450–500	75–90	100	10
Wood (1965)	500–705	450–645	75–105	100–120	~10
Corillion (1975)	500–700	450–650	75–100	100–120	~10
Hollerbach and Krassavina (1983)	500–590	440–480	80–90	120–130	10
Dąmbska and Karpiński (1954)	500–550	Up to 450	–	–	–
Krause (1997)	500–700	450–650	75	100	8–10
Urbaniak and Gąbka (2014)	500–760	450–640	–	–	–
Total range	**500–760**	**440–650**	**75–105**	**100–130**	**8–10**

Length in the first column excludes the coronula; its length and width are given separately, all values for length and width in μm

Table 76.1 does not contain values for specimens determined as "*T. stenhammariana*", which show a consistent tendency for smaller oogonia (Wood, 1965: length 465–555 μm; width 420–465 μm, coronula length 52–60 μm, coronula width ~94 μm; Corillion (1975): length 550–550 μm; width 420–450 μm, coronula length 52–60 μm, coronula width ~90μm).

Oospores: *Tolypella nidifica* is richly fertile, producing amazing numbers of oospores (e.g. Torn et al., 2015). Although sometimes described as ellipsoid (Hollerbach and Krassavina, 1983), the oospores are mostly subspherical with a flattened basal pole (Fig. 76.3b, c, e; Mouronval et al., 2015; Nowak et al., 2018; Vedder, 2004), showing the mark of the bipartite basal plate (Fig. 76.3f). Oospore colour is mostly reported as various kinds of dark-red, with only a few authors reporting black oospores. The oospore membrane is consistently described as smooth, often with complementary characters as granulated (pustular) ornamentation (Fig. 76.3b–d) (Table 76.2).

Antheridia: The adaxially inserted antheridia are solitary on a short pedicel. Data about antheridia size vary considerably between different authors with a tendency for lower diameters to be reported from northern/eastern regions, pointing to geographic differences (Table 76.3). Very low minimum values are reported by Migula (1889–1897: 300 μm) and Urbaniak and Gąbka (2014: 320 μm), overlapping with the range of *T. glomerata*. In Mediterranean beach lakes, maximum antheridia diameters up to 625 μm have been detected (pers. comm. R. Becker; own data). As with the oogonia, the antheridia of specimens determined as *T. stenhammariana* show a consistent tendency for being much smaller (Wood, 1965: c. 405 μm; Corillion, 1975: c. 400 μm; Langangen, 2014: 250–350 μm); they are not included in Table 76.3.

Bulbils: None.

76.2 Phenology and Reproduction

Tolypella nidifica is considered to be an annual species with very fast maturation. However, its development starts throughout the growing season and consequently plants can be found from March to October (November). To be reliably detected, investigations, in southern and central Europe, should be carried out before mid-July, because later in the year *T. nidifica* might be supressed by competition (especially filamentous algae, N. Stewart pers. comm.) or have even finished its life cycle (Hamann and Becker, 2016). Fructification has been reported from March to October (November), depending on the region. Along the Mediterranean, fructification seems to start earlier than in the Baltic Sea, where ripe oospores occur between June and September, again with geographical differences. N. Stewart (pers. comm.) reported ripe oospores in early October in the far north of Scotland (Orkney) and Langangen et al. (2002) until October/November from herbarium material from Finland. Winter et al. (1996) showed that the formation of mature oospores was inhibited at salinities exceeding 12 PSU.

76.3 Habitat Conditions and Ecology

Habitat type: *Tolypella nidifica* inhabits all kinds of coastal waters (coastal lagoons, bays, gulfs, fjords, estuaries; see e.g. Angus et al., 2016; Bazzichelli and Abdelahad, 2009; Becker, 2019; Bučas et al., 2019; Kiørboe, 1980; Martin, 2001; Morozova-Vodjanitzkaja, 1936; Olsen, 1944; Stewart et al., 2010) as well as other brackish

Fig. 76.3 *Tolypella nidifica* oospores. **a** Light micrograph of an oospore-membrane; **b–f** SEM photographs; **b–c** Oospore with variable membrane surface from smooth to granulated in different degree even within the same oospore; **d** Granulated (pustular) ornamentation of the membrane; **e**, **f** Side view and basal view showing the bipartite basal plate, formed by the two turning cells. Photographs: **a**-Klaus van de Weyer; **b**, **c**, **d**-Roman Romanov; **e**, **f**-Michelle T. Casanova

Table 76.2 Oospore morphometry of *Tolypella nidifica*

Author	Length	Width	Number of striae	Colour	Membrane
Migula (1889–1897)	390–480	340–450	7–8	Dark reddish brown	Smooth, with few indistinct dots on the ridges
Hy (1913)	up to 450–480	–	–	–	Almost smooth
Groves and Bullock-Webster (1920)	400–475	350–450	9	Wine red	Smooth or sometimes with scattered tubercles
Dąmbska and Karpiński (1954)	390–480	340–450	7–8	Brown–red to blackish	Smooth
Wood (1965)	400–475	350–450	7–8 (9)	Deep wine red or reddish brown	Smooth/granular/ vermiculate; occasionally tuberculate/ grumous especially near edge of fossa
Corillion (1975)	400–475	350–450	8–9	Brown–red to dark-red	Smooth; slightly tuberculated near the ridges
Hollerbach and Krassavina (1983)	385–485 (540)	340–430 (470)	7–8	Brownish–yellow to dark brown	Smooth or with some tubercles
Krause (1997)	400–500	350–450	7–9	Black or dark burgundy	–
Ray et al., (2001)	300–360	–	6–9	–	Fossa with pit-like depressions
Urbaniak (2004)	400–500	350–450	–	Dark wine-red, often almost black	–
Vedder (2004)	325–450	275–375	–	Brown to black	–
Bazzicheli and Abdelahad (2009)	400–500	350–450	7–9	Reddish–brown	–
Dilger (2010)	(250)310–430(540)	(210)250–380(400)	(5)6–8(9)	Black	–
Urbaniak et al., (2012)	361–456	320–441	6–8	Black, dark brown or brownish	Smooth/slightly pustular ornamentation
Urbaniak and Gąbka (2014)	360–450	300–425	–	Yellow–brown to black	–
Mouronval et al. (2015)	400–470	350–450	7–8(9)	Wine red to brown red even black	Smooth/pitted/ pitted only near the ridges
Total range	**(250)310–500(540)**	**(210)250–450**	**(5)6–9**		

All values for length and width in μm

sites like beach lakes (including ones located in salt marshes), ditches, temporary brackish pools and inland brackish water sites (Bryant and Stewart, 2011; Corillion, 1957, 1986; Hy, 1913; Thobor et al., 2000). It clearly prefers less exposed sites and is mainly found in sheltered conditions as long as suitable fine substrate is present. However, moderate wave action, removing the finest fractions of sediment or dispersing algal covering, or sheltering under *Fucus* fronds with sweeping effects on algal competition (N. Stewart pers. comm.), seems to favour the species occurrence

Table 76.3 Antheridia morphometry of *Tolypella nidifica*

Author	Diameter (μm)
Migula (1889–1897)	300–450
Groves and Bullock-Webster (1920)	550
Wood (1965)	(350)450–550
Corillion (1975)	500–550
Hollerbach and Krassavina (1983)	350–450
Krause (1997)	450–550
Urbaniak and Gąbka (2014)	320–450
Becker (pers. comm., own data)	Up to 625
Total range	**300–550(625)**

(Bučas et al., 2019; Herkül et al., 2018; Martin et al., 2010; Stewart et al., 2010; Torn et al., 2015). The species has also been observed in hard bottom communities, where it grows on patches of fine sediment accumulated between boulders (Martin et al., 2010; Urbaniak, 2004).

Sediment: Although requiring fine substrates, *Tolypella nidifica* clearly prefers sandy types. Bučas et al. (2019) could even show a positive correlation between coverage of *T. nidifica* and grain size. So, it is one of the species occurring among mixed substrates, as is common for example on the Estonian coast, resulting in co-occurrence with, e.g. fucoid algae (Martin et al., 2010). Nevertheless, the species can tolerate a broad range of sediment types ranging from flocculent mud (Angus et al., 2016) via various types of silt (c.f. Bučas et al., 2016) and muddy sediments (Herkül et al., 2018; Kuprijanov et al., 2017; Torn et al., 2015) in addition to the preferred sandy ones (Torn and Martin, 2003, Hamann and Becker, 2016). Outside the Baltic Sea region, Corillion (1957) reported an occurrence on calcareous clay.

Nutrient conditions: In general, most of the sites for the species are characterised as mesotrophic to slightly eutrophic without an indication that nutrient conditions themselves are decisive for the species' occurrence. Compiling the available data for Germany, Doege et al. (2016) report the following ranges: total phosphorus 0.01–0.11 mg/l; total nitrogen content 0.08–0.18 mg/l, nitrate nitrogen content <0.0226–0.043 mg/l and ammonium nitrogen content <0.003–0.103 mg/l. All other data published so far for this species (e.g. Kiørboe, 1980, Martin, 2001) fall well within the above ranges. However, there are no data available about the physiological limits of the species. Absence/presence studies, as asked for by Doege et al. (2016), are still lacking.

Water chemistry: Salinity seems to be the main factor for *Tolypella nidifica*, restricting vegetative occurrences mainly between 4 and 18(20) PSU (Olsen, 1944). This range covers also the reports from outside the Baltic Sea (e.g., Angus et al., 2016; Stewart et al., 2010). However, at a few places lower (0.5 PSU, Båstrup-Spohr et al., 2013) as well as higher (30 PSU, Olsen, 1944; 25 PSU, Martin, 2001) salinities have been measured at *T. nidifica*-sites. Interpretation of these values must take into account that salinity can abruptly vary at typical *T. nidifica* habitats, so extreme values may reflect exceptional events (as occurring after, e.g. rainfall, drought or flooding) rather than characterising the normal conditions required for growth and successful reproduction. Few data are available for pH conditions, but Corillion (1975) reports a rather narrow range between 7.3 and 8 for French occurrences. A broader range, reflecting at least in part diurnal variability caused by eutrophication, was reported by Hamann and Becker (2016) for Baltic Sea stands with a maximum pH of 8.9.

Depth range: Being a heliophilous species, *Tolypella nidifica* often grows just beneath the water surface (e.g. Corillion, 1975; Hamman and Becker, 2016). However, it reaches an astonishing 22–24 m depth at clear sites in the almost cloudless Crimean region (Morozova-Vodjanitzkaja, 1936); comparable depth maxima are also reported for the Eastern Baltic Sea (25 m for Tvärminne, Finland, by Urbaniak (2004) and 20 m for the Estonian coast by Martin et al., 2010). The upper limit is dictated by water level changes, sites with pronounced water level fluctuations can exhibit population maxima at intermediate depths between 0.5 and 1.2 m (e.g. Curonian Lagoon: Bučas et al., 2019, Rame Bay: Torn and Martin, 2003). Except for these record depths, the lower limit of Baltic Sea *T. nidifica* stands are at around 6 m (Urbaniak, 2004) with germlings being found a little deeper (at up to 7 m, Martin et al., 2010). An interesting bimodal depth pattern with a shallow water maximum within the first metre and a second one below around 1.5 m, has been observed for several Baltic occurrences (Urbaniak, 2004).

Associated species: Preferring sheltered brackish habitats, *T. nidifica* is mostly associated with salt-tolerant charophytes (mainly *Chara canescens, C. aspera, C. baltica)* and higher plants (mostly *Zostera marina, Ruppia maritima, Stuckenia pectinata, Zannichellia palustris*). Co-occurrence with other salt-tolerant species of the above groups (e.g. *Lamprothamnium papulosum, Tolypella hispanica, Chara connivens, C. tomentosa, C. horrida, Ranunculus baudotii, Ruppia cirrhosa, Zostera noltei*) are less frequent due to the restricted range of distribution or rareness of the respective species. At inland brackish-water sites with pronounced ion anomalies, a number of freshwater species were reported to co-occur with *T. nidifica* (e.g. *Chara virgata, C. contraria, C. galioides, C. vulgaris, Nitellopsis obtusa*). Co-occurrence with marine macroalgae (e.g. fucoids and *Ulva* spp.) mainly depends on the availability of mixed substrates, providing fine sediment for *T. nidifica* as well as at least stable gravel for the marine macrophytes.

With respect to phytosociology, *T. nidifica* belongs to the *Tolypelletum nidificae* Kornas 1959. It is assigned to the *Charion canescentis* F. Fukarek 1961 and included in the *Charetalia canescentis* Fukarek ex Krause, 1997 (Felzines and Lambert, 2012, 2016).

76.4 Variability

Along the Baltic Sea, *T. nidifica* shows little discernible variability, but considering data from Europe there is morphometric variability related to the ecological conditions of the respective sites. Consequently, some authors distinguished between several varieties and forms which differ with respect to quantitative characters such as size, internode and branchlet length etc.—all being regarded as site-specific alterations of a common basic morphology and therefore not discussed here. An exception might be "*T. stenhammariana*" which differs with respect to oogonia and antheridia size but, except for this, clearly exhibits the decisive characters of *T. nidifica*. However, as outlined in Table 78.4, regional differences in chromosome numbers between Baltic and Italian as well as French *T. nidifica* (the latter determined as *T. nidifica* subsp. *occidentalis*) point to either (A) geographical differences in chromosome number or, alternatively, (B) problems in discrimination between *T. glomerata* and *T. nidifica* (see following section).

76.5 Identification and Risk of Confusion

Tolypella nidifica belongs to a group of species with rounded tips to the end-cell of the branchlets. This group includes *T. normaniana*; for reliable distinction between *T. normaniana* and *T. nidifica* see Chap. 77 (*Tolypella normaniana,* this volume). The greatest difficulty, however, is to distinguish between *T. salina, T. nidifica*, and *T. glomerata*.

Chromosome numbers (Table 78.4 in this volume) show a distinction between the 3 species. However, a search for distinctive characters commonly applied and accepted by all authors has revealed little evidence so far because of inconsistencies and confusions over the characters separating these species. The following hints for determination consequently reflect the authors' opinion after checking fresh as well as herbarium material from the Atlantic and the Baltic Sea area and may not fully agree with literature data based on diverging concepts (see "remarks", below).

The polyembryony and the persistent proembryonic filament, already detectable when the plant is half buried in the sediment, is a character often cited to distinguish *T. salina* from *T. nidifica* (as well as *T. glomerata*). However, it is not a reliable character, because polyembryony as well as persistent proembryonic filaments may appear in *T. nidifica* also (own observation). A safer distinction between *T. nidifica* and *T. salina* uses gametangia characters. *T. salina* oogonia are smaller (220–520 μm width) and have fewer helical stripes (6–8) than *T. nidifica* (400–650 μm width, 8–10 helical stripes). They also differ with respect to the colour of the oogonia (*T. salina*: orange to pale brown; *T. nidifica*: dark reddish brown). Antheridia diameters overlap and may serve only as an additional character; *T. salina* (420–750 μm) and *T. nidifica* (250–550 μm). Another good indication is given by the antheridia's pedicels: whereas most antheridia of *T. salina* are stalked, pedicels of *T. nidifica* are almost exclusively short. When ripe, the oospore characters also help to distinguish the two species. *T. salina* oospores are only slightly smaller than *T. nidifica* ones with a large range of overlap; however, the number of striae (5–7 slightly prominent ones for *T. salina* in contrast to (6) 7–9 for *T. nidifica*) and colour (light brown to chestnut brown for *T. salina* in contrast to various shades of dark-red for *T. nidifica*) give indications for correct determination. The oospore membrane structure serves as an ultimate proof. The smooth membrane with the subtly granulated-vermiculated edges of the ridges of *T. salina* oospores is different from the smooth surface with slightly pustular ornamentation of *T. nidifica* oospore membranes. Another helpful character are the accessory branchlets, which often appear at the base of fertile whorls of *T. nidifica* but have not been observed for *T. salina*. In any case the habitat characteristics should be regarded carefully. *T. salina* can withstand far higher salinities (and even become fertile in them) compared to *T. nidifica* and, in contrast to *T. nidifica*, it requires dry periods for the production of ripe oospores.

Distinction between *T. nidifica* and *T. glomerata*, which develops larger antheridia when growing in brackish water compared to freshwater individuals, is extremely difficult because of overlapping salinity ranges. The only consistent difference found so far is, that *T. glomerata* is characterised by being dull whereas *T. nidifica* is always shiny (K. van de Weyer pers. comm.).

76.6 Distribution

See Fig. 76.4.

Tolypella nidifica can be found almost all over Europe. The main distribution centre is in the Baltic Sea region (and Norway) where the species is found all along the coastline in less exposed/sheltered bays. Inland occurrences are pretty rare and restricted to a few inland brackish water sites. In the Mediterranean and North Atlantic regions, the species is less abundant, probably because of the lack of habitats with suitable salinity conditions; in contrast to the brackish Baltic Sea, the salinity of lagoons and inlets in these regions can easily exceed the mesohaline range that *T. nidifica* requires for the formation of ripe oospores (see Steinhardt and Selig, 2011; Winter et al., 1996). The species has been found only in a very few places outside Europe and therefore must be regarded as a European species that has been unintentionally introduced to some places outside the continent.

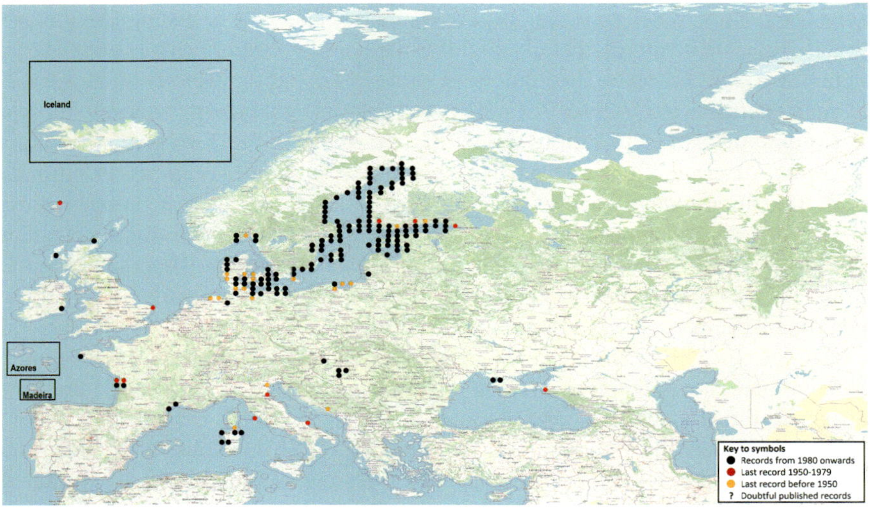

Fig. 76.4 European distribution of *Tolypella nidifica*. For worldwide distribution see Korsch (2018). Black dots are records from 1980 on, red dots from 1950 to 1979 and yellow dots before 1950; unconfirmed/doubtful published records (if applicable) are marked with "?" on top. Map prepared by Heiko Korsch

76.7 Status and Threats

The recent status of *Tolypella nidifica* is difficult to estimate because of its ephemeral character. In its Baltic region centre of distribution (including Norway), a decline in occurrence has been reported for all countries, most of them regarding the species as threatened. In this region, eutrophication and habitat loss are regarded as the main threats (Urbaniak, 2004), with eutrophication often acting via the overgrowing of stands by filamentous algal mats (e.g. Langangen, 2007a). On the other hand, *T. nidifica* is very abundant in the diaspore bank of Baltic Sea coastal ecosystems, where astonishing numbers of intact and viable oospores have been found in places where the species has not occurred for at least several years (Nowak et al., 2018; Selig et al., 2009). Probably this ephemeral small species has often been overlooked in Baltic Sea macrophyte monitoring programmes which focus on the outer shores, often excluding the shallow areas completely. A trend of loss of stands along the Swedish west and south coast (but not at other areas of the Swedish Baltic Sea coast) was verified by targeted monitoring of occurrences known from herbarium records by Blindow (2000a, 2000b). In any case the Baltic countries have a special responsibility for the worldwide protection of *T. nidifica* and most of them have classified the species respectively in their Red Lists (Becker, 2016, see also Chap. 9).

Distribution outside the Baltic region is scattered and restricted to marginal habitats, making it almost impossible to give a sound statement about the species' status. Declines mainly due to habitat loss are reported from the Mediterranean (including the Balkans) region as well as the UK and inland brackish water sites (e.g. Blaženčić and Blaženčić, 2002; Blaženčić et al., 2006, Hamann and Becker, 2016) resulting in a respective protection status (see Chap. 9). In Britain, eutrophication is thought to have caused the loss of one site and is a continuing threat at its Irish site. Salinity changes, including from sea level rise are also concerns in Britain and Ireland.

76.8 Remarks

As outlined above, the distinction between the European members of sect. *Obtusifolia*, especially *T. nidifica*, *T. salina* and *T. glomerata*, remains difficult because of conflicting opinions about the value of certain characters. The introduction of new varieties and forms such as *T. nidifica* subsp. *occidentalis* and *T. glomerata* f. *littorea* (French Atlantic coast and Mediterranean coast) has made the problem even worse. The remarks of different authors show the complexity: For Hy (1913) *T. glomerata* f. *littorea* includes various intermediates between *T. glomerata* and *T. nidifica*. For Corillion (1957) *T. nidifica* is a polymorphous plant, which is not without links to *T. glomerata* (especially *f. littorea*). For Corillion and Guerlesquin (1959) *T. salina*, by its main characters, recalls a type of littoral form and is placed by the authors under various names, either with *T. glomerata* or with *T. nidifica*. Consequently, a number of herbarium records have turned out to be misidentifications and a re-check especially of older material from Western Europe (especially material originating before the introduction of *T. salina*) is highly recommended.

All these difficulties have led charophytologists to wonder whether to set aside *T. glomerata* f. *littorea* and *T. nidifica* subsp. *occidentalis* and to merge *T. salina* with *T. nidifica*. However, pending more results from genetic analysis, chromosome number differences as well as some morphological differences observed directly on fresh and herbarium material of authenticated origin give enough evidence to keep the taxa separated at species level (see also Chap. 78).

Acknowledgements We thank Michelle T. Casanova, providing helpful material for characterisation of oospores as well as Nick Stewart and Irmgard Blindow for their helpful comments during preparation of the manuscript. Roman Romanov provided data and photographs of herbarium material, which is gratefully acknowledged.

References

Angus, S., Scott, S., Darwell, A., & Stewart, N. F. (2016). Bird's nest stonewort. Version 1.1. In: M. J. Gaywood, P. J. Boon, D. B. A. Thompson, & I. M. Strachan (Eds.), *The species action framework handbook* (220–227 pp.). Scottish Natural Heritage.

Båstrup-Spohr, L., Dahl-Nielsen, J., & Sand-Jensen, K. (2013). Kransnålalger rummer mange truede arter. *Urt, 37*, 66–70. [in Danish].

Bazzichelli, G., & Abdelahad, N. (2009). *Alghe d'acqua dolce d'Italia. Flora analitica delle Caroficee* . Sapienza Università di Roma. Ministero dell'Ambiente e della Tutela del territorio e del mare. Roma, Centro Stampa Università, Roma. [in Italian].

Becker, R. (2016). Gefährdung und Schutz von Characeen. In *Arbeitsgruppe Characeen Deutschlands, Armleuchteralgen: Die Characeen Deutschlands* (pp 149–191). Springer. [in German].

Becker, R. (2019). The Characeae (Charales, Charophyceae) of Sardinia (Italy): Habitats, distribution and conservation. *Webbia, 74*, 83–101.

Blaženčić, J., & Blaženčić, Ž. (2002). Rare and threatened species of Charophytes (Charophyta) in Southeast Europe. *Phytologia Balcanica, 8*, 315–326.

Blaženčić, J., Stevanović, B., Blaženčić, Z., & Stevanović, V. (2006). Distribution and ecology of charophytes recorded in the west and central Balkans. *Cryptogamie, Algologie, 27*, 311–322.

Blindow, I. (2000a). Distribution of Charophytes along the Swedish coast in relation to salinity and eutrophication. *International Review of Hydrobiology, 85*, 707–717.

Blindow, I. (2000b). Kransalger i Skåne. *Länsstyrelsen i Skåne Län, Rapportserien Skåne i Utveckling, 99*, 1–34. [in Swedish].

Bryant, J. A., & Stewart, N. F. (2011). Phylum Chlorophyta. Order Charales. In D. M. John, B. A. Whitton, & A. J. Brook (Eds.), *The freshwater algal flora of the British Isles. An identification guide to freshwater and terrestrial algae* (2nd ed., pp. 742–765). Cambridge University.

Bučas, M. A., Šiaulys, A., & Sinkevičienė, Z. (2016). Assessment of a simple hydroacoustic system for the mapping of macrophytes in extremely shallow and turbid lagoon. *Aquatic Botany, 134*, 39–46.

Bučas, M., Sinkevičienë, Z., Kataržytė, M., Vaičiūtė, D., Petkuvienė, J., Stragauskaitė, V., & Ilginė, R. (2019). How much can the occurrence and coverage of charophytes in an estuarine lagoon (Curonian Lagoon) be explained by environmental factors? *Estuarine, Coastal and Shelf Science, 16*, 128–138.

Corillion, R. (1957). Les Charophycées de France et d'Europe Occidentale. *Bulletin de la Société Scientifique de Bretagne, 32*(Hors série I & II), 1–499. [in French].

Corillion, R. (1975). *Flore et végétation du Massif Armoricain. T. IV. Flore des Charophytes (Characées) du Massif Armoricain et des contrées voisines d'Europe occidentale.* Jouve, Paris. [in French].

Corillion, R. (1986). Les Characées de la Palud de Treguennec (Finistère) et leur végétation. Evolution récente et état actuel (1985). *Bulletin de la Société D'études Scientifiques de L'anjou, 12*, 71–91. [in French].

Corillion, R., & Guerlesquin, M. (1959). Premières observations cytotaxonomiques sur le genre *Tolypella* (Charophycées). *Bulletin de la Société D'études Scientifiques D'angers, 2*, 167–170. [in French].

Dąmbska, I., & Karpiński, J. (1954). *Ramienice. Klucz do oznaczania gatunków krajowych.* Państwowe Wyda, wnictwo Naukowe, Warszawa. [in Polish].

Dilger, M. (2010). Merkmale zur Erkennung von *Nitella*- und *Tolypella*-Arten der Characeaen [sic] über Oosporen aus Sedimenten oder von Pflanzen. *Rostocker Meeresbiologische Beiträge, 23*, 40–59. [in German].

Doege, A., van de Weyer, K., Becker, R., & Schubert, H. (2016). Bioindikation mit Characeen. In Arbeitsgruppe Characeen Deutschlands, *Armleuchteralgen: Die Characeen Deutschlands* (pp. 97–138). Springer. [in German].

Felzines, J.-C., & Lambert, E. (2012). Contribution au prodrome des végétations de France : Les Charetea fragilis F. Fukarek 1961. *Le Journal de Botanique de la Société Botanique de France, 59*, 133–188. [in French].

Felzines, J.-C., & Lambert, E. (2016). Contribution au prodrome des végétations de France : Modification de la structure syntaxonomique des Charetea et compléments. *Le Journal de Botanique de la Société Botanique de France, 74*, 41–55. [in French].

Groves, J., & Bullock-Webster, G. R. (1920). *The British Charophyta.* Ray Society.

Hamann, U., & Becker, R. (2016). *Tolypella nidifica.* In Arbeitsgruppe Characeen Deutschlands, *Armleuchteralgen: Die Characeen Deutschlands* (pp. 388–397). Springer. [in German].

Herkül, K., Torn, K., & Möller, T. (2018). The environmental niche separation between charophytes and angiosperms in the northern Baltic Sea. *Botany Letters, 165*, 115–127.
Hollerbach, M. M., & Krassavina, L. K. [Голлербах, М. М. & Л. К. Красавина]. (1983). *Определитель пресноводных водорослей СССР. 14. Charophyta [Determination key of freshwater algae of USSR]*. Nauka. [in Russian].
Hy, F., 1913. Les Characées de France. *Bulletin de la Société Botanique de France, 60*(sup 1), 1–47. [in French].
Kiørboe, T. (1980). Distribution and production of submerged macrophytes in Tipper Grund (RingkobingFjord, Denmark), and the impact of waterfowl grazing. *Journal of Applied Ecology, 17*(3), 675–687.
Korsch, H. (2018). The worldwide range of the Charophyte species native to Germany. *Rostocker Meeresbiologische Beiträge, 28*, 45–96.
Krause, W. (1997). Charales (Charophyceae) Süßwasserflora von Mitteleuropa. In H. Ettl, G. Gärtner, H. Heynig, & D. Mollenhauer (Eds.), *Süßwasserflora von Mitteleuropa 18*. Fischer. [in German].
Kuprijanov, I., Herku, K., & Kotta, J. (2017). Ecological niche differentiation between native and non-native shrimps in the northern Baltic Sea. *Aquatic Ecology, 51*, 389–404.
Langangen, A. (2007a). *Charophytes of the Nordic countries*. Saeculum ANS, Oslo.
Langangen, A. (2007b). *Kransalger og deres forekomst i Norge: med korte beskrivelser av viktige norske lokaliteter*. Saeculum. [in Norvegian].
Langangen, A., 2014. Charophytes (Charales) from Samos and Ikaria (Greece) collected in 2013 and report on some localities in Skiathos (Greece). *Flora Mediterranea*, 139–151.
Langangen, A., Koistinen, M., & Blindow, I. (2002). The charophytes of Finland. *Memoranda Societatis pro Fauna Et Flora Fennica, 78*, 17–48.
Martin, A. (2001). *The ecology and palaeoecology of the charophyte Lamprothamnium papulosum in U.K. Coastal Lagoons*. PhD Thesis, University College, London.
Martin, G., Möller, T., Kotta, J., Daunys, D., Jermakovs, V., Bucas, M et al. (2010). Benthic Marine habitats of Eastern Baltic Sea. *Estonian Marine Institute Report Series, 15*.
Migula, W. (1889–1897). Fünfter Band: Die Characeen. In A. Fischer, E. Fischer, F. Hauck, G. Limpricht, C. Luerssen, W. Migula, H. Rehm, P. Richter, & G. Winter (Eds.), *Dr. L. Rabenhorst's Kryptogamenflora von Deutschland, Oesterreich und der Schweiz* (ed. 2). Eduard Kummer, Leipzig. [in German].
Moore, J. A. (1986). *Charophytes of Great Britain and Ireland*. BSBI Handbook 5. Botanical Society of the British Isles, London.
Morozova-Vodjanitzkaja, N. (1936). Phytobenthos of the Karkinitzky Gulf (Scientific meeting, Oct. 20, 1933). *Works of the Sevastopol Biological Station, T. V*: 219–232. [in Russian].
Mouronval, J. B., Baudouin, S., Borel, N., Soulié-Märsche, I., Klesczewski, M., & Grillas, P. (2015). *Guide des Characées de France méditerranéenne*. Office National de la Chasse et de la Faune Sauvage, Paris. [in French].
Müller, O. F. (1777). *Abbildungen der Pflanzen, welche in den Königreichen Dännemark und Norwegen, in den Herzogthümern Sleswig, Holstein und Oldenburg wild wachsen, zu Erläuterung des unter dem Titel Flora Danica, ... Band 5*. Martin Hallager, Kopenhagen. [in German].
Nowak, P., Steinhardt, T., von Ammon, U., Rohde, H., Schoor, A., Holzhausen, A., Schaible, R., & Schubert, H. (2018). Diaspore bank analysis of Baltic coastal Waters. *Botany Letters, 165*, 159–173.
Olsen, S. (1944). Danish Charophyta. Det Kongelige Danske Videnskabernes Selskab. *Biologiske Skrifter, 3*(1), 1–240, pl. 1–2. [in Danish].
Ray, S., Pekkari, S., & Snoeijs, P. (2001). Oospore dimensions and wall ornamentation patterns in Swedish charophytes. *Nordic Journal of Botany, 21*, 207–224.
Selig, U., Steinhardt, T., & Schubert, H. (2009). Interannual variability of submerged vegetation in a brackish coastal lagoon on the southern Baltic Sea. *Ekológia, 28*, 412–423.
Sinkevičienė, Z. (2004). Charophyta of the curonian lagoon. *Botanica Lithuanica, 10*, 33–57.

Steinhardt, T., & Selig, U. (2011). Influence of salinity and sediment resuspension on macrophyte germination in coastal lakes. *Journal of Limnology, 70*, 11–20.
Stewart, N., Darwell, A., & Scott, S. (2010). *Bird's nest stonewort Tolypella nidifica in Loch an t-Sruith Mhòir and Loch an Dùin, North Uist*. Report for Scottish Natural Heritage.
Thobor, C., Weigel, H.-P., & Rode, M. (2000). *Stehende Gewässer im Land Bremen. Zustand, Nutzung, Gewässergüte, Ökologie*. Bund für Umwelt und Naturschutz, Landesverband Bremen. [in German].
Torn, K., Kovtun-Kante, A., Herkül, K., Martin, G., & Mäemets, H. (2015). Distribution and predictive occurrence model of charophytes in Estonian waters. *Aquatic Botany, 120*, 142–149.
Torn, K., & Martin, G. (2003). Changes in the distribution of charophyte species in enclosed seabays of western Estonia. *Proceedings of the Estonian Academy of Sciences, Biology and Ecology, 52*, 134–140.
Urbaniak, J. (2004). *Tolypella nidifica*. In H. Schubert, & I. Blindow (Eds.), *Charophytes of the Baltic Sea* (pp. 223–233). A. R. G. Gantner.
Urbaniak, J., & Gąbka, M. (2014). *Polish Charophytes. An illustrated guide to identification*. Uniwersytet Przyrodniczy we Wroclawiu. [in Polish].
Urbaniak, J., Langangen, A., & van Raam, J. (2012). Oospore wall ornamentation in the genus *Tolypella* (Charales, Charophyceae). *Journal of Phycology, 48*, 1538–1545.
Vedder, F. (2004). Morphologie und Taxonomie rezenter und subfossiler Characeen-Oosporen aus der Ostsee. *Rostocker Meeresbiologische Beiträge, 13*, 43–54. [in German].
Winter, U., Soulié-Märsche, I., & Kirst, G. O. (1996). Effects of salinity on turgor pressure and fertility in *Tolypella* (Characeae). *Plant, Cell and Environment, 19*, 869–879.
Wood, R. D. (1965). *Monograph of the Characeae*. J. Cramer, Weiheim.

Chapter 77
Tolypella normaniana

Anders Langangen

Scientific name: *Tolypella normaniana* (Nordst.) Nordst., Acta Univ. Lund.25(4): 15. 1889

≡ *Nitella normaniana* Nordst. in Norman, Norske Videnskabers Selsk. Skr. 19de Aarhundr. 5: 334. 1868
Lectotype (designated here by Langangen): [Norway: Nordland: Beiarn] Ad littoral maris fundi intimi sinus Bejerfjord Nordlandiæ loco sub accessu astus inundato, sub recess emerso/1866–8-1/J.M. Norman (S A6842).

≡ *Tolypella nidifica* (O. F. Müll.) A. Braun f. *normaniana* (Nordst.) Wahlman 1875: 22

≡ *Tolypella nidifica* (O. F. Müll.) A. Braun subsp. *normaniana* (Nordst.) A. Braun & Nordst., Abh. Königl. Akad. Wiss. Berlin 1882: 95. 1882

Local names: NORWEGIAN: Fjæreglattkrans; RUSSIAN: Толипелла Нормана; SWEDISH: Fjordrufse.

77.1 Morphology[1]

See Figs. 77.1 and 77.2.

Habitus: Small green plants, under water or on tidal flats, often reddish as it is normally richly fertile. Often, but not always, forming heads 0.5–12 mm in diameter.

[1] See Langangen (1994) for further details of the morphology of this species.

A. Langangen (✉)
Natural History Museum, P.O. Box 1172 Blindern, 0318 Oslo, Norway
e-mail: langangen@hotmail.com

H. Schubert et al. (eds.), *Charophytes of Europe*,
https://doi.org/10.1007/978-3-031-31898-6_77

Fig. 77.1 Specimen of *Tolypella normaniana* from Mosjøen. Protonemal tips to 30 mm long. Each head is richly fertile. Photograph: Anders Langangen

Plants are found both encrusted and unencrusted. One oospore can give rise to as many as 12 plants (see Figs. 77.1 and 77.2).

Size: *Tolypella normaniana* is small, up to 10 cm long, but mostly shorter (around 2–3 cm). This length includes the whole plant to the tip of the branchlets and the protonemal process.

Main axis: The main axis is 150–600 μm in diameter. The internode between the protonemal whorl and the first whorl is 0.5–26 mm long; subsequent internodes are much shorter.

Cortex, spine-cells, stipulodes: Absent.

Branchlets: Whorls 1–6; only the first, protonemal whorl has 1–6 sterile, undivided branchlets, and 4–22 fertile branched branchlets which are up to 40 mm long. The other 1–5 whorls have only fertile branched branchlets. The fertile branchlets are normally curved inwards towards the main axis. The internode between the protonemal whorl and the first whorl varies from 0.5 to 16 mm in length. There is one terminal process produced at the protonemal whorl which is much longer than the heads and up to 90 mm long. This process appears to be a prolongation of the protonema and is often of similar diameter or wider than the branchlets and sometimes as wide as the protonemal internode.

Bract-cells, bracteoles, bractlets: Absent.

Sexuality: *Tolypella normaniana* is monoecious; gametangia are conjoined at the main axis nodes and fertile branchlet nodes.

Fig. 77.2 Habitus of two different specimens of *Tolypella normaniana*. **a** Specimen from Mosjøen with 8 plants from the same protonema. The germinated oospore is still fastened to the plant. Total length of the plant is 15 mm. **b** Specimen from Sørfjorden with 4 plants. The right plant is best developed. It has three whorls in addition to the protonemal whorl. Total length is 23 mm. Drawings: reproduced from Langangen (1994)

Oogonia: 300–450 μm long (including coronula) and 170–400 μm wide, with 7–8 helical stripes. Some oogonia have a stalk-cell up to 250 μm long.

Oospore: The oospore is dark brown to black, 230–300 μm long and 230–300 μm wide, with 6–8 sharp ridges. There are two basal impressions and the fossae are c. 45 μm across. The oospore membrane is granulate.

Antheridia: Solitary or 2–4 together, 250–450 μm in diameter. When they are stalked, the stalk-cell is up to 600 μm long.

Bulbils: Absent.

77.2 Phenology and Reproduction

This species is annual and survives the winter as oospores. Germination is presumed to be in the spring but affected by the seasonal climate conditions. Mature antheridia and oogonia have been observed from July onwards with ripe oospores in August. Numerous oogonia can still be present in November shortly before the fjord habitat freezes over (Langangen, 1994).

77.3 Habitat Conditions and Ecology

Further details of the ecology of this species can be found in Langangen (1994).

Habitat type: *Tolypella normaniana* grows in estuarine fjords where the fjord-heads receive large quantities of freshwater from large rivers. The tidal flats which make up large areas where the alga lives are exposed to the air at low tide. On these foreshore flats the alga survives in small pools or on wet bottom sediment, fastened by its rhizoids.

Sediment: The species grows on loose clay bottom, usually with firm clay beneath.

Nutrient conditions: No information available.

Water chemistry: *Tolypella normaniana* is found in localities with nearly fresh to slightly brackish water at least in summer (0.1–1.3 psu). The salt content is considerably higher in the autumn (up to 7.5 psu). pH ranges from 7.6 to 8.8.

Depth range: At low tide, the species grows on exposed wet mud or in shallow pools, but the tidal range of some sites is up to 3 m.

Associated species: This species can grow as scattered plants or in dense mats. Competition from other species is very low, except possibly at one site where filamentous algae are abundant (*Cladophora* spp., *Zygnema* spp., *Spirogyra* spp., *Oedogonium* spp., *Percursaria percursa*, *Vaucheria medusa*, *Bulbochaete* spp., *Tribonema* spp. *Ulva (=Enteromorpha)* spp.). Other aquatic associates include *Chara aspera*, *Potamogeton filiformis*, *Pelvetia canaliculata* and *Ruppia maritima*, but a number of terrestrial phanerogams may also be present, particularly near the shore, including *Cochlearia officinalis* ssp. *norvegica*, *Eleocharis uniglumis*, *Pedicularis palustris* and *Carex paleacea*.

77.4 Variability

The morphological variability of *Tolypella normaniana* is high, especially concerning the size of the specimens, number of plants on each protonema, length of branchlets etc.

77.5 Identification and Risk of Confusion

This species has blunt/rounded branchlet tips, a feature which it shares with *T. glomerata*, *T. nidifica* and *T. salina*. As Migula (1889–1897) stated, the specific character is that there are no sterile whorls because most branchlets in the protonemal whorl are fertile. This is in contrast to *Tolypella nidifica* and *T. glomerata* where one or two whorls have only sterile branchlets. A permanent terminal protonemal projection (Vorkeimspitze) is found on all specimens observed.

77.6 Distribution

See Fig. 77.3.

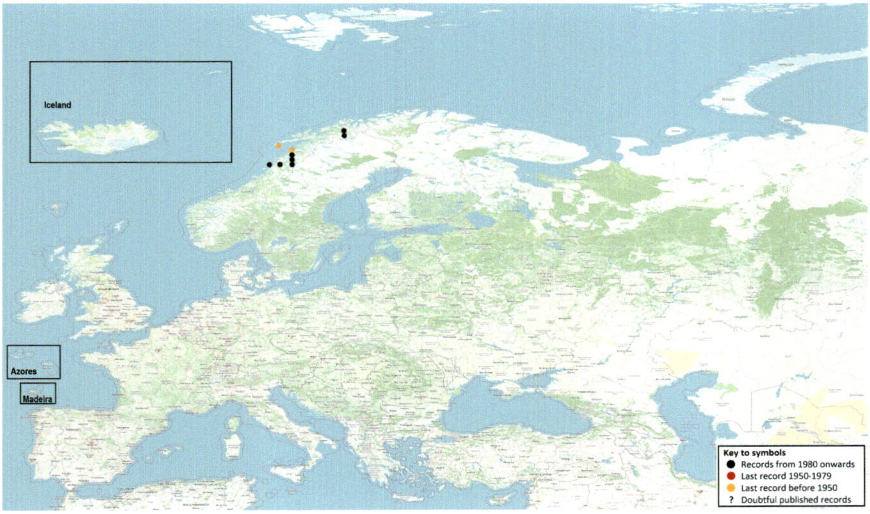

Fig. 77.3 Distribution of *Tolypella normaniana*. Black dots are records from 1980 on, red dots from 1950 to 1979 and yellow dots before 1950. Map prepared by Heiko Korsch

Tolypella normaniana is endemic to Norway and is found in only a few localities in the counties of Nordland and Troms (Langangen, 2011).

77.7 Status and Threats

The biggest threat for this species is from human activities in the form of infilling areas for roads and industrial use. The Red List status for *T. normaniana* in Norway is NT (Near Threatened). In the European Red List, the species is also assessed as NT (Near Threatened) (see Chap. 9).

References

Langangen, A. (1994). *Tolypella normaniana* Nordstedt, a little known charophyte from northern Norway. *Cryptogamie, Algologie, 15*, 221–236.

Langangen, A. (2011). Statusrapport for Nordlandsglattkrans *Tolypella normaniana*. Oslo. [in Norvegian].

Migula, W. (1889–1897). Fünfter Band: Die Characeen. In A. Fischer, E. Fischer, F. Hauck, G. Limpricht, C. Luerssen, W. Migula, H. Rehm, P. Richter, & G. Winter (Eds), *Dr. L. Rabenhorst's Kryptogamenflora von Deutschland, Oesterreich und der Schweiz* (ed. 2). Eduard Kummer. [in German].

Chapter 78
Tolypella salina

Elisabeth Lambert, Hendrik Schubert, and Didier Desmots

Scientific name: *Tolypella salina* Corill. in Rev. Algol. N. S. 5(3): 198. 1960

Holotype (Corillion 1960: 199): Gallia, prope Croix-de-Vie (Vendée)/24 Avril 1959 [squalm. 25. avril 1959]/[R. Corillion] (ANGUC, conserved in alcohol).

Nomenclatural note: Corillion (1960) indicated the 25th of April as collection date. This is probably a mistake as there are two alcohol bottles from the 24th and 27th 1959 in ANGUC, both marked as type by Corillion.

Local names: FRENCH: Tolypelle saline, Tolypelle des marais salants.

78.1 Morphology

See Figs. 78.1 and 78.2.

Habitus: *Tolypella salina* is generally small and scarcely branched. Depending on the ecological conditions, the colour can be "bright, grey, yellow or dark" green to brownish, becoming even darker at the end of the season. Often half buried in

E. Lambert (✉)
Faculté des Sciences, Département Biologie-Environnement, Université Catholique de l'Ouest (UCO), 3, Place André Leroy – BP 10808, 49008 Angers, Cedex 01, France
e-mail: lambert.elisa49@gmail.com

H. Schubert
Institut für Biowissenschaften, Universität Rostock, Albert-Einstein-Straße 3, 18051 Rostock, Germany
e-mail: hendrik.schubert@uni-rostock.de

D. Desmots
L.P.O., Réserve Naturelle Nationale des marais de Müllembourg, 85330 Noirmoutier-en-l'Ile, France
e-mail: didier.desmots@lpo.fr

H. Schubert et al. (eds.), *Charophytes of Europe*,
https://doi.org/10.1007/978-3-031-31898-6_78

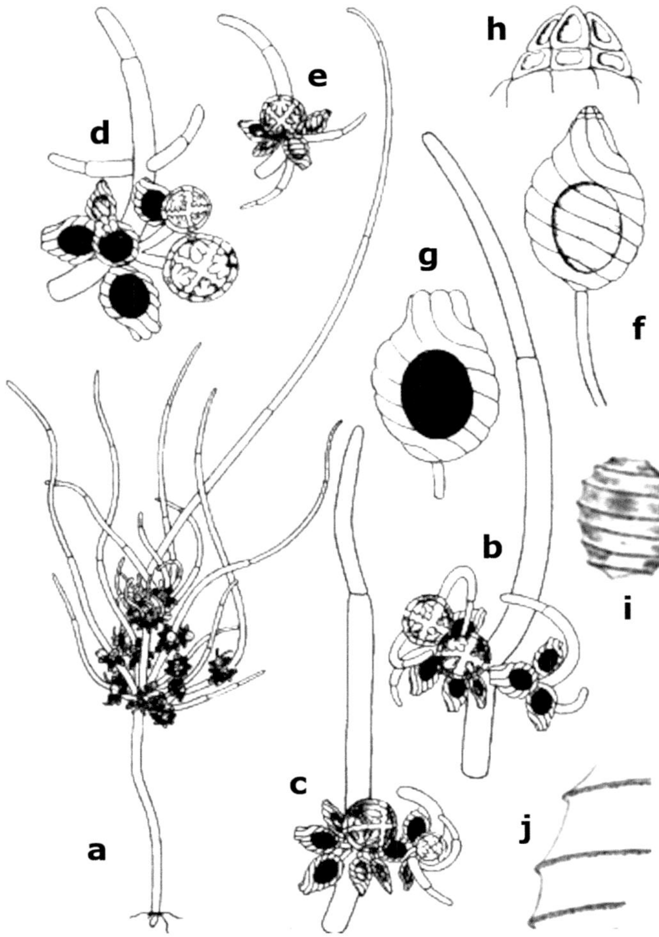

Fig. 78.1 *Tolypella salina*. **a** General appearance (4 cm total height); **b** Branchlets (5 mm total length); **c** Branchlet (4 mm total length); **d** Branchlet (4 mm length); **e** Branchlet (1.5 mm length), located at the base of a branchlet whorl; **f** Immature oogonium with coronula; **g** Mature oogonium without coronula; **h** Coronula; **i** Oospore (240 μm wide); **j** Oospore membrane with granulation at the ridge edges. Drawings: **a**, **b**, **c**, **d**, **e**, **f**, **g**, **h**—from Comelles (1986); **i**, **j**-from Corillion (1960)

the sediment, it is in this case only detectable by the long persistent pro-embryonic filament (or protonema), which emerges from the sediment and, at maturity, by the orange antheridia that manage to protrude. The vegetative parts comprise 1–3(6) small, very compact fertile heads, over which rises the long pro-embryonic filament. The axis of the primary and secondary rays is often very small (<3 mm) and the branchlet whorls seem completely agglomerated, giving the aspect of dense fertile heads. The vegetative apparatus of *T. salina* is slightly encrusted, in particular on the upper part of the pro-embryonic filament (Comelles, 1986).

Fig. 78.2 Habitus and habitat of *Tolypella salina*. **a** Cleaned individual, exhibiting polyembryony; **b** Typical appearance in the field; **c** Appearance of stands facing competition with filamentous algae; **d** Active salting basins on the Atlantic coast of France; **e–h** Life cycle of *T. salina*, starting with partly covered, but already antheridia-bearing plants (**e**), becoming fully fertile with developing oogonia (**f**), maturing and becoming fertilised at the onset of the dry period (**g**) and ripening when drying out, but still being wetted by moisture (**h**). Photographs: **a, d**-Didier Desmots; **b**, **c**, **e**, **f**, **g**, **h**-Elisabeth Lambert

Size: The species was originally described as mostly small, around 5 cm in height (sometimes up to 12 cm) (Corillion, 1960, 1975). However, size can reach (2)5–15 (20) cm without the pro-embryonic filament, depending on sufficient water depth (Lambert, 2012; Lambert et al., 2013; Mouronval et al., 2015).

Pro-embryonic filament: The whorls of branchlets are overtaken by a filament much longer and thicker than the branchlets. Its persistence is commonly observable throughout the growth period. Corillion (1960) described this pattern as follows: "The germinating oospore produces a relatively complicated pro-embryonic system. On the one hand, a very short main axis trunk divides laterally in the vicinity of the oospore into several proembryos (polyembryony which does not seem to have been previously reported in the *Charophyceae*); on the other hand, a rhizoid trunk of the same importance likely to split in the vicinity of the oospore without immediately giving birth to the delicate rhizoid typical of Characeae". The pro embryonic filament develops very quickly, when the plant is 1–2 cm high. It can reach 5.2–20 cm in large specimens in deeper water and has a dark green to brownish colour. The terminal pro-embryonic filament consists of 5–6 cells, whereas most other *Tolypella spp.* just have a 4-celled terminal filament (Corillion, 1960). Brownish pigmentation is visible from the pro-embryonic stages.

Main axis: The main axis is rather broad (0.25–0.50 mm to 1.1 mm in larger specimens). Internodes are small (0.5–3.5 mm), varying according to the site characteristics. In some cases, the internodes are shorter than the branchlets, resulting in dense spherical heads.

Cortex; spine-cells; stipulodes: Absent.

Branchlets: Whorls of fertile branchlets (usually 6, sometimes 8; 0.1–2.0 cm long) form very dense ovoid heads with a diameter of 5–7 mm. Branchlets are divided once or twice, slightly or strongly incurved. The primary rays are composed of 4–5 cells. Primary rays are 2–8 mm long and divide into 3–4 secondary rays at the first node. The secondary rays, consist of 3 to 4 cells and can be further divided into 2–3 celled tertiary rays (Corillion, 1960; Comelles, 1986; Lambert et al., 2013). Terminal segments are cylindrical-rounded and obtuse at their tips (sect. *Obtusifolia*). Short, indistinct and sterile branchlets can occur among the fertile ones (Cirujano et al., 2008; Comelles, 1986). Sterile branchlets were not observed in the type material "maybe due to the ecological conditions" (c.f. Corillion, 1960), but show up in the associated illustration.

Bract-cells; bracteoles; bractlets: None.

Sexuality: The species is monoecious, annual and vernal with early maturation. Gametangia can be observed as soon as the pro-embryo has developed its first whorl. Their disposition is typical of the genus *Tolypella*: Often quite numerous, the gametangia are present at the first node of the branchlets, but also at the bottom of the fertile whorls.

Oogonia: At the first nodes of the rays and also at the bottom of the whorl, oogonia ([2]4–6[8]) and antheridia (1 or 2, rarely 3) are developed. Oogonia are ovoid-subspherical, mostly stalked, sometimes sessile (Fig. 78.3). Spiral cells (usually 6 to 8 helical stripes) are sometimes slightly swollen at the top (Corillion, 1975). The coronula is deciduous (Comelles, 1986). Morphometric data of the oogonia are listed in Table 78.1.

Oospores: Mature oospores are light brown to chestnut brown; "reddish brown, bright, slightly golden brown under the optical microscope" (Comelles, 1986), with (5)6(7) slightly prominent ridges. Light microscopy also reveals that the edges of the ridges are subtly granulated-vermiculated, whereas the oospore membrane itself is smooth (Corillion, 1960; Urbaniak et al., 2012). SEM images show a smooth

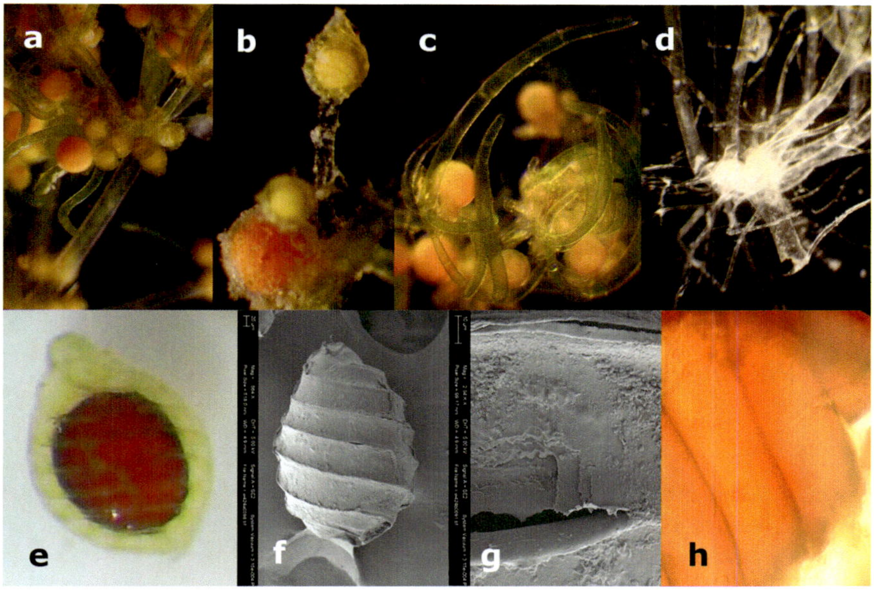

Fig. 78.3 Gametangia and oospores of *Tolypella salina*. **a** Fertile branchlets and rays (555 μm axis Ø); **b** Stalked oogonium and antheridium (510 μm Ø); **c** Fertile branchlet with central and adaxial antheridia (500 μm Ø) and rays with cylindrical-obtuse end cells; **d** Example of polyembryony; **e** Fertilised oogonium (435 μm length); **f**, **g** Oospore (**f**) and membrane surface ornamentation (**g**) both SEM pictures taken from type material; **h** Oospore membrane in light microscopy. Photographs: **a**, **b**, **c**, **d**, **e**, **h**-Elisabeth Lambert; **f**, **g**-Michelle T. Casanova

Table 78.1 Oogonia morphometry

Author	Length	Width	Coronula length and shape	Coronula width	Number of helical stripes
Corillion (1960, 1975)	400–620	430–480	78–117 Connivent to slightly spreading	120–156	6–8
Guerlesquin (1967)	400–700	429–480	78–117	120–156	–
Comelles (1986), Cirujano (1989), Cirujano et al. (2008)	400–650	360–490	75–120 Connivent	–	7–8
Lambert et al. (2013)	380–510	220–520	55–100 Connivent	–	6–7
Total range	**380–700**	**220–520**	**55–120**	**120–156**	**6–8**

Length in the first column is excluding the coronula, its length and width are given separately; all values for length and width in μm

membrane surface with more or less vermicular tubes adhering to it (Cirujano et al., 2008; Comelles, 1986; see also Fig. 78.3). The mother oospore often remains in situ, emptied of its content, throughout the development of the adult plant (Corillion, 1960). Table 78.2 gives an overview of the morphometry and colour of the oospores.

Antheridia: Antheridia are stalked or sessile, large and can be observed with the naked eye between the lateral oogonia. Table 78.3 gives an overview of the literature data on antheridia.

Bulbils: None.

78.2 Phenology and Reproduction

See Fig. 78.2.

Typically a halophytic species with fast maturation, occurring in saline pond complexes with regular drying periods (mineralisation), the species exhibits different phenology patterns depending on the ecological conditions of the site.

Atlantic French coast: In some saline stations the life cycle lasts from November to May. In November, the first plants appear; by the end of December the first antheridia develop; at the beginning of February the first oogonia appear and antheridia maturation starts, followed by ripening of the oogonia; around the end of March to the beginning of April the fertilised oogonia can be observed and the first mature oospores while the water level is decreasing. *T. salina* ends its maturation in the dried-up

Table 78.2 Oospore morphometry

Author	Length	Width	Number and shape of striae	Colour	Membrane
Corillion (1960, 1975)	273–366	258–312	5–7 Slightly prominent	Light to chestnut brown	Smooth membrane surface; subtle granulation on the prominent edges of the ridges
Guerlesquin (1967)	273–366	258–318	6	-	Smooth with tubercular ridges
Comelles (1986), Cirujano et al. (2008)	325- 370	295–330	7 Slightly prominent	Reddish brown, bright, slightly golden brown	Smooth or more or less vermicular tubes adhering to the smooth surface
Lambert et al. (2013) and own observ.	320–480(560)	200–300	6–7 Slightly prominent	Light to chestnut brown	Smooth with subtly granulated edges of the ridges
Mouronval et al. (2015)	270–370	260–330	(5)6–7 Slightly prominent	Chestnut brown	Smooth to slightly tuberculated-verrucate
Urbaniak et al. (2012)	-	-	4–6 Well developed, slightly flanged 9–13 μm long	-	Smooth oospore wall; ridges without any ornamentation
Casanova pers. comm.	-	-	-	Dry: brown-black Fresh: chestnut	Smooth, edges of the ridges subtly granulated-vermiculated
Total range	**270–460(560)**	**200–330**	**(4)5–7 Slightly prominent**		

All values for length and width in μm

Table 78.3 Antheridia morphometry

Author	Diameter and attachment (μm)
Corillion (1960, 1975), Guerlesquin (1967)	480–546 stalked
Comelles (1986), Cirujano (1989), Cirujano et al. (2008)	450–625, stalked
Lambert et al. (2013)	420–750, stalked
Mouronval et al. (2015)	> 600
Total range	**420–750, stalked**

basins, but still in moist conditions, often among filamentous algae. Antheridia can be observed still in mid-May together with mature oospores and the remains of the decayed axis and branchlets.

Plants in other basins or saline stations start their life cycle later (February/March), and end it in May. The first individuals, already with antheridia, are observed in February or March, depending on the basin's conditions. During March the oogonia are formed, and become mature in April. Decay occurs in May when the sites become dry.

On the Mediterranean French coast germination has already started in the autumn. Small plants can be found in January and antheridia are already formed in (February) March to April. Fructification reaches its peak in May and plants have disappeared by the beginning of June; just some few individuals can still be observed at the beginning of July.

This species can germinate in abundance in certain years and be completely absent the year after, with no clear explanation.

78.3 Habitat Conditions and Ecology

See Fig. 78.2.

Habitat type: This species grows in a variety of hyperhaline to mesohaline wetlands at shallow water depths, rapidly heating up in clear sunny weather conditions. Salinity is determined by the inflow of sea water or high salt content in the soil and is subject to pronounced seasonal and/or annual variation. Germination occurs at low salinity; salinity gradually increases during the plants' cycle while the pond or basin is drying up. Regular drying periods (mineralisation) are essential. The species can survive one year without a dry period but not more. *T. salina* seems to be the only *Tolypella* species that can tolerate hypersaline conditions. *T. salina* has a reproductive capacity that involves the development of fertilised oospores in conditions of higher salinity than other *Tolypella* species (example: end of March, salinity 12.8 g/l, presence of antheridia; end of April, salinity 15.5 g/l, oogonia observed; mid-May, salinity 21.8 g/l, ripe oospores observed; ripe oospores could be seen when salinity was 33.3 g/l.). Although there can be some turbid periods, the water of the basins containing *T. salina* populations is mostly clear. Conditions favourable for *T. salina* can be found especially in the salt pond complexes of the French Atlantic and Mediterranean coast and Italy (Trapani province, Sicily), with active salting basins (salterns) as well as abandoned salting basins. A detailed overview of the conditions in these salting basins is given by Lambert (2012). The species also occurs in coastal lagoons, separated from the ocean by a pebble bank (French Atlantic and Mediterranean coast, Sardinia, Adriatic coast, Greece), which dry out in spring/summer. Other suitable habitats for *T. salina* are temporary brackish waterbodies in endorheic areas under semi-arid conditions (inland Spain), and lagoons with short immersion periods, fed seasonally with brackish, strongly alkaline waters (French Mediterranean coast and inland Spain, Mouronval et al., 2009, 2015).

Sediment: *Tolypella salina* grows predominantly on hard sandy (sandy-clay), less frequently on clay substrates, with sometimes a fairly large amount of shell fragments. Substrates can be covered by certain amounts of smooth silt (0.5–1 cm thick). In most basins where *T. salina* has been observed, the estimated proportion of sand is generally greater than that of clay. Comparison of the basins has shown that the nature of the substrate is not enough in itself to explain the distribution of the species. The drying period is important for mineralisation resulting in low levels of organic matter.

Nutrient conditions: *Tolypella salina* is found in oligo- to mesotrophic stations. The species is able to tolerate a broad range of pH from 7.3 to over 9.9 (Lambert et al., 2013; Mouronval et al., 2015).

Water chemistry: During the growing period of *Tolypella salina*: variation of salinity from 18 to 80 g/l (salt harvest); from 11–20 to 40–66 g/l (recently abandoned) and from 4 to 24 g/l (old abandoned salting basins).

Depth range: *Tolypella salina* is a heliophilous species preferring shallow water with a flat bottom, in 2–40(70) cm depth with an optimum between 2 and 15 cm. Water depths are a little higher in winter and decrease slowly in spring.

Associated species: *Tolypella salina*, as a pioneer species, can be found sparsely without associates, half buried in the sediment. Even in denser stands it can form open vegetation. However, it is often associated with *Lamprothamnium papulosum* in open to fairly dense vegetation (2–20 cm tall). It is also observed with other halophilic *Characeae* (*Chara galioides, C. aspera,* Mediterranean Chara, *C. canescens, T. hispanica*) and transgressives of the *Ruppietalia maritimae* (*Althenia filiformis, A. orientalis, Zannichellia pedicellata, Ruppia maritima*). On the Mediterranean coast it has also been observed with the *Marchantiidae*: *Riella helicophylla*. *T. salina* participates in the *Charetalia canescentis* where it has sociological affinities with the *Lamprothamnietum papulosi* Corill. 1953. As a replacement for the invalid association *Tolypelletum salinae* Corill. 1960, the *Lamprothamnietum papulosi tolypelletosum salinae* Corill. ex Desmots, Lambert, Le Bail, Mouronval and Philippeau subass. has been integrated in the French *Charetea* synthesis (Felzines and Lambert, 2012) alongside the typicum *Lamprothamnietum papulosi* Corill. 1953.

78.4 Variability

There is some variability of size and shape (number of fertile whorls and the length of the plant) which are linked to ecological conditions.

78.5 Identification and Risk of Confusion

The largest specimens of *Tolypella salina* may resemble *Tolypella nidifica*. It can also be confused with *Tolypella glomerata*, which develops larger antheridia when growing in brackish water compared to freshwater individuals. However, *T. glomerata* is characterised by being dull, whereas *T. nidifica* and *T. salina* are always shiny (K. van de Weyer pers. comm.). The very large and persistent pro-embryonic filament as well as the polyembryony, irrespective of being not exclusive to *T. salina*, are good characters because of their constant presence in this species. Especially the lack of the large proembryonic filament can be seen as an indication that it is not *T. salina*, which is often half-buried and only detected by them. *T. nidifica* and *T. salina* can be separated by oogonia characters. *T. salina* oogonia are smaller (220–520 μm width) and have fewer helical stripes (6–8) than *T. nidifica* (400–650 μm width, 8–10 helical stripes). They also differ with respect to the colour of the oogonia (*T. salina*: orange to pale brown; *T. nidifica*: dark reddish brown). Antheridia diameters overlap and may serve only as an additional character; *T. salina* (420–750 μm mostly stalked) and *T. nidifica* (250–550 μm). When they are ripe, the oospore characters also help to distinguish the two species. *T. salina* oospores are only slightly smaller than *T. nidifica* ones with a large range of overlap; however, the number of striae (5–7 prominent ones for *T. salina* in contrast to (6) 7–9 for *T. nidifica*) and colour (brown-black to chestnut for *T. salina* in contrast to various shades of dark-red for *T. nidifica*) give clear indications for correct determination. The oospore membrane structure serves as an ultimate proof, especially the subtly granulated-vermiculated edges of the ridges of *T. salina* oospores, a character not observed on *T. nidifica* oospores. Additionally, the habitat characteristics should be taken into account. *T. salina* can withstand far higher salinities compared to *T. nidifica* and, in contrast to *T. nidifica*, requires dry periods for the production of ripe oospores. It is also a very early vernal species. While waiting for new genetic results, already the chromosome numbers published so far (Table 78.4) strongly support the species rank of *T. salina* by showing clear differences between the chromosome numbers of the 3 species: *T. salina* n = 50; *T. nidifica* n = c. 42; *T. nidifica* ssp. *occidentalis* n = c. 20 to c. 25; *T. glomerata* n = 15 to c. 20.

Table 78.4 Chromosomes numbers (n =) for *T. salina*, *T. nidifica* and *T. glomerata*

Country and authors	*Tolypella salina*	*Tolypella nidifica*	*Tolypella nidifica* subsp. *occidentalis*	*Tolypella glomerata*
France (Croix-de-Vie) Corillion and Guerlesquin (1959), Guerlesquin (1984)	50	-	-	-
Germany (Schleswig–Holstein) Lindenbein (1927)	-	c. 42	-	-
France (Oléron) Guerlesquin (1967, 1984)	-	-	c. 20	-
Italy (Maremma, Romagne) Guerlesquin (1967, 1984)	-	-	c. 25	-
France (Camargue) Corillion and Guerlesquin (1959), Guerlesquin (1967, 1984)	-	-	-	15
Italy (Maremma, Pouilles, Romagna, Venetia, Basilicata) Corillion and Guerlesquin (1963), Guerlesquin (1967, 1984)	-	-	-	c. 20
France (Aude, Somme) Guerlesquin (1967, 1984)	-	-	-	c. 20
Total range	**n = 50**	**n = c. 42**	**n = c. 20, c. 25**	**n = 15, c. 20**

78.6 Distribution

See Fig. 78.4.

The distribution of *T. salina* in Europe is rather limited, being mainly present on the French Atlantic and Mediterranean coasts in very salty to brackish wetlands (Beaubert et al., 2015; Borel, 2013; Citoleux et al., 2007; Desmots et al., 2011; Felzines and Lambert, 2012, 2016; Gernigon et al., 2014; Lafon and Lambert, 2020; Lambert et al., 2013). It is also observed in the same type of habitats in Greece, Sardinia & Sicily (Blaženčić et al., 2006a; Becker, 2019; Lambert et al., 2014; Romanov et al., 2019). In Spain, except for a few localities on the Atlantic coast, most of the occurrences are inland brackish water sites (Cirujano, 1989; Cirujano and Medina, 2002; Cirujano et al. 2013; Comelles, 1982, 1986; Espinar et al. 1997; García Murillo et al., 2006). A record from Britain (Bryant and Stewart, 2011), based on a specimen stored in L, has been redetermined as *T. glomerata* (K. van de Weyer pers. comm.).

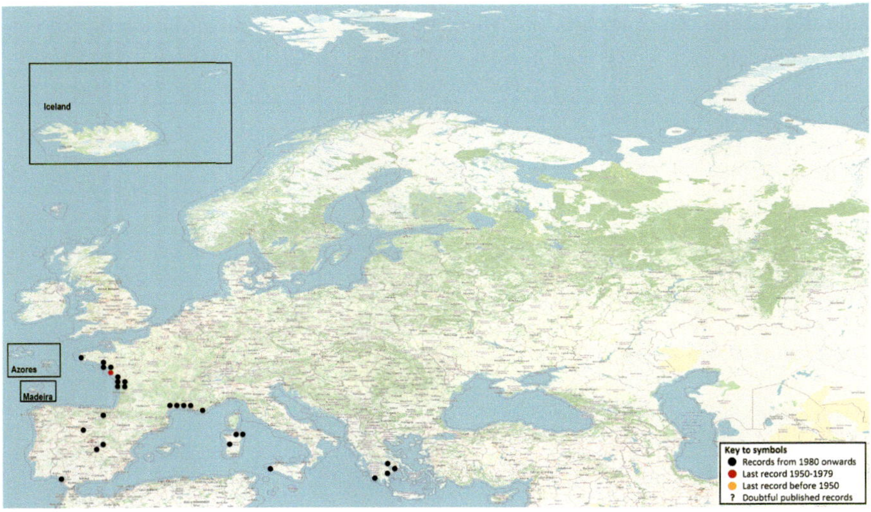

Fig. 78.4 European distribution of *Tolypella salina*. Black dots are records from 1980 on, red dots from 1950 to 1979 and yellow dots before 1950; unconfirmed/doubtful published records (if applicable) are marked with "?" on top. Map prepared by Heiko Korsch

78.7 Status and Threats

Tolypella salina is the only charophyte which is nationally protected in France (MEDDE, 2013). This was an important step towards the protection of certain vulnerable localities at the French littoral. In Spain *T. salina* is one of the rarest and most threatened charophyte species, being known only from a few locations. In the Eastern Adriatic, *T. salina* is considered as Critically Endangered (Blaženčić et al., 2006b). As this species typically lives in salting basins—active as well as abandoned ones—where the salinity regime depends on human activity, it is threatened by all actions that change the hydroregime, especially with respect to the required dry period. Some former salting basins have been reconstructed to provide warm water for shellfish, fish or worm aquaculture or, alternatively, *Salicornia* sp. cultivation, consequently making them unsuitable for *T. salina*. Other forms of habitat loss are complete drainage of former basins and wetlands for urban and industrial development as has occurred at the species *locus classicus*. It is important to promote the importance of the mosaic structure of salting areas with respect to the ecological conditions necessary for charophytes.

Acknowledgements We thank William Perez; Kenneth Karol and Petra Nowak for sharing the results of their work in progress about genetic differentiation between *T. salina, T. nidifica* and *T. glomerata* as well as Michelle T. Casanova for the work concerning the oospores from these species. Uwe Raabe provided specimens from Greece to E. Lambert for inspection, which is gratefully acknowledged as well as Nick Stewart's advice for amending the manuscript.

References

Beaubert, R., Desmots, D., Gernigon, J., Lambert, E., Lemesle, J.-C., & Micheau, A. (2015). *Macrophytes patrimoniales des lagunes de l'île de Ré.* Réserve Naturelle «Lilleau des Niges», LPO. [in French].

Becker, R. (2019). The Characeae (Charales, Charophyceae) of Sardinia (Italy): Habitats, distribution and conservation. *Webbia, 74*, 83–101.

Blaženčić, J., Stevanović, B., Blaženčić, Z., & Stevanović, V. (2006a). Distribution and ecology of charophytes recorded in the west and central Balkans. *Cryptogamie, Algologie, 27*, 311–322.

Blaženčić, J., Stevanović, B., Blaženčić, Ž., & Stevanović, V. (2006b). Red data list of Charophytes in the Balkans. *Biodiversity and Conservation, 15*, 3445–3457.

Borel, N. (2013). *Site des Salins d'Hyères. Nouvel état des lieux de la végétation aquatique.* Rapport d'étude. Conservatoire du Littoral; Communauté d'Agglomération Toulon Provence Méditerranée. [in French].

Bryant, J. A., & Stewart, N. F. (2011). Phylum Chlorophyta. Order Charales. In D. M. John, B. A. Whitton, & A. J. Brook (Eds.), *The freshwater algal flora of the British Isles. An identification guide to freshwater and terrestrial algae* (2nd ed., pp. 742–765). Cambridge University, Cambridge.

Cirujano, S., & Medina, L. (2002). *Plantas acuáticas de las lagunas y humedales de Castilla-La Mancha.* Junta de Comunidades de Castilla-La Mancha. [in Spanish].

Cirujano, S., Cambra, J., Sánchez Castillo, P. M., Meco, A., & Flor Arnau, N. (2008). *Flora ibérica. Algas continentales. Carófitos (Characeae).* Real Jardín Botánico, Madrid. [in Spanish].

Cirujano Bracamonte, S., Guerrero Maldonado, N., & García Murillo, P. (2013). The genus *Tolypella* (A. Braun) A. Braun in the Iberian Peninsula. *Botany Letters, 160*, 121–129.

Cirujano, S. (1989). *Tolypella salina* Corillion en la laguna de Carralogroño (Álava). *Anales del Jardín Botánico de Madrid, 45*, 547. [in Spanish].

Citoleux, J., Lambert, E., & Guerlesquin, M. (2007). Les Characées de la Baie d'Audierne: évolution du site au cours des 15 dernières années. *Conservatoire Botanique National de Brest, Echos du Réseau pour l'Inventaire et la Cartographie Armoricaine, 20*, 55–64. [in French].

Comelles, M. (1982). *Noves localitats i revisió de la distribució de les espècies de caròfits a Espanya.* Tesis de licenciatura. Facultad de Biología, Universidad Central de Barcelona. [in Spanish].

Comelles, M. (1986). *Tolypella salina* Corillion, carofícea nueva para España. *Anales del Jardín Botánico De Madrid, 42*, 293–298. [in Spanish].

Corillion, R. (1953). *Lamprothamnium papulosum* J. Groves . *Bulletin de la Société Scientifique de Bretagne, 28*, 33-41. [in French].

Corillion, R. (1960). *Tolypella salina*, sp. nov., Charophycée nouvelle des marais de Croix-de-Vie (Vendée). *Revue Algologique, Nouvelle Série, 5*, 198–207, pl. 21–22. [in French].

Corillion, R. (1975). *Flore et végétation du Massif Armoricain. T. IV. Flore des Charophytes (Characées) du Massif Armoricain et des contrées voisines d'Europe occidentale.* Jouve, Paris. [in French].

Corillion, R., & Guerlesquin, M. (1959). Premières observations cytotaxonomiques sur le genre *Tolypella* (Charophycées). *Bulletin de la Société d'Études Scientifiques d'Angers, N.S. 2*, 167–170. [in French].

Corillion, R., & Guerlesquin, M. (1963). Contribution à l'étude des végétations de Charophycées d'Italie péninsulaire (étage méditerranéen). *Bulletin de la Société Scientifique de Bretagne, 37*, 193–211. [in French].

Desmots, D., & Le Bail, J. (2011). Redécouverte de *Tolypella salina* Cor. 1960 en Pays de Loire. *Conservatoire Botanique National de Brest, Echos du Réseau pour l'Inventaire et la Cartographie Armoricaine, 24*, 31–36. [in French].

Espinar, J. L., Cirujano, S., & García Murillo, P. (1997). Contribución al conocimiento de los carófitos del parque nacional de Doñana. *Acta Botanica Malacitana, 22*, 209–211. [in Spanish].

Felzines, J.-C., & Lambert, E. (2012). Contribution au prodrome des végétations de France : Les Charetea fragilis F. Fukarek 1961. *Le Journal de Botanique de la Société Botanique de France, 59*, 133–188. [in French].

Felzines, J.-C., & Lambert, E. (2016). Contribution au prodrome des végétations de France : Modification de la structure syntaxonomique des Charetea et compléments. *Le Journal de Botanique de la Société Botanique de France, 74*, 41–55. [in French].

García Murillo, P., Fernández Zamudio, R., Cirujano, S., & Sousa, A. (2006). Aquatic macrophytes in Doñana protected area (SW Spain). An overview. *Limnetica, 25*, 71–80.

Gernigon, J., Desmots, D., & Lemesle, J. C. (2014). *Découverte de trois espèces de macrophytes aquatiques d'intérêt patrimonial sur l'ile de Ré. Etat des connaissances et perspectives.* Ed. RN Lilleau des Niges, LPO, MEDD. [in French].

Guerlesquin, M. (1967). *Recherches caryotypiques et cytotaxinomiques chez les Charophycées d'Europe occidentale et d'Afrique du Nord.* Thèse Université de Toulouse, Ed. Jouve, UCO, Angers. [in French].

Guerlesquin, M. (1984). Nombres chromosomiques et ploïdie chez les Charophytes. *Cryptogamie, Algologie, 5*, 115–126. [in French].

Lafon, P., & Lambert, E. (2020). État des connaissances sur les Charophytes d'Aquitaine et de Poitou-Charentes, pour un inventaire actualisé. *Bulletin de la Société Botanique du Centre-Ouest, 51*, 15–44. [in French].

Lambert, E. (2012). *Tolypella salina R. Cor., sur la façade atlantique. Contribution à l'amélioration des connaissances.* Université Catholique de l'Ouest, Angers; Fédération des Conservatoires Botaniques Nationaux. [in French].

Lambert, E., Desmots, D., Raabe, U., Le Bail, J., Lachaud, A., Grillas, P., Martinez, L., Mouronval, J. B., Lochon-Menseau, S., Michaud, H., & Borel, N. (2014). *Tolypella salina R. Cor., 1960; a species still overlooked.* Presentation 19th GEC, 11–14 Sept 2014, Vilnius, Lithuania.

Lambert, E., Desmots, D., Le Bail, J., Mouronval, J.-B., & Felzines, J.-C. (2013). *Tolypella salina* R. Cor. on the French Atlantic coast: Biology and ecology. *Acta Botanica Gallica, 160*, 107–119.

Lindenbein, W. (1927). Beitrag zur Cytologie der Charales. *Planta, 4*, 437–466. [in German].

MEDDE [Ministère de l'Environnement, du Développement Durable et de l'Ecologie] (2013). *Arrêté du 23 mai 2013 portant modification de l'arrêté du 20 janvier 1982 relatif à la liste des espèces végétales protégées sur l'ensemble du territoire national.* NOR: DEVL1308547A. J.O. 7 juin 2013. [in French].

Mouronval, J.-B., Olivier, A., & Galewski, T. (2009). *Les hydrophytes du salin du Caban.* Note technique de l'association Nacicca. [in French].

Mouronval, J. B., Baudouin, S., Borel, N., Soulié-Märsche, I., Klesczewski, M., & Grillas, P. (2015). *Guide des Characées de France méditerranéenne.* Office National de la Chasse et de la Faune Sauvage, Paris. [in French].

Romanov, R., Napolitano, T., van de Weyer, K., & Troia, A. (2019). New records and observations to the Characean flora of Sicily (Italy). *Webbia, 74*, 111–119.

Urbaniak, J., Langangen, A., & van Raam, J. (2012). Oospore wall ornamentation in the genus *Tolypella* (Charales, Charophyceae). *Journal of Phycology, 48*, 1538–1545.

Chapter 79
Taxon of Uncertain Affiliation *Chara filarszkyana* n. nov.

Roman Romanov and Attila Mesterházy

Scientific name: ***Chara filarszkyana*** **R. E. Romanov & Mesterházy n. nov. pro** ***Chara mucronata*** **Fil.**

≡ *Chara mucronata* Fil. in Mat. Term. Értes. 47: 671. 1930, n. ill. not *Chara mucronata* A. Braun in Ann. Sci. Nat., Bot. 1: 351. 1834.
Lectotype (designated here by R. E. Romanov & Mesterházy): Figs. 1–8 at page 673 in Filarszky in Mat. Term. Értes 47: 652–676. 1930.

Local names: No local names known.

79.1 Morphology

Figure 79.1, based on Filarszky (1930).

Habitus: The plant is bright green, unincrusted, small, growing as small bushes, variable in texture, resembling small tufted *Chara vulgaris*, bearing branchlets at the upper two thirds or half of the shoot.

Size: *Chara filarszkyana* is a small plant, 6–15 cm high.

Main axis: Internodes shorten towards the apex, moderately branched.

Cortex: Axial cortex is regularly triplostichous and isostichous, straight at the upper internodes, spirally twisted at the lowest internodes, sometimes diplostichous.

R. Romanov (✉)
Nišice Ujtin Potok b.b., Dobra Voda, 85356 Bar, Montenegro
e-mail: streptophytes@gmail.com

A. Mesterházy
Hunyadi Utca 55, H-9500 Celldomolk, Hungary
e-mail: amesterhazy@gmail.com

H. Schubert et al. (eds.), *Charophytes of Europe*,
https://doi.org/10.1007/978-3-031-31898-6_79

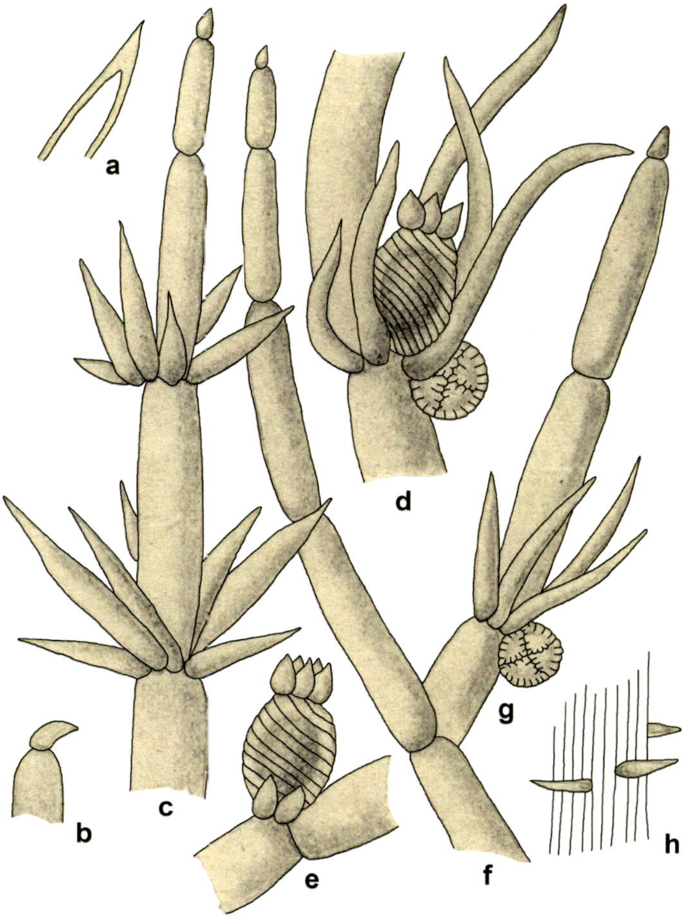

Fig. 79.1 Lectotype of *Chara filarszkyana*. **a** Stipulode ending with thickened cell wall at the tip; **b** End of branchlet with an abnormally bent small mucronate end cell; **c** Ecorticated sterile branchlet with two nodes; **d** Fertile node of ecorticate branchlet with conjoined gamethangia; **e** Branchlet node with oogonium; **f** Ecorticated nodeless (undifferentiated) six-celled branchlet; **g** Node with antheridium and ecorticated end of branchlet; **h** Triplostichous axis cortex with small spine-cells. Drawing: redrawn from Filarszky (1930)

Spine-cells: The spine-cells are solitary, sparse, straight, squarrose, thin, aculeate, long-pointed, with incrassed tip cell wall, variable in length, up to half of the main axis diameter.

Stipulodes: The stipulodes are short (undeveloped), papilliform, diplostephanous with extrastipulodes (?), but described as 2-3-stephanous with the third row not always visible. The length of stipulodes described in the protologue is in contradiction with the single incompletely illustrated stipulode (Fig. 79.1a).

Branchlets: The branchlets are 1–3 cm in length, up to 1 cm in the lower nodes, with one, two or even three nodes, with a three-celled nodeless end-segment with gradually narrowing cells, including a small or mucronate long-pointed end cell. The branchlets are illustrated as completely-ecorticate even for branchlets with two nodes, but the first, or first and second, or only the second segments, can be corticate according to the protologue. The first only, second only, or both first and second nodes are fertile, less often all three nodes are fertile. The nodeless branchlets are elongated, with five-(?) to six-cells. Upper branchlets envelop the apex.

Bract-cells: The bract-cells as illustrated are verticillate at sterile nodes and unilateral at fertile ones, but are described as unilateral, aculeate, long-pointed, variable in length, more or less unequal to each other in the same pair, from papilliform and quarter of the length of the oogonium to approximately 3-times longer.

Bracteoles: The bracteoles are more or less similar to the anterior bract-cells.

Sexuality: *Chara filarszkyana* is monoecious with conjoined and sejoined gametangia, somewhat protandrous, but only solitary antheridia form at the upper branchlets.

Oogonia: The oogonia are solitary, three-times larger than the antheridia, their spiral cells forming not less than 10 helical stripes and not swollen at the oogonium the apex. The coronula was illustrated as upright.

Oospores: No information is available.

Antheridia: The antheridia are solitary and octoscutate. No further information is available.

Bulbils: Not described.

79.2 Phenology and Reproduction

There is no information about phenology available. As this species was found in thermal springs, probably the seasons don't affect the phenology.

79.3 Habitat Conditions and Ecology

Habitat type: The habitat and environment are poorly known; up to date *C. filarszkyana* is known from water bodies associated with a thermal spring at the Római Baths in Budapest. Its recollection seems to be impossible because of serious transformation of the environment and disappearance of these water bodies (checked by A. Mesterházy). The species was collected in the outflow of thermal springs in Óbuda. There were many natural springs with 18-21 °C temperature at the Római-baths.

Sediment: Unknown.

Nutrient conditions and water chemistry: Total dissolved solids of thermal water is ca. 720 mg/l, because of its origin from Triassic-Eocene limestone. The lukewarm karst water consists of high $Ca(HCO_3)_2$ value and has slightly alkaline pH.

Depth range: It seems to be a shallow water species.

Associated species: The springs where *Chara filarszkyana* was collected were the richest site of charophytes in Hungary. Filarszky detected there 9 charophyte species with 40 forms. Besides *C. filarszkyana* the most interesting species were *C. crassicaulis* (*C. vulgaris* var. *crassicaulis*) and *C. papillosa*, which occurred in Hungary only in these springs (Boros, 1937). Regarding vascular plants, the site was associated with *Potamogeton coloratus*, *Persicaria amphibia*, *Stuckenia pectinata*, *Utricularia vulgaris*. The very rare *U. bremii* also occurred there (Filarszky, 1930).

79.4 Identification and Risk of Confusion

This species was not included in any manuals for European charophytes subsequent to its publication. The description of this species is incomplete mostly because of the missing traits of the gametangia and oospores, as well as their size. *Chara filarszkyana* can be recognised as a species from the subsection *Grovesia*. The combination of short papilliform stipulodes with somewhat elongated, solitary, pointed spine-cells is unusual for this subsection. Therefore, at the moment it seems to be better to keep this species separate from the similar *C. kokeilii* which is recognised by several authors as a gymnophyllous morphotype of *C. tenuispina*. *Chara filarszkyana* differs from *C. tenuispina* with short papillate stipulodes by the extrastipulodes (?), few nodes of the branchlets (1–2, rarely 3) and shorter bract-cells, and differs from *C. kokeilii* in having mostly 1–2 nodes on the branchlets, although its maximal number (rarely occurring 3) overlaps the known variability of *C. kokeilii*, which sometimes has two nodes at the lowest branchlets (Krause, 1990a, 1990b, 1997; Koletić et al., 2021). *Chara tenuispina* has a scattered distribution in slightly brackish shallow waterbodies in Hungary (A. Mesterházy ined.), whereas *C. kokeilii* was recorded once near Szeged by Filarszky (1930), but this record is not confirmed by herbarium specimens (A. Mesterházy ined.).

Further collections and detailed examination of the specimens are required to confirm the species rank of *C. filarszkyana*. The search of its specimens in BP by A. Mesterházy was unsuccessful. Some parts (incl. types of charophytes) of BP Herbarium collection were moved to Alsópetény before the siege of Budapest in 1945 and unfortunately the charophyte collection was destroyed there. Most probably the original material of *Chara filarszkyana* was included in this collection (Boros, 1947).

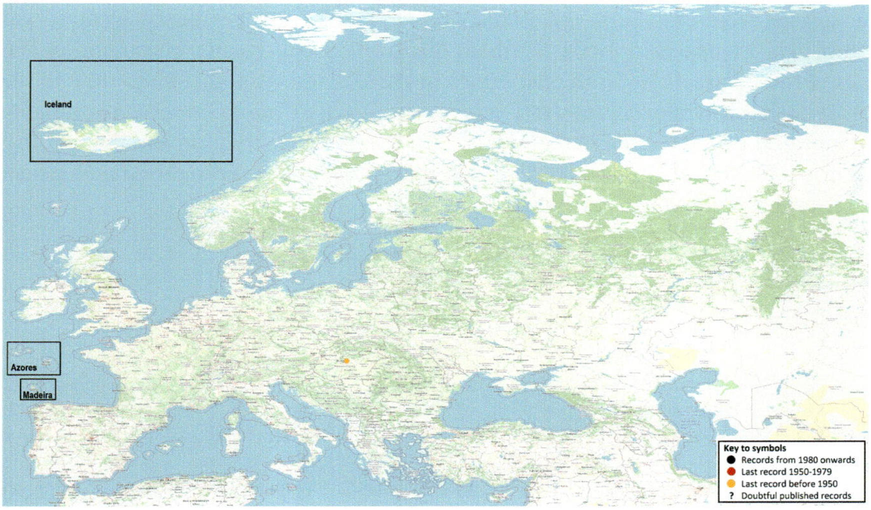

Fig. 79.2 Distribution of *Chara filarszkyana* in Europe. Black dots are records from 1980 on, red dots from 1950 to 1979 and yellow dots before 1950; unconfirmed/doubtful published records (if applicable) are marked with "?" superimposed. Map prepared by Heiko Korsch

79.5 Distribution

See the Fig. 79.2.

To date *C. filarszkyana* is known only from the northern territory of Budapest (Óbuda: Római-spa) (Filarszky, 1930). Authentic material (not found in BP) consists of: 1. Óbuda, Aquincum, a melegvizű csatornában [Budapest, Aquincum, in the warm water canal]/8 May 1905/N. Filarszky. 2. Tómedence kifolyásának legfelső részében, sebesen folyó vízben [Budapest, Óbuda thermal spring, in the upper part of the outflow of the lake basin, it flows rapidly in water]/18 May 1926/N. Filarszky. 3. Puskaporos-malom. m. [Budapest, Óbuda thermal spring, the old gunpowder mill, mixed with *Chara fragilis* Desv. f. *filiformis* Mig.]/20 October 1878/W. Steinitz.

79.6 Status and Threats

Chara filarszkyana has been assessed as EX (Extinct, European and Globally) in the European Red List (see Chap. 9). Currently the thermal springs of Óbuda supply the basins of the Római-spa, and these have been completely explored and managed. In ancient times, the Romans used the water of the springs to supply Aquincum, and later in the Middle Ages mills operated here. Despite regular use, the springs were still in natural conditions in the early twentieth century. The peat meadows along the springs were drained in the 1920s, and the lake together with the springs, was completely managed in 1958. By then, due to increasing water consumption and karst water abstractions from coal mines in the Dorog-Basin, the water flow of the springs had drastically decreased (Csemez et al., 1998). *Chara filarszkyana* probably disappeared from the area in the 1930s. Due to habitat destruction and the fact that the last collection is almost 100 years old, this species can be recognised as "Extinct (Ex)" according to IUCN categories.

References

Boros, Á. (1937). A magyarországi hévizek felsőbbrendű növényzete. *Botanikai Közlemények, 34*, 85–118. [in Hungarian].

Boros, Á. (1947). Szakosztályi ügyek az 1947 évi január hó 9én tartott 493, ülés. *Botanikai Közlemények, 44*, 79–80. [in Hungarian].

Csemez, A., Lorberer, Á., & Molnár, M. (1998). Mesél Óbuda földje. Óbuda-Békásmegyer természeti-táji értékei. *Guckler Károly Természetvédelmi Alapítvány*. [in Hungarian].

Filarszky, N. (1930). A Budai hőforrások nyílt vizeinek Chara-féléi és néhány más érdekesebb, ritkább algafaja. Matematikai Természettudományi Értesítö A. M. Tud. Akadémia III. *Osztályának Folyóirata, 47*, 652–674. [in Hungarian].

Koletić, N., Rimac, A., Šegota, V., Vuković, N., Blaženčić, J., & Alegro, A. (2021). Updated distribution of rare stonewort *Chara kokeilii* A.Braun (Charophyta) in Southeast Europe with environmental notes. *Cryptogamie, Algologie, 42*, 59–66. https://doi.org/10.5252/cryptogamie-algologie2021v42a5

Krause, W. (1990a). Über einen Fund der *Chara tenuispina* im Bodenseegebiet mit Blick auf die Gesamtverbreitung der Pflanze. *Carolinea, 48*, 31–36. [in German].

Krause, W. (1990b). Zum Neufund der *Chara kokeilii* A. Br. im Skutarisee. *Carinthia II, 180*, 705–713. [in German].

Krause, W. (1997). Charales (Charophyceae) Süßwasserflora von Mitteleuropa. In H. Ettl, G. Gärtner, H. Heynig, & D. Mollenhauer (Eds.), *Süßwasserflora von Mitteleuropa* Bd. 18. Fischer. [in German].

Chapter 80
Taxon of Uncertain Affiliation "Mediterranean Chara"

Irmgard Blindow, Ralf Becker, Roman Romanov, Hendrik Schubert, and Klaus van de Weyer

Previously included in: ***Chara baltica*** **(Hartm.) Bruzelius, Observ. Charae: 11. 1824**.

Taxonomic note: This taxon is under evaluation. Morphologically, it is clearly separated from *Chara baltica* (see below) and therefore, should not be attributed to this species. Further evaluation is needed to clarify the status of this taxon, its separation from other taxa, and its distribution range. Until a decision is reached about its taxonomic status, the taxon is attributed the name "Mediterranean Chara".

Local names: none.

I. Blindow (✉)
Biologische Station Hiddensee, Universität Greifswald, 18565 Kloster, Germany
e-mail: blindi@uni-greifswald.de

R. Becker
Am Schützenplatz 58, 26121 Oldenburg, Germany
e-mail: becker.r@posteo.de

R. Romanov
Nišice Ujtin Potok b.b., Dobra Voda 85356, Bar, Montenegro
e-mail: streptophytes@gmail.com

H. Schubert
Universität Rostock Institut für Biowissenschaften, Albert-Einstein-Straße 3, 18059 Rostock, Germany
e-mail: hendrik.schubert@uni-rostock.de

K. van de Weyer
lanaplan, Lobbericher Str. 5, 41334 Nettetal, Germany
e-mail: klaus.vdweyer@lanaplan.de

H. Schubert et al. (eds.), *Charophytes of Europe*,
https://doi.org/10.1007/978-3-031-31898-6_80

80.1 Morphology

See the Figs. 80.1, 80.2, 80.3 and 80.4.

Morphological details summarised below originate from Blume et al. (2009), Cirujano et al. (2008), Firbas and Al-Sabti (1995), Mouronval et al. (2015) and Zeneli and Kashta (2016) combined with data provided by M. Boegle and own observations, if not stated otherwise.

Habitus: Green to greyish green. Without or with weak encrustation.

Size: Medium-sized to large, 10–110 cm long.

Main axis: Diameter variable, about 0.6–2.0 mm.

Cortex: Diplostichous, rarely (partly) triplostichous and tylacanthous to isostichous, rarely with aulacanthous sections.

Spine-cells: Acute, 400–2500 μm, often variable in length even on the same individual plant, single or in clusters of 2–3, mainly pointing in the same direction or (somewhat) star-like. If in bunches, one of the spine-cells can be far longer than the other(s) which can be very short, or one of the spine-cells is very short compared to the other two.

Stipulodes: Acute, rarely ovoid, and arranged in two regular tiers of almost equal length, or the upper tier slightly longer than the lower, variable in length, from about 300 to 1100 μm long, shorter than or about as long as the axis diameter.

Branchlets: 8–10(11) branchlets in a whorl; branchlets consist of 4–9 segments, (1)2–6 corticate, 1–2(5) ecorticate, the penultimate segment somewhat inflated if ecorticate. The corticate part of the branchlet is usually far longer than the ecorticate part; in plants from Lake Preola, Sicily, cortication patterns were highly variable on the same individual plants, with only 1–2 corticate segments and the ecorticate part far longer than the corticate part in some whorls.

Bract-cells: Adaxial bract-cells are rather long, (460)600–2600 μm and up to 4 times as long as the branchlet diameter. In most collections, adaxial bract-cells are longer than abaxial which can be almost rudimentary, but (almost) verticillate bract-cells have been observed in about 1/3 of the collections checked.

Bracteoles: Long, in most cases far longer than the bract-cells, 2–3 times as long as the ripe oogonium, occasionally even longer, (1100)1900–2600(3800) μm long.

Sexuality: Monoecious, often richly fertile. Gametangia are conjoined on the 3-5 lowest branchlet nodes.

Oogonia: Coronula cells are spreading in some collections, not spreading in others (own data), convergent according to Cirujano et al. (2008). An overview of the morphometric data known so far is given in Table 80.1.

Fig. 80.1 Habitus of "Mediterranean Chara". **a**, **b** Field photographs of "Mediterranean Chara" from Li Cossi, Costa Paradiso, Sardinia; **c** Collection from Cala Fuili, Orosei, Sardinia with short to medium-sized bract-cells, stipulodes and spine-cells, showing the arrangement of gametangia; **d** Habitus of "Mediterranean Chara" from southern France; **e** Collection with long spine-cells, stipulodes and bract-cells from Albufera close to Valdivia, Spain, in B. Photographs and drawings: **a**, **b**—Klaus van de Weyer; **c**—Ralf Becker (from Becker, 2019); **d**—Drawing after Werner Krause, modified by Dagmar Wassong in van de Weyer and Schmidt (2018); **e**—Irmgard Blindow

Fig. 80.2 Details of "Mediterranean Chara". **a** Whorl of branchlets with short or moderately long ecorticate segments; **b** Fertile branchlet with verticillate bract-cells and conjoined gametangia; **c** Young oogonia; **d** Whorl with long ecorticate segments, stipulodes; **e** Nodal bulbil; **f** Main axis cortex. Photographs: Roman Romanov

Oospores: An overview of the morphometric data known so far is given in Table 80.2. Figures 80.3f and 80.4 depict the appearance of the oospores of this taxon. To what extent peculiar features as e.g. the prolonged apex and smooth oospore membrane are of discriminative value still needs to be studied.

Antheridia: Antheridia are orange to red. Mean values of antheridia diameter is around 500 μm (own data). An overview of the morphometric data known so far is given in Table 80.3.

Fig. 80.3 Details of "Mediterranean Chara" II. **a** Lago Preola, Sicily, locality with dense vegetation of "Mediterranean Chara"; **b** Collection of "Mediterranean Chara" during 2018; **c**, **d**, **e** Plants from Lake Preola under the stereomicroscope; **c** Acute stipulodes and rather long bract-cells; **d** Main axis with long, single spine-cells; **e** Cross-section through the axis showing tylacanthous cortication; **f** SEM image of oospore. Photographs: **a**, **b**—Irmgard Blindow, **c**, **d**, **e**, **f**—Michelle T. Casanova

Bulbils: Multicellular bulbils (modified main axis nodes) are not uncommon (Cirujano et al., 2008; own data; see Fig. 80.2). Mouronval et al. (2015) mention the occurrence of both main axis node and rhizoid bulbils.

Table 80.1 Oogonia morphometry

Author	Length	Width	Coronula length	Coronula width	Number of helical stripes
Firbas and Al-Sabti (1995)	920–1100	505–610	150–200	370–400	14–15
Cirujano et al. (2008)	875–1000	570–700	160–210	–	–
Zeneli and Kashta (2016)	800–900	500–600	180–230	330–360	–
Own data	580–1100	390–730	120–190	190–340	10–16
Total range	**580–1100**	**390–730**	**120–230**	**190–400**	**10–16**

Length in the first column excludes the coronula; its length and width are given separately, all values for length and width in μm

Table 80.2 Oospore morphometry

Author	Length	Width	Number of striae	Colour
Firbas and Al-Sabti (1995)	640–700	410–440	13	Black
Cirujano et al. (2008)	650–875	475–650	–	Black
Blume et al. (2009)	650–925	350–475	13–20	Black
Zeneli and Kashta (2016)	–	–	11–12	Black
Own data	670–730	390–510	–	Black
Total range	**640–925**	**350–650**	**11–20**	

All values for length and width in μm, own material and material presented by Blume et al. (2009) measured in dry state. For Blume et al. (2009), unpublished raw data were analysed

80.2 Phenology and Reproduction

Little is known about the phenology of the taxon. According to Mouronval et al. (2015), the life cycle can be both annual and perennial, the latter hibernating by both green parts and bulbils. Unripe oogonia and antheridia can still be found in August, ripe oospores as early as May. Plants collected in Sicily during September had ripe oogonia and oospores.

80.3 Habitat Conditions and Ecology

Habitat type: The taxon is found in brackish water. Almost all sites are close to the coast of the Mediterranean Sea, only a few from inland brackish water. The taxon is found in permanent, semi-permanent and temporary habitats such as brackish lakes,

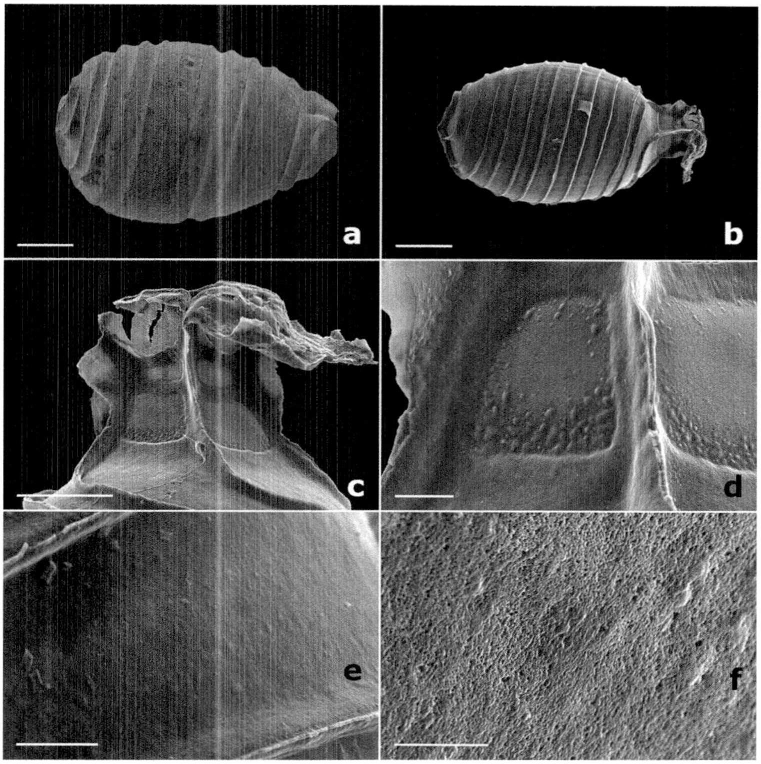

Fig. 80.4 SEM pictures of oospores of "Mediterranean Chara". **a** Gyrogonite (scale bar 200 μm); **b** Oospore, side view (scale bar 200 μm); **c** Apex of an oospore (scale bar 100 μm); **d** Detail of apex (scale bar 50 μm); **e** Smooth oospore-membrane surface (scale bar 20 μm); **f** Oospore-membrane in high resolution (scale bar 4 μm). Photographs: Roman Romanov

Table 80.3 Antheridia morphometry

Author	Diameter (μm)
Firbas and Al-Sabti (1995)	460–520
Cirujano et al. (2008)	470–600
Zeneli and Kashta (2016)	500–550
Own data	450–550
Total range	**450–600**

Own material was measured in dry state. No information about method is given by Cirujano et al. (2008), but most probably, wet or ethanol-fixed material was measured. No information about method is given in the other references river mouths, interdunal ponds, fish ponds, saltmarsh ditches and former stone quarries, as well as in periodically inundated dune depressions (Becker, 2019; Blaženčić et al., 2006a; Mouronval et al., 2015; Rodrigo and Alonso-Guillén, 2016; Zeneli and Kashta, 2016).

Sediment: The taxon was found on soft, sandy to silty sediments (Zeneli and Kashta, 2016; own data).

Nutrient conditions: In Cala Fuili, Sardinia, nitrate-N concentrations were 165 $\mu g\ L^{-1}$, total N 630 $\mu g\ L^{-1}$, PO_4-P 9 $\mu g\ L^{-1}$ and total P 20 $\mu g\ L^{-1}$ (R. Becker unpubl. data). In Lake Preola, Sicily, nitrate-N concentrations were below 1 $mg\ L^{-1}$, PO4-P concentrations below 30 $\mu g\ L^{-1}$ (Palumbo et al., 2018).

Water chemistry: The taxon was recorded at pH between 7.4 and 9.2 and salinities between 2 and 10 psu, occasionally at salinities as low as 0.8 psu or as high as 25–30 psu (Mouronval et al., 2015; Palumbo et al., 2018; Rodrigo and Alonso-Guillén, 2016; Zeneli and Kashta, 2016; van de Weyer unpubl. data). In Cala Fuili, Sardinia, Ca concentrations were 103 $mg\ L^{-1}$, Mg concentrations 175 $mg\ L^{-1}$ (R. Becker unpubl. data). In Lake Preola, Sicily, Ca concentrations were 98 $mg\ L^{-1}$, Mg concentrations 67 $mg\ L^{-1}$ (Palumbo et al., 2018).

Depth range: "Mediterranean Chara" has been found from very shallow water to about 5 m depth (Mouronval et al., 2015; Zeneli and Kashta, 2016; own data).

Associated species: The taxon was recorded together with other charophytes (*Tolypella glomerata*, *T. hispanica*, *Chara galioides*, *C. aspera*, *C. canescens*, *C. connivens*, *C. vulgaris*, *C. globularis*, *C. zeylanica*, *Nitella hyalina*), *Potamogeton coloratus*, *Stuckenia pectinata*, *Najas marina* ssp. *intermedia*, *Zannichellia* spp., *Ruppia* spp. (Guarino et al., 2019; Mouronval et al., 2015; Zeneli and Kashta, 2016; own data).

80.4 Variability

Like other taxa within the subsection *Hartmania*, "Mediterranean Chara" shows a rather high variability. Notably axis diameter, length of spine-cells, bract-cells and bracteoles as well as the number of oospore striae vary among populations.

80.5 Identification and Risk of Confusion

The taxon belongs to the subsection *Hartmania*, a complex of coarse, diplostichous *Chara* spp. that are hard to separate from each other. Although "Mediterranean Chara" has been determined as *Chara baltica* by several authors (e.g., Krause, 1997), it is morphologically closer to the cluster combining tylacanthous to isostichous monoecious taxa from freshwater and inland brackish water sites (see Chap. 15) than the cluster containing *Chara baltica*. It is now separated from *Chara baltica* on the basis of genetic differences (Boegle et al., 2010), failure of interfertility in crossing experiments (Blindow and Erichsen, 2010) and morphological differences such as smaller antheridia, acute and often longer spine-cells, stipulodes and bract-cells (own data). Oospores of "Mediterranean Chara" were morphometrically clearly separated from oospores of both *C. baltica* and *C. horrida* (Blume et al., 2009).

The determination of tylacanthous taxa of the subsection *Hartmania* is especially difficult in the Mediterranean area, where "Mediterranean Chara" overlaps morphologically with a number of other taxa. Most individuals are distinguished from *C. papillosa* by longer stipulodes, spine-cells, bract-cells and bracteoles, but short-spined individuals are hard to separate (see also Zeneli and Kashta, 2016). Mouronval et al. (2015) mention that both taxa are separated by ecological conditions with restriction of "Mediterranean Chara" to brackish and *C. papillosa* to freshwater habitats. However, *C. papillosa* can be found in both freshwater and inland brackish water sites (see Chap. 44). In crossing experiments, a plant conforming to *C. papillosa* collected from brackish water was interfertile with "Mediterranean Chara" (Blindow and Erichsen, 2010). "Mediterranean Chara" is separated from *C. aculeolata* by less dense and less clustered spine-cells. Romanov (2019) pointed out that the taxon may be confused with *Chara globata*, which, however, always has verticillate bract-cells. *C. globata* combines long bract-cells with short stipulodes and spine-cells, and is recognisable by its unique pattern of dense, "spiny" whorls and smooth axes (see Chap. 33). Long-spined individuals of "Mediterranean Chara" are hard to distinguish from *Chara corfuensis*. In this species, however, most collections have verticillate or near-verticillate bract-cells. Also *Chara affinis* is similar, but has often longer stipulodes and, especially, denser spine-cells.

As in *C. papillosa*, delicate, single-spined plants may be hard to separate from *C. contraria*.

80.6 Distribution

The taxon has been recorded from various countries around the Mediterranean Sea, almost exclusively from sites close to the coast of the Mediterranean Sea (see Fig. 80.5). Records are available from Spain, France, Italy, Slovenia, Albania, Greece. Several records from the French Atlantic coast have been attributed to this taxon. The taxon has also been reported from North Africa (Corillion, 1957, 1961; Corillion and Guerlesquin, 1971; García, 1993), but these records need to be confirmed. Records from freshwater highland lakes in the Maghreb (Muller et al., 2017) and from Egypt (Corillion and Guerlesquin, 1971) are possibly misidentifications (Romanov, 2019; see Chap. 33) (Fig. 80.5).

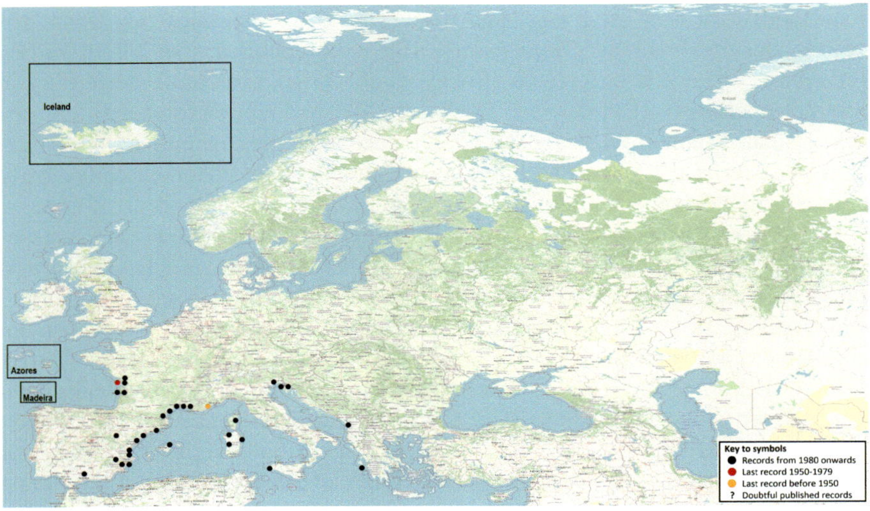

Fig. 80.5 European distribution of "Mediterranean Chara". Black dots are records from 1980 onwards, red dots last record from 1950 to 1979 and yellow dots last record before 1950; doubtful published records (if applicable) are marked with "?" superimposed. Map prepared by Heiko Korsch

80.7 Status and Threats

The taxon was classified as Critically Endangered in the Red List of the Balkans (Blaženčić et al. 2006b) but is not assessed in any other Red List in the Mediterranean area (Becker, 2016; see Chap. 9). As the taxon is under evaluation, it is attributed the status DD (Data Deficient) in the European Red List (see Chap. 9).

"Mediterranean Chara" is threatened by eutrophication, habitat destruction, tourism and changes in hydrological regimes (Becker, 2019; Krause, 1997). Becker (2019) proposes the implementation of action plans to protect threatened charophytes on Sardinia. Thereby, the taxon belongs to the group of brackish water charophytes which was proposed the highest priority for protection measures.

Acknowledgements We thank Michael Boegle for the contribution of data. Anders Langangen and Uwe Raabe kindly provided samples. Jelena Blaženčić and Angelo Troia gave valuable information.

References

Becker, R. (2016). Gefährdung und Schutz von Characeen. In Arbeitsgruppe Characeen Deutschlands, *Armleuchteralgen: Die Characeen Deutschlands* (pp. 149–191). Springer. [in German].

Becker, R. (2019). The Characeae (Charales, Charophyceae) of Sardinia (Italy): Habitats, distribution and conservation. *Webbia, 74*, 83–101.

Blaženčić, J., Stevanović, B., Blaženčić, Ž, & Stevanović, V. (2006a). Distribution and ecology of charophytes recorded in the west and central Balkans. *Cryptogamie, Algologie, 27*, 311–322.
Blaženčić, J., Stevanović, B., Blaženčić, Ž., & Stevanović, V. (2006b). Red data list of Charophytes in the Balkans. *Biodiversity and Conservation, 15*, 3445-3457.
Blindow, I., & Erichsen, I. (2010). Cross-fertilisation within the *Chara hispida* species complex. *Charophytes, 2*, 59–65.
Blume, M., Blindow, I., Dahlke, S., & Vedder, F. (2009). Oospore variation in closely related *Chara* taxa. *Journal of Phycology, 45*, 995–1002.
Boegle, M., Schneider, S., Schubert, H., & Melzer, A. (2010). *Chara baltica* Bruzelius 1824 and *Chara intermedia* A. Braun 1859. Distinct species or habitat specific modifications? *Aquatic Botany 93*, 195–201.
Cirujano, S., Cambra, J., Sánchez Castillo, P. M., Meco, A., & Flor Arnau, N. (2008). *Flora ibérica. Algas continentales. Carófitos (Characeae).* Real Jardín Botánico, Madrid. [in Spanish].
Corillion, R., (1957). Les Charophycées de France et d'Europe Occidentale. *Bulletin de la Société Scientifique de Bretagne, 32*(Hors série I & II), 1–499. [in French].
Corillion, R. (1961). Les stations à *Tolypella hispanica* Nordstedt (Charophycée) du territoire français. *Revue Algologique. Nouvelle Série, 6*, 33–37. [in French].
Corillion, R., & Guerlesquin, M. (1971). Notes phytogéographiques sur les Charophycées d'Egypte. *Revue Algologique, 2*, 177–191. [in French].
Firbas, P., & Al-Sabti, K. (1995). Cytosystematic studies on the charophyta in Slovenia. *Archives of Biological Science Belgrade, 47*, 49–58.
García, A. (1993). *Chara baltica* Bruz. (Charophyta) en Argentina. *Candollea, 48*, 297–303 [in Spanish].
Guarino, R., Marcenò, C., Ilardi, V., Mannino, A. M., & Troia, A. (2019). One *Chara* does not make *Charetea* in the Mediterranean aquatic vegetation. *Webbia, 74*, 139–147.
Krause, W. (1997). Charales (Charophyceae) Süßwasserflora von Mitteleuropa. In H. Ettl, G. Gärtner, H. Heynig, & D. Mollenhauer (Eds.), *Süßwasserflora von Mitteleuropa*. Fischer [in German].
Mouronval, J. B., Baudouin, S., Borel, N., Soulié-Märsche, I., Klesczewski, M., & Grillas, P. (2015). *Guide des Characées de France méditerranéenne.* Office national de la Chasse et de la Faune Sauvage. [in French].
Muller, S. D., Rhazi, L., & Soulié-Märsche, I. (2017). Diversity and distribution of Characeae in the Maghreb (Algeria, Morocco, Tunisia). *Cryptogamie, Algologie, 38*, 201–251.
Palumbo, V., Nicolosi, M., Pellerito, S., & Costa, N. (2018). *Piano operativo Acque Sotterranee, Task T.3.* Convenzione ARPA—DAR. [in Italian].
Rodrigo, M. A., & Alonso-Guillén, J. L. (2016). Charophyte flora in a Ramsar Mediterranean wetland (Albufera de València Natural Park, Spain) during the period 2007–2010. *Botanica Serbica, 40*, 205–215.
Romanov, R. (2019). New interesting records of charophytes (Charales, Charophyceae) from Eurasia and Africa. *Webbia, 74*, 159–166.
van de Weyer, K., & Schmidt, C. (2018). Bestimmungsschlüssel für die aquatischen Makrophyten (Gefäßpflanzen, Armleuchteralgen und Moose) in Deutschland: Band 1: Bestimmungsschlüssel, ed. 2. *Landesamt für Umwelt (LfU) Brandenburg, Potsdam, Fachbeiträge des LfU Brandenburg 119*, 1–180. [in German].
Zeneli, V., & Kashta, L. (2016). Some Charophyta (Charales) from coastal temporary ponds in Velipoja area (North Albania). *Journal of Environmental Science and Engineering, B5*, 69–77.

Chapter 81
Taxon of Uncertain Affiliation *Nitella opaca* var. *attenuata*

Michelle T. Casanova

Scientific name: ***Nitella opaca*** **C. Agardh ex Bruzelius) C. Agardh var.** ***attenuata*** **H. Groves & J. Groves in J. Bot. 19: 356 (1881)**

Neotype (designated here by Casanova): Stonyford Pond nr. Hythe, So. Hants, England/11.4.1882/H. G[roves] (BM 013738481).

Taxonomic note: In BM there is material of *N. opaca* var. *attenuata* collected from the type locality (Stonyford Pond nr. Hythe, So. Hants, England, BM013738481) which has oospores similar to those of *N. acuminata*. It is designated as a neotype here because the original material of *N. opaca* var. *attenuata* is missing (J. Wilbraham pers. comm.). The plant was described as having branchlets much more slender and longer than *N. opaca* with the secondary segments almost as long as the primary segments of the branchlets and resembling slender forms of *N. syncarpa*.

Local names: ENGLISH: Dark stonewort.

81.1 Morphology

See Fig. 81.1.

Habitus: *Nitella opaca* var. *attenuata* resembles a slender form of *N. syncarpa* (Groves and Groves, 1881) with an extremely slender and diffuse habit (Groves and Bullock-Webster, 1920).

Size: *Nitella opaca* var. *attenuata* is medium-sized (5–15 cm) but may reach up to 30 cm if growing in deeper water.

M. T. Casanova (✉)
273 Casanova Rd, Westmere, Victoria 3351, Australia
e-mail: m.casanova@nhm.ac.uk

H. Schubert et al. (eds.), *Charophytes of Europe*,
https://doi.org/10.1007/978-3-031-31898-6_81

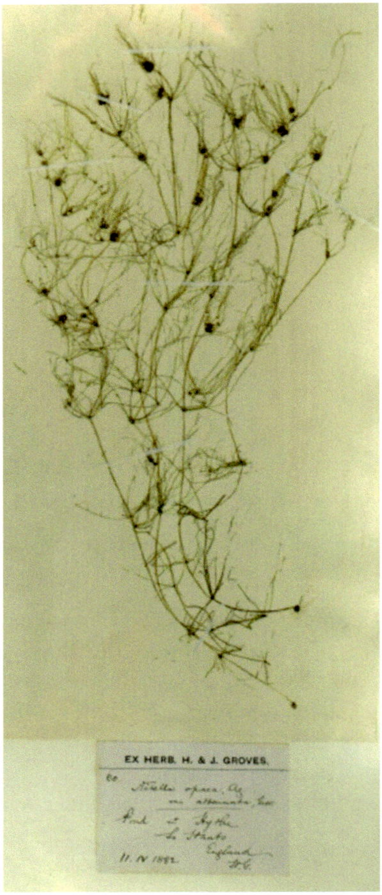

Fig. 81.1 *Nitella opaca* var. *attenuata*. Picture of the Topotype (BM013938474) Photograph: retrieved from https://data.nhm.ac.uk

Axis: Axis diameter is about 0.5 mm.

Sterile whorls: Sterile whorls consist of c. 6 long branchlets, which are 1 × divided into (1)2–4 secondary segments. The whorl diameter is 10–20 mm; secondary branchlet segments to 2 mm long (0.75″) and under 250 μm (0.1″) wide; (Groves and Groves, 1881).

Fertile whorls: Not described.

Dactyls: In *Nitella opaca* var. *attenuata* the dactyl tips are described as attenuate (rather than acute as in *N. opaca* var. *opaca*), and the illustration in Groves and Bullock-Webster (1920) shows a more elongate dactyl tip than the other examples illustrated.

Sexuality: *Nitella opaca* var. *attenuata* is not specifically described as monoecious or dioecious, but it is placed in *N. opaca* which is dioecious.

Oogonia: Oogonia are oval, with the two coronula cell rows of similar size, there are c. 8 stripes of the enveloping cells (convolutions) visible from the side.

Oogonia morphometry: No data are given for *Nitella opaca* var. *attenuata.*

Oospores: The oospore of *N. opaca* var. *attenuata* is described here as a slightly elongate sphere in side view, 340 × 280 μm (dry) with 7–8 striae with small flanges to 15 μm high, fossa width 30 to 65 μm (60 μm at mid-point), oospores without beak. The ornamentation is somewhat roughened and sparsely covered with flaccid, dentate projections 05–0.8 μm long (Fig. 81.2a–c).

Antheridia: Not described for *N. opaca* var. *attenuata.*

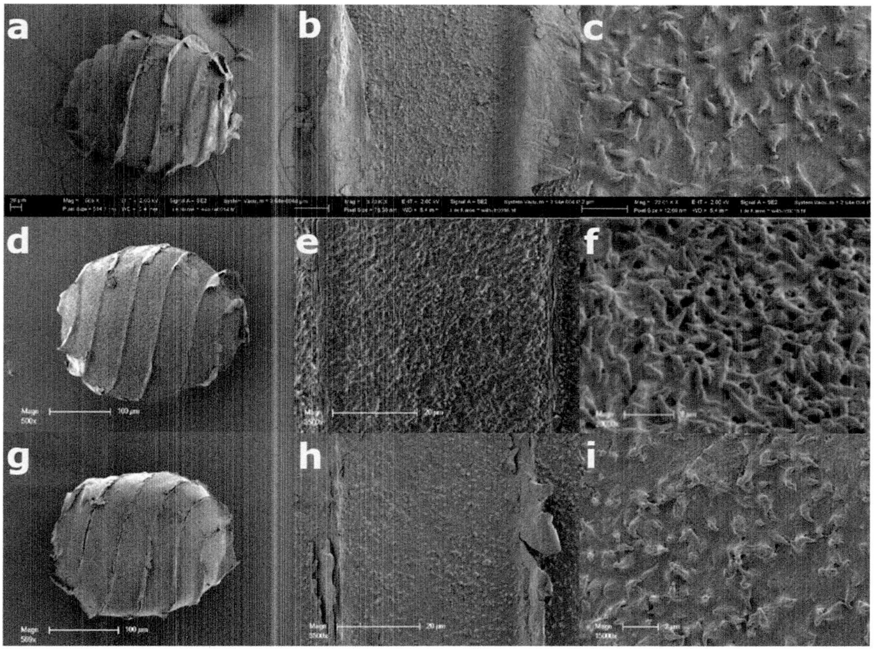

Fig. 81.2 Comparison of SEM-pictures of oospores of *Nitella opaca* var. *attenuata* (**a, b, c**), *N. acuminata* (**d, e, f**) and *N. acuminata* var. *belangeri* (**g, h, i**). Upper panels: Oospores of *Nitella opaca* var. *attenuata*. This material is not the type material, but was collected by H. Groves and came from the same locality as the type (Stonyford Pond near Hythe) in 1882 (the species was described in 1881 on the basis of material collected by H. Groves in 1878). **a** Side view of oospore; **b** View of the fossa between the striae; **c** Detailed view of fossa wall showing the less densely arranged, flaccid dentate ornamentation. Middle panels: Oospores of *Nitella acuminata* from Madagascar. This is not the type material which is from Reunion (Ille de Bourbon). **d** Side view of oospore; **e** View of the fossa between the striae; **f** Detailed view of fossa wall showing the densely arranged, flaccid dentate ornamentation. Lower panels: *Nitella acuminata* var. *belangeri* from the Philippines. **g** Side view of oospore; **h** View of the fossa between the striae; **i** Detailed view of fossa wall showing the less densely arranged, flaccid dentate ornamentation. Photographs: Michelle T. Casanova

Bulbils: None known.

81.2 Phenology and Reproduction

Material has been collected in April, May, August, September and October. Plants were not fertile when first collected in April, but oospores were available for examination from material collected in August. Plants found in 2016 are presumed to have germinated from the seed bank (Goodrich and Goldsmith, 2017).

81.3 Habitat Conditions and Ecology

Nitella opaca var. *attenuata* has been found in coastal ponds (Hythe, and Little Sea: note the Little Sea material has oospores typical of *Nitella opaca*), canals and lakes.

81.4 Variability

The variability within *Nitella opaca* var. *attenuata* is not known. Material with this name has been collected from several localities in England. The oospores of the plants from Little Sea are different (more like *N. opaca* var. *opaca*) from those on the material collected from the type locality.

81.5 Identification and Risk of Confusion

Nitella opaca var. *attenuata,* has single celled dactyls. When infertile it could be confused with *Nitella flexilis*, or infertile plants of *N. opaca* var. *opaca.* It is distinguished from those by the length of the dactyls and distinctive attenuate tips of the dactyls. It is possible that the designation of a variety of *N. opaca* is in error, because the oospores resemble those of *N. acuminata*. It is also possible that oospores became mixed with those of *N. acuminata* during processing for SEM.

81.6 Distribution

Nitella opaca var. *attenuata* occurs in sea-side ponds on the southern coast of Hampshire, Dorset, in Cornwall, and Frensham Great Pond in Surrey.

81.7 Status and Threats

The variety is apparently rare (records in BM date to 1950), and it has been rarely collected in the last 70 years (Goodrich and Goldsmith, 2017). It is possible that its habitat conditions have changed since then.

81.8 Remarks

Nitella opaca var. *attenuata* is assessed as a taxon of uncertain affiliation, largely due to a casual examination of two oospores (one from a plant from the type locality that resembles the oospore of *Nitella acuminata*: a species of Reunion, Indian Ocean Islands and India), and an oospore from a plant from Little Sea in Dorset that resembles other material called *N. opaca* from various localities (Fig. 81.2). It is possible that a mix up occurred in examination of the oospores, or that the specimen really is *Nitella acuminata* but the occurrence of *N. acuminata* is exotic and vagrant in England. A thorough review of all the specimens allocated to this variety is required.

References

Goodrich, S., & Goldsmith, B. (2017). *Heath Lake SSSI, Berkshire: Report on ecological lake monitoring 2016*. ECRC Research Report Number 178. University College London.

Groves, H., & Groves, J. (1881). Notes on the British Characeæ. *Journal of Botany, 18*, 353–356.

Groves, J., & Bullock-Webster, G. R. (1920). *The British Charophyta*. Ray Society.

Chapter 82
Taxon of Uncertain Affiliation
Nitella palhinhae

Roman Romanov

Scientific name: ***Nitella palhinhae*** **(A. G. Cunha) R. E. Romanov comb. nov**.
Basionym: *Tolypella palhinhae* A. G. Cunha in Bull. Soc. Portug. Sci. Nat. 12(8): 42. (1935).

Lectotype (Wood, 1965: 559): Lagoa dos Linhos, no Pinhal do Urso, ao S. da Fijueira da Foz [?], Portugal/May 1931/Dr. Goncalves da Cunha & L. Goncalves Sobrinho (LISU 45375).

Local names: Unknown.

82.1 Morphology[1]

See the Fig. 82.1.

Habitus: The plant is green, unencrusted, small, thin and flexible, richly (?) branched, consisting of many sprouts originating from the common base, with diffuse uncondensed single-furcated branchlets with tiny, inconspicuous heads at the tip (Cunha, 1935) or without them (Wood, 1965).

Size: Small plant, up to 16 cm high.

Main axis: Slender, up to 480 μ in diameter. The internodes appear to be always at least two times longer than the branchlets.

[1] Based on Cunha (1935), Corillion (1957a), Wood (1965).

R. Romanov (✉)
Nišice Ujtin Potok B.B., Dobra Voda, Bar 85356, Montenegro
e-mail: streptophytes@gmail.com

H. Schubert et al. (eds.), *Charophytes of Europe*,
https://doi.org/10.1007/978-3-031-31898-6_82

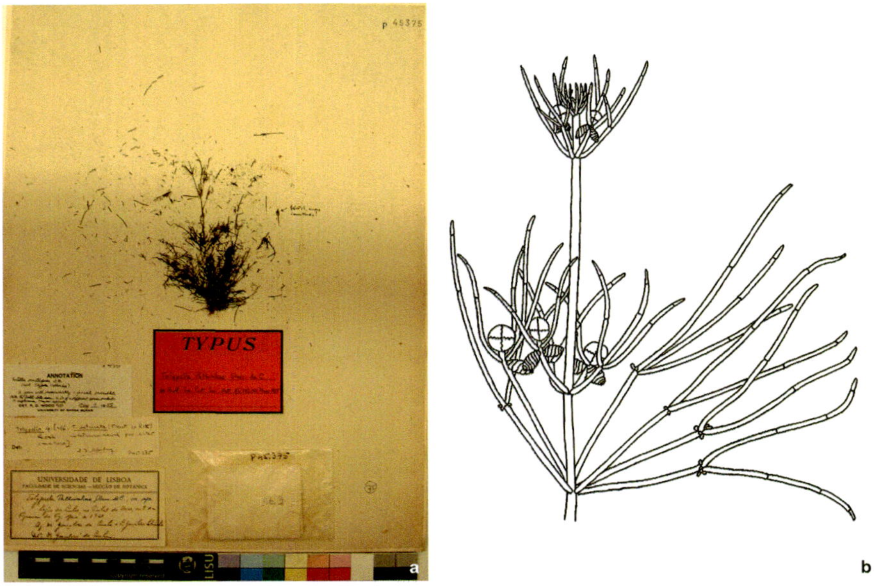

Fig. 82.1 *Nitella palhinhae*. **a** Type specimen (LISU); **b** Apical part. Photograph and drawing: **a**—Roman Romanov; **b**—Cunha (1935)

Branchlets: 5–6 in a whorl, 1(2?) furcate, 5–7 mm in length. Sterile branches are elongated, once furcate, with 2–4(5) dactyls. Fertile branches are usually much shorter. Primary segments are ca. 1/2–2/3 of the branchlet length. Dactyls are (1)2–4-celled, the end cell is short, conical, pointed (with a mucro).

Sexuality: Monoecious. The gametangia do not have surrounding mucus and are conjoined at the single furcation of the branchlet.

Oogonia: Solitary (?), mostly (?) geminate or triplicate, 450–550 μm in length and 300–400 μm in width, their spiral cells form 10 stripes in side-view and are not swollen at the oogonium apex. The coronula is persistent, ca. 25 μm in length and 60 μm in width, its cells are nearly equal in both rows.

Oospores: Dark yellow, 300 μm in length and 250 μm in width, with 6–7 flanged ribs.

Antheridia: Diameter ~500 μm, solitary, shortly stipitate, surrounded with 2–3 oogonia.

Bulbils: Unknown.

82.2 Phenology, Reproduction, Habitat Conditions and Ecology

Nitella palhinhae is probably a vernal species reproducing by oospores, its habitat conditions are poorly known. *N. palhinhae* was collected from a coastal lake, it seems to be a shallow water species.

82.3 Variability and Risk of Confusion

Corillion (1957a) suggested that *Tolypella palhinhae* could possibly be a regional variant of *Sphaerochara intricata (= Tolypella intricata)*, but *T. palhinhae* is here referred to *Nitella* because of the furcate rather than monopodial branchlets (Wood, 1965). Wood (1962; 1965) synonymised *T. palhinhae* with *N. ornithopoda*, and the decision was never disputed, but the former seems to be quite different from the latter, because it has diffuse once furcate branchlets without a central secondary segment, the first branchlet node is fertile, it has bigger gametangia that are geminate, and clustered oogonia as well as shortly stipitate antheridia. The habit of *N. palhinhae* is quite distinct because of the uncondensed long once-furcate branchlets. *N. palhinhae* is only known from the type locality. New specimens would be required for a detailed examination to confirm its species rank or merge it with another species.

82.4 Distribution

Nitella palhinhae is known from a single locality in Portugal (Cunha, 1935; LISU!, Fig. 82.2).

82.5 Status and Threats

Nitella palhinhae has been assessed as DD (Data Deficient) in the European Red List (see Chap. 9).

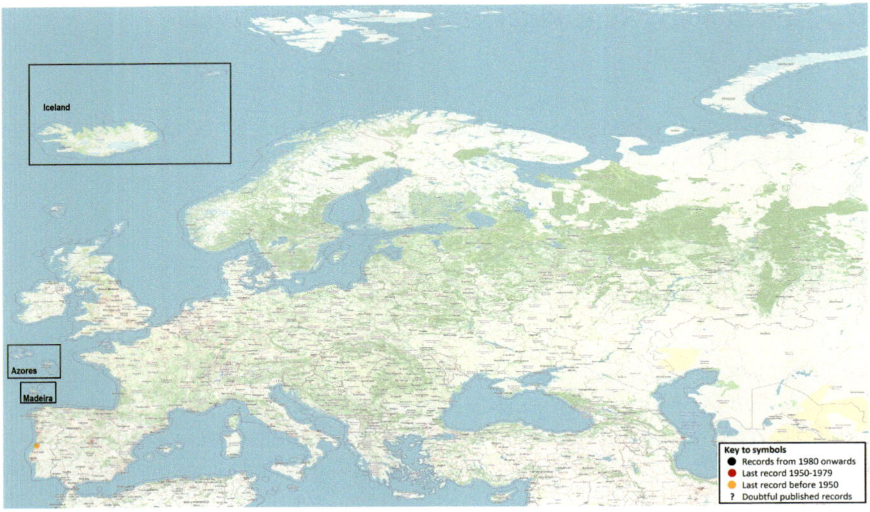

Fig. 82.2 European distribution of *Nitella palhinhae*. Black dots are records from 1980 on, red dots from 1950 to 1979 and yellow dots before 1950; unconfirmed/doubtful published records (if applicable) are marked with "?" superimposed. Map prepared by Heiko Korsch

Acknowledgements Ana Isabel Correia is kindly acknowledged for the encouragement and guidance during the study of the collection stored in LISU.

References

Corillion, R. (1957a). Les Charophycées de France et d'Europe Occidentale. *Bulletin de la Société Scientifique de Bretagne, 32* (Hors série I & II), 1–499. [in French].

Cunha, A. G. (1935). Quelques espèces nouvelles de Characées du Portugal. *Bulletin de la Société Portugaise des Sciences Naturelles, 12*, 41–58. [in French].

Wood, R. D. (1962). New combinations and taxa in the revision of Characeae. *Taxon, 11*, 7–25.

Wood, R. D. (1965). *Monograph of the Characeae*. J. Cramer.

Chapter 83
Addendum *Chara hydropithys*

Roman Romanov and Lefter Kashta

Scientific name: ***Chara hydropithys*** **Rchb., Handb. Gewächsk., ed. 2.: 1600. 1829**

Lectotype (Wood, 1965: 305): Surinam/1844/Weigelt (W).

Local names: RUSSIAN: Хара водяная сосенка; UKRAINIAN: Хара водяна сосонка

This species has been reported from a water body associated with a spring in the hills of Mesit, 7 km northeast of the city of Shkodër, Albania (Kashta, 1994, 2009). Its identification was based on Hollerbach and Krassavina (1983), but a voucher does not exist. Efforts to re-sample from the exact place where the plant was collected were unsuccessful (the waterbody does not exist anymore). Therefore, this species record can't be confirmed and *C. hydropithys* should be excluded from the list of European charophytes. The distribution area of *C. hydropithys* covers Africa, Southern and South-Eastern Asia, North and South America, mostly between 42°N and 30°S (Wood, 1965). The northernmost record in Eurasia is known from rice fields in Uzbekistan (Hollerbach and Krassavina, 1983; LE, checked by R. Romanov).

References

Hollerbach, M. M., & Krassavina, L. K. [Голлербах, М. М. & Л. К. Красавина]. (1983). *Определитель пресноводных водорослей СССР. 14. Charophyta [Determination key of freshwater algae of USSR 14. Charophyta]*. Nauka. [in Russian].

R. Romanov (✉)
Nišice Ujtin Potok b.b., Dobra Voda, Bar 85356, Montenegro
e-mail: streptophytes@gmail.com

L. Kashta
Research Center for Flora and Fauna, Faculty of Natural Sciences, University of Tirana, Place Mother Tereza, P.O. Box 183, Tirana, Albania

H. Schubert et al. (eds.), *Charophytes of Europe*,
https://doi.org/10.1007/978-3-031-31898-6_83

Kashta, L. (1994). Kontribut per algofloren e ujerave te embela te Shqiperise. *Buletini i Shkencave Natyrore, 1*, 63–67. [in Albanian].
Kashta, L. (2009). Flora harofite (Charophyta) e rajonit të Shkodrës. *Universiteti i Shkodrës "Luigj Gurakuqi" Buletini i Shkencave. Ser. Shkencave Natyrore, 59*, 97–106. [in Albanian].
Wood, R. D. (1965). *Monograph of the Characeae*. J. Cramer.

Chapter 84
Addendum *Chara socotrensis*

Nick Stewart

Scientific name: ***Chara socotrensis*** **Nordst. in Kuhn in Ber. Deutsch. Bot. Ges.1: 241. 1883**.

Lectotype (Wood, 1965: 281): Socotra. Im Bache bei Katheng/14. Mai 1881/G. Schweinfurth 690 (LD).

Local names: ENGLISH: Socotran Stonewort.

84.1 Morphology

See the Fig. 84.1.

Habit: Plants are glossy, mid-green to slightly olive-green, not to a little encrusted, with internodes often more or less as long as the branchlets but varying from 0.75–2.5 times.

Size: A slender to medium species, up to at least 30 cm tall.

Main axis: 550–680 μm diameter. Unbranched to very sparsely branched.

Cortex: More or less absent but rudimentary cortex is visible under high magnification, made up of chains of 3–4 minute globular cells 60–100 μm in diameter extending downwards from the axis nodes.

Spine-cells: Absent.

Stipulodes: Rudimentary to globular and up to 100(140) μm in diameter, occasionally the upper slightly elongate and ascending, very rarely up to 200 μm long, up to

N. Stewart (✉)
4 Hexton Road, Glastonbury BA6 8HL, United Kingdom
e-mail: nfstewart@freeuk.com

H. Schubert et al. (eds.), *Charophytes of Europe*,
https://doi.org/10.1007/978-3-031-31898-6_84

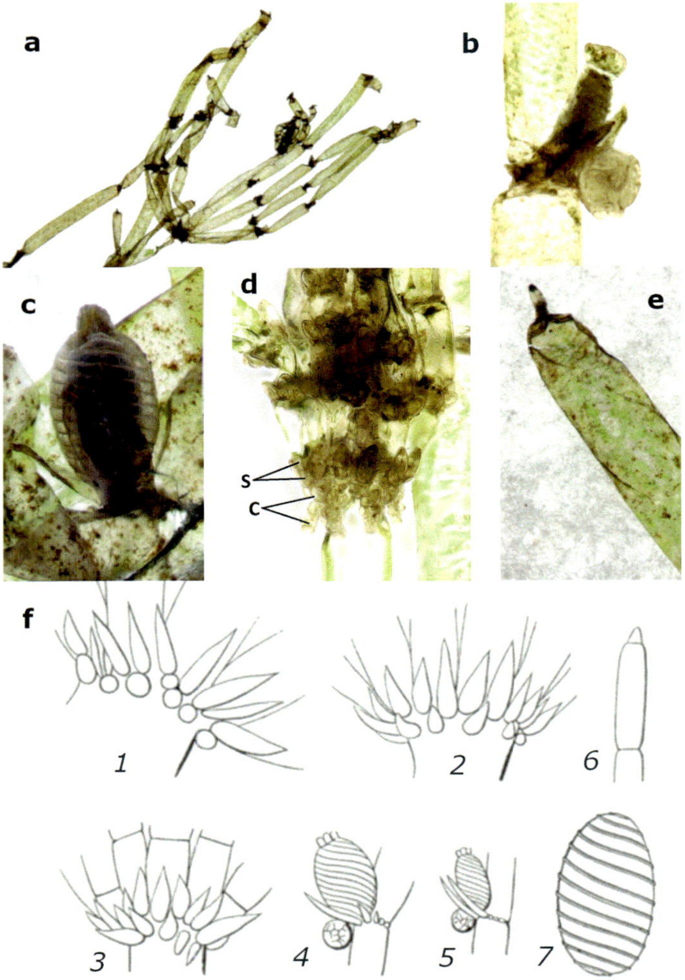

Fig. 84.1 *Chara socotrensis*. **a** Axis tip; **b** Branchlet node with gametangia; **c** Semi-mature oogonium; **d** Branchlet tip with end-cell forming apiculate tip; **e** Close-up of axis node showing two rows of minute stipulodes (S) and rudimentary cortical cells (C); **f** Illustrations from the type description (Nordstedt, 1883). 1–3 main axis nodes with stipulodes, 4–5 branchlet nodes with gametangia, 6 branchlet tip, 7 oospore. Photographs: Claudia Ferguson-Smyth, specimen from Lake of Menteith, Scotland, 2004, Hugh Dawson

twice as long as wide, obtuse. Their arrangement is unclear due to their small size but at least some are in two rows.

Branchlets: 6–8 in a whorl, ecorticate, to 4.5 cm long, 4–5 celled, constricted at the nodes, straight to flexuous. End-cell minute and forming an apiculate tip, to 200 μm long, acute to subacute, around a quarter of the width of the penultimate cell.

Sexuality: Monoecious. Gametangia at the lower 1–2 branchlet nodes, conjoined with a single antheridium and oogonium at each node.

Bract-cells: Present (with gametangia) at the lower 1–2 branchlet nodes, mostly globular but adaxial rarely slightly elongate up to twice as long as wide and obtuse.

Bracteoles: Two, 340–510 μm long, cylindrical below, gradually tapering above the middle, then narrowing abruptly to an obtuse or broadly acute tip, around half the length of the oogonia.

Bractlets: Absent.

Oogonia: Ovoid, 580–750 μm long (excluding coronula), 300–410 μm wide in Scottish specimens but oogonia not fully mature. Coronula 100–170 μm high, 170–210 μm wide, connivent to erect. Helical stripes indistinct due to immaturity, estimated at 10–11.

Oospores: Not mature. Estimated at c.650 μm long by c.400 μm wide, based on the most mature oogonia.

Antheridia: 300–400 μm diameter, based on dry, shrunken antheridia, likely to be 400–500 μm when fresh.

Bulbils: None recorded.

84.2 Phenology and Reproduction

Little is known about its phenology in its Scottish sites. It was recorded in August and September when the plants were fertile with antheridia and fairly well-developed but immature oogonia.

84.3 Habitat and Ecology

In its subtropical range it has been recorded from streams (Nordstedt, 1883), a calcareous swamp (Langangen, 2015) and a reservoir (Mondal et al., 2021). Both Scottish sites are large lowland mesotrophic to eutrophic lakes, well known for their rich aquatic flora. There, the species grew on sand (Loch Leven) in depths of 80–180 cm (both sites). One site (Loch Leven) has been badly impacted by nutrient enrichment in the past and although there have been improvements, this still remains a significant issue, with phytoplankton blooms occurring at some point in most summers. Mean TP concentrations were around 50–55 μg/l at the time that the species was recorded in the lake.

In Loch Leven it occurred in an open community with *Chara aspera*, *Nitella opaca* and *Potamogeton pusillus*. Other species occurring nearby included *Elodea canadensis*, *Tolypella glomerata*, *Callitriche hermaphroditica* and *Myriophyllum spicatum*. At Lake of Menteith it occurs with *Nitella opaca*, *Nitella translucens*, *Chara virgata*, *Isoetes lacustris*, *Fontinalis antipyretica*, *Potamogeton pusillus* and *Elodea canadensis*.

84.4 Variability

Specimens from both sites are similar but differ in several features from descriptions from within its natural range (see "remarks" section below).

84.5 Identification and Risk of Confusion

In Europe it is most likely to be confused with *Chara dissoluta* which also has only rudimentary cortex. However, the latter has a more matt appearance and has more obvious bundles of undeveloped cortical cells and stipulodes around the nodes. It also tends to have blunter (obtuse to broadly acute) bract-cells and branchlet tips. Other ecorticate species with *Chara*-like form and simple bracteate branchlets are *Chara braunii* and *Lamprothamnium* spp. Both of these have well-developed stipulodes in a single spreading row and *Chara braunii* also has a crown of several tiny cells at the tips of the branchlets.

84.6 Biogeography and Distribution

This species has been recorded from only a few localities in Yemen (Nordstedt, 1883), Somalia (Langangen, 2015), India (Mondal et al., 2021), Bolivia (Guerlesquin, 1981) and Brazil (Bueno et al., 2009). Other records are as forms of *Chara socotrensis* sensu Wood (1965), considered to be distinct species, with the possible exception of *C. socotrensis* var. *ebracteata* from NW India (Subramanian, 2005 ref. Gupta, 2012).

In Europe it has only been recorded as a presumed introduction in two lakes in central Scotland (Lake of Menteith and Loch Leven) (see Fig. 84.2). In both cases, it was recorded in 2004 in locally established populations. It is still present in Lake of Menteith but it has not been recorded since at Loch Leven.

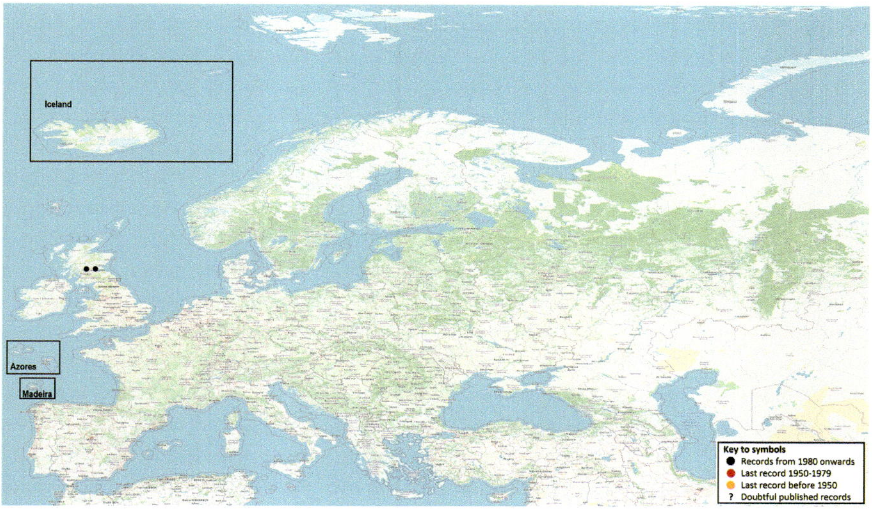

Fig. 84.2 European distribution of *Chara socotrensis*. Black dots are records from 1980 onwards, red dots last recorded from 1950 to 1979 and yellow dots last recorded before 1950; unconfirmed published records (if applicable) are marked with "?" superimposed. Map prepared by Heiko Korsch

84.7 Status and Threats

The presence of the species in Europe is presumed to have been as an introduction and it has not been refound at one of its sites, presumably having succumbed to the cold climate and/or the consequences of nutrient enrichment. The origin of the simultaneous introduction to two sites 55 kms apart is unclear. Both lakes are well-used by anglers and a speculative suggestion is that the plant was brought in from a trout hatchery when the lakes were stocked with fish.

84.8 Remarks

The description above is based on specimens from Scotland. These seem to fit the description of *Chara socotrensis* and were identified as this by Joop van Raam. However the status of this species along with the closely similar *Chara nuda* B. P. Pal from southern and south-eastern Asia (Gupta, 2012; Pal, 1929, 1932; Subramanien, 2002) needs review.

Chara *socotrensis* has been classified in the section *Protochara* of the subgenus *Charopsis* (e.g., Wood, 1965). The arrangement of the stipulodes in a single row (haplostephanous) is a key character in the distinction of sub-genus *Charopsis* from sub-genus *Chara*. On the type of *Chara socotrensis* the stipulodes are arranged rather irregularly and this has been interpreted in different ways by different authors. Nordstedt (1883) in his original description describes it as partly haplostephanous and partly diplostephanous and his illustrations show a regular bistipulate upper row and a more irregular unistipulate lower row (see Fig. 84.1f). However, Allen (1888) interprets these as a single tristipulate row. In contrast, Wood and Imahori (1964) and Wood (1965) describe and illustrate the stipulodes as being in a single unistipulate to bistipulate row, based on the same type material and one other specimen from Socotra Island. Langangen (2015) returns to Nordstedt's view of the species being partly haplostephanous and partly diplostephanous and found the same character in Somalian material. In contrast, all descriptions of *Chara nuda* describe the stipulodes as being in a single row, as also the records of *Chara socotrensis* from South America (Bueno et al., 2009; Guerlesquin, 1981) and India (Mondal et al., 2021). However, in most cases the stipulodes are so small that it is difficult to interpret the arrangement. Sources that indicate some elongation of the stipulodes (Bueno et al., 2009; Langangen, 2015; Nordstedt, 1883) all indicate the upper stipulodes as ascending and more or less appressed to the underside of the branchlets. This is the typical arrangement in diplostephenous species and not of haplostephanous species where the single row of stipulodes is patent, more or less bisecting the angle between the axis and the underside of the branchlets. It is clear, therefore, that *Chara socotrensis* has been misclassified and should be moved to the subgenus *Chara.*

In the Scottish material the stipulodes are so small that it is difficult to see how they are arranged. However, as part of the preparation of this account, sections of main axis were rewetted and photographed at higher magnification and these show, not only two rows of stipulodes but also some very reduced cortical cells (Fig. 84.1d).

Another complication is that the *Chara socotrensis /Chara nuda* complex has been recorded from very few localities that are widely scattered across the subtropical parts of the world (*Chara socotrensis* from Yemen, Somalia, India, Brazil and Bolivia and *Chara nuda* B. P. Pal from India and Myanmar). This raises the significant possibility that the complex is actually an agglomeration of several local but unrelated taxa that share the character of very reduced cortex. The presence in Myanmar of an intermediate taxon, *Chara denudata* f. *socotrensioides* R. D. Wood (Wood, 1965) which is very similar but has occasional elongate (but still very incomplete) cortical cells also fits with this possibility. If this is the case, then it follows that the Scottish material might also be an unusual form of a local taxon. Nearest in characters is *Chara dissoluta* (which is present in Ireland but not Britain) but as detailed under "risk of confusion", the two differ in appearance and a number of characters. Further study of the different populations is needed, supported by genetic studies similar to Casanova and Karol (2014).

A record of *Chara nuda* from Romania originated as a reference to "*Chara foetida* [var.] *nuda* A. Braun in litt." in Schur (1857). It was raised to species level by Fuss (1866) and subsequently cited by various sources as "*C. nuda* A. Braun" and most recently by Caraus (2017) as "*C. nuda* B. P. Pal". The name was never published by Braun and is a nomen nudum for a form of *Chara vulgaris* and unrelated to the taxon discussed here.

References

Allen, T. F. (1888). *The Characeæ of America. Part I.* New York.

Bueno, N. C., Bicudo, C. E. M., Biolo, S., & Meurer, T. (2009). Levantamento florístico das *Characeae* (Chlorophyta) de Mato Grosso e Mato Grosso do Sul, Brasil: *Chara. Revista Brasileira de Botanica, 32*, 759–774. [in Portuguese].

Caraus, I. (2017). Algae of Romania. A distributional checklist of actual algae. Version 2.4. *Studii si Cercetari Biologie, 7*, 1–1002.

Casanova, M. T., & Karol, K. G. (2014). A revision of *Chara* sect. *Protochara*, comb. et stat. nov. *(Characeae: Charophyceae). Australian Systematic Botany, 27*, 23–37.

Fuss, M. (1866). *Flora Transsilvaniae excursoria.* S. Filtsch, Cibinii (Hemanstadt).

Guerlesquin, M. (1981). Contribution à la connaissance des Characées d'Amérique du Sud (Bolivie, Equateur, Guyane française). *Revue d'Hydrobiologie Tropicale, 14*, 381–404. [in French].

Gupta, R. K. (2012). *Algae of India Volume 2. A checklist of Chlorophyceae, Xanthophyceae, Chrysophyceae* and *Euglenophyceae.* Botanical Survey of India, Ministry of Environment & Forests, Salt Lake, Kolkata. iii & 428, 8 pls.

Langangen, A. (2015). Some finds of charophytes from East-Africa (Zambia, Tanzania, Kenya and Somalia). *Acta Musei Nationalis Pragae B. Historia Naturalis, 71*, 239–248.

Mondal, K., Pal, T., & Keschr, J. P. (2021). On the morphology of five species of *Chara* (*Characeae, Charophyta* from West Bengal, India). *Nelumbo, 63*, 44–53.

Nordstedt, O. (1883). Characeae. *Berichte der Deutschen Botanischen Gesellschaft 1*, 238–243, Figs. 1–7.

Pal, B. P. (1929). A preliminary note on Burmese charophytes. *Journal of the Burma Research Society, 18*, 113.

Pal, B. P. (1932). Burmese Charophyta. *Journal of the Linnean Society of London, Botany, 49*, 47–92, 18 Figs.

Schur, F. (1857). Die Siebenbürgischen Characeen. *Oesterreichisches Botanisches Wochenblatt, 7*, 358–359. [in German].

Subramanian, D. (2002). *Monograph on Indian charophyta.* Bishen Singh & Mahendra Pal Singh, Dehra Dun, India.

Subramanian, D. (2005). More interesting Charophytes from Tamil Nadu. In J. P. Keshri, A. N. Kargupta, & G. Mustafa (Eds.), *Glimpses of Indian phycology* (pp. 61–74). Bishen Singh & Mahendra Pal Singh, Dehra Dun, India.

Wood, R. D. (1965). *Monograph of the Characeae.* J. Cramer, Weinheim.

Wood, R. D., & Imahori, K. (1964). *Iconograph of the Characeae.* J. Cramer, Weinheim.

Chapter 85
Addendum *Lamprothamnium succinctum*

Michelle T. Casanova

Scientific name: ***Lamprothamnium succinctum*** **(A. Braun in Asch.) R. D. Wood., Rev. Charac.: 340. (1965)**

≡ *Chara succincta* A. Braun in Asch. Oesterr. Bot. Zeitschr. 28: 257. 1878
Lectotypus (designated here by Casanova): In sehr salzigen Teichen bei den Reisfeldern in der kleinen Oase [Uah-el-Beharieh = Bahariyya, Ägypten] der libyschen Wüste bei dem Dorfe Sabu/1. Mai 1876/P. Ascherson (BM)

Nomenclatural remark: As original material has been found the neotype of Wood (1965: 344) is superseded.

85.1 Notes

Lamprothamnium succinctum has not been collected and vouchered in Europe, but the species is included here to provide comparison with other *Lamprothamnium* species in Europe, and to document its distinct morphology to prevent confusion with other species. Given the existing North African distribution of this species it is possible it will be found in the future as more saline habitats in the Mediterranean region are investigated. Because various specimens of *Lamprothamnium* have been mis-identified as *L. succinctum* in the past (e.g., Becker, 2019; Hotchkiss, 1965; Wood, 1965) the description here is based on the type material, the protologue, and one additional specimen from Libya (Harding-King, 1911: BM).

M. T. Casanova (✉)
273 Casanova Rd, Westmere, Victoria 3351, Australia
e-mail: m.casanova@nhm.ac.uk; mt.casanova@federation.edu.au

H. Schubert et al. (eds.), *Charophytes of Europe*,
https://doi.org/10.1007/978-3-031-31898-6_85

85.2 Morphology

Habit: In a dense lawn, drying out to a grey-green colour resulting from the coarse-grained, loose calcification (Figure 85.1)

Size: *Lamprothamnium succinctum* is small to medium-sized (12–30 cm). Plants have the size and structure of a sturdy *C. aspera* or a middle-sized *C. contraria.*

Main axis: Axis diameter is 0.7–1.1 mm; internodes are 1–2.5 cm long except in the apices.

Stipulodes: *Lamprothamnium succinctum* is haplostephanous, the stipulodes are arranged opposite the branchlets, however the single row of downward pointing stipulodes is often reduced or absent.

Branchlets: Whorls consist of 7–8 somewhat spreading, ecorticate branchlets 5–10 mm long, rarely to 20 mm long, up to c. 0.5 mm wide. Branchlets have usually 3 segments (rarely 4) with a short, mucronate end-cell, somewhat pointed (Figure 85.2). The basal branchlet cell is 2.5–3.5 mm long, sterile branchlets are absent from the type material.

Bract-cells: Rarely seen, occasionally one short one (to 0.5 mm long). Braun and Nordstedt (1882) record 5–6 at the first branchlet node without dimensions.

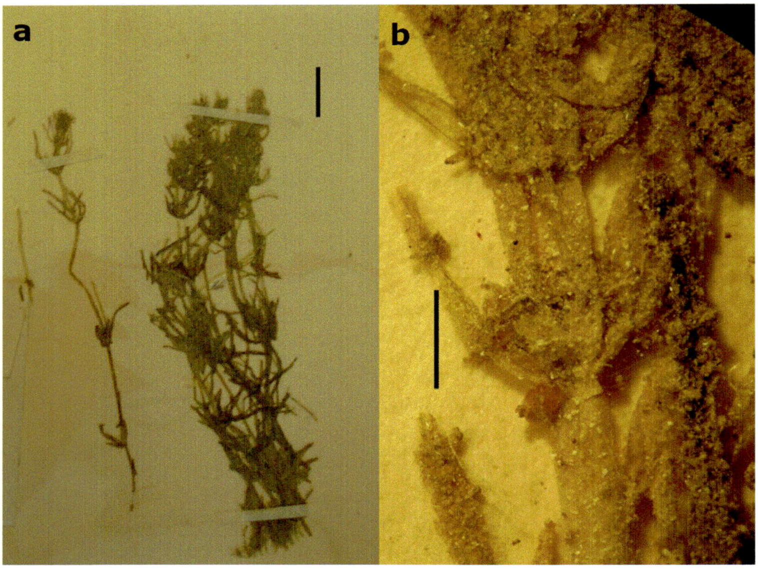

Fig. 85.1 Herbarium specimen of *L. succinctum* (BM: *P. Ascherton 659*). **a** General view of material, scale: 10 mm, **b** Close up of the base of a node, illustrating the oogonium external to the whorl of branchlets, and relatively sparse accessories, scale: 1 mm. Photographs: Michelle T. Casanova

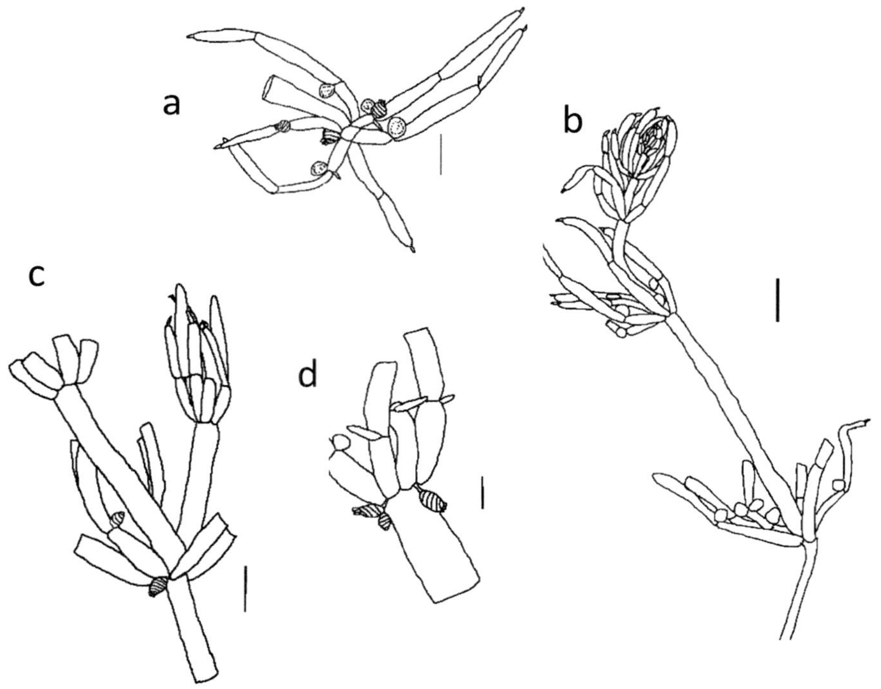

Fig. 85.2 Morphology of *L. succinctum*. **a** Whorl of branchlets illustrating placement of gametangia and 3–4-celled branchlets; **b** Axis and four branchlet whorls; **c** Base of a whorl with branchlets, a branch and oogonium external to the whorl; **d** Base of a whorl with few bract-cells on the branchlets and oogonia external to the whorl. All scales: 1 mm. Drawings: Michelle T. Casanova

In most species of *Lamprothamnium* they are usually verticillate where they are well-developed.

Sexuality: The species is monoecious, gametangia occur inside the base of the whorl and on the branchlets. There is occasionally an oogonium external to the base of the whorl. Oogonia and antheridia are conjoined or singular at the lowest 2–3 branchlet nodes.

Oogonia: Oogonia are apparently sessile, clustered inside and occasionally 1–2 outside the base of the whorl, and when occurring on the branchlets, sometimes conjoined with an antheridium, below or beside the antheridium, very rarely geminate, up to 780 μm long, coronula cells oval shaped, with 10–11 stripes.

Chromosomes: There is a single record for *L. succintum* from New Caledonia, but this is likely to be an erroneous identification. In that record (Hotchkiss, 1965) a count of $n = 42$ was measured on two separate specimens.

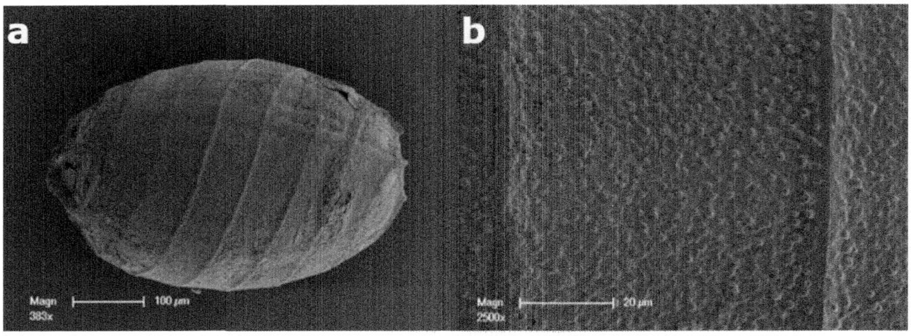

Fig. 85.3 Oospore of *L. succinctum* (BM: *P. Ascherton* 659). **a** Side view of oospore, apex to the left; **b** Detail of oospore wall showing verrucate surface of fossa. Images: Michelle T. Casanova

Table 85.1 Antheridia morphometry

Author	Diameter (µm)
Braun and Nordstedt (1882)	400–500
Measured on type (dry)	300

Oospores: Up to 420 µm × 350 µm with 8 striae of low ridges, black. The ornamentation is densely verrucate, the rounded verrucae c. 2 µm in diameter, 3–4 verrucae per 10 µm distance (Figure 85.3).

Gyrogonites: 10–11 stripes (Braun and Nordstedt, 1882).

Antheridia: Up to 300 µm in diameter (dry, recorded as 400–500 µm in protologue; Table 85.1), solitary or conjoined with the oogonia on the branchlets, rarely geminate on the branchlets. Never outside the base of the whorl. Never inside the base of the whorl.

Bulbils: Not known for this species.

85.3 Phenology and Reproduction

Not known for this species.

85.4 Habitat Conditions and Ecology

Habitat type: Known from very salty ponds in an oasis in Egypt's western desert (Braun and Nordstedt, 1882).

85.5 Variability

In the past any specimen of *Lamprothamnium* with gametangia external to the base of the whorl, and few bract-cells was identified as *L. succinctum* (Allen, 1888; García, 1994; García, 1999; García et al., 2002; Hotchkiss, 1965; Okazaki, 1996; Wood, 1965; Wood and Imahori, 1964). The separation of other species with basal gametangia, and the recognition of a greater diversity in the genus (Casanova, 2014) suggests that not all specimens with the above characters are definitively *L. succinctum*. The variability within reliably identified *L. succinctum* is therefore unknown.

85.6 Identification and Risk of Confusion

Lamprothamnium succinctum can be distinguished from *Lamprothamnium papulosum* by the wider axes and branchlets, the poor development of bract-cells on the branchlets, and by the presence of one or two oogonia external to the base of the whorl. The type material is from Libya. Vouchered specimens for Europe have not been seen for this study. Specimens from Australia, southern Africa and elsewhere (e.g., Hong Kong and New Caledonia) are not unequivocally *L. succinctum* as they either have many external gametangia, stipitate gametangia or antheridia external to the base of the whorl. These characters can be used to distinguish different species (Casanova, 2013). The oospore morphology of specimens identified as *L. succinctum* is currently being investigated.

85.7 Status and Threats

Lamprothamium succinctum is not currently recognised for the European flora, and its status in North Africa is unknown. On the basis of the lack of collections it is certainly rare. Brackish inland and coastal habitats for charophytes are usually under-valued by the general public. Many such habitats can be converted into salt-harvesting ponds, or drained, or used for coastal development. The status of saline ponds in Libya and North Africa in general is not known (Elbabour et al., 2017). In the Mediterranean region of Europe there is a large amount of coastal development and water use that could change the habitat parameters for this species.

85.8 Remarks

Braun and Nordstedt (1882) thought that this species (then determined as a *Chara*) was related to other species with gametangia at the base of the whorl (i.e. *C. corallina, C. australis* and *C. wallichii*). However, the similarity of *Lamprothamnium macropogon* (then cited as *Lychnothamnus macropogon*) was noted by those authors. The absence of antheridia at the base of the whorl was used in the key (Braun and Nordstedt, 1882: 17) and also noted by the collector (P. Ascherson) stating 'the lack of antheridia at the base of the whorl, the smaller oogonia and smaller dimensions' differentiates this species from other ecorticate species known to the experts at that time.

References

Allen, T. F. (1888). *The Characeæ of America*. Part I.

Becker, R. (2019). The Characeae (Charales, Charophyceae) of Sardinia (Italy): Habitats, distribution and conservation. *Webbia, 74*, 83–101.

Braun, A., & Nordstedt, C. F. O. (1882). Fragmente einer Monographie der Characeen. Nach den hinterlassen Manuscripten A. Braun's herausgegeben von Dr. Otto Nordstedt. *Abhandlungen der Königlichen Akademie der Wissenschaften zu Berlin für das Jahr, 1882*, 1–211. [in German].

Casanova, M. T. (2013). *Lamprothamnium* in Australia (Characeae, Charophyceae). *Australian Systematic Botany 26*, 268–290.

Casanova, M. T. (2014). A revision of *Chara* sect. *Charopsis* (Characeae, Charophyceae) in Australia, including specimens collected for Bush Blitz. *Australian Systematic Botany, 27*, 403–414.

Elbabour, M. M., Abdulsamad, E. O., & Mohamed, M. F. (2017). Wetlands along the Libyan Mediterranean Coastline: An integrated survey. *GeoLibya, 2*, 47–48.

García, A. (1994). Charophyta: Their use in paleolimnology. *Journal of Paleolimnology, 10*, 43–52.

Garcia, A. (1999). Quaternary charophytes from Salina del Bebedero Argentina: Their relationship with extant taxa and paleolimnological significance. *Journal of Palaeolimnology, 21*, 307–323.

García, A., Jones, B. G., Chenhall, B. E., & Murray-Wallace, C. V. (2002). The charophyte *Lamprothamnium succinctum* as an environmental indicator: A Holocene example from Tom Thumbs Lagoon, eastern Australia. *Alcheringa, 26*, 507–518.

Hotchkiss, A. (1965). Chromosome numbers in Characeae from the South Pacific. *Pacific Science, 19*, 31–37.

Okazaki, Y. (1996). Turgor regulation in a brackish water charophyte, *Lamprothamnium succinctum*. *Journal of Plant Research, 109*, 107–112.

Wood, R. D. (1965). *Monograph of the Characeae*. J. Cramer.

Wood, R. D., & Imahori, K. (1964). *Iconograph of the Characeae*. J. Cramer.

Chapter 86
Addendum *Nitella brachyteles*

Ivana Trbojević and Hendrik Schubert

Scientific name: ***Nitella brachyteles*** **A. Braun in Leonh. in Verh. Naturf. Vereins Brünn Abh. 2: 173. 1864**

Lectotype (designated here by Trbojevič & H. Schub. & T. Gregor): Algerie. Casele[?] de La Calle, Eaux stagnante/Jul. 1841/Durieu (BM 13828157, Ex Museo botanico Berolinensi).

≡ *Chara translucens* Pers. 1807 var. *brachyteles* (A. Braun) Feldm.-Maz. in Bull. Soc. Hist. Nat. Afrique N. 37: 72. 1946.

Taxonomic note: This species was described in Leonhardi (1864: 173) by A. Braun in a footnote (in German): [English translation] A similar species is the rare South-European and North-African *Nitella brachyteles*. It is intermediate in all characters between *Nitella translucens* and *N. mucronata*, also in the width of the main axis. The furcate apices in sterile branchlets are somewhat longer than in *N. translucens*. Braun (1868: 881–882) gave a thorough description of his new species and highlighted its intermediate position between *Nitella translucens* and *N. mucronata*. But it is closer to *N. translucens* by the only once furcate fertile branchlets with 2–3 oospores under each antheridium. He gave thorough information for two specimens from Algeria (both from La Calle) and more general information for a specimen from Corsica collected by Pouzols. Later authors (Corillion, 1975; Hollerbach and Krassavina, 1983; Hy, 1913; Migula, 1900) described the oospore membrane of *N. brachyteles*

I. Trbojević (✉)
University of Belgrade, Faculty of Biology, Institute of Botany and Botanical Garden "Jevremovac", Takovska 43, 11000 Belgrade, Serbia
e-mail: itrbojevic@bio.bg.ac.rs

H. Schubert
Universität Rostock, Institut für Biowissenschaften, Albert-Einstein-Straße 3, 18051 Rostock, Germany
e-mail: hendrik.schubert@uni-rostock.de

H. Schubert et al. (eds.), *Charophytes of Europe*,
https://doi.org/10.1007/978-3-031-31898-6_86

as finely punctuated (dotted), but it is uncertain which material they used for this statement.

Specimens deposited in BM (Fig. 86.1, left panel) were re-examined by us. The specimen on the left collected in 1839 by Bové in La Calle (El Kala, NE Algeria) and mentioned by Braun (1868) was considered by H. and J. Groves (as noted on the herbarium sheet) as a mixture of *Nitella brachyteles* and *N. virgata* (= *N. mucronata*). Wood (1965), examining material from the same collection stored in PC, describes it as "very poor material apparently a tangle of *N. tenuissima* and *N. mucronata*." An oospore from this BM specimen was investigated with SEM by M. T. Casanova (Fig. 86.2, upper row). The oospore membrane ornamentation (Fig. 86.2, upper row, right panel) is clearly fibrous, with thick, ropy fibres. Similar oospore membrane ornamentation is found in *N. gracilis, N. hyalina* and *N. confervacea* (John and Moore, 1987), all three species known from the El Kala region (Muller et al., 2017). However, the herbarium sheet apparently consists of a number of individuals, tangled and clogged, from more than one species, and we cannot claim the origin of the examined oospore (Fig. 86.2, upper row, right panel).

The specimen collected in 1841 by Durieu at the same location (near La Calle) and mentioned by Braun (1868) is selected by us as lectotype (Fig. 86.1, left panel, specimen on the right). Considering all the details it resembles a less furcate *N. mucronata*. The general habitus of plants couldn't be observed since the material is in poor condition, clogged, and hard to examine. The specific gametangia arrangement as described by Braun (1868), 2–3 oogonia under 1 antheridium, was observed in this specimen (see Fig. 86.2, lower row).

Apart from the two collections from La Calle mentioned by Braun (1868) and the one mentioned by him from Corsica, later authors claimed further records. Feldmann (1946) mentioned records from Maison-Carrée (collected by Trabut, today Al Harrach part of Algiers) and Oued Smar (specimen collected 1921 by L. Gauthier, deposited in NY, sterile). Part of the Corsica collection mentioned by Braun might have survived in LD, but this nineteenth century specimen bears only the annotation "*N. brachyteles* Corsica" (Fig. 86.1, lower row). Oospores from this specimen were examined by us, but unfortunately these were too immature for imaging.

Another specimen, deposited as *N. brachyteles* (NY), originates from Portugal (Fig. 86.1, right panel). It was collected by E. J. Mendes in 1954. Observing habitus and macroscopic morphological characteristics, this specimen does have qualities of *N. brachyteles*, as originally described by Braun—very similar to the previously discussed specimen from Corsica, it has once-divided fertile branchlets, with few oogonia at the furcation. An oospore from this specimen was collected for SEM, but unfortunately it was too immature for imaging (M. T. Casanova pers. communication).

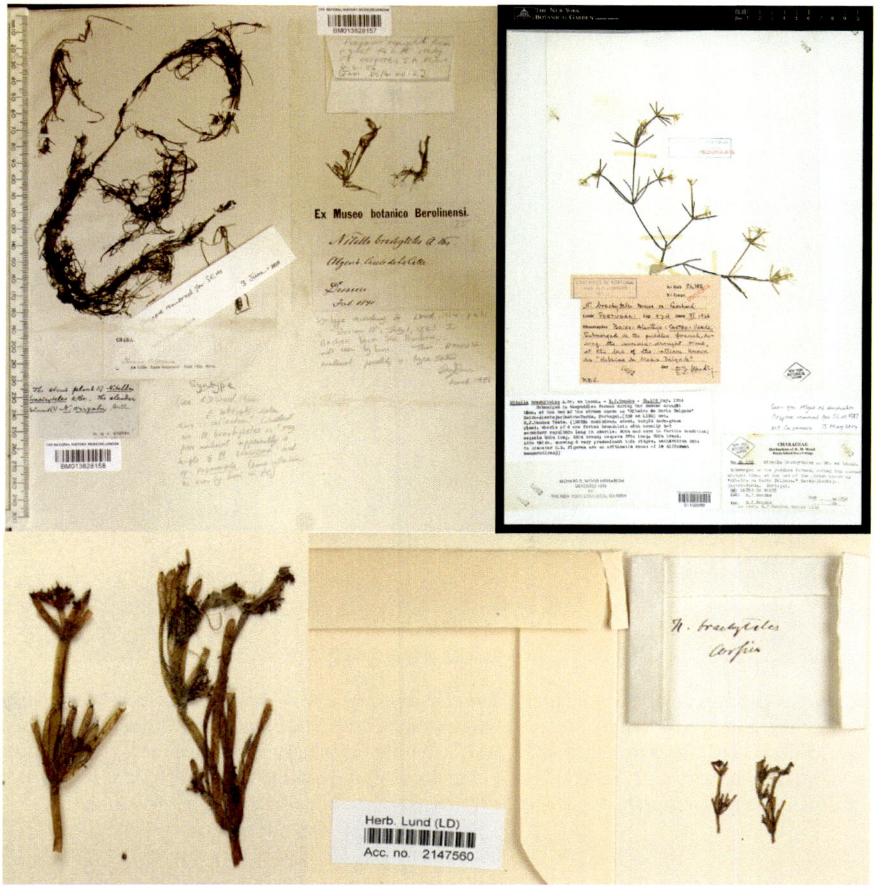

Fig. 86.1 *Nitella brachyteles*, herbarium material. Left panel upper row: Specimen collected near La Calle (Bové, June 1839, BM, left specimen), and specimen collected near La Calle (Durieu, July 1841, BM, right specimen). Right panel upper row: Specimen from Portugal (NY), Lower row: Specimen of *Nitella brachyteles* from Corsica, deposited in LD. Photographs: upper row left—Hendrik Schubert, upper row right—downloaded from http://sweetgum.nybg.org/images3/1478/173/01192289.jpg; lower row—Patrik Frödén

86.1 Conclusion

Morphological characters noted in the original description (Braun, 1868; Leonhardi, 1864) overlap between *N. mucronata* and *N. translucens*, and none of these characters clearly distinguishes *N. brachyteles*. It is possible that *N. brachyteles* is a synonym of *N. translucens* as stated by Muller et al. (2017), and previously discussed by Hy

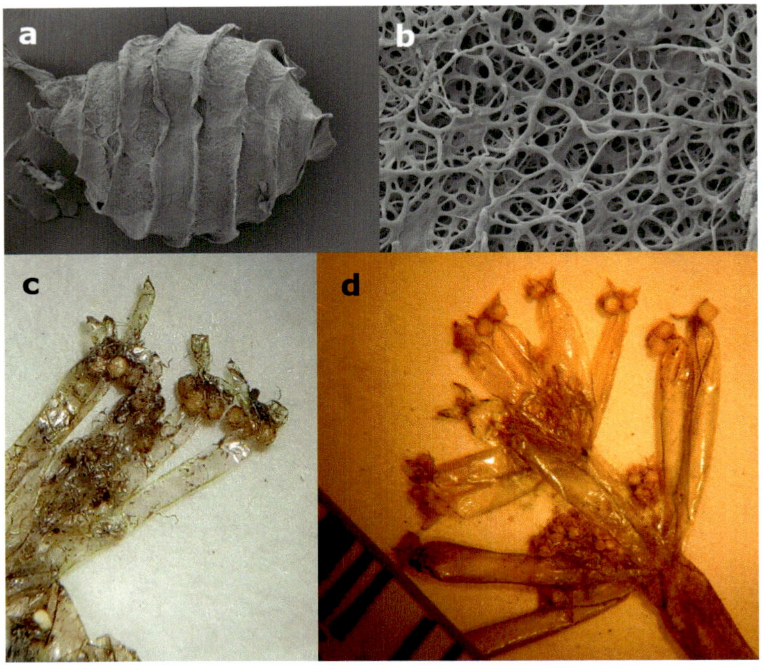

Fig. 86.2 Oospores and gametangia of *Nitella brachyteles*. **a** SEM image of the oospore; **b** SEM image of the fibrous oospore membrane ornamentation, both SEM images are from the specimen collected near La Calle (Bové, june 1839, BM); **c** Fertile branchlets of the specimen collected near La Calle (Durieu, July 1841, BM); **d** Fertile branchlets of the specimen from Portugal (NY, see also Fig. 86.1). Photographs: **a**, **b**, **d**—Michelle T. Casanova, **c**—Hendrik Schubert

(1914), particularly related to the variety *confervoides* (*N. translucens* var. *confervoides*). Feldmann (1946) also considered it as a variety of *N. translucens* (*N. translucens* var. *brachyteles*). The oospore membrane is a good specific indicator in *Nitella* species (see John and Moore, 1987; Mouronval et al., 2015; Urbaniak, 2009 and 2011) and as long as the oospore membrane of the lectotype is unknown, we abstain from a final decision.

References

Braun, A. (1868). *Die Characeen Afrika's*. Monatsberichte der Königlich Preussischen Akademie der Wissenschaften zu Berlin. [in German].

Corillion, R. (1975). *Flore et végétation du Massif Armoricain Tome IV Flore des Charophytes (Characées) du Massif Armoricain et des contrées voisines d'Europe occidentale*. Jouve, [in French].

Feldmann, G. (1946). Les Charophycées d'Afrique du Nord. *Bulletin de la Société d'Histoire de l'Afrique du Nord, 37*, 64–118. [in French].

Hollerbach, M. M. & Krassavina, L. K. [Голлербах, М. М. & Красавина, Л. К.], (1983). *Определитель пресноводных водорослей СССР. 14. Charophyta* [Determination key of freshwater algae of USSR]. Nauka. [in Russian].
Hy, F. (1913). Les Characées de France. *Bulletin de la Société Botanique de France, 60*, 1–47 [in French].
Hy, F. (1914). Les Characées de France. Note additionnelle. *Bulletin de la Société Botanique de France, 61*, 235–241. [in French].
John, D. M., & Moore, J. A. (1987). An SEM study of the oospore of some *Nitella* species (Charales, Chlorophyta) with descriptions of wall ornamentation and an assessment of its taxonomic importance. *Phycologia, 26*, 334–355.
Migula, W. (1900). Die Characeen Deutschlands, Oesterreichs und der Schweiz. In: Fischer, A., E. Fischer, F. Hauck, G. Limpricht, C. Luerssen, W. Migula, H. Rehm, P. Richter & G. Winter (Eds), *Dr. L. Rabenhorst's Kryptogamenflora von Deutschland, Oesterreich und der Schweiz* (2nd ed., Vol. 5). Eduard Kummer. [in German].
Mouronval, J. B., Baudouin, S., Borel, N., Soulié-Märsche, I., Klesczewski, M., & Grillas, P. (2015). *Guide des Characées de France méditerranéenne*. Office national de la Chasse et de la Faune Sauvage. [in French].
Muller, S. D., Rhazi, L., & Soulié-Märsche, I. (2017). Diversity and distribution of Characeae in the Maghreb (Algeria, Morocco, Tunisia). *Cryptogamie, Algologie, 38*, 201–251.
Urbaniak, J. (2009). Oospore variation in *Nitella gracilis* and *Nitella mucronata* (Charales, Charophyceae) from Poland. *Biologia, 64*, 252–260.
Urbaniak, J. (2011). A SEM study of the oospore wall ornamentation in Polish charophytes (Charales, Charophyceae) genera *Lychnothamnus*, *Nitella* and *Nitellopsis*. *Nova Hedwigia, 93*, 537–549.
Leonhardi, H. von (1864). Die bisher bekannten österreichischen Armleuchter-Gewächse besprochen von morphogenetischen Standpuncte. *Verhandlungen des naturforschenden Vereins in Brünn (Abhandlungen), 2*, 122–221. [in German].
Wood, R. D. (1965). *A revision of the Characeae. 1. Monograph of the Characeae*. J. Cramer.

Part III
Glossary and Index

Chapter 87
Glossary

John Bruinsma, Michelle T. Casanova, Dietmar Jäger, Elisabeth Lambert, Emile Nat, Hendrik Schubert, and Nick Stewart

87.1 Introduction

During the preparation of this book the authors soon realised that many terms are not well defined. Some of them, coming e.g. from leaf morphology, were applied to unicellular structures, others used as synonyms for structures which differed in their ontogeny. In order to sort out these problems, a short list of "problematic terms" rapidly evolved into a rather huge compendium, giving rise to an initiative to compile a comprehensive glossary. All specific terms dealt with in this book have been revised, defined and, in the case of confusing use between authors or disciplines,

Supplementary Information The online version contains supplementary material available at https://doi.org/10.1007/978-3-031-31898-6_87.

J. Bruinsma (✉)
Thorbeckelaan 24, NL 5694 CR Breugel, The Netherlands
e-mail: jhpbruinsma@gmail.com

M. T. Casanova
273 Casanova Rd., Westmere, VIC 3351, Australia
e-mail: m.casanova@nhm.ac.uk

D. Jäger
Herrenriedstrasse 4, 6845 Hohenems, Austria
e-mail: dietmar.jaeger@outlook.com

E. Lambert
Département Biologie-Environnement, Faculté des Sciences, Université Catholique de l'Ouest (UCO), 3, Place André Leroy—BP 10808, 49008 Angers, Cedex 01, France
e-mail: lambert.elisa49@gmail.com

H. Schubert et al. (eds.), *Charophytes of Europe*,
https://doi.org/10.1007/978-3-031-31898-6_87

a proposal for the future use of the term agreed upon after discussion. The printed glossary presented here is a compilation of terms that are used to describe (European) Charophytes. Included are the terms used in the present book and synonyms that were used in important Charophyte literature in the last century, for which we referred to Bryant and Stewart (2011); Groves and Bullock-Webster (1920); Moore (1986) and Wood and Imahori (1964, 1965). Many of the terms used to describe the (scanning electronic) microscopic characters of oospore-membranes are quite new to charology. Most of these terms are analogies of the terms which are used to describe pollen or surface characters: Faegri and Iverson (1964); Harris (1979).

The terms are illustrated by two kinds of pictures. The drawings are made by Dietmar Jäger. The SEM-pictures were made by Michelle T. Casanova. For SEM the oospores were prepared by cleaning and coating in gold, or gold–palladium and several different SEM machines were used over time (methods are outlined in Casanova, 2009).

The internet version of the glossary will be published which contains charophyte terms in Latin and in all European languages in which there is literature. At present these languages are available: Czech, Danish, Dutch, Estonian, French, German, Greek, Hungarian, Italian, Norwegian, Polish, Russian, Serbian, Slovakian, Spanish and Swedish. The following persons, mainly native speakers, contributed to the translations: Nadia Abdelahad, Király Angela, Lars Båstrup-Spohr, Jelena Blaženčic, Irmgard Blindow, Aurélie Boissezon, John Bruinsma, Michelle T. Casanova, Elena Chemeris, Luc Denys, Núria Flor Arnou, Király Gergeli, Tomáš Hauer, Alica Hindáková, Anja Holzhausen, Anders Langangen, Aleksandra Marković, Eva Papastergiadou, Mariusz Pełechaty, Jan Simons, Joep Spronk, Nick Stewart, Kaire Torn, Ivana Trbojević and Klaus van de Weyer.

87.2 Alphabetical List of Terms and Definitions

1-corticated	Synonym of haplostichous
2-corticated	Synonym of diplostichous

E. Nat
Dutch Information Center on Charophytes, Naturalis Biodiversity Center, P.O. Box 9517, 2300 RA Leiden, The Netherlands
e-mail: e.nat@kranswieren.nl

H. Schubert
Institut für Biowissenschaften, Universität Rostock, Albert-Einstein-Straße 3, 18059 Rostock, Germany
e-mail: hendrik.schubert@uni-rostock.de

N. Stewart
4 Hexton Road, Glastonbury BA6 8HL, UK
e-mail: nfstewart@freeuk.com

3-corticated Synonym of triplostichous

abaxial On the outer side of the branchlet, i.e. on the side away from the main axis

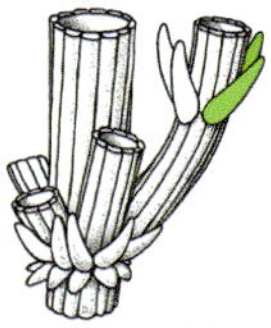

abscise (to-) To fall off

accessory branchlets The shorter and less developed branchlets. The term is used with heteroclemous *Nitella*-species

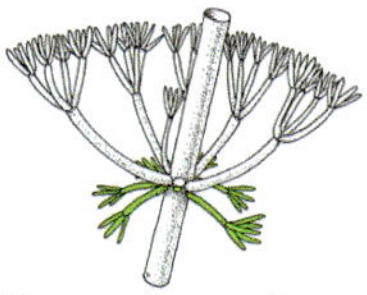

acuminate Tapered to a long point

acute Having a sharp-pointed tip with a less than 90 degree angle. Typically used for more abruptly pointed tips in contrast to acuminate for tapered acute tips

adaxial On the inner side of the branchlet, i.e. on the side nearest to the main axis

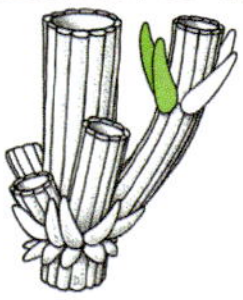

adventitious branchlets Synonym of accessory branchlets

agglutinate Clumped together

aggregate — Growing together. Authors may use 'aggregate' in a specific way, e.g. 'three or more together'

allantoid — Cylindrical(-elongated) and slightly curved with a rounded end; 'sausage-shaped'. It is also used for cylindrical, either straight or slightly curved and with a rounded or acute end

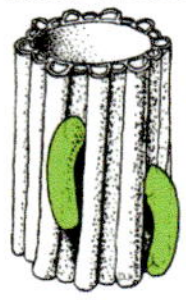

alopecuroid — Synonym of foxtail-like

alternate — (Of stipulodes) in between branchlets, not right below branchlet

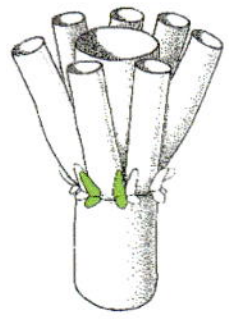

alveola/alveolate — A deep, regular, angular cavity, separated from each other by thin partitions. Alveolate: with alveolae, honeycombed

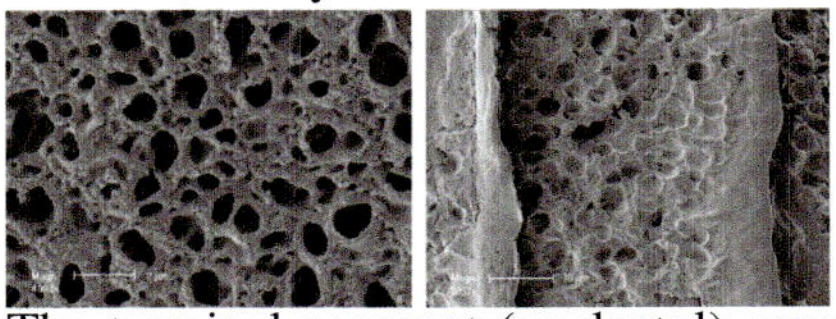

anarthrodactylous — The terminal segment (or dactyl) composed of one cell, especially in *Nitella*

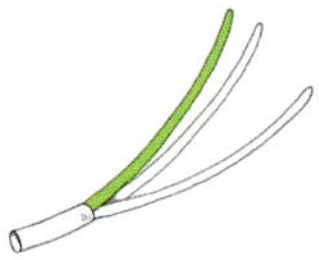

anastomosing — Intertwined fibres that join up in places

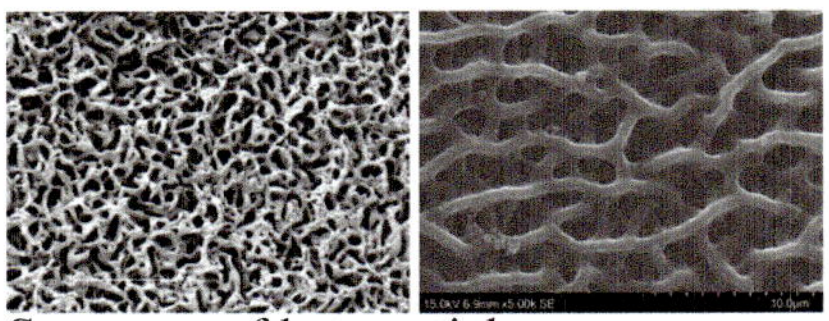

anisostichous — Synonym of heterostichous

anterior — Synonym of adaxial; also used for adaxial branchlet

antheridiocarp — (In charology) synonym of antheridium

antheridium (-dia) The male reproductive organ producing motile spermatozoids

antherozoids Synonym of spermatozoids

apiculate Abruptly narrowed into a small sharp point. See also mucronate

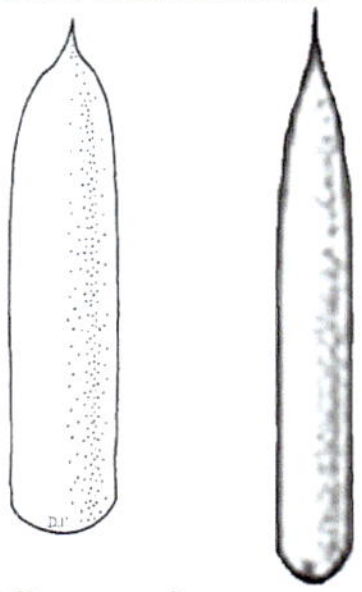

apomictic Generative reproduction without fusion of male and female gametes

arcuate Shaped like a bow; curved

armed (un-) Synonym of (not) spinose

arthrodactylous The terminal segment composed of more than one cell, especially in *Nitella*

articulation Synonym of segment

asperous	Rough and uneven 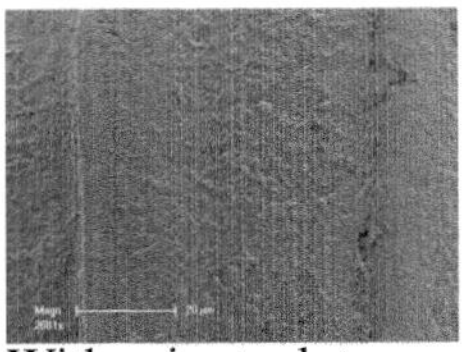
atomarious	With minute dots or points 
attenuate	A poorly defined term used differently by different authors and in general botany. Usually, but not exclusively, applied to leaf bases. In charophytes most often used to describe tips that are long-tapered, but here these are included within the range of acuminate
aulacanthous	Having the cortex of the main axis with the secondary cortical rows more prominent than the primary rows, so that the spine-cells may appear to grow in furrows 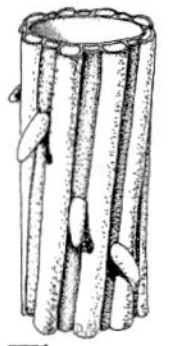
axis, main	The central part of a charophyte-plant, originating directly from a protonema or, alternatively, a bulbil. The main axis is of unlimited growth, lateral axes originate from its nodes 

barely visible spine(-cell) Synonym of papilla

basal cage Synonym of cage

basal impression Cellular impression at the bottom of oospores, usually pentagonal shaped in *Chareae* (on the left) or 1-3 impressions in *Nitelleae*, (on the right)

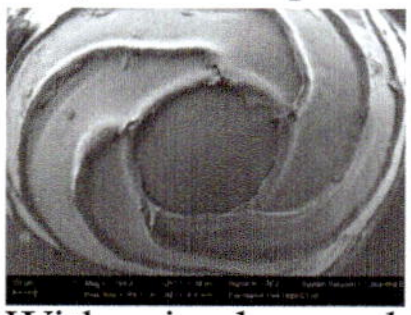
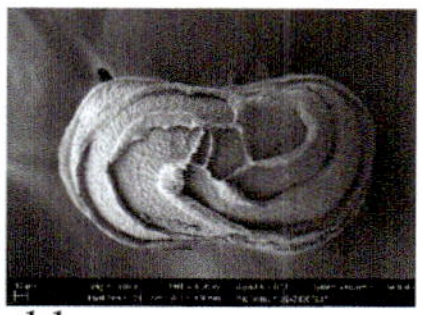

beaded With raised, rounded lumps

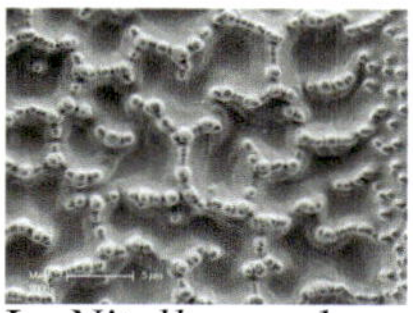

beak In *Nitella* an elongate extension of the oospore wall at the apex of the oospore

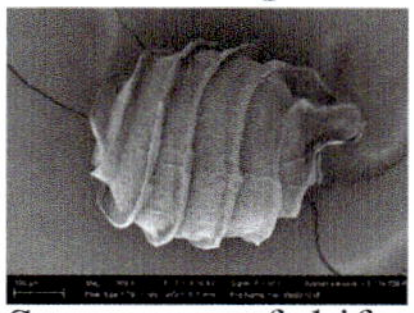
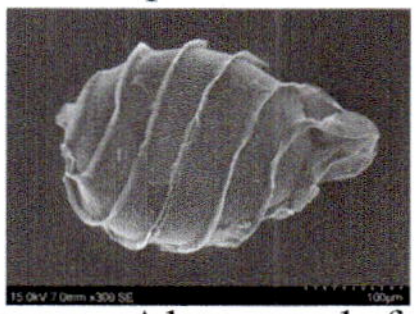

bifid Synonym of bifurcate. Also used for 'when a single cell is split length-wise' e.g. '*Nitella partita*'

bifurcate Forked in two

bistipulate Having two stipulodes per branchlet in the same whorl. The picture on the left is 'bistipulate and diplostephanous', the picture on the right is 'bistipulate and haplostephanous'

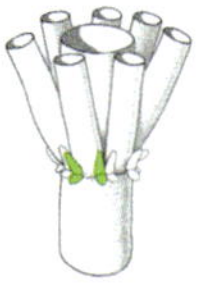
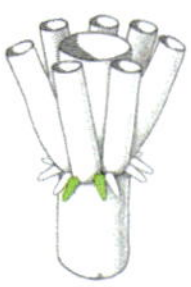

botuliform Synonym of allantoid

brachydactyl(ou)s Having dactyls much shorter than the penultimate ray of branchlets of *Nitella*, in other words, having abbreviated dactyls

bract Synonym of bractlet. In older literature it is also used for bract-cell

bract-cell Single-celled processes growing out from the peripheral cells of the branchlet-nodes

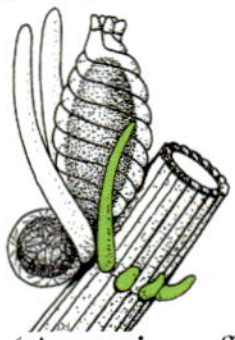

bracteole(s) (A pair of) single-celled processes originating from the basal node below the oogonium; one on each side of the oogonium

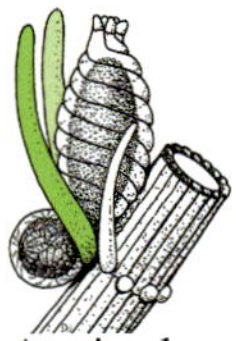

bractlet A single-celled process subtending the oogonium in female plants of dioecious species of *Chara*, replacing the antheridium

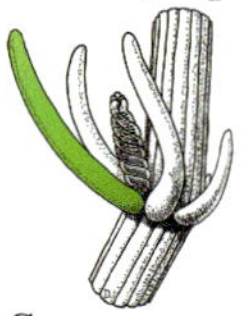

branch Synonym of lateral axis

branchlet The laterals of limited growth produced in whorls at the axial nodes

bridge An open raised element attached at both ends

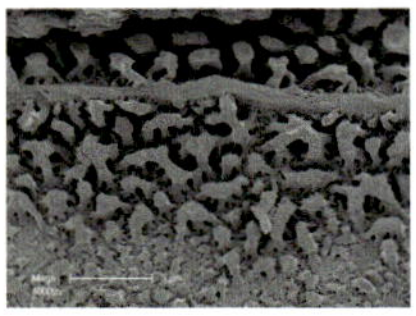

bulbil Enlarged cartilaginous nodal cells or agglomerations of starch-containing cells arising from nodal cells of the rhizoids and at the axis-nodes of some charophytes. Left: 2x *Nitellopsis obtusa*; Middle: *Chara fragifera*; Right: *Chara aspera*

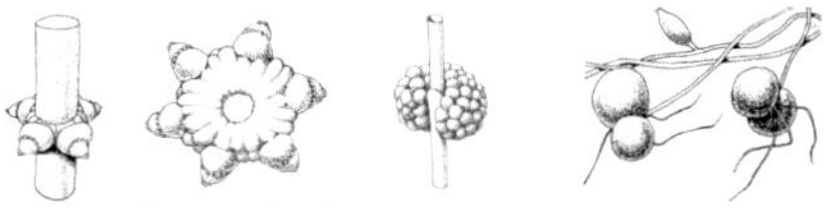

cage An extension of the oospore-wall around the edges and base of the basal cell of the oospore in some species of *Chara*, sometimes with windows

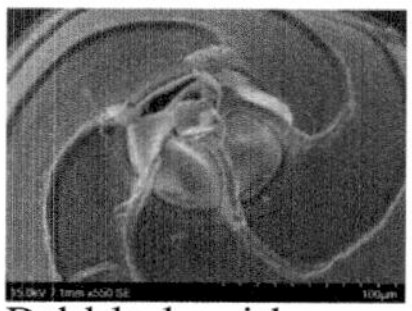
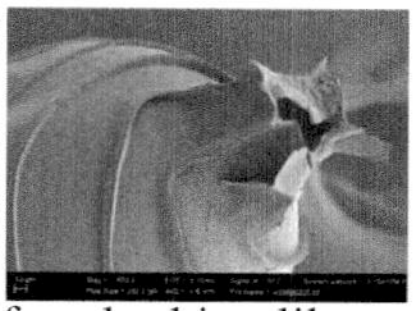

calculate Pebbled, with a surface looking like gravel

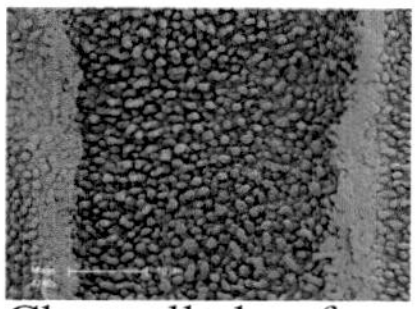

canaliculate Channelled or furrowed; longitudinally grooved

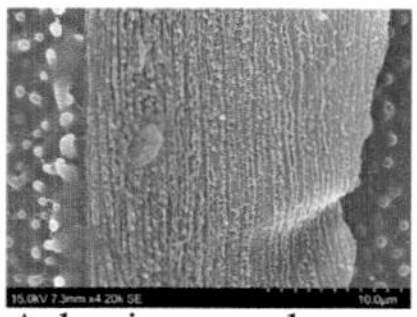

cancellate A lattice-work or grid; a series of enclosed cells or chambers

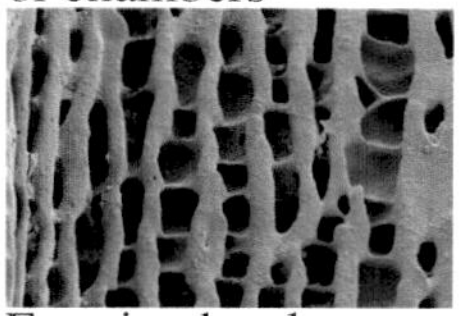

capitate Forming heads

capitulum (capitula) Synonym of capitulum-cells

capitulum-cells The final (innermost) cells of a manubrium; origin of the spermatogenic filaments

cariose Corroded; appearing worm eaten; with cavities or ulcerations

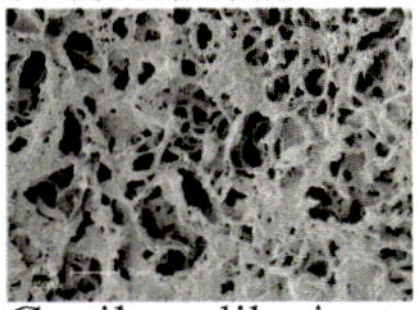

cartilaginous Cartilage-like in consistency, hard but easily cut with a knife, not green

catenate/catenulate Beads joined up in a chain. Catenulate: with small beads

cerebrellate Like the sinuous folds and creases on the surface of a brain

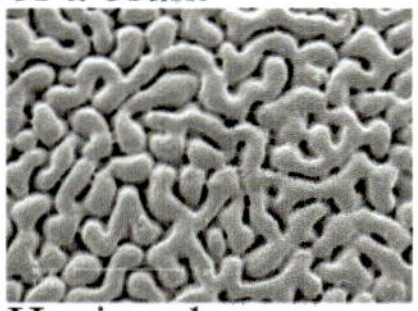

channelled Having deep grooves or channels

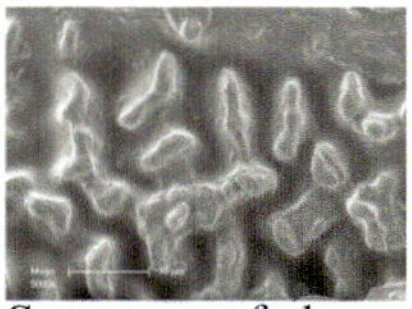

charad Synonym of charophyte

cicatrose A surface having indentations with elevated margins (varioles)

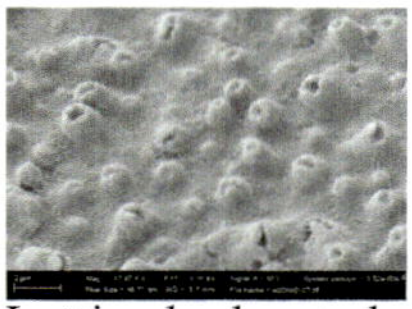

clathrate Latticed; elevated ridges meeting at right angles

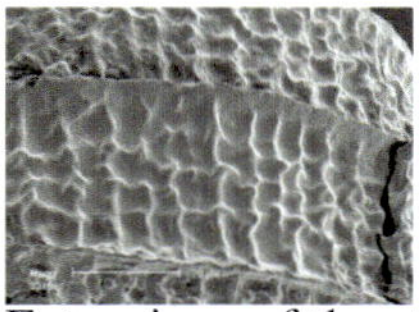

claws Extensions of the striae around the basal cell of the oospore in some species of *Chara*

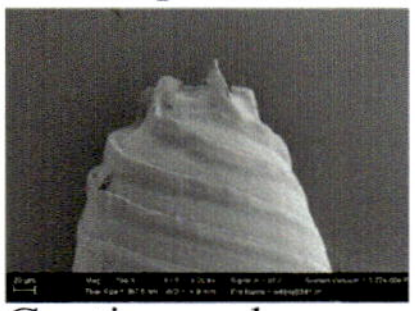

colliculate Continuously covered with low, rounded elevations

confluent Forming a cluster. E.g. heads of *Tolypella* and *Sphaerochara* species. Also: running into, aligning in the same direction, becoming continuous (in end-cells, c.f. mucronate)

confluent (with the penultimate cell) The base of the top-cell being as wide as the top of the penultimate cell

conjoined The antheridium and oogonium being adjacent at the same branchlet-node

connivent Converging, coming closely together at the tips, e.g. branchlets of *Chara connivens* (left), coronula-cells (right)

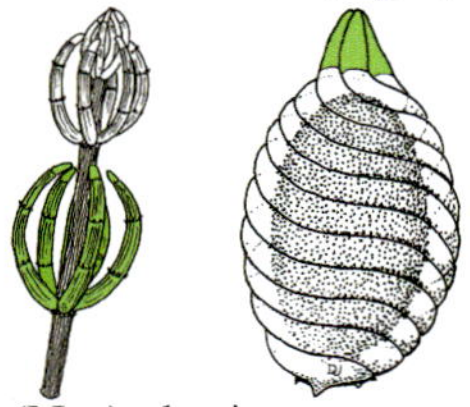

contiguous (non-) (Not) sharing a common border; (not) touching each other at the edges but not actually united

convolution(s) The number of rotations of the spiral cells around the oospore, so it can be determined by dividing the number of helical stripes by 5. However, the term is often mis-used as a synonym of helical stripes

coriaceous Leather-like in texture, with minute cracks like human skin

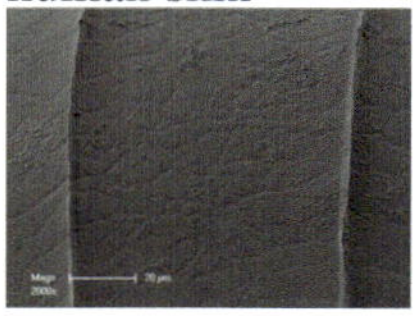

corona	The group of terminal cells or abbreviated dactyls on a branchlet-tip (e.g. *Chara braunii, Nitella translucens*) 
coronula	The small crown-like structure at the apex of the oogonium. Constructed by spiral-cells; one row in *Chareae* (left) and two in *Nitelleae* (right) 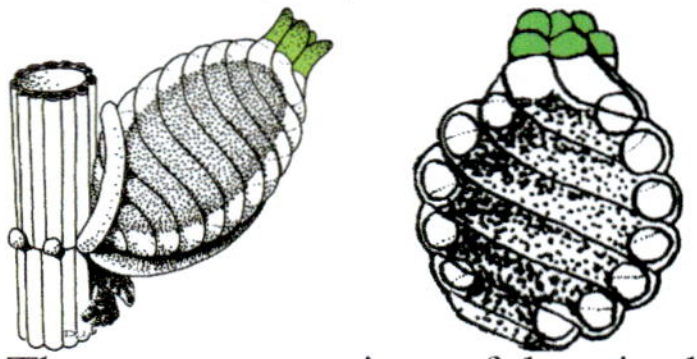
cortex/corticate	The outer covering of longitudinally-arranged cells in certain species of *Chara*, which gives the axis and/or branchlet a striped or ridged apperance. Corticate: having a cortex 
cortical cell(s)	Cells that make up the cortex 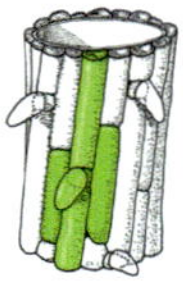
cortical row	A row of cortical cells 
corticated	Synonym of corticate
cortication	Synonym of cortex

corticinus With bark-like sculpturing or texture

costae/costate/costulate Longitudinal raised ribs or ridges. Costate: with costae. Costulate: with less prominent ribs or ridges than costate

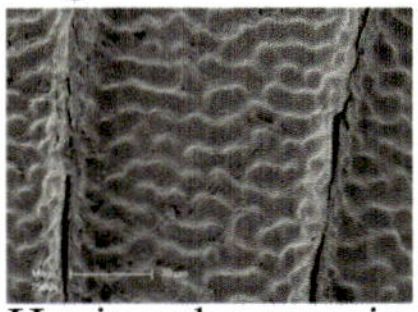

crenate/crenulate Having the margin notched with rounded teeth. Crenulate: finely crenate

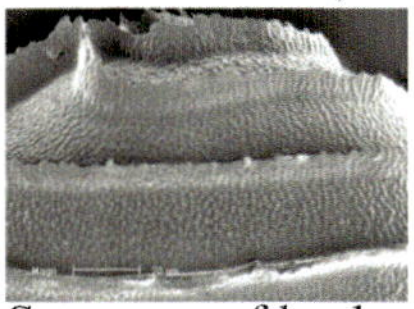

crest Synonym of beak

crust Synonym of encrustration

cusp(idate) (Narrowing to a) point

dactyl The terminal segment of the branchlet in the genus *Nitella*. It can be unicellulate (one cell), bicellulate (two cells, usually the end-cell minute and acute, but not always) or pluricellulate (more than two cells)

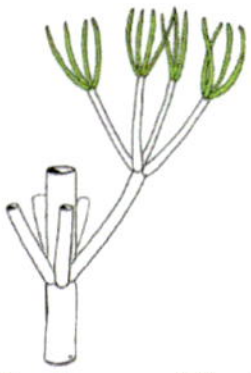

decumbent Lying with the extremity curving upwards

dendriform Synonym of dendroid

dendroid Tree-like. It is often used with the meaning of bushy

dentate With tooth-like or serrated projections

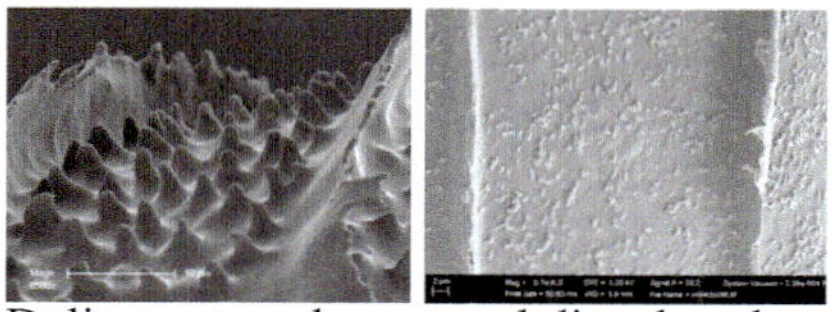

diaphanous Delicate, translucent and discolored

diffuse Loose, rather unorganised (e.g. of plants in deep water)

dioecious With gametangia produced on separate male and female plants

diplostephanous Having a double ring of stipulodes at the base of each whorl of branchlets

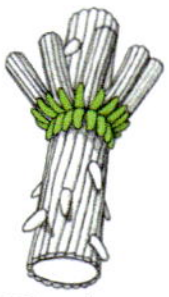

diplostichous Having a cortex arranged in alternate primary and secondary rows, there being two cortical rows corresponding to each branchlet

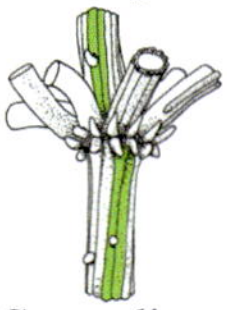

divergent, diverging Spreading

dorsal Synonym of abaxial

echinate/echinulate Thickly set or armed with short, stout spines or prickles; spiny like a hedgehog (left). Echinulate (right): minutely echinate

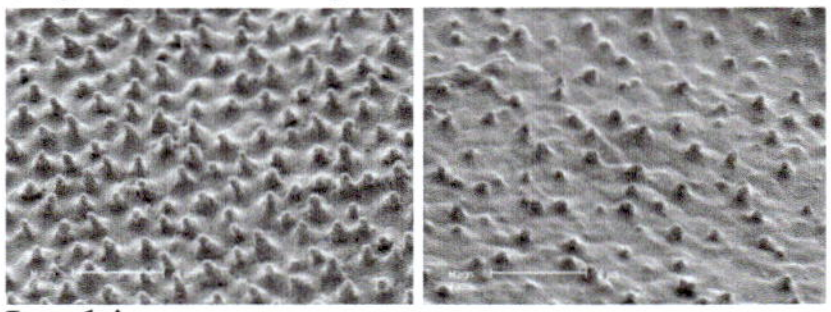

ecorticate Lacking a cortex

ecorticated Synonym of ecorticate

egg-cell Synonym of oosphere

embryonic branch Proembryo originating directly from a stem node instead of a germinating oospore. Translated from the original term “Zweigvorkeim” coined by Pringsheim 1864

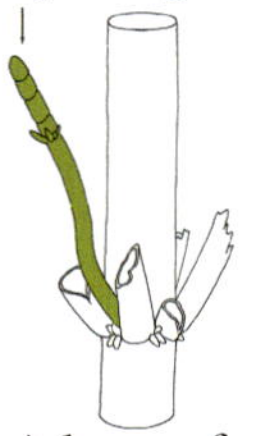

encrustation A layer of minerals around the plant or part of it

encrusted With an encrustration

end-cell In all charophytes: the terminial cell of the branchlet. It is the branchlet apical cell (see chapter 3 Ontogenesis) which has ceased further development

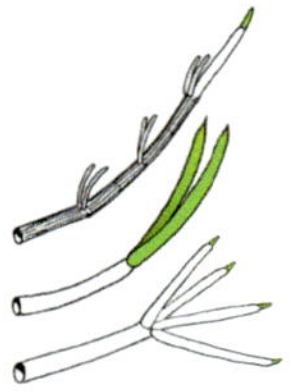

envelope-cell/enveloping cell Synonym of spiral-cell

equilateral Synonym of verticillate (bract-cells)

estephanous, (falsely) Synonym of (seemingly) haplostephanous

estipulodic, (falsely) (Seemingly) without stipulodes

exarate Ploughed; sculpted; furrowed; sulcate; an excavated surface in general

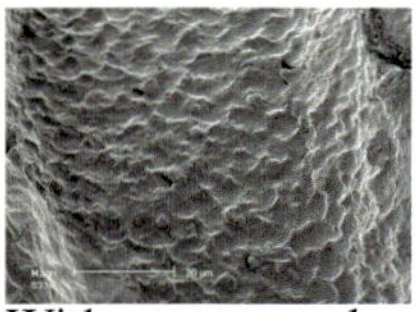

excavated With a scooped out depression; superficially, with a hollowed out area

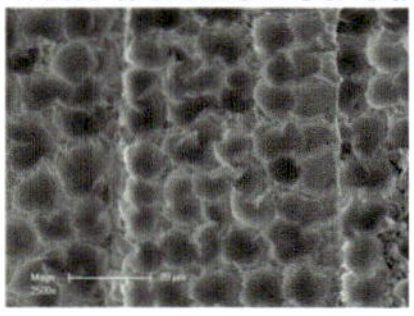

exsculptate With irregular, somewhat longitudinal depressions, as if carved or scooped out

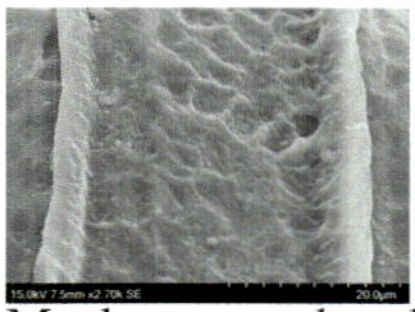

farinaceous Mealy or powdery looking

fascicled Synonym of fasciculate

fascicles/fasciculate (In) bunches or clusters

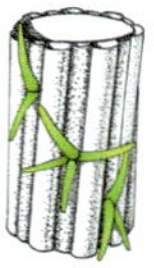

fatiscent With cracks, crevices, or similar openings

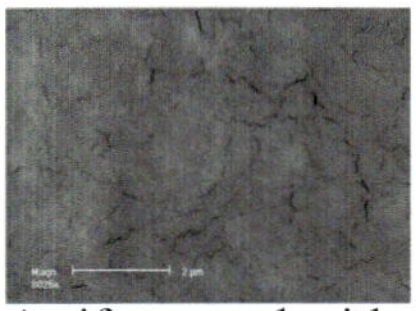

felted As if covered with felt

fenestrae/fenestrate Synonym of (having) windows

fibrous Invested with or made up of threads or fibres (i.e. very long flexible strands)

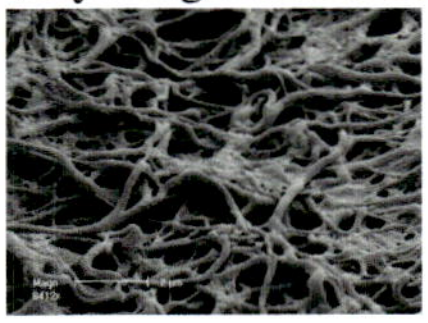

fimbriate Having a fringe or border of hairlike or finger-like projections

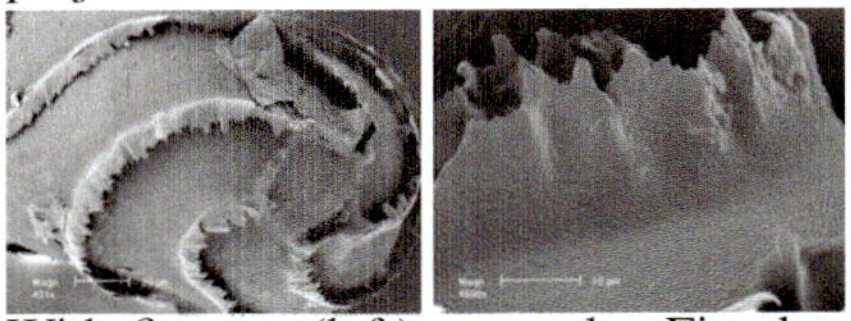

fissate/fissulate With fissures (left) or cracks. Fissulate (right): minutely cracked or fissured

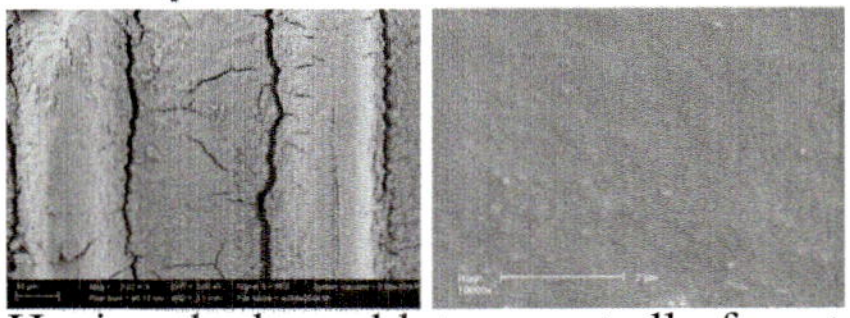

flabellate Having the branchlets repeatedly furcate

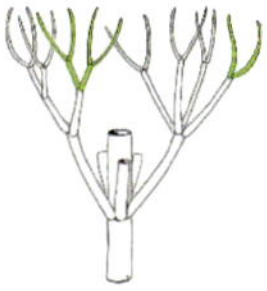

flange Attenuated extension of a stria that is thin and perpendicular to the surface of the oospore = a narrow, tall ridge

flask-cell The internodal cell beneath the antheridium, which grows into the antheridium. See chapter 3 Ontogenesis

fluted Having parallel grooves or flutes; channelled

foot-cell Synonym of flask-cell

form Purely habitat-dependent morphotype (e.g. low-light/high-light) which can be converted into another form by growth conditions

fossa (plural fossulae)/fossate/ fossulate When used in descriptions of the shape of the oospore: the space between the striae; synonym of helical stripe. When used as an ornamentation on oospores: with furrows (see illustration). Fossate: with fossae. Fossulate: with small fossae

fova/foveate/foveolate A shallow, rounded, regular depression or pit (upper left). Foveate (upper right): with fovae; foveolate (below): with small fovae

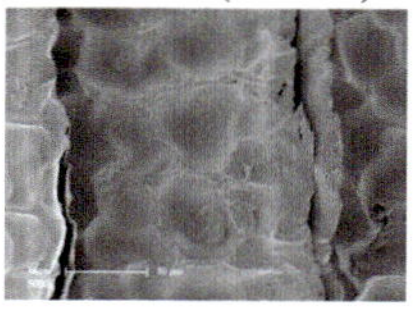

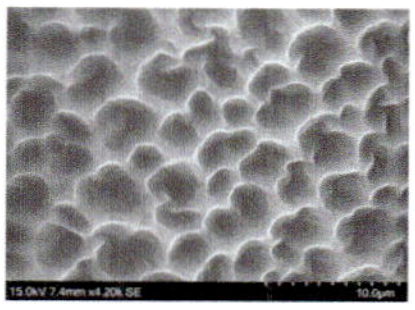

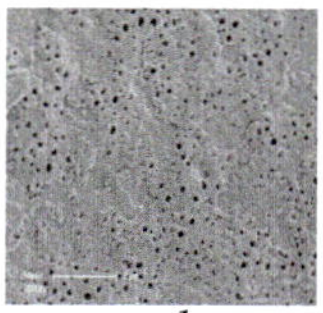

furcate Forked into more or less equal divisions

gametangial short branch Lateral of limited growth of a branchlet node, giving rise to the formation of gametangia. See chapter 3 Ontogenesis

gametangium (-gia) The sexual reproductive organ(s)

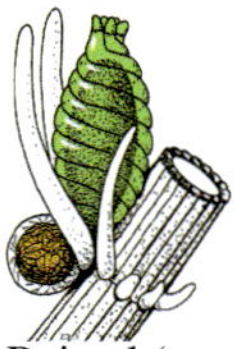

geminate Paired (gametangia, spines). E.g. in *Chara dissoluta* gametangia are solitary or geminate

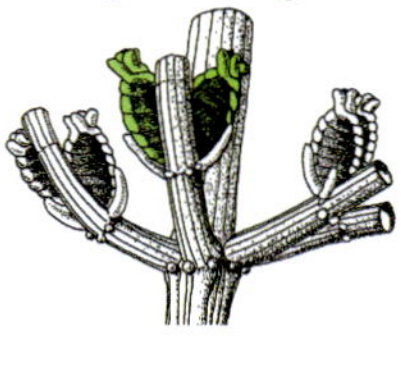

geniculate/geniculation Bent/bending like a knee

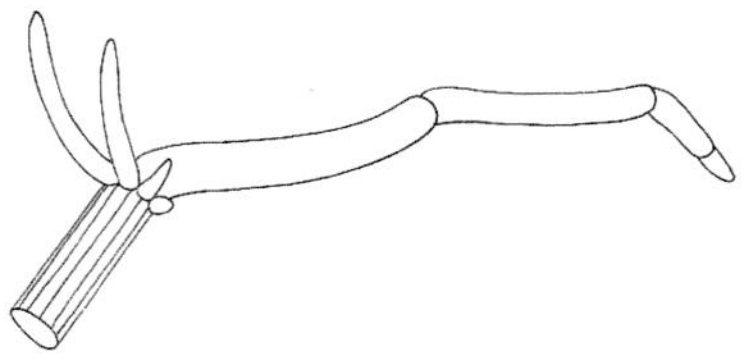

gibbous Inflated, swollen

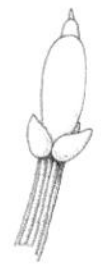

globule Synonym of antheridium

gloeocarpous Having the reproductive organs enveloped in mucus. See mucus

goffered With regular impressions, closely set, and separated by narrow ridges, waffling or honeycombs

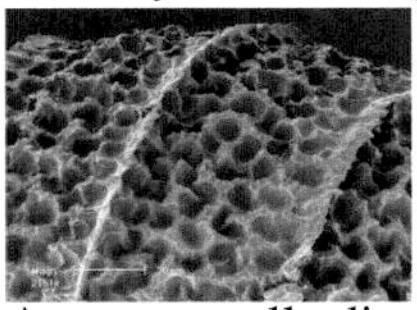

grain/granule/granulate A very small, discrete particle (left), approximately equal in all dimensions. Granule: a small grain. Granulate (right): covered with or made up of very small grains or granules

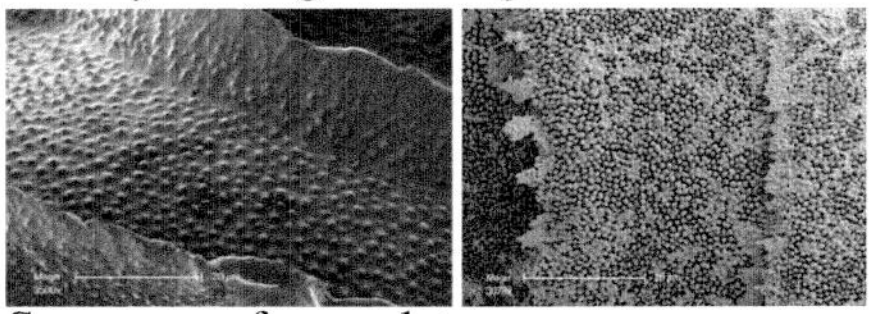

granolous Synonym of granulate

granulated Synonym of granulate

gristle Synonym of cartilage

grumous With clots

gymnocarpous Having the reproductive organs not enveloped in mucus

gymnophyllous Having naked branchlets, i.e. branchlets without a cortex in species of *Chara* where the main axes are corticate

gymnopodous Having the lowest internode of the branchlet ecorticate

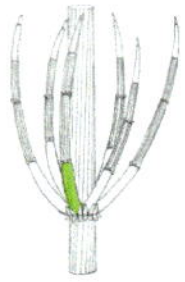

gyrogonite (Sub)fossil or recent female sex organ with a calcareous incrustation or the ovoid spiral-marked body that is the residue of the calcareous incrustation.

handle-cell Synonym of manubrium

haplostephanous Having a single ring of stipulodes at the base of each whorl of branchlets. The picture on the left is 'haplostephous, unistipulate', the picture on the right is 'haplostephous, bistipulate'

haplostichous Having a cortex with only primary rows, there being one cortical row corresponding to each branchlet

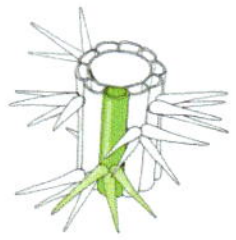

hatched Closely marked with short, transverse lines

head-cell (1) Synonym of capitulum. See chapter 3 Ontogenesis

head-cells (2) Cells originating at the exterior side of axis-cortex nodes and bearing the spines, especially in species with bundled spines

helical stripes The spaces between the striae, formed by the helical cells. They can be counted on both the oospore and the oogonium, but are usually used to describe the oogonium. In most older works they are called 'convolutions', but do not represent the number of times the helical cells rotate around the oospore

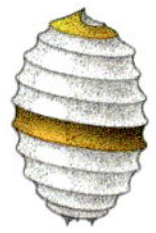

heteroclemous/heteroclemy (The phenomenon of) having branchlets of two kinds in the same whorl (the lower ring being homologous to stipulodes, see chapter 3 Ontogenesis)

heterophyllous Synonym of heteroclemous

heterostichous Having cortical cells of unequal diameter

hirsute Synonym of spinous

hispid With many spine-cells

holodactylous Synonym of anarthodactylous

homoclemous (The phenomenon of) having branchlets of the same form in the same whorl, antonym of heteroclemous

homophyllous In *Nitella*: having only one form of branchlet. Also written as 'homeophyllous'. Antonym of heterophyllous

incrustation/incrusted Synonym of encrustation/encrusted

internode Section between two nodes

irrorate Freckled or speckled; covered with minute spots or granules

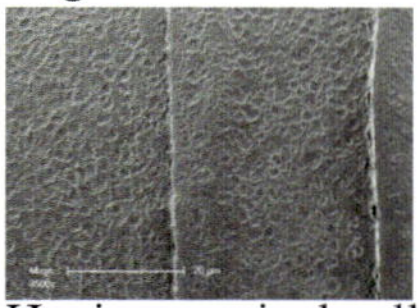

isostichous Having cortical cell(s) of about equal size

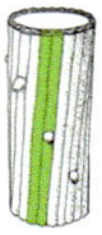

jungate Made up of interlocking angular elements

lacuna/lacunose Irregular, broad, shallow cavity (left). Lacunose (right): full of hollows or cavities; with scattered and irregular broad, shallow cavities

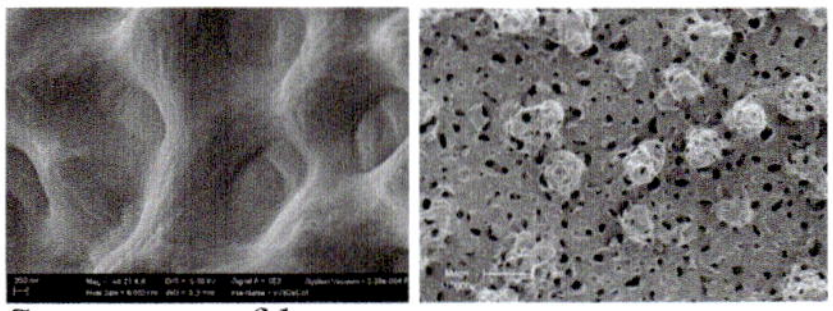

lacunate Synonym of lacunose

laevigate A smooth surface, sometimes somewhat shiny or polished; without elevations or depressions

lamella(e) — The layers of the membrane of the oospore. See chapter 3 Ontogenesis

lateral branch — Synonym of lateral axis

lateral axis — An axis of unlimited growth arising from a vegetative node and usually having whorls themselves

lateral axis origin cell — Omnipotent nodal cell able to develop into a side axis (instead of a primary cortex row) in *Chara* spp.

lateral shoot — Synonym of lateral axis

laterals — Synonym of branchlets

leaf(let) — Ambiguous terms, applied to single-celled outgrowths arising either from branchlet nodes or from gametangial nodes. See chapter 3 Ontogenesis

limb-cell — The first product of a division of an apical cell, which becomes further differentiated into a vegetative internodal cell and an omnipotent nodal cell, the latter one being able to develop into an apical cell again. See chapter 3 Ontogenesis

lime-shell — A hard shell surrounding the oospore, formed by a secretion of lime within the spiral-cells of the oogonium

lineate/lineolate — Longitudinally marked with raised or depressed parallel lines (left); with linear marks. Lineolate (right): finely lineate

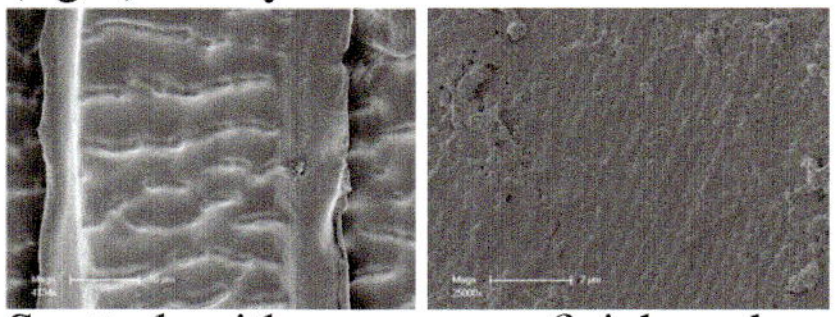

maculate — Spotted; with many superficial marks or spots

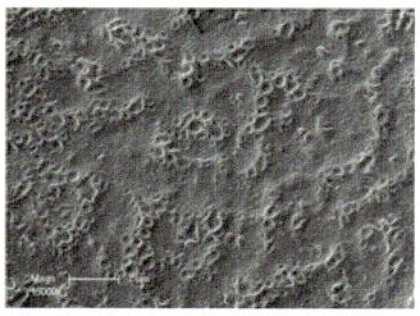

manubrium (-ia) Stalk-like cell(s) within the antheridium which support the capitula-cells (left). Manubrium of *Chara virgata* (right)

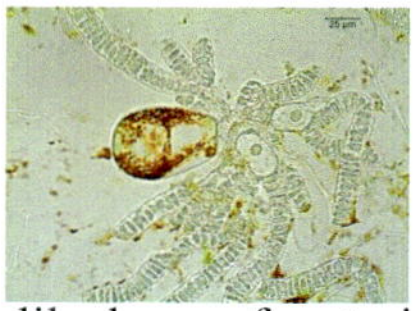

membrane Generally: a thin, tissue-like layer of material; in description of charophytes: synonym of oospore-membrane

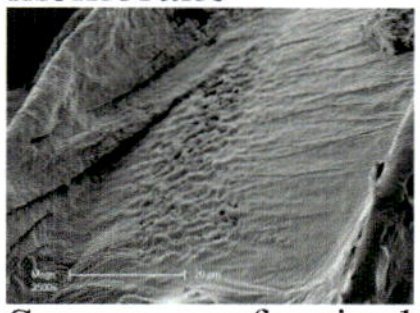

meshed Synonym of reticulate

microdactyl(ou)s Synonym of having a corona

monarthrodactylous Synonym of anarthrodactylous

moniliform Like a string of beads

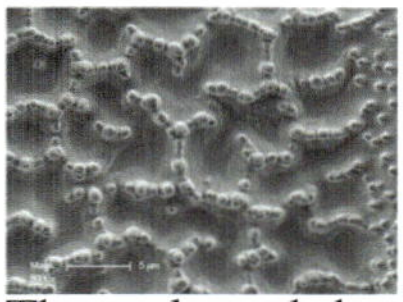

monoecious The male and the female gametangia produced on the same plant

monopodial Having a main axis of which the terminal bud continues to grow as a central leader shoot and the lateral branches remain subordinate

mottled With irregular patches of spots or patches of colour

mucilage Synonym of mucus

mucro/mucronate Mucro: a small, sharp point. Mucronate: abruptly narrowing into a small sharp point, comparable with 'apiculate'. European authors usually employ mucronate to describe species in which the point is a separate cell which is significantly narrower than the penultimate cell (e.g. *Nitella mucronata*)

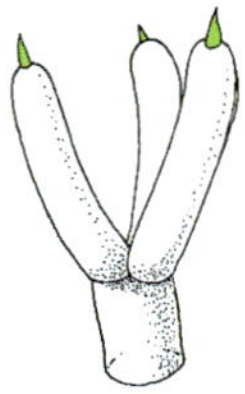

mucroniform Synonym of mucronate

mucus A slimy substance (produced around gametangia in several *Nitella* species)

net-like Synonym of reticulate

nitid Shiny or glossy; reflecting light

nodal cell A. s.l.: all cells a nodal complex is made of;
B. s.s.: the omnipotent cells originating from the differentiation of a limb cell. See chapter 3 Ontogenesis

node The point of an axis or branchlet at which laterals arise

nodule/nodulate A small rounded lump of matter distinct from its surroundings. Different from grains in being irregular shaped. Nodulate: with nodules

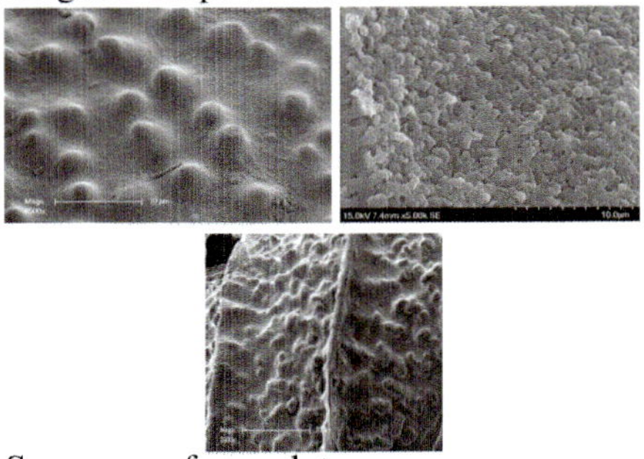

notate Synonym of maculate

nucule Synonym of oogonium

nuculus Synonym of oospore

nude Synonym of ecorticate

obtuse Having a tip with a 90 degree to 180 degree angle. Often obtuse tips are rounded, but not quite a semi-circle

octoscutate Antheridiae consisting of eight shield cells

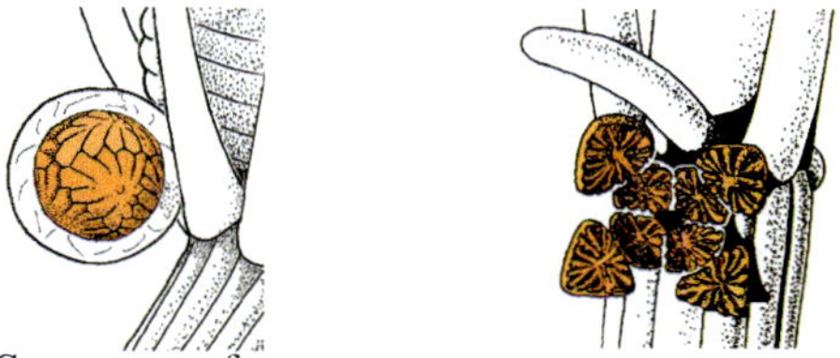

oocarp Synonym of oospore

oocyte Synonym of oosphere

oogonium (-nia) The female reproductive organ. *Chareae* (left) and *Nitelleae* (right)

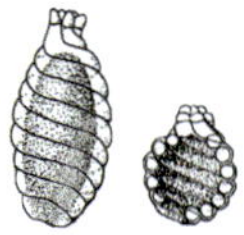

oosphere The egg-cell before fertilisation. See chapter 3 Ontogenesis

oosporangium Synonym of oogonium

oospore The fertilised egg-cell (zygote). In *Chara canescens* the oospore can be formed without fertilisation

oospore-membrane The thickened outer layer of the (usually fertilised) egg-cell formed by sporopollenin

oospore-shell The total of the outer envelopes, which may include calcite layers etc. as well

opposite (Of stipulodes) right below branchlet

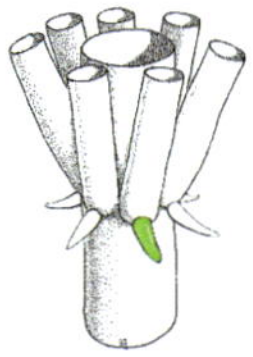

pachygyra With the surface of the oospore raised at the striae

papilla(leae)/papillus/papillate/ papillulate Of oospores: a small rounded protuberance, usually larger and more sparsely distributed than a grain (left). On the main axis of some *Chara* species: barely visible, small, nipplelike spine-cell (right). Papillate: with papillae. Papillulate: with small papillae

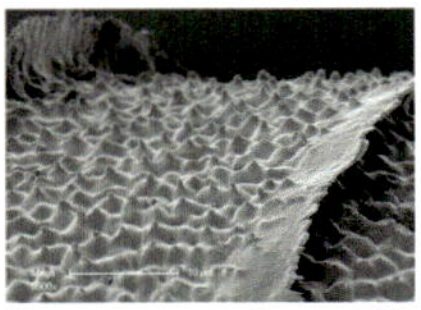

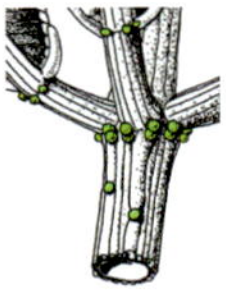

papilliform Synonym of papillate

papillose Synonym of papillate

parthenogenetic Producing viable oospores without fertilisation by male gametes

patent projecting more or less at a right angle. Used in the description of spine-cells

peltate More or less circular, with the stalk attached at a point on the underside

peripheral cell Synonym of secondary cortex-cell. See chapter 3 Ontogenesis

phloeopodous Synonym of gymnopodous

pillar Of oospores: a cylindrical protuberance

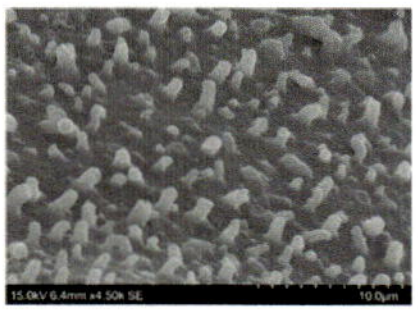

pitted Synonym of foveate

plasmodesmata Channels which traverse the cell walls, enabling transport and communication between neighbouring cells

plate Synonym of shield-cell. Also a term used for a variety of structures

plicate folded; with folds; impressed with striae to produce the appearance of having been folded or pleated

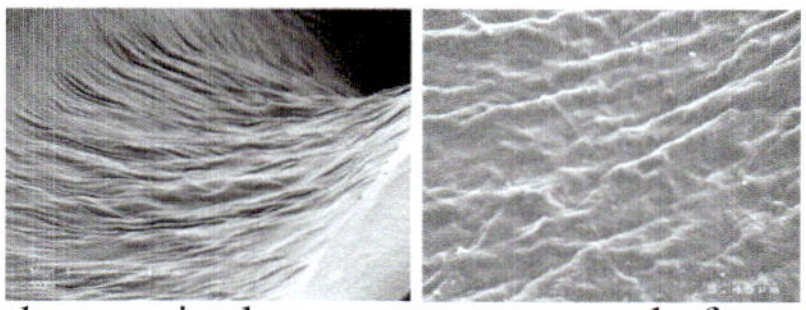

polyarthrodactylous the terminal segment composed of more than two cells

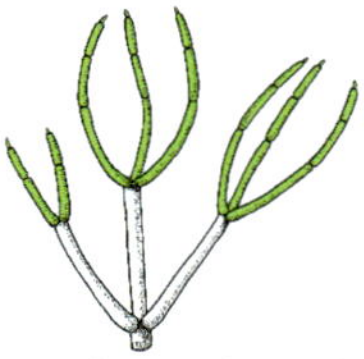

polyembryony The formation of more than one protonema from a single oospore as often observed in the genus *Tolypella*

pore/porate A minute opening/having pores

posterior Synonym of abaxial. Also used for abaxial branchlet

prickle Of oospores: a short, pointed projection on the oospore-wall

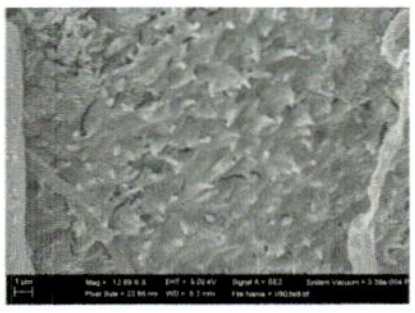

primary (cortical) cell(s) Cells that make up the primary row. See primary row

primary (cortical) row Row of alternating vegetative and nodal cortex cells; the nodal cells being the origin of secondary (vegetative) cells and spine-cells

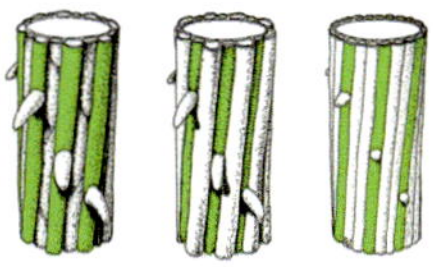

primary branchlets The fully developed whorl of branchlets of heteroclemous *Nitella* spp., able to bear gametangia. See heteroclemous

primary cortex Synonym of primary cortical row

primary rhizoid Rhizoid developing directly from a germinating zygote. See chapter 3 Ontogenesis

primary root Synonym of primary rhizoid

proembryo Synonym of protonema

proembryonic filament The terminal part of the proembryo above the last whorl of the proembryo. This last whorl gives rise to the vegetative plant (see chapter 3 Ontogenesis). About this whorl, the apical cell can further divide and release limb cell(s), which do not differentiate further to nodal complexes and internodes. Being inconspicuous and short-lived in most of the species, the pro-embryonic filament of some species, especially *Tolypella* and *Sphaerochara* spp. can develop to a persistent long and robust structure dominating the appearance of the whole plant and therefore being an important character for species discrimination

proembryonic whorl Whorl of branchlets of the prochara giving rise to the sexual plant. See chapter 3 Ontogenesis

protandrous Antheridia are produced earlier than the oogonia. A monoecious plant may appear first entirely male, then entirely female

protonema A pre-germ which originates directly from the oospore. From its only whorl-node the main shoot of a charophyte plant develops (see chapter 3 Ontogenesis)

punctuated Synonym of punctate

puncture/punctate/punctulate/puncticulate A small hole with depressed edges appearing to be the result of being pierced. Punctate: with punctures. Punctulate: finely punctate. Puncticulate: sparsely punctate with very fine, widely spaced punctures

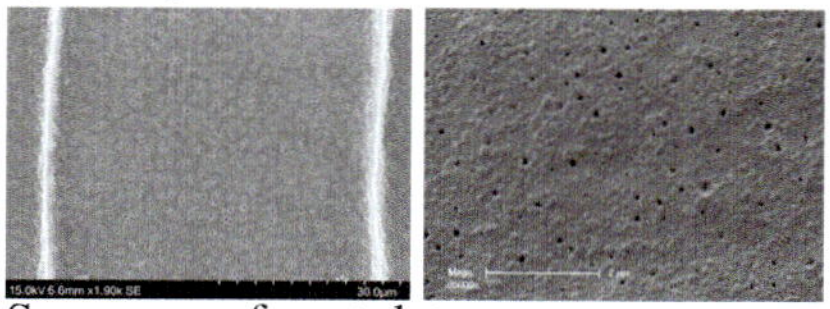

pustular/pusticular Synonym of pustulate

pustule/pustulate A deep pit or depression with raised edges, more protuberant than a variole. Pustulate: with many pustules

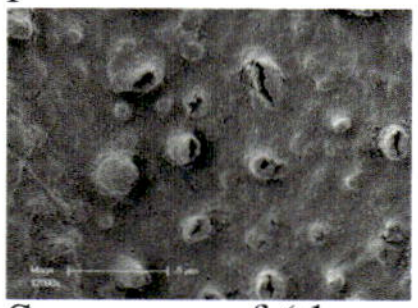

ramification Synonym of (the pattern of) branching

rastrate Covered as if with longitudinal scratches

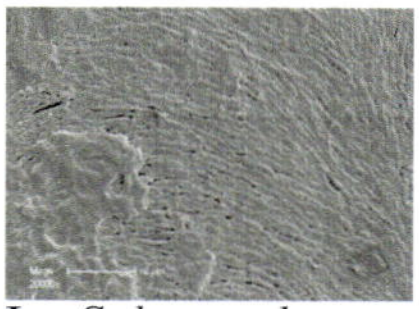

ray In *Sphaerochara* and *Tolypella*, the divisions of the branchlets, from base to end-cell; often described as primary ray for the dominant axis of the branchlets, secondary rays for lateral axes branching off the dominant axis and tertiary rays for laterals (if any) branching off the secondary rays. By contrast, also previously used in *Nitella* by some authors for the segments between furcations, including the dactyls. Blue = primary ray; yellow = secondary ray

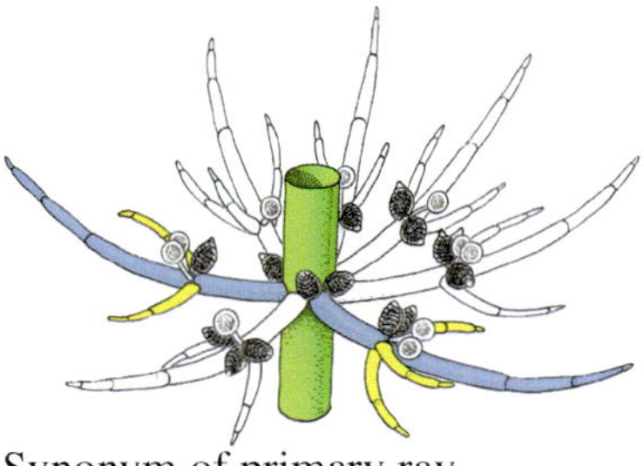

ray, first order Synonym of primary ray

ray, primary In *Sphaerochara* and *Tolypella*, the dominant axis of the branchlets, from base to end-cell. Also previously used in *Nitella* by some authors for the first (basal) segment of the branchlets from the base to the first furcation. Illustration see 'ray'

ray, quaternary In the branchlets of *Sphaerochara* and *Tolypella*, any lateral axis branching off the tertiary rays (from base to end-cell). Also previously used in *Nitella* by some authors for the fourth segment of the branchlets between the third and fourth furcations. (Rarely quinary ray was also a term used in some Australian and Asian species of *Nitella*)

ray, secondary In the branchlets of *Sphaerochara* and *Tolypella*, any lateral axes branching off the primary rays (from base to end-cell). Also previously used in *Nitella* by some authors for the second segment of the branchlets between the first and second furcations. Illustration see 'ray'

ray, second order Synonym of secondary ray

ray, tertiary In the branchlets of *Sphaerochara* and *Tolypella*, any lateral axis branching off the secondary rays (from base to end-cell). Also previously used in *Nitella* by some authors for the third segment of the branchlets between the second and third furcations

ray, third order Synonym of tertiary ray

ray, ultimate Synonym of dactyl

refracted Bent backwards at an acute angle

reticulated Synonym of reticulate

reticulately pitted Synonym of reticulate

reticulum/reticulate Network, like a fishing net. Reticulate: superficially net-like or made up of a network of lines

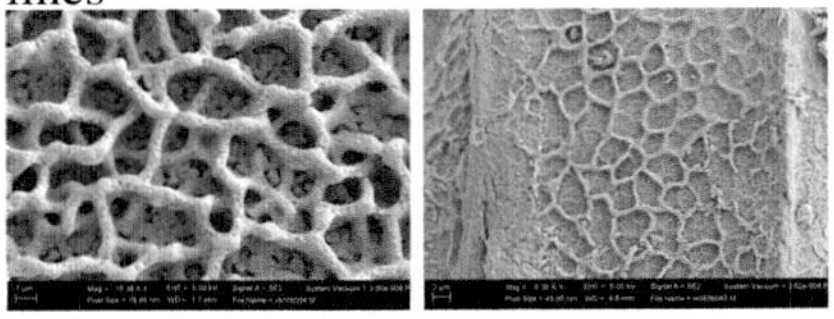

rhizoid Colourless, hair-like filaments growing from the plant base into the substrate, with a dual function of absorption and attachment

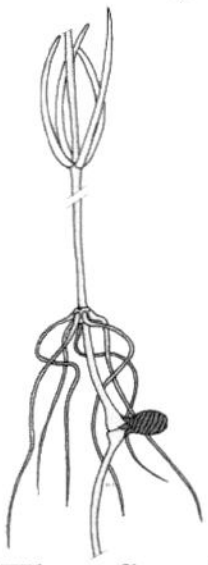

ribbon The flaccid extension of the striae on some species of *Chara*, formed between adjacent spiral-cells on the oospore

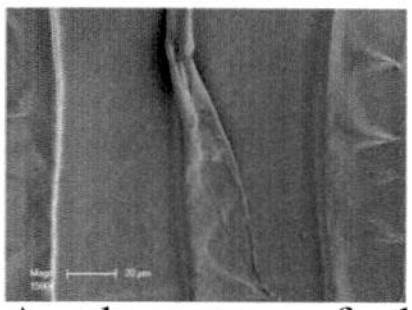

ridge A character of the striae: like a mountain ridge. Ridges can be sharp, broad, prominent, or extended into flanges

rimose/rimulose With minute, narrow and nearly parallel excavations (rimae) running into each other; chinky; resembling the cracked bark of a tree. Rimulose: minutely rimose

rivose/rivulose Marked with sinuate furrows, like rivulets, not running in a parallel direction: The picture is meant to indicate the valleys between the bumps. Rivulose: minutely rivose.

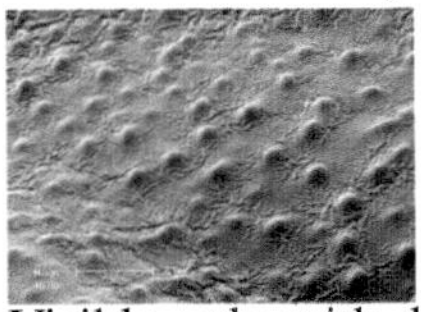

rudimentary (r. bract-cell, r. spine-cell, r. stipulode) Visible only with difficulty and on close examination. E.g. *Chara globularis* spine-cells are rudimentary

rugose/rugulose Wrinkled. Rugulose: minutely rugose; minutely wrinkled

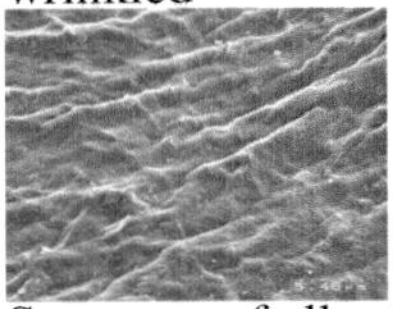

sausage-shaped Synonym of allantoid

scabrous/scabrid/scabridulous Rough; irregularly and roughly rugose; possessing short, sharp projections or wrinkles. Scabrid: sparsely scabrous. Scabridulous: finely scabrous; with fine and regular short, sharp, wrinkles and/or projections

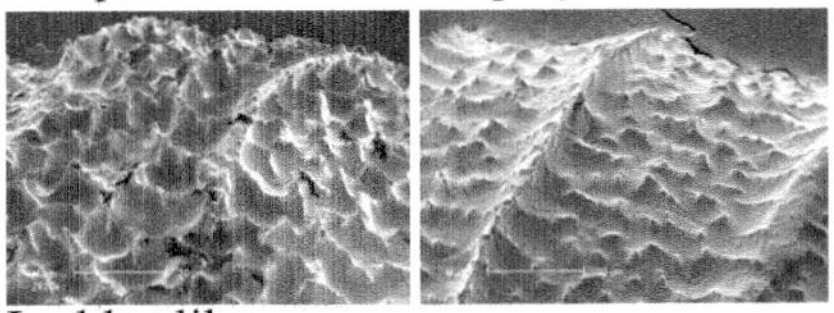

scalariform Ladder-like

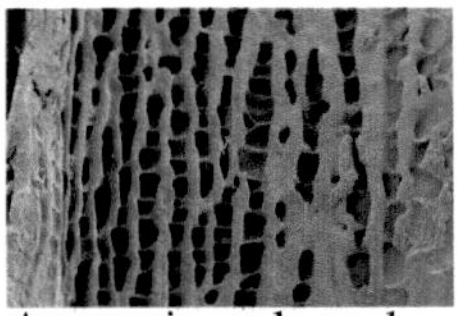

scarified Appearing clawed or scratched; furnished with fine, irregular grooves, coarser than aciculate

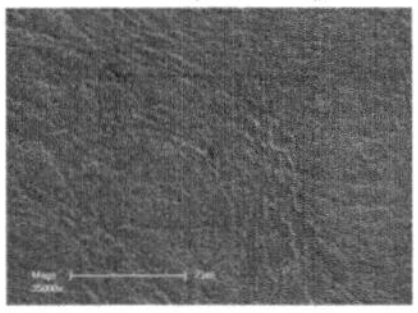

scutate/scutellate Covered with large, flat, scales. Scutellate: minutely scutate

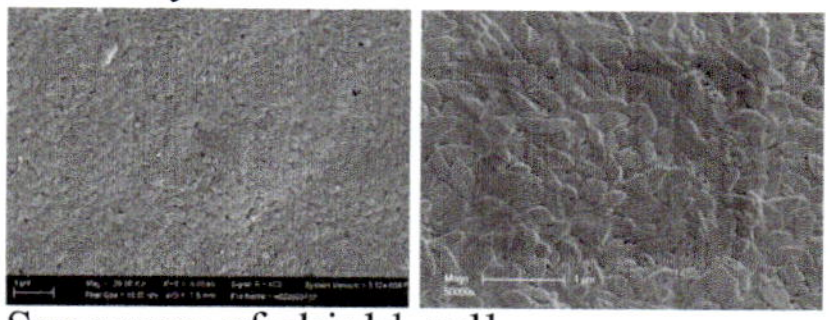

scute/scutum Synonym of shield-cell

secondaries/secondary cells Synonym of secondary cortical cells

secondary cortical cells The cells that make up the secondary cortical row

secondary cortical row Row of vegetative cortex cells, originating from cortex nodal cells, lying alongside the primary row. It is the row that does not bear the spine cells. In diplostichous species there is one secondary row, in triplostichous species there are two

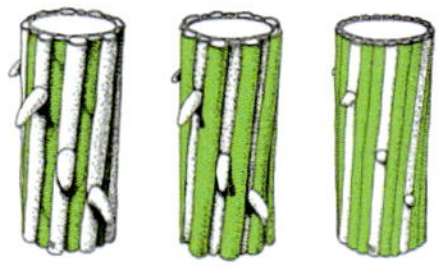

secondary branchlets Synonym of accessory branchlets

secondary cortex Synonym of secondary cortical row

secondary rhizoids Rhizoids originating from the rhizoidal node of the prochara; the term is often used in a broader sense, describing all rhizoids originating from nodes. See chapter 3 Ontogenesis

section A taxonomic rank between genus and species (but below subgenus, if also present). Sections are in use for organising very large genera. A botanist wanting to distinguish groups of species may prefer to create a taxon at the rank of section to avoid new combinations, which would need to be indexed in Index Kewensis

segment

Internode section of a branchlet. The part above the last branchlet node is called the "terminal segment", consisting either just of the branchlet apical cell (end-cell) or end cell and undifferentiated limb cells (see chapter 3 Ontogenesis). Other authors may use the term in a different way

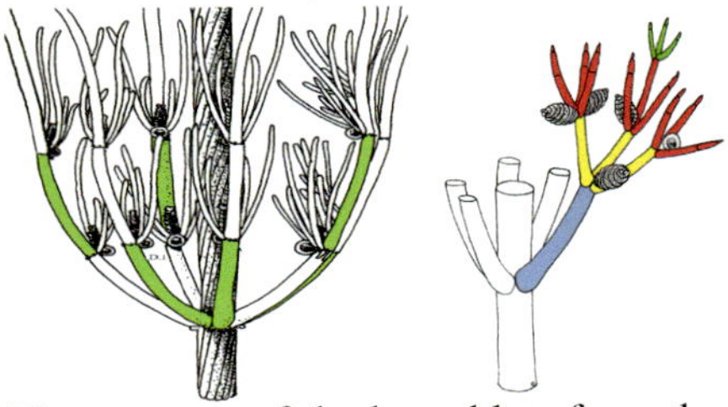

segment, first

The segment of the branchlets from the base to the first node

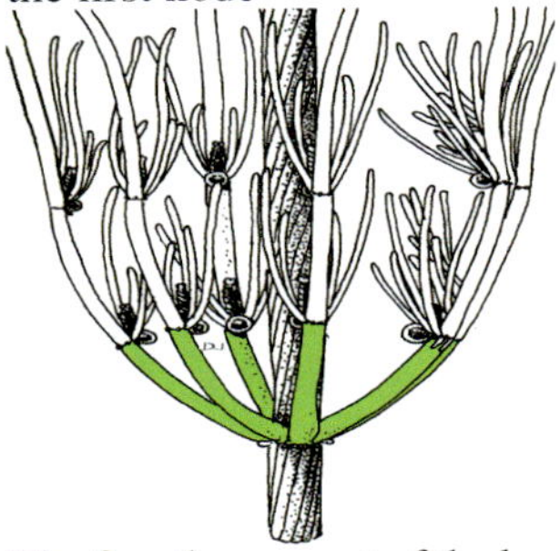

segment, fourth

The fourth segment of the branchlets between the third and fourth nodes

segment, primary

Synonym of first segment.

segment, quaternary

Synonym of fourth segment. Asian and Australian authors usually prefer 'quaternary segment' in describing *Nitella* and *Tolypella*

segment, second — The second segment of the branchlets between the first and second branchlet nodes

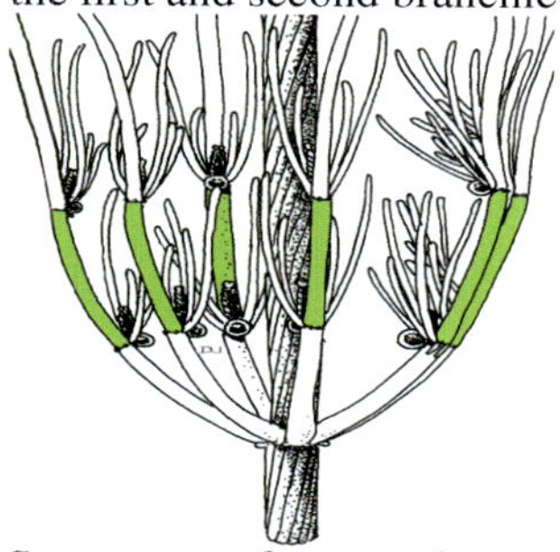

segment, secondary — Synonym of second segment. Asian and Australian authors usually prefer 'secondary segment' in describing *Nitella* and *Tolypella*

segment, terminal — The segment beyond the last node of the branchlet. It may consist of more than one cell. In *Nitelleae* the last segment is called 'dactyl'

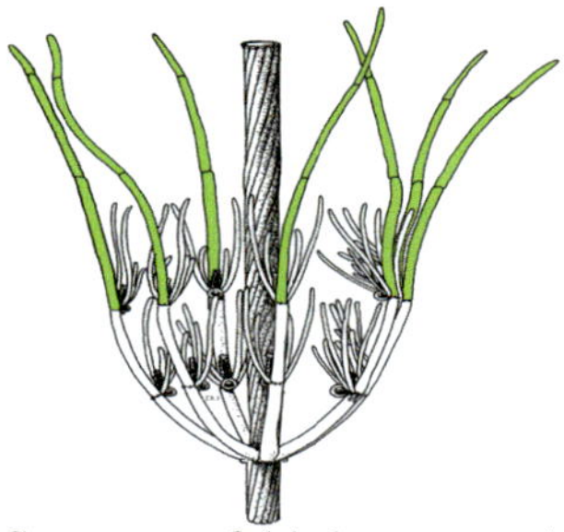

segment, tertiary — Synonym of third segment. Asian and Australian authors usually prefer 'tertiary segment' in describing *Nitelleae* and *Tolypella*

segment, third — The third segment of the branchlets between the second and third branchlet nodes, counting outwards from its base

sejoined — The antheridium and oogonium being produced at separate branchlet nodes of the same plant

senations — Lines arranged in parallel series

septum(-a) — The evolving cell wall in the process of division

series of cortex-cells — Synonym of row of cortical cells

shagreened Covered with a closely set roughness, like the rough-surfaced horse leather termed shagreen; like shark leather

shield Synonym of shield-cell

shield-cell Pigmented, plate-like cell with interlocking and wavy margins and a central round marking that make up the outer, protective layer of the antheridium

The picture shows seven shield cells. In position they are part of a 4- or 8-celled globe

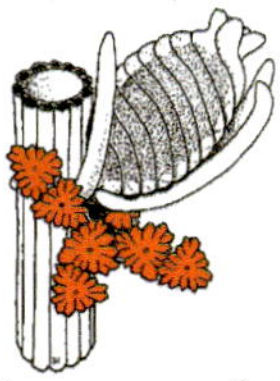

shieldlike cell Synonym of shield-cell

shoot Synonym of axis

shoot (main) Synonym of main axis

side branch Synonym of lateral branch

side shoot Synonym of lateral branch

side shoot origin cell Synonym of lateral branch origin cell

siliceate With the appearance of cobblestones

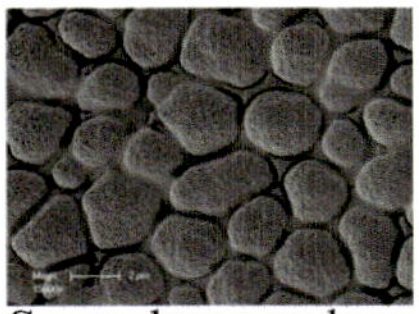

sinuate Strongly waved

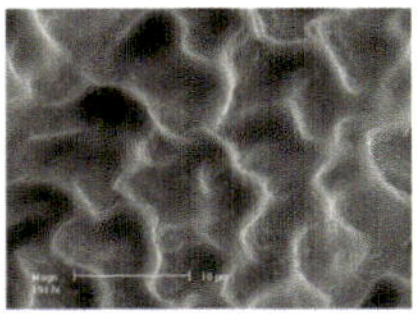

species A distinct combination of characters which, independent from the habitat that the individuals grow in, separates them from other "species"

spermatogenic filaments Series of cells originating from the capitulum, each being able to transform into a spermatozoid. See chapter 3 Ontogenesis

spermatozoids The minute motile bodies produced in the antheridium, presenting the male element

spherulate Provided with one or more rows of minute tubercles

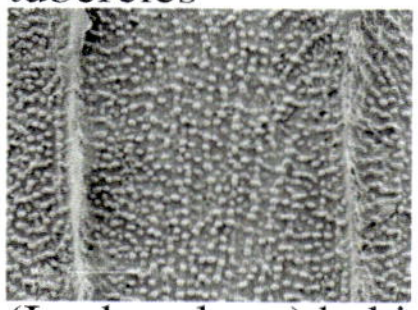

spicate/spike (In charology) habit elongate, not round or head-like (e.g. used to describe *Nitella* plants)

spine/spinose/spinulate/spiny Of cortex ornamentation: spine is a synonym of spine-cell. Of oospores (left): narrow, sharpened, elongate projection, longer than a prickle. Spinose (right): armed with thorny spines, more elongate than echinate. Spinulate: minutely spinose. Spiny: with spines

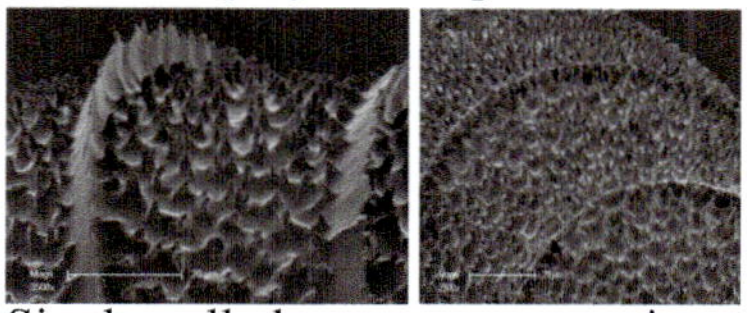

spine-cell Single-celled processes growing out from the node-cells of the stem-cortex

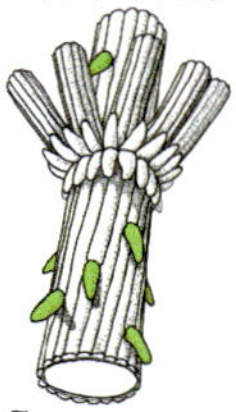

spiral ridges Synonym of striae

spiral-cells The cells surrounding the oogonium to form a protecting envelope

spongy With a texture similar to a marine sponge

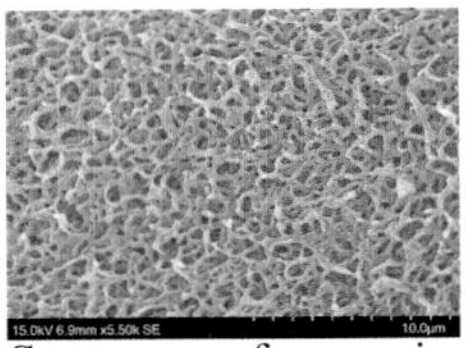

sporangium Synonym of oogonium

sporocarp Synonym of oospore

sporophyas Synonym of oogonium

sporophydium Synonym of oogonium

sporostegium Synonym of oogonium

squamate Synonym of squamose

squamose Rough with elevations; scurfy; rough with loose scales differing in direction or not parallel in direction

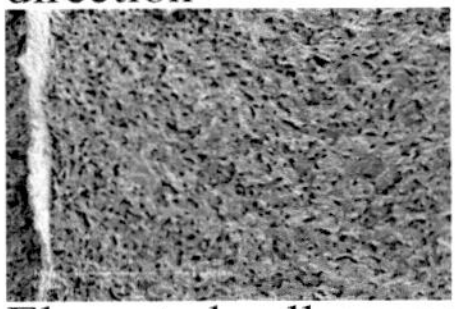

stalk/stalked Elongated cell connecting the oogonium to the node. Also antheridia can have stalks

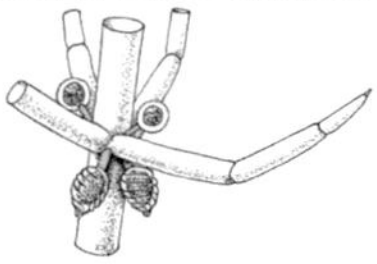

stellate Synonym of star shaped

stem Synonym of axis

stem (main) Synonym of main axis

stipe/stipitate Synonym of stalk(ed)

stipulae, (row of..), (series of..), (ring of..) Synonyms of whorl of stipulodes

stipule(s) Synonym of stipulode(s)

stipulode(s) In the *Chareae*: process of limited growth, usually single-celled, growing out from the base of the branchlet whorls

stipulodes (ring of..), (row of..), (series of..), (tier of..) Synonyms of whorl of stipulodes

stria(e) The stripes on the oospore that are formed at the intersections of the 5 spiral-cells

striate/striolate Marked with parallel, fine, longitudinal impressed lines or furrows. Striolate: minutely or finely striate

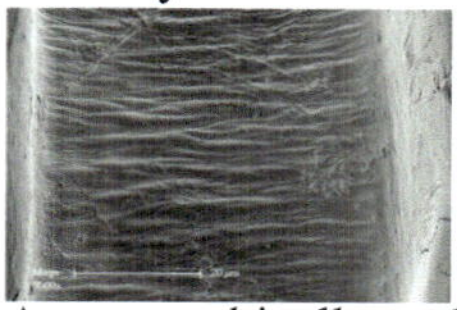

subspecies A geographically and/or climatically dependent variation of character expression well within the species range

subulate Broad at the base with a tapering point

sulci Synonym of fossulae

sulcus/sulcate A sinuate groove on a surface. Sulcate: with a sulcus

taxon A recognisable entity which may be separated from related entities at any level of the classificatory hierarchy

terete Circular in transverse section

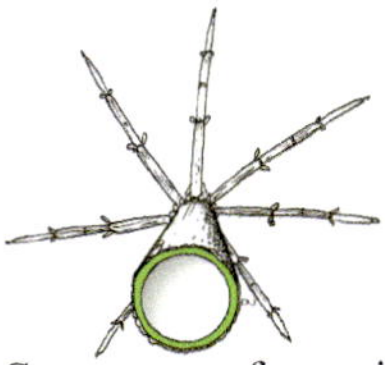

terminal process Synonym of terminal segment

tesselate Made up of squares like a chess board, either in sculpturing or in color

tetrascutate Antheridiae consisting of four shield cells

tier Synonym of whorl (e.g. of stipulodes)

tora/torose/torulose A knobby projection. Torose: superficially swelling in network of knots, knobs, or protuberances; knobby. Torulose: minutely torose

triplostichous The cortex having two secondary rows alternating with each primary row, there being three cortical rows corresponding to each branchlet

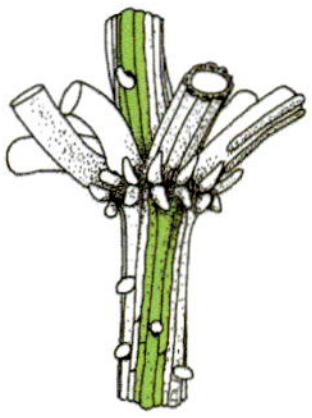

tubercle/tuberculate A small knob or button. Tuberculate: covered or furnished with rounded, projecting lobes; more projecting than granulate, papillate, or pustulate

tumulose With blister-like emanations larger than papillae and never with a terminal pore

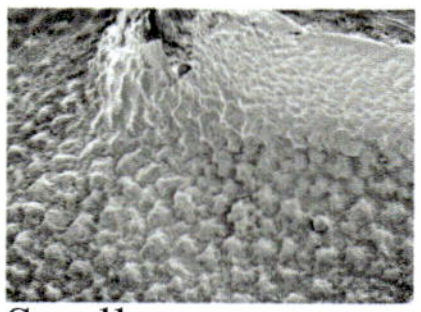

turgid Swollen

turning cell Sister cell(s) of the oocyte; comparable to an internodal cell, dissected from the gametangial apical cell when transforming to the oocyte. See chapter 3 Ontogenesis

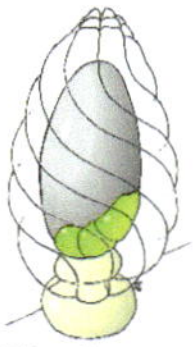

tylacanthous Having primary cortical cells that are more prominent than the secondary cortical cells, so that the spines may appear to sit on ridges

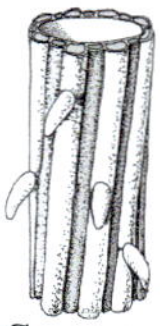

uncorticated Synonym of ecorticate

undose With undulating, broad, nearly parallel depressions running more or less into each other; wavy, resembling ripple marks on a sandy beach

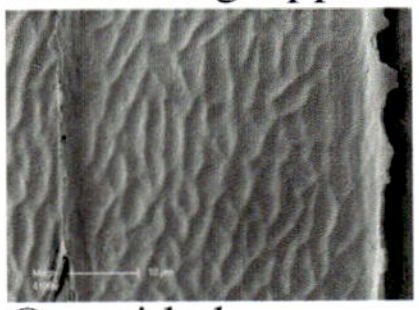

unilateral One-sided

unistipulate Having one stipulode per branchlet in the same whorl. Not to be confused with haplostephanous: having one ring of stipulodes at the base of each whorl of branchlets

variety Habitat selected genotype, exhibiting therefore a certain kind of expression of characters—which are inheritable, and possibly can (but not must!) be detected genetically and for which a discrete name is provided

variole/variolate A shallow pit or depression with raised edges (left). Variolate (right): pitted as if by smallpox; full of irregular indentations (varioles)

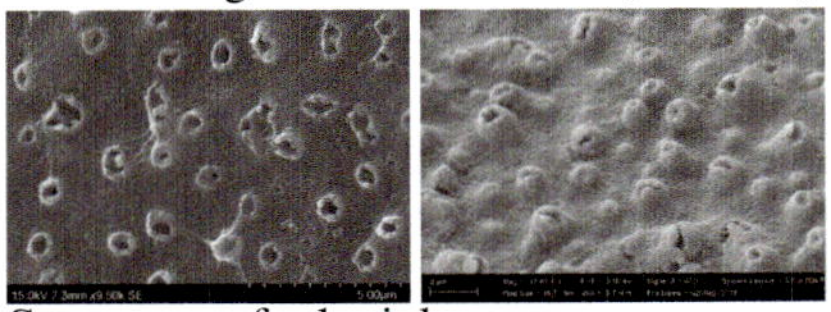

ventral Synonym of adaxial

vermicose/vermicular/vermiculated Synonyms of vermiculate

vermiculate With superficial, tortuous markings resembling the tracks of a worm

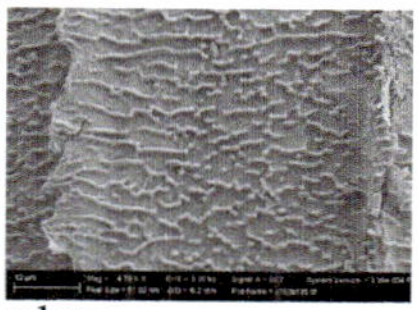

vermiferous Synonym of vermiculate

vermiform Synonym of vermiculate

verruca/verrucate Irregularly shaped lobe or wart-like protuberance. With irregularly shaped lobe or wart-like protuberance

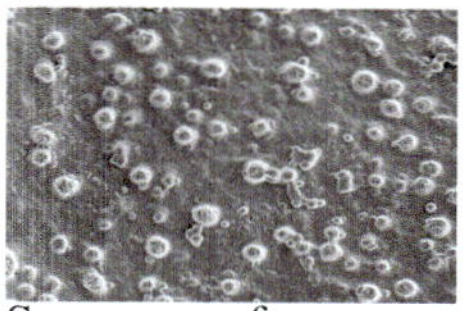
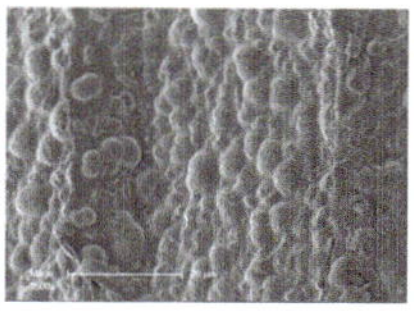

verrucose Synonym of verrucate

vertex-cell The outer node cell which arises during the process of the formation of the antheridia. It finally leads to the formation of the shield cell, the manubrium and the capitulum. For details of the process see Chapter 3 Ontogenesis

verticillate Arranged in whorls, equally long in all directions (e.g. antheridiae (left) around a branchlet, or adaxial and adaxial bracts of nearly equal length (right)

vestigial Rather undeveloped, rudimentary

vortex/vortices Synonym of whorl(s)

whorl Set of branchlets originating from the same node

whorl of stipulodes In *Chareae*: the whorl of stipulodes at the base of the branchlet whorl. Usually there are one or two whorls

window(s) A thinner, often transparent layer of oospore wall between claws at the base of the oospore of some species of *Chara*. The Illustration at the right hand is of windows in the lacunae of the wall of *Nitella*

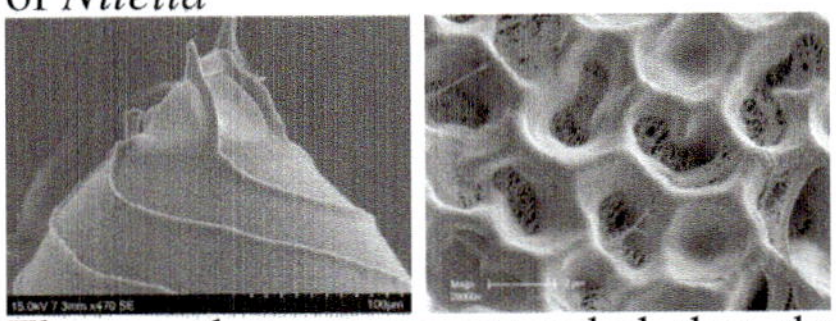

wing Flanges that are so extended that they would modify the movement of the oospore through the water. They are usually extended both at the apex (into a beak) and the base, obscuring the view of the basal cells

References

Bryant, J. A. (formerly Moore) & Stewart, N. F. (2011). Order Charales. In: D. M. John, B. A. Whitton, & A. J. Brook (Eds.), *The freshwater algal flora of the British Isles. An identification guide to freshwater and terrestrial algae* (ed. 2, pp. 742–765). Cambridge University Press.

Casanova, M. T. (2009). An overview of *Nitella* (Characeae, Charophyceae) in Australia. *Australian Systematic Botany, 22*, 193–218.

Faegri, K., & Iverson, J. (1964). *Textbook of pollen analysis*. With a chapter by H. T. Waterbolk. 2nd rev. ed. Blackwell Scientific Publications.

Groves, J., & Bullock-Webster, G. R. (1920). *The British Charophyta. Volume I. Nitelleae*. Ray Society, London.

Harris, R. A. (1979). A glossary of surface sculpturing. *Occasional Papers in Entomology, 28*, 1–31.

Moore, J. A. (1986). *Charophytes of Great Britain and Ireland*. BSBI-handbook No 5, Botanical Society of the British Isles, London.

Wood, R. D., & Imahori, K. (1964). *A revision of the Characeae, 2, Iconograph of the Characeae*. J. Cramer, Weinheim.

Wood, R.D., & (1965). *A revision of the Characeae. 1. Monograph of the Characeae*. J. Cramer, Weinheim.

Epithet Index

H. Schubert et al. (eds.), *Charophytes of Europe*, https://doi.org/10.1007/978-3-031-31898-6

Species Index

H. Schubert et al. (eds.), *Charophytes of Europe*, https://doi.org/10.1007/978-3-031-31898-6

D

E

F

G

H

O

P

GPSR Compliance

The European Union's (EU) General Product Safety Regulation (GPSR) is a set of rules that requires consumer products to be safe and our obligations to ensure this.

If you have any concerns about our products, you can contact us on ProductSafety@springernature.com

In case Publisher is established outside the EU, the EU authorized representative is:

Springer Nature Customer Service Center GmbH
Europaplatz 3
69115 Heidelberg, Germany

Batch number: 10756369

Printed by Printforce, the Netherlands